Industrial Engineering Terminology, Revised Edition

ANSI Standard Z94.0-2000

INDUSTRIAL ENGINEERING

TERMINOLOGY

REVISED EDITION

A revision of ANSI Z94.0-1989

An American National Standard
Approved 1998

Engineering and Management Press
Institute of Industrial Engineers
Norcross, Georgia

Library of Congress Cataloging-in-Publication Data

Industrial engineering terminology : a revision of ANSI Z94.0-1989 : an
   American National Standard approved 1998.-- Rev. ed.
         p.   cm.
   Includes bibliographical references and index.
   ISBN 0-89806-205-5
   1. Industrial engineering--Terminology--Standards--United States. I.
Title: Terminology. Industrial engineering. II. Institute of Industrial
Engineers (1981-). Engineering & Management Press.
   T56.24 .I53   1998
   670.42'014--dc21
                                                              2001024402

© 2000 Institute of Industrial Engineers. All Rights Reserved.

No part of this book may be reproduced in any form without written permission from the publisher.

Included in this volume are abbreviations, symbols, acronyms, functional designations, and letter combinations that are not official, nor are they comprehensive or definitive. They are intended only as a convenient reference supplement when so many aspects of industrial engineering are changing and expanding.

Additional copies may be obtained by contacting:
Member and Customer Support
Institute of Industrial Engineers
25 Technology Park
Norcross, GA 30092
770-449-0460
Fax: 770-441-3295
Email: cs@iienet.org
Quantity discounts available.

TABLE OF CONTENTS

INTRODUCTION vii

HOW TO USE THIS TERMINOLOGY BOOK xix

	Z94.1	Analytical Techniques & Operations Research
	Z94.2	Anthropometry & Biomechanics
	Z94.3	Computer & Information Systems
	Z94.4	Cost Engineering & Project Management
	Z94.5	Distribution & Marketing
	Z94.6	Employee & Industrial Relations
	Z94.7	Engineering Economy
	Z94.8	Facility Planning & Design
	Z94.9	Human Factors (Ergonomics) Engineering
	Z94.10	Management
	Z94.11	Manufacturing Systems
	Z94.12	Materials Processing
	Z94.13	Occupational Health & Safety
	Z94.14	Operations & Inventory Planning & Control
	Z94.15	Organization Planning and Theory
	Z94.16	Quality Assurance and Reliability
	Z94.17	Work Design and Measurement

INDEX 1-59

INTRODUCTION

Robert L. Williams, Ph.D., P.E., Chair
ANSI Z94 Committee on Industrial Engineering Terminology

MAKEUP OF THE Z94.0 STANDARD

This volume contains definitions of terms, symbols and acronyms that are associated with the broad field of industrial engineering. It is the result of the work of hundreds of leading practitioners and educators striving to present the best current usage of the language of the profession.

This is a revision of *Industrial Engineering Terminology*, the American National Standards Institute (ANSI) Standard Z94.0-1989, published in 1990. The original standard was published in 1972. This revision has been widely circulated for review and was approved as an American National Standard by ANSI in 1998.

The terminology is organized into seventeen sections as follows:

Z94.1 ANALYTICAL TECHNIQUES AND OPERATIONS RESEARCH
Z94.2 ANTHROPOMETRY AND BIOMECHANICS
Z94.3 COMPUTER AND INFORMATION SYSTEMS
Z94.4 COST ENGINEERING AND PROJECT MANAGEMENT
Z94.5 DISTRIBUTION AND MARKETING
Z94.6 EMPLOYEE AND INDUSTRIAL RELATIONS
Z94.7 ENGINEERING ECONOMY
Z94.8 FACILITY PLANNING AND DESIGN
Z94.9 HUMAN FACTORS ENGINEERING
Z94.10 MANAGEMENT
Z94.11 MANUFACTURING SYSTEMS
Z94.12 MATERIALS PROCESSING
Z94.13 OCCUPATIONAL HEALTH AND SAFETY
Z94.14 OPERATIONS AND INVENTORY PLANNING AND CONTROL
Z94.15 ORGANIZATION PLANNING AND THEORY
Z94.16 QUALITY ASSURANCE AND RELIABILITY
Z94.17 WORK DESIGN AND MEASUREMENT

In accordance with ANSI procedures all of these sections were reviewed and all were significantly revised and expanded. In revising Z94.1989, a new section on "Management" was added because of the long and close association between industrial engineering and management. In order to give emphasis to the areas, the new sections "Quality Assurance and Reliability" and "Analytical Techniques and Operations Research" evolved from the 1989 section entitled "Applied Mathematics." "Anthropometry and Biomechanics" resulted from a merging of two earlier sections as did "Occupational Health and Safety." And a few sections had title changes. Otherwise, the terminology is organized in the same manner as the 1989 standard that followed in large part the taxonomy used in the Institute of Industrial Engineers *AIIE Research Abstracts*. The section entitled "Cost Engineering and Project Management" was adapted from the terminology of the American Association of Cost Engineers and is hereby gratefully acknowledged.

Although this present edition of *Industrial Engineering Terminology* recognizes contemporary trends in industrial engineering, it is only appropriate to mention those leaders who, while not directly associated with this new edition (some having passed on), did lay the foundation for the work. Col. Clarence E. Davies (whose early association was with H. L. Gantt but whose later work was involved for more than half a century with ASME at the highest level). Professor David B. Porter, who also worked with Gantt, drew the first Gantt Progress Chart, and established the first collegiate course in motion study well over half a century ago. Henry N. Muller who was a main force behind the negotiation committee for the establishment of the terminology project as well as President of ASME, and Dr. Harold B. Maynard (MTM founder and author of ASME's 1955 publication, *Industrial Engineering Terminology*). These well-known leaders, together with AIIE Emeritus Executive Director Jack Jericho, Professor Delmar W. Karger (Emeritus Dean and Ford Foundation Professor at Rensselaer Polytechnic Institute) and Dr. William J. Jaffe (Emeritus Distinguished Professor, New Jersey Institute of Technology) were involved in standardizing industrial engineering definitions which ultimately resulted in the establishment of the Industrial Engineering Terminology project under the American National Standards Institute.

Every effort has been made to produce a clear and concise set of definitions for the terms. Some ambiguities and errors may regrettably have found their way into the terminology. Corrections and other suggestions for change should be sent to:

Dr. Timothy J. Greene, Chair
ANSI Z94 Committee on Industrial Engineering Terminology
c/o Executive Director
Institute of Industrial Engineers
25 Technology Park
Norcross, GA 30092
Email: terminology@iienet.org

PURPOSE AND SCOPE

A major goal of the terminology is to standardize usage of industrial engineering terms while reflecting the diversity of current usage. It is expected that more consistency of use in terminology will be achieved as the industrial engineering profession becomes accustomed to this standard and that a profession which is so committed to standards will itself submit to standard terminology. The terminology herein presented is an attempt to capture the current best usage of the terms.

In order to meet these objectives, invitations to join the committee were issued by the Institute of Industrial Engineers (IIE), Secretariat for the ANSI Z94 Committee on Industrial Engineering Terminology, to a wide variety of groups including professional societies, trade groups, and labor organizations. Those that elected to participate by appointing representatives are listed below.

American National Standards Committee Z94
Robert L. Williams, Chair
Maura Reeves, Secretary
James A. Bontadelli, Vice Chair

Standards Committee

AFL/CIO
  John L. Zalusky
  AFL/CIO
  Washington, DC

AMERICAN ARBITRATION ASSOCIATION
  Harvey Gittler, P.E.
  Oberlin, Ohio

AMERICAN ASSOCIATION OF COST ENGINEERS
  Robert C. Creese, Ph.D.
  West Virginia University
  Morgantown, West Virginia

Kenneth K. Humphreys, P.E., C.C.E.
American Association of Cost Engineers
Morgantown, West Virginia

AMERICAN COMPENSATION ASSOCIATION
Marc J. Wallace, Ph.D., CCP
University of Kentucky
Lexington, Kentucky

AMERICAN PRODUCTION AND INVENTORY CONTROL SOCIETY
Timothy J. Greene, Ph.D.
Oklahoma State University
Stillwater, Oklahoma

AMERICAN SOCIETY OF SAFETY ENGINEERS
Stuart R. Mirowitz, CIH, CSP
Paramus, New Jersey

AMERICAN SOCIETY FOR QUALITY CONTROL
Harrison Wadsworth, Jr., Ph.D., P.E.
Georgia Institute of Technology
Atlanta, Georgia

INSTITUTE OF INDUSTRIAL ENGINEERS
James A. Bontadelli, Ph.D., P.E.
Vice Chair of the Z94 Committee
University of Tennessee
Knoxville, Tennessee

Maura Reeves
Secretary of the Committee
Formerly of Institute of Industrial Engineers
Norcross, Georgia

Robert L. Williams, Ph.D., P.E.
Chair of the Z94 Committee
Ohio University
Athens, Ohio

PROJECT MANAGEMENT INSTITUTE
Philip C. Nunn
Management Technologies, Inc.
Troy, Michigan

SOCIETY OF MANUFACTURING ENGINEERS
Joseph I. ElGomayel, Ph.D.
Purdue University
West Lafayette, Indiana

HISTORY OF THE Z94 COMMITTEE

The Z94 Project, *Industrial Engineering Terminology*, was originally organized in 1966 under the procedures of the American National Standards Institute (formerly known as American Standards Association and then U.S. Standards Institute). After a lengthy period of preliminary discussions, the project was organized as a joint venture between the American Society of Mechanical Engineers and the American Institute of Industrial Engineers (now IIE)—with Dr. William J. Jaffe, Professor (now Emeritus Distinguished Professor) of Industrial Engineering, Newark College of Engineering (now NJ Institute of Technology), as Chairman, and Professor Delmar W. Karger (then Dean and now Emeritus Ford Foundation Professor of Management, School of Management, Rensselaer Polytechnic Institute) as Vice Chairman. Dr. Jaffe represented ASME and Professor Karger AIIE.

The committee was reorganized under the sole secretariat of the Institute of Industrial Engineers in 1982 with Dr. Robert L. Williams serving as Chair. Invitations were issued to other organizations to join the committee in order to help in the revision and development of the terminology. Dr. James A. Bontadelli was named Vice Chair and Mr. Gregory Balestrero was appointed Secretary. In revising Z94-1972, the ASME and AIIE suggested that *Z94.0 Industrial Engineering Terminology* appear in a single volume instead of the individual ones of the earlier Standard. However, to retain the advantages of separating the disciplines—see below (*Definitions*)—it was decided that all of the twelve previously issued single volumes (*Z94.1: Biomechanics; Z94.2: Cost Engineering*, etc.) appear, in their revised forms as Sections 1, 2, etc., respectively, in the new edition. Hence, Sections 1 through 12 were the revisions of the 1972 sections. The section entitled Applied Psychology was retained but reduced in scope since it was judged to be unrealistically comprehensive. As a result, new sections were added: Human Factors, Wage and Salary Administration, Anthropometry, Safety and Occupational Health and Medicine.

In revising Z94-1983, the need for a new section emphasizing the emerging field of integrated manufacturing was recognized. A section entitled Manufacturing Systems was added for that purpose. The section on Applied Psychology was eliminated because of the extensive duplication of terms with the section on Wage and Salary Administration. Terms in the deleted section were incorporated into the other sections as appropriate. The titles of two sections were changed. Wage and Salary Administration was changed to Employee and Industrial Relations to better reflect the current practice. For the same reason, Data Processing and Systems Design was changed to Computer and Information Systems.

In revising Z94.1989, a new section on "Management" was added because of its long and close association with industrial engineering. In order to give emphasis to the areas, "Quality Assurance and Reliability" and "Analytical Techniques and Operations Research" grew out of the 1983 section entitled "Applied Mathematics." "Biomechanics" and "Anthropometry" were merged into one section as were "Safety" and "Occupational Health and Medicine."

Seventeen subcommittees were organized to revise the terminology. The chairs of these working subcommittees and their respective sections are listed below.

SUBCOMMITTEES

Anita L. Callahan, Ph.D. Co-chair University of South Florida	ORGANIZATION PLANNING AND THEORY
Joseph I. ElGomayel, Professor Purdue University	MATERIALS PROCESSING
Lloyd English, P.E. American Association of Cost Engineers	COST ENGINEERING AND PROJECT MANAGEMENT
Mitchell Fein, P.E., C.M.C. Mitchell Fein, Inc.	WORK DESIGN AND MEASUREMENT
Gerald A. Fleischer, Professor University of Southern California	ENGINEERING ECONOMY
Paul E. Givens, Ph.D. Co-chair University of South Florida	ORGANIZATION PLANNING AND THEORY
Timothy J. Greene, Ph.D. Oklahoma State University	OPERATIONS AND INVENTORY PLANNING AND CONTROL

M. Susan Hallbeck, Ph.D., P.E. University of Nebraska, Lincoln	ANTHROPOMETRY AND BIOMECHANICS
William E. Hammer, Jr., CSP, President Hammer Management Consulting	COMPUTER AND INFORMATION SYSTEMS
L. Ken Keys, Ph.D., Dean Cleveland State University	MANAGEMENT
D. L. Kimbler, Ph.D., P.E. Clemson University	MANUFACTURING SYSTEMS
Alex Kirlik, Ph.D. Georgia Institute of Technology	HUMAN FACTORS (ERGONOMICS) ENGINEERING
Thomas A. Lacksonen, Ph.D., P.E. University of Wisconsin, Stout	FACILITY PLANNING AND DESIGN
Steven A. Lavender, Ph.D., (Coordinator) Rush-Presbyterian-St. Luke's Medical Center	HUMAN FACTORS/ANTHROPOMETRY
Kenneth D. Lawrence New Jersey Institute of Technology	DISTRIBUTION AND MARKETING
Soter G. Liberty, P.E., President S.G. Liberty Associates, Inc.	EMPLOYEE AND INDUSTRIAL RELATIONS
Jean-Paul Prentice, CCE Resource Optimization, Inc.	COST ENGINEERING AND PROJECT MANAGEMENT
Michael R. Taafe, Ph.D. University of Minnesota	ANALYTICAL TECHNIQUES AND OPERATIONS RESEARCH
John Talty, M.S., P.E., DEE National Institute for Occupational Safety and Health	OCCUPATIONAL HEALTH AND SAFETY
Harrison Wadsworth, Jr., Ph.D. Georgia Institute of Technology	QUALITY ASSURANCE AND RELIABILITY

The Z94 Committee is deeply indebted to the pioneering work of earlier committees and their leadership. Special acknowledgment must be given for the leadership of Dr. William J. Jaffe and Professor Delmar W. Karger, the chair and vice-chair, respectively, of the Z94 Committee from its inception through 1982.

DEFINITIONS [1]

The standard definitions for each section were derived by the respective Subcommittee. Upon adoption by the Subcommittee, the definitions were next submitted to all other subcommittees of Z94 for examination and suggestions. Revisions were often made, and the definitions adopted by the respective subcommittees were approved and adopted by the Z94 Standards Committee. Although the subcommittee members were chosen for their expertise, it was obvious that some definitions cut across all fields of IE, despite any attempt at dividing the entire IE discipline. Thus, for example, *planning* could not be the sole possession of the Organization Planning and Theory Subcommittee, but could be found in other sections as well (*e.g.*, Production Planning and Control). Again, the term *control* was a matter of concern to the Applied Mathematics Subcommittee (*e.g.*, Statistical Quality Control) as well as the Cost Engineering

Subcommittee. It was natural that, over the years, many supposedly common terms were used differently among the various divisions of the field. This should not be surprising, since even in the quantitative aspects the letter "c" could refer to c-charts (plotting number of defects) and to sampling plans where it represented the acceptance number. What was really surprising was the fact that in relatively few instances were there any real disagreements. However, in some instances, the term had become so ingrained that it was necessary to give both definitions. Again, in some cases, the "application" of the term was allowed to stand along with the definition.

Again, there was naturally the question as to the exact place where a definition belonged. This was especially true in cases where more than one subcommittee used the term. This may not be as simple as it sounds, for it is not limited to aspects where the same term is used in an exact sense. Consider, for example, the term *industrial engineering*, which might properly belong in any or all sections. However, because of the historical—as well as the continuing—association with Motion and Time Study, the term is found in the Work Measurement and Methods section. In all instances, the reader should bear in mind that the fundamental definition belongs to the basic subcommittee, and, in the event that another is given, the less general application is limited to use in a particular aspect of IE (*e.g.*, the Distribution and Marketing Subcommittee's application of Linear Programming—an Applied Mathematics Subcommittee term—to a marketing problem). However, a cross-reference between Z94 volumes is noted.

It must be admitted that this problem of determining where a definition belonged could have been easily avoided by grouping all definitions together—with no consideration as to a subdivision of the entire IE field. As tempting as that may have been, in view of the possible boundary questions, the purpose of breaking up the IE field into specific subdivisions and the resulting separate sections was for the convenience of the practitioner. Hence, the breakdown into seventeen separate sections. It must be admitted that even this is not without its impending problems in view of the fact that IE terms cut across the fictitious boundaries. To overcome this, an overall index was prepared, and, it is hoped this will prove useful in finding definitions in any specific category.

Although the basic responsibilities for the definitions are those of the particular subcommittee, in particular, and the Z94 Committee, in general, it is likely that the reader, particularly the practitioner, for whom these definitions are of more than academic interest, will have suggestions as to omissions, revisions, and additions. Such suggestions are not only invited, but solicited with advance appreciation. It will be helpful if these discerning readers and practitioners will contact Dr. Timothy J. Greene, Z94 Chair, c/o Executive Director, Institute of Industrial Engineers, 25 Technology Park, Norcross, GA 30092. Email: terminology@iienet.org

THE INDUSTRIAL ENGINEERING FIELD

The task of determining the exact limits of the IE field was not an easy one, despite the Committee's acceptance of the official IIE definition:

> Industrial Engineering is concerned with the design, improvement, and installation of integrated systems of people, materials, information, equipment, and energy. It draws upon specialized knowledge and skill in the mathematical, physical, and social sciences, together with the principles and methods of engineering analysis and design, to specify, predict and evaluate the results obtained from such systems.

Official recognition is, of course, exceedingly helpful, but in spite of all admonitions of strict adherence and interpretation, the field is a living and dynamic one. Even more important, terms are used by people and are at their mercy.

There is, perhaps, as far as the industrial engineer and his/her predecessors are concerned, no better example of the changing field than the statement made by the pioneer, Hoxie [2], who is probably best remembered as the one who was among the first to define the relationship between scientific and organized labor, and who insisted that "…Time and motion study, therefore, must be regarded as the chief cornerstone." Although hardly anyone would deny this as being the historical pattern, some today would seriously question so basic a role ascribed to this aspect of the field. Even the terms themselves—time and motion study and scientific management—are seldom used by practitioners at present. In fact, these are terms that are foreign to many of today's students, who may be unaware of the historical roots of IE.

If a pragmatic beginning is sought for industrial engineering, it can be found in the pioneering work of Taylor in his attempt to answer the question: What is a fair day's work? Taylor was fond of quoting President Theodore Roosevelt who insisted that "The conservation of our national resources is only preliminary to the larger question of national efficiency." Almost a hundred years ago, Taylor noted:

We can see our forests vanishing, our water-powers going to waste, our soil being carried by floods into the sea; and the end of our coal and iron is in sight. But our larger wastes of human effort, which go on every day through such of our acts are as blundering, ill-directed, or inefficient…

Hence, Taylor looked for the competently, well-trained employee and employer. "The principal object of management," he asserted, "should be to secure the maximum prosperity for the employer, coupled with the maximum prosperity of each employee." Initiative and incentive of the worker was only half the problem, management was the other half. Hence, he developed four principles: develop a science for each element of the employee's work to replace the old rule of thumb method; scientifically select, train, teach, and develop the worker; heartily cooperate so that all work is accomplished in accordance with the principles of the science developed; an almost equal division of work and responsibility between management and workers.

Early in his career it was Taylor's need to discover "some rule or law which would enable a foreman to know in advance how much of any kind of heavy laboring work one who was well-suited to the job ought to do in a day: that is, to study the tiring effect of heavy labor on a first-class man." Research on earlier work, indicated that two types of experiments had been made. Physiologists studied the endurance of the human animal; engineers, determined to discover what part of a horse-power a man-power was, had men lifting loads by turning winches, lifting weights, etc. Workers were studied by Taylor, and data were translated into foot-pounds. Using the concept of the product of force and distance could obviously not yield a relationship between foot-pounds of work and the tiring effect on the worker. (Clearly, a worker, standing still but holding a weight in his hands, was, according to the physicists, not working.) Convinced that there was a relationship somewhere in his data, Taylor assigned the task to Carl Barth, who derived a law, confined to that class of work in which the limit of a worker's capacity is reached because of being tired. This was the law of heavy laboring where the worker's strength is exerted by pushing, pulling, grasping. The worker could be under load only part of the day. Thus when pig iron was handled (with a pig weighing 92 pounds), the worker could be under load only 43 percent of the day, and free of load for 53 percent. Handling half a pig, the worker could be under load 58 percent of the day. As the load lightens, the time under load increases, until the load is so light, the worker can carry the load the entire day. At this point the law of endurance ceases to operate, and some other law must be found to indicate a worker's capacity for work.

Taylor, influenced by his early teachers—especially the celebrated American mathematics text writer, Wentworth—relied on the stopwatch. Hence, the pioneering work took place in what became known as *time study*. The Gilbreths, insisting on emphasizing the *method* of work, set the goal as *the one best way*; this became known as *motion study*. This schism has, over the years, subsided—but *motion and time study* marks the beginning of industrial engineering as we know it. It is altogether possible that the educators themselves—few, if any, of whom identify their respective departments as teaching scientific management—have added to the confusion of all, including students. Educators are prone to identify their IE departments as teaching solely industrial engineering. In fact, they often couple the IE with numerous other connotations. Thus, there are departments of Industrial Engineering and…Operations Research, Management Science, Management Engineering, Systems Engineering, Administrative Engineering, Engineering Management, to mention a few. To complicate matters further, they also teach, in addition to courses with the aforementioned titles, those in Industrial Management, Production Management, etc.

The IE educators are not alone in this endeavor to search for more encompassing appellations. The practitioners, themselves, have also engaged in this practice. Many, schooled in IE techniques, have identified themselves by titles quite similar to those that the educators have adopted: thus, they use the names Management Consultant, Management Engineer, Systems Engineer, and so on. Many believe, simply, that the adoption of the term Engineer, of course, should be limited to those licensed by the respective state boards. Many of these, in fact, simply use the term Professional Engineer, without further definition of their specialties. Naturally, their respective societies have followed suit, and these practitioners are found in divisions of the "founding" engineering societies, as well as in the "management," the "operations research," the "system" societies, in addition to the basic society for the IEs. In all of these societies, there has been, from time to time, strong movements for name changes. The reasons, like those in the educational institutions, have been manifold: the field is dynamic and not static, the splinter societies are attracting a great number of members merely by stressing these new fields, the name no longer identifies what the practitioner does, etc.

There is a sufficient history for this—in other fields as well. Probably the best example is the oldest of the engineering societies' fields—Civil Engineering. In many of the American schools of engineering, the "classical" title Civil Engineering Department is no more. If the name has not been changed completely, it has taken on companion titles—Sanitary Engineering, Urban Planning, Environmental Engineering, and others. Again, this is understandable, for the history of engineering shows a pattern of fragmentation, consolidation, and then further fragmentation and consolidation. Thus, in the United States, the American Institute of Civil Engineers, owes its foundation (1852) to those who felt that engineering had applications other than those with which the military was concerned. Moreover, this great interest and concern for nonmilitary (or civil) engineering was not limited to this country. The British Institution of Civil Engineers adopted quite early the famous definition of Thomas Tredgold (1882) and included it on its membership certificates [3]:

> A society for the general advancement of mechanical science and more particularly for promoting the acquisition of that species of knowledge which constitutes the profession of civil engineer, being the art of directing the great sources of power in nature for the use and convenience of man.

Hence, what began as a seeming fragmentation resulted in an all-encompassing objective.

Yet this was followed by more fragmentation. The engineers with specific interest in mining, for example, broke away and formed their own society which, incidentally, followed the fragmentation by a consolidation with the metallurgists to establish the American Institute of Mining and Metallurgical Engineers. The nonmilitary group soon experienced further fragmentation when the engineer with interests in machinery formed The American Society of Mechanical Engineers. Soon the electrical machine group broke away to form the American Institute of Electrical Engineers—only to experience similar fragmentation and consolidation with the radio engineers and the electronics group. Hence, the breaking away of the group of engineers and others who rallied around Frederick W. Taylor—himself a former ASME president—repeated the cycle that had become so common in engineering. In the first place, the "break" was never clear-cut. Although a few left ASME to join the Taylor Society, many retained dual membership—a practice that was continued when the Taylor Society combined with the Society of Industrial Engineers to form the Society for (the) Advancement of Management. In fact, when the AIIE came on the scene in 1948, many maintained triple membership.

This process continued as more and more specialized interest groups appeared—Quality Control, Operations Research, Management Science, Systems, etc. The IE practitioner, like the IE educator, often views these newer groups with forebodings. Unsure that his discipline has taken sufficient cognizance of this development, and fearful that many, in pursuit of this newer field, will forsake the older, he seeks to identify the old with the new. However, this search for the "best of all possible worlds" is not without its dangers, as well as its confusion.

Consider the use of the "systems" term. A "system," to quote Johnson, Kast, and Rosenzweig, authors of the *Theory and Management of Systems* [4], is an "organized or complex whole; an assemblage or combination of things or parts forming a complex of unitary whole." Thus, the data processing engineer, the electronics engineer, as well as the geographer, the clergyman, the anatomist, to mention but a few—can lay claim to data processing systems, communications systems, transportation systems, religious systems, anatomical systems, etc. The same is true of the industrial engineer. The quoted authors insist that, in input-output language, managers are essentially the agents who convert the inputs of disorganized resources of man (now identified as people), money and machinery, into outputs of useful goods and services. Thus management is decision systems, communication systems, control systems, etc.

This is not something new. In fact, it is almost impossible to discuss IE without reference to its historical roots, *scientific management* and Taylor. Here, too, matters seem a bit hazy, for even in reference to management itself, a babel of terms has persisted and continue to plague us. Even among English-speaking countries, there is confusion about the meaning of the term and its use.

Although the articulate may carefully distinguish among *management, rationalization, rationalisation, organization, organisation, administration,* the terms seem to be used almost interchangeably [5]. Although management has Latin roots (from Italian *maneggiare* = to handle horses; from Latin *manus* = hand), none of the other languages especially the Romance languages—use any form of the term. They advocate: *organisation scientifique* (thus, for example, *Comite International de l'Organisation Scientifique*), organizacao racional de trabalho, racionalizacion del travajo, organizzazione del lavore, Wirtschaflichkeit, Betriebslehre, Betriebsfuhrung, etc. Thus, the noun in the title of Fayol's classic, *Administration Industrielle et Generale,* was first—translated as *administration* by Conbrough (1929) and later *management* by Storr (1949). As a matter of fact, Urwick, not only noted the correctness of the Storr translation, but he also noted, unhappily, that the term management is popularly associated with definitions connotating cajolery and trickery on the one hand, but used more authoritatively, as an activity, a body of knowledge, a group of persons, etc. Just as there were problems of translating the title of Fayol's work into English, there were problems in the reverse process—viz., translating Taylor's title into French by Le Chatelier, who used the term *l'organisation scientifique du travail,* which essentially is the French equivalent for a labor union. As for the German term, it is not a happy choice, for *Betriebsfuhrung* is related to the German verb treiben—to drive.

The direct association of management with engineering—particularly in the United States—came early in the history of ASME, and when Taylor spoke of management—particularly shop management—the concern was a matter that had occupied the attention of the earlier ASME presidents. In fact, R.H. Thurston [6] the first ASME president, in his inaugural address, included social economy among the "Objects of the Society." Not only did H.R. Towne, in that "bench mark of scientific management," *The Engineer as Economist,* [7] include "shop management" as a field of interest, but, in an ASME session, substituting for President Coleman Sellars, [8] he suggested "industrial or economic questions" because ASME had, among its membership, "more than the Civil Engineers or the Mining Engineers men who are managers of labor, who are either owners or representatives of owners, and who, therefore, control capital." And, it is of special interest to note that (1920) the ASME council approved and adopted the *First Report of the Committee of Aims and Organizations* [8] (L.C. Marburg, Chairman), which included the statement: …that Industrial Engineering is a major subject for consideration by the Society and shall be placed on par with all major subjects. In his *Cost and Production Handbook*, L.P. Alford [9], codifier of management laws and principles, included under the all-inclusive term Industrial Engineering, the "laws" of Taylor, Fayol, and Alford, himself. Little wonder, then, that ASME in 1955, as an outgrowth of its Management Division's Work Standardization Committee (1948), sponsored an *IE Terminology* [10], which itself, has had its influence on this present *Terminology*.

Although it must be noted that the ASME never did establish an IE division, its IE partisans were mainly housed in its Management Division, again indicating some sort of concomitance between IE and Management. Nor was this the only instance. As already noted [11], some ASME members joined with some non-ASME members to form a group—at first informal, but later formal—known as the Society to Promote the Science of Management, which, after Taylor's death in 1915, became known as the Taylor Society. This society, in 1936, merged with the Society of Industrial Engineers—a group which, in 1917, probably as a measure for helping the war effort, was formed from the Western Efficiency Society (1910—to establish the Society for (the) Advancement of Management (SAM). SAM attracted a wide spectrum of members—some with ASME's Management Division, some with the National Association of Cost Accountants, some with the National Personnel Association, some unattached—to form an overall umbrella for most of the management practitioners. SAM's interests can be readily perceived from its *Glossary of Terms Used in Methods, Time Study and Wage Incentives* [12]. Among its other interests were problems relating to budgets and costing in production and distribution, performance in factories, offices, and stores, as well as operations in both profit and nonprofit institutions. In addition there were numerous other organizations specializing in all of these plus many more aspects of IE. Finally, in 1948, the American Institute of Industrial Engineers (AIIE) was established: its objectives were many and compelling, but perhaps the most important was the professional recognition of IE. AIIE published a terminology in 1965 to document terms in common usage [13]. This organization is now known as the Institute of Industrial Engineers (IIE).

The growth of AIIE has tended to emphasize the engineering aspects of the field, and, at the same time, has caused a continual re-examination and re-appraisal of the old concept of equating *IE with Management*. To date, no clear cut definitions have been recognized as far as the lay public is concerned, even though specific moves have been made by the professionals in identifying the field and acquainting the public with such notions. However, in many instances there have been concerted efforts to broaden the limits so as to encompass as much territory as possible. This has been especially true as new concepts and ideas have come into being: for example, operations research, bioengineering, systems engineering, to mention but a few. Many practitioners have found that such acquisition can be accomplished by considering themselves not only as adherents to an IE practice, but as advisers to management, in general. Many of the professional societies, fearful of either an exodus of members from their societies, or a failure to attract the new adherents who might seek independent societies devoted entirely to these "new frontiers" have attempted to solve the dilemma by establishing divisions in which specific interests may be pursued. Some societies have, in fact, considered name changes. This concern about territorial limits as well as the accompanying titles has long been a problem of the colleges which, of course, are among the first to sense the rise of the new branches.

Probably, of all the colleges, the most concerned have been the engineering colleges, in general, and the IE departments, in particular. Although accreditation is a major factor in all recognized colleges, it is particularly acute in the engineering college because of its relation to professional registration, a matter which is less critical in the liberal arts and the business colleges. In fact, a whole system of "common law" definitions have come about with *management* (without an identifying adjective) as the property of the business school, and *management* (accompanied by such terms as *engineering* and/or *industrial engineering*) as the concern of the engineering school, and *management* (with the single adjective *industrial*) as the subject matter of the technological institute. There is, of course, no hard and fast rule, but "IE" has been taken as belonging solely to the engineering school. However, this, despite the IIE definition, did not solve the boundary problem.

Historically [14], the first IE department was established at The Pennsylvania State—then, College, but now University, with Hugo Diemer at the head. He had been recommended by Taylor, who, in turn, had been asked by Beaver, a former Governor of Pennsylvania and President of the Penn State Board of Trustees, to suggest a "man who could teach mechanical engineering from the standpoint of production rather than machine performance" (1907). It is interesting to note that Diemer had, since 1905, been teaching courses in *shop management*, at the University of Kansas [15]. Individual courses were offered by Professor Diemer at the University of Kansas in 1901, by Professor Dexter Kimball at Cornell University in 1904 and by Professor Walter Rautenstrauch at Columbia in 1907 [16]. The University of Pennsylvania, on the other hand, was pioneering with its Wharton School, which had been established (1881) on the recommendation of James Wharton, in the belief that "management should be separated from ownership and developed as a profession." Because of the "scientific management" tradition associated with Taylor, it was clear that the engineering school's program included *management* courses. In fact, the management term appeared frequently in describing the courses and the departments, because many shied away from the IE term since it was associated by some with "efficiency" practitioners. And so, for a while, many departments named themselves with such titles as Administrative Engineering, Engineering Administration, Industrial Administration, Manufacturing Engineering, Manufacturing Management, Management Engineering, Commercial Engineering, etc. However, a basic group retained the IE appellation, which, eventually with accreditation programs, became the dominating title, especially with the establishment of AIIE and its promotion of a professional status.

Despite the title, management courses formed a most important part of the IE curriculum, and these attracted many students less technically oriented but anxious for positions in industry with high status and high wages. AIIE's attempt to solidify the professional engineering aspects, provided an external impetus, while the development of courses in mathematical and statistical applications served as an internal catalyst in strengthening the IE program. Soon the titles *Operations Research, Management Science*, and *Systems Engineering*, were added to the IE name.

These supplementary titles—used by the educators and the practitioners alike—have certainly not clarified matters, and both students and non-students are not too clear about the domain of IE. The defense, if any is needed, is quite simple: The IE field is broad and can and does encompass all such disciplines. Moreover, as each new vista appears, it attracts many, and so the consolidation—fragmentation cycle, noted above, begins all over again. Thus, in the World War II period, it was the application of statistical methods (especially in the field of quality and inspection), next came the more sophisticated mathematical and team approach (with operations research and management science), then the systems concept. In all probability, the next phase will be the application of biology and the life sciences.

IE, MANAGEMENT, ETC.

Through all these metamorphoses and connotations, one aspect seems to have prevailed from the beginning. This is *management*, for IE begins, as we know it, with Taylor—"the Father of Scientific Management." All of the pioneers were, in one way or another, concerned with it: Church (*The Science and Practice of Management*), Alford (*Laws of Management Applied to Manufacturing*), Cooke (*Spirit and Social Significance of Scientific Management*), Diemer (*Factory Organization and Administration*), the Gilbreths (*Primer of Scientific Management, The Psychology of Management*, etc.), Gantt (*Industrial Leadership*, etc.), Kimball (*Plant Management*), et al.

However, it must be admitted that no single definition of management has been accepted. The famous ASME Report of 1912 [17] dealt with the "new element," but did not offer any specific definition of *management*, scientific or otherwise. In fact, the very term *scientific management* was seemingly concocted by a distinguished group—Gantt, Gilbreth, Emerson, and others—to aid Brandeis in the celebrated Eastern Rate Case Hearings (1910), even though Taylor had used the term in a limited sense earlier (1903). The adjective was probably meant to indicate that management could be studied and practiced deliberately and systematically, and that it was not the result of capricious, rule-of-thumb, seat-of-pants, meanderings. However, the term (*scientific management*) has had limited use during approximately the last quarter of a century. On the other hand, the term, *management*, has been used extensively both as a noun and as an adjective especially in all fields associated with IE.

Numerous definitions of management have appeared, and even the briefest examination of a few of these will show the wide diversity the matter has aroused even among the professionals. [18] Thus,

Management is a broad term and covers almost all factors of an enterprise. [19]

Management proper is the function in industry concerned in the execution of policy, within the limits set up by administration and the employment of the organization for the particular objectives set before it. [20]

Management is the science of applied human effort…[21]

Management is involved in the avoidance of waste in any human effort. [22]

Management—the development of people, not the direction of things. [23]

Management is the art and science of organizing, preparing and directing the human effort applied to control the forces and utilize the materials of nature for the benefit of mankind. [24]

Management is the art of getting work done. [25]

To manage is to forecast and plan, to organize, to command, to coordinate and control. [26]

Of course, there are others, but the definition given in this Z94 Standard (see Z94.9 Organization Planning and Theory) is:

Management: (1) The process of utilizing material and human resources to accomplish designated objectives. It involves the activities of planning, organizing, directing, coordinating, and controlling. (2) That group of people who perform the functions described in (1), above.

This definition is basic. Certainly, it can be restated in many other ways, but retaining these fundamental concepts. Thus, it can be seen to be the process by which the execution of a given purpose is put into operation and supervised. Or, again, in another way, management may be defined as the process by means of which the purposes and objectives of a particular human group are determined, clarified, and effectuated.

Thus, Professor D.W. Karger can paraphrase his School of Management (Rensselaer Polytechnic Institute) Long Range Planning Document to say:

> Management refers to the systematic organization, allocation, and application of economic and human resources to bring about a controlled change; the role of management in these proceedings is one of organization and coordination of human effort in exploiting resources for the improvement of the organization involved.

With all these diverse "definitions" among the experts, it is not surprising to find that *management* takes on a wide variety of meanings among the laymen (including *cajolery, trickery,* and *totalitarianism*). However, even among the "experts" two other words—*administration* and *organization* have often been used interchangeably with it, especially administration. The interchangeable use, among many, of *management* and *administration* has been noted in this Z94 Standard, which also calls attention to a distinction (concerned with the goal-establishing, policy-formulation part of management). This concept is consistent with Sheldon and Spriegel, who point out some very interesting aspects of the two terms. According to Spriegel [27], business circles consider *management* the more inclusive term, while governmental agencies consider *administration* the more inclusive. To the business world, *administration* is usually equated with that part of management usually described as top management. Sheldon [18] puts it quite succinctly in considering *administration* as "the function in industry concerned with the determination of corporate policy."

Whether it is because, as Dionysius has said, "education is contact with manners," or, as Shakespeare has insisted, namely "that which we call a rose, by any other name would smell as sweet," the engineering college, in general, and the IE Department in particular, have adopted a great variety of names—not only to describe their courses, but to describe the IE Department itself. Others, of course, shy away from the IE connotation completely, and choose such names as a "Department of Administrative Engineering, Department of Engineering Administration," etc. Theoretically, *industrial management*, a term that had for years been used to describe some of the Industrial Engineering Department courses concerned with application of management to industrial problems, has been exiled to a strange limbo—a field that combines *management with business*, and "outside" the "true" engineering curriculum. It is paradoxical that it has been adopted by some of our foremost technological institutes—with great success. More often than not, the departure of *industrial management* from current IE curricula arises from a fear of a misunderstanding by engineering accreditation groups, as well as an inhibition on the part of purists who sense some contamination. On the other hand, in full realization of the need of such knowledge on the part of its students, quite frequently, the IE Department has adopted the material under some other name—probably being careful to add the term *engineering* to the title: *e.g., Management Engineering, Administrative Engineering.*

Because of the heavy mathematical orientation, *Operations Research (OR)* and *Management Science (MS)* have been quickly embraced as titles by the IE Department. In fact, so enthusiastic has been the adoption that many have given these equal billing with IE. (This has not been without problems, for the broad base of OR, for example, has caused it, in at least one great American engineering college, to move the subject from the IE Department to the Civil Engineering Department. In another, it was moved to the School of Commerce.) The breadth of the OR field allowed the IE Department to assume this discipline among its courses, while, on the other hand, OR provided a potent weapon for attacking IE problems.

The strength of OR can be readily observed from the definition given by Churchman, Ackoff, and Arnoff in their pioneering text [28]:

> Operations research in the most general sense can be characterized as the application of scientific methods, techniques, and tools to problems involving the operation of systems so as to provide those in control of the operations with the optimum solution of the problem.

With heavy emphasis on the multi-disciplined team and with a heavy orientation to a mathematical and statistical approach, OR, originally applied to military matters (hence its name), soon was useful in civilian matters. And the industrial engineer especially became interested in its application to management problems, particularly when they could be studied by mathematical models with optimization as a prime objective. Weinwurm [29] and others felt the "need for a thorough and detailed analysis of the decision making process…as a cause for the development of a management science"—with two approaches possible. One approach was "rationalistic with emphasis on building models," while the other insisted that the human element could not be neglected and took special cognizance of the fact that decisions were made by human beings—with the consequent result of a great need for the incorporation of "behavioristic and organization concepts." An integrated theory of organization soon took place (Simon) [30].

Thus, *IE*, whose relationship with *Management* continued, could not possibly overlook the development of OR and MS. And, hence, its association with these fields developed. But, this IE-Management association did not stop here. In more recent years management has come to be viewed in terms of *systems*. A *system* can be defined as…an organized or complex whole; an assemblage or combination of things or parts forming a complex of unitary whole [4]. The organization is, in this sense, a system wherein certain inputs are transformed to certain outputs—*i.e.*, Management is a system that converts men, money, materials and machines into goods and services. Moreover, management can thus be viewed as control systems, decision systems, and communications systems—domains claimed by managers and their IE advisers.

Little wonder, then, that IE, whose official definition includes such concepts as *systems, mathematical, physical,* and *social sciences, engineering analysis and design*, can take on, along with *management*, a term which has been with IE since its beginning, such appellations as *OR, MS* and *Systems*. Admittedly, the IE field is broad—exceedingly broad—and further and further horizons may be expected, and the consolidation-fragmentation cycle will, in all probability, continue. It is hopeful that this *Z94.0 Industrial Engineering Terminology* will—like the others that preceded and are bound to follow it—become a chart for delineating and defining the field.

REFERENCES AND NOTES

[1] The following sections (Definitions; The Industrial Engineering Field; IE, Management, etc.), were written by Dr. William J. Jaffe for earlier editions of the terminology and have been changed only in minor ways.

[2] R. R. Hoxie. Scientific management and labor welfare. *Journal of Political Economy*, XXIV, 1916, p.838. (C.F. Taylor said many years ago: "Time study is by far the most important element in scientific management." q.v. by A. H. Mogensen. *Common Sense Applied to Motion and Time Study*. New York: McGraw-Hill, 1932, p. 7.)

[3] A. C. Humphreys. *Lecture Notes on the Business Features of Engineering Practice*, 3rd ed. Hoboken, NJ: Stevens Institute of Technology, 1916, p. 7.

[4] R. A. Johnson, F. E. Kast, and J. E. Rosenzweig. *The Theory and Management of Systems*. New York: McGraw-Hill, 1963, p.4.

[5] L. M. Gilbreth and W. J. Jaffe. Management's past—a guide to its future. *Fifty Years Progress in Management 1910-1960*. ASME, 1960, p. 6.

[6] R. H. Thurston. President's inaugural address. *Transactions ASME*, 1, 1880, pp. 14-29.

[7] H. R. Towne. The engineer as economist. *Transactions ASME*, VII, 1886, pp. 428-432.

[8] W. J. Jaffe, L. P. Alford. *The Evolution of Modern Industrial Management*. New York University Press, 1957, pp. 29- 34.

[9] L. P. Alford (editor). *Cost and Production Handbook*. New York: Ronald, 1934.

[10] *Industrial Engineering Terminology*, ASME Standard 106, 1955.

[11] W. J. Jaffe. op. cit 111-117.

[12] *Glossary of Terms Used in Methods, Time Study and Wage Incentives*, Society for the Advancement of Management, 1952.

[13] R. L. Williams, et al. Industrial engineering terminology manual, *Journal of Industrial Engineering*, AIIE, XVI, No. 6, Nov-Dec 1965.

[14] W. J. Jaffe. op. cit 286-288.

[15] C. W. Lytle. Collegiate course for management. *Professional Society for Promotion of Engineering Education*, XXXIX, 1936, pp. 806-839.

[16] H. P. Emerson, D. C. E. Nachring. *Origins of Industrial Engineering, The Early Years of a Profession*. Norcross, GA: Institute of Industrial Engineers, 1988.

[17] The present state of the art of industrial management. ASME's 1912 Report, *Fifty Years Progress in Management 1910-1960*, pp. 293-323.

[18] E. H. Anderson and G. T. Schwenning. *The Science of Organization Production*. New York: Wiley, 1938, pp. 9-29.

[19] R. H. Lansburgh. *Industrial Management*, (2nd ed.) New York: Wiley, 1920, p. 4.

[20] O. Sheldon. *The Philosophy of Management*. London: Pitman, 1923, pp. 101 ff.

[21] A. H. Church. *The Science and Practice of Management*. New York: Engineering Magazine Co., 1914, p. 282.
[22] S. Webb. *The Works Manager Today*. New York: Longmans, Green, 1918, p. 5.
[23] B. W. Niebel. Letter to author.
[24] ASME Management Division 1921.
[25] L. P. Alford. Scientific industrial management. Paper No. 341, *World Engineering Congress*, Management Section, Tokyo, Oct. 30, 1929.
[26] H. Fayol. *General and Industrial Management, Storrs Translation*. London: Pitman, 1949, pp. 5-6.
[27] W. P. Spriegel. *Industrial Management*, (5th ed.) New York: Wiley, 1955, pp. 1, 8-10.
[28] C. W. Churchman, R. L. Ackoff, and E. L. Arnoff. *Introduction to Operations Research*. New York: Wiley, 1957, Chapter 1.
[29] E. H. Weinwurm. *Management Science. Fifty Years Progress in Management, 1910-1960*. ASME 1960, pp. 90-95.
[30] H. A. Simon. *Administrative Behavior*. New York: Macmillan, 147.

HOW TO USE THIS TERMINOLOGY BOOK

1. The terminology consists of seventeen sections: (1) Analytical Techniques and Operations Research, (2) Anthropometry and Biomechanics, (3) Computer and Information Systems, (4) Cost Engineering and Project Management, (5) Distribution and Marketing, (6) Employee and Industrial Relations, (7) Engineering Economy, (8) Facility Planning and Design, (9) Human Factors Engineering, (10) Management, (11) Manufacturing Systems, (12) Materials Processing, (13) Occupational Health and Safety, (14) Operations and Inventory Planning and Control, (15) Organization Planning and Theory, (16) Quality Assurance and Reliability, and (17) Work Design and Measurement. All terms in each of the respective sections are listed alphabetically. (However, if the term begins with a symbol, *e.g.*, λ-Criterion—the key word will be "lambda"—look under the word the symbol represents.)

2. In addition to the seventeen sections, there is an overall index—containing references to all the terms in each section. Hence, if you are not certain in which section you can find the term you seek, consult the overall index. The index is an alphabetical listing of every term from each section followed by the page number(s) where the definitions(s) can be located.

3. Although most terms owe their definitions to the writing of experts on the subcommittees and/or to conferences among the various Z94.0 subcommittees, some definitions come from accepted sources (books, societies, etc.). Hence, a Bibliography appears at the end of affected sections. In some sections—*e.g.*, Section 5 Distribution and Marketing—many terms are followed by a code number that refers to a specific source listed under the heading "Bibliography" or "References."

Z94.1 Analytical Techniques & Operations Research

Defined in this section are a set of terms that are in current use in the practice of teaching Operations Research. As with any such set of terms, the list is not exhaustive. However, we believe that the terms defined are characteristic of current practice, and span a significant space of modern Operations Research.

The previous edition, (1989) Industrial Engineering Terminology, is the basis for most of the definitions of this section. Many thanks to the 1989 committee for their work. A few new terms have been added for this edition and several terms had their definitions adjusted to reflect current practice or understanding.

Michael R. Taaffe,
Department of Operations and
Management Science
Graduate Faculty of Industrial Engineering
Graduate Faculty of Scientific Computation

Absolute Maximum (Minimum). The function f(x) defined over a set D is said to take on its absolute maximum (minimum) over D at a point x* in D if f(x) ≤ f(x*), (f(x) ≥ f(x*)), for every point x in D.

Accelerated Test. A test in which the applied stress level is chosen to exceed that stated in the reference conditions in order to shorten the time required to observe the stress response of the item, or magnify the response in a given time. To be valid, an accelerated test must not alter the basic modes and/or mechanisms of failure. [20]

Acceleration Factor. The ratio between the times necessary to obtain a stated proportion of failures for two different sets of stress conditions involving the same failure modes and/or mechanisms. [20]

Acceptance Sampling. Sampling inspection in which decisions are made to accept or not-accept product or service; also, the methodology that deals with the procedures by which decisions to accept or not-accept are based on the results of the inspection of samples. *Note:* In lot-by-lot sampling, acceptance and nonacceptance relate to individual lot. In continuous sampling, acceptance and non-acceptance relate to individual units, or to blocks of consecutive units, depending on the stated procedure.

Access, Random. When units in the waiting line of a queue are chosen for service by a random choice rule, the procedure is one of random access to service—or service by random access.

Accounting Prices. Internal prices assigned to company-owned resources as an aid to decision-making and control. In certain linear programming applications, the dual variables at optimality may be regarded as including a set of accounting prices for the various scarce resources. Accounting prices in the linear-programming format are called shadow prices (q.v.). [12]

Accuracy. Closeness of agreement between an observed value or test result and an accepted reference value. *Note:* The term accuracy, when applied to a set of observed values will be a combination of random components and a common systematic error or bias component.

Activity. (1) A decision variable whose level is to be computed in a programming problem. (2) A job in a network of jobs such as in a PERT network. [19]

Activity Analysis Problem. A linear programming problem whose constraints represent resource use and variables represent commodities to be manufactured. The i-th coefficient of an activity vector "j" represents the number of units of resources "i" required to produce a unit of commodity "j." The right-hand side coefficients indicate the amount of each resource available, while the objective function coefficients represent the profit per unit of activity. [15]

Activity Level. The value taken by a decision variable in an intermediate or final solution to a programming problem. [19]

Activity Redundancy. (See REDUNDANCY, ACTIVE.) [20]

Activity Vector. A column of the constraint matrix (q.v.) associated with a decision variable in a programming problem.

Adjacent Extreme Point Methods. Two extreme points of a convex set (q.v.) lying on the same boundary edge are called adjacent. All methods, such as the simplex method (q.v.) which move from one extreme point (q.v.) to an adjacent one are called extreme point methods. [9]

Admissible Basis. A set of m linearly independent activity vectors associated with a basic feasible solution (q.v.). (See BASIS)(BASIS SOLUTION.) [15]

Algorithm. An inductive and iterative mathematical technique for developing numerical solutions to certain classes of problems.

Alternate Optima. Distinct solutions to the same optimization problem. [9]

Analysis of Variance. A technique which subdivides the total variation of a set of data into meaningful component parts associated with specific sources of variation for the purpose of testing some hypothesis on the parameters of the model or estimating variance components.

Angular Matrix. A block-triangular matrix in which all blocks are zero except those down the diagonal and along the bottom row. That is, it is a matrix of the form

$$\begin{bmatrix} A_{11} & & & \\ & A_{22} & & \\ & & . & \\ A_{m1} & A_{m2} & . & A_{mm} \end{bmatrix}$$

where the Aij are submatrices (or "blocks") and all omitted entries are matrices consisting entirely of zeroes. Linear- programming problems with angular matrices lend themselves to special types of computational treatment, and in particular to that of decomposition. [12]

ARC. In a connected network with more than two nodes, an arc is a line connecting two nodes. It is often designated by an ordered pair (n_i, n_j), where n_i and n are the nodes of the network at the ends of the line.

ARITHMETIC AVERAGE. The sum of a set of sample values divided by the number of values in the set.

ARITHMETIC MEAN. The sum of a set of population values divided by the number of values in the set.

ARRIVAL RATE DISTRIBUTION, CONSTANT. The term constant applied to rates and times implies that a given rate or given time varies throughout its history under study in accordance with one probability density function. [12]

ARRIVAL RATE, MEAN. In a waiting line, is the expected number of arrivals occurring in a specified time unit.

ARRIVAL TIME, CONSTANT. (See arrival rate distribution, constant) [12]

ARTIFICIAL BASIS. A set of artificial vectors which may be used to initiate the simplex method (q.v.). [15]

ARTIFICIAL VARIABLES. Auxiliary variables introduced into the equality constraints, at the start of the simplex method of linear programming in order to provide an identity basis required to initiate the procedure. [12]

ARTIFICIAL VECTOR. The column vector associated with an artificial variable. It is a unit vector, with + 1 in the row where the artificial variable occurs and zero elsewhere. [12]

ASSEMBLY LINE BALANCING PROBLEM. An assembly line consists of a number of work stations. To assemble the product under consideration, a number of jobs must be performed subject to certain sequencing requirements concerning the order in which they are performed. Given the desired production rate of the product, management wishes to accomplish two objectives: assign jobs to balance the work among stations and minimize the total number of these stations. The times required to do each job are specified. The problem can be formulated as integer linear programming (q.v.). [18]

ASSIGNMENT PROBLEM. The problem of placing n elements into n cells, one element to a cell, at minimum cost, where the individual cost of putting element i in cell j is a given constant c_{ij}. The problem can be stated as a linear programming problem or a transportation problem. The problem is usually interpreted as one of assigning n persons to n jobs, such that the value of person i in job j is c_{ij}. The problem is then to determine the assignment with maximum total value. [19]

ATTRIBUTES INSPECTION. A term used to designate a method of inspection whereby units of product are examined to determine for each unit whether it does or does not conform to a requirement. Example: Go and Not- Go gaging of a dimension.
Note: inspection by attributes may be either one of two kinds: inspection whereby either the unit of product is classified simply as conforming or nonconforming, or the number of nonconformities in the units of product is counted, with respect to a given requirement or set of requirements.

AUGMENTED MATRIX. The coefficient matrix augmented by the column of right-hand-side constants; the same as the constraint matrix (q.v.). [19]

AVAILABILITY. The ability of an item to perform its designated function when required for use.

AVAILABILITY (ACHIEVED). The probability that an item when used and maintained under stated conditions in an ideal support environment will be in a satisfactory state at any given time. [20]

It may be expressed as:

$$A_a(t) = \frac{MTBM}{MTBM + \overline{M}}$$

where

MTBM = mean time between maintenance—is a function of both preventive and corrective maintenance requirements

M = mean active maintenance down time—a value which is a function of the repair times associated with corrective and preventive maintenance.

AVAILABILITY (INTRINSIC). The probability that an item is in satisfactory state at a stated instant of time (t) when it is used and maintained under stated conditions. It may be expressed as:

$$A_0(t) = \frac{MTBM}{MTBM + MDT}$$

MTBF = mean time between failures

MTTR = mean time to repair (excludes preventive maintenance downtime, supply downtime, and administrative downtime). [20]

Availability (Operation). The probability that an item is in a satisfactory state at a stated instant of time (t) when it is used and maintained under stated conditions in actual support environment. It may be expressed as:

$$A_0(t) = \frac{MTBM}{MTBM + MDT}$$

where

MTBM = mean time between maintenance

MDT = mean downtime (includes preventive maintenance, supply and administrative downtime). [2]

Average. (See ARITMETIC AVERAGE.)

Basic Feasible Solution. A basic solution to the m equations of the linear programming problem which also satisfies the non-negativity constraints. [19]

Basic Solution. Given a system of m simultaneous linear equation in "n" unknowns, $Ax = b(m < n)$ and with rank m. If an m x m nonsingular matrix is chosen from A, and if all the n-m variables associated with the remaining columns of this matrix are set equal to zero, the solution to the resulting system of equations is called a basic solution. [17]

Basic Variables. The m variables not initially set equal to zero in a basic solution (q.v.). [19]

Basis. A basis in e-dimensional Euclidean space is a set of n linearly independent vectors. In linear programming, a basis is an (m x m) nonsingular matrix composed of activity vectors, i.e., a basis matrix. [15]

Basis Inverse. The inverse of the basis matrix. [19]

Basis Matrix. (See BASIS) [19]

Basis Vector. A column of a basis matrix. [19]

Bias. A systematic error that contributes to the difference between a population mean of the measurements or test results and an accepted reference or true value.

Bidding, Competitive. Each firm seeking a particular contract must submit a sealed price and the firm which submits the lowest price is awarded the contract.

Birth-and-Death Process. A stochastic process which attempts to describe the growth and decay of a population the members of which may die or generate new individuals. The types mainly studied have relatively simple laws of reproduction and mortality. [22:29, 5:91, 13:402, 25:265]

Birth Process. A stochastic process describing the population of a system in which individual members may generate new members. The expression is often confined to the case where the variate (population) increases only by jumps of amount + 1, the probability of a jump from n to n + 1 in time dt being asymptotically $\lambda_n dt$ Here λ_n may also depend on t. (See also BRANCHING PROCESS)(POISON PROCESS) [22:29, 5:84, 13:402, 13:267]

Block Box. Describes a system (organism or mechanism) whose structure is unknown either because it cannot be observed or it is too complex to be understood.

Blemish. An imperfection that occurs with a severity sufficient to cause awareness but that should not cause any real impairment with respect to intended normal, or reasonably foreseeable, usage requirements.

Block Pivot. By introducing several columns at once into the basis, the notion of pivot element may be generalized to that of block pivot. [11]

Block-Triangular-Matrix. A matrix that can be written in the form

$$\begin{bmatrix} A_{11} & & & \\ & A_{22} & & \\ & & \cdot & \\ A_{m1} & A_{m2} & \cdot & A_{mm} \end{bmatrix}$$

where the Aij are submatrices or blocks and all entries not shown are submatrices consisting entirely of zeros. The Aij need not be square, but the vertical dimensions of all blocks in one row must coincide, as well as the horizontal dimensions of all blocks in one column. [12]

BOUNDED VARIABLE. A variable in a linear programming problem whose values are limited from above or below or both.

BOUNDED VARIABLE PROBLEM. A linear programming problem in which all or most of the variables have bounds of the type $l_j \leq x_j \leq b_j$. [15]

BRANCH AND BOUND. A method for solving mathematical programming problems based on partitioning the solution set into smaller subsets (branching) and establishing bounds on the value of the objective function for the subsets (bounding). A subset is excluded from further consideration when it is established that it contains no feasible solution having a value equal to or better than the current best solution.

BRANCHING PROCESS. A stochastic process describing the growth of a population in which the individual members may have offspring, the lines of descent "branching out" as new members are born. [22:32, 5:58, 13:272]

BURN-IN. The operation of items prior to their ultimate application intended to stabilize their characteristics and to identify early failures. [20]

CAPABILITY. A measure of the ability of an item to achieve mission objectives given the conditions during the mission. [28]

CAPACITATED TRANSPORTATION PROBLEM. A bounded variable transportation problem (q.v.). [15]

CATASTROPHIC FAILURE. (See FAILURE, CATASTROPHIC.) [20]

CATERER PROBLEM. A caterer requires for the next number of days r_j fresh napkins on the jth day. j = 1,2,...n. Two types of laundry service are available. A slow service which requires p days and costs b cents per napkin; a fast service requiring q days q < p but costing c cents per napkin c > b. Beginning with no usable napkins on hand or in the laundry, the caterer meets the demands by purchasing napkins at a cents per napkin. How does the caterer purchase and launder napkins to meet requirements so as to minimize the total cost of n days? [19]

CHAIN (IN A GRAPH). A sequence of arcs (i,i_1), (i_1,i_2), $(i_2,i_3)...(ik,j)$ connecting nodes i and j is called a chain. [11]

CHANCE CAUSES. Factors, generally numerous and individually of relatively small importance, which contribute to variation, but which are not feasible to detect or identify.

CHANCE CONSTRAINED PROGRAMMING. A method of treating the stochastic constraints in stochastic linear programming. A vector d_s with all components greater than 0 and less than or equal to 1 for all s in a given set S is defined. Then the stochastic linear program may be reformulated as

MIN cx

Subject to

Probability $\left\{ A^{(s)} x \geq b_s \right\} \geq d_s$ for all s in S; x ≥ 0

d_s is a vector chosen prior to solution of the change constrained problem. (See STOCHASTIC PROGRAMMING). [9]

CHANNELS, MULTIPLE. A waiting line is said to have multiple channels when there is more than one station at which service is provided.

CHARACTERISTIC. A property of items in a sample or population which, when measured, counted or otherwise observed, helps to distinguish between the items.

CHARACTERISTIC ROOT. The characteristic root of a square matrix A is a value such that $|\lambda - \lambda I| = 0$
where I is the identity matrix. For a p x p matrix there are, in general, p such roots. They are also known as Latent Roots and Eigenvalues.

The corresponding row-vectors u or column-vectors v for which
$$uA = \lambda u \text{ or } Av = \lambda v$$
are called characteristic vectors. [22]

CHECKOUT. Tests or observations of an item to determine its condition or status. [28]

CLUSTER SAMPLING. A method of sampling in which the population is divided into mutually exclusive aggregates (or clusters) of sampling units related in a certain manner. A sample of these clusters if taken at random and all the sampling units which constitute them are included in the sample.

COEFFICIENT MATRIX. The matrix of left-hand-side coefficients in a system of linear equations. It is to be distinguished from the matrix obtained by appending the right-hand side, which is called the "argumented matrix" of the system. [19]

COEFFICIENT OF LOSS OF SERVICE. The ratio of the average number of idle service stations (servers) to the total number of service stations (servers).

COLUMN VECTOR. One column of a matrix, or a matrix consisting of a single column. The elements of the column are interpreted as the components of the vector. [19]

COMPLEMENTARY SLACKNESS THEOREM. For symmetric primal and dual linear programs (q.v.) the following theorem holds: Whenever inequality occurs in the ith relation of the primal (or dual) constraints for an optimizing solution, then the ith dual (or primal) variable of an optimizing solution vanishes. Conversely, if the ith variable of the dual (or primal) system is positive, then the ith relation of the primal (or dual) system is an equality. [15]

COMPOSITE ALGORITHMS. For linear programming, if neither the basic solution nor the dual solution generated by its simplex multipliers, remain feasible, the corresponding algorithm is called composite. [19]

CONCAVE FUNCTION. Geometrically, a function whose graph is never below the chords joining any two points on the graph. A function f such that
$$\alpha_1 f(x_1) + \alpha_2 f(x_2) \leq f(\alpha_1 x_1 + \alpha_2 x_2)$$
for all $\alpha_1 \geq 0, \alpha_2 \geq 0, \alpha_1 + \alpha_2 = 1$ and x_1 and x_2 are in the convex domain of definition of f. A function for which the strict inequality always holds when $\alpha_1 > 0$ and $\alpha_2 > 0$ is called strictly concave. [19]

CONE. A set of vectors S is called a cone if for every vector X in S, λX is in S for all $\lambda \geq 0$. [15]

CONFIDENCE INTERVAL. An interval calculated from sample data and distribution parameters with a specified probability or confidence. To say that [a,b] is a $1 - \alpha$ confidence interval for the population parameter θ means that the a priori probability that the random interval [A,B] will contain H is $1 - \alpha$. Another interpretation would be that $(1-\alpha)(100)$ percent of such intervals calculated by different random samples of the same size would contain θ in the long run.

CONFIDENCE LEVEL. Probability that a particular value lies between an upper and a lower bound, the confidence limits.

CONFIDENCE LIMIT. The bounds of an interval. A probability can be given for the likelihood that the interval will contain the true value.

CONSTANT ARRIVAL RATE DISTRIBUTION. (See ARRIVAL RATE DISTRIBUTION, CONSTANT.)

CONSTANT FAILURE PERIOD. That period during which the failures of units of a particular item occurs at an approximately uniform rate. [20]

CONSTANT FAILURE RATE PERIOD. That possible period during which the failures occur at an approximately uniform rate. [20]

CONSTANT VECTOR. The right-hand side of a set of linear inequalities (equalities). By convention the linear relations of a linear programming problem are so arranged that all variables and their coefficients appear on the left, and the column of constants appears on the right. [19]

CONSTRAINT. An equation or inequality relating the variables in an optimization problem. [19]

CONSTRAINT MATRIX. In linear programming, the augmented matrix of the constraint equations. It is the matrix formed by the coefficient columns, or left hand sides, and the column of constants. [19]

CONSTRAINT QUALIFICATION. This qualification places restrictions on the nature of the vicinity of a point satisfying certain conditions for optimality. These restrictions insure that the conditions are indeed, properly related to optimality.

CONTROL CHART. A graphical method for evaluating whether a process is or is not in a "state of statistical control."

CONSUMER'S RISK. β For a given sampling plan, the probability of acceptance of a lot the quality of which has a designated numerical value representing a level which it is seldom desired to accept. Usually the designated value will be the Limiting Quality Level (LQL).

CONVEX COMBINATION. For the points $X_1, X_2, ..., X_n$ a convex combination is a point
$$X = \alpha_1 X_1 + \alpha_2 X_2 + ... + \alpha_n X_n$$
where
$$\alpha_i \geq 0 \text{ and } \Sigma_i \alpha_i = 1.$$

CONVEX CONE. A cone is convex if the positive sum of any two vectors in the cone is also in the cone.

CONVEX FUNCTION. Geometrically, a function whose graph is never above the chords joining two points on the graph. A function f such that

$\alpha_1 f(x_1) + \alpha_2 f(x_2) \geq f(\alpha_1 x_1 + \alpha_2 x_2)$ for all $\alpha_1 \geq 0, \alpha_2 \geq 0, \alpha_1 + \alpha_2 = 1$ and all x_1 and x_2 in the convex domain of definitions of f. [19]

Convex Hull. The convex hull of a set of points S is the set of all convex combinations of sets of points from S. [15]

Convex Polyhedral Cone. A cone generated by a convex polyhedron. [9]

Convex Polyhedron. The convex hull of a finite number of points. [15]

Convex Programming. Optimization of a convex function over a convex region. Linear programming and quadratic programming are special cases of convex programming.

Convex Set. Geometrically, a set of points that contains all the points on the line segment joining any two points of the set; *i.e.*, a set is convex if and only if each convex combination of any two points in the set is also in the set. [15]

Convolution. Let $F_1(x), F_2(x),...F_n(x)$ be a sequence of distribution functions. The distribution

$$f(x) = \int_{-\infty}^{\infty} dF_1(x_1) ... \int_{-\infty}^{\infty} dF_1(x - x_1 ... - x_{n-1})$$

is called the convolution of the distributions. The relation ship is sometimes written

$$F(x) = F_1(x)*F_2(x)* ... F_n(x)$$

If the associated variates are independent, F(x) is the distribution function of their sum.
[22:65, 5:8, 13:250, 25:404]

Correlation. The linear relationship between two or several random variables within a distribution of two or more random variables.

Cost Coefficient. The coefficient of a variable in the cost function of a linear program. The elements of any objective function are sometimes referred to as cost coefficients. [19]

Cost Function. The objective function of a cost minimization program. [19]

Cost Range. The range of values of a cost coefficient of a basic solution variable for which the current basis stays optimum. [19]

Cost Row. The row of objective function coefficients of a linear program matrix. [19]

Coupling Equations. When some of the constraints involve only the variables X and some others involve only the variables Y, the remaining equations involving both X and Y are called coupling equations. [19]

Cumulants. These are values that are given by the coefficients in the expansion of a power series formed from the logarithms of the characteristic function of a random variable.

Cumulative Distribution Function, F(x). A function giving, for every value x, the probability that the random variable X be less than or equal to x:

$$F(x) = P(X \leq x)$$

Cut. A partition of the nodes of a network into two sets such that the source node belongs to one set and the sink node to the other.

Cutting Plane. The hyperplane boundary of an added linear inequality constraint in a programming problem. It may be thought of as "cutting off" part of the original convex region of feasible solutions in order to form a new convex region. [11]

Cycling (Linear Programming). The repeating of a basic feasible solution during application of the Simplex Algorithm (q.v.). [15]

Debugging. The check out of equipment, complex items, or computer programs prior to use in order to detect design and/or operational difficulties.

Decision Function. A decision function is a rule of conduct which, at any stage of a sampling investigation, tells the statistician whether to take further observations or whether enough information has been collected, and in the latter case, what decision to make upon it. At each stage beyond the first the decision function is a function of the preceding observations. [22]

Decision Under Certainty. If a choice must be made between two or more actions, decision under certainty involves that realm where each action is known to lead invariably to a specific outcome.

Decision Under Risk. Decision under risk is concerned with that realm of choice where each action leads to one of a set of possible specific outcomes, each outcome occurring with a known probability.

DECISION UNDER UNCERTAINTY. Decision under uncertainty involves that realm of choice where each action or combination has as its consequence a set of possible specific outcomes, but where the probabilities of these outcomes are completely unknown or are not even meaningful.

DECOMPOSITION PRINCIPLE. A method for subdividing a large linear programming problem into smaller, more manageable linear programs, solving the smaller problems in the standard way, and then combining these solutions to solve the original problem. In particular linear programs which qualify for decomposition are those for which the variables can be separated into classes X, Y, Z, ... and the constraints into classes C_0, C_1, C_2, ... such that: The constraints in class C_0 can involve all the variables; constraints in class C_1 involve only the variables in X; the constraints in class C_2 involve only the variables Y; the constraints in class C_3 involve only the variables Z; and so on. The subsets of constraints, each involving only a part of the variables, are called subprograms; while the subset of constraints in C_0 is called the "master program. " [12]

DEFECT. A departure of a quality characteristic from its intended level or state that occurs with a severity sufficient to cause an associated product or service not to satisfy intended normal, or reasonably foreseeable, usage requirements.

DEGENERATE SOLUTION. For linear programs, a basic solution in which at least one basic variable is zero. [9]

DEGRADATION FAILURE. (See FAILURE, DEGRADATION.) [20]

DELAY. (See WAITING TIME.) [33]

DEPENDABILITY. A measure of the item operating condition at one or more points during the mission, including the effects of reliability, maintainability, and survivability, given the item condition(s) at the start of the mission. It may be stated as the probability that an item will (a) enter or occupy any one of its required operational modes during a specified mission, and (b) perform the functions associated with those operational modes. [28]

DEPENDENT VARIABLE. (See INDEPENDENT VARIABLE.)

DERATING. (1) Using an item in such a way that applied stresses are below rated values, or (2) The lowering of the rating of an item in one stress field to allow an increase in rating in another stress field. [28]

DESIGNATED IMPERFECTIONS. A category of imperfections which, because of their type and/or severity, are to be treated as an event for control purposes.

DETERMINISTIC MODEL. AS opposed to a stochastic model, a model which contains no random elements and for which, therefore, the future course of the system is determined by its state at present (and/or in the past).

DIET PROBLEM. A linear programming problem in which the constraints represent minimum daily requirements for specified nutrients and variables represent foods. The ith coefficient of the jth activity vector represents the amount of the ith nutrient in one unit of the jth food. The right-hand side coefficients indicate the daily requirement of each nutrient, while the objective function coefficients represents the cost per unit of each food. [15]

DIOPHANTINE PROGRAMMING. (See INTEGER LINEAR PROGRAMMING.) [19]

DISCRETE VARIABLE PROBLEM. (See INTEGER LINEAR PROGRAMMING.) [19]

DISTRIBUTION-FREE METHOD. A method, e.g., of testing a hypothesis or of setting up a confidence interval which does not depend on the form of the underlying distribution.

DOWNTIME. (See TIME, DOWN; Z94.17 WORK MEASUREMENT) [28]

DUAL LINEAR PROGRAMMING PROBLEMS. A pair of linear programs, called "primal" and "dual" of the following form:
Unsymmetric Case:
    Primal   Min cx
    subject to   AX=b
                         x $\geq$ 0

    Dual: Max Wb
    subject to   WA $\leq$ c

Symmetric Case:
    Primal   Min cx
    subject to   AX $\geq$ b
                         x $\geq$ 0

    Dual: Max Wb
    subject to   WA $\leq$ c
                         W $\geq$ 0   [15]

Dual Problem. (See DUAL LINEAR PROGRAMMING PROBLEMS.) [19]

Dual Simplex Algorithm. Starting with a dual feasible solution, the algorithm selects the most infeasible vector to leave the primal basis and then computes which vector must enter to maintain dual feasibility. As a consequence of this change of basis, other variables may become primal infeasible, but since the objective value changes monotonically toward optimum, the algorithm is finitely convergent. It is exactly the primal algorithm applied to the dual problem. [19]

Duality Theorems For Linear Programming. (1) Main theorem: If both the primal and dual problems have a finite optimum, then the optimum values of the objective functions are equal. (2) Corollary: If either problem has a feasible finite optimum, then so does the other, and the optimum values are equal. (3) Corollary: A feasible but unbounded solution to one problem implies no feasible solution for the other. (4) Corollary: No feasible solution to one problem implies that the other is either unbounded or infeasible. (5) Weak theorems of the alternative: A variable and its complementary slack are not both nonzero. (6) Strong theorem of the alternative: Among all alternate optima, at least one solution exists in which a variable and its complementary slack are not both zero, and the one of the pair that is zero in this solution is zero in all alternate solutions. [19]

Dynamic Programming. A method for optimizing a set of decisions which may be made sequentially. Characteristically, each decision may be made in the light of the information embodied in a small number of observable called state variables. The incurred cost for each period is a mathematical function of the current state and decision variables, while future states are functions of these variables. The aim of the decision policy is to minimize the total incurred cost, of equivalently the average cost per period. [19]

Early Failure Period. That possible early period beginning at some stated time and during which the failure rate is decreasing rapidly in comparison with the subsequent period.

Elementary Matrix. A unit or identity matrix except for one unique column or row. The inverse of a non singular elementary matrix is also an elementary matrix. [19]

Elimination Method. An algorithm for solving a set of m x m independent linear equations. (See GAUSSIAN ELIMINATION.) [19]

Equation. A proposition (or relation) that a given mathematical function shall be equal to another function (often a constant). [19]

Error Mean-Square. The residual or error sum-of-squares (q.v.) divided by the number of degrees of freedom on which the sum is based. It is an estimator of the residual or error variance. [22]

Error Sum of Squares. In the analysis of variance and covariance, and regression analysis, it is the sum of squares that remains after subtracting from the total sum of squares all sums of squares associated with the factors of the experiment. It also is referred to as the residual sum of squares.

Estimation. Inference about the numerical value of unknown population values from incomplete data such as a sample. If a single figure is calculated for each unknown parameter the process is called point estimation. If an interval is calculated within which the parameter is likely, in some sense, to lie, the process is called interval estimation. [22]

Event. An occurrence of some attribute. In strict probability terminology, it is a subset of the sample space.

Evolutionary Operation. Method of experimentation during production, whereby process variables are changed in a systematic manner and the results are observed. The changes are small in order to avoid the risk of off-standard product or other undesirable effects. Statistical analysis provides guides toward process improvement in the form of increased yield or improved quality.

Exponential Service Time. The service time is a random variable with an negative exponential distribution.

Extremal Problem. (See MASTER PROGRAM.)

Extreme Point. A point of a convex set which does not lie on a line segment joining any other two points of the set; also called a "vertex." [19]

Factor Analysis. A branch of multivariate analysis in which the observed variates x_i (i = 1,2,...,p) are supposed to be expressible in terms of a number m < p factors f_j together with residual elements.

Failure. The termination of the ability of any item to perform its required function under stated environmental conditions for a specified period of time. [3]

FAILURE ANALYSIS. The logical, systematic examination of an item or its diagram(s) to identify and analyze the probability, causes, and consequence of potential and real failures. [28]

FAILURE, CATASTROPHIC. Failures which are both sudden and complete. [20]

FAILURE, COMPLETE. Failure resulting from deviations in characteristic(s) beyond specified limits such as to cause complete lack of the required function.
Note: The limits referred to in this category are special limits specified for this purpose. [20]

FAILURE CRITERIA. Rules for failure relevancy such as specified limits for the acceptability of an item. [20]

FAILURE, DEGRADATION. Failures which are both gradual and partial. [20]

FAILURE, DEPENDENT. One which is caused by the failure of an associated item(s). Not independent. [28]

FAILURE, GRADUAL. Failures that could be anticipated by prior examination. [20]

FAILURE, INDEPENDENT. One which occurs without being related to the failure of associated items. Not dependent. [28]

FAILURE, INHERENT WEAKNESS. Failures attributable to weakness inherent in the item itself when subjected to stresses within the stated capabilities of that item. [20]

FAILURE MECHANISM. The physical, chemical or other process which results in a failure. [20]

FAILURE, MISUSE. Failures attributable to the application of stresses beyond the stated capabilities of the item. [20]

FAILURE MODE. The effect by which a failure is observed; for example, an open or short circuit condition, or a gain change. [20]

FAILURE, PARTIAL. Failures resulting from deviations in characteristic(s) beyond specified limits but not such as to cause complete lack of the required function. [20]

FAILURE, RANDOM. Any failure whose occurrence is unpredictable in an absolute sense but which is predictable only in a probabilistic or statistical sense. [28]

FAILURE RATE. The rate at which failures occur in a certain time interval; i.e., the probability that a failure per unit time occurs in the interval, given that a failure has not occured prior to the start of the interval. [28]

FAILURE RATE ACCELERATION FACTOR. The ratio of the accelerated testing failure rate understated reference test conditions and time period. [20]

FAILURE RATE, ASSESSED. The failure rate of an item determined as a limiting value or values of the confidence interval with a stated confidence level, based on the same data as the observed failure rate of nominally identical items.

FAILURE RATE, EXTRAPOLATED. Extension by a defined extrapolation or interpolation of the observed or assessed failure rate for durations and/or conditions different from those applying to the conditions of that observed or assessed failure rate.

FAILURE RATE, OBSERVED. The ratio of the total number of failures in a sample to the total cumulation observed time on that sample. The observed failure rate is to be associated with particular, and stated time intervals (or summation of intervals) in the life of the items, and with stated conditions.

FAILURE RATE, PREDICTED. For the stated conditions of use and the design considerations of an item, the failure rate computed from the observed, assessed or extrapolated failure rates of its parts. [20]

FAILURE, SECONDARY. Failure of an item caused either directly or indirectly by the failure of another item. [20]

FAILURE, SUDDEN. Failures that could not be anticipated by prior examination. [20]

FAILURE, WEAR-OUT. A failure which occurs as a result of deterioration processes or mechanical wear and whose probability of occurrence increases with time. [20]

FAIR GAME. In probability theory, a game consisting of a sequence of trials is deemed to be a "fair" game if the cost of each trial is equal to the expected value of the gain from each trial. In game theory, a game which with proper play neither adversary has an advantage.

FARKAS' LEMMA. If for every solution of $WA \leq 0$ it is also true that $Wb \leq 0$, then there exists a vector $X \geq 0$ such that $Ax = b$; and thus $(WA)X = Wb$. (If a linear homogeneous inequality $Wb \leq 0$ holds for all W satisfying a system of homogeneous inequalities $WA \leq 0$, then the inequality can be expressed as a non-negative combination of the inequalities of the system $WA \leq 0$.) [11]

FATHOM. A term used in branch and bound algorithms to indicate a node has been fully explored; i.e., it has been determined that the node cannot contain a solution better than the incumbent.

FEASIBLE BASIS. A basis yielding a basic feasible solution (q.v). [19]

FEASIBLE SOLUTION. A solution satisfying the constraints of a mathematical programming problem in which all variables are non-negative. [19]

FINITE AND INFINITE GAMES. A game is finite if each player has only a finite number of possible pure strategies; it is infinite if at least one player has an infinite number of possible pure strategies (e.g., a pure strategy might ideally consist of choosing an instant from a given interval of time at which to fire a gun). [21]

FIXED CHARGE PROBLEM. Nonlinear programming problem of the form.

$$\text{Minimize} \quad \Sigma f_j(x_j),$$
$$\text{Subject to} \quad AX \geq 0,$$
$$X \geq 0,$$
$$\text{Where } f_j(x_j) = c_j x_j + d_j \delta_j, d_j > 0,$$
$$\text{And } \delta_j = \begin{cases} 0 \text{ if } x_j = 0, \\ 1 \text{ if } x_j > 0. \end{cases}$$

The d_j are called fixed charges since d_j is incurred only if $x_j > 0$. [18]

FRACTIONAL PROGRAMMING. A class of mathematical programming problems in which the objective function is the quotient of linear functions. [19]

FRAME. The list of units, or items, accessible for test. Each unit has a serial number associated with it, actually or conceptually. If there are units in the population that are not covered by the frame, statistical inferences (estimates, confidence limits, etc.) refer to the frame, not the population. Generalizations from the frame to the population must be based on judgment. [3]

GAMBLER'S RUIN. The name given to one of the classical topics in probability theory. A game of chance can be related to a series of Bernoulli trials at which a gambler wins a certain predetermined sum of money for every success and loses a second sum of money for every failure. The play may proceed until his initial capital is exhausted and he is ruined. The statistical problems involved are concerned with the probability of the ruin of a player, given the stakes, initial capital and chances of success, and with such matters as the distribution of the length of play. There are many variations to this classical problem, which is closely associated with problems of the random walk and in the limiting case, Brownian motion, in particular, of sequential sampling. [22:117, 5.25, 13:334, 25:144]

GAME THEORY. The study of the following problem: if n players $P_1, P_2...P_N$ play a given game G, how must the ith player, Pi, play to achieve the most favorable result? Special two-person games can be solved by linear programming methods. More generally, the mathematical study of cooperative/competitive situations. [15]

GAUSSIAN ELIMINATION. A reduction method for systems of simultaneous linear equations, in which one of the equations is solved for one of the variables in terms of the remaining variables. When these expressions for the solved variable are substituted into the remaining equations, the result is an equivalent system with one less equation and one less variable. A Gaussian elimination step is exactly equivalent to a pivot step. It is a single change of basis and can be expressed functionally as premultiplying by the inverse of an appropriate elementary column matrix. Sufficient repetition of this procedure can yield the numerical solution in case of a nonsingular square system, and a solution of parametric form, (a linear function of a subset of the variables) if the number of variables exceeds the number of equations. [19]

GEOMETRIC SOLUTION. A graphic method of solving a linear programming problem, by plotting the halfplanes determined by the constraints and the lines of constant value for the objective function. [19]

GLOBAL OPTIMUM. A feasible solution (q.v.) which gives a value to the objective function at least as great (small) as any other in the feasible region. It is contrasted with a local optimum, which yields the best objective function value of all points in some subset of the feasible region. In convex programming problems a local optimum is a global optimum. (See ABSOLUTE MAXIMUM.) [19]

GOAL PROGRAMMING. A model and associated algorithm to minimize the absolute value of deviations from a set of values called goals subject to technological constraints. For example, one may have goals for profit, market share, and pollution limits.

GRADIENT OF A FUNCTION. A vector at a point, whose direction is the direction of most rapid change of some function f, and whose magnitude is the rate of change of f in that direction. [19]

GRADIENT METHODS. A term applied to algorithms which at each iteration in their use seek to improve the current function value by moving in the direction of the gradient of the function f(x) to be optimized. The application of this method in the presence of constraints gives rise to numerous forms depending upon the type of problem and the manner in which the method is modified to handle constraints. [18]

GRAPH (LINEAR). Compare network. A linear graph consists of a number of nodes or junction points, each joined to some or all of the others by arcs or lines. [11]

GRAPH, SIGNAL FLOW. (See SIGNAL FLOW GRAPH.)

HALF-SPACE. The set of points whose coordinates satisfy a linear inequality. [11]

HAZARD FUNCTION. The limit of the failure rate as the interval width approaches zero.

HOLDING TIME. Time that an item is not operational, so that it may be serviced.

HUNGARIAN METHOD. A combinatorial procedure designed for solving an assignment problem. It is based on a proof by the Hungarian mathematician Egervary. [11]

HYPERPLANE. The set of all points X in e-space whose coordinates $(x_1,...,x_n)$ satisfy a linear equation of the form:
$$\alpha_1 x_1 + ... + a_n x_n = b. [19]$$

IDENTITY MATRIX. A square matrix with all main diagonal elements unity and the remaining elements zero.

IMPLICIT ENUMERATION. A method (used in branch and bound algorithms) of checking all possible realizations of a binary vector without explicitly enumerating all possible cases.

IMPLICIT PRICES. Same as marginal values, shadow prices, dual variable levels, *i.e.,* numbers giving the incremental worth of a relaxation of one unit in the right-hand side of a constraint. [19]

INDEPENDENT EQUATIONS. A set of equations none of which can be expressed as a linear combination of the others. With linear equations, the condition for independence is that the matrix (coefficient columns) shall be nonsingular or, equivalently, have rank equal to the number of equations. [19]

INDEPENDENT TRIALS. The successive trials of an event are said to be independent if the probability of outcome of any trial is independent of the outcome of the others. The expression is usually confined to cases where the probability is the same for all trials. In the sampling of attributes, such a series of trials is often referred to as "Bernoullian Trials." It includes all the classical cases of drawing colored balls from urns with replacement after each draw, coin tossing, dice rolling, and the events associated with other games of chance. [22:136, 13:118, 25:95, 29:44]

INDEPENDENT VARIABLE. This term is regularly used in contradistinction to "dependent variable" in regression analysis. When a variate y is expressed as a function of variables x_1, x_2, etc., plus a stochastic term the x's are known as "independent variables." The terminology is rather unfortunate since the concept has no connection with either mathematical or statistical dependence. The usage is so convenient, however, and so common that strongly coordinated action would be necessary to change it. [22]

INEQUALITY. A proposition which relates the magnitudes of two mathematical expressions or functions A and B. Inequalities are of four types: A is greater than B (A > B); A is less than B (A < B); A is greater than or equal to B (A ≥ B); A is less than or equal to B (A ≤ B). The first two types are called "strict" and the last two "relaxed" or "weak." [19]

INEQUALITY RELATION. A constraint which is an inequality. [19]

INFEASIBLE BASIS. A basis corresponding to an infeasible solution. Postmultiplying the inverse of such a basis by the constant vector, one obtains a solution in which at least one variable value is negative. [19]

INITIAL BASIS. The set of column vectors associated with the basic variables for which a starting solution will be obtained in linear programming. It is often an identity matrix. [19]

INITIAL FEASIBLE BASIS. The set of column vectors associated with the basic variables of the first feasible solution to a linear programming problem. [19]

INITIAL SOLUTION. The solution used at the beginning of an iterative solution to a mathematical problem.

INPUT-OUTPUT COEFFICIENT. A coefficient of an activity vector of the activity analysis problem. The (i,j) coefficient represents the amount of resource i required to produce one unit of activity j. [19]

INSPECTION. Activities, such as measuring, examining, testing, gauging one or more characteristics of a product or service and comparing these with specified requirements to determine conformity.

INTEGER FORM. A cutting plan used to solve integer linear programs. [11]

INTEGER LINEAR PROGRAMMING. A linear programming problem with the added restriction that some or all of the variables are constrained to be integers. [19]

INTERARRIVAL TIME. The time between two successive arrivals. (Mean interarrival time: the arithmetic average of the actual interarrival times or the expected value of the interarrival time distribution; reciprocal of interarrival rate).

INTERINDUSTRY (INPUT-OUTPUT) ANALYSIS. An interpretation of the economy in terms of the interactions of industrial organizations. Also termed "Leontief Model." [19]

INVENTORY, ACTIVE. That group of items in a storage assigned an operational status. [28]

INVENTORY, INACTIVE. That group of items in storage being held in reserve for possible future commitment to the operational inventory. [28]

INVERSE OF A MATRIX. An inverse of the square matrix A is another matrix B of the same dimension such that AB = BA = I, where I is the identity matrix. [19]

INVERSION. An operation on a matrix yielding the matrix's inverse. [19]

ITEM, INTERCHANGEABLE. One which (1) possesses such functional and physical characteristics as to be equivalent in performance, reliability, and maintainability, to another item of similar or identical purpose; and (2) is capable of being exchanged for the other item a) without selection for fit or performance, and b) without alteration of the items themselves or of adjoining items, except for adjustment. [28]

ITEM, REPLACEABLE. One which is interchangeable with another item, but which differs physically from the original item in that the installation of the replaceable item requires operations such as drilling, reaming, cutting, filing, shimming, etc., in addition to the normal application and methods of attachment. [28]

ITEM, SUBSTITUTE. One which possesses such functional and physical characteristics as to be capable of being exchanged for another only under specified conditions or for particular applications and without alteration of the items themselves or of adjoining items. [28]

ITEM, UNIT. An object or quantity of material and/or service on which a set of observations can be made.

ITERATION. A single cycle of operations in a solution algorithm made up of a number of such cycles. [19]

KARMARKAR'S ALGORITHM. An algorithm proposed by Dr. Narendra Karmarkar to solve LP problems. The algorithm finds a feasible point and transforms the constraint space so that the feasible point is at the center of a sphere inscribed in the transformed constraint space. An new point is found by moving from the center of the sphere to the surface of the sphere by moving in a direction perpendicular to the objective hyperplane. The process is then repeated.

KARUCH-KUHN-TUCKER CONDITIONS. A set of conditions which are necessary and sometimes sufficient for a given feasible solution of a mathematical programming problem to yield optimum. [19]

KNAPSACK PROBLEM. If a_i is the weight of the ith object and b_i is its relative value, then the problem of outfitting a knapsack in order to maximize the total relative value and not exceed a weight limitation c is given by:

$$\text{Max} \sum_i b_i x_i$$

$$\text{subject to} \sum_i a_i x_i \leq c$$

where $x_i = 0$ or 1. It is an integer linear program. [11]

LAGRANGE MULTIPLIERS. Auxiliary variables $\lambda_1, \lambda_2, ..., \lambda_m$ used in optimizing a function $f(X) = f(x_1, x_2, ... X_N)$ subject to the m constraints, $\phi_1 = 0, .., \phi_m = 0$ on the x_i. The Lagrange function $F(X,\lambda) = f(X) + \sum \lambda_i \phi_i$ is formed and one set of (m + n) partial derivatives (for each χ_j and λ_i) are set equal to zero and solved for an optimizing set of χ_j and corresponding set of λ_i. [19]

LAPLACE TRANSFORM. If a function g(t) is related to a second function f(x) by the equation

$$g(t) = \int_0^\infty e^{-tx} f(x) dx$$

then g(t) is the Laplace transform of f(x).

In statistical theory it is more customary to use the Fourier transform, which has certain advantage over the Laplace form; e.g., if f(x) is a frequency function the Fourier transform always exists whereas the Laplace transform may not do so for real t. [22]

LEFT-HAND SIDE. The mathematical expression to the left of the equality or inequality sign in an equation or inequality. In linear programming, by convention, the left-hand side of each constraint is the complete linear function, while the right-hand side is the constant term. [19]

LEONTIEF SYSTEM. (See INTERINDUSTRY ANALYSIS.)

LEXICOGRAPHIC ORDERING. Dictionary ordering. [19]

LINEAR COMBINATION. An expression of the form

$$a_1 x_1 + a_2 x_2 + \ldots + a_n x_n$$

where the a_i are coefficients and the x_i are variables or vectors. [19]

LINEAR CONSTRAINT. A constraint which contains variables which are related in terms of linear combinations. [19]

LINEARLY DEPENDENT. A set of vectors {X} is linearly dependent if a set of numbers aj, not all equal to zero, can be found such that

$$a_1 X_1 + a_2 X_2 + \ldots + a_k X_k = 0. \text{ [19]}$$

LINEAR EQUATION. An equation whose left-hand side and right-hand side are both linear functions of the variables. Such an equation can always be put in the form

$$f(x, y, z,) = c,$$

where f is a linear function and c is a constant. [19]

LINEAR ESTIMATOR. An estimator which is a linear function of the observations. [22]

LINEAR FUNCTION. A function of the form

$$a_0 + a_1 x_1 + a_2 X_2 + \ldots + a_n x_n,$$

where the a_i are coefficients, not all 0, and the x_i are variables. The geometrical representation of a linear function is a straight line, plane, or hyperplane.

LINEAR INDEPENDENCE. A set of vectors (X_i) is linearly independent if the only set of numbers ai for which

$$a_1 X_1 + a_2 X_2 + \ldots + a_k X_k = 0$$
is $a_1 = a_2 = \ldots = a_k = 0$. [19]

LINEAR INEQUALITY. An inequality whose left-hand side and right-hand side are both linear functions of the variables. [19]

LINEAR MODEL. A model where each dependent variable is a linear function of independent variables.

LINEAR PROGRAMMING. The concept of expressing the interrelationship of activities of a system in terms of a set of linear constraints in nonnegative variables. A program, *i.e.*, values of the variables, is selected which satisfies the constraints and optimizes a linear objective function in these variables. [17]

LINEAR PROGRAMMING PROBLEM. The problem of minimizing or maximizing a linear function in n variables subject to m linear constraints, with the variables restricted to be nonnegative. Mathematically, we have

Min (max) cX   subject to
AX = b
X ≥ 0
with A an (mxn) matrix.

The constraints AX = b can also be given in terms of inequalities, *i.e.*, AX ≥ b, AX ≤ b or a combination of such constraints. [19]

LOCAL OPTIMUM. (See GLOBAL OPTIMUM.) [19]

LOOP (IN A GRAPH). A chain of arcs connecting node i to itself in a graph is called a loop (if the arcs are distinct, the loop is a simple loop). [11]

LOT TOLERANCE PERCENT DEFECTIVE (LTPD). Expressed in percent defective, the poorest quality in an individual lot that should be accepted.
Note: The LTPD is used as a basis for some inspection systems, and is commonly associated with a small consumer's risk. (See CONSUMER'S RISK.) [2]

MACHINE-ASSIGNMENT PROBLEM. Given m machines or machine types and n products or jobs; let d_{ij} be the time required to process one unit of product j on machine i, x_{ij} the number of units of j produced on machine i in tle coming time period, a_i the time available on machine i, b_j the number of units of j which must be completed, and c_{ij} the cost of processing one unit of product j on machinei. The machine-assignment problem thus reduces to finding nonnegative x_{ij} which will satisfy the constraints

$$\sum_{j=1}^{n} d_{ij} x_{ij} = a_i \quad i = 1,...,m$$

$$\sum_{i=1}^{m} x_{ij} = b_j \quad j = 1,...,n$$

and minimize the cost $Z = \sum_i \sum_j c_{ij} x_{ij}$. [17]

MAINTAINABILITY. A characteristic of design and installation which is expressed as the probability that an item will be retained in or restored to a specified condition within a given period of time, when the maintenance is performed in accordance with prescribed procedures and resources. [28]

MAINTENANCE. All actions necessary for retaining an item in or restoring it to a specified condition. [28]

MAINTENANCE, CORRECTIVE. The actions performed, as a result of failure, to restore an item to specified condition. [28]

MAINTENANCE, PREVENTIVE. The actions performed in an attempt to retain an item in a specified condition by providing systematic inspection, detection and prevention of incipient failure. [29]

MAN-FUNCTION. The function allocated to the human component of a system. [28]

MASTER PROGRAM. The part of a decomposed problem which expresses the extreme point solution of the subproblems as convex combinations and which also satisfy the constraints in the class C0 (the class of constraints which C0 can include all the variables). (See DECOMPOSTION PRINCIPLE.) [19]

MATHEMATICAL PROGRAMMING. The general problem of optimizing a function of several variables subject to a number of constraints. If the function and constraints are linear in the variables and a subset of the constraints restrict the variables to be nonnegative, we have a linear programming problem. [19]

MATRIX. A rectangular array of mn numbers arranged in m rows and n columns as follows:

$$\begin{bmatrix} A_{11} & . & . & A_{1n} \\ . & ... & ... & . \\ . & ... & ... & . \\ A_{m1} & . & . & A_{mn} \end{bmatrix}$$

MATRIX ELEMENT. One of the mn numbers of a matrix. [19]

MATRIX GAME. A zero-sum two-person game. The payoff (positive, negative or zero) of player one to player two when player one plays his i^{th} strategy and player two his j^{th} strategy is denoted by a_{ij}. The set of payoffs a_{ij} can be arranged into an m x n matrix. [19]

MAX-FLOW MIN-CUT THEOREM. A theorem which applies to a maximal flow network problem which states that the maximal flow from the source to the sink is equal to the minimal cut capacity of all cuts separating the source and sink nodes. [14]

MAXIMAL NETWORK FLOW PROBLEM. A problem involving a connected linear network in which goods must flow from an origin (source) to a final destination (sink) over the arcs of the network in such a fashion as to maximize the total flow that the network can support. The arcs are connected at intermediate nodes and each branch has a given flow capacity that cannot be exceeded. The flow into an intermediate node must equal the flow out of that node. [15]

MEAN. The expected value of a random variable.

MEAN LIFE. The arithmetic mean (q.v.) of the times-to-failure of the units of a given item. [20]

MEAN-MAINTENANCE-TIME. The total preventive and corrective maintenance time divided by the total number of preventive and corrective maintenance actions during a specified period of time. [28]

MEAN SERVICE RATE. The conditional expectation of the number of services completed in one time unit, given that service is going throughout the entire time unit.

MEAN TIME BETWEEN FAILURES(MTBF). For a particular interval, the total functioning life of a population of an item divided by the total number of failures within the population during the measurement interval. [28]

MEAN TIME TO FAILURE. This expression applies to components of a system and not to systems. The expected length of time that a component is in use in a system in operation from the moment the component is put into the system to the moment it fails. It is the expected length of life of the components. The meantime-to-failure is sometimes confused with the meantime-between-failures (MTBF) which applies to systems which experience subsystem or component failures which can be replaced or repaired, putting the system into operation again. The confusion occurs in part because for the Poisson failure distributions the two quantities have the same distribution.

MEAN TIME TO REPAIR (MTTR). The total corrective maintenance time divided by the total number of corrective maintenance actions during a given period of time. [28]

MEDIAN. A measure of central tendency with the number of data points above and below it equal; the 50th percentile value. [28]

MINIMAL COST FLOW PROBLEM. A network flow problem in which costs cij are the cost of shipping one unit of a homogeneous commodity from node i to node j. A given quantity F of the commodity must be shipped from the origin node to the destination at minimum cost. [19]

MINIMAX PRINCIPLE. A principle introduced into decision-function theory by Wald (1939). It supposes that decisions are taken subject to the condition that the maximum risk in taking a wrong decision is minimized. The principle has been critized on the grounds that decisions in real life are scarcely ever made by such a rule, which enjoins "that one should never walk under a tree for fear of being killed by its falling. " In the theory of games it is not open to the same objection, a prudent player being entitled to assume that his adversary will do his worst. [22]

MINIMAX THEOREM. Let X and Y represent the mixed strategy sets available to player I and player 2, respectively, in a matrix game with payoff matrix A. By definition.

$$\max_{X \in X} \min_{Y \in Y} E(X,Y) + \max_{X \in X} \min_{Y \in Y} XAY$$

and

$$\min_{Y \in Y} \max_{X \in X} E(X, Y) = \min_{Y \in Y} \max_{X \in X} XAY$$

where $X \geq 0$, $Y \geq O$ and the components of each X and Y sum to unity.

Theorem I. For any (mxn) matrix A the maximin and minimax values, as defined above, exist and are equal.

Theorem II. Every matrix game has a saddle point which defines an optimal strategy pair X* and Y* and a value of the game V* = E (X*, Y*). Analogous theorems hold, as well, for certain more complicated classes of games.

MISSION. The objective or task, together with the purpose, which clearly indicates the action to be taken. [28]

MIXED-INTEGER PROGRAMMING. Integer programs in which some, but not all, of the variables are restricted to integral values. [19]

MIXED STRATEGY. A mixed strategy for a player P is a vector $X=(x_1 x_2,...,x_m)$ of nonnegative numbers x_i such that

$$\sum_{i=1}^{m} x_i = 1$$

The elements x_i represent the frequencies with which the player selects his ith strategy.

MODE. The value(s) of a random variable such that the probability mass (discrete random variable) or the probability density (continuous random variable) has a local maximum for this value (or these values). *Note:* If there is one mode, the probability distribution of the random variable is said to be "unimodal"; if there is more than one mode the probability distribution is said to be "multimodal" (bimodal if there are two modes).

MODEL. A mathematical or physical representation of a system (q.v.) often used to explore the many variables (q.v.) influencing the system.

MONOTONE. A monotonic (or monotone) non-decreasing quantity is a quantity which never decreases (the quantity may be a function, sequence, etc., which either increases or remains the same, but never decreases). A sequence of sets El,E2,... is monotonic increasing if En is contained in En + I for each n. A monotonic (or monotone) non- increasing quantity is a quantity which never increases (the quantity may be a function, sequence, etc,. which either decreases or remains the same, but never increases). A sequence of sets, El,E2,... is monotonic decreasing if En contains En+ I for each n. A monotonic (or monotone) system of sets is a system of sets such that, for any two sets of the system one of the sets is contained in the other. A mapping of a topological space A onto a topological space b is said to be monotone if the inverse image of each point of B is a continuum. A mapping of an ordered set A onto an ordered set B is monotone provided x* precedes (or equals) y* whenever x* and y* are the images in B of points x and y of A for which x precedes y. [21]

Monotonic Increasing Function. A function f is monotonic increasing on an interval (a, b) if f(y) > f(x) for any two numbers x and y (of this interval) for which x < y. [21]

Monte Carlo Method. A simulation technique by which approximate evaluations are obtained in the solution of mathematical expressions so as to determine the range or optimum value. The technique consists of simulating an experiment to determine some probabilistic property of a system or population of objects or events by the use of random sampling applied to the components of the system, objects, or events.

Moving Average. An unweighted average of the latest n observations where the current observations has replaced the oldest of the previous n observations.

MTBF. (See MEAN TIME BETWEEN FAILURES.) [20]

Multicommodity Network Problem. A network problem in which more than one commodity can flow through the network at the same time. The capacity constraints on the arcs apply to the sum of the flows of the commodities. [19]

Multicriteria Optimization. A model and associated algorithm which attempts to find strategies which optimize several criterion measures instead of one. For example, there may be objectives and criterion measurements on economic, social, end environmental issues. The criterion measurements would be in different units.

Multiperson Game. A game where a number of players are each competitively striving towards a specific objective which any one player only partially controls.

Mutltiple Regression. A statistical procedure to determine a relationship between the mean of a dependent random variable and the given value of one or more independent variables.

Nash-Harsonyi Bargaining Model. A model of an n person bargaining game based on Nash's axioms of collective rationality, symmetry, linear invariance, and solution location.

N-Person Game. A multiperson game in which the success of (pay off to) each player depends not only on his actions but also on those of the other players.

Near-Optimum Solution. A feasible solution to a mathematical programming problem whose value of the objective function can be shown to be near the optimum value. [19]

Negative Definite or Semidefinite Quadratic Form. X^TAX is negative definite (semi-definite) if $-X^TAX$ is positive definite (semidefinite). [18]

Network Flow Problem. (See MAXIMAL NETWORK FLOW PROBLEM.)

Node. One of the defining points of a network; a junction point joined to some or all of the others by arcs. [9]

Nonbasic Variable. A variable in a solution to a linear programming problem obtained by the simplex method and whose value has been arbitrarily set to equal zero. [19]

Nonconformity. A departure of a quality characteristic from its intended level or state that occurs with a severity sufficient to cause an associated product or service not to meet a specification requirement.

Nondegeneracy Assumption. The assumption that all basic feasible solutions to a linear programming problem are nondegenerate. This assumption is convenient in some proofs which demonstrate the finiteness and convergence of the simplex algorithm. [15]

Nondegenerate Feasible Solution. A basic feasible solution in which all of the basic variables are positive. [19]

Nonegativity Constraint. A restriction that a variable can take on only positive or zero values, e.g., $x \geq 0$ [19]

Nonlinear Constraint. A constraint which contains variables whose relationship cannot be expressed in terms of linear combinations. [19]

Nonlinear Equation. An equation at least one of whose terms is a nonlinear function of the variables. [19]

Nonlinear Function. A function defined as something other than a sum of terms consisting of a constant times a single variable plus a final constant; *net,* not a linear function. [19]

NONLINEAR PROGRAMMING. An inclusive term covering all types of constrained optimization problems except those where the objective function and the constraints are all linear. Special types of nonlinear programming for which some theory has been developed are convex programming, concave programming, and quadratic programming. [19]

NONSINGULAR MATRIX. A square matrix whose determinant is not zero, and thus whose rank (number of linearly independent columns) is equal to its dimension (number of rows). Any nonsingular matrix has an inverse. [19]

NORTHWEST CORNER RULE. A procedure for determining a first basic feasible solution to a transportation problem. [15]

NT-HARD PROBLEMS. Problems for which the algorithms known to solve the problem do so in a time proportional to an exponential function of the value of a measure of problem size or complexity.

OBJECTIVE FUNCTION. The combination of the variables of a mathematical programming problem whose value is to be maximized or minimized subject to the constraints of the problem. For a linear programming problem, the objective function is a linear combination of variables. [19]

OBJECTIVE FUNCTION ELEMENT. One of the coefficients in the objective function. [19]

OBJECTIVE VALUE. The value of the objective function at a particular solution. [19]

OBSERVED VALUE. The particular value of a characteristic determined as a result of an observation or test. [3]

OPERABLE. The state of being able to perform the intended function. [28]

OPERATIONAL. Of, or pertaining to, the state of actual usage. [28]

OPERATIONS RESEARCH. An organized and systematic analysis of complex situations, such as arise in the activities of risk-taking organizations of people and resources. The analysis makes use of certain specific disciplinary methods, such as probability, statistics, mathematical programming, and queuing theory. The purpose of operations research is to provide a more complete and explicit understanding of complex situations, thus leading to an optimum performance of individuals, utilizing the resources available.

OPPORTUNITY COST. The money or other value sacrificed by choosing one course of action thus ruling out others.

OPTIMIZE. To maximize or minimize an objective function. [19]

OPTIMUM SOLUTION. A set of values of the variables which optimizes the objective function, subject to the constraints of the problem. [19]

ORDER-STATISTICS. When a sample of variate values are arrayed in ascending order of magnitude these ordered values are known as order-statistics. Examples are the smallest value of a sample and the median. More generally, any statistic based on order statistics in this narrower sense is called an order-statistic, *e.g.*, the range and the interquartile distance. [22]

PARAMETER. This word occurs in its customary mathematical meaning of an unknown quantity which may vary over a certain set of values. In statistics it most usually occurs in expressions defining frequency distributions (population parameters) or in models describing a stochastic situation *(e.g.,* regression parameters). The domain of permissible variation of the parameter defines the class of population or model under consideration. [22]

PARAMETRIC PROGRAMMING. A method for investigating the effect on an optimal linear programming solution of a sequence of proportionate changes in the elements of a single row or column of the matrix. Most commonly the method is applied to either the objective function row or the right-hand-side column. [19]

PARETO CURVE. An empirical relationship describing the number of persons y whose income is x, first advanced by Pareto (1879) in the form

$$y = Ax^{-(1+\alpha)} \quad 0 \leqq x \leqq \infty$$

The expression is now used to denote any frequency distribution of this form whether related to incomes or not. The variable x may be measured from some arbitrary value, not necessarily zero.

The coefficient A in the expression for the Pareto curve is generally referred to as the "Pareto Index." It affords evidence of the concentration of incomes, or, more generally, of the concentration of variate values in distributions of the Pareto type. [22]

PARETO OPTIMALITY (NON-DOMINATED SOLUTION). A solution to a multicriterion problem for which there exists no other feasible solution which has an equal or superior criterion value for each criterion.

PARTITION. To separate a linear program into related subprograms. More generally to divide a set into nonintersecting subsets. [19]

PAYOFF. The amount received by one of the players in a play of a game. [21]

PAYOFF FUNCTION. For a two-person zero-sum game, the payoff function M is the function for which M (x, y) (positive or negative) is the amount paid by the minimizing player to the maximizing player in case the maximizing player uses pure strategy x and minimizing player uses pure strategy y. [21].

PAYOFF MATRIX. A two dimensional array $\{P_{ij}\}$ in which the columns (or rows) represent the available strategies, the rows (or columns) represent the states of nature and the entries P_{ij} represent the outcomes (on some scale of measure) for the ith column and the jth row.

PERMANENTLY FEASIBLE SET (STOCHASTIC PROGRAMMING). Consider a linear program with random parameters. Here one considers only the convex set of those vectors X which will be feasible regardless of the subsequently observed parameters of the linear program. For each (A,b) the X's satisfying $AX = b$, $X \geq 0$ form a convex polyhedron. The set of permanently feasible X's are those that are elements of the intersection of these polyhedra where the intersection is overall A,b. [19]

PERSONNEL-ASSIGNMENT PROBLEM. (See ASSIGNMENT PROBLEM.) [15]

PERSONNEL SUBSYSTEM. A management concept which considers that functional part of a system which provides, through effective development and implementation of its various elements, the specified human performance necessary in the operation, maintenance, support, and control of the system in a specified environment.

PERTURBATION TECHNIQUES. The process of modifying slightly the right-hand-side coefficients of a linear programming problem to insure against the possibility of cycling. [15]

P-HARD PROBLEMS. Problem for which algorithms have been found that can be solved in a time proportional to a polynomial function of the value of a measure of the problem size or complexity.

PHASE I. The mathematical procedure used by the simplex algorithm, two-phase method, to find a first feasible solution to a linear programming problem. [19]

PHASE II. The mathematical procedure used by the simplex algorithm, two-phase method, to find an optimal basic feasible solution, given a first basic feasible solution. [19]

PHASE PROCESS. A renewal process where each independent and identically distributed interval-length distribution is the distribution of a phase random variable.

PHASE QUEUING MODEL. A queueing model in which either the arrival or service process or both is a phase process. This class of queueing model is a generalization of the so-called "method of stages," where the service-time or arrival-time distributions were either hyperexponential, Erland, or Coxian. Computational results for phase queueing models allow for numerical evaluation of complex models.

PHASE-TYPE RANDOM VARIABLE. The time-til absorption in a finite-state Markov Process having exactly one absorption state. Phase-type random variables are Markovian, thus their use in probability modeling allows for relatively easy computation in the form of small matrix inversions or numerical integration. Phase distributions are also dense in the set of distributions having support on the non-negative real-number line; meaning that phase distributions can take on any arbitrary shape. Their flexibility in shape and computational convenience has led to their adoption in a wide variety of probability models.

PIECEWISE LINEAR APPROXIMATION. The division of the domain of definition of a function into subregions, and the replacement of the function by some closefitting linear function in each subregion. [19]

PIVOT COLUMN. The column of the matrix containing the pivot element. In a linear programming iteration, it is the column associated with the entering variable (nonbasic variable picked to become basic). [19]

PIVOT ELEMENT. The element in a matrix found at the intersection of the pivot column and pivot row.

PIVOT ROW. The row of the matrix containing the pivot

element. In a linear programming iteration, it is the row associated with the departing variable (basic variable picked to become nonbasic). [19]

PIVOT STEP. A step consisting of a single transformation of the matrix in a pivotal method of reduction of a set of linear equations. [19]

PIVOTAL METHOD. One of the methods used in the solution of sets of linear equations, in which some particular coefficient plays a dominant role at each stage of the elimination process. One such method, which forms the base of the ordinary simplex method of linear programming, is called the method of multiplication and subtraction. A step in this method involves transforming the equations so that a designated variable X_s shall appear only in the rth equation and no others. To accomplish this, row r of the matrix is first divided by a_{rs} and suitable multiples of this row are subtracted from all other rows so that the coefficient of x_s in the other equations is zero. The coefficient a_{rs} appears in the computation of each new coefficient, it is called the pivot element. [12]

POLAR CONE. A cone in which all vectors make an angle of 90° or less with the generating vector.

POSITIVE DEFINITE (OR SEMIDEFINITE) QUADRATIC FORM. A quadratic form is called positive definite if

$$X^TAX > 0 \text{ for every } X \text{ except } X = 0.$$

It is called positive semidefinite if

$$X^TAX \geq 0 \text{ for every } X \text{ and there exist } X \neq 0 \text{ for which } X^TAX = 0. [17]$$

PRECISION. The closeness of agreement between randomly selected individual measurements or test results.

PREDICTED. Expected at some future date, on the basis of analysis of past experience. [28]

PREEMPTIVE PRIORITY (PREEMPTIVE SERVICE). A queue discipline in which the arrival of a higher priority while a lower priority is in service requires the return to the queue of the lower priority. There are two cases:

a. Repeat rule: The ejected item returns to service, having lost all service before ejection;

b. Preemptive resume rule: The ejected item resumes at the point of service where it was interrupted. [33]

PROBABILITY. A basic concept which may be taken either as expressing in some way a "degree of belief," or as the limiting frequency in an infinite random series. Both approaches lead to much the same calculus of probabilities. [22:226, 6:220, 25:18, 31:59]

PROCESS. Any set of conditions or causes which work together to produce a given result. [32]

PRODUCER'S RISK. For a given sampling plan, the probability of not accepting a lot the quality of which has a designated numerical value representing a level which it is generally desired to accept. Usually the designated value will be the Acceptable Quality Level (AQL).

PRODUCT FORM OF INVERSE (COMPUTING FORM). A computationally efficient (for sparse matrices) way of updating only the required portion of the inverse; often used in revised simplex (q.v.) codes due to its calculation and input-output characteristics. [19]

PROFIT RANGE. The interval $[c-\Delta_1, c + \Delta_2]$ where c is a profit coefficient in the objective function, and the two endpoints are points where a change of basis must first occur to maintain optimality. [19]

PROGRAMMING. The investigation of the structure and state of a system and the objective to be fulfilled in order to construct a statement of the actions to be performed, their timing, and their quantity (called a program or schedule) which will permit the system to move from a given status toward the defined objective. [11]

PROGRAMMING PROBLEMS. Programming problems are concerned with the efficient use or allocation of limited resources to meet desired objectives. [15]

PURE STRATEGY. A mixed strategy (q.v.) which has all frequencies equal to zero except one which equals unity. [11]

QUADRATIC FORM. A quadratic form is a numerical function of n variables which can be written

$$z = \sum_{i=1}^{n} \sum_{j=1}^{n} a_{ij} x_i x_j.$$

if $A = [a_{ij}]$, $X = (x_1, x_2 ... x_n)$

then in matrix notation

$$Z = X^TAX. \qquad [19]$$

QUADRATIC FUNCTION. A function which contains second-degree terms in the variables, *e.g.* $a + bx + cx^2$. [19]

QUADRATIC PROGRAMMING. Maximization (minimization), subject to linear constraints of an objective function which is a concave (convex) quadratic function. [19]

QUALITY. The totality of features and characteristics of a product or service that bear on its ability to satisfy stated or implied needs.

QUALITY ASSURANCE. All those planned or systematic actions necessary to provide adequate confidence that a product or service will satisfy given requirements for quality.

QUALITY CONTROL. The operational techniques and activities that are used to fulfill requirements of quality.

QUANTAL RESPONSE. The response of a subject to a stimulus is said to be quantal when the only observable—or the only recorded—consequence of applying the stimulus is the presence or absence of a certain reaction, *e.g.,* death. A quantal response may be expressed as a two-valued variate taking values 0 and 1. [22]

QUANTITATIVE DATA. A term usually used to describe data in which the variables concerned are quantities, as distinct from data derived from qualitative attributes.

QUEUE DISCIPLINE. The rule for selecting a unit in the waiting line for service.

QUEUING, MULTISTAGE. Involves two or more steps in a process, each of which involves waiting in a line. In this case, processing is done in sequence rather than in parallel.

QUEUING THEORY. The theory involving the use of mathematical models, theorems and algorithms in the analysis of systems in which some service is to be performed under conditions of randomly varying demand, and where waiting lines or queues may form due to lack of control over either the demand for service or the amount of service required, or both. Utilization of the theory extends to process, operation and work studies. (See source code 33.)

RANDOM. This word may be taken as representing an undefined idea, or, if defined, must be expressed in terms of the concept of probability. A process of selection applied to a set of objects is said to be random if it gives to each one an equal chance of being chosen. Generally, the use of the word "random" implies that the process under consideration is in some sense probabilistic. [22:237, 13:29]

RANDOM FAILURE. (See FAILURE, RANDOM.) [20]

RANDOM VARIABLE. A real-valued function that maps outcomes, which are elements of the sample space of a random experiment to the set of Real numbers. Distribution functions associated with random variables provide a probability associated with values of random variables.

RANDOM WALK. The path traversed by a particle which moves in steps, each step being determined by chance either in regard to direction or in regard to magnitude or both. Cases most frequently considered are those in which the particle moves on a lattice of points in one or more dimensions, and at each step is equally likely to move to any of the nearest neighboring points. The theory of random walks has many applications, e.g., to the migration of insects, sequential sampling and, in the limit, to diffusion processes. [22:239, 13:311, 25:393]

RANGE. The difference between the largest observed value and the smallest observed value. [2]

RANK OF A MATRIX. The maximum number of linearly independent rows of the matrix. Equivalently, the maximum number of linearly independent columns, or the size of the largest square submatrix with an inverse. [19]

RATING. The value of an item parameter which can be attained under specified conditions. [28]

READY RATE, OPERATIONAL (COMBAT). Percent of assigned items capable of performing the mission or function for which they were designed, at a random point in time. [28]

RECTILINEAR. A distance metric where movement is only allowed in two directions which are perpendicular to each other.

REDUNDANCY. In a system, the existence of more than one means for performing a given function.

REDUNDANCY, ACTIVE. That redundancy wherein all redundant items are operating simultaneously rather than being switched on when needed. [20]

REDUNDANCY, STANDBY. That redundancy wherein the alternative means of performing the function is inoperative until needed, and is switched on upon failure of the primary means of performing the function. [20]

REDUNDANT EQUATIONS. A set of equations, one of which may be expressed as a linear combination of the others. Such an equation can be omitted from the system without affecting the solution. [19]

RELIABILITY (OF AN ITEM, EXPRESSED NUMERICALLY). The probability that an item will perform a required function under stated conditions for a stated period of time. *Note:* This definition is used when defining the characteristic intended by such modified terms as *assessed reliability* and *predicted reliability*. [3]

RELIABILITY (GENERAL DEFINITION). Ability of an item to perform a required function under stated conditions for a stated period of time. [3]

RELIABILITY, HUMAN PERFORMANCE. The probability that man will accomplish all required human functions under specified conditions. [28]

RELIABILITY, INHERENT. The potential reliability of an item present in its design. [20]

REPAIR. (See MAINTENANCE, CORRECTIVE.) [28]

REPAIRABLE UNIT. (See UNIT, REPAIRABLE.) [20]

REPEATABILITY (MEASUREMENT). (See PRECISION.)

RESPONSE. The reaction of an individual unit to some form of stimulus. It may be reaction to a drug, as in big-assay, or the reaction to a request for information, as in sample surveys of human beings. [22]

RESTRICTION. An equation or inequality limiting the feasible range of variation of one or more variables. Also used for constraint. [19]

REVISED SIMPLEX METHOD (COMPUTING FORM). A variant of the simplex method, especially suitable for problems in which the number of variables is much larger than the number of equations. The method employs (1) an implicit inverse, "product form of the inverse," or (2) an explicit inverse of the current basis to compute an updated row or column of the tableau as required. The simplex multipliers comprise the profit row of the inverse; they are used to test for optimality and to select a new entering variable. [19]

RIGHT-HAND SIDE. The mathematical expression on the right of the equality or inequality sign in an equation or inequality. In linear programming, by convention, the right-hand side of each constraint is merely the constant term, with the complete linear function as the left-hand side. [19]

RIGHT-HAND SIDE ELEMENT. An element of the column vector comprising the right-hand sides of the constraints. [19]

RIGHT-HAND-SIDE RANGE. The interval $[b - \Delta_1, b + \Delta_2]$ where b is the original value of the right-hand-side element and the two endpoints are the first points where a change of basis must occur to maintain feasibility. Within this interval, the optimal objective value is a linear function of the right-hand-side element. [19]

SADDLEPOINT. A function $F(X, \lambda)$ is said to have a saddle point at the point (X_0, λ_0), if there exists an $\epsilon > 0$ such that for all X, $|X - X_0| < \epsilon$, and all λ, $|\lambda - \lambda_0| < \epsilon$, we have

$$F(X, \lambda_0) \leq F(X_0, \lambda_0) \leq F(X_0, \lambda) \qquad [18]$$

SADDLEPOINT OF A GAME. For any finite two-person zero-sum game, the elements a_{ij} of the payoff matrix satisfy the relation

$$\max_i(\min_j a_{ij}) \leq \min_j(\max_i a_{ij}).$$

If the sign of equality holds, then $\max_i(\min_j a_{ij}) = \min_j(\max_i a_{ij}) = v$ and there exist pure strategies i_0 and j_0 for the maximizing and minimizing players, respectively, such that if the maximizing player chooses i_0 then the payoff will be at least v no matter what strategy the minimizing player chooses, and if the minimizing player chooses j_0 then the payoff will be at most v no matter what strategy the maximizing player chooses. Thus

$$v = a_{i_0 j_0} = \max_i a_{i j_0} = \min_j a_{i_0 j}$$

In this case, the game is said to have a saddle point at (i_0, j_0). There might be more than one saddle point at which the value v is taken on. Similar statements hold for an infinite two-person zero-sum game, for which there might or might not be a saddle point. [21]

SAMPLE. A group of items, observations, test results, or portions of material, taken randomly from a larger collection of items, observations, tests results, or quantities of material, which serves to provide information that may be used as a basis for making a decision concerning the larger collection.

Sample Size. Number of units in a sample. When a sample of 10 units is taken from a population of 1,000 units, for example, we say that the sample size is 10.

Screening Test. A test or combination of tests intended to remove unsatisfactory items or those likely to exhibit early failures. [20]

Secondary Failure. (See FAILURE, SECONDARY). [20]

Sensitivity. The responsiveness of a solution to changes in the numerical values of the problem parameters. [19]

Sensitivity Analysis. An analysis of the effect on the solution of a mathematical problem, for example, as the parameters of the problem are varied. [11]

Sensitivity Testing. A general technique which attempts to determine a relation between quantal response and some stimulus by using a destructive test.

Separable Constraint. Let

$$g(x_1, x_2, ..., x_n) \leq 0$$

be the ith constraint in a mathematical programming formulation. Then, if $g(x_1, x_2, ..., x_n) = g_1(x_1) + g_2(x_2) + ... + g_n(x_n)$, the constraint is said to be separable. [18]

Separable Objective Function. When the objective function of a mathematical programming problem in n variables is capable of being written as a sum of n functions each of which is a function of only a single variable, the objective function is said to be separable. More precisely

$$Z = f(x_1, x_2, ..., x_n) = f_1(x_1) + f_2(X_2) + ... + f_n(x_n)$$

is separable. [18]

Separable Programming. A class of nonlinear programming problems in which each function appearing may be expressed as the sum of separate functions of single variables. When all the functions have been separated, it is possible to make a piecewise linear approximation to each one. For convex constraints, this is all that is required, since the LP algorithm will always choose the two nearest points surrounding the desired value, giving accuracy dependent only on the fineness of the approximation. For concave constraints, however, the ordinary LP algorithms will choose from the two most widely separated points, ignoring those in between and giving generally a most unsatisfactory accuracy. However, the separable programming algorithm assures that the points chosen are contiguous, and thus accurate. Of course, only a local optimum solution is assured.

Sequential Sampling. Sampling technique in which a succession of samples, of a particular size each, are chosen at random.

Service Rate Distribution, Constant. (See ARRIVAL RATE DISTRIBUTION, CONSTANT).

Service Time, Constant. (See ARRIVAL RATE DISTRIBUTION, CONSTANT).

Serviceability. A measure of the degree to which servicing of an item will be accomplished within a given time under specified conditions. [28]

Servicing. The replenishment of consumables needed to keep an item in operating condition, but not including any other preventive maintenance or any corrective maintenance. [28]

Service Rate, Mean. (See MEAN SERVICE RATE).

Set. Any well-defined collection of objects, things or symbols. To be well-defined, it must be possible to tell beyond doubt whether or not a given object, thing or symbol belongs to the collection being considered.

Shadow Prices. The values of the non-slack dual variables at optimality. This usage is appropriate in linear programming problems where the primary variables $x', ... x_n$ are quantities and where the right-hand side b_i of the typical constraint

$$\sum_{ij} a_{ij} x_j = b_i$$

is in the same unit and represents the quantity available system-wide of some particular scarce resource. Since at optimality the objective function z may be expressed as

$$z = b_1 w_1 + b_2 w_2 + ... + b_m w_m$$

where the w_i are the dual variables, we see that the profit accruing from the optimal usage of the resources $b_1, ..., b_m$ may be broken up into m imputed portions $b_i w_i$; and, in turn w_i may be regarded as an imputed unit value (shadow price) for b_i. The shadow price may be interpreted as a fictitious price that society gives to a commodity when the economy is operating in a state of perfect competition. [12]

Shortest Route Problem. Given N nodes numbered i = 1, 2,..., N and a set of numbers (d_{ij}) where d_{ij} = the distance required travel from the ith node to the jth node. Find the minimum distance route from node 1 to N. [19]

Signal Flow Graph. A graphical model of a system in which nodes represent the values of system variables and directed branches between the nodes represent functional relationships between the system variables.

Significance Testing. Statistical appraisal of the outcomes of sampling to note whether or not, at a certain level of risk, the results represent real effects or chance fluctuations of sampling and measurement.

Simplex. A simplex is an e-dimensional convex polyhedron having exactly n + 1 vertices. [15]

Simplex Algorithm. (See SIMPLEX METHOD).

Simplex Method. A computational procedure for solving a linear programming problem. The method embodies an algebraic algorithm, termed the simplex algorithm, which transforms by iterative steps a starting solution into an optimum solution (or determines that such a solution does not exist). [19]

Simplex Multipliers. The components of a vector $(\pi_1, \pi_2, ..., \pi_m)$ associated with the current solution of a linear programming problem. If the objective function is

$$\sum_i c_i x_i,$$

and the current basis is B, and the subscripts of the basic variables are $i_1, i_2, ..., i_m$, the simplex multipliers are defined by the equation

$$(\pi_1, \pi_2, ..., \pi_m) = (c_{i_1}, c_{i_2}, ..., c_{i_m}) B^{-1}$$

Simulation. The design and operation of a model of a system. Commonly implies the use of a computer program designed to accept selected inputs; to treat those inputs in a manner analogous to the way the real system operates, and to read out measurements of the status and change results of the programmed operations affected by those inputs.

Singular Matrix. A square matrix whose rank is less than its dimension and whose determinant is zero. A singular matrix has no inverse. [19]

Sink. That single node in a capacitated network at which all flow is assumed to terminate.

Slack. The difference between the larger and smaller members of an inequality. If $u \leq v$, the slack is defined as s = v - u. [12]

Slack Variable. An auxiliary variable introduced to convert an inequality constraint to an equation. [19]

Slack Vector. The column vector associated with a slack variable. It is a unit vector with + 1 or - 1 in the row in which the slack appears, and zeros elsewhere. [19]

Solution. In linear programming, a set of values of the variables which satisfy all the given constraints. If all of these values are nonnegative, the solution is called feasible; if one or more are negative, it is called nonfeasible. Solutions are otherwise classified as basic (number of variables with nonzero values less than or equal to the number of constraints) and nonbasic (number of nonzero variables greater than the number of constraints). [12]

Solution Level. The set of values taken by the variables in a solution.

Source. That single node in a capacitated network from which all flow is assumed to originate.

Sparse Vector or Matrix. A vector or matrix whose elements are mostly zeros. [19]

Standard Deviation. The most usual measure of dispersion of observed values or results expressed as the positive square root of the variance.

Standard Gamble. A gamble where one gains X_L with probability p or otherwise gains X_H with probability (1-p).

Standby Redundancy. (See REDUNDANCY, STANDBY). [20]

Starting Basis. The set of column vectors associated with the basic variables of a starting solution in linear programming. [19]

Starting Solution. The first solution used at the beginning of an iterative-type solution to an optimization problem.

State Probability In Queuing Models. The state of a system is specified by the number of units in the queue, waiting for service, the number of units in service, etc. The state probability is then the probability that the system is in a given state.

Stationary Process. A stochastic process $\{x_t\}$ is said to be strictly stationary if the multivariate distribution of

$$x_{t1+h}, x_{t2+h}, \ldots x_{tn+h}$$

is independent of h for any finite set of parameter values

$$x_{t1+h}, \ldots x_{tn+h}, t_1, t_2, \ldots t_n.$$

The process is said to be stationary in the wide sense if the mean and variance exist and are independent of t. [22:279, 5:4]

Statistic. A function of one or more random variables which does not depend upon any unknown parameter. Common statistics are the prior sample mean or variance of a random sample.

Statistical Quality Control. The application of statistical techniques to the control of quality.

Steady State. A physical condition in which the variables (q.v.) affecting a system (q.v.) are either invariant or periodic functions of time.

Steppingstone Method. A name for a special form of the simplex method used for solving the transportation problem. Given a feasible solution, the algorithm tells how to improve this solution by introducing a new source- destination allocation and reallocating among existing allocations so that the constraints remain satisfied. Those existing allocations whose values are modified in this way are called steppingstones. If no such reallocation can improve the solution, the current solution is optimal. [19]

Step Stress Test. A test consisting of several stress levels applied sequentially for periods of equal duration to a sample. During each period a state stress level is applied and the stress level is increased from one step to the next. [20]

Stochastic. The adjective "stochastic" implies the presence of a random variable; e.g., stochastic variation is variation in which at least one of the elements is a variate and a stochastic process is one wherein the system incorporates an element of randomness as opposed to a deterministic system. [22:279, 4:6, 29:161]

Stochastic Programming. A generalization of linear programming in which any of the unit costs, the coefficients in the constraint equations, and the right hand sides may be random variables. The aim of such programming is to choose levels for the variables which will minimize some function of the cost. (See PERMANENTLY FEASIBLE SET). [19]

Storage Life (Shelf Life). The length of time an item can be stored under specified conditions and still meet specified requirements. [28]

Strategy. The decision rule used for making the choice from available courses of action.

Strategy (Mixed). (See MINIMAX THEOREM).

Stratified Random Sampling. The process of selecting a simple random sample from each of the population strata. [26]

Stratum. A group of units from a population, a subpopulation, usually defined by relevant population characteristics. [26]

Strictly Concave Function. The negative of a strictly convex function. [19]

Strictly Convex Function. (See CONVEX FUNCTION).

Strictly Increasing Function. A function f is strictly increasing on an interval (a,b) if $f(y) > f(x)$ for any two numbers x and y (of this interval), for which $x < y$. [21]

Suboptimal. (1) Not yet optimal. (2) Optimal over a subregion of the feasible region. [19]

Supporting Hyperplane. Let W be a boundary point of a convex set D. Then $C^TX = Z$ is called a supporting hyperplane at W if $C^TW = Z$ and if all of D lies in one closed half-space produced by the hyperplane. For every boundary point W of D, there exists at least one supporting hyperplane at W. [18]

Surplus Variable. A slack variable when an inequality is of the "greater than or equal to" form.

Surveillance. (1) The continuing analysis and evaluation of records, methods, and procedures including the act of verification to assure conformance with technical requirements. (2) A system whereby supplies and equipment in storage are subjected to, but not limited to, cyclic, scheduled, and special inspection and continuous action to assure that material is maintained in ready for issue condition. [27]

Survivability. The measure of the degree to which an item will withstand hostile man-made environment and not suffer abortive impairment of its ability to accomplish its designated mission. [28]

SYMMETRIC PARAMETRIC PROGRAMMING. The simultaneous parameterization of the right-hand side and the objective function. This is useful in economic studies when both costs and requirements change as a linear function of a single parameter - for example, time. Another use is as a sort of primal-dual algorithm to go from a pseudo right-hand side and objective function (for which the starting basis is optimal and feasible) to true values of the right-hand side and objective function. In some problems, this approach requires far fewer iterations than a more conventional algorithm. [19]

SYSTEM. A set of objects with relations between the objects and their attributes.

SYSTEM EFFECTIVENESS. A measure of the degree to which an item can be expected to achieve a set of specific mission requirements, and which may be expressed as a function of availability, dependability and capability. [28]

SYSTEM, LINEAR. A system is said to be linear if an input $f_1(t)$ produces an output $g_1(t)$ and an input $f_2(t)$ produces an output $g_2(t)$ and if then an input $af_1(t) + bf_2(t)$ produces an output $ag_1(t) + bg_2(t)$ for all $f_1(t)$ and $f_2(t)$ and all constants a and b.

TABLEAU. The current matrix, with auxiliary row and/or columns, as it appears at an iterative stage in the standard simplex method.

TECHNOLOGY MATRIX. The coefficient matrix, or (a_{ij}), as distinguished from the costs coefficients (c_j) and the right-hand sides (b_j) of a linear programming problem.

TESTING. A means of determining the capability of an item to meet specified requirements by subjecting the item to a set of physical, chemical, environmental, or operating actions and conditions. [3:17]

TIME, ADJUSTMENT OR CALIBRATION. That element of maintenance time during which the needed adjustments or calibrations are made. [28]

TIME, ADMINISTRATIVE. Those elements of delay time that are not included in supply delay time. [28]

TIME, ALERT. That element of uptime during which an item is thought to be in specified operating condition and is awaiting a command to perform its intended mission. [28]

TIME (AS USED IN RELIABILITY DEFINITIONS). Refers to any duration of observations of the considered items—either in actual operation or in storage, readiness, etc., but excludes downtime due to a failure. *Note:* In definitions where "time" is used, this parameter may be replaced by distance, cycles, or other measures of life as may be appropriate. This refers to terms such as acceleration factor, wear-out failure, failure rate, mean life, mean-time-between-failures, mean-time-to-failure reliability, and useful life. [20]

TIME, CHECKOUT. That element of maintenance time during which performance of an item is verified to be in specified condition. [28]

TIME, CLEANUP. That element of maintenance time during which the item is enclosed and extraneous material not required for operation is removed. [28]

TIME DELAY. That element of downtime during which no maintenance is being accomplished on the item because of either supply delay or administrative reasons. [28]

TIME, DOWN (DOWNTIME). The element of time during which the item is not in condition to perform its intended function. [28]

TIME, FAULT CORRECTION. That element on maintenance time during which a failure is corrected by a) repairing in place; b) removing, repairing, and replacing; or c) removing and replacing with alike serviceable item. [28]

TIME, FAULT LOCATION. That element of maintenance time during which testing and analysis are performed on an item to isolate a failure. [28]

TIME, INACTIVE. That time during which an item is in reserve (in the inactive inventory). [28]

TIME, ITEM OBTAINMENT. That element of maintenance time during which the needed item or items are being obtained from designated organization stockrooms. [28]

TIME, MISSION. That element of uptime during which the item is performing its designated mission. [28]

TIME, MODIFICATION. The time necessary to introduce any specific change(s) to an item to improve its characteristics or to add new ones. [28]

Time, Preparation. That element of maintenance time needed to obtain the necessary test equipment and maintenance manuals, and set up the necessary equipment to initiate fault location. [28]

Time-Series. A time-series is a set of ordered random variables on a quantitative characteristic of an individual or collective phenomenon taken at different points of time.

Tolerance (Specification Sense). The total allowable variation around a level or state (upper limit minus lower limit), or the maximum acceptable excursion of a characteristic.

Transformation Matrix. An elementary matrix representing a single change of basis. It is the inverse of a matrix formed from an identity matrix (which is the present basis in terms of itself) by replacing one unit vector (representing the departing variable) with another vector (representing the entering vector). The entering variable is expressed in terms of the present (old) basis. [19]

Transformation of a Matrix. A change in the appearance of a matrix which leaves the solutions to the corresponding set of linear equations unchanged; a change of basis. [19]

Transient State. In contrast with steady state, a state before some time t such that there is a probability approaching zero that the system can return to that (transient) state.

Transportation Problem. A homogeneous product is to be shipped in the amounts $a_1, a_2,...a_m$ respectively from each of m shipping origins and received in amounts $b_1, b_2,...,b_n$ respectively by each of n shipping destinations. The cost of shipping a unit amount from the ith origin to the jth destination is c_{ij} and is known for all combinations (i,j). The problem is to determine the amounts x_{ij} to be shipped over all routes (i,j) so as to minimize the total cost of transportation. [15]

Transshipment Problem. A generalized transportation problem in which items may pass through intermediate nodes between source and destination. [11]

Traveling-Salesman Problem. A salesman is required to visit each of n cities, indexed by 1,...n. He leaves from a home base city indexed by 0, visits each of the n cities exactly once, and returns to city 0. It is required to find an itinerary which minimizes the total distance traveled by the salesman. [15]

Tree. A network having n nodes is a tree if it has (n - 1) arcs and no loops. [11]

Trim Problem. The problem of cutting rolls of paper or other material to meet the orders for a specified list of roll widths. The problem can be formulated as a linear programming problem. [19]

Unbounded Solution. Solution with some infinite or unbounded values, and therefore with an infinite optimum for the objective function. [19]

Unique Optimal Solution. The optimal solution in a problem which has one and only one optimal solution. [19]

Unit, Nonrepaired. An item which is not repaired after a failure.

Unit, Repairable. A unit which can be repaired after a failure.

Universe. The totality of the test of items, units, or measurements, etc., real or conceptual, that is under consideration.

Upper Bounded Linear Programming Problem. A linear programming problem in which all the variables x_j are restricted by a corresponding upperbound d_j, i.e., $x_j \leq d_j$. [15]

Useful Life. The period from a stated time, during which, under stated conditions, an item has an acceptable failure rate, or until an irreparable failure occurs. [20]

Utility. A numerical index representing relative preferences assigned among a group of objects, alternatives, prospects, or choices.

Utilization Parameter. In queuing theory, is the ratio between mean arrival and mean servicing rates.

Value of a Game. The number v (expected payoff) associated with any two-person zero-sum game for which the minimax theorem (q.v.) holds. [21]

Variable. Generally, any quantity which varies. More precisely, a variable in the mathematical sense, i.e., a quantity which may take any one of a specified set of values. It is convenient to apply the same word to denote nonmeasurable characteristics, e.g., "sex" is a variable in this sense since any human individual may take one of two "values," male or female. It is useful, but far from being the general practice, to distinguish between a variable as so defined and a random variable or variate (q.v.). [22:310, 7:422]

VARIANCE. A measure of the squared dispersion of observed values or measurements expressed as a function of the sum of the squared deviations from the population mean or sample average.

VARIATE. In contradistinction to a variable (q.v.) a variate is a quantity which may take any of the values of a specified set with a specified relative frequency or probability. The variate is therefore often known as a random variable. It is to be regarded as defined, not merely by a set of permissible values like an ordinary mathematical variable, but by an associated frequency (probability) function expressing how often those values appear in the situation under discussion. [22:312, 4:26, 13:1993]

VECTOR. An ordered set or array of quantities, usually n real numbers (x_1, x_2,...x). A vector arrayed horizontally is called a row vector; one arrayed vertically is called a column vector. For example, the rows of a matrix are row vectors, the columns of a matrix are column vectors. [19]

VENN DIAGRAM. A schematic representation of the universal set and its subsets. A rectangle is usually used to represent the universal set, points to designate elements of the set and circles to depict subsets of the universal set. The diagram is sometimes referred to as an Euler Diagram.

VERTEX. A point of a convex set which does not lie on a line segment joining any other two points of the set; also called an extreme point. Sometimes used as a synonym for node (q.v.). [19]

WAITING LINE (QUEUE). A line formed by units waiting for service.

WAITING TIME. The time elapsed while a unit is waiting for service.

WEAR OUT FAILURE. (See FAILURE, WEAR-OUT).

WEAR-OUT FAILURE PERIOD. That possible period during which the failure rates of units of a particular item are rapidly increasing due to deterioration processes. [20]

ZERO-SUM TWO-PERSON GAME (MATRIX GAME). A two-person game of strategy in which the amount lost by one player is won by the other. Such a game has an equivalent formulation as a linear programming problem. [15]

BIBLIOGRAPHY

[1] American Society for Quality Control, Standard A1-1971, 20 ANSI Z1.5-1971, *Definitions, Symbols, Formulas and Tables for Control Charts.* Milwaukee, WI: American Society for Quality Control.

[2] American Society for Quality Control, Standard A2-1978, ANSI Z1.6-1978, *Terms, Symbols, and Definitions for Acceptance Sampling.* Milwaukee, WI: American Society for Quality Control.

[3] American Society for Quality Control, Standard A3-1979, ANSI Z1.7-1979, *Quality Systems Terminology.* Milwaukee, Wl: American Society for Quality Control.

[4] Arley, N. and Buch, K.R. *Introduction to Theory of Probability and Statistics.* New York: John Wiley & Sons, 1950.

[5] Bailey, N.T.J. *The Elements of Stochastic Processes.* New York: John Wiley & Sons, 1964.

[6] Bizley, M.T.L. *Probability.* Cambridge, 1957.

[7] Brownlee, K.A. *Statistical Theory and Methodology in Science and Engineering.* (2nd ea.) New York: John Wiley & Sons, 1965.

[8] Burington, R.S. and May, D.C. *Handbook of Probability and Statistics.* Handbook Publishers, 1953.

[9] Charnes, A. and Cooper, W.W. *Management Models and industrial Applications of Linear Programming,* Volumes I & 11. New York: John Wiley & Sons, 1961.

[10] Cramer, H. *The Elements of Probability Theory.* New York: John Wiley & Sons, 1955.

[11] Dantzig, G.B. *Linear Programming and Extensions.* Princeton, NJ: Princeton University Press, 1963.

[12] *Decomposition Principles for Solving Large Structured Linear Programs.* Princeton, NJ: Mathematica, 1963.

[13] Feller, W. *An Introduction to Probability Theory and Its Applications, Volume 1.* (2nd ea.) New York: John Wiley & Sons, 1957.

[14] Ford, L.R. and Fulkerson, D.R. *Flows in Networks.* Princeton, NJ: Princeton University Press, 1962.

[15] Gass, S.I. *Linear Programming.* (2nd ea.) New York: McGraw-Hill, 1964.

[16] *Glossary and Abbreviations for Quality Assurance Program.* New Mexico: Quality Assurance Department, Sandia Corporation, 1962.

[17] Hadley, G. *Linear Programming.* Reading, MA: Addison-Wesley, 1962.

[18] Hadley, G. *Nonlinear and Dynamic Programming.* Reading, MA: Addison-Wesley, 1964.

[19] IBM. *An Introduction to Linear Programming.* White Plains, NY: IBM Corporation.

[20] International Electrotechnical Commission (EC), Technical Committee 56 (Reliability), Draft Copy, *Definitions of Reliability Terms.*

[21] James, G., and James, R.C. *Mathematics Dictionary, Multi-lingual Edition.* D. Van Nostrand Company, Inc., 1959 (1966 Reprint).

[22] Kendall, M.G.. and Buckland, W.R. *Dictionary of Statistical Terms.* (2nd ea.) New York: Hafner Publishing Company, 1960.

[23] McElrath, G.W., Astrachan, M., Leone, F.C., and Bocke, R.W. *Significance Tests,* four papers presented at the Twelfth Annual Convention of the American Society for Quality

Control, Boston, Massachusetts, May 26-28, 1958.

[24] Nemhauser, G.L. *Introduction to Dynamic Programming,* New York: John Wiley and Sons, 1966.

[25] Parzen, E. *Modern Probability Theory and Its Application.* New York: John Wiley & Sons, 1960.

[26] Stuart, A. *Basic Ideas of Scientific Sampling,* Number Four of Griffin's Statistical Monographs and Courses. New York: Hafner Publishing Company, 1962.

[27] United States Department of Defense, Mil.-Std.- 109A, 30 October 1961, *Quality Assurance Terms and Definitions.* Washington, DC: United States Government Printing Office, 1961.

[28] United States Department of Defense, Mill-Std.-721 B, 25 August 1966, *Definitions of Effectiveness Terms for Reliability, Maintainability, Human Factors, and Safety.* Washington, DC: United States Government Printing Office, 1966.

[29] Uspensky, J.V. *Introduction to Mathematical Probability.* New York: McGraw-Hill, 1937.

[30] Wilks, S.S. *Elementary Statistical Analysis.* Princeton, NJ: Princeton University Press, 1951.

[31] Wilks, S.S. *Annuals of Mathematical Statistics,* Vol. 12 (1941), p. 91 and Vol. 13 (1942), p. 400.

[32] Morse, P.M. *Queues, Inventories and Maintenance.* New York: John Wiley and Sons, Inc.

[33] Saaty, T.L. *Elements of Queuing Theory.* New York: McGraw-Hill, 1959.

[34] Bellman, R.E. *Dynamic Programming.* Princeton, NJ: Princeton University Press, 1957.

[35] Eisenhart, C. Realistic evaluation of the precision and accuracy of instrument calibration systems. *Journal of Research of the National Bureau of Standards,* 67C, No. 2, 161-187, 1963.

[36] Murphy, R.B. On the meaning of precision and accuracy. *Material Research and Standard,* Vol. l, No. 4, 264- 267, 1961.

Z94.2
Anthropometry & Biomechanics

Anthropometry

Editorial Note: Even the most casual reader will note that the data in this section are treated differently from those in the other sections; the reasons are many. First and foremost, a straight alphabetical arrangement would not be helpful. For example, why should "knuckle" follow "knee" when the reader is probably interested in more leg dimensions? Clearly a diagrammatic arrangement would be more helpful and logical. However, for the sake of consistency and helpfulness, the anthropometry terms can be found in alphabetical order in the overall index. Nevertheless, the reader should be alerted to the fact that when a term is indexed as being in the *Anthropometry* section, the reader should refer to at least one — if not all — of the three listings in this chapter: *Dimensional Terminology*, *Plate List*, and *Cross Reference List*. Also the reader should refer to the *Glossary*. Although these details are carefully delineated in the following Introduction, it would not be amiss to emphasize here that the numbers found in connection with the Plates also appear in the Dimensional Terminology. Thus, "knee-to-knee breadth, sitting (IV, 30)" can be found diagrammatically in Plate IV, etc. As for actual numerical dimensions, the reader should note the details in the various references in the *Bibliographies*. Again, of special interest to engineers, designers, inventors, et al. — each of the *Anthropometry* terms notes relevant applications: *e.g.*, "knuckle height... carrying handles... suitcases...". Additionally, information on the use of these dimensions in design applications has been updated and enhanced.

The work of the Anthropometry Subcommittee was made much easier due to the efforts of previous edition subcommittees. The subcommittee of the 1983 edition consisted of Charles L. Mauro, Mary A. Marrey, Patricia A. Moore, and Tado B. Warashina. The subcommittee for the 1989 revision consisted of William S. Marras (Chair), Richard J. Cruise, Anil Mital, and Nancy L. Philippart. Many of their definitions have been retained in this edition.

The present subcommittee, distinguished in its own right, has made important updates to this chapter. The Anthropometry Subcommittee included the following individuals:

Chairperson
M. Susan Hallbeck, Ph.D., P.E.
University of Nebraska — Lincoln

Subcommittee

Claire C. Gordon, Ph.D.
U.S. Army Natick Research,
Development and Engineering Center

Joe W. McDaniel, Ph.D.
U.S. Airforce Armstrong Laboratory

Mary Danz Reece, Ph.D.
Exxon Biomedical Corporation

Human Factors and Ergonomics Section Coordinator:
Steven A. Lavender, Ph.D.
Rush-Presbyterian-St. Luke's Medical Center, Chicago

INTRODUCTION

Anthropometry, the study of people in terms of their physical dimensions, has classically been performed by the physical anthropologist. More recently, in an attempt to create more efficient people-machine interfaces, engineers and designers have become increasingly aware of anthropometric terminology and dimensions.

This increased awareness on the part of practicing professionals has brought to light the need for a standard glossary of terms related to anthropometry as it is used in the fields of engineering and design. Unfortunately, most designers and engineers are not versed in medical terminology, nor are most anthropologists familiar with engineering. As a result, the majority of available references fail to communicate to both audiences. Therefore, the Z94.15 Committee was formed for the purpose of identifying those dimensions required in industry, translating their medical definitions, and compiling the resulting terms into a working glossary.

The main purpose of the standard is intended to be that of a technical reference. As such, the practicing professional should be able to consult the standard when encountering unfamiliar terms in texts or when devising a proper reference term for further definition of written or presented material.

A historical overview of the development of anthropometry is discussed in order to provide the reader with an awareness of the varying backgrounds from which anthropometry grew and also, as a result, an indication of the problems of application to the disciplines of industrial engineering and design.

A HISTORICAL OVERVIEW

The following work is intended to give the reader a reference source on the measurements most commonly taken in anthropometric research, as well as their relevance for equipment design. Since all human beings change, it is important to have a means of quantifying the variations of these physical changes. Anthropometry is the formal name for the techniques used to express quantitatively the form of the human body. Thus, anthropometry is the measurement of humans, whether living or dead, and consists primarily in the detailed, accurate, and consistent measurement of the dimensions of the human body (Montagu, 1960:3).

In 1920, the physical anthropologist, Ales Hrdlicka, summarized the purposes for taking anthropometric measurements. He stated that these measurements were taken for the purposes of industrial design, artistic expression of the human body, military purposes, medical, surgical and dental research and procedures, detection of bodily defects and their correction and forensic identification (Hrdlicka, 1920:8). His approach was narrowed considerably since he did admit later on in his treatise (ibid., 1920:8) that the "industrial and artistic systems are of little interest to us."

To practitioners of anthropometry and scholars in anatomy and physical anthropology, it soon became evident that standardization of measurements and measuring techniques was sorely needed and that this methodological disarray was very prejudicial to progress in the field. From 1870 to 1920, many international meetings were set up to discuss, refine, and standardize anthropometric techniques. Among the most significant of the conferences were the first and second craniometric conferences held in Munich and in Berlin in the years 1877 and 1880 respectively and the thirteenth General Congress of the German Anthropological Societies, held in 1882 at Frankfurt-on-Main.

It must be noted that the field of physical anthropology was concerned basically with anthropometric measurements until the early 1950's when the theoretical orientation of the field changed radically and interest in anthropometry was replaced by an ever-growing interest in human evolutionary processes.

Studies of movements of the body were undertaken in industry in this country during the second decade of the present century in order to improve per capita work output. During the 1920's, many management personnel realized the value of placing the work within easy reach of the workers and this led to a large amount of research geared to studying maximum and normal work areas dimensions. During the 1930's and the 1940's, a large amount of military anthropological research was done establishing effects of bodily dimensions and physiques on the design and use of such military equipment as apparel, aircraft, battleships, tanks, etc. During the same period O'Brien addressed the issue of fitting the masses through her work in adult and children garment sizing. These studies helped in synthesizing data from psychology, physiology, anthropology and medicine with engineering to form a field generally known as human engineering.

After World War II, research emphasizing the fitting of the machine to the human became more fully developed as both commercial and military outfits continued to perform research on body dimensions and work space requirements. The design of motor vehicles and mass transit facilities was reevaluated in light of this new orientation of human-machine design specifications.

Recently, it became necessary, as the technological revolution continued, to change the approaches, the types of data gathered, and the measurement instruments in anthropometric research. Dynamic anthropometry was established as it was necessary to develop a scheme of measurement, taking into account three-dimensional coordinate systems. The science of biomechanics (essentially the interdisciplinary study of the mechanical nature and behavior of biological materials) is now closely related to the fields of classical anthropometry and engineering anthropometry.

With the advent of the computer it is now possible to consider the effect of more than two measurements at a time through the use of multivariate analytic routines in which one does not have to determine dependent and independent variables, and can determine the interaction between and within a large amount of variables at the same time.

TYPES OF DIMENSIONS

As mentioned in the historical overview, different types of dimensions have been defined for different applications: These include static and dynamic anthropometry. Static dimensions may be subdivided into circumferences, lengths, skinfolds, and volumetric measurements. Dynamic dimensions include link measurements, center of gravity measurements, and body landmark locations. Static and dynamic anthropometry are also referred to by the names structural and functional anthropometry, terms which more explicitly express the body and its action. Static dimensions are taken with body parts held in fixed, standardized positions. These dimensions are easily obtainable but not so easily applied since design applications often involve the body in functional attitudes. Dynamic dimensions are taken with the body at work, in motion or in workspace attitudes. These measurements are more difficult to obtain with application limited to a particular workspace or movement studied. Functional dimensions account for interactions among body members. For example, the limit of functional arm reach is not due to arm length alone but is affected by shoulder movement, some trunk movement and the function to be performed by the hand.

Functional anthropometric models used in workspace design are statistical in that they describe the probabilistic location of body landmarks of a population of users in a given workspace.

Human body measurements, when taken from a sufficiently large sample, follow a normal distribution. Static and dynamic dimensions are generally given as percentiles (*e.g.* one can discuss a 10th percentile female popliteal height as well as a 10th percentile female eye location in a given workspace attitude): a measure which describes the proportion of a subject population less than or equal to that value in a given dimension. Percentiles provide a univariate basis for estimating the proportion of a population accommodated or inconvenienced by a specific design. Multivariate analysis techniques are becoming increasingly employed in anthropometric design as explained in subsequent sections.

Anthropometric Design Principles

Accommodation of anthropometric characteristics is important to the design of anything which must be operated or maintained by a human. This section will examine some of the design principles associated with anthropometric accommodation. The fundamental principle of design must be that all intended users of an item will be able to use it effectively and efficiently, and not be limited by their size.

There are five approaches to achieving anthropometric accommodation:

1. Select workers who fit the existing design
2. Custom fit each individual
3. Have (several) fixed sizes
4. Make it adjustable
5. Design for the extreme individual

Selecting workers who fit the existing design is appropriate when it is not practical to modify the design. This approach tends to be self-propagating because once a population of users becomes restricted, the restrictions tend to become institutionalized. It is usually less expensive in the future to continue to design to a limited population than to relax the restriction and expand the design range to be accommodated. Examples of situations where this is approach is used is in commercial and military aircraft. These cockpits accommodate a certain range of sitting heights to afford outside vision and a certain range of limb lengths to afford reach to controls. After several generations of aircraft were built for this population, it would be economically disadvantageous to change the characteristics of the pilot population.

Custom fitting the design to each individual user is a good approach when only a very small number of items are being produced. For example, if a one-of-a-kind race car is being built for a specific driver, that driver's size is the only one that requires accommodation. The small population of astronauts have always had custom-fitted space suits.

Having several fixed sizes is the approach used where (1) there is a large population of users, (2) the unit price is relatively small, and (3) adjustment is impractical. An example is shoes and clothing, which come in a variety of fixed sizes, and the user selects the one best matching his/her personal fit criterion. When using this approach, inventory schemes must be developed. A inventory scheme must define the number produced and stocked at each location for each size to accommodate the demand. Because size is approximately normally distributed, the demand for sizes near the mean will be greater than for sizes near the extremes. One common mistake is not empirically validating the inventory scheme, since the fit criteria applied by individuals may not match the theoretical criteria assumed by the designer.

Making the design adjustable is the approach used where (1) there is a medium-sized population of users or (2) adjustability is practical, or (3) the unit price is relatively large. In automobiles, for example, the high unit cost requires adjustability to fit any potential user. In an office chair, the unit cost is not great, but neither is the cost making chairs adjustable.

Designing for the extreme individual or dimension is appropriate where some limiting factor can define either a minimum or maximum value which will accommodate the population. One example of is locating an emergency stop button within the smallest reach. Designing for the extreme individual or dimension is a special case of approach 3 (several fixed sizes) or even of approach 4 (make it adjustable). Designing for the extremes of an unrestricted population would increase the standard 80 inch (2 meter) high doorway by 20 inches (0.5 meters) to accommodate the tallest people, who may be over 8 feet tall.

Usefulness of anthropometric measures for design. Designers must take care when using anthropometric data and selecting which measures to use. For the most part, anthropometric measures were not defined to be useful to designers, but rather for the ease and repeatability of making the measures. Because the human body is pliable and comes in a great range of sizes and proportions, measures were defined to make the best use of surface features and bony landmarks. Anthropometric measures are usually two dimensional (unrelated lengths) and do not describe functional abilities, such as reach envelopes. For this reason, the best use designers can make of anthropometric data is for scaling functional capabilities, and relating the characteristics and performance of specific individuals to populations.

Defining the population. The first and most critical step in anthropometric accommodation is defining the population to be accommodated. If the user population is small, or if only a small number of measures are needed, the designer can consider measuring the actual user population. If this is impractical, the next best practice is to find a matching population which has already been measured.

Selecting a correlated survey
To find a matching population, one must consider many factors, such as gender, age range, and race. The table below has a sampling of the richness and variety of existing surveys from which designers can choose.

In selecting a matching population, pay particular attention as to how the population was restricted. Military surveys generally have the greatest number of measures, some approaching 200 measures per individual. However, most military populations have restrictions on size and weight, and a usually have more limited age range than is found among civilian workers.

While this section contains a number of definitions of anthropometric measures, this is only a small subset of the hundreds available. The designer should also be aware that measures with the same name may have definitions which differ from one survey to another. So even if the measure has the same name, it is important to compare the actual definition to verify that it is the same.

General Design Requirements
Design and sizing should ensure accommodation, compatibility, operability, and maintainability by the user population. Generally, design limits should be based upon a range from the 5th percentile (small) female to the 95th percentile (large) male values for critical body dimensions. For any body dimension, the 5th percentile values means that five percent of the population will be equal to or smaller than that value, and 95 percent will be larger; conversely, the 95th percentile value indicates that 95 percent of the population will be equal to or smaller than that value, and 5 percent will be larger. Therefore, use of a design range from the 5th percentile female to 95th percentile male values will theoretically provide coverage for 95 percent of the user population for that dimension. (See FOOTNOTE).

[FOOTNOTE: If the population being accommodated were all the same gender, the design range of 5th percentile male to 95th percentile male values will theoretically provide coverage for 90 percent of the user population. However, in a mixed population, the design range from the 5th percentile female to 95th percentile male values will theoretically provide coverage for 95 percent, not 90 percent of the user population. The correct value is 95 percent because of the Distributive Law of Algebra: $mA + mB = m(A + B)$. In a mixed male/female population, let M be the number of men and F the number of women, then the total number of people is (M + F). Since no men are excluded by the 5th percentile female cutoff and no women are excluded by the 95th percentile male cutoff, then the following is true: Excluded are 5% of M plus 5% of F equals 5% of (M+F), which means 95 percent remain after the exclusion, not 90 percent]

TABLE OF MISCELLANEOUS CIVILIAN ANTHROPOMETRIC SURVEYS

Law Enforcement Officers	1975
Health and Nutrition, Education Survey	1974
Japanese Civilians	1974
Australian Female Pilots	1973
Airline Stewardesses	1971
Slovakian Civilians	1969
German Office Workers	1969
Czechoslovakian Lumbermen	1969
Swedish Industrial Workers	1969
Swedish Civilian Women	1969
Bantu Miners	1968
South Africans	1968
French Young Men	1968
Air Traffic Controllers	1965
National Health Exam Survey	1962
English Civilian Women	1957
Dutch Civilian Women	1951
Women in Dept. of Agriculture	1940

Safety considerations: Where the design is safety critical (that is, failure to perform could result in serious injury to personnel or equipment), it is recommended that design range from the 1st percentile female to 99th percentile male values be accommodated.

Univariate Standard scores, or Z scores. While some may be more familiar with Z scores than percentiles, they are, in fact, two ways of representing the same phenomenon, that is locating individual measures within a distribution. Anthropometrists prefer percentiles because they are easier to explain, and since anthropometric measures are approximately normally distributed, the two can be related using a table of areas under the normal curve. Percentiles have the additional advantage of being computed from cumulative frequencies, and do not require the distributions to be normal, so they also apply to strength measures, which are often not normally distributed. Percentiles are more straight forward, since they have the same sign, whereas Z scores have different signs above and below the mean of the distribution. To convert a percentile to a Z score, simply use the percentile as an area, and look up the Z score in the table. For example, a 95th percentile has an area of the mean (50th percentile) plus an additional 45 percent, which equates to a $Z = +1.645$. Similarly, a 5th percentile value is 45 percent below the mean and equates to a $Z = -1.645$.

Clothing allowances: Because the anthropometric data represents nude body dimensions taken in standard anthropometric postures, suitable allowances should be made for postural variation, light or heavy clothing, flying suits, helmets, boots, body armor, load-carrying equipment, protective equipment, and other worn or carried items.

Strength considerations: In most instances where strength is the parameter being accommodated, the design should not link strength and body size. The correlation between body size and strength is low, so a designer should not assume that small people are weak and big people are strong. Good design practice requires accommodating size and strength separately where both are relevant. Strength is task specific, so one should not apply a strength measured in one location and direction of force to a situation where the location or direction of force is different. Strength accommodation should also consider the duration of the exertion, that is, endurance.

Multivariate Accommodation

Multivariate Analysis. Univariate percentiles are not appropriate when two or more dimensions are simultaneously used as criteria for design (Moroney and Smith, 1972). When the operator or maintainer must simultaneously perform two or more different actions in such a way that they interact, multivariate techniques should be used to verify accommodation. For example, a vehicle operator must simultaneously have sufficient sitting height/seat adjustment to see over the forward instrument panel, sufficient arm length to reach all hand operated controls, and sufficient leg length to operate all foot pedals. To verify accommodation, individuals with extreme combinations of body proportions are used: tall torso with short limbs, short torso with long limbs, etc. (Hendy, 1990).

Non-additive Percentiles: There is no such thing as a 5th percentile person. However, there are people with one body dimension equal to a 5th percentile population value. Because of lack of proportionality, however, it is unlikely that any one individual will have more than a few dimensions close to 5th percentile. In practice, to exclude only the smaller 5 percent of a population when considering two or more simultaneous variables, the design cutoff value of each variable must be lowered to less than the 5th percentile value.

The percentile values reported in anthropometric surveys are suitable only for univariate accommodation and should not be used for designs where two or more dimensions are used simultaneously as design parameters. While the means of different measures of body parts can be added, the measures away from the means (at the tails of the distribution where the 5th and 95th percentile values occur) are not additive. Because of lack of perfect correlation of body dimensions, those people excluded by the cutoff of one variable are not the same as the people excluded by the cutoff of a second variable. In a simplistic example, the 95th percentile cutoff on one dimension could give 0.95 x 0.95 = 0.90 or about a 10 percent excluded at the large end of the distribution. If three variables are considered simultaneously, then 0.95 x 0.95 x 0.95 = 0.85, or about a 15 percent excluded at the large end of the distribution (Robinette and McConville, 1982: Moroney and Smith, 1972).

Methods for determining multivariate limits: There are several methods for evaluating multivariate accommodation. Where accommodation of a certain percentage of a population is desired, such as the central 95 percent, using the raw data from the original anthropometric survey, the critical limits can be found by making iterative eliminations with trial combinations of the critical dimensions until the desired percentage of population remains. A second method involves using a hypothetical sample of individuals having extreme dimensions on one or a combination of variables (Bittner, 1986). A dial-up data base is now available which can be used to calculate multivariate "test cases" for a number of workplace/crew station design applications. The data base is described in "User's Guide to the Anthropometric Database at the Computerized Anthropometric Research and Design (CARD) Laboratory," by J. C. Robinson, K. M. Robinette, and G. F. Zehner (1992).

Methods of verifying multivariate limits: One method involves testing with a group of subjects carefully selected so each has at least two or more of the required dimensions. Each of the subjects are tested in a mockup or the actual equipment. In practice, it is difficult to find subjects with the rare combinations of size and proportions desired. It may be necessary to use subjects with correct torso and arm dimensions to evaluate the seat-hand control accommodation, then use separation subjects with correct torso and leg dimensions to evaluate the seat-pedal accommodation. There are also techniques for "shimming" subjects to simulate the desired dimensions.

Computer models. Computer models can generate 3-D manikins which can be dimensioned to match all the subjects in a multivariate set. Whereas finding people with specific dimensions and unusual body proportions is a very difficult task, computer models allow the designer to easily define subjects of any size and proportion. These models can be used to evaluate "electronic mockups" as described in "Models for Ergonomic Analysis and Design: COMBIMAN and CREW CHIEF" (McDaniel, 1990) and in "The Development of Computer Models for Ergonomic Accommodation" (McDaniel, 1990).

Why 5th to 95th design criteria?

The traditional design range of from 5th to 95th percentile of the variable is a trade-off between the necessity of not considering everyone in the design range and the practicality of including almost everyone.

The range of human size is much greater than most people realize, with stature of adults ranging from 2 to 8 feet and the weight of adults ranging from 20 to 1,000 pounds. Most of us will never even meet people with these extremes of body size, so the designer should not be forced to consider them when designing a product.

Figure 1 is a plot of a hypothetical dimension (Y axis) to be accommodated against percentiles (X axis) for a cumulative-normal distribution. While the normal distribution has tails which theoretically go to out to infinity, the range of percentiles in this figure go from the 0.05 percentile (that is, 5 in 10,000) to the 99.95 percentile. Thus, the graph is constructed to that the range of Y, which goes from 0 to 100, covers the frequency range of the central 99.9 percent of the population, or all but 1 in 1,000.

Now, compare the extremes of the distribution with the 5th percentile measures (which equals 25 on the Y axis) and the 95th percentile measure (which equals 75 on Y axis). This shows that the included range in Y between the 5th and 95th percentile is 50 units or half the Y range. In other words, accommodating the central 90 percent (5th to 95th percentile) costs half as much as accommodating the whole range.

Examining the slope of the curve, one can also see that it turns and gets very steep beyond the 5th to 95th percentile points. So the 5th and 95th percentile marks have been chosen as the most cost effective design range. It excludes a very small part of the population, and most of those excluded can still cope with the design, since people have some ability to stretch or stoop.

SKINFOLDS

The problem of the quantification of body builds (somatotyping) has been investigated for many years. One of the earliest attempts at quantification of the bodily physique was undertaken by J. Matiegka (1921) who broke down the human physique into four (4) components: gross body weight, skeletal, dermal, and muscular components. Matiegka's study was important in that he pointed to a direction of considering the human body as capable of being measured in a functionally oriented, "dynamic" fashion.

Unfortunately, Matiegka's article was long ignored in anthropometric circles. Using more sophisticated mathematical models, Tanner, Healy and Whitehouse (1959) reconfirmed with slight modifications Matiegka's original observations by performing a factor analysis of body build and finding that there are four orthogonal (independent) factors: skeletal frame size, skeletal width, muscle width, and fat width (Tanner et al, 1959:91). (Factor analysis is a multivariate analytical procedure which identifies the independent factors that explain most of the variation in a sample in decreasing order of importance.)

It must be noted that the measurement of body composition is an important advance from the grossly inappropriate height-weight tables that are currently enjoying great popularity. Taking someone's weight and comparing the resulting deviation from the corresponding "ideal weight" is almost biologically meaningless. One does not have any idea of the amount of muscle versus the amount of fat in the body. The visual system of "body typing" developed by W. C. Sheldon in 1940 is far from desirable when scrutinizing the human body in a functional frame of reference.

We shall briefly consider here the role of the skinfold as an indicator of "fatness vs. leanness", *i.e.*, as an estimator to the individual's muscularity. The relevance of this procedure to apparel design will then become apparent. However, even though at present, the skinfolds' primary area of application would be apparel design and for biomedical and physiological studies, this does not negate the potential for use in the very near future, especially with regard to dynamic anthropometric studies.

Definition and Methodology. The procedure known as "skinfold" is used basically to determine the amount of sub-cutaneous fat. It must be noted that the thickness of the skin is relatively constant throughout the body (Montagu, 1960:85). However, the pattern of fat deposition is highly variable both within and between individuals and populations. As a result, the choice of the site for the taking of a skinfold measurement is critically important. At present, there are two (2) primary sites where comparable data on skinfolds can be extracted.

Triceps skinfold: This is one of the least culturally objectionable and most readily accessible sites on the body. Individuals of both sexes in most cultures can be measured at this site with the least amount of tension and conflict. The site is located at the back of the right or left (it is important to be consistent with the choice of side) upper arm over the triceps muscle at a level approximately midway between the tip of the scapular acromial process and the elbow. The forearm is conventionally flexed at a 90° angle. Once the site is located, the subject then lowers the forearm so that the arm hangs freely as the measurement is taken by the investigator.

Subscapular skinfold: The subscapular skinfold is taken with the subject standing freely. The skin is lifted most readily along a line at about a 45° angle, going medially upwards and laterally downwards. At this site, the thickness of the subcutaneous adipose tissue layer is fairly consistent.

Other sites: A variety of other sites, such as the midaxillary line, chest, abdomen, and legs have been proposed, but the variability encountered with the amount of subcutaneous adipose tissue, as well as cultural discomforts, make these sites less desirable.

Summary: In order to assure comparability of skinfold measurements, the following factors should be borne in mind:

(1) the "skin" should be lifted by firmly grasping the fold between thumb and forefinger;

(2) the width of the skin enclosed between the fingers should be minimal, but still yielding a well defined fold;

(3) the depth of the caliper placement should be the minimal distance from the crest where a true fold, with surfaces approximately parallel to each other, occurs;

(4) the side chosen (right or left) and site locations should be consistent at all times; and

(5) the caliper should consistently be read approximately three seconds after the skinfold is lifted to standardize the effects of compression that inevitably occur.

Finally, note that upper arm circumference minus skin and subcutaneous adipose tissue (skinfold) is a gross measure of the muscularity of the subject.

INTENDED USE OF THE STANDARD

Static dimensions, skinfolds and some dynamic dimensions are defined in this standard. The static dimensions which are included are those frequently referred to in design and engineering and have a well established military data base. A few of the dimensions also have an established civilian data base. There are many dimensions not included, but these are available from the secondary bibliographic sources listed. The standard is not intended to be a complete listing of available anthropometric dimensions. Several functional dimensions used in workspace design are also included.

Dimensions specific to safety in the design of furniture and toys for children have not been included, but appropriate references are included in the bibliography.

Each definition is cross-referenced to a drawing which illustrates the anatomical or body surface landmarks which comprise the dimension.

Sample

POPLITEAL HEIGHT, SITTING (III, 25)
Dimension Name (Plate Number, Dimension Number)

A Cross-Reference list of Dimension Numbers with Dimension Names has also been included since the reader may not immediately know the dimension name by inspection of the plates. In referring to the drawings, locate the number corresponding to the dimension in which you are interested; then, refer to the Cross-Reference List to locate the name of the dimension.

A Glossary of terms is included following the Cross-Reference List for explanation of anatomic terms. The reader may find that referring to the drawings in conjunction with use of the glossary explanation will clarify the term.

The drawings included represent only the adult population since bone maturation is completed by then. The focus of this reference is toward design for the adult population. A male skeleton has been used in the drawings, since, in most cases, the dimensions are similarly taken for male and for female, although the actual data bases are distinct, statistically speaking. For data taking in children's populations, some dimensions would require redefinition inasmuch as certain of the bony landmarks are not well-developed and thus are not palpable.

BIBLIOGRAPHY

If the reader needs further information on actual data bases or methods of taking anthropometric data, he or she should consult the annotated bibliography. The bibliography is subdivided into two sections: primary and secondary. The primary bibliography lists those sources from which the dimensions in this standard were selected. In some cases, the dimension definition has not been changed at all; changes were made only in those instances where it was necessary for clarity, *i.e.*, by providing language understandable by the layperson. The drawings, however, in this standard do include the skeleton as a reference for the dimensions whereas the figures shown in the primary sources did not provide skeletal reference.

The secondary bibliography lists sources of static and dynamic anthropometry which include a more extensive listing of dimensions as well as a data base specific to children's dimensions. Other sources included present methodological information and practical evaluations of the use of static and dynamic anthropometry.

DYNAMIC/FUNCTIONAL DIMENSIONS

These dimensions are workspace specific. General definitions are provided, but application requires knowledge of relevant parameters of workspace as well as body attitude in the workspace.

FUNCTIONAL ARM REACH CONTOURS. Statistical envelopes which describe functional arm reach at various lateral angles of certain percentages of a representative subject population for restrained and unrestrained conditions. Specific derivative of this model is control reach contours in which reach to specific control locations by a representative population of users is determined.

FUNCTIONAL LEG REACH CONTOURS. Statistical envelops which describe functional leg reach at various pedal heights of certain percentages of a representative subject population.

EYE CONTOURS. Statistical envelopes (elliptical in shape) which describe where eyes of certain percentages of representative user population are located in workspace environment. Model can be developed with or without head movement.

HEAD CONTOURS. Statistical envelopes which describe location below which certain percentages of representative user population are located in workspace environment. This definition for head contour can be generalized to other body landmarks (*e.g.* knees, elbows, shoulders, etc.).

SLEEP ENVELOPE. Statistical contours which describe height, breadth and length dimensions within which certain percentages of a representative user population can assume specific sleep postures (*e.g.*, prone or fetal).

STATIC DIMENSIONS

ARM REACH FROM WALL (II, 8). The horizontal distance from the wall to the tip of the longest finger, (usually the middle finger). To measure, the subject stands erect, with heels, buttocks, and shoulders (interscapular region) pressed against a wall, with the right arm and hand extended forward horizontally and maximally in the sagittal plane. To ensure that both shoulders are equally pressed against the wall, extend both arms. By convention, the right arm is then measured. The relevant forward distance reachable by the fingertips.

BICANTHIC DIAMETER (IX, 46). The maximum distance from the lateral point of the junction of the upper and lower eyelids of the right eye to the most lateral point of the junction of the upper and lower eyelids of the left eye. The relevant application is the greatest lateral distance required for proper eye protection and optical equipment, *e.g.*, eye goggles, ophthalmic instruments.

BUTTOCK DEPTH (II, 11). The horizontal distance between the buttocks and the abdomen at the level of the maximum protrusion of the buttocks. To measure, the subject stands erect. The relevant application is fore and aft space at buttock level while the subject is standing, *e.g.*, turnstile design.

BUTTOCK-KNEE LENGTH (III, 20). The horizontal distance from the most posterior point on the buttocks to the most anterior point on the knee. To measure, the subject sits erect with each knee at a right angle, with the upper leg parallel to the floor, and feet flat on the floor. The relevant application is for establishing the distance between the seat back and objects located in front of the knees. It is a static, minimum or clearance dimension, *e.g.*, airplane seat spacing.

BUTTOCK-LEG LENGTH (III, 23). The horizontal distance from the most posterior point on the buttocks to the base of the heel. To measure, the subject sits erect on the floor with the knee fully extended and the ankle at 90°. The relevant application is clearance for the outstretched leg, *e.g.*, wheelchair design, examination tables.

BUTTOCK-POPLITEAL LENGTH (III, 26). The horizontal distance from the plane of the most posterior point on the buttocks to the back of the lower leg at the knee. To measure, the subject sits erect with the knees at right angles, with the upper leg parallel to the floor, and feet flat on the floor. The relevant application is for determining maximum seat depth. Too long a seat will severely discommode short-legged operators, *e.g.*, chair depth.

CHEEK BREADTH (VIII, 41). The maximum distance between the most laterally situated points on the zygomatic arches (cheekbones) (Bizygomatic breadth). The relevant application is the greatest distance between the outermost points of the cheekbones, *e.g.*, helmet design.

CHEST BREADTH (I, 2). On males, the horizontal frontal distance across the chest at nipple level, and on females, at the level where the fourth rib meets the breastbone. To measure, the subject stands erect, breathes normally, and has the arms hanging naturally at the sides. The dimension is the maximal measurement at quiet inspiration. The relevant application is for determining maximum lateral space available at chest level, *e.g.*, clothing.

CHEST DEPTH (II, 9). On males, the horizontal distance from front to back of the chest at nipple level and on females, above the breasts, at the level where the fourth rib meets the sternum or breastbone. To measure, the subject stands erect, breathing normally. The dimension is the maximal measurement at quiet inspiration. The relevant application is fore and aft space available at chest level.

EAR BREADTH (IX, 54). With the subject's head oriented in the Frankfort plane, the distance between the points where an imaginary line drawn perpendicular to the long axis of the external ear meets the most anterior point and the most posterior point of the external ear. The relevant application is the maximum breadth of the ear, *e.g.*, earphone design.

EAR HEIGHT (IX, 48). With the subject's head oriented in the Frankfort plane, the distance between the highest point of the incurved rim of the external ear and the lowest point on the inferior border of the ear lobe. The relevant application is the maximum height of the ear, *e.g.*, earphone design.

ELBOW-TO-ELBOW BREADTH (IV, 28). The maximum horizontal distance between the lateral surfaces of the elbows. To measure, the subject sits erect with the upper arms vertical and lightly touching the sides, and the forearms extended horizontally. A relevant application is, passage-way design.

ELBOW HEIGHT (I, 6). The vertical distance from the floor to the radiale (the depression at the elbow formed where the bones of the upper arm and forearm meet). To measure, the subject stands erect with the arms hanging naturally at the sides. The relevant application is work or rest surface height, the vertical distance between the floor and table tops, *e.g.*, desks and workbenches used in the standing position.

ELBOW REST HEIGHT, SITTING (III, 22). The vertical distance from the sitting surface to the bottom of the right elbow. To measure, the subject sits erect with the upper arm vertical at the side, forearm at a right angle to the upper arm. The relevant application is for determining the vertical distance between the seat surface and the top surface of an elbow rest, *e.g.*, arm rest in seating design.

EXTRACANTHIC DIAMETER (IX, 47). The distance between the lateral (outer) and the medial (inner) corners of each eye. This measurement should be taken on both the right and left eyes. The relevant application is the greatest width of the eye, *e.g.*, binocular equipment, telescopes.

EYE HEIGHT, SITTING (III, 16). The vertical distance from the sitting surface to the lateral (outer) corner of the eye (ectocanthus). To measure, the subject sits erect and looks straight ahead. The relevant application is seated eye level, or the vertical distance between the seat surface and the eye, *e.g.*, location of primary functions, *i.e.*, control panels, observation windows, cockpit design, vehicular design.

EYE HEIGHT, STANDING (II, 13). The vertical distance from the floor to the lateral (outer) corner of the eye (ectocanthus). To measure, the subject stands erect and looks straight ahead. The relevant application is eye level, or the vertical distance from the floor to the eye position which affords the best visual field to the standing subject, *e.g.*, standing control panels for the placement of primary visual functions.

FACIAL LENGTH (IX, 51). The distance between the midpoint on the forehead and the lowest median point on the lower border of the mandible. The relevant application is the greatest distance from the top point of the forehead to the outermost point of the chin, *e.g.*, full face masks, helmets.

FINGER LENGTH (V, 31). The length of the right middle finger (digit 3) is the distance from the finger tip to the lower crease on the palmar side (Metacarpal-phalangeal joint crease). The relevant application is the longest finger length, *e.g.*, gloves.

FOOT BREADTH (VII, 36). The maximum horizontal distance across the foot. To measure, the subject stands with weight equally distributed on both feet. The relevant application is for determining lateral space available for the foot, *e.g.*, shoe size, brake pedal width, foot control width.

FOOT LENGTH (VII, 37). The maximum horizontal distance parallel to the long axis of the foot, from the back of the heel to the tip of the longest toe. To measure, the subject stands with weight equally distributed on both feet. The relevant application is for determining fore and aft space available for the foot, *e.g.*, shoe size, brake pedal length, foot control length.

FOREARM-HAND LENGTH (III, 17). The horizontal distance from the tip of the right elbow to the tip of the longest finger. To measure, the subject sits erect with the upper right arm vertical at the side and the forearm, hand and fingers extended horizontally. The relevant application is for determining the maximum fingertip reach from a fixed elbow point, *e.g.*, hand control location, arm rest length.

FRANKFORT PLANE (IX, 55). The landmark called the Frankfort plane is a standard plane for orientation of the head. It is established by a line passing through the tragion (approximately the earhole) and the lowest point of the eye socket (infraorbitale). The relevant application is as a landmark for various dimensions.

HAND BREADTH AT METACARPAL (V, 32). The maximum breadth across the hand where the fingers join the palm. To measure, the right hand is extended straight and stiff with the fingers held together. The relevant application is for determining the breadth available for palm, with fingers extended, *e.g.*, handle widths.

HAND LENGTH (V, 33). The distance from the wrist crease (palmar side) to the middle fingertip of the right hand extended straight on the arm. To measure, the right hand and lower arm is extended straight and stiff with the fingers held together. The relevant application is for determining maximum fingertip reach from the wrist, *e.g.*, gloves.

HAND THICKNESS AT METACARPAL III (VI, 35). The maximum distance between the back and palm surfaces of the hand at the knuckle (metacarpal-phalangeal joint) of the middle finger where it joins the palm of the right hand when the fingers are extended. To measure, the right hand is extended straight and stiff with the fingers held together. The relevant application is the space available for the flat palm. More space is required for the whole hand than the palm alone, since the muscle masses at the bases of the thumb and the little finger, (the thenar and hypothenar eminences) increase hand thickness, *e.g.*, handles, refrigerator doors.

HEAD BREADTH (VIII, 38). The maximum horizontal head breadth above the ears, at right angles to the mid-sagittal plane. The location of this dimension is highly variable. To measure, the subject is sitting erect, looking forward. The relevant application is the breadth across the head or the lateral space available for the head above the ears, *e.g.*, headrests, earphones, helmets.

HEAD LENGTH (VIII, 43). The distance between the most anterior point on the forehead (between the brow ridges) and the most posterior point, on the back of the head, in the midline, To measure, the subject is sitting erect, looking forward. The relevant application is head length or the space available between the front and the back of the head, *e.g.*, head gear.

HEAD LENGTH, MAXIMUM (VIII, 42). The distance between the most anterior point of the nose and the most posterior point on the back of the head, in the midline. To measure, the subject sits erect looking forward. The relevant application is maximum head length, or the space available between the tip of the nose and the back of the head, *e.g.*, face guards.

HIP BREADTH, SITTING (IV, 29). The maximum horizontal distance across the hips when seated. To measure, the subject sits erect, knees and ankles supported at right angles; knees and heels together. The relevant application is the space available across the hips, *e.g.*, seat breadth.

HIP BREADTH, STANDING (I, 3). The maximum horizontal distance across the hips. To measure, the subject stands erect, with heels together. The relevant application is the lateral space available at the level of the hips, while the person is standing, *e.g.*, turnstile aisles.

INTERPUPILLARY (VIII, 39). The distance between the centers of the pupils. To measure, the subject sits erect and looks straight ahead. The relevant application is binocular spacing or the separation distance between binocular eyepieces, *e.g.*, sighting stations, stereoscopic photogrammetry, microscopy, etc.

JAW BREADTH (VIII, 40). The straight-line distance between the right and left angles where the body of the mandible and the ascending ramus meet (Bigonial breadth). To measure, the subject's head is tilted slightly upwards; the angle of the mandible is usually palpated both anteriorly and inferiorly from the lowest point of the ear lobes. The relevant application is the greatest distance on the outermost points of the jaw, *e.g.*, partial face masks, chin straps.

JAW HEIGHT, TOTAL (IX, 53). The distance in the midline or sagittal plane between the point at which the nasal septum merges with the upper cutaneous tip (subnasal) and the lowest median point on the lower border of the mandible (menton). The relevant application is the greatest distance between the bottom of the nose and the lowest point on the chin, *e.g.*, face masks.

KNEE HEIGHT, SITTING (III, 24). The vertical distance from the floor to the uppermost point on the knee. To measure, the subject sits erect with his knees at right angles. The relevant application is the space available above the knees, *e.g.*, car dashboards, knee clearance in table heights.

KNEE-TO-KNEE BREADTH, SITTING (IV, 30). The maximum horizontal distance across the lateral surfaces of the knees. To measure, the subject sits erect with his knees at right angles and touching lightly. The relevant application is the space available for determining knee clearance, *e.g.*, cockpit clearance.

KNUCKLE HEIGHT (I, 7). The vertical distance from the floor to the largest knuckle of the middle finger, where the finger meets the palm (metacarpal-phalangeal joint of digit 3). To measure, the subject stands erect, palm flat against the side of the thigh. The relevant application is for determining the maximum permissible vertical distance between the base of an object and the underside of an attached carrying handle, *e.g.*, suitcases.

LOWER ARM LENGTH (I, 4). The distance from the elbow to the wrist joint when the arm is hanging down and the palm facing inward. To measure, the subject stands erect with the arms hanging freely at the sides. The relevant application is for determining the length of the lower arm, *e.g.*, clothing design.

MAXIMUM BODY BREADTH (I, 1). The maximum breadth across the body including the arms. To measure, the subject stands erect with arms hanging relaxed at the sides. The relevant application is for determining passage clearances, *e.g.*, door width.

MAXIMUM BODY DEPTH (II, 12). The maximum horizontal distance between the vertical lines tangent to the most anterior and posterior points on the trunk. To measure, the subject stands erect with the arms at the sides. The relevant application is as a minimum clearance dimension of the body, *e.g.*, passage clearance.

MOUTH BREADTH, MAXIMUM (IX, 50). The distance between the two most lateral points on the mouth when the lips are joined together and the mouth is in a relaxed state. The relevant application is the greatest distance from the outermost edges of the lips across the mouth, *e.g.*, the design of mouthpieces.

NASAL BREADTH (IX, 49). With the subject's head tilted slightly upwards, the maximum transverse distance between the most laterally situated points on the wings of the nose. The relevant application is the greatest distance between the outer edges of the lower part of the nose, *e.g.*, nose plugs, eye glass frames.

NASAL HEIGHT (IX, 52). With the subject's head tilted slightly upwards, the distance between sellion and the point where the nasal septum merges with the upper cutaneous lip (subnasale landmark). The relevant application is the greatest distance from the bridge of the nose to the underside of the nose, e.g., face masks.

NOSE BRIDGE HEIGHT (VIII, 45). The distance between the sellion and the lowest median point on the lower border of the mandible or jaw (Menton-Sellion length). The relevant application is the greatest distance between the bridge of the nose and the lowest edge of the chin, e.g., full face masks.

POPLITEAL HEIGHT, SITTING (III, 25). The vertical distance from the floor to the underside of the thigh immediately behind the knee. To measure, the subject sits erect with knees at right angles and the bottom of the thighs and the back of the knees barely touching the sitting surface. The relevant application is the measurement of the vertical distance between the floor and the highest point on the front of the sitting surface, e.g., seat height, i.e., secretarial chairs.

SHOULDER BREADTH (IV, 27). The maximum horizontal distance across the lateral surfaces of the shoulders. To measure, the subject sits erect with upper arms touching the sides and the forearms extended horizontally. The relevant application is for measuring the space available across the shoulders, e.g., shoulder pads.

SHOULDER ELBOW LENGTH (III, 18). The uppermost point on the lateral edge of the shoulder to the bottom of the elbow. The subject sits erect with his upper arm vertical at his side and the forearm making a right angle with it. The relevant application is the distance between objects located at or above the shoulder as well as at or below the elbow. It is a minimum clearance dimension, e.g., accessories and equipment hung from the shoulder.

SHOULDER HEIGHT, SITTING (III, 19). The vertical distance from the sitting surfaces to the uppermost point on the lateral edge of the shoulder with the subject sitting erect. The relevant application is for determining the vertical distance between the seat surface and objects located at or above the shoulder, e.g., seat-back design.

SHOULDER HEIGHT, STANDING (I, 5). The vertical distance from the floor to the uppermost point on the lateral edge of the shoulder with the subject standing erect. The relevant application is the total distance from the floor to the shoulder, e.g., standing control position, shelf height.

SITTING HEIGHT (III, 15). The vertical distance from the sitting surface to the top of the head. To measure, the subject sits erect, looking straight ahead, with the knees at right angles. The relevant application is the greatest distance between the seat and the top of the head, e.g., furniture design.

STATURE (II, 14). The vertical distance from the floor to the top of the head (vertex). To measure, the subject stands erect and looks straight ahead. The relevant application is the maximum height of an individual, e.g., the minimum standing clearance in the workspace.

THIGH CLEARANCE HEIGHT, SITTING (III, 21). The vertical distance from the sitting surface to the top of the thigh at its intersection with the abdomen. The subject sits erect with his knees at right angles. The relevant application is the maximum distance between the seat and the top of the thigh, e.g., table and chair height relationships.

UNDEREYE HEIGHT (VIII, 44). With the subject's head in the Frankfort plane, the distance in the midline or sagittal plane between the lowest point on the inferior border of the orbit as palpated through the skin and the top of the head. The relevant application is the greatest distance between the top of the head and the lower orbit of the eye, e.g., headgear, goggles.

WAIST DEPTH (II, 10). The horizontal distance between the back and abdomen at the level of the greatest lateral indentation of the waist; (if this is not apparent, at the level at which the belt is worn). To measure, the subject stands erect with his abdomen relaxed. The relevant application is for determining the fore and aft space at abdominal level; clearances between the operator's back and control stick, steering wheel, and work surfaces are examples.

WRIST-THUMBTIP LENGTH (V, 35). The distance from the wrist crease (palmar side) to the tip of the thumb. The relevant application is the length of the thumb from the wrist crease, e.g., the limiting factor in pinch length and gloves.

CROSS-REFERENCE LIST

Plate Number	Dimension Number	Dimension Name
I	1	Maximum Body Breadth
	2	Chest Breadth
	3	Hip Breadth, Standing
	4	Lower Arm Length
	5	Shoulder Height, Standing
	6	Elbow Height
	7	Knuckle Height
II	8	Arm Reach from Wall
	9	Chest Depth
	10	Waist Depth
	11	Buttock Depth
	12	Maximum Body Depth
	13	Eye Height, Standing
	14	Stature
III	15	Sitting Height
	16	Eye Height, Sitting
	17	Forearm-Hand Length
	18	Shoulder-Elbow Length
	19	Shoulder Height, Sitting
	20	Buttock-Knee Length
	21	Thigh Clearance Height, Sitting
	22	Elbow Rest Height, Sitting
	23	Buttock-Leg Length
	24	Knee Height, Sitting
	25	Popliteal Height, Sitting
	26	Buttock-Popliteal Length
IV	27	Shoulder Breadth
	28	Elbow-to-Elbow Breadth
	29	Hip Breadth, Sitting
	30	Knee-to-Knee Breadth, Sitting
V	31	Finger Length
	32	Hand Breadth at Metacarpal
	33	Hand Length
	34	Wrist-Thumbtip Length
VI	35	Hand Thickness at Metacarpal III
VII	36	Foot Breadth
	37	Foot Length
VIII	38	Head Breadth
	39	Interpupillary Distance
	40	Jaw Breadth
	41	Cheek Breadth
	42	Head Length, Maximum
	43	Head Length
	44	Undereye Height
	45	Nose Bridge Height
IX	46	Bicanthic Diameter
	47	Extracanthic Diameter
	48	Ear Height
	49	Nasal Breadth
	50	Mouth Breadth, Maximum
	51	Facial Length
	52	Nasal Height
	53	Jaw Height, Total
	54	Ear Breadth
	55	Frankfort Plane

2-14   INDUSTRIAL ENGINEERING TERMINOLOGY

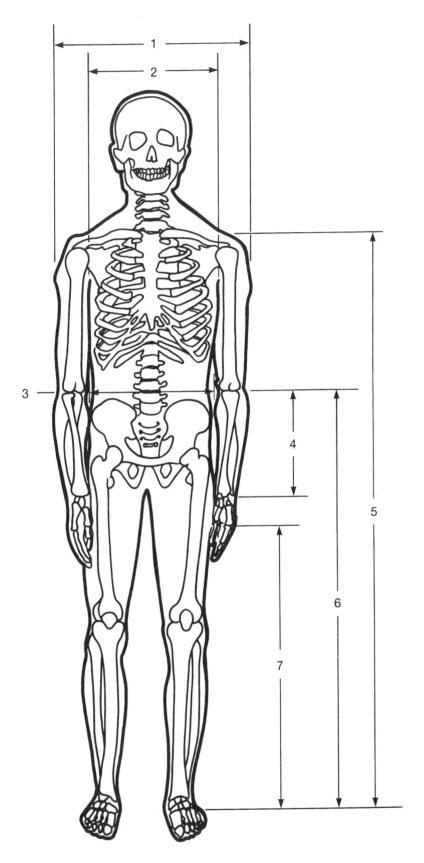

PLATE NO. I

Anthropometry & Biomechanics 2-15

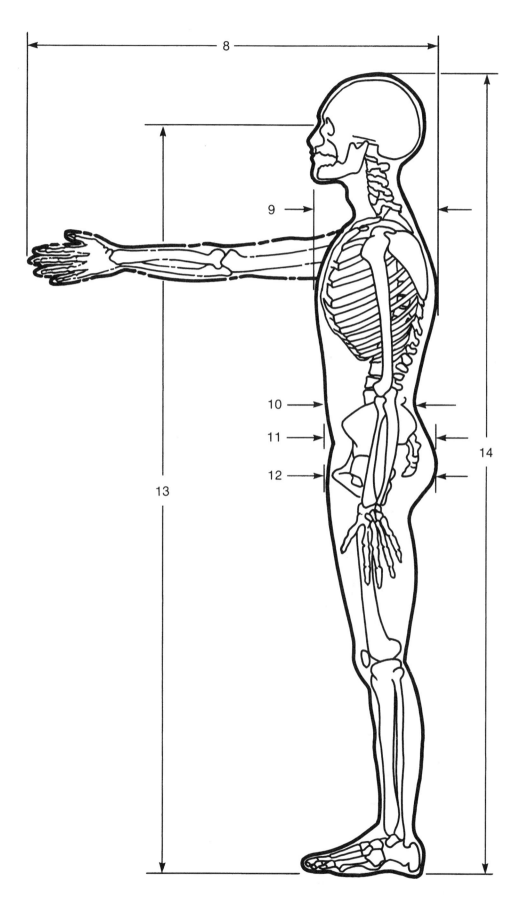

PLATE NO. II

2-16 INDUSTRIAL ENGINEERING TERMINOLOGY

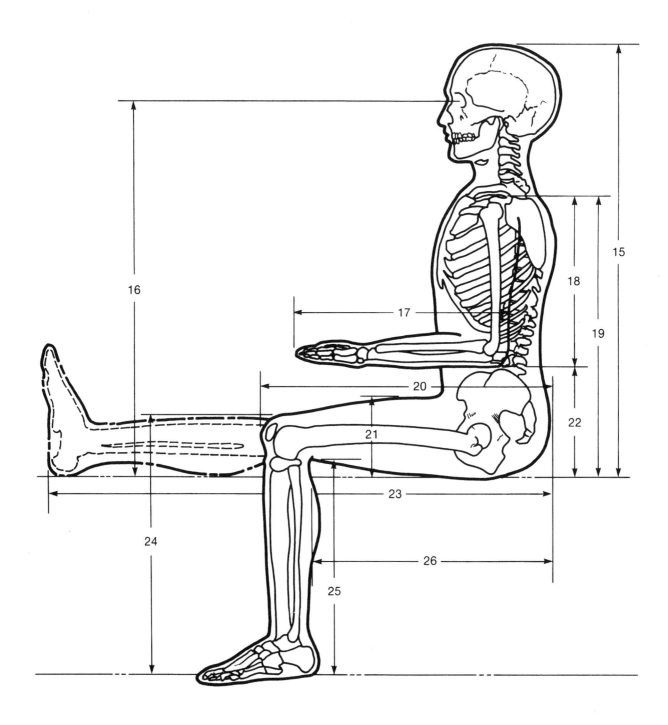

PLATE NO. III

Anthropometry & Biomechanics 2-17

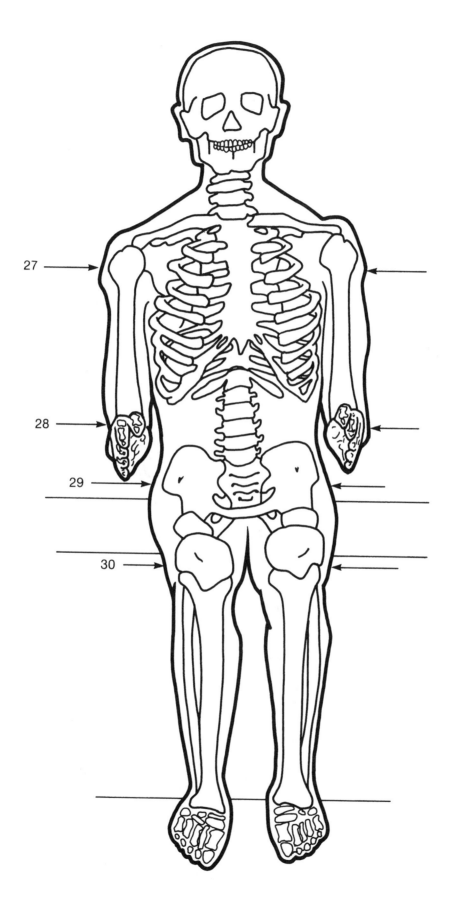

PLATE NO. IV

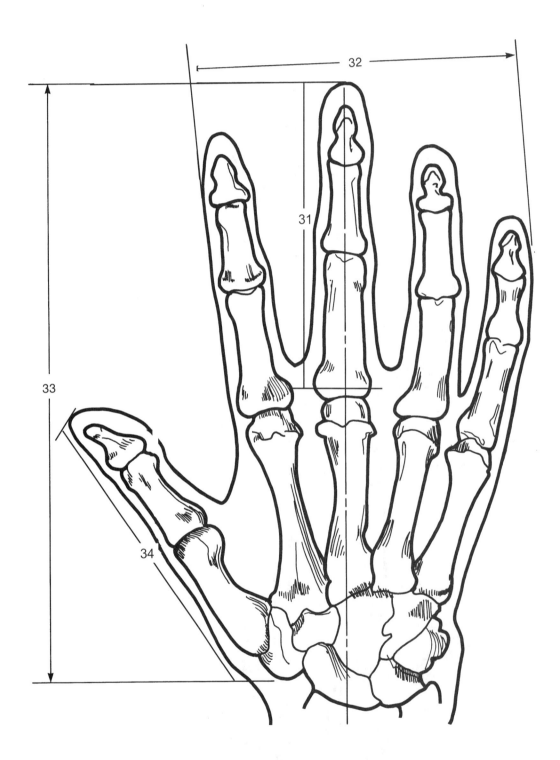

PLATE NO. V

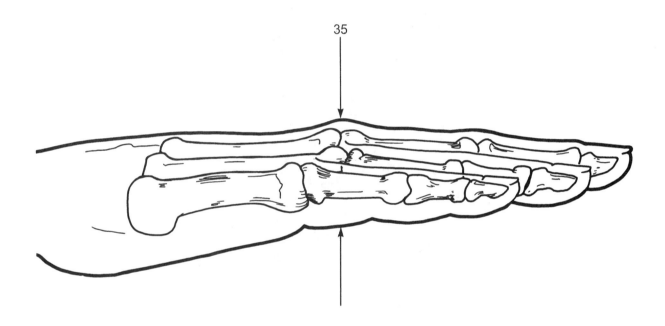

PLATE NO. VI

2-20 INDUSTRIAL ENGINEERING TERMINOLOGY

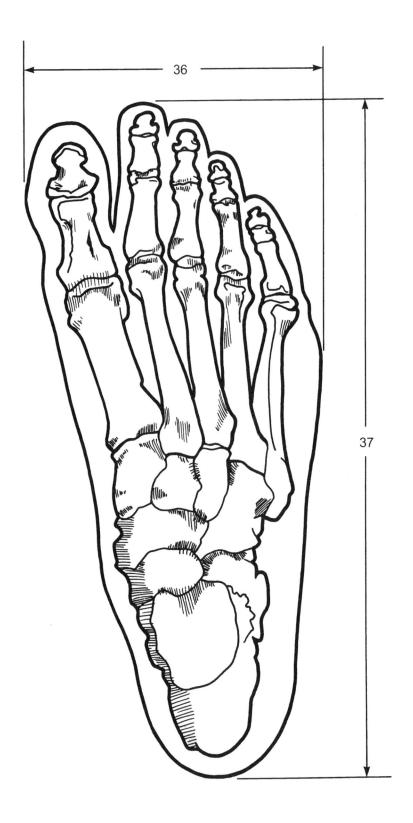

PLATE NO. VII

Anthropometry & Biomechanics 2-21

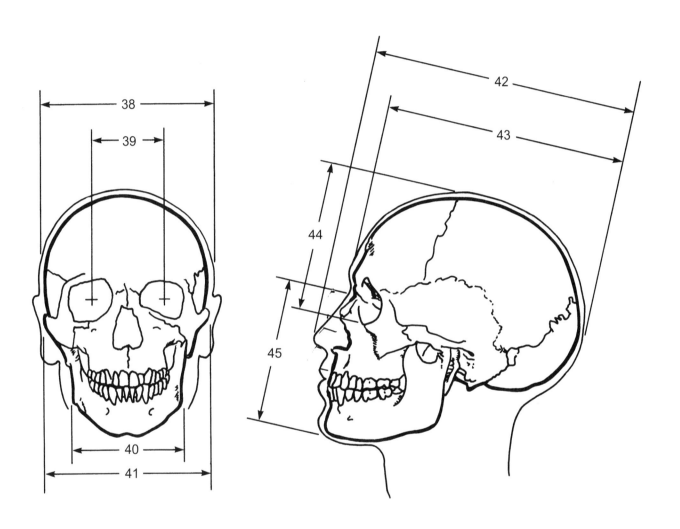

PLATE NO. VIII

INDUSTRIAL ENGINEERING TERMINOLOGY

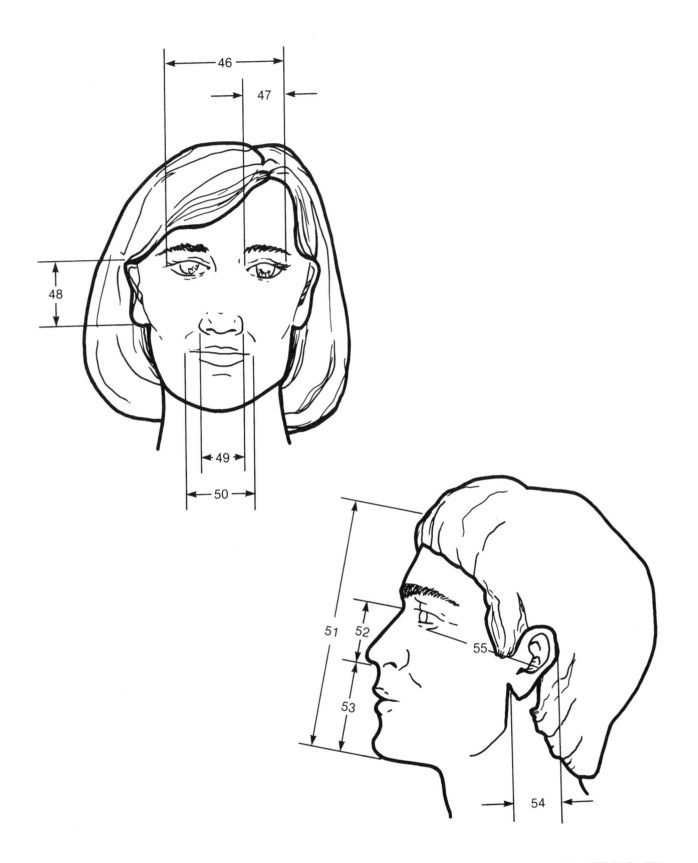

PLATE NO. IX

GLOSSARY

ANTERIOR. Toward the front or ventral side of the body; *e.g.*, the breast is on the anterior chest wall.

ASCENDING RAMUS. Refers to that part of the mandible or jaw which forms an angle with the main body of the jaw.

CORONAL. Frontal; any plane dividing the body into anterior (or ventral) and posterior (or dorsal) portions at right angles to the sagittal plane.

CUTANEOUS. Relating to the skin.

ECTOCANTHUS. The external corner of the eye.

FRANKFORT PLANE. The Frankfort plane is a standard plane for orientation of the head. It is established by a line passing through the tragion (approximately the earhole) and the lowest point of the eye socket (infraorbitale). The Frankfort plane can be approximated by having the subject stand so that the natural line of sight is horizontal.

GLABELLA. The most anterior part of the forehead between the brow ridges in the mid-sagittal plane.

HYPOTHENAR. The fleshy mass at the medial side of the palm.

HYPOTHENAR EMINENCE. The elevation on the medial side of the palm produced by the short muscles of the little finger.

INFERIOR. Lowermost or below; the foot is inferior to the ankle.

INFRAORBITALE. The lowest point of the eye socket.

INTERPUPILLARY. Between the pupils.

INTERSCAPULAE. Between the scapulae (shoulder blade, q.v.).

LATERAL. Toward the side; opposite of medial.

MANDIBLE. Jaw bone.

MEDIAL. Relating to the middle or center; nearer to the median or mid-sagittal plane.

MENTON. The inferior point on the mandible in the mid-sagittal plane; bottom of the chin.

METACARPAL. Relating to the metacarpus (plural, metacarpi), which are five bones of the hand between the carpus (wrist bones) and the phalanges.

MID-SAGITTAL. The plane vertically dividing the body through the midline into symmetrical right and left halves.

NASAL SEPTUM. The thin wall dividing the nasal cavity into halves; it is composed of a central supporting skeleton covered on each side by a mucous membrane.

OCCIPUT. The prominence at the lower back of the skull.

PALPATE. To examine by feeling and pressing with the palms of the hands and the fingers.

POPLITEAL. Relating to the posterior surface of the knee. The popliteal point is the hollowed out region of the leg directly behind the knee where the bottom of the thigh and the top of the calf intersect.

POSTERIOR. Toward the back or dorsal side of the body; the vertebral column is posterior to the digestive tract.

PROTUBERANCE. An outgrowth, a swelling, a knob.

SAGITTAL. Any plane parallel to the mid-sagittal line vertically dividing the body into right and left portions.

SCAPULA. Shoulder blade; a large triangular flattened bone lying over the ribs, posteriorly on either side, articulating laterally with the clavicle and the humerus.

SELLION. The point of the deepest depression of the nasal bones at the top of the nose.

SUBNASALE. The point of intersection of the groove of the upper lip with the inferior surface of the nose in the mid-sagittal plane.

THENAR. The fleshy mass on the lateral side of the palm; the base of the thumb.

THENAR EMINENCE. The swelling on the lateral part of the palm of the hand caused by the short muscle of the thumb.

TRAGION. The superior point on the juncture of the cartilaginous flap (tragus) of the ear with the head.

TRANSVERSE. Horizontal; any plane dividing the body into superior and inferior portions.

Vertex. The highest point on the head when the head is in the Frankfort plane; top of the head.

Zygomatic Arch. Cheekbone.

PRIMARY BIBLIOGRAPHY

Air Standardization Coordinating Committee. "A Basis for Common Practices and Goals in the Conduct of Anthropometric Surveys", *Air Standard 61/83*, September, 1991. The Air Standard provides standard practice and definitions for the conduct of anthropometric surveys such as: posture definition, measuring techniques, accuracy, definition of standard body marks, and definition of a set of anthropometric measurements and background information that should be collected in an anthropometric survey. It was a primary source of the dimensions and definitions used in this document.

Bittner, A. "A Family of Manikins for Workstation Design", *NAEC-TR-2100-07B*, 1986. This paper describes a method involving the use of a hypothetical sample of individuals having extreme dimensions on one or a combination of variables.

Damon, A., Stoudt, H.W. and McFarland, R.A. *The Human Body in Equipment Design*. Cambridge: Harvard University Press, 1966. A primary source from which dimensions were chosen for the standard. The chapter on "Anthropometry and Human Engineering" presents data on 37 dimensions, the data base from which is available military and civilian studies including the National Health Survey data of 1960-1962. Explanations accompanying the dimensions include examples applicable to specific equipment design. General discussion provides guidelines for selecting minimum or maximum values and a consideration of factors affecting individual variability. Other dimension types discussed and illustrated are circumferences, functional dimensions, and center of gravity measurements. In addition to the chapter on anthropometry are sections on biomechanics and equipment design, control design, hand and foot controls, and seat design.

Hendy, K.C. "Aircrew/Cockpit Compatibility: A Multivariate Problem Seeking a Multivariate Solution," *AGARD Conference Proceedings No. 491*, April 1990. This paper describes situations where the operator or maintainer must simultaneously perform two or more different actions in such a way that they interact and suggests that multivariate techniques be used to verify accommodation. This paper explains one method to verify accommodation, individuals with extreme combinations of body proportions are used: tall torso with short limbs, short torso with long limbs, etc.

Hertzberg, H.T.E. "Chapter 11: Engineering Anthropology" in *Human Engineering Guide to Equipment Design*, edited by Harold P. Van Cott and Robert G. Kinkade. Washington, D.C.: U.S. Government Printing Office, Superintendent of Documents, 1972. A primary source from which dimensions were chosen for inclusion in the standard, "Engineering Anthropology" presents tabular data of male and female, military and civilian studies as well as international data (where available) for 39 static dimensions. The use of anthropometry is clarified with specific guidelines for the choice of percentiles for a given design problem. Variability of military and national groups is discussed. Data for functional dimensions as well as the discussion of factors affecting muscle capacity give the reader a perspective for interpretation and use of data.

McDaniel, J. W. "Models for Ergonomic Analysis and Design: COMBIMAN and CREW CHIEF", In *Computer-Aided Ergonomics: A Researchers Guide* (Karwowski, W., Genaidy, A. M., and Asfour, S. S., Eds.). Taylor and Francis, London, 1990. This paper discusses COMBIMAN and CREW CHIEF computer models which can generate 3-D manikins. These models can be dimensioned to match all the subjects in a multivariate set. These models can be used to analyze, design, and evaluate "electronic mockups" of workplaces.

McDaniel, J. W. "The Development of Computer Models for Ergonomic Accommodation," In *Workplace Equipment and Tool Design*, (Mital, A. and Karwowski, W., Eds.). Elsevier, Amsterdam, 1990. This paper discusses the development of computer models which can be dimensioned in three-dimensions to match all the subjects in a multivariate set. These models can be used to generate and evaluate "electronic mockups".

Montagu, M.F. and Ashley, A. *A Handbook of Anthropometry*. Springfield: Charles C. Thomas, 1960. A primary source from which definitions of head dimensions were chosen for inclusion in the standard, this volume is a detailed compendium on the theory, practice, and methods of anthropometry. It considers the cephalometry and somatometry of living humans, the craniometry of skeletal populations, postcranial osteometry, developmental assessment of the dentition, the measurement of body composition, and an unusually detailed section on references and directories for further information.

Moroney, W. F., and Smith, M. J., "Empirical Reduction in Potential User Population as the Result of Imposed Multi-variate Anthropometric Limits," Naval Aerospace Medical Research Laboratory, *NAMRL - 1164*, September, 1972. This paper discusses why univariate percentiles are not appropriate when two or more dimensions are simultaneously used as criteria for design. They explain that percentile values reported in anthropometric surveys are suitable only for univariate accommodation and should not be used for designs where two or more dimensions are used simultaneously as design parameters; that while the means of different measures of body parts can be added, the measures away from the means (at the tails of the distribution where the 5th and 95th percentile values occur) are not additive.

Robinette, K. and McConville, J. "An Alternative to Percentile Models", *SAE Technical Paper Series No. 810217*, 1982. This report demonstrates the seriousness of the problems associated with the use of percentiles, and describes and compares an alternative regression approach for representing human body size variability.

Robinson, J.C., Robinette, K.M., and Zehner, G.F. "User's Guide to the Anthropometric Database at the Computerized Anthropometric Research and Design (CARD) Laboratory", *AL-TR-1992-0036*, Feb. 1992. A dial-up database is now available which can be used to calculate multivariate "test cases" for a number of work-place/crew station design applications with a user's guide.

SECONDARY BIBLIOGRAPHY

Abraham, S. *Preliminary Findings of the First Health and Nutrition Examination Survey*, United States, 1971- 1972: *Anthropometric and Clinical Findings*. Rockville, Maryland: Health Resources Administration, National Center for Health Statistics, April 1975. This document presents preliminary findings collected on a probability sample of U.S. population by age, sex, race, and income level. Data are presented on selected anthropometric measurements of children 1-17 years of age, obesity in adults 20-74 years of age, and clinical signs of possible nutritional deficiency for persons 1-74 years of age. The particular contribution to anthropometric data is the inclusion of data by yearly intervals on height, weight, triceps skinfold, and subscapular skinfold for ages 1-17.

Bass, W. *Human Osteology*. Columbia, Missouri: Archeological Survey, 1971. This volume is a primer in skeletal anatomy and in the anthropometry of the human skeleton. It gives a thorough introduction to the theory and practice of anthropometry but it emphasizes the bony landmarks and is geared to the measurement, aging, sexing, and cataloguing of skeletal material found in archeological sites.

Brozek, J. *Body Measurements and Human Nutrition*. Detroit: Wayne University Press, 1956. This edited volume considers the effect that varying degrees of nutritional status has on a battery of anthropometric measurements. Most of the samples are drawn from the armed forces, yet, this volume also starts to consider the effects of nutritional status on the anthropometry of women.

Chapanis, A. (ed.). *Ethnic Variables in Human Factors Engineering*. Baltimore: The Johns Hopkins University Press, 1975. A diverse reference on human factors, the book presents 18 essays from papers presented at a NATO symposium, three of which specifically deal with anthropometry: "Population differences in dimensions, their genetic basis and their relevance to practical problems of design" by D.F. Roberts; "Anthropometric measurements on selected populations of the world" by Robert M. White; and "International anthropometric variability and its effects on aircraft cockpit design" by Kenneth W. Kennedy. In particular, the article by White includes statistics on weight, stature, sitting height, and chest circumference for a number of populations of different countries. Data important in the design of international products.

Churchill, E. and McConville, J.T.. *Sampling and Data Gathering Strategies for Future USAF Anthropometry*. Springfield, Virginia: National Technical Information Service, 1976. A methodology is presented for tailoring a data gathering study to the specific design needs of a project as well as a discussion of the different types of variability, necessary to the determination of the data gathering approach needed. Types of variability discussed are intra-individual, inter-individual, and secular variability. The appendix includes a list of all the anthropometric dimensions available in the Aerospace Medical Research Lab Data Bank.

Clauser, C.E., Tucker, P.E., McConville, J.T., Churchill, E., Laubach, L.L., and Reardon, J.A. *Anthropometry of Air Force Women*. Wright-Patterson Air Force Base, Ohio: Aerospace Medical Research Laboratory, 1972. A total of 137 dimensions are included with graphs representing the actual distributions of the population surveyed. Dimension types include body lengths and heights, circumferences, weights, skinfolds, breadths and depths, body surface distances, as well as measures of the head, the hands, and the feet. Correlation coefficients which establish the ratio of certainty by which one dimension can be determined from another are developed for all variables as well as regression equations for those variables with a moderately large intercorrelation.

Dempster, W.T. *Space Requirements of the Seated Operator: Geometrical, Kinematic, and Mechanical Aspects of the Body with Special Reference to the Limbs*.

Wright-Patterson Air Force Base, Ohio: Wright Air Development Center, United States Air Force, 1955. This study explores the relationships between joint "centers" and joint ranges of motion based on a subject population representing a cross section of somatotypes. Links are defined, as a result of this study, as a proportion of the corresponding static segment length. Links are useful to the designer and engineer in determining workspace layout as a function of potential ranges of motion. Specifications for construction of drafting board mannequins were worked out for seated postures.

Diffrient, N., Tilley, A.R. and Bardagjy, J.C. *Humanscale 1/2/3*. Cambridge, The MIT Press, 1974. Humanscale consists of three pictorial selectors with rotary dials correlating anthropometric data with age and body size. Included are body and link measurements, seat and table design measurements, and some data on wheelchair users, the handicapped and the elderly. Although the information is presented in an easily usable form, there is a drawback in this method of presentation. Since for any given dimension, there have been numerous surveys conducted, attention needs to be given to the criteria by which one selects the data of one survey over another. Humanscale presents no discussion on why or by what criteria they selected the data which is presented on the rotary selectors.

Dreyfuss, H. *The Measure of Man*. New York: Whitney Library of Design, 1967. This portfolio of anthropometric data is accompanied by design specifications, bibliography, charts, and two life-size standing human figures. It provides male and female, seated and standing dimensions.

Gordon, C. C., Bradtwiller, B., Churchill, T., Clauser, C. E., McConville, J. T., Tebbetts, I., Walker, R. A. *Anthropometric Survey of U.S. Army Personnel: Methods and Summary Statistics*. Natick/TR-89/044. Natick, M.A.: U.S. Army Natick, R, D, and E Center, 1989. Most recent human engineering data collected on U.S. populations; extensive descriptions, photos of landmarks and measurements; appendix indicating usefulness of various dimensions for specific engineering applications.

Hrdlicka, A. *Anthropometry*. Philadelphia: The Wistar Institute of Anatomy and Biology, 1920. This volume is another detailed primer on the measurement of humans. It goes into great detail on how to measure humans, anthropometric landmarks, anthropometric instruments, as well as a section on aging and sexing skeletal material.

Lockhart, R.D., Hamilton, G.F. and Fyfe, F.W. *Anatomy of the Human Body*. Philadelphia: J.B. Lippincott Co., 1959. This volume is an authoritative source book on the structure and function of the human body. It first considers the skeleton in great detail, then, it considers each part of the human body in an integrated, functional manner.

Robinette, K.M., and Fowler, J. *An Annotated Bibliography of United States Air Force Engineering Anthropometry—1946 to 1988*. AAMRL-TR-88-013. Wright Patterson A.F.B., Ohio: Armstrong Aerospace Medical Research Laboratory. Bell, N. A., Donelson, S. M., and Wolfson, E. *An Annotated Bibliography of U.S. Army Natick Anthropology (1947-1991)*. Natick/TR-91/044. Natick, M.A.: U.S. Army Natick, R, D, and E Center.

Roebuck, J.A., Kroemer, K.H.E. and Thomson, W.G. *Engineering Anthropometry Methods*. New York: John Wiley & Sons, 1975 Methods for both measuring and applying data on human body dimensions and strength to engineering design are presented which the engineer and designer may use in the conceptual development of systems where applicable data is not already available. A detailed discussion of the use of statistical theory in the development of anthropometric data is covered. A section on applications methodology discusses the role of models and mock-ups of many types, for example, drawings, mannequins, models of reach envelopes, and mathematical models in the design evaluation process. A detailed bibliography documenting examples of methods presented and a correlation coefficient table relating 96 anthropometric dimensions included in the appendices.

Sinclair, D. *Human Growth After Birth*, 5th Edition. London: Oxford University Press, 1989. This is a volume written for the medical scientist or the medical student on aspects of post-natal growth of humans. It covers the growth of the individual from birth until maturity with regards to height and weight, tissues, organ systems, indices of maturity, and ontogenetic changes in shape and posture. It then considers factors affecting growth, repair, and maturation.

Snyder, R.G. Chaffin, D.B. and Schutz, R.K. *Link System of the Human Torso*. Wright-Patterson Air Force Base, Ohio: Aerospace Medical Research Laboratory, United States Air Force, 1972. Vector prediction equations are developed by use of anthropometry, photogrammetry, radiography and cineflouroscopy of 28 male volunteer subjects for location of cervical, thoracic, and lumbar vertebrae relative to given surface landmarks. This information is useful in the development of work-space envelopes. This kind of information is extremely beneficial in the dynamic evaluation of man-machine interface problems. Included also are anthropometric dimensions and skinfolds of 28 subjects used in the study.

Snyder, R.G., Spencer, M.L. Owings, C.L. and Schneider, L.W. *Anthropometry of U.S. Infants and Children*. Detroit: Society of Automotive Engineers, Inc., 1975. This reference presents in tabular and graphic form the results of anthropometric data collection on 4024 infants and children from birth to 13 years of age, in half-year intervals. Methods of data collection and analysis are also described.

Stoudt, H.W., Damon, A., McFarland, R.A. and Roberts, J.

Skinfolds, Body Girths, Biacromial Diameter and Selected Anthropometric Indices of Adults: United States, 1960-1962. Rockville, Maryland; Health Resources Administration, National Center for Health Statistics, 1970. This survey includes the following data by age in 10-year intervals (except for the intervals 18-24 years and 75-79 years) and by sex; right-arm and infrascapular skinfolds; right arm, waist, and chest girths; sum of skinfolds; ponderal index; ratios of sitting height erect to stature, chest girth to stature, and biacromial diameter to stature. The equations for predicting each of the physical measurements from height, weight, and age, and the interrelation of all the measurements is summarized graphically and analyzed.

Stoudt, H.W., Damon, A., McFarland, R.A. and Roberts, J. *Weight, Height, and Selected Body Dimensions of Adults: United States, 1960-1962*. Rockville, Maryland: Health Resources Administration, National Center for Health Statistics, 1965. Data on the civilian population with a sample size of 3091 men and 3581 women, age 18 through 79, includes: height, weight, sitting height (normal and erect), knee height, popliteal height, elbow rest height, thigh clearance height, buttock-knee length, buttock-popliteal length, elbow-to-elbow breadth, and seat breadth. Results of large-scale studies done in earlier years on college populations and others are compared.

Webb Associates. *Anthropometric Source Book, Vol. I: Anthropometry for Designers, NASA* 1024, National Aeronautics and Space Administration, Washington, DC 1978. A source book was designed to provide NASA, NASA Contractors, the aerospace industry, government agencies, and a wide variety of industrial users in the civilian sector with a comprehensive, up-to-date tabulation of anthropometric data. Specifically, it is tailored to meet the needs of engineers engaged in the design of equipment, habitability areas, workspace layouts, life-support hardware, and clothing for the NASA Space Shuttle/Spacelab program. The intent was to provide the designer not only with dimensional data but with underlying anthropometric concepts and their application to design.

Z94.2
Anthropometry & Biomechanics (continued)

Biomechanics

The Biomechanics Subcommittee acknowledges the efforts of previous subcommittees in establishing a substantial base for this chapter. Our task was facilitated greatly by these previous subcommittee members. Many of the terms developed by these people have been retained because we could not improve them. There are too many previous committee members to mention here. However, they are identified in previous editions of this book.

The present Biomechanics Subcommittee members have updated and revised many of the terms relating to biomechanics. They have made substantial constristutions, which cannot be diminished by the use of previous editions of this chapter. The subcommittee consists of the following members.

Chairperson
Sudhakar L. Rajulu, Ph.D
Lockheed Engineering & Sciences Co.

Subcommittee

Jeff Poliner, M.S.
Lockheed Engineering & Sciences Co.

Lara Stoycos, M.S.
Lockheed Engineering & Sciences Co.

Human Factors and Ergonomics Section Coordinator:
Steven A. Lavender, Ph.D.
Rush-Presbyterian-St. Luke's Medical Center, Chicago

ABDOMINAL WALL. The covering of the abdominal cavity composed of skin, fatty tissue, muscles and fibrous structures. Extends from the rib cage to the pelvis. The muscles of the abdominal wall assist in respiration and in the maintenance of posture. The workplace should not mechanically press against the abdominal wall.

ABDUCTION. Deflection of a limb away from a sagittal plane (q.v.). A basic element of motions inventory. (See ADDUCTION.)

ABDUCTORS. Muscles which, by contraction, effect abduction of a limb. E.g., the deltoid (q.v.) abducts the arm,

ACOUSTIC STIMULUS. Sound signal which indicated that an individual is to initiate, maintain, or terminate an activity.

ACROMION. A bony outcrop of the shoulder blade which forms a joint with the collarbone and is attached to the deltoid and trapezius muscles. It can be felt through the skin in front of the shoulder. It is used as an anatomical reference point (q.v.) in anthropometry (q.v.).

ACTION POTENTIAL. Electric activity produced in nerve, muscle or other excitable tissue during activity. Once the threshold of stimulation is reached, the action potential is triggered and the "all or none law" (q.v.) applies. Most electrophysiological recordings are produced by conditioned signals evoked originally by action potentials, e.g., electrocardiograms, electromyogram.

ACTIVATING RECEPTORS. End-organs of nerves which trigger a specific response reaction upon receipt of a stimulus. (See SENSORY END ORGANS.)

ADDUCTION. Movement of a limb towards the mid-sagittal plane (q.v.). (See ABDUCTION.)

ADDUCTOR. Muscle which by contraction causes adduction of a limb. E.g., pectoralis major adducts the arm and the abductor magnus adducts the thigh.

AEROBIC METABOLISM. Physiological "combustion" of body fuels with oxygen in muscle during exertion. This is a highly efficient form of energy release available for long periods of time in light and medium work. During heavy work aerobic muscle metabolism is preceded by anaerobic (q.v.) muscle metabolism as aerobic metabolism takes several minutes to be initiated.

AFFERENT. Conveying a stimulus or a body fluid inwardly towards a biological processing station (See EFFERENT.)

AGONIST. (See PRIME MOVERS.)

AIRWAY. The path that air of other respiratory gases take in going from the mouth or nostrils, to the alveoli.

AIRWAY RESISTANCE. The resistance, usually measured as a pressure drop, which must be overcome for air to flow through the airway.

ALKALOSIS. A pathological condition in the body of excessive base or below normal acids. One occupationally important cause is hyperventilation (q.v.), which may result from anxiety, drugs or disease or from bad environmental factors such as high temperature, high concentration of solvents or too rapid or heavy a work rate. Physiological response is evidenced in respiratory rate, cardiac rate and kidney function, or it may cause fainting.

ALL OR NONE LAW. Nerve fibers and associated muscle fibers respond entirely to a stimulus or do not respond at all. In muscle, strength of contraction is governed by the number of fibers stimulated and the frequency of stimulation; it is not determined by the individual fiber, which, once stimulated, cannot be stopped halfway in its contractile element, but is limited by its elasticity.

ANAEROBIC METABOLISM. An energy generating process within the body that occurs without the presence of oxygen. Occurs when energy demand of a task exceeds energy available by aerobic metabolism (q.v.) due to inadequate supply of oxygen in a muscle This results in a state of "oxygen debt" where body fats are "burned" without the use of oxygen. This is a low efficiency process because the anaerobic combustion products must be removed from the body by a secondary aerobic process occurring after the activity has finished. Hence, man pants after heavy exercise. Anaerobic activity should be minimized in task design. There must be a payback for any anaerobic effort.

ANALYTIC WORKPLACE DESIGN. Based on established biomechanical and behavioral concepts including the known operating characteristics of man. Produces a workplace situation well within the range of human capacity and does not generally require modification improvement, or preliminary experimental "mock-ups."

ANATOMICAL POSITION. Subject standing erect against a wall with feet parallel and touching, arms adducted (q.v.) and supinated (q.v.), with palm faced forward. Used as a reference posture in anatomical description.

ANATOMICAL REFERENCE POINT. A prominent structure or feature in anatomy which can be located and described by visual inspection or palpation of the body surface. Movements and postures are often defined by description of the relative position and displacement of various anatomical reference points with respect to each other. Also known as anatomical landmark.

ANATOMY. The discipline dealing with the geometrical and topographical description of the structures of the body. It includes description of the structures such as bones, muscles, blood vessels and nerves, including the shape and dimensions (gross anatomy) as well as description of the finer features of human tissues visible only through the microscope (histology). Since structure determines function, a knowledge of anatomy as the structural basis of human performance is indispensable for the proper understanding of human function. It is both static and dynamic.

ANATOMY OF FUNCTION. A subdiscipline of kinesiology (q.v.) which describes changes in the configuration of limbs during the performance of task. Different from "functional anatomy" (q.v.) which relates to physiological function.

ANGLE OF ABDUCTION. Angle between the longitudinal axis of a limb and a sagittal plane (q.v.).

ANGULAR ACCELERATION. The time rate of change of the angular velocity.

ANGULAR VELOCITY. The time rate of change of an angular measure.

ANISOTROPIC. Indicating nonuniform directional physical properties. Body tissues such as bone are anisotropic. The strength of bone differs in different directions. Muscle also has highly directional properties.

ANOXIA. Lack of oxygen in the blood stream or tissue cell. Although the result of many mechanisms, the ultimate effect frequently is defective function of many sensory and motor functions. Produced environmentally by high altitude, chemically by pollutants and drugs, or locally by overexertion. May result in death.

ANTAGONIST. A muscle opposing the action of another muscle. An active antagonist is essential for control and stability of action by a prime mover (q.v.) E.g., the biceps are antagonist ti the pronator during forearm pronation.

ANTHROPOMETRIC TABLES. An arrangement of tabulations and dimensional drawings stating human body measurements. Good quality tables stress data in ranges of dimensions ascribed to certain percentiles of the population rather than listing of mean values. Anthropometric tables are essential to analytical workplace design (q.v.). (See ANTHROPOMETRY.)

ANTHROPOMETRY. The study of people in terms of their physical dimensions. It is the formal name for the technique used to express quantitatively the form of the human body.

ANTHROPOMORPHIC. Shaped like man or man's limbs. Mechanical manipulators and bionic devices often have anthropomorphic features. Cosmetic prostheses are anthropomorphic.

APPLIED ANATOMY. A subdiscipline of anatomy dealing with problems involving physical (as opposed to physiological) function of body systems. The term is sometimes also used to indicate the application of anatomical principles to specific fields of human activity such as surgical anatomy, diagnostic anatomy, anatomy of work.

APONEUROSIS. A fibrous sheet of connective tissues which serves as attachment of muscles or muscle fibers at origin or insertion on bone; an expanded tendon. It is normally pressure resistant and is a preferred load-bearing site. E.g., the seat back should support the aponeurosis of the latissimus dorsi (q.v.). It does not have the elastic properties of the muscles attached to it.

ARM-TOOL AGGREGATE. The arm holding and manipulating a tool, acting with the tool as an integral biomechanical unit. In estimating physiological work, the arm-tool aggregate mass must be considered.

ARTHRITIS. Inflammation of joint structures. Such inflammation may or may not cause pain, and may result in joint destruction and/or limitation of motion. Important cause of partial occupational disability, manipulative limitations, and reduced capacity for head scanning, and increases with aging.

ARTICULATION. Refers to the junction of bones and ligaments which allow body motion. Commonly known as a joint. e.g., knee, elbow. The wrist is the articulation between forearm and hand. (See JOINT.)

ASTHENIA. Lack or loss of strength and energy. When pertaining to somatotype (body build), refers to condition of very slender, feeble-appearing person.

Atrophy. Wasting away of body tissues or organs from disuse, poor nourishment, or from disease, leaving decreased mass of muscle.

Available Motions Inventory. The nature and quality of motions which are available from man in the performance of a specific task. It may be limited by physiological or anatomical deficiencies of individuals.

Axis of Rotation. The true line about which angular motion takes place at any instant. Not necessarily identical with anatomical axis of symmetry of a limb nor necessarily fixed. Thus, forearm rotates about an axis which extends obliquely from lateral side of elbow to a point between the little finger and ring finger. The elbow joint has a fixed axis maintained by circular joint surfaces, but the knee has a moving axis as its cam-shaped surfaces articulate. Axis of rotation of tools should be aligned with true limb axis of rotation. Systems of predetermined motion times often specify such axes incorrectly. Almost all joints move through multicentric axes.

Axis of Thrust. Line along which thrust can be transmitted safely. In the forearm it coincides with the longitudinal axis of the radius. Tools should be designed to align with this axis. Ulnar or radial deviation which produces misalignment causes bending stress acting on the wrist.

Basic Grasp. One of five fundamental means of prehension (q.v.). All activities of grasping and manipulation involve the basic grasps or combinations of them. They are: contact grasp, instrument grasp, power grasp, tripodal grasp and wraparound grasp (q.v.).

Benchmark. A reference value against which performance is compared. It may be physiological (e.g., "normal" heartbeat is 60-80 beats per minute) or functional (e.g., "normal" walking speed in industrial work is 2 miles per hour; a worker can assemble 10 parts in an hour). A "normal" electrocardiogram has an acceptable shape; "normal" acceleration signatures (q.v.) of limit movements are characteristically recognizable. Often used in rating. The effectiveness of evaluation procedures in biomechanics often depends on the intelligent selection of bench marks.

Bicipital. Relating to the biceps muscle, or the groove in which its tendon lies.

Bifurcation. A division into two parts or branches; the bifurcations of the arterial tree.

Biocontrol System. A mechanical system controlled by biological signals; e.g., electromyographically controlled prostheses; e.g., a device alerting truck drivers who are about to fall asleep by means of a bell operated by encephalographic patterns. A biocontrol system should make use of signals of unambiguous meaning which cannot be accidentally generated.

Biomechanical Hypothesis. An a priori assumption about the operating characteristics of man in a work situation based on functional anatomy. Thus, it might be hypothesized that a task which requires ulnar deviation will limit the rotation of the wrist.

Biomechanical Profile. Electromyographic and biomechanical data recorded simultaneously during motion against a resistance. Included as displacement signature (q.v.), which indicates range and pattern of motion; velocity signature (q.v.), which provides an index of strength and speed of motion; and acceleration signature (q.v.), which shows the quality of motion and is an index of control over precision. Electromyogram of muscle masses involved gives an index of their sequencing and coordination. The profile permits objective evaluation of changes in functional capacity resulting from modifications of man-equipment interfaces. (See ELECTROMYOGRAPHY, BIOMECHANICS.)

Biomechanics. The study of the human body as a system operating under two sets of laws: the laws of Newtonian mechanics and the biological laws of life.

Bionic Device. Man made device resembling a biological structure in form, mode of operation and function, e.g., space suits, manipulators for radioactive devices, flap-wing aircraft.

Biotaxis. Contact with living things. One of the ecological stress vectors. Includes both contact with fellow man, i.e., sociotaxis (q.v.) and with other organisms, e.g., microbiotaxis (q.v.).

Bitrochal Seat. Seat designed to fit bitrochanteric width. Anthropometric term roughly synonymous with hip breadth. This dimension is obtained by measurement with calipers applied to the greater trochanter of both femurs in standing position. Important in seat design.

Blood Pressure. The pressure exerted by blood against containing vascular elements, tissues or the heart chamber, measured in millimeters of mercury above atmospheric pressure. In biomechanics the product of systolic blood pressure and pulse rate, under non-stressful conditions, is considered a good index of physiological work stress and/or the mechanical output of the heart. Together with the pulse beat, it is used as an index of fitness in work. Blood pressure is dependent on the energy of the heart action, the elasticity of the walls of the arteries, and the volume and viscosity of the blood. The maximum (systolic) pressure occurs near the end of the strike of the heart. The minimum (diastolic) pressure occurs late in the ventricular diastole. Note: systolic is derived from the Greek (systole) meaning drawing together or contraction and refers to the contraction or period of contraction of the heart, especially that of the ventricle. Diastolic is from the Greek (diastole) meaning drawing asunder or expansion and refers to the dilation (or dilation period) of the heart, especially of the ventricles, which coincides with the interval between the second and first heart sounds.

BODY-LOAD AGGREGATE. In lifting and manual materials handling, the combined weight of the load manipulated and the body segments involved in the task.

BRACHIALIS MUSCLE. Short, strong muscle originating at lower end of humerus (q.v.) and inserting into ulna (q.v.). Operates at mechanical disadvantage, powerful flexor of forearm, employed when lifting.

CAPITULUM OF HUMERUS. A smooth hemispherical protuberance at the distal end of the humerus (q.v.) articulating with the head of the radius. Irritation caused by pressure between the capitulum and head of the radius is called tennis elbow. (See RADIOHUMERAL JOINT.)

CARBON DIOXIDE. The gaseous product of oxidation or aerobic metabolism (q.v) CO_2 (chemical symbol) is transported in the plasma of venous blood to the lungs for discharge. Its partial pressure, pCO_2, controls many body functions including respiratory rate and cardiac rate.

CARDIAC RATE. Heart beats per minute (BPM). The heart can change its output by increasing its stroke volume or its cardiac rate, where the latter is the most effective and most efficient method. Heart rate varies widely with age, sex, environmental stress, state of health, etc. Therefore, the term "normal" heart rate is meaningless. However, ratio of working cardiac rate over resting cardiac rate is one of the measures of work tolerance and work stress.

CARPAL TUNNEL. A passage in the wrist through which important blood vessels and nerves pass to the hand from the forearm. The carpal tunnel is comprised of the concave surfaces of the palmar carpal bones covered by the transverse carpal ligament. The carpal tunnel acts as the conduit for the flexor tendons, the median nerve, and the median artery. Ulnar or radial deviation cause misalignment of the carpal tunnel and irritation of structures passing through it. (See CARPAL TUNNEL SYNDROME.)

CARPAL TUNNEL SYNDROME (CTS). A common affliction of assembly workers caused by compression of the median nerve in the carpal tunnel. The symptoms of carpal tunnel syndrome are burning, numbness, tingling, and pain in the thumb, index and long fingers and in the lateral half of the palm. The thenar eminence may be emaciated. Symptoms are generally felt at night, and may wake the individual from sleep. The pain may radiate to the arm and up into the shoulder following the median nerve path. These symptoms are produced by an entrapment of the median nerve in the carpal tunnel. CTS results from disease either from small or restricted carpal tunnel which compromises the contents, or an enlargement of one of the structures in the canal. CTS is correlated with vibratory hand tool usage, jobs requiring considerable use of the hands, such as assembly work. Results in reduced manipulative skills, particularly if thumb is involved and often reduces work output. Marie and Foix in 1913 first recognized the surgical methods of decompression of the medial nerve to prevent paralysis of the thenar muscles. Medical awareness began in the 1930's, when surgical techniques were first refined. In the 40's and 50's, general recognition surfaced.

CARTILAGINOUS PLATES. Flattened masses of cartilage which act as supporting structures. Intervertebral discs (q.v.) are separated from the vertebrae by cartilaginous plates.

CENTER OF GRAVITY. Equilibrium point of a supported body where all its weight is concentrated (See CENTER OF MASS).

CENTER OF MASS. That point at the exact center of an object's mass; often called the center of gravity (See CENTER OF GRAVITY).

CENTER OF ROTATION. A point around which circular motion is described.

CEREBELLUM. The "small brain" located below the cerebral hemispheres. It exerts a regulatory influence of muscular activity and is concerned with balance, posture, and muscular coordination. It is a subdivision of the mentencephalen, which grows upward and outward from the axis of the brain stem to hide the more inferior portions not covered by the cerebral hemispheres.

CEREBRUM. The anterior-superior part of the brain which governs voluntarily coordinated activities of the body, e.g., sensory-motor functions and associative activities. It is the main portion of the brain, occupying the upper part of the cranial cavity. Its two hemispheres, united by the corpus callosum (also known as commissura magna cerebri), form the largest part of the human nervous system.

CHEMOTAXIS. Relating to contact with a chemical environment. A primary consideration of environmental health, e.g., air pollution. One of the ecological stress vectors.

CHRONOCYCLEGRAM. The photographic record made by a chronocyclegraph, (q.v.).

CHRONOCYCLEGRAPH. An instrument for measuring the pathway taken by a body reference point or an object during the performance of a task as part of the work cycle. It consists of a flickering light attached to an anatomical reference point (q.v.) and photographic recording of the moving light. The chronocyclegraph enables an investigator to determine exact speed, relative speeds and directions as well as changes in motion such as may be caused by fatigue.

CLAVICLE. The collarbone. The sole bony connection between the arm and the trunk. It is easily fractured in falls broken by an extended arm.

CONDITIONED REFLEX. Patterned reflex response to an external stimulus developed as a result of either habituation or training. Easier to learn than to unlearn, e.g., stop on red, go on green. (See SIMPLE REFLEX.)

CONDUCTIVE DEAFNESS. Impaired hearing resulting from interference with, or injury to, the mechanisms for the conduction of sound waves into the inner ear. This transmission takes place either through the external auditory canal and the structures of the middle ear (air conduction) or through vibrations applied to the bones of the skull (bone conduction). This type of hearing loss can be contrasted with Boilermaker's Deafness which is due to perceptive deafness (q.v.).

CONTRACTILE TISSUE. Tissue which is capable of shortening in response to physiological or environmental stimuli, E.g., muscle.

CONTACT GRASP. Action of the hand when pushing, e.g., a coin over a flat surface. One of the five basic grasps of the hand. Not a grasp in the literal sense. The index finger is generally used. Overuse may contribute to high stress concentration on base of distal phalanx and physical damage.

CORONAL PLANE. Any plane which divides the body into anterior and posterior portions. It is perpendicular to the sagittal and transverse planes (q.v.) and is also known as the frontal plane.

CORONOID FOSSA. A depression in the lower end of the front of the humerus in which the coronoid process of the ulna lies when the arm is bent.

CORRELATIVE KINESIOLOGY. That branch of kinesiology (q.v.) which quantitatively relates myoelectric activity (q.v.) with resultant movement. The basis of rational workplace design (q.v.) for minimal fatigue.

COVERT LIFTING TASK. An operation which may not involve the handling of a load but which exhibits all of the biomechanical characteristics ascribed to an overt lifting task (q.v.). Its principal characteristic resides in the existence of a bending moment acting on the vertebral column for reasons other than load lifting, e.g., excessive heel height may cause postural changes which impose bending moments on the lumbar spine. Side-stepping may impose a lateral bending moment. Reaching forward constitutes covert lifting task because torque of the arm at the shoulder joint is transferred to the vertebral column. (See LIFTING TASK).

CUMULATIVE TRAUMA. The development of trauma or work strain from repeated or continuous application of work stress which for short periods of time or single applications would not be harmful. An important factor to consider in workplace and tool design. This does not include delayed onset muscle soreness.

DECELERATION. Rate of slowing down a motion. A negative acceleration. Very significant in biomechanics in the control of end points of a motion. Adequate coordination and voluntary control requires accurate deceleration. Many drugs (e.g., muscle relaxants) interfere with deceleration control and thus destroy coordination.

DELTOID. The muscle of the shoulder, responsible for extending the arm sideways, and for swinging the arm at the shoulder. Overuse of the deltoid muscle may cause fatigue or pain in the shoulder.

DEMANDED MOTIONS INVENTORY. The nature and quality of motions which are required to perform a specific task. The demanded motions inventory may be limited by constraints imposed by the working environment including machinery and tools. Also used to denote the nature and sequence of motions demanded by a specific task design (See AVAILABLE MOTIONS INVENTORY.)

DE QUERVAIN'S DISEASE. A stenosing tenosynovitis involving the abductor pollicis longus and extensor pollicis brevis tendons of the thumb. This disease appears as a weakness that affects the gripping of objects. A pain over the radial styloid process is felt. One occupational cause is the forceful gripping of the type found in clothes wringing.

DIAPHYSIS. The central portion or shaft of a long bone.

DISTAL. Away from the central axis of the body; opposite of proximal.

DOPPLER EFFECT. An apparent signal frequency change caused by relative motion of transmitter and observer. Characteristically, a light color change or sound pitch change. Sometimes used in work measurement to investigate motion patterns. Current devices are not always reliable in the response ranges of interest.

DORSIFLEXION. Bending upwards around an axis. For example, when the ankle flexion takes place upwardly it is called dorsiflexion and the downward flexion is called plantar flexion.

DYNAMIC MOMENT. A force-distance relationship dealing with dynamic (resulting from inertia and acceleration) forces as opposed to static (resulting from gravitational accelerations only) forces. An example of dynamic moment is that experienced during a lifting task. At the instant of lift-off (load moves off platform), the acceleration of the load is caused by the dynamic moment on the lumbo-sacral joint. The consequences of this phenomenon are not always clearly predictable.

DYNAMIC WORK. Muscle work performed when one end of a muscle moves with respect to the other end and external movement is produced. (See ISOMETRIC WORK).

DYNAMOMETER. Apparatus for measuring force or work output external to a subject. Often used to compare external output with associated physiological phenomena (electromyography, spirometry, etc.) to assess physiological work efficiency.

ECCENTRIC CONTRACTION. Increase of tension within a muscle while lengthening. For example, the brachialis exerts a force resisting the pull of gravity when extending a flexed forearm slowly.

ECHOGRAPHY. The use of ultrasonic energy for imaging internal organs and tissues. Same as sonography. (See ULTRASONICS.)

ECOLOGICAL STRESS VECTOR. Any vector in the environment producing a physiological response in man. Such response may or may not be pathological. Major stress vectors are: climatotaxis, biotaxis, mechanotaxis, chemotaxis, particulotaxis (q.v.).

EFFECTIVE TEMPERATURE. A measure of "warmth as related to comfort." Combines environmental factors, humidity, true temperature, and air velocity to provide a scale of subjective comfort. Two experimental scales of the American Society of Heating and Ventilating Engineers pertain to normally-dressed and partially-dressed people. A comfortable effective temperature is significant in maintaining high productivity and good quality control. (Corrected effective temperature (CET) is often used in place of effective temperature.)

EFFERENT. Conveying away from a biological processing station, specifically by conveying nervous impulses from a neural center to a muscle. (See AFFERENT.)

ELASTIC LIMIT OF TISSUES. The level of physical deformation caused by the application of a force beyond which tissue damage occurs and beyond which the tissue will not return to its original function when the force is removed.

ELECTRICAL-SILENCE. Absence of a measurable action potential in a biological structure. A zero recorder pen deflection should ideally occur during electromyography of a non-contracting muscle. In practice, true electrical-silence cannot be observed unless random noise signals are filtered out.

ELECTROCARDIOGRAPH. A recording galvanometer which produces an analog representation of the electrical activity of the conductive tissues of the heart, i.e., an electrocardiogram (ECG or EKG). Useful in the diagnosis of heart disease and assessment of work stress in health and disease.

ELECTRODE. A device used to sense the electric potential of body tissues or to transmit such a potential. It can be surface-mounted as a skin electrode, internally mounted as a needle electrode, or as a very fine wire inserted into the muscle. An effective electrode minimizes contact resistance and junction potentials. (Used in myography, cardiography, and encephalography.)

ELECTROMYOGRAPHIC KINESIOLOGY. The analysis and evaluation of human motion patterns by means of electromyography. Useful in establishing optimal motion patterns through workplace and tool design.

ELECTROMYOGRAPHY. The technique of recording the electrical potential generated by muscle activity. Surface electromyography will provide a measure of muscular sequence and physiological effort associated with a particular task.

ELECTROPHYSIOLOGICAL APPARATUS. Research and evaluation devices which sense bioelectrical signals from heart, muscles, brain, etc. The sensed signal is amplified and displayed in the form of an analogue or digital readout. Examples: electromyography, electrocardiography, polygraph, (q.v.).

EMPIRICAL WORKPLACE DESIGN. Design based on operating characteristics and mechanical needs of equipment or on previous experience with similar work situations. Must normally proceed through a number of "improvements" to adapt to the needs of the worker.

ENDOCHTHON. Innate, originating from within. In hand-operation of a lever, the power is endochthonous to the operator. (See EXOCHTHON.)

ENVIRONMENTAL ENGINEERING. Application of engineering principles to create and maintain surroundings which are favorable for the sustenance of lift.

ENVIRONMENTAL INPUTS. Often used as synonym for ecological stress vector (q.v.). More accurately, the following factors to which the worker responds either physiologically or behaviorally; economic, social, managerial, mechanical, climatic, psychological. The resulting outputs are economical (in the form of production), emotional (in the form of employee behavior), and physiological (in the form of employee health and fatigue).

ENVIRONMENTAL STRESS VECTOR. Synonym for ecological stress vector (q.v.).

EPICONDYLE. A raised area (bump) on the condyle of a bone, from which muscles originate. E.g., the flexor muscles of the fingers originate from the medial epicondyle of the humerus.

EPICONDYLITIS. Inflammation or infection in the general area of an epicondyle, e.g., tennis elbow.

EPIPHYSIS. The ends of a long bone. Joint surfaces are always part of the epiphysis.

EQUILIBRIUM. State of a system or a body in which internal change does not occur. In mechanics, the sums of all forces and moments are zero.

ERECTOR SPINAE MUSCLES. Large back muscle which originate on the sacrum and lumbar vertebral bodies and insert on the rib cage and thoracic vertebrae. The muscle is comprised of three components: 1) spinalis, 2) longissimus, and 3) ilio-costalis. The erector spinae are the primary muscles recruited during lifting activities.

ERGONOMICS. The application of a body of knowledge (life sciences, physical sciences, engineering, etc.) dealing with the interactions between man and his total working environment, such as atmosphere, heat, light and sound, as well as all tools and equipment of the workplace.

ERGONOMIC ANALYSIS. Application of the principles of ergonomics to study in detail all of the specific elements which are pertinent to a man-equipment or man-task interface.

ETIOLOGY. The study of causes of disease. In biomechanics, etiology of trauma or work strain is complicated because the site of such trauma is often remote from the focus of work stress. Industrial hygiene experience is a valuable aid to determining causes of workplace-generated trauma.

EXOCHTHON. Outside of and not inherent to a body. Operation of pneumatic or electrical tools requires management by the operator of power which is exochthonous to his system, i.e., not generated by him. (See ENDOCHTHON.)

EXTENSION. Straightening of a curve or angle; the position of the joints of the extremities and back when one stands at rest, or the direction of motion that tends to restore this position; the opposite of flexion.

EXTENSOR MUSCLE. A muscle which when active increases the angle between limb segments, e.g. the muscles which straighten the knee or elbow, open the hand or straighten the back.

EXTENSOR RETINACULUM. Also known as transverse dorsal ligament. A membranous band of fibers at the back of the hand which forms a tunnel through which the extensor tendons of the fingers pass. The retinaculum acts as a guide and prevents the tendons from bowstringing when the wrist is hyper-extended.

Extensor Tendon. Connecting structure between an extensor muscle and the bone into which it inserts. Examples are the hard, longitudinal tendons found on the back of the hand when the fingers are fully extended. (See TENDON, EXTENSOR MUSCLE, INSERTION.)

External Mechanical Environment. The man-made physical environment, e.g., equipment, tools, machine controls, clothing. Ant: internal (bio)mechanical environment (q.v.). external working environment, The environment external to the body at the workplace. Of particular importance in ergonomics because of the effect of surroundings on work performance. It includes immediate environment such as tool handles, clothing, temperature, humidity, pressure, and composition of atmosphere. (See GILBRETHIAN VARIABLES, INTERNAL (BIO)MECHANICAL ENVIRONMENT.)

Extrinsic. Anatomical term referring to a component, usually a muscle that originates outside of the structure on which it acts, e.g. relating to vision, the extrinsic muscles of the eye are located in the orbit and move the eyeballs. The intrinsic muscles (q.v.) are located within the eyeball and operate the iris and lens. Ant: intrinsic.

Eye Scanning. Binocular scanning of the visual field by movement of the eyeballs alone and without head rotation through use of the extrinsic muscles of the eye. Eye scanning is a rapid and easy process, but can be effectively performed only with a central visual cone of 60 degrees or loss of depth perception will result.

Fatigue. That physiological state characterized by a lessened capacity for work and/or the inability to maintain consistent work quality levels. May be due to intensive mental or physical activity or depletion of physiological reserves.

Feedback System. A control system in which information from the controlled element is returned to the controlling element. The controlling element then modulates its requirements of the controlled element according to performance requirements. In biomechanics, muscle coordination by visual or tactile signals are examples of feedback systems. Improper performance by a worker may be caused by interference in his feedback system (disease) or improper operation of his sensors (optical illusion). Homeostatic systems are examples of physiological feedback systems. (See HOMEOSTASIS.)

Femur. The long bone of the upper leg (thigh). The upper end (head) of the femur articulates with the pelvis to form the hip joint, the lower end (condyles) articulates with the tibia to form the knee joint. The hip and knee joints often are the site of arthritis (q.v.), which may make walking painful and limited (elderly workers).

Fibula. The smaller bone of the lower leg (shank). Extending from just below the knee to the ankle joint, it is lateral and posterior to the tibia (q.v.), which is the main load-bearing bone. The fibula is attached to the tibia by the interosseus membrane and provides a site for muscle attachment. The lower end (lateral malleolus) forms the lateral side of the ankle joint itself, is essential to ankle stability, and often is injured by twisting forces.

Fifth Lumbar Vertebra. The lowest vertebral element in the lumbar region, in the small of the back immediately above the sacrum. Because the flexible lumber section is adjacent to the rigid sacral and pelvic structure, the fifth lumbar vertebra and adjacent discs are subjected to high stress, especially in lifting tasks. Trauma due to excessive stress at this site can cause lower back pain and pain in the legs. (See LUMBOSACRAL ANGLE, HERNIATED DISC.)

Fixed Linkage Mechanism. Linkage formed between skeletal elements of man and machine in a man-machine system. Movement of any link produces movement in other links, e.g., machine operator and drill press. Configuration of fixed linkage systems determines biomechanical efficiency of work, e.g., seat height at drill press.

Flexion. Movement at a joint whereby the angle between two bones connected at a joint is changed in a manner tending to approximate the body segments it connects. Motion of this type is produced by contraction of flexor muscles.

Flexor Muscle. A muscle which, when contracting, changes the angle between two limb or body segments in a manner tending to approximate the segments it connects. The principal flexor of the elbow is the brachialis muscle. Flexors of the fingers and the wrist are the large muscles of the forearm originating at the elbow. (See EXTENSOR MUSCLE.)

Flexor Retinaculum. Touch fibrous ligament forming the ceiling of the carpal tunnel (q.v.). Also called transverse carpal ligament, this structure prevents bowstringing of the flexor tendons at the wrist.

Force Platform (Plate). (See REACTANCE PLATFORM.)

Force-Time. Applied force multiplied by time of application. Used to quantify isometric work (q.v.) and has the dimensions of linear impulse, for short time duration phenomenon.

Functional Anatomy. Study of the body and its component parts, taking into account those structural features related directly to physiological function. E.g. functional anatomy of elbow defines relative flexor force developed by biceps and brachialis at various degrees of flexion.

Frequency. The number of repeated occurrences in a given time interval. The frequency of electrical or mechanical activity is usually specified as cycles per second (c.p.s.) or hertz (Hz).

Gait Analysis. Study of human locomotion. Analyses can be kinematic, kinetic, electromyographic, or some combination of these techniques. Used in the design of ramps and lower limb prostheses, evaluation of load-carrying situations, and in defining neurological or structural abnormalities of gait. Factors examined include muscular activity and coordination, ground reaction forces, joint and limb segment positions, forces and accelerations, and postural changes.

Gangrene. A condition in which a localized area of body tissue dies and decays. Gangrene occurs in many forms, but primarily refers to death or necrosis of tissue, as caused by loss of blood flow to a specific area (e.g. arterial injury or occlusion, frostbite, etc.), infection by bacteria (e.g., gas gangrene associated with traumatic wounds), or both. Diabetic workers are more susceptible to gangrene than others due to circulatory insufficiency associated with the disease. Also, paraplegic workers are subject to decubitus ulcers over the buttocks which are associated with a lack of sensation which predisposes the workers to unrelieved sitting pressures which in turn occludes capillary circulation long enough to cause tissue death.

Gilbrethian "Systems Concept". The original notion calling for matching of the internal physiological and biomechanical environment of the worker with his external physiological and mechanical environment. First postulated by Frank B. and Lillian M. Giibreth in 1911 by their setting forth comprehensive list of interacting variables affecting the worker, his environment, and his motions requirements. (See GILBRETHIAM VARIABLES.)

Gilbrethiam Variables. System of three sets of variables intrinsic to every task. They are variables of the worker relating to anatomic and psychological factors: variables of the environment including physical and economic factors: and variables of motion including effort and movement patterns. The variables are used as a basic tool in work system analysis and design.

Glenoid Cavity. Glenoid cavity is used commonly with reference to the socket of the shoulder joint formed by the scapula (q.v.) to receive the humerus (q.v.). Syn: joint socket.

Gluteus Maximus. Large muscle of the rump which acts across the hip joint and extends the thigh, i.e., brings the thigh from the position occupied in sitting to the position in standing. It is an important muscle for walking, lifting and standing. Electromyogram of the gluteus maximus is a good index of the relative muscle activity during lifting.

Goniometer. An instrument by which angles of joint rotation are measured, either statically or during movement of the joint. Important in the evaluation of industrial workplace layouts and in the analysis of the body in motion (See REFLEX GONIOMETRY.)

Grasp Reflex. A reflex which is both basic and conditioned. It is initiated by stimulation of the median nerve (q.v.) sensory feedback area in the hand. Surgeons need a well conditioned grasp reflex so that when surgical tools are being transferred, the surgeon can concentrate on the patient. However, in the machine shop people have lost fingers by reflexively grasping moving saw blades, etc. (See PREHENSILE.)

Grooving. Designing a tool with grooves on the handle to accommodate the fingers of the user. Considered, a bad practice, because of the great variation in the size of workers hands. Grooving interferes with sensory feedback. Intense pain may be caused by grooving to the arthritic hand.

Ground Reaction Force. A gravitational force produced by the weight of an object against the surface on which it lies.

Hamate Bone. One of the wrist bones which possesses a hook projecting toward the palm of the hand on the side opposite the thumb. This hook provides one attachment of the covering of the carpal tunnel (q.v.) flexor retinaculum (q.v.) through which the flexor tendons of the fingers pass; the hook helps hold these tendons in place. (See PALMAR ARCH, ULNAR NERVE.)

HAMSTRINGS. Muscles located at the back of the thigh which are employed in bending the knee and erecting the trunk. They are important in lifting tasks where bent knees are employed and in tasks requiring operation of controls with the leg.

HEAD OF RADIUS. Proximal end of the radius bone (q.v.) which is in contact with the humerus (q.v.) at the elbow joint is biomechanically significant in the operation of many hand tools.

HEAT ACCLIMATIZATION. Physiological adjustment to temperature changes of the external working environment (q.v.). Such adjustments include change in heart rate, respiratory rate, transpiration and perspiration and peripheral circulation. The effect of these changes is to maintain a constant core-temperature and other physiologic parameters essential to the well-being and efficient work performance of the individual.

HEAT STROKE. Illness caused by exposure to excessively high temperatures resulting in overload and breakdown of the physiological control system permitting acclimatization. Symptoms are headache, dizziness, confusion, hot dry skin, collapse and coma, high fever, and increased pulse rate. Heat stroke may occur where work space is not properly cooled and ventilated and especially where work situations involve intense heat, i.e., steel mills, roasters, etc. Prolonged and repeated exposure can cause serious brain damage and/or death. Clothing (such as impermeable synthetics) which prevent adequate heat dissipation can cause heat stroke. In other cases, symptoms of prostration similar to shock with collapse, cool, damp skin and subnormal temperature may occur.

HERNIA. Also known as rupture. Failure or weakness in the tissue wall which contains an organ or other structure or the protrusion of the organ through the area of weakness. Herniation of the groin, in which intestines may protrude through a mechanically weak area in the abdominal wall, is industrially significant because of the work limitations which it imposes. Good workplace layout and careful attention to posture will minimize restrictions imposed on the worker.

HERNIATED DISC. Also known as "slipped disc", this term commonly refers to herniation of the nucleus pulposus or protrusion of the interior soft part of the intervertebral disc through the outer, fibrous containing layer which fails following degeneration or excessive stress. The condition may cause pressure on spinal nerves, with resulting low back and sciatic pain down the leg, disability and often partial paralysis. Low back trauma and/or overload has a causal relationship.

HINGE JOINT. A single degree of freedom joint such as that encountered at the elbow (humero-ulnar joint) where motion in only one plane is possible. The elbow, as a whole, has two degrees of freedom due to the humero-radial and radioulnar joints which permit rotation.

HOMEOSTASIS. The state of equilibrium or constancy of physiological conditions in the living body, e.g., chemical equilibrium, fluid content, blood pressure. Also used to describe the physiological mechanism by which equilibrium is maintained in the body, e.g., regulation of blood pressure, blood chemical factors, body temperature, etc.

HOMOLOGOUS MOTION. A motion produced by one set of muscles which substitutes for a motion which could be produced by another set of muscles. In workplace design, the muscles required for a particular task can be chosen by the workplace layout. Thus, a lateral humeral rotation (q.v.) or its homologous motion, a shoulder swing, is exercised depending on seat height or table height. Homologous motions which have the same output may produce substantially different work stress.

HUMAN ENGINEERING. Application of anatomical, biomechanical, psychological, physiological to the design of work situations. The objective is to optimize human performance in the monitoring and operation of tools and equipment. Sometimes referred to as Human Factors Engineering. (See ERGONOMICS.)

HUMERAL ROTATION. Rotation of the arm about the long axis of the humerus (q.v.) as contrasted with circumduction or shoulder swing. Humeral rotation with forearm flexed produces wrist displacement in the plane of the forearm, probably with less physiological work than would be required to achieve the same position by shoulder abduction.

HUMERUS. The bone of the upper arm which starts at the shoulder joint, the head of the humerus articulates with the glenoid cavity. (q.v.) and ends in the humeral condyles within the elbow joint. Muscles which move the upper arm, forearm and hand are attached to this bone.

Hyperventilation. "Over breathing" (a state in which there is an increased amount of air entering the pulmonary alveoli). A condition of prolonged, rapid and/or deep breathing which leads to excessive expulsion of residual CO_2 from the blood via the lungs. Conducive to alkalosis (q.v.). Often produced by excessive physical work load and/or heat stress or apprehension.

Hypoxia. Synonym anoxia, condition in which there is an oxygen deficiency in organs and tissues, i.e., less than a normal amount. Mechanisms which cause hypoxia include insufficient partial pressure of oxygen in air (i.e., high altitude), insufficient pulmonary ventilation, or exposure to toxic gases (i.e., carbon monoxide) or chemicals. Impeded circulation, caused for example, by the pressure of a tool on the hand, can reduce oxygen delivery to tissues resulting in a local area of hypoxia.

Idiopathic. Refers to a disease of unknown cause, especially one arising in the body without an identifiable extrinsic cause. Diseases caused by toxic industrial agents tended to be considered idiopathic until causes were found.

Iliac Crest. The upper rounded border of the lateral aspect of the pelvic bone above the hip joint. No muscles cross the iliac crest and it lies immediately below the skin. It is an important anatomical reference point (q.v.) because it can be felt through the skin. Seat backrests should clear the iliac crest.

Impairment. A dysfunction in man in which a body part (or parts) is incapable of performance within established and accepted standards of "normal" performance.

Improvement Approach to Workplace Design. (See WORKPLACE LAYOUT.). Syn: empirical workplace design.

Index of Work Tolerance. Any measure indicative of the length of time during which an individual can perform a specific task with necessary efficiency and at the same time experience desirable levels of physiological and emotional well-being.

Industrial Hygiene. That science and art devoted to the recognition, evaluation and control of those environmental factors or stresses, arising in or from the workplace, which may cause sickness, impaired health and well-being, or significant discomfort and inefficiency among workers. (Adapted from *Fundamentals of Industrial Hygiene*, National Safety Council.)

Infrared Photography. A technique whereby only the infrared radiation of the spectrum is photographed by using infrared sensitive film and infrared transmitting filters. Infrared radiation varies with the temperature of the radiator. Therefore, infrared photography is used to indicate local temperature variations in human subjects which can be caused by local inflammation, ischemia, or peripheral vasoconstriction. This technique is useful in evaluating the effects of hand tools and hand-operated machinery on soft tissues. (See THERMOGRAPH.)

Input Variables of External Environment. Six motiva-tional vectors in human performance, economic, social, managerial, biomechanical, climatic, and behavioral.

Insertion. The anatomic point of attachment of a muscle to the bone which moves when the muscle contracts. The insertion is usually via a tendon or aponeurosis (q.v.). Insertion is at the distal (q.v.) end of a muscle. (See ORIGIN (MUSCLE))

Integrated Surface Myogram. A recording of the level of electrical activity (firing rate) associated with a muscle contraction. Sensed by electrodes mounted on the skin and bracketing the belly of the muscle under consideration. The ISM represents a summation of all the number of potential changes generated at any instant by the contraction of the muscle mass. It is proportionate to the number of fibers contracting at any time and indicates the coordination, sequencing and level of muscular activity involved in specific body maneuvers. An important index of work stress, and training and aptitude for the performance of a task.

Internal (Bio)Mechanical Environment. The muscles, bones and tissues of the body, all of which are subject to the same Newtonian force as external objects in their interacting with other bodies and natural forces. When designing for the body one must consider the forces that the internal mechanical environment must withstand.

Interphalangeal Joints. The finger or the toe joints. The thumb has one interphalangeal joint, the fingers have two interphalangeal joint each.

Intervertebral Discs. Fibro-cartilaginous pads which separate the bodies of the vertebrae. Bending of the spine requires deformation of the disc which are susceptible to injury during incorrect lifting.

Intrinsic Muscle. Muscles located within an anatomical entity which contribute to its function, e.g., intrinsic muscles of the hand which are used for the fine manipulation, as opposed to extrinsic muscles which lie in the forearm but are used to move the fingers and wrist. Intrinsic muscles are smaller and weaker and fatigue earlier than extrinsic muscles.

ISCHIAL TUBEROSITY. A rounded projection of the ischium (q.v.). It is a point of attachment for several muscles involved in moving the femur (q.v.) and the knee. It can be affected by improper design of chairs and by situations involving trauma to the pelvic region. When seated, pressure is borne at the site of the ischial tuberosities. Chair design should provide support to the pressure projection of the ischial tuberosity through the skin of the buttocks.

ISCHEMIA. Lack of blood flow. Loss of sufficient replacements to maintain normal metabolism (q.v.) in the cells. Caused by blockage in the circulatory system or failure of the cardiac system. Blockage may be by internal biological agents, such as arterial wall deposits, or by external environmental agents, such as poorly designed tools or workplace which press against arteries and occlude them. Depending on the degree of ischemia, numbness, fatigue and tingling may be evidenced in the limbs. At the workplace, loss of precision in manipulation may lead to reduced efficiency, poor quality and the possibility of accidents.

ISCHIUM. The lowest component bone of the hip complex. Important biomechanically as the main support of the body in seating. (See ISCHIAL TUBEROSITY.)

ISOINERTIAL. Human muscle force applied to a constant mass in motion.

ISOKINETIC. Human muscle force exerted during constant velocity of motion.

ISOMETRIC WORK. Referring to a state of muscular contraction without movement. Although no work in the "physics" sense is done, physiologic work (energy utilization and heat production) occurs. In isometric exercise, muscles are tightened against immovable objects. In work measurements isometric muscular contractions must be considered as a major factor of task severity. (See DYNAMIC WORK, NEGATIVE WORK, WORK.)

JOINT. Vernacular term for articulation (q.v.) between two bones which may permit motion in one or more planes. Joints may become the sites for work-induced trauma (see TENNIS ELBOW, etc.) or other disorders (e.g., arthritis).

JOINT REACTION FORCE. The internal reaction force acting at the contact surfaces when a joint in the body is subjected to internal loads, external loads, or is moved.

KINEMATICS. The branch of mechanics that deals with motion of a body without reference to force or mass.

KINEMATIC CHAIN. A combination of body segments connected by joints (q.v.) which, when operating together, provide a wide range of motion for the distal (q.v.) element. A single joint only allows rotation, but kinematic chains, by combining joints, enable translational motion to result from the rotary motions of the limb segments. Familiarity with the separate rotary motions and their limitations is necessary for comprehension of the characteristics of the resultant motion. By combining joints whose axes are not parallel, the kinematic chain enables a person to reach every point within his span of reach (q.v.)

KINESIOLOGY. The study of human movement in terms of anatomy of function.

KINESIOMETER. A device which measures parameters of motion such as displacement, velocity and acceleration, either as vector sums, or resolved into component parts. Used to evaluate the effect of changes in the workplace such as chair height, lighting, task elements, drugs, antihistamines, etc., on the motions patterns of the worker.

KINETICS. The branch of mechanics that deals with the motion of a body under the action of given forces.

KNEE SWITCH. A control device which is operated by lateral movement of the knee when seated, requiring operation of the abductor muscles of the leg. In the seated position the leg cannot rotate but the abductors operate with strength, precision and control. Knee switches are used to control sewing machines, etc.

LACTIC ACID. A product of anaerobic metabolism (q.v.) of muscle which is cleared from the muscle tissues after the activity which produced it has ceased. Presence of lactic acid causes fatigue and muscular pain and contributes to condition of acidosis (q.v.). Lactic acid is removed from the body by a subsequent aerobic metabolism (q.v.). Work schedules should provide periods of recovery following heavy exertion.

LAMBERT SURFACE. An illuminated reflecting surface whose brightness appears equal at any angle of observation. A theoretical ideal for a workplace surface.

Lateral Displacement. Movement of a limb or body seg-ment away from the mid-sagittal plane (q.v.). Movement of legs (side-stepping) or arms sideways are examples of lateral displacement. (See MEDIAL DISPLACEMENT.)

Lateral Transfer. Personnel reassignment in a company organization to an equivalent position in a collateral subdivision. When lateral transfers are actively sought by employees, organizational or environmental problems should be suspected. Minor discomforts such as sore backs or elbows may be the cause of requests for lateral transfers.

Latissimus Dorsi. A large flat muscle of the back which originates from the spine of the lower back and inserts into the humerus (q.v.) at the armpit. It adducts the upper arm, and when the elbow is abducted, it rotates the arm medially and bings the shoulder back to an anatomically neutral . It is actively used in operating equipment such as the drill press where a downward pull by the arm is required.

Lesion. A wound, injury, or unnatural change in tissue texture. Describes local manifestation of disease whether traumatic or not.

Lifting Task. Any industrial task which applies a moment to the vertebral column. It includes one or more of the following elements: sagittal bending moment, lateral bending moment, torsional moment, inertial moment, isometric component, negative component, dynamic component, and frequency of lift.

Lifting Torque. The most reliable measure of lifting stress. The product of load and distance from a fulcrum within the vertebral column created by a lifting task. This torque must be balanced by an opposing one in the musculoskeletal system.

Ligament. A tough fibrous band or loop of tissue which connects bones and supports body tissues. The geometry of joint movement depends on the arrangement of the ligaments. The biomechanist must be familiar with arrangement of ligaments in designing workplace and tasks for the industrial worker.

Light Task. Any job which requires a physiological output rate of from 0.01 to 0.025 horsepower. This range assumes that general working conditions are sufficient to maintain an 18 percent to 20 percent rate of worker efficiency (the metabolic conversion of fuel into useful energy). This definition is applicable only when large muscle groups are used.

Limb-Load Aggregate. The total load of lifting to be considered in biomechanical task analysis. Where the load is small, the physiological effort is primarily dependent on limb weight. Such an effort is a covert lifting task (q.v.). (See ARM-TOOL AGGREGATE.)

Linkage. (See FIXED LINKAGE MECHANISM.)

Locomotion. Movement of the body, limbs or an anatomical reference point (q.v.) from one place to another.

Locomotor System. Anatomic structure used in locomotion (q.v.) consisting of musculoskeletal system (q.v.). Interaction of the musculoskeletal system and the nervous system. The performance of any motor task involves the use of the locomotor system.

Lordosis. Curvature in the sagittal plane (q.v.) of the cervical and lumbar regions of the spine. Normal lordosis is a prerequisite of safe materials handling. Lordotic configuration changes when the stress equivalent (q.v.) of the loads handled exceeds safe limits. Bone disease may also affect lordosis. In the industrial environment, lordotic changes caused by improper heel heights may cause discomfort and inefficiency in workers, especially women.

Lordotactic. Mechanically leaning against the lumbar spine.

Lumbar Spine. Lowest section of the spinal column or vertebral column immediately above the sacrum (q.v.). Located in the small of the back and consisting of five large lumbar vertebrae (q.v.), it is a highly stressed area in work situations and in supporting the body structure.

Lumbar Vertebrae. The five vertebrae located in the lumbar spine. They are the largest in cross section and the strongest of the vertebrae.

Lumbosacral Angle. Angle between the back of the lumbar spine and the sacrum. Correct angle is essential to postural integrity and safety and affects walking speed. Frequently distorted by improper shoe heel height and poorly designed backs of working chair.

Lumbosacral Joint. The joint between the fifth lumbar vertebra and the sacrum. Often the site of spinal trauma because of large moments imposed by lifting tasks. This joint allows the greatest motion of the joints in the lumbar spine.

Lumen of the Transverse Canal. Hole in the vertebral column formed by the junction of semicircular notches of successive vertebrae through which nerves emerge. Injury from the stress of poor lifting technique, mechanical impact or biological changes of aging, may cause compression of the nerve, thereby causing partial or total disability, paralysis and pain.

Man-Environment Interface. Region of contact of man with his environment in the performance of a work task. It can include areas of exposure between the worker and the parameters which affect output performance levels such as climate, temperature, and illumination.

Man-Equipment Interface. Areas of physical or perceptual contact between man and equipment. The design characteristics of the man-equipment interface determine the transfer of information and motor skill. Poorly designed interfaces with the workplace, machines, tools, fixtures and the workplace may lead to localized trauma (e.g., calluses) or fatigue.

Man-Equipment Task System. System where the motor skills of man are matched to those demanded by equipment in such a manner as to facilitate the effective performance of a task. Input-output relations in these systems (i.e. efficiency) are specific and are functions of task design.

Manipulation. To operate through the use of the hands. Most industrial tasks involve motor skills of manipulation.

Manipulative Skill. The ability to move objects with reference to location in space, speed of movement and control of movement. Manipulative skill depends on intrinsic muscles (q.v.) of the hand and eye-hand coordination.

Man-Task System. Simple system involving man performing a given task, usually involving only perception and motor skills.

Mechanical Analogue. Mechanical model or representation of another type of system (i.e., physiological, anatomical, etc.) which responds to inputs with outputs corresponding to those in the real system.

Mechanoreceptor. Sensory end organs receptive to mechanical stimuli, e.g., pressure sensors in the skin. Impairment of mechanoreceptors can lead to degraded quality of work and increased risk at the workplace because of insufficient sensory feedback (q.v.). Control in manipulation of tools or objects depends on effective mechanoreceptors.

Mechanotaxis. Contact with a mechanical environment consisting of forces (pressure, moment), vibration, etc. One of the ecological stress vectors. Improper design of the mechanotactic interface may lead to instantaneous trauma, cumulative pathogenesis, or death.

Medial Displacement. Movement of a limb or body segment toward the mid-sagittal plane (q.v.). Bringing the legs or arms together are examples of medial displacement. (See LATERAL DISPLACEMENT.)

Medial Popliteal Nerve. Important nerve at the back of the leg in the hollow of the knee. (See popliteal region.) Pressure on the medial popliteal nerve caused by poor seat design may lead to inadequate blood flow to the feet.

Median Nerve. A major nerve controlling the flexor muscles (q.v.) of the wrist and hand. Tool handles and other objects to be grasped should make good contact with the sensory feedback (q.v.) area of this nerve located in the palmar surface of the thumb, index, middle, and part of ring finger. It is the median nerve which is compressed in carpal tunnel syndrome.

Metabolic Cost. The amount of energy consumed as the result of a specific activity, generally stated in calorie (q.v.) units. Can be measured by spirometry (q.v.) or respiratory gas exchange. A helpful measure of task severity. (See METABOLISM, LIGHT TASK.)

Metabolic Rate. Rate of energy consumption required for performance of body activities. Basal metabolic rate is the minimum rate at which vital body functions can be maintained. Units are in calories per unit time. Metabolic rate can be determined indirectly by oxygen uptake or carbon dioxide discharge and is a good index of work strain of steady state work when work is medium-heavy or heavy, and large muscle masses are involved.

Metabolism. The physiological combustion process. Conversion of food stuffs into body tissue, energy, and waste products. A continuous complex process influenced by the energy demands of the body as well as its state of health, temperature, emotion, etc. (See AEROBIC METABOLISM, METABOLIC RATE, METABOLIC COST.)

Metacarpal Bones. The five bones in the hand which connect the carpal (wrist) bones with the phalanges (knuckles). The metacarpals form the structural base of the palm of the hand.

MICROBIOTAXIS. Contact with microorganisms. Exposure to an environment of certain microbes. One of the ecological stress vectors. A form of biotaxis (q.v.).

MICROMOTION FILM. Motion picture or video tape made of a specific job or class or work. Each component of the task is portrayed on a single frame for independent study. With such films exposed at fast speeds (frames/ sec), it is possible to break down complex motion patterns into their most basic elements for therblig determination and efficiency analysis. In biomechanics this technique is used to identify short duration acceleration patterns and reflex responses, e.g., during lifting, which are too fast for visual observation.

MID-SAGITTAL PLANE. A reference plane formed by bisecting the human anatomy so as to have a right and left aspect. Human motor function can be described in terms of movement relative to the mid-sagittal plane. (See TRANSVERSE PLANE, CORONAL PLANE.)

MOMENT OF FORCE (TORQUE). The effectiveness of a force to produce rotation about an axis measured by the product of the force and perpendicular distance from the line of force to the axis. (SI unit is Nm.) Popularly it refers to the twisting or turning effect of a force. Moments, like forces, have both internal and external effects upon the body in which they are acting. The external effect of a moment on a body is to change or attempt to change the angular or rotational velocity of the body. The internal effects of moments are to cause a state of strain. A moment is necessary to produce angular acceleration. (See TORQUE.)

MOMENT CONCEPT. The concept based on theoretical and experimental bases that lifting stress depends on the bending moment exerted at susceptible points of the vertebral column rather than depending on weight alone.

MOMENT OF INERTIA. The inertial resistance to a change in rotational motion which must be overcome by a torque (q.v.), in order to produce angular acceleration. It is a function of the distribution of mass about the axis of rotation.

MOTIONS INVENTORY. (See AVAILABLE MOTIONS INVENTORY, DEMANDED MOTIONS INVENTORY.)

MOTIONS PATHWAY. The locus of travel of an anatomical landmark, often a "knuckle," in moving from one point of the workplace to another. It includes the elemental increments of "reach" and "move" as defined by the various predetermined motion time systems (q.v.) and can also include such motor therbligs as "position," "examine," "hold," etc. motion time methods system, Abbreviated MTM. (See PREDETERMINED MOTION TIME SYSTEM.)

MOTOR NERVE. A nerve also known as an efferent nerve which conducts impulses from the brain or spinal cord to muscle.

MUSCULOSKELETAL SYSTEM. The combined system of muscles and bones which comprise the internal biomechanical environment.

MYOGRAPHY. (See ELECTROMYOGRAPHY.)

NAVICULAR BONE. Synonym for scaphoid (q.v.) bone.

NEGATIVE WORK. Work performed with the assistance of gravity as in the lowering of an object or a body segment. Its muscular effort mainly stresses the control of the load rather than the power required in lifting. Its effort is usually rated at from 1/3 to 1/2 of that of positive work (q.v). One of the elements of a lifting task (q.v.).

NEUROPATHY. Pathology of a nerve trunk of varied etiology. Pain, paralysis, muscle atrophy, numbness, and loss of reflexes may occur in areas supplied by the involved nerve. Neuropathy may be due to injury, exposure to harmful chemicals, infection, compression of a nerve, or poisoning by heavy metals. It may also result from a generalized disease such as diabetes.

NON-POSITIVE TACTILE STIMULI. The cessation of feedback signal from the tactile sensors to the brain. For example, if one is sliding an object along a surface toward a hole without looking at it, a non-positive tactile stimulus will be obtained when the object falls into the hole indicating that the task has been successfully completed.

NYSTAGMOGRAM. A recording of the quick rhythmic oscillations of the eyeballs usually known as saccadic eye movements. These oscillations are typically involuntary and can be either horizontal, vertical or rotary. The recordings are a good indicator of the effects of rotation on postural controls as maintained by the semicircular canals and also on the effectiveness of eye scanning (q.v.) as opposed to head scanning (q.v.).

OCCUPATIONAL ECOLOGY. Study of the worker, his environment and the interaction of worker with environment. The occupational ecologist is interested in matching man and environment for optimal ergonomic efficiency (q.v.) and minimal disturbance to the environment.

OCCUPATIONAL MEDICINE. The medical specialty concerned with the epidemiology, prevention and treatment of disease commonly observed in industrial environments. These include the diseases and/or accidents which are primarily associated with particular occupations.

OCULOGRAM. Recording of overall eye movement patterns, obtained by electromyographic observation of the rectus lateralis muscles of both eyes. Instantaneous readout of the position of sight focus within the visual field can be shown if instruments are properly calibrated.

OLECRANON FOSSA. A depression in the back of the lower end of the humerus in which the ulna (q.v.) bone rest when the arm is straight.

OLECRANON PROCESS. The elbow bone. A part of the ulna which makes up the bony point of the elbow, best felt when the arm is bent. An important anatomical reference point (q.v.).

ORIGIN (MUSCLE). The anatomic point of attachment of the non-moving end of a muscle to a bone. When the muscle contracts it brings an associated body segment toward the origin. The origin is at the proximal (q.v.) end of the muscle.

ORTHOAXIS. The true anatomical axis about which a limb rotates as opposed to the assumed axis. The assumed axis is usually the most obvious or geometric one, while the orthoaxis is less evident and can only be referenced by the use of anatomical landmarks.

ORTHOCENTRE. The instantaneous anatomical center about which a limb rotates. It varies with the angle of rotation and can fall outside the physical limits of the joint involved.

ORTHOSIS. A device applied externally or internally, to control or enhance human limb movement, or prevent bone deformity. May take the form of splints, or self-help devices for the disabled.

OSCILLOGRAPH. A electrical recording device which yields a permanent trace of an electrical signal. The recording may be produced by pen and ink on standard paper, pressure stylus on waxed paper, heated stylus on heat-sensitive paper, or a light beam on light-sensitive paper. A good oscillographic recorder will not distort the signal which is being recorded. Used in electromyography (q.v.), cardiography (q.v.), etc.

OVERLOAD PRINCIPLE. The principle which states that when a system is saturated by the requirements of a performance task, it fails to function entirely. Two pertinent examples of this principle are found in the functioning of muscle groups under excess load and the ability to process information when the information load becomes too large.

OVERT LIFTING TASK. Lifting and manipulation of substantial loads as opposed to the covert lifting task (q.v.). The stresses which are generated are resisted by the vertebral column and the erector spinae muscles of the trunk and back.

OXYGEN DEBT. Quantitative expression of the amount of oxygen necessary for physiological combustion and elimination of the by-products of anaerobic metabolism (q.v.).

PALMAR ARCH. Blood vessel in the palm of the hand from which the arteries supplying blood to the fingers are branched. Pressure against the palmar arch by poorly designed tool handles may cause ischemia (q.v.) of the fingers and loss of tactile sensation (q.v.) and precision of movement.

PARTICULOTAXIS. Contact with particulate matter. One of the ecological stress vectors. Particulotaxis at the workplace has caused numerous industrial diseases such as silicosis, which is commonly called potter's consumption or stonemason's disease, berylliosis, which results from exposure to beryllia dust, and black lung disease found among coal miners. Ventilation, electrostatic percipitators, proper work clothes, makes, etc., can effectively abate the dangers of particulotaxis.

PATHOCUMULUS. Trauma resulting from repetitive application of work stress.

PATHOLOGICAL PROCESS. Any process which causes temporary or permanent changes in physiological function or anatomical structure resulting in a state of disease. Pathological processes may be triggered or exacerbated by poorly designed man-task interfaces.

PERIPHERAL HEMODYNAMICS. The study of blood flow phenomena in the outer regions of the body. Because work tasks involve the extremities, knowledge of peripheral hemodynamics is necessary to understand the interaction of internal and external environments (q.v.)

PHALANX. Colloquially known as the knuckle, any of the long bones of the fingers or toes. Frequently used as anatomical reference points (q.v.) in work analysis.

PHARMACOKINESIS. Motions caused by the administration of drugs.

PHRENIC NERVE. Motor nerve to the diaphragm, originating from the vertebral column in the neck and transversing the thoracic cavity. Important in diaphragmatic respiration. Can be irritated in the neck by a poorly designed seat harness. Repeated pressure surges in the abdominal cavity caused by movement of the back required by improper seat height are transmitted to the diaphragm and may cause pain to be projected to the shoulder.

PHYSIOLOGICAL OPTIMAL ALIGNMENT. Alignment of the principal axis of rotation of a task with the orthoaxis of a limb. For most tasks, the physiological optimal alignment is not a fixed position due to variations between the individuals.

PHYSIOLOGICAL RESPONSE. Any of the body's reactions (hormonal, electrochemical, muscular, or nervous, etc.) to any internal or external stimulus. Physiological response to external stress is synonymous with work strain (q.v.).

PHYSIOLOGY. The study of the biological, biochemical and biophysical functions of living organisms. It is one of the basic disciplines in the practice of biomechanics.

PIEZOELECTRIC EFFECT. A property exhibited by certain crystalline substances in which the application of mechanical stress produces a redistribution of electric charge and the application of a voltage produces a mechanical deformation. This property is exhibited by bone and other tissue.

PINCH GRASP. Also known as key grasp. One of the lesser used grips of the hand. Involves the thumb and the near side of the index finger at the second phalanx (q.v.). Not a natural grip, it must be learned and requires high levels of manual dexterity and tactile discrimination. Used to apply large forces to small objects because of the strength of the thumb.

PIVOT JOINT. A joint in which motion is limited to rotation about an axis perpendicular to the contact surface. The atlas and axis, the two uppermost vertebrae, form such a joint which accommodates rotation of the head. Attempts at movement other than pure rotation at such joints may be dangerous or cause discomfort.

PLANTAR FLEXION. The motion about the ankle joint which raises the heel from the ground and points the toe, e.g., standing on toes, operating a gas pedal.

PLAY FOR POSITION. One of the fundamental motions or the therbligs at the workplace. The pre-positioning of an object for subsequent operation. Mechanical guides aid in playing for position.

PNEUMOTACHOGRAPH. An instrument used to record the rate of breathing and pulmonary ventilation (q.v.). Designed by Wolff, it is commonly used in work physiology, especially in agriculture, forestry, and foundry work.

POLYGRAPH. Recording instrument used to monitor and detect changes in physiological systems (e.g., sweat rate, heart rate, respiratory rate, etc.) simultaneously. It is generally used for work physiological measurements but its popularity is derived from its association with lie-detecting tests.

POSITIVE WORK. Work performed by a person in applying a force through a distance. Muscular effort consists of generating the energy to perform work in the true engineering sense. Work stress (q.v.) and strain (q.v.) (muscular shortening) are necessary corollaries of positive work. It is one of the elements of a lifting task (q.v.). (See NEGATIVE WORK, ISOMETRIC WORK.)

POSTPRANDIAL. Literally after meal. Its importance in the industrial workplace is in understanding the effect of digestion upon work performance. Basically, the consumption of food in substantial quantities causes a shift in the flow of blood away from the skeletal muscles and peripheral organs and toward the viscera, severely limiting metabolic activity of the skeletal muscles. Also, after heavy meals a lower level of alertness and a greater accident proneness may exist. Indication of pre or post prandial state of subject should be included in work-task study.

POWER GRIP. One of five basic grasps of the hand. It consists of the fingers wrapping around the gripped object with the thumb placed against it. This allows use of the strong opposing muscles of the thumb and the combined strength of the finger flexor muscles. Used in hammering operations, with special pliers, and handbrake applications. The tool designer should be aware of the strength available from this grip.

PREDETERMINED MOTION TIME SYSTEM. Any scheme useful in the prediction of performance times of industrial work tasks. It analyzes all motions into elemental components whose unit times have been computed according to such factors as length, degree of muscle control required, precision, strength, etc. Time standards of several of these are used as bench mark (q.v.) levels for normal performance in biomechanics and work physiology. For these purposes they are divided into systems: derived from taxonomy and kinesiology; non-taxonomic and kinesiological; taxonomic but non-kinesiological.

PREHENSILE. Adapted for taking hold. Usually refers to grasping motions of the hand. The prehensile and manipulative (q.v.) ability varies considerably with levels of training and cultural and social background. Performance of workers on a given task may differ for these reasons. (See GRASP REFLEX.)

PREHENSION. Refers to the ability to grasp or take hold. (See PREHENSILE.)

PRESSURE SENSOR. One of the sensory end organs (q.v.) responsive to a pressure stimulus. Important for tactile control (q.v.). Feedback from pressure sensors between fingers enables the worker to moderate his grasp. Interference with pressure sensor function, as with heavy gloves or shoes, will impair touch perception and may cause faulty grasp or insufficient foot pedal control.

PRIME MOVERS. The muscles which produce or maintain a specific motion or posture. Also called agonist, they are assisted by synergists (q.v.) and opposed by antagonists (q.v.). Prime mover activities may be preceded by action of a trigger muscle initiating the movement. Knowledge of the prime movers is essential in workplace design and in electromyographic evaluation of task severity.

PRONATION. Rotation of the forearm in a direction to face the palm downward when the forearm is horizontal or backward when the body is in anatomical position (q.v.). An important element of industrial demanded motions inventory (q.v.), it is performed by muscles whose efficiency is a function of arm position. (See RANGE OF FOREARM PRONATION AND SUPINATION.)

PROPRIOCEPTION. The sensing of one's location relative to the external environment. An important sense for maintenance of balance and for orienting one's self for performing work tasks. An impaired proprioceptive sense can cause industrial accidents or faulty performance where controls are located outside the visual field.

PROSTHESIS. The replacement of a body part. The replacement may be for a limb, a blood vessel, an organ, or a skeletal member. It may be partially or completely non-functional when external, although it is usually cosmetic. (See ORTHOSIS.)

PROXIMAL. Describing that part of a limb, body segment, or muscle which is closest to the point of attachment. The elbow is proximal to the wrist which is proximal to the fingers. Generally refers to distance from trunk.

PSYCHOSOMATIC RESPONSE. A physical or physiological response resulting from psychological stimulus. Often a defensive reaction. Important to understand in the practice of human engineering (q.v.) and in modifying sociotaxis (q.v.) in ergonomic analysis (q.v.).

PULMONARY VENTILATION. Commonly referred to as breathing. Volume of respiratory gases passing in and out of the lungs per unit time. Used in spirometry (q.v.) as in index of work stress (q.v.).

PULSE RATE RATIO. The ratio of pulse rate after or during exercise to pulse rate at rest. Typically used as a measure of heart function, the pulse ratio is also used as a measure of work stress.

RADIAL DEVIATION. Movement of the hand which deflects its longitudinal axis toward the radius (q.v.). It causes the head of radius (q.v.) to press against the capitulum of humerus (q.v.) and may lead to irritation known as "tennis elbow." Tool design should minimize radial deviation. Strength of grasp is diminished in radial deviation.

RADIAL NERVE. One of the main nerves of the arm providing motor stimuli to extensor muscles of the forearm, wrist and fingers. Receives sensory feedback from back of forearm and hand.

RADIOCARPAL JOINT. Commonly known as the wrist, this joint provides all movements between the radius-ulna and hand. Because of its flexibility, the radiocarpal joint is often abused in adapting to the use of improperly designed tools.

RADIOHUMERAL JOINT. Part of the elbow which is thrust bearing. The site of tennis elbow (q.v.). Composed of the head of radius (q.v.) and capitulum of humerus (q.v.). (See RADIAL DEVIATION.)

RADIO-ULNAR JOINTS. The mechanism whereby the radius and ulna rotate about one another during pronation (q.v.) and supination (q.v.). At the elbow, the head of radius pivots in a notch in the ulna and is contained by an encircling annular ligament, thus forming a journal bearing. The bones are maintained in relative position by a ligamentous sheet connecting their central sections. During pronation and supination the distal end of the radius glides around the head of the ulna and rotates on its own axis. Workplace design should assure proper wrist and hand alignment during supination and pronation to avoid trauma. (See RADIAL DEVIATION, TENNIS ELBOW.)

RADIUS. The long bone of the forearm in line with the thumb. It is the active element in the forearm during pronation (q.v.) and supination (q.v.). It also provides the forearm connection of the wrist joint. (See RADIOCARPAL JOINT, RADIOHUMERAL JOINT, ULNA.)

RANGE OF FOREARM PRONATION AND SUPINATION. The total useful rotational capability of the wrist about the longitudinal axis of the forearm. Very important in industrial tasks involving screwing actions, twisting motions, etc. Range (approximately 180 degrees) is greatest if no other deflection (ulnar or radial deviation) occurs at the same time.

RATIONAL WORKPLACE DESIGN. Application, a priori, of principles of anatomy, physiology, and systems engineering, in optimizing task and performance efficiency. (See ANALYTIC WORKPLACE DESIGN, IMPROVEMENT APPROACH TO WORKPLACE DESIGN, AND EMPIRICAL WORKPLACE DESIGN.)

RAYNAUD'S DISEASE. Constriction of the blood vessels of the hand from cold temperature, emotion, or unknown cause. Afflicts women predominantly and affects both hands simultaneously. Primary Raynaud's disease is believed to exist in 15% of the general population, and is hereditary and nonoccupational in nature. Occupational, or secondary Raynaud's Syndrome, is also referred to as vibration white finger, and is associated with excessive vibration with use of heavy, vibrating, and reciprocating hand tools, and working in cold temperatures. Hands become cold, blue and numb and lose fine prehensile (q.v.) ability. On recovery hands become red accompanied by burning sensation. Easily confused with one-sided numbness and tingling caused by poor tool design and resulting pressure. (See SCALENUS ANTICUS SYNDROME.)

REACTANCE PLATFORM (FORCE PLATFORM). A balance system for measuring forces or accelerations of a supported body. In biomechanics the force platform is used to measure the forces associated with movement and the accelerations associated with these forces. Used as inputs along with kinematic data in biomechanical models to determine joint loading during lifting or gait analyses. Shifts in the center of gravity of the total body can also be determined.

REACTIONS INVENTORY. The available physiological, kinesiological or psychological responses of an individual to a stimulus or set of stimuli. Reactions will vary with time and depend on the condition (preparedness, fatigue, etc.) of the individual.

RECTUS ABDOMINUS. Long strap-like muscle originating in the pubic region and running vertically to insert in the lower ribs and sternum. By contracting, it flexes the vertebral column and compresses the abdominal wall. It functions as the antagonist (q.v.) to the sacrospinalis (Erector Spinae (q.v.)). The rectus abdominis, by contracting can break the force of a blow to protect the abdominal viscera. However, workplace design should avoid repeated impact of the rectus abdominis.

REFERRED PAIN. Pain sensed at a place distant from its true origin (e.g., pain originating in stomach may be felt high in the lumbar region; e.g., pressure on the front of the abdomen may produce pain in the shoulder). Unawareness of referred pain patterns is frequent cause of incorrect identification of loci of work stresses.

REFLEX. (See SIMPLE REFLEX.)

REFLEX GONIOMETRY. Quantification of reflexive response by measurement of angular displacement, velocity and/or acceleration of a limb about a joint. Used in the evaluation of drug effects as related to safety at the workplace.

REHABILITATION. Treatment to restore a loss of capacity by retraining, medication, corrective exercise, or substitution of new faculties to replace those lost or damaged. Includes physical, emotional and technological rehabilitation.

RESPIRATION. Commonly called breathing. The composite process of internal respiration wherein an exchange of gases between circulatory fluids and cells occurs and external respiration which covers the gas exchange between blood and the air in the lungs. Respiration must occur to sustain life and must be adequate for efficient workplace function, i.e., the task should not interfere with normal breathing. Inadequate internal respiration is one of the causes of muscular fatigue. (See ANOXIA, ISCHEMIA, ANAEROBIC METABOLISM.)

RESPIRATORY QUOTIENT. The ratio of carbon dioxide volume exhaled to volume of oxygen up-take. Respiratory quotient varies considerably with type of foodstuff being metabolized and is increased by hyperventilation and acidosis. In steady work, the respiratory quotient is an index of work strain.

Resting Metabolism. Physiological combustion process usually expressed in units of oxygen consumed per unit time in a person at rest while seated or standing in normal position at the workplace. Increase in metabolic rate (q.v.) during specific activities above resting metabolism is the metabolic cost (q.v.) of the activity.

Rotation. Motion in which all points describe circular arcs above an immovable line or axis.

Rotator Cuff. A group of muscles surrounding the shoulder joint, Because of the many components of available movement at the shoulder, it is susceptible to self-generated trauma caused by actions of the muscles of the rotator cuff. In the industrial setting, rapid reversals of movement in the shoulder should be minimized to avoid trauma or fatigue which may result.

Sacrospinalis Muscle. The long muscle of the back with origin (q.v.) in the sacrum and insertion (q.v.) in the thoracic and lumbar vertebrae. Its primary function is to keep the spine erect. A key muscle to be monitored during electromyographic studies of whole body lifting tasks since its response is characteristic of the total effort being expanded. Operates at a poor mechanical advantage and should not be over-stressed in task design or damage to vertebral column or spinal nerves (q.v.) may result. (See ERECTOR SPINAE MUSCLES.)

Sacrum. A triangular, slightly curved bone at the base of the spine. The joint between sacrum and pelvis and between sacrum and lumbar spine are frequent sites of lower back trauma.

Sagittal Plane. A plane from back to front vertically dividing the body into right and left portions. Important in anthropometric definitions. Mid-sagittal plane is a sagittal plane symmetrically dividing the body.

Saphenous Veins. Two major superficial veins in the leg. Long saphenous vein passes approximately from the front of the medial ankle to the grain. Short saphenous vein located at back of leg passes from ankle to the popliteal region (q.v.). Both veins are frequently sites of varicosity. Good chair design eliminates pressure against short saphenous vein, thus avoiding interference with venous return and minimizing risk of irritation and possible general inflammation of vein.

Scalenus Anticus Syndrome. Numbness, pain and tingling in fingers and hand, often in the area of ulnar nerve (q.v.) distribution. Differs from Raynaud's disease (q.v.) in that if may appear on one side only. Caused by a squeezing of large nerves and blood vessels supplying the arm between the scalenus muscles in the neck or sometimes between scalenus muscle and ancillary rib. Occupationally significant because it may manifest itself in some persons only when the arm is abducted or when weights are lifted.

Scaphoid. One of the bones of the hand. Located at the base of the palm on the thumb side. It bears against the radius (q.v.) and is directly on the axis of thrust from the hand to the arm. Its proximal surface is an important moving element of the wrist joint involved in flexion, extension, adduction and abduction. Because of its location, the scaphoid is vulnerable to disability through fracture from impact and thrust loads. Also the name of a bone in the tarsal region of the foot. (See SCAPHOID TUBERCLE.)

Scaphoid Tubercle. Protuberance of scaphoid bone (q.v.). Important as one of the anchor points of the flexor retinaculum (q.v.). An important anatomical reference point (q.v.) for locating the wrist joint.

Scapula. The shoulder blade. Large, triangular flat bone; it forms the socket for the proximal joint of the humerus and constitutes at the same time the place of the large muscle of the shoulder and back. This bone positions and stabilizes the shoulder joint. Backrests of chairs should clear the lowest point of the scapula. A bony ridge on the scapula serves as an important anatomical reference point (q.v.)

Semilunar Notch. A wide but shallow crescent-shaped indentation found on the upper end of the ulna. It articulates with the trochlea (q.v.) of the humerus and forms the bearing of the hinge joint during extension and flexion of the forearm. The orthoaxis of flexion of the forearm passes through the center of curvature of the semilunar notch.

Semispinalis Capitis. Muscle of the neck, antagonist to sternomastoid muscle (q.v.). Important element in head scanning (q.v.) and postural control.

Sesarnoid Bone. Small rounded bone enclosed and/or enveloped by a tendon between tendon and joint structures (e.g., kneecap) and may protect joint (e.g., sesamoid bone at base of big toe). In optimal posture the center of mass of the body is located above the sesamoid bone of the big toe.

SHEAR. A loading mode in which a load is applied parallel to the surface of the structure, causing internal angular deformation or slip.

SIGNATURE. Record of a characteristic pattern by a subject (e.g., the electrocardiogram is a signature). Motions pathways (q.v.) or their time derivatives which are repeatable are signatures indicative of an individual's motor skills, state of health or fatigue. (See VELOCITY SIGNATURE, ACCELERATION SIGNATURE, CHRONOCYCLEGRAPH.)

SIMPLE REFLEX. Automatic motor reaction to an external stimulus where the pathway from sensory end organ to muscle short-circuits the brain and passes through the spinal cord only. The knee jerk and the Achilles tendon reflex are examples of simple reflexes which maintain integrity of posture. (See CONDITIONED REFLEX.)

SKELETAL CONFIGURATION. The general arrangement of body parts, such as muscles and bones, which may be necessary in the performance of tasks. Skeletal configuration largely determines the biomechanical efficiency of performance.

SOCIOTAXIS. Contact with fellow man either individually or in groups under circumstances which triggers behavioral reactions. One of the ecological stress vectors. Improper design of sociotactic interfaces may cause emotional work stress, while good sociotactic interfacing will promote emotional well-being and high motivation at work.

SOMATOTYPING. System for classifying body builds by grouping them into distinctive types. These classifications are loosely correlated with personality traits, and may be helpful in identifying proneness to disease and accident.

SONOGRAPHY. The use of ultrasonic energy for imaging internal organs and tissues. Same as echography (See ULTRASONICS.)

SPAN OF REACH. Boundaries of normal reach with the tip of the index finger around the shoulder joint. Often assumed to be a sphere; however limitation of freedom of skeletal configurations make some points within this sphere out of reach.

SPASM. Involuntary contraction of isolated bundles of muscle or muscle groups. Generally caused by local chemical imbalance resulting from fatigue, local ischemia or trauma or other reasons. Cannot be consciously controlled and is a potential cause of accidents in the workplace. Must not be confused with spasticity (q.v.). Because the conditions causing spasm can be induced by enforced postural rigidity or by poorly designed tools, the industrial engineer must be aware of the etiology (q.v.) of spasm.

SPASTICITY. Heightened reactivity of affected muscle groups. Results in increased resistance to passive stretch and increased reaction of tendon reflexes. Caused by interference with cerebral or cortical control of musculature. Do not confuse with spasm (q.v.). Important in evaluation of disabled workers in rehabilitative effort. (See SIMPLE REFLEX.)

SPECIFIC ACOUSTIC IMPEDANCE. A material property that is equal to the product of the density and the speed of sound in that material. A sound wave will be reflected from the interface of two materials when each has a different specific acoustic impedance.

SPIKE. Recording of an isolated electrophysiological event of less than 25 millisecond duration. It appears as a sharp V or inverted V.

SPINAL NERVE. One of the 31 pairs of nerves originating from the spinal cord. It leaves the vertebral column (q.v.) by two roots and contains both motor and sensory components (q.v.). At the point of exit between two vertebrae the spinal nerve is susceptible to compression trauma which may cause pain, paralysis or numbness at points distal (q.v.) to the site of the lesion.

SPINE. Also called the backbone. The vertebral column (q.v.) which is the central skeletal structure of the body. Also a sharp projection of any bone.

SPINOUS PROCESS OF THE VERTEBRA. A bony rearward extension of the vertebra. It provides extended surfaces for muscle attachment, and for stabilizing adjoining vertebral segments. The spinous process is that part of the vertebra which can be felt through the skin at the back. Often used as anatomical reference points (q.v.). The spinous process of the seventh cervical vertebra is an example of an important anatomical reference point.

SPIROMETRY. Measurement of gas volumes moved during breathing. Used as an index of work stress (q.v.).

Also used in indirect inference of metabolic rates (q.v.). In industry, a portable gasometer or a pneumotachograph (q.v.) is commonly used.

STATIC WORK. Syn: isometric work (q.v.).

STERNOCLEIDOMASTOID MUSCLES. A pair of muscles connecting the breastbone and collar bones to the mastoid processes of the lower skull behind the ears, which provide support for the head. When operating together, the right and left sternocleidomastoid pull the head forward and the back of the skull downward, and when operating singly each turns the head to the opposite side. They oppose the semispinalis (q.v.) muscles and stabilize the head. In the workplace the worker's head position should be near vertical to minimize activity of the semi-spinal and sternocleidomastoid muscles. The sternocleidomastoid is also functionally important in head scanning (q.v.).

STRAIN. (1) in mechanics, a measure of deformation, either elongation, contraction or angular (shear) deformation. (2) in medicine, any injury involving overextension, compression or twisting of a muscle, ligament or joint. (See WORK STRAIN.)

STRAIN PROPAGATION. Internal transmission of the response reaction in the body to external stress. Propagation may be by means of mechanical or biological processes such as hormonal, neural (nerves), or circulatory mechanisms. Because of strain propagation resultant trauma is often evidenced at locations remote from the point of stress application, e.g., scalenus anticus syndrome (q.v.).

STRAIN SYNTHESIS. Artificial simulation of work strain by combination of work stresses. (1) In mechanics, a force per unit are either external or internal to the body. Normal stresses are tensile or compressive. Tangential stresses are called shear stresses. (2) In medicine, any force or other stimulus applied to or acting on an individual such as excessive lifting force, excessive noise or emotional upset. (See WORK STRESS.)

STRESS EQUIVALENT. In biomechanics, the quantitative relationship between physiological outputs or stresses of different dimensions generated by work stress or work strain. Thus, both isometric work with the dimensions of linear impulse, and dynamic work with conventional dimensions of work, can be compared through the resultant work strain expressed in terms of metabolic activity which under such circumstances serves as the stress equivalent. In lifting, light bulky objects exert the same moment as heavy dense objects and the recorded height of the integrated electromyogram (q.v.) is the stress equivalent of each task expressed in foot-pounds of torque on the spine per millivolt of mean myogram height.

STRESS TRANSMITTAL. Mode of transfer of external force from the man-equipment interface (q.v.) into distant points of the body. If the level of stress transmittal is low, then excessive work stress is manifested by superficial lesion (q.v.) as blisters or calluses. If the level of transmittal is high, then cumulative pathogenesis (q.v.) may produce pathological lesions in anatomical structure, remote from the point of work stress application, e.g., tennis elbow caused by poor machine control design, e.g., broken collarbone caused by breaking a fall with outstretched hand.

STIFFNESS. A measure of resistance offered to external loads by a specimen or structure as it deforms.

SUPINATION. Rotation of the forearm about its own longitudinal axis bringing the thumb side of the hand from a position next to the body to one away from it when the arm is in anatomical position (q.v.). Supination tends to turn the palm upward when the elbow is flexed 90 degrees and the forearm is horizontal and forward. Supination is an important element of available motions inventory (q.v.) for industrial application, particularly where tools such as screwdrivers are used. Efficiency in supination depends on arm position. Workplace design should provide for elbow flexion at 90 degrees. (See PRONATION, RANGE OF FOREARM PRONATION AND SUPINATION.)

SYNAPSE. Junction of a nerve cell with another. In the simplest case of a reflex arc the synapse joins the sensory (receptor) fiber with the motor fiber to stimulate muscle action. (See SIMPLE REFLEX, SENSORY NERVE, MOTOR NERVE.)

SYNERGIST. A muscle which acts to assist a prime mover (q.v.) in performing a specific action. A muscle may act as synergist in one action and as antagonist (q.v.) in another action to the same prime mover. Work tasks should be designed to use synergists effectively.

SYNDROME. A complex of symptoms which identifies a physiological or pathological condition. (See SCALENUS ANTICUS SYNDROME, CARPAL TUNNEL SYNDROME.)

SYNOVIAL FLUID. Fluid which provides lubrication to the joints. Secreted by membranes of the joint and contained by surrounding tissues, it is very viscous and an effective lubricant. An excess or lack of fluids resulting from work stress trauma or disease will cause swelling and pain in the joint and may limit the range of motion. (Also known as synovia.)

Synovial Structures. All of those elements concerned with generation and/or use of synovial fluid (q.v.). Generally located at joints or in tendon sheaths wherein the tendon is lubricated by synovial fluid. When synovial structure is damaged, a state of disease may develop, e.g., tenosynovitis (q.v.), tendinitis, synovitis, or arthritis.

Tangential. Relating to a straight line that is the limiting through a fixed point.

Task Load. A measure of task severity. Expressed as a stress equivalent (q.v.) which is biomechanically a more accurate measure of task load than measure of metabolic activity, heart rate or pulmonary ventilation.

Tendinitis. Inflammation of tendon (including tendon sheath) (see TENOSYNOVITIS)

Tendon. Fibrous end section of muscle. It attaches to bone at the area of application of tensile force. When its cross section is small, stresses in the tendon are high, particularly because the total force of many muscle fibers is applied at the single terminal tendon. The site of many industrial diseases caused by trauma, biomechanically improper movements, or failure of lubrication, the tendon must be protected in tool and workplace design. (See TENOSYNOVITIS.)

Tennis Elbow. Sometimes called lateral epicondylitis. An inflammatory and/or degenerative reaction of tissues in the lateral elbow region. In an industrial environment it may follow a single, violent effort or repetitive subsymptomatic strain in supination (q.v.) against resistance (as in screwdriving) or violent extension of the wrist with hand pronated. Can frequently be avoided by assuring that the axis of rotation of a tool or machine control coincides with the orthoaxis (q.v.) of rotation of the forearm.

Tenosynovitis. A disease of the tendon sheath, commonly occurring in the fingers, hand, and wrist. The tendon sheath accumulates excessive amounts of synovial fluid with extremely repetitious actions, for example 1500 to 2000 repetitions per hour, which leads to swelling and inflammation of the tendon sheaths. Often associated with continual ulnar deviation (q.v.) of the wrist during rotational movements (e.g., screw driving) or by other overwork and trauma. In industry, extensor sheath inflammation is more frequent. Work tolerance is reduced because of pain during wrist and finger movement.

Thermal Environment. The surroundings in which a worker is situated which affect his ability to reject or absorb heat. It includes the local temperature, humidity and air velocity. Presence of radiating surfaces also affect the thermal environment. It is significant in the workplace for its effect on work rate, quality control, accidents and physiological well-being (See EFFECTIVE TEMPERATURE.)

Thermal Stress. Deviation of an individual's core temperature from it's normal working range caused by extremes in the thermal environment. Can cause changes in body temperature, body chemistry and function of heart, lungs, kidneys and other organs. Influences ability to evaluate sensory feedback as well as many work skills.

Thermodynamic Efficiency. In biomechanics, the ratio of the work performed to the energy consumed by the human body. This is commonly estimated at 20% unless man goes into oxygen debt (q.v.). It is computed by dividing the external work performed (e.g., the number of bricks lifted multiplied by weight and height) by the energy consumption of the human body during the task. Important index for effort rating. (See METABOLISM.)

Thermograph. A device for producing graphic representation of temperature distribution on the body surface. Commonly an infrared sensor connected to a suitable display device, e.g., video projector, with either graded black-white shades or discrete colors to indicate different temperature levels. Thermographs can accurately indicate local hot spots when such spots are not otherwise readily visible. An important tool in locating excessive work strain and in the maintenance of industrial health.

Theoretical Biomechanics. The analytical study of the human body in relation to its immediate mechanical environment. The principles of classical mechanics and the life sciences are employed in the formation of models which predict the effect of mechanical stress on the body. These models can be used as guides in the design of man-equipment interactions such as the workplace and for predictive hypotheses of performance efficiency. In situations where experimentation with man is not feasible, theoretical biomechanics is used to evaluate and predict potentially destructive task situations, e.g., strong vibrations, highway collisions.

TIBIA. The larger of the two bones of the leg below the knee; also referred to as the shinbone. The tibia is the major weight-bearing bone of the leg and transmits forces between the ankle and the knee joint.

TISSUE RHEOLOGY. Study of the deformation and flow of live or excised tissues under external stress. Usually evaluated as a function of time. Can be used to evaluate industrial hardware to avoid stress concentration and instantaneous or cumulative trauma.

TONUS. Commonly known as muscle tone. The tension inherent in muscle which is nominally at rest. A minimal tonus condition is characteristic of healthy muscle. Flabbiness and poor postural control result from poor tonus. Exercise and regular activity aid in maintaining healthy tonus which contributes to worker efficiency.

TORQUE. Refers to the moment of force (q.v.) exerted around any joint during the performance of a task. A moment, or torque, is the product of the perpendicular component of force exerted and the distance from the point of application to the fulcrum or "pivot point" (the joint). Isometric muscle torque which does not result in motion causes high joint forces with physiologic effects dependent on the duration and magnitude of the torque and the joint configuration. E.g., in lifting an object and holding it elevated, torque is exerted by the paraspinal muscles with resultant heavy loads on the lumbar vertebrae. Formerly measured in foot-pounds, inch-ounces, but now in Newton-meters (Nm). If a force, F, acts to produce rotation about a center at a distance, d, from the line in which the force acts, the force has a torque, $T = F*d$. (See MOMENT OF FORCE (TORQUE))

TRANSVERSE LIGAMENT. There are several transverse ligaments (q.v.) in the body. Usually identified by the local anatomical structure with which they are associated. E.g. flexor retinaculum.

TRANSVERSE PLANE. An anatomical reference plan parallel with the ground (i.e., horizontal) when considering a man in anatomical position (q.v.). Also that plane at right angles to the long axis of an isolated organ. (See SAGITTAL PLANE, CORONAL PLANE.)

TRAPEZIOUS MUSCLE. Broad flat muscle of the upper back. Maintains posture and is an important moving and activating muscle of the shoulder blade. Tasks requiring the worker to hold elbows high will cause fatigue in the trapezius muscle and loss of postural stability. Often the cause of neck pain in these work situations. Important muscle in elevating the arms.

TRAUMA. An injury or wound, generally caused by a physical agent. Cuts, bruises and abrasions are examples of trauma, but trauma may be present even though it is not visible, e.g., strained muscle. The causes of trauma must be anticipated in workplace design or tool design. Protective devices and special clothing (work shoes, gloves) are used to avoid trauma. (See CUMULATIVE TRAUMA).

TRAUMATIC ARTHRITIS. Inflammation, pain and swelling of a joint caused by the application of excessive forces at the joint. Can be triggered by a direct blow or sudden strain and leads to restricted motion. Task design should avoid repeated mild trauma or prolonged strain which can cause traumatic arthritis or exacerbate pre-existing arthritic conditions.

TRAUMATIC FIBROSITIS. Inflammation in muscle and increase in fibrous tissues in muscle caused by application of excessive force or by disease. Can result from direct blow or repeated mild trauma at the workplace. Weakness and disability are consequences.

TREMOR. A rapid involuntary alternation of motion of a part of the body. May occur at rest, or only on initiation of movement (intention tremor). Directly caused by alternate rapid contractions of muscle groups and their antagonists (q.v.). Originates from fatigue, disease of nerves or muscles, injury, or response to environment, e.g., shivering from the cold. Presence of tremor disables the worker in precision tasks and increases the risk of accidents.

TRIGGER FINGER. Also known as snapping finger. A condition of partial obstruction in flexion or extension (q.v.) of a finger. Once past the point of obstruction, movement is eased. May occur without apparent cause. A symptom of tenosynovitis (q.v.). (stenosing tenosynovitis creptians). Caused by thickening of a tendon or a localized reduction of the tendon sheath. In the workplace, flexing against strong antagonists and flexing of the distal phalanx (q.v.) without middle phalanx movement is suspected. In addition, tool handles with sharp edges that create local pressure points and localized stress on a particular tendon have been associated with trigger finger. Tool handles should be designed for trigger operation by multiple fingers thereby relieving the stress on a single digit.

TROCHLEA. Bobbin-shaped outcrop of bone at distal end of humerus (q.v.), forming axis shaft of elbow hinge joint, i.e., humero-ulnar joint.

TRIPODAL GRASP. One of the five basic grasps of the hand.

Also called manipulative grasp. Object is retained by thumb, index finger an middle finger, which forms a tripod with object possibly in contact with palm. Delicate rotational control is available to the worker for fine screwdriver manipulation, etc.

TUBEROSITY. Large rough outgrowth on a bone which serves as a point of attachment for tendons. It may be used as a convenient anatomical landmark (q.v.), particularly with respect to design of seats and equipment controls.

TUBULAR BONE. Construction of the long bones of the body, e.g., humerus. Provides maximum strength/weight properties while providing centrally located and protected site for blood vessels, nutrients, and bone marrow. Workplace design should avoid imposition of loads with torsional and transverse components.

ULNA. One of the two bones of the forearm. It forms the hinge joint at the elbow and does not rotate about its longitudinal axis. It terminates at the wrist on the same side as the little finger. Task design should not impose thrust loads through the ulna.

ULNAR DEVIATION. A position of the hand in which the wrist is bent toward the little finger. Ulnar deviation is a poor working position for the hand and causes nerve and tendon damage. It reduces the useful range of pronation and supination (q.v.) by approximately 50 percent, and work performed in ulnar deviation proceeds at low efficiency. Hand tool design should avoid ulnar deviation.

ULNAR NERVE. One of three large nerves of the hand. It supplies muscles of fine manipulation (q.v.) located in the hand and provides sensory feedback (q.v.) via the skin of the little finger and part of the ring finger. Because the ulnar nerve enters the hand immediately below the skin, it is susceptible to damage from poorly designed tool handles which apply pressure to the palm on the side of the little finger.

ULTRASONICS. The use of high frequency vibrations for cleaning, heating tissues, dispersing solid mass, and imaging objects by reflection (1) Ultrasonic Cleaning The use of high frequency vibrations in a solvent bath to loosen and disperse surface contaminants. Typical frequencies for this application are 20-50 kHz. Also used in dentistry to remove deposits from tooth structure. (2) Ultrasonic Therapy The use of high frequency mechanical vibrations to transmit energy into and through body tissue to heat or disperse material. Typical frequency for this application is about one MHz. Ultrasonic phacoemulsification for removal of eye cataracts. Ultrasonic disintegration for pulverization of bladder calculi. (3) Ultrasonic Imaging The use of high frequency vibrations to obtain a visual image of internal organs and soft tissue, most often using pulse echo techniques. Typical frequencies for this application are 1-15 MHz. Examples are echocardiology and echoencephalography.

ULTRASONIC TRANSDUCER. A device to convert electrical energy to mechanical energy, using piezo-electric materials such as quartz, barium titanate or lead metaniobate.

UNIVOCAL. Having a unique association. A relationship wherein one single specific effect can only be produced by one single specific cause, e.g., an ulnar claw can be produced only by damage to the ulnar nerve (q.v.).

VASOCONSTRICTION. Decrease in the cross-sectional area of blood vessels. This may result from contraction of a muscle layer within the walls of the vessels or may be the result of mechanical pressure. Reduction in blood flow results. Vasoconstriction can be controlled by the nervous system and is influenced by certain drugs. Improper design of the man equipment interface (q.v.) may cause vasoconstriction, i.e., poor seat design, improper tool handles.

VELOCITY SIGNATURE. An analogue recording which shows the velocity history of a body part in motion. Signature size and shape are indices of strength and motor skills of an individual. Repeatability of velocity signatures in indicative of status of health, fatigue and/or training. An important component of the biomechanical profile (q.v.).

VERTEBRA. One of the irregularly shaped bones of the spine (q.v.). The anterior portion, known as the vertebral body, is the load bearing component of the stucture. The posterior portion provides muscle attachment points, longitudinal passage of the spinal cord, exits for the spinal nerves (q.v.), and stabilizing surfaces for adjoining vertebrae which also regulate motion and are load bearing under certain circumstances. (See SPINOUS PROCESS OF THE VERTEBRA, VERTEBRAL COLUMN).

VERTEBRAL COLUMN. Bony structure commonly known as the spine. Provides the primary supporting structure of the trunk of the body, and houses and protects the nerves of the spinal cord. Extends from the head to the coccyx. Consists of 24 bones (vertebrae) separated by shock-absorbing inter-vertebral discs (q.v.). Trauma to the vertebral column (shock, twisting, compression, fracture, or stretching) can cause severe pain and/or disability. Curvature of the vertebral column is an important factor in the design of seats. (See vertebra, spine, spinous process.) Note: Because of the irregularities (especially at the lower end of the spinal column), texts may differ in their count of the number of vertebrae. E.g., Dorland's Illustrated Medical Dictionary defines, vertebrae as "any of the thirty-three bones of the spinal column, comprising seven cervical, twelve thoracic, five lumbar, five sacral, and four occygeal vertebrae." Blakiston also lists 33. In lay terms, the five regions of the spinal column are: cervical (neck), thoracic (chest), lumbar (lower back), sacral and coccygeal (both in the pelvis). The first three regions have separate and movable vertebrae throughout life, but, in the adult, the sacral and coccygeal vertebrae become fused to form the sacrum and coccyx. This fusion and, hence, resulting irregularity in the lower regions give rise to what may seem to be contradictions in texts to lay readers. Again, there may also seem to be confusion at the other (upper) end, depending on whether the uppermost bone is considered part of the skull or spinal column. Thus, it may be noted that the 35th Edition (British) of Gray's Anatomy lists 31 vertebrae (8 cervical, 12 thoracic, 5 lumbar, 5 sacral, and 1 coccygeal). Thus, orthopedists are quite definite when they note, for example "slipping of L5 over S1" (referring to the "fifth lumbar" and "first sacral") probably one of the most common of lower back ailments. Fifth lumbar vertebra (q.v.).

Work. (1) In mechanics, the product of force and distance through which the force acts. (2) In medicine, a measure of physical or mental effort expended in any task. Physical work results in metabolic activity locally and/or systemically. In biomechanics, physical work is classified as positive work (q.v.), negative work (q.v.) or isometric work (q.v.). Isotonic work, used in physiology, means under a constant tension.

Work Measurement. In biomechanics, the application of work physiological methods to the measurement of task severity and/or conjointly with the construction of a biomechanical profile (q.v.) for a task or individual in the evaluation of the compatibility of the available and demanded motions inventories (q.v.).

Work Metabolism. Physiological energy consumption ascribable to the performance of a specific task in excess of resting metabolism (q.v.). Usually expressed as an index of energy consumed per unit time, e.g., calories per minute. In industry, working metabolism is obtained by analysis of respiratory gas exchange.

Work Physiology. Application of principles of physiology to ergonomics (q.v.) and industrial engineering in design of tasks and workplace. Concerned with moderation of metabolic cost (q.v.), measurement and prevention of harmful work strain (q.v.). Uses physiological measurements of gas exchange, pulmonary ventilation (q.v.), metabolism (q.v.) and heart rate to measure the effects of workplace-related environmental stress vectors.

Work Strain. The natural physiological response reaction of the body to the application of work stress (q.v.). The locus of the reaction may often be remote from the point of application of work stress. Work strain is not necessarily traumatic but may appear as trauma when excessive, either directly or cumulatively, and must be considered by the industrial engineer in equipment and task design. Thus, increase of heart rate is non-traumatic work strain resulting from physical exertion, but tenosynovitis (q.v.) may represent pathological work strain resulting from undue work stress on a tendon and/or its sheath.

Work Stress. Biomechanically, any external force acting on the body during the performance of a task. It always produces work strain (q.v.). Application of work stress to the human body is the inevitable consequence of performance of any task and is, therefore, only synonymous with "stressful work conditions" when excessive. Work stress analysis is an integral part of task design.

Work Tolerance. Span of time during which a worker can effectively perform a task without rest period at levels of physiological and emotional well-being acceptable to the individual and without pathogenesis. Work tolerance is affected by environmental inputs (q.v.), workplace layout (q.v.), and state of health.

Workplace Layout. The designed arrangement of a work situation comprising man, materials, equipment and the physical environment of the workplace. Optimally it should be the result of deliberate task design which considers the implications of ecological stress vectors on performance efficiency, occupational safety and health.

Wraparound Grasp. One of five basic grasps of the hand. The grasped object is held against the palm by the fingers wrapped around it with the thumb opposing the index finger (holding a broomstick). The thumb is of minimal importance in this grasp, which is an innate aptitude of man. It requires almost no learning to apply. A straight wrist is needed for successful application. Flexion of the wrist "breaks" the strength of the grasp and may lead to accidents at the workplace.

BIBLIOGRAPHY

Astrand, P. and Rodahl, K. *Textbook of Work Physiology*. McGraw-Hill Book Company, 1970.

Barham, J.N. and Wooten, E. P. *Structural Kinesiology*. Macmillan Company, 1973.

Barnes, R.M. *Motion and Time Study*. Sixth Edition. John Wiley & Sons, Inc., 1968.

Basmajian, J.V. *Grant's Method of Anatomy*. Ninth Edition. Williams and Wilkins Company, 1975.

Brunnstrom, S. *Clinical Kinesiology*. Third Edition, F.A. Davis Company, 1972.

Chaffin, D. B. and Anderson, G. B., *Occupational Biomechanics*, Wiley-Interscience, 1984.

Clarke, D.H. *Exercise Physiology*. Prentice-Hall, Inc., 1975.

Dorland, W.A. *Dorland's Illustrated Medical Dictionary*. Twenty-fifth Edition. W.B. Saunders Company, 1974.

Downey, J.A. and Darling, R.C. *Physiological Basis of Rehabilitation Medicine*. W.B. Saunders Company, 1971.

Easton, D.M. *Mechanisms of Body Function*. Second Edition. Prentice-Hall Inc., 1974.

Frankel, V. H., and M. Nordin, *Basic Biomechanics of the Skeletal System*, Lea & Febiger, Philadelphia, 1981.

Fung, Y.C., Perrone, N. and Anliker, M. *Biomechanics: Its Foundations and Objectives*. Prentice-Hall Inc., 1972.

Grandjean, E. *Fitting the Task to the Man*. Second Edition. Taylor & Francis Ltd., 1963.

Guyton, A.C. *Textbook of Medical Physiology*. Fifth Edition. W.B. Saunders Company, 1976.

Hunter, D. *The Diseases of Occupation. Fifth Edition*. Little, Brown and Company, 1975.

Jacob, S.W. and Francone, C.A. *Structure and Function in Man. Third Edition*. W.B. Saunders Company, 1974.

Karpovich, P.V. and Sinning, W.E. *Physiology of Muscular Activity. Seventh Edition*. W.B. Saunders Company, 1971.

Katz, B. *Nerve, Muscle, and Synapse*. McGraw-Hill Book Company, 1966.

Konz, S., *Work Design: Industrial Ergonomics*, Grid Publishing Inc., 1983.

Maynard, H.B. *Industrial Engineering Handbook. Third Edition*. McGraw-Hill Book Company, 1971.

Mountcastle, V.B. *Medical Physiology. Thirteenth Edition*. C.V. Mosby Company, 1974.

Murrell, K.F.H. *Human Performance in Industry*. Reinhold Publishing Corp., 1965.

Niebel, B.W. *Motion and Time Study*. Sixth Edition. Richard D. Irwin, Inc., 1976.

Putz-Anerson, V. *Cumulative Trauma Disorders: A manual for musculoskeletal diseases of the upper limbs*. Taylor and Francis, London, 1988.

Rasch, P.J. and Burke, R.K. *Kinesiology and Applied Anatomy*. Fifth Edition. Lea and Febiger, 1974.

Tichauer, E.R. *Biomechanics of Lifting*. Report RD-3130-MPO-69 for Dept. of Health, Education and Welfare. Institute of Rehabilitation Medicine, February 15, 1970.

Tichauer, E.R. *Ergonomics: The State of the Art*. American Industrial Hygiene Association Journal. Vol. 28. March 1967.

Tichauer, E.R. *Gilbreth Revisited*. ASME Publication 66-WA/BHF-7. November 27, 1965.

Tichauer, E.R. *Human Capacity, A limiting Factor in Design*. Institution of Mechanical Engineers Publication P3764, July 1963.

Tichauer, E.R. *More Bricks, Less Sweat*. ASME Publication. Frank Gilbreth Centennial, 1969.

Tichauer, E. R., *The Biomechanical Basis of Ergonomics*, Wiley-Interscience, New York, 1978.

White, A. A., III, and M. M. Panjabi, *Clinical Biomechanics of the Spine*, Lippincott, Philadelphia, 1978.

U.S. Department of Health, Education, and Welfare. *The Industrial Environment ItsEvaluation and Control*. Government Printing Office, 1973.

Z94.3
Computer & Information Systems

The goal of the subcomittee was to include computer and information systems terms commonly used by industrial engineers. The subcommittee reviewed a number of established sources in determining the terms and definitions to be included in this section. The primary sources utilized were the American National Standard for Information Systems (X3.172 - 1990) published by the American National Standards Institute in 1990. The secretariat was the Computer and Business Equipment Manufacturers Association. The subcommittee used definitions from the Dictionary for Information Systems for terms included in this section to the extent possible. Exceptions were made to be consistent with current accepted usage in the field of computer and information systems and by industrial engineers. While additional words could have been included, the members of the subcommittee believe those included reflect current usage by industrial engineers. Overly simple terms and unnecessary variations of basic terms were excluded. The subcommittee is keenly aware of the continuous creation of new terms and rapid change in usage of terms associated with the field of computer and information systems. However, considerable effort was made to provide a comprehensive and current reference of industrial engineering usage of these terms.

Chairman
William E. Hammer, Jr., CSP
President
Hammer Management Consulting

Subcommittee

Richard M. Morris, III
President
R.M. Morris and Associates

Andrew J. Saunders, CMFGT
Information Services Analyst
The Duriron Company, Inc.

ACCESS TIME. (1) (ANSI/ISO) The time interval between the instant at which an instruction control unit initiates a call for data and the instant at which delivery of the data is completed. Access time equals latency plus transfer time. In some countries, the concept of access time is identical to that of latency. (2) (ANSI/ISO) Deprecated term for cycle time.

ACCURACY. (1) (ANSI/ISO) A quality of that which is free of error. (2) (ANSI/ISO) A qualitative assessment of freedom from error, with a high assessment corresponding to a small error.

ADDRESS. (1) (ANSI/ISO) A character or group of characters that identifies a register, a particular part of storage, or some other data source or destination. (2) (ANSI/ISO) To refer to a device or a data item by its address.

AI. (See ARTIFICIAL INTELLIGENCE.)

ALGOL (ALGORITHMIC LANGUAGE). (ANSI) A language used to express computer programs by algorithms.

ALGORITHM. (ANSI/ISO) A finite set of well-defined rules for the solution of a problem in a finite number of steps.

ALIAS. (ANSI) An alternate label. For example, a label and one or more aliases may be used to refer to the same data element or point in a computer program.

ALPHANUMERIC. (ANSI) Pertaining to a character set that contains letters, digits, and usually other characters such as punctuation marks. *Syn*: alphameric.

ANALOG. (1) (ANSI) Pertaining to data consisting of continuously variable physical quantities. (2) (ANSI) Contrast with digital, discrete.

ANALYST. (ANSI/ISO) A person who defines problems and develops algorithms and procedures for their solution.

AND. (ANSI/ISO) A logic operator having the property that if P is a statement, Q is a statement, R is a statement..., then the *and* of P,Q,R... is true if all statements are true, false if any statement is false. P AND Q is often represented by P·Q, PQ, P^Q. *Syn*: logical multiply.

APPLICATION. (1) (ANSI) A particular kind of work that a user performs on a computer; e.g., a payroll application, an airline reservation system. (2) (ANSI) A shortened form of application program with information provided by the user.

APPLICATIONS GENERATOR. Software that generates an application program with information provided by the user.

ARGUMENT. (1) (ANSI/ISO) An independent variable. (2) (ANSI/ISO) Any value of an independent variable, *e.g.*, a search key; a number identifying the location of an item in a table.

ARITHMETIC OPERATION. (ANSI) An operation that follows the rules of arithmetic.

ARRAY. (ANSI) An arrangement of elements in one or more dimensions. (PROGRAMMABLE LOGIC ARRAY.)

ARTIFICIAL INTELLIGENCE. (1) (ANSI) The capability of a device to perform functions that are normally associated with human intelligence, such as reasoning, learning, and self-improvement.

ASCII (AMERICAN NATIONAL STANDARD CODE FOR INFORMATION INTERCHANGE). (ANSI) The standard code, using a coded character set consisting of 7-bit coded characters (8-bits including parity check), used for information interchange amount data processing systems, data communication systems, and associated equipment. The ASCII set consists of control characters and graphic characters.

ASSEMBLER. (ANSI/ISO) A computer program used to assemble. *Syn*: assembly program.

ASSEMBLY LANGUAGE. (ANSI/ISO) A computer-oriented language whose instructions are usually in one-to-one correspondence with computer instructions and that may provide facilities such as the use of macroinstructions. *Syn*: computer-dependent language.

ASYNCHRONOUS OPERATION. (1) (ANSI/ISO) An operation that occurs without a regular or predictable time relationship to a specified event, *e.g.*, the calling of an error diagnostic routine that may receive control at any time during the execution of a computer program. *Syn*: asynchronous working. (2) (ANSI) A sequence of operations in which operations are executed out of time coincidence with any event.

ATTRIBUTE. A property possessed by a record or entity. Two examples of attributes of the department entity are its name and its budget.

AUTOMATIC DATA PROCESSING (ADP). (ANSI) Data processing by means of one or more devices that use common storage for all or part of a computer program and also for all or part of the data necessary for execution of the program; that execute user-written or user-designated programs; that perform user-designated symbol manipulation, such as arithmetic operations, logic operations, or character-string manipulations; and that can execute programs that modify themselves during their execution. Automatic data processing may be performed by a stand-alone unit or by several connected units.

AUXILIARY OPERATION. (ANSI) An off-line operation performed by equipment not under control of the processing unit.

AUXILIARY STORAGE. (1) (ANSI) Storage that is available to a processor only through input/output channels. (2) In a microcomputer, storage that is not memory; e.g. storage on diskettes, on streaming tapes, or on magnetic tape cartridges.

BACKBONE. This form of network is the organization's central information path. In the application of a backbone network, hub locations' telecommunications capabilities are sized to provide effective distribution of information to all the organization's locations.

BACKGROUND PROCESSING. (1) (ANSI) The execution of lower priority computer programs when higher priority programs are not using the system resources.

BACKUP. A copy of disk or tape files used for recovery if the operational files are lost or damaged.

BAR CODE. (ANSI/ISO) A code representing characters by sets of parallel bars of varying thickness and separation that are read optically by transverse scanning.

BASIC. (BEGINNER'S ALL-PURPOSE SYMBOLIC INSTRUCTION CODE). (ANSI) A procedural algebraic language originally designed for ease of learning with a small instruction repertoire.

BATCH PROCESSING. (1) (ANSI/ISO) The processing of data or the accomplishment of jobs accumulated in advance, in such a manner that the user cannot further influence its processing while it is in progress. (2) (ANSI) The processing of data accumulated over a period of time. (3) (ANSI) Loosely, the execution of computer programs serially. (4) (ANSI) Pertaining to the technique of executing a set of computer programs such that each is completed before the next program of the set is started. (5) (ANSI) Pertaining to the sequential input of computer programs or data.

BAUD. (1) (ANSI) A unit of signaling speed equal to the number of discrete conditions or signal events per second. For example, one baud equals one-half dot cycle per second in Morse code, one bit per second in a train of binary signals, and one 3-bit value per second in a train of signals each of which can assume one of eight different states. (2) (ANSI) In asynchronous transmission, the unit of modulation rate corresponding to one unit interval per second, *i.e.*, if the duration of the unit interval is 20 milliseconds, the modulation rate is 50 baud.

BCD. (See BINARY-CODED DECIMAL NOTATION.)

BENCHMARK PROBLEM. (1) (ANSI) A problem used to evaluate the performance of hardware or software or both. (2) (ANSI) A problem used to evaluate the performance of several computers relative to each other, or a single computer relative to system specifications.

BINARY. (1) (ANSI/ISO) Pertaining to a selection, choice, or condition that has two possible different values or states. (2) (ANSI/ISO) Pertaining to a fixed radix numeration system having a radix of two.

BINARY-CODED DECIMAL NOTATION (BCD). (ANSI/ISO) A binary-coded notation in which each of the decimal digits is represented by a binary numeral, *e.g.*, in binary-coded decimal notation that uses the weights 8, 4, 2, 1, the number "twenty-three" is represented by 0010 0011 (compare its representation 10111 in the pure binary numeration system). *Syns*: binary-coded decimal code, binary-coded decimal representation, coded decimal notation.

BINARY DIGIT (BIT). (1) (ANSI/ISO) In binary notation, either of the characters, 0 or 1.

BINARY SEARCH. (ANSI/ISO) A dichotomizing search in which, at each step of the search, the set of items is partitioned into two equal parts, some appropriate action being taken in the case of an odd number of items.

BIT. (See BINARY DIGIT.)

BIT MAPPING. The process of defining a pattern of bits stored in memory and used to display images on a video display screen and/or to print images on a printer.

BLOCK. (1) (ANSI/ISO) A string of records, a string of words, or a character string, formed for technical or logical purposes are treated as an entity. (2) (ANSI) A collection of contiguous records that are recorded as a unit, and the units are separated by interblock gaps. (3) (ANSI) A group of bits or digits that are transmitted as a unit and that may be encoded for error-control purposes. (4) (ANSI) In programming languages, a subdivision of a program that serves to group related statements, delimit routines, specify storage allocation, delineate the applicability of labels, or segment parts of the program for other purposes. In FORTRAN, a block may be a sequence of statements: in COBOL, it may be a physical record.

BOOLEAN OPERATION. (1) (ANSI/ISO) Any operation in which each of the operands and the result take one of two values. (2) (ANSI/ISO) An operation that follows the rules of Boolean algebra.

BOOTSTRAP. (1) (ANSI/ISO) A set of instructions that cause additional instructions to be loaded until the complete computer program is in storage. (2) (ANSI) A technique or device designed to bring itself into a desired state by means of its own action, *e.g.*, a machine routine whose first few instructions are sufficient to bring the rest of itself into the computer from an input device. (3) (ANSI) That part of a computer program used to establish another version of the computer program. (4) (ANSI/ISO) To use a bootstrap.

BUBBLE SORT. (ANSI) An exchange sort in which the sequence of examination of pairs of items is reversed whenever an exchange is made. *Syn*: sifting sort.

BUS. (ANSI) One or more conductors used for transmitting signals or power.

BYTE. (ANSI/ISO) A binary character string operated upon as a unit and usually shorter than a computer word.

CACHE MEMORY. (ANSI/ISO0 A special buffer storage, smaller and faster than main storage, that is used to hold a copy of instructions and data in main storage that are likely to be needed next by the processor, and that have been obtained automatically from main storage.

CALL. (1) (ANSI/ISO) The action of bringing a computer program, a routine, or a subroutine into effect, usually by specifying the entry conditions and jumping to an entry point. (2) (ANSI) In data communication, the action performed by the calling party, or the operations necessary in making a call, or the effective use made of a connection between two stations. (3) (ANSI/ISO) In computer programming, to execute a call. (4) (ANSI) To transfer control to a specified closed subroutine.

CATHODE RAY TUBE DISPLAY. (ANSI) A device that presents data in visual form by means of controlled electron beams.

CELL. The intersection of a row and column in a matrix or spreadsheet.

CHANNEL. (1) (ANSI/ISO) A means of one-way transmission. A channel may be provided; e.g., by frequency or time division multiplexing. Syn: data transmission channel. (2) (ANSI) A path along which signals can be sent, *e.g.*, data channel, output channel. (2) (ANSI) The portion of a storage medium that is accessible to a given reading or writing station, *e.g.*, track, band. (3) (ANSI/ISO)

CHARACTER. (1) (ANSI/ISO) A member of a set of elements upon which agreement has been reached and that is used for the organization, control, or representation of information. Characters may be letters, digits, punctuation marks, or other symbols.

CHECK DIGIT. (ANSI) A check key that consists of a single digit.

CLIENT-SERVER. A version of cooperative processing in which networked computers act as servers and/or clients. Clients are requesters of information. Servers are repositories of information or coordinators of requests. Systems on the network may act as both clients and servers when this is appropriate.

CLOCK. (1) (ANSI/ISO) A device that generates periodic, accurately spaced signals used for such purposes as timing, regulation of the operations of a processor, or generation of interrupts. (2) (ANSI) (See MASTER CLOCK.) (3)

COBOL (See COMMON BUSINESS ORIENTED LANGUAGE.)

CODE. (1) (ANSI/ISO) A set of rules that maps the elements of one set, the coded set, onto the elements of another set, the code element set. Syn: coding scheme. (2) (ANSI) A set of items, such as abbreviations, that represent the members of another set. (3) (ANSI/ISO) To represent data or a computer program in a symbolic form that can be accepted by a processor. (4) (ANSI) To write a routine.

COLLISION. (ANSI/ISO) An unwanted condition that arises from concurrent transmissions on a channel and that results in garbled data.

COM. (ANSI) Computer Output Microfilming.

COMMON BUSINESS ORIENTED LANGUAGE (COBOL). (ANSI) A programming language designed for business data processing.

COMPILE. (1) (ANSI/ISO) To translate a computer program expressed in a problem-oriented language into a computer-oriented language. (2) (ANSI) To prepare a machine language program from a computer program written in another programming language by making use of the overall logic structure of the program, or generating more than one computer instruction for each symbolic statement, or both, as well as performing the function of an assembler.

COMPILER. (ANSI/ISO) A computer program used to compile. *Syn*: compiling program.

CONCURRENT OPERATION. (1) (ANSI/ISO) Pertaining to processes that take place within a common interval of time during which they may have to alternately share common resources; e.g. several programs are concurrent when they are executed by multiprogramming in a computer having a single instruction control unit. (2) (ANSI) Contrast with simultaneous.

CONSOLE. (ANSI) A part of a computer used for communication between the operator or maintenance engineer and the computer.

CONTENTION. (1) (ANSI/ISO) In a local area network, a situation in which two or more data stations are allowed by the protocol to start transmitting concurrently and thus risk collision. (2) (ANSI/ISO) A condition that arises when two or more data stations attempt to transmit at the same time over a shared channel, or when two data stations attempt to transmit at the same time in two-way alternate communications.

COOPERATIVE PROCESSING. A system in which applications work together cooperatively, without respect to location or system. For example, a workstation might initiate a transaction and the software, recognizing that it needs information stored elsewhere, would retrieve it from the appropriate computer system. Client/Server is one approach to this.

COPY. (1) (ANSI/ISO) To read data from a source, leaving the source data unchanged, and to write the same data elsewhere on a data medium that may differ from that of the source; e.g. to copy a file from a magnetic tape onto a magnetic disk. (2) (ANSI/ISO) The reproduction of selected recorded text from memory or from a recording medium to another recording medium.

CPU. (ANSI) Central Processing Unit.

CRT DISPLAY. (See CATHODE RAY TUBE DISPLAY.)

CURSOR. (ANSI/ISO) A movable, visible mark used to indicate the position on which the next operation will occur on a display surface.

CUT AND PASTE. (ANSI/ISO) In text processing, a function that enables the user to designate a block of text and to move it from one point to another within a document or into another document. Syn: block move.

CYBERNETICS. (ANSI) The branch of learning that brings together theories and studies on communication and control in living organisms and in machines.

DATA. (1) (ANSI/ISO) A representation of facts, concepts, or instructions in a normalized manner suitable for communication, interpretation, or processing by humans or by automatic means. (2) (ANSI) Any representations such as characters or analog quantities to which meaning is or might be assigned.

DATA BASE. (1) (ANSI/ISO) A collection of interrelated data, often with controlled redundancy, organized according to a schema to serve one or more applications; the data are stored so that they can be used by different programs without concern for the data structure or organization. A common approach is used to add new data and to modify and retrieve existing data.

DATA BASE MANAGEMENT SYSTEM. (DBMS). (1) (ANSI) An integrated set of computer programs that collectively provide all of the capabilities required for centralized management, organization, and control of access to a database that is shared by many users. (2) (ANSI) A computer-based system used to establish, make available, and maintain the integrity of a database, that may be invoked by nonprogrammers or by application programs to define, create, revise, retire, interrogate, and process transactions; and to update, back up, recover, validate, secure, and monitor the database.

DATA BASE MANAGER. A shortened version of data base management system that is commonly used by microcomputer users.

DATA DIRECTORY. (1) (ANSI) An inventory that specifies the source, location, ownership, usage, and destination of all of the data elements that are stored in a database. (2) (ANSI) A subset of a data dictionary/directory that has the functions of (1).

DBMS. (See DATA BASE MANAGEMENT SYSTEM.)

DECISION TABLE. (1) (ANSI/ISO) A table of all contingencies that are to be considered in the description of a problem, together with the actions to be taken for each set of contingencies.

Digital Computer. (ANSI) A computer that consists of one or more associated processing units and peripheral equipment and that is controlled by internally-stored programs. A computer may be a stand-alone unit or may consist of several interconnected units.

Digitize. (ANSI/ISO) To express or represent in a digital form data that are not discrete data, *e.g.*, to obtain a digital representation of the magnitude of a physical quantity from an analog representation of that magnitude.

Direct Access. (1) (ANSI/ISO) The capability to obtain data from storage devices or to enter data into a storage device in a sequence independent of their relative position, by means of addresses that indicate the physical location of the data. (2) (ANSI) Pertaining to the organization and access method that must be used for a storage structure in which locations of records are determined by their keys, without reference to an index or to other records that may have been previously accessed.

Directory. (See DATA DIRECTORY.)

Disk. (See DISKETTE, FLEXIBLE DISK, MAGNETIC DISK.)

Disk Drive. (1) (ANSI/ISO) A mechanism for moving a disk pack or a magnetic disk and controlling its movement.

Diskette. (ANSI/ISO) A flexible magnetic disk enclosed in a protective container. *Syn*: flexible disk.

Disk Operating System. The software that controls or manages the basic functions of a computer.

Document. (ANSI) A medium and the information recorded on it that generally has permanence and that can be read by humans or machines.

DOS. (See DISK OPERATING SYSTEM.)

Dot Matrix Printer. *Syn*: matrix printer.

Double Precision. (ANSI/ISO) The use of two computer words to represent a number in accordance with required precision.

Duplex Transmission. (ANSI/ISO) Data transmission in both directions at the same time.

EBCDIC (See EXTENDED BINARY-CODED DECIMAL INTERCHANGE CODE.)

Edit. (ANSI/ISO) To prepare data for a later operation. Editing may include the following: the rearrangement or the addition of data; the deletion of unwanted data; format control; code conversion; and the application of standard processes such as zero suppression.

Editor Program. (ANSI) A computer program designed to perform such functions as the rearrangement, modification, and deletion of data in accordance with prescribed rules.

Electronic Data Interchange (EDI). EDI is the electronic transfer of business information from one independent computer system to another, using agreed-upon standards for terms and formats of documents. The most common application of this method is the translation of a purchase order on a customer's system into a sales order on a vendor's system. However, many other types of transactions are possible including notification of shipment, invoicing, cash transfers, etc. Standards have normally been adapted by industry segments and this has inhibited some advancement in use of this tool. The success of EDI relies on rethinking business processes, not speeding up old bureaucratic procedures. The two key standards are X.12 and EDIFACT.

Electronic Mail. (1) (ANSI) The use of a computer to transmit correspondence between workstations. (2) This facility of communication can be provided via a single computer system, an owned network, or a more public value-added network. The principle is to send messages, not to a place as in a telephone, but to particular people. Messages sent are placed in a repository until the receiving party picks them up. Sending and receiving can conceivable be done from any location with the proper equipment and data communication capability. Recent definition of a standard, X.400, has aided the implementation of this tool in business.

Electrostatic Plotter. (ANSI/ISO) A raster plotter that uses a row of electrodes to fix the ink electrostatically on the paper.

Encode. (1) (ANSI/ISO) To convert data by the use of a code or a coded character set in such a manner that reconversion to the original form is possible. Encode is sometimes loosely used when complete reconversion is not possible.

Equivalence Operation. (1) (ANSI/ISO) The dyadic Boolean operation whose result has the Boolean value 1 if and only if the operands have the same Boolean value. *Syn*: if-and-only-if operation.

Erasable Programmable Read-only Memory

(EPROM). (ANSI/ISO) A programmable read-only memory that can be erased by a special process and reused. Syn: reprogrammable read-only memory.

Execute. (ANSI/ISO) To perform the actions specified by a program or portion of a program.

Exit. (1) (ANSI/ISO) To execute an instruction or statement within a portion of a program in order to terminate the execution of that portion. Such portions may include loops, routines, subroutines, modules. (2) (ANSI/ISO) Any instruction in a computer program, in a routine, or in a subroutine after the execution of which control is no longer exercised by that computer program, that routine, or that subroutine.

Expert System. (ANSI) In artificial intelligence, a functional unit for solving problems in a particular field of knowledge by drawing inferences from a knowledge base acquired through human experience. Syn: Knowledge-based system.

Extended Binary-Coded Decimal Interchange Code (EBCDIC). (ANSI) A coded character set consisting of 8-bit coded characters.

FIFO (first-in-first-out). (ANSI) A queuing technique in which the next item to be retrieved is the item that has been in the queue for the longest time.

File. (1) (ANSI/ISO) A set of related records treated as a unit, *e.g.*, in stock control, a file could consist of a set of invoices. (2) (ANSI) (inverted file, master file, transaction file.)

Flag. (1) (ANSI/ISO) An indicator or parameter that shows the setting of a switch. Syn: switch indicator. (2) (ANSI) Any of various indicators used for identification purposes; e.g., a word mark. (3) (ANSI) A character that signals the occurrence of some condition, such as the end of a word. (4) Syn: Sentinel.

Flexible Disk. (ANSI/ISO) A flexible magnetic disk enclosed in a protective container. Syn: Floppy Disk.

Floating-Point Representation System. (ANSI/ISO) A numeration system in which a real number is represented by a pair of distinct numerals, the real number being the product of the fixed-point part, one of the numerals, and a value obtained by raising the implicit floating-point base to a power denoted by the exponent in the floating-point representation, indicated by the second numeral.

Floppy Disk. (See flexible disk.)

Flowchart. (ANSI/ISO) A graphical representation in which symbols are used to represent such things as operations, data, flow direction, and equipment, for the definition, analysis, or solution of a problem. Syn: flow diagram.

Font. (ANSI) A family or assortment of characters of a given size and style, *e.g.*, 9-point Bodoni modern.

Format. (ANSI/ISO) The arrangement or layout of data in or on a data medium. (2) (ANSI) In a programming language, a language construct that specifies the rules for transformation between internal and character representations of data objects. (3) (ANSI) In text processing, and arrangement or layout of text.

FORTRAN. (formula translation). (ANSI) A programming language primarily used to express computer programs by arithmetic formulas.

Function. (1) (ANSI/ISO) A mathematical entity whose value, *i.e.*, the value of the dependent variable, depends in a specified manner on the values of one or more independent variables, not more than one value of the dependent variable corresponding to each permissible combination of values from the respective ranges of the independent variables. (2) (ANSI) A specific purpose of an entity, or its characteristic action. (3) (ANSI) In data communication, a machine action such as carriage return or line feed.

Global. (ANSI) Pertaining to that which is defined in one subdivision of a computer program and used in at least one other subdivision of that computer program.

Graphic. (ANSI/ISO) A symbol produced by a process such as handwriting, drawing, or printing. *Syn*: graphic symbol.

Half-duplex. (ANSI) In data communication, pertaining to an alternating, one way at a time, independent transmission.

Halfword. (ANSI) A contiguous sequence of bits or characters which comprises half a computer word and is capable of being addressed as a unit.

Hard Copy. (ANSI/ISO) In computer graphics, a permanent copy of a display image that is portable and can be read directly by human beings, *e.g.*, a display image that is recorded on paper.

Hardware. (1) (ANSI/ISO) Physical equipment as opposed to programs, procedures, rules, and associated documentation. (2) (ANSI) Contrast with software.

Hash Total. (ANSI) The result obtained by applying an algorithm to a set of data for checking purposes, *e.g.*, a summation obtained by treating data items as numbers.

Heuristic Method. (ANSI) A method of solving problems that consists of a sequence of trials yielding approximate results, with control of the progression toward an acceptable final result; e.g. the method of successive approximations.

Hierarchical Model. (1) (ANSI) A data model whose pattern of organization has the form of a tree structure. (2) (ANSI) A data model that provides a tree structure for relating data elements, where each node of the tree corresponds to a group of data elements or a record type, and may have only one superior node.

High-level Language. (ANSI/ISO) A programming language that does not reflect the structure of any one given computer or that of any given class of computers. High level languages are primarily designed for, and are syntactically oriented to, particular classes of problems.

Icon. (ANSI) A symbol, displayed on a screen, that enables a user to select an action, the object of the action, or both.

Idle Time. (ANSI/ISO) That part of operable time during which a functional unit is not operated.

If-and-only-if Operation. (See EQUIVALENCE OPERATION.)

If-then Operation. (ANSI/ISO) *Syn*: Implication.

Illegal Character. (ANSI) A character or combination of bits that is not valid according to some criteria, *e.g.*, with respect to a specified alphabet, a character that is not a member.

Impact Printer. (ANSI/ISO) A printer in which printing is the result of mechanical impacts.

Index. (1) (ANSI/ISO) A reference of integer value, or an expression that yields an integer value, that identifies the position of a data item with respect to some other data item. (2) (ANSI) A table or list of the contents of a storage medium, file, document, or database, together with keys or references for locating the contents. (3) (ANSI) A symbol or a numeral array of similar quantities. (4) (ANSI) In micrographics, a guide for locating information on a roll of microfilm using targets, flash cards, lines, bars, or other optical codes. (5) (ANSI) To prepare a list as in (2). (6) (ANSI) To move a machine part to a predetermined position, or by a predetermined amount, on a quantized scale.

Information. (1) (ANSI/ISO) The meaning that is currently assigned to data by means of the conventions applied to that data. (2) (ANSI) In a conceptual schema language, any kind of knowledge about things, facts, or concepts of a universe of discourse that is exchangeable among users.

Initialize. (1) (ANSI) To set counters, switches, addresses, or contents of storage to zero or other starting values at the beginning of, or at prescribed points in, the operation of a computer routine.

Initial Program Loader (IPL). (ANSI/ISO) A bootstrap loader used in a computer to load that part of an operating system needed to load the remainder of the operating system.

Ink Jet Printer (ANSI/ISO). A non-impact printer in which the characters are formed by projecting a jet of ink onto paper.

Input. (1) (ANSI/ISO) Pertaining to a device, process, or channel involved in the reception of data by a computer or by any of its components. (2) (ANSI) An input state, or a sequence of states.

Input-output (I/O). (1) (ANSI) Pertaining to input, output, or both.

Insert. (1) (ANSI) To introduce data between previously stored items of data. (2) (ANSI) In text processing, to introduce new characters or text within previously recorded text. The text is automatically rearranged to accommodate the addition.

Instruction. (ANSI/ISO) In a programming language, any expression that specifies one operation and identifies its operands, if any.

Instruction Set. (ANSI/ISO) The set of the instructions of a computer, of a programming language, or of the programming languages in a programming system.

Interface. (1) (ANSI/ISO) A shared boundary between two functional units, defined by functional characteristics, common physical interconnection characteristics, signal characteristics, and other characteristics, as appropriate. The concept involves the specification of the connection of two devices having differnet functions. (2) (ANSI) A point of communication between two or more processes, persons, or other physical entities.

Internal Sort. (1) (ANSI) A sort performed within internal storage. (2) (ANSI) A sort program or a sort phase that sorts two or more items within main storage.

Interoperability. (1) (ANSI) The capability of two or more systems to exchange and use information. (2) The ability to communicate across previously incompatible aggregations of hardware and software with a minimum of human intervention. It is usually achieved by specialized tools placed at network junctures. Interoperability has proven to be an achievable tool.

Interpreter. (1) (ANSI/ISO) A computer program used to interpret. *Syn*: interpretive program. (2) (ANSI/ISO) A device that prints on a punched card the characters corresponding to hole patterns punched in the card.

Interrupt. (ANSI/ISO) A suspension of a process, such as the execution of a computer program, caused by an event external to that process, and performed in such a way that the process can be resumed. *Syn*: interruption.

I/O. (See INPUT-OUTPUT.)

IPL. (See INITIAL PROGRAM LOADER.)

ISO. (ANSI) The International Organization for Standardization

Iterative Operation. (ANSI/ISO) The repetition of the algorithm for the solution of a set of equations with successive combinations of initial conditions or other parameters; each successive combination is selected by a subsidiary computation based on a predetermined set of iteration rules. Iterative operation is usually used to permit solution of boundary value problems or for automatic optimization of system parameters. *Syn*: automatic sequential operation.

JCL. (See JOB CONTROL LANGUAGE.)

Job. (ANSI/ISO) A unit of work that is defined by a user and that is to be accomplished by a computer. Loosely, the term "job" is sometimes used to refer to a representation of a job. This representation may include a set of computer programs, files, and control statements to the operating system.

Job Control Language (JCL). (ANSI) A problem-oriented language designed to express statements in a job that are used to identify the job or describe its requirements to an operating system.

Joining. The process of merging data from two files (tables) by combining records of one file with records of the other file that have the contents of specified fields in common. The combined records may exist in memory or constitute a new file.

Joy Stick. (ANSI/ISO) In computer graphics, a lever with at least two degrees of freedom that is used as an input device, normally as a locator.

K. When referring to storage capacity, two to the tenth power, 1024 in decimal notation.

Key. (1) (ANSI/ISO) An identifier within a set of data elements. (2) (ANSI/ISO) One or more characters, within a set of data, that contains information about the set, including its identification. (3) In a record, a data element whose value is unique for each occurrence of the record and is used to identify or locate the record in a database management system. (4) (ANSI) On a keyboard, a manually actuated mechanism that performs a specific operation or causes the printing of a particular character.

Keyword. (1) (ANSI) In a programming language, a token that is usually specified by the language, and that uniquely characterizes a statement, or part of the statement; e.g. In some languages IF designates an IF-statement. (2) (ANSI) One of the predefined words of an artificial language. (3) (ANSI) *Syn*: descriptor, in the context of information retrieval.

Knowledge-Based System. (See EXPERT SYSTEM.)

Label. (1) (ANSI/ISO) An identifier within or attached to a set of data elements. (2) (ANSI/ISO) In computer languages, an identifier that names a statement.

LAN. (See LOCAL AREA NETWORK.)

LAN gateway. (ANSI/ISO) A functional unit that connects a local area network with another network using different protocols. The network to which a local area network is connected may be another local area network, a public data network (PDN), or another type of network.

Language. (1) (ANSI/ISO) A set of characters, conventions, and rules, that is used for conveying information. (See ALGOL, ASSEMBLY LANGUAGE, HIGH-LEVEL LANGUAGE, JOB CONTROL LANGUAGE, NATURAL LANGUAGE, PROCEDURE-ORIENTED LANGUAGE, PROGRAMMING LANGUAGE, SOURCE LANGUAGE, SYMBOLIC LANGUAGE.)

Laser printer. (ANSI/ISO) A nonimpact printer that creates, by means of a laser beam directed on a photosensitive surface, a latent image which is then made visible by a toner and transferred and fixed on paper. Syn: laser beam printer.

LIBRARY. (1) (ANSI) A file or a set of related files, *e.g.*, in stock control, a set of inventory control fields. (2) (ANSI) A repository for dismountable recorded media, such as magnetic disk packs and magnetic tapes.

LIFO (LAST-IN-FIRST-OUT). (ANSI) A queuing technique in which the next item to be retrieved is the item most recently placed in the queue.

LIGHT PEN. (ANSI/ISO) A light-sensitive pick device that is used by pointing it at the display surface.

LINEAR PROGRAMMING (LP). (1) (ANSI/ISO) In operations research, a procedure for locating the maximum or minimum of a linear function of variables that are subject to linear constraints. *Syn*: linear optimization.

LINE PRINTER. (1) (ANSI/ISO) A device that prints a line of characters as a unit.

LINKAGE EDITOR. (ANSI/ISO) A program for creating a load module from one or more object modules or load modules, or by resolving cross-references among the object modules, and possibly by relocating elements. *Syn*: linker.

LIST. (1) (ANSI/ISO) An ordered set of items of data. (2) (ANSI) To print or otherwise display items of data that meet specified criteria.

LOAD. (ANSI/ISO) (1) To enter data or programs into storage or working registers. (2) (ANSI) To insert data values into a database what previously contained no occurrences of data.

LOCAL AREA NETWORK (LAN). (ANSI) A data network, located on a user's premises, within a limited geographic region. Communication within a local area network is not subject to external regulation; however, communication across the network boundary may be subject to some form of regulation. A local area network does not use store-and-forward techniques.

LOGICAL RECORD. (ANSI) A record independent of its physical environment. Portions of the same logical record may be located in different physical records, or several logical records or parts of logical records may be located in one physical record.

LOOP. (ANSI/ISO) A sequence of instructions that may be executed repeatedly while a certain condition prevails. In some implementations, no test is made to discover whether the condition prevails until the loop has been executed once.

MACHINE CODE. (ANSI/ISO) *Syn*: computer instruction code.

MACRO. (See MACROINSTRUCTION.)

MACROINSTRUCTION. (ANSI/ISO) An instruction in a source language that is to be replaced by a defined sequence of instruction in the same source language and may also specify values for parameters in the replaced instructions. *Syn*: macro statement.

MAGNETIC DISK. (ANSI/ISO) A flat circular plate with a magnetizable surface layer, on one side or both sides of which data can be stored.

MAGNETIC STORAGE. (ANSI) A storage device that utilizes the magnetic properties of certain materials.

MAINFRAME. (1) A computer that is relatively large and provides the capability to perform applications requiring large amounts of data. These computers are more expensive than microcomputers and minicomputers. Due to the rapid development in computer technology, the specific description of what constitutes a mainframe computer has and will continue to change. (2) (ANSI) A mainframe computer, usually one to which other computers are connected in order to share its resources and computing power.

MANAGEMENT INFORMATION SYSTEM (MIS). (ANSI) The total flow of information within an enterprise that supports the decision-making functions of management at all organizational levels of the enterprise.

MARK SENSING. (ANSI/ISO) The electrical sensing of conductive marks usually recorded manually on a nonconductive data medium.

MASK. (ANSI/ISO) A pattern of characters that is used to control the retention or elimination of portions of another pattern of characters.

MASTER CLOCK. (ANSI/ISO) A clock whose main function is to control other clocks.

MATHEMATICAL MODEL. (ANSI) A mathematical representation of a process, device, or concept.

MATRIX. (1) (ANSI/ISO) A rectangular array of elements, arranged in rows and columns, that may be manipulated according to the rules of matrix algebra. (2) (ANSI) By extension, an array of any number of dimensions. (3) (ANSI) In computers, a logic network in the form of an array of input leads and output leads with logic elements connected at some of their intersections.

Matrix Printer. (ANSI/ISO) *Syn*: dot matrix printer.

Mean Time Between Failures. (MTBF). (ANSI/ISO) For a stated period in the life of a functional unit, the mean value of the lengths of time between consecutive failures under stated conditions.

Memory. (1) (ANSI/ISO) All of the addressable storage space in a processing unit and other internal memory that is used to execute instructions. (2) (ANSI/ISO) Main storage, when used in reference to calculators, microcomputers, and some minicomputers.

Menu. (ANSI) A list of options displayed on video screen from which the user selects an action to be performed.

Merge. (ANSI/ISO) To combine the items of two or more sets that are each in the same given order into one set in that order.

Microcomputer. (ANSI) A computer system whose processing unit is a microprocessor. A basic microcomputer includes a microprocessor, storage, and input/output facility, which may or may not be on one chip.

Microfiche. (ANSI) A sheet of microfilm capable of containing microimages in a grid pattern, usually containing a title that can be read without magnification.

Microprocessor. (1) (ANSI/ISO) A processor whose elements have been miniaturized into one or a few integrated circuits. (2) (ANSI) An integrated circuit that accepts coded instructions at one or more terminals, executes the instructions received, and delivers signals describing its status. The instructions may be entered, integrated, or stored internally.

MIS. (See management information system.)

Mnemonic Symbol. (ANSI/ISO) A symbol chosen to assist the human memory, *e.g.*, an abbreviation such as "mpy" for "multiply."

Modem. (ANSI/ISO) A functional unit that modulates and demodulates signals. One of the functions of a modem is to enable digital data to be transmitted over analog transmission facilities. Modem is a contraction of modulator-demodulator.

Monitor. (1) (ANSI/ISO) A functional unit that observes and records selected activities for analysis within a data processing system. Possible uses are to indicate significant departures from the norm, or to determine levels of utilization of particular functional units. (2) (ANSI) Software or hardware that observes, supervises, controls, or verifies the operation of a system.

Mouse. (ANSI/ISO) In computer graphics, a hand-held locator operated by moving it on a flat surface. A mouse generally contains a control ball or a pair of wheels.

Move. (ANSI/ISO) *Syn*: transfer.

Multiplexer. (1) (ANSI/ISO) In process control, a device that combines several input signals into a single output signal in such a manner that each of the input signals can be recovered. (2) (ANSI) A device capable of interleaving the events of two or more activities or capable of distributing the events of an interleaved sequence to the respective activities.

Multiprocessing. (1) (ANSI/ISO) A mode of operation that provides for parallel processing by two or more processors of a multiprocessor. (2) (ANSI) The simultaneous execution of two or more computer programs or sequences of instructions by a computer. (3) Loosely, parallel processing.

Multiprogramming. (ANSI/ISO) A mode of operation that provides for the interleaved execution of two or more computer programs by a single processor.

Multitasking. (ANSI/ISO) A mode of operation that provides for the concurrent performance, or interleaved execution of two or more tasks.

Natural Language. (ANSI/ISO) A language whose rules are based on current usage without being explicitly prescribed.

Network. An arrangement of nodes and interconnecting branches.

Network Model. (ANSI) A data model that consists of a modified tree structure that permits all of the root record to have multiple owner records.

Node. (1) (ANSI/ISO) In a network, a point where one or more functional units interconnect transmission lines. (2) (ANSI) The representation of a state or an event by means of a point on a diagram. (3) (ANSI) In a tree structure, a point at which subordinate items of data originate.

Nonimpact printer. (ANSI/ISO) A printer in which

printing is not the result of mechanical contacts with the printing medium.

NONLINEAR PROGRAMMING. (ANSI/ISO) In operations research, a procedure for locating the maximum or minimum of a function of variables that are subject to constraints, when either the function or the constraints, or both, are nonlinear. *Syn*: nonlinear optimization.

NUMERICAL. (ANSI) Pertaining to data or to physical quantities that consist of numerals.

NUMERICAL CONTROL (NC). (ANSI/ISO) Automatic control of a process performed by a device that makes use of numerical data usually introduced as the operation is in progress.

OBJECT CODE. (ANSI) Output from a compiler or assembler which is itself executable machine code or is suitable for processing to produce executable machine code.

OCR. (See OPTICAL CHARACTER RECOGNITION.)

OFFLINE. (ANSI/ISO) Pertaining to the operation of a functional unit when not under the direct control of a computer.

ONLINE. (1) (ANSI/ISO) Pertaining to the operation of a functional unit that is under the direct control of a computer. (2) (ANSI) Pertaining to a user's ability to interact with a computer. (3) (ANSI) Pertaining to the user's access to a computer via a user terminal.

OPEN SHOP. (1) (ANSI) Pertaining to the operation of a computer facility in which most productive problem programming is performed by the problem originator rather than by a group of programming specialists. The use of the computer itself may also be described as open shop if the user/programmer also serves as the operator.

OPERATING SYSTEM. (ANSI/ISO) Software that controls the execution of programs. An operating system may provide services such as resource allocation scheduling, input/output control, and data management. Although operating systems are predominately software, partial or complete hardware implementations are possible.

OPTICAL CHARACTER RECOGNITION (OCR). (ANSI/ISO) Character recognition that uses optical means to identify graphic characters.

OUTPUT. (1) (ANSI/ISO) Pertaining to a device, process, or channel involved in the production of data by a computer or by any of its components. (2) (ANSI/ISO) An output state, or sequence of states. (3) (ANSI) Information retrieved from a functional unit or from a network, usually after some processing.

OVERFLOW. (ANSI/ISO) In a calculator, the state in which the calculator is unable to accept or process the number of digits in the entry or in the result.

OVERLAY. (1) (ANSI/ISO) One of several segments of a computer program that, during execution, occupy the same area of main storage, one segment at a time. (2) (ANSI) To load an overlay.

PACKET SWITCHING. (ANSI/ISO) The process of routing and transferring data by means of addressed packets so that a channel is occupied only during the transmission of a packet; upon completion of the transmission, the channel is made available for the transfer of other packets.

PADDING. (1) (ANSI/ISO) A technique that incorporates fillers in data. (2) (ANSI) A technique that incorporates fillers into data. (2) (ANSI) A technique used to fill a field, record, or block with dummy data, usually zeroes or spaces.

PAGE PRINTER. (1) (ANSI/ISO) A printer that prints one page as a unit, *e.g.*, COM printer; xerographic printer.

PAGING. (1) (ANSI/ISO) The transfer of pages between real storage and auxiliary storage. (2) (ANSI) An allocation technique by which main storage is divided into page frames. A computer program need not be located in contiguous page frames in order to be executed.

PARALLEL COMPUTER. (1) (ANSI) A computer that has multiple arithmetic or logic units that are used to accomplish parallel operations or parallel processing. (2) (ANSI) Contrast with serial computer.

PARALLEL PROCESSING. (1) (ANSI) The concurrent or simultaneous execution of two or more processes in a single unit. (2) (ANSI) Contrast with serial processing.

PARAMETER. (ANSI/ISO) A variable that is given a constant value for a specified application and that may denote the application.

PARITY BIT. (ANSI) A binary digit appended to a group of binary digits to make the sum of all the digits, including the appended binary digit, either odd or even, as predetermined.

PASSWORD. (ANSI/ISO) A character string that enables a user to have full or limited access to a system or to a set of data.

PATTERN RECOGNITION. (ANSI/ISO) The identification of shapes, forms, or configurations by automatic means.

PERIPHERAL UNIT. (ANSI) With respect to a particular processing unit, any equipment that can communicate directly with that unit. *Syn*: peripheral device.

PIXEL. (ANSI/ISO) The smallest element of a display surface that can be independently assigned color or intensity. *Syn*: picture element. (PEL)

PL/I. (ANSI) A programming language designed for use in a wide range of commercial and scientific computer applications.

PLOTTER. (ANSI/ISO) An output unit that presents data in the form of a two-dimensional graphic representation.

POLLING. (1) (ANSI/ISO) On a multipoint connection, the process whereby data stations are invited one at a time to transmit. (2) (ANSI) Interrogation of devices for purposes such as to avoid contention, to determine operational status, or to determine readiness to send or receive data.

PORTABILITY. (1) (ANSI) The ability to transfer data from one system to another without being required to recreate or reenter data descriptions or to significantly modify the application being transported. (2) (ANSI) The ability of software or of a system to run on more than one type or size of computer or under more than one operating system. (3) (ANSI) Syn: transportability.

POSTPROCESSOR. (ANSI/ISO) A computer program that effects some final computation or organization.

PRECISION. (1) (ANSI/ISO) A measure of the ability to distinguish between nearly equal values; e.g., four-place numerals are less precise than six-place numerals; nevertheless, a properly computed four-placer numeral may be more accurate than an improperly computed six-place numeral. (2) (ANSI) The degree of discrimination with which a quantity is stated; e.g., a three-digit numeral discriminates among 1000 possibilities.

PREPROCESSOR. (ANSI/ISO) A computer program that effects some preliminary computation or organization.

PRINTER. (ANSI/ISO) An output unit that processes a durable record of data in the form of a sequence of discrete graphic characters belonging to a predetermined character set.

PRINT WHEEL. (ANSI/ISO) A rotating disk that presents characters at a single print position.

PROCEDURAL LANGUAGE. (See PROCEDURE-ORIENTED LANGUAGE.)

PROCEDURE-ORIENTED LANGUAGE. (ANSI) A problem-oriented language that facilitates the expression of a procedure as an explicit algorithm, *e.g.*, FORTRAN, ALGOL, COBOL, PL/I. *Syn*: imperative language.

PROCESS CONTROL. (ANSI/ISO) Automatic control of a process, in which a computer system is used to regulate usually continuous operations or processes.

PROGRAM. (1) (ANSI/ISO) A sequence of instructions suitable for processing. Processing may include the use of an assembler, a compiler, an interpreter or another translator to prepare the program for execution; the instructions may include statements and necessary declarations. (2) (ANSI/ISO) To design, write, and test computer programs. (3) (ANSI) In programming languages, a set of one or more interrelated modules capable of being executed. (4) Loosely, a routine. (5) Loosely, to write a routine.

PROGRAMMABLE READ-ONLY MEMORY (PROM). (ANSI/ISO) A storage device that, after being written once, becomes a read-only memory.

PROGRAMMER. (ANSI) A person who designs, writes, and tests computer programs.

PROGRAMMING. (ANSI/ISO) The designing, writing, and testing of programs.

PROGRAMMING LANGUAGE. (ANSI/ISO) An artificial language designed to generate or to express programs.

PROM. (See PROGRAMMABLE READ-ONLY MEMORY.)

PROTOCOL. (ANSI/ISO) A set of semantic and syntactic rules that determines the behavior of functional units to achieve communication.

RAM. (See RANDOM-ACCESS MEMORY.)

RANDOM ACCESS. (1) (ANSI) An access mode in which specific logical records are obtained from, or placed into, a mass storage file in a nonsequential manner. (2) (ANSI) Pertaining to the organization and access method for a storage structure in which locations of records are determined by a randomizing or hashing algorithm applied to the values of their keys so that the random numbers thus generated serve as addresses of the records. (3) (ANSI) Deprecated term for direct access.

Random Access Memory (RAM). (ANSI) High speed read/write memory with an access time that is the same for all storage locations.

Raster Graphics. (ANSI/ISO) Computer graphics in which a display image is composed of an array of pixels arranged in rows and columns.

Read. (ANSI/ISO) To acquire or to interpret data from a storage device, from a data medium, or from another source.

Read-Only Memory (ROM). (ANSI/ISO) A memory in which data, under normal conditions, can only be read.

Real Time. (1) (ANSI/ISO) Pertaining to the processing of data by a computer in connection with another process outside the computer according to time requirements imposed by the outside process. This term is also used to describe systems operating conversational mode and processes that can be influenced by human intervention while they are in progress.

Real-time. (1) (ANSI/ISO) A mode of operation of a data processing system when performing real-time jobs. *Syn*: real-time operation. (2) (ANSI/ISO) The manipulation of data that are required or generated by some process while the process is in operation. Usually the results are used to influence the process, and perhaps related processes, while it is occurring.

Record. (1) (ANSI/ISO) A group of related data elements treated as a unit. (2) (ANSI) A named and usually ordered collection of zero or more data items and data aggregates that represent the occurrence of a set of data values that describe the attributes of a particular entity. (3) (ANSI) In programming languages, an aggregate that consists of data objects, each of which may be uniquely referenced by its own identifier.

Record Layout. (ANSI/ISO) The arrangement and structure of data or words in a record including the order and size of the components of the record.

Record Length. (ANSI/ISO) Syn: record size.

Record Size. (ANSI/ISO) The number of characters or bytes in a record. Syn: record length.

Reentry Point. (ANSI/ISO) The address or the label of the instruction at which the computer program that called a subroutine is reentered from the subroutine.

Refresh. (ANSI/ISO) The process of repeatedly producing a display image on a display surface so that the image remains visible.

Register. (ANSI/ISO) A storage device that has a specified storage capacity.

Relative Address. (ANSI/ISO) An address calculated as a displacement from a base address.

Relocatable Program. (ANSI/ISO) A computer program that is suitable for dynamic relocation.

Remote Job Entry (RJE). (ANSI/ISO) Submission of a job through an input unit that has access to a computer through a data link.

Reprogrammable Read-only Memory. (ANSI/ISO) Syn: erasable programmable read-only memory.

Reserved Word. (ANSI) In programming languages, a keyword whose definition is fixed by the programming language and which cannot be changed by the user. In Ada and COBOL all keywords are reserved words.

Reset. (1) (ANSI/ISO) To cause a counter to take the state corresponding to a specified initial number. (2) (ANSI/ISO) To put all or part of a data processing device back into a prescribed state.

Resident. (ANSI/ISO) Pertaining to computer programs that remain on a particular storage device.

Resource Allocation. (ANSI/ISO) The assignment of the facilities of a computer system for the accomplishment of jobs, *e.g.*, the assignment of main storage, input-output units, or files.

Response Time. (ANSI/ISO) The elapsed time between the end of an inquiry or demand on a computer system and the beginning of the response, *e.g.*, the length of time between an indication of the end of an inquiry and the display of the first character of the response at a user terminal.

Return. (ANSI) In programming languages, within a procedure, a language construct that designates a statement that ends the execution of the procedure.

Reusable Program. (ANSI/ISO) A computer program that may be loaded once and executed repeatedly, subject to the requirements that any instructions that are modified during its execution are returned to their initial states and that its external program parameters are preserved unchanged.

RJE. (See REMOTE JOB ENTRY.)

ROM. (See READ-ONLY MEMORY.)

ROUTINE. (1) (ANSI/ISO) A program, called by another program, that may have some general or frequent use.

RPG. (ANSI) Report Program Generator.

RUN. (1) (ANSI/ISO) A performance of one or more jobs. (2) (ANSI/ISO) A performance of one or more programs.

SCAN. (ANSI) To examine every reference or every entry in a file routinely as part of a retrieval scheme. (2) (ANSI) To examine sequentially, part by part.

SCROLLING. (ANSI/ISO) Moving a window vertically or horizontally in order to view data not otherwise visible within the boundaries of a display screen or window.

SEARCH. (1) (ANSI/ISO) The examination of one or more data elements of a set for one or more elements that have a given property. (2) (ANSI) To examine a set of items for one or more having a given property.

SEARCH KEY. (1) (ANSI/ISO) A key used for data retrieval. (2) (ANSI) In a record, a data item that represents one of the data values of the range allowed for a particular attribute of an entity.

SECTOR. (ANSI/ISO) A predetermined angular part of a track or a band on a magnetic drum or magnetic disk that can be addressed.

SEQUENCE. (1) (ANSI/ISO) A series of items that have been sequenced. (2) (ANSI) An arrangement of items according to a specified set of rules, *e.g.*, items arranged alphabetically, numerically, or chronologically. (3) (ANSI) To place items in an arrangement in accordance with the order of the natural numbers. (4) (ANSI/ISO) Deprecated term for order.

SEQUENTIAL ACCESS. (1) (ANSI/ISO) The capability to enter data into a storage device or data medium in the same sequence as the data are ordered, or to obtain data in the same order as they were entered. (2) (ANSI) Syn: serial access.

SERIAL. (1) (ANSI/ISO) Pertaining to a process in which all events occur one after the other; e.g., the serial transmission of the bits of a character according to the CCITT V25 protocol. (2) (ANSI) Pertaining to the sequential or consecutive occurrence of two or more related activities in a single device or channel. (3) (ANSI) Pertaining to the sequential processing of the individual parts of a whole, such as the bits of a character or the characters of a word, using the same facilities for successive parts. (4) (ANSI) Contrast with parallel.

SERIAL PRINTER. (ANSI/ISO) *Syn*: character printer.

SERIAL SORT. (ANSI) A sort that requires only sequential access to the items in a set.

SET. (1) (ANSI/ISO) A finite or infinite number of objects, entities, or concepts, that have a given property or properties in common. (2) (ANSI/ISO) To cause a counter to take the state corresponding to a specified number. (3) (ANSI/ISO) To put all or part of a data processing device into a specified state. (4) (ANSI) In the CODASYL model, a set of records that represents a hierarchical relationship between its owner record and the member records of the set.

SIGNAL. (1) (ANSI/ISO) A variation of a physical quantity, used to convey data. (2) (ANSI/ISO) A time-dependent value attached to a physical phenomenon and conveying data.

SIMULATION. (ANSI/ISO) The representation of selected characteristics of the behavior of one physical or abstract system by another system: e.g., the representation of physical phenomena by means of operations performed by a computer system; or the representation of operations of a data processing system by those of another data processing system.

SINGLE PRECISION. (ANSI/ISO) Characterized by the use of one computer word to represent a number in accordance with the required precision.

SOFTWARE. (1) (ANSI) Programs, procedures, rules, and any associated documentation pertaining to the operation of a system. (2) (ANSI) Contrast with hardware.

SORT. (1) (ANSI/ISO) To segregate items into groups according to specified criteria without necessarily ordering the items within each group. (2) (ANSI) To arrange a set of items according to keys which are used as a basis for determining the sequence of the items, *e.g.*, to arrange the records of a personnel file into alphabetical sequence by using the employee names as sort keys.

SORT KEY. (1) (ANSI) A key used as a basis for determining the sequence of items in a set. *Syn*: sequencing key.

SOURCE LANGUAGE. (ANSI/ISO) A language from which statements are translated.

SOURCE PROGRAM. (1) (ANSI/ISO) A computer program expressed in a source language.

SPOOLING. (ANSI/ISO) The use of auxiliary storage as a buffer storage to reduce processing delays when transferring data between peripheral equipment and the processors of a computer. The term is derived from the expression "simultaneous peripheral operation on line."

SPREADSHEET. (ANSI) A worksheet arranged in rows and columns, in which a change in the contents of one cell can cause electronic recomputation of one or more cells, based on user-defined relations among the cells.

STATEMENT. (1) (ANSI/ISO) In a programming language, a language construct that represents a set of declarations or a step in a sequence of actions. (2) (ANSI) In computer programming, a symbol string or other arrangement of symbols. (3) (ANSI/ISO) Deprecated term for instruction.

STORAGE. (1) (ANSI/ISO) The retention of data in a storage device. (2) (ANSI/ISO) The action of placing data into a storage device. (3) (ANSI) A storage device. *Syn*: memory. (4) (ANSI) Any medium in which data can be retained.

STRING. (1) (ANSI/ISO) A sequence of elements of the same type, such as characters, considered as a whole.

STYLUS. (ANSI/ISO) A pointer that is operated by placing it in a display space or a table, *e.g.*, a light pen, sonic pen, voltage pencil.

SUBROUTINE. (1) (ANSI/ISO) A sequenced set of statements that may be used in one or more computer programs, and at one or more points in a computer program. The execution of a subroutine is usually invoked by a call. (2) (ANSI) A routine that can be part of another routine.

SUBROUTINE CALL. (ANSI) The language construct, in object coding, that performs the call function.

SUBSCRIPT. (ANSI/ISO) A symbol that is associated with the name of a set to identify a particular subset or element of the set.

SUPERVISORY PROGRAM. (1) (ANSI/ISO) A computer program, usually part of an operating system, that controls the execution of other computer programs and regulates the flow of work in a data processing system. (2) (ANSI) *Syn*: executive program, supervisor.

SYMBOL. (1) (ANSI/ISO) A conventional representation of a concept or a representation of a concept upon which agreement has been reached. (2) (ANSI) A representation of something by reason of relationship, association, or convention.

SYMBOLIC LANGUAGE. (ANSI) A programming language that expresses addresses and operation codes of instructions in symbols convenient to humans rather than to machine language.

SYNCHRONOUS TRANSMISSION. (ANSI/ISO) Data transmission in which the time of occurrence of each signal representing a bit is related to a fixed time base. (2) (ANSI) Contrast with asynchronous transmission.

SYNTAX. (1) (ANSI) The relationships among characters or groups of characters, independent of their meanings or the manner of their interpretation and use. (2) (ANSI) The structure of expressions in a language. (3) (ANSI) The rules governing the structure of a language. (4) (ANSI) The relationship among symbols.

SYSGEN. (See SYSTEM GENERATION.)

SYSTEM. (ANSI/ISO) People, machines, and methods organized to accomplish a set of specific functions.

SYSTEM GENERATION (SYSGEN). (ANSI/ISO) The process of selecting optional parts of an operating system and of creating a particular operating system tailored to the requirements of a data processing installation.

TABLE. (1) (ANSI/ISO) An array of data items, each of which may be unambiguously identified by means of one or more arguments. (2) (ANSI) A collection of data elements, each of which may be uniquely identified by a label, by its position relative to the other items, or by some other means. *Syn*: dictionary.

TABLE LOOKUP. (ANSI/ISO) A procedure for obtaining the value corresponding to an argument from a table of values.

TABLET. (ANSI/ISO) A special flat surface with a mechanism for indicating positions thereon, normally used as a locator.

TASK. (ANSI/ISO) In a multiprogramming or multiprocessing environment, one or more sequences of instructions treated by a control program as a unit of work to be accomplished by a computer.

TELECOMMUNICATION. (ANSI/ISO) The transmission of signals over long distances, such as by telegraph, radio, or television.

TEMPLATE. A model based on spreadsheet software in which all of the formulas and output formats are in place.

TERMINAL. (ANSI) A point in a system or communications network at which data can either enter or leave.

TEXT. (1) (ANSI) In ASCII and data communication, a sequence of characters treated as an entity, if preceded by one start-of-text character and terminated by one end-of-text character, respectively. (2) (ANSI) In text processing, the information that consists of symbols, words, phrases, sentences, paragraphs, and tables that are to be printed or displayed.

TEXT PROCESSOR. (ANSI) A device with associated software or a computer program that allows a user to do text processing. Syn: word processor.

THROUGHPUT. (ANSI/ISO) A measure of the amount of work performed by a computer system over a period of time, *e.g.*, number of jobs per day.

THUMB WHEEL. (ANSI/ISO) In computer graphics, a wheel, movable about its axis, that provides a scalar value. A pair of thumb wheels can be used as a locator.

TIME OUT. (ANSI) An interval of time after which an enforced event occurs. The time out can be prevented by an appropriate signal.

TIME SHARING. (1) (ANSI/ISO) An operating technique of a computer system that provides for the interleaving in time of two or more processes in one processor. (2) (ANSI) The concurrent use of a device by a number of users.

TRACE. (ANSI) A record of the execution of a computer program; it exhibits the sequences in which the instructions were executed.

TRANSLATOR. (ANSI/ISO) A computer program that translates from one language into another language and in particular from one programming language into another programming language. *Syn*: translating program.

TRANSMISSION. (1) (ANSI) The sending of data from one place for reception elsewhere.

TRIPLE PRECISION. (ANSI/ISO) Characterized by the use of three computer words to represent a number in accordance with required precision.

TRUNCATION. (ANSI) The deletion or omission of a leading or of a trailing portion of a string in accordance with specified criteria.

TRUTH TABLE. (1) (ANSI/ISO) An operation table for a logic operation. (2) (ANSI) A table that describes a logic function by listing all possible combinations of input values and indicating, for each combination, the true output values.

TURING MACHINE. (ANSI) A mathematical model of a device that changes its internal state and reads from, writes on, and moves a potentially infinite tape, all in accordance with its present state, thereby constituting a model for computer-like behavior.

TURNAROUND TIME. (ANSI/ISO) The elapsed time between submission of a job and the return of the complete output.

UTILITY PROGRAM. (ANSI/ISO) A computer program in general support of the processes of a computer, *e.g.*, a diagnostic program, a trace program, a sort program. *Syn*: service program.

VALIDATION. (ANSI) Tests to determine whether an implemented system fulfills its requirements.

VARIABLE. (1) (ANSI) A quantity that can assume any of a given set of values. (2) (ANSI) In a conceptual schema language, a term that refers to unspecified, indeterminate entities in the universe of discourse.

VARIABLE-LENGTH RECORD. Pertaining to a file in which the records need not be uniform in length.

VECTOR. (1) (ANSI/ISO) A quantity usually characterized by an ordered set of scalars. (2) (ANSI) Contrast with scalar.

VERIFY. (1) (ANSI) To determine whether a transcription of data or other operation has been accomplished accurately. (2) (ANSI) To check the results of data entry.

VIRTUAL STORAGE. (ANSI/ISO) The storage space that may be regarded as addressable main storage by the user of a computer system in which virtual addresses are mapped into real addresses. The size of virtual storage is limited by the addressing scheme of the computer system and by the amount of auxiliary storage available, and not by the actual number of main storage locations.

WINDOW. (1) (ANSI/ISO) In computer graphics, a predefined part of a virtual space. (2) (ANSI) A portion of a display surface in which display images pertaining to a particular application can be presented. Different applications can be displayed simultaneously in different windows.

Word. (ANSI/ISO) A character string or a bit string considered to be an entity for some purpose.

Word Processor. (See TEXT PROCESSOR.)

Wraparound. (1) (ANSI/ISO) Forcing that part of an image that lies outside an edge of a display space to be displayed at the opposite edge of that space. (2) (ANSI) Syn: word wrap.

Write. (ANSI/ISO) To make a permanent or transient recording of data in a storage device or on a data medium.

Zerofill. (ANSI/ISO) To fill unused storage locations with the representation of the character denoting zero.

Zooming. (ANSI/ISO) The progressive scaling of an entire display image to give the visual impression of movement of all or part of a display group toward or away from an observer.

SOURCE DESIGNATIONS

(ANSI) - Indicates definitions adopted by American National Standard for Information Systems (X3.172-1990) published by the American National Standards Institute in 1990. The secretariat was the Computer and Business Equipment Manufacturers Association.

(ANSI/ISO) - Same as above but has also been adopted by the International Standardization Organization.

Z94.4
Cost Engineering & Project Management

This section contains terms selected from STANDARD COST ENGINEERING TERMINOLOGY, AACE Standard No. 105-90 (draft), 1990, a document produced by the American Association of Cost Engineers (AACE). By definition, Cost Engineering incorporates the professional disciplines of cost control, cost estimation, business planning, management science, profitability analysis, project management, value engineering; and planning and scheduling

To avoid excessive duplication, many definitions that the reader may expect to find in this chapter will be found in other sections of this terminology, — especially Section 7, Engineering Economy and Section 1, Analytical Techniques and Operational Research. The reader should also consult the overall index.

Editing for this section, including selection of the terms from STANDARD COST ENGINEERING TERMINOLOGY, was done by the chair of the Z94 Committee, Dr. Williams. Comments concerning omissions and errors should be directed to him. The Z94 Committee on Industrial Engineering Terminology gratefully acknowledges the contributions of the American Association of Cost Engineers for generously allowing their terminology to be used. Mention must also be made of individual members of the American Association of Cost Engineers who contributed to this effort:

American Association of Cost Engineers
209 Prarie Ave., Suite 100
Morgantown, WV 26505
Telephone: 800-858-2678
Fax: 304-291-5728
e-mail: info@aacei.org
URL: http://www.aacei.org

Chairman

Jean-Paul Prentice, CCE
 Resource Optimization Inc.
e-mail: jproi8000@aol.com

Subcommittee

William R. Barry, CCC
ASEC Corporation

Robert C. Creese, Ph.D.
Department of Industrial Engineering
West Virginia University

Fred R. Douglas
Texaco, Inc. (retired)

Kenneth Humphreys, Ph.D.
Past Executive Director
American Association of Cost Engineers

Barry G. McMillan
Staff Director - Education and Certification
American Association of Cost Engineers

Robert Templeton
M.W. Kellog Company (retired)

Acceleration. Conduct by the owner or the engineer (either in a directed or constructive manner) in which a contractor is required to complete performance of the contracted scope of work earlier than scheduled. A Directed Acceleration occurs when the owner formally direct such accelerated completion. A Constructive Acceleration generally occurs when a contractor is entitled to an excusable delay; the contractor requests a time extension from the owner; the owner declines to grant a time extension or grants one in an untimely manner; the owner or the engineer either expressly orders completion within the original performance period or implied in a clear manner that timely completion within the original performance period if expected; and the contractor gives notice to the owner or the engineer that the contractor considers this action an acceleration order.

Acceptance, Final (Partial). The formal action by the owner accepting the work (or a specified part thereof), following written notice from the engineer that the work (or specified part thereof) has been completed and is acceptable subject to the provisions of the contract regarding acceptance.

Access To The Work. The right of the contractor to ingress and egress, and to occupy the work site as required to reasonably perform the work described in the contract documents. An example of denial of access to the work would be on the segment of a sewer installation project where no easements or work limits are indicated, but the contractor is ordered, after contract award, to conduct operations within a narrow work corridor necessitating different or unanticipated construction methods (eg, use of sheeting).

Accountability. Answerable, but not necessarily charged personally with doing the work. Accountability cannot be delegated but it can be shared.

Account Code Structure. The system used to assign summary numbers to elements of the work breakdown and account numbers to individual work packages.

Account Number. A numeric identification of a work package. An account number may be assigned to one or more activities. Syn. Shop Order Number.

Accounts Payable. The value of goods and services rendered on which payment has not yet been made. (See TAXES PAYABLE.)

Accounts Receivable. The value of goods shipped or services rendered to a customer on which payment has not yet been received. Usually includes an allowance for bad debts.

Activity. A basic element of work, or task that must be performed in order to complete a project. An activity occurs over a given period of time. (See WORK ITEM.)

Activity Code. Any combination of letters, numbers, or blanks which describes and identifies any activity or task shown on the schedule. Syn.: Activity Identifier.

Activity Description. A concise explanation of the nature and scope of the work to be performed, which easily identifies an activity to any recipient of the schedule.

Activity Duration. The length of time from start to finish of an activity, estimated or actual, in working or calendar time units.

Activity Identifier. (See ACTIVITY CODE.)

Activity Splitting. Dividing (ie, splitting) an activity of stated scope, description and schedule into two or more activities which are rescoped and rescheduled. The sum of the split activities is normally the total of the original.

Activity Times. Time information generated through the CPM calculation that identifies the start and finish times for each activity in the network.

Activity Total Slack. The latest allowable end time minus earliest allowable end time. The activity slack is always greater than or equal to the slack of the activity ending event.

Acts of God. (1) an extraordinary interruptions by a natural cause, as a flood or earthquake, or the usual course of events that experience, foresight or care cannot reasonable foresee or prevent; (2) an event in nature over which neither the owner nor the contractor has any control.

Actual Completion Date. The calendar date on which an activity was completed. (See ACTUAL FINISH DATE.)

Actual Cost. The actual expenditures incurred by a program or project.

Actual Cost of Work Performed. The direct costs actually incurred and the direct costs actually recorded and assigned in accomplishing the work performed. These costs should reconcile with the contractors incurred cost ledgers when they are audited by the client.

Actual Finish Date. The calendar date on which the activity was actually completed. It must be prior to or equal to the data date. The remaining duration of this activity is zero.

ACTUAL START DATE. The calendar date on which work actually began on an activity.

ADDENDA. Written or graphic instruments issued prior to the date for opening of bids which may interpret or modify the bidding documents by additions, deletions, clarification, or corrections.

ADM. (See ARROW DIAGRAMING METHOD.)

ADMINISTRATIVE EXPENSE. The overhead cost due to the nonprofit-specific operations of a company. Generally includes top management salaries and the costs of legal, central purchasing, traffic, accounting, and other staff functions and their expenses for travel and accommodations.

AGENT. A person authorized to represent another (the principal) in some capacity. The agent can only act within this capacity or "scope of authority" to bind the principal. Agency agreements can be oral or in writing.

AGREEMENT. The written agreement between the owner and the contractor covering the work to be performed: other contract documents are attached to the agreement and made a part thereof as provided therein.

ALLOWANCES. Additional resources included in estimates to cover the cost of known but undefined requirements for an individual activity, work item, account or subaccount.

AMBIGUITY. An uncertainty in the meaning of provisions of a contract, document or specification. Mere disagreement about the meaning of a provision does not indicate an ambiguity. There must be genuine uncertainty of meaning based on logical interpretation of the language used in the contract. Generally, ambiguities in contract are construed against the drafter of the agreement.

AMENDMENT. A modification of the contract by a subsequent agreement. This does not change the entire existing contract but does alter the terms of the affected provisions or requirements.

AMORTIZATION. (1) As applied to a capitalized asset, the distribution of the initial cost by periodic charges to operations as in depreciation. Most properly applies to assets with indefinite life; (2) The reduction of a debt by either periodic or irregular payments; (3) A plan to pay off a financial obligation according to some prearranged schedule .

ANALYSIS. The examination of a complex whole and the separation and identification of its constituent parts and their relationships.

ANNUALLY RECURRING COST. Those costs that are incurred in a regular pattern each year throughout the study period.

ANNUITY. (1) An amount of money payable to a beneficiary at regular intervals for a prescribed period of time out of a fund reserved for that purpose; (2) A series of equal payments occurring at equal periods of time.

ANTICIPATORY BREACH. A specific refusal by the contractor to perform within the terms of the contract documents before performance is due; or a clear indication that the contractor is unable or unwilling to perform.

APPLICATION FOR PAYMENT. The form furnished by the owner or the engineer which is to be used by the contractor in requesting progress of final payments and which shall contain an affidavit, if required, in the general or supplementary conditions. The application for payment includes all supporting documentation as required by the contract documents.

APPROVE. To accept as technically satisfactory by person or persons in authority. The approval may still require confirmation by someone else at a higher level of authority for legal or commercial considerations.

ARBITRATION. A method for the resolution of disputes by an informal tribunal in which a neutral person or persons with specialized knowledge in the field in question renders a decision on the dispute. An arbitrator may grant any award which is deemed to be just and equitable after having afforded each party full and equal opportunity for the presentation of the case. Arbitration does not strictly follow the rules of evidence and discovery procedures found in litigation. Arbitration may be conducted under the auspices of an organization (eg, the American Arbitration Association) which is available as a vehicle for conducting an arbitration.

ARROW. The graphic representation of an activity in the ADM network. One arrow represents one activity. The tail (see I-NODE) of the arrow represents the start of the activity. The head (see J-NODE) of the arrow represents the finish. The arrow is not a vector quantity and is not drawn to scale. It is uniquely defined by two event codes.

ARROW DIAGRAM. A network (logic diagram) on which the activities are represented by arrows between event nodes.

Arrow Diagramming Method. A method of constructing a logical network of activities using arrows to represent the activities and connecting them head to tail. This diagramming method shows the sequence, predecessor and successor relationships of the activities.

As-Built Schedule. The final project timetable, which depicts for each activity actual start and completion date, actual duration, costs, and consumed resources.

Assessed Value. That value entered on the official assessors records as the value of the property applicable in determining the amount of taxes to be assessed against that property.

Authorized Work. Activity that has been approved to proceed by the client. The cope may or may not be well defined at the time authorized; it is usually defined by contract.

Backcharge. A cost caused by defective or deficient work by the contractor deducted from or used to offset the amount due to the contractor.

Backup. Supporting documents for an estimate or schedule including detailed calculations, descriptions of data sources, and comments on the quality of the data.

Backward Pass. Calculation of the latest finish time and latest start time for all uncompleted network activities or late time for events in the ADM and PDM methods. It is determined by working from the final activity and subtracting durations from uncompleted activities.

Base Date. (See BASE TIME.)

Base Period (Of a Given Price Index). Period for which prices serve as a reference for current period prices; in other words, the period for which an index is defined as 100 (if expressed in percentage form) or as 1 (if expressed in ratio form).

Base Point For Escalation. Cost index value for a specific month or an average of several months that is used as a basis for calculating escalation.

Base Time. The date to which all future and past benefits and costs are converted when a present value method is used (usually the beginning of the study period). *Syn.*: Base Date.

Battery Limit. Comprises one or more geographic boundaries, imaginary or real, enclosing a plant or unit being engineered and/or erected, established for the purpose of providing a means of specifically identifying certain portions of the plant, related groups of equipment, or associated facilities. It generally refers to the processing area and includes all the process equipment, and excludes such other facilities as storage, utilities, administration buildings, or auxiliary facilities. The scope included within a battery limit must be well-defined 80 that all personnel will clearly understand it. (See OFF-SITES)

Beginning Event. An event that signifies the beginning of an activity. Syn. Predecessor Event; Preceding Event; Starting Event.

Beginning Network Event. The event that signifies the beginning of a network (or subnet).

Beginning (Start) Node of Network. A node at which no activities end, but one or more activities begin.

Benchmark Indexes. For most manufacturing and all mining industries, indexes reflecting changes in output between census years.

Beneficial Occupancy. Use of a building, structure, or facility by the owner for its intended purpose (functionally complete), although other contract work, nonessential to the function of the occupied section, remains to be completed. (See SUBSTANTIAL COMPLETION.)

Bid. To submit a price for services; a proposition either verbal or written, for doing work and for supplying materials and/or equipment.

Bid Bond. A bond that guarantees the bidder will enter into a contract on the basis of his/her bond.

Bidder. The individual, partnership, or corporation, or combination thereof, acting directly or through an authorized representative, formally submitting a bid directly to the owner, as distinct from a sub- bidder, who submits a bid to a bidder.

Bid Security. Security is provided in connection with the submittal of a bid to guarantee that the bidder, if awarded or offered the contract, will execute the contract and perform the work. The requirements for the bid security are usually designated in a specific section of the bidding documents. The bid security is payable to the owner (usually around 5% of the total bid price) in the form of either a certified or bank check or a bid bond issued by a surety satisfactory to the owner. The bid security of the successful bidder is usually retained until the bidder has executed the agreement and furnished the required contract security, whereupon the bid security is returned. Bid security of the other bidders is returned after the bid opening.

BID SHOPPING. An effort by a prime contractor to reduce the prices quoted by subcontractors and/or suppliers, by providing the bid price to other subcontractors or suppliers in an attempt to get the other subcontractors or suppliers to underbid the original price quoted. The reverse of this situation is when subcontractors try to get a better price out of a prime contractor. This is known as Bid Peddling.

BIDDING DOCUMENTS. The advertisement for bids, instructions to bidders, information available to bidder, bid form with all attachments, and proposed contract documents (including all addenda issued prior to receipt of bids).

BIDDING REQUIREMENTS. The advertisement for bids, instructions to bidders, supplementary instructions and all attachments therein, information to bidders and all attach-ments therein, and bid form and all attachments therein.

BLACK BOX. Describes a system (organism or mechanism) whose structure is unknown either because it cannot be observed or it is proprietary, classified or too complex to be understood.

BLANKET BOND. A bond covering a group of persons, articles, or properties.

BLS. Bureau of Labor Statistics.

BLS PERIODICALS
- *CPI Detailed Report*, issued monthly
- *Current Wage Developments*, issued monthly
- *Employment and Earnings*, issued monthly
- *Monthly Labor Review,* issued monthly
- *Occupational Outlook Quarterly,* issued quarterly
- *Producers Prices and Price Indexes*, issued monthly (previously *Wholesale Price Index*)

BONDS. Instruments of security furnished by the contractor and/or surety in accordance with the contract documents. The terms contract security refers to the payment bond, performance bond and those other instruments of security required in the contract documents.

BOND, BID. A bond that is executed in connection with the submittal of a bid and which guarantees that the bidder, if awarded or offered the contract, will execute the contract and perform the work. The bidding documents sometimes include a specific form for submittal of the bid bond and may be used to satisfy the requirement for bid security as defined in the bidding documents.

BOND, PAYMENT. A bond that is executed in connection with a contract and which secures the payment of all persons applying labor and material in the prosecution of the work provided for in the contract.

BOND PERFORMANCE. A bond that is executed in connection with a contract and which secures the performance and fulfillment of all undertakings, covenants, terms, conditions, and agreements contained in the contract.

BONUS-PENALTY. A contractual arrangement between a client and a contractor wherein the contractor is provided a bonus, usually a fixed sum of money, for each day the project is completed ahead of a specified schedule and/or below a specified cost, and agrees to pay a similar penalty for each day of completion after the schedule date or over a specified cost up to a specified maximum either way. The penalty situation is sometimes referred to as liquidated damages.

BOOK VALUE (NET). (l) Current investment value on the books calculated as original value less depreciated accruals; (2) New asset value for accounting use; (3) The value of an outstanding share of stock of a corporation at any one time, determined by the number of shares of that class outstanding.

BREACH OF CONTRACT. Failure, by either the owner or the contractor, without legal excuse, to perform any work or duty owed to the other person .

BREAK-EVEN CHART. A graphic representation of the relation between total income and total costs for various levels of production and sales indicating areas of profit and loss.

BREAK-EVEN POINT. (1) In business operations, the rate of operations output, or sales at which income is sufficient to equal operating costs or operating cost plus additional obligations that may be specified: (2) The operating condition, such as output, at which two alternatives are equal in economy; (3) The percentage of capacity operation of a manufacturing plant at which income will just cover expenses.

BREAKOUT SCHEDULE. This jobsite schedule, generally in bar chart form is used to communicate the day-to-day activities to all working levels on the project as directed by the construction manager. Detail information with regard to equipment use, bulk material requirements, and craft skills distribution, as well as the work to be accomplished, forms the content of this schedule. The schedule is issued on a weekly basis with a two to three-week look ahead from the issue date. This schedule generally contains from 25 to 100 activities.

Budget. A planned allocation of resources. The planned cost of needed materials is usually subdivided into quantity required and unit cost. The planned cost of labor is usually subdivided into the workhours required and the wage rate (plus fringe benefits and taxes).

Budget Cost of Work Performed (BCWP). The sum of the budgets for completed portion of in-process work, plus the appropriate portion of the budget for level of effort and apportioned effort for the relevant time period BCWP is commonly referred to as "earned value".

Budget Cost of Work Scheduled (BCWS). The sum of the budgets for work scheduled to be accomplished (including work-in- process), plus the appropriate portion of the budgets for level of effort and apportioned effort for the relevant time period.

Bulk Material. Material bought in lots. These items can be purchased from a standard catalog description and are bought in quantity for distribution as required. Examples are pipe (nonspooled), conduit, fittings, and wire.

Burden. In construction, the cost of maintaining an office with staff other than operating personnel. Include also federal, state and local taxes, fringe benefits and other union contract obligations. In manufacturing, burden sometimes denotes overhead.

Burden of Proof. The necessity of proving the facts in a dispute on an issue raised between the owner and the contractor. In a claim situation, the burden of proof is always on the person filing the claim. This is true whether the contractor is claiming against the owner, or the owner is making a claim against the contractor.

Business Planning. The determination of financial, production and sales goals of a business organization; and the identification of resources, methods, and procedures required to achieve the established objectives within specified budgets and timetables.

Calendar. Time schedule of project activities. l The calendar identifies working days, holidays, and the length of the working day in time units and/or shifts.

Calendar Range. The span of the calendar from the calendar start date through the calendar end date. The calendar start date is unit number one. The calendar range is usually expressed in years.

Calendar Unit. The smallest time unit of the calendar that is in use to estimate activity duration. This unit is generally in hours, shifts, days, or weeks. Syn. Time Unit.

Calendar Start Date. The date assigned to the first unit of the defined calendar; the first day of the schedule.

Capacity Factor. (l) the ratio of average load to maximum capacity; (2) the ratio between average load and the rated capacity of the apparatus; (3) the ratio of the average actual use to the rated available capacity. Also called Capacity Utilization Factor.

Capital Budgeting. A systematic procedure for classifying, evaluating, and ranking proposed capital expenditure for the purpose of comparison and selection, combined with the analysis of the financing requirements.

Capital, Direct. (See DIRECT COST (1).)

Capital, Fixed. The total original value of physical facilities which are not carried as a current expense on the books of account and for which depreciation is allowed by the Federal Government. It includes plant equipment, building, furniture and fixtures, transportation equipment used directly in the production of a product or service. It includes all costs incident to getting the property in place and in operating condition, including legal costs, purchased patents, and paid-up licenses. Land, which is not depreciable, is often included. Characteristically it cannot be converted readily into cash.

Capital, Indirect. (See INDIRECT COSTS (1).)

Capital, Operating. Capital associated with process facilities inside battery limits.

Capital, Sustaining. The fixed capital requirements to (l) maintain the competitive position of a project throughout its commercial life by, improving product quality, related services, safety, or economy, or (2) required to replace facilities which wear out before the end of the project life.

Capital, Total. Sum of fixed and working capital.

Capital, Venture. Capital invested in technology or markets new at least to the particular organization.

Capital, Working. The funds in addition to fixed capital and land investment which a company must contribute to the project (excluding startup expense) to get the project started and meet subsequent obligations as they come due. Working capital includes inventories, cash and accounts receivable minus accounts payable. Characteristically, these funds can be converted readily into cash. Working capital is normally assumed recovered at the end of the project.

Cash Costs. Total cost excluding capital and depreciation spent on a regular basis over a period of time, usually one year. Cash costs consist of manufacturing cost and other expenses such as transportation cost, selling expense, research and development cost or corporate administrative expense.

Cash Flow. The net flow of dollars into or out of a project. The algebraic sum, in any time period, of all cash receipts, expenses, and investments. Also called cash proceeds or cash generated. The stream of monetary (dollar) values — costs and benefits — resulting from a project investment.

Cash Return, Percent of Total Capital. Ratio of average depreciation plus average profit, to total fixed and working capital, for a year of capacity sales. Under certain limited conditions, this figure closely approximates that calculated by profitability index techniques where it is defined as the difference, in any time period, between revenues and all cash expenses, including taxes.

Causation. An explanation or description of the facts and circumstances that produce a result, the cause and effect for which the contractor claims entitlement to compensation from the owner under the contract.

Chain Index. An index which globally measures the price change of a range of commodities.

Change. Alteration or variation to a scope of work and/or the schedule for completing the work.

Change, Cardinal. Work that is beyond the scope of that specified in the contract and consequently unauthorized. The basic tests for a cardinal change are whether the type of work was within the contemplation of the parties when they entered into the contract and whether the job as modified is still the same basic job.

Change, Constructive. An act or failure to act by the owner or the engineer that is not a directed change, but which has the effect of requiring the contractor to accomplish work different from that required by the existing contract documents.

Change in Scope. A change in objectives (either in quality or quantity of the specifications and/or material), work plan, or schedule that results in a material difference from the terms of an approval to proceed previously granted by higher authority. Under certain conditions (normally so stated in the approval instrument), a change in resource applications may constitute a change in scope.

Change Order. A document requesting a scope change or correction. It must be approved by both the client and the contractor before it becomes a legal change to the contract.

Change, Unilateral. (See MODIFICATION, UNILATERAL.)

Change in Sequence. A change in the order of work initially specified or planned by the contractor. If this change is ordered by the owner and results in additional cost to the contractor, the contractor may be entitled to recovery under the changes clause.

Changed Conditions. (See DIFFERING SITE CONDITIONS.)

Chart of Accounts. (See CODE OF ACCOUNTS.)

Chemical Engineering Plant Cost Index. An index tailor-made specifically for chemical plant construction, composed of many subindexes for the various components of a chemical plant.

Claim. A written statement requesting additional time and/or money for acts or omissions during the performance of the construction contract. The contract must set forth the facts and circumstances for which the owner or the engineer is responsible to be entitled to additional compensation and/or time.

Code of Accounts. A systematic numeric method of identifying various categories of costs incurred in the progress of a job; the segregation of engineering, procurement, fabrication, construction, and associated project costs into elements for accounting purposes. Syn. Chart of Accounts.

Commitments. The sum of all financial obligations made, including incurred costs and expenditures as well as obligations, which will not be performed until later.

Commodity. In price index nomenclature, a good and sometimes a service.

Completed Activity. An activity with an actual finish date.

Composite Price Index. An index which globally measures the price change of a range of commodities.

Conceptual Schedule. A conceptual schedule is similar to a proposal schedule except it is usually time-scaled and is developed from the abstract design of the project. This schedule is used primarily to give the client a general idea of the project scope and an overview of activities. Most conceptual schedules contain between 30 and 200 activities.

CONFLICT IN PLANS AND SPECIFICATIONS. Statements or meanings in the contract documents (including drawings and specifications) that cannot be reconciled by reasonable interpretation on the part of the contractor and which may require the owner to provide an interpretation between alternatives.

CONSENT OF SURETY. An acknowledgment by a surety that its bond, given in connection with a contract, continues to apply to the contract as modified; or, at the end of a contract, permission from the surety to release all retainage to the contractor.

CONSTANT BASKET. A set of goods and services with quantities fixed in relation to a given time period, used for computing composite price indexes.

CONSTANT BASKET PRICE INDEX. Price index which measures price change by comparing the expenditures necessary to provide the game set of goods and services at different point in time.

CONSTANT DOLLARS. Dollars of uniform purchasing power exclusive of general inflation or deflation. Constant dollars are tied to a reference year.

CONSTANT UTILITY PRICE INDEX. A composite price index which measures price change by comparing the expenditures necessary to provide substantially equivalent sets of goods and services at different points in time.

CONSTRAINT. (See RESTRAINT.)

CONSTRAINT DATE. (See PLUG DATE.)

CONSTRUCTION COST. The sum of all costs, direct and indirect, inherent in converting a design plan for material and equipment into a project ready for start-up, but not necessarily in production operation; the sum of field labor, supervision, administration, tools, field office expense, materials, and equipment.

CONSTRUCTION MANAGEMENT. Project management as applied to construction.

CONSUMABLES. Supplies and materials used up during construction. Includes utilities, fuels and lubricants, welding supplies, worker's supplies, medical supplies, etc.

CONSUMER PRICE INDEX. A measure of time-to-time fluctuations in the price of a quantitatively constant market basket of goods and services, selected as representative of a special level of living.

CONTINGENCY. An amount added to an estimate to allow for changes that experience shows will likely be required. May be derived either through statistical analysis of past project costs or by applying experience from similar projects. Usually excludes changes in scope or unforeseeable major events such as strikes, earthquakes, etc.

CONTRACTOR. A business entity that enters into contract to provide goods or services to another party.

CONTRACT COMPLETION DATE. The date established in the contract for completion of all or specified portions of the work. This date may be expressed as a calendar date or as a number of days after the date for commencement of the contract time is issued.

CONTRACT DATE. Any date specified in the contract or imposed on any project activity or event that impacts the activity/project schedule. *Syn.* Scheduled Date.

CONTRACT DOCUMENTS. The agreement, addenda (which pertain to the contract documents), contractor's bid (including documentation accompanying the bid and any post-bid documentation submitted prior to the notice of award) when attached as an exhibit to the agreement, the bonds, the general conditions, the supplementary conditions, the specifications and the drawings as the game are more specifically identified in the agreement, together with all amendments, modifications and supplements issued pursuant to the general conditions on or after the effective date of the agreement.

CONTRACT PRICE. The monies payable by the owner to the contractor under the contract documents as stated in the agreement.

CONTRACT, "READ AS WHOLE". Reading an entire contract document, instead of reading each clause in the contract in isolation. If a clause is ambiguous and can be interpreted in more than one way, the meaning that conforms to the rest of the document is usually the accepted meaning.

CONTRACT TIME. The number of days within which, or the dates by which, the work, or any specified part thereof, is to be completed.

CONTRACT WORK BREAKDOWN STRUCTURE (CWBS). (See WORK BREAKDOWN STRUCTURE.)

CONTRACTS. Legal agreements between two or more parties, which may be of the types enumerated below:

1. In **COST PLUS** contracts the contractor agrees to furnish to the client services and materials at actual cost, plus an agreed upon fee for these services. This type of contract is employed most often when the scope of services to be provided is not well defined.

    a. *Cost Plus Percentage Burden and Fee* — the client will pay all costs as defined in the term of the contract, plus "burden and fee' at a specified percent of the labor costs which the client is paying for directly. This type of contract generally is used for engineering services. In contracts with some governmental agencies, burden items are included in indirect cost.

    b. *Cost Plus Fixed Fee* — the client pays costs as defined inthe contract document. Burden on reimbursable technicallabor cost is considered in this case as part of cost. Inaddition to the costs and burden, the client also pays afixed amount as the contractor's "fee".

    c. *Cost Plus Fixed Sum* — the client will pay costs defined by contract plus a fixed sum which will cover "non-reimbursable" costs and provide for a fee. This type of contract is used in lieu of a cost plus fixed fee contract where the client wishes to have the contractor assume some of the risk for items, which would be reimbursable under a Cost Plus Fixed Fee type of contract.

    d. *Cost Plus Percentage Fee* — the client pays all costs, plus a percentage for the use of the contractor's organization.

2. **FIXED PRICE** types of contract are ones wherein a contractor agrees to furnish services and material at a specified price, possibly with a mutually agreed upon escalation clause. This type of contract is most often employed when the scope of services to be provided is well defined.

    a. *Lump Sum* — contractor agrees to perform all services asspecified by the contract for a fixed amount. A variationof this type may include a turn-key arrangement where the contractor guarantees quality, quantity and yield on a process plant or other installation.

    b. *Unit Price* — contractor will be paid at an agreed upon unit rate for services performed. For example, technical work-hours will be paid for at the unit price agreed upon. Often field work is assigned to a subcontractor by the prime contractor on a unit price basis.

    c. *Guaranteed Maximum (Target Price)* — a contractor agrees to perform all services as defined in the contract document guaranteeing that the total cost to the client will not exceed a stipulated maximum figure. Quite often, thesetypes of contracts will contain special share-of-the-savingarrangements to provide incentive to the contractor tominimize costs below the stipulated maximum.

    d. *Bonus-Penalty* — a special contractual arrangement usually between a client and a contractor wherein the contractor is guaranteed a bonus, usually a fixed sum of money, for each day the project is completed ahead of a specified schedule and/or below a specified cost, and agrees to pay a similar penalty for each day of completion after the schedule date or over a specified cost up to a specified maximum either way. The penalty situation is sometimes referred to as liquidated damages.

CONTROL. Management action, either preplanned to achieve the desired result or taken as a corrective measure prompted by the monitoring process.

CORRECTION PERIOD. The period of time within which the contractor shall promptly, without cost to the owner and in accordance with the owners written instructions, either correct defective work, or if it has been rejected by the owner, remove it from the site and replace it with nondefective work, pursuant to the general conditions.

COST. The amount measured in money, cash expended, or liability incurred, in consideration of goods and/or services received.

COST ACCOUNTING. The historical reporting of disbursements and costs and expenditures on a project. When used in conjunction with a current working estimate, cost accounting can assist in giving the precise status of the project to date.

COST ANALYSIS. A historical an/or predictive method of ascertaining for what purpose expenditures on a project were made and utilizing this information to project the cost of a project as well as costs of future projects. The analysis may also include application of escalation, cost differentials between various localities, types of buildings, types of projects, and time of year.

Cost Approach. One of the three approaches in the appraisal process. Underlying the theory of the cost approach is the principle of substitution, which suggests that no rational person will pay more for a property than the amount with which he/she can obtain, by purchase of a site and construction of a building without undue delay, a property of equal desirability and utility.

Cost and Schedule Control Systems Criteria (C/SCSC). Established characteristics that a contractor's internal management control system must possess to assure effective planning and control of contract work, costs, and schedules.

Cost Category. The name and number, or both, of a function, hardware, or other significant cost category for which costs are to be summarized.

Cost Control. The application of procedures to monitor expenditures and performance against progress of projects or manufacturing operations; to measure variance from authorized budgets and allow effective action to be taken to achieve minimum costs.

Cost Engineer. An engineer whose judgment and experience are utilize in the application of scientific principles and techniques to problems estimation; cost control; business planning and management science; profitability analysis; and project management, planning and scheduling.

Cost Estimation. The determination of quantity and the predicting or forecasting, within a defined scope, of the costs required to construct and equip a facility, to manufacture goods, or to furnish a service. Costs are determined utilizing experience and calculating and forecasting the future cost of resources, methods, and management within a scheduled time frame. Included in these costs are assessments and an evaluation of risks and uncertainties. Cost estimation provides a basis for feasibility studies, business planning, budget preparation, and cost and schedule control.

Cost Index (Price Index). A number which relates the cost of an item u at a specific time to the corresponding cost at some arbitrarily specified time in the past. (See PRICE INDEX.)

Cost of Capital. A term, usually used in capital budgeting, to express as an interest rate percentage the overall estimated cost of investment capital at a given point in time, including both equity and borrowed funds.

Cost of Lost Business Advantage. The cost associated with loss of repeat business and/or the loss of business due to required resources and costs.

Cost of Ownership. The cost of operations, maintenance, follow-on logistical support, and end item and associated support systems. Syn. Operating and Support Costs.

Cost of Quality. Consists of the sum of those costs associated with: (a) cost of quality conformance, (b) cost of quality nonconformance, (c) cost of lost business advantage.

Cost of Quality Conformance. The cost associated with the quality management activities of appraisal, training, and prevention.

Cost of Quality Nonconformance. The cost associated with deviations involving rework and/or the provision of deliverables that are more than required.

Cost Value. (See FUNCTIONAL WORTH.)

Cost of Living Index. In modern usage, a price index based on a constant utility concept as opposed to a constant basket concept.

Criteria. A document that provides objectives, guidelines, procedures, and standards to be used to execute the develop-ment, design, and/or construction portions of a project.

Critical Activity. Any activity on a critical path.

Critical Path. Sequence of jobs or activities in a network analysis project such that the total duration equals the sum of the duration of the individual jobs in the sequence. There is no time leeway or slack (float) in activity along critical path (ie, if the time to complete on or more jobs in the critical path increase, the total production time increases). It is the longest time path through the network.

Critical Path Method. A scheduling technique using arrow, precedence, or PERT diagrams to determine the length of a project and to identify the activities and constraints on the critical path.

Criticality. A measure of the significance or impact of failure of a product, process, or service to meet established requirements.

Crude Materials. Includes products entering the market for the first time which have not been fabricated or manufactured but will be processed before becoming finished goods (eg, steel scrap, wheat, raw cotton). Syn: Raw Materials

Current Cost Accounting. A methodology prescribed by the Financial Accounting Board to compute and report financial activities in constant dollars.

Current Dollars. Dollars of purchasing power in which actual priceshare stated, including inflation or deflation. In the absence of inflation or deflation, current dollars equal constant dollars.

Current Period (of a Given Price Index). Period for which prices are compared to the base period prices.

Custom in the Industry. An established practice in a particular industry in the general area. It may be used to show the practice to be followed in a particular circumstance.

Damages, Actual. The increased cost to one party resulting from another party's acts or omissions affecting the contract but not incorporated into a contract modification.

Damages, Liquidated. An amount of money stated in the contract as being the liability of a contractor for failure to complete the work by the designated time(s). Liquidated damages ordinarily stop at the point of substantial completion of the project or beneficial occupancy by the owner.

Damages, Ripple. (See IMPACT COST.)

Data Date (DD). The calendar date that indicates when the project has been updated.

Date for the Commencement of the Contract Time. The date when the contract time commences to run and on which the contractor shall start to perform the contractor's obligations under the contract documents.

Deceleration. The opposite of acceleration. A direction, either expressed or implied, to slow down job progress.

Declining Balance Depreciation. Method of computing depreciation in which the annual charge is a fixed percentage of the depreciated book value at the beginning of the year to which the depreciation applies. Syn: Percent on Diminishing Value

De-escalate. A method to convert present-day costs or costs of any point in time to costs at some previous date via applicable indexes.

Defect. A deviation of a severity sufficient to require corrective action.

Defective. An adjective which, when modifying the work, refers to work that is unsatisfactory, faulty or deficient, or does not conform to the contract documents, or does not meet the requirements of any inspection, reference standard, test or approval referred to in the contract documents, or has been damaged prior to the engineer's recommendation of final payment (unless responsibility for the protection thereof has been assumed by the owner at substantial completion in accordance with the contract documents).

Defective Specifications. Specifications and/or drawings which contain errors, omissions, and/or conflicts, which affect or prevent the contractors performance of the work.

Defect, Latent. A defect in the work which cannot be observed by reasonable inspection.

Defect, Patent. A defect in the work which can be observed by reasonable inspection.

Deflation. An absolute price decline for a commodity; also, an operation by means of which a current dollar value series is transformed into a constant dollar value series (ie, is expressed in "real terms using appropriate price indexes as deflators).

Delay, Compensable. Any delay beyond the control and without the fault or negligence of the contractor resulting from the owner-caused changes in the work, differing site conditions, suspensions of the work, or termination for convenience by the owner.

Delay, Concurrent. Two or more delays in the same time frame or which have an independent effect on the end date. The owner/engineer and the contractor may each be responsible for delay in completing the work. This may bar either party from assessing damage against the other. This may also refer to two or more delays by the same party during a single time period.

Delay, Excusable. Any delay beyond the control and without the fault or negligence of the contractor or the owner, caused by events or circumstances such as, but not limited to, acts of God or of the public enemy, acts of intervenors, acts of government other than the owner, fires, floods, epidemics, quarantine restrictions, freight embargoes, hurricanes, tornadoes, labor disputes, etc. Generally, a delay caused by an excusable delay to another contractor is compensable when the contract documents specifically void recovery of delay costs.

DELAY, INEXCUSABLE. Any delay caused by events or circumstances within the control of the contractor, such as inadequate crewing, slow submittals, etc., which might have been avoided by the exercise of care, prudence, foresight, or diligence on the part of the contractor.

DELAY, NONPREJUDICIAL. Any delay impacting a portion of the work within the available total float or slack time, and not necessarily preventing completion of the work within the contract time.

DELAY, PREJUDICIAL. Any excusable or compensable delay impacting the work and exceeding the total float available in the progress schedule, thus preventing completion of the work within the contract time unless the work is accelerated.

DELIVERABLE. A report or product of one or more tasks that satisfy one or more objectives and must be delivered to satisfy contractual requirements.

DEMAND FACTOR. (1) The ratio of the maximum instantaneous production rate to the production rate for which the equipment wag designed; (2) the ratio between the maximum power demand and the total connected load of the system.

DEMOGRAPHIC INDEX. Cost indexes developed to deal with geographic cost differences.

DEMURRAGE. A charge made on cars, vehicles, or vessels held by or for consignor or consignee for loading or unloading, for forwarding directions or for any other purpose.

DEPLETION. (1) A form of capital recovery applicable to extractive property (eg, mines). Depletion can be on a unit-of-output basis related to original or current appraisal of extent and value of the deposit. (Known as percentage depletion.) (2) lessening of the value of an asset due to a decrease in the quantity available. Depletion is similar to depreciation except that it refers to such natural resources as coal, oil, and timber in forests.

DEPRECIATED BOOK VALUE. The first cost of the capitalized asset minus the accumulation of annual depreciation cost charges.

DEPRECIATION. (1) Decline in value of a capitalized asset; (2) a form of capital recovery applicable to a property with a life span of more than one year, in which an appropriate portion of the asset's value is periodically charged to current operations.

DETAILED ENGINEERING. The detailed design, drafting, engineering, and other related services necessary to purchase equipment and materials and construct a facility.

DETAILED SCHEDULE. A schedule which displays the lowest level of detail necessary to control the project through job completion. The intent of this schedule is to finalize remaining requirements for the total project.

DEVELOPMENT COSTS. Those costs specific to a project, either capital or expense items, which occur prior to commercial sales and which are necessary determining the potential of that project for consideration and eventual promotion. Major cost areas include process, product, and market research and development.

DEVIATION. A departure from established requirements. A deviation in the work product may be classified as an imperfection, nonconformance, or defect, based on its severity in failing to meet or unnecessarily exceed the requirements.

DEVIATION COSTS. The sum of those costs, including consequential costs such as schedule impact, associated with the rejection or rework of a product, process, or service due to a departure from established requirements. Also may include the cost associated with the provision of deliverables that are more than required.

DIFFERENTIAL PRICE ESCALATION RATE. The expected percent difference between the rate of increase assumed for a given item of cost (such as energy), and the general rate of inflation.

DIFFERING SITE CONDITIONS. Subsurface or latent physical conditions at the site differing materially from those conditions indicated in the contract documents or unknown physical conditions at the site, of an unusual nature, differing materially from conditions normally encountered and generally recognized as inherent in work of the nature provided for in the contract.

DIRECT COST. (1) In construction, cost of installed equipment, material and labor directly involved in the physical construction of the permanent facility. (2) in manufacturing, service and other non-construction industries, the portion of operating costs that is generally assignable to a specific product or process area. Usually included are:
  a. Input Materials
  b. Operating, Supervision, and Clerical Payroll

c. Fringe Benefits
d. Maintenance
e. Utilities
f. Catalysts, Chemicals and Operating Supplies
g. Miscellaneous (Royalties, Services, Packaging, etc.)

Definitions of the above classifications are:

a. *Input Material* — raw materials which appear in some form as a product. For example, water added to resin formulation is an input material, but sulfuric acid catalyst, consumed in manufacturing high octane alkylate, is not.

b. *Operating, Supervision, and Clerical Payroll* — wages and salaries paid to personnel who operate the production facilities.

c. *Fringe Benefits* — payroll costs other than wages not paid directly to the employee. They include costs for: (l) Holidays, vacations, sick leave; (2) Federal old age insurance; (3) Pensions, life insurance, savings plans, etc.

In contracts with some governmental agencies these items are included in indirect cost.

d. *Maintenance Cost* – expense incurred to keep manufacturing facilities operational. It consists of: (1) Maintenance Payroll Cost; (2) Maintenance Materials and Supplies Cost.

Maintenance materials which have a life of more than one year are usually considered capital investment in detailed cash flow accounting.

e. *Utilities* — the fuel, steam, air, power and water which must be purchased or generated to support the plant operation.

f. *Catalysts, Chemicals and Operating Supplies* — materials consumed in the manufacturing operation, but not appearing as a product. Operating supplies are a minor cost in process industries and are sometimes assumed to be in the maintenance materials estimate; but in many industries, mining for example, they are a significant proportion of direct cost.

g. *Miscellaneous*. (l) Costs paid to others for the use of a proprietary process. Both paid-up and "running" royalties are used. Cost of paid-up royalties are usually on the basis of production rate. Royalties vary widely, however, and are specific for the situation under consideration. (2) Packaging Cost – material and labor necessary to place the product in a suitable container for shipment. Also called Packaging and Container Cost or Packing Cost. Sometimes considered an indirect cost together with distribution costs such as for warehousing, loading and transportation. (3) Although the direct costs described above are typical and in general use, each industry has unique costs which fall into the "direct cost" category. A few examples are equipment rental, waste disposal, contracts, etc.

DISINFLATION. A downward trend in inflation rates, effected by weak or declining demand. It may well portend deflation.

DISPERSION. The scattering of values from the mean.

DISPUTE. A disagreement between the owner and the contractor as to a question of fact or contract interpretation which cannot be resolved to the mutual satisfaction of the parties.

DISRUPTION. An action or event that hinders a party from proceeding with construction. If such disruption is caused by owner or engineer action (or failure to act), the contractor may be entitled to recover any resulting costs.

DISTRIBUTABLES. The field portion of a construction project that can be associated with any specific account. Includes the field nonmanual staff, field office, office supplies, temporary construction, utilities, small tools, construction equipment, weather protection, snow removal, lost time, labor burden, etc. When completion cost reports are prepared, the distributable costs may be distributed across the direct accounts.

DISTRIBUTION. The broad range of activities concerned with efficient movement of finished products from the end of the production line to the consumer; in some cases it may include the movement of raw materials from the source of supply to the beginning of the production line. These activities include freight transportation, warehousing, material handling, protective packaging, inventory control, plant and warehouse site selection, order processing, market and sales forecasting, customer service, attendant management information systems; and in some cases, buying activities.

DRAWINGS, PLANS. The drawings, plans or reproductions thereof, which show location, character, dimensions, and details of the work to be performed and which are referred to in the contract documents.

Dummy Activity. An activity, always of zero duration, used to show logical dependency when an activity cannot start before another is complete, but which does not lie on the same path through the network. Normally, these dummy activities are graphically represented as a dashed line headed by an arrow and inserted between two nodes to indicate a precedence relationship or to maintain a unique numbering of concurrent activities.

Dummy Start Activity. An activity entered into the network for the sole purpose of creating a single start for the network.

Durable Goods. Generally, any producer or consumer goods whose continuous serviceability is likely to exceed three years (eg, trucks, furniture).

Duration. The time required to accomplish an activity. (See ACTIVITY DURATION.)

Earliest Expected Completion Date. The earliest calendar date on which the completion of an activity work package or summary item occurs.

Early Event Time (EV). The earliest time at which an event may occur.

Early Finish Time (EF). The earliest time at which an activity can be completed; equal to the early start of the activity plus its remaining duration.

Early Start Time (ES). The earliest time any activity may begin as logically constrained by the network for a specific work schedule.

Early Work Schedule. Predicated on the parameters established by the proposal schedule and any negotiated changes, the early work schedule defines reportable pieces of work within major areas. The format is developed into a logic network including engineering drawings, bid inquiries, purchase orders, and equipment deliveries, and can be displayed as a time-phased network. The detail of this schedule concentrates on projected engineering construction issue drawings released and equipment deliveries. The activities of the early part of construction are more defined than in the proposal or milestone schedule.

Earned Value. The periodic, consistent measurement of work performed in terms of the budget planned for that work. In criteria terminology, earned value is the budgeted cost of work performed. It is compared to the budgeted cost of work scheduled (planned) to obtain scheduleperformance and it is compared to the actual cost of work performed to obtain cost performance.

Earned Value Concept. The measurement at any time of work accomplished (performed) in terms of budgets planned for that work, and the use of these data to indicate contract cost and schedule performance. The earned value of work done is quantified as the budgeted cost for work performed (BCWP) compared to the budgeted cost for work scheduled (BCWS) to show schedule performance and compared to the actual cost of work performed (ACWP) to indicate cost performance.

Earned Value Reports. Cost and schedule performance reports that are part of the performance measurement system. These reports make use of the earned value concept of measuring work accomplishment.

Earnings Value. The present worth of an income producer's probable future net earnings, as prognosticated on the basis of recent and present expense and earnings and the business outlook.

Economic Return. The profit derived from a project or business enterprise without consideration of obligations to financial contributors and claims of others based on profit.

Economic Value. The value of property in view of all its expected economic uses, as distinct from its value in view of any particular use. Also, economic value reflects the importance of a property as an economic means to an end, rather than as an end in itself.

Economy. The cost or profit situation regarding a practical enterprise or project as in economy study, engineering economy, and project economy.

Effective Date of the Agreement. The date indicated in the agreement on which it becomes effective, but if no such data is indicated, the date on which the agreement is signed and delivered by the last of the two parties to sign and deliver.

Elementary Commodity Groups (Elementary Groups). The lowest level of goods and services for which a consistent set of value weights is available.

Ending Event. The event that signifies the completion of all activities leading to that event.

Ending Node of Network (ADM). A node where no activities begin, but one or more activities end.

End Network Event. The event that signifies the end of a network.

ENDOWMENT. A fund established for the support of some project or succession of donations or financial obligations.

ENGINEER (IN CONTRACTS). The individual, partnership, corporation, joint venture, or any combination thereof, named as the engineer in the agreement who will have the rights and authority assigned to the engineer in the contract documents. The term "the engineer" means the engineer or the engineer's authorized representative.

EQUITABLE ADJUSTMENT. A change in the contract price and/or time to compensate the contractor for expense or delay incurred due to the actions or lack of action of the owner or the owner's representatives or other occurrences, or to compensate the owner for contract reductions. The objective of an equitable adjustment is to put the contractor on the same relative financial position after the change as before the change.

EQUIVALENT SETS OF COMMODITIES. Sets of commodities which provide the same total satisfaction to a given group of consumers (without necessarily being identical).

ERROR. Any item or activity in a system that is performed incorrectly, resulting in a deviation, eg, design error, fabrication error, construction error, etc. An error requires an evaluation to determine what corrective action is necessary.

ERRORS AND OMISSIONS. Deficiencies, usually in design or drafting, in the plans and specifications that must be corrected in order for the facility to operate properly. Errors in plans and specifications are normally items that are shown incorrectly, while omissions are normally items that are not shown at all.

ESCALATION. The provision in actual or estimated costs for an increase in the cost of equipment, material, labor, etc, over that specified in the purchase order or contract due to continuing price level changes over time.

ESCALATOR CLAUSE. Clause contained in collective agreements, providing for an automatic price adjustment based on changes in specified indices.

ESTEEM VALUE. (See FUNCTIONAL WORTH.)

ESTIMATE, COST. An evaluation of all the costs of the elements of a project or effort as defined by an agreed-upon scope. Three specific types based on degree of definition of a process industry plant are:

1. *Order of Magnitude Estimate* – an estimate made without detailed engineering data. Some examples would be: an estimate from cost capacity curves, an estimate using scaleup or down factors, and as approximate ratio estimate. It is normally expected that an estimate of this type would be accurate within plus 50 percent or minus 30 percent.

2. *Budget Estimate* – Budget in this case applies to the owner's budget and not to the budget as a project control document. A budget estimate is prepared with the use of flow sheets, layouts and equipment details .It is normally expected that an estimate of this type would be accurate within plus 30 percent or minus 15 percent.

3. *Definitive Estimate* – as the name implies, this is an estimate prepared from very defined engineering data. The engineering data includes as a minimum, nearly complete plot plans and elevations, piping and instrument diagrams, one line electrical diagrams, equipment data sheets and quotations, structural sketches, soil data and sketches of major foundations, building sketches, and a complete set of specifications. This category of estimate covers all types from the minimum described above to the maximum definitive type which would be made from "approved for construction" drawings and specifications. It is normally expected that an estimate of this type would be accurate within plus 15 percent and minus 5 percent.

ESTIMATE-TO-COMPLETE. The estimated workhours, costs, and time and/or v materials required to complete a work package or summary item (includes applicable overhead unless only direct costs are specified).

EVENT. An identifiable single point in time on a project. Graphically, it is represented by a node. An event occurs only when all work preceding it has been completed. It has zero duration.

EVENT NAME. An alphanumeric description of an event.

EVENT NUMBER. A numerical description of an event for computation and identification.

EVENT SLACK. The difference between the latest allowable date and the earliest date for an event.

EVENT TIMES. Time information generated through the network analysis calculation, which identifies the start and finish times for each event in the network.

EXCHANGE VALUE. (See FUNCTIONAL WORTH.)

EXEMPT. Employees exempt from federal wage and hours guidelines.

EXPANSION. Any increase in the capacity of a plant facility or unit, usually by added investment. The scope of its possible application extends from the elimination of problem areas to the complete replacement of an existing facility with a larger one.

EXPECTED BEGIN DATE. Begin date assigned to a specific activity. Syn: Target Start Date.

EXPENSE. Expenditures of short-term value, including depreciation, as opposed to land and other fixed capital. For factory expense. (See PLANT OVERHEAD.)

EXTRAPOLATION. To infer from values within an observed interval, or to project or extend beyond observed data.

EXPECTED DURATION. The length of time anticipated for a particular activity in the PERT method or in arrow or precedence diagramming methods (ADM, PDM).

EXPECTED ELAPSED TIME. Statistically weighted time estimates or a single knowledgeable estimate for activity duration. If a weighted or mean time estimate, it incorporates an optimistic (a) most likely (m) and pessimistic (b) estimate for the work to be accomplished.

FACTORY EXPENSE. (See PLANT OVERHEAD.)

FAIR VALUE. That estimate of the value of a property that is reasonable and fair to all concerned, after every proper consideration has been given due weight.

FEE. The charge for the use of ones services to the extent specified in the contract.

FIELD COST. Engineering and construction costs associated with the construction site rather than with the home office.

FIELD LABOR OVERHEAD. The sum of the cost of payroll burden, temporary construction facilities, consumables, field supervision, and construction tools and equipment.

FIELD ORDER. A written order issued by the engineer to the contractor which orders minor changes in the work but which does not involve an adjustment in the contract price or the contract time.

FIELD SUPERVISION. The cost of salaries and wages of all field supervisory and field support staff personnel (excluding general foreman), plus associated payroll burdens, home office overhead, living and travel allowances, and field office operating costs.

FIFO (FIRST IN, FIRST OUT). A method of determining the cost of inventory used in a product. In this method, the costs of materials are transferred to the product in chronological order. Also used to describe the movement of materials. (See LIFO.)

FINANCIAL LIFE. (See VENTURE LIFE.)

FINISHED GOODS. Commodities that will not undergo any further processing and are ready for sale to the user (eg, apparel, automobiles, bread).

FIRST COST. Costs incurred in placing a facility into service, including but not limited to costs of planning, design, engineering, site acquisition and preparation, construction, purchase, installation, property taxes paid and interest during the construction period, and construction-related fees. Syn: Initial Investment Cost; Initial Cost.

FIRST EVENT NUMBER. The number of the first event in time for a work package or summary item. This event number defines the beginning of the work package or summary item in relation to the network.

FIXED COSTS. Those costs independent of short term variations in output of the system under consideration. Includes such costs as maintenance; plant overhead; and administrative, selling and research expense. For the purpose of cash flow calculation, depreciation is excluded (except in income tax calculations).

FIXED PRICE CONTRACT. Contract where the contractor agrees to furnish services and material at a specified price, possibly with a mutually agreed-upon escalation clause. This type of contract is most often employed when the scope of services to be provided is well defined.

FLOAT. (l) In manufacturing, the amount of material in a system or process, at a given point in time, that is not being directly employed or worked upon. (2) in construction, the cushion or slack in any noncritical path in a network planning system. *Syn*: Slack, Path Float.

FORECAST. An estimate and prediction of future conditions and events based on information and knowledge available at the time of the forecast.

FORWARD PASS. (1) In construction, network calculations which determine the earliest start/earliest finish time (data) of each activity. (2) in manufacturing, often referred to as forward scheduling, a scheduling technique where the scheduler proceeds from a known start date and computes the completion date for an order usually proceeding from the first operation to the last.

FRACTILE. A selected portion of a distribution of values (eg, quartile).

FRAG NET. A portion or fragment of a CPM network usually used to illustrate changes to the whole network.

FREE FLOAT (FF). The amount of time that the completion of an activity may exceed its scheduled finish time without increasing the start time of any succeeding activity.

FREE HAUL. The distance every cubic yard of excavated material is entitled to be moved without an additional charge for haul.

FRINGE BENEFITS. Employee welfare benefits, ie, expenses of employment such as holidays, sick leave, health and welfare benefits, retirement fund, training, supplemental union benefits, etc.

FUNCTION. An expression of conceptual relationships useful in model formulations (eg, productivity is a function of hours worked).

FUNCTIONAL REPLACEMENT COST. The current cost of acquiring the same service potential as embodied by the asset under consideration.

FUNCTIONAL USE AREA. The net usable area of a building or project exclusive of storage, circulation, mechanical, and similar types of space.

FUNCTIONAL SYSTEM. An assembly of parts or components and/or subsystems having one primary end use in the project. It should be noted that secondary and tertiary uses for functional systems are common.

FUNCTIONAL WORTH. The lowest overall cost for performing a function. Four types are as follows:

Cost Value — the monetary sum of labor, material, burden, and all other elements of cost required to produce an item or provide a service.

Esteem Value — the monetary measure of the properties of a product or service, which contribute to desirability or salability but not to required functional performance.

Exchange Value — the monetary sum at which a product or service can be traded.

Use Value – the monetary measure of the necessary functional properties of a product or service that contribute to performance.

GENERAL PURPOSE INDEX. A broad-based index designed to reflect general changes in the economy (eg, Gross National Expenditures Implicit Price Index).

GENERAL TERMS AND CONDITIONS. That part of a contract, purchase order, or specification that is not specific to the particular transaction but applies to all transactions.

GENERAL OVERHEAD. The fixed cost in operation of a business. General overhead is also associated with office, plant, equipment, staffing, and expenses thereof, maintained by a contractor for general business operations. The costs of general overhead are not specifically applicable to any given job or project. (See OVERHEAD.)

GENERAL REQUIREMENTS. Distributables and field costs.

GIVEN YEAR. The year or period selected for comparison, relative to the base year or base period.

GROSS AREA. Generally, the sum of all the floor or slab areas of a project that are enclosed by the exterior skin of the building.

GROSS NATIONAL PRODUCT (GNP). The total national output of goods and services at the market prices for the stated year.

GUIDELINE. A document that recommends methods to be used to accomplish an objective.

HAMMOCK. An aggregate or summary activity spanning the nodes of two or more activities and reported at a summary management level.

HANGER. A beginning or ending node not intended in the network (a break in a network path).

HAUL DISTANCE. The distance measured along the center line or most direct practical route between the center of mass of excavation and the center of mass finally placed. It is the average distance material is moved by a vehicle.

HEDGE. In master production scheduling, a quantity of stock used to protect against uncertainty in demand. The hedge is similar to safety stock, except that a hedge has the dimension of timing as well as amount.

Highest and Best Use. The valuation concept that requires consideration of all appropriate purposes or uses of the subject property in order to determine the most profitable likely utilization.

Holding Time. Time that an item is not operational so that it may be serviced.

Home Office Cost. Those necessary costs involved in the conduct of everyday business, which can be directly assigned to specific projects, processes, or end products, such as engineering, procurement, expediting, legal fees, auditor fees inspection, estimating, cost control, taxes, travel, reproduction, communications, etc.

Ideal Index. The geometric mean of the Laspeyres index and the Paasche index.

Idle Equipment Cost. The cost of equipment that remains on site ready for use but is placed in a standby basis. Ownership or rental costs are still incurred while the equipment is idle.

Impact Cost. Added expenses due to the indirect results of a changed condition, delay, or changes that are a consequence of the initial event. Examples of these costs are premium time, lost efficiency, and extended overhead.

Imperfection. A deviation that does not affect the use or performance of the product, process, or service. In practice, imperfections are deviations that are accepted as-is.

Imposed Date. A date externally assigned to an activity that establishes the earliest or latest date in which the activity is allowed to start or finish.

Imposed Finish Date. A predetermined calendar date set without regard to logical considerations of the network, fixing the end of an activity and all other activities preceding that ending node.

Impossibility. An inability to meet contract requirements because it was in fact physically impossible to do so (Actual Impossibility).

Impracticability. Inability to perform because of extreme and unreasonable difficulty, expense, injury, or loss involved. This is sometimes considered Practical Impossibility.

Imputation (of Price Movement). The assignment of known price change to a certain commodity on the basis of the assumed similarity of price movement.

Income. Used interchangeably with profit. Avoid using Income instead of Sales Revenue. (See PROFIT.)

Incremental Costs (Benefits). The additional cost (benefit) resulting from an increase in the investment in a project. Syn: Marginal Cost (Benefit).

Indirect Costs. (1) in construction, all costs which do not become final part of the installation, but which are required for the orderly completion of the installation and may include, but are not limited to, field administration, direct supervision, capital tools, startup costs, contractor's fees, insurance, taxes, etc.; (2) In manufacturing, costs not directly assignable to the end product or process, such as overhead and general purpose labor, or costs of outside operations, such as transportation and distribution. Indirect manufacturing cost sometimes includes insurance, property taxes, maintenance, depreciation, packaging, warehousing and loading. In government contracts, indirect cost is often calculated as a fixed percent of direct payroll cost.

Individual Price Index. An index which measures the price change for a particular commodity and which may be computed as the ratio of its prices at two points in time.

Inefficiency. Level of production or performance that is legs than that achieved under normal working conditions. Some of the causes that may lead to inefficient performance are changes, delays, and differing site conditions.

Inflation. A rise in the general price level, usually expressed as a percentage rate.

Initial Cost. (See FIRST COST.)

Initial Investment Cost. (See FIRST COST.)

I-Node (ADM). The node signifying the start of the activity (the tail of the arrow).

In-Place Value. Value of a physical property, ie, market value plus costs of transportation to site and installation.

In-Progress Activity. An activity that has been started but is not completed on the reporting date.

Input-Output Analysis. A matrix which provides a quantitative framework for the description of an economic unit. Basic to input-output analysis is a unique set of input-output ratios for each production and distribution process. If the ratios of input per unit of output are known for all production processes, and if the total output are known for all production processed, and if the total production of each end product of the economy, or of the section being studied is known, it is possible to compute precisely the production levels required at every intermediate stage to supply the total sum of end products. Further, it is possible to determine the effects at every point in the production process of a specified change in the volume and mix of end products.

INTANGIBLES. (1) in economy studies, conditions or economy factors that cannot be readily evaluated in quantitative terms as in money; (2) in accounting, the assets that cannot be reliably evaluated (eg, goodwill).

INTERFACE ACTIVITY. An activity connecting an event in one subnetwork with an event in another subnetwork, and representing a logical or imposed interdependence between them.

INTERFACE NODE. A common node for two or more subnets representing logical interdependence.

INTERFERENCE. Conduct that interrupts the normal flow of operations and impedes performance. A condition implied in every construction contract is that neither party will do anything to hinder the performance of the other party.

INTERIM DATES. Dates established in the contract designating the start or the completion of designated facilities or features of a facility. Interim dates are also referred to as Intermediate Access or Intermediate Completion Dates.

INTERMEDIATE EVENTS. Detailed events and activities, the completion of which are necessary for and lead to the completion of a major milestone.

INTERMEDIATE MATERIALS. Commodities that have been processed but require further processing before they become finished goods (eg, fabric, flour, sheet metal).

INTERMEDIATE NODE. A node where at least one activity begins and one activity ends.

INVENTORY. Materials, products in process, and finished products required for plant operation or the value of such material and other supplies, ie, for those maintenance, catalyst, chemicals, spare parts.

INVESTMENT. The sum of the original costs or values of the items that constitute the enterprise; used interchangeably with capital; may include expenses associated with capital outlays such as mine development.

INVESTMENT COST. Includes first cost and later expenditures that have substantial and enduring value (generally more than one year) for upgrading, expanding, or changing the functional use of a facility, product, or process.

ITEM. A commodity designated and defined specifically for direct price, observation.

J-NODE (ADM). The node signifying the finish of the activity (the head of the arrow).

JOB OVERHEAD. The expense of such items as trailer, toilets, telephone, superintendent, transportation, temporary heat, testing, power, water, cleanup, and similar items possibly including bond and insurance associated with the particular project.

JUDGEMENTAL SAMPLING. A procedure of selecting the sample which is based on specific criteria established by sample designers. The selection of priced items and outlets is not a probability sample; that is, it is not based on random chance.

KEY ACTIVITY. An activity that is considered of major significance. A key activity is sometimes referred to as a milestone activity.

LABOR BURDEN. Taxes and insurances the employer is required to pay by law based on labor payroll, on behalf of or for the benefit of labor. (In the US these are federal old age benefits, federal unemployment insurance tax, state unemployment tax, and workers compensation).

LABOR COST, MANUAL. The salary plus all fringe benefits of construction workers and general labor on construction projects and labor crews in manufacturing or processing areas which can be definitely assigned to one product or process area or cost center.

LABOR COST, NON-MANUAL. In construction, normally refers to field personnel other than crafts and includes field administration and field engineering.

LABOR FACTOR. The ratio between the workhours actually required to perform a task under project conditions and the workhours required to perform an identical task under standard conditions.

LADDERING. A method of showing the logic relationship of a set of several parallel activities with the arrow technique.

LAG. Specified time increment or delay between the start or completion of an activity and the start or completion of a successor activity.

LAG RELATIONSHIP. The four basic types of lag relationships between the start and/or finish of a work item and the start and/or finish of another work item are:

1. Finish to Start
2. Start to Finish
3. Finish to Finish
4. Start to Start

LASPWYRES-TYPE PRICE INDEX (STRICT APPELLATION). A composite index founded on a Constant Basket which is taken from the base period of this index.

LATE FINISH (LF). The latest time an activity may be completed without delaying the project finish date.

LATENT CONDITION. A concealed, hidden, or dormant condition that cannot be observed by a reasonable inspection.

LATEST EVENT TIME (LET). The latest time an event may occur without increasing the project's scheduled completion date.

LATE START. The latest time at which an activity can start without lengthening the project.

LATEST REVISED ESTIMATE. The sum of the actual incurred costs plus the latest estimate-to-complete for a work package or summary item as currently reviewed and revised, or both (including applicable overhead where direct costs are specified).

LAWS AND REGULATIONS. Laws, rules, regulations, ordinances, codes and/or orders.

LEAD. A PDM constraint introduced before a series of activities to schedule them at a later time.

LEARNING CURVE. A graphic representation of the progress in production effectiveness as time passes. Learning curves are useful planning tools, particularly in the project oriented industries where new products are phased in rather frequently. The basis for the learning curve calculation is the fact that workers will be able to produce the product more quickly after they get used to making it.

LETTER OF CREDIT. A vehicle that is used in lieu of retention and is purchased by the contractor from a bank for a predetermined amount of credit that the owner may draw against in the event of default in acceptance criteria by the contractor. Also applies when an owner establishes a line of credit in a foreign country to provide for payment to suppliers of contractors for goods and services supplied.

LEVEL FINISH SCHEDULE (SF). The date when the activity is scheduled to be completed using the resource allocation process. Level finish is equal to the level start plus duration except when split.

LEVEL FLOAT. The difference between the level finish and the imposed finish date.

LEVELIZED FIXED-CHARGE RATE. The ratio of uniform annual revenue requirements to the initial investment, expressed as a percent.

LEVEL OF EFFORT (LOE). Support effort (eg, vendor liaison) that does not readily lend itself to measurement of discrete accomplishment. It is generally characterized by a uniform rate of activity over a specific period of time.

LEVEL START SCHEDULE (SS). The date the activity is scheduled to begin using the resource allocation process. This date is equal to or later in time than early start.

LEVERAGE (TRADING ON EQUITY). The use of borrowed funds or preferred stock in the intent of employing these senior funds at a rate of return higher than their cost in order to increase the return upon the investment of the residual owners.

LIFE. (1) physical: that period of time after which a machine or facility can no longer be repaired in order to perform it design function properly. (2) service: the period of time that a machine or facility will satisfactorily perform its function without a major overhaul. (See VENTURE LIFE; STUDY PERIOD; ECONOMIC LIFE).

LIFE CYCLE. (See STUDY PERIOD; LIFE).

LIFE-CYCLE COST (LCC) METHOD. A technique of economic evaluation that sums over a given study period the costs of initial investment (less resale value), replacements, operations (including energy use), and maintenance and repair of an investment decision (expressed in present or annual value terms).

LIFO (LAST IN, FIRST OUT). A method of determining the cost of inventory used in a product. In this method, the costs of material are transferred to the product in reverse chronological order. LIFO is used to describe the movement of goods. (See FIFO.)

LINE OF CREDIT. Generally an informal understanding between the borrower and the bank as to the maximum amount of credit that the bank will provide the borrower at any one time.

LINKING PROCEDURE. A procedure by which a "new" series of indexes is connected to an "old" series in a given link period, generally because of a change in baskets. Actually, indexes of the new series with link period as time base are multiplied by the old index for the link period as the given period. (See SPLICING TECHNIQUE).

LOAD FACTOR. (1) a ratio that applies to physical plant or equipment average load/maximum demand, usually expressed as a percentage. It is equivalent to percent of capacity operation if facilities just accommodate the maximum demand; (2) the ratio of average load to maximum load.

LOAD LEVELING. The technique of averaging, to a workable number, the amount or number of people working on a given project or in a given area of a project at a particular point in time. Load leveling is a benefit of most scheduling techniques and is necessary to insure a stable use of resources. Syn: Workpower Leveling.

LOCAL COST. In foreign work, the cost of local labor, equipment taxes, insurance, equipment, and construction materials incorporated in a construction project, with local currencies. This includes the finishing of imported goods using local labor and materials, the cost of transforming imported raw or semi-finished products using local labor and plant facilities and the marketing of locally produced products.

LOCATION FACTOR. An estimating factor used to convert the cost of an identical plant from one location to another. This factor takes into consideration the impact of climatic conditions, local infrastructure, local soil conditions, safety and environmental regulations, taxation and insurance regulations, labor availability and productivity, etc.

LOGICAL RESTRAINT. A dummy arrow or constraint connection that is used as a logical connector but that does not represent actual work items. It is usually represented by a dotted line, and is sometimes called a dummy because it does not represent work. It is an indispensable part of the network concept when using the arrow diagramming method of CPM scheduling.

LOOP. A path in a network closed on itself passing through any node or activity more than once, or, a sequence of activities in the network with no start or end.

LOSS OF PRODUCTIVITY/EFFICIENCY. (See INEFFICIENCY).

LOT BATCH. A definite quantity of some product manufactured under conditions of production that are considered uniform.

LOT SIZE. The number of units in the lot.

LUMP-SUM. The complete in-place cost of a system, a sub-system, a particular item, or an entire project. Lump-sum contracts imply that no additional charges or costs will be assessed against the owner. (See FIXED PRICE CONTRACT).

MAINTENANCE AND REPAIR COST. The total of labor, material, and other related costs incurred in conducting corrective and preventative maintenance and repair on a facility, on its systems and components, or on both. Maintenance does not usually include those items that cannot be expended within the year purchased. Such items must be considered as fixed capital.

MAJOR COMPONENTS. Part of the aggregation structure of a price index (eg, a CPI can be subdivided into major components of food, housing, clothing, transportation, health and personal care, recreation, reading and education, tobacco and alcohol).

MAJOR MILESTONE. The most significant milestones in the project's life or duration, representing major accomplishments or decision points; usually associated with the first breakdown level in the work breakdown structure.

MAJOR SYSTEM ACQUISITION PROJECTS. Those projects that are directed at and are critical to fulfilling a mission, entail the allocation of relatively large resources, and warrant special management attention.

MANUFACTURING COST. The total of variable and fixed or direct and indirect costs chargeable to the production of a given product, usually expressed in cents or dollars per unit of production, or dollars per year. Transportation and distribution costs, and research, development, selling and corporate administrative expenses are usually excluded. (See also OPERATING COSTS).

MARGINAL ANALYSIS. An economic concept concerned with those incremental elements of costs and revenue which are associated directly with a specific course of action, normally using available Current costs and revenue as a base and usually independent of traditional accounting allocation procedures.

MARGINAL COST (BENEFIT). (See INCREMENTAL COSTS (BENEFIT).)

MARKET VALUE. The monetary price upon which a willing buyer and a willing seller in a free market will agree to exchange ownership, both parties knowing all the material facts but neither being compelled to act. The market value fluctuates with the degree of willingness of the buyer and seller and with the conditions of the sale. The use of the term market suggest the idea of barter. When numerous sales occur on the market, the result is to establish fairly definite market prices as the basis of exchanges.

Mark Up. As variously used in construction estimating, includes such percentage applications as general overhead, profit, and other indirect costs. When mark-up is applied to the bottom of a bid sheet for a particular item, system, or other construction price, any or all of the above items (or more) may be included, depending on local practice.

Material Cost. The cost of everything of a substantial nature that is essential to the construction or operation of a facility, both of a direct or indirect nature. Generally includes all manufactured equipment as a basic part.

Material Difference. A change that is important to the performance of the work or that will have a measurable influence or effect on the time, cost of, or procedures for the work under the contract.

Maximum Out-of-Pocket Cash. The highest year-end negative cash balance during project life.

Mechanical Completion. Placing a fixed asset in service. Mechanical completion is an event.

Merit Shop. (See OPEN SHOP).

Method of Performance. The manner in which the specified product or objective is accomplished, which is left to the discretion of the contractor unless otherwise provided in the contract. If the owner or the engineer orders the contractor to modify the construction procedure, this constitutes a change in method. If the imposition of this modification results in additional cost to the contractor, the contractor may be entitled to recovery under the changes clause.

Milestone. An important or critical event and/or activity that must occur when scheduled in the project cycle in order to achieve the project objective(s).

Milestone Flag. A numeric code that may be entered on an event to flag the event as a milestone.

Milestone Level. The level of management at which a particular event is considered to be a key event or milestone.

Milestone Report. An output report at a specified level showing the latest allowable date, expected date, schedule completion date, and the lack for the successor event contained on each activity or event name flagged as a milestone at the level specified.

Milestone Schedule. A schedule comprised of key events or milestones elected as a result of coordination between the clients and the contractor's project management. These events are generally critical accomplishments planned at time intervals throughout the project and used as a basis to monitor overall project performance. The format may be either network or bar chart and may contain minimal detail at a highly summarized level.

Misrepresentation. Inaccurate factual information furnished by either party to a contract, even if done unintentionally.

Mitigation of Damages. To take all possible measures to avoid damage and delay and, if not avoidable, to reduce or lessen the extra costs incurred due to occurrence of the event.

Model Pricing. The techniques of using verbal, symbolic, or analog models to depict cost relationships, and the form which they take. Mathematics and digital computers are basic analytical tools for model pricing.

Modification, Bilateral. An agreement negotiated by and entered into by both parties for a modification of the existing contract terms of a mutually agreed time or price adjustment.

Modification, Unilateral. A modification to the contract issued by the owner without the agreement of the contractor as to the time or price adjustment.

Monitoring. Periodic gathering, validating and analyzing various data on contract status to determine any existing or potential problems. Usually one accomplishes this through use of the data provided in contractor reports on schedule, labor, cost and technical status to measure progress against the established baselines for each of these report areas. However, when deemed necessary, on-site inspection and validation and other methods can be employed.

Monthly Guide Schedule. A detailed two-month schedule used to detail the sequence of activities in an area for analysis or to plan work assignments. This schedule is usually prepared on an "as needed" basis or within a critical area. Syn: Short-Term Activities.

Month-to-Month Price Index. A price index for a given month with the preceding month as the base period.

Most Likely Time Estimate. The most realistic estimate of the time an activity might consume.

Moving Average. Smoothing a time series by replacing a value with the mean of itself and adjacent values.

MULTIPLE FINISH NETWORK. A network that has more than one finish activity or finish event.

MULTIPLE START NETWORK. A network that has more than one start activity or event.

MULTIPLE STRAIGHT-LINE DEPRECIATION METHOD. A method of depreciation accounting in which two or more straight line rates are used. This method permits a predetermined portion of the asset to be written off in a fixed number of years. One common practice is to employ a straight line rate which will write off 3/4 of the cost in the first half of the anticipated service life; with a second straight line rate to write off the remaining 1/4 in the remaining half life.

NEGLIGENCE. Failure to exercise that degree of care in the conduct of professional duties that should be exercised by the average, prudent professional, practicing in the same community under similar circumstance. Under this concept, an architect/engineer is not liable for errors of judgment, but only for a breach of duty to exercise care and skill.

NET AREA. When used in building construction, it is the area, exclusive of encroachment by partitions, mechanical space, etc., which is available for circulation or for any other functional use within a project.

NET BENEFITS (SAVINGS). The difference between the benefits and the costs — where both are discounted to present or annual value dollars.

NET PROFIT. Earnings after all operating expenses have been deducted from net operating revenues for a given period.

NET PROFIT, PERCENT OF SALES. The ratio of annual profits to total sales for a representative year of capacity operations. An incomplete measure of profitability, but a useful guidepost for comparing similar products and companies. Syn.: Profit Margin.

NET PURCHASES (CONCEPT OF). According to this concept, any proceeds from the sale in the reference year of a used commodity belongs to a given elementary group and are subtracted from the expenditure reported on commodities in that elementary group.

NETWORK. A logic diagram of a project consisting of the activities and events that must be accomplished to reach the objectives, showing their required sequence of accomplishments and interdependencies.

NETWORK ANALYSIS. Technique used in planning a project consisting of a sequence of activities and their interrelationship within a network of activities making up a project. (See CRITICAL PATH.)

NETWORK PLANNING. A broad generic term for techniques used to plan complex projects using logic diagrams (networks). Two of the most popular techniques are ADM and PDM.

NODE. The symbol on a logic diagram at the intersection of arrows (activities). Nodes identify completion and/or start of activities. (See EVENT.)

NONCASH. A term frequently used for tangible commodities to be used from inventory and not replaced.

NONDURABLE GOODS. Goods whose serviceability is generally limited to a period of less than three years (such as perishable goods and semidurable goods).

NONEXEMPT. Employees not exempt from federal wage and hours guidelines.

NONWORK UNIT. A calendar-specified time unit during which work will not be scheduled.

NOTICE OF AWARD. The written notice of acceptance of the bid by the owner to a bidder stating that upon compliance by the bidder with the conditions precedent enumerated therein, within the time specified, the owner will sign and deliver the agreement.

NOTICE TO PROCEED. A written notice issued by the owner to the contractor authorizing the contractor to proceed with the work and establishing the date for commencement of the contract time.

OBJECTIVE EVENT. An event that signifies the completion of a path through the network. A network may have more than one objective event.

OBSOLESCENCE. The condition of being out of date. A loss of value occasioned by new developments which place the older property at a competitive disadvantage. A factor in depreciation; (2) a decrease in the value of an asset brought about by the development of new and more economical methods, processes, and/or machinery; (3) the loss of usefulness or worth of a product or facility as a result of the appearance of better and/or more economical products, methods or facilities.

OFFSITES. General facilities outside the battery limits of process units, such as field storage, service facilities, utilities, and administrative buildings.

Omission. Any part of a system, including design, construction and fabrication, that has been left out, resulting in a deviation. An omission requires an evaluation to determine what corrective action is necessary.

On-Stream Factor. The ratio of actual operating days to calendar days.

Open Shop. An employment or project condition where either union or non-union contractors or individuals may be working. Open shop implies that the owner or prime contractor has no union agreement with workers. Also referred to as merit shop.

Operating Costs. The expenses incurred during the normal operation of a facility, or component, including labor, materials, utilities, and other related costs. Includes all fuel, lubricants, and normally scheduled part changes in order to keep a subsystem, system, particular item, or entire project functioning. Operating costs may also include general building maintenance, cleaning services, taxes, and similar items. (See MANUFACTURING COST.)

Optimistic Time Estimate. The minimum time in which the activity can be completed if everything goes exceptionally well. It is generally held that an activity would have no more than one chance in a hundred of being completed within this time.

Optimum Plant Size. The plant capacity which represents the best balance between the economics of size and the cost of carrying excess capacity during the initial years of sales.

Organizational Codes. Numerical or alphabetized characters that the user specifies for the system to associate with a particular activity for sorting purposes.

Original Duration. The initial accepted estimate of an activity duration used in the original baseline schedule.

Over-Haul. The distance in excess of that given as the stated haul distance to transport excavated material.

Overhead. A cost or expense inherent in the performing of an operation, ie, engineering, construction, operating or manufacturing, which cannot be charged to or identified with a part of the work, product or asset and, therefore, must be allocated on some arbitrary base believed to be equitable, or handled as a business expense independent of the volume of production. Plant overhead is also called factory expense.

Over (Under) Plan. The planned cost to date minus the latest revised estimate of cost to date. When planned cost exceeds latest revised estimate, a projected underplan condition exists. When latest revised estimate exceeds planned cost, a projected overplan condition exists.

Overrun (Underrun). The value for the work performed to date minus the actual cost for that game work. When value exceeds actual cost, an underrun condition exists. When actual cost exceeds value, an overrun condition exists.

Owner. The public body or authority, corporation, association, firm or person with whom the contractor has entered into the agreement and for whom the work is to be provided.

Paasche-Type Price Index. A composite index founded on a fixed basket which is taken from the current period of this index.

Partial Utilization. Placing a portion of the work in service for the purpose for which it is intended (or a related purpose) before reaching substantial completion for all the work.

Path. The logically continuous series of connected activities through a network.

Path Float. (See FLOAT.)

Payback Method. A technique of economic evaluation that determines the time required for the cumulative benefits from an investment to recover the investment cost and other accrued costs. (See simple payback period.)

Payoff (Payback) Period. (See PAYOUT TIME.)

Payout Time. The time required to recover the original fixed investment from profit and depreciation. Most recent practice is to base payout time on an actual sales projection. Syn: Payoff Period. (See SIMPLE PAYBACK PERIOD.)

Payroll Burden. Includes all payroll taxes, payroll insurances, fringe benefits, and living and transportation allowances.

PDM. (See PRECEDENCE DIAGRAM METHOD.)

PDM Arrow. A graphical symbol in PDM networks used to represent the lag describing the relationship between work items.

PDM Finish to Finish Relationship. This relationship

restricts the finish of the work item until some specified duration following the finish of another work item.

PDM FINISH TO START RELATIONSHIP. The standard node type of relationship as used in ADM where the activity of work item may start just as soon as another work item is finished.

PDM START TO FINISH RELATIONSHIP. The relationship restricts the finish of the work item until some duration following the start of another work item.

PDM START TO START RELATIONSHIP. This relationship restricts the start of the work item until some specified duration following the start of the preceding work item.

PERCENT COMPLETE. A comparison of the work completed to the current projection of total work. The percent complete of an activity in a program can be determined by inspection of quantities placed as workhours expended and compared with quantities planned or workhours planned. Other methods can also be used.

PERCENT ON DIMINISHING VALUE. (See DECLINING BALANCE DEPRECIATION.)

PERFORMANCE BOND. A bond that guarantees the work will be completed in accordance with the contract documents. The bond also assures the owner that the contractor will fulfill all contractual and financial obligations.

PERFORMANCE MEASUREMENT BASELINE. The time-phased budget plan against which contract performance is measured. It is formed by the budget assigned to scheduled work elements and the applicable indirect budgets. For future effort not planned in detail, the performance measurement baseline also includes budgets assigned to higher level CWBS elements and undistributed budget. It will reconcile to the contract budget base. It equals the total allocated budget legs management reserve.

PERT. An acronym for Project Evaluation Review Technique which is a probabilistic technique, used mostly by government agencies, for calculating the most likely durations for network activities. Most recently, however, the term PERT has been used as a synonym for CPM.

PESSIMISTIC TIME ESTIMATE. The maximum time required for an activity under adverse conditions. It is generally held that an activity would have no more than one chance in a hundred of exceeding this amount of time.

PHASED CONSTRUCTION. As most commonly used today, implies that construction of a facility or system or sub-system commences before final design is complete. Phased Construction is used in order to achieve beneficial use at an advanced date.

PHYSICAL PROGRESS. The status of a task, activity, or discipline based on pre-established guidelines related to the amount or extent of work completed.

PLAN. A predetermined course of action over a specified period of time which represents a projected response to an anticipated environment in order to accomplish a specific get of adaptive objectives.

PLANNED COST. The approved estimated cost for a work package or summary item. This cost when totaled with the estimated costs for all other work packages results in the total cost estimate committed under the contract for the program or project.

PLANNING. The determination of a project's objectives with identification of the activities to be performed, methods and resources to be used for accomplishing the task, assignment of responsibility and accountability, and establishment of an integrated plan to achieve completion as required.

PLANNING PACKAGE. A logical aggregation of work within a cost account, normally the far term effort that can be identified and budgeted in early baseline planning, but which will be further defined into work packages, LOE, or apportioned effort.

PLANT OVERHEAD. Those costs in a plant that are not directly attributable to any one production or processing unit and are allocated on some arbitrary basis believed to be equitable. Includes plant management salaries, payroll department, local purchasing and accounting, etc. *Syn.*: Factory Expense.

PLUG DATE. A date assigned externally to an activity that establishes the earliest or latest date when the activity is scheduled to start or finish. Syn.: Constraint Date.

PRECEDENCE DIAGRAM METHOD (PDM). A method of constructing a logic network using nodes to represent the activities and connecting them by lines that show logic relationships.

PRECEDING EVENT. (See BEGINNING EVENT.)

PRECONSTRUCTION CPM. A plan and schedule of the construction work developed during the design phase preceding the award of contract.

Predecessor Activity. Any activity that exists on a common path with the activity in question and occurs before the activity in question.

Predecessoor Event. (See BEGINNING EVENT.)

Preferential Logic. The contractor's approach to sequencing of the work over and above those sequences indicated in or required by the contract documents. Examples include equipment restraints, crew movements, form reuse, special logic (lead/lag) restraints, etc. factored into the progress schedule instead of disclosing the associated float time.

Preliminary CPM Plan. CPM analysis of the construction phase made before the award of contracts to determine a reasonable construction period. (See PRECONSTRUCTION CPM).

Preliminary Engineering. Includes all design-related services during the evaluation and definition phases of a project.

Prevention. Quality activities employed to avoid deviations; includes such activities as quality systems development, quality program development, feasibility studies, quality system audits, contractor/subcontractor evaluation, vendors/suppliers of information/materials evaluation, quality orientation activities, and certification/qualification.

Price. The amount of money asked or given for a product (eg, exchange value). The chief function of price is rationing the existing supply among prospective buyers.

Price Index. The representation of price changes, which is usually derived by dividing the current price for a specific good by some base period price. (See COST INDEX.)

Price Relatives. The ratio of the commodity price in a given period to it price in the base period.

Pricing. The observation and recording (collecting) of prices of commodities.

Pricing, Forward. An estimation of the cost of work prior to actual performance. It is also known as Prospective Pricing. Pricing forward is generally used relative to the pricing of proposed change orders.

Pricing, Retrospective. The pricing of work after it has been accomplished

Primary Classification. The classification of commodities by "commodity type."

Procurement. The acquisition (and directly related matters) of equipment, material, and non-personal services (including construction) by such means as purchasing, renting, leasing (including real property), contracting, or bartering, but not by seizure, condemnation, or donation. Includes preparation of inquiry packages, requisitions, and bid evaluations; purchase order award and documentation; plus expediting, in-plant inspection, reporting, and evaluation of vendor performance.

Productivity. Relative measure of labor efficiency, either good or bad, when compared to an established base or norm as determined from an area of great experience. Alternatively, productivity is defined as the reciprocal of the labor factor.

Profit.

Gross Profit — earnings from an on-going business after direct costs of goods sold have been deducted from sales revenue for a given period.

Net Profit — earnings or income after subtracting miscellaneous income and expenses (patent royalties, interest, capital gains) and federal income tax from operating profit.

Operating Profit — earnings or income after all expenses (selling, administrative, depreciation) have been deducted from gross profit.

Profit Margin. (See NET PROFIT, PERCENT OF SALES.)

Profitability. A measure of the excess income over expenditure during a given period of time.

Profitability Analysis. The evaluation of the economics of a project, manufactured product, or service within a specific time frame.

Profitability Index (PI). The rate of compound interest at which the company's outstanding investment is repaid by proceeds for the project. All proceeds from the project, beyond that required for interest, are credited, by the method of solution, toward repayment of investment by this calculation. Also called discounted cash flow, interest rate of return, investors method, internal rate of return. Although frequently requiring more time to calculate than other valid yardsticks, PI reflects in a single number both the dollar and the time values of all money involved in a project. In some very special cases, such as multiple changes of sign in cumulative cash position, false and multiple solutions can be obtained by this technique.

PROGRAM. An endeavor of considerable scope and enduring in nature as opposed to a project; usually representing some definable portion of the basic agency mission and defined as a line item in the agency budget.

PROGRAM MANAGER. An official in the program division who has been assigned responsibility for accomplishing a specific get of program objectives. This involves planning, directing and controlling one or more projects of a new or continuing nature, initiation of any acquisition processes necessary to get project work under way, monitoring of contractor performance and the like.

PROGRESS. Development to a more advanced stage. Progress relates to a progression of development and, therefore, shows relationships between current conditions and past conditions. In networking, progress indicates activities have started or completed, or are in progress. (See STATUS.)

PROGRESS TREND. An indication of whether the progress rate of an activity or of a project is increasing, decreasing, or remaining the game (steady) over a period of time.

PROJECT. An endeavor with a specific objective to be met within the prescribed time and dollar limitations and which has been assigned for definition or execution.

PROJECT CONTROL. The ability to determine project progress and status as it relates to the selected schedule.

PROJECT DURATION. The elapsed duration from project start date through v project finish date.

PROJECTED FINISH DATE. The current estimate of the calendar date when an activity will be completed.

PROJECTED START DATE. The current estimate of the calendar date when an activity will begin.

PROJECTED UNDERRUN (OVERRUN). The planned costs minus the latest revised estimate for a work package or summary item. When planned cost exceeds the latest revised estimate, a projected underrun condition exists. When the latest revised estimate exceeds the planned cost, a projected overrun condition exists.

PROJECT FINISH DATE (SCHEDULE). The latest scheduled calendar finish date of all activities on the project.

PROJECTION. An extension of a series, or any get of values, beyond the range of the observed data.

PROJECT LIFE. (See ECONOMIC LIFE.)

PROJECT MANAGEMENT. The utilization of skills and knowledge in coordinating the organizing, planning, scheduling, directing, controlling, monitoring and evaluating of prescribed activities to ensure that the stated objectives of a project, manufactured product, or service, are achieved.

PROJECT MANAGER. An individual who has been assigned responsibility and authority for accomplishing a specifically designated unit of work effort or group of closely related efforts established to achieve stated or anticipated objective, defined tasks, or other units of related effort on a schedule for performing the stated work funded as a part of the project. The project manager is responsible for the planning, controlling, and reporting of the project.

PROJECT NETWORK ANALYSIS (PNA). A group of techniques based on the network project representation to assist managers in planning, scheduling, and controlling a project.

PROJECT OFFICE. The organization responsible for administration of the project management system, maintenance of project files and document, and staff support for officials throughout the project life cycle.

PROJECT PHASE. The major phases of a project, which include preplanning, design, procurement, construction, start-up, operation, and final disposition.

PROJECT PLAN. The primary document for project activities. It covers the project from initiation through completion.

PROJECT START DATE (SCHEDULE). The earliest calendar start date among all activities in the network.

PROJECT SUMMARY WORK BREAKDOWN STRUCTURE (PSWBS). A summary WBS tailored by project management to the specific project, and identifying the elements unique to the project.

PROJECT TIME. The time dimension in which the project is being planned.

PROPOSAL SCHEDULE. The first schedule issued on a project; accompanies either the client's request or the contractor's proposal.

PROPOSED BASE CONTRACT PRICE. The sum total of the individual total price amounts for items of work designated as base bid items listed on the schedule of prices on the bid form (excluding alternates, if any).

Proposed Change Order. The form furnished by the owner or the engineer which is to be used (1) by the owner, when signed by directive authorizing addition to, deletion from, or revision in the work, or an adjustment in contract price or contract time, or any combination thereof; (2) by the owner, when unsigned, to require that the contractor figure the potential effect on contract price or contract time of a proposed change, if the proposed change is ordered upon signing by the owner; (3) by the contractor, to notify the owner that in the opinion of the contractor, a change is required as provided in the applicable provisions of the contract documents. When signed by the owner, a proposed change order may or may not fully adjust contract price or contract time, but is evidence that the change directed by the proposed change order will be incorporated in a subsequently issued change order following negotiations as to its effect, if any, on contract price or contract time. When countersigned by the contractor, a proposed change order is evidence of the contractors acceptance of the basis for contract adjustments provided, except as otherwise specifically noted.

Proposed Confined Contract Price. The sum total of bidders proposed base contract price and all of the individual total price amounts for items of work designated as alternate bid items listed on the schedule of prices for alternate bid items on the bid form (excluding all additional alternates, if any).

Prudent Investment. That amount invested in the acquisition of the property of an enterprise when all expenditures were made in a careful, businesslike, and competent manner.

Punchlist. A list generated by the owner, architect, engineer, or contractor of items yet to be completed by the contractor. Sometimes called a but list ("but" for these items the work is complete).

Pure Price Change. Change in the price of a particular commodity which is not attributable to change in its quality or quantity.

Qualification Submittals. Data pertaining to a bidders qualifications which shall be submitted as set forth in h instructions to bidders.

Quantity Ratio. A ratio which measures, for a given commodity, its quantitative shift between alternative baskets.

Quantity Survey. Using standard methods measuring all labor and material required for a specific building or structure and itemizing these detailed quantities in a book or bill of quantities.

Quantity Surveyor. In the United Kingdom, contractors bidding a job receive a document called a bill of quantities, in addition to plans and specifications, which is prepared by a quantity surveyor, according to well-established rules. To learn these rules the quantity surveyor has to undergo five years of technical training and must pass a series of professional examinations. In the United Kingdom a quantity surveyor establishes the quantities for all bidders, and is professionally licensed to do so.

Real Dollars. (See CONSTANT DOLLARS.)

Real Estate. This reefers to the physical land and appurtenances, including structures affixed thereto. In some states, by statute, this term is synonymous with real property.

Real Property. Refers to the interests, benefits, and rights inherent in the ownership of physical real estate. It is the bundle of rights with which the ownership of real estate is endowed.

Reasonableness Standard. Costs that do not exceed the amount incurred by a prudent contractor or those costs which are generally accepted. Some factors on which reasonableness is based are recognition of the costs as ordinary and necessary and restraints imposed by law, contract terms, or sound business practices.

Rebasing. Conversion of a price index from one time base to another.

Remaining Available Resources. The difference between the resource availability pool and the level schedule resource requirements. Its computed from the resource allocation process.

Remaining Duration. The estimated work units needed to complete an activity as of the data date.

Remaining Float (RF). The difference between the early finish and the late finish.

Rental (Leased) Equipment Cost. The amount which the owner of the equipment (lessor) charges to a lessee for use of the equipment. The best evidence of such costs are rental invoices that indicate the amount paid for leasing such equipment.

Replacement. A facility proposed to take the place of an existing facility, without increasing its capacity, caused either by obsolescence or physical deterioration.

Replacement Cost. (l) the cost of replacing the productive capacity of existing property by another property of any type, to achieve the most economical service, at prices as of the date specified; (2) facility component replacement and related costs, included in the capital budget, that are expected to be incurred during the study period.

Replacement Value. That value of an item determined by re-pricing the item on the basis of replacing it, in new condition, with another item that gives the same ability to serve, or the game productive capacity, but which applies current economic design, adjusted for the existing property's physical deterioration.

Reproduction Cost. The cost of reproducing substantially the identical item or facility at a price level as of the date specified.

Reprogramming. A comprehensive re-planning of the efforts remaining in the contract resulting in a revised total allocated budget which exceeds the contract budget base.

Repudiation. (See ANTICIPATORY BREACH.)

Required Completion Date. The required date of completion assigned to a specific activity or project.

Required Return. The minimum return or profit necessary to justify an investment. It is often termed interest, expected return or profit, or charge for the use of capital.

Requirement. An established requisite characteristic of a product, process, or service. A characteristic is a physical or chemical property, a dimension, a temperature, a pressure, or any other specification used to define the nature of a product, process, or service.

Resale Value. The monetary sum expected from the disposal of an asset at the end of its economic life, its useful life, or at the end of the study period.

Reschedule. (1) in construction, the process of changing the duration and/or dates of an existing schedule in response to externally imposed conditions or progress. (2) in manufacturing, the process of changing order or operation due dates, usually as a result of their being out of phase with when they are needed.

Research Expense. Those continuing expenses required to provide and maintain the facilities to develop new products and improve present products.

Reserve Stock. (See SAFETY STOCK.)

Resident Engineer. The authorized representative of the engineer who is assigned to the site or any part thereof whose duties are ordinarily set forth in the contract document and/or the engineers agreement with the owner.

Resource. Any consumable, except time, required to accomplish an activity.

Resource allocation process (RAP). The scheduling of activities in a network with the knowledge of certain resource constraints and requirements. This process adjusts activity level start and finish dates to conform to resource availability and use.

Resource Availability Date. The calendar date when a resource level becomes available to be allocated to project activity.

Resource Availability Pool. The amount of resource availability for any given allocation period.

Resource Code. The code for a particular labor skill, material, equipment type; the code used to identify a given resource.

Resource Description. The actual name or identification associated with a resource code.

Resource Histogram. A graphic display of the amount of resource required as a function of time on a graph. Individual, summary, incremental, and cumulative resource curve levels can be shown. Syn.: Resource Plot.

Resource Limited Scheduling. A schedule of activities so that a pre-imposed resource availability level (constant or variable) is not exceeded in any given project time unit.

Resource Plot. (See RESOURCE HISTOGRAM.)

Responsible Organization. The organization responsible for management of a work package.

Responsibility. Originates when one accepts the assignment to perform assigned duties and activities. The acceptance creates a liability for which the assignee ig held answerable for and to the assignor. It constitutes an obligation or accountability for performance.

Restraint. In externally imposed factor affecting the scheduling of an activity. The external factor may be a resource, such as labor, cost or equipment, or, it can be a physical event that must be completed prior to the activity being restrained. Syn.: Constraint.

RETENTION. Usually refers to a percent of contract value (usually 5 or 10 percent) retained by the purchaser until work is finished and testing of equipment is satisfactorily completed.

RETIREMENT OF DEBT. The termination of a debt obligation by appropriate settlement with the lender. It is understood to be in full amount unless partial settlement is specified.

REVISION. In the context of scheduling, a change in the network logic, activity duration, resources availability or resources demand which requires network recalculation and drawing correction(s).

RISK. The degree of dispersion or variability around the expected or "best" value which is estimated to exist for the economic variable in question, eg, a quantitative measure of the upper and lower limits which are considered reasonable for the factor being estimated.

ROYALTIES. Payments a company receives to allow others to use a design or concept the company has researched and developed to commercialization. Generally, one of two types: (1) paid-up royalties where a lump sum payment is made, and (2) running royalties where continuous payments are made, usually based on actual production or revenues.

SALES. Orders booked by customers.

SALES ANALYSIS (OR RESEARCH). A systematic study and comparison of sales for consumption data along the lines of market areas, organizational units, products or product groups, customers or customer groups, or such other units as may be useful. (MARKET RESEARCH.)
Typical analyses would include:

Promotion Evaluation,
Quota Assignment, and
Territory Assignment.

SALES FORECAST. A prediction or estimate of sales, in dollar or physical units, for a specified future period under a proposed marketing plan or program and under an assumed get of economic and other forces outside the unit for which the forecast is made. The forecast may be for a specified item of merchandise or for an entire line.

SALES PROFILE. The growth or decline of historical or forecast sales volume, by years.

SALES PRICE. The revenue received for a unit of a product. Gross sales price is the total amount paid. Net sales are gross sales less returns, discounts, freight and allowances. Plant netbacks are net sales less selling, administrative and research expenses. Syn.: Selling Price.

SALES REVENUE. Revenue received as a result of sales, but not necessarily during the same time period.

SALVAGE VALUE. (l) the cost recovered or which could be recovered from a used property when removed, sold, or scrapped; (2) the market value of a machine or facility at any point in time (normally an estimate of an assets net market value at the end of its estimated life); (3) the value of an asset, assigned for tax computation purposes, that is expected to remain at the end of the depreciation period.

SCHEDULE. The plan for completion of a project based on a logical arrangement of activities, resources available, imposed dates or funding budget.

SCHEDULED COMPLETION DATE. A date assigned for completion of activity or accomplishment of an event for purposes of meeting specified schedule requirements.

SCHEDULED DATE. (See CONTRACT DATE.)

SCHEDULED VARIANCE. The difference between projected start and finish dates and actual or revised start and finish dates.

SCHEDULED EVENT TIME. In PERT, an arbitrary schedule time that can be v introduced at any event but is usually only used at a certain milestone or the last event.

SCHEDULE VARIANCE. The difference between BCWP and BCWS. At any point in time it represents the difference between the dollar value of work actually performed (accomplished) and that scheduled to be accomplished.

SCHEDULING. The assignment of desired start and finish times to each activity in the project within the overall time cycle required for completion according to plan.

SCHEDULING RULES. Basic rules that are spelled out ahead of time so that they can be used consistently in a scheduling system.

SCOPE. Defines the equipment and materials to be provided, and the work to be done, and is documented by the contract parameters for a project to which the company is committed.

SCOPE CHANGE. A deviation from the project scope originally agreed to in the contract. A scope change can consist of an activity either added to or deleted from the original scope. A contract change order is needed to alter the project scope.

SEASONAL COMMODITIES. Commodities which are normally available in the market-place only in a given season of the year.

SEASONAL CARIATION. That movement in many economics series which tends to repeat itself within periods of a year.

SECONDARY FLOAT (SF). Is the same as the Total Float, except that it is calculated from a schedule date et upon an intermediate event.

SECULAR TREND. The smooth or regular movement of a long-term time series trend over a fairly long period of time.

SELLING EXPENSE. The total expense involved in marketing the products in question. This normally includes direct selling costs, advertising, and customer service.

SELLING PRICE. (See SALES PRICE.)

SENSITIVITY ANALYSIS. A test of the outcome of an analysis by altering one or more parameters from an initially assumed value(s).

SENTIMENTAL VALUE. A value associated with an individuals personal desire, usually related to a prior personal relationship.

SERVICEABILITY. A measure of the degree to which servicing of an item will be accomplished within a given time under specified conditions.

SERVICING. The replenishment of consumables needed to keep an item in operating condition, but not including any other preventive maintenance or any corrective maintenance.

SERVICE WORTH VALUE. Earning value, assuming the rates and/or prices charged are just equal to the reasonable worth to customers of the services and/or commodities sold.

SHIFTING BASE. Changing the point of reference of an index number series from one time reference period to another.

SHOP DRAWINGS. All drawings, diagrams, illustrations, schedules and other data which are specifically prepared by or for the contractor to illustrate some portion of the work and all illustrations, brochures, standard schedule, performance charts, instructions, diagrams and other information prepared by a supplier and submitted by the contractor to illustrate material or equipment for some portion of the work.

SHOP ORDER NUMBER. (See ACCOUNT NUMBER.)

SHORT-TERM ACTIVITIES. (See MONTHLY GUIDE SCHEDULE.)

SHUTDOWN POINT. The production level at which it becomes less expensive to close the plant and pay remaining fixed expenses out-of-pocket rather than continue operations; that is, the plant cannot meet its variable expense. (BREAKDOWN POINT.)

SIC CODE. The Standard Industrial Classification of the office of Management and Budget, which provides the framework for the industry-sector index classification scheme. Product indexes are aggregated to five-digit product classes and four-digit industries. Industry indexes can be aggregated to three- and two-digit levels as well.

SIGNIFICANT VARIANCES. Those differences between planned and actual performance which exceed established thresholds and which require further review, analysis and action.

SIMPLE PAYBACK PERIOD (SPP). The time required for the cumulative benefits from an investment to pay back the investment cost and other accrued costs, not considering the time value of money.

SITE PREPARATION. An act involving grading, landscaping, installation of roads and siding, of an area of ground upon which anything previously located had been cleared so as to make the area free of obstructions, entanglements or possible collisions with the positioning or placing of anything new or planned.

SLACK. (See FLOAT.)

SLACK PATHS. The sequences of activities and events that do not lie on the critical path or paths.

SLACK TIME. The difference in calendar time between the scheduled due date for a job and the estimated completion date. If a job is to be completed ahead of schedule, it is said to have slack time; if it is likely to be completed behind schedule, it is said to have negative slack time. Slack time can be used to calculate job priorities using methods such as the critical ratio. In the critical path method, total slack is the amount of time a job may be delayed in starting without necessarily delaying the project completion time. Free slack is the amount of time a job may be delayed in starting without delaying the start of any other job in the project.

SPECIFICATION, DESIGN (PRESCRIPTIVE). A design specification providing a detailed written and/or graphic presentation of the required properties of a product, material, or piece of equipment, and prescribing the procedure for its fabrication, erection, and installation.

SPECIFICATION, PERFORMANCE. A statement of required results, verifiable as meeting stipulated criteria, and generally free of instruction as to the method of accomplishment.

SPECIFICATIONS. Written directions regarding the quality of materials and the nature of the workmanship for a job. Specifications may be written directly on the drawings, or presented in a separate document.

SPLICING TECHNIQUE. One of the procedures used for maintaining the continuity of a price index series in the case of substituted items (and/or replaced outlets). The basic assumption underlying the technique is that, at a given point in time, the relative difference prices between the replaced and replacing items (and/or outlets) reflects the difference in respective qualities. In effect, the splicing technique is analogous to, and may be considered a particular case of, the linking procedure.

SPOT MARKET PRICE INDEX. Daily index used as a measure of price movements of 22 sensitive basic commodities whose markets are to be presumed to be among the first to be influenced by changes in economic conditions. It serves as one early indicator of impending changes in business activity.

STAGE OF PROCESSING. A commodity's intermediate position in the value- added channel of production.

STANDARD DEVIATION. The most widely used measure of dispersion of a frequency distribution. It is calculated by summing squared deviations from the mean, dividing by the number of items in the group and taking the square root of the quotient.

STANDARD INDUSTRIAL CLASSIFICATION (SIC CODE). A classification system of the Office of Management and Budget which provides the framework for, the industry-sector index classification scheme. Product indexes are aggregated to five-digit product classes and four-digit industries. Industry indexes can be aggregated to three- and two-digit levels as well.

> Example: industry code - 3443 – fabricated platework
>
> product code - 80201 – carbon steel tanks and vessels

STANDARD NETWORK DIAGRAM. A predefined network intended to be used v more than one time in any given project.

STARTING EVENT. (See BEGINNING EVENT.)

STARTUP. That period after the date of initial operation, during which the unit is brought up to acceptable production capacity and quality within estimated production costs. Startup is the activity that commences on the date of initial activity that has significant duration on most projects, but is often confused (used interchangeably) with date of initial operation.

STARTUP COSTS. Extra operating costs to bring the plant on stream incurred between the completion of construction and beginning of normal operations. In addition to the difference between actual operating costs during that period and normal costs, it also includes employee training, equipment tests, process adjustments, salaries and travel expense of temporary labor, staff and consultants, report writing, post-startup monitoring and associated overhead. Additional capital required to correct plant problems may be included. Startup costs are sometimes capitalized.

STATUS. The condition of the project at a specified point in time relative to its plan. An instantaneous snapshot of the then current conditions. (See PROGRESS.)

STATUS LINE. A vertical line on a time-scaled schedule indicating the point in time (date) on which the status of the project is reported. Often referred to as the time now line. (See DATA DATE.)

STATUSING. Indicating on the schedule the most current project status. (See UPDATING.)

STOCK AND BOND VALUE. A special form of market value for enterprises, which can be owned through possession of their securities. Stock and bond value is the sum of (1) the par values in dollars of the different issues of bond multiplied by the corresponding ratios of the market price to the par value, and (2) the number of shares of each issue of stock multiplied by the corresponding market price in dollars per share.

Stop Work Order. (See SUSPENSION OF WORK, DIRECTED.)

Straight-Line Depreciation. Method of depreciation whereby the amount to be recovered (written off) is spread uniformly over the estimated life of the asset in terms of time periods or units of output.

Study Period. The length of time over which an investment is analyzed. Syn: Life Cycle, Time Horizon.

Subcontract. Any agreement or arrangement between a contractor and any person (in which the parties do not stand in the relationship of an employer and an employee) and where neither party is the owner.

Subcontractor. An individual, partnership, corporation, joint venture or other combination thereof having a direct contract with the contractor or with any other subcontractor for the performance of a part of the work at the site.

Subindex. A price index for a sub-aggregate of a given basket of commodities.

Subnet. The subdivision of a network into segments usually representing some form of sub-project; a portion of a larger network generally for a unique area of a project. (See FRAG NET.)

Substantial Completion. Work (or a specified part thereof) which has progressed to the point where in the opinion of the engineer, as evidenced by the engineers definitive certificate of substantial completion, it is sufficiently complete, in accordance with the contract documents, so that the work (or specified part) can be utilized for the purposes for which it is intended; or if there be no such certificate issued, when final payment is due in accordance with the general conditions. Substantial completion of the work, or specified part thereof, may be achieved either upon completion of pre-operational testing or startup testing, depending upon the requirements of the contract documents. The terms Substantially Complete and Substantially Completed as applied to any work refer to substantial completion thereof.

Subsystem. An aggregation of component items (hardware and software) performing some distinguishable portion of the function of the total system of which it is a part. Normally, a subsystem could be considered a system in itself if it were not an integral part of the larger system.

Successor Activity. Any activity that exists on a common path with the activity in question and occurs after the activity in question.

Successor Event. The event that signifies the completion of an activity. (See J-NODE (ADM))

Summary Item. An item appearing in the work breakdown structure.

Summary Network. A summarization of the CPM network for presentation purposes. This network is not computed.

Summary Number. A number that identifies an item in the work breakdown structure.

Sum-of-Digits-Method. A method of computing depreciation in which the amount for any year is based on the ratio: (years of remaining life)/(1+2+3...+n), n being the total anticipated life. Also known as sum-of-the-years-digits method.

Sunk Cost. A cost that has already been incurred and which should not be considered in making a new investment decision.

Superior Knowledge. (See MISREPRESENTATION.)

Supplementary Conditions. The part of the contract documents which amends or supplements the general conditions.

Supplier. A manufacturer, fabricator, distributor or vendor.

Surety. A bonding company licensed to conduct business which guarantees the owner that the contract will be completed (Performance Bond) and that subcontractors and suppliers will be paid (Payment Bond).

Suspension of Work, Constructive. An act or failure to act by the owner, or the owners representative, which is not a directed suspension of work or work stoppage, but which has the effect of delaying, interrupting, or suspending all or a portion of the work.

Suspension of Work, Directed. Actions resulting from an order of the V owner to delay, interrupt, or suspend any or all portions of the work for a given period of time, for the convenience of the owner.

System. A collection of hardware (equipment and facilities) and related software (procedures, etc.) designated to perform a unique and useful function. A system contains everything necessary (except personnel and materials or supplies) to perform its defined function.

Systems Studies. The development and application of methods and techniques for analyzing and assessing programs, activities and projects to review and assess efforts to date and to determine future courses and directions. These studies include cost/ benefit analysis, environmental impact analysis, assessment of the likelihood of technical success, forecasts of possible futures resulting from specific actions, and guidance for energy program planning and implementation.

Take-Off. Measuring and listing from drawings the quantities of materials required in order to price their cost of supply and installation in an estimate and to proceed with procurement of the materials.

Tangibles. Things that can be quantitatively measured or valued, such as items of cost and physical assets.

Target Date. The date an activity is desired to be started or completed; either externally imposed on the system by project management or client, or accepted as the date generated by the initial CPM schedule operation.

Target Reporting. A method of reporting the current schedule against some established base line schedule and the computations of variances between them.

Target Start Date. (See EXPECTED BEGIN DATE.)

Task. The smallest unit of work planned. It must have an identifiable start and finish and usually produces some recognizable results.

Task Monitor. The individual assigned the monitoring responsibility for a major effort within the program.

Taxes Payable. Tax accruals due within a year.

Temporary Construction Cost. Includes costs of erecting, operating, and dismantling impermanent facilities, such as offices, workshops, etc., and providing associated services such as utilities.

Termination. Actions by the owner, in accordance with contract clauses, to end, in whole or in part, the services of the contractor. Termination may be for the convenience of the owner or for default by the contractor.

Terms of Payment. Defines a specific time schedule for payment of goods and services and usually forms the basis for any contract price adjustments on those contracts that are subject to escalation.

Third Party Claim. A claim against either or both the owner or the contractor by members of the public, or other parties, usually for property damage or personal injury.

Tied Activity. An activity that must start within a specified time or immediately after its predecessors completion or start.

Time Extension. An increase in the contract time by modification to complete an item of work. Time extension may be granted under the corresponding provisions in the general conditions. An excusable delay generally entitles a contractor to a time extension.

Time Horizon. (See STUDY PERIOD.)

Time-limited scheduling. The scheduling of activities so predetermined resource availability pools are not exceeded unless the further delay will cause the project finish to be delayed. Activities can be delayed only until their late start date. However, activities will begin when the late start date is reached, even if resource limits are exceeded. Networks with negative total float time cannot be processed by time-limited scheduling.

Time Now Line. The point in time that the network analysis is based upon. May or may not be the data date. (See STATUS LINE.)

Time of the Essence. A contract requirement that completion of the work within the time limits in the contract is essential. Failure to do so is a breach for which the injured party is entitled to damages.

Time Scaled CPM. A plotted or drawn representation of a CPM network where the length of the activities indicates the duration of the activity as drawn to a calendar scale. Float is usually shown with a dashed line as are dummy activities.

Time Unit. (See CALENDAR UNIT.)

Total Cost Bidding. A method of establishing the purchase price of movable equipment; the buyer is guaranteed that maintenance will not exceed a set maximum amount during a fixed period and that the equipment will be repurchased at a set minimum price when the period ends.

Total Float (TF). The amount of time (in work units) that an activity may be delayed from its early start without delaying the project finish date. Total float is equal to the late finish minus the early finish or the late start minus the early start of the activity.

TRACKING. A form of monitoring applied to projects. The measurements are expected to change according to the planned progress.

TRANSFER PRICE. A term used in economic analysis in the mineral processing industries; used to assign a value to raw materials when the same company does the mining and processing; usually equal to the fair market value.

TURNOVER RATIO. The ratio of annual sales to investment. Inclusion of working capital is preferable, but not always done. Turnover ratio is considered by some to be reasonable basis for a guesstimate of facilities cost, for new products similar to existing products. It ranges around 1.0 for many chemical plants. The product of turnover ratio and profit margin on sales gives a return-on-investment measure.

UNCERTAINTY. Unknown future events which cannot be predicted quantitatively within useful limits, eg, accidents which destroy invested facilities, a major strike, a competitor's innovation which makes the new product obsolete.

UNDERGROUND FACILITIES. All pipelines, conduits, ducts, cables, wires, utility access ways, vaults, tanks, tunnels or other such facilities or attachments, and any encasements containing such facilities which have been installed under-ground to furnish any of the following services or materials: electricity, gases, steam, liquid petroleum products, telephone or other communications, cable television, sewage and drainage removal, traffic or other control systems or water.

UNIT COST. Dollar per unit of production. It is usually total cost divided by units of production, but a major cost divided by units of production is frequently referred to as a unit cost; for example, the total unit cost is frequently subdivided into the unit costs for labor, chemicals, etc.

UNJUST ENRICHMENT DOCTRINE. The belief in law that one person should not be allowed to profit or enrich himself or herself unfairly at the expense of another person.

UNUSUALLY SEVERE WEATHER. Adverse weather which, at the time of year in which it occurred, is unusual for the place of contract performance. No matter how severe or destructive, if the weather is not unusual for the particular time and place, the contractor is not entitled to relief. Unusual or normal weather does not mean ideal weather or the best weather that can be expected; rather it means the normal weather pattern, both good and bad, that could be reasonably anticipated in a particular area. The normal weather pattern is generally that based on the record of the prior ten years unless the contract documents provide for a different period.

UPDATING. The regular review, analysis, evaluation, and reporting of progress of the project, including re-computation of an estimate or schedule. (See STATUSING.)

UNION. A group of workers who organize together for the purpose of negotiating wage rates, working conditions and fringe benefits.

USEFUL LIFE. The period of time over which an investment is considered to meet its original objective.

USE VALUE. (See FUNCTIONAL WORTH.)

VALUATION OR APPRAISAL. The art of estimating the fair-exchange value of specific properties.

VALUE, ACTIVITY. That portion of the contract price which represents a fair value for the part of the work identified by that activity.

VALUE ADDED BY DISTRIBUTION. The portion of the value of a product or service to the consumer or user which results from distribution activities. This value includes such components as time utility and place utility.

VALUE ADDED BY MARKETING. That portion of the value of a product or service to the consumer or user which results from marketing activities. This value includes such components as price reduction through economies of scale and buyer awareness of more desirable innovations in products or services.

VALUE OF WORK PERFORMED TO DATE. The planned cost for completed work.

VALUE EFFECTIVE. Generally used to describe decisions which have a cost impact; value-effective decisions tend to optimize the value received for the decision made and to maximize return on investments.

VALUE ENGINEERING. A practice function targeted at the design itself, which has as its objective the development of design of a facility or item that will yield least life-cycle costs or provide greatest value while satisfying all performance and other criteria established for it.

VALUE ENGINEERING COST AVOIDANCE. A decrease in the estimated overall cost for accomplishing a function.

VALUE ENGINEERING COST REDUCTION. A decrease in the

committed established overall cost for accomplishing a function.

VALUE ENGINEERING JOB PLAN. An aid to problem recognition, definition, and solution. It is a formal, step-by-step procedure followed in carrying out a value engineering study.

VARIABLE COSTS. Those costs that are a function of production, eg, raw materials coats, by-product credits, and those processing costs that vary with plant output (such as utilities, catalysts and chemical, packaging, and labor for batch operations).

VARIANCE. An cost control, the difference between actual cost or forecast budget cost.

VARIATION IN ESTIMATED QUANTITY. The difference between the quantity estimated in the bid schedule and the quantity actually required to complete the bid item. Negotiation or adjustment for variations are generally called for when an increase or decrease exceeds 15 percent.

VENTURE LIFE. The total time span during which expenditures and/or reimbursements related to the venture occur. Venture life may include the research and development, construction, production and liquidation periods. (See FINANCIAL LIFE.)

VENTURE WORTH. Present worth of cash flows above an acceptable minimum rate, discounted at the average rate of earnings.

VERTICAL EVENT NUMBERING. Assigning event numbers in vertical order.

WAGE RATE. The hourly, daily or weekly cost of a person who works for wages, eg, mechanics, laborers, steamfitters.

WEIGHTS. Numerical modifiers used to infer importance of commodities in an aggregative index.

WORK. Any and all obligations, duties, responsibilities, labor, materials, equipment, temporary facilities, and incidentals, and the furnishing thereof necessary to complete the construction which are assigned to, or undertaken by the contractor, pursuant to the contract documents. Also, the entire completed construction or the various separately identifiable parts thereof required to be furnished under the contract documents. Work is the result of performing services, furnishing labor, and furnishing and incorporating materials and equipment into the construction, all as required by the contract documents.

WORK BREAKDOWN STRUCTURE (WBS). A product-oriented family tree division of hardware, software, facilities and other items which organizes, defines and displays all of the work to be performed in accomplishing the project objectives.

1. *Contract Work Breakdown Structure (CWBS)* — the complete WBS for a contract developed and used by a contractor in accordance with the contract work statement. It extends the PSWBS to the lowest level appropriate to the definition of the contract work.

2. *Project Summary Work Breakdown Structure (PSWBS)* — a summary WBS tailored by project management to the specific project with the addition of the elements unique to the project.

WORK BREAKDOWN STRUCTURE ELEMENT. Any one of the individual items or entries in the WBS hierarchy, regardless of level.

WORK DIRECTIVE CHANGE. A written directive to the contractor, issued on or after the effective date of the agreement and signed by the owner and recommended by the engineer ordering an addition, deletion or revision in the work, or responding to differing or unforeseen physical conditions or emergencies under which the work is to be performed as provided in the general conditions. A work directive change may not change the contract price or the contract time, but is evidence that the parties expect that the change directed or documented by a work directive change will be incorporated in a subsequently issued change order following negotiations by the parties as to its effect, if any, on the contract price or contract time.

WORKHOUR. An analysis of planned versus actual staffing of the project used to determine work progress, productivity rates, staffing of the project, etc.

WORK-IN-PROCESS. Product in various stages of completion throughout the factory, including raw material that has been released for initial processing and completely processed material awaiting final inspection and acceptance as finished product or shipment to a customer. Many accounting systems also include semi-finished stock and components in this category. Syn: In-Process Inventory.

WORK ITEM. The precedence notation equivalent of an activity. (See ACTIVITY.)

WORK PACKAGE. A segment of effort required to complete a specific job such as a research or technological study or report, experiment or test, design specification, piece of hardware, element of software, process, construction drawing, site survey, construction phase element, procurement phase element, or service, which is within the responsibility of a single unit within the performing organization. The work package is usually a functional division of an element of the lowest level of the WBS.

WORKPOWER LEVELING. (See LOAD LEVELING.)

WORK SITE. The area designated in the contract where the facility is to be constructed.

WORK UNIT. A unit of time used to estimate the duration of activities.

WORTH. The worth of an item or groups of items, as in a complete facility, is determined by the return on invesment compared to the amount invested. The worth of an item is dependent upon the analysis of feasibility of the entire item or group or items under discussion (or examination.)

WRITTEN AMENDMENT. A written amendment of the contract documents, signed by the owner and the contractor on or after the effective date of the agreement and normally dealing with the non-engineering or non-technical rather than strictly work-related aspects of the contract documents.

YEAR-TO-YEAR PRICE INDEX. A price index for a given year with the preceding year as the base period.

Z94.5
Distribution & Marketing

Marketing is the revenue generating activity of a corporation. As companies in the United States have increased their production capacity, many companies have found themselves with the capacity to produce more product than they can sell. In such a surplus economy, marketing becomes very important to the success of the firm. Marketing generally deals with all the activities that move the product from the producer to the consumer and includes such activities as: product development, product design, market research, pricing, transportation, promotion, sales, advertising, public relations, and distribution.

With the development of surplus production capacity and surplus products, many companies have adopted what is called the marekting concept, whichis to design products and services based on the needs,wants, and preferences of consumers.

Interest in marketing has expanded out of the business world into many non-business organizations. Organizations and individuals such as charities, political candidates, art groups, hospitals and government agencies have incorporated marketing into their organizational activities.

In recent years, marketing has taen on strong quantitative orientations. Statisticians, mathematicians, economists, and social scientists have been attracted to do research on marketing problems.

The definitions in this section are from the American Marketing Association's Marketing definitions, as well as from several marketing text books and marketing reference books.

> Michael D. Geurts
> Professor of Marketing
> Marriott School of Management
> Brigham Young University
> Provo, Utah
>
> Kenneth D. Lawrence
> Professor of Management and Marketing Science
> School of Management
> New Jersey Institute of Technology
> Newark, NJ 07102

Adoption Process. The steps in a consumer's decision-making process through which a product passes on its way to acceptance, i.e., awareness, interest, evaluation, trial, and adoption.

Advertising. (1) Any paid form of non-personal presentation and promotion of ideas, goods, or services by an identified sponsor. It involves the use of such media as the following:
    Magazine and newspaper space
    Motion pictures
    Outdoor (posters, signs, skywriting, etc.)
    Direct mail
    Novelties (calendars, blotters, etc.)
    Radio and television
    Cards (car, bus, etc.)
    Catalogues
    Directories and references
    Programs and menus
    Circulars
This list is intended to be illustrative, not inclusive.

(2) A non-personal sales presentation, set at a predetermined level, aimed at an audience within a specific period of time, and paid for by an identifiable sponsor. *Comment.* Advertising is generally but not necessarily carried on through mass media. While the postal system is not technically considered a "paid" medium, material distributed by mail is definitely a form of presentation that is paid for by the sponsor. For kindred activities see "Publicity" and "Sales Promotion."

Advertising Evaluation. A group of techniques designed to reveal the relative effectiveness of various advertising strategies and tactics in the light of competitive activity. (See MARKETING RESEARCH.)

Allowances. Discounts that are given to final consumers, customers, or channel members for doing "something" or accepting less of "something."

Area Sampling. A market research survey technique which uses geography, such as blocks in a city, as the basis for selecting a random sample test population.

Automatic Selling. The retail sale of goods or services through currency operated machines activated by the ultimate-consumer buyer.
    Comment. Most, if not all, machines now used in automatic selling are coin operated. There are reports, however, of promising experiments with such devices that may be activated by paper currency; machines that provide change for a dollar bill are already on the market. Some service stations are in operation which accept debit cards or credit cards.

Available Stock. Amount of finished product in inventory that has not been committed to cover orders which have been placed.

Average Cost. Obtained by dividing total cost by the related quantity (i.e., the total quantity which causes the total cost).

Average Fixed Cost. A figure obtained by dividing total fixed costs by the associated quantity.

Average Inventory. In a simple inventory system, this is the sum of half the lot size plus reserve stock; or half the maximum number of units in stock.

Average Revenue. A figure obtained by dividing total revenue by the associated quantity.

Average Variable Cost. Total variable costs divided by the relevant quantity.

Backhaul. Use of a common carrier or vehicle from a private fleet for a return load to source after completion of a shipment.

Bait Pricing. Pricing a product below normal to entice a customer into the store–and then trying to get the customer to trade up to a higher-quality product or brand.

Base Inventory Level. The normal aggregate inventory level made up of the aggregate lot size inventory plus the aggregate safety stock inventory but not taking into account the anticipation inventory that will result from the production plan. The base inventory should be made known before the production plan is made. Essentially, it is that inventory level necessary to minimize the probability of running out of stock–i.e., buffer stock q.v.).

Benefit Segmentation. The breaking down of a market into groups based upon the benefits purchased, the values received, and the needs or wants matched.

Blind Tests. Consumer preference tests in which care is exercised to avoid identification of the products or brands involved.

Branch House (Manufacturer's). An establishment maintained by a manufacturer detached from the headquarters establishment and used primarily for the purpose of stocking, selling, delivering, and servicing his product.

Branch Office (Manufacturer's). An establishment maintained by a manufacturer, detached from the headquarters establishment and used for the purpose of selling his products or providing service.
Comment. The characteristic of the branch house that distinguishes it from the branch office is that it is used in the physical storage, handling, and delivery of merchandise; otherwise the two are identical.

Branch Store. A subsidiary retailing business owned and operated at a separate location by an established store.

Brand. Identification in the form of a name, term, sign, symbol, or design, or a combination of them which is intended to identify the goods or services of one seller or group of sellers and to differentiate them from those of competitors.
Comment. A brand may include a brand name, a trade mark, or both. The term brand is sufficiently comprehensive to include practically all means of identification except perhaps the package and the shape of the product. All brand names and all trade marks are brands or parts of brands but not all brands are either brand names or trade marks. Brand is the inclusive general term. The others are more particularized.

Brand Insistence. The final stage of the brand acceptance process, in which consumers refuse to accept substitutes and search for the desired brand.

Brand Management. Product management as applied to a specific class of goods identified by name as the product of a single firm or manufacturer.

Brand Recall. Extent to which consumers can remember and state the brand name used in a previously seen and/or heard advertisement.

Brand Recognition. The first stage in the brand acceptance process. At this point the consumer is simply familiar with the existence of a particular product.

Brand Switching. Sequential choices by consumers of different brands having the same or similar characteristics and functions.

Break-even Analysis. A financial evaluation of the profit potential of alternative prices; more specifically the quantity at which total revenue is equal to the sum of the fixed costs and the variable costs associated with that quantity.

Broker. (1) An agent who does not have direct physical control of the goods in which he deals but represents either buyer or seller in negotiating purchases or sales for his principal. (2) An agent wholesaler who specializes in certain products and functions by bringing together buyers and sellers.
Comment. The broker's powers as to prices and terms of sale are usually limited by his principal. The term is often loosely used in a generic sense to include such specific business units as free-lance brokers, manufacturer's agents, selling agents, and purchasing agents.

Buffer Stock. That quantity of an item of inventory held in stock for absorbing expected variations in usage between the time reorder action is initiated and the first part of the new order is received in stock.

Buyer's Market. An economic condition in which an abundance of goods exceeds the demand for the goods.

Cash-and-Carry Wholesaler. One who sells only for cash; credit and delivery services are not available.

Cash Cows. Strategic business units with high relative market shares in low-growth markets; managerial action is normally to "milk" these.

Cash Discount. A percentage of the invoice amount that the seller will allow the buyer to deduct in return for specifically prompt payment.

Category Killer. A retail store that carries a very large assortment of merchandise and is able to buy at low prices and dominate competitors.

Census. The collection of data for market research purposes.

Chain Stores. A group of stores owned and operated by one company. Each carries the same merchandise, but operates in different geographical areas.

Chain Store System. A group of retail stores of essentially the same type, centrally owned and with some degree of centralized control of operation. The term Chain Store may also refer to a single store as a unit of such a group.
Comment. According to the dictionary, two may apparently be construed to constitute a "group."

Channel Alignment. The coordination of efforts in the form of pricing, transportation, inventory planning and ownership between upstream and downstream sites in the supply chain.

CODE DATING. Identifying attributes of a product – especially date of manufacture, lot number, and recommended usage date – by a system of symbols (letters, numbers, or words).

COMMISSION MERCHANT. An agent wholesaler who has physical control over the merchandise and earns commissions from the sales of goods handled.

COMMON CARRIERS. Transporters which maintain regular schedules and accept goods from any shipper.

COMPARATIVE ADVERTISING. Advertising which makes direct comparisons of the product being promoted with those of competitors.

COMPETITIVE BIDDING. A situation in which several suppliers submit price quotations based on the buyer's specifications for a product or service.

CONCENTRATED MARKETING. A marketing program in which one market is singled out for intense, exclusive treatment.

CONCEPT TESTING. The process of evaluating a new product idea prior to its actual physical development.

CONJOINT MEASUREMENT (OR TRADEOFF ANALYSIS). Methods for measuring the relative importance of various combinations of product attributes in terms of overall consumer satisfaction.

CONSOLIDATED METROPOLITAN STATISTICAL AREAS (CMSAs). Sprawling giant urban areas that include two or more adjoining SMSAs.

CONSUMER BEHAVIOR. The general manner in which individuals reach decisions related to the selection, purchase, and use of goods and services.

CONSUMERS' COOPERATIVE. A retail business owned and operated by ultimate consumers to purchase and distribute goods and services primarily to the membership–sometimes called purchasing cooperatives.
Comment. The Consumers' Cooperative is a type of cooperative marketing institution. Through federation, retail units frequently acquire wholesaling and manufacturing institutions. The definition confines the use of the term to the cooperative purchasing activities of ultimate consumers and does not embrace collective buying by business establishments or institutions.

CONSUMER GOODS. Products made expressly for use by the final consumer, as opposed to those made for resale or for further use in the manufacture of other goods.

CONSUMERISM. A movement in which consumers demand that marketers pay more attention to their needs and wants as well as to product quality and service.

CONSUMER PANELS. Groups of consumers that provide information on a continuing basis.

CONSUMER PRODUCT SAFETY COMMISSION [CPSC]. A government agency that is charged with protecting consumers from unsafe products.

CONSUMER SURPLUS. The difference to consumers between the value of a purchase and the price they pay.

CONTRACT CARRIERS. Transporters who are willing to work for anyone for an agreed sum and for any length of time.

CONVENIENCE GOODS. Those consumers' goods which the customer usually purchases frequently, immediately, and with the minimum of effort in comparison and buying. Examples of merchandise customarily bought as convenience goods are; tobacco products, soap, newspapers, magazines, chewing gum, small packaged confections, and many food products.
Comment. These articles are usually of small unit value and are bought in small quantities at any one time, although when a number of them are bought together as in a supermarket, the combined purchase may assume sizable proportions in both bulk and value. The convenience involved may be in terms of nearness to the buyer's home, easy accessibility to some means of transport, or close proximity to places where people go during the day or evening, for example, downtown to work.

COOPERATIVE MARKETING. The process by which independent producers, wholesalers, retailers, consumers, or combinations of them act collectively in buying or selling or both.

CORRECTIVE ADVERTISING. Advertising required by the Federal Trade Commission to correct previous deceptive advertising.

COST-PLUS. A pricing system in which the cost of the product or service is used as a base to which a profit factor is added.

COUNTERADVERTISING. A Federal Trade Commission plan under which consumer groups can advertise against the sale of a product they consider harmful or uneconomical.

COUPON. A sales promotion device providing a purchase incentive in the form of a price reduction when it is presented with the specified product at the checkout counter and/or mailed to a redemption agency.

CRITICAL PATH SCHEDULING OR CRITICAL PATH METHOD (CPM). A network planning technique used for planning and controlling elements in a project. By showing each of these elements and associated completion time requirements, the "critical path" can be determined. The critical path identifies those elements that actually control the lead time of the project. CPM uses network diagrams with precedence constraints. It is easily run on a computer.

CUMULATIVE QUANTITY DISCOUNTS. These apply to purchases over a given period and the discount usually increases as the amount purchased increases.

CUSTOMER SERVICE LEVEL. A measure of how rapidly and dependably a firm can deliver what customers want; generally measured as a percentage of line items filled "from stock" or within scheduled lead time.

CYCLE STOCK. One of the two main components of any item of inventory, the cycle stock is the most active part, i.e., that which depletes gradually and is replenished cyclically. (Another part of the item inventory is the safety stock which is a cushion of protection against uncertainty in the demand or in the replenishment lead time. The stock level necessary to prevent shortage during ordering, producing, and shipping of replacement work.)

CYCLE TIME. The time it takes in a physical distribution process to complete a process.

DEALER. A firm that buys and resells merchandise at either retail or wholesale.
Comment. The term is naturally ambiguous. For clarity, it should be used with a qualifying adjective, such as "retail" or "wholesale."

DEMAND CURVE. A graphic representation of the quantity of a product or service demand at various price levels.

DEMOGRAPHICS. Periodic measures relating to people in specific geographical areas—including income, living conditions, family size occupations, ethnic backgrounds, and educational levels (e.g., census tract data).

DEPARTMENT STORE. A large retailing business unit which handles a wide variety of shopping and specialty goods, including women's ready-to-wear and accessories, men's and boy's wear, piece goods, small wares, and home furnishings, and which is organized into separate departments for purposes of promotion, service and control. Examples of very large department stores are Macy's, New York, J.L. Hudson Co. of Detroit, Marshall Field & Co. of Chicago, and Famous, Barr of St. Louis. Two well-known smaller ones are Bresee's of Oneonta, New York, and A.B. Wycoff of Stoudsburg, Penn.
Comment. Many department stores have become units of chains, commonly called "ownership groups," since each store retails its local identity, even though centrally owned. The definition above stresses three elements: large size, wide variety of clothing and home furnishings, and departmentization. Size is not spelled out in terms of either sales volume or number of employees, since the concept keeps changing upwards.

DIFFERENTIAL MARKETING. The use of different marketing programs for separate segments of the market.

DIFFUSION. The process by which new products are accepted by consumers. It begins with innovator use and proceeds through early adapters, early majority, late majority, and laggards.

DIRECT MARKETING. The process of selling to consumers by excluding middlemen such as retailers or wholesalers.

DISCRIMINATORY ANALYSIS. Refers to statistical methods used in classification problems. Given that an individual may have emanated from one of K populations, the major problem is to allocate it to the correct population with minimum error, usually based on multiple measurements on the individual and a prior set of similar measurements on individuals whose origin is known.

DISPOSABLE INCOME. Personal income remaining after the deduction of taxes on personal income and compulsory payments, such as social security levies.
Comment. This is substantially the Department of Commerce concept.

DISTRIBUTION. The broad range of activities concerned with efficient movement of finished products from the end of the production line to the consumer; in some cases it may include the movement of raw materials from the source of supply to the beginning of the production line. These activities include freight transportation, warehousing, material handling, protective packaging, inventory control, plant and warehouse site selection, order processing, market and sales forecasting, customer service, and attendant management information systems; in some cases it may include buying activities. In which case it is more properly called Logistics.

Distribution Center. Intermediate warehouse(s) where products from different sources are assembled for shipment to specific customer locations.

Distribution Warehouses. Facilities for product storage and reshipment.°Used to facilitate the rapid movement of goods when trading areas are remote from point of manufacture.

Distributive Lag Effect of Advertising. The effect that advertising has on each period in the future in addition to the current period effect.

Diversification. The process of adding different types of products to an already existing line of products.

Drop Shipper. A merchant who sells products which are shipped directly to the buyer by the manufacturer.

Dual Distribution. Occurs when a manufacturer uses several competing channels to reach the same target market— perhaps using several middlemen and selling directly himself.

Early Adopters. Adopters who are well respected by their peers and usually high in opinion leadership.

Economic Order Quantity. The optimum quantity of a product as determined by balancing the cost to carry inventory against the cost of ordering (and/or set-up).

Economies of Scale. As a company produces larger numbers of°a°particular product, the cost for each of these products goes down.

Elastic Demand. The quantity demanded would increase enough to increase total revenue if price were decreased (and vice versa if price were increased).

Elastic Supply. The quantity supplied increases at a greater rate than the increases in price.°

Electronic Data Interchange. The electronic transmission of standard business documents in a predefined format from one company computer application to its trading partners computer.

Exclusive Dealing Contract. A contract which prohibits middlemen (or intermediaries) from handling competitive products, except where such action would have the effect of reducing competition or creating a monopoly.

Exclusive Distribution. A situation in which a manufacturer grants exclusive distribution rights to an intermediary in a particular territory.

Exclusive Outlet Selling. That form of selective selling whereby sales of an article or a service or brand of an article to any one type of buyer are confined to one retailer or wholesaler in each area, usually on a contractual basis.
Comment.° This definition does not include the practice of designating two or more wholesalers or retailers in an area as selected outlets.°While this practice is a form of Selective Selling, it is not Exclusive Outlet Selling.°The term does not apply to the reverse contractual relationship in which a dealer must buy exclusively from a supplier.

Factor. (1) A specialized financial institution engaged in selling their accounts receivable and lending on the security of inventory.° (2)A type of commission house which often advances funds to the consignor, identified chiefly with the raw cotton and naval stores trades.
Comment.° The type of factor described in (1) above operates extensively in the textile field but is expanding into other fields.

Factoring. A specialized financial function whereby producers, wholesalers, and retailers sell their accounts receivable to financial institutions, including factors and banks, often on a non-recourse basis.
Comment.° Commercial banks as well as factors and finance companies engage in this activity.

Fair Trade. Retail resale price maintenance imposed by suppliers of branded goods under authorization of state and federal laws.
Comment.° This is a special usage of the term promulgated by the advocates of resale price maintenance and bears no relation to the fair practices concept of the Federal Trade Commission; nor is it the antithesis of unfair trading outlawed by the antitrust laws.

FIFO, First-In, First-Out. An inventory valuation method, in which costs of material are transferred in chronological order of receipt.°

Fixed Costs. Costs which do not vary with changes in output; e.g., rent, depreciation, insurance, etc.

FIXED ORDER INTERVAL SYSTEMS. is a an approach to ordering products occurs at predetermined intervals.

F.O.B. ORIGIN. A term stating that the price of goods quoted does not include shipping charges, which are the buyer's responsibility and which normally gives the buyer the right to specify the shipper.

FRANCHISE. An arrangement by which individuals can sell the products of a manufacturer under terms specified by the manufacturer.

FREIGHT FORWARDER. An intermediary who consolidates the shipments of several companies to effect cost saving with carload lots.

FREQUENCY DISTRIBUTION. A specification of the way in which frequencies of members of a population are distributed according to the values of the variates which they exhibit.

FULL-FUNCTION WHOLESALER. An intermediary who provides services such as storage, delivery, credit, return privileges, and marketing intelligence for his or her retail customers.

GENERAL MERCHANDISE STORES. Retail stores, such as department and discount stores, which carry a wide range of merchandise in depth.

HORIZONTAL CHANNEL INTEGRATION. Combines institutions at the same level of operations under one management. An organization may integrate horizontally to merging with other organizations at the same level in the marketing channel.

IMPULSE ITEMS. Products that are bought with little thought or effort, and which are usually placed near the store's cash registers.

INDUSTRIAL GOODS. Goods which are destined to be sold primarily for use in producing other goods or rendering services as contrasted with goods destined to be sold primarily to the ultimate consumer. They include equipment (installed and accessory), component parts, maintenance, repair and operating supplies, raw materials, fabricating materials.
Comment. The distinguishing characteristics of these goods is the purpose for which they are primarily destined to be used, in carrying on business or industrial activities rather than for consumption by individual ultimate consumers or resale to them. The category also includes merchandise destined for use in carrying on various types of institutional enterprises. Relatively few goods are exclusively industrial goods. The same article may, under one set of circumstances, be an industrial good, and under other conditions a consumers' good.

INFORMATION SHARING. Means that all parties involved share information on point-of-sale data and forecasts on this data only.

INSTITUTIONAL ADVERTISING. Advertising for the purpose of promoting a concept, or the goodwill of a company or organization.

JOBBER. This term is widely used as a synonym of "wholesaler" or "distributor;" derived from their practice of buying from manufacturers in "job lots."
Comment. The term is sometimes used in certain trades and localities to designate special types of wholesalers. This usage is especially common in the distribution of agricultural products. The characteristics of the wholesalers so designated vary from trade to trade and from locality to locality. Most of the schedules submitted to the Bureau of the Census by the members of the wholesale trades show no clear line of demarcation between those who call themselves jobbers and those who prefer to be known as wholesalers. Therefore, it does not seem wise to attempt to set up any general basis of distinction between the terms in those few trades or markets in which one exists. There are scattered examples of special distinctive usage of the term "Jobber." The precise nature of such usage must be sought in each trade or area in which it is employed.

JUST-IN-TIME. An approach to materials management in which the product arrives just as they are needed for production or sales.

LABEL. The identifying portion of a package which generally contains the brand name, name of the manufacturer or distributor, product ingredients, and suggested uses or safeguards.

LAGGARDS. Those that prefer to do things the way they have been done in the past and are very suspicious of new ideas.

LIFE STYLE. The way in which people live their lives in career, social, and consumer terms.

LIFO, LAST-IN-FIRST-OUT. An inventory valuation method, in which costs are transferred in reverse chronological order.

LIMITED-LINE STORE. A retail store which offers a complete selection of a narrow line of products.

LIST PRICE. The full price that is quoted subject to appropriate trade discounts to a potential customer; it is generally recommended by the manufacturer.

LOSS LEADER. A product of known or accepted quality priced at a loss or no profit for the purpose of attracting patronage to a store.
Comment. This term is peculiar to the retail trade—elsewhere the same item is called a "leader" or a "special."

MAIL ORDER HOUSE (RETAIL). A retailing business that receives its orders primarily by mail or telephone and generally offers its goods and services for its sale from a catalogue or other printed material.
Comment. Other types of retail stores often conduct a mail order business, usually through departments set up for that purpose, although this fact does not make them mail order houses. On the other hand, some firms that originally confined themselves to the mail order business now also operate chain store systems.

MANUFACTURER'S AGENT. An agent who generally operates on an extended contractual basis; often sells within an exclusive territory; handles non-competing but related lines of goods; and possesses limited authority with regard to prices and terms of sale. He may be authorized to sell a definite portion of his principal's output.
Comment. The manufacturer's agent has often been defined as a species of broker. In the majority of cases this seems to be substantially accurate. It is probably more accurate in seeking to define the entire group not to classify them as a specialized type of broker but to regard them as a special variety of agent since many of them carry stocks. The term "Manufacturer's Representative" is sometimes applied to this agent. Since this term is also used to designate a salesman in the employ of a manufacturer, its use as a synonym for "Manufacturer's Agent" is discouraged.

MARGINAL COST. The change in total cost resulting from the production of one additional unit.

MARGINAL REVENUE. The change in total revenue resulting from the sale of an additional unit.

MARKET. (1) The aggregate of forces or conditions within which buyers and sellers make decisions that result in the transfer of goods and services. (2) The aggregate demand of the potential buyers of a product or service under given conditions (such as price, availability, buyer awareness). In either case, the definition usually implies a specific geographical area.

MARKET DELINEATION. The process of determining potential purchasers and their identifying characteristics.

MARKET GROWTH. The second stage in the product life cycle when a new product's sales are growing fast.

MARKET POTENTIAL (ALSO MARKET OR TOTAL MARKET). A calculation of maximum possible sales opportunities for all sellers of a good or service during a stated period.

MARKET SEGMENTATION. The act of dividing or partitioning a market into distinct groups of potential buyers where each such group might require a separate product and/or marketing mix.

MARKET SHARE. The portion of a market controlled by a particular producer. Reaching a specified share of market is often stated as a goal of a marketing plan.

MARKET TESTING. The marketing for test purposes of a new or improved product and/or a new or improved advertising or sales promotion plan. Typically, market testing is done within a limited and self-contained territory. Tools of marketing research are used in interpreting the test.

MARKETING. The performance of business activities that direct the flow of goods and services from producer to consumer or user.
Comment. The task of defining Marketing may be approached from at least three points of view. (1) The "legalistic" of which the following is a good example: "Marketing includes all activities having to do with effecting changes in the ownership and possession of goods and services." It seems obviously of doubtful desirability to adopt a definition which throws so much emphasis upon the legal phases of what is essentially a commercial subject. (2) The "economic" examples of which are: "That part of economics which deals with the creation of time, place, and possession utilities." "That phase of business activity through which human wants are satisfied by the exchange of goods and services for some valuable consideration." Such definitions are apt to assume somewhat more understanding of economic concepts than are ordinarily found in the market place. (3) The "factual or descriptive" of which the definition suggested by the committee is an example. This type of definition

merely seeks to describe its subject in terms likely to be understood by both professional economists and business men without reference to legal or economic implications. This definition seeks to include such facilitating activities as marketing research, transportation, certain aspects of product and package planning, and the use of credit as a means of influencing patronage.

MARKETING CHANNELS. The paths taken by a product on its way to the consumer.

MARKETING CONCEPT. A guiding-light philosophical orientation which holds that the key to achieving organizational goals consists in determining the needs, wants and desires of target markets, and matching those needs/wants/desires to an integrated marketing mix so as to deliver the target markets' desired satisfactions more effectively and efficiently than competitors while maximizing profits.

MARKETING COST ACCOUNTING. The branch of cost accounting which involves the allocation of marketing costs according to customers, marketing units, products, territories, or marketing activities.

MARKETING COST ANALYSIS. The study and evaluation of the relative profitability or costs of different marketing operations in terms of customers, marketing units, commodities, territories, or marketing activities.
Comment. Marketing Cost Accounting is one of the tools used in Marketing Cost Analysis.

MARKETING FUNCTION. A major specialized activity or group of related activities performed in marketing.
Comment. There is no generally accepted list of marketing functions, nor is there any generally accepted basis on which the lists compiled by various writers are chosen. The reason for these limitations is fairly apparent. Under this term students of marketing have sought to squeeze a heterogeneous and non-consistent group of activities. Some of them are broad business functions with special marketing implications; others are peculiar to the marketing process. The function of assembling is performed through buying, selling, and transpor-tation. Assembling, storage, and transporting are general economic functions; selling and buying are more nearly individual in character. Most of the lists fail sadly to embrace all the activities a marketing manager worries about in the course of doing his job.

MARKETING INFORMATION SYSTEMS. Methods, either manual or computerized, for systematic gathering, editing, storage, and retrieval of data relevant to the marketing of goods and services. Such systems may include methods for summarizing and analyzing the data for either short-range or long-range purposes.

MARKETING MANAGEMENT. The planning, direction and control of the entire marketing activity of a firm or division of a firm, including the formulation of marketing objectives, policies, programs and strategy, and commonly embracing product development, organizing and staffing to carry out plans, supervising marketing operations, and controlling marketing performance.
Comment. In most firms the man who performs these functions is a member of top management in that he plays a part in determining company policy, in making product decisions, and in coordinating marketing operations with other functional activities to achieve the objectives of the company as a whole. No definition of his position is included in this report because there is no uniformity in the titles applied to it. He is variously designated Marketing Manager, Director of Marketing, Vice President for Marketing, Director or Vice President of Marketing and Sales, General Sales Manager.

MARKETING MIX. The blending of the elements of product planning, distribution strategy, promotion, and price to meet the needs of a specific market.

MARKETING MODEL. Any representation which aids understanding of one or more characteristics of a market. Models take many forms, varying from concrete versions, such as physical models of equipment and facilities, to abstract versions, such as mathematical or statistical models of share interchanges among competing brands.

MARKETING PLANNING. The work of setting up objectives for marketing activity and of determining and scheduling the steps necessary to achieve such objectives.
Comment. This term includes not only the work of deciding upon the goals or results to be attained through marketing activity but also the determination in detail of exactly how they are to be accomplished.

MARKETING POLICY. A course of action established and disseminated in order to obtain consistency of marketing decisions and operations under recurring and essentially similar circumstances.

MARKETING RESEARCH. The systematic gathering, recording, and analyzing of data about problems relating to the marketing of goods and services. Such research may be undertaken by impartial agencies or by business firms, or their agents. Marketing research is an inclusive term which includes various subsidiary types: (1) market analysis, of which product potential is a type, which is the study of the size, location, nature, and characteristics of markets. (2) sales analysis (or research), which is the systematic study and comparison of sales (or consumption) data. (3) consumer research, of which motivation research is a type, which is concerned chiefly with the discovery and analysis of consumer attitudes, reactions, and preferences. (4) advertising research, of which advertising evaluation is a type, which is aimed toward improving the use of non-personal aids to marketing, primarily the mass media. The techniques of operation research and statistics are often useful in marketing research.

MARK UP. The amount of money added to cost to determine the selling price.

MERCHANDISING. The planning and supervision involved in marketing the particular merchandise or service at the places, times, and prices and in the quantities which will best serve to realize the marketing objectives of the business.
Comment. This term has been used in a great variety of meanings, most of them confusing. The usage recommended by the Committee adheres closely to the essential meaning of the word. The term is most widely used in the sense in the wholesaling and retailing trades. Many manufacturers designate this activity as Product Planning or Management and include in it such tasks as selecting the article to be produced or stocked and deciding such matters as the size, appearance, form, packaging, quantities to bought or made, time of procurement, and price lines to be offered.

MOTIVATION RESEARCH. A group of techniques developed by the behavioral scientists which are used by marketing researchers to discover factors influencing marketing behavior.
Comment. These techniques are widely used outside the marketing sphere, for example, to discover factors influencing the behavior of employees and voters. The Committee has confined its definition to the marketing uses of the tool. Motivation Research is only one of several ways to study marketing behavior.

MULTIDIMENSIONAL SCALING. A statistical technique that looks at consumer perceptions of several product features and compares product positions based on features sometimes called perceptional mapping.

MULTILEVEL MARKETING. A method of marketing through independent distributors who both sell product and recruit other distributors. The recruiting distributor receives a commission on sales made by several levels of recruits below him. Sometimes call a pyramid or network marketing,

NATIONAL BRAND. A manufacturer's or producer's brand, usually enjoying wide territorial distribution.
Comment. The usage of the terms National Brand and Private Brand in this report, while generally current and commonly accepted, is highly illogical and nondescriptive. But since it is widespread and persistent, the Committee embodies it in this report.

NETWORK ANALYSIS. Technique used in planning a project consisting of a sequence of activities and their interrelationship within a network of activities making up a project. (See CRITICAL PATH SCHEDULING/METHOD--CPM.)

NETWORK PLANNING. A broad generic term for techniques used to plan complex projects. Two of the most popular techniques are PERT and CPM.

NON-METRIC SCALING. Methods for constructing a spatial configuration of objects whose rank order of distances best reproduces the rank order of dissimilarities among the objects from original data.

NON-PARAMETRIC STATISTICS. Methods of statistical inference that do not depend upon assumptions about an underlying probability distribution.

OLIGOPOLY. A market characterized by relatively few sellers and which is difficult to penetrate by others.

OPEN DATING. The marking of perishable or semiperishable goods to show the last possible day they should be sold.

OPINION LEADER. An individual whose opinion is respected by and who influences others in a group.

ORDER LEAD TIME. The average time lapse between placing the order and receiving it.

PERCEPTION. The attachment of meaning to information received through the five senses.

PERCEPTUAL MAPPING. Methods for measuring consumer perceptions of a product relative to competitive products based on a given set of product attributes.

PHYSICAL DISTRIBUTION. The process by which goods are moved from the manufacturer, through the various intermediaries on to the ultimate customer, including shipping, warehousing inventory control as well as order processing and customer service functions.

PLANNED OBSOLESCENCE. The practice of manufacturing products with a predictably limited life due to changes in styles and/or physical wear-out.

PLANT SITE SELECTION. Identification, measurement, and comparison of total costs for providing sets of goods or services to all buyers from existing and contemplated sources. Objectives of such analyses are to optimize total costs while maintaining appropriate customer service. Short term analyses are used to influence operating decisions (rescheduling and distribution); long term analyses are used to influence capital investment decisions (adding capacity within existing plants or establishment of new plants).

POINT-OF-PURCHASE ADVERTISING. Displays, signs, and demonstrations which promote a product at a time and place close to the point of sale.

POSITIONING. A marketing strategy which concentrates on a specific market segment by attempting to relate a particular product to its competitors.

PREMIUM. A bonus given without charge when a product is purchased.

PRICE CUTTING. Offering merchandise or a service for sale at a price below that recognized as usual or appropriate by its buyers and sellers.
Comment. One obvious criticism of this definition is that it is indefinite. But that very indefiniteness also causes it to be more accurately descriptive of a concept which is characterized by a high degree of indefiniteness in the mind of the average person affected by price cutting. Traders' ideas of what constitutes price cutting are so vague and indefinite that any precise or highly specific definition of the phenomenon is bound to fail to include all its manifestations. If you ask a group of traders in a specific commodity to define price cutting, you will get as many conflicting formulas as there are traders. But if you ask those same traders at any particular time whether selling at a certain price constitutes price cutting, you will probably get a considerable degree of uniformity of opinion. It is precisely this condition which the definition is designed to reflect.

PRICE LEADER. A firm whose pricing behavior is generally followed by other companies in the same industry.
Comment. The price leadership of a firm may be limited to a certain geographical area, as in the oil business, or to certain products or groups of products, as in the steel business.

PRIMARY DATA. Original information collected for a specific market research study.

PRIVATE BRANDS. Brands sponsored by merchants or agents as distinguished from those produced and distributed by manufacturers or producers.
Comment. This usage is thoroughly illogical, since no seller wants his brand to be private in the sense of being secret and all brands are private in the sense that they are special and not common or general in use. But the usage is common in marketing literature and among traders. Therefore the Committee presents it in this report.

PROBABILITY DISTRIBUTION. A distribution giving the chance of value x as a function of x; or more generally, the probability of joint occurrence of a set of variables $x_1, x_2, ..., x_p$ as a function of those quantities.

PRODUCERS' COOPERATIVE MARKETING. That type of cooperative marketing which primarily involves the sale of goods or services of the associated producing membership. May perform only an assembly or brokerage function but in some cases, notably milk marketing, extends into processing and distribution of the members' production.

PRODUCT ADVERTISING. Advertising whose major purpose is to sell a product.

PRODUCT LIFE CYCLE. The stages a new product idea goes through from beginning to end.

PRODUCT LINE. A group of products that are closely related either because they satisfy a class of need, are used together, are sold to the same customer groups, are marketed through the same type of outlets or fall within given price ranges. Example, carpenters' tools.
Comment. Sub-lines of products may be distinguished, such as hammers or saws, within a Product Line.

PRODUCT MANAGEMENT. The planning, direction, and control of all phases of the life cycle of products, including the creation or discovery of ideas for new products, the screening of such ideas, the coordination of the work of research and physical development of

products, their packaging and branding, their introduction on the market, their market development, their modification, the discovery of new uses for them, their repair and servicing, and their deletion.
Comment. It is not safe to think of Product Management as the work of the executive known as the Product Manager, because the dimensions of his/her job vary widely from company to company, sometimes embracing all the activities listed in the definition and sometimes being limited to the sales promotion of the products in his/her care.

PRODUCT MIX. The composite of products offered for sale by a firm or a business unit.
Comment. Toothpaste is a product. The 50 cent tube of toothpaste is an item. Toothpastes and powders, mouthwashes, and other allied items compose an oral hygiene product line. Soaps, cosmetics, dentifrices, drug items, cake mixes, shortenings and other items may comprise a product mix if marketed by the same company.

PRODUCT POTENTIAL. A class of studies designed to express the potential market for a product in such terms as the personal characteristics of individual consumers or the demographic characteristics of market areas.

PRODUCT RECALL. Return of product to manufacturer or local dealer for reprocessing, repair, or destruction because of failure to meet specifications or due to a potential safety defect.

PRODUCT OBJECTIVE. (OR PROFIT FORECAST) A goal set by product management for profit to be achieved on a specified item of merchandise or on an entire line, based upon a sales volume and price forecasts and a total cost analysis.

PROJECT MANAGEMENT. This includes the areas of project evaluation and scheduling plus project coordination and control. Useful techniques in project management include the concepts of cash flow and present worth, decision tree analysis, critical resource analysis and critical path scheduling.

PROMOTIONAL ALLOWANCE. A grant of money made by a manufacturer to those in the distribution channel to help promote a product.

PROMOTION EVALUATION. A group of techniques designed to reveal the relative effectiveness of various promotion strategies and tactics which may include consideration of competitive activity.

PROMOTIONAL PRICE. A price created specifically as part of a selling strategy, e.g., "Buy one and get one free."

PUBLIC RELATIONS. The communications and other relationships a firm has with its various audiences (i.e., customers, stock holders, employees, government, and neighbors of the plant).

PULLING STRATEGY. A plan to build user demand for a product that distribution channels will be forced to meet.

PURCHASING POWER. (BUYING POWER) The capacity to purchase possessed by an individual buyer, a group of buyers, or the aggregate of the buyers in an area or a market.

PUSHING STRATEGY. A marketing method directed to the channels of distribution, rather than the end user.

QUANTITY DISCOUNT. A price reduction offered when large quantities of a product are ordered; can be at one time or cumulative over a period.

QUALIFYING DIMENSIONS. The dimensions which are relevant to a product-market.

QUALITATIVE RESEARCH. Seeks in-depth, open-ended responses.

QUANTITATIVE RESEARCH. seeks structured responses that can be summarized in numbers.

REFERENCE GROUPS. The groups of people with whom a person identifies.

REORDER POINT. The inventory level that signals the need to place a new order.

RESALE PRICE MAINTENANCE. Control by a supplier of the selling prices of his branded goods at subsequent stages of distribution by means of contractual agreement under fair trade laws or other legal devices.

RESEARCH DESIGN. An all-encompassing plan used to conduct a market research project.

RESIDENT BUYER. An agent who specializes in buying, on a fee or commission basis, chiefly for retailers.
Comment. The term as defined above, is limited to agents residing in the market cities who charge their retail principals fees for buying assistance rendered, but there are resident buying offices that are owned by out-of-town stores and some that are owned coop-

eratively by a group of stores. The former are called private offices and the latter associated offices. Neither of them should be confused with the central buying office of the typical chain, where the buying function is performed by the office directly, not acting as a specialized assistant to store buyers. Resident Buyers should also be distinguished from apparel merchandise brokers who represent competing manufacturers in the garment trades and have as customers out-of-town smaller stores in search of fashion merchandise. These brokers are paid by the manufacturers to whom they bring additional business, on a percentage of sales basis.

RETAILER. A merchant, or occasionally an agent, whose main business is selling directly to the ultimate consumer.
Comment. The retailer is to be distinguished by the nature of his sales rather than by the way he procures the goods in which he deals. The size of the units in which he sells is an incidental rather than a primary element in his character. His essential distinguishing mark is the fact that his typical sale is made to the ultimate consumer.

RETURN ON INVESTMENT (ROI). A figure used to measure selling success, which is derived by multiplying the rate of profit (net profit/sales) by turnover (sales/investment).

RESPONSE FUNCTION. A mathematical equation that measures the changes in sales as a result of changes in various marketing variables such as price, advertising and sale promotions

SAFESTOCK. The amount of extra inventory that is kept to guard against stockouts of items.

SALES ANALYSIS. The study of sales figures for the purpose of reviewing, improving, or correcting a marketing situation. Sales information is broken down into individual components and examined as it relates to other factors operating within the marketing mix.

SALES BUDGET. The part of the marketing budget which is concerned with planned dollar and unit sales and planned costs of personal selling during a specified future period.

SALES FORECAST. An estimate of sales, in dollars and/or physical units for a specified future period under a proposed marketing plan or program and under an assumed set of economic and other forces outside the unit for which the forecast is made. The forecast may be for a specified item of merchandise or for an entire line.
Comment. Two sets of factors are involved in making a Sales Forecast; (1) those forces outside the control of the firm for which the forecast is made that are likely to influence its sales, and (2) changes in the marketing methods or practices of the firm that are likely to affect its sales. In the course of planning future activities, the management of a given firm may make several sales forecasts each consisting of an estimate of probable sales if a given marketing plan is adopted or a given set of outside forces prevails. The estimated effects that several marketing plans may have on Sales and Profits may be compared in the process of arriving at that marketing program which will, in the opinion of the officials of the company, be best designed to promote its economic welfare.

SALES MANAGER. The executive who plans, directs, and controls the activities of the sales staff.
Comment. This definition distinguishes sharply between the manager who conducts the personal selling activities of a business unit and his superior, the executive, variously called Marketing Manager, Director of Marketing, Vice President for Marketing, who has charge of all marketing activities. The usage of this form of organization has been growing rapidly during recent years.

SALES PLANNING. That part of the Marketing Planning work which is concerned with making sales forecasts, devising programs for reaching the sales target, and deriving a sales budget.

SALES PROMOTION. In a specific sense, those marketing activities, other than personal selling, advertising, and publicity, that stimulate consumer purchasing and dealer effectiveness, such as display, shows and exhibitions, demonstrations, and various non-recurrent selling efforts not in the ordinary routine. (2.) In retailing, all methods of stimulating customer purchasing, including personal selling, advertising, and publicity.
Comment. This definition includes the two most logical and commonly accepted usages of this much abused term. It is the suggestion of the Committee that insofar as possible, the use of the term be confined to the first of the two definitions given above.

SALES QUOTA. A projected volume of sales assigned to a marketing unit for use in the management of sales efforts. It applies to a specified period and may be expressed in dollars and/or in physical units of various product lines.

Comment. The quota may be used in checking the efficiency or stimulating the efforts of or in remunerating individual salesmen or other personnel engaged in sales work. A quota may be for a salesman, a territory, a department, a branch house, a wholesaler or retailer, or for the company as a whole. It may be different from the sales figure set up in the sales budget. Since it is a managerial device, it is not an immutable figure inexorably arrived at by the application of absolutely "exact" statistical formulas.

SALES TERRITORY. A geographic area which is the responsibility of one salesperson or several working in a coordinated effort.

SAMPLING. The process of selecting a representative number of people from a given universe for the purpose of market research (or of products for quality assurance).

SELECTIVE DISTRIBUTION. The use of a small, but carefully selected number of retailers to handle a product line.

SELLING. The personal or impersonal process of assisting and/or persuading a prospective customer to buy a commodity or a service or to act favorably upon an idea that has commercial significance to the seller.

Comment. This definition includes advertising, other forms of publicity, and sales promotion as well as personal selling.

SELLING UP. A sales technique used to induce a customer to buy a better and more expensive product than was originally being considered.

SERVICES. Activities benefits, or satisfactions which are offered for sale or are provided in connection with the sale of goods. Examples are amusements, hotel service, electric service, transportation, the services of barber shops and beauty shops, repair and maintenance service, the work of credit rating bureaus. This list is merely illustrative and no attempt has been made to make it complete. The term also applies to the various activities such as credit extension, advice and help of sales people, delivery, by which the seller serves the convenience of his customers.

SHOPPING CENTER. A cluster of retail stores, centrally located to provide easy access for a large number of people.

SHOPPING GOODS. Those consumers' goods which the customer in the process of selection and purchase characteristically compares on such bases as suitability, quality, price and style. Examples of goods that most consumers probably buy as Shopping Goods are: millinery, furniture, dress goods, women's ready-to-wear and shoes, automobiles, and major appliances.

Comment. It should be emphasized that a given article may be bought by one customer as a Shopping Good and by another as a Specialty or Convenience Good. The general classification depends upon the way in which the average or typical buyer purchases.

SKIMMING PRICE POLICY. Trying to sell the top of the demand curve at a high price before aiming at more price-sensitive customers.

SHRINKAGE. A term the refers to missing inventory at a retail store. The missing inventory can be the result of shoplifting, employee theft or accounting mistakes.

SLOTTING FEE. A fee paid by manufacturers for a better location on a retailers shelf. A type of rental fee paid for use of shelf space in a retail store.

SPECIALTY GOODS. Those consumers' goods with unique characteristics and/or brand identification for which a significant group of buyers are habitually willing to make a special purchasing effort. Examples of articles that are usually bought as Specialty Goods are: specific brands and types of fancy foods, hi-fi components, certain types of sporting equipment, photographic equipment, and men's suits.

Comment Price is not usually the primary factor in consumer choice of specialty goods although their prices are often higher than those of other articles serving the same basic want but without their special characteristics.

STANDARD INDUSTRIAL CLASSIFICATION (SIC) CODES. A multi-digit coding system which identifies type of business, by logical groups and related subdivisions, and is used in market and research planning, as well as for government industrial census information.

STANDARDIZATION. The determination of basic limits or grade ranges in the form of uniform specifications to which particular manufactured goods may conform and uniform classes into which the products of agriculture and the extractive industries may or must be sorted or assigned.

Comment. This term does not include Grading which is the process of sorting or assigning units of a commodity to the grades or classes that have been established through the process of Standardization. Some systems of standardization and grading for agricultural products are compulsory by law.

STATUS. A psychological term which describes an individual's relative position within a specific group.

STOCKOUTS. Are the shortages of a product resulting from carrying too few product items in inventory.

STOCK TURNOVER. The number of times an average inventory is turned over within a given year.

STRATEGIC BUSINESS UNIT (SBU). An organizational unit within a larger company which focuses its efforts on some product-markets and is treated as a separate profit center.

STRATEGIC (MANAGEMENT) PLANNING. The managerial process of developing and maintaining a match between the resources available to an organization and its market opportunities.

STRIP MALL. A small mall that is usually several stores in a row.

SUBLIMINAL ADVERTISING. The attempt to influence people by presenting a stimulus (advertising message) below the threshold of recognition.

SUBOPTIMIZATION. An operations research term describing a problem solution that is best from a narrow (but not overall) point of view.

SUPERMARKET. A department store for food products and related items which operates as a self-service unit, and often competes with other types of stores by offering lower prices on certain consumer items.

SUPPLY CHAIN MANAGEMENT. Is the logistics of managing the pipeline of goods from contracts with suppliers and receipt of incoming material, control of work-in-process and finished goods inventories in the plant, to contracting the movement of finished goods through the channel of distribution.

SURVEY. A market research study which is conducted by asking respondents specific questions in order to obtain information on attitudes, motives, and opinions. Such studies can be conducted face-to-face, by telephone, or through the mail.

SWEETHEARTING. A method of stealing from a retail store when the checkout worker does not charge a shopper for all products.

TARGET RETURN OBJECTIVES. The long- and short-term profit goals sought by a firm, usually stated as a percentage of sales or investment.

TERRITORY ASSIGNMENT. A geographical area assigned to a sales unit for use in managing marketing efforts or to a producing unit for use in managing distribution efforts.

TOTAL AVAILABLE MARKET. The non-captive portion of the total market.

TRADE DISCOUNT. A reduction from the list price given to an organization in the marketing channel for services that would otherwise have to have been performed by the manufacturer.

TRADEMARK. A brand with the benefit of legal protection. A trademark includes graphic content as well as the brand name.

TRADING AREA. A district whose size is usually determined by the boundaries within which it is economical in terms of volume and cost for a marketing unit or group to sell and/or deliver a good or service.
Comment. A single business may have several trading areas: for example, the Trading Area of Chicago's Marshall Field for its retail store business is different from that for its catalogue business.

TRAFFIC MANAGEMENT. The planning, selection, and direction of all means and method of transportation involved in the movement of goods in the marketing process (and the related material purchase efforts).
Comment This definition is confined to those activities in connection with transportation that have to do particularly with marketing and form an inseparable part of any well-organized system of distribution. It includes control of the movement of goods in vehicles owned by the marketing concern as well as by public carrier. It does not include the movement of goods within the warehouse of a producer or within the store of a retail concern.

TRANSPORTATION. The moving of a product item from where it is made to where it is purchased and will be used.

ULTIMATE CONSUMER. One who buys and/or uses goods or services to satisfy personal or household wants rather than for resale or for use in business, institutional, or industrial operations.
Comment. The definition distinguishes sharply between Commercial or Industrial Users and Ultimate

Consumers. A firm buying and using an office machine, a drum of lubricating oil, or a carload of steel billets is an Industrial User of those products, not an Ultimate Consumer of them. A vital difference exists between the purposes motivating the two types of purchases which in turn results in highly significant differences in buying methods, marketing organization, and selling practices.

UNDIFFERENTIATED MARKETING. A condition in which a firm produces one product and attempts to make it fit all potential markets with a single marketing mix.

UNIT PRICING. Items priced in terms of a standard unit of measure, e.g., pints, quarts, gallons, pounds, square feet, tons, etc.

UNIVERSAL PRODUCT CODE (UPC). The 10 digit, all-numeric, code for representing grocery products. The first five digits uniquely identify the grocery manufacturer and the last five assigned by the manufacturer to identify his products. It is frequently incorporated into the package by means of a bar code which can be read electronically.

USAGE RATE. The rate at which a product's inventory is consumed or sold per period of time.

VALUE ADDED BY DISTRIBUTION. That portion of the value of a product or service to the consumer or user which results from distribution activities. This value includes such components as time utility and place utility.

VALUE ADDED BY MARKETING. That portion of the value of a product or service to the consumer or user which results from marketing activities. This value includes such components as price reduction through economies of scale and buyer awareness of more desirable innovations in products or services.

VARIABLE COSTS. Costs that change with varying production levels; it is generally conceived as being "fixed" per unit of product.

VERTICAL CHANNEL INTEGRATION. Combines two or more stages of the channel process under one management. This may occur when one member of a marketing channel purchases the operation of another member or simply performs the function of another member, eliminating the need for the intermediary as a separate entity.

VERTICAL MARKETING SYSTEM. Distribution systems that are professionally managed, centrally controlled, and created to attain operating economies that would be impossible on a catch-as-catch-can basis.

WAREHOUSE SITE SELECTION. Selecting a site for a warehouse the number of alternatives to be considered is large because of potential intermediate storage between sources and buyers. (See DISTRIBUTION.)

WARRANTY. A guarantee given to the buyer that covers a fixed period of time and insures that the manufacturer or the retailer will replace a product or provide a full or prorated refund if the product is defective, or fails to meet the conditions stated by the manufacturer.

WHOLESALER. A business unit which buys and resells merchandise to retailers and other merchants and/or to industrial, institutional, and commercial users but which does not sell insignificant amounts to ultimate consumers. In the basic materials, semi-finished goods, and tool and machinery trades merchants of this type are commonly known as "distributors" or "supply houses."
Comment. Generally these merchants render a wide variety of services to their customers. Those who render all the services normally expected in the wholesale trade are known as Service Wholesalers; those who render only a few of the wholesale services are known as Limited Function Wholesalers. The latter group is composed mainly of Cash and Carry Wholesalers who do not render the credit or delivery service, Drop Shipment Wholesalers who sell for delivery by the producer direct to the buyer, Truck Wholesalers who combine selling, delivery, and collection in one operation, and Mail Order Wholesalers who perform the selling service entirely by mail. This definition ignores or minimizes two bases upon which the term is often defined; first, the size of the lots in which wholesalers deal, and second, the fact that they habitually sell for resale. The figures show that many wholesalers operate on a very small scale and in small lots. Most of them make a significant portion of their sales to industrial users.

ZONE PRICING. A system of uniform prices quoted on shipments anywhere within a given geographical area. To the extent that transportation (freight) charges are the primary cause of price differentials, manufacturers can offer "freight equalized" prices which match those quoted by competitors located in (or closer to) the zone.

BIBLIOGRAPHY

E. Jerome McCarthy, William D. Perreault, Jr., *Basic Marketing*. 8th edition.

Schewe, Smith. *Marketing concepts and applications*. 2nd edition.

Marketing Definitions. Compiled by Ralph S. Alexander and the Committee on Definitions of the American Marketing Association.

Herbert F. Holtje. *Theory and Problems of Marketing. Distribution and Marketing*.

Z94.6
Employee & Industrial Relations

In the first edition of Industrial Engineering Terminology, the terms and definitions comprising the field of industrial relations, were included in the section called Applied Psychology. In the second edition (1982), a section covering Wage and Salary Administration was added.

In the third edition (1989), the Z94 Committee voted to combine the two sections under the title Employee and Industrial Relations. This change reflected the more commonly understood and generally used terms in referring to this organizational function as the industrial relations department, the employee relations department or the personnel department. Today, in some enterprises, such departments are often known as the Human Resources Department. Nevertheless the functions are the same and are included in the definitions within this section.

Although the administration of the wage and salary functions are the responsibility of industrial relations departments, the Z94 Committee voted to include the Wage and Salary Administration terminology in the Employee and Labor Relations section because of the involvement of industrial engineering in the design and control of incentive plans, training, human factors (ergonomics), work place environment, job analysis, job evaluation, and participation in handling grievances concerning work standards, productivity and safety.

The previous, and third edition (1989) of Industrial Engineering Terminology formed the basis for the work of this subcommittee in reviewing the contents to determine what terms are now archaic and no longer needed to be included and, more important, what terms have come into the language since the last edition.

In addition, the subcommittee reviewed and used definitions set forth by the American Arbitration Association in its book Labor Arbitration, the American Compensation Association in its pamphlet Glossary of Compensation Terms, and in the glossary set forth by R.S Schuler, N.J. Buetell and S.A. Youngblood in their publication, Effective Personnel Management (West Publishing Co. third edition, 1989) The Z94 subcommittee wishes to express its appreciation to these organizations for their help and assistance and permission to include some of their definitions in this section.

Other publications referred to include Glossary of Personnel Management and Industrial Relations Terms published by the Society for Advancement of Management in 1959, and the Glossary of Current Industrial Relations and Wage Terms, Bulletin Number 1438 published by the United States Department of Labor.

The members of the Employee and Industrial Relations Subcommittee, involved in the preparation of this section of Industrial Engineering Terminology, consisted of representatives from business, labor and academia and are listed below:

Chairman
Soter G. Liberty, P.E.
President, S.G. Liberty Associates Inc.
Adjunct Professor
Lawrence Tech University
Director, IIE Society for Eng. & Mgmt.
    Systems
Retired Chrysler Executive.

Subcommittee

Dr. Adnan Aswad Ph.D
Professor
Industrial & Manufacturing
Systems Engineering
University of Michigan

Irving Bluestone
Professor of Labor Studies
Wayne State University/Detroit MI.
Retired UAW Vice President

Stanley Harris, Ph.D., CMfgE, CEI
Associate Professor
College of Enbineering
Lawrence Technological University

Dolores Infante
Ergonomics Coordinator
Industrial Engineering
GM Tech Center
General Motors

Douglass V. Koch SPHR
Associate Provost
Human Resources Consultant
Lawrence Technological University

John E.S. Moffat E.A.
Industrial Engineer
and Tax Consultant
Washington, D.C.

Irvin Otis P.E.
IIE Area II Vice President
President, Otis & Associates
Adjunct Professor Central Michigan
Retired Chrysler Executive

Abandon. The act of shutting down a plant or facility for an indefinite period of time; a period during which operations are temporarily suspended; differs from a close down in which an operation is completely closed with no plans for reopening. (Specific to the steel industry.)

Abandonment. The dollar loss recognized as a tax deduction when a taxpayer irrevocably discards a depreciable asset with the intention of neither reusing it nor reselling it.

Ability. Demonstrable knowledge or skill. Ability includes aptitude and achievement.

Ability To Pay. A concept sometimes expressed in collective bargaining related to the economic base upon which the cost of wages and/or benefits is to be borne. Uses the effects of wage levels on costs to an organization (rather than as income to employees) and, therefore, helps determine whether or not an organization can afford a specific wage and benefit level.

Absence. The act of employees on a company's active payroll who are away from their work with or without the employer's prior knowledge or authorization.

Absenteeism. Failure of workers to report to work when scheduled. Often applied to unjustified failure to report to work.

Absentee Rate. A ratio indicating the number of man-days or man-hours lost to the total number of available man-days of employment during some base period; usually one month.

Accession. The hiring of a new employee, or the rehiring, appointment, or reinstatement of a former employee.

Accident. Any unintentional event which causes injury, death or property damage.

Accident and Sickness Benefits. Regular payments to workers who lose time from work due to off-the-job disabilities through accident or sickness. Usually insured and part of a private group health and insurance plan financed in whole or in part by the employer. (See HEALTH AND INSURANCE PLAN, TEMPORARY DISABILITY INSURANCE, WORKER'S COMPENSATION, SICK LEAVE.)

Accidental Death and Dismemberment Benefits. An extra lump-sum payment made under many life insurance plans for loss of life, limb, or sight as a direct result of an accident. Coverage may be for occupational and non-occupational accidents. (See WORKER'S COMPENSATION.)

Accident Proneness. The tendency of some workers, because of peculiarities in intelligence, coordination, temperament, or other physical and mental characteristics, to become victims of accidents or possibly the cause of accidents to others.

Accountability. The total obligation that subordinates have to render to their superiors an account concerning the degree to which assigned responsibilities have been or have not been met and carried out.

Accrual Accounts. Accounts set up within an accounting system to provide funds for an obligation at some future date.

Accrual of Benefits. The process, method or formula normally based on various factors such as length of service, hours worked, or level of responsibility which is used to determine either when benefits become legally enforceable claims or increase in number, value, or percentage.

Achievement. An accomplishment of value or importance in relation to a standard.

Across the Board Increase. An identical pay increase (the amount or percent) given to every employee. Sometimes known as a general increase.

Active Employees. Employees at work, as distinguished from retired or laid-off employees. (Includes Employees on sick leave.)

Activity Plateau. A level of performance established informally, resulting from past practice, or by design such as under the application of a wage incentive plan.

Actual-hours-worked. The number of hours worked in a pay period.

Actuary. A person normally trained in mathematics, statistics, legal accounting methods, and the principles of sound operation of insurance, annuities, and pension plans, who employs life expectancy projections, financial projections, and related data in the funding and management of such plans.

ADA. (See AMERICANS WITH DISABILITIES ACT).

Additive, Hourly (add-on, adder). A fixed sum of

money that is added to incentive earnings for work upon which an operator has no opportunity to earn incentive wages. Wages paid outside incentive earnings or the base for incentive earnings. (Initially designed to separate out COLA payments).

ADMINISTRATIVE LABOR. In the steel industry, any labor, whether paid hourly or salary, which is not direct labor or indirect labor. In the automotive industry, all labor residing outside of the production or manufacturing plants, is considered as General & Administrative (G&A) i.e., corporate and division headquarter locations. In the production (manufacturing) plants all labor is either direct or indirect.

ADMINISTRATION. (1) (noun) That body of individuals in an organization accountable and responsible for the performance of the functions of executive leadership. (2) (verb) Performing the administrative functions of planning, organizing, motivating and controlling the activities of an organization so that it may effectively achieve its goals.

ADMINISTRATIVE SERVICES ONLY (ASO). Claims services arrangement provided by insurance carriers to employers with self-insured health and disability benefit plans.

ADMINISTRATOR. (1) A person responsible for the performance of specific administrative duties. (2) The person or organization (frequently the sponsor) specifically designated by the terms of the instrument under which a pension or welfare plan operates to direct the plan.

ADVANCE NOTICE. In general, an announcement of an intention to carry out a certain action, given to an affected or interested party in sufficient time to prepare for it, as in informing a union of planned changes in production methods or plant shut-down, notifying a worker that he or she will be laid-off on a certain date, and notifying management of the union's intention to terminate or modify a collective bargaining agreement on its expiration date. (See PAY-IN-LIEU-OF NOTICE.)

ADVANCE ON WAGES. Refers to any practice by which employees are entitled to draw wages or salaries in advance of actual work performance or prior to the normal pay date for work already completed.

AFFIRMATIVE ACTION. Recruitment to accelerate achievement of a balanced work force according to non-discriminatory guidelines as to gender, ethnic, and/or racial minorities.

AFL-CIO (AMERICAN FEDERATION OF LABOR AND CONGRESS OF INDUSTRIAL ORGANIZATIONS). Federation of autonomous national and international unions created by the merger of the American Federation of Labor (AFL) and the Congress of Industrial Organizations (CIO) in December 1955. More than 75 percent of union members in the United States come within the orbit of the AFL-CIO through their membership in affiliated unions. The initials AFL-CIO after the name of the union indicate that the union is an affiliate.

AGE DISCRIMINATION AND EMPLOYMENT ACT (ADEA) OF 1967 (AMENDED, 1978). This law makes non-federal employees between 40 and 70 a protected class relative to their treatment (in pay, benefits, employment and other personnel actions). Most direct impact on compensation programs was to preclude mandatory retirement prior to 70 (in most states) and to mandate group insurance coverage for employees over 60.

AGENCY SHOP. Provision in a collective bargaining agreement that assesses all bargaining unit employees who do not join the union to pay a fixed amount monthly, usually the equivalent of union dues, as a condition of employment to help defray the union's expenses in acting as a bargaining agent. Under some arrangements, the payments are allocated to the union's welfare fund or to a recognized charity. May operate in conjunction with a modified union shop. (See UNION SHOP, EXCLUSIVE BARGAINING RIGHTS.)

AGREEMENT. Written contract between an employer (or an association of employers) and a union (or unions), usually for a definite term, defining conditions of employment (wages, hours, vacations, holidays, overtime payments, working conditions, benefits, etc.), rights of workers and union, and procedures to be followed in settling disputes or handling issues that arise during the life of the contract. (See COLLECTIVE BARGAINING.) (BARGAINING AGREEMENT.)

ALCOHOLISM PROGRAM. A program provided by an employer, a union, or both to assist employees in rehabilitation from alcoholism. The service may be supplied directly by the employer or by an outside service agency.

ALLOWANCE. A pay or work time adjustment to compensate an employee for job fatigue, unavoidable delays, personal needs, and rest.

ALLOWED TIME. The basic time established for the performance of a task increased by appropriate allowances.

Amalgamated Craft Union. A craft union formed by two or more trade unions of the same general make-up in order to consolidate related trades.

American Arbitration Association. Private nonprofit organization established to aid professional arbitrators in their work through legal and technical services, and to promote arbitration as a method of settling commercial and labor disputes. Provides lists of qualified arbitrators to unions and employers on request.

Americans with Disabilities Act (ADA). This Act, enacted July 26, 1992, contains requirements concerning employment of people with handicap ping conditions. The definition of handicapped has been broadened significantly. Under this Act even applicants or workers who are temporarily disabled may request accommodation, consequently more careful scrutiny must be given to including, rather than excluding, persons with handicaps for job opportunities.

Annual Bonus. Usually a lump sum payment made in addition to an employee's normal salary or wage. May be based on performance (individual or company).

Annual Earnings. The total amount of compensation received for services by a worker during the year, including wages, salaries, bonuses, overtime pay, and incentive earnings. The total annual earnings of a worker may be the result of work performed for a single employer or a number of employers in a given year, and includes profit sharing.

Annual Improvement Factor (productivity). Introduced in the 1948 agreement between the General Motors Corporation and the United Automobile Workers, and altered over the years, for wage increases granted automatically each contract year, in addition to the cost-of-living adjustment. The provision is prefaced with the following words which set it apart from ordinary deferred wage increases: "the annual improvement factor provided herein recognizes that a continuing improvement in the standard of living of employees depends upon technological progress. It further recognizes the principle that to produce more with the same amount of human effort is a sound economic and social objective." (See DEFERRED WAGE INCREASES.)

Annual Wage or Employment Guarantee. An arrangement under which an employer guarantees workers a minimum amount of wages or hours of work during a year.

Annuity. A fixed payment to a specific person at stated intervals either for a definite number of years or for life. Usually associated in industrial relations with a pension plan. (See PENSION PLAN.)

Anti-racketeering Law. Federal law making it a felony to obstruct, delay, or affect interstate commerce by robbery or extortion, e.g., Hobbs Act (1934).

Anti-strikebreaker Law. Federal law prohibiting the interstate transportation of strikebreakers, e.g., Byrens Act (1936).

Anxiety. A state of apprehension or uneasiness related to fear. The object of anxiety is ordinarily less specific than the object of fear.

Appendix. The final portion of a labor contract which may include wage schedules, details of insurance plans, and other material not usually included in the main body of an agreement or labor contract.

Applicant. A person who seeks and applies for employment in an organization.

Application Form (blank). A prepared set of questions arranged to provide an applicant with the means to set forth in writing, pertinent personal experience, and educational information to serve as a basis for his/her selection or rejection by an employer or agent.

Apprentice. A person who enters into an agreement to learn a skilled trade and to achieve a journeyman status through supervised training and experience, usually for a specified period of time. Practical training is supplemented by related technical off- the-job instruction.

Apprentice Rate. The schedule of wage rates applicable to workers being given formal apprenticeship training for a skilled job, in accordance with set standards. The rate schedule is usually established in such a manner as to permit the gradual achievement of the minimum journeyman wage rate.

Aptitude. Ability to develop requisite performance skills.

Aptitude Test. An actual or simulated trial to determine qualification or fitness to perform a task or job. A method to determine aptitude.

Arbitration (voluntary, compulsory, advisory). Method of settling labor-management disputes through recourse to an impartial third party, whose decision is usually final and binding. Arbitration is voluntary when both parties agree to submit disputed issues to arbitration, and compulsory if required by law. (A court order to carry through a voluntary arbitration agreement is not generally considered compulsory arbitration.) Advisory arbitration: As provided in federal government agreements, arbitration without a final and binding award.

ARBITRATOR. An impartial third party to whom disputing parties submit their differences for decision (award). The arbitrator may be involved in "issue" arbitration as well as "grievance" arbitration. An adhoc arbitrator is one selected to act in a specific case or a limited group of cases. A permanent arbitrator is one selected to serve for the life of the contract or a stipulated term. (See IMPARTIAL CHAIRMAN.)

AREA DIFFERENTIAL. Additional compensation paid for similar jobs performed in one geographical location as compared to another. Also called area rate differential. (See INTERCITY DIFFERENTIAL.)

AREA WAGE SURVEY. Wage surveys conducted by the Bureau of Labor Statistics of the U.S. Department of Labor in Standard Metropolitan Statistical Areas (SMSAs).

ASSESSMENT. A charge levied by a union on each member for a purpose not covered by the regular dues. Assessments may be either one-time or periodic charges.

ASSOCIATION AGREEMENT. An agreement negotiated and signed by an employer's association, on behalf of its members, with a union or unions. (See MULTI-EMPLOYER BARGAINING.)

ATTENDANCE BONUS. Payment or other type of reward (e.g., a day off) for employees whose record of daily reporting for work, without absences, meets certain standards of excellence.

ATTENTION. The focusing of perception leading to heightened awareness of a limited range of stimuli.

ATTITUDE. An orientation toward or away from some object, concept, or situation; readiness to respond in a predetermined manner to the object, concept, or situation.

ATTITUDE SURVEY. A device for appraising employee morale by determining workers' feelings or opinions toward their job, supervisor, management policies and practices, or the company in general.

ATTRITION ARRANGEMENT. A process of relying upon voluntary quits, deaths, and retirements to reduce a company's labor force over time instead of resorting to dismissal of workers.

AUTHORITY. The right to influence or command thought, opinion or behavior, e.g., the right of an executive to evaluate performance or to prepare and issue directions with respect to the plans, procedures, processes, policies, practices, or other factors relating to particular functions or activities, wherever they may be performed within the organization.

AUTHORIZATION CARD. A statement signed by the worker authorizing a union to act as the representative in dealings with management, and/or authorizing the company to deduct union dues from his/her pay. (See CHECK-OFF.)

AUTOMATIC INCREASES. Additional pay resulting from predetermined policy. (See AUTOMATIC PROGRESSION.)

AUTOMATIC PROGRESSION. Increases in job pay or classification resulting from predetermined formula or policy.

AUTOMATIC RETIREMENT. (See COMPULSORY RETIREMENT.)

AUTOMATIC WAGE ADJUSTMENT. Automatically increasing or decreasing wages in accordance with some specific plan.

AUTOMATIC WAGE PROGRESSION. Automatically increasing wages after specified periods of service (also known as length-of-service increases).

AUTOMATION. Highly mechanized processes designed to deliver consistent production through the integration of various mechanisms to produce a finished item with relatively few or no worker operations; usually includes electronic computing controls, and robots. Advanced automation systems involve self-regulating machines (feedback) that can perform highly precise sequential operations. In common practice, the term is often used in reference to any type of advanced mechanization; more specifically, it is often associated with cybernetics.

AVERAGE EARNED RATE. An hourly rate arrived at by dividing hours worked into the equivalent earnings paid for a calendar quarter for use in the next quarter. Excludes payments not considered to be earnings.

AVERAGE EARNINGS. Individual or group earnings over a

period of time divided by the number of time or production units in the period under consideration.

Average Hourly Earnings Exclusive of Overtime Payments. Average hourly earnings from which premium payments for overtime work have been eliminated. Also a measure of average hourly earnings published by the Bureau of Labor Statistics in which gross average hourly earnings in manufacturing are adjusted statistically to eliminate the influence of premium overtime payments at time and one half the regular rate of pay after a fixed number (presently, 40) hours of work a week. The adjustment does not compensate for other forms of overtime payment nor for other types of premium pay. (See AVERAGE STRAIGHT-TIME HOURLY EARNINGS, GROSS AVERAGE HOURLY EARNINGS.)

Average Incentive Earnings. The amount of money earned while performing under an incentive plan divided by the number of hours worked under the incentive plan.

Average Straight-Time Hourly Earnings. Average wages earned per hour excluding premiums for overtime and shift assignment.

Average Time. Mathematical mean of selected work cycle or elemental times.

Avoidable Delay. Delay controllable by a worker and therefore not allowed in the job standard.

Award. The decision of an arbitrator in a dispute. In labor arbitration, the arbitrator's reasons are generally expressed in the form of a written opinion which accompanies the award.

Award a Job. The act of assigning a job to an employee where a bidding process is used. After all the bids on the job have been reviewed, the successful candidate is then awarded the job.

Back Pay. Delayed payment of compensation for work performed and arising from arbitration awards or a grievance procedure regarding particular rates, errors in computation of pay, improper layoff, unwarranted discharge or current legal interpretation of wage legislation. Includes contract settlement made retroactive.

Back-to-work Movement. Return of some or all striking workers to their jobs before the strike is ended.

Band Width. The maximum length of work day from which an employee can choose the hours he or she will work.

Bank. The storage of parts. Bank withdrawals are reported at a later date in order to control incentive earnings at a particular time or to control efficiency performance at that time for non-incentive operations. Considered a falsification of production records in many companies, therefore, a practice subject to disciplinary action.

Bargaining. The process by which persons or groups with partly conflicting and partly harmonious interests try to agree on a procedure for dividing available resources. Bargaining is likely to occur when each person controls resources desired by the other and a range of agreements can be made that will benefit both persons.

Bargaining Agent. Union designated by appropriate government agency, such as the National Labor Relations Board, or recognized voluntarily by the employer, as the exclusive representative of all employees in the bargaining unit for purposes of collective bargaining.

Bargaining Agreement. (See AGREEMENT.)

Bargaining Rights. Legally recognized right of unions to represent workers in dealings with employers.

Bargaining Unit. A group of employees in a craft, department, plant, firm or industry recognized by the employer or group of employers or designated by an authorized agency such as the National Labor Relations Board, as appropriate for representation by a union for purposes of collective bargaining.

Base Pay. Minimum rate of compensation for a given task when performed by a fully qualified employee. May differ from rate paid during training or probationary periods.

Base Points. Term used in job analysis referring to minimum point score value for a given evaluated factor.

Base Rate. The amount of pay set for a given job or classification based on a period of time, exclusive of bonus, premium or differential amounts which may be added for specific reasons.

Base Salary. (See BASE PAY.)

Base Wage Rate. (See BASE PAY.)

Basic Piece Rate. A set payment for a unit of production exclusive of additional allowances.

Beginner Rate. Compensation rate during initial probationary period or while learning a new job.

Behavioral Sciences. The sciences concerned with the behavior of men and women in organizations, especially social anthropology, psychology, and sociology, including aspects of biology, economics, political science, history, philosophy, and other fields.

Benchmark. A standard with characteristics so detailed that other classifications can be compared as being above, below, or comparable to it. Most frequently in job evaluation, benchmark refers to a job, or group of jobs, used for making comparisons either within the organization or to comparable jobs outside the organization.

Benchmark Evaluation Method. A job evaluation technique in which the pay rate for a job under study is established by comparison with pay rates for jobs selected as standards or benchmarks.

Benchmark Job. A job or task accepted as a gauge for comparison of other jobs or tasks.

Benchwork Job. A task performed at a table, bench or fixture of similar configuration.

Benefit. Compensation other than direct wages or salary. Usually includes holiday and vacation pay; health, disability and life insurance; social security and unemployment compensation; and pension contributions paid by the employer.

Benefit Limitations. The minimum and maximum restrictions placed on a benefit.

Bereavement Pay. (See FUNERAL LEAVE PAY.)

Binding Arbitration. An agreement between the parties to arbitration in which the arbitrator's award is final and binding upon both parties. Labor arbitration awards may be appealed to the courts if one of the parties to the arbitration believes that the arbitrator exceeded his/her authority under the contract.

Blue-collar Workers. Term for manual workers, usually those employed in production, maintenance, and related occupations, and paid by the hour or on an incentive basis. (See WHITE-COLLAR WORKER.)

Board of Arbitration. A board usually consisting of three members, a union representative, a company representative, and an independent representative chosen by the union and management, who join together to hear a labor dispute.

Board of Inquiry. A board appointed by the President of the United States, under the Labor Management Relations Act, to examine and report on the facts and positions of the parties in a "national emergency" dispute. The term is often used for any board set up by a public agency to investigate a labor dispute. (See FACT-FINDING BOARD.)

Bogey. A target level of performance usually based on minimum acceptable productivity.

Bogus. A term used in the printing industry to designate typesetting work which is not needed for printing but which is required by the collective bargaining agreement.

Bonafide Occupational Qualification (BFOQ). An exception to the restrictions of Title VII of the Civil Rights Act (1964) regarding discrimination on the basis of sex, religion, and national origin. Under certain conditions an employer may require a person of specific sex, national origin, or religious affiliation to staff certain jobs. The intent of this provision is to specify that there are certain jobs for which race, sex, or religion may be legitimate qualifications.

Bonus. Extra compensation paid for a specific reason. Unlike premium amount of compensation, may be nonrecurring and established by a wide variety of methods.

Bonus Determinant. A factor which establishes the earnings potential under a bonus plan.

Bonus Earnings. Additional compensation over and above base earnings paid for specific reasons such as high productivity, low waste or defects, low absenteeism, employment anniversary, end of year, company profitability, etc.

Bonus Plan. A formula or plan used to determine bonus eligibility and payment.

Bonus Restriction. Limits placed on an incentive plan so that earnings come from extra operator effort rather than from technological improvements.

Book Member. (See UNION MEMBER.)

Bootleg Wages. The wages above those at the prevailing rate or the union's scale which an employer may pay in a tight labor market to hold or attract employees. May also refer to wages at rates below the prevailing or union rate which an employee may accept in order to obtain employment. (See KICK-BACK.)

Bottom Out. A point in incentive plans where the incentive performance falls below the 100% of standard level and where the operator is guaranteed at least 100% of base rate. The point in which incentive plans are considered to have deteriorated beyond a level of usefulness. A term commonly used in the steel industry.

Boycott. Efforts by a union, usually in collaboration with other unions, to discourage the purchase, handling, or use of products of an employer with whom the union is in dispute (primary boycott). When such action is extended to another company doing business with the employer involved in the dispute, it is termed a secondary boycott (which is an illegal activity). (See HOT-CARGO CLAUSE.)

Brainstorming. A problem-solving conference technique. Participants announce ideas as fast as possible and no one may criticize or evaluate them while the meeting is in session. The objective of brain storming is to make public the maximum number of possible problem solutions, useful as well as useless ones.

Break-in Allowance. A pay or time allowance granted to an employee when learning a new job.

Break in Service. The time between when an employee leaves a company and later returns and is eligible for certain benefits.

Break Time. (See REST PERIOD.)

Bridge Benefits. (See SURVIVORS' BENEFITS.)

Brief. A written statement in support of a party's position which is submitted to an arbitrator or court either before or after the hearing.

Bug. (See UNION LABEL.)

Bumping. Practice that allows a senior employee (in seniority ranking or length of service) to displace a junior employee in another job or department during a layoff or reduction in force.

Bureau of Labor Statistics (BLS). U.S. Department of Labor.

Bureau of National Affairs, Inc. (BNA). A commercial non-governmental organization engaged in providing various types of reports and services dealing with industrial relations and labor affairs.

Business Agent. Generally a full-time paid employee or official of a local union whose duties include day-to-day dealing with employers and workers, adjustment of grievances, enforcement of agreements, and similar activities.

Business Unionism ("bread-and-butter" unionism). Union policy that places primary emphasis on securing higher wages and better working conditions for its members through collective bargaining rather than through political action or radical reform of society. The term has been widely used to characterize the objectives of the trade union movement in the United States.

Buy-out. In labor negotiations, the act of offering employees a lump sum money settlement to surrender employment rights.

Bylaws. Generally, provisions supplementing charters or constitutions of unions or other organizations, setting forth the rules for the organization.

Cafeteria Benefits. A benefit plan in which employees have a choice of benefits within some dollar limit. Usually a common core benefits package is required (specific minimum levels of health, disability, retirement, and death benefit) plus a group of elective programs from which the employee may select a set dollar amount. Additional coverage may be available through employee contributions.

Callback Pay. Pay, usually at premium rates, received by a worker called back to duty after completing his or her regular assignment.

Call-in Pay. The amount of pay guaranteed to a worker who is called to work on a day on which he or she otherwise would have not reported, and finds no work available or is not given a full or half shift's employment. Call-in pay may be higher than the amount of reporting pay, and may be provided for at premium rates on specified premium days, such as Saturdays, Sundays, and holidays. (See REPORTING PAY.)

Cancelled Rate. A rate of pay or production which has been changed or cancelled because of changes in method, material, or because an error has been found in the rate.

Capped Rate. A control limit placed on certain incentive standards which control the top earnings level. Can be a self-imposed level of production which employees will not exceed or a control designed into the incentive plan.

Card Check. A procedure whereby signed union authorization cards are checked against a list of workers in a prospective bargaining unit to determine if the union has majority status. The employer may recognize the union on the basis of this check without the necessity of a formal election. Often conducted by an outside party, e.g., a respected member of the community. (See AUTHORIZATION CARD.)

Casual Workers. Workers who have no steady employer, but who shift from employer to employer. Also used in longshoring to refer to workers not regularly attached to a particular work group. Sometimes applied to temporary employees.

Catastrophe Insurance. (See MAJOR MEDICAL EXPENSE BENEFITS.)

Cell Manning. Similar to team manning in which all of the operators in a group are able to perform all of the operations in the group.

Central Labor Council. An AFL-CIO organization formed by association of local unions in a community or other geographical area, to further union interests and activities. Also known as City Central Body.

Certification. Formal designation by a government agency, such as the National Labor Relations Board, of the union selected by the majority of the employees in a supervised election to act as exclusive bargaining agent for all employees in the bargaining unit.

Changeover Time. The time required to modify or replace an existing facility or workplace, usually including both teardown time for the existing condition and setup for the new condition.

Chapter. (See LOCAL UNION.)

Charade. Action during negotiations to achieve a certain position through pretense or bluffing.

Charge. (NRLB) A written statement of alleged unfair practices filed with the NLRB by management, labor, or any aggrieved employee under the provisions of the Labor-Management Relations Act.

Charter. Written authorization to establish a union or a subordinate or affiliated body.

Check-off. The practice whereby the employer, by agreement with the union, regularly withholds from the wages of union workers assessments and dues, and transmits these funds to the union. Under the Labor Management Relations Act of 1947, the employer must receive from each employee a written assignment which shall not be revocable for a period of more than one year, or beyond the termination date of the agreement, whichever occurs sooner.

Civil Rights Act. Under Title VII of this federal act (1964), employers, unions, and employment agencies are required to treat all persons equally, regardless of race, color, religion, sex, or national origin, in all phases of employment, including hiring, promotion, compensation, firing, apprenticeship, job assignments, and training. An Equal Employment Opportunity Commission was created to assist in carrying out this section of the act.

Civil Service Reform Act. Title VII of this act was passed in 1978 and is known as the most significant change in federal personnel administration since the passage of the Civil Service Act pub 1883. It deals with unfair labor practices in the federal government.

Classification. A category or group in which a job is located on basis of pay rate, competence or training level, seniority or other job identification criteria.

Classification Act Employees. Federal government employees, typically professional, administrative, technical and clerical employees, whose salary rates and certain other conditions of employment are determined by Congress. (See WAGE BOARD EMPLOYEES.)

Classification Change. A grade or title change in classification based on reevaluation or change in duties and responsibilities of a particular position.

Classification Index. A numerical value assigned to a job classification which establishes the relationship of that job classification to others.

Classification Method of Job Evaluation. A method which compares jobs on a whole job basis. Predefined class descriptions are established for a series of job classes and a job is placed in whichever classification best describes it.

Class of Positions. A group of positions, regardless of

location, that are alike enough in duties and responsibilities to be called the same descriptive title, to be given the same pay scale under similar conditions, and to require substantially the same qualifications.

CLASS RATE. Pay applicable to a given grouping of jobs or employees.

CLEANUP TIME. Paid time allowed to workers to clean their workplaces or tools or to wash up before leaving the plant at the close of the workday or for lunch. Also known as Washup time. (See CLOTHES CHANGING TIME.)

CLOCK-OUT (OFF). The act of punching off a job on a time clock for a variety of reasons such as, machine breakdown, waiting for work or shift end.

CLOSED SHOP. Form of union security provided in an agreement which binds the employer to hire and retain only union members in good standing. The key distinction between a closed shop and a union shop lies in the hiring restriction, a restriction prohibited by the Labor Management Relations Act, 1947. Legal closed shops may be found outside the scope of this act (which applies to employers and employees in industries affecting interstate commerce) and outside of states with "right-to-work" laws.

CLOSED UNION. A union which bars new members or makes membership acquisition difficult (e.g., by very high initiation fees) in order to protect job opportunities for its present members, or for other reasons. Some unions accept only sons of present members. (See OPEN UNION.)

CLOTHES CHANGING TIME. Time allocated within the paid workday for changing from street wear to working clothes, or from working clothes to street wear, or both. (See CLEANUP TIME.)

CLOTHING ALLOWANCE. An allowance granted by an employer to those employees who are required to buy special clothing such as uniforms, safety shoes, and other safety garments, in connection with the performance of their work.

CODETERMINATION. A system of governing an organization where the employees in addition to management help run the company through union or worker representation on the board of directors.

CODES OF ETHICAL PRACTICES. Rules adopted by the AFL-CIO in 1956-57, setting standards of behavior for unions and their officers.

COFFEE BREAK. (See REST PERIOD.)

COLA. (See COST OF LIVING ADJUSTMENT).

COLLECTIVE BARGAINING. Method whereby representatives of the employees (the union) and employer determine the conditions of employment through direct negotiation, normally resulting in a written contract setting forth the wages, hours, and other conditions to be observed for a stipulated period (e.g., two years). (See AGREEMENT.)

COMMISSION. Compensation for services rendered in arranging a transaction, or be a percentage of the transaction amount.

COMMISSION EARNINGS. Compensation to sales personnel based on a percentage of value of sales. Commission earnings may be in addition to a guaranteed salary or may constitute total pay. Sales personnel on straight commission usually have a fixed drawing account which is balanced against actually realized commission earnings at specific periods. (See DRAWING ACCOUNT.)

COMMON LABOR. General term used to designate unskilled workers performing heavy labor. In specific plants, may refer to unskilled workers not assigned to a particular job. The latter use is probably now the more frequent one.

COMMON LABOR RATE. The hourly rate paid for physical or manual labor of a general character and simple nature requiring no special training or skill and requiring little or no previous experience.

COMMUNICATION. The passing of information and understanding from a person to another person or a group.

COMMUNITY WAGE SURVEY. A general term used to describe a survey designed to reveal the structure and level of wages within a particular geographic area for a given industry or, more typically for broad categories of industry. (See AREA WAGE SURVEY.)

COMPANY UNION. A derogatory charge leveled against a union suspected of being an ineffectual employee representative.

COMPANYWIDE BARGAINING. (See MULTIPLANT BARGAINING.)

COMPARABLE RATE. A rate paid for work agreed or determined to be comparable within a plant or occupations with similar characteristics in other industries.

COMPARABLE WAGES. Equal pay for equal work.

COMPARABLE WORTH. The doctrine that men and women who perform work of the same "inherent value" should receive comparable compensation, excepting allowable differences (for example, seniority plan, merit plan, production-based pay plans, or different locations).

COMPA-RATIO. The ratio of an actual pay rate (numerator) to the midpoint for the respective pay grade (denominator). Compa-ratios are used primarily to compare an individual rate of pay to the midpoint of some other control point of the structure. It is most frequently used as an index of a person's relationship to the structure. In addition, compa-ratios can be calculated for a group of people, a department, or an entire organization.

COMPENSABLE INJURY. A work injury for which compensation indemnity benefits are payable to the injured worker or his or her beneficiary under worker's compensation laws.

COMPENSATION. The total of value received for the performance of a job. Sometimes used to encompass the entire range of wages and benefits, both current and deferred, which workers receive out of their employment.

COMPENSATORY TIME OFF. A compensation plan for overtime work which gives the employee paid time off from duty in lieu of overtime pay. The Supreme Court ruled such plans unconstitutional for certain federal or government agencies, requiring full overtime rates for overtime worked.

COMPETITIVE WAGE. The wage within a given labor market required to balance the demand and supply for a particular labor type that is required by a company to maintain a competitive price position with other firms in the same industry.

COMPLAINT. (NLRB) A formal paper issued by the NLRB, stating the alleged unfair labor practice and the basis for the Board's jurisdiction, in order to start an unfair labor practice hearing.

COMPRESSION. Reduction of the pay differentials for jobs having different responsibility and skill requirements. May apply either to actual pay or pay ranges.

COMPULSORY ARBITRATION. (See ARBITRATION.)

COMPULSORY RETIREMENT. Involuntary separation from employment in a company upon reaching a specified age. In precise pension terms, a distinction is usually made between compulsory and automatic retirement. The age of compulsory retirement is that point at which a worker loses the right to decide whether he should retire or continue on his job. The age of automatic retirement is the age beyond which no employee may continue to work under the terms of the pension plan. In other words, an employee may work beyond the compulsory retirement age if the employer consents, but automatic retirement rules out the option on both sides. (See AUTOMATIC RETIREMENT.)

CONCESSION BARGAINING. A negotiating pattern in which a union agrees to certain reductions in pay and/or benefits in order to assist a company in becoming more competitive.

CONCILIATION. (See MEDIATION.)

CONFLICT. The simultaneous presence of opposing or mutually exclusive impulses, desires, or tendencies.

CONFORMITY. A general term referring to adherence to a group norm concerning beliefs, values, attitudes, or behavior. Such adherence may reflect a variety of underlying psychological processes.

CONSTANT SHARING PLAN. A series of incentive plans in which the percent earnings of operators above a base level is either greater than or less than the percent increase in production. An incentive plan with a straight line incentive pay curve which does not pass through the (0/0) point. Most common constant sharing plans pay less than 1:1 incentive bonus; usually a 50-50 sharing plan in which the employees receive a 1% bonus for each 2% increase in production.

CONSTANT TOTAL COST PLAN. An incentive plan with a pay curve that keeps the total cost per piece (labor plus overhead) at a constant rate.

CONSTANT UNIT LABOR COST PLAN. An incentive plan with a pay curve that keeps the labor cost per piece a constant.

CONSULTATION. An obligation on the part of management to consult the union on particular issues (e.g., contracting-out) in advance of taking action is frequently provided by agreements. What consultation actually means in each situation is what the parties want it to mean. In general, the process of consultation lies between notification to the union, which may amount simply to providing information, and negotiation, which implies agreement on the part of the union before the action can be taken.

CONSUMER PRICE INDEX (CPI). A government index, issued monthly by the Bureau of Labor Statistics, which measures the average change in prices of goods and services purchased by urban wage- earner and clerical-worker families. (See COST OF LIVING ADJUSTMENT, ESCALATOR CLAUSE.)

CONTINUOUS BARGAINING COMMITTEES (INTERIM COMMITTEES). Committees established by management and union in a collective bargaining relationship to keep the agreement under constant review, and to discuss possible contract changes, long in advance of the contract expiration date. May provide for third- party participation. (See HUMAN RELATIONS COMMITTEES, CRISIS BARGAINING.)

CONTINUOUS OPERATIONS. Necessary plant operations (powerhouse, maintenance, plant protection, including some manufacturing processes, etc.) that must continue to function on a 24-hour, 7-day basis. (See CONTINUOUS PROCESS)(ROUND-THE-CLOCK OPERATIONS.)

CONTINUOUS PROCESS. A process which, once begun, must continue without interruption for a long period, making the use of multiple shifts necessary.

CONTRACT. (See AGREEMENT.)

CONTRACT BAR. A denial of the request for a representation election, based on the existence of any agreement. Such an election will not be conducted by the National Labor Relations Board if there is in effect a written agreement which is binding upon the parties, has not been in effect for more than a "reasonable" time, and its terms are consistent with the National Labor Relations Act.

CONTRACTING-OUT (SUBCONTRACTING; FARMING OUT). Practice of having certain steps in a manufacturing process, plant maintenance, or other work functions performed by outside contractors.

CONTRACT WAGE PAYMENT. An arrangement whereby the worker contracts to perform a specific job for a predetermined amount of compensation.

CONTRIBUTORY PENSION PLAN. A pension plan for the benefit of the employee under which the cost is shared by both the employer and the employee. (See NON-CONTRIBUTORY PENSION PLAN.)

CONTROL (WAGE AND SALARY). The administration of compensation plans in accordance with set predetermined policy.

CONTROL POINT. The point within a salary range representing the desired average or median pay for a job or group of jobs at a given time.

COOLING-OFF PERIOD. A period of time that must elapse before a strike or lockout can begin or be resumed, by agreement or by law. The term derives from the hope that the tensions of unsuccessful negotiations will subside in time and that a work stoppage will be averted. (See NATIONAL EMERGENCY DISPUTE.)

CORRELATION. The relationship between total point score value of a job, as determined by job analysis and pay grade.

COST OF LIVING ADJUSTMENT (COLA). An adjustment of wages or salaries in accordance with a formula taking into consideration changes in the cost of living as measured by an appropriate index of the retail prices of goods and services that enter into the consumption of low- or moderate-income families. (See ESCALATOR CLAUSE.)

COST OF LIVING ALLOWANCE. Regular cents-per-hour or percentage payments made to workers through the operation of escalator clauses or other types of cost-of-living allowances not incorporated into base rates. Also known as float.

COUNCIL OF ECONOMIC ADVISERS. This small federal government agency, established under the terms of the Employment Act of 1946, advises the President on economic developments, appraises government economic growth and stability, and assists in the preparation of the President's annual economic report to the Congress. (See GUIDEPOSTS.)

CPI. (See CONSUMER PRICE INDEX.)

CRAFT. Usually, a skilled occupation requiring a thorough knowledge of processes involved in the work, the exercise of considerable independent judgment, usually a high degree of manual dexterity, and, in some instances, extensive responsibility for valuable product or equipment.

CRAFT UNION. A labor organization which limits member ship to workers having a particular craft or skill or working at closely related trades, for example, electricians, machinists, or plumbers. (See SKILLED TRADES.)

CREDITED SERVICE. Years of employment counted for retirement, severance pay, seniority, etc. The definition of a credited year of service varies among companies and plans.

CREDIT UNION. A financial institution voluntarily organized and operated by a group of individuals having a common interest; providing for deposits to and withdrawals from a common fund with the purpose of extending credit to members at low, regulated interest rates and to encourage thrift.

CREDIT UNION DEDUCTIONS. Automatic deductions made by an employer from an employee's pay and paid to the credit union to satisfy the employee's repayment of a loan. Done by agreement among the employer, the employee, and the credit union.

CREEP (CREEPING METHOD CHANGES). Changes in the method of performing a job over a long period of time so that the exact date when a change took place cannot be determined.

CRISIS BARGAINING. Term used to characterize collective bargaining taking place under the shadow of an imminent strike deadline, as distinguished from extended negotiations in which both parties have ample time to present and discuss their positions.

CRITICAL INCIDENTS. Those events which are particularly conducive to success or failure in performing any task.

CYCLE TIME. (1) The time required to perform one complete sequence of job activities. (2) A period of time within which a round of regularly recurring events is completed.

DAILY/DAY RATE. For a worker hired on a daily rate basis, the rate of pay is normally expressed as a rate for a standard number of work hours per day; the hourly rate, the quotation of a daily rate normally excludes premiums that may be paid for late-shift work or overtime hours, as well as bonuses for special conditions of work or for other reasons unrelated directly to production.

DAY WORK. Usually refers to work for which pay is based on time as contrasted to pay based on output.

DEADHEADING PAY. Payment to a transportation worker for traveling without a load.

DEAD TIME. (See DOWNTIME) (DELAY TIME)

DEAD WORK. A term used in mining, referring to nonproductive work, including the removal of rock, debris, and other waste matter from the product mined.

DEATH BENEFIT. Payment, usually a lump sum of money, provided to a worker's beneficiary, in the event of death. May be provided by a pension plan or another type of employer-sponsored welfare plan, or by a union to its members. (See LIFE INSURANCE PLAN.)

DECERTIFICATION. Withdrawal by a government agency, such as the National Labor Relations Board (NLRB), of a union's official recognition as exclusive bargaining representative.

DECERTIFICATIONS ELECTION. An election conducted by the National Labor Relations Board (NLRB) to remove a union from representation if the employees currently represented by the union vote, by a majority of those voting, to do so.

DECISION. A determination arrived at after review, investigation, interpretation, or consideration of facts or information.

DECISION MAKING. The response to a need or stimulus by means of acquiring and organizing information to yield alternate courses of action; and selecting one course of action from the alternatives.

DEDUCTIBLE. That portion of insured expense which the insured must pay before the plan's benefits begin.

DEFERRED BONUS PAYMENT. A plan in which bonus earned in one period is paid out in one or more subsequent periods.

DEFICIT REDUCTION ACT OF 1984. Congress legislated new restrictions on the depreciation deductions and investment tax credits allowed to taxpayers for automobiles, airplanes, computers and other tangible properties used in a trade or business and/or in the production of income.

DEFERRED COMPENSATION. Money or other financial rewards earned in one period and paid in a subsequent period or periods. Deferred compensation can be salary deferrals, bonus deferrals, or supplementary retirement payments. The earnings are paid at a later date usually when the individual is in a lower tax bracket. No use of the funds can be made by the recipient until the agreed-upon transfer date. Postponed earnings designated for future income declaration under a qualified plan for postponing income tax on earnings until future years.

DEFERRED WAGE INCREASES. Wage changes that do not become effective until a specified future date.

DEFINED BENEFIT PENSION PLAN. A pension plan which specifies the benefits or the methods of determining the benefits but not the level or rate of contribution. Contributions are determined actuarially on the basis of the benefits expected to become payable.

DEFINED CONTRIBUTION PLAN. A procedure for investment of set contribution amounts for purposes of retirement, the value of which, at time of retirement, being dependent upon the market value of the investments made.

DELAY. An interruption of a specified sequence of work activity.

DELAY ALLOWANCE. A time adjustment for unavoidable interruption of work activity.

DELAY TIME. (See DOWNTIME.)

DEMOTION. An employee transfer to a lower paying and/or responsibility job.

DEPRESSED (INCENTIVE OR STANDARD). An incentive operation in which the performance is below the 100% level and where the operator is guaranteed at least 100% of the base rate.

DICTIONARY OF OCCUPATIONAL TITLES. A comprehensive compilation of job descriptions and titles published by the U.S. Department of Labor.

DIFFERENTIAL PIECE RATE. Plan under which piece rates vary at different levels of output.

DIFFERENTIAL PIECEWORK. A wage incentive plan using two or more piece rates. One rate is paid for production of a fixed quantity in a given time, and a higher rate(s) is paid for production beyond the initial level. Also known as Taylor differential piecework plan. (See MULTIPLE PIECE RATE PLAN.)

DIFFERENTIAL SKILL. Those objectively measurable factors (skills) which are required for a position, and, thereby, define that position as uniquely different from another position. The skills demonstrated by one employee as compared to those of any other employee.

DIFFERENTIAL TIME PLAN. A wage incentive plan that pays a higher base wage rate as production levels increase. A series of step bonuses for increased production.

DIRECT COMPENSATION. As opposed to indirect compensation, all forms of compensation before required (FICA, income taxes) and/or authorized (Community Chest, 401k, insurance premiums) deductions that involve direct and immediate payment to the individual, often including a base wage or salary payment.

DIRECT LABOR. (1) Work which is readily chargeable to or identifiable with a specific product. (2) Work performed on a product or service that advances the product or service towards its ultimate specifications or objectives. (3) Any labor whose cost is directly allocated to a product.

DIRECT LABOR BUDGET. A projection correlated with future time periods of total direct labor required to meet forecast work or production.

DIRECT OBSERVATION. A job analysis technique which involves direct observation of incumbent(s) actually performing the job.

DIRTY MONEY. Extra pay for especially disagreeable work.

DISABILITY. Any injury or illness, temporary or permanent, which prevents a worker from carrying on his/her usual occupation. (See PERMANENT AND TOTAL DISABILITY.)

DISABILITY PAY. Compensation for time away from work due to work-related injury or illness.

DISABILITY RETIREMENT. Retirement because of physical inability to perform the job. (See PERMANENT AND TOTAL DISABILITY.)

DISAFFILIATION. Withdrawal of a local union from membership in a national or international union.

DISCHARGE. Dismissal of a worker from employment for cause.

DISCIPLINARY INTERVIEW. The right of an employee who has been charged with wrong-doing; it precedes the assessment of discipline or discharge. It is a tool used to determine if the disciplinary action contemplated is appropriate or necessary.

DISCIPLINE. (1) (noun) A force resulting from training and education which causes an individual or group to react on the basis of close cooperation, mutual understanding, and knowledge of regulations, customs, habits, traditions, or mores of the group. (2) (verb) To apply penalties for undesirable behavior.

Discrimination. (1) Term applied to a prejudice against or unequal treatment of workers in hiring, employment, pay or conditions of work, because of race, national origin, creed, color, sex, religious beliefs, age, union membership or activity, or any other characteristic not related to ability or job performance. (2) A more general meaning is "to distinguish between or among".

Dismissal Compensation. (See SEVERANCE PAY.)

Dispute. Disagreement between union and management. (LABOR DISPUTE).

Dividend. Shareholder's earnings from corporate profits based upon stock ownership (number of shares owned).

Division of Labor. Differentiating of work into specialized jobs or positions.

Double Back Pay. Penalty reimbursement to discharged employees. Specified by Labor Reform Act of 1977 (NLRA) as penalty for discharge of employees for protected activities during a union organization campaign.

Double Dipping. Receiving retirement pension payments from two or more employers.

Double Time. Premium rate for overtime work, amounting to twice the employee's regular pay rate for each hour worked.

Downgrading/Downgrade. The lowering of a particular job in scope, authority, responsibility, degree of difficulty and skills required, with a possible reduction in wage or salary.

Downtime. Time during which an operation is halted for such causes as: lack of material, machinery breakdown, improper tooling, power loss, etc.

Draw. A designated amount against which expenses/wages may be charged and payment withdrawn.

Drawing Account. An account from which advances from commissions are paid.

Dual Pay. A system of wage payment used by railroads under which employees are paid on a mileage or hourly basis. A standard mileage is defined as a basic day, usually eight hours, for the purpose of determining the daily rate. Wages are computed on the number of hours or miles, whichever yields the greater compensation to the employee.

Dual Unionism. A charge (usually a punishable offense) leveled at a union member or officer who seeks or accepts membership or position in a rival union or otherwise attempts to undermine a union by helping its rival.

Dues. Refers to money collected from union members by a union as a cost of belonging to the union. The amount of union dues is established by the members or their elected representatives at the union's convention.

Early Retirement Age (pension plans). The age when an employee is first permitted to retire and to elect either immediate or deferred receipt of income. If payments begin immediately, they are generally paid in a reduced amount. Company consent to the election may or may not be required.

Earned Hours. The product of pieces or quantity produced times the times standard, usually in hours. Used as the basis for measuring labor productivity, computing incentive earnings, or measuring efficiency (performance).

Earned Rate. Used to denote pay rate for job performed, as in the case of job transfer.

Earnings. Remuneration for services performed during a specific period of time. The term invariably carries defining words, e.g. hourly, daily, weekly, annual, average, gross, straight-time hourly earnings. Since a statistical concept is usually involved in the term and its variations, the producers and users of earnings figures have an obligation to define them. The Bureau of Labor Statistics, in its wage surveys, defines straight-time earnings to exclude premium pay (for overtime and for work on weekends and holidays) and shift differentials. Compensation is a term sometimes used to encompass the entire range of wages and benefits, both current and deferred, which workers receive out of their employment.

Earnings Efficiency. Actual earnings, normally calculated in terms of earned hours, compared to standard or expected earnings.

Economic Opportunity Act. A 1964 act "to mobilize the human and financial resources of the Nation to combat poverty in the United States," which includes the work-training program (administered by the U.S. Department of Labor), is directed to encourage young unemployed persons (age 16-21, inclusive) to stay in school or obtain job experience that would prepare them for meaningful work careers.

ECONOMIC SUPPLEMENTS. A type of compensation for employees including things like pensions, vacations, paid holidays, sick leave, health and insurance, and supplemental unemployment benefits.

ECONOMIC STRIKES. Union-authorized strikes to bring about changes in wages, hours, or working conditions, usually associated with an impasse in contract negotiations.

EDUCATIONAL ASSISTANCE. Reimbursement of all or a share of tuition and expenses for an employee's studies at a training institution not directly associated with the employee's firm.

EFFICIENCY. A measure of work performance determined by comparing actual work done with the standard amount of work for this job determined a "normal." The ratio of a standard performance time to an actual performance time is usually expressed as a percentage.

EFFORT. Physical and mental exertion/energy required to perform a given job or task. A criterion or job factor of a job evaluation plan.

EFFORT RATING. Evaluated worker performance. In work measurement, an estimate of observed performance, normally expressed as a percentage of "normal" performance. (See LEVELING, NORMALIZE.)

ELEMENT. A basic segment of a job or task identified by a distinct start/stop breakpoint, e.g., grasp, or release.

ELEMENT TIME. Time allowed to perform a specified part of a job or task at normal effort.

ELISA TEST. The most commonly used and the least expensive test to determine whether an individual is infected by HIV, a precursor of AIDS.

EMERGENCY BOARD. A board appointed by the United States President, whenever a labor dispute threatens to seriously interrupt interstate commerce, to investigate and report on the conditions underlying the dispute within a specified time period. After the board reports, no change may be made in the conditions underlying the dispute, except by agreement of the parties. This is the last formal step in the procedures regarding contract disputes.

EMPLOYEE. General term for an employed wage earner or salaried worker.

EMPLOYEE ASSISTANCE PROGRAMS (EAP). Programs specifically designed to assist employees with chronic personal problems (for example: marital dysfunctions, alcohol or drug abuse) that hinder their job performance, attendance, and corporate citizenship.

EMPLOYEE BENEFITS. Provisions made available to an employee by an employer, e.g. health insurance, vacations, holidays, schooling allowance, etc. Such benefits may be paid totally by the employer or the costs may be shared.

EMPLOYEE BUY-OUT. A program whereby employees of a company band together to buy the company from the owners (stockholders) and then operate it as the new owners.

EMPLOYEE CLASSIFICATION. A designation given to an employee based upon some selected basis for differentiation such as pay grade, skill or trade, length of employment, membership of a work unit, etc.

EMPLOYEE CONTRIBUTIONS. Any contributions to a benefit plan made by employees.

EMPLOYEE INVOLVEMENT PROGRAMS. Programs set up by an employer, frequently in cooperation with a union, to allow employees to be involved in the management of various operations or jobs.

EMPLOYEE PACING. Condition in which the pace or rate at which the employee works is determined by the employee and not the machine as under machine pacing.

EMPLOYEE POLYGRAPH PROTECTION ACT (1988). Prohibits most employers from polygraph testing without reasonable suspicion.

EMPLOYEE SERVICES. A wide range of employer provisions at no cost to the employee or made available at discounted or token prices.

EMPLOYEE SERVICES AND PERQUISITES. A form of indirect compensation that varies depending on employee type and organization to offset the pressures associated with working (for example, day care) or used to symbolize a status differential eg. Company paid memberships to country, athletic, and social clubs.

EMPLOYEE STOCK OWNERSHIP PLAN (ESOP) (LEVERAGE ESOP). A plan in which a company borrows money from a financial institution using its stock as security or collateral for the loan. Principal and interest loan repayments are tax deductible. With each loan repayment, the lending institution releases a certain amount of stock being held as security. The stock is then placed into an Employee Stock Ownership Trust (ESOT) for distribution at no cost to all employees. The employees receive the stock upon retirement or separation from the company.

EMPLOYER. General term for any individual, government agency, corporation, or other operating group, which hires workers (employees). The terms "employer" and "management" are often used interchangeably when there is no intent to draw a distinction between owners and managers.

EMPLOYERS' ASSOCIATION. Voluntary membership organization of employers established to deal with issues common to the group. It may be formed specially to handle industrial relations and to negotiate with a union or unions. (See ASSOCIATION AGREEMENT.)

EMPLOYMENT CONTRACT. A legal agreement concerning period of employment, earnings, conditions of employment and duties of the employee and the employer.

EMPLOYMENT DIRECTOR (MANAGER). A specialist who recruits, investigates, selects, and recommends the hiring of personnel to meet the requirements of the firm or agency.

EMPLOYMENT SECURITY. (See JOB SECURITY.)

ENTRANCE RATE. The hourly rate a worker receives upon being hired.

ENTRY LEVEL. Basic or lowest job in a line of progression in which employees entering that line of progression first accumulate seniority rights.

ENTRY RATE. Compensation initially paid new or transferred employees during a probationary or training period.

EQUAL EMPLOYMENT OPPORTUNITY COMMISSION (EEOC). A commission of the Federal Government charged with enforcing the provisions of the Civil Rights Act of 1964. In addition the commission is charged with enforcing the provisions of the Equal Pay Act of 1963 as it pertains to sex discrimination in pay.

EQUAL PAY ACT OF 1963. An amendment to the Fair Labor Standards Act of 1938, prohibiting pay differentials on jobs that are essentially equal in terms of skill, effort, responsibility, and working conditions, except when they are the result of bonafide seniority, merit or production-based pay systems or other-job related factor other than gender.

EQUAL PAY FOR COMPARABLE WORK. This doctrine is much broader than the equal pay doctrine. In effect this doctrine moves beyond the four compensable factors embodied in the Equal Pay Act of 1963 and specifies that so long as the work performed by women and men in question is comparable, charges of sex discrimination in pay can be brought under Title VII of the 1964 Civil Rights Act. The U.S. Supreme Court in Gunther vs. County of Washington allowed plaintiffs to bring suit on sex discrimination in pay outside the narrow confines of equal work and under the broader confines of equal pay for comparable work. This principle exists midway between the extremes of the equal pay for equal work doctrine and the comparable worth doctrine. (See EQUAL PAY FOR EQUAL WORK.)

EQUAL PAY FOR EQUAL WORK. The doctrine embodied in the Equal Pay Act of 1963, intended to protect women from sex discrimination in pay. The Equal Pay for Equal Work doctrine says that women may not be paid less than men on jobs that are equal in terms of four compensable factors: equal effort, equal skill, equal responsibility, and equal working conditions. The Equal Pay for Equal Work doctrine is a very narrow definition of pay equity across the sexes.

EQUITY. Anything of value earned through the provision or investment of something of value. In the case of compensation, an employee earns equity interest through the provision of labor on a job. So defined, equity is often used as a fairness criterion in compensation: People should be paid according to their contributions. Also, one of the four basic principles in productivity gainsharing plans. For banking and accounting, equity means the capital accruing to ownership that results after the subtractions of liabilities from assets.

EQUITY FUNDING. The funding of a portion of a retirement plan investment in equity. Equity funding affects the employer's contributions but not the employees' benefits.

EQUIVALENT WORKERS. A method used to calculate the total number of employees that would be required if all operators worked 40 hours a week for 52 weeks a year. The total actual hours worked divided by the product of 40 hours times 52 weeks a year.

ERISA. The Employee Retirement Income Security Act, passed by Congress in 1974, legislated the manner for administering pensions, pension plans, and other service-related benefits. Compliance of ERISA legislation is the responsibility of the Department of Labor, while the administration of taxation on benefits is the responsibility of the Internal Revenue Service.

ERTA. The Economic Recovery Act of 1981 enabled employees to establish Individual Retirement Accounts. The 1981 Act also accelerated tax deductions for individuals and businesses on the depreciation of real property and tangible personal property. The rules for these tax shelters were amended successively by the Tax Reform Act of 1986, the Omnibus Budget Reconciliation Act of 1989 and the Unemployment Compensation Amendments of 1992.

ESCALATOR CLAUSE. A provision in a union agreement allowing for the adjustment of wages in accordance with specific changes in the cost of living as measured by an appropriate index, the price of production materials or some other agreed-upon criterion. (See COST OF LIVING ADJUSTMENT.)

ESCAPE CLAUSE. General term signifying release from an obligation. One example is found in maintenance- of-membership arrangements which give union members an "escape-period" during which they may resign from membership in the union without forfeiting their jobs.

ESOP. (See EMPLOYEE STOCK OWNERSHIP PLAN.)

EVALUATED RATE. The base pay for a job determined by an analytical procedure for comparing jobs and setting rates of pay.

EVALUCOMP. A job evaluation plan for office personnel, technicians, professional and scientific personnel, and all levels of management. Developed and offered by the American Management Association, Evalucomp differs from traditional job evaluation systems in that no effort is made to construct an internal hierarchy of jobs on the basis of job characteristics. Rather an extensive description of each job is provided to allow for matching along with Evalucomp wage survey data.

EXCLUSIVE BARGAINING RIGHTS. The right and obligation of a union designated as majority representative to bargain collectively for all employees, including nonmembers, in the bargaining unit.

EXECUTIVE BOARD. Constitutional administrative body composed of elected officials and other elected or appointed members generally responsible for overseeing operations and establishing operating policies.

EXECUTIVE COMPENSATION. Money and fringe benefits paid to an executive for services performed. Executive compensation may include money, deferred financial payment, car, club membership, stock options and other perquisites.

EXECUTIVE COMPENSATION PLAN. Any of a number of special pay arrangements to motivate and reward key management employees. Includes incentive bonuses, stock plans, insurance deferred compensation, and perquisites.

EXEMPT EMPLOYEE/JOB. Any employee or job excluded from the Wage and Hour Law. Includes executive, administrative, professional, or outside salespersons who meet certain criteria covering earnings, direction of work force, exercise of discretion and independent judgment, advanced knowledge in a field of science or learning, or selling outside the employer's place of business.

EXEMPT JOB. A job not subject to the provisions of the Fair Labor Standards Act with respect to minimum wage and overtime. Exempt employees include most professionals, administrators, executives, and outside sales representatives.

EXPECTED EARNINGS (LEVEL). Under an incentive plan, the level at which employees are targeted to earn. Incentive plans are designed to allow a properly skilled operator working with the necessary effort to earn 125% or 133% (or some other value) above the base rate. This target level is the expected earnings rate or level.

EXPENSE ACCOUNT. An account of reimbursable expenses paid or incurred by an employee in connection with the performance of services such as transportation, meals, and lodging while away from home.

EXPERIENCE CURVE. (See LEARNING CURVE.)

EXPERIENCED EMPLOYEE. A worker who has satisfactorily performed a job or task for a lengthy period of time.

EXPERIENCE RATING. (1) Process of basing tax rates or insurance premiums on the employer's own record; as in worker's compensation, unemployment insurance, and commercially insured health and insurance programs. (2)The process of classifying individuals according to previous experience and/or qualifications.

EXPERT WITNESS. A person with special skills or experience in a profession or technical area who testifies in a legal proceeding.

EXPIRATION DATE. Formal termination date established in a collective bargaining agreement.

EXTENDED LEAVE PLAN. A plan allowing workers to take extended, unpaid leave without loss of job or seniority for specific reasons.

EXTENDED VACATION PLAN. A plan providing extra-long paid vacations to qualified (long service) workers at regular intervals supplementing an annual paid vacation plan.

FACT-FINDING BOARD. A group of individuals appointed by a governing body to investigate, assemble, and report the facts in a labor dispute, sometimes with authority to make recommendations for settlement. (See BOARD OF INQUIRY.)

FACTOR COMPARISON. A job evaluation method comparing and evaluating jobs for purposes of pay determination based on selected key factors or attributes common to the jobs being compared.

FACTOR SCALES. The point ranges assigned to the various factors (skill, effort, responsibility, etc.) in a job evaluation plan which defines the importance of the factor in the overall plan.

FACTOR WEIGHT. A weight indicating the relative importance of a compensable factor in a job evaluation system.

FAIR DAY'S PAY. The level of pay for a job for which a standard level of work performance is expected. Equated to a "fair day's work." Normally the base rate or guaranteed pay rate of a job. The "normal" or "100%" base of a wage incentive plan.

FAIR DAY'S WORK. Standard level of work performance for the base or guaranteed pay rate "fair day's pay."

FAIR EMPLOYMENT PRACTICE LAWS. Laws that forbid discrimination in hiring, promotion, discharge, or conditions of employment on the basis of race, creed, color, national origin, and in some cases, sex and age. Such laws exist at federal, state, and local levels. (See CIVIL RIGHTS ACT.)

FAIR LABOR STANDARDS ACT. The 1936 federal law which prohibits oppressive child labor and established a minimum hourly wage and premium overtime pay for hours in excess of a specific level (now time and one-half after 40 hours per week) for most workers engaged in interstate commerce. The minimum wage and the coverage of the act have been modified several times since enactment. Also known as FLSA; wage-hour law.

FAMILY AND MEDICAL LEAVE ACT (FMLA). This act, effective August 5, 1993, requires private-sector employers, with 50 or more employees, to provide up to 12 weeks of unpaid leave in any 12 month period, with guaranteed reemployment, for the birth, adoption or placement of a child, care of a family member with a serious health condition and an employee with a serious health condition. Eligible employees must have worked approximately 25 hours a week for one year.

FARMING-OUT. (See CONTRACTING-OUT.)

FATIGUE. A decline in performance due to physiological or psychological tiring.

FATIGUE ALLOWANCE. Percentage of time added to the normal time to allow for excessive fatigue.

"FAVORED NATIONS" CLAUSE. A provision indicating that one party to the agreement (employer or union) should have the opportunity to share in more favorable terms negotiated by the other party with another employer or union.

FEATHERBEDDING. A practice, working rule, or agreement provision which limits output or requires employment of excess workers thereby preserving unnecessary jobs.

FEDERAL LABOR RELATIONS AUTHORITY (FLRA). The Authority with responsibility for adjudicating unfair labor practices and other labor disputes within the federal government. (Similar to the NLRB)

FEDERAL MEDIATION AND CONCILIATION SERVICE (FMCS). An independent U.S. Government agency which provides mediators to assist the parties engaged in interstate commerce and involved in negotiations, or labor disputes, in reaching a settlement; provides lists of suitable arbitrators on request; and engages in various types of "preventive mediation." Mediation services are also provided by several state agencies.

FEDERATION. Association formed to promote common interests.

FINAL OR FINAL AVERAGE PAY (PENSION BENEFIT FORMULA). A formula which bases benefits on the credited earnings of an employee at or during a selected number of years (typically 3-5) immediately preceding retirement.

FINANCIAL INCENTIVE. Monetary compensation other than base pay to encourage and reward productivity above standard work performance.

FINANCIAL INCENTIVE PLAN. A method or system for providing financial incentives for increased worker productivity. (See PRODUCTIVITY GAINSHARING (PGS))

FINISH-GO-HOME BASIS OF PAY. A practice under which employees are permitted to go home after completing a specific work assignment generally considered a standard day's work.

FINK. Also applies to an informer or "company spy" (See STRIKEBREAKER).

FIRST AID CASE ACCIDENTS. An accident in which the person injured receives medical treatment, usually at the company's medical center and/or at the company's expense, and is permitted to return to his/her regular job within 24 hours after the injury.

FIXED BENEFIT RETIREMENT PLAN. A type of defined benefit pension plan providing retirement benefits in a fixed amount or at a fixed percentage.

FIXED SHIFT. The type of shift on which a group of workers maintains the same schedule of hours week after week, rather than rotating time-of-day assignments periodically with other groups.

FLAGGED RATE. (See RED CIRCLE RATE.)

FLAT BENEFIT RETIREMENT PLAN. A defined benefit pension plan providing benefits unrelated to earnings. An example would be one that specifies a certain amount of money per month per year of service.

FLAT INCREASE. A given adjustment in compensation similarly applied to all employees within a group. May also be called an across-the-board increase.

FLAT RATE. A fixed compensation for a job task.

FLEXIBLE BENEFITS PROGRAM. (See CAFETERIA BENEFITS.)

FLEXTIME. Work schedule arrangement in which an employee is allowed flexibility within certain time constraints to choose individual or group working hours.

FLOOR UNDER WAGES. (See MINIMUM WAGE.)

FLSA. (See FAIR LABOR STANDARDS ACT.)

FMLA. (See FAMILY AND MEDICAL LEAVE ACT.)

FOREIGN SERVICE PREMIUM. Additional compensation made available to employees assigned to positions in another country.

FORM AGREEMENT. Uniform agreement signed by individual members of an employers' association and often by employers in the same line of work but outside the association. (See STANDARD AGREEMENT.)

FREE RIDERS. A term applied by unions to nonmembers who, because of being in the bargaining unit, share in benefits resulting from union activities without paying dues.

FREQUENCY RATE OF ACCIDENTS. The number of disabling occupational injuries per million man-hours worked.

FRINGE BENEFITS. Benefits supplemental to wages received by workers, at a cost to employers. Among these benefits, commonly designated as "fringe" are paid holidays, paid vacations, pensions, child care, tuition refund, and insurance benefits (life, accident, health, hospitalization, and medical). Compensation other than direct wages or salary.

FRONT END. A compensation arrangement covering a set period of time in which the greater portion or rate of compensation occurs during the early part of the time period.

FULL CREW LAW OR RULES. Laws or regulations of several states which require a minimum number of workers having specified skills for each railway train, e.g., engineer, firemen, conductors, brake men, and flagmen.

FULL-TIME EARNINGS. Earnings received for working a regular schedule of hours over a stated period of time, e.g. a day, week, month, et. al.

FULL-TIME WORKER RATE. A rate paid to a full-time worker as distinguished from that paid to a part-time, or temporary worker.

FUND. (noun) Money and investments set aside in a separate account to take care of the payment of pensions, supplemental unemployment benefits, strike benefits, etc. (verb) To set up a (fund); to set aside adequate reserves. Also known as trust fund.

Funeral Leave Pay. Pay to the worker for time lost because of the death and funeral of a designated member of his family. Also known as bereavement pay.

Gainsharing. A financial incentive plan in which the savings from improved productivity (savings gained) are shared between the company and the employees. (See PROFIT SHARING, PRODUCTIVITY GAINSHARING (PGS))

Garnishment of Wages. The practice of legally attaching the wages of a debtor and collecting the debt directly via his employer.

General Wage Changes. Wage adjustments which affect large numbers of workers in a similar manner at the same time.

Geographic Differentials. Pay differentials established for the same job based on differences in cost of living between or among two or more geographical areas.

Ghost Employees. Employees on the payroll who do not actually report for work; frequently used as a conduit for illegal payments to various individuals or groups.

Giveaways. Benefits offered by a company and not asked for by the union or the employees.

Givebacks. Negotiated elimination from a future labor agreement of past benefits or wages. Also known as concessions.

Going Rate (Going Wages). Usual or prevailing compensation for a job or job classification. (See PREVAILING RATE.)

Golden Handcuffs. Employee benefits and related incentives that tend to hold an individual in continued employment by an organization.

Golden Parachute. A special financial protection plan for key executives in the event of an unfriendly takeover by another firm.

Gonveal. Illegal return of a portion of earnings by a worker to an employer or supervisor as a tribute for retaining his job.

Goon. Slang term for a person hired by either management or union during a labor dispute to make trouble and intimidate the opposition by violence or the threat of violence.

Grade Description. (See JOB DESCRIPTION) (POSITION DESCRIPTION.)

Grandfathering. Maintaining of benefits or other advantages for designated individuals in specific instances because of seniority or prior circumstance. (See RED CIRCLE RATE.)

Graveyard Shift. The shift beginning around midnight and ending in the morning. (See SHIFT.)

Green Circled Rate. An individual rate that is being paid below the minimum rate of the established pay range for that person's job or pay grade. The rate is green circled to assure that a speeded up adjustment is made to bring the person into the rate range.

Grievance. Any complaint or expressed dissatisfaction by an employee in connection with job, pay, or other aspects of employment. An alleged violation of the contract. Whether it is formally recognized and handled as a "grievance" depends on the scope of the grievance procedure.

Grievance Committee. (See SHOP COMMITTEE.)

Grievance Procedure. A formal plan, specified in union agreements, which provides a channel for the adjustment of grievances through discussions at progressively higher levels of authority in management and union, often culminating in arbitration if necessary. Formal grievance procedure plans may also be found in nonunion companies.

Grievance Steps. The specified steps in a grievance procedure by which a grievance dispute moves from one level of authority to the next higher level.

Gross Average Hourly Earnings. A measure of hourly wages obtained by dividing total compensation, including premium payments for overtime and late-shift work, as well as recurrent production and nonproduction bonuses, prior to payroll deduction for taxes, social security, or other purposes for a given payroll period by man-hours worked plus hours paid for sick leave, holidays and vacations. Total compensation computed monthly on an industry basis for comprehensive groups of manufacturing and nonmanufacturing industries by the Bureau of Labor Statistics in its employment and payroll reporting program.

GROSS AVERAGE WEEKLY EARNINGS. A measure of weekly wages obtained by dividing total compensation prior to payroll deductions for taxes, social security payments, and other purposes for a given weekly payroll period by total employment or, alternatively, by multiplying average hourly earnings by average weekly hours. Computed monthly on an industry basis for comprehensive groups of manufacturing and nonmanufacturing industries by the Bureau of Labor Statistics in its employment and payroll reporting program.

GROUP ANNUITY PLAN. Pension plan underwritten and administered by an insurance company.

Trusteed Plan. Noninsured; contributions deposited with bank, trust company, or board of trustees, who administers the program.

Deposit Administration. Insurance company assumes role of trustee, as above.

Money Purchase Plan. Fixed contributions to the worker's account, the pension is thus determined by the amount contributed. (See RETIREMENT PLAN, GROUP ANNUITY PLAN, ANNUITY.)

GROUP BONUS. A bonus payment based on the performance of a group of workers operating as a unit.

GROUP INCENTIVE (PLAN). An incentive plan in which individual worker earnings are based upon a share of the total incentive earnings reflecting the combined performance of the entire group.

GROUP JOBS. Jobs where several employees work in concert to accomplish a particular task. Members of the group may or may not be skilled in all of the jobs within the group. (See TEAM MANNING.)

GROUP PAYMENT. Compensation amount paid to employees comprising an identifiable work unit. May be paid as incentive earnings.

GROUP PIECEWORK PLAN. A method of compensating all workers in a designated group on the basis of total group piecework earnings.

GUARANTEE. The minimum time for a job upon which compensation for work is calculated. (MINIMUM GUARANTEE.)

GUARANTEED ANNUAL WAGE. Minimum assured compensation under plan providing guaranteed annual employment.

GUARANTEED BASE RATE. Minimum pay rate for specific tasks or for work performed under special circumstances.

GUARANTEED EMPLOYMENT. A plan providing a minimum number of days employment in a work year.

GUARANTEED HOURLY RATE. An assured rate of pay.

GUARANTEED RATE. The rate of pay assured to an incentive worker.

GUARANTEED STANDARD. An assured rate of pay based upon quantity of work produced.

GUARANTEED TIME. Rate of hourly or weekly pay guaranteed to a worker under an incentive system. May differ from base rate. The term is sometimes used for weekly wage or employment guarantees (e.g., a guarantee of 36 hours work for employees called to work on the first day of the workweek).

GUARANTEED WAGE PLAN. An arrangement by which an employer guarantees or assures employees a specific amount of wages. (See ANNUAL WAGE OR EMPLOYMENT GUARANTEE, WAGE ADVANCE PLAN.)

GUARANTEED WAGE RATE. An assured rate of pay for a given job or classification. Usually applied per hour worked.

GUARANTEE OR TRIAL RATE. A minimum guaranteed rate that remains in effect during the time that trial runs are made on new work and a trial or temporary piece rate is in effect. The level of these guaranteed rates is usually higher than plant minimum job or base rates, and is related to past earnings of the individual or the group of workers affected. (See TEMPORARY RATE.)

GUIDELINE METHOD OF JOB EVALUATION OR JOB PRICING. A method which involves "pricing" a job to a pre-determined salary structure (by selecting the range midpoint closest to the job price) and then realigning the hierarchy on the basis of relative internal job worth.

GUIDEPOSTS. Standards by which unions and business leaders, and the general public, can appraise particular wage and price decisions.

HALO EFFECT OR HALO ERROR. The tendency, and error, to rate an employee on several dimensions based on how well they perform on just one dimension. Typically this tendency reveals itself in the attribution of favorable characteristics to a person who is already favorably evaluated and unfavorable characteristics to a person who is already unfavorably evaluated.

HANDICAPPED WORKER RATE. A lower rate of pay for a worker whose efficiency is impaired because of physical or mental disabilities. Under the Federal Fair Labor Standards Act, rates below the legal minimum wage may be established for disabled workers in accordance with regulations issued pursuant to section 14 of the act. (See SUBSTANDARD RATE.)

HANDICAPPED WORKERS. Workers whose capacities or earning abilities are impaired by physical or mental disability.

HARDSHIP ALLOWANCE PREMIUM. Hardship allowance of premium bears no relation to the work to be done nor living costs but is paid in recognition of extraordinarily difficult living conditions, great physical demands, or notably unhealthy conditions or environment.

HAY SYSTEM. A point factor system that evaluates jobs with respect to "know-how," "problem-solving," and "accountability." It is used primarily for salaried jobs.

HAZARD PAY. Extra payment for work performed while exposed to abnormal or unsafe conditions.

HAZARDS. Danger or risks encountered by employees in the work environment, or resulting from conditions of employment.

HEALTH AND INSURANCE PLAN. A program of providing financial protection to workers and families against death, illness, accidents, and other risks, in which the costs are borne in whole or in part by the employer. One or more of the following major benefits may be provided for the worker and dependents: life insurance plan, accidental death and dismemberment benefits, accident and sickness benefits, surgical and medical benefits, and major medical expense benefits.

HEALTH CENTER. A clinic administered by a union or by employer(s) and union(s) where members and their families may receive medical examinations and treatment free or at a nominal charge.

HEALTH MAINTENANCE ORGANIZATION (HMO). Prepaid group medical service organization emphasizing preventative health care. Defined in the Health Maintenance Organization Act of 1973 as "an organized system for the delivery of comprehensive health maintenance and treatment services to voluntarily enrolled members for a pre-negotiated, fixed periodic payment." Subject to meeting certain standards and conditions specified in the act and associated regulations, HMO's must be offered to participants in group health plans as an alternative choice for coverage.

HEARING. The presentation of an arbitration case or the gathering of information regarding an issue.

HIGH TASK. Performance of an average experienced operator working at an incentive pace, over an eight-hour day under safe conditions, without undue or cumulative fatigue. Often stated as a percentage above normal performance. (See NORMAL PERFORMANCE, LOW TASK.)

HIGH TIME. The time workers are engaged in a job high above ground, and, thus, dangerous or uncomfortable, as in construction. Sometimes also applied to work below ground level with extra dangers or discomforts for the worker. (See HAZARD PAY.)

HIRING. The procurement and employment of personnel to fill current or future position vacancies.

HIRING HALL. An office maintained by a union, or jointly by employers and a union, for referring workers to jobs or for the actual hiring operation. Common in maritime industry.

HIRING RATE. (See ENTRANCE RATE.)

HIT-THE-BRICKS. Slang for "to go on strike."

HOLDBACK PAY. Any wages withheld from an employee's pay other than that required by law or an employee benefit program.

HOLIDAY PAY. Pay to workers, typically at regular rates, for holidays not worked. Payments are often provided at premium rates for work on such days

HOME LEAVE. A periodic leave in the country of domicile for overseas staff members and their dependents.

HOMEWORK. Production of industrial goods by workers in their homes from materials supplied by the employer.

HORIZONTAL UNION. A union which includes only workers in a single craft or skill, or closely related skills, such as carpenters, electricians, etc., usually cutting across industry lines. Use of term is declining. (See CRAFT UNION.)

Hospitalization Benefits. Plan that provides workers, and in many cases, their dependents, with hospital room and board (e.g., semi-private room) or cash allowances towards the cost of such care for a specified number of days plus the full cost of specified services. Usually part of a more inclusive health insurance program. (See HEALTH AND INSURANCE PLAN.)

Hot-cargo Clause. An agreement provision stipulating that employees covered by the agreement cannot be required to handle or use goods shipped from, or bound to, an employer who is involved in a strike with a union.

Hourly Rate. The rate of pay expressed in terms of dollars-and-cents-per-hour. Hourly rates are normally basic rates i.e., exclusive of extra payments for shift work and overtime, and exclusive of production or non-production bonus payments. May also mean "earned rate per hour" under incentive methods of wage payment.

Hourly Rate Wage Plan. A plan for paying an employee a fixed amount per hour regardless of production.

House Organ(s). A company magazine or publication containing news and articles concerned with employees, company plans, programs and products, and primarily of interest to employees, customers, and stockholders.

Housing Allowance. A differential paid to adjust for differences between local housing costs and the costs for similar housing at other employer locations.

Human Capital Theory. A branch of labor economics proposing that the investment one is willing to make to enter an occupation is related to the returns one will earn over time in the form of compensation.

Human Relations. Term applied to a broad area of managerial effort and research dealing with the social and psychological relations among people at work; practical application includes improving personal relationships, reducing friction, improving organization, and thereby enhancing effectiveness.

Human Relations Committees. Continuing committees of employees and management, often called quality circles or participative management teams set up to study problems, and to make recommendations for improvements. In some instances participative teams are empowered to implement the decisions made.

Human Resources Department. A term replacing personnel department, industrial relations department, or employment department, emphasizing performance improvement and personnel development.

Human Resources Programs. Plans of the human resources department for the development and training of personnel.

Hundred Percent Incentive. A wage incentive plan which, unlike a sharing plan, permits employees to receive the entire monetary value of direct labor savings resulting from above standard performance.

Idle Time. Time during which a worker is not working; includes occurrences when machine controlled elements in a task exceed the time required for "internal" manual operations.

Impartial Chairman. An arbitrator employed jointly by a union and employer, usually for a definite term, to serve as the impartial party on a tripartite arbitration board and to decide all disputes or specific kinds of disputes arising during the life of the contract. The functions of an impartial chairman often expand with experience and the growing confidence of the parties, and he/she frequently acts alone in deciding a case. (UMPIRE.)

Impasse. Point at which a deadlock is reached in labor negotiations.

Improshare. Improved productivity through sharing; a gainsharing incentive pay program for employees based on improved productivity. (See GAINSHARING)

Improvement Factor. Wage earner benefits resulting from increased productivity. A term used in an agreement negotiated by the United Automobile Workers (C.I.O.) with the General Motors Corporation in May, 1948, describing an annual increase in wages of a stipulated amount during the life of the agreement.

Incentive. A reward and worker inducement for above normal effort and continued performance above standard.

Incentive Earnings. Actual earnings less base pay for actual hours worked under an incentive system. (See INCENTIVE PAY.)

Incentive Opportunity. Earnings potential over and above a base wage for productivity which exceeds an established performance standard.

Incentive Pace. Performance of a qualified worker at a speed above that established as standard or normal for a task or job.

Incentive Pay. Extra pay for work performed in excess of a set quantity or "standard."

Incentive Performance. Worker output in excess of a set predetermined standard of productivity. (See Incentive pace.)

Incentive Plan. A method for compensating work performed over and above an established level or standard.

Incentive Plan, Group. (See Group incentive plan.)

Incentive Plan, Piecework. An earnings system based upon the number or quantity of units produced.

Incentive Plan, Standard Hour. An earnings system based on pay hours earned above actual clock hours worked.

Incentive Rate. (1) The term "incentive rate" may apply to pay per unit for production above a predetermined minimum standard of output, or a ratio of management-labor sharing of labor cost savings resulting from the operation of an incentive system. (2) Rate upon which compensation for producing a given quantity of work is based (differs from day work).

Incentive Wage System. General term for methods of wage payment which relate earnings of workers to their actual production, individually or as a group. (See Group incentive plan, piecework.)

Incidence rate. An explicit formula for determining the number of accidents and diseases per year by the number of employee exposure levels. Required by OSHA.

Income. Any remuneration paid by an employer which is perceived by the employee to have utility.

Income Policy. A guideline for setting wages.

Income Security Plans. An employer financed agreement to provide guaranteed weekly payments to laid-off workers.

Incumbent. A person occupying and carrying out a job.

Independent Union. Term applied to local, national, and international unions not affiliated with the AFL-CIO.

Indexing. An automatic wage adjustment to reflect changes in cost of living, or other economic indices of the economy.

Indirect Compensation. Money paid by an employer for an employee to a third party such as hospitalization, life insurance, disability insurance and other fringe benefits; or benefits derived by an employee from such services provided for by an employer. Also known as fringe benefits or supplemental compensation.

Indirect Labor. Labor that is necessary to support the manufacture of a product or the provision of a service but does not directly enter into transforming the material into the product or performing the service.

Individual Equity. A fairness criterion that directs employers to set wage rates for individual employees (people on the same job, in the simplest case) according to individual variation in merit.

Individual Job Rate Structure. (See Individual rate.)

Individual Rate. Compensation based, in a loose way, upon the job being done, or related to the training, ability, skill, and bargaining power of the individual worker. In many establishments, there is no formal wage structure (either job rates or rate ranges), and the rates paid are known as individual rates or individual pay rates. The term "individual rate" is also used to indicate the rate actually received by the individual worker, as distinguished from the job rate shown in the rate structure maintained by an employer.

Individual Retirement Account (IRA). A tax-deferred retirement savings plan.

Individual Wage Determination. The method of arriving at the actual wage to be paid an individual considering such factors as individual performance, wage history, position in a range and experience.

Industrial Democracy. Labor participation with management in making decisions relating to the operation of the organization.

Industrial Relations Center. A part of a university dedicated to research, training, and education in the field of union/management interaction. These centers may also be called institutes or schools.

Industrial Relations. General term covering matters of mutual concern to employers and employees and includes the relationships, formal and informal, between employer and employees or their representatives; an area of specialization within an organization.

INDUSTRIAL UNION. A union that represents all or most of the production, maintenance, and related workers, both skilled and unskilled, in an industry or company. May also include office, sales, and technical employees of the same companies. (See CRAFT UNION.)

INDUSTRY-WIDE BARGAINING. Negotiations between an employers' group and a union resulting in an agreement covering a substantial part of an industry, e.g. all class one railroads. The term "industry-wide" does not necessarily imply nationwide coverage. It is usually safe to assume that in whatever way "industry" is used, it does not include every establishment.

INEQUITY. A real or alleged variation, generally applied to wage rates or benefits that are lower than those prevailing elsewhere in the plant, company or industry for the same or comparable work.

INEQUITY ADJUSTMENT. Pay adjustment for inequity.

INFLATION. A rise in the general level of prices as measured by changes in the Consumer Price Index.

INFORMATIONAL PICKETING. Picketing designed to inform the public that a labor dispute exists between workers and their organization.

INITIATION FEE. Payment required of a worker when he/she joins a union.

INJUNCTION. Court order restraining one or more persons, corporations, or unions from performing some act which the court believes would result in irreparable injury to property or other rights. Also known as labor injunction.

INTEGRATED SALARY STRUCTURE. A compensation system that brings together theories, facts, discoveries, principles, and considerations of many fields and disciplines involving sound, meaningful, and effective policy recommendations which anticipate changing conditions.

INTERCITY DIFFERENTIAL. Differences in prevailing wage levels among a group of cities; such differences are measured by rates for comparable occupations and industries on a city to city basis. (See AREA DIFFERENTIAL.)

INTERIM EARNINGS. Income earned after an employee is discharged which may be deducted from an award of back pay upon reinstatement by an arbitrator.

INTERNAL EQUITY. A fairness criterion that directs an employer to set wage rates that correspond to the relative value of each job to the organization.

INTERVIEW. A question and answer method for gathering information about a job and a job candidate involving a person who is knowledgeable about the job and the candidate. (See DISCIPLINARY INTERVIEW.)

INTRINSIC REWARDS. Rewards that are associated with the job itself, such as the opportunity to perform meaningful work, complete cycles of work, see finished products, experience variety, and receive feedback on work results.

IRA. Individual Retirement Account. (See ERTA, INDIVIDUAL RETIREMENT ACCOUNT (IRA).)

JOB. A work assignment, a task or related series of tasks, a position classification or quantity of work. May be used in connection with either wage or salary activities.

JOB ANALYSIS. Systematic study of a position leading to a detailed specification of qualification and performance requirements for purposes of wage (or salary) administration, employee selection or training, determination of skill transfer opportunity, comparison and consolidation of positions or organizational design.

JOB BANKS. Places where computerized listing of jobs and their characteristics are maintained. These banks are generally associated with public employment agencies. (For "company" job banks see protected employees.)

JOB BURNOUT. A specific set of symptoms brought on by severe or chronic stress directly related to the career rather than personal difficulties. Related symptoms are chronic fatigue, low energy, irritability, and negative attitude toward job and self.

JOB BREAKDOWN. A job analysis detailing various elements pertaining to skill requirements, training and experience, mental and physical demands, working conditions, hazards exposure, and requirements of responsibilities of performance.

JOB CHARACTERISTIC. An attribute of a particular work function related to skill, experience, responsibility, effort or working conditions.

Job Class. A designation given to jobs grouped by some criteria such as pay rate, skill level, or permanency. (See JOB CLASSIFICATION.)

Job Classification. An arrangement of jobs in an establishment or industry into a series of categories, each of which is based on progressively higher requirements in terms of skill, experience, training, and similar considerations resulting in a grouping of occupations where distinctions between jobs are clear and sharp. Job descriptions are often used as a basis for classification. (See JOB EVALUATION.)

Job Cluster. A grouping of jobs resulting from relatively equal job evaluation scores or related within an organization and linked together by technology, organization, or custom, e.g. a wide cluster could include all factory jobs; narrower clusters would be assembly jobs and maintenance jobs.

Job Combination. The practice, usually as a cost improvement, of combining two jobs into one job by eliminating some of the duties from one or both jobs and combining the remaining duties. The new job may be in a higher, lower, or the same labor grade.

Job Comparison. A listing of key factors of each job being compared based upon their quantifiable values in order to permit relative analysis.

Job Content. Reference to basic work function elements and scope of duties involved in a given job (or position) primarily those pertaining to skills, mental and physical demands, and responsibilities.

Job Content Evaluation Method. Analysis techniques that concentrate on the actual work activities involved in determining relative value.

Job Cycle. Time necessary to perform an operation before having to start it all over again.

Job Definition. (See JOB DESCRIPTION, JOB SPECIFICATION.)

Job Description. An established written summary statement of a position which describes its functional requirements. It may include such specifics as equipment or tools used, physical and mental skills, training, working conditions, duties, responsibilities, location, and designation of supervisor.

Job Enlargement. Horizontally broadening work responsibilities, i.e., adding more of the same type of duties requiring the same skills.

Job Enrichment. Vertically broadening work responsibilities include inspection, planning, etc. in contrast to the horizontal expansion of job duties implicit in job enlargement. (See JOB ENLARGEMENT.)

Job Evaluation. Rating jobs to determine their position in a job hierarchy through assignment of points or use of some other systematic rating method for essential job requirements such as skill, experience, and responsibility. Job evaluation is widely used in the establishment of wage rate structures and in the elimination of wage inequalities. (See JOB CLASSIFICATION, LABOR GRADE.)

Job Evaluation Committee. A committee whose membership is charged with the responsibility of a) selecting a work function analysis system and b) carrying out or overseeing the system operation. Job evaluation committee members usually represent all major constituencies within the organization.

Job Evaluation Manual. Organized documents defining policies, procedures, standards, work functions and the resulting analysis.

Job Evaluation Point Method. A method of job evaluation wherein factors are measured according to a numerical index, the sum total of these numerical values becoming the relative measure of value (and pay rate) to other jobs.

Job Factor. A defined characteristic or attribute of a job common to other jobs and suitable for basing a comparison of value or relative worth between jobs.

Job Family. Jobs involving work of the same nature but requiring different skill and responsibility levels.

Job Feedback. The degree to which the job itself provides the worker information about how well the job is being performed.

Job Freezing. Deferral of permanent assignment for a set period of time or until the person designated for the job meets qualification requirements.

Job Grade. (See PAY GRADE.)

Job Instruction Training (JIT). A systematic technique for on the-job training consisting of four steps (l) careful selection and preparation of trainer and trainee for the learning experience to follow; (2) full explanation and demonstration by the trainer ; (3) a trial on-the-job performance by the trainee; and (4) a thorough feedback session highlighting job performance and job requirements.

Job Levels. Pay grade or pay level assigned to a work function.

Job Matching. An essential function in effective recruiting that entails fitting the qualifications of people to the requirements of the job.

Job Posting. An announcement of job openings for the purpose of soliciting bids for promotion or transfer from employees.

Job Pricing. The practice of establishing wage rates for positions within the organization, usually combining judgments regarding market value and internal evaluation results.

Job Profile. A major component or feature of a job matching system that contains the descriptions of jobs that are available.

Job Ranking. A method of job evaluation wherein various positions are set in a compensation level array based on relative appraisal of job factors.

Job Rate. The established pay for a given position. (See MINIMUM JOB RATE, STANDARD RATE.)

Job Rating. (See JOB EVALUATION.)

Job Relatedness. Refers to selection tests and qualifications being related to an employee's potential success on the job. If a test or qualification is shown to be job related, a disparate impact charge can be defended.

Job Scope. The range of duties, activities, responsibilities and authority involved in the proper performance of a job. (See JOB DESCRIPTION.)

Job Security. (1) (Explicit.) In certain labor contracts, the employment level to be maintained in the plant over the life of the contract is stipulated and is, therefore, explicit. The maintenance of employment by a specific clause in a contract or agreement. (2) (Implicit.) The practice, in certain labor contracts, where the union will relinquish certain benefits in exchange for a surmised level of continuing employment. But because the level of employment is a matter of discussion and not a guarantee, the maintenance of employment (job security) is said to be implicit.

Job Sharing. The redesign of a full-time job resulting in two or more employees holding responsibility for a single job, dividing time, salary or wages, and fringe benefits.

Job Specification. Necessary capabilities for satisfactory activity performance. (See JOB DESCRIPTION.)

Job Title. A label for a work function or assignment.

Joint Action Process (See UNION-MANAGEMENT COOPERATION.)

Joint and Survivor Option. A pension plan provision under which the pensioner may receive a reduced benefit with a guarantee that, if he/she dies while his/her beneficiary is living, payments of a predetermined proportion of the reduced benefit will be continued to this beneficiary for life. Also known as survivor's option.

Joint Bargaining. Two or more unions negotiating an agreement with a single employer or two or more employers negotiating an agreement with a single union.

Joint Board (or council). Delegate body composed of representatives of locals of a single national or international union in a particular area, working together to further the interests of the union. When more than one union is involved, the term "trade council" may be used.

Joint Lockout. (See LOCKOUT.)

Joint Rate Setting. The process of management and labor establishing wages. The extent of labor participation in the actual process of rate setting varies from industry to industry, and from establishment to establishment

Journeyman. A fully qualified trade craftsman who has successfully completed an apprentice program.

Journeyman Rate. The rate of pay for a fully qualified worker in a skilled trade or craft, who has completed an apprenticeship or equivalent training. Some journeymen may receive rates above or below union scale.

Jurisdiction. The union right to represent workers within specified occupations, industries, or geographical boundaries.

Jurisdictional Dispute. Conflict between two or more unions over the organization of a particular establishment or whether a certain type of work should be performed by members of one union or another.

Jurisdictional Strike. A work stoppage resulting from a jurisdictional dispute.

Jury-duty Pay. Pay or allowance granted to an employee for working time spent on jury duty, usually in addition to fees paid by the court.

Just Cause. Good or fair reasons for discipline. This term is commonly used in agreement provisions safeguarding workers from unjustified discharge or other disciplinary action.

Keogh Retirement Plan. An individual retirement account designed for self-employed persons.

Key Employee Plan. Supplementary bonus or incentive provisions designed to compensate department heads and supervisors as opposed to officer/ executive levels, for improved or measured performance.

Key Job. (See BENCHMARK JOB.)

Key Man Insurance. Insurance that is designed to protect the business against possible death of an employee with unique and largely irreplaceable skills and knowledge. The business pays the premium and normally is the beneficiary.

Kick-back. A practice by which an employer or representative arranges with his workers for a return of a part of their wages, established by union contract or by law, as a condition of employment. A federal Anti-kick-back Law was enacted in 1934, prohibiting kickbacks by workers employed on public construction work or on any work financed wholly or in part by federal funds. (See BOOTLEG WAGES.)

Labor. Various meanings include 1) human effort aimed at the production and/or distribution of goods and services; 2) one of the four factors of production; 3) an organized labor movement or union.

Labor Area. Geographical area from which workers may be recruited. (See LABOR MARKET AREA.)

Labor Class (LABOR GRADE, SALARY GRADE, JOB GRADE, JOB CLASS). A series of rate steps (single rate/2-rate range) in the wage structure usually determined by job evaluations or by negotiations so that different jobs or positions of the same relative worth fall into the same grouping and are therefore, paid at the same rate.

Labor Cost. The cost of human resources that is a part of the cost of goods or services. Usually refers to those wages which go directly into the product, but may include the indirect support functions.

Labor Dispute. (See DISPUTE.)

Labor Economics. The study of wages, employment and the operation of the labor market in general.

Labor Force. In census terms, all persons age 16 or over, employed or unemployed and actively seeking work. Total labor force includes members of the Armed Forces; Civilian labor force excludes them. Term is often used to designate total employment of a particular company or industry. Also known as work force.

Labor Force Participation Rate. The proportion of the population 16 years old and older that is employed or unemployed and actively seeking work.

Labor Grade. One of a series of rate steps (single rate or rate ranges) in the wage rate structure of an establishment. An outcome typically of some form of job evaluation in which various occupational classifications are rated on the basis of such requirements as skill, experience, training, education requirements and working conditions. Occupations are grouped into a limited number of steps so that occupations of approximately equal "value" or "worth" fall into the same level. (See JOB CLASSIFICATIONS, JOB EVALUATION.)

Labor Injunction. (See INJUNCTION.)

Labor-management Committee. A committee of labor and management representatives focusing on improving labor-management relations and joint problem solving.

Labor Management Relations Act. Refers especially to the Taft-Hartley Act (1947) a federal law, amending the National Labor Relations Act (Wagner Act), 1935, which, among other changes, defined and made illegal a number of unfair labor practices by unions. It preserved the guarantee of the right of workers to organize and bargain collectively with their employers and retained the definition of unfair labor practices as applied to employers. The act does not apply to employees in a business or industry where a labor dispute would not affect interstate commerce. Other major exclusions are: employees subject to Railway Labor Act, agricultural workers, government employees, domestic servants, and supervisors. Amended by Labor-Management Reporting and Disclosure Act of 1959. (See NATIONAL LABOR RELATIONS ACT, NATIONAL LABOR RELATIONS BOARD, UNFAIR LABOR PRACTICE, SECTION 14 (B), LABOR MANAGEMENT RELATIONS ACT, 1947.)

Labor-Management Reporting And Disclosure Act Of 1959, (Landrum-Griffin Act). This federal law was designed "to eliminate or prevent improper practices on the part of labor organizations, employers," etc. Its seven titles include a bill of rights to protect members in their relations with unions; regulations of trusteeships; standards for elections; and fiduciary responsibility of union officers. The Labor Management Relations Act, 1947, was amended in certain respects by this act. Among other changes, hot-cargo clauses in contracts were forbidden, except for apparel and construction industries. Restrictions were placed on secondary boycotts and picketing.

Labor Market Area. A geographical territory from which workers may be recruited, surrounding a concentration of establishments. Usually a metropolitan area, consisting of a central city and its suburbs. Also a Labor Area.

Labor Mobility. The extent to which workers can, are willing, or do move from job to job, employer to employer, or place to place to find employment.

Labor Movement. General term usually applied to organized labor; its growth, structure, and activities.

Labor Productivity. The rate of output of employees per unit of time measured against a standard or expected rate of output per man-hour.

Labor Relations. In an organization it is concerned with the relationship between the employees and the organization. In a unionized organization, the working relationship between the union representatives and the management representatives.

Labor-saving Ratio. The mathematical relationship of the unit labor cost of one method to the unit labor cost of another, usually an improved method compared to an existing method.

Labor Supply (labor supply curve). The minimum wage necessary to attract a given number of employees or level of employment. Economists most often consider labor supply as a function or line of points defining this wage for a series of employment levels.

Labor Turnover (turnover). Movement of workers into and out of employment in a company or industry through hiring, layoffs, recall, quits, etc. Labor turnover rates are usually expressed as the number of accessions and separations during a given period per 100 employees.

Landrum-Griffin Act. (See LABOR MANAGEMENT REPORTING AND DISCLOSURE ACT OF 1959.)

Layoff (reduction in force). Involuntary separation from employment for a temporary or indefinite period, without prejudice, that is, resulting from no fault of the workers. Although "layoff" usually implies eventual recall, or at least an intent to recall workers to their jobs, the term is occasionally used for separations plainly signifying permanent loss of jobs, as in plant shutdowns.

Layoff Allowance. (See SEVERANCE PAY.)

Layoff Costs/Benefits. The benefits provided to an employee while on lay-off and the costs to the company for supplying such benefits. Also called trailing costs.

Layoff Pay. Benefits received because of temporary or permanent termination. (See SEVERANCE PAY.)

Learner. A person acquiring knowledge, generally for the purpose of performing a job satisfactorily. (See APPRENTICE.)

Learner Rates. The schedule of wages applicable to workers inexperienced in the job for which they are employed, during their period of training. The schedule of rates is usually established in such a manner as to permit the gradual achievement of the minimum job rate as the learner develops competence on the job. Under the Fair Labor Standards Act, an employer may be permitted to employ learners in a specific plant at a wage lower than the legal minimum, whenever employment of learners at such a lower rate is believed necessary to prevent curtailment of employment opportunities. Hearings are held by the administrator to determine under what limitations as to wages, time, number, proportion, and length of service, special certificates authorizing the employment of learning at subminimal rates may be issued to an employer for certain occupations in the plant facility.

Learner's Allowance. A special allowance attached to an incentive standard to encourage new employees or employees on new operations or products to reach incentive level performance. It is often used with group incentives to avoid penalizing the senior members of the group while the new employee(s) are "coming up-to-speed."

Learner's Certificates. Certificates issued by the U.S. Department of Labor, under provisions of the Fair Labor Standards Act of 1938, enabling employers to pay rates below the statutory minimum to learners, messengers, apprentices, and handicapped workers so as not to curtail opportunities for their employment.

LEARNING CURVE, MANUFACTURING PROGRESS FUNCTION (MPF). A graph or mathematical formula plotting the course of production during a period in which an employee is learning a job; the vertical axis plots a measure of proficiency while the horizontal axis represents some measure of practice or experience (time). Also known as an experience curve, a quantitative model relating the factors of time, productivity improvement and cost, applied to predict future costs under conditions of production volume increases. Essentially, a linear relationship on log-log paper reflecting an equivalent percentage reduction as production quantities double.

LEAVE OF ABSENCE. Generally, an unpaid excused period away from work, without loss of job or seniority.

LEGALLY REQUIRED BENEFITS. Insurance programs to which employers must contribute for employees according to law. Includes social security, unemployment compensation, worker's compensation, state temporary disability insurance, and special programs for railway workers.

LEGAL PLAN, PREPAID. An employer and/or a union agreement with a law firm or group of lawyers to supply legal services to workers at no or minimum employee cost. Some unions, such as the UAW, have a group of lawyers who supply legal services to their members at no or minimum cost. Such services are usually supplied for civil actions as opposed to criminal actions. Also called prepaid legal services. When supplied by an employer, it is considered a fringe cost.

LEVELING. A procedure for rating work performance in order to adjust observed average elemental times used to establish a labor standard and/or standard data. (See NORMALIZE.)

LEVEL INCOME OPTION. (See SOCIAL SECURITY ADJUSTMENT OPTION.)

LEVEL OF ASPIRATION. A goal that an individual sets as something he/she expects or strives to achieve. Reaching the goal is interpreted by him/her as success, falling short as failure.

LIFE INSURANCE PLAN. Group term insurance coverage for employees, paid for in whole or in part by the employer, providing a lump-sum payment to a worker's beneficiary in the event of death. (See HEALTH AND INSURANCE PLAN, DEATH BENEFIT.)

LINE OF PROGRESSION (LOP). A designated sequence of related job classifications through which an employee may progress for purposes of promotion or demotion in cases of layoff or work cutback. Frequently called a "promotion schedule."

LIVING DOCUMENT. This term, as used by unions, expresses the belief that the terms of an agreement, particularly a long-term agreement, should be subject to review and renegotiation by the parties if conditions change or unforeseen events come about, despite the absence of a reopening clause.

LIVING WAGE. A term used normally in collective bargaining to mean the minimum compensation under current economic conditions which is perceived by workers as necessary to sustain a particular standard of living.

LOCAL UNION. Labor organization comprising the members of a union within a particular area or establishment, which has been chartered by, and is affiliated with a national or international union. Also known as local; chapter; lodge.

LOCKOUT. A temporary withholding of work or denial of employment to a group of workers by an employer during a labor dispute in order to compel a settlement at or close to the employer's terms. A joint lockout is such an action undertaken at the same time by a group of employers. Technically, the distinction between a strike and a lockout depends on which party actually initiates the stoppage. One, however, can develop into the other. (See WORK STOPPAGE) (JOINT LOCKOUT.)

LONGEVITY PAY. Wage or salary adjustments based on length of service.

LONG TERM BONUS. Usually a form of deferred compensation that establishes an income stream in the form of a bonus over time.

LONG TERM CONTRACT. A multiple-year wage contract or collective bargaining agreement as contrasted to a one-year agreement.

LONG-TERM DISABILITY. Wages or salary paid by an employer or a third party to workers who are ill or injured and unable to work for long periods of time.

LOOSE RATE. A job rate which produces a higher payment than would be paid for similar work performed under correctly applied or maintained work methods and related standards. Runaway rate.

Loose Standard. (See LOOSE RATE.)

Low Task. Performance level normal for unmeasured or unstandardized tasks; also called unmeasured daywork pace.

Lump Sum Award. One time payment of a bonus based upon an exceptional performance. A one time payment made to an employee for a variety of reasons such as an arbitration award or the granting of a patent or a method improvement.

Lump Sum Increase. The practice of giving employees an annual or periodic increase as a lump sum rather than in regular paychecks. If an employee was granted a $1,000 per year increase, he/she would receive the full amount at one time (less applicable taxes) rather than $83.33 per month.

Machine-controlled Time. The time during a work cycle in which production is entirely dependent on machine performance.

Machine Downtime. Time during a work shift when a machine is inoperable.

Machine Hour. A unit of production used to measure capacity and utilization of machines based upon standard hourly performance expectations.

Machine Idle Time. Machine down time or time when equipment is unproductive during a work cycle or work shift.

Machine pacing. A condition under which the machine determines how fast the work must be done; therefore, the pace at which the employee works is determined by the machine.

Maintenance of Benefits. A clause in a collective bargaining agreement requiring benefits to be maintained at the current level for the term of the contract.

Maintenance-of-membership Clauses. An arrangement provided for in a collective bargaining agreement whereby employees who are members of the union at the time the agreement is negotiated, or who voluntarily join the union subsequently, must maintain their membership for the duration of the agreement, or possibly, a shorter period, as a condition of continued employment. (See UNION SECURITY.)

Major Medical Expense Benefits. Plan designed to insure workers against the heavy medical expenses resulting from catastrophic or prolonged illness or injury. If the benefit supplements the benefit payable by a basic health insurance plan (hospital, medical, or surgical), it is called a "supplementary" plan, otherwise, it is called a "comprehensive" plan. (See DEDUCTIBLE, HEALTH AND INSURANCE PLAN.) (CATASTROPHE INSURANCE)

Make-up Pay. Allowances given by employers to incentive workers to make up differences between actual earnings and earnings at guaranteed rates or statutory minimum rates. At times, the term is also associated with the practice of permitting employees to earn a full week's wages by making up for lost time.

Make-work. (See FEATHERBEDDING.)

Management. Term applied to the employer and representatives, or to corporation executives who are responsible for the administration and direction of an enterprise. (See EMPLOYER.)

Management By Objectives (MBO). An employee development technique which consists of a process in which a supervisor and an employee or group of employees jointly identify, establish, and measure achievement against common goals for the employee(s). The unique feature of management by objectives is that the process requires involvement of employees in the setting of performance goals.

Management Prerogatives. As used in union-management relationships, this term is applied to rights reserved to management, which may be expressly noted as such in a collective bargaining agreement, usually including the right to schedule production, to determine the process of manufacture, to maintain order and efficiency, to hire, etc.

Mandatory Bargaining Subjects. Items which under the National Labor Relations Act must be negotiated if either party wishes to do so.

Man-days of Strike Idleness. A key measure of strike activity reflecting working time lost because of strikes. The figures on strike idleness do not include secondary idleness-that is, the effects of a work stoppage on other establishments or industries whose employees may be made idle as a result of material or service shortages.

Man-hour. A unit of productivity equivalent to the output of one worker performing at standard normal pace for one hour.

Man-hour Output. A productivity measure in which numerical quantity produced is divided by hours applied to the production of that quantity.

Man-minute. A unit of productivity used in some incentive plans and equivalent to one worker performing at the standard normal pace for one minute. This is the basis for the Bedaux incentive plan, i.e., the 60B unit is a normal hour.

Manning Table. A listing of the positions and the number of workers, both direct and indirect, authorized in the operation of a business. In a production plant a variable manning table may be used to authorize the manpower required at different production levels.

Manpower. General term used to designate the number of persons required in operating a specific organizational entity.

Manual Workers. (See BLUE-COLLAR WORKERS.)

Market Comparison Study. Survey of wages paid by other companies for like or similar services.

Marginal Productivity Theory of Wages. An economic theory that seeks to explain wage determination based upon contribution of the last or marginal worker to total production.

Market Pricing. A wage and salary setting policy that sets the rates to be paid for a job to the organization's best estimate of the going wage rate in the external marketplace for that job. It is a process that defines a job's worth solely by the going rate in the labor market.

Market Rates. The employer's best estimate of the wage rates that are clearing in the external labor market for a given job or occupation.

Master Agreement. A single or uniform collective bargaining agreement covering a number of plants of a single employer or the members of an employer's association. (See MULTIPLANT BARGAINING, MULTI-EMPLOYER BARGAINING.)

Mastercraft. The job created by the reduction of several skilled crafts such as electrician, carpenter, millwright, plumber, welder, etc. into smaller groups so that a skilled craftsman, after appropriate training, can be called on to perform several crafts. The combining of several craft skills into one higher skilled craft.

Maternity Benefits. Term applied to health and insurance plan benefits payable to an employee because of pregnancy and childbirth, and for hospital, surgical, and medical benefits related thereto. In some plans, men are also paid for time away from work in helping to care for the new born child.

Maternity Leave. Leave granted to undergo and recover from childbirth. In some firms, husbands are granted paternity leave.

Maturity Comparisons. The consideration given education, experience, age, seniority, performance, and other factors in determining an individual's wages. (See INDIVIDUAL WAGE DETERMINATION.)

Maturity Curve. The chart of curves which relate an individual's years of experience and level of responsibility to compensation within a given profession or field as opposed to compensation rates within a particular organization.

Mealtime. (See PAID LUNCH PERIOD.)

Measured Day Work. A plan using standards to measure performance but which does not pay daily or weekly incentive earnings; some plans do include the adjustment of the employee's base hourly rate on a quarterly basis.

Mediation. A procedure in which a central third party assists union and management negotiators in reaching a voluntary agreement. An attempt by a third party to help in the settlement of a dispute between employer and union through suggestion, advice, or other methods of stimulating agreement. In contrast to an arbitrator, a mediator has no authority to impose terms on the parties. Most of the mediation in the United States is under taken through federal and state mediation agencies. (See FEDERAL MEDIATION AND CONCILIATION SERVICE)

Mediator. Term used to designate a person who under takes mediation of a dispute. In practice synonymous with Conciliator. (See FEDERAL MEDIATION AND CONCILIATION SERVICE.)

Mediation/Arbitration (Med/Arb). A technique whereby an arbitrator attempts to mediate a dispute but, if unsuccessful, makes a binding decision.

Medical Benefits. Plan which provides workers, and in many cases their dependents, with specified medical care (other than that connected with surgery) or a cash allowance toward the cost of doctors' visits. Generally part of a health and insurance program. (See HEALTH AND INSURANCE PLAN, HEALTH CENTER.)

MEMBER. An employee who has joined a union paying the required initiation fee and periodic dues is considered to be a "member-in-good-standing." (See UNION MEMBER.)

MEMBER-IN-GOOD-STANDING. (See UNION MEMBER.)

MERIT BUDGET. Allocation of funds to identified groups of employees for merit award use.

MERIT INCREASE. An increase in the wage rate of an individual worker on the basis of performance or service. This is widely used as a method of advancing workers within established rate ranges, sometimes in conjunction with a provision for automatic increases over part of the range. Merit increases may be administered informally at the discretion of the employer, or provision may exist for the periodic review of the performance of employees for granting of merit increases. (See WAGE REVIEW).

MERIT PAY. Compensation over and above an employee's pay rate based upon a plan for rewarding performance meeting a set criteria for excellence.

MERIT POOL. A fund from which merit payments are allocated. May be based upon a formula for accumulation reflecting improvement in operating profit, loss reduction, productivity, or other criteria.

MERIT PROGRESSION. A formula for progressing a person through a wage structure according to performance, time, or some other individual equity basis.

MERIT RATING. A formalized system of appraising employee performance according to an established group of factors. Usually the appraisal is annual or semiannual and is by immediate supervision. Factors frequently considered include quality of work, quantity of work, reliability, adaptability, initiative, attitude, etc.

MERIT RATING INCREASES. (See MERIT INCREASE.)

METHODS ANALYSIS. A systematic and critical examination of existing and proposed ways for performing a specific task, operation or activity.

METHODS ENGINEER. A person who, by specialized training, analyzes and designs work methods and procedures. (Formerly motion study engineer.)

METHODS ENGINEERING. Analysis and design of work methods and procedures. (Formerly identified with motion study.) (See MOTION STUDY.)

METHODS STUDY. Detailed study of work aimed at establishing the optimum or best way to perform the desired task. (Formerly identified with motion study.) (See MOTION STUDY.)

ME-TOO CLAUSE. A clause in a labor contract or agreement that grants to one union any better terms won by another union in subsequent negotiations with the same employer.

MIDPOINT. The salary or wage midway between the minimum and maximum rates of the range.

MIDPOINT PROGRESSION (MIDPOINT SEPARATION). The slope of the salary schedule from midpoint to midpoint expressed as a percent of the lower midpoint. Progression may be constant throughout the schedule or may increase at higher salary grade levels.

MIGRATORY WORKERS. Persons whose principal income is earned from temporary employment (usually in farming) and who, in the course of a year, move one or more times, often through several states.

MILITARY LEAVE. Excused leave of absence for military service, reserve training, National Guard duty, etc. Time lost may be paid for by the employer in whole or in part.

MINIMUM FUNDING. The least amount of reserve investment necessary to back an employee welfare plan.

MINIMUM GUARANTEE. (See GUARANTEE.)

MINIMUM JOB RATE. The minimum rate of pay for workers on a given job. The minimum rate may be either a single or the minimum of a rate range. Union rates or union scales are usually minimum job rates. Normally, entrance rates, probationary rates, or learner rates fall below the minimum job rates.

MINIMUM MANNING. The minimum number of workers who can be assigned to operate a particular machine, operation, or assembly line where manning can be varied.

MINIMUM PLANT RATE. Normally the minimum rate of pay for experienced workers in the lowest-paid job in the establishment. The term may, however, mean different things in plants with different organized wage structures. In some plants, the term refers to the rate for the lowest-paid production job, although lower rates may exist for such jobs as common labor or janitor. In some plants the so-called minimum rate may actually be a hiring or probationary rate. (See ENTRANCE RATE.)

MINIMUM RATE. There are several kinds of minimum rates, those that are applicable to specific jobs and those that are applicable to entire establishments. Normally those that are applicable to specific jobs are called minimum job rates and those applicable to entire establishments are called minimum plant rates. In addition, there are several varieties of guaranteed minimum rates usually applicable to individual jobs under wage incentive systems.

MINIMUM TIME. The least time wherein a task may be reasonably expected to be accomplished.

MINIMUM WAGE. Rates of wages, established legally or through collective bargaining, below which workers cannot be employed. The Fair Labor Standards Act establishes the legal minimum wage to be paid to workers engaged in interstate commerce, unless such workers are covered by state laws which provide for higher minimum wages. Minimum rates are also established by collective bargaining and are applicable to individual plants, or to groupings of plants within an area or within an industry.

MODIFIED UNION SHOP. (See UNION SHOP.)

MONEY-PURCHASE PLAN. (See PENSION PLAN.)

MONEY WAGES (NOMINAL WAGES). Amount of cash or money actually received as compared to the real wage which measures purchasing power.

MONITORSHIP. Supervision or surveillance of a union by an outside party, usually for a limited time, imposed by order of a court or parent union organization.

MONTHLY LABOR REVIEW. A monthly magazine devoted to general economic and labor matters, issued by the Bureau of Labor Statistics, United States Department of Labor.

MOONLIGHTING. Practice of maintaining regular employment with more than one employer. Term refers to employee practice of taking employment in a secondary job during evenings or nights. Reverse arrangement is sometimes termed "sunlighting."

MOTION STUDY (MOTION ANALYSIS). The study of the basic manual movements and procedures associated with the accomplishment of work. Historically associated with Gilbreth's work, in contrast to time study, associated with Taylor.

MOTION-TIME ANALYSIS. Motion analysis with time consideration.

MOVING ALLOWANCE. (See RELOCATION ALLOWANCE.)

MTM (METHODS TIME MEASUREMENT). A widely available system of predetermined times for basic motions used to set work standards and/or labor time estimates.

MULTI-EMPLOYER BARGAINING. Collective bargaining between a union or unions and a group of employers, usually represented by an employer association, resulting in a uniform or master agreement.

MULTIPLANT BARGAINING. Collective bargaining between a company and the union or unions representing workers in more than one of its plants, usually resulting in a master agreement. If all or most plants are involved, the term "companywide" is appropriately used. (See COMPANYWIDE BARGAINING.)

MULTIPLE PIECE RATE PLAN. A wage incentive plan using two or more piece rates. One rate is paid for production of a fixed quantity in a given time, and a higher rate(s) is paid for production beyond the initial level. (See DIFFERENTIAL PIECEWORK.)

MULTIPLE TIME PLAN. A wage incentive plan that pays a higher base wage rate as production levels increase. A series of step bonuses for increased production. (See DIFFERENTIAL TIME PLAN.)

NATIONAL EMERGENCY DISPUTE. Term used in the Labor Management Relations (Taft-Hartley) Act to designate an actual or threatened strike or lockout which may imperil the national health or safety. If such a possibility exists in the opinion of the President, he or she may appoint a board of inquiry to investigate the issues in the dispute. Upon receiving a report from the board, the President may direct the Attorney General to petition the appropriate district court for a time specified injunction, during which the board of inquiry and the National Labor Relations Board have certain functions to perform. If no settlement is reached during this cooling-off period, the parties are free to resume their dispute, and the President may recommend appropriate action to the Congress.

NATIONAL INSTITUTE OF OCCUPATIONAL SAFETY AND HEALTH (NIOSH). A federal agency created to aid in the research, dissemination, and education of health and safety issues and provides expertise to organizations in need of such services.

NATIONAL LABOR RELATIONS ACT (WAGNER ACT OF 1935). Basic federal act guaranteeing workers the right to organize and bargain collectively through representatives of their own choosing. Also defined "unfair labor practices" as regards employers. Amended by the Labor Management Relations Act, 1947 and the Labor-Management Reporting and Disclosure Act of 1959.

NATIONAL LABOR RELATIONS BOARD (NLRB). Agency created by the National Labor Relations Act, 1935, and continued through subsequent amendments, whose functions are to define appropriate bargaining units, to hold elections to determine whether a majority of workers want to be represented by a specific union or no union, to certify unions to represent employees, to interpret and apply the act's provisions prohibiting certain employer and union unfair practices, and otherwise to administer the provisions of the act. (See LABOR MANAGEMENT RELATIONS ACT, 1947.)

NATIONAL MEDIATION BOARD. Agency established by the Railway Labor Act, 1926, to provide aid in settling disputes between railway and airline companies and unions over union representation, negotiation of changes in agreements, and interpretation of agreements reached through mediation.

NATIONAL RAILROAD ADJUSTMENT BOARD. Federal agency established in 1934 which functions as a board of arbitration, handing down final and binding decisions on disputes arising out of grievances, or the application and interpretation of agreements.

NATIONAL UNION. Ordinarily, a union composed of a number of affiliated local unions. The Bureau of Labor Statistics, in the union directory, defines a national union as one with agreements with different employers in more than one state, or an affiliate of the AFL-CIO, or a national organization of government employees.

NEGOTIATION. (See COLLECTIVE BARGAINING.)

NIGHT SHIFT. (See SHIFT.)

NON-CONTRIBUTORY BENEFIT PLAN. A plan in which the employer pays the entire premium or the full cost of building up a fund from which to pay benefits.

NON-CONTRIBUTORY PENSION PLAN. A pension plan for the benefit of the employee under which the entire cost is borne by the employer. (See CONTRIBUTORY PENSION PLAN.)

NON-EXEMPT EMPLOYEES. All employees covered by the Wage and Hour Law and not exempted from the requirements of that law.

NON-FINANCIAL INCENTIVE (PLAN). All influences, other than financial, which tend to stimulate employees to higher levels of productivity, such as promotion, training, competition with others, recognition, praise, and, at times, reprimands. Programs to achieve the above.

NON-FINANCIAL REWARDS. The rewards given an employee for his/her services which have no direct financial value. Such rewards might include recognition for certain achievements by letter, postings, certificates, or announcements.

NON-PERFORMANCE PAY. Payment in recognition of certain events or occurrences such as service, anniversary, or generally observed holidays.

NON-PRODUCTION BONUS. An extra payment to employees which depends on factors other than the output of the individual worker, such as profit-sharing, safety, attendance, and Christmas bonuses. (See BONUS PLAN, PRODUCTION BONUS.)

NON-PRODUCTIVE LABOR. Term sometimes applied to indirect labor.

NONQUALIFIED PENSION PLANS. Pension plans that do not come under the purview of ERISA. In addition, under these plans, the employer's contribution is not tax deductible, and the employee is taxed on contributions.

NON-QUANTITATIVE JOB EVALUATION. Any method of job evaluation that does not employ quantitative methods in the valuing of jobs. Examples of non-quantitative job evaluation methods include classification and job ranking.

NORMALIZE. To adjust observed data to a standard base. (See LEVELING.)

NORMAL OPERATION. (See Z94.17 WORK DESIGN AND MEASUREMENT.)

NORMAL PACE. Work pace associated with normal or standard production. (See NORMAL PERFORMANCE.)

NORMAL PERFORMANCE. The work output of a qualified employee working with standardized work procedures, excluding time for allowances.

NORMAL RETIREMENT. (See RETIREMENT.)

NORMAL RETIREMENT AGE. In technical terms, the earliest age at which a worker under a pension plan may retire of his/her own accord and receive the full amount of benefits to which he/she is entitled under the normal benefit formula of the plan. In most plans, and under social security, this age is 65 years. Historically, it is the age "set" by Prince von Bismarck.

NORMAL WORK RATE. (See NORMAL PERFORMANCE.)

NO-STRIKE, NO-LOCKOUT CLAUSE. Provision in a collective bargaining agreement through which the union agrees not to strike and the employer agrees not to lock out employees for the duration of the contract. These pledges may be hedged by certain qualifications, e.g., the union may strike if the employer violates the agreement.

OASDI. (See OLD AGE, SURVIVORS, AND DISABILITY INSURANCE BENEFITS.)

OCCUPATION. A generalized job or family of jobs common to many industries and areas.

OCCUPATIONAL DIFFERENTIALS. Relatively stable differences in wage rates between or among occupations.

OCCUPATIONAL RATE. Rates (single or ranges) that are designated for particular occupations in an establishment, area, or industry. Generally, these rates are formal rates, and are paid to any worker who is qualified to perform the work of the occupation.

OCCUPATIONAL WAGE. (See OCCUPATIONAL RATE.)

OCCUPATIONAL WAGE RELATIONSHIP. The relationship of wage rates among occupations representative of a range of duties, skills, and responsibilities. Relationships may be analyzed within an individual plant, a community or region, or on an industry basis.

OFF-THE-BOOKS. Refers especially to work performed but not recorded officially in order to escape scrutiny especially by financial, union, safety, tax and/or governmental committees, and/or agencies.

OLD-AGE SURVIVORS AND DISABILITY INSURANCE BENEFITS (OASDI). Retirement income and survivors' and disability payments available to eligible workers covered by federal social security legislation.

OPEN-END-AGREEMENT. Collective bargaining agreement with no definite termination date, usually subject to reopening for negotiations or to termination at any time upon proper notice by either party.

OPEN PAY SYSTEM. A compensation system in which information about wage ranges-at the extreme, even individual employee wage levels-is made public.

OPEN SHOP. Term commonly applied to an establishment with a policy of not recognizing or dealing with a labor union. Term may sometimes be applied to an organized establishment where union membership is not a condition of employment. (See UNION SECURITY.)

OPEN UNION. A union that will admit any qualified person to membership usually upon payment of reasonable initiation fees. (See CLOSED UNION.)

OPERATION. A job or task consisting of multiple work elements.

OPERATION ANALYSIS CHART. A special form (chart) used to analyze an operation and show all important factors, such as material, shipment, time, labor, etc., affecting the operation either for instructional purpose or to assist in improving the operation.

OPINION. A written document in which an arbitrator sets forth the reasons for the award. In most labor cases the parties want the arbitrator to explain the reasoning in order to give them some guidance for similar situations that may arise under the contract.

ORGANIZATIONAL DEVELOPMENT (OD). Attempts at improving organizational and individual behavior and performance through training and development efforts. Specific techniques may include group exercises, management by objectives (MBO), sensitivity training, and group discussion.

ORGANIZATIONAL STRESS. A sociopsychological work environment component characterized by organizational changes, work overload, poor supervision, unfair salaries, job insecurity, and physical insecurity all producing uncertainty.

ORGANIZER. Employee of a union or federation (usually paid but sometimes a volunteer) whose duties include recruiting new members for the union, assisting in forming unions in nonunion companies, assisting in campaigns for recognition, etc. (UNION ORGANIZER.)

ORIENTATION PROGRAMS. Activities used by employers, or jointly with the unions, to help familiarize new employees with the work environment and the culture of the firm.

OUTLAW STRIKE. (See STRIKE.)

OUT-OF-LINE RATE. A job rate which is higher or lower than that determined as proper for the work involved based upon a systematic evaluation of a number of related jobs or in relation to a job pay regression line established through job evaluation. (See RED CIRCLE RATE.)

OUT-OF-LINE REQUEST. A request by a laid off union employee for any union position the employee is qualified to perform.

OUT-OF-WORK BENEFITS. Usually, payments made by a union (and possibly supplemented by the company) to unemployed members.

OUTPUT PER MAN-HOUR. A measure of labor productivity which is a portion of total factory productivity.

OVERTIME. Work performed in excess of basic workday or work-week, as defined by law, collective bargaining agreement, or company policy. Sometimes applied to work performed on Saturdays, Sun days, and holidays at premium rates. Under the Fair Labor Standards Act, non-exempt employees must be paid one and one half times their normal wage rates for all hours worked in excess of forty in any work week.

OVERTIME PAY. (1) Pay at higher or equal to regular job rates for work performed at times other than the regular work shift or work day. (2) Payment at premium rates for work performed in excess of or at times other than specified workday or work week.

OVERTIME PREMIUM PAY. Payment of wages at premium rate for time worked beyond the regular hours of employment established by union agreement, employer or industry practice, or law. In the United States, payment is typically made at one and a half times the regular rate of pay. Higher premium rates are found to a limited extent. (See PREMIUM RATE OF PAY.)

PACE. Rate of output or performance compared to an accepted standard. May be expressed quantitatively in terms of units per time or in terms of percent relative to standard.

PACE RATING. A worker's speed of performance as compared with normal pace. (See PERFORMANCE RATING.)

PACE SETTER. A worker who is better than average on a particular job, and whose production is used by the employer as a standard for measuring the amount of work which can be done in a given period of time.

PACKAGE SETTLEMENT. A term used to describe a combination of benefits received by workers as a result of collective bargaining. A package may include wage increases and other benefits of monetary value, such as insurance, paid holidays, paid vacations, and sick leave. The term generally implies that a specific amount of increase was to be applied partly to rates of pay, and partly to the financing of the related benefits.

PACT. (See AGREEMENT.)

PAID ABSENCE ALLOWANCE. Payment for lost working time available to workers for various types of leave not otherwise compensated for, e.g., excused personal leave.

PAID HOLIDAYS. Holidays are days of special religious, cultural, social, or patriotic significance on which work or business ordinarily ceases. Paid holidays are those established by agreement or by company policy, for which workers receive their full daily pay without working. (See HOLIDAY PAY, UNPAID HOLIDAYS.)

PAID LUNCH PERIOD. Time allowed for eating lunch (or the midshift meal on late shifts), commonly 20-30 minutes, counted as part of the paid workday. Usually practiced where employees cannot leave their workplace for meals (e.g., coal mining). Agreements sometimes also require a company to furnish meals when workers remain in the plant for overtime work.

PAID VACATIONS. Excused leave of absence of a week or more, with full pay, granted to workers annually for purposes of rest and recreation. Paid vacations are provided in private industry by collective bargaining agreements or company policy, not by law. Vacations are frequently graduated by length of service. (See EXTENDED VACATION PLAN.)

PAPER LOCALS. Local unions which exist only "on paper" (charter) with no actual membership.

PARTICIPATION POINT. Established level of job performance at which employee is eligible for additional earnings or compensation.

Part-time Employee. Worker employed on a temporary or regular basis for a workweek substantially shorter than the scheduled week for full-time employees. Usually under 25 hours per week in order to minimize obligation for fringe benefits. (State law often determines how many hours constitutes "Part-time").

Part-time Worker Rate. A rate paid to a part-time temporary, or contingent worker, as distinguished from that paid to a regular or full-time worker. Part-time rates may be equal to, or lower or higher than, regular or full-time rates. During periods of ample labor supply, part-time rates are usually lower, but may become equivalent or higher when the labor market is tight because of keen competition for such help. Retail trade establishments and restaurants are among the industries dependent on part-time or temporary help to carry on their normal functions.

Past Practice. Existing practices in the plant or company, sanctioned by use and acceptance, that are not specifically included in the collective bargaining agreement, except, perhaps, by reference to their continuance.

Past Service. Under a pension plan, years of employment or credit service prior to the establishment of the plan or a change in the plan's benefits.

Pattern Bargaining. Term applied to follow-the-leader negotiating practices in an industry.

Pay Adjustment. A general revision of pay raises. The adjustment may be either across-the-board, such as cost of living or COLA increases, or spot adjustments for revisions in prevailing wage rates.

Pay-As-You-Go. (See UNFUNDED PLAN.)

Pay Compression. (See COMPRESSION.)

Pay Curve. (See WAGE CURVE.)

Pay Grade. One of the classes, levels or groups into which jobs of the same or similar value are grouped for compensation purposes. All jobs in a pay grade have the same pay range: maximum, minimum, and midpoint.

Pay Grade Overlap. The degree to which adjacent pay grades in a structure overlap. Numerically, the percentage overlap between two adjacent pay grades.

Pay Grade Width. (See SALARY RANGES.)

Pay-in-lieu-of notice. Where employers are required to provide advance notice of layoff and fail to do so, agreements often require the employer to pay workers for the full notice period as a penalty.

Payment By Result. Refers to any method of wage payment where the amount of the wage depends upon the amount of output. The term applies to straight piecework or other types of incentive systems. The production to which wages are related may be the output of an individual worker or the output of a group of workers.

Payment Certain Guarantee. (See PERIOD CERTAIN OPTION.)

Pay Plan. A schedule of pay rates or ranges and a list showing the assignments of each class in the classification plan to one of the rates or ranges. May extend to rules of administration and the benefit package.

Pay Range. (See SALARY RANGES.)

Payroll Deduction. A deduction from an employee's gross earnings made by the employer for social security, unemployment insurance, federal income tax, local government payroll tax, union dues, special union assessments, group insurance premiums, savings plans, etc.

Payroll Period. The established frequency with which workers are paid in a particular industry, regard less of the time to which the rate applies.

Payroll Tax. Taxes levied by the government and paid by employers, employees, or both, creating funds from which employees receive retirement, unemployment, or other benefits. Also may refer to employer contributions, based on fixed percentages of total payroll, to union or other private health and welfare and vacation funds, and to payroll taxes levied by cities.

Pay Steps. The levels within a pay range.

Pay Structure. A tabulation of wage scales for groupings or various jobs. May be applied to arrangement of wage levels or adjustments within a given job or compensation category.

Pay Survey. The gathering of data on wages and salaries paid by other employers for selected key classes of jobs or benchmark jobs.

Pay Trend Line. A line fitted to a scatter plot that treats pay as a function of job values. The most common technique for fitting a pay trend line is regression analysis.

PEG POINT. An occupational rate for a key unskilled, semi-skilled, or skilled job establishing differentials within the wage structure. The term first used by the National War Labor Board in its decisions on wages in the cotton textile industry in 1945 (see 21 W.L.B.882) and thereafter applied to the wage structure through collective bargaining.

PEL. Paid education leave.

PENALTY RATE. An extra rate which is paid for hazardous jobs, late-shift work, or for overtime. (See PREMIUM RATE OF PAY.)

PENSION. The amount of money usually paid at regular intervals to an employee who is retired from a company and is eligible under a pension plan to receive such benefits.

PENSION ADMINISTRATOR. As defined by the Employment Retirement Income Security Act (ERISA) of 1974, the person or organization (frequently the sponsor) designated by the terms of the instrument under which a pension or welfare plan operates. Also called plan administrator.

PENSION BENEFIT GUARANTEE CORPORATION (PBGC). A Federal corporation set up in the Labor Department, similar to the Federal Deposit Insurance Corporation (FDIC) which guarantees vested pensions (defined benefits). Insurance premiums are paid by employers with covered pension programs.

PENSION LIABILITY. The financial liability assumed by an organization in the administration of a pension plan. Under the Employee Retirement Income Security Act (ERISA), organizations cannot abandon pension plans without conforming to specific government regulations and, therefore, have a definite financial liability.

PENSION PARACHUTE. A special arrangement in certain pension plans that allows excess money in the plan to accrue to the beneficiaries and paid to said beneficiaries when an unfriendly takeover of a company becomes imminent.

PENSION PLAN. Any plan whose primary purpose is to provide regular payments for a determined period of time, usually life, to employees upon retirement. Additional benefits are often provided. The term private pension plans is often used to distinguish voluntary plans from the social security system. If the employee shares in the cost, the plan is contributory; if the cost is borne entirely by the employer, the plan is noncontributory.

PENSION TRUST FUND. A fund consisting of money contributed by the employer and, in some cases, the employee to provide pension benefits. Contributions are paid to a trustee who invests the money, collects the interests and earnings, and disperses the benefits under the terms of the plan and trust agreement.

PER CAPITA TAX. Regular payments made on the basis of a paid-up membership count by a local union to its national organization, or by a national union to a federation, to finance the activities of the parent organization. Amount usually set by union constitution.

PERCENTAGE INCREASE. The percent increase of wage or salary rates from the original rate as opposed to a dollar increase added to wage and salary rates.

PERFORMANCE. The degree with which an employee applied skill and effort to an operation or task as measured against an established standard. Standard time divided by actual time expressed as a percentage.

PERFORMANCE APPRAISAL. Supervisory or peer analysis of work performance. May be made in connection with wage and salary review, promotion, transfer, or employee training.

PERFORMANCE EFFICIENCY. A measure of work performance determined by comparing actual work done with the standard amount of work for this job determined as "normal."

PERFORMANCE RATING. Observation of worker performance to determine productivity in terms of standard or normal.

PERFORMANCE RATING FACTOR. The ratio of a standard performance time to an actual performance time, usually expressed as a percentage.

PERFORMANCE RATING SCALE. Correlation of performance with a numerical index usually designating standard or normal performance as 100.

PERFORMANCE STANDARD. (See STANDARD PERFORMANCE.)

PERFORMANCE STANDARDS. Task or behavioral standards established as goals to be achieved by an employee and providing the basis for performance appraisal.

PERK (PERQUISITE). Employer-paid benefits or expenses of employees extending beyond normal fringe benefits and paid compensation. (See also PERQUISITE)

PERIOD CERTAIN OPTION. Provision in a pension plan under which the pensioner may elect to receive a reduced benefit for life, on the condition that if he dies before receiving a specified number of payments the balance is continued to the beneficiary. A guarantee of a specified number of payments may be a standard plan provision, in which case it is called a payment certain guarantee.

PERMANENT AND TOTAL DISABILITY. Inability of a worker to perform any job owing to physical or mental impairment which is expected to be of long- continued and indefinite duration. The existence of the impairment must be certified by a physician, under the Social Security Act and most private pension plans, in order to qualify for benefits. Mental disabilities may be excluded by some pension plans.

PERMANENT ARBITRATOR. An arbitrator who is selected to serve under the terms of a collective bargaining agreement for a specified period of time or for the life of the contract. The duties of the permanent arbitrator are defined in the contract.

PERMANENT PART TIME. Fixed arrangements for regular employees to work fewer than five days per week or forty hours per week.

PERMANENT PIECE RATE. A rate established for a piecework job calculated to yield an appropriate level of earnings and based, generally, on experience with trial rates for the job assignment; such rates are expected to persist until basic conditions change.

PERMANENT UMPIRE. (See PERMANENT ARBITRATOR.)

PERMIT CARD. Card issued by a union to a nonmember which permits the recipient to accept or retain employment on a temporary basis in a union shop or on a union job. Also known as work permit.

PERQUISITE. Relates to the furnishing by employers of food, lodging and other payments in kind to workers in addition to monetary compensation thus, waitresses are generally allowed a certain number of meals, depending upon the length of shift; board and lodging are usually supplied to workers in lumber camps and in some cases to farm labor. Often called perks. (See EMPLOYEE SERVICES AND PERQUISITES, PERK)

PERSONAL LEAVE. Excused leave for reasons important to the individual worker, but not otherwise provided for, e.g., getting married. May be included as part of vacation allowance.

PERSONAL RATE. (See GREEN CIRCLED RATE, RED CIRCLE RATE.)

PERSONAL TIME. An allowance usually included in a work standard as a percentage to provide time for the personal needs of the worker during the workday.

PERSONNEL RELATIONS. (See LABOR RELATIONS.)

PHANTOM STOCK, ALSO CALLED "SHADOW STOCK". An involved financial compensation technique in which a company buys stock for a particular employee without the use of that employee's name and at an agreed upon time turns the proceeds or the stock itself over to the employee or an amount equal to the worth of the transactions if the company elected not to actually purchase the stock. Either way, the name of the employee is not involved in the transactions until payment is made. A form of deferred compensation.

PHANTOM STOCK PLANS. The actual programs or procedures used to execute the purchase of phantom stocks.

PHYSICAL ABILITIES TEST. Tests used to measure the physical suitability of job applicants to determine whether the applicant can fulfill the established performance standards.

PHYSICAL WORK ENVIRONMENT. Composed of the building, chairs, equipment, machines, lights, noise, heat, chemicals, toxins and other factors associated with occupational accidents, diseases, fatigue, physical or psychological stress, and the consequent loss of productivity.

PICKETING. Patrolling near employer's place of business by union members (pickets) to publicize the existence of a labor dispute, persuade workers to join the union or the strike, discourage customers from buying or using employer's goods or service, etc.

Informational picketing. Picketing directed toward advising the public that an employer does not employ members of, or have a contract with, a union.

Organization picketing. Picketing carried on for the purpose of persuading employees to join the union or authorize the union to represent them.

Recognitional picketing. Picketing to compel the employer to recognize the union as the exclusive bargaining agent for employees.

PIECE RATE. Under an incentive wage system, the predetermined amount paid to a worker for each unit of output. Rates may be based on individual or group output. (See INCENTIVE RATE.)

PIECE RATE PLAN. A method of wage payment based on piece rates. (See PIECEWORK RATE.)

PIECE SCALE. (See PRICE LIST.)

PIECEWORK. Production for which payment is based on rates per unit of output. Method of wage payment based on the number of units produced, or any work for which piece rates are paid.

PIECEWORK EARNINGS. Wages received for work produced while on piece rate assignment.

PIECEWORK PLAN. A method of wage payment based upon a set amount of pay earned for each unit of production. (See PIECE RATE PLAN.)

PIECEWORK RATE. A set amount of pay for a unit of production or worker output.

PIECEWORK WITHOUT BASE GUARANTEE. A piece work pay arrangement wherein the amount of pay strictly reflects the amount of production or output with the extreme of no pay for no production.

PINK COLLAR WORKER. Term used to identify the influx of women to the industrial work force. Clearly, a "pink collar" can be "white" or "blue." (See BLUE COLLAR WORKERS, WHITE COLLAR WORKER.)

PLANT CLOSING ACT (1988). Requires 60 days notice of plant or office closing for employers with more than 100 employees.

P.M. An incentive payment to sales personnel in retail trade to push and sell items on which the margin of profit is large, to dispose of slow moving items, or to clean out old stock. Also referred to as premium money, push money.

POINT. An increment of value applied to a job factor when measured in connection with a point factor job evaluation plan.

POINT METHOD. (See JOB EVALUATION POINT METHOD.)

POINT PLAN. (See JOB EVALUATION POINT METHOD.)

POINT SYSTEM. (See JOB EVALUATION POINT METHOD.)

PORTABILITY. A pension plan feature that allows participants to change employers without changing the source from which benefits (for both past and future accruals) are to be paid.

PORTAL-TO-PORTAL. (1) Payment for travel to and from a work-site from the point at which an employee enters the plant or arrives at the place of employment. (2) Payment for travel time from a point of entry to the job site and return.

POSITION. A job or work classification. A collection of related tasks, duties, responsibilities, and authority.

POSITION CLASS. A group designation for various positions based normally on compensation rate and job duties, but which may reflect other considerations (i.e., exempt or non-exempt).

POSITION DESCRIPTION. (See JOB DESCRIPTION.)

POSITION GUIDE. A position description usually expanded to encompass a definition of working relation ships with other individuals or groups, general responsibilities and overall aspects of working procedure.

POSTING. (See JOB POSTING.)

POTENTIAL EARNINGS. Projected level of compensation reasonably expected from a high level of performance.

POWER. A person's control in a relationship is the capacity to influence others in that relationship.

PREFERENTIAL HIRING. Agreed-upon arrangement whereby the employer gives preference in hiring to union members, to applicants with previous training and experience in the industry, to workers displaced from another plant or from another part of a particular plant, or by order of the National Labor Relations Board to "employees found to be discriminatorily discharged."

PREMIUM. (1) Payment over and above a regular job rate for various reasons such as production performance above standard, or for working overtime or on other than regular shifts. (2) Compensation at greater than regular rate. Usually directly based on the hourly rate time during which work is performed or quantity of work produced as opposed to bonus.

PREMIUM MONEY. (See P.M.)

PREMIUM OVERTIME PAY. (See OVERTIME PAY, OVERTIME PREMIUM PAY.)

PREMIUM RATE OF PAY. An extra rate paid for overtime, work on late shifts, holidays and Sunday work, or for work in particularly dangerous or unpleasant occupations. The term is also used in reference to extra rates paid to employees, usually because of exceptional ability or skill in the occupation. (See OVERTIME PREMIUM PAY, PENALTY RATE, SKILL DIFFERENTIAL.)

PREPAID LEGAL PLAN. (See LEGAL PLAN, PREPAID.)

PREVAILING RATE. Typically, the predominant or more common rate paid to a group of workers, usually with reference to specific occupations in an industry or labor market area. In actual application, the term "prevailing rate" is used in a variety of ways. Some of the variations arise from differences in the concept of the geographic unit or industry that is pertinent to a particular situation. Also called prevailing wage. In some situations, notably locality wage surveys, measures of prevailing rates relate to the arithmetic mean, or to the median. In view of these variations, the use of the term "prevailing rate" requires specific mention of the area, occupation, industry, rate, and type of quantitative measure involved to have definite meaning.

PREVAILING WAGE LAW. Federal act requiring the payment of prevailing wage rates in the locality on construction, alteration, or repair of public works performed under contract with the federal government, e.g., Davis-Bacon Act (1931) amended in 1964 to include certain payments for fringe benefits as part of the prevailing rate.

PREVENTIVE MEDIATION. A function of the Federal Mediation and Conciliation Service involving the development of procedures by union and management designed to anticipate and to study potential problems. This may take the form of early entry into labor disputes before a strike threatens.

PRICE LIST. A listing of piece prices or rates to be paid by a company or a group of companies making similar products. In unionized establishments, price lists are established typically upon agreement between the union and the employer.

PRIVACY RIGHTS. The Privacy Act of 1974, applies to federal agencies, and its purpose is to protect all individuals from unauthorized disclosure of personal information by government employees. The Act protects an employee's right on the job through the establishment of rules for the confidential verification of references in selection and employment decisions.

PROBATIONARY PERIOD. Usually a stipulated period of time during which a newly hired employee is on trial prior to establishing seniority or otherwise becoming a regular employee. Sometimes used in relation to discipline, e.g., a period during which a regular employee, guilty of misbehavior, is on trial. Usually 60 or 90 days.

PROBATIONARY RATE. The rate of pay for a new employee during the trial period of employment. The probationary rate is usually lower than the minimum rate for that job. May apply to other employees on new assignments or new plants. (See ENTRANCE RATE.)

PRODUCTION. The generation of goods and services.

PRODUCTION BONUS. A bonus payment directly related to the output of an individual worker or group of workers. Usually paid for production in excess of quota or for the completion of a job in less than standard time. The bonus may be a flat amount paid for all production above standard, or it may increase in various proportions as production increases. May be non-monetary such as time off. (See BONUS, INCENTIVE PAY, NONPRODUCTION BONUS.)

PRODUCTION RATE. The unit volume of output during a set time unit such as x pieces per hour.

PRODUCTION STANDARDS. Usually, expected output of a worker or group of workers, consistent with the quality of workmanship, efficiency of operations, and the reasonable working capacities of normal operators. An expectation of output for a worker used as a base to evaluate performance, production rate, and/or pay.

PRODUCTION WORKERS. Usually, employees directly involved in manufacturing or operational processes, as distinguished from supervisory, sales, executive, and office employees. The term "production and related workers" as used in federal government statistics is usually specifically defined for survey purposes.

PRODUCTIVE LABOR. Term sometimes applied to direct labor.

PRODUCTIVE TIME. Period when worker is performing assigned task. Sometimes called "up time" as opposed to "down time" when work is delayed.

PRODUCTIVITY. The relationship between the amount or volume of output or service provided and the resources contributing to the actual production of the output or services. The ratio of output to total input; each of these may be measured in physical terms or in financial terms. The ratio of actual production to standard production-applicable to either an individual worker or group of workers. Units per manhour or manhours per unit.

PRODUCTIVITY BARGAINING. Negotiation of contractual obligations to increase productivity expressed in a collective bargaining agreement in exchange for increases in employee compensation or benefits. May have form of givebacks (q.v.).

PRODUCTIVITY FACTOR. (See ANNUAL IMPROVEMENT FACTOR.)

PRODUCTIVITY GAINSHARING (PGS). A plan based on an identification by the employees with the goals of the enterprise, b) involvement of the employees in developing and administering the plan, c) equity in the distribution of the future gains between the employees and the company, and d) managerial competence in the direction and control of the business. The plan involves developing an adequate measure of overall productivity, computing changes in productivity levels, and distributing the financial gains to all participants. (See GAINSHARING, RUCKER PLAN, SCANLON PLAN.)

PRODUCTIVITY MEASUREMENT. A gauge of productivity using real or physical volume factors used for promoting productivity and making budget and long-term projections.

PROFIT SHARING. Any procedure in which employees receive in addition to their regular pay or wages, a share of the profits of the business. A disbursement of business profits paid in addition to regular compensation usually based on a specific formula and in recognition of employee contributions to improved business performance.

PROFIT SHARING PLAN. The plan and/or procedures governing the distribution of a portion of an organization's profit to employees. The plan may be based on a single factor, such as a percentage of the profits, or on multiple factors such as profit, length of service, or position. Different amounts may be distributed among different groups of employees such as hourly, salaried non- exempt, exempt, and executive.

Deferred profit sharing plan. Share of profits set aside in a fund to be distributed at some later date, usually when the employee retires (a form of retirement plan).

PROFIT SHARING TRUST. In a deferred profit sharing plan, the money due employees under a given formula is invested via a trust fund and the income and principle is distributed to the employees at time of separation. The investment vehicle is referred to as a profit sharing trust.

PROGRESSION. Movement to a higher level of work or to a higher level of pay in same grade. (See LINE OF PROGRESSION.)

PROGRESSION, AUTOMATIC. (See AUTOMATIC INCREASES.)

PROGRESSION SYSTEM. Wage progression.

PROMOTION. The assignment of an employee to a job in a higher job classification or pay grade.

PROMOTIONAL PAY. (1) Additional payment usually in the form of a bonus for stimulating sales or for accomplishing specific sales program objectives aimed at increasing sales performance. (2) An increase in base wages as a result of work performance or contracted obligation.

PROMOTION INCREASE. An increase in a salary or wage rate accrues to a person because of a promotion to a higher level job.

PROTECTED EMPLOYEES. Protected status for union employees who have been rendered excess or redundant as a result of contractual specified occurrences such as productivity improvements, certain volume reductions et al. Employees in this status are not laid-off and may be redeployed, with full pay and benefits, to non-traditional activities including community work. (This protective status varies with each company.)

PROTEST PRICE. In some industries, notably pottery and women's dresses, piece rates on new work are determined on the basis of previously developed time elements. A worker may not be able to earn an appropriate amount under such estimated time allowances and piece rates. If he/she does not earn enough, he/she enters a protest but continues to work at these rates until a review is made and new rates are set. Any adjustment in rates is usually retroactive to the time of protest, or the time the worker was started on the new work.

PURCHASING POWER. Measure of level of compensation in terms of equivalent goods and services for which compensation may be exchanged at current market prices.

PUSH MONEY. An incentive payment for certain types of employees. (See P.M.)

PYRAMIDING. (1) Double payment of overtime rates for overtime work which may result from paying both daily and weekly overtime rates for same hours of work; sometimes applied to any premium added to another premium rate. (2) Multiple pay premium resulting in increased earnings.

QUALIFIED OPERATOR. A worker judged to have met certain criteria to perform a given job.

QUALIFIED STOCK OPTION. A stock option granted after December 31, 1963, which meets certain IRS codes and allows participants to take advantage of such tax benefits as capital gains which are not allowed under other options.

QUALITY BONUS. Additional payment for maintaining or attaining specified levels of excellence of work output.

QUALITY CIRCLE. A technique frequently used in Japanese companies and adopted by some American companies in which groups of workers, usually under the guidance of a facilitator or trainer, seek to find ways to resolve quality problems, improve quality, and increase production. Can be used with both office and factory workers. Considered a form of participative management.

QUALITY OF WORKLIFE PROGRAM. Programs developed by management, often in cooperation with employees and/or employee groups (unions) and usually administered by the personnel department (or jointly with the union) designed to improve the social and psychological needs of employees. Such programs are designed to recognize employees as total human beings as opposed to the concept of labor as a commodity and, therefore, to improve the workplace in terms of human needs for recognition, motivation, praise, appreciation, etc.

QUANTITATIVE JOB EVALUATION. Job evaluation systems that involve the use of numerical indices and analyses in the estimation of job value. The most common quantitative methods of job evaluation are the factor comparison method and the point factor method.

QUICKIE STRIKE. (See STRIKE.)

QUIT. Voluntary termination of employment initiated by the employee, as distinguished from dismissal or layoff which are involuntary.

QUOTA HIRING. Employment of persons in accordance with a predetermined plan aimed at changing the composition or past characteristics of a work force or group.

QUOTAS. Incentive systems frequently involve the establishment of quotas to be set as objectives to be obtained by employees. A given level of earnings is guaranteed for the achievement of a quota and bonuses for incentives are created for those employees who exceed the quotas.

RAIDING (NO-RAIDING AGREEMENT). Term applied to a union's attempt to enroll members belonging to another union or already covered by a collective bargaining agreement negotiated by another union, with the intent to usurp the other union's bargaining relationship.

RAISE. The actual increase in any salary over a period of time, usually one year.

RANK AND FILE. Members of an organization, exclusive of officers, managers, and supervisors. In unions, refers to all members who are not officers and/or members of a committee or elected officials.

RANKING METHOD OF JOB EVALUATION. The simplest form of job evaluation. A whole job, job to job comparison, resulting in an ordering of jobs from highest to lowest.

RATE. Pay for work based on a set unit of time or productivity.

RATE CHANGE. An adjustment in a production standard.

RATE CUTTING. Term applied to a reduction by management of established incentive or time wage rates in the absence of comparable changes in job content, or any actions by companies in reducing wages.

RATE EROSION. A gradual development of rate incorrectness resulting from gradual changes in job conditions without compensating adjustments in the job standard. (See LOOSE RATE.)

Rate Range. A range of rates for the same job, with the specific rates of individual workers within the range determined by merit, length of service, or a combination of various concepts of merit and length of service. Rate ranges may be set up with various degrees of formality and more or less rigid rules respecting the position within the range at which new workers are hired and the rules concerning their automatic or nonautomatic advancement to the maximum rate. The range may be expressed as a spread from a set minimum to a set maximum rate.

Rate Setting. (1) The establishment of pay per unit for incentive work. (2) The establishment of a standard time.

Ratification. Formal approval of a newly negotiated agreement by vote of the union members affected.

Rating. (See LEVELING, NORMALIZE) (SPEED RATING.)

Rating Scale. A graphical means whereby employee qualities such as ambition, initiative, personality, performance, and potential may be evaluated against some predetermined standard. (Some times referred to as a Traditional Rating Scale).

Real Wages. Real wages are represented by the goods and services typically consumed by workers that can be purchased with money wages, i.e., real wages are an expression of the purchasing power of money wages. Over periods of time, changes in real wages are obtained by dividing indexes of money wages by an appropriate index of consumers' prices.

Recall. Process of bringing laid-off employees back to work, usually based on the same principles that govern order of layoff in inverse order (e.g., last worker laid off is first to be recalled). In union affairs, recall is a procedure for removing (disciplining) an officer by means of a member ship vote.

Reclassification. Change in compensation grade or level of a given job or position resulting from changes in job duties, responsibilities, or working conditions.

Recognition. (See UNION RECOGNITION.)

Recruitment. Exploiting the sources of supply for applicants and providing potential employees.

Red Circle Rate. Rate of pay higher than the contractual, or formally established, rate for a job. The special rate is usually attached to the incumbent worker, not to the job as such. This procedure is commonly used to protect long-service workers from a decline in earnings through no fault of their own. Also known as out-of-line rate, personalized rate, flagged rate.

Reduction In Force. Layoff, or by attrition.

Reevaluation. The process of review of evaluated jobs for the purpose of determining changes which may lead to a new job rate.

Referendum. Process by which all members of a union vote, usually as individuals, for the election of officers, changes in union constitution, etc. as distinguished from decision-making through delegates assembled in convention.

Regional Differential. Differences in wage levels among several broad geographic subdivisions especially in the United States.

Regression Line. The line of best fit among an array of job scores in a job evaluation analysis designating a trend or relationship between the jobs.

Regular Employee. Usually, a full-time employee who has fulfilled formal or informal probationary requirements, as distinguished from seasonal, part-time, probationary, and temporary employees.

Regular Rate. The rate of pay received by a worker for all hours of work performed at straight-time rates. Also refers to the rate of pay at which a worker is predominantly engaged when he/she is subject to assignments at varying rates.

Rehire. (noun) A former employee returned to his/her job as a new employee. (verb) To reemploy a worker previously separated.

Reinstatement. Restoration of an improperly discharged employee to his/her former job and to the rights formerly held by him/her.

Relevant Labor Market. The geographical area from which a company obtains a large portion of its work force for a given occupational group.

Relief Time. Time during which a worker is permitted to leave the workplace, usually for personal needs, with the place being taken by a substitute when necessary. (See REST PERIOD) (SPELLOUT.)

Relocation Allowance. Payment to employees of all or part of their expenses in moving to a new location, or a fixed allowance to be used for this purpose. (MOVING ALLOWANCE.)

Relocation Reimbursement. Payment for travel, meals, lodging and related expenses in connection with an employee's move to another location.

Reopening Clause. Clause in a collective bargaining agreement stating the time or the circumstances under which negotiations can be requested, prior to the expiration of the contract. Reopenings are usually restricted to wage issues and, perhaps, other specified economic issues, not to the contract as a whole. (See WAGE REOPENING.)

Reporting Pay. The amount of pay guaranteed to a worker who reports for work at the usual hour, without notification to the contrary, and finds no work available or is not given a job. Typically, pay for a minimum number of hours at regular rate is provided for in union agreements. (See CALL-IN PAY.)

Representation Election. Election conducted to determine by a majority vote of the employees in an appropriate unit, which, if any, union is desired as their exclusive representative. These elections are usually conducted by the National Labor Relations Board or by state labor relations agencies. (See BARGAINING UNIT.)

Residency Period. The time an employee must perform a given job in order to have established seniority rights to that job or to have adequately prepared for progression to an advanced job for which the lesser job is a prerequisite.

Residual Rights. The residual rights doctrine gives management the benefit of the doubt concerning rights and powers on which a contract is silent.

Responsibility. The obligation of an employee to perform an assigned task to the best of his/her ability and in accordance with the directive of the supervisor and/or management.

Rest Period. Brief interruption in the workday, usually of 5 to 15 minutes duration, during which the worker rests, or takes refreshments without loss of pay. (See RELIEF TIME) (COFFEE BREAK, BREAK TIME.)

Restricted Job. Work limited by conditions beyond the control of the worker such as machine speed or process cycle time.

Retirement. Withdrawal from working life or from a particular employment because of age, disability, etc., with an income. Normal retirement is retirement for age, presently at age 65. Early retirement is retirement prior to the normal retirement age. Disability retirement is retirement prior to the normal retirement age because of poor health or injury disabling the worker. Special early retirement-extra early retirement benefits provided under specified circumstances, e.g., involuntary separation. (See PENSION PLAN, SOCIAL SECURITY ACT.)

Retirement Plan. (See PENSION PLAN.)

Retraining (retraining program). Development of new skills for workers through a definite program, so that they are able to qualify for new or different work. A training program in which experienced employees are given additional training as a result of work rotation, technological changes, new and more difficult assignments, change in the nature of jobs, and reassignment of displaced, old or handicapped workers.

Retroactive Pay. Past wages due for work performed during a prior time period such as may result from an adjustment in rates of compensation or settlement of a pay claim. Often called back pay.

Revenue Reconciliation Act of 1993. In addition to raising individual tax rates, the Act extended the limitations on deductibles. For nonresidential real property placed in service on or after May 13, 1993, the 1993 Act extended from 31.5 to 39 years the recovery period (depreciation period) and reduced the annual tax deduction for such tax payers. Starting in 1994, the 1993 Act ends deductibility for club dues and limits to 50 per cent the portion of business-meal and entertainment costs; further, the tax law eliminates moving expense deductions for house-hunting trips, meals and temporary living expenses.

Right-to-Work Law. Term applied to state legislation which prohibits any contractual requirement that a worker join a union in order to get or keep a job, thus banning provisions in agreements requiring employees to become and remain union members (otherwise permissible under the Labor Management Relations Act). The law prohibits union shop agreements.

Role-playing. A method for teaching principles affecting interpersonal relations by having subjects assume various roles during interaction.

Rolling. (See BUMPING.)

Rotating Shift. The system of rotating the crews where two or more shifts are worked in an establishment. This system is designed to distribute day and night work on an equal basis among the various workers. In some industries, where 7-day operations are common, the work schedules may be arranged so that workers are given different days off in each week. (See FIXED SHIFT, SPLIT SHIFT, SWING SHIFT.)

Round-the-clock Operations. (See CONTINUOUS OPERATIONS.)

Royalty. In relation to wages, the payment to union health and welfare funds, such as those benefiting members of the United Mine Workers and the American Federation of Musicians, although the term is not the official designation for such payments. In these cases, the application of the term stems, at least in part, from the fact that employer contributions are based on tons of coal mined and number of musical records produced. For some types of professional workers, such as musicians, singers, and writers, payment for work is frequently based on a percent of sales of the final product (book, article, or song). Such payments are referred to as royalties.

Rucker Plan. A productivity gainsharing plan based on the ratio of total labor costs to the value added revenue; i.e., sales, outside purchases. (See GAINSHARING.)

Runaway Rate. A standard production rate which is considered "loose." A standard which yields very high earnings when compared with earnings for a task of similar difficulty.

Runaway Shop. Term used by unions to characterize a business establishment which moves to evade union or state labor laws, or to reap a competitive advantage from low wage standards in another area, dismissing all or most of its regular employees in the process.

Runoff Election. A second election conducted after the first produces no winner according to the rules. If more than two contenders were in the first contest, the runoff may be limited to the two highest. (See REPRESENTATION ELECTION)

Safety. The prevention and reduction of the causes of accidents or job related illness.

Safety Education. The process of presenting information and reminders to employees to enable them to gain understanding and desired attitudes concerning a respectful appreciation for the dangers, hazards, or unsafe practices in the work environment with the means to avoid or prevent them.

Safety Engineering. That phase of engineering which deals with the prevention of industrial accidents and is primarily concerned with redesigning machinery, to eliminate accident producing hazards, to invent safety devices, and to design production and maintenance conditions for maximum safety conditions.

Salary. The fixed monetary compensation paid to an employee either weekly, monthly, or yearly (rather than by the hour) for services rendered; usually based on a certain minimum of hours per period. Generally applies to non-production, non- routine or supervisory jobs exempt from the provisions of the Fair Labor Standard Act although an employee can be paid a salary and still be covered by said act.

Salary Administration. The methods, procedures, and policies used to control and guide various salary programs and plans.

Salary and Commission. (See COMMISSION EARNINGS.)

Salary Budget. An amount or pool of money set aside for paying salaries during some period. Salary budgets must be taken into account when planning adjustments to a wage structure and when planning individual employee adjustment.

Salary Compression. Occurs when new employees are hired in at pay rates (salary) close to, equal to, or even greater than the pay rates of present employees in the same or similar positions. Most prevalent during periods of high inflation.

Salary Curve. A curve, often linear, on a graph describing the relationship of jobs, either in points or grades, to the compensation for each of the jobs, often with job values in points or grades along the horizontal axis and dollar compensation along the vertical.

Salary Differential. The difference between the rate of compensation from one position to another anywhere on the salary curve.

Salary Guide. A listing of compensation amounts for a given occupation group in order of progression based upon an application criterion for advancement, such as time in position or level of job qualification.

Salary Level Change. The increase in average salary from one period to the next stated in terms of a percent.

SALARY PLUS BONUS. A type of salary compensation plan in which an employee is paid a base salary not tied to performance and a bonus over and above the base salary based upon a formula measuring performance. The bonus is usually based upon some percent of base salary.

SALARY PLUS COMMISSION. A type of salary compensation plan in which an employee is paid a base salary not tied to performance and a set dollar amount or a percent dollar amount of the value of all units sold or produced above a given level.

SALARY PROFILE. A tabulation showing numbers of persons at various compensation levels in a given job, occupation, organization, geographical location, or other grouping.

SALARY RANGES. High and low designations of compensation for a given job or position. May incorporate designation of pay progression amounts from low to high applicable on basis of set formula or policy.

SALARY RATE. For workers hired on a weekly, monthly, or annual basis, the rate of pay is normally ex pressed in terms of dollars per week, month, or year. Workers employed on a monthly, or annual salary basis may actually be paid monthly, semi- monthly, or more frequently. Usually, the length of the work week is specified and a policy is established for compensation in the event that longer or shorter hours than a full week are worked.

SALARY SCATTERGRAM. An arrangement in graphical or matrix form of numbers of employees at each designated compensation amount or grade.

SALARY STRUCTURE (WAGE STRUCTURE). The structure of pay grades and ranges established for jobs within an organization. The salary structure may be expressed in terms of pay grades, job evaluation points, or policy lines.

SALES COMPENSATION. Any form of compensation paid to sales representatives. Sales compensation formulas usually attempt to establish direct incentives for sales outcomes such as the establishment of commissions as a percent of sales.

SAVINGS PLAN. (See THRIFT PLAN.)

SCAB. (See STRIKEBREAKER.)

SCALE. (See UNION RATE.)

SCANLON PLAN. A productivity gainsharing plan based on the ratio of selected labor costs to sales revenues adjusted for returns and allowances and changes in inventory levels. The Scanlon Plan involves much employee participation, predating quality circles with most of the same techniques. In effect, employees share in organization profits as a result of contributing and cooperating to attain higher productivity.

SCATTERGRAM. A mathematical technique designed to display a "picture" of a relationship between two variables. A scattergram is a plotting of coordinates simultaneously representing x and y scores for a number of observations. Usually used in establishing wage rates for job evaluation plans. The horizontal axis (x) shows job evaluation points or grades and the vertical axis (y) shows dollar amounts.

SEASONAL EMPLOYMENT. Employment during part of the year only, arising out of the seasonal character of an industry. Agricultural, cannery, construction, and lumber workers are examples of workers subject to seasonal employment. Seasonal unemployment, unemployment caused by seasonal or climatic conditions or by customs which make full time employment impossible in some industries.

SECONDARY BOYCOTT. (See BOYCOTT.)

SECTION 14 (B), LABOR MANAGEMENT RELATIONS ACT, 1947. This section of the Taft-Hartley Act provides the opening through which states may enact "right-to-work" laws. It reads as follows: "Nothing in this act shall be construed as authorizing the execution or application of agreements requiring member ship in a labor organization as a condition of employment in any State or Territory in which such execution or application is prohibited by State or Territorial law." (See RIGHT-TO-WORK LAW.)

SELECTION. The choice of the best qualified personnel for employment from the recruited.

SEMI-SKILLED LABOR. Employees whose work requires some manipulative ability but is limited to a fairly definite work routine. They are expected to exercise only limited independent judgment, and training is usually short with no formal apprenticeship.

SENIORITY. Term used to designate an employee's status relative to other employees, as in determining order of promotion, layoff, vacations, etc. Straight seniority-seniority acquired solely through length of service. Qualified seniority-other factors such as ability considered with length of service. Departmental or unit seniority-seniority applicable in a particular section of a plant rather than in the entire establishment. Plantwide or companywide seniority-seniority applicable throughout the plant or company. Seniority list-individual workers ranked in order of seniority. (See SUPER SENIORITY.)

SENIORITY RIGHTS. Employment rights, privileges, and status gained by employees in relation to others on the basis of length of service.

SENSITIVITY TRAINING. A group process that is designed to create an opportunity for individuals to expose their behavior and to give as well as receive feedback. Members frankly say how they react to certain behaviors so that each gains an awareness of the way his or her behavior affects others.

SEPARATION PAY. (See SEVERANCE PAY.)

SEPARATIONS. Termination of temporary or permanent employment initiated by the employee or employer.

SERVICE FEE. Fee required by unions to be paid by non members applying for employment in union hiring halls, as a condition of referral to employment.

SERVICE, LENGTH OF. Total amount of time an individual has worked for an employer is measured by the employer's standards, union contracts, or customs and traditions and used as a basis for computing seniority.

SEVERANCE PAY. Monetary allowance paid by employer to displaced employees generally upon permanent termination of employment with no chance of recall, but often upon indefinite layoff with recall rights intact. Plans usually graduate payments by length of service. Also known as dismissed pay or allowance, termination pay, separation pay, layoff allowance. In some cases, severance pay formulas may be employed to create incentives for a person to leave prior to retirement.

SEVERITY RATE OF ACCIDENTS. The total of time charges or occupational injuries per million man-hours of exposure. Time charges include actual calendar days of disability resulting from temporary total injuries and scheduled changes for deaths and permanent disabilities.

SEXUAL HARASSMENT. Physical violation or verbal abuse of a sexual nature of employees.

SHAPEUP. System of hiring work gangs from a group of workers assembled to seek employment. Used in longshore work in some ports, and in the hiring of migratory farm workers.

SHARE-THE-WORK. (See WORK SHARING.)

SHARING PLAN. Payment to employees of part of productivity gains over and above an established goal or standard. An incentive plan. (See PRODUCTIVITY GAINSHARING, RUCKER PLAN, AND SCANLON PLAN.)

SHESTIL. Term used to describe an additional initiation fee required by some craft unions to permit part-time employees to accept full-time employment. This practice has been outlawed by the Taft-Hartley Act.

SHIFT. A term applied to a work period where two or more groups of workers are employed at different hours during the operating time of an establishment; e.g., an establishment may operate two shifts of eight hours each or 16 hours a day. In some industries, the term "trick", "turn", or "tour" is used instead of shift. (See FIXED SHIFT, ROTATING SHIFT, SPLIT SHIFT, SWING SHIFT.)

SHIFT DIFFERENTIAL. Added compensation to workers who are employed on a work schedule other than the regular daytime schedule. Shift differentials may be paid in a number of ways: (1) a fixed amount per hour above the rate paid on the regular day shift; (2) a percentage over earnings at the regular day shift rates; (3) shorter hours with full daily pay, or; (4) both shorter hours and additional monetary compensation above full daily pay.

SHIFT PREMIUM. (See SHIFT DIFFERENTIAL.)

SHOP COMMITTEE. Group of workers selected by fellow employees, usually union members, to represent them in their dealings with management. Special types are: grievance committee, negotiating committee.

SHOP RULES. (See WORK RULES).

SHOP STEWARD. A local union's representative in a plant or department selected by union members (or sometimes appointed by the union) to carry out union duties, adjust grievances, collect dues and solicit new members. Usually a fellow employee.

SHORT-TERM DISABILITY. A part of worker compensation insurance.

Short-term Income Protection (unemployment insurance). State administered programs which provide financial protection for workers during periods of joblessness. These plans are wholly financed by employers except in Alabama, Alaska, and New Jersey where there are provisions for relatively small employee contributions.

Short Workweek Benefit. As part of a supplemental unemployment benefit plan, payment to worker for the difference between a specified level of weekly hours and the hours actually worked or paid for.

Shut-down. The stoppage of activity in an establishment pending settlement of a labor dispute, installation of new equipment, repairs, lack of material, financial troubles, etc.

Sick Leave. Period of time during which a worker may be absent without loss of job or seniority if unable to work because of illness or accident. A paid sick leave provides for full or partial pay for such absence, usually up to a stipulated maximum. Sick leave plans differ from accident and sickness benefits principally in that the former cover shorter periods of absence and are uninsured. (See ACCIDENT AND SICKNESS BENEFITS.)

Significant Risks. An enforcement tool superseding the no-risk standard as a result of a court decision in Industrial Union Department, AFL-CIO v. American Petroleum Institute, 1980. This standard implied that OSHA cannot demand compliance with the no-risk standard if the organization in question can show its existing exposure level to harmful agents is below a threshold assessment as determined by OSHA.

Single Company Union. An independent or unaffiliated union of employees of one company, usually with no formal ties to any other labor organization.

Single Rate. A rate which is the same for all workers on the same job or in the same job classification, and under which the individual worker on a job receives the same rate during the entire time that he/she is holding the job. The single rate usually is paid to experienced workers in jobs requiring varying degrees of skill. Learners or apprentices may be paid according to rate schedules which start below the single rate and permit the workers to achieve the full job rate over a period of time. In the less skilled jobs, the rates for beginners and experienced workers may be identical because the period of time necessary to become familiar with all phases of the work is relatively short.

Sitdown Strike. (See STRIKE.)

Skill. A factor of comparison in a job evaluation plan. Relates to manual and mental dexterity. Developed through practice, training and experience.

Skill Differential. Differences in wage rates paid to workers engaged in occupations requiring varying levels of skill in work performance. May also refer to differentials in rates of workers in the same occupation, higher rates being paid to those who usually perform the more complex tasks. (See PREMIUM RATE OF PAY.)

Skilled Labor. Employees who have mastered one of the traditional crafts, usually through an apprenticeship, and who possess a thorough comprehensive knowledge of the job, have the ability to exercise considerable independent judgement, and the capability of assuming responsibility.

Skilled Trades. A subunit in an industrial union's local made up of craft employees. They frequently have the right to a separate ratification vote on new contracts and serve on labor-management committees dealing with contracting-out.

Skills Inventory. A planning tool that indicates all people currently employed by the organization and classifies them according to their skills, job assignments, age, sex, and other factors relevant to human resource planning. The device is employed primarily as a way of classifying internal labor supplies for human resource planning. (See MANNING TABLE).

Sliding Scale Wage Plan. A wage which links wages directly to profits, prices, sales, or output.

Slotting. The act of folding a job into a category, classification, or any other class defined by a job evaluation system. Slotting usually involves the comparison of a job to some standard or bench mark job.

Slowdown. A deliberate reduction of output without an actual strike in order to force concessions from an employer. (See STRIKE.)

Social Security Act, 1935. Federal law establishing a national social insurance program. The law provides for: old-age, survivors' and disability benefits (an all-federal program); public assistance to the aged, the blind, and to needy families; and unemployment insurance (both federal- state programs). The coverage and other provisions have been modified several times since enactment.

SOCIAL SECURITY ADJUSTMENT OPTION. Pension plan provision under which a worker eligible for an early retirement benefit may elect to get a larger plan benefit than is actually due up to the time his social security benefit is payable, and a smaller benefit thereafter, so that a level income is maintained throughout retirement. (LEVEL INCOME OPTION.)

SOCIAL SECURITY OFFSET. Under some pension plans, the amount of social security benefits to which a retiring worker is entitled that is to be deducted from the private plan benefit, as computed, to obtain the actual benefit payable. The offset or deduction may be all or part of the social security benefit.

SOCIOPSYCHOLOGICAL CONDITIONS. Conditions in the workplace that lead to perceived stress and low quality of work life.

SOCIOPSYCHOLOGICAL WORK ENVIRONMENT. The non physical parts of the work environment including such things as relationships with supervisors, company policies, structure of the organization, organizational changes, uncertainty, conflicts, and relationship with co-workers.

SPAN OF CONTROL. A concept referring to the range of supervision or the number of subordinates which an executive can direct and control with maximum efficiency.

SPECIAL PERMIT RATE. A rate paid to a union worker who comes from another city and is employed under a special permit because of local labor shortages. The rate received is the same as that paid to a permanent worker in the area. In the unionized brewery industry, this term refers to the rates paid to special workers who are temporarily employed during the peak summer period. These rates are usually lower than those received by "regular" union workers.

SPEED RATING. (See LEVELING, PERFORMANCE RATING.)

SPEEDUP. Workers' term for conditions which force them to increase effort or production within a given time without a compensating increase in earnings. (See STRETCHOUT.)

SPELL. A work period.

SPELLHAND. An extra person or persons added to a group and becoming a permanent part of the group so as to provide relief to all members of the group without waiting for a specific person to supply such relief time.

SPELLOUT. (See RELIEF TIME.)

SPENDABLE EARNINGS. In general, the monthly earnings of workers less various amounts deducted for taxes and other purposes from payrolls; hence, "spend able earnings" may be identified broadly with "take-home" earnings or disposable income. Term also used in the sense of the earnings available for private spending or saving, but this usage would include certain types of deductions (e.g., union dues) as spendable earnings. "Net spendable average weekly earnings" is a series developed by the Bureau of Labor Statistics in which Federal Social Security and Income Taxes are deducted from gross average weekly earnings for workers with a specified number of dependents. (See TAKE-HOME PAY.)

SPLIT SHIFT. The daily working time that is not continuous, but split into two or more working periods. "Split shifts" are usually found in industries such as local transportation, which is affected by peak or rush periods at various times of the day. (See FIXED SHIFT, ROTATING SHIFT, SHIFT, SWING SHIFT.)

SPREAD-THE-WORK. (See WORK SHARING.)

STANDARD. An established norm against which various measurements are compared. An established norm of productivity defined in terms of units of output per set time or in standard time. The time allowed to perform a specific job including quantity of work to be produced.

STANDARD AGREEMENT. Collective bargaining agreement prepared by a national or international union for use by, or guidance of, its local unions, designed to produce standardization of practices within the union's bargaining relationships. (See FORM AGREEMENT.)

STANDARD ALLOWANCE. An established time provided for compensation for anticipated circumstances. Frequently expressed as a percentage of the base normal time to cover personal, fatigue, and delays (P,F, and D). (See ALLOWANCE.)

STANDARD ALLOWED TIME. A unit time value for the accomplishment of a task inclusive of adjustments or allowances for various circumstances.

STANDARD DATA. A structured collection of normal time values for elements of tasks or jobs.

STANDARD HOUR. The time earned for producing an amount of work equal to the production quota for one hour.

Standard Hour Plan. A method or formula of compensation based on earned standard hours credited on the basis of output during a given time period.

Standard Hours Produced. The total output of a worker during a given time period expressed in terms of standard allowed time and calculated as the sum of earned time for all work produced during the period.

Standard Metropolitan Statistical Area (SMSA). Geographic division used by the Bureau of the Census and the Social and Economic Statistics Administration of the Department of Commerce. There are a set number of SMSA's in the U.S. and Puerto Rico, the exact number changing over time. Each one consists of one or more entire counties that meet standards pertaining to population and metropolitan character. Federal studies of the population characteristics of SMSA's can be helpful in analyzing information about an organization's labor market.

Standard Performance. Level of performance required to accomplish a standardized task or job in the standard time established.

Standard Rate. A basic rate of pay established for an occupation in a plant, industry, or community through collective bargaining, company regulation, or by law. May also refer to established rates for services rendered in a community in connection with maintenance and repair of automobiles, appliances, buildings, etc.

Standard Time. A unit time value for the accomplishment of a work task or job.

Standard Time Data. A structured collection of time values for elements of work performed under standard conditions at normal pace. (See STANDARD DATA.)

Starting Rate. Rate of pay during probationary period, during period of training, or for a period of time following a job transfer. Entrance rate.

Status Pay. A form of a non-financial reward, such as office furnishings or office size, given to an employee by virtue of position within the organization.

Step Bonus. A plan of compensation under which payments are provided in increasing progression for increased productivity or accomplishment of work-related goals.

Step Rates. Fixed levels between the minimum rates for an occupation in wage progression system. (See AUTOMATIC PROGRESSION) (WAGE PROGRESSION.)

Stint. Shift.

Stock Bonus. Stock in a company awarded to an employee by the company, usually for a particular reason such as personal performance or corporate performance, and paid for by the company.

Stock Option. The privilege granted to certain employees to purchase shares of the company's stock as of a given date and price at some time in the future when, it is assumed, the stock will be selling on the market at a higher price.

Stock Option Discount. The difference between the quoted market price for a given stock and the price quoted to an employee at the time a stock option is granted.

Stock Option Exercise. The actual purchase of shares of stock at the option price after a predetermined date.

Stock Option Forfeit. The election by an employee not to exercise his/her rights to purchase stock under the option plan. Usually occurs when the market price falls below the option price.

Stock Option Holding Period. The time a participant in a stock option plan must hold purchased stock in order to qualify for the special tax treatments (capital gains) available. Usually refers to qualified stock option plans. Under laws governing qualified plans, participants had to hold stock for three years to take advantage of capital gains and exercise options within five years.

Stock Option (non-qualified). A stock option granted to employees that has fewer restrictions than a qualified plan. The option may be priced below market value, the time duration may be longer than five years, and more recently lower-priced options may be exercised before higher-priced options still outstanding. A nonqualified option is not eligible for capital gains treatment and, when exercised, becomes taxable income to the employee.

Stock Option Plans. Plans allowing employees and/or officers the privilege of purchasing company stock at a certain price on certain dates. The formal policies and procedures that outline how stock options will be administered. The usual plan would define participation, amounts of stock, prices, time limits, and methods for exercising rights.

Stock Option Term. The period of time in which a participant has the right to purchase stock. The maximum for qualified plans is five years.

Stock Purchase Plan. (1) (General) A plan enabling employees to purchase shares of stock in a company, with or without employer contribution, generally under more favorable terms than those available on the open market. The formal policies and procedures that outline how employees may purchase stock and the method of administering the plan. (2) (Non-qualified Stock Purchase Plan) A plan that is, in effect, a management stock purchase plan. It allows senior management or other key personnel to purchase stock in the business. There are, however, certain restrictions: a) the stockholder must be employed for a certain period of time, b) the business has the right to buy back the stock, c) stockholders cannot sell the stock for a defined period of time. (3) (Qualified) A program under which employees buy shares of the company's stock, with the company contributing a specific amount for each unit of employee contribution. Also stock may be offered at a fixed price (usually below market) and paid for in full by the employees. Benefits are distributed in stock of the employing company.

Stock Warrant. An option to purchase common stock at a fixed price intended to be favorable in the future. Considered a common stock equivalent. May be issued by a company to an employee or purchased by anyone in an open market.

Straight Commission. Compensation, usually for sales or service personnel, based on a predetermined percentage of the value of sales or services rendered.

Straight Salary. Compensation based on a fixed amount for a given period of time (per hour, per day, per week, per month, or per year).

Straight Time. Time worked at regular rate, as distinguished from overtime, without regard to output. (See EARNINGS, OVERTIME.)

Straight Time Rate. Rate of pay applicable to regular time work.

Stretchout. Term used by workers when they are required to tend more machines or assume additional duties within a given time without a corresponding increase in earnings. (See SPEEDUP.)

Strike (wildcat, outlaw, quickie, slowdown, sympathy, sit-down, general walkout). Temporary stoppage of work by a group of employees (not necessarily members of a union) to express a grievance, enforce a demand for changes in the conditions of employment, obtain recognition, or resolve a dispute with management.

Quickie Strike. A spontaneous or unannounced strike or strike of very short duration. (Sometimes referred to as a walkout).

Sympathy Strike. Strike of workers not directly involved in a dispute, but who wish to demonstrate worker solidarity or bring additional pressure upon the company involved.

Sitdown Strike. Strike during which workers stay inside the plant or work-place, but refuse to work or allow others to do so.

General Strike. Strike involving all organized workers in a community or county. (Rare in the United States.)

Strike Benefits. Union payments made to members who are on strike.

Strikebreaker. Worker or person on hire who accepts employment or continues to work in a plant where an authorized strike is in process, filling the job of a striker and knowingly assisting in defeating the strike. (See ANTI-STRIKEBREAKER LAW, FINK.) (SCAB)

Strike Deadline. Time set by the union for beginning a strike if a satisfactory settlement is not reached. Typically, this is at midnight of the last day of the contract term or the start of the next day's first shift.

Strike Fund. Money allocated by a union or set aside in a separate account to pay strike benefits and to defray other expenses of strikes.

Strike Insurance. Payment by companies in an association to a fund, or for the purchase of insurance, to reimburse a struck member company for lost business.

Strike Notice. Formal notice of an intention to strike, presented by the union to the employer, or to the appropriate federal government agency, e.g., the Federal Mediation and Conciliation Service.

Strike Vote. Vote conducted among members of a union to determine whether or not a strike should be called.

Typically a "given" action in contract negotiations in order to show solidarity.

Struck Work or Goods. Goods produced or services performed by a firm while a strike of its employees is in process. (See HOT-CARGO CLAUSE.)

Style Development Rate. Similar to temporary, experimental, or trial rate. The term is used in the hosiery manufacturing industry and relates to work on new styles for which no piece rates have yet been set. Generally, hourly rates are paid on such work. Usually, these hourly rates average close to the workers' previous piece-rate earnings. The style development rates are in effect for a specified time and are then replaced by new piece rates.

SUB. Income security plan. (See SUPPLEMENTAL UNEMPLOYMENT BENEFIT PLAN).

Subcontracting (contracting-out, farming out). Practice of having certain steps in a manufacturing process, plant maintenance, or other work functions performed by outside contractors, using their own work forces. Limitations on what work can be subcontracted is often covered in union contracts.

Subminimal Rate. A rate below the minimum established for an occupation, establishment, industry, or area by union agreement, law, or policy. Such rates may be paid to learners, probationary, substandard or special permit workers.

Subsistence Allowance. A payment to a worker for expenses covering meals, lodging, and transportation while in a traveling status for his employer. Such allowances may be based on a fixed amount for meals and lodging plus other expenditures or on the actual expenses incurred for all items. There also are cases where institutional workers (e.g., nurses) receive a subsistence allowance for living outside the institution, since free room and board are incorporated into the wage structure.

Substandard Rate. A rate of pay below the prevailing or standard level for a worker whose efficiency is impaired because of physical or mental handicaps. The term is also used to refer to rates below federal or state minimum wage levels or below prevailing levels for an occupation in an industry or area. (See HANDICAPPED WORKER RATE.)

Suggestion System. Plan whereby employees' ideas that may increase efficiency or quality, improve operations and/or safety, are channeled to the attention of management; usually combined with a system of rewards for acceptable ideas.

Sunlighting. Secondary employment during daytime hours (See MOONLIGHTING).

Superannuated Rate. A rate of pay below the prevailing level for a worker above a certain age. Such rates are frequently allowed in union agreements. At times, the agreement requires the employment of a certain ratio of older workers at superannuated rates. Superannuated workers with long service are sometimes retained in an employed status because of their economic need; also their services are sometimes sought during periods of labor shortages.

Superannuated Workers. Term sometimes applied to employees who are unable to perform their jobs, or any job, at the normal level because of advanced age and its attendant infirmities.

Supercraft. (See MASTERCRAFT.)

Super Seniority. A position on the seniority list ahead of what the employee would acquire solely on the basis of length of service or other general seniority factors. Usually such favored treatment is reserved to union stewards or other workers entitled to special consideration in connection with layoff and recall to work.

Supervision. (1) The function of directing the work of others to insure performance in accordance with plans and instructions. (2) A collective term for those engaged in this function.

Supervisor. An individual within the management group who is responsible for the work assignments and work results of operative employees.

Supervisory Incentive. An incentive plan designed to financially reward supervisors over and beyond their base pay for meeting predetermined objectives for productivity and costs of the operations under their supervision.

Supplemental Compensation. Money paid an employee over and above the agreed-upon base rate usually for exceeding predetermined targets or achieving or exceeding productivity standards. Can be either commission or bonus.

Supplemental Unemployment Benefit Plan. An agreement negotiated between Ford Motor Company and the United Automobile Workers in 1955 and later extended to other companies. Such plans provide regular weekly payments to laid-off workers receiving state unemployment insurance, through funds financed by the employer. Additional pay given to laid-off employees over and above the state supplied unemployment insurance payments.

SUPPLEMENTARY BENEFITS. (See FRINGE BENEFITS.)

SUPPLEMENTS TO WAGES AND SALARIES. As defined by the U.S. Department of Commerce for national income purposes: "Supplements to wages and salaries is the monetary compensation of employees not commonly regarded as wages and salaries. It consists of employer contributions to private pension and welfare funds, compensation for injuries, director's fees, pay of the military reserve, and a few other minor items of labor income." Term sometimes used more broadly to refer to all supplements to basic wage or salary rates.

SURGICAL BENEFITS. Plans which provide workers, and in many cases, their dependents, with specified surgical care or a cash allowance toward the cost of such care, usually in accordance with a schedule of surgeon's fees. Often referred to as Blue Shield. Generally part of a health and insurance program. (See HEALTH AND INSURANCE PLAN.)

SURVIVOR PROTECTION. Compensation paid to survivors of an employee. Most commonly provided by group life insurance plans.

SURVIVORS' BENEFITS (TRANSITION BENEFITS, BRIDGE BENEFITS, WIDOW'S ALLOWANCE). Payments to dependents of employees who die prior to retirement, financed in whole or in part by the employer. May be in the form of payments for a fixed period (e.g., 24 months) supplementing regular life insurance benefits, a benefit for life out of a pension program, a lump-sum payment, etc.
SURVIVOR'S OPTION. (See JOINT AND SURVIVOR OPTION.)

SUSPENSION. Form of disciplinary action of a temporary nature, as in removing a worker from his/her job for a stipulated time with the consequent loss of pay as punishment, or in removing a union official from office until his affairs are checked or put into order.

SWEATSHOP. Term of contempt applied to an establishment employing workers for long hours at low wages under unfavorable working conditions.

SWEETHEART AGREEMENT (CONTRACT). A derogatory term for a union contract exceptionally favorable to a particular employer, in comparison with other contracts, implying less favorable conditions of employment than could be obtained under a legitimate collective bargaining relationship.

SWING SHIFT. An extra shift of workers required in establishments where continuous or seven-day operations are scheduled, to provide the other crews with days off. The "swing crew" usually rotates among all of the other shifts. Also, refers to the practice of one of three rotating shifts staying on the job through two shift periods, thus "swinging" the shifts into their new assignments. (See FIXED SHIFT, ROTATING SHIFT, SPLIT SHIFT.)

SYMPATHY STRIKE. (See STRIKE.)

SYNTHETIC TIME STANDARD. A time standard developed from data compiled from prior study or research rather than from a study of the actual job to which the standard is to be applied.

TAFT-HARTLEY ACT. (See LABOR MANAGEMENT RELATIONS ACT.)

TAKE-HOME PAY. Typically, earnings for a payroll period, less required deductions. Represents net earnings. (See SPENDABLE EARNINGS.)

TAKE-HOME WAGES. (See TAKE-HOME PAY.)

TARGET. In incentive systems, a rate or standard is set with the objective of making it possible for a worker or a group of workers to earn a specific percentage above the base rate. The expected earnings to which an incentive standard (including a piece work rate) is geared.

TASK. The standard quantity of production expected under normal conditions. Usually designated as the point above which incentive pay is earned under a wage incentive plan.

TASK AND BONUS PLAN. An incentive plan wherein a set compensation amount is given an employee for output above a predetermined level or task.

TAX CREDIT ESOP. A plan designed to meet conditions specified by the Economic Tax Recovery Act of 1981, permitting employers tax credits for the value of the stock.

TAX REFORM ACT STOCK OWNERSHIP PLAN (TRASOP). A special ESOP that came into existence with the Tax

Reform Act of 1976. This act grants businesses the right to recover a specified amount of "start-up" and administrative costs. Employees have full and immediate vesting to all shares placed in their accounts.

TAX EQUITY AND FISCAL RESPONSIBILITY ACT, 1982 (TEFRA). A comprehensive tax law which imposed greater restrictions on employee benefit plans.

TAX SHELTER. A device for restructuring compensation, or investments, with the principal objective of avoiding current taxes and postponing payment of such taxes. For compensation purposes, usually refers to a plan for postponing declaration of earnings for tax purposes to a future period of time when tax rate for an individual is expected to be lower. (See FRTA)

TEAM MANNING. The establishment of a group of workers to operate a particular machine or assembly line or to complete a particular task.

TECHNOLOGICAL UNEMPLOYMENT. Displacement of workers caused by or attributed to the installation of labor-saving methods or machines.

TEFRA. The Tax Equity and Fiscal Responsibility Act of 1982, sharply cut the maximum pension benefit and contribution limits for qualified pension plans.

TEMPORARY DISABILITY INSURANCE. Provision presently enacted into law in certain states providing payments for a limited period of time to workers suffering loss of wages due to sickness or disability incurred off the job. In some states, employers may substitute privately insured coverage meeting certain standards. (See WORKER'S COMPENSATION, ACCIDENT AND SICKNESS BENEFITS.)

TEMPORARY EMPLOYEE. An employ hired for a specific job on a temporary basis. The "temporary" status is often used to better evaluate a prospective employ prior to granting a permanent status. (See PROBATIONARY PERIOD).

TEMPORARY RATE. A rate or standard set tentatively on new or changed work. When new or changed work is started under an incentive plan, it is often not known whether or not correct rates or standards can be properly set for the tasks involved, temporary rates are established. These rates are later revised and are made permanent when found to be satisfactory. Sometimes they are called "experimental" or "trial" rates. (See GUARANTEED RATE, PROTEST PRICE.)

TEMPORARY STANDARD. (See TEMPORARY RATE.)

TERMINATION. Separation from an organization due to resignation, layoff, retirement, or dismissal for cause (stealing, insubordination, incompetence, etc.)

TERMINATION PAY. (See SEVERANCE PAY.)

THEORY X. Based on the assumption that workers dislike work, wish to avoid responsibility, like to be told what to do, lack ambition, and show little concern or loyalty toward their organization. Theory X managers use firm direction and strict control over their people with close surveillance and threats of punishment. (Theory accredited to Douglas McGregor).

THEORY Y. Based on the assumption that people seek responsibility, and become committed to objectives through the satisfaction of self-actualization. That they have the capacity to demonstrate imagination and creativity in solving organizational problems. Theory Y managers delegate authority and permit greater participation of workers in policy and decision making in the areas that affect them. (Theory accredited to Douglas McGregor).

THEORY Z. Emphasizes the importance of adapting managerial strategies and tactics to the regional or national culture and life-styles of employees. It stresses job security and the opportunity to gain recognition through employee participation in planning, problem solving, decision making, and quality awareness. (Theory accredited to William Ouchi).

THRIFT PLAN. An arrangement under which payroll deductions are made, with the worker's consent, for investment and saving, to which the employer contributes. Much of the total fund is subject to tax-deferral. The accumulated amounts (contributions plus earnings on the contributions) become available to the worker, usually after certain conditions are met. (SAVINGS PLAN.)

TIGHT RATE. A time standard under an incentive plan in which it is difficult, when applying the appropriate effort and skill, to reach the expected incentive earnings. Theoretically, a rate in which insufficient time is allowed for the performance of an operation under the conditions of the incentive plan in use.

TIGHT STANDARD. (See TIGHT RATE.)

TIME AND ONE-HALF. Premium rate consisting of one and one-half times the employee's regular rate.

TIME RATE. Compensation rate based on hours worked, i.e., hourly wage rate. (See TIME WAGE.)

TIME RATE EARNINGS. Compensation based on time rate multiplied by the amount of time of the work period.

TIME STANDARD. Designated time for the accomplishment of a specified task or job.

TIME STUDY. A work measurement technique consisting of a careful time measurement of the task with a time measuring instrument which has been adjusted for any observed variance from normal effort of pace and to allow adequate time for unavoidable or machine delays, rest to overcome fatigue, and personal needs.

TIME WAGE. Compensation based on hours worked rather than quantity produced.

TIN HANDCUFFS. Relatively small monetary or stock inducements to entice employees to stay with the company.

TIN PARACHUTES. Relatively small amounts of monetary compensation given to employees dismissed as a consequence of a merger or acquisition.

TIP. A gratuity given by a customer or patron in recognition of satisfactory personal service or through custom. Tips are considered as compensation by the Bureau of Internal Revenue, thus constituting taxable income. Employers are required to withhold taxes based on an estimate of the expected amount of tips received by each employee.

TONNAGE RATE. Pay for a unit of work applicable to incentive workers and common in such industries as coal mining and basic iron and steel, where output for important categories of workers can be measured on a tonnage basis.

TOOL ALLOWANCE. Allowance to an employee, paid by employer, as reimbursement for the cost of tools and upkeep, where the employee furnishes his/her own tools or is responsible for their maintenance. An allowance in a standard to cover tool maintenance such as tool sharpening.

TOOL MAINTENANCE TIME. (See ALLOWED TIME.)

TOTAL CASH COMPENSATION. The total cash payment made to an individual in the form of compensation in a given year.

TOTAL COMPENSATION. The complete pay package for employees including all forms of money, benefits, and services and in-kind payments.

TOTAL PAY. The net amount paid to an expatriate employee consisting of the sum of his or her base salary, expatriation premium and allowances less the tax equalization factor.

TOUCH LABOR. Work directly applied to material or product. Refers to those workers actually in physical contact with product during manufacture.

TOUR OF DUTY. (See SHIFT.)

TRADE COUNCIL. (See JOINT BOARD.)

TRADE UNION. (See UNION.)

TRAILING COSTS. (See LAYOFF COSTS/BENEFITS.)

TRAINEE. The term "trainee" applies to workers who receive formal training for occupations requiring a limited degree of skill. The training may include some classroom work. A trainee differs from a learner in that a learner does not receive formal training but learns the job through actual performance, under supervision.

TRANSFER CARD. Card issued by a local union to a member in good standing, certifying eligibility to join another local of the same union in a different location.

TRANSFER PAY. Compensation for period of temporary reassignment or for period of time subsequent to permanent reassignment.

TRANSITION BENEFITS. (See SURVIVORS' BENEFITS.)

TRAVEL TIME. The time spent traveling to and from a designated point and place of work. Such travel includes portal-to-portal in mining, deadheading on railroads, and out-of-town work performed by building tradesmen, mechanics, musicians, etc.

TRICK. (See SHIFT.)

TRUSTEE. A person, bank, or trust company who administers and takes responsibility for a trust fund, or a person who is a member of a board of trustees. (See FUND.)

TRUSTEESHIP. In union affairs, the taking of direct control of a local union by the national or international union, generally to correct mismanagement or illegal practices on the part of local officers, to prevent secession, or to strengthen the local. Control of the local is returned to members or officers after the cause for trusteeship has been corrected. (See MONITORSHIP.)

TRUST FUND. (See FUND.)

TUITION PAYMENT PLAN. Plan providing for payment by the employer of a part or all of the costs of job-related training courses undertaken by employees.

TURN. (See SHIFT.)

TURNOVER. (See LABOR TURNOVER.)

UMPIRE. (See IMPARTIAL CHAIRMAN.)

UNAFFILIATED UNION. (See INDEPENDENT UNION.)

UNEMPLOYMENT INSURANCE (UNEMPLOYMENT COMPENSATION). Joint federal-state program, established in 1935 under the Social Security Act and subject to the standards set forth in the Federal Unemployment Tax Act, under which states administered funds obtained through payroll taxes provide payments to eligible unemployed persons for specified periods of time. Levels of benefits and tax rates are established by each state. Generally, excluded groups include among others, railroad workers (covered by Railroad Unemployment Insurance Act), agricultural workers, state and municipal employees, and workers in nonprofit institutions. The federal part of the program is administered by the United States Department of Labor.

UNEMPLOYMENT PAY. Compensation to a person no longer employed.

UNFAIR LABOR PRACTICE. Action by either an employer or union which violates the provisions of national or state labor relations act, such as refusal to bargain in good faith.

UNFAIR LABOR PRACTICE STRIKE. A strike caused, at least in part, by an employer's unfair labor practice.

UNFAIR LIST. Union list of employers designated as unfair to organized labor.

UNFUNDED PLAN. A plan (e.g., pension plan) under which benefits are paid, like wages, directly from an employer's general assets, often as a payroll item, as distinguished from a fund irretrievably segregated from the general assets of the firm and separately administered. Some pension plans are funded in part and unfunded with regard to certain benefits. (PAY-AS-YOU-GO.)

UNION. An organization with the legal authority to negotiate with the employer on behalf of the employees and to administer the ensuing agreement. Unions are voluntary organizations and need no license from the government to operate. Also known as labor union, labor organization.

UNION AGREEMENT. (See AGREEMENT.)

UNION CONTRACT. (See AGREEMENT.)

UNION CONVENTION. An assembly of delegates meeting periodically to act on union problems, elect officers, and determine policies. The convention is typically the chief governing body of the union in constitutional terms.

UNION DUES. Fee paid periodically, usually monthly, by members of a union, typically as a condition of continued membership. Each union sets its own dues requirement. (See CHECK-OFF.)

UNION LABEL (BUG). Tag, imprint, or design attached to an article as evidence that it was produced by union labor.

UNION LEAVE. Paid or unpaid, but excused, leave for union representatives, shop stewards, etc. to attend to union business, e.g., participating in union conventions.

UNION-MANAGEMENT COOPERATION. Voluntary joint participation of union and management in solving problems such as production and safety, or in engaging in certain outside activities, such as community or charitable work. The term is usually reserved to joint actions outside of the process of collective bargaining itself.

UNION MEMBER. A union member may be defined in broad terms as a worker who has met the union's qualification for membership, has joined the union, and has maintained his membership rights. Each union usually determines its own qualifications. In general, dues-paying members are those who pay dues to the union on a regular basis. Members in good standing include dues-paying members and members exempted for various reasons (unemployed, on strike, ill, etc.) but carried on the union rolls as full-fledged members. Book members are those listed on the union rolls, dues-paying or not.

Union Organizer. (See ORGANIZER.)

Union Rate. An hourly rate, usually a single rate for an occupation or trade, established by agreement reached through collective bargaining. A union rate or scale is usually the minimum rate that may be paid to qualified persons in the job; there are usually no restrictions prohibiting the employer from paying higher rates.

Union Recognition. Employer acceptance of a union as the representative of his employees, the first step in the establishment of a collective bargaining relationship.

Union Scale. (See UNION RATE.)

Union Security. Protection of a union's status by a provision in the collective bargaining agreement establishing a closed shop, union shop, agency shop, or maintenance of membership arrangement. In the absence of such provisions, employees in the bargaining union are free to join or support the union at will, and, thus, in union reasoning, are susceptible to pressures to refrain or to the inducement of a "free ride".

Union Shop. Provision in a collective bargaining agreement that requires all employees to become members of the union within a specified time after hiring (typically 30 days), or after a new provision is negotiated, and to remain members of the union as a condition of continued employment.

Modified union shop. Variation of the union shop. Certain employees may be exempted, e.g., those already employed at the time the provision was negotiated who had, up until then, not joined the union.

Union Steward. (See SHOP STEWARD.)

Unit of Production. A measured quantity of output. May be based on variables such as physical count, weight, or accumulated standard time.

Unlicensed Personnel. Seamen not required to have an official license, such as deckhands, stewards, firemen, etc. as distinguished from licensed masters, mates, and engineers.

Unpaid Holidays. Holidays observed by an establishment only to the extent of providing premium pay for work on that day. Paid time off not provided. (See PAID HOLIDAYS.)

Unrestricted Job. A job in which the output of production is limited only by the ability of the worker to produce.

Upgrading. The process of advancing of workers to jobs having greater skill and/or experience requirements and commanding higher rates of pay. Workers may be upgraded without the necessary skill and/or experience when workers are not available to fill such jobs. Under those conditions the worker learns the new job on the basis of on- the-job training. May also result from re-evaluation of that job.

Urban Wage Rate Index. Series maintained by the Bureau of Labor Statistics, beginning in 1943, to measure the movement of wage rates in urban areas in manufacturing industry groups, and selected nonmanufacturing industries.

U.S. Department of Labor. The department was established by an Act of Congress in 1913 to "foster, promote, and develop the welfare of the wage earners of the United States, to improve their working conditions, and to advance their opportunities for profitable employment." The Department also has important functions in the field of international labor affairs. The major bureaus and offices of the Department include: Bureau of Apprenticeship and Training, Bureau of Employment Security, Bureau of Labor Statistics, Bureau of Labor Standards, Wage and Hour and Public Contracts Divisions, Women's Bureau, Bureau of International Labor Affairs, Office of Manpower, Automation and Training, Office of the Solicitor, Neighborhood Youth Corps, Labor-Management Services Administration, and Manpower Administration. The Secretary of Labor heads the Department.

Utilityman (utility operator or worker). A worker who can perform a variety of jobs and is used to do such work on an "as needed" basis, e.g., an operator who relieves on an assembly line.

Vacation Pay. Payment for a period of time received by workers for vacation purposes. The time period frequently varies with length of service. During busy times or in a tight labor market, workers may be given the option of accepting vacation pay in lieu of time off.

Validity. The degree to which a predictor or criterion measures what it purports to measure or demonstrates the job-relatedness of a test by showing how well an applicant will perform based on the test predictions

VEBA. Voluntary Employees' Beneficiary Association intended to provide life, sickness, accident or other benefits to members. Usually established jointly by a company and a union.

VERTICAL LOADING. Adding duties to a job that are different from those that already in the job and that require different skills, knowledge, and abilities.

VERTICAL UNION. (See INDUSTRIAL UNION.)

VESTIBULE TRAINING. A method particularly effective for training a large number of workers, usually unskilled or semiskilled in the processes of the job, using machines and tools, directly related to the job, set up in a replica inside or outside the plant.

VESTING (VESTED RIGHTS). As commonly used in connection with pension plans, a guarantee to a worker of equity in the plan, based on contributions by the employer, should employment terminate before the employee becomes eligible for retirement. The worker usually must meet specified minimum age and service requirements for qualification. The vested worker receives the pension earned when he/she reaches retirement age, wherever he/she is then employed.

VOCATIONAL REHABILITATION ACT OF 1973. Section 503 of this act is intended to promote job access in employment for handicapped people. All employers with contracts or grants from the federal government in excess of $10,000 and who employ 50 or more people must develop formal affirmative action plans in regard to handicapped people.

VOCATIONAL TRAINING. Training for workers supplied by employers, schools, or state or federal government agencies to teach specific skills such as auto mechanics, brick laying, carpentry, and various factory skills.

W2 FORM. The form a company gives an individual at year's end which reports the individual's total earnings reported to the government, and shows the total of all items withheld for various purposes, including federal, state, local and FICA taxes.

WAGE. Payment for work performed. Usually applied to blue collar work as opposed to salary applied to white-collar work.

WAGE ADVANCE PLAN. Advancing of wages in workweeks of short duration under plans obligating employers to maintain weekly wages up to a specified minimum level. Wages must be repaid during later weeks in which regular or longer hours are worked. No repayment is required unless the employer provides sufficient work to enable the advance to be repaid. (See ANNUAL WAGE OR EMPLOYMENT GUARANTEE, GUARANTEED WAGE PLAN.)

WAGE AND PRICE CONTROLS. An occasional program in which wages and prices are established through government regulation.

WAGE AND SALARIES. As defined by the U.S. Department of Commerce for national income purposes: "Wage and salaries consist of the monetary remuneration of employees commonly regarded as wages and salaries, inclusive of executives' compensation, commissions, tips, and bonuses, and of payments in kind which represent income to the recipients." More generally, this term refers to remuneration to individuals for productive effort.

WAGE AND SALARY ADMINISTRATION. The managing and supervision of the wage structure of an employer. It involves the application of wage and salary adjustments, according to established policies, and the analysis of data such as cost of living, prices, wage and salary surveys, which have a direct bearing on the wage structure and are used in wage negotiations. May also involve the establishment of new rates through job evaluation, job analysis, and time studies.

WAGE AND SALARY RECEIPTS. As defined by the U.S. Department of Commerce for national income purposes: "Wage and salary receipts are equal to wages and salaries less employee contributions for social insurance, except that retroactive wages are counted when paid rather than when earned."

WAGE AND SALARY STRUCTURE. An arrangement in an ordered manner of compensation rates reflecting relationships between jobs. May also include supplemental benefits.

WAGE AND SALARY SURVEY. The collection of data and the preparation of information and statistics covering prevailing rates of pay for given jobs throughout an industry, community, or competitive labor market.

WAGE ARBITRATION. The referral of wage disputes between employers and unions to an arbitrator or Board of Arbitration. The arbitrator's award or decision is customarily binding upon both parties. Arbitration is usually voluntary, both parties having agreed to refer the dispute to a third party for a decision. (See WAGE MEDIATION.)

WAGE ASSIGNMENT. A voluntary transfer by a worker of some of his earned wages or commissions to another party or parties. Such assignments may be used for payment of purchased goods and debts, purchase of savings bonds, and payment of union dues and assessments.

Wage Award. A decision or judgement regarding wages handed down by an arbitrator, board, or other authoritative agency in a dispute between management and labor.

Wage Board Employees. Federal government employees, typically manual workers, whose rates of pay are determined on the basis of prevailing rates for comparable work in the area, as distinguished from classification act employees.

Wage Curve (wage and salary curve). Salary curve, a curve, often linear, on a graph describing the relationship of jobs, either in points or grades, to the compensation for each of the jobs, often with job values in points or grades along the horizontal axis and dollar compensation along the vertical.

Wage Deductions. A sum withheld by the employer from an employee's earnings to pay certain specific and authorized charges against an employee's income such as taxes, union dues, credit union, and insurance.

Wage Determination. The process of establishing wage rates and wage structures through collective bargaining, arbitration, individual employer determination, etc. The process may involve comparisons with rates paid by other firms, the use of job evaluation, or other techniques. The term is also applied to findings, orders, or decisions of wage regulatory bodies such as minimum wage boards.

Wage Differential. Differences in wages among occupations, industries, or areas. Historical wage differentials, to which frequent reference is made, are those which have existed over long periods of time.

Wage Drift. Term generally used to describe the differential change in average earning levels over time as measured against negotiated changes. The difference between the level of actual earnings, which is influenced by many factors, and the level at which earnings would be if formal general wage changes alone are taken into account, is thus likened to an upward drift.

Wage Escalation. (See escalator clause.)

Wage Guide. (See salary guide.)

Wage Guidelines. Standards used to measure compliance of wage increases with established policy objectives.

Wage-hour Law. (See fair labor standards act.)

Wage Incentive. Payment for performance over a predetermined level or standard.

Wage Incentive Payment. That portion of compensation reflecting earnings above base pay for productivity in excess of standard or normal under a wage incentive plan.

Wage Incentive Plan. A defined procedure or method for providing workers opportunity for extra compensation above base pay for productivity in excess of a set standard.

Wage Inequality. An unadjusted disparity between rates of workers whose duties and responsibilities are similar or identical. Wage inequalities can be considered either on an intraplant or interplant basis. The elimination of wage inequalities is often accomplished through job review or the adoption of job evaluation plans. (See wage inequity.)

Wage Inequity. An unadjusted relationship or disparity between the wage rates of workers or of job classifications. (See wage inequality.)

Wage Leaders. An employer or group of employees accepted as trend setters in a labor market.

Wage Leadership. The influence exercised by the wage settlements reached by a large firm or group of firms on other settlements in an industry or labor market. "Follow-the-leader" wage adjustments appear to be particularly significant in some industries. May also relate to a policy adopted by a firm maintaining a position of wage leadership in an industry or area.

Wage Level. The level of wages received by workers in an occupation, establishment, industry, or area. Wage levels are generally indicated by average rates.

Wage Mediation. The entrance of a disinterested third party into a wage dispute in an effort to effect a settlement. Unlike arbitration, the mediator merely makes recommendations and assists the disputant parties in reaching a settlement. (See wage arbitration.)

Wage Movement. (1) Changes in wage levels over a period of time. (See wage drift.) (2) An increase or decrease in the wage level being paid a job or an occupation in a wage market.

Wage Plan. A plan covering the manner in which employees will be paid. A collection of such plans that may include such devices as pay for time worked, for production (an incentive plan), or a combination of base wages and/or incentive or bonus payments. (See WAGE AND SALARY ADMINISTRATION.)

WAGE PATTERN. Identifiable trends or similarities in job compensation.

WAGE POLICY. The formalized practice of an establishment or industry relating to elements of wages, such as wage rate scales, shift differentials, overtime provisions, non-production bonuses, automatic increments, paid holidays, paid vacations, pensions, and insurance benefits. In a broader sense, criteria for wage adjustments are stated in terms of objectives (e.g., stabilization, rising standard of living, etc.) or in terms of prevailing economic conditions.

WAGE PROFILE. (See SALARY PROFILE.)

WAGE PROGRESSION. (See AUTOMATIC PROGRESSION.)

WAGE RATE. The monetary compensation for a given unit of time or effort by which a worker's pay is calculated. There are several kinds of wage rates, related to the system of wage payment used in an establishment. The principal kinds are hourly rates, daily rates, weekly rates, monthly rates, annual rates, and various kinds of incentive rates.

WAGE RATE BRACKET. In the administration of wartime wages stabilization policy by the National War Labor Board, the term referred to a range of "sound and tested going rates" for an occupation in a labor market area. The minimum of the range or bracket, the most important point in actual wage administration, was frequently set at the first substantial cluster of rates in a wage distribution. The minimum of the bracket with that point up to which the War Labor Board would permit adjustments in interplant inequity cases.

WAGE REOPENING. A provision or clause in a union agreement permitting the question of wages to be reopened for negotiation before the expiration of the agreement, usually at a set time.

WAGE REVIEW. A periodic review of the performance of workers to determine or select those who deserve merit increases, or advancement to higher paying jobs. (See MERIT INCREASE.)

WAGE STRUCTURE. The sum total of the various elements and considerations that characterize a specific rate schedule in an establishment, industry, area, or country as a whole. Typical of such elements are: a) relationship between rates of occupations of different skills; b) relationship between rates of pay for men, women, and workers of different races and color in the same occupations; c) provision for extra pay for late shift work, overtime, hazardous, unpleasant, or unhealthful work; d) interarea and interregional variations in rates of pay; e) methods of pay; f) provisions for lunch and rest periods; and g) supplementary benefits, such as vacations, insurance, sick leave, and holiday provisions.

WAGE SURVEY. (See WAGE AND SALARY SURVEY.)

WAGNER ACT. The major comprehensive labor code enacted in 1935 with the intent by Congress to restore equality of bargaining power between labor and management . Also known as the National Labor Relations Act (NLRA).

WAITING TIME. (See DOWNTIME, IDLE TIME.) (DEAD TIME, DELAY TIME)

WALKOUT. A strike.

WALSH-HEALY PUBLIC CONTRACTS ACT OF 1936. A federal law requiring certain employers holding federal contracts for the manufacture or provision of materials, supplies, and equipment to pay industry-prevailing wage rates.

WASH-UP TIME. Also known as clean-up time. A paid allowance for personal needs at end of work shift.

WASTE BONUS. Compensation for performance in reducing scrap, damage, material losses, or sub-standard work.

WATCH. The shift or work period for employees in certain industries. Particularly used in the military. (See SHIFT.)

WELFARE PLAN. An arrangement which provides various health and other employee interest benefits-including death benefits. The term is used also in connection with various employee benefit plans.

WELLNESS PROGRAM. A program for reducing employee illness, accidents, and medical claims, by establishing fitness centers sponsoring health screenings and wellness education. These programs monitor the overall health of employees and provide the means for reducing risks through systematic exercise and life style modification eg. annual physicals, comprehensive health risk assessment, smoking and alcohol abeyance programs, weight reduction, blood pressure control, drug rehabilitation, stress management, exercise programs, and other health instructive programs.

WHIPSAWING. Term applied to a union tactic of negotiating with one employer at a time, using each negotiated gain as a lever against the next employer. Usually within the same industry.

WHISTLEBLOWER PROTECTION ACT OF 1989 (WPA). This act was established to protect the rights of federal employees who make disclosures of corruption, fraud, waste, abuse, and unnecessary government expenditures. The Act also established the Office of Special Counsel to receive and investigate allegations, prevent reprisals, issue petitions for corrective action, file complaints and make recommendations for disciplinary action.

WILDCAT OR OUTLAW STRIKE. A strike not sanctioned by the union officers and one which usually violates the agreement. (See STRIKE).

WHITE COLLAR UNION. An organization of individuals in office, technical, professional, or minor supervisory work assignments, for the purpose of improving status, professional standards, hours, working conditions, and compensation. Used in comparison with blue collar unions or unions of hourly, factory workers.

WHITE-COLLAR WORKER. Usually associated with the description of an employee on the office, clerical, sales, semi-technical and/or professional staffs. More specifically, the term is used to describe an employee not "on production work." Usually, it is regarded as the antonym of blue-collar worker (q.v.).

WITHHOLDING. A portion of an employee's gross wage withheld either by law, regulation, or consent for allocation and payment by the employer to its properly allocated source.

WORK. Generally, the expenditure of physical and mental energy for the purpose of accomplishing a particular task. The application of various skills and effort to the accomplishment of a given task or series of tasks. A job. The tasks or duties that enable one to earn money. A place where industrial labor is carried out.

WORK CONTENT. The sum total of operations, methods, activities and procedures required to accomplish one unit of a given task or series of tasks.

WORK CYCLE. One complete sequence of operations, activities, and procedures required to accomplish one unit of a given task.

WORK FORCE. (See LABOR FORCE.)

WORK HOURS ACT. Legislation applicable to federal public work or any work in which the federal government is involved and which provides an 8-hour workday, a 40-hour workweek and overtime at 1-1/2 times basic rate.

WORK MEASUREMENT. A generic term used to refer to the determination or setting of time standards for the performance of standardized tasks using recognized industrial engineering techniques, such as time study, standard data, work sampling, or predetermined motion time systems. It includes the assessment of actual effort exerted versus the real effort required to accomplish the task.

WORK RULES. Workplace requirements established to maintain discipline, prevent injuries and accidents, and to maintain productivity. Work rules deal with such items as relief periods, coffee breaks, absence, tardiness, day-to-day conduct in the plant-operations, hygiene, and safety procedures. Work rules may be set forth in collective bargaining agreements and in some union constitutions. Also known as working rules or shop rules.

WORKER PARACHUTES. A special form of protection plan (cash payments and/or guarantee of future employment) for workers in the event of mergers with another company. (See GOLDEN PARACHUTE.)

WORKER PARTICIPATION CLAUSE. A union-management contract clause which spells out the areas or functions in which workers will assist or participate in the management of the company.

WORK SAMPLING STUDY. A procedure for observing work activities at random intervals of time from which statistical inference may be made relative to the entire scope of work involved.

WORK SHARING. Spreading available work among all employees in a group in order to prevent or reduce the extent of a layoff.

WORK SIMPLIFICATION. An organized approach to improvement of worker performance involving training, implementation of improvements and follow up.

WORK STANDARD (STANDARD, STANDARD TIME). An establlished measurement base, usually a unit time value stated as time per unit of production or per task, for the accomplishment of a work task as determined by the proper application of appropriate work measurement techniques. The time allowed for the performance of a task by a qualified, trained, and experienced employee when working at an average pace (normal), with appropriate allowances.

WORK STOPPAGE. Any temporary halt to work. It may be initiated either by the employees or by the employer. The term has been adopted by the Bureau of Labor Statistics to replace the terms strike and lockout.

WORK STUDY. The techniques of methods study and work measurement employed to ensure the best possible use of human and material resources in carrying out a specific activity.

WORK TICKET. A specification of work required. An authorization to proceed with a particular job. A work order.

WORK TIME. The time a worker or process is productively involved with a job. May also be used to designate a period of time such as a work shift.

WORK WEEK. The scheduled number of working hours in any seven day period that an employee is required to work.

WORKER'S COMPENSATION. Also called workmen's compensation. A system of insurance required by state law and financed by the employer through private carrier or state government agency which provides payment to workers or their families for occupational illness or injuries that causes an employee to miss work or incur other costs, usually medical, related to the illness or injury. Also covers payments to families in the event of fatal injuries.

WORKING CONDITIONS. The environment under which work is performed with reference to heat or cold, light, noise, ventilation, dust, vibration, fumes, moisture, and exposure to hazards and injury or illness producing conditions.

WORKING FORCE (WORK FORCE). (See LABOR FORCE.)

WORKING RULES. (See WORK RULES.)

WORKLOAD. Amount of work to be performed by an employee or a group of employees, or output expected, in a given period of time.

WORKPLACE. The location at which a worker performs his or her job.

WORKPLACE LAYOUT. Specification or design showing positions of workers, extent of working area, contents and arrangement of workplace, location of materials and equipment.

ZERO-BALANCE REIMBURSEMENT ACCOUNT (ZEBRA PLAN). A flexible spending plan in which no money is set aside initially, but employees are reimbursed for some expenses as they occur.

Z94.7
Engineering Economy

This section consists of two major elements. **Part A** consists of definitions and symbols of the most commonly used parameters as well as a set of functional forms for those compound interest factors frequently encountered in the engineering economy literature. These can be summarized in two exhibits. **Part B** is a glossary of technical terms used in engineering economy.

Exhibit I, definitions and symbols used for parameters, is very similar to Table 1 in the 1982 edition of *Industrial Engineering Terminology*. Significant changes incorporated into this new version include:

(1) The symbols for the number of compounding periods (N) and the number of compounding subperiods per period (M) are capitalized. This is consistent with our basic scheme: all amounts are capitalized and all rates are lower case letters.

(2) Four new symbols have been added: the periodic rate of increase or the geometric gradient (g), the periodic "inflation" rate (f), salvage value (S), and end-of-period cash flow (A.).

Exhibit II is a summary of the functional forms of the compound interest factors, including factor names and appropriate algebraic expressions. These are classified into four groups:

	Compounding of Interest	
Cash Flow	**End of Period**	**Continuous**
Discrete	Group A	Group B
Continuous	—	Groups C, D

The continuous compounding cases, Groups B, C and D, are based on the assumption that interest is compounded continuously during the period at nominal interest rate r, where the effective and nominal rates are related by $r = ln(1+i)$. More precisely, assuming that the interest period is divided into M subperiods of equal duration, and further assuming that interest is compounded at the end of each subperiod at rate r/M, then the effective rate per period (i) is related to the nominal rate per period (r) by $i = (1+r/M)^M - 1$. At the limit, as M approaches ∞, then i approaches $\exp(r)-1$.

Chairman
Gerald A. Fleischer
Professor
University of Southern California

Subcommitee

Richard S. Leavenworth
Professor Emeritus
University of Florida

Wolter J. Fabrycky
Professor
Virginia Polytechnic Institute and State University

Exhibit I

DEFINITIONS AND SYMBOLS USED FOR PARAMETERS

Ref. No.	Definition of Parameter	Symbol
1.	Cash Flow at end of period j.	A_j
2.	End-of-period cash flows (or equivalent end-of-period values) in a uniform series continuing for a specified number of periods.	A
3.	Amount of money (or equivalent value) flowing continuously and uniformly during each and every period, continuing for a specified number of periods.	$\overline{A}$
4.	Future sum of money. The letter "*F*" implies future (or equivalent future value).	F
5.	Uniform period-by-period increase or decrease in cash flows (or equivalent values); the arithmetic gradient.	G
6.	Number of compounding subperiods per interest period.*	M
7.	Number of compounding periods.	N
8.	Present sum of money. The letter "*P*" implies present (or equivalent present value). Sometimes used to indicate initial capital investment.	P
9.	Amount of money (or equivalent value) flowing continuously and uniformly during a given period.	$\overline{P}$ or $\overline{F}$
10.	Salvage (residual) value of capital investment.	S
11.	Rate of price level increase or decrease per period; an "inflation" or "escalation" rate.	f
12.	Uniform rate of cash flow increase or decrease from period to period; the geometric gradient.	g
13.	Effective interest rate per interest period* (discount rate), expressed as a percent or decimal fraction.	i
14.	Nominal interest rate per interest period*, expressed as a percent or decimal fraction.	r

*Normally, but not always, the interest *period* is taken as one *year*. Subperiods, then, would be quarters, months, weeks, etc.

Exhibit II

FUNCTIONAL FORMS OF COMPOUND INTEREST FACTORS*

Ref. No.	Name of Factor	Algebraic Formulation	Functional Format
Group A. All cash flows discrete: end-of-period compounding			
1.	Compound Amount (single payment)	$(1+i)^N$	$(F/P, i, N)$
2.	Present Worth (single payment)	$(1+i)^{-N}$	$(P/F, i, N)$
3.	Sinking Fund	$\dfrac{i}{(1+i)^N - 1}$	$(A/F, i, N)$
4.	Capital Recovery	$\dfrac{i(1+i)^N}{(1+i)^N - 1}$	$(A/P, i, N)$
5.	Compound Amount (uniform series)	$\dfrac{(1+i)^N - 1}{i}$	$(F/A, i, N)$
6.	Present Worth (uniform series)	$\dfrac{(1+i)^N - 1}{i(1+i)^N}$	$(P/A, i, N)$
7.	Arithmetic Gradient to Uniform Series	$\dfrac{(1+i)^N - iN - 1}{i(1+i)^N - i}$	$(A/G, i, N)$

Engineering Economy   7-3

Exhibit II (continued)

FUNCTIONAL FORMS OF COMPOUND INTEREST FACTORS*

Ref. No.	Name of Factor	Algebraic Formulation	Functional Format
8.	Arithmetic Gradient to Present Worth	$\dfrac{(1+i)^N - iN - 1}{i^2(1+i)^N}$	(P/G, i, N)
9.	Geometric Gradient to Present Worth (for i≠g)	$\dfrac{1 - (1+g)^N(1+i)^{-N}}{i - g}$	(P/A, g, i, N)

Group B. All cash flows discrete: continuous compounding at nominal rate r per period.

Ref. No.	Name of Factor	Algebraic Formulation	Functional Format
10.	Continuous Compounding Compound Amount (single payment)	e^{rN}	(F/P, r, N)
11.	Continuous Compounding Present Worth (single payment)	$e^{-rN} = \exp(-rN)$	(P/F, r, N)
12.	Continuous Compounding Present Worth (uniform series)	$\dfrac{e^{rN} - 1}{e^{rN}(e^r - 1)}$	(P/A, r, N)
13.	Continuous Compounding Sinking Fund	$\dfrac{e^r - 1}{e^{rN} - 1}$	(A/F, r, N)
14.	Continuous Compounding Capital Recovery	$\dfrac{e^{rN}(e^r - 1)}{e^{rN} - 1}$	(A/P, r, N)
15.	Continuous Compounding Compound Amount (uniform series)	$\dfrac{e^{rN} - 1}{e^r - 1}$	(F/A, r, N)

Exhibit II (continued)

FUNCTIONAL FORMS OF COMPOUND INTEREST FACTORS*

Ref.No.	Algebraic Name of Factor	Formulation	Functional Format

Group C. Continuous, uniform cash flows: continuous compounding (payments during one period only)

16.	Continuous Compounding Present Worth (single, continuous payment)	$\dfrac{i(1+i)^{-N}}{\ln(1+i)}$	$(F/\overline{P}, i, N)$
17.	Continuous Compounding Compound Amount (single, continuous payment)	$\dfrac{i(1+i)^{N-1}}{\ln(1+i)}$	$(F/\overline{P}, i, N)$

Group D. Continuous, uniform cash flows: continuous compounding (payments during a continuous series of periods)

18.	Continuous Compounding Sinking Fund (continuous, uniform payments)	$\dfrac{\ln(1+i)}{(1+i)^N - 1}$	$(\overline{A}/F, i, N)$
19.	Continuous Compounding Capital Recovery (continuous, uniform payments)	$\dfrac{(1+i)^N \ln(1+i)}{(1+i)^N - 1}$	$(\overline{A}/P, i, N)$
20.	Continuous Compounding Compound Amount (continuous, uniform payments)	$\dfrac{(1+i)^N - 1}{\ln(1+i)}$	$(F/\overline{A}, i, N)$
21.	Continuous Compounding Present Worth (continuous, uniform payments)	$\dfrac{(1+i)^N - 1}{(1+i)^N \ln(1+i)}$	$(P/\overline{A}, i, N)$

*See Exhibit I for definitions of symbols used in this exhibit.

Accounting Life. The period of time over which the amount of asset cost to be depreciated, or recovered, will be allocated to expenses by accountants.

Actual dollars. Cash flow at the time of the transaction.

Alternative, Contingent. An alternative which is feasible only if some other alternative is accepted. The opposite of a mutually exclusive alternative.

Alternative, Economic. A plan, project, or course of action intended to accomplish some objective that has or will be valued in monetary terms.

Alternative, Independent. An alternative such that its acceptance has no influence on the acceptance of other alternatives under consideration.

Alternative, Mutually Exclusive. An alternative such that its selection rules out the selection of any other alternatives under consideration.

Amortization. (1) a) As applied to a capitalized asset, the distribution of the initial cost by periodic charges to expenses as in depreciation. Most amortizable assets have no fixed life; b) The reduction of a debt by either periodic or irregular payments. (2) A plan to pay off a financial obligation according to some prearranged program.

Annual Cost. The negative of annual worth. (EQUIVALENT UNIFORM ANNUAL COST.)

Annual Equivalent. In time value of money, one of a sequence of equal end-of-year payments which would have the same financial effect when interest is considered as another payment or sequence of payments which are not necessarily equal in amount or equally spaced in time.

Annual Worth. A uniform amount of money at the end of each and every period over the planning horizon, equivalent to all cash flows occurring over the planning horizon when interest is considered.

Annuity. (1) An amount of money payable to a beneficiary at regular intervals for a prescribed period of time out of a fund reserved for that purpose. (2) A series of equal payments occurring at equally spaced periods of time.

Annuity Factor. The function of interest rate and time that determines the amount of periodic annuity that may be paid out of a given fund. (See CAPITAL RECOVERY FACTOR.)

Annuity Fund. A fund that is reserved for payment of annuities. The present worth of funds required to support future annuity payments.

Annuity Fund Factor. The function of interest rate and time that determines the present worth of funds required to support a specified schedule of annuity payments. (See PRESENT WORTH FACTOR, UNIFORM GRADIENT SERIES.)

Apportion. In accounting or budgeting, the process by which a cash receipt or disbursement is divided among and assigned to specific time periods, individuals, organization units, products, projects, services, or orders.

Bayesian Statistics. (1) classical - The use of probabilistic prior information and evidence about a process to predict probabilities of future events. (2) subjective - The use of subjective forecasts to predict probabilities of future events.

Benefit-Cost (Cost-Benefit) Analysis. An analysis technique in which the consequences on an investment evaluated in monetary terms are divided into separate categories of costs and benefits. Each category is then converted into an annual equivalent or present worth for analysis purposes.

Benefit-Cost Ratio. A measure of project worth in which the equivalent benefits are divided by the equivalent costs.

Benefit-Cost Ratio Method. (See BENEFIT-COST ANALYSIS.)

Book Value. The original cost of an asset or group of assets less the accumulated book depreciation.

Break-Even Chart. (1) A graphic representation of the relation between total income and total costs (sum of fixed and variable costs) for various levels of production and sales indicating areas of profit and loss. (2) Graphic representation of a figure of merit as a function of a specified relevant parameter.

Break-Even Point. (1) The rates of operations, output, or sales at which income will just cover costs. Discounting may or may not be used in making these calculations. (2) The value of a parameter such that two courses of action result in an equal value for the figure of merit.

Capacity Factor. (1) The ratio of current output to maximum capacity of the production unit. (2) In electric utility operations, it is the ratio of the average load carried during a period of time divided by the installed rating of the equipment carrying the load. (See DEMAND FACTOR, LOAD FACTOR.)

CAPITAL. (1) The financial resources involved in establishing and sustaining an enterprise or project. (2) A term describing wealth which may be utilized to economic advantage. The form that this wealth takes may be as cash, land, equipment, patents, raw materials, finished products, etc. (See INVESTMENT, WORKING CAPITAL.)

CAPITAL BUDGETING. The process by which organizations periodically allocate investment funds to proposed plans, programs, or projects.

CAPITALIZED ASSET. Any asset capitalized on the books of account of an enterprise.

CAPITALIZED COST. (1) The present worth of a uniform series of periodic costs that continue indefinitely (hypothetically infinite). Not to be confused with capitalized expenditure. (2) The present sum of capital which, if invested in a fund earning a stipulated interest rate, will be sufficient to provide for all payments required to replace and/or maintain an asset in perpetual service.

CAPITAL RECOVERY. (1) Charging periodically to operations amounts that will ultimately equal the amount of capital expended. (2) The replacement of the original cost of asset plus interest. (3) The process of regaining the new investment in a project by means of setting revenues in excess of the economic investment costs. (See AMORTIZATION, DEPLETION, AND DEPRECIATION.)

CAPITAL RECOVERY FACTOR. A number, which is a function of time and interest rate, used to convert a present sum to an equivalent uniform annual series of end-of-period cash flows. (See ANNUITY FACTOR.)

CAPITAL RECOVERY WITH RETURN. The recovery of an original investment with interest. In the public utility industry frequently this is referred to as the revenue requirements approach.

CASH FLOW. The actual monetary units (*e.g.*, dollars) passing into and out of a financial venture or project being analyzed.

CASH FLOW DIAGRAM. The illustration of cash flows (usually vertical arrows) on a horizontal line where the scale along the line is divided into time period units.

CASH FLOW TABLE. A listing of cash flows, positive and negative, in a table in order of the time period in which the cash flow occurs.

CHALLENGER. In replacement analysis, a proposed property or equipment which is being considered as a replacement for the presently owned property or equipment (the defender). In the analysis of multiple alternatives, an alternative under consideration which is to be compared with the last acceptable alternative (the defender). (See MAPI METHOD.)

COMMON COSTS. In accounting, costs which cannot be identified with a given output of products, operations, or services. Expenditures which are common to all alternatives.

COMPOUND AMOUNT. (1) The equivalent value, including interest, at some stipulated time in the future of a series of cash flows occurring prior to that time. (2) The monetary sum which is equivalent to a single (or a series of) prior sum(s) when interest is compounded at a given rate.

COMPOUND AMOUNT FACTOR(S). Functions of interest and time which, when multiplied by a single cash flow (single payment compound amount factor) or a uniform series of cash flows (uniform series compound amount factor) will give the future worth at compound interest of such single cash flow or series.

COMPOUNDING, CONTINUOUS. A compound interest assumption in which the compounding period is of infinitesimal length and the number of periods is infinitely great. A mathematical concept that is conceptually attractive and mathematically convenient for dealing with frequent (*e.g.*, daily) compounding periods within a year.

COMPOUNDING, DISCRETE. A compound interest assumption in which the compounding period is of specified length such as a day, week, month, quarter year, half year, or year.

COMPOUNDING PERIOD. The time interval between dates (or discrete times) at which interest is paid and added to the amount of an investment or loan. Usually designates the frequency of compounding during a year.

COMPOUND INTEREST. (1) The type of interest that is periodically added to the amount of investment (or loan) so that subsequent interest is based on the cumulative amount. (2) The interest charges under the condition that interest is charged on any previous interest earned in any time period, as well as on the principal.

CONSTANT DOLLARS. An amount of money at some point in time, usually the beginning of the planning horizon, equivalent in purchasing power to the actual dollars necessary to buy the good or service. Actual dollars adjusted for relative price change.

COST-BENEFIT ANALYSIS. (See BENEFIT-COST ANALYSIS.)

COST EFFECTIVENESS ANALYSIS. An analysis in which the major benefits may not be expressed in monetary terms. One or more effectiveness measures are substituted for monetary values resulting in a trade off between marginal increases in effectiveness versus marginal increases in costs.

COST OF CAPITAL. A term, usually used in capital budgeting, to express as an interest rate percentage the overall estimated cost of investment capital at a given point in time, including both equity and borrowed funds.

CUTOFF RATE OF RETURN (HURDLE RATE). The rate of return after taxes that will be used as a criterion for approving projects or investments. It is determined by management based on the supply and demand for funds. It may or may not be equal to the minimum attractive rate of return (MARR) but is at least equal to the estimated cost of capital.

CURRENT DOLLARS. (See ACTUAL DOLLARS.)

DECISION THEORY. With reference to engineering economy, it is a branch of economic analysis devoted to the study of decision processes involving multiple possible outcomes, defined either discretely or on a continuum, and deriving from the theory of games and economic behavior and probabilistic modeling.

DECISION TREE. In decision analysis, a graphical representation of the anatomy of decisions showing the interplay between a present decision, probability of chance events, possible outcomes and future decisions, and their results or payoffs.

DECISIONS UNDER CERTAINTY. From the literature of decision theory, that class of problems wherein single estimates with respect to cash flows and economic life (complete information) are used in arriving at a decision among alternatives.

DECISIONS UNDER RISK. From the literature of decision theory, that class of problems in which multiple outcomes are considered explicitly for each alternative and the probabilities of the outcomes are assumed to be known.

DECISIONS UNDER UNCERTAINTY. From the literature of decision theory, that class of problems in which multiple outcomes are considered explicitly for each alternative but the probabilities of the outcomes are assumed to be unknown.

DEFENDER. In replacement analysis, the presently owned property or equipment being considered for replacement by the most economical challenger. In the analysis of multiple alternatives, the previously judged acceptable alternative against which the next alternative to be evaluated (the challenger) is to be compared.

DEFLATING (BY A PRICE INDEX). Adjusting some nominal magnitude, *e.g.*, an actual dollar estimate, by a price index. It may be required in order to express that magnitude in units of constant purchasing power. (See ACTUAL DOLLARS, INFLATING, CONSTANT DOLLARS.) (CURRENT DOLLARS.)

DEFLATION. A decrease in the relative price level of a factor of production, an output, or the general price level of all goods and services. A deflationary period is one in which there is (or is expected to be) a sustained decrease in price levels.

DEMAND FACTOR. (1) The ratio of the current production rate of the system divided by the maximum instantaneous production rate. (2) The ratio of the average production rate, as determined over a specified period of time, divided by the maximum production rate. (3) In electric utility operations, it is the ratio of the maximum kilowatt load demanded during a given period divided by the connected load. (See CAPACITY FACTOR, LOAD FACTOR.)

DEPLETION. (1) An estimate of the lessening of the value of an asset due to a decrease in the quantity available for exploitation. It is similar to depreciation except that it refers to such natural resources such as coal, oil, and timber. (2) A form of capital recovery applicable to properties such as listed above. Its determination for income tax purposes may be on a unit of production basis, related to original cost or appraised value of the resource (known as cost depletion), or based on a percentage of the income received from extracting or harvesting (known as percentage depletion).

DEPLETION ALLOWANCE. An annual tax deduction based upon resource extraction. (See DEPLETION.)

DEPRECIATION. (1) a) Decline in value of a capitalized asset; b) A form of capital recovery, usually without interest, applicable to property with two or more years' life span in which an appropriate portion of the asset's value periodically is charged to current operations. (2) A loss of value due to physical or economic reasons. (3) In accounting, depreciation is the allocation of the book value of this loss to current operations according to some systematic plan. Depending on then existing income tax laws, the amount and timing of the charge to current operations for tax purposes may differ from that used to report annual profit and loss.

DEPRECIATION, ACCELERATED. Depreciation methods accepted by the taxing authority which write off the value (cost) of an asset usually over a shorter period of time (*i.e.*, at a faster rate) than the expected economic life of the asset. For example, the Accelerated Cost Recovery System (ACRS) introduced in the U.S. in 1981 and modified in later years.

DEPRECIATION ALLOWANCE. An annual income tax deduction, and/or charge to current operations, of the original cost of a fixed asset. The income tax deduction may not equal the charge to current operations. (See DEPRECIATION.)

DEPRECIATION BASIS. In tax accounting, the cost or otherwise determined value of a group of fixed assets, including installation costs and certain other expenditures, and excluding certain allowances. The depreciation basis is the amount which by law, and/or acceptance by the taxing authority, may be written off for tax purposes over a period of years.

DEPRECIATION, DECLINING BALANCE. A method of computing depreciation in which the annual charge is a fixed percentage of the depreciated book value at the beginning of the year to which the depreciation charge applies.

DEPRECIATION, MULTIPLE STRAIGHT-LINE. A method of depreciation accounting in which two or more straight-line rates are used. This method permits a predetermined portion of the asset to be written off in a fixed number of years. One common practice is to employ a straight-line rate which will write off 3/4 of the cost in the first half of the anticipated service life with a second straight-line rate used to write off the remaining 1/4 in the remaining half life. (See DEPRECIATION, STRAIGHT LINE.)

DEPRECIATION, SINKING FUND. (1) A method of computing depreciation in which the periodic charge is assumed to be deposited in a sinking fund that earns interest at a specified rate. The sinking fund may be real but usually is hypothetical. (2) A method of depreciation where a fixed sum of money regularly is deposited at compound interest in a real or hypothetical fund in order to accumulate an amount equal to the total depreciation of an asset at the end of the asset's estimated life. The depreciation charge to operations for each period equals the sinking fund deposit amount plus interest on the beginning of period sinking fund balance.

DEPRECIATION, STRAIGHT-LINE. A method of computing depreciation wherein the amount charged to current operations is spread uniformly over the estimated life of an asset. The allocation may be performed on a unit of time basis or a unit of production basis or some combination of the two.

DEPRECIATION, SUM-OF-YEARS-DIGITS. A method of computing depreciation wherein the amount charged to current operations for any year is based on the ratio: (years of remaining life)/$(1 + 2 + 3 + ... + n)$, n being the estimated life.

DETERIORATION. A reduction in value of a fixed asset due to wear and tear and action of the elements. It is a term used frequently in replacement analysis.

DEVELOPMENT COST. (1) The sum of all the costs incurred by an inventor or sponsor of a project up to the time that the project is accepted by those who will promote it. (2) In international literature, activities in developing nations intended to improve their infrastructure.

DIRECT COST. A traceable cost that can be segregated and charged against specific products, operations, or services.

DISCOUNTED CASH FLOW. (1) Any method of handling cash flows over time, either receipts or disbursements, in which compound interest and compound interest formulae are employed in their analytical treatment. (2) An investment analysis which compares the present worth of projected receipts and disbursements occurring at designated times in order to estimate the rate of return from the investment or project. (See RATE OF RETURN, PROFITABILITY INDEX.)

DISCOUNT RATE. (See INTEREST RATE, DISCOUNTED CASH FLOW.)

DOLLARS, CONSTANT (REAL DOLLARS). Dollars (or some other monetary unit) of constant purchasing power independent of the passage of time. In situations where inflationary or deflationary effects have been assumed when cash flows were estimated, those estimates are converted to constant dollars (base year dollars) by adjustment by some readily accepted general inflation/deflation index. Sometimes termed *actual* dollars, although this term also is used to describe current dollar values. (See ACTUAL DOLLARS, INFLATING, DEFLATING.) (CURRENT DOLLARS.)

DOLLARS, CURRENT (THEN CURRENT DOLLARS). Estimates of future cash flows which include any anticipated changes in amount due to inflationary or deflationary effects. Usually these amounts are determined by applying an index to base year dollar (or other monetary unit) estimates. Sometimes termed *actual dollars*, although this term also is used to describe constant dollar values. (See CONSTANT DOLLARS, INFLATING, DEFLATING.)

EARNINGS VALUE (EARNING POWER OF MONEY). The present worth of an income producer's estimated future net earnings as predicted on the basis of recent and present expenses and earnings and the business outlook.

ECONOMIC LIFE. The period of time, extending from the date of installation to the date of retirement from the intended service, over which a prudent owner expects to retain an equipment or property so as to minimize cost or maximize net return. (See LIFE.)

ECONOMY. (1) The cost or net return situation regarding a practical enterprise or project, as in economy study, engineering economy, or project economy. (2) A system for the management of resources. (3) The avoidance of (or freedom from) waste in the management of resources.

EFFECTIVE INTEREST. (See INTEREST RATE, EFFECTIVE.)

EFFECTIVENESS. In engineering economy, the measurable consequences of an investment not reduced to monetary terms; *e.g.*, reliability, maintainability, safety.

ENDOWMENT. A fund established for the support of some project or succession of donations or financial obligations.

ENDOWMENT METHOD. As applied to an economy study, a comparison of alternatives based on the present worth or capitalized cost of the anticipated financial events.

ENGINEERING ECONOMY. (1) The application of economic or mathematical analysis and synthesis to engineering decisions. (2) A body of knowledge and techniques concerned with the evaluation of the worth of commodities and services relative to their costs and with methods of estimating inputs.

EQUIVALENT UNIFORM ANNUAL COST (EUAC). (See ANNUAL COST.)

ESTIMATE. A magnitude determined as closely as it can be by the use of past history and the exercise of sound judgment based upon approximate computations, not to be confused with offhand approximations that are little better than outright guesses.

EXCHANGE RATE. The rate at a given point in time at which the currency of one nation exchanges for that of another.

EXPECTED YIELD. In finance, the ratio of the expected return from an investment, usually on an after-tax basis, divided by the investment.

EXTERNAL RATE OF RETURN. A rate of return calculation which takes into account the cash receipts and disbursements of a project and assumes that all net receipts (cash throwoffs) are reinvested elsewhere in the enterprise at some stipulated interest rate. (See RATE OF RETURN, INTERNAL RATE OF RETURN.)

FAIR RATE OF RETURN. The maximum rate of return which an investor owned public utility is entitled to earn on its rate base in order to pay interest and dividends and attract new capital. The rate, or percentage, usually is determined by state or federal regulatory bodies.

FIRST COST. The initial investment in a project or the initial cost of capitalized property including transportation, installation, preparation for service, and other related initial expenditures.

FIXED ASSETS. That tangible portion of an investment in an enterprise or project comprising land, buildings, furniture, fixtures, and equipment with an expected life greater than one year.

FIXED COST. Those costs which tend to be unaffected by changes in the number of units produced or the volume of service given.

FUTURE WORTH. (1) The equivalent value at a designated future date based on the time value of money. (2) The monetary sum, at a given future time, which is equivalent to one or more sums at given earlier times when interest is compounded at a given rate.

GOING-CONCERN VALUE. The difference between the value of a property as it stands possessed of its going elements and the value of the property alone as it would stand at completion of construction as a bare or inert assembly of physical parts.

GOOD-WILL VALUE. That element of value which inheres in the fixed and favorable consideration of customers arising from an established well-known and well-conducted business. This is determined as the difference between what a prudent business person is willing to pay for the property and its going-concern value.

Gradient Factors. A group of compound interest factors used for equivalence conversions of arithmetic or geometric gradients in cash flow. In general use are the arithmetic gradient to uniform series (gradient conversion) factor, the arithmetic gradient to present worth (gradient present worth) factor, and the geometric gradient to present worth factor.

Increment Cost (Incremental Cost). The additional (or direct) cost which will be incurred as the result of increasing output by one unit more. Conversely, it may be defined as the cost which will not be incurred if the output is reduced by one unit. (2) The variation in output resulting from a unit change in input. (3) The difference in costs between a pair of mutually exclusive alternatives.

Indirect Cost. Nontraceable or common costs which are not charged against specific products, operations, or services but rather are allocated against all (or some group of) products, operations, and/or services by a predetermined formula.

Inflating (By A Price Index). The adjustment of a present or base year price by a price index in order to obtain an estimate of the current (or then current) price at future points in time. (See DEFLATING, CONSTANT DOLLARS, ACTUAL DOLLARS) (CURRENT DOLLARS.)

Inflation. A persistent rise in price levels, generally not justified by increased productivity, and usually resulting in a decline in purchasing power. Sometimes the term is used interchangeably with escalation. However, this latter term more often is restricted to the differential increase in a price relative to some specific index of general changes in price levels. (See DEFLATION.)

Intangibles. (1) In economy studies, those elements, conditions or economic factors which cannot be evaluated readily or accurately in monetary terms. (2) In accounting, the assets of an enterprise which cannot reliably be values in monetary terms (*e.g.*, goodwill). (See IRREDUCIBLES.)

Interest. (1) The monetary return or other expectation which is necessary to divert money away from consumption and into long term investment. (2) The cost of the use of capital. It is synonymous with the term time value of money. (3) In accounting and finance, a) a financial share in a project or enterprise; b) periodic compensation for the lending of money.

Interest Rate. The ratio of the interest accrued in a given period of time to the amount owed or invested at the start of that period.

Interest Rate, Effective. The actual interest rate for one specified period of time. Frequently the term is used to differentiate between nominal annual interest rates and actual annual interest rates when there is more than one compounding period in a year.

Interest Rate, Market. The rate of interest quoted in the market place which includes the combined effects of the earning value of capital, the availability of funds, and anticipated inflation or deflation.

Interest Rate, Nominal. (1) The interest rate for some period of time which ignores the compounding effect of interest calculations during subperiods within that period. (2) The annual interest rate, or Annual Percentage Rate (APR), frequently quoted in the media.

Interest Rate, Real. An estimate of the true earning rate of money when other factors, especially inflation, affecting the market rate have been removed.

Internal Rate of Return. A rate of return calculation which takes into account only the cash receipts and disbursements generated by an investment, their timing, and the time value of money. (See RATE OF RETURN, DISCOUNTED CASH FLOW, EXTERNAL RATE OF RETURN.)

Investment. (1) As applied to an enterprise as a whole, the cost (or present value) of all the properties and funds necessary to establish and maintain the enterprise as a going concern. The capital tied up in an enterprise or project. (2) Any expenditure which has substantial and enduring value (generally more than one year) and which is therefore capitalized. (See FIRST COST.)

Investor's Method. A term most often used in the valuation of bonds. (See RATE OF RETURN, INTERNAL RATE OF RETURN, DISCOUNTED CASH FLOW.)

Irreducibles. Those intangible conditions or economic factors which cannot readily be reduced to monetary terms; *e.g.*, ethical considerations, esthetic values, or nonquantifiable potential environment concerns.

Leaseback. A business arrangement wherein the owner of land, buildings, and/or equipment sells such assets and simultaneously leases them back under a long term lease.

Life. (1) ECONOMIC: That period of time after which a machine or facility should be retired from primary service and/or replaced as determined by an engineering economy study. The economic impairment may be absolute or relative. (2) PHYSICAL: that period of time after which a machine or facility can no longer be repaired or refurbished to a level such that it can perform a useful function. (3) SERVICE: that period of time after which a machine or facility cannot perform satisfactorily its intended function without major overhaul.

LIFE CYCLE COST. The present worth or equivalent uniform annual cost of an equipment or project which takes into account all associated cash flows throughout its life including the cost of removal and disposal.

LOAD FACTOR. (1) Applied to physical plant or equipment, it is the ratio of average production rate for some period of time to the maximum rate. Frequently, it is expressed as a percentage. (2) In electric utility operations, it is the average electric usage for some period of time divided by the maximum possible usage. (See CAPACITY FACTOR, DEMAND FACTOR.)

MAPI METHOD. A procedure for equipment replacement analysis developed by George Terborgh for the Machinery and Allied Products Institute. It uses a fixed format and provides charts and graphs to facilitate calculations. A prominent feature of this method is that it includes explicitly an allowance for obsolescence.

MARGINAL COST. (1) The rate of change of cost as a function of production or output. (2) The cost of one additional unit of production, activity, or service. (See INCREMENTAL COST, DIRECT COST.)

MATHESON FORMULA. A title for the formula used for declining balance depreciation. (See DEPRECIATION, DECLINING BALANCE.)

MAXIMAX CRITERION. In decision theory, probabilities unknown, a rule that says choose the alternative with the maximum of the maximum returns identified for each alternative.

MAXIMIN CRITERION. In decision theory, probabilities unknown, a rule that says choose the alternative with the maximum of the minimum returns identified for each alternative. Also called a maximum security level strategy or Wald's strategy.

MINIMAX CRITERION. In decision theory, probabilities unknown, a rule that says choose the alternative with the minimum of the maximum costs identified for each alternative. Also called a maximum security level strategy.

MINIMAX REGRET CRITERION. In decision making under uncertainty, a rule that says choose the alternative with the least potential net return or cost regret.

MINIMIN CRITERION. In decision theory, probabilities unknown, a rule that says choose the alternative with the minimum of the minimum costs identified for each alternative.

MINIMUM ATTRACTIVE RATE OF RETURN. The effective annual rate of return on investment, either before or after taxes, which just meets the investor's threshold of acceptability. It takes into account the availability and demand for funds as well as the cost of capital. Sometimes termed the minimum acceptable return. (See COST OF CAPITAL, CUTOFF RATE OF RETURN.)

MINIMUM COST LIFE. (See ECONOMIC LIFE.)

MULTIPLE RATES OF RETURN (MULTIPLE ROOTS). A situation in which the structure of a cash flow time series is such that it contains more than one solving internal rate of return.

NOMINAL INTEREST. (See INTEREST RATE-NOMINAL.)

OBSOLESCENCE. (1) The condition of being out-of-date. A loss of value occasioned by new developments which place the older property at a competitive disadvantage. A factor in depreciation. (2) A decrease in the value of an asset brought about by the development of new and more economical methods, processes, and/or machinery. (3) The loss of usefulness or worth of a product or facility as the result of the appearance of better and/or more economical products, methods, or facilities.

OPPORTUNITY COST. The cost of not being able to use monetary funds otherwise due to that limited resource being applied to an "approved" investment alternative, and thus not being available for investment in other income-producing alternatives. Sometimes expressed as a rate.

PAYBACK PERIOD. (1) Regarding an investment, the number of years (or months) required for the related profit or savings in operating cost to equal the amount of said investment. (2) The period of time at which a machine, facility, or other investment has produced sufficient net revenue to recover its investment costs.

PAYBACK PERIOD, DISCOUNTED. Same as payback period except the period includes a return on investment at the interest rate used in the discounting.

PAYOFF PERIOD. (See PAYBACK PERIOD.)

PAYOFF TABLE. A tabular presentation of the payoff results of complex decision questions involving many alternatives, events, and possible future states.

Payout Period. (See PAYBACK PERIOD.)

Perpetual Endowment. An endowment with hypothetically infinite life. (See CAPITALIZED COST, ENDOWMENT.)

Planning Horizon. (1) A stipulated period of time over which proposed projects are to be evaluated. (2) That point of time in the future at which subsequent courses of action are independent of decisions made prior to that time. (3) In utility theory, the largest single dollar amount that a decision maker would recommend be spent. (See UTILITY.)

Present Worth (Present Value). (1) The monetary sum which is equivalent to a future sum or sums when interest is compounded at a given rate. (2) The discounted value of future sums.

Present Worth Factor(s). (1) Mathematical formulae involving compound interest used to calculate present worths of various cash flow streams. In table form, these formulae may include factors to calculate the present worth of a single payment, of a uniform annual series, of an arithmetic gradient, and of a geometric gradient. (2) A mathematical expression also known as the present value of an annuity of one. (The present worth factor, uniform series, also is known as the annuity fund factor.)

Principal. Property or capital, as opposed to interest or income.

Profitability Index. An economic measure of project performance. There are a number of such indexes described in the literature. One of the most widely quoted is one originally developed and so named (the PI) by Ray I. Reul, which essentially is based upon the internal rate of return. (See DISCOUNTED CASH FLOW, INVESTOR'S METHOD, RATE OF RETURN.)

Promotion Cost. The sum of all expenses found to be necessary to arrange for the financing and organizing of the business unit which will build and operate a project.

Rate of Return (Internal Rate of Return). (1) The interest rate earned by an investment. (2) The interest rate at which the present worth equation (or the equivalent annual worth or future worth equations) for the cash flows of a project or project increment equals zero. (3) As used in accounting, often it is the ratio of annual profit, or average annual profit, to the initial investment or the average book value.

Rate of Return, External. A rate of return calculation which employs one or more supplemental interest rates to produce equivalence transformations on a portion or all of the cash flows and then solves for rate of return on that equivalent cash flow series.

Real Dollars. (See CONSTANT DOLLARS.)

Replacement Policy. A set of decision rules for the replacement of facilities that wear out, deteriorate, become obsolete, or fail over a period of time. Replacement models generally are concerned with comparing the increasing operating costs (and possibly decreasing revenues) associated with aging equipment against the net proceeds from alternative equipment.

Replacement Study. An economic analysis involving the comparison of an existing facility and one or more facilities with equal or improved characteristics proposed to supplant or displace the existing facility.

Required Return. The minimum return or profit necessary to justify an investment. Often it is termed interest, expected return or profit, or charge for the use of capital. It is the minimum acceptable percentage, no more and no less. (See COST OF CAPITAL, CUTOFF RATE OF RETURN, MINIMUM ATTRACTIVE RATE OF RETURN.)

Retirement of Debt. The termination of a debt obligation by appropriate settlement with the lender. The repayment is understood to be in full amount unless partial settlement is specified.

Risk. (1) Exposure to a chance of loss or injury. (2) Exposure to undesired economic consequences.

Risk Analysis. Any analysis performed to assess economic risk. Often this term is associated with the use of decision trees. (See DECISIONS UNDER RISK, DECISION TREE.)

Salvage Value. (1) The cost recovered or which could be recovered from a used property when removed from service, sold, or scrapped. A factor in appraisal of property value and in computing depreciation. (2) Normally, an estimate of an asset's net market value at the end of its estimated life. In some cases, the cost of removal may exceed any sale or scrap value; thus net salvage value is negative. (3) The market value of a machine or facility at any point in time.

Sensitivity. The relative magnitude of decision criterion change with changes in one or more elements of an economy study. If the relative magnitude of the criterion exhibits large change, the criterion is said to be sensitive; otherwise it is insensitive.

Sensitivity Analysis. A study in which the elements of an engineering economy study are changed in order to test for sensitivity of the decision criterion. Typically it is used to assess needed measurement or estimation precision and often it is used as a substitute for more formal or sophisticated methods such as risk analysis.

Service Life. (See LIFE.)

Simple Interest. (1) Interest that is not compounded, *i.e.*, is not added to the income-producing investment or loan. (2) Interest charges under the condition that interest in any time period is only charged on the principal. Frequently interest is charged on the original principal amount disregarding the fact that the principal still owing may be declining through time. (See INTEREST RATE-NOMINAL.)

Sinking Fund. (1) A fund accumulated by periodic deposits and reserved exclusively for a specific purpose, such as retirement of a debt or replacement of a property. (2) A fund created by making periodic deposits (usually equal) at compound interest in order to accumulate a given sum at a given future time usually for some specific purpose.

Sinking Fund Deposit Factor. (See SINKING FUND FACTOR.)

Sinking Fund Factor. The function of interest rate and time that determines the periodic deposit required to accumulate a specified future amount.

Study Period. The length of time that is presumed to be covered in the schedule of events and appraisal of results. Often it is the anticipated life of the project under consideration, but may be either longer or (more likely) shorter. (See LIFE, PLANNING HORIZON.)

Sunk Cost. A cost which, since it occurred in the past, has no relevance with respect to estimates of future receipts or disbursements. This concept implies that, since a past outlay is the same regardless of the alternative selected, it should not influence a new choice among alternatives.

Time Value of Money. (1) The cumulative effect of elapsed time and the money value of an event, based on the earning power of equivalent invested funds and on changes in purchasing power. (2) The expected compound interest rate that capital should or will earn. (See INTEREST.)

Traceable Costs. Cost elements which can be identified with a given product, operation, or service. (See DIRECT COST, MARGINAL COST.)

Uncertainty. (1) That which is indeterminate, indefinite, or problematical. (2) An attribute of the precision of an individual's or group's precision of knowledge about some fact, event, consequence, or measurement.

Uniform Gradient Series. A uniform or arithmetic pattern of receipts or disbursements increasing or decreasing by a constant amount in each time period. (See GRADIENT FACTORS.)

Utility. (1) In economics, a process of evaluating factor inputs and outputs in quantitative units (*i.e.*, Utiles) in order to arrive at a single measure of performance to assist in decision making. (2) In economic analysis, a measured preference among various choices available in risk situations based on the decision making environment, the alternatives being considered, and the decision maker's personal attitudes.

Utility Function. A mathematically derived relationship between utility, measured in utiles, and quantities of money and/or commodities or attributes based on a decision maker's attitudes and preferences.

Valuation or Appraisal. The art and science of estimating the fair exchange monetary value of specific properties.

Variable Cost. A cost which tends to fluctuate according to changes in the number of units produced. (See MARGINAL COST.)

Working Capital. (1) That portion of investment represented by current assets (assets that are not capitalized) less the current liabilities. The capital necessary to sustain operations as opposed to that invested in fixed assets. (2) Those funds, other than investments in fixed assets, required to make the enterprise or project a going concern.

Yield. In evaluating investments, especially those offered by lending institutions, the true annual rate of return to the investor. In bond valuation, the annual dividend of a bond divided by the current market price and usually expressed as a percent. (See EXPECTED YIELD.)

BIBLIOGRAPHY

Au, T. and Au, T.P. *Engineering Economics for Capital Investment Analysis*. Newton, Mass: Allyn and Bacon, 1983.

Bierman, H., Jr. and Smidt, S. *The Capital Budgeting Decision*. (6th ed.) New York: MacMillian Publishing Company, 1984.

Blanchard, B.S. *Design and Manage to Life Cycle Cost*. Portland, OR: M/A Press, 1978.

Blank, L. and Tarquin, A. *Engineering Economy*. (3rd ed.) New York: McGraw-Hill, Inc., 1989.

Bussey, L.E. *The Economic Analysis of Industrial Projects*. Englewood Cliffs, NJ: Prentice-Hall, 1978.

Canada, J.R. and White, J.A. *Capital Investment Decision Analysis for Management and Engineering*. (2nd ed.) Englewood Cliffs, NJ: Prentice-Hall, 1980.

Commonwealth Edison Company. *Engineering Economics*. Commonwealth Edison Company, 1975.

DeGarmo, P.E., Sullivan, W.G. and Bontadelli, J.A. *Engineering Economy*. (8th ed.) New York: MacMillian Publishing Company, 1984.

Emerson, R.C. and Taylor, W.R. *An Introduction to Engineering Economy*. (2nd ed.) Cardinal Publishers, 1979.

Fabrycky, W.J. and Thuesen, G.J. *Economic Decision Analysis*. (3rd ed.) Englewood Cliffs, NJ: Prentice-Hall, Inc., 1984.

Fleischer, G.A. *Engineering Economy: Capitol Allocation Theory*. PWS Kent, 1984.

Frost, M.J. *How to Use Cost Benefit Analysis in Project Appraisal*. (2nd ed.) New York: Halsted Press, 1975.

Grant, E.L., Ireson, G.W. and Leavenworth, R.S. *Principles of Engineering Economy*. (8th ed.) New York: John Wiley & Sons, 1990.

Greynolds, E.B., Aronofsky, J.S. and Frame. R.J. *Financial Analysis Using Calculators: Time Value of Money*. New York: McGraw-Hill Book Company, 1980.

Holloway, C.A. *Decision Making Under Uncertainty, Models and Choices*. Englewood Cliffs, NJ: Prentice-Hall, Inc., 1979.

Humphreys, K.K. and Katell, S. *Basic Cost Engineering*. New York: Marcel Dekker, Inc., 1981.

Jelen, F.C. and Black, J.H. *Cost and Optimization Engineering*. (2nd ed.) New York: McGraw-Hill, 1983.

Jones, B.W. *Inflation in Engineering Economic Analysis*. New York: John Wiley & Sons, 1982.

Kurtz, M. *Handbook of Engineering Economics*. New York: McGraw-Hill Book Company, 1984.

Mallik, A.K. *Engineering Economy with Computer Applications*. Mahomet, IL: Engineering Technology, Inc., 1979.

Morris, W.T. *Engineering Economic Analysis*. Reston, VA: Reston Publishing Company, 1976.

Newnan, D.G. *Engineering Economic Analysis*. (3rd ed.) Engineering Press, Inc., 1988.

Peters, M.S. and Timerhaus, K.D. *Plant Design and Economics for Chemical Engineers*. (3rd ed.) McGraw-Hill Book Company, 1980.

Riggs, J.L. and West T.M. *Engineering Economics*. (3rd ed.) New York: McGraw-Hill Book Company, 1986.

Rose, I.M. *Engineering Investment Decision*. New York: Elsevier Scientific Publishing Company, 1976.

Sassone, P.G. and Schaffer, W.A. *Cost-Benefit Analysis: A Handbook*. Orlando, FL: Academic Press, 1978.

Sepulveda, J.A., Souder, W. and Gottfried, B.S. *Schaum's Outline: Engineering Economics*. New York: McGraw-Hill Book Company, 1984.

Steiner, H.M. *Public and Private Investments: Socioeconomic Analysis*. New York: John Wiley & Sons, Inc., 1980.

Stevens, G.T. Jr. *The Economic Analysis of Capital Investments for Managers and Engineers*. Reston, VA: Reston Publishing Company, 1983.

Swalm, R.O. and Lopez-Leautaud, J.L. *Engineering Economic Analysis: A Future Wealth Approach*. New York: John Wiley & Sons, 1984.

Taylor, G.A. *Managerial and Engineering Economics*. (3rd ed.) D. Van Nostrand Company, 1980.

Thuesen, G.J. and Fabrycky, W.J. *Engineering Economy*. (7th ed.) Englewood Cliffs, NJ: Prentice-Hall, Inc., 1989.

Z94.8 Facility Planning & Design

The facility planning & design section contains terms in the areas of facility layout, material handling and storage equipment, bar coding, and warehousing. This revision of facility planning terminology is base upon the 1989 edition. A minor number of terms were deleted, a significant number of term definitions were changed, a significant number of terms were added, and a bibliography was added. Past subcommittee members, whose names are listed in the previous editions, are thanked for their development of a solid base of definitions. Similarly, the efforts of the present subcommittee members are appreciated for their work in contributing to the expanded set of facility planning definitions.

Chairman

Thomas A. Lacksonen, Ph.D., P.E.
Assistant Professor
Department of Industrial and Manufacturing -
Systems Engineering
Ohio University

Subcommittee

Mark H. Baldwin
Director, Warehouse & Distribution Systems Div.
Remstar International Inc.

Sunderesh S. Heragu, Ph.D.
Assistant Professor
Decision Sciences and Engineering Systems
Rensselear Polytechnical Institute

Russell D. Meller, Ph.D.
Assistant Professor
Department of Industrial and Systems Engineering
Auburn University

John Raab
Vice President, Marketing
Rapistan-Demag Corporation

Acceleration Resistance. Inertial reaction opposing increase in speed of a vehicle.

Accessory. Any supplemental device for an industrial truck that cannot be classified as an attachment.

Accident. Any unplanned occurrence which may result in damage to property or equipment, injury or illness to persons, and/or adversely affects an activity or function.

Accordion Roller Conveyor. A roller conveyor with a flexible latticed frame which permits variation in length. (See ROLLER CONVEYOR.)

Accumulator Conveyor. Any conveyor designed to permit accumulation of packages, objects, or carriers. May be roller, live roller, roller slat, belt, power-and-free, or tow conveyors. (See MINIMUM PRESSURE ACCUMULATING CONVEYOR.) Syn. Accumulation Conveyor

Address. (1) A character or group of characters that identifies a register, a particular part of storage, or some other data source or destination. (2) To refer to a device or an item of data by its address.

AGV Routing. The act of determining the specific paths over which an automated guided vehicle must travel in a given time period and programming the controller to execute the travel path.

AIM. Automatic Identification Manufacturers, Inc.

Airbill. A nonnegotiable shipping document used by U.S. airlines for domestic air freight. It contains shipping instructions, a description of the commodity and applicable transportation charges. May be used as through documents for coordinating other transportation.

Air Cargo. In the United States, goods transported by air mail, air freight or air express. In other countries, air freight only.

Aircraft Accident. Any accident that occurs between the time the aircraft engine(s) is started with the intent of flight until the aircraft comes to rest with engine(s) stopped upon completion of flight.

Air-float Chain Conveyor. A chain and roller conveyor in which the chain is supported by an inflatable air hose to control and minimize line pressure.

Air Freight. Air cargo service for the transport of goods not classified as air mail or air express.

Air Freight Forwarder. An agent who collects, ships and distributes cargo under his or her own tariff usually consolidating the cargo into a larger unit that is tendered to the airline. Clearing of cargo through Federal Customs may be handled by the agent.

Air Mail. Mail service having priority over all other air cargo.

Air Parcel Post. Air mail weighing more than eight ounces and not exceeding 70 pounds and 100 inches in combined length and girth. In common practice, Air parcel post is most widely used for packages of five pounds or less.

Air Waybill. A nonnegotiable shipping document used for air freight.

Aisle. That space used to accommodate the movement of people, material, firefighting apparatus and/or equipment.

Aisle Captive AS/RS. A person-on-board or automated storage/retrieval device that is designed or restricted to travel within a specific aisle.

Aisle to Aisle S/R Device. A person-on-board or automated storage/retrieval device that is able and allowed to travel from aisle to aisle.

Aisle Transfer Car. A machine or vehicle for transferring an S/R machine from aisle to aisle, which normally runs on a rail or rails.

Alternate Capacity. A capacity, other than basic capacity, that adjusts for variations such as load center, fork height, and application of an attachment.

Approved Industrial Truck. A truck that is listed or approved for fire safety purposes for the intended use by a nationally recognized testing laboratory using nationally recognized testing standards.

Apron. A series of pans or plates pivotally attached to one another to form the conveying medium of a conveyor.

Apron Conveyor. A conveyor in which a series of pans or plates, attached to a chain or pivotally attached to one another, form the moving bed.

Area. Length X width space required for a department or facility.

Areaway. An open subsurface space, exterior and adjacent to the building, used to admit light or air, or as a means of access.

ARM CONVEYOR. A conveyor consisting of an endless belt, one or more chains, to which are attached projecting arms, or shelves, for handling packages or objects in vertical or inclined path.

ARM TO COLUMN CONNECTOR. A bracket affixed to the arm, which is used to attach a complete arm to a column.

ASME TYPE LP GAS CYLINDER. A fuel container for liquefied petroleum gas made and inspected under the ASME Boiler and Pressure Vessel Code, Section VIII, for Unfired Pressure Vessels. Where used for storage of an industrial truck's fuel supply, it may be permanently affixed to the vehicle or removable.

ASPECT RATIO. (1) The ratio of height-to-width of a bar code symbol. A code twice as high as wide would have an aspect ratio of 2; a code twice as wide as high would have an aspect ratio of 1/2, or 0.5. (2) Maximum of L/W ratio and W/L ratio.

AS/RS RACK. A rack with bar codes and other special attachments and surfaces that can be used by a storage /retrieval device.

ASSEMBLY SERVICE (AIR TRANSPORT). A service under which an airline consolidates shipments from many shippers and transports them as one shipment to one receiver.

ATTIC. The space between the ceiling framing of the top story and the roof framing.

AUTOMATED ELECTRIFIED MONORAIL (AEM). A conveyor system on which each individual vehicle is powered by its own on-board electric drive unit.

AUTOMATED STORAGE AND RETRIEVAL SYSTEM (AS/RS). A combination of equipment and controls which handles, stores, and retrieves materials with precision, accuracy and speed under a defined degree of automation. Systems vary from relatively simple, manually controlled order-picking machines operating in small storage structures to giant, computer-controlled storage/retrieval systems totally integrated into the manufacturing and distribution process. Vertical heights of these latter systems can exceed 100 feet.

AUTOMATIC COUPLER. A trailer connection that may be connected without manual actuation.

AUTOMATIC DISPENSER. A sortation system in which small items are manually stacked up on the exterior of an A frame in their respective columns and dispensed by a computer-controlled mechanism.

AUTOMATIC GUIDED VEHICLE SYSTEM (AGVS). A vehicle equipped with automatic guidance equipment, either electromagnetic or optical. Such a vehicle is capable of following prescribed guide paths and may be equipped for vehicle programming and stop selection, blocking, and any other special functions required by the system.

BACK CROSS BRACE. Diagonal members forming a rigid cross between the back post ties on the uprights at the end of the drive-in rack bay.

BACK DIAGONAL BRACE. A diagonal member connecting back post ties at the end of the drive-in rack bay.

BACK POST TIES. Horizontal ties across the last posts at the end of the drive-in rack bay.

BAG CONVEYOR, DOUBLE HELICAL. Closely spaced parallel tubes with right and left hand helical threads rotating in opposite directions on which bags or other objects are carried while being conveyed.

BALCONY. An elevated platform projecting from an exterior or interior wall, supported by columns or the wall, and enclosed by a railing.

BAR. The dark element of a printed symbol.

BAR CODE. An array of rectangular bars and spaces that are arranged in a predetermined pattern following unambiguous rules in a specific way to represent elements of data, which are referred to as characters.

BAR CODE DENSITY. The number of characters that can be represented in a lineal inch.

BAR CODE LABEL. A label that carries a bar code and is suitable to be affixed to an article.

BAR CODE READER. A device used to identify and read a bar code symbol.

BAR LENGTH. The bar dimension perpendicular to the bar width.

BAR WIDTH. The thickness of a bar measured from the edge closest to the symbol start character to the trailing edge of the same bar.

BASE (CANTILEVER RACK). The bottom member of a rack in contact with the floor and permanently affixed to the column. May also be affixed to the floor and used as a load bearing arm. In free standing racks, the base usually extends beyond the length of the arms. Purpose of the base is to prevent the rack from tipping over.

BASE (PORTABLE METAL STACKING RACK). Consists of structural cold-formed or tubed members joined together into a square or rectangular load carrying perimeter to which posts or sockets are attached, when required. The four basic styles are as follows:

Base No. 1—base is constructed with two center structural members or intermediate framing inside the basic rack for allowing 4-way entry of light truck forks.

Base No. 2—base is constructed with six structural members or intermediate framing for allowing 4-way entry for racks that are wider than the length of the lift truck forks.

Base No. 3—base is constructed with X-type structural members or intermediate frame for 4-way entry of lift truck forks.

Base No. 4—base is constructed *without* corner sockets, but with open style intermediate bracing for load only. This style for transportation by crane with special tongs or slings.

BASE GUIDE RAIL. Member attached to the outside edge of the base and used to align a row or pair of complete frames. Used also as a truck guide.

BASEMENT. Any story having at least 1/2 of its volume below the adjacent finished grade.

BATCH PICKING. An order picking method in which an operator is responsible for picking multiple orders to completion in the same work cycle.

BATCH PROCESS. An industrial manufacturing method in which one or several units are produced at a time, in contrast to continuous process (q.v.).

BATTERY (BAT.)—AIEE (60.06.010). A battery consists of two or more cells electrically connected for producing electric energy. (Common usage permits this designation to be applied also to a single cell used independently. In the following definitions, unless otherwise specified, the term battery will be used in this dual sense.)

BATTERY-ELECTRIC TRUCK. An electric truck in which the power source is a storage battery.

BAY. (1) The space defined by a pair of upright frames, the overall depth, and the clear height between shelves. (2) A cubic space with limits normally defined by functional or physical constraints; as example:

Structural bay. The space defined by four columns.

Storage bay. The space defined by the size of a block of material stored within it.

Truck bay. The space defined by the area required to park a truck.

Service bay. The space defined by the area required to park and service a vehicle.

BAY MARKER. A marker either on the floor, overhead or on posts, numbered, and/or lettered to designate a bay in a storage area.

BEAM CLAMP. A type of suspension fitting used to support tracks from overhead structure.

BEAMS. A horizontal load-carrying member on each end of which is a connector used in tying together two upright frames.

BEARING BRACE. The member of an upright frame with full length floor contact.

BEARING PLATE. A metal plate affixed to the bottom of a post for load distribution, permitting lagging or bolting to the floor.

BELT CONVEYOR. An endless fabric, rubber, plastic, leather, or metal belt operating over suitable drive, tail end and bend terminals and over belt idlers or slider bed for handling bulk materials, packages, or objects placed directly upon the belt.

BELT CONVEYOR, CLOSED. Moving, endless, flexible belt, or belts, which may be formed into a tubular shape by joining of edges, and which are opened while in motion to receive load, closed to convey or elevate, and opened to discharge.

BELT CONVEYOR, FLAT. A belt conveyor in which the carrying run of the conveyor belt is supported by flat belt idlers or by a flat surface.

BELT CONVEYOR, MAGNETIC. An inclined belt conveyor operating over a slider bed containing permanent magnets for transporting ferrous metal parts against gravitational pull.

BELT CONVEYOR, MULTIPLE CORD. A belt conveyor composed of two or more spaced strands of Vee, double Vee, or round belts.

BELT CONVEYOR, MULTIPLE RIBBON. A belt conveyor having a conveying surface of two or more spaced strands of narrow conveyor belts.

Belt Conveyor, Portable. A portable conveyor in which a belt is used as the conveying medium.

Belt Conveyor, Troughed. A belt conveyor with the conveyor belt edges elevated on the carrying run to form a trough by conforming to the shape of the troughed carrying idlers or other supporting surface.

Belt Feeder. A belt conveyor adapted for feeder service. (See FEEDER; CONVEYOR TYPE FEEDER.)

Bidirectional Path. A material flow path over which material flow takes place in two opposite directions.

Bidirectional Read. The ability to read data successfully whether the scanning motion is left to right or right to left.

Bidirectional Symbol. A bar code symbol that permits reading in complementary (2) directions.

Bill of Lading. A receipt given by the carrier or agent for goods received for shipment. (See STRAIGHT BILL OF LADING, ORDER BILL OF LADING.)

Binder. A strip of burlap, heavy paper board, thin lumber or similar material placed between layers of containers to keep the stack together.

Bin Storage. An area of a plant or facility devoted entirely to loose item or open stock placement.

Blending Conveyor. (See MIXING CONVEYOR, PADDLE TYPE.)

Block. A storage term denoting a rectangular area within a facility or building in which items-objects are densely located, and in which only the exterior faces of the area may be conveniently reached.

Blocked Space (air transport). A program of reduced rates and reserved space granted a shipper who assures a minimum of freight (15,000 lbs. a week, for example) between the same two cities for a given period (90 days, for example).

Block, Leg or Column. Deck spacers used to make full four-way or all-way entry pallets.

Block Stacking. The action of putting objects into a block pattern in a storage area up to five tiers in height.

Boom. A cantilever member, with a load-carrying hook at the free end.

Boom Conveyor. Any type of conveyor mounted on a boom.

Booster Conveyor. Any power conveyor used to regain elevation lost in gravity roller or wheel conveyor lines.

Bottom Discharge Bucket Conveyor. A conveyor for carrying bulk materials in a horizontal path consisting of an endless chain to which roller-supported, cam-operated conveyor buckets are attached continuously.

Brace. Stiffening member(s) connecting two posts in an upright frame, i.e., diagonal, cross, or horizontal.

Bridge Crane. (See OVERHEAD CRANE.)

Bridge Plate. A plate, usually of metal, used to span the space between a freight car or truck and the loading dock.

Bridge Rail. The rail supported by the bridge girders on which the trolley travels.

Bucket Conveyor. Any conveyor in which the material is carried in a series of buckets. (See BUCKET ELEVATOR; GRAVITY DISCHARGE BUCKET CONVEYOR; PIVOTED BUCKET CONVEYOR.)

Bucket Elevator. A conveyor for carrying bulk materials in a vertical or inclined path, consisting of endless belt, chain or chains, to which elevator buckets are attached, the necessary head and boot terminal machinery and supporting frame or casing. (See INTERNAL DISCHARGE BUCKET ELEVATOR; POSITIVE DISCHARGE BUCKET ELEVATOR.)

Bucket Elevator, Gravity Discharge. (See GRAVITY DISCHARGE BUCKET CONVEYOR.)

Bucket Elevator, Pivoted. (See PIVOTED BUCKET CONVEYOR.)

Bucket Loader. A form of portable, self-feeding, inclined bucket elevator for loading bulk materials into cars, trucks, or other conveyors. (See BUCKET ELEVATOR; PORTABLE CONVEYOR.)

Building. A roofed structure for the shelter, support, or enclosure of persons or property.

Bulk Stock. Full and unbroken containers of supplies. Bulk storage can involve piles of raw material (stone), tanks (oil), silos (sand), and similar on-site methods of containing homogeneous commodities.

Bulk Storage. The area within a facility or warehouse devoted to the placement of large, greater than loose-issue quantities of items, or in which each single item is too large to be placed in a bin storage location.

Bumper Pad. Molded or laminated cushions fastened to a dock to prevent damage to it from a truck bumping against the dock while parking.

Cab. An equipment-mounted enclosure for operators from which motion of equipment may be controlled.

Cab Controlled. Equipment controlled from an operator's cab.

Cable-screw Conveyor. A one-way or closed-circuit conveyor of which the propelling medium is a flexible torque-transmitting cable of which helical (screw) threads are an integral part. Loads or load carriers engage the thread and advance a distance equal to one pitch for each revolution of the cable screw.

Cableway. A cable, or rope-supported system in which the material handling carriers are not detached from the operating mechanism and the travel is wholly within the span.

Cableway, Slack Line. A cable or rope-supported system in which the supporting cable is adjusted in length to provide the lifting function of the unit.

Caged Storage. The area within a facility or warehouse which has been set aside for pilferable items and is afforded a measure of security by means of a partition or other enclosure, usually wire screening.

Camber (girder). The slight vertical curve given to girders to compensate partially for deflection due to load.

Canopy. A roofed projection from the wall of a building, either exterior or interior, without enclosures below its edges, and providing protection from above.

Cantilever Rack. A rack consisting of load arms cantilevered from column(s); typically used for storing long products such as tubing, bars, pipes, etc.

Cantilever Truck. A self-loading counterbalanced or noncounterbalanced truck, equipped with cantilever load engaging means, such as forks.

Capacity. The maximum pounds, or the maximum load in pounds at a given load center, that a truck can safely transport and/or stack to a specified height.

Captive Container. A container whose use cycle remains within an enterprise (private, corporate, or government).

Cargo. The product shipped in an aircraft, railroad car, ship or truck.

Cargo Container Capacity. The inside usable cubic volume of the container.

Cargo Container Displacement. The outside cubic volume of the container. (Cargo Container Length x Cargo Container Width x Cargo Container Height.)

Cargo Container Height. The vertical overall outside dimension of the container.

Cargo Container Length. The horizontal overall outside dimension usually parallel to the longitudinal axis of carrier equipment.

Cargo Container Width. The horizontal overall outside dimension at right angles to the container length.

Carriage. (1) A support structure for forks or attachments, generally roller mounted, traveling vertically within the mast of cantilever truck or stacker crane. (2) That part of an S/R machine by which a load is moved in the vertical direction.

Carrier. (1) A person or firm engaged in the business of transporting goods. (2) An assembly usually with four or more wheels which will run on monorails or underhung crane girders, and which will support a load or a hoist and load. (3) The product carrying levels in horizontal and vertical carousels. (See COMMON CARRIER).

Carrier Head. A two-wheeled assembly used with load bars to form a carrier or an end truck.

Carousel Conveyor. A continuous platform or series of spaced platforms which move in a circular or elliptical horizontal path. Note: The term "carousel" has been applied to other forms of conveyors such as car type and pallet type.

Carousel, Horizontal. A modular storage and retrieval device providing carriers and shelves which rotates around a vertical axis, presenting items to be picked or stored at a workstation.

Carousel, Vertical. A modular storage and retrieval device providing carriers and shelves which rotates around a horizontal axis, presenting items to be picked or stored at a workstation opening.

Car Type Conveyor. A series of cars attached to and propelled by an endless chain or other linkage running on the horizontal or a slight incline.

Catwalk. An elevated service platform or walkway constructed to permit access to equipment, controls or other devices not frequently used. The catwalk generally is designed to support only itself and the weight of people and repair parts required for access to the aforementioned items.

CEILING TIE. A member tying the rack to the overhead building structure.

CELLULAR LAYOUT. Layout of machines, workstations and support services in separate groups such that each group or cell is dedicated to the manufacture of a specific family of parts. Machines, workstations and support services in a cell are typically dissimilar in their processing capabilities and parts belonging to a part family are almost entirely processd in their corresponding cell.

CENTER CONTROL TRUCK. A powered industrial truck in which the driver's position is located near the longitudinal center of the truck.

CHAIN CONVEYOR. Any conveyor in which one or more chains act as the conveying element. A British term for trolley conveyor. (See DRAG CHAIN CONVEYOR, ROLLING CHAIN CONVEYOR, SLIDING CHAIN CONVEYOR, TOW CONVEYOR.)

CHAIN ELEVATOR. (See BUCKET ELEVATOR.)

CHAIN FEEDER. (See CONVEYOR TYPE FEEDER.)

CHARACTER. (1) A single group of bars and spaces that represents an individual number, letter, punctuation mark or other graphic fonts. (2) A graphic shape representing a letter, numeral or symbol. (3) A letter, digit, or other symbol that is used as part of the organization, control, or representation of data. A character is often in the form of a spatial arrangement of adjacent or connected strokes.

CHARACTER DENSITY. The dimension, in linear inches, required to encode one character.

CHARACTER READING. Reading of alpha or numeric characters, and/or symbols, by optical means.

CHARACTER SET. Those characters available for encoding purposes.

CHARGED AND DRY. Charged and dry is the condition of a storage battery when it is assembled with charged plates, dry separators, and with no electrolyte. Charged and dry batteries can be used as soon as they are filled but often are given an initial charge after filling and before use.

CHARGED AND WET. Charged and wet is the condition of a storage battery when it is filled with electrolyte and is fully charged.

CHECK DIGIT OR CHECK CHARACTER. A digit or character included within a symbol whose value is based mathematically on other characters included in the symbol. It is used for the purpose of performing a mathematical check to ensure the accuracy of the read.

CHUTE CONVEYOR. A conveyor typically used to transport material from a higher floor to a lower floor via enclosed or open slides.

CICMHE. College-Industry Council on Material Handling Education of the Material Handling Institute.

CITY TERMINAL SERVICE (AIR TRANSPORT). Shipments accepted or delivered by airlines at the terminals of cartage agents or other designated in-town terminals, at lower rates than those charged for the door-to-door airport pick-up and delivery service.

CLAMP. An attachment which uses a clamping action to engage the load from opposite sides. (See CLAMP ARMS.)

CLAMP ARMS. Auxiliary members that bolt or are welded to clamp arm supports. Various types are available for clamp(s) to suit them for specific loads, such as bag, bale, drum, carton, paper roll, tire, etc.

CLOSED CIRCUIT CONVEYOR. An arrangement of a conveyor or conveyors capable of moving material through all portions of a circuit, and returning the undistributed portion to the starting point.

CLASS-BASED STORAGE POLICY. A storage policy in which incoming items are classified depending upon the level of S/R activity they generate.

CLASSIFICATION AND CODING SYSTEM. A computer software system that allows part designers to classify a part based on its physical and operating characteristics (shape, material, tolerance requirement, etc.) and assign an alphanumeric code.

CLOSET. A small room designed to provide limited storage.

CODE. A set of unambiguous rules specifying the way in which data may be represented.

CODE READER OR SCANNER. A device that examines a spatial pattern, one part after another, and generates analog or digital signals corresponding to the pattern.

Collapsible Container. A container, the major components of which can be disassembled and later reassembled for use.

Collector, Shoe. The portion of a collector that makes contact by sliding on the conductor bar.

Collector, Wheel. The portion of a collector that makes contact by rolling on the conductor.

Collectors. Electrical contacting devices providing a path for current flow from stationary conductors to moving equipment.

Collision Avoidance Rules. Rules programmed into the controller of an AGV to ensure it will not collide with other vehicles or moving objects in the shop-floor.

Column. (1) A single vertical member of the cantilever rack that may have a means for adjusting arms on preset centers. (2) Steel beam or post used to support the roof in a facility.

Column Spacing. The distance between consecutive columns in a facility.

Commodity Code. A system for identifying a given commodity by a number.

Commodity Rate. A charge applicable on a specific commodity between certain specified points.

Common Carrier. A carrier engaged in the business of transporting goods and/or persons for compensation and without discrimination.

Compartment (battery). The chamber on a truck in which a storage battery is housed and mounted.

Conductors. Bar or wire used to transmit an electrical current.

Conductors, Enclosed. Conductors enclosed in a nonconducting material to minimize accidental contact with the conductor.

Connecting Carrier. A carrier maintaining a direct physical connection with one or more carriers.

Connector (battery). An item for quick connection composed either of male and female halves or two halves that differ only in method of mounting. Generally the two halves are known as charging plug and charging receptacle.

Connector Lock. A safety device used to lock the rail support connector to the post.

Consolidating Point or Area. A point at which less than carload, or less than truckload, shipments are brought together to be reforwarded as a carload or truckload.

Container. A receptacle such as a bag, barrel, drum, box, crate or package used in storage or shipment of a commodity to provide protection from physical damage or contamination of the commodity. (See DEMOUNTABLE CARGO CONTAINER)

Container Corner Fittings. The fittings located at the corners of the container, which provide means for handling, supporting and/or securing the container.

Container Load. A shipment sufficient in size to "fill" a container, either by cubic measurement or weight, depending upon governing tariff to meet the provided minimums.

Container Marking. Bar codes, numbers, nomenclature, or symbols stamped or painted on, or otherwise affixed to, items or containers for purposes of identification.

Continuous Code. A bar code or symbol where the space between characters ("inter-character gap") is part of the code.

Continuous Process. An operation wherein raw materials enter into the process in an unbroken stream and proceed to a final form, in contrast to batch processing (q.v.).

Contract Carrier. A person or firm, not a common carrier, that transports goods and/or persons for compensation under private or special contract agreements.

Control Bars. Conductor bars that carry the control current.

Controlled Velocity Roller Conveyor. A roller conveyor having means to control the velocity of objects being conveyed. (See ROLLER CONVEYOR.)

Controllers. Electronic, mechanical, or electromechanical devices by means of which an operator or computer controls the speed, direction or other action of motor driven equipment or systems.

Control, Single Speed. A drive control system providing one speed operation in either direction.

CONTROL, VARIABLE SPEED. A drive control system providing more than single-speed operation in either direction.

CONTROL, VOLTAGE. The voltage impressed on the control device(s).

CONVEYOR. A horizontal, inclined or vertical device for moving or transporting bulk materials, packages or objects in a path predetermined by the design of the device and having points of loading and discharge fixed, or selective; included are skip hoists and vertical reciprocating and inclined reciprocating conveyors; typical exceptions are those devices known as industrial trucks, tractors and trailers, tiering machines (truck type), cranes, hoists, monorail cranes, power and hand shovels, power scoops, bucket drag lines, platform elevators designed to carry passengers or the elevator operator, and highway or rail vehicles.

CONVEYOR-ELEVATOR. A conveyor that follows a path, part of which is substantially horizontal or on a slope less than the angle of slide of the material and part of which is substantially vertical or on a slope steeper than the angle of slide.

CONVEYOR TYPE FEEDER. Any conveyor such as apron, belt, chain, flight, pan, oscillating, screw, or vibrating, adapted for feeder service.

CORNER MARKER. A marker used at aisle intersections to prevent workers and equipment from bumping into and/or damaging the stacks.

CORROSION. (1) The oxidation of metals. (2) Chemical or electrochemical deterioration by reaction. (3) "Rust" in ferrous metals.

CORROSION PREVENTION. The protection given metallic items by covering surfaces subject to corrosion to prevent contact with water, moisture, vapor, acids, or other contaminating substances or storage in dehumidified air. It includes careful cleaning of all or part of the items, the application of a barrier of oil, grease, or moisture-vaporproof paper to exclude air and moisture, and the covering of the barrier to provide for its protection.

CORROSION PREVENTIVE. Any agent such as oil, plastic, paint, wrap or other surface treatment of metals whose primary function is to prevent corrosion. This may exclude atmosphere by means of a continuous film, or may direct corrosion to another element (cathodic protection).

COUNTERBALANCED FRONT SIDELOADER TRUCK. A self-loading high lift counterbalanced truck (equipped with a fixed or tiltable elevating mechanism) capable of transporting and tiering a load in both the counterbalanced forward position and any location up to and including 90° from the longitudinal centerline of the truck, which has the capability of traversing the load laterally.

COUNTERBALANCED TRUCK. A truck equipped with load engaging means wherein all the load during normal transporting is external to the polygon formed by the wheel contacts thus requiring a counterbalance to avoid truck tipping.

COUPLER HEIGHT. The vertical distance from floor to the effective horizontal center line of coupler.

CRANE. A machine for lifting, and/or moving a load with the hoisting mechanism an integral part of the machine. Cranes may be traveling, portable, or fixed type. (See BRIDGE, FIXED, JIB, PORTABLE, TOWER, OR TRAVELING CRANES.)

CRANE ATTACHMENT. A device that can be attached to a heavy-duty truck to make it a mobile crane.

CRANE BAR (PORTABLE METAL STACKING RACK). (See TIERING GUIDE.)

CRANE BRIDGE (TOP-RUNNING). The girders, trucks, end ties, footwalk (if any), and drive mechanism which carries the trolley.

CRANE DRIVE, INDIVIDUAL. Separate drives mounted on individual end trucks.

CRATE. A rigid shipping container of framed construction joined together with nails, bolts or any equivalent method of fastening. The framework may or may not be enclosed with sheeting.

CRAWLER CRANE. A portable crane mounted on a base equipped with crawler treads.

CROSS AISLE. (1) An aisle at right angles to main aisles, used for the movement of goods, equipment, and personnel. (2) The space used to accommodate secondary movement of people, material, and/or equipment, and intersecting another aisle.

CROSS BAR CONVEYOR. A conveyor having endless chains supporting spaced cross members from which materials are hung or festooned while being processed.

CROSS-DOCKING. Products are received, immediately broken into smaller shipments, then shipped without the products having gone into storage.

Cross Stacking. The placing of one layer of containers at right angles to those just below to increase the stability of the stack.

Cross Tie. Cross layers of supplies as in cross stacking, except that only an occasional layer is crossed, and not every other one.

Cube. Cube expresses volume and is the product of length, width and depth. A common measure of storage space utilization is termed "cube utilization."

Cube-per-Order Index (COI). The ratio of the item's storage space requirement (volume) to the number of S/R transactions for that item.

Cube per Order Index Storage Policy. A storage policy in which each item is allocated warehouse space based on its cube-per-order index. Items are located nearest the I/O or P/D point in a non-decreasing order of their COI.

Curved Belt Conveyor. A unit load belt conveyor usually operating horizontally through 90° or 180° turns.

Cushion Tire. A solid tire having a high rounded profile in cross section.

Dead Loads. The loads on a structure that remain in a fixed position relative to the structure.

Decentralized Storage. Storage that is distributed throughout a facility and is not centralized in one location.

Deck. The horizontal load-carrying or load-bearing surface of a pallet. Top deck—Load-carrying surface. Bottom deck—Load-bearing surface.

Deck (pallet storage rack). A solid top or platform supported by beams.

Deck (portable metal stacking rack). The platform attached to the base.

Deckmember. A surface element used in the construction of a pallet deck.

Deckmember Chamfer. A beveled edge on the side of the bottom deckmember for the purpose of easing the entry and exit of pallet truck load wheels.

Deck Opening. Any void in the deck caused by spacing of surface elements.

Deck Spacer. A structural member which supports the top deck and/or separates the top and bottom decks.

Declared Value (air transport). The value of the shipment declared by the shipper. In air freight, declared value on most shipments has historically been 50 cents a pound or 50 dollars, whichever is greater, assumed unless the shipper declares a higher value.

Decoder Logic. The electronic package which receives the signals from a scanner, performs the algorithm to interpret the signals into meaningful data and provides the interface to other devices.

Dedicated Storage Policy. A storage policy in which each item is assigned to a specific set of storage spaces.

Deep Lane Storage. Storage depth which is greater than two loads deep on one or both sides of an aisle.

Deferred Rate (air transport). Rate for deferred service in return for the greater flexibility given the airlines in determining when to move such freight. The rates are usually lower than general commodity rates.

Deferred Service (air transport). A service under which delivery is deferred for a specified number of days (to the third or fourth day, for example).

Dehumidify. To dry out or absorb moisture by means of machinery, baking, ventilating, or by the use of desiccant.

Delivery Table. (1) A conveyor which transports material from the discharge of a machine; (2) A table to which a chute discharges.

Demountable Box. A steel case with a cover in which a battery is assembled for use on a truck that does not have its own integral battery compartment.

Demountable Cargo Container. A reusable weatherproof cargo container for shipping or storing materials as a unit that can be secured to carrier equipment.

demountable cargo container—standard group i. A demountable cargo container of rigid construction, rectangular configuration, with end doors, in nominal lengths of 10, 20, 30 and 40 feet and width and height of 8 feet. Containers can be stacked and are normally positioned longitudinally on carrier equipment.

demountable cargo container—standard group ii. A demountable cargo container of rigid construction, rectangular configuration, with end doors, in nominal lengths of 6 2/3 feet and 5 feet, with a width of 8 feet and optional heights of 8 feet or 6 feet 10 inches, capable of being stacked, supported and mounted on 6-inch side runners for either platform or forklift handling. Containers are normally positioned transversely on carrier equipment.

DEMURRAGE. A penalty charge assessed by carriers for detention of cars, ships or trucks by shippers or receivers of goods beyond a specified free time.

DEPTH OF FIELD. The distance between the maximum and minimum plane in which a symbol can be read.

DERRICK CRANE. A fixed crane consisting of a mast or equivalent member held at the head by guys or braces, with or without a boom, for use with a hoisting mechanism.

DESICCANT. A substance which by virtue of its physical structure and/or chemical nature, absorbs water; it is typically used within a sealed package or enclosed space.

DESTINATION INSPECTION. The inspection performed at the receiving point of the consignee of material to ascertain whether the shipment is in conformance with purchase specifications; frequently called receiving or incoming inspection.

DETERIORATION. Any impairment of quality, value, or usefulness, including damage caused by erosion, corrosion, combustion, and contamination.

DIESEL-ELECTRIC TRUCK. An electric truck in which the power source is a diesel engine-driven generator.

DIMENSIONAL WEIGHT. The density, or weight per cubic foot of a shipment.

DIMENSIONAL WEIGHT RULE. A practice applicable to low density shipments under which the transportation charges are based on a dimensional weight rather than upon actual weight.

DISCHARGED AND WET. Is the condition of a storage battery when it is filled with electrolyte and is discharged. An initial charge is required before it is ready for use.

DISCONNECT DEVICE, MAIN LINE. A device used to open the electric circuit.

DISCRETE CODE. A bar code or symbol where the space between characters ("Inter-character gap") is not part of the code.

DISCRETE MANUFACTURING SYSTEM. A system in which the manufacturing process is not continuous

DISPATCH RULES. Rules programmed into the controller of a discrete material handling device or used manually to control when, where and how the device is to be released into the material flow network.

DISTRIBUTION POINTS. Depots, warehouses, or other warehousing activities having the mission for the receipt, storage, and issue of goods, normally for a given geographical area.

DISTRIBUTION SERVICE. A service under which a distributor accepts and transports one shipment from one shipper and separates the shipment into a number of parts at the destination and distributes them to many receivers.

DIVERTER. An automated fixture on a slat conveyor capable of moving in a direction along the length of the slat. It is used to direct items from one conveyor line to another.

DOCK. A horizontal platform designed to place the floor of a building level with the bed of a vehicle.

DOCKBOARD. A portable or fixed device for spanning the gap or compensating for difference in level between loading platforms and carriers.

DOCK LEVELER. An adjustable rectangular platform of approximately highway truck width built into the dock edge used to compensate for the difference between varying truck bed heights and delivery platforms of a facility.

DOCK SHELTER. A stationary or retractable, rigid or flexible shield that allows goods from a truck to be transported into the building with little or no exposure to outside elements.

DOLLY. A wheeled manual transporting device with an open horizontal platform for low volume transportation.

DOUBLE ARM. Horizontal load member affixed to the column for loading on either side of the column.

DOUBLE DEEP STORAGE. Loads that are stored two deep on each side of an aisle.

DOUBLE LEG BUCKET ELEVATOR. A bucket elevator having the carrying and return runs enclosed in separate casings between the head and boot. (See BUCKET ELEVATOR.)

Double Leg En Masse Conveyor. An en masse conveyor or elevator in which the carrying and return runs are operated in separated parallel and adjacent casings.

Drag Chain Conveyor. A conveyor having one or more endless chains which drag bulk materials through a trough. (See SLIDING CHAIN CONVEYOR.)

Drag Conveyor, Portable. (See PORTABLE DRAG CONVEYOR.)

Drawbar Pull, Maximum. The maximum towing force in pounds at a specified coupler height that a vehicle will develop on a level surface based upon a prescribed coefficient of friction existing between the driving wheel(s) and the supporting surface when the load is being moved at a uniform rate of not less than 1 percent of the top level travel speed.

Drawbar Pull, Rated Normal. The greatest sustained towing force in pounds at a specified coupler height that a vehicle will develop on a level surface and within a given duty cycle without exceeding the allowable continuous temperature rating for the components.

Drive Girder. The girder on which the bridge drive machinery is mounted.

Drive-in Rack. A rack consisting of upright frames, rails, and ties permitting a vehicle to enter the structure from one side only to pickup up or deposit pallets on continuous rails. The structure dictates first-in, last-out storage.

Drive-thru Rack. A rack consisting of upright frames, rails, and ties permitting a vehicle to enter the structure from opposite sides to pick up or deposit pallets on continuous rails. The structure permits first-in, first-out storage.

Driving Head. A motor-driven carrier head, the load-bearing wheels of which are driven.

Drop Section. A section of track that can be lowered out of alignment with a stationary track.

Drum Grooves. Helical grooves cut in the periphery of the drum to guide the hoisting rope.

Drum Handler. A clamp to engage a drum(s). Drum handlers are available which engage the drum(s) in various ways and positions. Certain models may also manipulate drum(s).

Drum Support. A unit composed of a pair of front to back members placed between a pair of beams used to support cylindrical objects.

Dual Command Cycle. The time between actuation of the S/R machine and completion of a cycle in which one load is stored and another load is retrieved and the S/R machine is ready to start a new cycle.

Dummy (battery). Generally of wood, is employed in lead batteries to fill the space left in a tray when the specified number of cells is less than the tray's capacity.

Dumper. An attachment for dumping containers. Various types are available to suit container and operational requirements.

Dunnage. Any material such as boards, planks, blocks, plastics, or metal bracing that is used in transportation and storage to support and secure supplies, to protect them from damage, or for convenience in handling.

Dynamic Lowering. A method of control by which the hoist motor is so connected in the lowering direction, that when it is overhauled by the load, it acts as a generator and forces current either through the resistors or back into the line, thus acting as a brake or speed limit.

Economy Service (air transport). (See DEFERRED SERVICE.)

Edgemember. A deckmember assembled at right angles to and at the extreme ends of stringers or stringer members.

El Conveyor. A trough type roller or wheel conveyor consisting of two parallel rows of rollers or wheels set at a 90° included angle, with one row providing a sloped carrying surface and the other acting as guard. (See ROLLER CONVEYOR; WHEEL CONVEYOR.)

Electrically Baffled. Wired in such a way that electric current is cut off from motor-driven equipment approaching units, which are not properly set for passage of the equipment.

Electric Tractor. An industrial tractor in which the principal energy is transmitted from power source to motor(s) in the form of electricity and can be either battery, gas-electric, diesel-electric or tethered-electric.

Electric Truck. A truck in which principal energy is transmitted from power source to motor(s) in the form of electricity.

Elevating Conveyor. Any conveyor used to discharge material at a point higher than that at which it was received. Term is specifically applied to certain underground mine conveyors.

Emergency Exit. An exit used only by people in case of an emergency. There are building codes that must be adhered to when locating.

Empty Pallet Stacker. A device for receiving empty pallets and automatically stacking them for re-use or shipment.

Enclosed Dock. A dock enclosed from the top, or top and two sides, to provide minimal protection to goods and handling personnel from exposure to outside elements.

Enclosed Track Trolley Conveyor. A trolley conveyor where the propelling medium such as chain, cable or other linkage and the trolleys or load-carrying wheels are supported and completely enclosed by a tubular type track. (See TROLLEY CONVEYOR.)

Encoded Area. The total lineal dimension consumed by all characters of a code pattern including start/stop and data.

End Control Truck. A powered industrial truck in which the driver's position is located at the end opposite the load end of the vehicle.

End Frames (PORTABLE METAL STACKING RACK). A completely removable unit making up one side of the rack. Consists of two posts with appropriate bracing separating the posts so as to nest with two corners of the base rack.

End Stop. A device located at the end of a track or crane bridge to prevent the carrier or crane from leaving the track or crane bridge.

End Tie. A structural member other than the end truck which connects the ends of the girders to maintain the squareness of the bridge.

End Trucks. The assemblies consisting of truck frame and wheels which support the crane girders.

En Masse Conveyor. A conveyor comprised of a series of skeleton or solid flights on an endless chain or other linkage which operates in horizontal, inclined, or vertical paths within a closely fitted casing for the carrying run. Bulk material is conveyed and elevated in a substantially continuous stream with a full cross section of the casing.

Escalator. A passenger conveyor in which the passenger-carrying surfaces usually form stairs where travel is on a slope.

Expendable Container. A container intended to be discarded after one cycle of use.

Extendable Conveyor. (1) For bulk materials; conveyor is usually of troughed design and may be lengthened or shortened while in operation. Commonly used in underground mine conveyor work. (2) For packaged materials, objects, or units; conveyor may be one of several types including roller, wheel, or belt conveyors. Construction is such that the conveyor may be lengthened or shortened within limits to suit operating needs.

Facility. A combination of all of the physical elements required to manufacture a product, or to provide a service.

Factory Mutual Listing (FM). A list of truck models approved as complying with the requirements of and used by the mutual group of insurance companies to determine the insurance premium rate for various areas of operation.

Fastener. Any device or medium used to hold together the components of an assembled pallet.

FCL. (See FULL CONTAINER LOAD.)

Feeder. A conveyor or other mechanism adapted to control the rate of delivery of bulk material, packages or objects or a device that controls, separates, or assembles objects.

Feeder and Catchers Table. A pair of reversible conveyors, entry and exit, which provide for repeat feeding of metal being processed through a rolling mill.

Field. Any group of characters defined as a unit of information. This differs from a line in that one line may contain several fields.

Field Separator. A printed mark or symbol that identifies fields to scanner.

Fire Accident. Any instance of unintentional ignition, uncontrolled burning, and/or explosion of combustible materials.

First Floor. The floor of a story which is closest to grade.

First Read Rate. The percentage representing the number of successful reads per 100 attempts to read a particular symbol.

FIXED CRANE. Cranes that are nonmobile. (See BRIDGE, JIB, PORTABLE, TOWER, OR TRAVELING CRANES.)

FIXED PLATFORM TRUCK. An industrial truck equipped with a load platform and not capable of self-loading.

FIXED POSITION LAYOUT. A type of layout in which the processing equipment move to the job which is at a stationary location, as in shipyards, construction sites, aircraft manufacturing and servicing, and so on.

FIXTURE CONVEYOR. Usually a slat or apron conveyor on which are mounted "pedestals" or fixtures for mounting loads such as engine blocks, etc.

FLAT TOP CONVEYOR. Special slat conveyors which form a continuous top and are supported by chain rollers.

FLAT WIRE CONVEYOR BELT. A belt composed of flat metal strips formed in a series of lateral, rectangular loops held together by lateral wires passing through holes in the flat strips to provide articulation similar to a conveyor chain and arranged to be driven by sprockets.

FLIGHT. (1) Plain or shaped plates suitably made for attachment to the propelling medium of a flight conveyor. (2) A term applied to any conveyor in a tandem series.

FLIGHT CONVEYOR. A conveyor comprised of one or more endless propelling media, such as a chain, to which flights are attached, and a trough through which material is pushed by the flights.

FLOOR. The walking level of a story.

FLOOR CONTROLLED UNITS. Motor propelled units which are controlled by an operator on the floor by means of a control suspended from the overhead equipment.

FLOOR CONVEYOR. Any of several types of conveyors using chain, cable, or other linkage mounted near or flush with the floor for the purpose of assembling, or finishing built-up products and subassemblies. (See CAR TYPE CONVEYOR; PALLET TYPE CONVEYOR; SLAT CONVEYOR.)

FLOOR LOAD RATING. The uniformly distributed live load that can safely be supported by a floor, expressed in pounds per square foot of floor space.

FLOW ANALYSIS. The detailed examination of the progressive travel, either of people or material, from place to place and/or from operation to operation. The examination consists of questioning the reason for the existence of every aspect of the operation, or the travel; and, also, the examination seeks reasons for determining how the progressive travel, or the operation, may be changed or modified to achieve the utmost economies in both time and material, all other things being equal.

FLOWCHART. A tabular arrangement of data depicting the number of material handling trips performed per time unit for a pair of planning departments.

FLOW DIAGRAM. A representation of the location of activities or operations and the flow of materials between activities on a pictorial layout of a process. Usually accompanied by a flow process chart.

FLOW LINE. (See LINE OF FLOW.)

FLOW PATH. The travel path of material as it flows through the manufacturing or service process.

FLOW PROCESS CHART. A graphic, symbolic representation of the work performed, or to be performed, on a product as it passes through some or all of the stages of a process.

Typically, the information included in this type of chart concerns the quantity, distance moved, type of work done (by symbol with explanation), and equipment used. A composite of the process charts of individual components, plan assembly and related operations. Work times may also be included. Process symbols generally used are:

○ *operation*: A subdivision of a process that changes or modifies a part, material, or product, and is done essentially at one workplace location.

➡ *move*: Change in location of a part, material, or product from one workplace to another.

☐ *inspection*: Comparison of observed quality or quantity of product with a quality or quantity standard.

▽ *storage*: Keeping a product, material, or part protected against unauthorized removal.

D *delay*: An event which occurs when conditions (except those which intentionally change the physical or chemical characteristics of the part, material, or product) do not permit or require immediate performance of the next planned step.

The above symbols are ASME Standard Symbols, 101.

FLOW SHOP. A manufacturing enterprise devoted to producing products in a system with processes that are dedicated by product family.

FLOW RACK. A rack in which items are replenished from one end of the rack and gradually move to the other end on sloping wheels (gravity flow rack) or powered rollers where a lift truck, person or other device picks up the items.

FLUID COUPLING. A hydrodynamic drive which transmits power without ability to change torque. (Torque ratio is unity for all speed ratios.)

FLUME CONVEYOR. A channel through which a liquid flows and conveys materials.

FLUSH DOCK. A dock that is located flat against the plane of the building wall.

FOLDING CONTAINER. A container having hinged components that can be folded with all parts and/or hardware attached.

FORK ADAPTER. Quickly detachable attachment that mounts on a fork to adapt the truck to handle specific loads.

FORK ADJUSTER. A manual or power actuated accessory to facilitate lateral positioning of forks.

FORK HEIGHT. The vertical distance from the floor to the load-carrying surface adjacent to the heel of the forks with mast vertical, and in the case of reach trucks, with the forks extended.

FORKLIFT OPENING. (See FORK POCKETS.)

FORKLIFT TRUCK. A high lift self-loading truck, equipped with load carriage and forks for transporting and tiering loads.

FORK OPENING. An opening in a pallet to allow insertion and withdrawal of a lifting device.

FORK POCKETS. Openings in container base for inserting the tines of a fork-lifting device.

FORKS. Horizontal tine-like projections, normally suspended from the carriage, for engaging and supporting loads

FORK, SAFETY. A mechanical device which prevents passage of carriers when the elements of interlocking equipment are not interlocked.

FORK SPACING.

Maximum inside fork spacing—the nominal distance between the inside edges of adjustable forks at their greatest separated position.

Maximum outside fork spacing—the nominal distance between the outside edges of adjustable forks at their greatest separated position.

Minimum inside fork spacing—the nominal distance between the inside edges of adjustable forks at their least separated position.

Minimum outside fork spacing—the nominal distance between the outside edges of adjustable forks at their least separated position.

FORWARDING AREA. An area of the warehouse where items are accumulated just prior to shipping.

FOUR-WAY CONTAINER. A container whose configuration permits retrieving or discharging from adjacent right angle directions in the same horizontal plane.

FRAGILE. Delicate, weak or easily damaged goods.

FRAME (CANTILEVER RACK). One upright complete with arms.

FREE LIFT. The attainable lift of the carriage from the extreme lowered position before the stated overall lowered height of the mast is exceeded by any standard part of the forks, mast or carriage assemblies.

FREE STANDING RACK. Is a structure composed of one or more racks without external bracing, but which may be lagged to the floor.

FREE STANDING RACK STRUCTURE. A rack structure installed inside a building of conventional construction, supported only by the floor and not supported or attached to any building structure.

FREIGHT. A method or service for transporting cargo by air, land, or water. Cargo carried by transportation lines. The cost for such transportation.

FREIGHT FORWARDER. A person or firm engaged in the business of collecting, consolidating, shipping, and distributing goods acting on behalf of the shipper and attending to the necessary details of shipping, insuring, and documenting of goods. Services may also include the preparation of consular invoices and the clearing of goods through customs.

Front to Back Members (safety spreaders). Formed structural shapes placed between two beams to form a single shelf unit. Among common types are welded, bolted, or drop-in-place.

Welded. Members welded at both ends to a pair of beams.

Bolted. Members bolted at both ends to a pair of beams.

Drop-in-place. Members placed between a pair of beams.

Full Container Load—FCL. Where the load carried in a container equals one of the two operating maxima—in weight or volume.

Gantry Crane. A traveling crane similar to an overhead crane, except that the bridge for carrying the hoisting mechanism is rigidly supported on two or more legs running on fixed rails or other runway.

Gas-electric Truck. An electric truck in which the power source is a gasoline or LP gas engine-driven generator.

Gateway City (air transport). That city at which the aircraft first lands upon arrival in the USA and/or the last city from which the aircraft takes off when leaving the USA.

Gauge of Double Girder Cranes. The center-to-center distance between two crane girders.

Gauger Table. The combination of a conveyor and any mechanism to stop and gauge the cutting length at a shear in a metal processing line.

General Commodity Rate (air transport). Published air freight rates for all commodities except those for which specific rates have been filed.

Girder Crane. Horizontal beam of the crane bridge which supports the trolley and is supported by the end trucks.

Grade Clearance. The maximum grade change that the vehicle will clear empty and loaded without abnormal mechanical contact between the floor and the vehicle.

Grade Resistance. The component of gravitational force affecting movement up an inclined plane and acting at center of gravity of a vehicle in a direction parallel to the surface on which the wheels are supported. It is numerically equal to the product of gross vehicle weight and the sine of the angle of incline.

Gravity Discharge Bucket Conveyor. A conveyor using gravity discharge buckets attached between two endless chains and which operate in suitable troughs and casings in horizontal, inclined and vertical paths over suitable drive, corner and take-up terminals.

Gravity Hopper. A surge bin attached to the tail end of a portable conveyor which utilizes the conveyor belt as a live bottom.

Gross Weight. The weight of a container plus its contents.

Guided Rail (drive-in, drive-thru racks). A combination of rail and rail guide.

Hand. (1) The right hand or left hand of a conveyor is determined by facing the direction in which the material is flowing. In the case of a reversible conveyor, the hand is determined when the material is flowing toward the drive end. (2) In a screw conveyor a right-hand screw when rotated clockwise will move material toward the observer; a left-hand screw when rotated clockwise will move material away from the observer.

Hand Truck. A general group of nonpowered industrial trucks for operation by a pedestrian.

Hanger Rods. Steel rods which, together with other suspension fittings, are used to suspend equipment from the supporting structure.

Hazardous Material or Operation. Any material or operation which has the potential for producing environmental hazards, personal injury and/or property damage.

Heel (cantilever rack). The back part of the base of a single sided upright.

Height Above Deck (portable metal stacking rack). The clear vertical opening from the deck surface to the underside of the target.

Height of Scan. The maximum vertical scanning dimension of a moving scanner at a specific distance from the face of the scanner.

Helium Neon Laser. The type of laser most commonly used in bar code scanners.

High Lift Platform Truck. A self-loading truck equipped with a load platform intended primarily for transporting and tiering loaded skid platforms.

High Lift Truck. A self-loading truck equipped with an elevating mechanism designed to permit tiering.

High Rise AS/RS. A warehouse with an AS/RS of height 50 to 120 feet.

Hinged Feeder. One which vertically reciprocates one end of an adjoining hinged horizontal conveyor to provide synchronization with the movement of a vertical conveyor during the period required to permit transfer of objects from the vertical to the horizontal conveyor; or from the horizontal to the vertical conveyor.

Hoist. Mechanism for lifting and lowering loads.

Honeycombing. The practice of removing merchandise in pallet-load quantities where the space is not exhausted in an orderly fashion. This results in inefficiencies due to the fact that the received merchandise may not be efficiently stored in the space which is created by the honeycombing.

Hook Approach. The minimum horizontal distance between the center of the runway rail and the hook.

Horizontal Bay Clearance. The internal dimension between upright posts parallel to the aisle.

Horizontal Rail Clearance. The dimension between the two nearest edges of the rails.

Hugger Belt Conveyor. Two belt conveyors whose conveying surfaces combine to convey loads up steep inclines or vertically.

Hydraulic Conveyor. A conveyor in which water jets form the conveying medium for bulk materials through pipes and troughs.

ICC Type LP Gas Cylinder. A fuel container for liquefied petroleum gas made and inspected under Interstate Commerce Commission regulations. When used for storage of an industrial truck's fuel supply, it may be permanently or removably affixed to the vehicle.

ICHCA. International Cargo Handling Coordination Association.

Idler Sheave. A sheave used to equalize tension in opposite parts of a rope.

Imbalanced Load, Carousel. The maximum load difference tolerated in a carousel, as measured between the two parallel sides, under which the carousel can continue to operate within its normal duty cycle.

Impact Factor. An allowance for additional load resulting from the dynamic effect of the live load.

In Bond Cargo. Cargo that moves under the carrier's bond from the gateway city until released by customs at an inland point.

Inclined Reciprocating Conveyor. A reciprocating power- or gravity-actuated unit that receives only inanimate objects on a track, roller conveyor or other form of carrying surface not designed to carry passengers or the operator. These units operate on inclines of less than 90° to the horizontal and usually at less than 45°. This type of conveyor is never designed to carry an operator or passengers and no person may be permitted to ride on it. No controls for movement of the carrier may be located on the carrier nor within reach of any person who might be standing on the carrier. For carriers operating on steeper than stairway slopes further safeguards are necessary and the construction should follow the provisions given for vertical reciprocating conveyors. (See VERTICAL RECIPROCATING CONVEYOR.)

Indexing. Controlled spacing or feeding.

Industrial Tow Tractor. A powered industrial truck designed primarily to draw one or more non-powered trucks, trailers or other mobile loads.

Industrial Truck. A wheeled vehicle, primarily for the movement of objects or materials, and usually associated with manufacturing, processing, or warehousing, but not including vehicles intended primarily for earth-moving or over-the-road hauling.

Input and Output (I/O) Point (See PICK-UP AND DELIVERY STATION).

Intercharacter Gap. The space between two adjacent bar code characters.

Interleaved Bar Code. A bar code in which characters are paired together using bars to represent the first character and spaces to represent the second.

Interline. The movement of a shipment via two or more carriers.

Interlocking Mechanism. A mechanical device to lock together the adjacent ends of two cranes or a crane and spur truck to permit the transfer of carriers from one crane or track to the other.

INTERLOCKS, ELECTRICAL. Electrical devices in the starter circuit that prevent a short circuit when opposite controls are operated at the same time.

INTERLOCKS, MECHANICAL. Mechanical devices that prevent operation of opposite controls at the same time.

INTERMEDIATE PACKAGE. An interior container that contains two or more unit packages of identical items.

INTERNAL COMBUSTION ENGINE TRUCK. A truck in which the power source is a gasoline, LP gas, or diesel engine.

INTERNAL COMBUSTION TRACTOR. An industrial tractor in which the power source is gasoline, LP gas, or diesel engine.

INTERNAL DISCHARGE BUCKET ELEVATOR. A bucket elevator having continuous buckets abutting hinged, or overlapped and designed for loading and discharging along the inner boundary of the closed path of the buckets. (See BUCKET ELEVATOR.)

INTERNAL RIBBON CONVEYOR. A trunnion-supported revolving cylinder the inner surface of which is fitted with continuous or interrupted ribbon fighting.

INVENTORY. (1) The material maintained in a facility for further processing and/or sale and shipment to a customer. (2) The physical count of items of stock, raw material, work-in-process, or finished goods located within a depot, warehouse, plant, or other storage or warehousing activity.

INVERTED POWER AND FREE CONVEYOR. A Power-and-Free Conveyor in which the two conveyor lines - powered and free - are at ground level and the loads are above the conveyor lines.

ITEM-TO-PERSON AS/RS. An AS/RS in which the items come from their storage locations to the end of an aisle where an operator or a transportation mechanism sends the items to their point of use.

JIB CRANE. A fixed crane with a cantilevered bridge supported from a stationary vertical support.

JOB SHOP. A manufacturing enterprise devoted to producing parts or products in a system with processes that are not dedicated by product family.

JOINT RATE. A single through rate on cargo moving via two or more carriers.

KNOCKDOWN CONTAINER. (See COLLAPSIBLE CONTAINER.)

LABORATORY. The space and resources assigned to perform analysis, research, testing or experimental activities.

LANYARD. A rope suitable for supporting one person. One end is fastened to a safety belt or harness and the other end is secured to a substantial object or a lifeline.

LAYOUT. (1) A scaled two-dimensional drawing or scaled three dimensional model created by hand or computer or using templates, icons and other display devices. (2) Physical arrangement of machines, workstations and support services in a manufacturing facility.

LEAD STORAGE BATTERY—AIEE (60.08.011). A lead storage battery is a storage battery the electrodes of which are made of lead and the electrolyte consisting of a solution of sulfuric acid. (These cells have a nominal voltage of 2.0.)

LESS-THAN-CARLOAD (LCL). That weight which is the minimum required for the assessment of normal carload rates. These carload (CL) rates can be further reduced by exceeding the related minimum weight, using multiple cars, or contracting for a unit train (such as coal shipments from mines to utility plants.)

LESS-THAN-TRUCKLOAD. A shipment that requires a fraction of a vehicle's capacity (both in terms of volume and weight).

LEVEL. Pertains to the storage height of a row of arms, or base or shelf, on the same horizontal plane. For example: a third level would indicate the third load or storage plane from the floor.

LIFELINE. A rope, suitable for supporting one person, to which a lanyard or safety belt (or harness) is attached.

LIFT. The vertical travel of the lifting device.

LIFTING MAGNET. An electromagnetic device for attracting and holding ferrous materials for hoisting.

LIFT-OUT LINKS OR TRACK OPENERS. Sections of monorail or runway track arranged to lift or swing out of the line of track to permit door operation.

LIFT SECTION. A section of track which can be raised out of alignment with a stationary track.

LIFT SPEED. The vertical velocity of the lifting mechanism.

Lift Table. A stationary or mobile platform that can be raised or lowered or tilted to provide working space at various angles and heights for a variety of applications.

Lift Truck. An industrial truck equipped with a lifting device, e.g., clamps, push-pull attachment, or forks.

Light Pen. A hand held scanning wand which is used as a contact bar code reader.

Limit Switch. A device actuated by the motion of another object to alter an electric circuit.

Line Item. A single item of inventory separately identified, but not limited by quantity.

Line Layout. A type of layout suitable for mass production in which machines, workstations and support services are arranged along a straight, U-shaped, L-shaped, S-shaped, or Z-shaped line in a sequence corresponding to the operational sequence of the dominant part(s).

Line of Flow. The route of travel for people and/or material during a process or procedure.

Liquefied Petroleum Gas (LP gas) (NBFU #58). A fuel which is composed predominantly of any of the following hydrocarbons, or mixtures of them: propane, propylene, butanes (normal butane or isobutane), and butylenes.

Live Load. The load being moved not including the weight of equipment being used to effect the movement.

Load Axle. The vehicle axle nearest the load.

Load Backrest Extension. A device extending vertically from the fork carriage frame.

Load Bar. A load-carrying member between carrier heads or carriers.

Load Block. The assembly of hook, swivel, bearing, sheaves, pins and frame suspended by the hoisting ropes.

Load Carrying Flange. The lower flange of the track on which the load bearing wheels roll.

Load Centering Device. A horizontal mechanism to ensure loads are centered when being transferred between two conveyor lines that are in angular relation to one another.

Load Center. The horizontal longitudinal distance from the intersection of the horizontal load-carrying surfaces and vertical load-engaging face of the forks (or equivalent load positioning structure) to the center of gravity of the load.

Loaded. When not otherwise specified, "loaded" is understood to mean the condition when the truck is handling the equivalent of a symmetrical cubic load.

Load Height. Maximum overall dimension from the bottom of a pallet or load module to the top of the load.

Loading Platform. An area of a warehouse or other facility where merchandise is received or shipped. The height of the structure is usually designed to conform to truck bed height on the truck side of the facility, and to rail car bed height on the rail car facility.

Load Length. (1) The length of a load being handled by a forklift truck. It is measured from the heel of the forks to the front of the load. (2) Maximum overall dimension of a pallet or load module and load in the direction perpendicular to the length of the aisle.

Load Moment. In counterbalanced trucks, reach trucks, and sideloaders, the nominal moment produced by the load tending to overturn the truck.

Load Moment Constant. In a counterbalanced truck, the longitudinal horizontal dimension from the overturning axis to the vertical load-engaging face with mast vertical.

Load Overhang. The amount of load projection past the base of a pallet or load module in any direction.

Load Width. Maximum overall dimension of a pallet or load module and load in the direction parallel to the aisle.

Locator System. A record that shows the exact location of items stored within a storage activity.

Locomotive Crane. A portable crane mounted on a base or car equipped for travel on a railroad track which may be self-propelled or propelled by an outside source.

Loft. A large room or space above the first floor, generally enclosed only by the exterior walls of the building. A loft can be either the top story of any type of building or all of the upper stories of a multistoried industrial building, and generally is intended for division into one large space only.

Looping Table. A roller, wheel, or other suitable type of conveyor located between two metal strip processing units over which the strip can loop when the forward machine slows down.

Lowering Conveyor. Any type of vertical conveyor for lowering objects at a controlled speed. (See ARM CONVEYOR; SUSPENDED TRAY CONVEYOR; INCLINED RECIPROCATING CONVEYOR; VERTICAL RECIPROCATING CONVEYOR.)

Low-lift Pallet Truck. A self-loading industrial truck equipped with wheeled forks of dimensions to go between the top and bottom boards of a double-faced pallet, and having wheels capable of lowering into spaces between the bottom boards so as to raise the pallet off the floor for transporting.

Low-lift Platform Truck. A self-loading industrial truck equipped with a load platform intended primarily for transporting loaded skid platforms.

Low-lift Truck. A self-loading truck equipped with an elevating mechanism designed to raise the load only sufficiently to permit horizontal movement. (Popular types are low lift platform truck and pallet truck.)

Low Solid Tire. A solid tire having a low flat profile in cross section.

Machine Layout. Detailed positions of machines, workstations, and support services in a factory.

Magazine. A type of warehouse usually covered with earth or depressed, where ammunition or other hazardous material is stored.

Magnetic Controller. A device or system of devices having all basic functions operated by electromagnets.

Main Aisle. An aisle wide enough to permit the easy flow of people, material, firefighting apparatus and/or equipment in two directions, generally running the length of the facility.

Main Floor. The floor of a story in which the principal entrance of a building is located.

Manifest. A document giving the description of aircraft, car, ship, or truck cargo.

Manlift. A passenger conveyor consisting of a vertical, endless belt with projecting steps and handles on its outer surface for transporting passengers between different elevations.

Man-on-board. A storage and retrieval concept whereby materials are accessed by taking the operator to the materials on board an S/R machine.

Marine Accident. Any grounding, fire, explosion, or other accident involving a vessel while afloat.

Marine Leg. A self-feeding bucket elevator with a means for lowering into the hold of a vessel. (See BUCKET ELEVATOR.)

Mark Sensing. Machine reading of marks by use or the conductive properties of the marks themselves.

Mass Production. A manufacturing operation in which a few high volume products are handled.

Mast. The support member providing the guideways permitting vertical movement of the carriage. It is usually constructed in the form of channels or similar sections providing the supporting pathway for the carriage.

Master Switch (crane). A manually operated device which dominates the operation of contactors and auxiliary devices of an electric circuit.

Mast Vertical (fork truck with tilting mast). Mast is considered vertical at any given fork height when adjusted so that the intersection of load-carrying surfaces and forward faces of forks or their equivalent is the same horizontal distance from the load axle as it is when at the height of the mast pivot.

Material Flow Network. The set of paths used by material handling devices for transporting material in a factory.

Material Handling Assignment. The act of assigning a specific material handling device to a specific material transfer request.

Material Handling System Selection. The act of selecting the types and quantity of one or more types of material handling devices from a candidate list to handle all the material transfers in a factory for a specified time period, usually a few years.

Materials Handling Equipment. Such items as forklift trucks and attachments, towing tractors, automatic guided vehicles, overhead or stacker cranes, robots, pallet trucks, warehouse trailers, and conveyor systems, used in storage and handling operations.

Matte Feeder. An extremely heavy duty type of apron feeder consisting of thick steel flights attached to a solid mat of chain links supported, in turn, by closely spaced rollers.

MAXIMUM FORK HEIGHT. The fork height attainable in fully raised position when loaded.

MAXIMUM LIFT. The lift from the extreme lowered position of the lifting device to the fully elevated position when loaded.

MEASUREMENT CARGO. A cargo on which transportation charges are based on measurement of dimensions.

MECHANICAL LOAD BRAKE. An automatic type of brake used for controlling loads in the lowering direction when simple reversing control is used. The brake remains released as long as the motor is driving in the down direction but automatically sets to provide braking action when the load begins to overhaul the motor.

MEMORANDUM TARIFF. Nonofficial documents which contain rule and rate information extracted from official tariffs and published by carriers.

MERGING FEEDER. One which consolidates the movement of objects from two or more lanes into a single lane.

MERRY-GO-ROUND CONVEYOR. (See CAROUSEL CONVEYOR.)

METERING CONVEYOR. In package handling, a power conveyor whose speed and motion are controlled to release units at a predetermined rate.

MEZZANINE. A platform within a story between the floor of the story, and the floor or roof next above it; elevated to a height which permits movement of people both below, and on the mezzanine; and generally having an area of not more than 1/3 the area of the floor immediately below the mezzanine.

MHA. The Material Handling Institute of America.

MICRO LOAD AS/RS. A fully enclosed AS/RS with narrow aisles that have just enough room for an S/R machine that can handle very small loads (e.g., electronic components, memory chips, etc).

MINI-LOAD. A storage and retrieval concept whereby materials are accessed by bringing the container to the operator where he/she processes less than unit load quantities. The term is typically used in small parts applications and/or where the weight of the container does not exceed 750 pounds.

MINIMUM PRESSURE ACCUMULATING CONVEYOR. A conveyor designed to minimize buildup of pressure between adjacent packages or cartons.

MINIMUM UNDERCLEARANCE. The vertical dimension from the lowest point on the loaded or empty vehicle, excluding wheel assemblies, to the vehicle supporting plane.

MIXING CONVEYOR, PADDLE TYPE. A conveyor consisting of one or more parallel paddle conveyor screws.

MIXING CONVEYOR, SCREW TYPE. A screw conveyor consisting of one or more conveyor screws, ribbon flight, cut-flight, or cut-and-folded flight conveyor screws with or without auxiliary paddles.

MOBILE RACK. A rack with wheels at the base that permit movement of the entire rack structure along tracks.

MODULAR DRAWER. Drawers with storage compartments that are modular in design and contained in a storage cabinet, providing high density storage.

MONORAIL. An overhead track on which a powered or unpowered carrying device moves equally spaced, top running or underhung carriers.

MOTOR BRAKE. A mechanical device which applies braking action to stop the rotation of the armature.

MOTORIZED HAND/RIDER TRUCK. A dual purpose truck that is designed to be controlled by a walking or a riding operator.

MOTORIZED HAND TRUCK. A truck that is designed to be controlled by a walking operator.

MOTOR VEHICLE ACCIDENT. An accident involving any powered vehicle designed primarily for on-the-road operations. This excludes equipment designed primarily for off-the-road operations such as material handling equipment, amphibious vehicles when afloat, tractors, etc.

MOVING BEAM BAR CODE READER. A device which dynamically searches for a bar code pattern by sweeping a moving optical beam through a field of view.

MULTIPLE FLIGHT CONVEYOR SCREW. A conveyor screw having two or more conveyor screw flights mounted at equal intervals around the pipe or shaft. The axial distance between adjacent flights is equal to the pitch of the conveyor screw flight divided by the number of flights.

MULTIPLE FONT. A scanner having the capability to

recognize more than one type font, but only one at a time.

MULTIPLE SCREW FEEDER. A series of conveyor screws installed side by side, usually in a flat bottom bin.

MULTIPLE STRAND CONVEYOR. (1) Any conveyor which employs two or more spaced strands of chain, belts, or cords as the load-supporting medium. (2) Any conveyor in which two or more strands are used as the propelling medium connecting pans, pallets, etc.

MULTI-ROW LAYOUT. A layout of machines, workstations, and support services along two or more well-defined rows.

NARROW AISLE TRUCK. A self-loading truck primarily intended for right angle stacking in aisles narrower than those normally required by counterbalanced trucks of the same capacity.

NATURAL FREQUENCY VIBRATING CONVEYOR. A vibrating conveyor in which the rate of free vibration of the trough on its resilient supports is approximately the same as the rate of vibration induced by the driving mechanism.

NESTING TARGET (PORTABLE METAL STACKING RACK). A formed metal piece (female form) affixed to the top of the post designed to retain the plug on the leg.

NET WEIGHT. The weight of the contents, not including the container (gross weight less tare weight).

NICKEL-IRON STORAGE BATTERY. A storage battery in which the positive active material consists of oxides of nickel, the negative active material consists of a mixture of iron and iron oxides and the alkaline electrolyte is usually an aqueous solution of potassium hydroxide. (The cells have a nominal voltage of 1.2.)

NONPERISHABLES. Items of inventory which do not spoil or deteriorate readily.

NON-READ. (See NO-READ.)

NONRIGID TIE. A member that connects two single racks maintaining a constant space, but does not add to the rigidity of the assembly.

NONTELESCOPING MAST. A mast in which the support member or members providing the guideways for vertical movement of the fork carriage do not move vertically with respect to the truck.

NO-READ. The absence of data at the scanner output after an attempted scan due to no code, defective code, scanner failure or operator error.

NORMAL AND PROPER USAGE. Operation of the equipment with a program of regular maintenance in accordance with generally accepted practices and within the rated capacity and service classification for which it was specified and designed.

NOTCHED STRINGER. A stringer which has openings cut out for insertion of pallet lifting devices.

OBSTACLE. A general term used to describe any unusable part of a facility.

OFFICE. A room in which the administrative functions of a facility are conducted or a service is supplied; generally having a high lighting level, low noise level, and finished floors, ceilings and walls.

OFFICE LAYOUT. Physical arrangement of offices, conference rooms, rest-rooms and other support services in a service facility. (See LAYOUT).

OPEN DOCK. A dock open from all sides with no protection from external elements.

OPTICAL CHARACTER READER. An information processing device that accepts prepared forms and converts data from them to computer output media via Optical Character Recognition (OCR).

OPTIONAL FEATURE. Any substitution for the standard attachments, accessories or components or any deviation from standard operational methods.

ORDER BATCHING. The act of combining several orders into one group for convenience in picking.

ORDER BILL OF LADING. A negotiable document by which a carrier acknowledges receipt of freight and contracts for its movement. The surrender of the original order bill of lading, properly endorsed is required by transportation lines upon delivery of the freight, in accordance with the terms of the bill of lading.

ORDER COLLATION. The gathering together of items in a specific order.

ORDER PICKER TRUCK, HIGH LIFT. A high-lift truck controllable by the operator stationed on a platform movable with the load-engaging means and intended for (manual) stock selection. The truck may be capable of self-loading and/or tiering.

Order Picking. The selection of normally less-than-unit load quantities of material for individual orders.

Order Picking and Stacking Truck. A self-loading industrial truck, primarily intended for right-angle stacking and order picking, equipped with an operator control platform that moves vertically on the mast with the load-engaging means.

Order Picking Sequence. The sequence in which items in an order are picked.

Order Picking Truck. A manually loaded industrial truck equipped with both a load platform, and an operator control platform movable as a unit on a mast.

Original Package. In storage operations, the original package is the first package to be applied to a specified quantity of items for the purpose of providing protection from corrosion or physical damage.

Oscillating Conveyor. A vibrating conveyor having a relatively low frequency and large amplitude of motion. (See vibrating conveyor.)

Oscillating Feeder. (See conveyor type feeder.)

Outrigger. A stabilizing structural member extending beyond the main body of the vehicle.

Overall Depth. The horizontal front-to-back dimension of a rack machine or other device.

Overall Height. The vertical dimension from the floor to the topmost point of the upright frame of a machine or device.

Overall Lowered Height (OALH). The maximum vertical dimension from the vehicle supporting plane (the ground) to the extreme top point of the mast with the fork carriage in the fully lowered position and unloaded. (Not to be confused with overall vehicle height.)

Overall Vehicle Height. The minimum height of the vehicle measured from the supporting plane.

Overall Width. The horizontal left-to-right dimension of a machine or device.

Over-and-under Conveyor. Two endless chains or other linkage between which carriers are mounted and controlled so that the carriers remain in an upright and horizontal position throughout the complete cycle of the conveyor.

Overhead Conveyor. (See trolley conveyor; power-and-free conveyor.)

Overhead Crane. A traveling crane with a movable bridge running on the top surface of rails of an overhead fixed runway structure and carrying a movable or fixed hoisting mechanism. (Syn. Bridge Crane)

Overhead Trolley Conveyor. (See trolley conveyor.)

Packaging. Application or use of appropriate wrappings, cushioning, interior containers and complete identification marking, up to but not including the exterior shipping container.

Packing. Application or the use of exterior shipping containers and assembling of items or packages therein, together with necessary blocking and bracing or cushioning, weatherproofing exterior, strapping, and marking of shipping containers.

Paddle Conveyor. (See mixing conveyor, paddle type.)

Paddle Conveyor Screw. A conveyor screw in which paddles are pitched and positioned to form the material propelling means.

Paddle Mixer. (See mixing conveyor, paddle type.)

Paddle Washer. A conveyor consisting of one or two parallel inclined paddle conveyor screws in a conveyor trough having a receiving tank and an overflow weir at the lower end and a discharge opening at the upper end.

Pallet. A horizontal platform device used as a base for assembling, storing and handling materials and/or products in a unit load, usually consisting of 3 stages, with upper and lower dockboards.

Pallet, All-way Entry. A pallet whose configuration permits retrieving or discharging from any direction in the same horizontal plane.

Pallet, Captive. A pallet whose use cycle remains within a single enterprise (private, corporate, or government).

Pallet Dispenser. A device for feeding empty pallets as required by a pallet loader or other machine or operation.

Pallet, Double Deck. A pallet having top and bottom decks.

Pallet, Double Entry. (See pallet, two-way entry.)

Pallet, Double Faced. (See PALLET, DOUBLE DECK.)

Pallet, Double Wing. A pallet whose top and bottom decks protrude beyond the outer edges of the deck spacers.

Pallet, Euro. Standardized pallet for the European Economic Community

Pallet, Expendable. A pallet intended to be discarded after one cycle of use.

Pallet, Flush. A pallet whose decks do not protrude beyond the deck spacers.

Pallet Frame. (See PALLET STORAGE SHELF)

Pallet, Full Four-way Entry. Full four-way pallets permit four-way insertion and withdrawal by forks of industrial trucks or by load wheeled forks of low-lift pallet truck.

Pallet Height. The total vertical dimension between the outer surfaces of the top and bottom decks.

Palletizer. A device to manually or automatically palletize cases coming from a production line via conveyors or other material handling devices.

Pallet Jack. A manual or battery powered device for limited lifting, maneuvering and transporting a pallet through short distances.

Pallet Length. The first dimension stated in designating a pallet size. Based on common practice among pallet manufacturers and users, the first dimension is defined as follows:

one-way, two-way, and partial four-way entry pallets—by the length of the stringer.

four-way entry and all-way entry pallets having two or more deck members at the top deck—by the length of the stringer.

four-way and all-way entry pallets having a single deck member and a solid top deck.

single deck or double deck reversible—longest dimension of the pallet.

double deck nonreversible—the horizontal dimension at right angles to the length of bottom deck member.

Palletless Loader. An automatic machine consisting of synchronized conveyors and mechanisms to receive objects and arrange them automatically into a unit in accordance with a predetermined pattern.

Pallet Load Detierer. A device for separating tiered pallet loads.

Pallet Loader. An automatic or semiautomatic machine, consisting of synchronized conveyors and mechanisms to receive objects from a conveyor(s) and place them onto pallets according to a prearranged pattern.

Pallet Load Tierer. A device for stacking full pallet loads for storage.

Pallet, Noncaptive. A pallet whose use cycle extends through one or more enterprises (private, corporate, or government) and may include a common carrier service.

Pallet, Nonreversible. A pallet having only one load-carrying surface.

Pallet, Partial Four-way Entry. Partial four-way pallets permit four-way insertion and withdrawal by forklift trucks and restrict the load wheeled forks of low-lift pallet trucks to two-way entry.

Pallet, Reusable. A pallet intended for multiple cycles of use.

Pallet, Reversible. A pallet having top and bottom decks capable of carrying a load.

Pallet, Single Deck. A pallet having only one deck.

Pallet, Single Faced. (See PALLET, SINGLE DECK.)

Pallet, Single Wing. A pallet whose top deck only protrudes beyond the outer edges of the deck spacers.

Pallet Size. The horizontal dimensional length and width of the top deck of a pallet.

Pallet Stacker. (See EMPTY PALLET STACKER.)

Pallet, Stacking. A pallet having a superstructure of vertical members (fixed, removable or collapsible), to support a superimposed load.

Pallet Storage Rack. A structure composed of two or more upright frames, beams, and connectors, for the purpose of supporting palletized materials in storage. Among the common methods of assembly are welded, bolted, or clipped.

Pallet Storage Shelf. At least two parallel beams in the same horizontal plane with connectors in place between upright frames for supporting pallets in storage.

Pallet, Take-it-or-leave-it. A pallet configuration that permits two choices of handling: (1) retrieving the pallet and unit load together, or (2) retrieving the unit load and leaving the pallet.

Pallet-to-post Clearance. The dimension between the outside edge of the pallet to the inside edge of the post on either side of the pallet.

Pallet Truck. A self-loading, low lift truck equipped with wheeled forks of dimensions to go between the top and bottom boards of a double faced pallet and having wheels capable of lowering into spaces between the bottom boards so as to raise the pallet off the floor for transporting.

Pallet Truck Openings. Openings in the bottom deck of a pallet to allow the load wheels of a low-lift pallet truck to bear on the floor.

Pallet, Two-way Entry. A pallet whose configuration permits insertion and withdrawal of a lifting device from opposite directions along the same horizontal axis.

Pallet Type Conveyor. A series of flat or shaped wheelless carriers propelled by and attached to one or more endless chains or other linkage.

Pallet Unloader. An automatic machine consisting of synchronized conveyors and mechanisms to disassemble a pallet load and discharge the objects singly.

Pallet Width. The horizontal dimension at right angles to the pallet length.

Pallet, Wing. A pallet whose deck(s) boards protrude beyond the outer edges of the deck spacers.

Pan Conveyor. A conveyor comprised of one or more endless chains or other linkage to which overlapping or interlocking pans are attached to form a series of shallow, open-topped containers. Some are called apron conveyors.

Pan Feeder. (See CONVEYOR TYPE FEEDER.)

Parcel Post Air Freight. Consolidation of a number of parcel post packages (with destination postage affixed by the shipper) for shipment as air freight to the postmaster at another city for subsequent delivery within local postal zones or beyond.

Parking Brake. A device to prevent the movement of a stationary vehicle.

Passenger Conveyor. A conveyor for transporting people who enter and leave the conveyor by walking or stepping and who either stand, sit or walk on the conveyor while being transported.

Paulin. Sheet of canvas or other materials usually treated to make it resistant to moisture and chemicals, used as a protective covering (tarpaulin).

Penthouse. An enclosed floor above the main roof of the building, generally substantially smaller than the area of the main floor of the building, and generally used to house either building or process equipment or in some cases, small habitable rooms, for specialized functions.

Percent (%) Grade. The measure of the rate of ascent of an inclined plane numerically equal to the vertical rise divided by the horizontal length, multiplied by 100.

Personal Injury. Any bodily harm sustained by a person as the result of an accident.

Personal Protective Equipment. Any device utilized by the individual to prevent personal injury.

Personnel Aisle. An aisle used only as pedestrian routes for access to doors or to special interior areas.

Person-to-Item AS/RS. An AS/RS in which the storage racks are stationary and a person or machine goes to the storage location of items and picks them.

Pick List. A list that contains items from one or more orders to be picked.

Pickstation. The primary established location where an order picker stands or is assigned. For pickers who move, this may also be referred to as their "pick zone." A location for miniload and multiple modular AS/RS devices to deliver their products.

Pick-up and Delivery Service (air transport). An optional service for the surface transport of shipments from shipper's door to originating air terminal and from the air terminal of destination to receiver's door. Pick-up service is provided upon shipper's request. Delivery service is provided automatically by the air carrier at an additional charge unless the shipper requests otherwise. PU&D is normally performed within a 25-mile radius of the airport. The latter is commonly referred to as the terminal area. (For service beyond the terminal area see TRUCK/AIR/TRUCK.)

Pick-up and Delivery Stations (P&D stations). A location at which a load entering or leaving storage is supported in a manner suitable for handling by material handling equipment (also called transfer station, I/O station, feed/discharge station, etc.)

Pick-Up/Drop-Off Point. (See pick-up and delivery station).

Piggy-back. The transportation of truck trailers and containers on railroad cars.

Pilot Plant. A facility devoted to the production of initial lots, or to the continuous production of small quantities of a product for the purpose of experimenting with its design, composition, production methods, and/or equipment used.

Pins. Fasteners such as rivets, belts, lags or other material used in assembling rack frames.

Pin Type Slat Conveyor. Two or more endless chains to which crossbars are attached at spaced intervals, each having affixed to it a series of pointed rods extending in a vertical plane on which work is carried. Used principally in spraying or washing operations where the least amount of area of the product is contacted.

Pit. A depression below the adjacent floor or grade level, generally open above.

Pivoted Bucket Conveyor. A conveyor using pivoted buckets attached between two endless chains which operate in suitable guides or casing in horizontal, vertical, or inclined, or a combination of these paths over drive, corner, and take-up terminals. The buckets remain in the carrying position until they are tipped or inverted to discharge.

Plain Chain Conveyor. (See sliding chain conveyor.)

Planning Department. Collection of workstations to be grouped together during the facilities planning process.

Planograph. A storage plan usually in two dimensions, on paper, which indicates the storage location of merchandise within a warehousing facility. Its primary function is to achieve the most effective utilization of space within the warehouse complex.

Plant Layout. A block layout of a factory, showing the relative location of departments, but not the machine locations or the material handling aisles.

Plant Location Analysis. The compilation and evaluation of data necessary to render a decision concerning site selection.

Platform. An area that is raised in relation to one or more adjacent areas.

Platform Truck, Powered. A dolly that is powered electrically or by gasoline.

Platform Truck, Unpowered. (See dolly).

Plug (portable metal stacking rack). A formed, cast, or forged metal piece (male form) affixed to the bottom of each leg designed to fit the nesting target on the post.

Pneumatic Chute or Spout. A chute or spout in which air is introduced through the bottom to facilitate movement of bulk material down a slight decline.

Pneumatic Conveyor. An arrangement of tubes or ducts through which bulk material or objects are conveyed in a pressure and/or vacuum system.

Pocket Belt Conveyor. (See pocket conveyor.)

Pocket Conveyor. A continuous series of pockets formed of a flexible material festooned between cross-rods carried by two endless chains or other linkage which operate in horizontal, vertical, or inclined paths.

Portable Conveyor. Any type of transportable conveyor, usually having supports which provide mobility.

Portable Cranes. Cranes having a revolving superstructure with power plant, operating machinery and boom mounted on a fully mobile carriage not confined to a fixed path. (See crawler, locomotive, or truck crane.)

Portable Drag Conveyor. A portable conveyor upon which endless drag chains are used as the conveying medium. Sometimes applied to a portable flight conveyor. (See drag chain conveyor.)

Portable Metal Stacking Racks. A structure constructed of welded tubing or structural shapes or a combination of both. The rack bed is rectangular with or without short legs on each corner which allow truck forks to penetrate under the bed. Rack legs can be fixed or removable. Some have one or two removable side frames rather than single legs. Decking, when required, can be of wood or metal. When legs or frames are in place, one rack rests on top of another by means of interlocking male and female socket arrangements or tiering guides. The racks can be customized to hold anything from small stampings (by adding wire mesh sides and bottoms) to diesel engines.

Port of Entry. A harbor or airport, designated by the government, at which goods or persons are admitted into the country.

Port-of-origin Cargo Clearance. Customs formalities can be handled at certain originating cities in the USA rather than delaying such procedures until the export cargo reaches a gateway city.

Positive Discharge Bucket Elevator. A spaced bucket type of elevator in which the buckets are maintained over the discharge chute for a sufficient time to permit free gravity discharge of bulk materials. (See BUCKET ELEVATOR.)

Post. A single upright member, a part of an upright frame that may have provision for vertical adjustment of rails and beams.

Power-and-free Conveyor. A conveying system wherein the load is carried on a trolley or trolleys which are conveyor propelled through part of the system and may be gravity or manually propelled through another part. This arrangement provides a means of switching the free trolleys into and out of adjacent lines. The spur or subsidiary lines may or may not be powered.

Power Bars. Conductor bars which carry motor operating current.

Power Conveyor. Any conveyor which requires power to move its load.

Power Curve. A power driven conveyor for moving unit loads around a horizontal curve.

Power Driven Roller Conveyor. (See ROLLER CONVEYOR, LIVE.)

Powered Industrial Truck. A mobile, power-driven truck used to carry, push, pull, lift, stack or tier material.

Preprinted Symbol. A symbol which is printed in advance of application either on a label or on the article to be identified.

Preservation. Application or use of adequate protective measures to prevent deterioration resulting from exposure to atmospheric conditions during shipment and storage.

Preservatives. Material used to protect an item against deterioration.

Process Layout. A layout in which machines, workstations and support services with similar processing capabilities are grouped together.

Product Layout. A layout in which machines, workstations and support services to produce a product or family of products are grouped together.

Property Damage. The total cost to restore the property to the condition that existed prior to the accident.

Push-Back Rack. A rack in which items are loaded on a rail-guided carrier from the front of the rack and the previously stored loads are pushed back one position to the rear to make room for the new load. When a load is removed, the remaining loads advance forward due to a gentle slope in the racks.

Push Button Station (Crane). A means by which an operator at a fixed or moving location may actuate push buttons to control powered equipment.

Push Diverter. A computer-controlled arm-like mechanism that can push items into or away from one conveyor line to another or a temporary hold area.

Pusher Bar Conveyor. Two endless chains cross-connected at intervals by bars or rotatable pushers, which propel the load along the bed or trough of the conveyor.

Pusher Chain Conveyor. One or more endless chains with attachments which propel or retard the movement of packages, objects, trucks, dollies, or cars along stationary wood, metal, or roller beds, troughs, rails, or tracks.

Push-Pull. A lift truck attachment designed to load, transport and unload unit loads stacked on slip sheets

Queue. A line formed by loads or items while waiting for processing.

Quiet Area. A clear space, containing no dark marks, which precedes the start character of a symbol and follows the stop character.

Rack Supported Building Structure. A complete and independent load storage system in which the storage rack is the basic structural system.

Radio Controlled. Motor-propelled units which are controlled by an operator by means of a remote radio transmitter.

Radio Frequency (RF) Identification. An electronic method of assigning a piece of information to a product, process, or person with a transponder (tag) and reading the information with a reader.

Rail. A horizontal load-carrying member running the full depth of the rack. Within each bay, two rails are used at each level.

Rail and Support Height. The combined height in inches of the guide rail and support bracket in a drive-thru rack.

Rail Guide. A device attached to the rail for alignment of pallets.

Rail Projection. The horizontal dimension from the post to the outside edge of the rail assembly.

Rail Span. The length of rail between the centerline of the uprights.

Rail Support. A member between the rail and the upright post.

Rail Support Connector. A device connecting the rail support to the post.

Rail Tie. A member used to connect a pair of rails at or near the end of the rail assembly.

Ram. An attachment comprising a cantilever member, extending horizontally forward from a lift truck carriage, which is used to handle hollow core, cylindrical or similar loads, such as rugs, coils, pipe, etc.

Ramp. A sloping aisle or roadway connecting two different levels, generally not exceeding 45 degrees.

Random Storage Policy. A storage policy in which an incoming item is stored in any available location.

Rated Load (crane). The maximum load the equipment is designed to handle safely.

Reach Truck. A self-loading truck, generally high lift, having load-engaging means mounted so it can be extended forwardly under control to permit a load to be picked up and deposited in the extended position and transported in the retracted position.

Receiving. The function of accepting, recording, and reporting merchandise into a facility.

Receiving Area. The space reserved for the receiving function.

Reciprocating Beam Conveyor. One or more parallel reciprocating beams with tilting dogs or pushers arranged to progressively advance objects.

Removable Posts (portable metal stacking racks). A load-carrying column that is inserted at each corner of the base by a simple mechanical means. A nesting target or plug is permanently affixed to the top of the column.

Replenishment. Re-stocking primary picking locations from reserve storage locations.

Resistors, Ballast (crane or carrier). Electrical resistance added in the motor circuit to provide reduced starting current for gradual acceleration.

Restricted Articles (air transport). Certain articles (explosives, for example) which are either excluded from air cargo entirely or subject to stringent requirements on volume and packaging. A restricted articles tariff, giving full details on such articles, is published by Air Tariff Publishers, Inc. for the airlines.

Retarder. Any device used to slow the rate of travel of bulk material or objects on a conveyor.

Retarding Conveyor. Any conveyor used to retard the rate of movement of bulk materials, packages or objects where the slope is such that the conveyed material tends to propel the conveying medium.

Reusable Container. A container intended for multiple cycles of use.

Rewarehousing. Rearrangement of supplies from one storage area to another within the same storage activity.

Ribbon Belt Conveyor. (See BELT CONVEYOR, MULTIPLE RIBBON.)

Ribbon Conveyor. (See RIBBON FLIGHT SCREW CONVEYOR.)

Ribbon Conveyor, Internal. (See INTERNAL RIBBON CONVEYOR.)

Ribbon Flight. A conveyor screw flight proportioned to provide a space between it and the internal supporting pipe or shaft when mounted thereon.

Ribbon Flight Conveyor Screw. A conveyor screw in which the conveyor screw flight is of the ribbon flight type.

Ribbon Flight Screw Conveyor. A screw conveyor having a ribbon flight screw. (See SCREW CONVEYOR.)

RIDER TRUCK. A truck that is designed to be controlled by an operator from either a standing or sitting position.

RIGHT ANGLE STACK AISLE WIDTH. The aisle width necessary to place in location the maximum dimension load perpendicular to the aisle with a given vehicle.

RIGID BACK-TO-BACK TIE. A member connecting back-to-back racks, maintaining a constant space and adding rigidity to the complete assembly.

RIGID CONTAINER. A container with components permanently assembled.

RIGID POST (PORTABLE METAL STACKING RACK). A load-carrying column permanently attached to the corner of the base. A nesting target or plug is permanently affixed to the top of the column.

ROLLER CONVEYOR. A series of rollers supported in a frame over which objects are advanced manually, by gravity, or by power.

ROLLER CONVEYOR, HERRINGBONE. A roller conveyor consisting of two parallel series of rollers having one or both series skewed.

ROLLER CONVEYOR, HYDROSTATIC. A section of roller conveyor having rollers suitably weighted with liquid to control the velocity of objects being conveyed.

ROLLER CONVEYOR, LIVE. A series of rollers over which objects are moved by the application of power to all or some of the rollers. The transmitting medium is usually belting or chain.

ROLLER CONVEYOR, SHOCK ABSORBING. (See ROLLER CONVEYOR, SPRING MOUNTED.)

ROLLER CONVEYOR, SKEWED. A roller conveyor having a series of rollers skewed to direct objects laterally while being conveyed.

ROLLER CONVEYOR, SPEED-UP. A powered roller conveyor operating at a higher speed than its feeder to create space between unit loads.

ROLLER CONVEYOR, SPOOL TYPE. A roller conveyor in which the rollers are of conical or tapered shape with diameter at ends of roller larger than at the center.

ROLLER CONVEYOR, SPRING MOUNTED. (1) A roller conveyor where the ends of each roller are supported on springs. (2) A section of roller conveyor supported on springs.

ROLLER CONVEYOR, TROUGHED. A roller conveyor having two rows of rollers set at an angle to form a trough over which objects are conveyed.

ROLLER SPIRAL. An assembly of curved sections of roller conveyor arranged helically and over which objects are lowered by gravity.

ROLLER TABLE. A table with a surface consisting of a series of rollers.

ROLLING CHAIN CONVEYOR. A conveyor consisting of one or more endless roller chains on which packages or objects are carried on the chain rollers. The speed of transportation is double that of the chain speed.

ROLLING RESISTANCE. Force, resisting tractive effort, caused by friction losses. It is preferably expressed in pounds, or as pounds per thousand pounds of vehicle weight, or as a percent of gross vehicle weight.

ROLLWAY SKID. A roller conveyor with brakes to prevent the rollers from turning except in one direction.

ROOM. An enclosed space within a building.

ROPE AND BUTTON CONVEYOR. A series of buttons or flights attached to an endless wire rope or cable for the purpose of conveying or retarding the movement of bulk material or objects along a stationary trough.

ROTARY SWITCH. A switch with a movable inner frame containing straight and/or curved sections of track. The inner frame can be rotated to align these sections of track with other tracks for routing carriers from one track to another.

ROTARY TABLE FEEDER. A rotating horizontal, circular table to which material flows from a round bin or hopper opening and from which it is discharged by a plow.

ROTARY VANE FEEDER. A rotor of cylindrical outline with radial, spaced plates or vanes rotating on a horizontal axis for controlling the flow of bulk materials.

ROTATOR. A power-actuated attachment which provides rotary motion about an axis that is usually perpendicular to the face of the carriage.

ROUGH TERRAIN TRUCK. A wheeled type truck designed primarily as a fork truck, but which may be equipped with attachments. This truck is intended for operation on unimproved natural terrain as well as the disturbed terrain of construction sites.

Row. A continuous series of racks.

Runners. Integral rails or skids on which demountable containers are supported.

Running Sheave. A sheave which rotates as hook is raised or lowered.

Runway. The rails, beams, brackets and framework on which the crane operates.

Runway Conductors. The conductors mounted on or parallel to the runway which carry current to the crane.

Safety Belt. A device, usually worn around the waist, which, by reason of its attachment to a lanyard and lifeline or a structure, will prevent a worker from falling.

Safety Work Surface. A surface intended to reduce the possibility of a foot slippage.

Saw-Tooth Dock. A particular dock layout requiring trucks to enter and exit the dock area at an angle to the building wall where the docks are located.

Scan. (1) The search for a symbol which is to be optically recognized. (2) A search for marks to be recognized by the recognition unit of an optical scanner.

Scan Area. The area intended to contain a symbol.

Scanning Curtain. The effective reading area (width x height) of a moving beam scanner, which is equal to its depth-of-field and height-of-scan at a specific operating range.

Scheduling. Assignment of material transfer requests (jobs) to material handling devices.

Scoop. An attachment to handle loose or bulk material by a shoveling action. Can be manually or power manipulated and dumped.

Screw Conveyor. A conveyor screw revolving in a suitably shaped stationary trough or casing fitted with hangers, trough ends and other auxiliary accessories.

Screw Conveyor, Rotating Casing. A screw conveyor in which a tubular casing rotates.

Screw Conveyor, Vertical. A screw conveyor that conveys in a substantially vertical path.

Screw, Cut-and-folded Flight. A conveyor screw with a section or sections of each pitch cut and folded back.

Screw, Cut Flight. A conveyor screw with a section or sections notched from each pitch.

Screw, Double Flight. A conveyor screw having two conveyor screw flights mounted 180° apart on the pipe or shaft. The axial distance between adjacent flights is equal to one-half the pitch of the conveyor screw flight.

Section. (1) A non-permanent or permanent physical division within a facility, and may be delineated by the structural features of the facility such as columns, partitions, firewalls, etc. (2) A single column of storage racks or shelves.

Selective Rack. A rack that provides complete access to every pallet load on it.

Self-feeding Conveyor. Any conveyor so arranged as to feed itself automatically without the necessity of using a separate feeder.

Self-feeding Portable Conveyor. Any power-propelled conveyor designed to advance into a pile of bulk material, thereby automatically feeding itself.

Self Loading. The capability of a powered industrial truck to pick up, carry, and deposit its load without the aid of external handling means.

Semi-gantry Crane. A traveling crane with one end of the bridge supported on one or more legs running on fixed rails or other runway and the other end of the bridge supported by a truck running on an elevated rail or runway.

Serial Code. A bar code symbol typically used with a fixed beam scanner where the scanning action is caused by the motion of the symbol past the scanning head. The bits of the symbol are evaluated one at a time (serially) as the symbol passes.

Service Aisle. An aisle necessary to provide access to interiors of stacks for inventory, inspection, or for protective processing.

Shaft. (1) A vertical space within a building, generally extending through more than one story, usually enclosed, created to house vertical functions. (2) A passageway through the ground, generally constructed to provide for movement of men or material from one level to another.

Shaker Conveyor. (See oscillating conveyor.)

Shared Storage Policy. A storage policy in which

items with relatively high transaction rates are stored in spaces near the I/O point and items with relatively low transaction rates are stored farther away from the I/O point.

Shelf. A horizontal supporting surface above floor level within a rack or storage section. When carousel carriers are subdivided horizontally, the sub-division surfaces are also referred to as shelves.

Shelf Connector. A device used at each end of a beam to connect the beam to an upright frame.

Shipping. The function of recording, reporting, and sending merchandise to the consumer.

Shipping Area. The space reserved for the shipping function.

Ship's Conveyor Elevator. A deck-mounted pocket or suspended tray conveyor arranged to be lowered into a ship's hold to varying depths for the purpose of loading or unloading packages and objects. Not to be confused with a marine leg.

Shop. The space and resources assigned to a functional group engaged in making or repairing a variety of articles.

Shrinkwrap. A method of unitizing a load by placing a plastic sheet over the load and applying heat and suction so that the load is tightly wrapped by the plastic film.

Shuttle. The load supporting mechanism on the carriage of an S/R machine which provides for movement of loads into or out of storage compartments and P & D stations.

Shuttle Conveyor. Any conveyor such as a belt, chain, apron, screw, etc., in a self-contained structure movable in a defined path parallel to the flow of the material.

Side Loader. A self-loading truck, generally high lift, having load engaging means mounted in such a manner that it can be extended laterally under control to permit a load to be picked up and deposited in the extended position and transported in the retracted position on the truck bed.

Side Loading Attachment. An attachment used to pick up loads and deposit them at right angles to the longitudinal axis of a lift truck. (Primarily intended for narrow aisle operations.)

Side-Mounted Operator Compartment. An operator compartment a portion of which is located outside of a longitudinal line connecting between outermost front and rear tires.

Side-pull En Masse Conveyor. An arrangement of horizontal closed circuit conveyor in which the tension element is at one side and above the moving stream of material.

Side Pusher Conveyor. A trolley conveyor with arms cantilevered to the side to push free trolleys on parallel tracks or with pushers to engage the cantilevered arms of free trolleys.

Side Shifter. A device to permit lateral movement of the load-engaging means on a lift truck.

Silo. (1) A tall cylindrical building, usually used for storing bulk commodities. (2) A tall cylindrical building in the ground such as that used for storing missiles.

Single Arm. Horizontal load member affixed to a column for single side loading.

Single Command Cycle. The time between actuation of the S/R machine and completion of a cycle in which one load is stored or retrieved and the S/R machine is ready to start a new cycle.

Single Deep Storage. Loads which are stored one deep on each side of an aisle.

Single Rack. A pallet storage rack with only two upright frames and one or more shelves.

Single-Row Layout. A layout of machines, workstations, and support services along a well defined row.

SITC. Standard International Trade Classification. (See COMMODITY CODE. See also STANDARD INTERNATIONAL TRADE CLASSIFICATION.)

Skate-wheel Conveyor. A wheel conveyor making use of series of skate wheels mounted on common shafts or axles, or mounted on parallel spaced bars on individual axles. (See WHEEL CONVEYOR.)

Skew Table. A live roller conveyor having its rollers skewed for the purpose of moving objects laterally against a guide member. It may have a fixed or an adjustable guide member. (See ROLLER CONVEYOR, SKEWED.)

Skid Support. A pair of upturned channels placed front to back across a pair of shelf beams.

Skip Hoist. A bucket or car operating up and down a defined path receiving, elevating, and discharging bulk materials.

SKU. Stock Keeping Unit.

Slat Conveyor. A conveyor employing one or more endless chains to which nonoverlapping, noninterlocking, spaced slats are attached.

Slave Pallets. A handling base or container on which a unit load is supported and which normally is captive to an AS/RS.

Slider Bed. A stationary surface on which the carrying portion of a conveyor belt slides.

Sliding Chain Conveyor. One or more endless chains sliding on tracks on which packages or objects are carried.

Sliding Type Switch. A switch with a movable inner frame containing straight and/or curved sections of track. The inner frame can be slid to align these sections of track with other tracks for routing carriers from one track to another.

Sling. A flexible means of connecting the load and the hoisting device.

Slope Conveyor. Usually a troughed belt conveyor used for transporting coal or ore through an inclined passage to the surface from an underground mine. (See APRON CONVEYOR, BELT CONVEYOR, FLIGHT CONVEYOR.)

Solid Tire. A tire, such as "Cushion Tire" or "Low-Solid Tire," made of resilient solid material.

Sortation System. An automated conveyor system with diverters used for sorting items in a warehouse.

Sorting Conveyor. A conveyor which receives mixed unit loads and discharges them to segregated spaces or conveyors in response to an automatic dispatch control. Operator attention is usually required to introduce the dispatch signal into a memory system.

Space. One or more areas.

Span. Center to center distance between runways.

Spark Enclosing Equipment. Auxiliary units which completely enclose the electrical equipment on the vehicle to prevent the emission of electric sparks. It includes the enclosure of the generator, electric motors, contactors, and protection of certain other electrical equipment.

Spindle Conveyor. A chain-on-end conveyor in which the chain pins extend in a vertical plane and are usually of enlarged diameter in that portion above the chain on which special revolvable fixtures can be rotated for the purpose of spraying or drying. Outboard rollers or sliding shoes support the chain and products.

Spine Layout. A type of layout with a central aisle (spine) from which branch secondary aisles usually perpendicular to the central aisle.

Spot. The placing of a truck or boxcar where it is required for loading or unloading.

Spur Track. A fixed track arranged to interlock with an adjacent crane girder to permit passage of carriers from the spur track to the crane and vice versa.

Squaring Shaft. A driven shaft which transmits torque to drive wheels operating on two or more tracks.

Stability. A truck's resistance to overturning. (The recognized method of measurement is the "Tilting Platform Tests" covered in Appendix A of ASA Safety Code B56.1-1959 plus additions shown in Section 4 of ITA Recommended Practices Manual.)

Stacker Conveyor. (1) A conveyor adapted to piling or stacking bulk materials, packages or objects. (See APRON CONVEYOR; BELT CONVEYOR; FLIGHT CONVEYOR; PORTABLE CONVEYOR; SLAT CONVEYOR.) (2) A fixed or pivotally mounted boom conveyor. (3) With a blending system the stacker operates over the stocking conveyor in a manner similar to a wing tripper to build layered piles or beds of material parallel to the stocking conveyor. (See BOOM CONVEYOR; PORTABLE CONVEYOR.)

Stair. A series of steps allowing walking access from one level to another.

Stairway. One or more flights of stairs and the necessary landings and platforms connected therewith to form a continuous and uninterrupted passage from one floor to another or to the roof or to the ground.

Standard International Trade Classification (SITC). A numerical code developed by the United Nations and adopted by the U.S. scheduled airlines and airlines of most other countries as a basis for numerical identification of commodities moving in airfreight. (See COMMODITY CODE.)

Start/Stop Character. A bar code character that provides the scanner with start and stop reading instructions as well as code orientation. The start character is normally at the left-hand end of a picket fence oriented code and adjacent to the most significant character. The stop character is normally at the right-hand end of a picket fence oriented code and adjacent to the last significant character.

STATION. An established location in which something or someone stands or is assigned to stand or remain. This space includes that area necessary for the operation, material being processed or worked on, and the auxiliary equipment necessary to perform the assigned function.

STEEL BELT CONVEYOR. A belt conveyor using a steel band belt as the conveying medium.

STOCK NUMBER. Number assigned to an item, principally to identify that item for storage and issue purposes.

STORAGE MODULE. (1) Those items such as pallets, containers, boxes, etc., containing, holding or constituting the unit load. (2) A modular AS/RS device.

STORAGE/RETRIEVAL (S/R) MACHINE. A machine operating on floor or other mounted rail(s) used for transferring a load from storage compartment to a P & D station and from a P & D station to a storage compartment. The S/R machine is capable of moving a load both vertically and parallel with the aisle and laterally placing the load in a storage location.

STOREHOUSE. A place or space for storing goods.

STORY. The space between the surface of two successive floors in a building, or between the top floor and the ceiling or underside of the roof frame.

STOPPING TOLERANCE (CAROUSEL). The maximum deviation expected when a rotating carousel is stopped under maximum imbalanced load and maximum unit-load conditions.

STORAGE POLICY. A set of rules used by warehouse managers to determine where incoming items are to be stored in a warehouse and how many storage spaces must be assigned to each.

STRADDLE TRUCK. A self-loading outrigger type industrial truck, generally high-lift, for picking up and handling loads between its outrigger arms.

STRAIGHT BILL OF LADING. A nonnegotiable document by which a carrier acknowledges receipt of freight and contracts for its movement. The surrender of the original straight bill of lading is not required by the transportation line upon delivery of the freight except when necessary for the purpose of identifying the consignee.

STRAIGHT CHUTE. A sloped chute designed to transfer bulk materials, packages, or objects in a straight line from point of entry to point of discharge.

STRAIGHT-IN STRAIGHT-OUT DOCK. A particular dock layout in which the docks are constructed in a way that allows trucks to enter and exit the dock area in a straight line perpendicular to the building wall where the dock is located.

STRAPPING. A reinforcing or securing of crates, boxes, bales or bundles by encircling with round or flat banding material (metal or fabric).

STRAP SLOT. A small opening in the stringer or on the underside of top deckmember to facilitate use and prevent movement of unit load fastening material.

STRETCHWRAP. A method of unitizing a load by stretching and wrapping a plastic sheet tightly over the load.

STRINGER. A continuous longitudinal deck spacer.

STRINGERMEMBER. A subsurface element on full four-way entry wood pallets beneath and at right angles to the deckmember but above the block, leg or post spacers.

SUPPORTS, CONDUCTOR. Mechanical supports for conductor bars.

SUSPENDED TRAY CONVEYOR. A vertical conveyor having one or more endless chains with suitable pendant trays, cars or carriers which receive objects at one elevation(s) and deliver them to another elevation(s).

SUSPENDED TRAY ELEVATOR. (See SUSPENDED TRAY CONVEYOR.)

SUSPENDED TRAY LIFT. (See SUSPENDED TRAY CONVEYOR.)

SWITCH, MONORAIL. A device with a moving section or sections of track which can be moved to permit passage of a carrier from an incoming track to one of various outgoing tracks. (See ROTARY OR SLIDING TYPE SWITCH.)

SYMBOL. A combination of characters including start/stop characters and check characters, as required, which form a complete scannable entity.

SYMBOL DENSITY. The number of characters per lineal inch.

SYMBOL LENGTH. The length of the symbol measured from the beginning of the quiet area adjacent to the start character to the end of the quiet area adjacent to a stop character.

TAG LINE. An electrical conductor system employing flexible cable or cables as the conductors.

TARE. Gross weight less weight of contents.

TARIFF (AIR TRANSPORT). A document filed with the Civil Aeronautics Board setting forth applicable rules, rates and charges for the movement of goods.

TARIFF (MOTOR CARRIER). Documents published and filed with I.C.C., containing rates, fares, rules and regulations of common carriers.

TARIFF (RAILROAD). A schedule of rates charged with governing rules and regulations. (Part I of I.C.C. Act.)

TELESCOPING CONVEYOR. A conveyor, the length of which may be varied by telescoping frame members. (See EXTENDABLE CONVEYOR.)

TELESCOPING MAST. A multiple mast wherein one member is stationary and the other(s) movable vertically with respect to the stationary member and supporting the fork carriage in its vertical movement.

TETHERED-ELECTRIC TRUCK. An electric truck in which the power source is remote from the vehicle and connected by a flexible cable.

TIER. (1) A horizontal layer of a column, row or stack. Tiers are numbered in the order of their stacking from the bottom up. (2) A set of storage locations having a common elevation in an AS/RS.

TIERER, PALLET LOAD. (See PALLET LOAD TIERER.)

TIERING. The process of placing one load above another.

TIERING GUIDE (PORTABLE METAL STACKING RACK). A formed metal piece to serve a dual purpose (1) as a lifting device for crane handling, (2) as a tiering guide.

TILT. Longitudinal angular displacement of mast structure, forks, or carriage.

TOE (CANTILEVER RACK). The front edge of an extended base on a single or double sided upright.

TOE BOARD. A vertical projection above a platform or walkway to contain objects and to define the walkway for personnel.

TONGUE SWITCH. A switch that contains one straight section of track or tongue, pivoted at one end. The tongue can be swung to various positions to connect this section of track to other tracks for transfer of carriers from one track to another.

TOP CROSS BRACES. Diagonal members in the form of a rigid cross connecting the uppermost portion of the rack assembly.

TOP DIAGONAL BRACE. A diagonal member connecting uppermost portions of rack assemblies.

TOP HORIZONTAL BRACE. A member tying two posts together parallel to the depth of the bay at the top of the rack.

TOP-RUNNING CRANE. A traveling crane with a moveable bridge running on the upper flanges of an overhead fixed runway structure and carrying a movable or fixed hoisting mechanism.

TOP TIE. A member connecting rack posts at the top of the structure.

TORQUE CONVERTER TRANSMISSION. A hydrodynamic drive which transmits power with ability to change torque. (Torque ratio is a function of speed ratio.)

TORQUE, FULL LOAD. The torque necessary for a motor to produce its rated horsepower at full-load speed.

TOW CONVEYOR. An endless chain supported by trolleys from an overhead track or running in a track at the floor with means for towing floor supported trucks, dollies, or carts.

TOWER CRANE. A crane with one or two uprights that can move horizontally along tracks and may also be capable of rotating around the upright. It is usually used in open air applications where heavy, bulky material must be handled, e.g., shipyards and construction sites.

TRACTIVE EFFORT. The motive force, exerted at the circumferences of the driving wheels at the point of contact with the ground.

TRACTIVE EFFORT, REQUIRED. Tractive effort needed to propel a vehicle and any trailers being towed by it, while overcoming (1) rolling resistance of all wheels, including all wheel bearing friction, (2) grade resistance, and (3) inertia of vehicle and any trailers being towed by it.

TRACTOR, MOTOR DRIVEN. A motor driven unit, supported from track wheels and propelled by a friction driving wheel or wheels. It is normally attached to, and drives, a carrier or crane end truck.

TRAFFIC CONTROL. A mechanical or electrical mechanism to prevent collision of objects as they merge from two conveyor lines into a single line.

TRAMWAY. A system in which carriers are supported by a track cable and in which the movement of carriers is continuous over one or more spans.

TRANSFER LIFT SECTION. A section of track which can be moved vertically out of alignment with one stationary track into alignment with another stationary track at a different elevation in order to transfer carriers from one track to the other.

TRAVELING CRANES. Cranes that follow a fixed path. (See OVERHEAD, GANTRY, SEMI-GANTRY, WALL CRANES.)

TRAVEL SPEED. The velocity of a vehicle in a horizontal plane.

TRAVEL TIME MODELS. Mathematical models that estimate the time taken by an S/R device for an S/R transaction.

TRAY. A captive support container used to hold and transport stored products in a vertical lift module or mini-load.

TRAY (BATTERY). A tray of a storage battery is a support or container for one or more storage cells.

TROLLEY CONVEYOR. A series of trolleys supported from or within an overhead track and connected by endless propelling means such as chain, cable, or other linkage with loads usually suspended from the trolleys.

TROLLEY (TOP RUNNING). The unit including the hoisting mechanism which travels on the bridge rails.

TROLLEY (UNDERHUNG). (See CARRIER.)

TRUCK. (See POWERED INDUSTRIAL TRUCK.)

TRUCK/AIR/TRUCK SERVICE. The surface movement of air freight to and from airports and origin and destination points beyond the usual 25-mile radius of pick-up and delivery service.

TRUCK CRANE. A portable crane mounted on an automotive vehicle equipped with a power plant for travel.

TRUCK LEVELER. A hydraulically operated ground level platform over which the rear wheels of a truck or trailer may be driven for the express purpose of raising or lowering the truck body in order to compensate for varying truck bed and delivery platform heights.

TUNNEL. A below grade passageway.

TURNING CENTER. The center of the circle described by the truck moving in a turn.

TURNING RADIUS (INSIDE). Under conditions described for turning radius (outside), the distance from center of the smallest circle to the nearest projection of the truck. Primarily used to establish operational clearance.

TURNING RADIUS (OUTSIDE). Half the diameter of the largest circle described by the outermost projection of the truck when driving slowly empty or loaded under its own power with the steering mechanism at the optimum steering angle, both forward and reverse, and to the left and the right.

TURNTABLE. (1) A device with a movable inner frame containing a straight section of track. The inner frame can be rotated, with a loaded carrier on it, to align the section of track with other tracks for the transfer of carriers from one track to another. (2) A device used to rotate cartons or other items so they can be fed from a side conveyor line to a main conveyor line or vice-versa.

TWIST CONVEYOR. An el conveyor in which the carrying surface and guard gradually exchange their functional duties.

TWO-WAY CONTAINER. A container whose configuration permits retrieving or discharging from opposite directions along the same horizontal axis.

U-LINE LAYOUT. A type of layout in which the machines and workstations are arranged in the shape of a U permitting an operator to closely monitor several machines or workstations.

UNDERCLEARANCE (PORTABLE METAL STACKING RACK). The dimension from the floor to the underside of the base frame.

UNDERDRIVE. A crane or carrier arrangement which propels the equipment by means of a driving wheel pressing on the underside of the track.

UNDER-RUNNING CRANE. A traveling crane with a moveable bridge moving on the bottom flanges of an overhead fixed runway structure and carrying a moveable or fixed hoisting mechanism.

UNDERVOLTAGE PROTECTION. A device operative on the reduction or failure of voltage which interrupts the main circuit and maintains this condition.

UNDERWRITERS' LABORATORIES LISTING (UL). A list of truck models approved as complying with the requirements of and used by the Stock group of insurance companies to determine the premium rate for insurance coverage for various areas of operation.

UNIDIRECTIONAL PATH. A material flow path over which material flow takes place in one direction only.

UNIT LOAD. Any load configuration handled as a single item.

UNIT OF ISSUE. The quantity of an item issued from storage, such as each, number, dozen, gallon, pair, pound, ream, set, yard, pallet load.

UNIVERSAL PRODUCT CODE (UPC). The voluntary standard bar code symbol for retail food packages in the United States.

UPRIGHT (CANTILEVER RACK). A column with the supports necessary to keep it in vertical alignment. The six basic types are as follows:

(Type I) single sides with extended base—column with the arms and the base extended in one direction only.

(Type II) double sided with extended base—column with arms and base extended from two sides of the column in the same vertical plane.

(Type III) imbedded—column with the bottom of the column imbedded into the floor.

(Type IV) pinned—column affixed by pins to the ceiling and floor.

(Type V) bottom pinned—column with the column affixed to the floor and the top of the column affixed to another column in another row by an aisle tie.

(Type VI) free standing—column with 3 or more extended base legs to make column rigid without being permanently affixed to the floor.

UPRIGHT FRAME. An assembly comprised of posts, horizontal braces, and/or cross braces and diagonal braces in one plane.

VALUATION CHARGES. Transportation charges assessed shippers whose shipments have declared values higher than the historically assumed value to 50 cents a pound or 50 dollars, whichever is greater.

VALUE-ADDED SERVICES (IN A WAREHOUSE). Special services (other than merely filling a customer's order), performed in a warehouse as required by the customer.

VARIABLE REACH LIFT TRUCK. A high lift truck with the additional capability of extending and retracting the forks (and load) in a longitudinal direction.

VAULT. A room designed to provide security of contents.

VERTICAL BAR CODE. A code pattern presented in such orientation that the overall coded area from start to stop is perpendicular to the horizon. The individual bars are in an array appearing as rungs of a ladder.

VERTICAL BELT CONVEYOR. A vertical conveyor in which sections of articulated slat conveyor apron form rigid platforms for vertical movement in continuous flow. The platforms are flexible in but one direction and they assume a vertical position on the noncarrying run to minimize space requirements.

VERTICAL CHAIN CONVEYOR, OPPOSED SHELF TYPE. Two or more vertical elevating-conveying units opposed to each other. Each unit consists of one or more endless chains whose adjacent facing runs operate in parallel paths. Thus each pair of opposing shelves or brackets receive objects (usually dish trays) and deliver them to any number of stations.

VERTICAL CLEARANCE. The dimension from the floor to the underside of the rail assembly; or from the topside of the rail assembly to the underside of the next rail assembly.

VERTICAL LIFT MODULE. A modular storage and retrieval device similar to a mini-load, which stores product vertically on special trays, retrieving them and presenting these trays of items to be picked at a pickstation.

VERTICAL RECIPROCATING CONVEYOR. A reciprocating power or gravity actuated unit which receives only inanimate objects on a track, roller, conveyor, or power conveyor forming the bed of the carrier and transmits these inanimate objects vertically from one elevation to another. This type of conveyor is never designed to carry an operator or passengers and no person may ever be permitted to ride upon it. The carrier must never be furnished with flooring of any type which might provide safe footing for any person. The conveyor shaft must be enclosed and where manual loading and unloading is employed, doors at each operating level must be installed and so interlocked with the carrier that doors at any level can be opened only when the carrier has stopped at that level and carrier cannot be moved until such open doors are closed. No controls for movement of carrier may be located on carrier, or within the enclosure, or within reach of any person standing on the carrier. For further reference see the current B 20 Safety Code for "Conveyors & Related Equipment."

VIBRATING CONVEYOR. A trough or tube flexibly supported and vibrated at a relatively high frequency and small amplitude to convey bulk material or objects. (See OSCILLATING CONVEYOR.)

VIBRATING CONVEYOR, BALANCED. A vibrating conveyor having a secondary mass operating in a direction opposite to the movement of the troughed pan for the purpose of minimizing the dynamic force normally directed into the supports.

VIBRATING CONVEYOR, NATURAL FREQUENCY. A vibrating conveyor in which the rate of free vibration of the trough on its resilient support is approximately the same as the rate of vibration induced by the driving mechanism.

WALKIE STACKER. A power operated stacker (controlled by an operator walking behind it) that allows a pallet to be lifted, stacked and transported.

WALKING BEAM CONVEYOR. A conveyor employing a multiple arrangement of walking beams with associated rails and/or rollers.

WALL CRANE. A traveling crane having a jib with a movable or fixed hoisting mechanism and operating on a runway attached to the side walls or columns of a building.

WALL TIE. A member connecting the rack to a wall.

WAND SCANNER. A hand held scanning device used as a contact bar code or OCR reader.

WAREHOUSE. A building where wares, and/or goods, are received, stored, and shipped.

WAREHOUSE LAYOUT. Layout of the AS/RS, storage racks, and other support services in a warehouse.

WATER MUFFLER. A special muffler utilizing a water trap to prevent the emission of sparks or flames from the exhaust system.

WAVE PICKING. An order picking method in which a designated group (wave) of orders are fragmented into multiple pickstations and picked simultaneously. Typically the fragments of orders are consolidated into single entities prior to shipping.

WAYBILL. A document prepared by the transportation line at the point of origin giving a description of the shipment and shipping instructions and sent with the shipment or mailed to the agent at the waybill destination or transfer point.

WHEEL BASE. Distance from center to center of outermost wheels.

WHEEL CONVEYOR. A series of wheels supported on axles in a frame over which objects are moved manually or by gravity. (See SKATE-WHEEL CONVEYOR.)

WHEEL, DRIVING. Wheel to which tractive effort is applied to cause equipment to move.

WHEEL LOAD. The load on any wheel with the trolley and lifted load (rated capacity) positioned on the bridge to give maximum loading.

WHEEL SPIRAL. Helically wound, curved sections of wheel conveyor that lower objects by gravity.

WICKET CONVEYOR. A conveyor comprised of two or more endless chains connected by cross bars to which vertical rods are attached at spaced intervals. The cross bars are also provided with spaced projections at level of same to form, in effect, a continuous carrying surface through which the loads cannot fall. Used for handling such products as painted sheets, wall, or composition board, etc., on edge through dryers or bake ovens.

ZONE PICKING. An order picking method in which an operator is responsible for picking only the items in an order that are within his/her zone.

BIBLIOGRAPHY

Askins, R.G. and C.R. Standridge (1993), *Modeling and Analysis of Manufacturing Systems*, Wiley, New York, NY.

Heragu, S.S. (1997), *Facilities Design*, PWS Publishing Company, Boston, MA

Advanced Material Handling, Material Handling Institute, Inc., Charlotte, NC

Sule, D.R. (1994), *Manufacturing Facilities Location, Planning, and Design,* 2nd ed., PWS Publishing Company, Boston, MA

Tompkins, J.A., J.A. White, Y.A. Bozer, E.H. Frazelle, J.M.A. Tanchoco, J. Trevino (1996), *Facilities Planning*, 2nd ed., New York, NY.

Z94.9 Human Factors (Ergonomics) Engineering

The Human Factors Subcommittee's task was made considerably easier due to the work of previous subcommittees. Earlier editions of this book served as a substantial base for this latest revision. The names of previous subcommittee members can be found in previous editions of this book. We appreciate their efforts.

The Human Factors Subcommittee has put forth considerable effort in revising and updating this chapter. Subcommittee members consisted of the following members:

Chairperson:
Alex Kirlik, Ph.D.
School of Industrial and Systems Engineering
Georgia Institute of Technology

Subcommittee:

Arthur D. Fisk, Ph.D.
School of Psychology
Georgia Institute of Technology

Richard Henneman, Ph.D.
NCR Human Interface Technology Center

HUMAN FACTORS AND ERGONOMICS SECTION COORDINATOR:

Steven A. Lavender, Ph.D.
Department of Orthopedic Surgery
Rush-Presbyterian-St. Luke's Medical Center, Chicago

Abduction. Movement of a limb away from midline axis of body.

Absolute Pitch. Refers to the ability of a person to identify the pitch of a pure tone without any external reference.

Absolute Threshold. That stimulus value which marks the transition between no response and response on the part of an observer; occasionally used analogously to describe a qualitative change in the nature of the response, as the "threshold of feeling" between sound and pain at intense levels of acoustic stimulation; or the "pricking pain threshold" between perceived warmth and burn. Values obtained for the threshold depend on the statistical properties of the psychophysical method used.

Accessibility Score. A rating of the passage space available for a person to reach his/her work station based on the width of the passageway. The rating is multiplied by the number of people affected and averaged across people.

Accident. Any event caused by human situational and environmental factors or combinations of those factors which interrupt the work process and which may or may not result in injury, death, property damage, or other undesired events but which has the potential to do so.

Accommodation. The adjustment of a sense organ to receive an impression distinctly. In vision, the ability of the eye to focus on objects at varying distances effected by the gradual thickening and increasing curvature of the crystalline lens as objects become closer.

Accuracy. A human performance measure indicating the correctness of behavior as measured against an objective or normative standard.

Acoustic Scattering. The irregular reflection, refraction, or diffraction of a sound in many directions.

Acoustic (or auditory) Stimulus. The waveform of air pressure at the eardrum.

Acoustic Trauma. Injury to the ear caused by an intense auditory stimulus resulting in a certain degree of temporary or permanent hearing loss.

Activity Analysis. A structural listing of the operations, behaviors, or tasks that make up a job or work. Data gathered typically includes the sequence, frequency, and importance of activities and the time devoted to each, as well as their interrelationships. Various procedures and formats are used, according to the purpose of the analysis. (See ACTIVITY SAMPLING.)

Activity Sampling. Performing an activity analysis by observing only portions or segments of the ongoing work process, with observation periods being scheduled in advance according to a sampling rationale. (See ACTIVITY ANALYSIS.)

Acuity. A measure of the ability to resolve or discriminate sensory stimulation.

Adaptation, General. An adjustment internal to a behaving system to maintain stable or more successful relationships with an external environment.

Adaptation, Sensory. The maintenance of sensory effectiveness under changing stimulation (positive adaptation) or the reduction of sensory responsiveness with continued stimulation (negative adaptation); for example, maintenance of visual acuity under reduced illumination, or reduced responsiveness to an unpleasant odor.

Adaptive Control. Form of system control in which the control law changes as a function of time-varying dynamics of the controlled system.

Adduction. Movement of a limb toward midline axis of body.

Advisory Light. A signal indicating safe or normal performance, operation of essential equipment, or to alert an operator for routine action purposes.

Advisory System. Form of human operator aiding in which automation, such as an expert system, provides recommended courses of action to the human operator.

Affordance. A description of an object in terms of the actions it makes available.

Afterimage. A visual image or other sense impression that persists after the stimulus is no longer operative. Visual after-images are complex perceptions that can vary in hue, brightness, saturation, shape, pattern, texture, focus, latency, duration and developmental sequence.

Alarm. Device indicating (typically undesired) deviation of the state of a controlled system from a nominal or expected value.

ALPHA. Probability of Type I error; i.e., the probability of incorrectly rejecting a null hypothesis.

ALPHA TESTING. First phase of software evaluation involving limited product release.

ALPHANUMERIC DISPLAY. Display consisting solely of numbers and letters.

ANALOG DISPLAY. Display format in which a quantity is represented by a physical dimension of a display; e.g., a clock.

ANECHOIC ROOM. A room in which boundaries absorb almost all incident sound resulting in virtual free field conditions.

ANIMATED-PANEL DISPLAY. Diagrammatic representation of components of a system hardware and/or simple semi-functioning models of such components.

ANOXIA. Lack of oxygen in the blood stream or tissue cell. Although the result of many mechanisms, the ultimate effect frequently is defective function of many sensory and motor functions. Produced environmentally by high altitude, chemically by pollutants and drugs, or locally by overexertion. May result in death.

ANTHROPOMETRY. Classically, the study of people in terms of physical dimensions. (1) The measurement of static and dynamic features or characteristics of the human body. Typically included are linear and arc dimensions, mass and volume. (2) The techniques used to quantitatively express the form and dimensions of the body. (See Z94.2 ANTHROPOMETRY & BIOMECHANICS.)

APPARENT MOTION. A sensory or perceptual illusion in which stationary stimuli are perceived to be moving.

APPLICATION. Computer software dedicated to a particular task.

ARTICULATION INDEX (AI). A numerically calculated measure of the intelligibility of speech in the presence of background noise. It is generally based on the speech/noise ratio for each of twenty frequency bands. This ratio is weighted, multiplied by .05, and summed over the twenty bands to yield the AI which ranges from 0.0 (voice communication nearly impossible) to 1.0 (excellent communication).

ARTIFICIAL HORIZON. A pictorial display used in aircraft to provide information as to the orientation of the plane with respect to the horizontal, indicating to the pilot whether he/she is flying straight and level.

ARTIFICIAL INTELLIGENCE. The study of the use of computers to perform tasks that are consensually agreed to display aspects of intelligence, rationality, or adaptation.

ARTIFICIAL REALITY. (See VIRTUAL ENVIRONMENT.)

ATTENTION. Allocation of sensory, perceptual and cognitive functionality to stimuli or information processing tasks.

ATTENTION, DIVIDED. Allocation of attention among two or more concurrent tasks.

ATTENTION, SELECTIVE. Ability to focus on a desired set of information or a task, to the exclusion of the remaining set of information or competing tasks.

AUDIOGRAVIC ILLUSION. Errors in auditory localization which accompany an error in the perception of body position.

AUDIOGYRAL ILLUSION. Errors in auditory localization following rapid rotation of the body in the absence of visual cues.

AUDIOMETER. Instrument used to assess human auditory sensitivity as a function of frequency. Tones of varying frequencies and intensities are presented through earphones to a subject who indicates the point at which a given tone just becomes, or ceases to be, audible.

AUDITORY AFTER-EFFECT. A phenomenon lasting several seconds in which familiar sounds are modulated following listening to rapid, high-intensity pulses for about one minute.

AUDITORY FATIGUE. A temporary increase in auditory threshold due to a previous auditory stimulus.

AUDITORY FLUTTER. A wavering auditory sensation resulting from the periodic interruption of a continuous sound at a sufficiently slow rate.

AUDITORY FUSION. A phenomenon in which a series of short duration primary sounds with successive arrival times at the ear(s) produce the sensation of a single secondary sound.

AUDITORY LATERALIZATION. The determination by a subject that the apparent direction of a sound is either left or right of the frontal-medial plane of the head.

AUDITORY LOCALIZATION. The determination by a subject of the apparent direction and/or distance of a sound.

AUTOKINETIC EFFECT. An illusion of the apparent movement of a stationary point light source presented in a dark stimulus field in the absence of a visual frame of reference.

AUTOMATION. The use of electromechanical or information technology to perform autonomous functions.

AUTOMATICITY. A psychological construct denoting a combination of information processing characteristics, including processing that is fast, reliable, unlimited by short term memory capacity, robust to increase in task difficulty, and difficult to modicy once learned.

AVAILABILITY (2). A cognitive heuristic in which the likelihood of an event is judged by the ease with which examples of the event can be generated or recalled.

AVAILABILITY, MEASURE OF. The ratio of the total time a system is capable of performing its function to the total time that there is a requirement for its operation.

"AVERAGE MAN" CONCEPT. An oversimplified method of describing the characteristics of a varied population based on mean measurements. When more than one measurement is used the "average man" disappears.

BACKLASH. Tendency for control system response to be reversed when a control movement is stopped.

BACKWARD CHAINING. An automated or human reasoning strategy in which a sequence of actions are sought to transform a problem from a goal state to an initial or given state.

BANDPASS. Transmission of a limited signal frequency range.

BANDWIDTH. Range of frequencies to which a system is sensitive.

BANG-BANG CONTROL. Discrete system control rule in which only maximum positive and negative control inputs are used.

BAR GRAPH. Display format in which a quantity is represented by the length of a bar.

BASE RATE. A population mean frequency; specifically, the prior odds in a Bayesian evidential reasoning task.

BASE RATE INSENSITIVITY. Human tendency to place less than objectively correct weights on base rate information in evidential reasoning. Highly influenced by contextual factors.

BEATS. Periodic variations in sound resulting from hearing two equally intense simple tones of slightly different frequencies. As the difference between the frequencies is increased, a train of subjective experiences is produced, beginning with waxing and waning and progressing to throbbing (beats), then changing to rattling or roughness. That stage lasts until the two frequencies are separated enough that the subjective interaction disappears and two distinct tones are heard.

BEGINNING SPURT. A transient burst of activity at the beginning of the work period.

BETA. Probability of Type II error; i.e., the probability of incorrectly accepting the null hypothesis.

BETA TESTING. Second phase of software evaluation involving limited product release.

BINOCULAR PARALLAX. The lack of correspondence between the images which fall on each retina due to the interpupillary distance between the two eyes; a cue to depth. (See PARALLAX.)

BIOMECHANICS. Application of mechanical principles and methods of analysis to the structure and movements of the body and body parts. Frequently-studied aspects include the range, strength, endurance, and speed of movements and responses to such physical forces as acceleration and vibration. (See Z94.2 ANTHROPOMETRY & BIOMECHANICS.)

BIONICS. A study of living systems or organisms with the intention of deriving technological knowledge applicable to the design and performance of artificial systems.

BIT. An information processing unit. (See Z94.3 COMPUTER AND INFORMATION SYSTEMS.)

BLIND-POSITIONING REACTIONS. Movements that require the localization of objects in space without the aid of visual cues.

BRIGHTNESS. The amount of light emitted by, or reflected from, a surface in a specified direction. Measured as light intensity per unit projected area. Also referred to as luminance.

BRIGHTNESS CONTRAST. The relationship between the brightness of an object, Bo, and the brightness of the object's background, Bb. It is expressed as the ratio of the difference to the background.

$$\frac{Bo - Bb}{Bb} \times 100$$

Brightness Control. Potentiometer for adjusting display luminance.

BTPS Conditions. The temperature and pressure of gas in the lungs at body temperature (37°C) and saturated with water (47 torr or 6.266134kPa) at ambient pressure. (See STPD conditions.)

Candela (CD). The luminous intensity, in the direction of the normal, of a blackbody surface 1/100,000 m² in area, at the temperature of solidification of platinum (2042 K) under a pressure of 101,325 N/m². (See Candle.)

Candle. (new unit). 1/60 of the intensity of 1 cm² (100 mm²) of a blackbody radiator at the temperature of solidification of platinum (2042 K). To understand this, a brief history may be in order: the first light standard was the *standard candle*, that was a candle of spermacel of which six weighed one pound with the candle burning at a rate of 6 grains per hour. Since this spermacel candle was not of sufficiently constant illuminating power to be of much scientific value, other standards were used.

In the USA, the *harcourt pentane lamp*, in which air was drawn over the pentane liquid, with the mixture burned in a standard burner and the flame adjusted to a definite height, with corrections made when atmospheric conditions varied from normal, became the standard. The illuminating power of such a lamp was equal to about 10 standard candles and was accepted as an international standard (defined as illuminating power equal to 1/10 of that of the Harcourt Lamp).

In Germany, the *hefner lamp* (constructed after a standard pattern and burning amyl acetate) was the standard and equal to 0.9 International Candles.

In France, the *carcel lamp* in which coiza oil was burned was the standard, and the Carcel Standard was 9.62 International Candles. For practical purposes, to measure the "quantity of light," the intensity of illumination produced by a standard candle at a distance of 1 m was adopted and called a candlemeter. Analogously, there was the foot candle, with 1 candlemeter = 0.093 foot candles.

Candlepower. Light intensity of a light source. Also "candle" and a "lumer/steradian." A source of one candle intensity at the center of a sphere emits a total radian flux of 12.57 ($=4\pi$) lumens. The total surface of the sphere is 12.57 steradians; thus one candle provides an intensity of 1 lumen/steradian.

Ceiling Effect. Characteristic of behavioral data due to performance reaching its upper limit.

Central Hearing Loss. Hearing impairment following damage in the auditory pathways or the auditory areas of the brain.

Channel Capacity. By analogy to engineering usage, the concept of a theoretical upper limit on the amount of information that human beings can receive and process, usually within a given time. The term is most frequently applied to the sensory modalities and/or the central nervous system.

Character Height. Vertical distance, measured by physical length or number of visual picture elements (pixels), of a displayed alphanumeric symbol.

Character Width. Horizontal distance, measured by physical length or number of visual picture elements (pixels), of a displayed alphanumeric symbol.

Chemoreceptor. An end organ of the nervous system which triggers neural signaling activity in response to chemical stimuli, *e.g.*, taste buds, olfactory end organs.

Chernoff Display. Display of multidimensional data using a facial representation.

Choice Reaction Time. Amount of time required to make different responses to different stimuli. The time increment above what is needed for a response to a single stimulus is sometimes taken as a measure of the time to process the cognitive choice. (See reaction time.)

Clo Unit. A measure of the thermal insulation provided by clothing.
$$1 \text{ clo} = \frac{.155 \text{m}^2 \cdot \text{deg.C}}{\text{Watt}}$$

Click. Depress and release a button on a mouse or other computer input device.

Closed Loop. System in which some function of system output is used as system input.

Cognition. Higher mental activities typically involving the use of stored information or knowledge.

Cognitive Engineering. The analysis and design of systems performing cognitive functions. Also, the application to cognitive psychology to the design of such systems.

Command. Input to a computer or control system.

Command Line Interface. Computer interface in which commands are entered as strings of alphanumeric characters.

COMMAND LANGUAGE. The elements and syntax of a set of computer commands.

COMPATIBILITY. How consistently the spatial movement or conceptual relationships of stimuli and responses meet human expectations. With reference to controls and displays, the naturalness of the movement of a control compared to what is displayed to the operator.

COMPENSATORY DISPLAY. In tracking tasks, a display on which either the input signal (target) or the output signal (cursor) is fixed and the other moves; the operator's task in manipulating the controls is to eliminate or minimize system error by superimposing the cursor onto the target.

COMPUTER GRAPHICS. Manipulation and display of pictorial or graphical data on a computer display.

COMPUTER-AIDED INSTRUCTION. Use of computer technology to support learning and education.

COMPUTER SUPPORTED COOPERATIVE WORK (CSCW). Use of computer technology to support communication and collaboration in a group work environment.

CONCEPT TRAINERS. Aids used when the concepts to be learned are too complex to be absorbed in a few trials from verbal descriptions, and when the principles to be used in task performance can best be simulated by physical objects and real actions such as miniature representation of control components, and check points of a complex electronic system.

CONDUCTIVE HEARING LOSS. Hearing impairment due to interference with the transmission of sound waves through the external or middle ear.

CONES. Visual receptor cells found most densely toward the center of the retina. They are responsible for sharp vision and color discrimination, but cease functioning at low light intensities.

CONSISTENT MAPPING. An invariant relationship between stimuli and required response.

CONSONANCE. The phenomenon in which tones presented simultaneously produce a blended or pleasant sensation.

CONTINGENCY ANALYSIS. The part of a task analysis performed to identify non-routine situations with which a system may have to deal, in order to determine any special human performance required by such events.

CONTRAST. The brightness relationships of two non-specular surfaces adjacent to each other when the illumination of both the objects and the immediate surroundings are the same.

CONTROL. (1) A device, usually mechanical, electrical, or electronic, which directs the action of some mechanism or produces some change in the operation of a process or system. (2) The activity of bringing a system to a desired state, or maintaining a desired system state in the presence of external disturbances.

CONTROL LAYOUT. The spatial organization of controls at an interface or workplace.

CONTROL SYSTEM. A system performing a control function.

CONTROLLED SYSTEM. A system under the control of humans or automation.

CONTROL-DISPLAY RATIO. The ratio of movement of a control device to the movement of a display indicator.

CONTROL-FORCE CURVE. A function describing the relationship between the amount of force applied by the operator and the resulting displacement of a control.

CONVERGENCE. Coordinated directing of both eyes at the same point so as to obtain single vision by bringing the image of the object to the fovea of each eye.

COPPER-HARPER-SCALE. Ordinal scale used to indicate task difficulty or workload level.

CORIOLIS EFFECT. The difference in effective G force at different distances from the rotation axes in a short-axis rotating environment. When limbs or head move across a G force gradient, misperceptions of body orientation (tilting, rotation) or of limb position may be experienced, with concomitant physiological effects such as nausea or vertigo.

CREW STATION. Work Station within a vehicle cockpit.

CRITICAL FLICKER FREQUENCY. The minimum number of alternations per second at which a regularly intermittent light ceases to appear as flickering and fuses to yield a perception of steady light. Also called critical fusion frequency or flicker-fusion frequency.

CRITICAL INCIDENT TECHNIQUE. An approach to job analysis which calls for factual description of specific events or behavior. The technique is also applicable to other situations, such as accident investigation in which reports of errors or near-accidents may be collected.

Cue. A stimulus probabilistically related to a judgment or action.

Cursor. A moving element on a display. On an instrument, it is used to show position discrepancy or error in reference to another "target" element. On a CRT display, it is used to show row and column positions or to indicate where the next activity will occur.

Cut and Paste. A text editing task involving the selection and removal of a text block, followed by insertion of the block at a new point in the text.

Cybernetics. Study of communication and control functions common to both living and engineered systems with the emphasis on attempting to understand organisms through making analogies to machines. ("I made up the word... from the Greek...'Steersman'...from the same root as 'governor' and refers to control and communication in the animal and the machine." - Norbert Wiener.)

Dark Adaptation. A process of neuro-chemical changes whereby the visual system becomes much more sensitive to light.

Data Entry. Task in which information is entered into a computer.

Data Glove. Input device consisting of a glove with sensors detecting finger movements.

Dazzle. Extreme brightness of direct or reflected light that causes difficulty in seeing.

Dead Band. Loci of control positions, usually around the center position of a knob or lever, for which there is no control output.

Deadman Control. Mechanism designed to remove or reduce hazardous energy to a safe level in the absence of the kind of force exerted by a conscious operator.

Debug. Removal of software errors.

Decibel. A logarithmic scale expressing quantities of sound, or electrical power, signal attenuation and gain. Measurement is relative to a specified reference level.

Decision Making. Selection of action from a set of alternatives.

Decision Time. Delay attributable to cognitive processing of input data.

Delphi Method. A procedure for obtaining the consensus of experts in areas such as long-range forecasts by successive iterations of questioning interspersed with feedback to each expert on the others' opinions and supporting reasons.

Depth Perception. Visual discrimination along the dimension of distance from an observer, using monocular cues (*e.g.*, interposition, accommodation, size) and/or binocular cues (convergence, stereopsis).

Describing Function. Mathematical model of human performance, especially in tracking tasks such as aircraft piloting.

Detectability. Quality of a signal, display, or stimulus that affects the probability of its presence being perceived.

Dialog Box. A display window for entering computer commands.

Dichotic. Pertaining to differential stimulation of the two ears.

Diotic. Pertaining to the stimulation of the two ears by identical stimuli.

Difference Limen. Difference threshold–the amount by which two supraliminal stimuli must differ if this difference is to be perceived in an exactly specified portion (usually 50 or 70%) of trials. (See LIMEN.)

Direct Manipulation. Computer interface in which commands are entered by actions upon displayed objects.

Display. The presentation of output data from a device or system in a form designed for human perception through vision, audition, or other sensory modality.

Display Layout. The spatial organization of displays at an interface or workplace.

Distributed Control. System arrangement in which control activity is distributed among multiple controllers.

Distributed Decision Making. System arrangement in which decision making activity is distributed among multiple decision makers.

Double Click. Two clicks in rapid succession, as a method of computer input.

Dynamic Display. Display containing information that is updated autonomously.

Dynamic System. A system whose state changes autonomously as well as through controller activity.

Dynamic Visual Acuity. The measure of visual acuity used when a target moves with respect to an observer. (See VISUAL ACUITY.)

Dynamometer. A device for measuring the strength of a muscular contraction; typically a hand-grip device with a scale showing squeeze force.

Earcon. Auditory display symbol, in analogy to visual icon.

Echo. Computer interface in which commands are reflected on the display.

Ecological. Pertaining to the environment for a particular organism.

Ecological Psychology. Study of the interaction of organisms and environments.

Ecological Validity. Originally a measure of cue validity in a judgement task, now the degree of correspondence between laboratory conditions and the target environment to which laboratory results are intended to generalize.

Edit. Alter data in a computer file, as in wordprocessing.

Emergent Feature. A perceptible feature on a graphical display whose existence owes to a relational property between primitive display elements.

Empty Field Myopia. The normal accommodation of the eye for near, rather than far, vision when viewing a homogeneous field such as an empty sky.

Energy Expenditure. The amount of energy used during activity or rest. Usually expressed in calories per unit time.

Engineering Psychology. One of several terms used to define approximately the same discipline. Other terms are human factors, human factors engineering, and ergonomics. Ergonomics is used predominantly outside the U.S.A.; the others predominantly within. The aim of the discipline is the evaluation and design of facilities, environments, jobs, training methods and equipment to match the capabilities of users and workers. Exact definition of these terms or the scope of work is improbable. (See ERGONOMICS.)

Ergometer. A device for measuring the energy output of human muscular work; for example, a stationary bicycle.

Ergonomics. The application of a body of knowledge (life sciences, physical science, engineering, psychology, etc.) dealing with the interactions between man and the total working environment, such as atmosphere, heat, light and sound, as well as all tools and equipment of the workplace.

Error, Constant. Deviation from true value or established standard due to factors remaining unchanged during the series of measurements.

Error, Variable. Deviation from true value or established standard due to random factors that affect each measurement separately.

Etiology. Study of the causes of disease; in Human Factors, study of the causes of human error over its time course.

Expert. An individual who has acquired special skill in, or knowledge of, a particular subject through professional training or practical experience.

Expert System. A computer program typically containing knowledge derived from human expertise in a particular area of knowledge.

Face Validity. Apparent relevance or appropriateness of a measure.

Fail Safe. Characteristic of system that avoids injury to humans and/or catastrophic damage when an element malfunctions.

Fatigue. Decreased ability to perform and/or feelings of tiredness following effort.

Fault. Abnormal condition in a controlled system, possibly leading to system failure.

Fault Localization. Systematic procedure for identifying malfunctioning element of a system.

Fault Tree Analysis. A logical approach to identify the probabilities and frequencies of events in a system that are most critical to uninterrupted and safe operation. This analysis may include failure mode and effect analysis (determining result of component failure interactions towards system safety) and Techniques for Human Error Prediction (THERP).

Feedback. Data on the performance of a system, equipment, group, or individual for use in modifying performance.

Fidelity. (1) The degree to which a system's output

duplicates the input. (2) The faithfulness with which a simulation represents features of a system.

FIELD OF VIEW. The solid angle within which an optical sensor (such as the human eye) provides useful data.

FIELD STUDY. Form of scientific investigation in which activities in actual work environments are observed.

FLASH BLINDNESS. Temporary loss of vision for low-luminance objects following brief exposures to intense light.

FLICKER. Rapid yet perceptible changes in display luminance.

FLICKER FUSION. The tendency of an oscillating or flickering sensory input signal to be observed as continuous. At a certain threshold frequency (typically above 6 Hz for vision), the perceiving apparatus, (e.g., eye, ear, tactile or other sensory areas) is no longer able to separate the individual pulses and instead perceives one continuous signal. This effect makes motion pictures seem to run smoothly. Flicker fusion caused, for instance, by low frequency fans or objects on fast conveyor belts passing before a light source, may lead to errors such as seeing a moving saw blade apparently at rest, and leading to accidents (See STROBOSCOPE.)

FLIGHT DECK. Workstation in an aircraft cockpit.

FLIGHT SIMULATOR. A ground-based device containing a representation of an aircraft cockpit with a computer driving functional displays and controls. It may be used for training, for research, or for engineering development.

FLIGHT TRAINER. A ground-based device containing a representation of an aircraft cockpit, used for pilot training. It need not have the fidelity of a flight simulator.

FLYBAR (FLYING BY AUDITORY REFERENCE). A system providing turn, bank, and air speed information via auditory signal, rather than with conventional flight instruments.

FOOT CANDLE. An obsolete measure of the intensity of illumination of a surface, or, the flux density of light incident on a surface. A flux of 1 lumen per square foot of surface is defined as 1 foot candle of illumination. It is no measure of the light reflected from a surface and, therefore, no "index of visibility" of an object to the worker. (See BRIGHTNESS CONTRAST, VISUAL ACUITY, REFLECTANCE, FOOT-LAMBERT.)

FOOT-LAMBERT. An obsolete measure of the illumination flux density reflected from a surface. Used in measurement of the effectiveness of workplace lighting, the foot-lambert is the only accurate index of light actually perceived by the eye: equals $1/\pi$ candles per square foot. (See REFLECTANCE.)

FORCE JOYSTICK. Control in which input corresponds to applied force.

FORCE-PACED. Experimental or actual task in which a performer's activity is paced by external events.

FORWARD CHAINING. An automated or human reasoning strategy in which a sequence of actions are sought to transform a problem from an initial or given state to a goal state.

FUNCTION ALLOCATION. Stage of system design in which specific functions are assigned to individuals or automated systems.

GAIN. (1) An increase in signal power in transmission from one point to another. Gain is usually expressed in decibels and is widely used to denote transducer gain. (2) An increase or amplification. (3) More generally, the sensitivity or degree of control provided by any manual adjustment.

GANGED CONTROLS. Controls which are grouped or "stacked" on a single axis so as to reduce crowding on panels of limited dimensions. For example, two control knobs attached to concentric shafts.

GLARE. A sensation produced by a luminance in the visual field that is sufficiently greater than the luminance to which the eyes are adapted. Glare may cause annoyance, discomfort, disability, distraction or a reduction in visibility.

GLASS COCKPIT. Aircraft flightdeck characterized by the use of multi-function computer displays rather than single-sensor, single-indicator (SSSI) instruments.

GOMS. A method of task analysis describing activity in terms of goals, operators, methods, and selection rules. Often used in the description of human-computer interaction.

GO, NO-GO DISPLAY. A visual display which provides only two alternate choices of information (e.g., on-off, start-stop, etc.).

GRAPHICAL DISPLAY. Computer display containing graphical or pictorial elements.

GRAPHICAL USER INTERFACE (GUI). Interface based on graphical displays supporting direct manipulation.

GRAYOUT. A temporary condition in which vision is hazy, restricted, or otherwise impaired, owing to insufficient oxygen.

GROUND. The unfocused surroundings and interstices of a figure or object, perceived as lying beyond and not belonging to the figure or object, *e.g.*, the background in a painting. Figure and ground are sometimes reversible, as when an interwoven black-white pattern may appear either as a white figure on a black background, or vice versa.

G-TOLERANCE. A tolerance in a person or animal, or in a piece of equipment, to an acceleration of a particular value and direction.

HAWTHORNE EFFECT. An experimental confound resulting from the act of observation changing the essential characteristics of the phenomenon observed.

HEAD SCANNING. Scanning of the visual field through head movement combined with eye movement. Head scanning is normally a slow process which, when performed too frequently or rapidly, may induce giddiness and lead to errors in visual judgment. Good workplace design minimizes the need for head scanning.

HEAD UP DISPLAY (HUD). Mode of display in which information is presented on a transparent surface (e.g., a windshield) along the human's natural line of sight.

HEART RATE. A physiological measure that is used as an estimate of job stress, work load, or environmental stress. Usually expressed in beats per minute.

HEAT STRESS. Additional physiological load induced by working in a hot environment. The effects of heat stress include increased heart rate and increased body temperature, and may also include increased sweat rate and feelings of fatigue. Severe heat stress can result in heat exhaustion.

HEAT STRESS INDEX. Any of several methods that attempt to combine some or all of the interactive effects of temperature, humidity, air velocity, work load, and clothing into a single measure predictive of physiological effect.

HELMET MOUNTED DISPLAY (HMD). Mode of display in which information is presented on a helmet visor along the human's natural line of sight.

HELP SYSTEM. On-line user assistance available within many application programs.

HEURISTIC. A useful but not completely reliable rule or strategy, in either human or automated reasoning.

HIERARCHICAL MENU SYSTEM. An interface consisting of menus interconnected in a tree-like structure.

HIGHLIGHT. Method of display in which a rare feature is used to capture a user's attention.

HUE. The attribute of color determined primarily by the wavelength of light entering the eye. Spectral hues range from red through orange, yellow, green, and blue to violet.

HUMAN ENGINEERING. One of several terms used to define approximately the same discipline. Other terms are human factors, human factors engineering, and ergonomics. Ergonomics is used predominantly outside the U.S.A.; the others predominantly within. The aim of the discipline is the evaluation and design of facilities, environments, jobs, training methods and equipment to match the capabilities of users and workers. Exact definition of these terms or the scope of work is improbable. (See ERGONOMICS.)

HUMAN ERROR. Human behavior with undesirable consequences.

HUMAN FACTORS. One of several terms used to define approximately the same discipline. Other terms are human engineering, human factors engineering, and ergonomics. Ergonomics is used predominantly outside the U.S.A.; the others predominantly within. The aim of the discipline is the evaluation and design of facilities, environments, jobs, training methods and equipment to match the capabilities of users and workers. Exact definition of these terms or the scope of work is improbable. (See ERGONOMICS.)

HUMAN OPERATOR. A person who participates in some aspect of operation or support of a system and its associated equipment and facilities. (Generally refers to one who operates equipment as opposed to one who maintains the equipment.)

HUMAN TRANSFER FUNCTION. The mathematical description of a human operator's outputs (*e.g.* control movement) in a tracking task as a function of inputs (display indications), described in terms of a linear differential equation.

HUMAN-COMPUTER INTERACTION (HCI). The communication between a computer and a human user, or the scientific study of this phenomenon.

HUMAN-MACHINE SYSTEM. A functioning system comprised of a machine and at least one individual.

HYPERTEXT. Format of database organization in which data elements are linked together in a tangled hierarchy, allowing the user to easily navigate between related elements.

HYPOXIA. An insufficiency of the oxygen supply to living tissues relative to their existing requirement, such as may occur with reduced atmospheric pressure at high altitudes. The central nervous system and heart are most vulnerable: marked cardiovascular and behavioral symptoms may develop over a period of minutes, and include disorientation, dizziness, dimness of vision, and breathlessness, progressing to unconsciousness and collapse ("blackout").

ICON. A graphical display symbol used on a system interface.

IMPULSE. (1) The product of a force and the time during which the force is applied. (2) Psychology; human response based more on emotion than cognition.

IMPULSE NOISE. Noise generated in discrete energy bursts, which has a characteristic wave shape of its own.

INCREMENTAL TRANSFER EFFECTIVENESS. A measure of effectiveness that compares successive increments of time spent in one training task with successive increments saved in subsequent training.

INDICATOR. An instrument or device for displaying information; *e.g.* over-limit, location, or speed. Indicators may be mechanical (symbolic or pictorial), electrical (warning lights), or electronic (cathode-ray tube or screen).

INFORMATION. In information theory, a purely quantitative property of an ensemble of items that enables categorization or classification of some or all of them. The amount of information in an ensemble (symbolized by H) is measured by the average number of operations (statements, decisions, tests, etc.) needed to effect categorization of the items.

INFORMATION THEORY. An interdisciplinary study dealing with the transmission of messages or signals, or the communication of information. It draws upon communications theory (which includes much from physics and engineering), linguistics, psychology, and sociology.

INGRESS. Access for entering an area, such as an operating or passenger station within a vehicle or work area.

INJURY. Generally any physical harm or damage to a person. A disabling injury is one which prevents a person from performing his regular job for a full day beyond the day of the accident which produced the injury. In OSHA usage, an injury may or may not involve loss of time from work.

INPUT. (1) The path through which information is applied to any device. (2) The means for supplying information to a machine. (3) Information or energy entering into a system. (4) The quantity to be measured, or otherwise operated upon, which is received by an instrument. Also called input signal.

INPUT DEVICE. Equipment, such as a mouse or keyboard, allowing an individual to enter commands or data into a computer system.

INTEGRATED CONTROLLER. A control device which combines more than one aspect of an operation (*e.g.*, control of steering, acceleration and braking in a single joystick).

INTEGRATED DISPLAY. A visual display which combines several related parameters into a single function format.

INTERACTIVE. A feature of a task indicating that human actions alter external task conditions.

INTERFACE. (1) A common boundary between two parts of a system, whether material or non-material. (2) The physical boundary between an operator and the equipment used, *e.g.*, control, display, seat, etc.

INTERLOCK. A feature of a device which precludes its operation in the event that specified conditions are not met.

INTERNATIONAL SYSTEM OF UNITS. (See SI UNITS.)

JOB ANALYSIS. A study based upon measurement, observation, and interviews to determine and identify the duties, tasks and functions involved, together with the skills, knowledge, and responsibilities required. (See Z94.6 EMPLOYEE AND INDUSTRIAL RELATIONS.)

JOB ENLARGEMENT. Extending the parameters of the job being performed by an employee in order to make the work more psychologically rewarding. A job may be extended in two dimensions-horizontal and vertical. Horizontal job enlargement can take the form of a greater variety of tasks on which the employee works, an increased number of tasks, and job rotation. Vertical job enlargement, or job enrichment, usually takes the form of more planning, more controlling, or more team participation on the part of the job incumbent. The additional controlling aspects may include self-pacing as well as inspection. This form of enlargement involves building motivators into the job. (See Z94.15 ORGANIZATION PLANNING & THEORY.)

JOYSTICK. A control or input device consisting of a movable lever with at least one degree of freedom.

JUST NOTICEABLE DIFFERENCE. The least amount of a stimulus which, added to or subtracted from a standard stimulus, produces a detectably different experience. Also called just perceptible difference, least noticeable difference, minimal change.

KEYCLICK. An audible signal indicating that a keypress has occurred.

KEYBOARD. An input device consisting of spatially organized alphanumerical keys.

KEYPAD. An input device consisting of numerical keys and, typically, simple arithmetic function keys.

KINESTHESIS. The sense that yields knowledge of the movements of the body or of its members; awareness of movement due to the mechanical stimulation of special receptors in the muscular tissue, tendons, and joints.

KNOWLEDGE OF RESULTS. Feedback of information about performance.

LANDOLT RING. A ring with a small gap at one point, used to test visual acuity by having observer report orientation of the gap.

LATERAL. Of or pertaining to the side or lateral axis; directed, moving or located along, or parallel to, the lateral axis.

LEARNING. A relatively permanent change in personality (including cognitive, affective, attitudinal, motivational, behavioral, experiential, and the like), reflecting a change in performance usually brought about by practice although it may arise from insight or other factors, including memory.

LEARNING CURVE. A function indicating learning rate. When graphed, typically time or practice trial is plotted on the abscissa and a performance measure is plotted on the ordinate.

LEARNING, TRIAL-AND-ERROR. The process in which a person or animal, having no already established response to the requirements of the task, responds at first to only the task's general features with a wide variety of acts, then gradually eliminates the responses that prove unsatisfactory and repeats with increasing frequency those that prove satisfactory.

LEARNING, WHOLE VS. PART. In "whole learning," all of the material to be learned is worked through in successive repetitions from first to last; in "part learning" (or "piecemeal learning"), the material is broken into smaller segments to be learned separately and then combined into the whole.

LIFE SUPPORT. An area of human factors which focuses on health promotion, biomedical aspects of safety, protection, sustenance, escape, survival, and recovery of personnel.

LIGHT ADAPTATION. A process of neuro-chemical changes whereby the visual system becomes less sensitive to light.

LIGHT FLUX. Rate at which a source emits light energy (evaluated in terms of visual effect) and expressed in lumens (lm) (q.v.).

LIGHT PEN. An input device shaped like a pen which emits energy to a particular location on a light sensitive display.

LIKERT SCALE. An ordinal rating scale, typically consisting of 5 or 7 items, labelled from strongly agree to strongly disagree.

LIMEN. Threshold; a psychophysical concept denoting the lowest detectable intensity of any sensory stimulus.

LINK ANALYSIS. An analysis of the visual, auditory, and tactual links between human and machine or between one human and another involved in an operation. Primary objectives are determination of the importance of links, frequency of their use, and their adequacy.

LOAD. The demand made by inputs on performance.

LOAD, SENSORY. The number and variety of stimuli to which responses must be made. For example, if the operator must discriminate among several different classes of visual stimuli, the load on the visual system is greater than if discrimination is of only one type or of limited types.

LOCALIZATION, AUDITORY. The capability of an observer to identify the position of a sound source with reference to himself.

LOUDNESS. The attribute of auditory sensation by which sounds may be ordered on a scale extending from soft to loud. The unit of loudness is the sone. Often a weighting network is used to determine loudness (*i.e.* a weighted scale for noise measurement).

LOUDNESS CONTOUR. A graph of sound pressure level versus frequency, showing the sensitivity of the ear.

LOUDNESS LEVEL. The (judged) loudness level of a sound, in phons, is numerically equal to the median sound pressure level, in decibels relative to 20 micronewtons per square meter, of a 1000 hertz reference tone presented to subjects facing the sound source and judged by the subjects in a number of trials to be as loud as the sound under test.

LUMEN. A unit of luminous flux equal to the luminous flux radiated into a unit solid angle (steradian; sr) from a point source having a luminous intensity of 1 candela (cd).

SI symbol: ℓ m; formula: cd $\cdot$ sr

LUMINOUS FLUX. The total visible energy emitted by a source per unit of time is called the "total luminous flux from the source." The unit of flux, the LUMEN (lm) is the flux emitted in a unit solid angle (steradian) by a point source of one candela luminous intensity. A uniform point source of one candela intensity thus emits 4 lumens.

MACRO. A user-programmed set of input commands that can be executed as a single command.

MAIN MENU. The initial, or highest level, menu in a hierarchy.

MANUAL. (1) Operated by people rather than automatically. (2) Documentation giving instructions for the assembly, operation, or repair of a system or software application.

MASKING. Masking is the amount by which the threshold of audibility of a sound is raised by the presence of another (masking) sound. The unit customarily used is the decibel.

MEAN TIME BETWEEN FAILURE (MTBF). Availability formulation expressing the dependence of availability on reliability. Total system operating time divided by the number of system failures during that time period.

MEDIAL PLANE. The medial plane is a vertical plane that divides the body into left and right. It is perpendicular to the frontal plane.

MEMORY. (1) Recall and recognition of anything previously learned or experienced. (2) The component of a computer, control system, or the like, designed to provide ready access to data or instructions previously recorded.

MENTAL MODEL. Psychological construct describing an internal representation of a system used for predictions, control, and fault diagnosis.

MENTAL WORKLOAD. A measure of demands for cognitive activity or cognitive effort during the performance of a task.

MENU. A display of command options or data elements in a software application.

MENU BREADTH. The amount of options per menu screen in a menu system.

MENU DEPTH. The number of levels in a menu system.

MENU SYSTEM. A typically hierarchical arrangement of menus within a software application.

MENU NAVIGATION. The process of moving through a menu system.

METABOLIC RESERVES. The energy source stored in chemical form, such as carbohydrates, that can be efficiently mobilized and utilized by the body, particularly for muscular activity and work beyond the normal level of activity of an individual.

METABOLISM. The sum of all the physical and chemical processes by which living organized substance is produced and maintained; also the transformation by which energy is made available for the uses of organisms.

METHOD OF ADJUSTMENT. A psychophysical method used primarily to determine thresholds; in this procedure the subject varies some dimension of a stimulus until that stimulus appears equal to or just noticeably different from a reference stimulus.

METHOD OF CONSTANT STIMULI. A psychophysical method used primarily to determine thresholds; in this procedure a number of stimuli, ranging from rarely to almost always perceivable (or rarely to almost always perceivably different from some reference stimulus), are presented one at a time. The subject responds to each presentation: "yes-no," "same-different," "greater than - equal to - less than," etc.

METHOD OF EQUAL SENSE DISTANCES. A psychophysical

method used primarily to scale sensations; in this procedure the subject adjusts a set of stimuli until the elements of the set appear to be equidistant along some dimension.

Method Of Limits. A psychophysical method used primarily to determine thresholds; in this procedure some dimension of a stimulus, or of the difference between two stimuli, is varied incrementally until the subject changes his response.

Method of Magnitude Estimation. A psychophysical method used primarily to scale sensations; in this procedure the subject assigns to a set of stimuli numbers which are proportional to some subjective dimension of the stimuli.

Method Of Magnitude Production. A psychophysical method used primarily to scale sensations; in this procedure the subject adjusts a stimulus along some dimension until the magnitude of the stimulus appears equal to some specified magnitude.

Method Of Paired Comparisons. A psychophysical method used primarily to scale responses; in this procedure stimuli are presented in pairs to a subject who compares them along some dimension.

Method Of Rank Order. A psychophysical method used primarily to scale sensations; in this procedure stimuli are presented to a subject and he orders them along some dimension.

Method Of Ratio Production. A psychophysical method used primarily to scale sensations; in this procedure the subject adjusts a stimulus along some dimension until that stimulus appears to be a specific fraction or multiple of a reference stimulus.

Method Of Single Stimuli. A psychophysical method used primarily to scale sensations; in this procedure stimuli are presented singly to a subject who rates them along some dimension.

Mets. A functional energy expenditure classification, using multiples of the resting metabolic value. Thus, 2 mets is two times resting metabolism.

Minimum Separable Acuity. Smallest space between two lines that can be discriminated as a gap. It is measured in terms of the angle subtended by the gap, measured at the eye.

Mnemonic. A technique or designed coding system to aid in memory storage or retrieval.

Mode. The internal state of a system which determines how controls or input commands are interpreted.

Mode Error. Selection of action inappropriate for the current system mode.

Monitor. To observe, listen in on, keep track of, or exercise surveillance over by any appropriate means, as, to monitor radio signals; to monitor the flight of a rocket by radar; to monitor a landing approach.

Motion And Time Study. A systematic study of work systems, which have the purposes of: (1) developing a preferred system and method (usually one with lowest cost); (2) standardizing this system and method; (3) determining the time required by a qualified and properly trained person working at a normal pace to perform a specific task or operation; (4) assisting and training a worker in the preferred method. Motion study (or methods design), finding the preferred method of doing work. Time study (or work measurement), determining standard time for performing a specific task. Taylor used the term time study almost indiscriminately—including what the Gilbreths called motion study—much to their chagrin, especially when he took time studies without first studying the "one best way." (See Z94.17 Work Design & Measurement.)

Motion Parallax. The apparent difference in rate of movement of two objects actually moving at the same velocity but at different distances from the observer.

Motor Skill. The ability to achieve the more or less complicated adjustments of hands, fingers, legs, feet, or other parts of the body in an integrated, smoothly flowing sequence resulting in the performance of some act.

Mouse. An input device moved by the hand along a flat surface to move a cursor on a display.

Multimedia. Information display presenting multiple formats or stimulating multiple sensory modalities.

Multitasking. The performance of multiple tasks simultaneously.

Negative Transfer. A mental process occurring when past learning interferes with new learning. This can occur when the stimuli in the new situation are similar to the stimuli in the old situation, but different responses are required. It can also occur when the stimuli in a new situation are different from the stimuli in the old situation, but the responses are the same.

Neuromuscular. Pertaining jointly to nerves and muscles, as neuromuscular junction.

Night Vision. Vision which occurs in faint light, or after dark adaptation. Sometimes called twilight or scotopic vision. Hues and saturations cannot be distinguished.

Noise. (1) Unwanted signal(s) that typically interfere with wanted signals. Noise may be audible (voice communications), or visible (radar, hard copy). (2) Unwanted energy (or the voltage produced), usually of random character, present in a transmission system due to any cause.

Noise Criterion (NC) Curves. Any of several versions (SC, NC, NCA, PNC) of criteria used for rating the acceptability of continuous indoor noise.

Numerical Display. Format of information display containing only numerical data.

Nystagmus. An involuntary oscillation of the eyeballs, especially occurring as a result of eye fixations and stimulations of the inner ear during rotation of the body.

Open Loop. A control system to which there is no self-correcting action, operating without feedback, or with only partial feedback.

Operations Research. A mathematical modeling procedure aimed at discovering how, under the necessary conditions, requirements, and constraints, a system can best accomplish what it is designed or expected to do. Beginning with a clear statement of objectives, it involves identifying the relevant variables, expressing them in quantitative terms, measuring their fluctuations in given situations, and discovering how they interact. The system may be a machine, a social organization, or a combination of human and machine.

Operator. An individual performing the functions of system control, monitoring, diagnosis, or repair.

Operator Inputs. In a human-machine system, information received (sensed) by the operator from instructions or displays or directly from the environment.

Operator Outputs. In a human-machine system, the action taken by the operator; the result of a decision based on the input, *e.g.*, manipulation of controls and verbal communication.

Operator Overload. Condition in which an operator is required to perform more functions than he/she is able to handle effectively within given time limitations, such as identifying signals, processing information, making decisions, etc.

Optimum-location Principle (equipment design). The principle of arrangement such that each display and control should be placed in its optimum location in terms of some criterion of use (convenience, accuracy, speed, force to be applied, etc.).

Output device. Any device such as a display or printer, used to communicate the output of a computer to a user.

Oxygen Consumption. Rate of oxygen use of organisms, tissues, or cells.

Parallax. The difference in the apparent direction or position of an object when viewed from different points. (See BINOCULAR PARALLAX.)

Part Task Simulation. That type of simulation which gives the subject an opportunity to learn selected aspects of the total task.

Pattern Recognition. A human or machine process associated with identifying or classifying a set of input stimuli.

Percentile Threshold. Relates to the collective visual acuity of a group, stating the percentile probability of detection of a small object. Object size is expressed by its subtended angle at the eye. An important parameter in visual inspection for quality control. (See VISUAL ACUITY.)

Perception. (1) Perception is the process by which sensations are interpreted. (2) The awareness of external objects, qualities, or relations, which ensues directly upon sensory processes.

Perceptive Deafness. Deafness from conditions involving the cochlear structures or auditory nerve. Persons so affected may be deaf only to high frequency sounds, or, alternatively, only to low frequencies. Perceptive deafness is considered to be occupational since it often occurs in persons working in noisy environments. Also called boilermaker's deafness. (See CONDUCTIVE HEARING LOSS.)

Perceptual Augmentation. A type of display aiding in which enriched perceptual information is provided to a performer.

Perceptual Overload. Input of too much information.

The condition in which a person receives too much sensory information. The nervous system, on becoming saturated, is unable to process or understand the information received. The result of a perceptual overload is no reaction or response.

PERCEPTUAL-MOTOR TASK. A task involving an overt movement response to a nonverbal stimulus situation; the response is determined by the organization of sensory cues within the individual.

PERCEPTUAL SKILL. Relative ability to detect and interpret information received through the sensory channels; contrast with motor skill.

PERFORMANCE DECREMENT. A decrease in human proficiency often attributable to operator overload, stress, or fatigue, and characterized by increasing errors, misjudgments, omission of task elements, reduced intensity of effort, etc.

PERFORMANCE MEASURES. Objective and subjective measures developed to evaluate personnel effectiveness; objective measures include productivity, job knowledge tests, job performance samples, proficiency tests and checklists; subjective techniques include peer ratings, supervisory ratings, and self-ratings.

PERIPHERAL VISION. "Side" vision, primarily sensed by the rod cells. May produce an orienting reflex whereby the head turns and the visual object is focused on the fovea for optimal perception.

PERSONNEL SUBSYSTEM (PS). The segment of system management which provides the human performance necessary to operate, maintain, support, and control the system in its operational environment.

PHON. The unit of loudness level of sound, numerically equal to the sound pressure level in decibels, relative to 20μ N/m^2 x (20μ Pa) of a simple 1000 Hz tone judged by listeners to be equivalent in loudness. *Compare* sone.

PHOTOPIC ADAPTATION. The decreased visual sensitivity to light, sometimes manifested by decreased perceived brightness of a fixed stimulus.

PHOTOPIC VISION. Vision associated with levels of illumination (luminance) about 30 cd/m^2 or higher, characterized by the ability to distinguish colors and small details. Also called foveal vision.

PHOTORECEPTORS. Sensory end organs (q.v.) of the optic nerve. Elements of the retina called rods and cones (q.v.). They contain biochemicals which operate in the interpretation of perceived light. The industrial designer should be aware of the differences in these end organs. Rods are black/white discriminating, cones are color sensitive. Rods are effective in the peripheral visual field and color perception is limited to the central visual field.

PIE CHART. A circular, radially divided, graphical display indicating proportional data.

PILOT STUDY. An initial evaluation of the procedures of an experimental study prior to conducting the full study.

PIXEL. The smallest element of an image that can be individually processed in a video display system.

POPULATION STEREOTYPE. A consistent set of expectations within a specific social or cultural group. Distinguished from compatibility in that stereotypes are learned expectations, such as read meaning danger, up meaning more.

POP UP MENU. An overlay menu in a software application, activated by a command input.

POSITIONING MOVEMENT, PRIMARY. The first movement an operator makes in positioning a control; movement which carries the body member approximately to the point of aim; also called "gross adjustment" or "slewing."

POSITIONING MOVEMENT, SECOND CORRECTIVE. That part of the positioning movement made to bring the body member into exact relation with the point of aim; also called "fine adjustment."

POSITION, TRACKING. A tracking task in which movement of the operator's control is associated with a direct displacement or movement of the tracking indicator, a linear relationship is implied between system error and control movement.

PRACTICE EFFECT. A change in performance due to repetitive experience at a task.

PREDICTOR DISPLAY. Shows the operator some aspect of what will happen to the system in the future. Such displays may be pursuit or compensatory, symbolic or pictorial; and they may predict the vehicle output, the system input, or both. The distinguishing feature is that they provide advanced information that allows the operator to anticipate future requirements for control movements.

PRESENCE. The degree to which experience in a virtual environment is felt as real.

PREVIEW CONTROL. Control involving the use of a predictor display.

PROACTIVE INHIBITION. The state or process hypothesized to account for the lessened ease of learning of the later members of a series following learning of an earlier member.

PROBLEM SOLVING. The generation of a set of actions to accomplish a goal.

PROBLEM SPACE. A representation of a problem solving task indicating the set of problem states and the actions which cause transitions between states.

PROCESS CONTROL. The control of a continuous state production system, such as industrial, chemical, and energy conversion processes.

PROPRIOCEPTION. The sensing of one's location relative to the external environment. An important sense for maintenance of balance and for orienting one's self for performing work tasks. An impaired proprioceptive sense can cause industrial accidents or faulty performance where controls are located outside the visual field.

PROPRIOCEPTOR. Any receptor sensitive to the position and movement of the body and its members, including (1) receptors in the vestibule of the inner ear and in the semicircular canals sensitive to the orientation of the body in space, and to bodily rotation, and (2) receptors in the muscles, tendons, and joints giving rise to kinesthetic sensations.

PROTOTYPING TOOLS. Software applications allowing rapid construction and testing of an initial design for an interface.

PSYCHOMETRICS. The measurement of psychological processes through application of mathematical and statistical techniques.

PSYCHOMOTOR ABILITY. Of or pertaining to muscular action ensuing directly from a mental process, as in the coordinated manipulation of aircraft or spacecraft controls.

PSYCHOMOTOR TASK. A task which involves precise coordination of a sensory or ideational process and a motor activity, *e.g.*, a tracking task.

PSYCHOPHYSICAL METHODS. Standardized procedures for presenting stimulus material to a subject for judging and for recording the results. Originally developed for determining functional relations between physical stimuli and correlated sensory responses, but now used more widely.

PSYCHOPHYSICAL QUANTITY. A physical measurement, as a threshold; dependent on human attributes or perception.

PSYCHOPHYSICS. A branch of psychology that deals with the quantitative relationship between physical and psychological events.

PULL DOWN MENU. A menu at the top of a computer display, extending vertically, activated by a command input.

PURKINJE EFFECT. The response of the human eye which makes it less sensitive to lights of longer wavelengths under conditions of decreased illumination, *e.g.*, red appears darker at night than blue having the same brightness under photopic conditions.

PURSUIT TRACKING. A task in which the subject is required to keep a response marker in line with a moving target.

QUANTITATIVE DISPLAY. A display which provides numerical values as opposed to one in which only qualitative information is provided.

QUICKENING. A form of display aiding in a manual control task, in which displayed error is computed as a combination of present error, velocity, and acceleration. The display thus indicates the likely future state of the system in the absence of control.

RADIALE HEIGHT (ELBOW HEIGHT). Vertical distance from the floor to radiale, the depression at the elbow between the bones of the upper arm (humerus) and forearm (radius); subject standing erect, arms hanging naturally at sides. (See Z94.2 ANTHROPOMETRY & BIOMECHANICS.)

RATE TRACKING. A task in which the operator moves a control in such a way as to keep an indicator in line with a moving object, thus measuring the object's speed of motion; involves an exponential relationship between system error and control movement. (See POSITION, TRACKING; TRACKING.)

REACTION TIME (RT). The interval between application of a stimulus and the beginning of the subject's response (sometimes measured to response completion).

REACTION TIME, COMPLEX. The time required to react when a discrimination needs to be made; *e.g.*, a response is made to one, but not to others, of two or more expected signals (discrimination reaction time), or a different response is specified for each kind of stimulus (choice reaction time).

Reaction Time, Simple. The time required to make a predetermined response as quickly as possible to pre-arranged signal; a combination of sensing time and response time.

Real Time. The absence of delay, except for the time required for the transmission by electromagnetic energy, between the occurrence of an event or the transmission of data, and the knowledge of the event, or reception of the data at some other location.

Receiver-Operating-Characteristic. A receiver-operating-characteristic is a graphical summary of the performance of a detector. Detection probability is plotted on the ordinate and false alarm probability is plotted on the abscissa. These are conditional probabilities, that is, the probability that the condition (signal present or signal absent) is true. A family of non-intersecting curves is often plotted with either constant detectability index or constant signal-to-noise ratio as the parameter. Each curve shows how detection probability and false-alarm probability vary monotonically as a function of the decision criteria (operating point) of the detector.

Receptor. A sensory nerve ending or end-organ in a living organism that is sensitive to physical or chemical stimuli.

Recognition. The psychological process in which an observer so interprets stimuli received from an object that he forms a correct conclusion as to the nature of that object.

Red-green Blindness. A common form of partial color blindness, or dichromatism, in which red and green stimuli are confused because they are seen as various saturations and brightnesses of yellow, blue, or gray.

Redout. The condition occurring under negative g in which objects appear to have red coloration due to uncertain causes, possibly venous congestion of engorged eyelids.

Redundancy. The existence of more than one means of accomplishing a given function. Each means of accomplishing the function need not necessarily be identical. (Active)—that redundancy wherein all redundant items are operating simultaneously rather than being switched on when needed. (Sometimes referred to as parallel redundancy.) (Standby)—the redundancy wherein the alternative means of performing the function is inoperative until needed and is switched on upon failure of the primary means of performing the function.

Reflectance. The ratio of the reflected incident light which on striking a surface is reflected. Although lighting at the workplace is important, reflectance is more significant because it provides the visual input to the worker. Reflectance is expressed as a percentage of incident flux density.

Refractory Period. A brief period following stimulation of a nerve or muscle during which it is unresponsive to a second stimulus; may be absolute (no response) or relative (response only if the stimulus is very strong.)

Rehearsal. The psychological process of repeating information in working memory for maintenance.

Reinforcement. The action of a symbolic or concrete reward upon a response so that the response is strengthened (positive reinforcement); or, the action of symbolic or physical punishment on a response so that the response is weakened or replaced by an escape or avoidance response (negative reinforcement).

Reliability. The probability that a system will perform a required function under specified conditions for a specified period of time or at a given point in time.

Reliability (psychological test). The complex property of a series of observations or measuring instruments that makes possible the obtaining of similar results upon repetition.

Remote Handling. Process by which manipulative skills are transferred from the proximity of the human operator to a more distant area by means of mechanically or electronically linked remote-control systems.

Remote Indicating. Of an instrument, displaying indications at a point remote from its sensing element, often by electrical or electronic means.

Representativeness. A cognitive heuristic in which the probability that a item belongs to a particular class is judged by the similarity of the item to the prototype member of the class.

Resolving Power. The capability of the eyes to see two objects viewed simultaneously as two distinct objects; the capability to perceive as distinct two objects, in close proximity, casting images on the retina.

Respiratory Quotient. Quotient of the volume of CO_2 produced, divided by the volume of O_2 consumed by an organism, an organ, or a tissue during a given period of time.

Response. (1) Physiological—the muscular contraction, glandular secretion, or other activity of an organism which results from stimulation. (2) Psychological—a behavioral action, most commonly motor or verbal, that occurs following an external or internal stimulus.

Response Time. The time required for a human operator to make a given response after a stimulus is perceived. Under simple conditions (*e.g.*, pushing a button), it may be a few hundredths of a second.

Retina. The inner liner of the eye which contains optic end organs. Composed of rods, cones (q.v.) and photosensitive cells, it translates light energy into nervous impulses.

Retinal Disparity. The difference which exists between the images formed in the right and left eyes when a solid object is viewed binocularly.

Retinal Field. The extended mosaic of the rod and cone receptor elements of the retina, which forms something of an anatomical correlate of the stimulus field.

Retinal Illuminance. The illuminance of the retina, the usual units being the troland and the lux.

Retinal Rivalry. Alternation of sensations first from one eye and then from the other, when the two eyes are simultaneously stimulated by different colors or figures. Also called binocular rivalry. Contrast with binocular fusion, in which the two impressions are fused into a single impression.

Rhodospin. A substance found in the rods of the dark-adapted eye, which bleaches rapidly on exposure to light, and is converted to opsin and retinal.

Rods. Photosensitive cells of the retina which are specifically reactive to dim light. They are black-and-white sensitive only, and their proper function is required in poorly illuminated or night-vision applications.

Root Mean Square Pressure of Fundamental Speech Sounds. A method of relating the measurements of individual sounds as uttered, by making the intervals of time over which the squared pressures are averaged correspond directly to the intervals during which the specific vowels and consonants are spoken. Values are expressed in decibels relative to the r.m.s. pressure of the weakest speech sound (the initial consonant, "th," of "thin").

Rotation (zero g). The controlled movement of a free-floating orbital worker to achieve a desirable work position after translation from one location in space to another; a positioning maneuver required in order to perform maintenance or other duties inside or outside an orbiting space vehicle.

Saccadic Movements. Sudden movement of the eyes from one fixation point to another.

Safety. Freedom from those conditions which can cause injury or death to personnel and damage to or loss of equipment or property.

Saturation. Extent to which a chromatic color differs from a gray of the same brightness, measured on an arbitrary scale from 0% to 100% (where 0% is gray).

Scintillation. (1) Generic term for rapid variations in apparent position, brightness, or color of a distant luminous object viewed through the atmosphere. (2) A flash of light produced in a phosphor by an ionizing event. (3) On a radar display, a rapid apparent displacement of the target from its mean position. Also called target glint or wander. This includes but is not limited to shift of effective reflection point on the target.

Scotoma. A blind or partially blind area in the visual field.

Scotopic Adaptation. Like dark adaptation, but with more explicit reference to the part played by the rod system of the retina.

Scotopic Vision. Vision which occurs in faint light, or after dark adaptation. Sometimes called twilight or night vision. Hues and saturations cannot be distinguished.

Screen Dump. Print the contents of a computer display.

Scroll. Either vertically or horizontally advance through the contents of a computer display.

Seat Reference Point (SRP). The point at which the midlines of the seat and backrest intersect.

Secchi Disk. A white disk which, when submerged to varying depths, aids in determining the color and depth of light penetration in the sea.

Self-Paced. Experimental or actual task in which a performer's activity is pace solely by the performer.

Sensation. Subjective response or any experience aroused by stimulation of a sense organ.

Sensing Time. The time required for a human operator to become aware of a signal.

Sensor. The nerve endings or sense organs which receive information from the environment, from the organism, or from both.

Sensory End Organs. Receptor organs of the sensory nerves located in the skin or other tissues. Each end organ can sense only a specific type of stimulus. Primary stimuli are heat, cold, or pressure, each requiring different end organs. Knowledge of end organ distribution is of importance to the safety engineer. For example, there are a few heat receptors on the outer surface of the forearm, so that the skin may be severely burned before heat is sensed.

Sensory Feedback. Signals perceived by sense organs (*e.g.*, eye, ear) to indicate quality or level of performance of an event triggered by voluntary action. On the basis of sensory feedback information, decisions may be made permitting or not permitting an event to run its course; enhancing or decreasing activity levels.

Sensory Nerve. A nerve (also called afferent nerve) which conducts stimuli from sensory end organs (q.v.) which change physical information (such as pressure, temperature, light) into nerve impulses, to the spinal cord or the brain where cognition occurs and/or the stimulus is converted to a motor signal for transmission via a motor nerve (q.v.)

Sequence-Of-Use Principle (equipment design). The principle of arrangement that controls and displays should be so positioned that those used in sequence would be physically arranged in order of operation.

Service Test Model. A model used to determine the characteristics, capabilities, and limitations of a piece of equipment or a complete system under either simulated or actual service operational conditions.

Servomechanism. A control system which brings a dynamic system to a desired state by operating upon an error signal.

Shade. Any color darker, *i.e.*, of lower lightness, than median gray.

Shake Table. A test device for determining the effects of vibration on human or animal subjects or on equipment.

Shape Coding. Varying the shape of controls to make them distinctive. Shape coding is effective both visually and actually, that is, the difference can be both seen and felt. The shape of a control should suggest its purpose, and the shape should be distinguishable not only with the naked hand, but also with gloves.

Short-Term Memory. The storage of recently received inputs for a period of seconds or minutes.

Shoulder-Elbow Distance. From top of acromion to tip of elbow, measured when subject is sitting erect, with upper arm vertical, forearm horizontal.

Side Tone. The signal from the talker's microphone usually returned to the talker via his earphones. This feedback signal is called the side tone. The talker's speech can be manipulated by varying the side tone.

Signal Detection Theory. A psychophysical model which views human perception as a process in which decisions based on uncertain sensory information are made under conditions of risk.

Simulation. A set of test conditions designed to duplicate field operation and usage environments.

Simulator. Any machine or apparatus that simulates a desired condition or set of conditions, such as a flight simulator.

Sitting Height. The vertical distance from the sitting surface to the top of the head. The subject sits erect, looking straight ahead, with knees at right angles.

Situation Awareness. The psychological activities involved in attaining a veridical internal representation of the environment.

SI Units. Abbreviation of Le Systeme International d'Unites (International System of Units).

Skinfold Measurement For Estimating Body Fat. Permits reasonably close estimation of body fat. Calipers are used to exert a pressure of $10g/mm^2$ over an area of 20 to 40 mm^2. Measurements are made over the right triceps muscle and below the inferior angle of the right scapula.

Snow-blindness. A temporary abnormality of the color sense, in which all objects are tinged with red. Caused by long-continued exposure to very bright light, as in Arctic exploration, on glaciers, in telescopic observation of the sun, watching welding operations, etc.

Somatotyping. A nearly obsolete means of numerical classification of various human body types. Endomorphy: soft, round form with loose, flabby tissue. Mesomorphy: massive, solid form with cubical head, heavy muscles. Ectomorphy: slender limbs and body, with slight head.

Sone. Unit of loudness; a simple tone of 1000 Hz 40 dB above a listener's threshold produces a loudness of one sone.

Space Myopia (empty field myopia). A disturbance of the visual accommodative mechanism due to insufficiently sharp detail, making it difficult to detect small objects in the relatively unstructured visual environment of space.

Spectrum Colors. The series of saturated colors normally evoked by photopic stimulation of the retina with radiant energy of continuously differing single wavelengths through the visible range.

Specular Reflection. Reflection in which the reflected radiation is not diffused; reflection as from a mirror. Also called regular reflection, simple reflection.

Speech Articulation Index. The method of estimating the intelligibility of speech by measuring differences (in decibels) between speech and noise levels at different frequency bands.

Speech Interference Level (SIL). Used as a gross basis for comparing the relative effectiveness of speech in noise. It is usually the simple numerical average of the decibel level of noise in three octave bands, namely, those with centers at 500, 1000, and 2000 Hz.

Speech Recognition. An automated process of recognizing spoken language.

Speech Synthesis. An automated process of generating spoken language.

Spin Table. A round platform on which human or animal subjects can be placed in various positions and rapidly rotated in order to simulate and study the effects of prolonged tumbling at high rates. Complex types of tumbling can be simulated by mounting the spin table on the arm of a centrifuge.

Squatting Height. The height from the top of the subject's head to the floor when the subject balances on toes, body erect.

Squeeze. Squeeze in diving is due to the effect of increasing external pressure upon the ears and sinuses, the face plate or the swim suit, uncompensated by an equal increase in pressure from within.

Stabilization (zero g). A condition required during translation and after arrival of a free-floating orbital worker at a desired position inside or outside an orbiting space vehicle. A mechanism is necessary to stabilize the worker against accelerations that may cause visual disorientation, labyrinthine reactions, and movement away from the work area.

Standard Observer. A hypothetical observer with a visual response mechanism possessing the colorimetric properties defined by the 1931 ICI tables of the distribution coefficients, x, y, z, and the trichromatic coefficients, x, y, z, of the equal energy spectrum. The y coefficients of the equal energy spectrum are the relative luminosity values defining the standard observer for photometry.

Standing Knee Height. Vertical distance from top of kneecap to floor when subject is standing.

Static Display. A display containing information that does not change.

Stature. The vertical distance from the floor to the top of the head. The subject stands erect and looks straight ahead.

Stereoscopic Acuity. Ability to sense three-dimensional aspect of physical space. A function of independent image inputs, $i.e.$, viewing with two eyes. Stereoscopic acuity also depends on brightness contrast and distance from observed objects.

Sternum Height. Vertical distance from lower tip of sternum to floor when subject is standing.

Stimulus. Energy, external or internal, which excites a receptor.

Stimulus Generalization. An aspect of learning characterized by the spreading of a response learned in association with one stimulus to other stimuli which are similar but not identical to the first.

STPD Conditions. Standard temperature and pressure, dry. These are conditions of a volume of gas at 0°C, at 760 torr (101.32472 kPa) without water vapor. A STPD volume of a given gas contains a known number of moles of that gas. (See BTPS CONDITIONS.)

Stress. (1) The force per unit area on a body that tends to produce a deformation. (2) The effect of a physiological, psychological, or mental load on a biological organism which may produce fatigue and degrade proficiency.

Stroboscope. An optical instrument for observing moving bodies by making them visible intermittently. If the motion is cyclic, such as rotation or vibration, an optical illusion is created of the moving body being stationary or moving at a slower rate. (See FLICKER FUSION.)

STROPHOSPHERES. Defines the region in a person's fixed workplace that is common to the hand motions made with various hand manipulations.

SUBJECT. A member of a specified population from whom an experimenter obtains response data for specific stimulus variables. The data obtained are often generalized to that population.

SUBJECTIVE EXPECTED UTILITY (SEU). A decision making model which assumes a decision maker will enumerate the set of feasible alternatives and select the alternative that maximizes expected utility.

SUPERVISORY CONTROL. A mode of control in which a human operator or team monitors and manages the operation of automated or semi-automated systems which control a system.

SURROUND BRIGHTNESS. The brightness of the area immediately adjacent to an area of visual work.

SYSTEM. A composite of equipment, skills, and techniques (including all related facilities, equipment, material, services, and personnel) that is capable of performing and/or supporting an operational role.

SYSTEM ANALYSIS. Identification of the dynamic relationships between elements of a system.

SYSTEM ENGINEERING. The study and planning of a system so that the relationships of various parts of the system are fully established before designs are committed, or to look for improvements in existing systems.

TABULAR DISPLAY. Display of alphanumeric and other symbols in a row and column format.

TACHISTOSCOPE. An experimental device for controlling the exposure of visual stimulus material, *e.g.*, symbols.

TACTILE. Of or relating to the sense of touch. Tactual.

TACTILE CONTROL. Control regulated through tactile sensory feedback (q.v.). It may be reinforced by designing equipment controls to incorporate distinctive texture patterns (*e.g.*, knurling) or shapes to prevent error. Other means of tactile control are the initiation of movement with a vibratory cue and termination through a positive stop. (See TACTILE SENSE, TACTILE STIMULUS.)

TACTILE SENSE. The sense of touch. Experienced through four types of mechanoreceptors (q.v.) located throughout the body. Significant source of sensory feedback (q.v.). Improper clothing (thick gloves) may impair tactile sensation and cause inaccurate discrimination of tactile signals. (See TACTILE CONTROL, TACTILE STIMULUS.)

TACTILE STIMULUS. An input originating in the touch mechanoreceptors (q.v.) as sensory end organs (q.v.). The input is converted into a nerve signal for afferent transmission along the sensory nerves (q.v.) to the brain or spinal cord, whence an efferent signal is transmitted along a motor nerve (q.v.) causing muscular reaction. Tactile stimuli in the workplace replace other cognitive judgment and are conducive to quick error-free control action. (See TACTILE SENSE, TACTILE CONTROL.)

TARGET. (1) Any object, point, etc., toward which something is directed. (2) An object which reflects a sufficient amount of a radiated signal to produce an echo signal on detection equipment.

TARGET ACQUISITION. The process of optically, mechanically, or electronically orienting a tracking system in direction and range to lock on a target.

TARGET DISCRIMINATION. (1) Perceiving a desired signal within a background of noise. (2) Distinguishing multiple signals.

TASK. A group of related job elements performed within a work cycle and directed toward a goal; a composite of the discriminations, decisions, and motor activities required of an individual in accomplishing a unit of work.

TASK ANALYSIS. An analytical process employed to determine the specific behaviors required of human components in a human-machine system. It involves determining, on a time base, the detailed performance required of human and machine, the nature and extent of their interactions, and the effect of environmental conditions and malfunctions. Within each task, behavioral steps are isolated in terms of the perceptions, decision, memory storage, and motor outputs required, as well as the errors which may be expected. The data are used to establish equipment design criteria, personnel and training requirements, etc.

TASK CODING. A numerical and/or alphabetical index sometimes used during task analysis to indicate in relative terms: a) how demanding a task is with respect to perceptual, judgmental, and motor skills (Skill Coding); b) how critical the correct performance of the task is to system performance or mission success (Task Criticality); c) whether personnel are already doing the task or whether it is new (Task Newness).

Task, Continuous. Those tasks which involve a continuously changing response to a continuously changing stimulus. The operator compares the desired output with the error signal which is fed back and makes continuous adjustments accordingly; *e.g.*, maintaining a given vehicle attitude by manual control movements in response to instrument indications.

Task Element. A basic unit of behavior, made up of the smallest logically definable set of perceptions, decisions, and responses which the human being is required to perform in completing a task; *e.g.*, a single action identified by a specific signal on a specified indicator or display, a specific control actuation, and a feedback signal or response adequacy.

Task Equivalence. The condition existing when two tasks have sufficient elements in common to permit an individual to transfer from one to the other at the same level of skill without additional training or experience.

Tasks, Procedural. Those tasks which involve stepwise, discontinuous, or all-or-none responses to discrete perceptual cues. The operator observes a displayed condition and performs a single response, or observes a series of conditions and performs a series of separate nonrepeated responses; *e.g.*, turning on deicing equipment when icing conditions are indicated.

Teleoperation. The remote control of a teleoperator such as a robot, typically accomplished via transmitted video information.

Threshold, Absolute Visual. The minimum intensity of light that can be seen by the human eye after complete dark adaptation, generally considered to be about 1/1,000,000,000 of a lambert in the periphery. Theoretically, as few as six quanta of light reaching the retina have a good probability of yielding a visual sensation.

Threshold Contrast. The smallest difference between luminances (or brightnesses) that is perceptible to the human eye under specified conditions of adaptation luminance and target visual angle. Also called contrast threshold, liminal contrast. *Compare* threshold illuminance. Psychophysically, the existence of a threshold contrast is merely a special case of the general rule that for every sensory process there is a corresponding lowest detectable intensity of stimulus, *i.e.*, a limen.

Threshold, Difference. The minimum difference between two stimuli that can be perceived as different; statistically, that difference which is perceptible 50% of the time.

Threshold (Limen). The statistically determined point at which a stimulus is just barely adequate to elicit a specified organismic response (absolute threshold) or at which it differs enough from another stimulus to elicit a different response (difference or differential threshold).

Threshold, Masked Differential. The least amount of change of a stimulus necessary to elicit a response when the stimulus is obscured (masked) by another stimulus presented simultaneously; applies to brightness, loudness, etc.

Threshold Of Audibility. For a specified signal, the minimum effective sound pressure level of the signal that is capable of evoking an auditory sensation in a specified fraction of the trials. The characteristics of the signal, the manner in which it is presented to the listener, and the point at which the sound pressure level is measured must be specified. Also called threshold of detectability.

Threshold Of Detectability. Threshold of audibility.

Threshold Of Discomfort. In acoustics, for a specified signal, the minimum effective sound pressure level of that signal which, in a specified fraction of the trials, will stimulate the ear to a point at which the sensation of feeling becomes uncomfortable. The term applies similarly for other senses.

Threshold Of Pain. A specified stimulus signal, the minimum level of that stimulus which, in a specified fraction of trials, will stimulate a sense organ to a point at which the discomfort gives way to definite pain that is distinct from mere nonnoxious feeling of discomfort.

Threshold Shift. A threshold shift is a change in the threshold of audibility for an ear at a specified frequency. The amount of threshold shift is customarily expressed in decibels.

Threshold, Terminal. A maximum stimulus that will produce a given type of sensory experience or elicit a given kind of response.

Through-put Capacity. The output of thoroughly processed items from a machine.

TIMBRE. That attribute of auditory sensation by which a listener discriminates between two sounds of similar loudness and pitch, but of different tonal quality.

TIME AND LINK ANALYSIS. An analysis of the number and duration of activities assigned to each potential element of a system. It provides an overview of what the human component is expected to do at each point in an operation, an outline of the process of information exchange among operators and between operators and equipment, and data useful in the determination of crew size for each major component of the system.

TIME-LINE ANALYSIS. An analytical technique for the derivation of human performance requirements which provides an indication (by means of a time-line chart) of the functional and temporal relationships among tasks, as well as the behavior and task loadings for any given combination of tasks.

TIME-SHARING. The division of an operator's perceptual, decision-making, or response time among activities or tasks which must be performed at or about the same time. For example, flight control movements must be time-shared with engine control adjustments, communications, etc; in general, tasks which must be time-shared in the operational situation should also be time-shared during training.

TINT. Any color lighter, *i.e.*, of higher lightness, than median gray. May imply weak saturation as well as relatively high lightness.

TONAL GAP. A gap is a frequency band of low auditory sensitivity above and below which the sensitivity is better.

TONAL ISLAND. A tonal island is a frequency region bonded above and below by tonal gaps.

TONALITY (TONAL). Tonality is the phenomenon in which tones spaced an octave apart sound more alike than those spaced along any other interval.

TONE. (1) Physics—a sound wave capable of exciting an auditory sensation having pitch. (2) Psychology—a sound sensation having pitch.

TOUCH SCREEN. An input device consisting of a screen sensitive to either the position or pressure of an object pressed against the screen.

TRACKBALL. An input device consisting of a ball resting in a socket, which can be rotated by a user to control cursor position.

TRACKING. The process by which an operator continually attempts to minimize some measure of the difference between a desired and an actual output of a system; an activity that involves following a target moving in two- to three-dimensional space with reference to the observer or equipment.

TRAINING. The totality of instructions, planned circumstances, and directed activity by which personnel acquire and/or strengthen new concepts, knowledge, skills, strength habits, or attitudes which will enable them to perform assigned duties with maximum reliability, efficiency, uniformity, safety, and economy.

TRANSFER OF LEARNING. The phenomena through which information acquired in one situation is applied or "transferred" to some other situation, most usually an operational one.

TRANSILLUMINATION. (1) Indirect illumination used on console panels and utilizing edge- and back-lighting techniques on clear, fluorescent, or sandwich-type plastic materials. (2) Transmitting light through material to enhance seeing quality deviations.

TRANSMITTANCE. The ratio of luminous flux transmitted through a medium to the incident flux, expressed as a percentage. "Selective transmittance" refers to the transmission of particular wavelengths of light through a transparent or translucent medium, such as a red filter.

TREMOR. Shaking, trembling, or oscillation which may result from natural impairment or from excitement, emotion, cold, or an effort to maintain a part of the body in a fixed position. It can be measured by the distance or the number of departures from a fixed path or position in a given time. Fatigue or efforts to prevent tremor may increase it; strong support of the body or part of the body affected, or the application of friction, tend to decrease the trembling.

TRIAL. A trial is a sequence of events during an experimental session in which a subject (or subjects) is presented a stimulus set and the resulting response set is recorded.

TRIANOMALY. Rare type of trichromatism in which an abnormally large portion of blue stimulus is required in a blue-green mixture to match a given cyan.

TRICHROMATIC THEORY. A color theory based upon the facts that all hues may be derived from the mixture of two or more of three primaries.

TRICHROMATISM. Form of vision yielding colors which require in general three independently adjustable primaries (such as red, green, and blue) for their duplication by stimulus mixture.

TRITANOPIA. Form of dichromatism in which reddish blue and greenish yellow stimuli are confused. Tritanopia is a common result of retinal disease, but in rare cases may be inherited. Sometimes called blue blindness.

TRUNK. Human body torso.

TURNAROUND TIME. That element of maintenance time needed to service or check out an item.

ULTRASONIC. In acoustics, of or pertaining to frequencies above those that affect the human ear, *i.e.*, more than 20,000 Hz.

UNBURDENING. A method of easing the human task in tracking by using computers to perform mathematical integrations; or, the mechanization of a system to reduce physical effort on the part of the operator.

UPTIME. That element of active time during which a system element is either alert, reacting, or performing.

USABILITY. A measure of the ease with which an individual can either operate, or learn how to operate, a system.

USABILITY TESTING. Empirical testing to measure the usability of a system.

VALIDITY (TEST). The degree to which a test measures what it was designed to measure, estimated by means of coefficient of correlation between test scores and a criterion measure, such as actual on-the-job performance; empirical validity.

VALUE. (1) The dimension of the Munsell system of color which corresponds most closely to lightness. (2) Numerical quantity. (3) Worth, as in value engineering.

VARIED MAPPING. A variable relationship between stimuli and desired response.

VERBAL MEDIATION. The technique of verbalizing instructions or directions so as to emphasize cue-response relationships in learning or performing a task; used especially during the early stages of learning.

VERBAL PROTOCOL ANALYSIS. A method of behavioral measurement involving the collection and examination of the verbal behavior of a performer.

VERTIGO. Sensation of dizziness or whirling, in which objects, though stationary, appear to move in various directions, and the person finds it difficult to maintain an erect posture. It may result from overstimulation of the sense organs, from changes in the blood supply of the brain, or from disease. "Aviator's vertigo," caused by conflict between visual and gravitational cues, results in a disturbance in the pilot's orientation with respect to the earth.

VIEWING ANGLE. The angle formed by a line from the eye to the viewed surface.

VIGILANCE. An activity or type of research involving continuous watch and the ability of individuals to detect signals of varying frequencies during these watches.

VISUAL ACUITY. The ability to recognize objects or sets of objects by eye as distinct, specific and separate. Depends on illumination, object size, distance from object to observer, object configuration and brightness of contrast. Expressed quantitatively as the ratio of distance at which the observer recognizes to the distance at which a normal eye recognizes. 20/100 indicates that a person recognizes at 20 feet a sight chart which can be seen by the normal eye at 100 feet. Visual acuity of individuals should be matched to the needs of their task assignments.

VISUAL FIELD DEFECT. A condition of impaired eyesight where the eye cannot see portions of the area in its field of vision. Blind spots of various size and position exist. Visual field defects are sufficiently common to warrant consideration in work placement. Not usually detected in eye examination for visual acuity (q.v.).. Should be suspected whenever a person appears to be accident-prone.

VIRTUAL ENVIRONMENT. Artificial, typically computer generated, stimulation of the senses to give the subjective impression of existence within a particular environment.

VIRTUAL IMAGE. An image that cannot be shown on a surface but is visible, as in a mirror.

VISIBILITY. The capacity of radiant energy, within a certain range of wavelengths, to excite a visual receptor process and thereby evoke the phenomenon of brightness.

VISION, FOVEAL. Visual sensations or perceptions due to stimulation of the fovea centralis, or center of the retina.

VISION, PERIPHERAL. Visual sensations or perceptions due to stimulation of the outlying portions of the retina.

VISION, PERSISTENCE OF. The tendency of visual excitation to outlast the stimulus, or more generally the tendency of changes in visual sensory response to lag behind changes in the stimulus.

VISUAL ACUITY. The ability to perceive detail at various distances. More specifically, it can be expressed as the reciprocal of the visual angle, in minutes of arc, that is subtended by the smallest detail that can be visually discriminated.

VISUAL ANGLE. The angle subtended by an object in the visual field at the nodal point of the eye. The angle determines the size of the image on the retina. Objects of different sizes or distances have the same-sized image on the retina if they subtend the same angle.

VISUAL ADAPTATION. Adjustive change in visual sensitivity due to continued visual stimulation. Three recognized types are: (1) scotopic or dark adaptation, (2) photopic or light adaptation, and (3) chromatic or color adaptation.

VISUAL FIELD. That part of space that can be seen when head and eyes are motionless, (or) the totality of visual stimuli which act upon the unmoving eye at a given moment.

VISUAL PHOTOMETRY. A subjective approach to photometry, wherein the human eye is used as the sensing element; to be distinguished from photoelectric photometry.

VISUAL RANGE. The distance, under daylight conditions, at which the apparent contrast between a specified type of target and its background becomes just equal to the threshold contrast of an observer; to be distinguished from the night visual range. Also called daytime visual range.

VISUAL SPACE. This term, like visual field, refers to the extended world as perceived by means of the eyes but is commonly used in a more generic and abstract way in discussions of the perception of distance and length, of depth or distance away from the retina, and of form or figure in two and three dimensions.

VOLUME (TONAL). Volume is that attribute of auditory sensation in terms of which sounds may be ordered on a scale extending from spatially small to spatially large.

WAIST DEPTH. The horizontal distance between the back and abdomen at the level of the greatest lateral indentation of the waist (if this is not apparent, at the level at which the belt is worn).

WARM-UP TIME. Time measured from the application of power to an operable system to the instant when the system is capable of functioning in its intended manner.

WARNING. A signal, statement, or label indicating the existence of hazardous conditions.

WATCHKEEPING. That task in which the operator monitors some condition such as the status of automatic equipment.

WEAK COLOR. A color of low saturation.

WEBER-FECHNER LAW. An approximate psychophysical law relating degree of response or sensation of a sense organ and the intensity of the stimulus.

WEBER'S LAW. A law that states that for each sense parameter the ratio between the just noticeable difference and the stimulus intensity remains constant.

WEIGHTLESSNESS. (1) A condition in which no acceleration, whether of gravity or other force, can be detected by an observer within the system in question. (2) A condition in which gravitational and other external forces acting on a body produce no effect, either internal or external on the body. (3) The absence of gravitational pull on a body or object. (4) Condition in free fall or in space beyond the earth's gravitational field; may be simulated by air-bearing devices or during parabolic flight of high-performance aircraft. (5) A condition that is characterized by the absence of gravity. For clarity, it has been proposed that the term zero g or null g be used to refer to the physical state of an object or body and that the term weightlessness be used to refer to the physiological and psychological experience of living organisms under zero g conditions.

WHITE NOISE. (1) Randomly fluctuating noise. (2) The noise that is heard when many sound waves of different lengths are combined so that they reinforce or cancel one another in a non-uniform fashion. (3) A noise for which the spectrum density is substantially independent of frequency range. (4) A sound or electromagnetic warehouse spectrum is continuous and uniform as a function of frequency.

WHITEOUT. An atmospheric and surface condition in the arctic in which no object casts a shadow, the horizon being indiscernible, and only very dark objects being seen.

WINDCHILL. A commonly used scale for expressing the severity of cold environments, indicates the effects of wind velocity and temperature combined.

WORK. (1) Expression for human effort measured in physical units, or specific output results. (2) General description of task.

WORK CYCLE. The total series of actions and events which characterizes or describes a work assignment or single operation.

WORK PACE. The rate at which a task or activity is done. Work may be paced externally, as by machine or mission rates, or self-paced by the worker.

WORK PHYSIOLOGY. The study of bodily responses to physical and mental effort. Includes measures of stress on the cardiovascular, respiratory and musculoskeletal systems in particular.

WORK REST CYCLE. The ratio of work time to non-work time. The larger the ratio the more opportunity there is for fatigue.

WORKSPACE. The physical area in which an individual performs some duty or task.

WORK SPACE LAYOUT. A design of work area or work station to include provisions for seating, physical movement of human operators, operational maintenance, and other factors permitting adequate person-to-person contact and man-machine interaction.

WORK STUDY. Analysis of work methods, techniques, and procedures.

WYSIWYG. A feature of a computer display, in which the displayed image exactly corresponds to the printed image (What You See Is What You Get).

YAW. (1) The rotational or oscillatory movement of an aircraft, rocket, or the like about a vertical axis. The amount of this movement is the angle of yaw. (2) To cause to rotate about a vertical axis.

YAW AXIS. A vertical axis through an aircraft, rocket, or similar body, about which the body yaws. Also called a yawing axis.

YAWING MOVEMENT. A movement that tends to rotate an aircraft, etc., about a vertical axis.

YES-NO EXPERIMENT. An experiment in which each trial contains a single observation interval and the subject judges whether or not a stimulus was presented during that interval.

BIBLIOGRAPHY

Air Force Systems Command. *Design Handbook*, Series 1-0, General; AFSC DH 1-1, General Index and Reference. (3rd ed.) December, 1973.

American National Standards Institute. *Psychoacoustical Terminology*. ANSI S3.20-1973.

Anastasi, A. *Fields of Applied Psychology*. New York: McGraw-Hill, 1964.

Army Missile Command. *Military Standard: Human Engineering Design Criteria for Military Systems, Equipment, and Facilities*, (MIL-STD-1472A) Redstone Arsenal, Alabama. 1970.

Bard, P. *Medical Physiology*. (10th ed.) St. Louis: Mosby, 1956.

Bartels, H., Dejours, P., Kellogg, R.H., and Mead, J. Glossary on respiration and gas exchange. *Journal of Applied Physiology*. (April 1973):34(4).

Bilodeau, E.A. *Principles of Skill Acquisition*. New York: Academic Press, 1969.

Blesser, A. *Systems Approach to Bio-medicine*. New York: McGraw-Hill, 1969.

Chapanis, A. *Man-Machine Engineering*. Belmont, CA: Wadsworth Publishing Co., 1965.

Chapanis, A. *Research Techniques in Human Factors Engineering*. Baltimore: Johns Hopkins, 1959.

Chapanis, A., Garner, W.R., and Morgan, C.T. *Applied Experimental Psychology: Human Factors in Engineering Design*. New York: Wiley, 1949.

Damon, A., Stoudt, H.W., and McFarland, R.A. *The Human Body in Equipment Design*. Cambridge, MA: Harvard University Press, 1966.

DeGreene, K.B. *Systems Psychology*. New York: McGraw-Hill, 1970.

Department of Defense. *Definitions of Effectiveness Terms of Reliability, Maintainability, Human Factors and Safety*. MIL-STD-721B, 25 August 1966.

Firenze, R.J. *The Process of Hazard Control*. Dubuque, IA: Kendall/Hunt, 1978.

Fogel, L.J. *Biotechnology: Concepts and Applications*. Englewood Cliffs, NJ: Prentice-Hall, Inc., 1963.

Gagne, R.M., and Fleishman, E.A. *Psychology and Human Performance*. New York: Holt, Rinehart and Winston, 1959.

Gagne, R.M. (ed.) *Psychological Principles in System Development*. New York: Holt, Rinehart and Winston, 1962.

Kubokawa, C. *Databook for Human Factors Engineers*. Man Factors, Inc., 1969.

McCormick, E.J. *Human Factors Engineering*, (3rd ed.) New York: McGraw-Hill, 1970.

Meister, D., and Rabideau, G.F. *Human Factors Evaluation in System Development*. New York: Wiley-Interscience, 1965.

Morgan, C.T., Chapanis, A., Cook, J.S. III, and Lund, M.W. (Eds.) *Human Engineering Guide to Equipment Design*. U.S. DOD Joint Services Steering Committee. New York: McGraw-Hill, 1963.

Murrell, K.F.H. *Human Performance in Industry*. (2nd ed.) New York: Reinhold, 1969.

Parsons, H.M. *Man-Machine System Experiments*. Baltimore: The Johns Hopkins Press, 1972.

Post Office Department. *The Post Office Challenge to Industry*. Section VI. *Contractors Guide to Terminology*, 1967.

Sahakian, W.S. *Introduction to the Psychology of Learning*. Chicago: Rand-McNally, 1976.

Shurtleff, D. *Design Problems in Visual Displays*. ESD-TR-66-299, 1966.

Stevens, S.S. (ed.) *Handbook of Experimental Psychology*. New York: Wiley, 1951.

Tiffin, J., and McCormick, E.J. *Industrial Psychology*. (5th ed.) Englewood Cliffs, N.J.: Prentice-Hall, 1965.

VanCott, H.P. and Kinkade, R.G. *Human Engineering Guide to Equipment Design*. Washington, D.C.: American Institute for Research, 1972. For sale by Superintendent of Documents, U.S. Gov. Printing Office.

Woodson, W.E., and Conover, D.W. *Human Engineering Guide for Equipment Designers*. (2nd ed.) Berkeley, CA: University of California Press, 1970.

Woodworth, R.S., and Schlosberg, H. *Experimental Psychology*. (Rev. ed.) New York: Holt, 1954.

Z94.10
Management

In the 1989, Industrial Engineering Terminology Revised Edition, reference is made on page XVI to Management as: Management refers to the systematic organization, allocation, and application of economic and human resources to bring about a controlled change: the role of management in these proceedings is one of organization and consideration of human effort in exploiting resources for the improvement of the organization involved [Professor D. W. Karger taken from his School of Management (Rensselaer Polytehnic Institute) Long Range Planning Document].

While the Management subcommittee has reviewed much literature and a number of texts published over the past several years, as part of its normal education mission, the most complete (and recent) reference is the 1991 text; *Management*, by Kathryn M. Bartol, University of Maryland College Park; and David C. Martin, American University, McGraw-Hill, Inc., New York, N.Y., (1991), 813 pages.

This text therefore represents the major source of the start of our first Management Terminology Section.

Chairman

L. Ken Keys, Ph.D.
Professor and Chairman
Industrial and Manufacturing Systems
 Engineering Department
Louisiana State University, Baton Rouge

Subcommittee

Charles Parks, Ph.D., P.E.
Associate Professor
Industrial and Manufacturing Systems
 Engineering Department
Louisiana State University, Baton Rouge

Paul Givens, Ph.D.
Professor and Chairman
Industrial and Management Systems
University of South Florida, Tampa

Anita Callahan, Ph.D.
Assistant Professor
Industrial and Management Systems
University of South Florida, Tampa

ABILITY. The capacity to perform; along with effort, it is a key determinant of performance.

ACCEPTABLE QUALITY LEVEL (AQL). A predetermined standard against which random samples of produced materials are compared in acceptance sampling.

ACCEPTANCE SAMPLING. A statistical technique used in quality control that involves evaluating random samples from a group, or "lot," of produced materials to determine whether the lot meets acceptable quality levels.

ACCEPTANCE THEORY OF AUTHORITY. A theory connected with administrative management approach which argues that authority does not depend as much on "persons of authority" who *give* orders as on the willingness of those who *receive* the orders to comply.

ACCOMMODATION. A conflict-handing mode that involves solving conflicts by allowing the desires of the other party to prevail.

ACCOUNTABILITY. The requirement to provide satisfactory reasons for significant deviations from duties or expected results.

ACHIEVEMENT-ORIENTED. A leader behavior identified in path-goal theory that involves setting challenging goals, expecting subordinates to perform at their highest level, and conveying a high degree of confidence in subordinates.

ACTION PLAN. Plans that identify day-to-day activities in support of the goals of the organization.

ACQUIRED-NEEDS THEORY. A content theory of motivation (developed by David C. McClelland) which argues that our needs are acquired or learned on the basis of our life experience.

ACQUISITION. The purchase of all or part of one organization by another, and a means of implementing growth strategies.

ACTION RESEARCH. A method used in the diagnosis phase of organizational development that places heavy emphasis on data gathering and collaborative diagnosis before action is taken.

ACTIVE LISTENING. The process in which a listener actively participates in attempting to grasp the facts and the feelings being expressed by the speaker.

ACTIVITY. A work component to be accomplished, represented by an arrow on a PERT network diagram or bar on a GANTT Chart.

AD HOC COMMITTEE. A committee created for a short term purpose.

ADAPTIVE MODE. An approach to strategy formulation that emphasizes taking small incremental steps, reacting to problems rather than seeking opportunities, and attempting to satisfy a number of organizational power groups.

ADHOCRACY. The structural configuration in Mintzberg's typology characterized by various forms of matrix departmentalization, expertise dispersed throughout, low formalization, and emphasis on mutual adjustment.

ADJOURNING. A stage of group development in which group members prepare for disengagement as the group nears successful completion of its goals.

ADMINISTRATIVE MANAGEMENT. An approach within classical management theory that focuses on principles that can be used by managers to coordinate the internal activities of organizations.

ADMINISTRATIVE PROTECTIONS. A type of trade barrier in the form of various rules and regulations that make it more difficult for foreign firms to conduct business in a particular country.

ADVERSE IMPACT. The effect produced when a job selection rate for a protected group is less than 80 percent of the rate for the majority group.

ADVERTISING. An approach to influencing the environment involving the use of communications media to gain favorable publicity for particular products and services.

AFFIRMATIVE ACTION. Any special activity undertaken by employers to increase equal employment opportunities for groups protected by federal equal employment opportunity laws and related regulations.

AFFIRMATIVE ACTION PLAN. A written, systematic plan that specifies goals and timetables for hiring, training, promoting, and retraining groups protected by federal equal employment laws and related regulations.

AGGREGATE PRODUCTION PLANNING. A primary operating system used in operations management that is concerned with planning how to match supply with product or service demand over a time horizon of about 1 year.

ALTERNATIVE WORK SCHEDULES. Schedules based on adjustments in the normal work schedule rather than in the job content or activities.

AMORAL MANAGEMENT. An approach to managerial ethics that is neither immoral nor moral but, rather, ignores or is oblivious to ethical considerations.

ANCHORING AND ADJUSTMENT. A decision-making bias that involves the tendency to be influenced by an initial figure, even the information is largely irrelevant.

ANTIFREELOADER ARGUMENT. An argument that indicates that since businesses benefit from a better society, they should bear part of the costs by actively working to bring about solutions to social problems.

APPLICATION BLANK. A form used widely as a selection method that contains a series of inquires about such issues as an applicant's educational background, previous job experience, physical health, and other information that may be useful in assessing an individual's ability to perform a job.

APPLICATION SOFTWARE PACKAGES. Special purpose computer programs available for sale or lease from commercial sources.

ARTIFICIAL INTELLIGENCE. A field of information technology aimed at developing computers that have humanlike capabilities, such as seeing, hearing, and thinking (i.e. reasoning and learning from experience).

ASSESSMENT CENTER. A controlled environment used to predict the probable managerial success of individuals mainly on the basis of evaluations of their behaviors in a variety of simulated situations, thereby facilitating selection.

ASSET MANAGEMENT RATIOS. Financial ratios that measure how effectively an organization uses its resources.

AUTHORITY. The right to make decisions, carry out actions, and direct others in matters related to the duties and goals of a position.

AUTOCRATIC. The behavioral style of leaders who makes unilateral decisions, dictates work methods, limits worker knowledge about goals, the next step to be performed, and sometimes gives feedback that is punitive.

AUTONOMOUS WORK GROUP. Another name for a self-managing team.

AUTONOMY. A core job characteristic involving the amount of discretion allowed in determining schedules and work methods for achieving the required output.

AVAILABILITY. A decision-making bias that involves the tendency to judge the likelihood of an occurrence on the basis of the extent to which other like instances or occurrences can easily be recalled.

AVOIDANCE. A conflict-handing mode that involves ignoring or suppressing a conflict in the hope that it will either go away or not become too disruptive.

BACKORDER. Orders for goods which cannot be filled immediately but which will be honored as more goods as service slots become available.

BALANCE OF PAYMENTS. An international economic element that is an account of goods and services, capital loans, gold, and other items entering and leaving a country, and a factor influencing the ability of an organization to conduct international business in that country successfully.

BALANCE OF TRADE. The difference between a country's exports and imports, and, generally, the most critical determinant of a country's balance of payments.

BALANCE SHEET. A financial statement that depicts an organization's assets and claims against those assets at a given point in time.

BALANCE SHEET BUDGET. A financial budget that forecasts the assets, liabilities, and shareholders' equity at the end of the budget period.

BANKRUPTCY. A defensive strategy in which an organization that is unable to pay its debts can seek court protection from creditors and from certain contact obligations while it attempts to regain financial stability.

BARGAINING. A method of formulating goals to match agreements for the exchange of goods or services with other organizations.

BATCH PROCESSING. An arrangement whereby data is accumulated and then processed as a whole.

BCG GROWTH-SHARE MATRIX. A portfolio approach involving a four-cell matrix (developed by the Boston Consulting Group) that compares various businesses in an organization's portfolio on the basis of relative market share and market growth rate.

BEHAVIOR MODIFICATION. The use of techniques associated with reinforcement theory.

BEHAVIORAL DISPLACEMENT. A condition that is a side effect of poorly designed and/or excessive controls in which individuals engage in behaviors that are encouraged by controls and related reward systems even though the behaviors actually are inconsistent with organizational goals.

BEHAVIORAL SCIENCE. An approach that emphasizes *scientific research* as the basis for developing theories about human behavior in organizations that can be used to develop practical guidelines for managers.

BEHAVIORAL VIEWPOINT. A perspective on management that emphasizes the importance of attempting to understand the various factors that affect human behavior in organizations.

BEHAVIORALLY ANCHORED RATING SCALES (BARS). Performance appraisal scales containing sets of specific behaviors that represent gradations of performance used as common reference points (or anchors) for rating employees on various job dimensions.

BELONGINGNESS NEEDS. The needs in Maslow's hierarchy that involve the desire to affiliate with and be accepted by others.

BENEFITS. Forms of compensation beyond wages for time worked, including various protection plans, services, pay for time not worked, and income supplements.

BILL OF MATERIAL (BOM). An input to MRP systems that consists of a listing of all components, including partially assembled pieces and basic parts, that make up an end product.

BOTTOM-UP BUDGETING. A process of developing budgets in which lower-level and middle managers specify their budgetary needs and top management attempts to accommodate them to the extent possible.

BOUNDARY SPANNING. An approach to influencing an environment that involves creating roles within the organization that interface with important elements in the environment.

BOUNDED RATIONALITY. A concept that suggests the ability of managers to be perfectly reasonable in making decisions is limited by such factors as cognitive capacity and time constraints.

BRAINSTORMING. A technique for enhancing group creativity that encourages group members to generate as many novel ideas as possible on a given topic without evaluating them.

BREAK-EVEN ANALYSIS. A quantitative technique based on a graphic model that helps decision makers understand the relationships among sales volume, costs, and revenues in an organization.

BUDGETING. The process of stating in quantitative terms, usually dollars, planned organizational activities for a given period of time.

BUFFERING. A method of adapting to environmental fluctuations that involves stockpiling either inputs into or outputs from a production or service process.

BUREAUCRATIC CONTROL. A managerial approach that relies on regulation through rules, policies, supervision, budgets, schedules, reward systems, and other administrative mechanisms aimed at ensuring that employees exhibit appropriate behaviors and meet performance standards.

BUREAUCRATIC MANAGEMENT. An approach within classical management theory that emphasizes the need for organizations to operate in a rational manner rather than relying on the arbitrary whims of owners and managers.

BUSINESS-LEVEL STRATEGY. A type of strategy that concentrates on the best means of competing within a particular business while also supporting the corporate-level strategy.

BUSINESS PLAN. A document written by the prospective owner or entrepreneur that details the nature of the business, the product or service, the customers, the competition, the production and marketing methods, the management, the financing, and other significant aspects of the proposed business venture.

CAFETERIA STYLE BENEFIT PLAN. A benefit plan in which employees can choose from a wide range of alternative benefits, tailoring them to their particular situations.

CAPACITY ARGUMENT. A theory which states that the private sector, because of its considerable economic and human resources, must make up for recent government cutbacks in social programs.

CAPACITY PLANNING. A primary operating system used in operations management that is concerned with determining the people, machines, and major physical resources, such as buildings, that will be necessary to meet the production objectives of the organization.

CAPACITY REQUIREMENTS PLANNING. A technique for determining what personnel and equipment are needed to meet short-term production objectives.

CAPITAL EXPENDITURES BUDGET. A type of budget that involves a plan for the acquisition or divestiture of major fixed assets, such as land, buildings, or equipment.

CAPITALIST ECONOMY. An economy in which economic activity is governed by market forces and the means of production are privately owned by individuals.

CAREER. A sequence of work-related activities and experiences that span the course of a person's life.

CAREER MANAGEMENT. The continuing process of setting career goals, formulating and implementing strategies for reaching the goals, and monitoring the results.

CARRYING, OR HOLDING, COST. An inventory cost comprised of the expenses associated with keeping an item on hand (such as storage, insurance, pilferage, breakage).

CASH BUDGET. A financial budget that projects future cash flows arising from cash receipts and disbursements by the organization during a specified period.

CENTRAL PROCESSING UNIT. The main memory and processing section of a computer.

CENTRALIZATION. A vertical coordination method that addresses the extent to which power and authority are retained at the top organizational levels.

CEREMONIAL. A system of rites performed in conjunction with a single occasion or event.

CHAIN OF COMMAND. The unbroken line of authority that ultimately links each individual with the top organizational position through a managerial position at each successive layer in between.

CHANGE. Any alteration of the status quo.

CHANGE AGENT. An individual with a fresh perspective and a knowledge of the behavioral sciences who acts as a catalyst in helping the organization approach old problems in new or innovative ways and thus plays a key role in OD interventions.

CHARISMA. A leadership factor that comprises the leader's ability to inspire pride, faith, and respect; to recognize what is really important; and to articulate effectively a sense of mission or vision that inspires followers.

CLAN CONTROL. A managerial approach that relies on values, beliefs, traditions, corporate culture, shared norms, and informal relationships to regulate employee behavior and facilitate the reaching of organizational goals.

CLASSICAL VIEWPOINT. A perspective on management that emphasizes finding ways to manage work and organizations more effectively.

CLIENTS. Those individuals and organizations that purchase an organization's products and/or services.

CLOSED SYSTEM. A system that does little or no interacting with its environment and receives little feedback.

COERCIVE POWER. Power that depends on the ability to punish others when they do not engage in desired behaviors.

COGNITIVE RESOURCE THEORY. A theory, which is a major revision and extension of Fiedler's contingency model, that considers the additional factor of a leader's cognitive resource use in predicting performance.

COGNITIVE RESOURCES. Factors that involve the intellectual abilities, technical competence, and job relevant knowledge that leaders bring to their jobs.

COGNITIVE THEORIES. Theories that attempt to isolate the thinking patterns that we use in deciding whether or not to behave in a certain way.

COLLABORATION. A conflict-handling mode that strives to resolve conflict by devising solutions that allow both parties to achieve their desired outcomes.

COMMAND GROUP. A formal group consisting of a manager and all the subordinates who report to that manager.

COMMUNICATION. The exchange of messages between

people for the purpose of achieving common meanings.

COMMUNICATION CHANNELS. Various patterns of organizational communication flow that represent potential established conduits through which managers and other organization members can send and receive information.

COMMUNICATION NETWORK. The pattern of information flow among task group members.

COMPENSATION. Wages paid directly for time worked, as well as more indirect benefits that employees receive as part of their employment relationship with an organization.

COMPETITION. A conflict-handling mode that involves attempting to win a conflict at the other party's expense.

COMPETITIVE ADVANTAGE. A significant and specific edge one enjoys over a business rival.

COMPETITIVE ADVANTAGE OF NATIONS. The concept that environmental elements within a nation can foster innovation in certain industries, thereby increasing the prospects of success of home-based companies operating internationally within those industries.

COMPETITORS. The element of task environment that includes other organizations that either offer or have a high potential of offering rival products or services.

COMPLACENCY. A condition preventing effective decision making in which individuals either do not see the signs of danger or opportunity or ignore them.

COMPRESSED WORKWEEK. An alternative work schedule whereby employees work four 10-hour days or some similar combination, rather than the usual five 8-hour days.

COMPROMISE. A conflict-handling mode that aims to solve conflict issues by having each party give up some desired outcomes in order to get other desired outcomes.

COMPUTER-AIDED DESIGN (CAD). A system on which CIM systems rely that uses computers to geometrically prepare, review, and evaluate product designs.

COMPUTER-AIDED MANUFACTURING (CAM). A system on which CIM systems rely that uses computers to design and control production processes.

COMPUTER-ASSISTED INSTRUCTION (CAI). A training technique in which computers are used to lessen the time necessary for training by instructors and to provide additional help to individual trainees.

COMPUTER-BASED INFORMATION SYSTEMS (CBISs). Information systems that involve the use of computer technology.

COMPUTER-INTEGRATED MANUFACTURING (CIM). The computerized integration of all major functions associated with the production of a product.

COMPUTER VIRUS. A small program, usually hidden inside another program, that replicates itself and surfaces at a predetermined time to cause disruption and possibly destruction.

CONCENTRATION. A growth strategy that focuses on guiding the growth of a single product or service or a small number of closely related products or services.

CONCEPTUAL SKILLS. Key management skills related to the ability to visualize the organization as a whole, discern interrelationships among organizational parts, and understand how the organization fits into the wider context of the industry, concurrent enterprise, community, and world.

CONCURRENT CONTROL. A control type based on timing involving the regulation of ongoing activities that are part of the transformation process to ensure that they conform to organizational standards.

CONCURRENT ENGINEERING. (C.E., also called Concurrent Enterprise). A systematic approach to the integrated, simultaneous design of products and their related processes, including manufacture and support. This approach is intended to cause the developers, from the outset, to consider all elements of the product life cycle from conception through disposal, including quality, cost, schedule, and user requirements. Also often called Simultaneous Engineering.

CONFLICT. A perceived difference between two or more parties that results in mutual opposition.

CONSIDERATION. The degree to which a leader builds mutual trust with subordinates, respects their ideas, and shows concern for their feelings.

CONSTRAINTS. Conditions that must be met in the course of solving a linear programming problem.

Contingency Planning. The development of alternative plans for use in the event that conditions evolve differently than anticipated, rendering original plans unwise or unfeasible.

Contingency Theory. A viewpoint that argues that appropriate managerial action depends on the particular parameters of situation.

Continuous-Process Production. A type of technology in which products are made through a uninterrupted operation.

Control System. A set of mechanisms, established as part of the control process, that are designed to increased the probability of meeting organizational standards and goals.

Controlling. The management function that is aimed at regulating organizational activities so that actual performance conforms to expected organizational standards and goals.

Convergent Thinking. A way of thinking related to creativity in which an individual attempts to solve problems by beginning with a problem and attempting to move logically to a solution.

Co-opting. An approach to influencing the environment that involves absorbing key members of important environmental elements into the leadership or policymaking structure of an organization.

Corporate Culture. A term sometimes used for organizational culture.

Corporate-Level Strategy. A type of strategy that addresses what businesses the organization will operate, how the strategies of those businesses will be coordinated to strengthen the organization's competitive position, and how resources will be allocated among the businesses.

Corporate Philanthropy. Corporate contributions for charitable and social responsibility purposes.

Corporate Social Responsibility. A term often used to refer to the concept of organizational social responsibility as applied to business organizations.

Corporate Social Responsiveness. A term used to refer to the concept of organizational social responsiveness as applied to business organizations.

Corridor Principle. A principle which states that the process of beginning a new venture helps entrepreneurs visualize other opportunities that they could not envision or take advantage of until they started the initial venture.

Cost Leadership Strategy. A generic business-level strategy outlined by Michael E. Porter that involves emphasizing organizational efficiency so that the overall costs of providing products and services are lower than those of competitors.

Creativity. The cognitive process of developing an idea, concept, commodity, or discovery that is viewed as novel by its creator or a target audience.

Crisis Problem. A type of problem in managerial decision making involving a serious difficulty that requires immediate action.

Critical Path. The path in a PERT network that will take the longest to complete.

Customer Divisions. A form of divisional structure involving divisions set up to service particular types of clients or customers.

Customers. Those individuals and organizations that purchase an organization's products and/or services.

Cybernetic Control System. A self-regulating control system that, once it is put into operation, can automatically monitor the situation and take corrective action when necessary.

Data. Unanalyzed facts and figures.

Data Base. A set of data organized efficiently in a central location so that it can serve a number of information system applications.

Data Base Management System. The software that allows an organization to build, manage, and provide access to its stored data.

Debt Capital. A type of financing available to entrepreneurs that involves a loan to be repaid, usually with interest.

Debt Management Ratios. Financial ratios that assess the extent to which an organization uses debt to finance investments, as well as the degree to which it is able to meet its long-term obligations.

DECENTRALIZATION. A vertical coordination method that addresses the extent to which power and authority are delegated to lower levels.

DECIDING TO DECIDE. A response in which decision makers accept the challenge of deciding what to do about a problem and follow an effective decision-making process.

DECISION MAKING. The process through which manages identify organizational problems and attempt to resolve them.

DECISION MATRIX. (See PAYOFF TABLE.)

DECISION SUPPORT-SYSTEM (DSS). A computer-based information system that supports the process of managerial decision making in situations that are well structured.

DECISION TREE. A quantitative decision-making aid based on a graphic model that displays the structure of a sequence of alternative courses of action and usually shows the payoffs associated with various paths and the probabilities associated with potential future conditions.

DECODING. The process, in communication, of translating symbols into meaning.

DEFENSIVE AVOIDANCE. A condition preventing effective decision making in which individuals either deny the importance of a danger or an opportunity or deny any responsibility for taking action.

DEFENSIVE STRATEGIES. Strategies (sometimes called *retrenchment strategies*) that focus on the desire or need to reduce organizational operations usually through cost and/or asset reductions.

DELEGATION. A means of vertical coordination that involves the assignment of part of a manager's work to others along with both the responsibility and the authority necessary to achieve expected results.

DELPHI METHOD. A method of technological or qualitative forecasting that uses a structured approach to gain the judgments of a number of experts on a specific issue relating to the future.

DEMOCRATIC. The behavioral style of leaders who tend to involve the group in decision making, let the group determine work methods, make overall goals known, and use feedback as an opportunity for helpful coaching.

DEPARTMENTALIZATION. An aspect of organization structure involving the clustering of individuals into units and of units into departments and larger units in order to facilitate achieving organizational goals.

DEPENDENT DEMAND INVENTORY. A type of inventory consisting of raw materials, components, and subassemblies that are used in the production of an end product or service.

DESCRIPTIVE DECISION-MAKING MODELS. Models that attempt to document how managers actually *do* make decisions.

DEVELOPED COUNTRIES. A group of countries that are characterized by high levels of economic or industrial development and that include the United States, western Europe, Canada, Australia, New Zealand, and Japan.

DEVIL'S ADVOCATES. Individuals who are assigned the role of making sure that the negative aspects of any attractive decision alternatives are considered and whose involvement in a group helps avoid group-think.

DIALECTICAL INQUIRY. A technique used to help avoid group-think that involves approaching a decision situation from two opposite points of view.

DIFFERENTATION. The extent to which organizational units differ from one another in terms of the behaviors and orientations of their members and their formal structures; also used to refer to the tendency for open systems to become more complex.

DIFFERENTIATION PARADOX. The idea that although separating efforts to innovate from the rest of the organization increases the likelihood of developing radical ideas, such differentiation also decreases the likelihood that the radical ideas will ever be implemented.

DIFFERENTIATION STRATEGY. A generic business-level strategy outlined by Michael E. Porter that involves an attempt to develop products and services that are viewed as unique in the industry.

DIRECT CONTACT. A means of facilitating lateral relations that involves communication between two or more persons at similar levels in different work units for purposes of coordinating work and solving problems.

DIRECT INTERLOCK. A situation in which two companies have a director in common.

Direct Investment. A means of entering international markets involving the establishment of operating facilities in a foreign country.

Directive. A leader behavior identified in path-goal theory that involves letting subordinates know what is expected of them, providing guidance about work methods, developing work schedules, identifying work evaluation standards, and indicating the basis for outcomes or rewards.

Discretionary Expense Center. A responsibility center whose budgetary performance is based on achieving its goals by operating within predetermined expense constraints set through managerial judgment or discretion.

Dissatisfies. A type of factor which figures in the two-factor theory of motivation that is largely associated with the work environment (such as working conditions and supervision) and that can influence the degree of worker dissatisfaction.

Distinctive Competence. An organizational strength that is unique and not easily matched or imitated by competitors.

Distributed Processing. An approach to controlling information system resources in which computers are distributed to various organizational locations where they serve the organization's local and/or regional needs and are interconnected for electronic communication.

Divergent Thinking. A way of thinking related to creativity in which an individual attempts to solve problems by generating new ways of viewing the problem and seeking novel alternatives.

Diversification. A growth strategy that entails effecting growth through the development of new areas that are clearly distinct from current businesses.

Divestiture. A defensive strategy that involves an organization's selling or divesting of a business or part of a business.

Divisional Structure. A type of departmentalization in which positions are grouped according to similarity of products, services, or markets.

Divisionalized Form. The structural configuration in Mintzberg's typology characterized by divisional departmentalization, a strong management group at the division level, high formalization within divisions, and an emphasis on standardized outputs.

Domain Shifts. Changes in the mix of products and services offered so that an organization will interface with more favorable environmental elements.

Downsizing. A method of increasing organizational efficiency and effectiveness that involves significantly reducing the layers of middle management, expanding spans of control, and shrinking the size of the work force.

Downward Communication. Vertical communication that flows from a higher level to one or more lower levels in the organization.

Driving Forces. Forces studied in force-field analysis that involve factors that pressure *for* a particular change.

Econometric Models. Explanatory models based on systems of simultaneous multiple regression equations involving several predictor variables that are used to identify and measure relationships or interrelationships that exist in the economy.

Economic Element. The part of the mega-environment that encompasses the systems of producing, distributing, and consuming wealth.

Economic Order Quantity (EOQ). An inventory control method developed to minimize order holding costs, while avoiding stockout costs.

Effectiveness. A dimension of organizational performance involving the ability to choose and achieve appropriate goals.

Efficiency. A dimension of organizational performance involving the ability to make the best use of available resources in the process of achieving goals.

Effort-Performance Expectancy. A component of expectancy theory that concerns our assessment of the probability that our efforts will lead to the required performance level.

Electronic Data Processing (EDP). The transformation of data into meaningful information through electronic means.

Electronic Mail System. The mail system that allows high-speed exchange of written messages through the use of computerized text-processing and communications networks.

Electronic Monitoring. An issue in job-related stress and health that involves the use of computers to continually assess employee performance.

Employee Assistance Program (EAP). A program through which employers help employees overcome personal problems that are adversely affecting their job performance.

Employee-Centered. A leadership approach in which managers channel their main attention to the human aspects of subordinates' problems and to the development of an effective work group dedicated to high performance goals.

Employee Involvement Teams. Small groups of employees who work on solving specific problems related to quality and productivity, often with stated targets for improvement.

Employment at Will. The legal principle that holds that either employee or employer can terminate employment at any time for any reason.

Employment Test. A selection method involving the assessment of a job applicant's characteristics through paper-and-pencil responses or simulated exercises.

Encoding. The process, in communication, of translating the intended meaning into symbols.

End User. The same as a user.

End-User Computing. The development and/or management of information systems by users.

Enlightened Self-Interest Argument. An argument that holds that businesses exist at society's pleasure and that, for their own legitimacy and survival, businesses should meet the expectations of the public regarding social responsibility.

Entrepreneur. An individual who creates a new enterprise.

Entrepreneurial Mode. An approach to strategic management in which strategy is formulated mainly by a strong visionary chief executive who actively searches for new opportunities, is heavily oriented toward growth, and is willing to make bold decisions or to shift strategic rapidly.

Entrepreneurial Team. A special type of project team comprising a group of individuals with diverse expertise and backgrounds who are brought together to develop and implement innovative ideas aimed at creating new products or services or significantly improving existing ones.

Entrepreneurship. The "creation of new enterprise".

Enterprise Integration. A concept that encompasses the entire scope of a business, characterized by: (1) optimization across the total organization based on common goals; (2) balancing the needs, capabilities, and relationships of people, processes, and tools; (3) shared data, processes, and tools to perform processes; (4) universal availability and access of data, processes, and tools; and (5) storage and retrieval of data, rather than transfer of data.

Environmental Complexity. A factor affecting the level of environmental uncertainty that involves the number of elements in an organization's environment and their degree of similarity.

Environmental Dynamism. A factor affecting the level of environmental uncertainty that involves the rate and predictability of change in the elements of an organization's environment.

Environmental Munificence. The extent to which the environment can support sustained growth and stability.

Environmental Uncertainty. A condition of the environment in which future conditions affecting an organization cannot be accurately assessed and predicted.

Equity Capital. A type of financing available to entrepreneurs that usually requires that the investor be given some form of ownership in the venture.

Equity Theory. A cognitive theory of work motivation which argues that we prefer situations of balance, or equity, which exist when we perceive the ratio of our inputs and outcomes to be equal to the ratio of inputs and outcomes for a comparison other.

ERG Theory. An alternative (proposed by Clayton Alderfer) to Maslow's hierarchy of needs theory which argues that there are three levels of individual needs.

Escalation Situations. Situations that signal the strong possibility of escalating commitment and accelerating losses.

ESTEEM NEEDS. The needs in Maslow's hierarchy related to a two-pronged desire to have a positive self-image and to have our contributions valued and appreciated by others.

ETHICS. A set of standards and code of conduct that define what is right, wrong, and just in human actions.

ETHNOCENTRIC ORIENTATION. An approach to international management (also known as home-country orientation) in which executives assume that practices that work in the headquarters or home country must necessarily work elsewhere.

EVENT. An indication of the beginning and/or ending of activities in a PERT network.

EXCHANGE RATE. An international economic element that is the rate at which one country's currency can be exchanged for another country's currency.

EXECUTIVE SUPPORT SYSTEM (ESS). A computer-based information system that supports decision making and effective functioning at the top levels of an organization.

EXISTENCE NEEDS. The needs in ERG theory that include the various forms of material and physiological desires, such as food and water, as well as such work-related forms as pay, fringe benefits, and physical working conditions.

EXPATRIATES. Individuals who are not citizens of the countries in which they are assigned to work.

EXPECTANCY THEORY. A cognitive theory of workmotivation (originally proposed by Victor H. Vroom) which argues that we consider three main issues before we expend the effort necessary to perform at a given level.

EXPECTED VALUE. The sum of payoffs times; the respective probabilities for a given alternative.

EXPENSE BUDGET. An operating budget that documents expected expenses during the budget period.

EXPERT POWER. Power that is based on the possession of unique knowledge that is valued by others.

EXPERT SYSTEM (ES). A computer-based system that applies the substantial knowledge of an expert to help solve problems in a specific area.

EXPLANATORY, OR CAUSAL MODELS. Methods of quantitative forecasting that attempt to identify the major variables that are related to or have caused particular past conditions and then use current measures of those variables (predictors) to predict future conditions.

EXPORTING. A means of entering international markets that involves an organization's making a product in the home country and sending it overseas.

EXPROPRIATION. The seizure of a foreign company's assets by a host-country government.

EXTERNAL AUDIT. A financial audit involving the review and verification of the fairness of an organization's financial statements that is conducted by an independent auditor.

EXTERNAL ENVIRONMENT. The major forces outside the organization that have the potential of significantly impacting on the likely success of products or services.

EXTINCTION. A type of reinforcement in behavior modification that involves withholding previously available positive consequences associated with a behavior in order to *decrease* that behavior.

EXTRINSIC REWARDS. Rewards, such as bonuses, awards, or promotions, that are provided by others.

FACILITIES. The land, buildings, equipment, and other major physical inputs that substantially determine productive capacity, require time to alter, and involve significant capital investment.

FEEDBACK. A term that generally denotes information about the results of actions; used in conjunction with systems theory to signify information about results and organizational status relative to the environment; as applied to the job characteristics model of job design, specifies core job characteristic associated with the degree to which the job provides for clear, timely information about performance results; also used to indicate an element in the communication process that refers to the basic response of the receiver to the interpreted message in the communication process.

FEEDBACK CONTROL. A control type based on timing that involves regulation exercised after a product or service has been completed in order to ensure that the final output meets organizational standards and goals.

FEEDFORWARD CONTROL. A control type based on timing that focuses on the regulation of inputs to ensure that they meet the standards necessary for the transformation process.

FIELDER'S CONTINGENCY MODEL. A situational leadership approach originally develop by Fred Fiedler and his associates.

FINANCIAL BUDGETS. Plans that outline how an organization is going to acquire its cash and how it intends to use the cash.

FINANCIAL STATEMENT. A summary of a major aspect of an organization's economic status which is used as a fiscal control technique.

FINISHED-GOODS INVENTORY. The stock of items that have been produced and are awaiting sale or transit to a customer.

FIRST-LINE MANAGERS. Managers at the lowest level of the hierarchy who are directly responsible for the work of operating (nonmanagerial) employees.

FIRST-LINE SUPERVISORS. The same as first-line managers.

FIVE COMPETITIVE FORCES MODEL. A model developed by Michael E. Porter and used in the examination of an organization's task environment that offers an approach to analyzing the nature and intensity of competition in a given industry in terms of five major forces.

FIXED-INTERVAL SCHEDULE OF REINFORCEMENT. A type of partial reinforcement schedule based on a pattern in which a reinforcer is administered on a fixed time schedule, assuming that the desired behavior has continued at an appropriate level.

FIXED POSITION LAYOUT. A type of facilities layout having a production configuration in which the product or client remains in one location and the tools, equipment, and expertise are brought to it, as necessary, to complete the productive process.

FIXED-RATIO SCHEDULE OF REINFORCEMENT. A type of partial reinforcement schedule based on a pattern in which a reinforcer is provided after a fixed number of occurrences of the desired behavior.

FLAT STRUCTURE. A structure that has few hierarchical levels and wide spans of control.

FLEXIBLE MANUFACTURING SYSTEM (FMS). A manufacturing system on which CIM systems rely that uses computers to control machines and the production process automatically so that different types of parts or product configurations can be handled on the same production line.

FLEXTIME. An alternative work schedule that specifies certain core hours when individuals are expected to be on the job and then allows flexibility in starting and quitting times as long as individuals work the total number of required hours per day.

FOCUS STRATEGY. A generic, business-level strategy outlined by Michael E. Porter that entails specializing by establishing a position of overall cost leadership, differentiation, or both, but only within a particular portion or segment of an entire market.

FORCE-FIELD ANALYSIS. A method that involves analyzing the two types of forces that influence any proposed change; driving forces and restraining forces.

FORMAL COMMUNICATION. Vertical and horizontal communication that follows paths specified by the official hierarchical organization structure and related task requirements.

FORMAL GROUP. A group officially created by an organization for a specific purpose.

FORMALIZATION. A method of vertical coordination that addresses the degree to which written policies, rules, procedures, job descriptions, and other documents specify what actions are (or are not) to be taken under a given set of circumstances.

FORMING. A stage of group development in which group members attempt to assess the ground rules that will apply to a task and to group interaction.

FRAMING. A decision-making bias that involves the tendency to make different decisions depending on how a problem is presented.

FRANCHISE. A continuing arrangement between a franchiser and a franchise in which the franchiser's knowledge, image, manufacturing or service expertise, and marketing techniques are made available to the franchises in return for the payment of various fees or royalties and conformity to standard operating procedures.

FRANCHISEE. An individual who purchases a franchise and, in the process, is given an opportunity to enter a new business hopefully with an enhanced chance of success.

Free Riders. Individuals who engage in social loafing, thus benefiting from the work of the group without bearing their proportional share of the cost involved.

Friendship Group. An informal group that evolves primarily to meet employee social needs.

Frustration Regression Principle. A principle incorporated into ERG theory which states the if we are continually frustrated in our attempts to satisfy a higher-level need, we may cease to be concerned about that need.

Functional Audit. A technique for evaluating internal strengths and weaknesses that involves an exhaustive appraisal of an organization and / or its individual businesses conducted by assessing the important positive and negative attributes of each major functional area.

Functional Group. A formal group consisting of a manager and all the subordinates that report to that manager.

Functional-Level Strategy. A type of strategy that focuses on action plans for managing a particular area within a business in a way that supports the business-level strategy.

Functional Managers. Managers who have responsibility for a specific, specialized area of the organization and supervise mainly individuals with expertise and training in that specialized area.

Functional Structure. A type of departmentalization in which positions are grouped according to their main functional (or specialized) area.

Futurists. Individuals useful in social forecasting who track significant social and other trends in the environment and attempt to predict their impact on the organization.

Gainsharing. A compensation system in which employees throughout an organization are encouraged to become involved in solving problems and are then given bonuses tied to organizationwide performance improvements.

Game Theory. A quantitative technique for facilitating decision making in situations of conflict among two or more decision makers seeking to maximize their own welfares.

Gantt Chart. A type of planning and control model developed by Henry L. Gantt that relies on a specialized bar chart showing the current progress on each major project activity relative to necessary completion dates.

Garbage-Can Model. A nonrational model of managerial decision making stating that managers behave in virtually a random pattern in making nonprogrammed decisions.

GE Business Screen. A portfolio approach involving a nine-cell matrix (developed by General Electric with McKinsey & Company) that is based on long-term industry attractiveness and on business strength.

General Managers. Managers who have responsibility for a whole organization or a substantial subunit that includes most of the common specialized areas within it.

Geocentric Orientation. An approach to international management (also known as *world orientation*) whereby executives believe that a global view is needed in both the headquarters of the parent company and its various subsidiaries and that the best individuals, regardless of home-or host-country origin, should be utilized to solve company problems anywhere in the world.

Geographic Divisions. A form of divisional structure involving divisions designed to serve different geographic areas.

Globalization. A strategy aimed at developing relatively standardized products with global appeal, as well as at rationalizing operations throughout the world.

Goal. A major planning component that is a future target or end result that an organization wishes to achieve.

Goal Commitment. A critical goal setting element that involves one's attachment to, or determination to reach, a goal.

Goal Incongruence. A condition in which there are major incompatibilities between the goals of an organization member and those of the organization, the effects of which can be reduced by the establishment of standards.

Government Agencies. The element of the task environment that includes agencies that provide service and monitor compliance with laws and regulations at local, state or regional, and national levels.

GRAND STRATEGY. A master strategy that provides the basic strategic direction at the corporate level.

GRAPEVINE. Another term for informal communication.

GRAPHIC RATING SCALES. Performance appraisal scales that list a number of rating factors, including general behaviors and characteristics, on which an employer is rated by the supervisor.

GROUP. Two or more interdependent individuals who interact with and influence each other in collective pursuit of a common goal.

GROUP COHESIVENESS. A major group process factor that concerns the degree to which members are attracted to a group, are motivated to remain in the group, are motivated to remain in the group, and are mutually influenced by one another.

GROUP MAINTENANCE ROLES. Roles that do not directly address a task itself but, instead, help foster group unity, positive interpersonal relations among group members, and development of the ability of members to work effectively together.

GROUP TASK ROLES. Roles that help a group develop and accomplish its goals.

GROUP TECHNOLOGY. The classification of parts into families (groups of parts or products that have some similarities in the way they are manufactured) so that members of the same family can be manufactured on the same production line.

GROUPTHINK. A phenomenon of group decision making in which cohesive groups tend to seek agreement about an issue at the expense of realistically appraising the situation.

GROWTH-NEED STRENGTH. The degree an individual needs personal growth and development on the job.

GROWTH NEEDS. The needs in ERG theory that impel creativity and innovation, along with the desire to have a productive impact on our surroundings.

GROWTH STRATEGIES. Grand strategies that involve organizational expansion along some major dimension.

HACKERS. Individuals who are knowledgeable about computers and who gain unauthorized entry to, and sometimes tamper with, computer networks and files of organizations with which they have no affiliation.

HALO EFFECT. The tendency to use a general impression (based on a few characteristics) of an individual to judge other characteristics of that same individual.

HAND OF GOVERNMENT. A view of corporate social responsibility which argues that the interests of society are best served by having the regularity hands of the law and the political process, rather than the invisible hand, guide the results of corporations' endeavors.

HAND OF MANAGEMENT. A view of corporate social responsibility which states that corporations and their managers are expected to act in ways that protect and improve the welfare of society as a whole, as well as advance corporate economic interests.

HARDWARE. Physical computer equipment, including the computer itself and related devices.

HARVEST. A defensive strategy that entails minimizing investments while attempting to maximize short-run profits and cash flow, with the long-run intention of exiting the market.

HAWTHORNE EFFECT. The possibility that individuals singled out for a study may improve their performance simply because of the added attention they receive from the researches, rather than because of any specific factors being tested.

HAWTHORNE STUDIES. A group of studies conducted at the Hawthorne plant of the Western Electric Company during the late 1920s and early 1930s whose results ultimately led to the human relations view of management.

HIERARCHY OF NEEDS THEORY. A content theory of motivation (developed by Abraham Maslow) which argues that individual needs form a five-level hierarchy.

HORIZONTAL COMMUNICATION. Lateral or diagonal message exchange either within work-unit boundaries, involving peers who report to the same supervisor, or across work-unit boundaries, involving individuals who report to different supervisors.

HORIZONTAL COORDINATION. An aspect of organization structure involving the linking of activities across departments at similar levels.

HUMAN RESOURCE MANAGEMENT (HRM). The management of various activities designed to enhance the effectiveness of an organization's work force in achieving organizational goals.

HUMAN RESOURCE PLANNING. The process of determining future human resources needs relative to an organization's strategic plan and devising the steps necessary to meet those needs.

HUMAN SKILLS. Key management skills associated with a manager's ability to work well with others both as a member of a group and as a leader who gets things done through others.

HUBRID STRUCTURE. A type of departmentalization that adopts parts of both functional and divisional structures at the same level of management.

HYGIENE FACTORS. A type of factor which figures in the two-factor theory of motivation that is largely associated with the work environment (such as working conditions and supervision) and that can influence the degree of worker dissatisfaction.

IDEA CHAMPION. An individual who generates a new idea or believes in the value of a new idea and supports it in the face of numerous potential obstacles.

IMMORAL MANAGEMENT. An approach to managerial ethics that not only lacks ethical principles but is actively opposed to ethical behavior.

IMPORT QUOTA. A type of trade barrier in the form of a limit on the amount of a product that may be imported over a given period of time.

INCOME STATEMENT. A financial statement that summarizes the financial results of company operations over a specified time period, such as a quarter or a year.

INCREMENTAL MODEL. A nonrational model of managerial decision making stating that managers make the smallest response possible that will reduce the problem to at least a tolerable level.

INCREMENTALIST APPROACH. An approach to controlling an innovation project to controlling an innovation project that relies heavily on clan control but also involves a phased set of plans and accompanying bureaucratic controls that begin at a very general level and grow more specific as the project progresses.

INCUBATOR. An organization whose purpose is to nurture new ventures in their very early stages by providing space (usually at a site housing other new ventures as well), stimulation, support, and a variety of basic services, often at reduced fees.

INDEPENDENT DEMAND INVENTORY. A type of inventory consisting of end products, parts used for repairs, and other items whose demand is tied more directly than that for dependent demand inventory items to market issues.

INDIRECT INTERLOCK. A situation having environmental influence in which two companies each have a director on the board of a third company.

INDIVIDUALISM-COLLECTIVISM. A cultural dimension in Gert Hofstede's framework for analyzing societies that involves the degree to which individuals concern themselves with their own interests and those of their immediate family as opposed to the interests of a larger group.

INDIVIDUALIZED CONSIDERATION. A leadership factor that involves delegating projects to help develop each follower's capabilities, paying personal attention to each follower's needs, and treating each follower as an individual worthy of respect.

INFORMAL COMMUNICATION. Communication that takes place without regard to hierarchical or task requirements.

INFORMAL GROUP. A group established by employees, rather than by the organization, in order to serve group members' interests or social needs.

INFORMAL LEADER. An individual, other than the formal leader, who emerges from the group as a major influence and is perceived by group members as a leader.

INFORMATION. Data that has been analyzed or processed into a form that is meaningful for decision makers.

INFORMATION CENTER. A centrally located group of hardware, software, and information system professionals which is dedicated to assisting users with information system development and which can be used to manage end-use computing.

INFORMATION POWER. Power that results from access to and control over the distribution of important information about organizational operations and future plans.

INFORMATION SYSTEM. A set of procedures designed to collect (or retrieve), process, store, and disseminate information to support planning, decision making, coordination, and control.

INFRASTRUCTURE. The highways, railways, airports, sewage facilities, housing, educational institutions, recreation facilities, and other economic and social amenities that signal the extent of economic development in an area and constitute an economic factor that influences the ability of organization to conduct business in that area successfully.

INITIATING STRUCTURE. The degree to which a leader defines his or her own role and the roles of subordinates in terms of achieving unit goals.

INNOVATION. A new idea applied to initiating or improving a process, product, or service.

INPUTS. The components of an organizational system that include the various human, material, financial, equipment, and informational resources required to produce goods and services.

INSTITUTIONAL POWER. A need for power in which individuals focus on working with others to solve problems and further organizational goals.

INTEGRATION. The extent to which there is collaboration among departments that need to coordinate their efforts.

INTELLECTUAL STIMULATION. A leadership factor that involves offering new ideas to stimulate followers to rethink old ways of doing things, encouraging followers to look at problems from multiple vantage points, and fostering creative breakthroughs in obstacles that had seemed insurmountable.

INTEREST GROUP. An informal group created to facilitate employee pursuits of common concern.

INTERLOCKING DIRECTORATES. A situation in which organizations have board members in common either directly or indirectly.

INTERNAL AUDIT. A financial audit involving a review of both financial statements and internal operating efficiency that is conducted by members of the organization.

INTERNAL ENVIRONMENT. The general conditions that exist within an organization.

INTERNATIONAL BUSINESS. Profit-related activities conducted across national boundaries.

INTERNATIONAL ELEMENT. The element of the mega-environment that includes the developments in countries outside an organization's home country that have the potential of impacting on the organization.

INTERNATIONAL MANAGEMENT. The process of organizing, leading, and controlling in organizations engaged in international business.

INTERVENTIONS. Change strategies developed and initiated with the help of a change agent.

INTRAPRENEURS. Individuals who engage in entrepreneurial roles inside organization.

INTRAPRENEURSHIP. The process of innovating within an existing organization.

INTRINSIC REWARDS. Rewards that are related to our own internal experiences with successful performance, such as feelings of achievement, challenge, and growth.

INVENTORY. A stock of materials that are used to facilitate production or to satisfy customer demand.

INVENTORY MODELS. Quantitative approaches to planning the appropriate level for the stocks of materials needed by an organization.

INVESTMENT CENTER. A responsibility center whose budgetary performance is based on return on investment.

INVISIBLE HAND. A classical view of corporate social responsibility which holds that the entire social responsibility of a corporation can be summed up as "make profits and obey the law".

IRON LAW OF RESPONSIBILITY. A law connected with the enlightened self-interest argument which states that "in the long run, those who do not use power in a manner that society considers responsible will tend to lose it".

ISSUES MANAGEMENT. The process of identifying a relatively small number of emerging social concerns of particular relevance to the organization, analyzing their potential impact, and preparing an effective response.

ITEM COST. An inventory cost that is the price of an inventory item itself.

JAPANESE MANAGEMENT. An approach that focuses on aspects of management in Japan that may be appropriate for adoption in the United States.

JOB ANALYSIS. A key activity in human resources planning that involves the systematic collection and recording of information concerning the purpose of a job, its major duties, the conditions under which it is performed, the contacts with others that performance of the job requires, and the knowledge, skills, and abilities needed for performing the job effectively.

JOB-CENTERED. A leadership approach in which leaders divide work into routine tasks, determine work methods, and closely supervise workers to ensure the methods are followed and productivity standards are met.

JOB CHARACTERISTICS MODEL. A model developed to guide job enrichment efforts that include consideration of core job characteristics, critical psychological states, and outcomes.

JOB DEPTH. An aspect of job design that addresses the degree to which individuals can plan and control the work involved in their jobs.

JOB DESCRIPTION. A statement of the duties, working conditions, and other significant requirements associated with a particular job.

JOB DESIGN. The specification of task activities associated with a particular job.

JOB ENLARGEMENT. A job design approach that involves the allocation of a wider variety of similar tasks to a job in order to make it more challenging.

JOB ENRICHMENT. A job design approach that upgrades the job-task mix in order to increase significantly the potential for growth, achievement, responsibility, and recognition.

JOB EVALUATION. A systematic process of establishing the relative worth of jobs within a single organization in order to help establish pay differentials among jobs; it is the foundation of most major compensation systems.

JOB POSTING. An internal recruiting practice whereby information about job vacancies is placed in conspicuous places in an organization, such as on bulletin boards or in organizational newsletters.

JOB ROTATION. A job design approach that involves periodically shifting workers through a set of jobs in a planned sequence.

JOB SCOPE. An alternative work schedule in which two or more people share a single full-time job.

JOB SIMPLIFICATION. A job design approach whereby jobs are configured so that jobholders have only a small number of narrow activities to perform.

JOB SPECIFICATION. A statement of the skills, abilities, education, and previous work experience that are required to perform a particular job.

JOINT VENTURE. An agreement involving two or more organizations that arrange to produce a product or service jointly.

JUDGMENTAL FORECASTING. A type of forecasting that relies mainly on individual judgments or committee agreements regarding future conditions.

JURY OF EXECUTIVE OPINION. A method of judgmental forecasting in which organization executives hold a meeting and estimate, as a group, a forecast for a particular item.

JUST-IN-TIME (JIT) INVENTORY CONTROL. An approach to inventory control that emphasizes having materials arrive just as they are needed in the production process.

KANBAN. A subsystem of the JIT approach involving a simple parts-movement system that depends on cards and containers to pull parts from one work center to another.

KINESIC BEHAVIOR. A category of nonverbal communication that includes body movements: gestures, facial expressions, eye movements, and posture.

LA PROSPECTIVE. A method of technological or qualitative forecasting that addresses a variety of possible futures by evaluating major environmental variables, assessing the likely strategies of other significant actors, devising possible counter strategies, developing ranked hypotheses about the variables, and formulating alternative scenarios that do not greatly inhibit freedom of choice.

LABOR-MANAGEMENT RELATIONS. The process through which employers and unions negotiate pay, hours of work, and other conditions of employment; sign a contract governing such conditions for a specific period of time; and share responsibilities for administering the resulting contact.

LABOR SUPPLY. The element of the task environment that consists of those individuals who are potentially employable by an organization.

Laissez-Faire. Behavioral style of leaders who generally give the group complete freedom, provide necessary materials, participate only to answer questions, and avoid giving feedback.

Large-Batch and Mass Production. A type of technology in which products are manufactured in large quantities, frequently on an assembly line.

Lateral Relations. An approach to horizontal coordination that involves coordinating efforts through communicating and problem solving with peers in other departments or units, rather than referring most issues up the hierarchy for resolution.

Law of Effect. A concept on which reinforcement theory relies heavily, which states that behaviors having pleasant or positive consequences are more likely to be repeated and that behaviors having unpleasant or negative consequences are less likely to be repeated.

Leadership. The process of influencing others toward the achievement of organizational goals.

Leading. The management function that involves influencing others to engage in the work behaviors necessary to each organizational goals.

Leading Indicators. Explanatory models based on variables that tend to be correlated with the phenomenon of major interest but also tend to occur in advance of that phenomenon.

Legal-Political Element. The part of the mega-environment that includes the legal and governmental systems within which an organization must function.

Legitimate Power. Power that stems from a position's placement in the managerial hierarchy and the authority vested in the position.

Less Developed Countries (LDCs). A group of countries, often called the "third world", that consists primarily of relatively poor nations characterized by low per capita income, little industry, and high birthrates.

Liaison Role. A role to which a specific individual is appointed to facilitate communication and resolution of issues between two or more departments, thereby facilitating lateral relations.

Licensing. An agreement in which one organization gives limited rights to another to use certain of its assets, such as expertise, patents, copyrights, or equipment, for an agreed-upon fee or royalty.

Life Cycles. Predictable stages of development.

Line Authority. The authority that follows the chain of command established by the formal hierarchy.

Line Position. A position that has authority and responsibility for achieving the major goals of the organization.

Linear Programming (LP). In Applied Mathematics, a quantitative tool for planning how to allocate limited or scarce resources so that a single criterion or goal (often profits) is optimized.

Linking Pin. An individual who provides a means of coordination between command groups at two different levels by fulfilling a supervisory role in the lower-level command group and a subordinate role in the higher-level command group.

Liquidation. A defensive strategy that entails selling or dissolving an entire organization.

Liquidity Ratios. Financial ratios that measure the degree to which an organization's current assets are adequate to pay current liabilities (current debt obligations).

Local Area Networks (LANs). Interconnections (usually by cable) through which end-user computing can be managed that allow communications among computers in a single building or within close proximity.

Logical Office. The concept that portable microcomputers allow an individual's office to be anywhere the individual is, rather than being restricted to one specific location.

LPC Orientation. A personality trait in Fiedler's contingency model that is measured by the least preferred coworker (LPC) scale.

Machine Bureaucracy. The structural configuration in Mintzberg's typology characterized by functional departmentalization, a strong group of technical specialists, high formalization, and emphasis on standardization of work.

Management. The process of achieving organizational goals through engaging in the four major functions of planning, organizing, leading, and controlling.

MANAGEMENT BY EXCEPTION. A control principle associated with comparing performance against standards which suggests that managers should be informed of a situation only if control data show a significant deviation from standards.

MANAGEMENT BY OBJECTIVES (MBO). A process through which specific goals are set collaboratively for the organization as a whole and every unit and individual within it, the goals are then used as a basis for planning, managing organizational activities, and assessing and rewarding contributions.

MANAGEMENT BY WANDERING AROUND (MBWA). A practice whereby managers frequently tour areas for which they are responsible, talk to various employees, and encourage upward communication.

MANAGEMENT INFORMATION SYSTEM (MIS). A computer-based information system that produces routine reports and often allows on-line access to current and historical information needed by managers mainly at the middle and first-line levels; the name also given to the field of management that focuses on designing and implementing computer-based information systems for use by management.

MANAGEMENT OF TECHNOLOGY (MOT). The linking of engineering, science, and management disciplines to plan, develop, and implement technological capabilities to shape and accomplish the strategic and operational objectives of an organization.

MANAGEMENT SCIENCE. An approach within the quantitative management viewpoint that is aimed at increasing decision effectiveness through the use of sophisticated mathematical models and statistical methods.

MANAGERIAL ETHICS. Standards of conduct and moral judgment used by managers of organizations in carrying out their business.

MANAGERIAL INTEGRATOR. A separate manager who is given the task of coordinating related work that involves several functional departments and who facilitates lateral relations.

MANUFACTURING RESOURCE PLANNING (MRP II). A computer-based information system that integrates the production planning and control activities of basic MRP systems with related financial, accounting, personnel, engineering, and marketing information.

MARKET CONTROL. A managerial approach that relies on market mechanisms to regulate prices for certain clearly specified goods and services needed by an organization.

MASCULINITY-FEMININITY. A cultural dimension in Geert Hofstede's framework for analyzing societies that involves the extent to which a society emphasizes traditional male values such as assertiveness, competitiveness, and material success rather than traditional female values such as passivity, cooperation, and feelings.

MASTER PRODUCTION SCHEDULE (MPS). A schedule that translates the aggregate plan into a formalized production plan encompassing specific products to be produced or services to be offered and specific capacity requirements over a designated time period.

MATERIALS REQUIREMENTS PLANNING (MRP). A primary operating system used in operations management which consists of a computer-based inventory system that develops materials requirements for the goods and services specified in the master schedule and initiates the procurement actions necessary to acquire the materials when needed.

MATRIX STRUCTURE. A type of departmentalization that superimposes a horizontal set of divisional reporting relationships onto a hierarchical functional structure.

MECHANISTIC CHARACTERISTICS. The likely characteristics of firms operating in a stable environment, such as high centralization of decision making, many rules and regulations, and mainly vertical communication channels.

MEDIUM. The method used in the communication process to convey the message to the intended receiver.

MEGA-ENVIRONMENT. The segment of the external environment that reflects the broad conditions and trends in the societies within which an organization operates.

MENTOR. An individual who contributes significantly to the career development of a junior colleague or a peer.

MERGER. The combining of two or more companies into one organization and thus a means of implementing growth strategies.

MESSAGE. The encoding-process outcome, which consists of verbal and nonverbal symbols that have been developed to convey meaning to the receiver.

Middle Managers. Managers beneath the top levels of the hierarchy who are directly responsible for the work of other managers below them.

Mission. The organization's purpose or fundamental reason for existence.

Mission Statement. A broad declaration of the basic, unique purpose and scope of operations that distinguishes the organization from others of its type.

Mode. An indication of the beginning and/or ending of activities in a PERT network.

Modeling. A component of social learning theory that involves observing and attempting to imitate the behaviors of others.

Monitoring Methods. Methods of quantitative forecasting that provide early warning signals of significant changes in established patterns and relationships so that managers can assess a possible impact and plan responses if necessary.

Moral Management. An approach to managerial ethics that strives to follow ethical principles and precepts.

Morphological Analysis. A method of technological or qualitative forecasting that focuses on predicting potential technological breakthroughs by breaking the possibilities into component attributes and evaluating various attribute combinations.

Motivation. The force that energizes behavior, gives direction to behavior, and underlies the tendency to persist.

Motivators. A type of factor that figures in the two-factor theory of motivation that relates mainly to the content of the job (such as the work itself and feelings of achievement) and that can influence the degree of worker satisfaction.

Multifocal Strategy. A strategy aimed at achieving the advantages of worldwide integration whenever possible, while still attempting to be responsive to important national needs.

Multinational Corporation (MNC). An organization that engages in production or service activities through its own affiliates in several countries, maintains control over the policies of those affiliates, and manages from a global perspective.

Multiple Control Systems. Systems that use two or more of the feed-forward, concurrent, and feedback control processes and involve several strategic control points.

National Responsiveness Strategy. A strategy of allowing subsidiaries to have substantial latitude in adapting products and services to suit the particular needs and political realities of the countries in which they operate.

Natural Selection Model. (See POPULATION ECOLOGY MODEL.)

Need for Achievement (NACH). The desire to accomplish challenging tasks and achieve a standard of excellence in one's work.

Need for Affiliation (NAFF). The desire to maintain warm, friendly relationships with others.

Need for Power (NPOW). The desire to influence others and control one's environment.

Needs Analysis. An assessment of an organization's training needs that is developed by considering overall organizational requirements, tasks associated with jobs for which training is needed, and the degree to which individuals are able to perform those tasks effectively.

Negative Entropy. The ability of open systems to bring in new energy in the form of inputs and feedback from the environment in order to delay or arrest entropy.

Negative Reinforcement. A type of reinforcement in behavior modification, aimed at *increasing* a desired behavior, that involves providing a noxious stimuli so that an individual will engage in the desired behavior in order to stop the noxious stimuli.

Negative Synergy. The result that occurs when group process losses are greater than any gains achieved from combining the forces of group members.

Negotiating Contracts. An approach to influencing the environment that involves seeking favorable agreements on matters of importance to the organization.

Network. A management process element consisting of a set of cooperative relationships with individuals whose help is needed in order for a manager to function effectively.

Network Diagram. A diagram constructed as a step in setting up PERT that constitutes a graphic depiction of the interrelationships among the activities in a project.

Neutralizers. Situational factors that make it *impossible* for a given leader behavior to have an impact on subordinate performance and/or satisfaction.

New Venture. An enterprise that is in the process of being created by an entrepreneur.

New Venture Teams. Temporary task forces or teams made up of individuals who have been relieved of their normal duties in order to develop a new process, product, or program.

New Venture Units. Either separate divisions or specially incorporated companies created for the specific purpose of developing new products or business ideas and initiatives.

Newly Industrialized Countries (NICs). Countries within the LDGs that are emerging as major exporters of manufactured goods, including such nations as Hong Kong, Taiwan, and South Korea.

Noise. Any factor in the communication process that interferes with exchanging messages and achieving common meaning.

Nominal Group Technique (NGT). A technique for enhancing group creativity that integrates both individual work and group interaction within certain ground rules.

Noncrisis Problem. A type of problem in managerial decision making involving an issue that requires resolution but does not simultaneously have the importance and immediacy characteristics of a crisis.

Noncybernetic Control System. A control system that relies on human discretion as a basic part of its process.

Nonprogrammed Decisions. Managerial decisions for which predetermined decision rules are impractical because the situations are novel and/or ill-structured.

Nonrational Escalation. The tendency to increase commitment to a previously selected course of action beyond the level that would be expected if the manager followed an effective decision-making process; also called *escalation phenomenon*.

Nonrational Models. Models of managerial decision making which suggest that information-gathering and -processing limitations make it difficult for managers to make optimal decisions.

Nonverbal Communication. Communication by means of elements and behaviors that are not coded into words.

Normative Decision-Making Models. Models of decision making that attempt to prescribe how managers *should* make decisions.

Normative Leadership Model. A situational leadership theory model (designed by Vroom and Yetton) that helps leaders assess important situation factors that affect the extent to which they should involve subordinates in particular decisions.

Norming. A stage of group development in which group members begin to build group cohesion, as well as develop consensus about norms for performing a task and relating to one another.

Norms. A major group process factor involving expected behaviors sanctioned by the group that regulate and foster uniformity in member behaviors.

Not-For-Profit Organization. An organization whose purposes focus on issues other than making profits.

Object Language. A category of nonverbal communication that involves the communicative use of material things, including clothing, cosmetics, furniture, and architecture.

Objective Function. A mathematical representation of the relationship to be optimized in a linear programming problem.

Office Automation System (OAS). A computer-based information system aimed at facilitating communication and increasing the productivity of managers and office workers through document and message processing.

Ombudsperson. Usually an executive operating outside the normal chain of command whose job is to handle issues involving employee grievances and warnings about serious ethical problems.

On-Line Processing. An arrangement whereby data can be accessed and processed immediately.

One-Way Communication. The communication that results when the communication process does not allow for feedback.

Open System. A system that operates in continual interaction with its environment.

Operating Budget. A type of budget involving a statement that presents the financial plan for each responsibility center during the budget period and reflects operating activities involving revenues and expenses.

Operational Control. A control type concerning mainly lower-level managers that involves overseeing the implementation of operating plans, monitoring day-to-day results, and taking corrective action when required.

Operational Goals. Targets or future end results set by lower management that address specific measurable outcomes required from the lower levels.

Operational Plans. The means devised to support implementation of tactical plans and achievement of operational goals; such plans are usually developed by lower management in conjunction with middle management.

Operations Management. The management of the productive processes that convert inputs into goods and services; the name also given to the function or field of expertise that is primarily responsible for managing the production and delivery of an organization's products and services.

Operations Research. Another name commonly used for management science.

Opportunity Problem. A type of problem in managerial decision making involving a situation that offers a strong potential for significant organizational gain if appropriate actions are taken.

Orchestrator. A high-level manager who articulates the need for innovation, provides funding for innovating activities, creates incentives for middle managers to sponsor new ideas, and protects idea people.

Ordering Cost. An inventory cost that is comprised of the expenses involved in placing an order (such as paperwork, postage, and time).

Organic Characteristics. The likely characteristics of firms operating in a highly unstable environment, such as decentralization of decision making, few rules and regulations, and both hierarchical and lateral communication channels.

Organization. Two or more persons engaged in a systematic effort to produce goods or services.

Organization Chart. A line diagram that depicts the broad outlines of an organization's structure.

Organization Design. The process of developing an organization structure.

Organization Structure. The formal pattern of interactions and coordination design by management to link the tasks of individuals and groups in achieving organizational goals.

Organizational Cultural Change. An intervention involving the development of a corporate culture that is in synchronization with organizational strategies and other factors, such as structure.

Organizational Culture. A system of shared values, assumptions, beliefs, and norms that unite the members of an organization.

Organizational Development (OD). A charge effort that is planned, focused on an entire organization or a large subsystem, managed from the top, aimed at enhancing organizational health and effectiveness, and based on planned interventions.

Organizational Problems. Discrepancies between a current state or condition and what is desired.

Organizational Social Responsibility. The obligation of an organization to seek actions that protect and improve the welfare of society along with its own interests.

Organizational Social Responsiveness. A term that refers to the development of organizational decision processes whereby managers anticipate, respond to, and manage areas of social responsibility.

Organizational Termination. The process of ceasing to exist as an identifiable organization.

Organizing. The management function that focuses on allocating and arranging human and non-human resources so that plans can be carried out successfully.

Ossification. A condition that may occur in stage V (resource maturity) of small-business growth that is characterized by lack of innovation and avoidance of risk.

Outputs. The components of an organizational system that include the products, services, and other outcomes produced by the organization.

Overconfidence. The tendency to be more certain of judgments regarding the likelihood of a future event than one's actual predictive accuracy warrants.

Overcontrol. The limiting of individual job autonomy to such a point that it seriously inhibits effective job performance.

Panic. A reaction preventing effective decision making in which individuals become so upset that they frantically seek a way to solve a problem.

Paralanguage. A category of nonverbal communication that involves vocal aspects of communication that relate to how something is said rather than to what is said.

Partial-Factor Productivity. A productivity approach that measures organizational productivity by considering the total output relative to a specific input, such as labor.

Participative. A leader behavior identified an path-goal theory that is characterized by consulting with subordinates, encouraging their suggestions, and carefully considering their ideas when making decisions.

Path-Goal Theory. A situational leadership theory that attempts to explain how leader behavior impacts the motivation and job satisfaction of subordinates.

Pay Survey. A survey of the labor market to determine the current rates of pay for benchmark, or key, jobs which is used to address the issue of external equity (of compensation).

Payoff. The amount of decision-maker value associated with a particular decision alternative and future condition in a payoff table.

Payoff Table. A quantitative decision-making and consisting of a two-dimensional matrix that allows a decision maker to compare how different future conditions are likely to affect the respective outcomes of two or more decision alternatives.

Perception. The process that individuals use to acquire information from the environment.

Perceptual Defense. The tendency to block out or distort information that one finds threatening or that challenges one's beliefs.

Performance Appraisal. The process of defining expectations for employee performance; measuring, evaluating, and recording employee performance relative to those expectations; and providing feedback to the employee.

Performance-Outcome Expectancy. A component of expectancy theory that is our assessment of the probability that our successful performance will lead to certain outcomes.

Performing. A stage of group development in which energy is channeled toward a task.

Personal Power. A form of need for power in which individuals want to dominate others for the sake of demonstrating their ability to wield power.

PERT. The program evaluation and review technique consists of breaking down a project into a network of specific activities, mapping out their sequence and interdependencies, and necessary completion times and dates.

Physiological Needs. The needs in Maslow's hierarchy that are required for survival, such as food, water, and shelter.

Plan. The means devised for attempting to reach a goal.

Planned Change. Change that involves actions based on a carefully thought-out process for change that anticipates future difficulties, threats, and opportunities.

Planning. The management function that involves setting goals and deciding how best to achieve them.

Planning Mode. An approach to strategy formulation involving systematic, comprehensive analysis, along with integration of various decisions and strategies.

Planning Staff. A small group of individuals who assist top-level managers in developing the various components of the planning process.

Plant Closings. A generic term that refers to shutting down operations at a factory or nonfactory site either permanently or for an extended period of time (an issue connected with a corporation's responsibility to employees).

Point Factor Method. A job evaluation approach in which points are assigned to jobs on the basis of the degree to which the jobs contain selected compensable factors.

Policy. A standing plan that provides a general guide specifying the broad parameters within which organization members are expected to operate in pursuit of organizational goals.

Political Risk. The probability of the occurrence of political actions that will result in loss of either enterprise ownership or significant benefits from conducting business.

Polycentric Orientation. An approach to international management (also known as *host-country orientation*) whereby executives view host-country cultures and foreigners as difficult to fathom and, therefore, believe that the parts of the organization located in a given host country should be staffed by local individuals to the fullest extent possible.

Pooled Interdependence. A type of technological interdependence in which units operate independently but their individual efforts are important to the success of the organization as a whole.

Population Ecology Model. A view of the organization-environment interface that focuses on populations, groups or organizations and argues that environmental factors cause organizations with appropriate characteristics to survive and others to fail.

Portfolio Strategy Approach. A corporate-level strategy approach that involves analyzing an organization's mix of businesses in terms of both individual and collective contributions to strategic goals.

Positive Synergy. The force that results when the combined gains from group interaction are greater than group process losses.

Power. The capacity to affect the behavior of others.

Power Distance. A cultural dimension in Geert Hofstede's framework for analyzing societies that involves the degree to which individuals in a society accept differences in the distribution of power as reasonable and normal.

Procedure. A standing plan that involves a prescribed series of related steps to be taken under certain recurring circumstances.

Process Consultation. An intervention concerned with the interpersonal relations and dynamics operating in work groups.

Process Layout. A type of facilities layout having a production configuration in which the processing components are grouped according to the type of function that they perform.

Product Divisions. A form of divisional structure involving divisions created to concentrate on a single product or service or at least a relatively homogeneous set of products or services.

Product Layout. A type of facilities layout having a production configuration in which the processing components are arranged in a specialized line along which the product or client passes during the production process.

Product/Market Evolution Matrix. A portfolio approach involving a 15-cell matrix (developed by Charles W. Hofer) in which businesses are plotted according to the business unit's business strength, or competitive position, and the industry's stage in the evolutionary product/market life cycle.

Product Life Cycle. The stages through which a product or service goes: (1) conceptual (2) technical feasibility, (3) development, (4) commercial validation and production processes, (5) full scale promotion, (6) product support.

Productivity. An efficiency concept that gauges the ratio of outputs relative to inputs into a productive process.

Professional Bureaucracy. The structural configuration in Mintzberg's typology characterized by functional or hybrid departmentalization, a strong group of professionals operating at the lower levels, low formalization, and emphasis on standardization of skills.

Profit Budget. An operating budget that focuses on the profit to be derived from the difference between anticipated revenues and expenses.

Profit Center. A responsibility center whose budgetary performance is measured by the difference between revenues and costs, in other words, profits.

Profitability Ratios. Financial ratios that help measure management's ability to control expenses and earn profits through the use of organizational resources.

PROGRAM. A comprehensive single use plan that coordinates a complex set of activities related to a major nonrecurring goal.

PROGRAM EVALUATION AND REVIEW TECHNIQUE (PERT). A network planning method for managing large projects.

PROGRAMMED DECISIONS. Managerial decisions made in routine, repetitive, well-structured situations through the use of predetermined decision rules.

PROJECT. A single-use plan that coordinates a set of limited-scope activities that do not need to be divided into several major projects in order to reach a major nonrecurring goal.

PROJECT MANAGERS. Managers who have responsibility for coordinating efforts involving individuals in severaldifferent organizational units who are all working on a particular project.

PROJECTION. The tendency of an individual to assume that others share his or her thoughts, feelings, and characteristics.

PROSPECT THEORY. A theory explaining certain decision-making biases which posits that decision makers find the prospect of incurring an actual loss more painful than giving up the possibility of a gain.

PROTOTYPE. A working model of a product, process, or system developed to test the product, process or system functions and features. It is usually hand made.

PROTOTYPING. A means of developing a system that involves building a rough, working model of all or parts of a proposed system for purposes of preliminary evaluation and further refinement.

PROXEMICS. A category of nonverbal communication that involves the influence of proximity and space on communication.

PUBLIC AFFAIRS DEPARTMENT. A permanent department, used as a mechanism for facilitating an organization's internal social response that coordinates various ongoing social responsibilities and identifies and recommends policies for new social issues.

PUBLIC RELATIONS. An approach to influencing the environment involving the use of communication media and related activities to create a favorable overall impression of the organization among the public.

PUNISHMENT. A type of reinforcement in behavior modification that involves providing negative consequences in order to *decrease* or discourage a behavior.

PURCHASING. A primary operating system used in operations management that is involved with acquiring necessary goods or services in exchange for funds or other remuneration.

QUALITATIVE FORECASTING. A type of forecasting aimed primarily at predicting long-term trends in technology and other important aspects the environment.

QUALITY. The totality of features and characteristics of a product or service that bear on its ability to satisfy stated or implied needs.

QUALITY CIRCLE (QC). A small group of employees who meet periodically to solve quality problems related to their jobs.

QUANTITATIVE FORECASTING. A type of forecasting that relies on numerical data and mathematical models to predict future conditions.

QUESTIONABLE PAYMENTS. An international social responsibility issue concerning business payments that raise significant ethical questions of right or wrong either in the host nation or in other nations.

QUEUING MODELS. A quantitative planning technique based on mathematical models that describe the operating characteristics of situations in which service is provided to persons or units waiting in line.

RATIO ANALYSIS. A financial control technique that involves determining and evaluating financial ratios.

RATIONAL MODEL. A model of managerial decision making which suggests that managers engage in completely rational decision processes, ultimately make optimal decisions, and possess and understand all information relevant to their decisions at the time they make them.

RATIONALIZATION. The strategy of assigning activities to those parts of organization, regardless of their location, that are best suited to produce the desired results and then selling the finished products where they are likely to yield the best profits.

RATIONING. A method of adapting to environmental fluctuations that involves limiting access to a product or service that is in high demand.

RAW MATERIALS INVENTORY. The stock of parts, ingredients, and other basic inputs to a production or service process.

REACTIVE CHANGE. Change that occurs when one takes action in response to perceived problems, threats, or opportunities.

REALISTIC JOB PREVIEW. A technique used during the recruiting process in which the job candidate is presented with a balanced view of both the positive and the negative aspects of the job and the organization.

RECEIVER. The person with whom the message is exchanged in the communication process.

RECIPROCAL INTERDEPENDENCE. A type of technological interdependence in which one unit's outputs become inputs to the other unit and vice versa.

RECRUITMENT. An activity in the staffing process that involves finding and attempting to attract job candidates who are capable of effectively filling job vacancies.

REFERENCE CHECKS. A selection method involving attempts to obtain job-related information about job applicants from individuals who are in a position to be knowledgeable about the applicants' qualifications.

REFERENT POWER. Power that results from being admired, personally identified with, or liked by others.

REGRESSION MODELS. Explanatory models based on equations that express the fluctuation in the variable being forecasted in terms of fluctuations among one or more other variables.

REINFORCEMENT THEORY. A theory of motivation that argues that our behavior can be explained by consequences in the environment and, therefore, that is not necessary to look for cognitive explanation.

RELATEDNESS NEEDS. The needs in ERG theory that address our relationships with significant others, such as families, friendship groups, work groups, and professional groups.

REORDER POINT (ROP). The inventory level at which a new order should be place.

REPLACEMENT CHART. A partial organization chart showing the major managerial positions in an organization, current incumbents, potential replacements for each position, and the age of each person listed on the chart.

REPRESENTATIVENESS. A decision-making has that involves the tendency to be overly influenced by stereotypes in making judgements about the like of occurrences.

RESERVATIONS. Organizational units that devote full time to the generation of innovative ideas for future business.

RESOURCE DEPENDENCE. An approach to controls that argues that managers need to consider controls mainly in areas in which they depend on others in resources necessary to reach organizational goals.

RESOURCE DEPENDENCE MODEL. A view of the organization-environment interface that highlights organizational dependence on the environment for resources and argues that organizations attempt to manipulate the environment to reduce that dependence.

RESPONSIBILITY. The obligation to carry out duties and active goals related to a position.

RESPONSIBILITY CENTER. A subunit headed by a manager who is responsible for achieving one or more goals.

RESTRAINING FORCES. Forces studied in force-field analysis that involve factors that pressure *against* a change.

RESTRUCTURING. A method of increasing organizational efficiency and effectiveness that involves making changes in organization structure, often includes reducing management levels and changing components of the organization through divestiture and/or acquisition, as well as shrinking the size of the work force.

REVENUE BUDGET. An operating budget that indicates anticipated revenues.

REVENUE CENTER. A responsibility center whose budgetary performance is measured primarily by its ability to generate a specified level of revenue.

REVITALIZATION. The renewal of the innovative vigor of organizations sought in the elaboration-of-structure stage of the organizational life cycle.

REWARD POWER. Power that is based on the capacity to control and provide valued rewards to others.

RISK. The possibility, characteristic of decisions made under uncertainty, that a chosen action could lead to losses rather than the intended results.

RITE. A relatively elaborate, dramatic, planned set of activities intended to convey cultural values to participants and, usually, an audience.

ROLE. A major work group input that involves a set of behaviors expected of an individual who occupies a particular position in a group; also used to denote a management process element consisting of an organized set of behaviors associated with a particular office or position.

ROMANCE OF LEADERSHIP. The phenomenon of leadership possibly being given more than its due credit for positive results.

ROUTING, OR DISTRIBUTION MODELS. Quantitative methods to assist managers in planning the most effective and economical approaches to distribution problems.

RULE. A standing plan that is a statement spelling out specific actions to be taken or not taken in a given situation.

SAFETY NEEDS. The needs in Maslow's hierarchy that pertain to the desire to feel secure, and free from threats to our existence.

SALES BUDGET. An operating budget that indicates anticipated revenues.

SALES-FORCE COMPOSITE. A method of judgmental forecasting that is used mainly to predict future sales and typically involves obtaining the views of various salespeople, sales managers, and/or distributors regarding the sales outlook.

SATISFACTION-PROGRESSION PRINCIPLE. A principle incorporated into ERG theory which states that satisfaction of one level of need encourages concern with the next level.

SATISFACTION MODEL. A nonrational model of managerial decision making stating that managers seek alternatives only until they find one that looks *satisfactory*, rather than seeking the optimal decision.

SATISFIERS. Motivating factors which figure in the two-factor theory of motivation that relate mainly to the content of the job (such as the work itself and feelings of achievement).

SCENARIOS. Outlines of possible future conditions, including possible paths the organization could take that would likely lead to these conditions.

SCHEDULES OF REINFORCEMENT. Patterns of rewarding that specify the basis for and timing of positive reinforcement.

SCIENTIFIC MANAGEMENT. An approach within classical management theory that emphasizes the scientific study of work methods in order to improve worker efficiency.

SELECTION. An activity in the staffing process that involves determining which job candidates best suit organizational needs.

SELECTION INTERVIEW. A relatively formal, in-depth conversation conducted for the purpose of assessing a candidate's knowledge, skills, and abilities, as well as providing information to the candidate about the organization and potential jobs.

SELF-ACTUALIZED NEEDS. The needs in Maslow's hierarchy that pertain to the requirement of developing our capabilities and reaching our full potential.

SELF-CONTROL. A component of social learning theory involving our ability to exercise control over our own behavior by setting standards and providing consequences for our own actions.

SELF-EFFICACY. The belief in one's capabilities to perform a specific task.

SELF-MANAGING TEAM. A work group given responsibility for a task area without day-to-day supervision and with authority to influence and control both group membership and behavior.

SELF-ORIENTED ROLES. Roles that are related to the personal needs of group members and often negatively influence the effectiveness of a group.

SELF-SERVING BIAS. The tendency to perceive oneself as responsible for successes and others as responsible for failure.

SEMANTIC BLOCKS. The blockages or difficulties in communication that arise from word choices.

SEMANTIC NET. The network of words and word meanings that a given individual has available for recall.

SENDER. The initiator of the message in the communication process.

SEQUENTIAL INTERDEPENDENCE. A type of technological

interdependence in which one unit must complete its work before the next until in the sequence can begin work.

SHAPING. A technique associated with positive reinforcement that involves the successive rewarding of behaviors that closely approximate the desire response until the actual desired response is made.

SIMPLE STRUCTURE. The structural configuration in Mintzberg's typology characterized by functional departmentalization, a strong concentration of power at the top, low formalization, and emphasis on direct supervision.

SIMULATION. A quantitative planning technique that uses mathematical models to imitate reality.

SINGLE-USE PLANS. Plans aimed at achieving a specific goal that, once reached, will most likely not recur in the future.

SITUATIONAL LEADERSHIP THEORY. A contingency theory (developed by Paul Hersey and Ken Blanchard) based on the premise that leaders need to alter their behaviors depending on the readiness of followers.

SITUATIONAL THEORIES. Theories of leadership that take into consideration important situational factors.

SKILL-BASED PAY. A compensation system in which employees' rates of pay are based on the number of predetermined skills that the employees have mastered.

SKILL VARIETY. A core job characteristic involving the extent to which the job entails a number of activities that require different skills.

SKILLS INVENTORY. A data bank (usually computerized) containing basic information about each employee that can be used to assess the likely availability of individuals for meeting current and future human resources needs.

SLACK. Latitude about when various activities can be started on the noncritical paths in a PERT network without endangering the completion date of the entire project.

SLACK RESOURCES. A means of facilitating horizontal coordination that involves a cushion of resources that aids adaptation to internal and external pressures, as well as initiation of changes.

SMOOTHING. A method of adapting to environmental fluctuations that involves taking actions aimed at reducing the impact of fluctuations, given the market.

SOCIAL AUDIT. A systematic study and evaluation of the social, rather than the economic, performance of an organization.

SOCIAL FORECASTING. The process of identifying social trends, evaluating the organizational importance of those trends, and integrating these assessments into the organization's forecasting program.

SOCIAL INFORMATION-PROCESSING APPROACH. A job design approach arguing that individuals often form impressions of their jobs from socially provided information, such as comments by supervisors and coworkers.

SOCIAL LEARNING THEORY. A theory of motivation having aspects of both cognitive and reinforcement theories which argues that learning occurs through the continuous reciprocal interaction of our behaviors, various personal factors, and environmental forces.

SOCIAL LOAFING. The tendency of individuals to expend less effort when working in groups than when working alone.

SOCIAL SCANNING. The general surveillance of various elements in the task environment to detect evidence of impending changes that will affect the organization's social responsibilities.

SOCIALIST ECONOMY. An economy in which the means of production are owned by the state and economic activity is coordinated by plan.

SOCIOCULTURAL ELEMENT. The element of the mega-environment that includes the attitudes, values, norms, beliefs, behaviors, and associated demographic trends that and characteristic of a given geographic area.

SOFTWARE. The set of programs, documents, procedures, and routines associated with the operation of a computer system that makes the hardware capable of its various activities.

SOLDIERING. A work practice whereby workers deliberately work at less-than-full capacity, the observation of which led Taylor to develop his scientific management approach.

SPAN OF MANAGEMENT, OR SPAN OF CONTROL. A means

of vertical coordination involving the number of subordinates who report directly to a specific manager.

SPONSOR. A middle manager who recognizes the organizational significance of an idea, helps obtain the necessary funding for development of the innovation, and facilitates its actual implementation.

STABILITY STRATEGY. A grand strategy that involves maintaining the status quo or growing in a methodical, but slow manner.

STAFF POSITION. A position whose primary purpose is providing specialized expertise and assistance to line positions.

STAFFING. The set of activities aimed at attracting and selecting individuals for positions in a way that will facilitate the achievement of organizational goals.

STANDARD COST CENTER. A responsibility center whose budgetary performance depends on achieving its goals by operating within standard cost constraints.

STANDING COMMITTEE. A permanent task group of individuals charged with handling recurring matters in a narrowly defined subject area over an indefinite, but generally lengthy, period of time.

STANDING PLANS. Plans that provide ongoing guidance for performing recurring activities.

START-UP. A new firm or venture started from scratch by an entrepreneur.

STATISTICAL PROCESS CONTROL. A statistical technique employed in quality control that uses periodic random samples taken during actual production to determine whether acceptable quality levels are being met or production should be stopped for remedial action.

STEREOTYPING. The tendency to attribute characteristics to an individual on the basis of an assessment of the group to which the individual belongs.

STOCKOUT COST. An inventory cost that involves the economic consequences of running out stock (such as loss of customer goodwill and possibly sales).

STORMING. A stage of group development in which group members frequently experience conflict with one another as they locate and attempt to resolve differences of opinion regarding key issues.

STORY. A narrative based on true events, which sometimes may be embellished to highlight the intended value.

STRATEGIC BUSINESS UNIT (SBU). A distinct business, with its own set of competitors, that can be managed reasonably independently of other businesses within the organization.

STRATEGIC CONTROL. A control type concerning mainly top managers that involves monitoring critical environmental factors that could affect the viability of strategic plans, assessing the effects of organizational strategic actions, and ensuring that strategic plans are implemented as intended.

STRATEGIC CONTROL POINTS. Performance areas chosen for control because they are particularly important in meeting organizational goals.

STRATEGIC GOALS. Broadly defined targets or future end results set by top management.

STRATEGIC MANAGEMENT. A process through which managers formulate and implement strategies geared to optimizing strategic goal achievement, given available environmental and internal conditions.

STRATEGIC PLANS. Detailed action steps such plans are developed by top management in consultation with the board of directors and middle management.

STRATEGIES. Large-scale action plans for interacting with the environment in order to achieve long-term goals.

STRATEGY FORMULATION. The part of the strategic management process that involves identifying the mission and strategic goals, conducting competitive analysis, and developing specific strategies.

STRATEGY IMPLEMENTATION. The part of the strategic management process that focuses on carrying out strategic plans and maintaining control over how those plans are carried out.

SUBSTITUTES. Situational factors that make leadership impact not only *impossible* but also unnecessary or negate their effectiveness.

SUBSTITUTES FOR LEADERSHIP. An approach that attempts to specify some of the main situational factors likely to make leader behaviors unnecessary or to negate their effectiveness.

Sunk Costs. Costs that, once incurred, are not recoverable and should not enter into considerations of future courses of action.

Superordinate Goals. Major common goals that require the support and effort of all parties and on which the manager is sometimes able to refocus individuals or groups in conflict, thus reducing or resolving the conflict.

Suppliers. The element of the task environment that includes those organizations and individuals that supply the resources an organization needs to conduct its operations.

Supportive. A leader behavior identified in path-theory that entails showing concern for the status, well-being, and needs of subordinates; doing small things to make the work more pleasant; and being friendly and approachable.

SWOT Analysis. A method of analyzing an organization's competitive situation that involves assessing organizational strength (S) and weaknesses (W), as well as environmental opportunities (O) and threats (T).

Symbol. An object, act, event, or quality that serves as a vehicle for conveying meaning.

Symbolic Processes. Components of social learning theory involving the various ways that we use verbal and imagined symbols to process and store experiences in representational forms that can serve as guides to future behavior.

Synectics. A technique for enhancing creativity that relies on analogies to help group members look at problems from new perspectives.

Synergy. A major characteristic of open systems; the ability of the whole to become more than the sum of its parts.

System. A set of interrelated parts that operate as a whole in pursuit of common goals.

Systems Development Life Cycle. A series of stages that are used in the development of most medium- and large-size information systems.

Systems Theory. A view of management based on the notion that systems.

Tactical Control. A control type concerning mainly middle managers that focuses on assessing the implementation of tactical plans at department levels, monitoring associated periodic results, and taking corrective action as necessary.

Tactical Goals. Targets or future end results usually set by middle management for specific departments or units.

Tactical Plans. The means charted to support implementation of the strategic plan and achievement of tactical goals; such plans are developed by middle management, sometimes in consultation with lower management.

Takeover. A form of acquisition involving the purchase of a controlling share of voting stock in a publicly-traded company and thus a potential reason for organizational termination.

Tall Structure. A structure that has many hierarchical levels and narrow spans of control.

Tariff. A type of trade barrier in the form of a customs duty or tax levied mainly on imports.

Task Environment. The segment of the external environment made up of the specific outside elements with which an organization interfaces in the course of conducting its business.

Task Force. A temporary group usually formed to make recommendations on a specific issue.

Task Group. A formal group that is created for a specific purpose that supplements or replaces work normally done by command groups.

Task Identity. Core job characteristics involving the degree to which a job allows the completion of a major identifiable piece of work, rather than just a fragment.

Task Significance. A core job characteristic involving the extent to which the worker sees the job output as having an important impact on others.

Team. A temporary or ongoing group whose members are charged with working together to identify problems, form a consensus about what should be done, and implement necessary actions in relation to a particular task or organizational area.

Team Building. An intervention aimed at helping work groups become effective at task accomplishment.

TECHNICAL SKILLS. Key management abilities that reflect both an understanding of and a proficiency in a specialized field.

TECHNOLOGICAL ELEMENT. The part of the mega-environment that reflects the current state of knowledge regarding the production of products and services.

TECHNOLOGICAL FORECASTING. A type of forecasting aimed primarily at predicting long-term trends in technology and other important aspects of the environment.

TECHNOLOGICAL INTERDEPENDENCE. The degree to which different parts of the organization must exchange information and materials in order to perform their required activities.

TECHNOLOGICAL TRANSFER. The transmission of technology from those who possess it to those who do not.

TECHNOLOGY. The knowledge, tools, equipment, and work techniques used by an organization in delivering its product or service.

TECHNOSTRUCTURAL ACTIVITIES. An intervention involving activities intended to improve work technology and/or organization structure.

TELECOMMUTING. A form of working at home that is made possible by using computer technology to remain in touch with the office.

THEORY Z. A concept that combines positive aspects of American and Japanese management into a modified approach aimed at increasing U.S. managerial effectiveness while remaining compatible with the norms and values of American society and culture.

THIRD-PARTY INTERVENTION. A technique concerned with helping individuals, groups, or departments resolve serious conflicts that may relate to specific work issues or may be caused by suboptimal interpersonal relations.

TIME-SERIES METHODS. Methods of quantitative forecasting that use historical data to develop forecasts of the future.

TOP-DOWN BUDGETING. A process of developing budgets in which top management outlines the overall figures and middle and lower-level managers plan accordingly.

TOP MANAGERS. Managers at the very top levels of the hierarchy who are ultimately responsible for the entire organization.

TOTAL-FACTOR PRODUCTIVITY. A productivity approach which measures organizational productivity by considering all the inputs involved in producing outputs.

TOTAL QUALITY CONTROL (TQC). A quality control approach that emphasizes organizationwide commitment, integration of quality improvement efforts with organizational goals, and inclusion of quality as a factor in performance appraisals.

TRADE ASSOCIATIONS. Organizations that are composed of individuals or firms with common business concerns and which have environmental influence.

TRAINING AND DEVELOPMENT. A planned effort to facilitate employee learning of job-related behaviors in order to improve employee performance.

TRAITS. Distinctive internal qualities or characteristics of an individual, such as physical characteristics, personality characteristics, skills and abilities, and social factors.

TRANSACTION-PROCESSING SYSTEM (TPS). A computer-based information system that executes and records the day-to-day routine transactions required to conduct an organization's business.

TRANSACTIONAL LEADERS. Leaders who motivate subordinates to perform at expected levels by helping them recognize task responsibilities, identify goals, acquire confidence about meeting desired performance levels, and understand how their needs and the rewards that they desire are linked to goal achievement.

TRANSFORMATION PROCESSES. The components of an organizational system comprising the organization's managerial and technological abilities that are applied to convert inputs into outputs.

TRANSFORMATIONAL LEADERS. Leaders who motivate individuals to perform beyond normal expectations by inspiring subordinates to focus on broader missions that transcend their own immediate self-interests, to concentrate on intrinsic higher level goals rather than extrinsic lower-level goals, and to have confidence in their abilities to achieve the extraordinary missions articulated by the leader.

TURNAROUND. A defensive strategy design to reverse a

negative trend and restore the organization to appropriate levels of profitability.

Two-Factor Theory. A content theory of motivation (developed by Frederick Herzberg) which argues that potential rewards fit into two categories, hygiene factors and motivators, each having distinctly different implications for employee motivation.

Two-Way Communication. The communication that results when the communication process explicitly includes feedback.

Uncertainty. An aspect of nonprogrammed decisions that is a condition in which the decision maker must choose a course of action without complete knowledge of the consequences that will follow implementation.

Uncertainty Avoidance. A cultural dimension in Geert Hofstede's framework for analyzing societies that involves the extent to which members of a society feel uncomfortable with and try to avoid situations that they perceive as unstructured, unclear, or unpredictable.

Undercontrol. The granting of autonomy to an employee to such a point that the organization loses its ability to direct the individual's efforts toward achieving organizational goals.

Unions. Employee groups formed for the purpose of negotiating with management about conditions relating to their work.

Unit and Small-Batch Production. A type of technologyin which products are custom-produced to meet customer specifications or meet customer specifications or they are made in small quantities primarily by craft specialists.

Upward Communication. The vertical flow of communication from a lower level to one or more higher levels in the organization.

User. An individual, other than an information system professional, who is engaged in the development and/or management of computer-based information systems.

Valence. A factor in expectancy theory that is our assessment of the anticipated value of various outcomes or rewards.

Validity. A concept underlying the use of various selection methods that concerns the degree to which a measure actually assesses an attribute that it is designed to measure.

Variable-Interval Schedule of Reinforcement. A type of partial reinforcement schedule based on a pattern in which a reinforcer is administered on a varying, or random, time schedule that *averages* out to a predetermined time frequency.

Variable-Ratio Schedule of Reinforcement. A type of partial reinforcement schedule based on a pattern in which a reinforcer is provided after a varying, or random, number of occurrences of the desired behavior in such a way that the reinforcement pattern *averages* out to a predetermined ratio of occurrences per reinforcement.

Venture Team. A group of two or more individuals who band together for the purpose of creating a new venture.

Verbal Communication. The written or oral use of words to communicate.

Vertical Communication. Communication that involves a message exchange between two or more levels of the organizational hierarchy.

Vertical Coordination. The linking of activities at the top of the organization with those at the middle and lower levels in order to achieve organizational goals.

Vertical Integration. A growth strategy that involves the production of inputs previously provided by suppliers or through the replacement of a customer role by disposing of one's own outputs.

Vicarious Learning. A component of social learning theory involving our ability to learn new behaviors and/or assess their probable consequences by observing others.

Video Teleconferencing. The holding of meetings with individuals in two or more locations by means of closed-circuit television.

Waiting-In-Line Models. A quantitative planning technique based on mathematical models that describe the operating characteristics of queuing situations in which service is provided to persons or units waiting in line.

Whistle-Blower. An employee who reports a real or perceived wrongdoing under the control of his or her employer to those who may be able to take appropriate action.

WHOLLY OWNED SUBSIDIARY. An operation on foreign soil, by means of which organizations can conduct business internationally, that is totally owned and controlled by a company with headquarters outside the host country.

WIDE AREA NETWORKS. Networks through which end-user computing can be managed that provide communications among computers over long distances, usually through the facilities and services of public or private communications companies.

WORK AGENDA. A management process element composed of a loosely connected set of tentative goals and tasks that a manager is attempting to accomplish.

WORK-IN-PROCESS INVENTORY. The stock of items that are currently being transformed into a final product or service.

WORK SPECIALIZATION. The degree to which the work necessary to achieve organizational goals is broken down into various jobs.

WORLDWIDE INTEGRATION STRATEGY. A strategy aimed at developing relatively standardized products with global appeal, as well as rationing operations throughout the world.

ZERO-BASE BUDGETING (ZBB). A budget approach in which responsibility centers start with zero in preparing their budget requests and must justify the contributions of each of their activities to organizational goals.

ZERO DEFECTS. A quality mentality that the total quality control approach attempts to achieve in which the work force strives to make a product or service conform exactly to desired standards.

BIBLIOGRAPHY

Aldag, Ramon J., and Stearns, Timothy M., *Management*, 2nd Edition, South-Western Publishing Co., Cincinnati, Ohio, 1991, 800 pgs.

Bartol, Kathryn M., and Martin, David C., *Management*, McGraw-Hill Inc., Pub., New York, N.Y., 1991, 813 pgs.

Betz, Frederick, *Management of Technology*, Prentice-Hall, Inc., Englewood Cliffs, New Jersey, (1990), 268 pgs.

Daft, Richard L., *Manageameant*, The Dryden Press, Pub., Chicago, Ill., 1988, 787 pgs.

Dunham, Randall B., and Pierce, John L., *Management*, Scott, Foresman and Co., Pub., Glenview, Ill, 1989, 877 pgs.

Hanna, Allistair M., and Rubenstein, Albert H., Co chairman, *Research On The Management of Techology, Unleashing the Hidden Advantage, NRC,* National Academy Press, Washington D.C., 1991, 40 pgs.

Kuper, George H., "The IE and EI", *International Industrial Engineering Conference,* Detroit, Michigan, May 22nd, 1991.

LeFevre, E. Walter, "*Engineering Stages of New Product Development*, NSPE Publication #3018, 1990, National Society of Professional Engineers, Alexandria, Virginia, 32 pgs.

Robbins, Stephen P., *Management,* 3rd Edition, Prentice-Hall, Pub., Englewood Cliffs, New Jersey, 1991, 732 pgs.

Stoner, James A.F., and Freeman, Edward R., *Management*, 4th Edition Prentice-Hall, Inc., Pub., Englewood Cliffs, New Jersey, 1989, 796 pgs.

Winner, Robert I., Pennell, James P., Bertrand, Harold E., and Shusarczuk, Maika M.C. *The role of Concurrent Engineering in Weapons Systems Acquisition*, IDA Report R-338, December 1988, Institute For Defense Analysis, Alexandria, Virginia, 175 pgs.

Z94.11 Manufacturing Systems

This section appeared for the first time in the previous edition of this standard. The study and use of manufacturing systems have continued to evolve. The initial version described a move from study of processes to application of analytical techniques. More recently, and reflected in this edition, manufacturing systems are viewed in new ways, and older models have become more formalized. The success of just-in-time production has led to extensions to the original theories. The more formal definitions of just-in-time systems have spawned studies of manufacturing as lean and agile systems. Advances in computing and the application of computing power in manufacturing have brought new possibilities as well, both for the analytical power available and for the application of artificail intelligence and related methods in manufacturing systems. Fortunatly, many of the older terms continue to appllyas manufacturing systems continue to grow in complexity and sophistication.

Chairman

D. L. Kimbler, Ph.D., P.E.
Department of Industrial Engineering
Clemson University

Subcommittee

J. Temple Black, Ph.D.
Department of Industrial Engineering
Auburn University

Robert P. Davis, Ph.D., P.E.
Department of Industrial Engineering
Clemson University

Richard A. Wysk, Ph.D.
Department of Industrial and
  Manufacturing Engineering
Pennsylvania State University

The subcommittee wishes to acknowledge the assistance of Mr. Ravidar Patil, Clemson University, and Auburn University students in IE482 (spring 1998) for their assistance. We also recognize the continuing valuable contributions of the other participants in the original development of this section: Adnan Aswad, Ph.D., Ray Cole, P.E., Fred Choobineh, Ph.D., P.E., Edward L. Fisher, Ph.D., P.E., T. C. Chang, Ph.D.

ABSOLUTE. A coordinate system in which each location is completely specified by its distance from the origin.

ACCURACY. (1) Quality, state, or degree of conformance to a recognized standard. (2) Difference between the actual response and the target position desired or commanded of an automatic control system.

ACTIVE ACCOMODATION. Integration of sensors, control and robot motion to achieve alteration of robot's preprogrammed motions in response to felt forces.

ACTUATOR. A transducer that converts electrical, hydraulic, or pneumatic energy to effect motion.

ADAPTIVE CONTROL. A control method in which control parameters are continuously and automatically adjusted in response to measured process variables to achieve near-optimum performance.

AGILE (MANUFACTURING). Manufacturing system design, control and execution with change and adaptability in mind; the ability of a system to adapt to changes in demand, product, and technology.

ALGORITHM. A prescribed set of well defined rules or processes for the solution of a problem in a finite number of steps.

ALLOCATE. To assign a resource for use in a job performing a specific task.

ALTERNATE ROUTING. An alternate method or sequence of performing an operation, a series of operations, or a complete routing. The alternate is generally used because of a machine breakdown or an excessive overload on the machines or work centers specified in the primary routing. An operation may be replaced by either a single alternate operation or a sequence of operations.

ALTERNATE WORK CENTER. A work center that can be used in case of breakdowns or overloads in the primary work center.

ANALOG DATA. Data represented in a continuous form, as contrasted with digital data represented in a discrete (discontinuous) form.

ANALOG-TO-DIGITAL CONVERTER (A/D, ADC). A hardware device which senses an analog signal and converts it to a representation in digital form.

ANDON. A device or signal board that signals the operational status of a process.

APPLICATION PROGRAM. Computer program devised for a specific task.

ARCHITECTURE. Preset, physical, and logical arrangement of a system.

ARM. A manipulator comprising an interconnected set of links and powered joints, which supports or moves a hand or end effector.

ASSEMBLY. (1) The fitting together of fabricated parts into a complete machine structure or unit. (2) A group of subassemblies and/or parts which are put together to form a single unit.

ASSEMBLY ROBOT. A robot designed, programmed, or dedicated to putting together parts into subassemblies or complete products.

ASYNCHRONOUS. Not related through repeating time patterns.

AUDIT. A survey or methodical investigation for gathering information.

AUTOMATED ASSEMBLY. Assembly by means of operations performed automatically by machines. A computer system may monitor the production and quality levels of the assembly operations.

AUTOMATED GUIDED VEHICLE (AGV). A self-controlled vehicle that follows specified paths in a plant floor to move material, tools, and other items. Although most systems are directed (guided) through a set of predefined (fixed) paths, new guidance systems can plan paths and control the vehicle dynamically.

AUTOMATED PROCESS PLANNING. Creation of process plans, with either partial or total computer assistance.

AUTOMATED PROGRAMMED TOOL (APT). A high level computer assisted programming language used for numerical control of machine tools.

AUTOMATIC. Pertaining to a process or device that, under specified conditions, functions without intervention by a human operator.

AUTOMATED STORAGE AND RETRIEVAL SYSTEM (AS/RS). Computer controlled high-density rack system and access device for rapid storage and retrieval of parts and tools.

Automation. (1) The implementation of processes by automatic means. (2) The theory, art, or technique of making a process more automatic. (3) The investigation, design, development, and application of methods of rendering processes automatic, self-moving, or self-controlling.

Autonomation. The automatic control of quantity and quality through the separation of human and machine work.

Available. Not yet allocated (firm). An available item may be on hand or in the process of being manufactured or delivered.

Availability. (1) The proportion of time that a machine is in service and capable of work. (2) Total time less downtime and maintenance time.

Backlash. Free play in a power transmission system such as a gear train, resulting in a characteristic form of hysteresis.

Backward Scheduling. A scheduling technique where the schedule is computed starting with the due date for the order and working backward to determine the required start date. (See FORWARD SCHEDULING.)

Balancing. In the Toyota Production System, the coordination of manufacturing and subassembly cells in producing the daily quantity demanded by final assembly.

Bang-Bang Control. Control achieved by a command to the actuator that any time tells it either to operate in one direction or the other with maximum energy.

Bang-Bang-Off Control. Control achieved by a command to the actuator which at any time tells it either to operate in one direction or the other with maximum energy or to do nothing.

Base. (1) The platform or structure to which a robot err is attached. (2) The end of a kinematic chain of arm links and joints opposite to that which grasps or processes external objects.

Batch Manufacturing. The production of parts in discrete runs, or batches. Batch operations can be interspersed with other production operations or runs of other parts.

Batch Production. Non-continuous processing of unlike parts. Contrast with mass production.

Bit. An acronym for binary digit; the smallest unit of information in the binary numbering system, represented by the digits 0 and 1.

Blank. A raw material piece, ready for a subsequent operation.

Blanket Routing. A routing that lists a group of operations needed to produce a family of items. The items may have small differences in size, but they use the same sequence of operations. Specific times or tools for each individual item can be included. (See GROUP TECHNOLOGY).

Bottleneck. A facility, process, department, etc., that limits production capacity. For example, a machine or work center where jobs arrive faster than they leave.

Buffer. A temporary storage in front of or following a process or work station.

Cable Drive. Transmission of power from an actuator to a remote mechanism by means of flexible cable and pulleys.

Calibration. (1) The act of determining, marking, or rectifying the capacity or scale graduations of a measuring instrument or replicating machine. (2) Determination of the deviation from standard so as to ascertain the proper correction factors.

CAMAC. Computer-assisted measurement and control.

Capacity. The highest sustainable output rate that can be achieved with the current product specifications product mix, worker effort, plant, and equipment.

Capacity Loading. Work center loading where work will be rescheduled into other time periods if capacity is not available for it in the required time period.

Capacity Planning. The function of setting the limits or levels of manufacturing operations in the future, consideration being given to sales forecasts and the requirements and availability of people, machines, materials, and money.

Capacity Requirements. The projected future production capacity needs expressed in terms of people, machines, and facilities.

CAPITAL TOOLING. Jigs, fixtures, dies, and ancillary support tooling that require tool design and construction.

CARTESIAN COORDINATE SYSTEM. A coordinate system whose axes or dimensions are three intersecting perpendicular straight lines and whose origin is the intersection.

CELL. Collection of manufacturing operations and machines consisting of a number of work stations, operators, specialized tooling, materials-transport mechanisms and decouplers to make a family of parts.

CELL CONTROL. A module in the control hierarchy that controls a cell. The cell-control module is controlled by a center control module, if one exists. Otherwise it is controlled by the factory-control level.

CELLULAR MANUFACTURING. Organization of manufacturing equipment into groups according to function and inter-machine relationships to process a family of parts.

CENTER. Manufacturing unit consisting of a number of cells and the materials transport and storage buffers that interconnect them.

CENTER CONTROL. A module of the control hierarchy that controls a center. The center-control module is controlled by the factory-controlled level.

CHAIN DRIVE. Transmission of power from an actuator to a remote mechanism by means of flexible chain and mating-toothed sprocket wheels.

CHANGE ORDER. A formal notification that an order must be changed in some form. This can either result from a changed date or specification by the customer, an engineering change, or a change in inventory requirement date.

CLASSIFICATION. A process in which items are separated into groups based on the presence or absence of characteristic attributes.

CLOCK RATE. The speed (frequency) at which the processor operates, as determined by the rate at which words or bits are transferred through internal logic sequences.

CLOSED LOOP CONTROL. Control achieved by a closed feedback loop, i.e., by measuring the degree to which actual system response conforms to desired system response and utilizing the difference to drive the system into conformance.

CLOSED LOOP SYSTEM. A system in which the output or some result of the output is measured and fed back to the control for comparison with the input.

CODING. A process of establishing symbols to be used for meaningful communications.

COMMAND LANGUAGE. A source language consisting primarily of procedural operators, each capable of invoking a function to be executed.

COMMON PARTS. Parts which are used in two or more products or models.

COMMUNICATIONS LINK. Any mechanism for the transmission of information, usually electrical.

COMPENSATION. Logical operations employed in a control scheme to counteract dynamic lags or otherwise to modify the transformation between measured signals and controller output to produce prompt stable response.

COMPLIANCE. (1) The quality or state of bending or deforming to stresses within the elastic limit. (2) The amount of displacement per unit of applied force. (3) Slight mechanical motion to accommodate a fit or interference with motion.

COMPONENT. An inclusive term used to describe a subassembly or part that goes into higher level assemblies.

COMPOSITE COMPONENT. A hypothetical component which contains all the features that exist in a part family.

COMPOSITE ROUTING. A routing that lists a group of operations which are needed to produce a family of items, but which are not used for all items. The operations used depend on the characteristics of each particular item. (See group technology).

COMPUTED PATH CONTROL. A control scheme wherein the path of the manipulator end point is computed to achieve a desired result in conformance to a given criterion, an acceleration limit, a minimum time, etc.

COMPUTER-AIDED DESIGN (CAD). Any system that uses a computer to assist in the creation or modification of a design, typically including drafting.

COMPUTER-AIDED DESIGN/COMPUTER-AIDED MANUFACTURING (CAD/CAM). A technology of using computers to perform certain functions in design and manufacturing where a data base is shared by both functions.

COMPUTER-AIDED ENGINEERING (CAE). The use of computers in all aspects of engineering design, analysis and manufacturing functions.

COMPUTER-AIDED MANUFACTURING (CAM). The effective utilization of computer technology in the management, control, and operations of the manufacturing facility through either direct or indirect computer interface with the physical and human resources of the company.

COMPUTER-AIDED PROCESS PLANNING (CAPP). The use of computers in the preparation of process plans for a piece part or assembly. CAPP approaches are generally classified as variant and generative.

COMPUTER-INTEGRATED MANUFACTURING (CIM). The use of computers in all aspects of manufacturing, with integration of function and control in a hierarchy of computer systems.

COMPUTER NETWORK. A complex consisting of two or more interconnected computing units.

COMPUTER NUMERICAL CONTROL (CNC). The operation of machine tools and other processing machines by a series of coded instructions, executed by a computer with interface supporting the machine tool functions.

CONCURRENT ENGINEERING. Philosophy and approach to product design that promotes interactive design and manufacturing efforts to develop product and process as nearly to simultaneously as possible.

CONFIGURATION CONTROL. A means of ensuring that the product being built and shipped corresponds to the product ordered and designed.

CONFIGURATION MANAGEMENT. A system, usually software, for implementing configuration control through coordination of design and production documents.

CONTACT SENSOR. A device capable of sensing mechanical contact of the hand or some other part of a robot with an external object.

CONTINUOUS PATH CONTROL. A discriminable control scheme whereby the inputs or commands specify every point along a desired path of motion.

CONTROLLER. A device that uses predetermined procedures or rules to control a machine.

COORDINATE MEASURING MACHINE (CMM). An automatic inspection machine consisting of a contact probe and a means of positioning it in a three-dimensional space relative to the surfaces of a geometric shope to be inspected.

COORDINATED AXIS CONTROL. (1) Control wherein the axes of a robot arrive at their respective end points simultaneously, giving a smooth appearance to the motiom. (2) Control wherein the motions of the axes are such that the end point moves along a pre-specified type of path (line, circle, etc.) Also called end-point control.

COUNTER. (1) In relay-panel hardware, an electro-mechanical device which can be wired and preset to control other devices according to the total cycles of one on and offfunction. (2) In programmable controllers (PC), a logical analog of the electro-mechanical device internal to the processor.

CRITICAL ITEMS. Items that have a lead time longer than the normal planning span time, or items whose scarcity may impose a limit on production.

CRITICAL RATIO. A dynamic priority technique where job priorities are recalculated at regular intervals based on the progress the job has made through production and the current need date.

CRITICAL RATIO SCHEDULING. The sequencing of jobs in the queue of a work center in accordance with their critical ratio priorities.

CRITICAL WORK CENTER. (1) A work center that is working close to its maximum capacity or where a bottleneck (overload) occurs. (2) A work center that processes the work of an important part of the plant or product line, or one where a breakdown would be critical, or one that consists of a machine with unique characteristics for which an alternate is not available.

CUTTER PATH. The moving path of a machining cutter in relation to a work-piece.

CYLINDRICAL COORDINATE SYSTEM. A procedure used to specify a point on a cylinder in three dimensions, two linear and one angular.

CYCLE TIME. The span of time necessary to produce an item on one machine, or the period of time from starting one machine operation to starting another (in a pattern of continuous repetition).

DAMPING. (1) The absorption of energy, as in viscous damping of mechanical energy and resistive damping of electrical energy. (2) A property of a dynamic system which causes oscillations to die out and makes the response of the system approach a constant value.

DATA COLLECTION. The act of bringing data from one or more points to a central point.

DATA ENTRY. The process of coding or reading a single data device - such as a card or badge reader, numeric keyboard or rotary switch - for storage and later use.

DATA HIERARCHY. A data structure consisting of sets and subsets such that every subset of a set is of lower rank than the data of the set.

DATA LOGGING. Recording of data about events that occur in time sequence.

DEAD BAND. A range within which a nonzero input causes no output.

DECOUPLER. An element in a manufacturing cell that assists the flow of material in the cell or performs other functions to assist the operators or robots.

DEGREE OF FREEDOM. One of a limited number of ways in which a point or a body may move or in which a dynamic system may change, each way being expressed by an independent variable and all required to be specified if the physical state of the body or system is to be completely defined

DELIVERY CYCLE. The time from the receipt of the customer order to the time of the shipment of the product or of the supplying of the service, also called order cycle.

DERIVATIVE CONTROL. Control scheme whereby the actuator drive signal is proportional to the time derivative of the difference between the input, desired output, and the measured actual output.

DESIGN. The process of developing a detailed description that allows a product or process to be built or reproduced.

DIGITAL. The representation of numerical quantities by means of discrete numbers. It is possible to express in binary digital form all information stored, transferred, or processed by dual-state conditions, e.g., onloff openlclosed, octal and BCD values.

DIGITAL-TO-ANALOG CONVERTER (D/A, DAC). A device which transforms digital data into analog data.

DIRECT COST. Cost factors that are assignable to individual units of production, such as labor in assembly or materials in a product.

DIRECT DIGITAL CONTROL (DDC). Use of a computer to provide the computations for the control functions of one or multiple control loops used in process control operations.

DIRECT NUMERICAL CONTROL (DNC). The use of an external shared computer for distribution of part program data via data lines to more than one remote machine tool.

DISASSEMBLY. Operation in which a single unit is separated into two or more parts or subassemblies.

DISCRETE. Pertaining to distinct elements or to representation by means of distinct elements, such as characters.

DISCRETE PARTS MANUFACTURING. A process that produces distinct items.

DISPATCHING. (1) Selecting and sequencing of available jobs to be run at individual work stations and the assignment of these jobs. (2) The act of scheduling a task for execution.

DISTAL. Away from the base, toward the end effector of the arm.

DISTRIBUTED NUMERICAL CONTROL. A computer numerical control system in which a microprocessor used as a resident controller for a robot or machine tool also interacts with a host computer or other controllers.

DOWNLOAD. Transfer of files andlor programs from a computer to a local computerlcontroller.

DOWNTIME. The time interval during which a device is inoperative.

DRAGGING. In computer graphics, the translation of a selected display item along a path defined by a graphic input device.

Drift. The tendency of a system's response to move gradually away from the desired response.

Due Date. The latest calendar date at which an operation, or part, or order is to be completed.

Duplex. Data transmission in two directions. Full duplex describes two data paths which allow simultaneous data transmission in both directions. Half-duplex describes one data path which allows data transmission in either of two directions, but only one direction at a time.

Duty Cycle. The fraction of time during which a device or system will be active or at full power.

Dynamic Accuracy. (1) Deviation from true value when relevant variables are changing with time. (2) Difference between actual position response and position desired or commanded of an automatic control system as measured during motion.

Dynamic Model. A representation that deals with changing factors over time

Empirical Data. Data originating in or based on observation or experience.

Encoder. A type of transducer commonly used to convert angular or linear position to digital data

End Effector. An actuator, gripper, or mechanical device attached to the wrist of a manipulator by which objects can be grasped or otherwise operated on.

End Item. End product or the highest level of assembly shown by the bill of material

End-Point Control. Any control scheme in which only the motion of the manipulator end point may be commended and the computer can command the actuators at the various degrees of freedom to achieve the desired result.

End-Point Rigidity. The resistance of the hand, tool, or end point of a manipulator arm to motion under applied force

Enterprise Resource Planning (ERP). A complex computer system for controlling the scheduling and material flow in a manufacturing system.

Ergonomics. The study of human aspects of processes, especially behavioral, cognitive, and physiological.

Expediting. (1) The function of searching out and correcting conditions accounting for discrepancies between planned and actual performance. (2) The "rushing" or changing of production orders which must be furfilled in less than normal lead time.

Explosion. An extension of a bill of materials into the total of each of the components required to manufacture an assembly or subassembly quantity.

Extension. Orientation or motion toward a position where the joint angle between two connected bodies is 180 degrees.

External Sensor. A sensor for measuring displacements, forces, or other variables in the environment external to the robot.

Expert System. Intelligent computer program that uses knowledge and inference rules and procedures to solve problems that require significant human expertise for their solution.

Fabrication. (1) A term used to distinguish production operations for components as opposed to assembly operations. (2) Processing of natural or synthetic materials for desired mod)fication of shape and properties.

Factory. Manufacturing unit consisting of a number of centers and the materials transport, storage buffers, and communications that interconnect them.

Factory Control. A module in the control hierarchy that controls a factory. Factories are controlled by management personnel and policies.

Feedback Data. Data describing the result of a previous decision or action and used to determine actual status and deviation from a plan, so as to initiate corrective action.

Fixed-Stop Robot. A robot with a stop-point control but no trajectory control. That is, each of its axes has a fixed limit at each end of its stroke and cannot stop except at one or the other of these limits.

Fixture. A device to hold and locate a workpiece during inspection or production operations.

Flexible Manufacturing Cell (FMC). A smaller version of an FMS having only one or two CNC machines usually served by a robot for material handling.

Flexible Manufacturing System (FMS). An integrated system of process machines, material handling, and computer control system.

Flexible Workstation. An automated production system usually consisting of a CNC machining center and/or specialized machines, typically linked with a tool store and material handling system through industrial robots.

Flexion. Orientation or motion toward a position where the joint angle between two connected bodies is small.

Floorstock. Inventory issued to the plant in excess of immediate requirements.

Floor-To-Floor Time. The total standard time elapsed for picking up a part, loading it into a machine, carrying out operations, and unloading it (back to the floor, bin, pallet, etc.); generally applies to batch production.

Flowline. An arrangement of manufacturing machines and processes in which parts or products move sequentially from one station to the next.

Flowshop. A manufacturing system where machines are placed in sequence of the processes needed to produce the item.

Force Sensor. A sensor capable of measuring the forces and torques exerted by a robot at its wrist. Such sensors usually contain six or more independent sets of strain gauges plus amplifiers. Computer processing (analog or digital) converts the strain readings into three or-thogonal torque readings in an arbitrary coordinate system. When mounted in the work surface, rather than the robot's wrist, such a sensor is often called pedestal sensor.

Fuzzy Logic Control. Adaptive control technique using concepts of fuzzy logic that allow manipulation of linguistic variables, rather than numerical variables.

Forward Scheduling. A scheduling technique where the scheduler proceeds from a known start date and computes the completion data for an order usually preceding from the first operation to the last. (See BACKWARD SCHEDULING).

Frame Buffer. An electronic device capable of storing a digitized image in a digital memory for later readout and processing.

Gantry Robot. A robot that has a wrist or arm suspended from a bridge-like frame, the latter having axes that supply linear displacement to the wrist or arm.

Generative Process Planning. A process planning function which synthesizes a plan for a part by reasoning from first principles. Contrast with variant process planning.

Grey-Scale Picture. A digitized image in which the brightness of the pixels can have more than two values, typically, 128 or 256; requires more storage space and much more sophisticated image processing than a binary image, but offers potential for improved visual sensing.

Gripper. (1) A manipulator hand. (2) A device by which a robot may grasp and hold external objects.

Group Technology. A philosophy of manufacturing in which parts, products, and/or machines may be groupea functionally based on shared characteristics.

Group Technology Code (GT Code). The classification code to represent features of a part.

Hand. A device attached to the end of a manipulator arm that has a mechanism for closing jaws or other means to grasp objects.

Harmonic Drive. A gearing system in which one gear rotates within another, the two differing slightly in number of teeth, found in SCARA systems.

Heuristic. Pertaining to exploratory methods of problem solving in which solutions are discovered by evaluation of the progress made toward the final result. Contrast with algorithm.

Hierarchical Control. A distributed control technique in which the controlling processes are arranged in a hierarchy.

Homogeneous Transform. In robotics, a square transform matrix developed for a particular class of robots, used in computing a new spatial position and orientation vectors of a robot joint or work-piece.

Host. Common term for a computer which serves as a supervisor or file server to one or more automated machines.

Human Factors. The study of human aspects of systems

focusing primarily on cognitive and sensory processes. (See ERGONOMICS).

HYDRAULIC MOTOR. An actuator consisting of interconnected valves or pistons that convert high-pressure hydraulic or pneumatic fluid into mechanical shaft rotation.

HYSTERESIS. (1) The lagging of a physical response of a body behind its cause. (2) The asymmetry of the force/ displacement relationship in one direction compared to that of another direction.

IDLE TIME. (1) That part of available time during which the hardware is not teeing used. (2) The time during which a machine or operator is nonproductive because of lack of work.

INDIRECT COST. Cost factors that must be allocated to a group of parts or products because there is no close association between the cost and the product, such as building utilities or cost in the purchasing and personnel departments.

INDIRECT MATERIALS. Raw materials which become part of the final product but in such small quantities that their cost is not applied directly to the product. Instead, their costs become a part of manufacturing overhead.

INDUCTION MOTOR. An alternating-current motor in which torque is produced by the reaction between a varying or rotating magnetic field that is generated in stationary field magnets and the current that is induced in the coils or circuits of the rotor.

IN-PROCESS INVENTORY. Products in various stages of completion throughout the factory.

INPUT/OUTPUT (I/O). (1) Pertaining to either input or output, or both. (2) A general term for the equipment used to communicate with a computer. (3) The data involved in such communication. (4) The media carrying the data for input/output.

INPUT/OUTPUT MODULE. The printed circuit board or electronic assembly that is the termination for field wiring of I/O devices.

INTEGRAL CONTROL. Control scheme whereby the signal that drives the actuator equals the time integral of the difference between the input (desired output) and the measured actual output.

INTEGRATED COMPUTER-AIDED MANUFACTURING (ICAM). A research and development program sponsored by the U.S. Air Force to advance the state of manufacturing technology.

INTEGRATED DECISION SUPPORT SYSTEM (IDSS). A term that refers to a computer-software-based support system that integrates a number of decision-making tools in a single package.

INTEGRATED INFORMATION SUPPORT SYSTEM (IISS). A term that encompasses the acquisition and use of information to make effective decisions in a CAM environment.

INTELLIGENT ROBOT. A robot that can make sophisticated decisions and behavioral choices through its sensing and recognizing capabilities.

INTELLIGENT SYSTEM. Use of automation and artificial intelligence to reduce safety risks and implement cognitive capabilities, such as speech understanding, logic deduction, picture understanding, reasoning with expert knowledge, problem solving, and decision making in machines.

INTERACTIVE. Processing of data on a two-way basis, and with human intervention providing redirection or processing in a predetermined manner.

INTERFACE. A shared boundary. An interface may be a mechanical or electrical connection between two devices, a portion of computer storage accessed by two or more programs, or a device for communication to or from a human operator.

INTERLOCK. (1) A device to prevent a machine from initiating further operations until some condition or set of conditions is furfilled. (2) To prevent a machine or device from initiating further operations until the operation in process is completed.

INTERMITTENT PRODUCTION. A production system in which the productive units are organized according to function. The jobs pass through the functional department departments in lots and each lot may have a different routing.

INTERNAL SENSOR. A sensor for measuring displacements, forces, or other variables internal to a robot.

INTERRUPT. A signal that temporarily suspends the normal sequence of operations of a computer or controller.

Jig. A device that holds and locates a workpiece and also guides, controls, or limits one or more cutting tools. (See FIXTURE).

Job Assignment. Assignment of an employee to a machine or team, or of a job to a machine, employee, or team.

Job Shop. A discrete parts manufacturing facility characterized by a high mix of products of relatively low volume production in batch lots where work is carried out in functional areas made up of like manufacturing processes.

Job Status. The stage of activity toward a defined task or responsibility at any given time. May be a measurement against scheduled requirements of an overall task or plan.

Joint. (1) Rotary or linear articulation. (2) Axis of rotational or translational (sliding) degree-of-freedom of manipulator arm.

Joint Mode. Robot motion mode in which the arm motions are defined by, and motion occurs in, joint angles.

Joint Space. The vector that specifies the angular or translational displacement of each joint of a multidegree-of-freedom linkage relative to a reference displacement for each such joint.

Joystick. A movable handle that a human operator may grasp and rotate to a limited extent in one or more degrees of freedom and whose variable position or applied force is measured, resulting in commands to a control system.

Justification. The process of analyzing the benefits, primarily economic, in a system.

Just-In-Time (JIT). A synonym for the Toyota Production System or lean production where the manufacturing system is a linked-cell design that uses Kanban for material control.

Kaizen. Quality and process improvement process involving a formal program of gradual and continuing improvement, originated by Masaaki Imai.

Kanban. A production inventory and scheduling system pioneered by Toyota that tends to minimize work in process by pulling materials from components to final assembly where a fixed number of parts are allowed in each stage.

Ladder Diagram. An industry standard scheme for representing control logic relay systems.

Lead Screw. A precision machine screw which, when turned, drives a sliding nut or mating part in translation.

Lean Production. (1.) Manufacturing activities performed with a specific objective of minimizing inventory related costs and activities. (2) A term developed by MIIT researchers to describe companies which have implemented the Toyota Production System. (See JUST-IN-TIME).

Learning Control. A control scheme whereby experience is automatically used to provide for better future control decisions than those of the past.

Level Production. Scheduling final assembly so the demand for components and subassemblies is even on a daily basis.

Limit Switch. An electrical switch positioned to be actuated when a certain motion limit occurs, thereby deactivating the actuator causing that motion.

Linear Array Camera. A television camera (usually solid-state) with an aspect ratio of 1:n; today, n is typically 128, 256, or 512.

Linearity. (1) The degree to which an input/output relationship is proportional. (2) The degree to which a motion intended to be in a straight line conforms to a straight line.

Line Balancing. (1) The assignment of tasks to work stations or allocation of work stations to tasks so as to minimize the number of work stations and to minimize the total amount of unassigned time at all stations. (2) A technique for determining a fairly consistent flow of work through an assembly line at the planned line rate.

Line Synchronization. The ability to synchronize the operation of an industrial robot with a moving production line so that variations in line speed are automatically compensated for.

Load Capacity. The maximum weight or mass of material that can be handled by a machine or process without failure.

Load Deflection. (1) The difference in position of some point on a body between a non-loaded and an externally loaded condition. (2) The difference in position of a manipulator hand or tool, usually with the arm extended, between a non-loaded condition (other than gravity) and an externally loaded condition. Either or both static and dynamic (inertial) loads may be considered.

LOAD LEVELING. The procedure of moving operations or orders in such a manner as to smooth, in accordance with the capacity, manpower, and machine requirements over part or all of the planning horizon.

LOCATING SURFACES. Machined surfaces on apart that are used as reference surfaces for precise locating and clamping of the part in a fixture.

LOCAL AREA NETWORK. Data communication network that interconnects workstations, computers, machine tools, robots, and/or other digital devices.

LOT SIZE. The number of units within the same order.

MACHINE CENTER. A group of similar machines which can all be considered together for purposes of loading. (See WORK CENTER.)

MACHINED SURFACE. A group of faces on a part that can be machined by a single process.

MACHINE TOOL. A powered machine used to shape a part, typically by the action of a tool moving in relation to the work-piece.

MACHINING CENTER. A numerically controlled machine tool, such as a milling machine, capable of performing a variety of operations such as milling, drilling, tapping, reaming, and boring. Usually also included are arrangements for storing 10 to 100 tools and mechanisms for automatic tool change.

MACHINING PROCESS. Any particular machining operation viewed as an indivisible activity for planning purposes.

MAINTENANCE, Any activity intended to eliminate faults or to keep hardware or programs in satisfactory working condition, including tests, measurements, replacements, adjustments, and repairs.

MANIPULATOR. A mechanism typically consisting of a series of segments, jointed or sliding relative to one another, for the purpose of grasping and moving objects, usually in several degrees of freedom. It may be remotely controlled by a computer or by a human.

MANUFACTURING. A series of interrelated activities and operations involving the design, materials selection, planning, production, quality assurance, management, and marketing of discrete consumer and durable goods.

MANUFACTURING AUTOMATION PROTOCOL (MAP). A communication standard specifically tailored for the needs of automation and communication in manufacturing.

MANUFACTURING PLANNING. The function of setting the limits or levels of manufacturing operations in the future, consideration being given to sales forecasts and the requirements and availability of personnel, machines, materials, and finances.

MANUFACTURING SYSTEM. A complex arrangement of machines, transportation elements, tooling, people, storage buffers, and other items that are used together for manufacturing; coordinated to facilitate the flow of materials and information to coordinate inputs, processes and outputs, and characterized by measurable parameters.

MASS PRODUCTION. A manufacturing system where the industry sales volume is established and production rates are independent of individual orders.

MASTER-SLAVE MANIPULATOR. A class of teleoperator having geometrically isomorphic master and slave arms. The master is held and positioned by a person; the slave duplicates the motions, sometimes with a change of scale in displacement or force.

MATERIAL CONTROL. The function of maintaining and accounting for a constantly available supply of raw materials, purchased parts and supplies that are required for the production of products.

MATERIAL HANDLING. The processes and systems th transfer and manage the transfer of goods from one place to another.

MATERIALS PLANNING. The planning of requirements fo components based upon requirements for higher level assemblies.

MATHEMATICAL MODEL. A symbolic representation of, process, device, or concept.

MATRIX-ARRAY CAMERA. A television camera (usually solid-state) with an aspect ration of n:m, where neither n nor m is 1.

Mean Time Between Failure (MTBF). (1) The average time that a device will operate before failure. (2) The mean of the failure time distribution.

Mean Time To Repair (MTTR). The average time the device is expected to be out of service after failure.

Mechanization. A well-developed production system basis which traditionally does not involve a computer or adaptive controls, although sensors, electronic and electric controls are common, and performs automatic work-piece handling and tooling.

Microcomputer. A computer that is constructed using a single microprocessor CPU as the basic element.

Microprocessor. A basic element of a central processing unit that is a single integrated circuit. A microprocessor has a limited instruction set which is usually expanded by microprogramming. A microprocessor requires additional circuits to become a suitable central processing unit.

Model. An approximate representation of a process o system that attempts to relate (usually mathematically some or all of the variables in the system in such a wa that an increased understanding of the system is attained.

Move Time. The actual time that a job spends in transit from one operation to another in the shop.

Multiplexing. The time-shared scanning of a number of data lines into a single channel. Only one data line is enabled at any instant.

Net Load Capacity. The additional weight or mass of; material that can be handled by a machine or process without failure over and above that required for; container, pallet, or other device that necessarily accompanies the material.

Noise. (1) A spurious, unwanted, or disturbing signal. (2) A signal having energy over a wide range of frequencies.

Numerical Control (NC). A technique that provides prerecorded information in a symbolic form representing the complete instructions for the operation of a machine.

Off-Line. (1) Pertaining to devices not under direct control of the central processing unit. (2) Operation where the CPU operates independently of the time base of input data or peripheral equipment.

Online. (1) Pertaining to equipment or devices under direct control of the central processing unit. (2) Operation where input data is fed directly from measuring devices into the CPU or MCU. (3) In teleprocessing, a system in which the input data enters the computer directly from the point of origin and/or in which data is transmitted directly to where it is used.

Open-Loop Control. Control achieved by driving control actuators with a sequence of preprogrammed signals without measuring actual system response and closing the feedback loop.

Operation. A defined action, i.e., the act of obtaining a result from one or more elements.

Optical Character Recognition (OCR). The machine identification of printed characters through use of light sensitive devices.

Optical Scanner. (1) A device that scans optically and usually generates an analog or digital signal. (2) A device that optically scans printed or written data and generates their digital representations.

Optimal Control. A control scheme whereby the system response to a commanded input is optimal according to a specified objective function or criterion of performance, given the dynamics of the process to be controlled and the constraints on measuring.

Optimization. The analysis of a system to determine parameters that make it perform best in some sense, such as minimum cost.

Output. (1) The finished result of an activity. (2) Data that has been processed from an internal storage to an external storage.

Overshoot. The degree to which a system response, such as change in reference input, goes beyond the desired value.

Pallet. (1) Device that serves as a standardized conveyance for the part in an automated work environment. (2) Material handling device used for loading and unloading components.

Pan. (1) Orientation of a view, as with video camera, in azimuth. (2) Motion in the azimuth direction.

Part. A material item which is used as a component and is not an assembly.

PART CLASSIFICATION. A coding scheme, typically involving four or more digits, that specifies a discrete product as belonging to a part family. (See GROUP TECHNOLOGY.)

PART FAMILY. A set of discrete products that can be produced by the same sequence of machining operations. This term is primarily associated with group technology.

PART HANDLING SYSTEM (PHS). Network including carts, pathways, conveyors, chains, and storage areas for parts, fixtures, and pallets.

PART ORIENTATION. The angular displacement of a product being manufactured relative to a coordinate system.

PART PROGRAM. The set of instructions and descriptions necessary to guide a machining operation.

PASSIVE ACCOMODATION. Compliant behavior of a robots end point solely in response to forces exerted on it.

PATH. In robotics, the space curve traced by the end effector.

PATTERN RECOGNITION. Description or classification of pictures or other data structures into a set of classes or categories.

PAYLOAD. The maximum weight or mass of a material that can be handled satisfactorily by a machine or process in normal and continuous operation.

PEDESTAL ROBOT. A robot mounted on a single fixed support.

PERIPHERAL EQUIPMENT. Units which may communicate with a computer or programmable controller, but are not part of it, e.g., teletype, cassette recorder, CRT terminal, tape reader, etc.

PICK AND PLACE. Simple transfer motion of robots.

PICK AND PLACE ROBOT. A simple robot, often with only two or three degrees of freedom, that transfers items from place to place by means of point-to-point moves. Little or no trajectory control is available. Often referred to as a bang-bang robot.

PITCH. (1) An angular displacement up or down as viewed along the principal axes of a body having a top side, especially along its line of motion. (2) The axial displacement of successive threads of a screw.

PLANETARY DRIVE. A gear reduction arrangement consisting of a sum spur gear, two or more planetary spur gears, and an internally toothed ring gear.

PLAYBACK ACCURACY. (1) Difference between a position command recorded in an automatic control system and that actually produced at a later time when the recorded position is used to execute control. (2) Difference between actual position response of an automatic control system during a programming or teaching run and that corresponding response in a subsequent run.

POINT-TO-POINT CONTROL. A control scheme whereby the inputs or commands specify only a limited number of points along a desired path of motion. The control system determines the intervening path segments.

POKA-YOKE. An approach to minimizing mistakes or defects by modifying a system to reduce the possibility of their occurring; originated by Dr. Shigeo Shingo.

POLAR COORDINATE SYSTEM. A coordinate system in two variables, an angle of rotation and distance from origin, usually as applied to points in a plane. Two coordinates specify a point on a circle.

POLLING. A technique by which each of the devices sharing a communications line or network is periodically interrogated to determine whether it requires servicing.

POSITION CONTROL. Control system in which the input (desired option) is the position of some body.

POSITION ERROR. In a servomechanism that operates a manipulator joint, the difference between the actual position of that joint and the commanded position.

POSTPROCESSOR. Program or computer function necessary in some numerical control systems to customize generic programs to run on a specific machine.

PRECISION. The standard deviation or root-mean-squared deviation of values around their mean.

PRIORITY SCHEDULING RULE. A technique for assigning an objective priority sequence number that can then be utilized for scheduling of production jobs.

PROCESS. A systematic sequence of operations to produce a specified result.

PROCESS CONTROL. Systems for automation of continuous operations. This is contrasted with numerical control, which provides automation of discrete operations.

PROCESS DETAILING. Planning for the details of machining. Process detailing includes cutting parameter selection, cutter path generation, etc.

PROCESS INSTRUCTION. The part of a process plan which specifies the operations on a part and the sequences of these operations, with alternative operations and routings wherever feasible.

PROCESSOR. (1) In hardware, a data process. (2) In software, a computer program that includes the compiling, assembling, translating, and related functions for a specific programming language, e.g., Cobol processor, Fortran processor.

PROCESS PLAN. A detailed plan for the production of a piece part or assembly. It includes a sequence of steps to be executed according to the instructions in each step and consistent with the controls indicated in the instructions.

PROCESS PLANNING. (1) The act of preparing a process plan for the fabrication or assembly of parts. Planning may be performed manually or with computer assistance. (2) National Bureau of Standards (NBS) Planning of all the required operations for making a single component.

PROCESS SEQUENCING. Determining a suitable order of processes for manufacturing a part.

PRODUCT GROUP. A group of products having common classification criteria.

PRODUCTION. The converting of a raw material into a finished product.

PRODUCTION CONTROL. The function of directing or regulating the orderly movement of goods through the entire production cycle from the requisitioning of raw materials to the delivery of the finished product.

PRODUCTION CYCLE TIME. The elapsed time to produce a product.

PRODUCTION LEVELS. The quantity of production usually expressed in units, or some other broad measure.

PRODUCTION MONITORING. The checking of the status and progress of production activities.

PRODUCTION PLAN. Information required to produce a desired system output, including part numbers, quantities, tools, machines, and schedules.

PRODUCTION RATES. The quantity of production, usually expressed in units/hour.

PRODUCT MIX. (1) The distribution of the various items in a production plan. (2) The combination of individual product types and the volume produced that make up the total production volume.

PROGRAM SCAN. The time required for a programmable controller processor to execute all instructions in the program.

PROGRAMMABLE CONTROLLER (PC). A solid state control system which has a user programmable memory for storage of instructions to implement specific functions such as I/O control logic, timing, counting, arithmetic, and data manipulation.

PRONATION. Orientation or motion toward a position with the back, or protected side, facing up or exposed.

PROPORTIONAL CONTROL. Control scheme whereby the signal that drives the actuator equals the difference between the input (desired output) and measured actual output.

PROPORTIONAL-INTEGRAL-DERIVATIVE CONTROL (PID). Control scheme whereby the signal that drives the actuator equals the weighted sum of the difference, time integral of the difference, and time derivative of the difference between the input and the measured actual output.

PROTOCOL. The rules for controlling data communications between devices in computer systems or computer networks.

PROXIMITY SENSOR. A device that senses the presence of an object and/or measures how far away it is. Proximity sensors work on the principles of triangulation of reflected light, lapsed time for reflected sound, and others.

PULL SYSTEM. In JIT systems, the downstream process pulls (withdraws) the needed materials from the upstream process in the needed quantities at the correct time.

QUALITY. (1) The characteristics of a product that affect its ability to satisfy customer needs, whether stated or implied. (2) A state of a product which is free of defects.

QUALITY CONTROL. Activities and processes used to regulate the level of quality.

QUEUE. A waiting line formed by items in a system waiting for service.

QUEUEING. The process and study of waiting lines.

RATE CONTROL. Control system in which the input is the desired velocity of the controlled object.

RATIONALIZE. To simplify and organize to allow further use and analysis, as in putting parts into a family based on group technology principles.

REAL TIME. Pertaining to computation performed while the related physical process is taking place so that results of the computation can be used in guiding the physical process, without the process response dynamics being changed.

RECORD-PLAYBACK ROBOT. A manipulator for which the critical points and desired trajectories are stored in sequence by recording the actual values of the joint position encoders of the robot as it is moved under operator control. To perform the task, these points are played back to the robot servo system.

REDUNDANCY. Duplication of information or devices in order to improve reliability.

REGISTER. A computer or other electronic storage area set aside to hold a particular quantity, such as motor registers used to keep a running account of arm motion.

RELATIVE. A coordinate system in which each location is defined by its incremental distance from a previous position.

RELIABILITY. (1) The likelihood that a machine will be operable, specified as a percent or as a time until breakdown. (2) The probability that a device will function without failure over a specified time period or amount of usage.

REMOTE CENTER COMPLIANCE (RCC). A compliant device used to interface a robot or other mechanical workhead to its tool or working medium. The RCC allows a gripped part to rotate about its tip or to translate without rotating when pushed laterally at its tip. The RCC thus provides general lateral and rotational float and greatly eases robot or other mechanical assembly in the presence of errors in parts.

REPEATABILITY. In robotics, the error involved when a robot is taught a point and then is commanded to return to that point.

RESOLUTION. (1) The least interval between two adjacent discrete details that can be distinguished from one another. (2) The smallest increment of distance that can be read and acted upon by an automatic control system. (3) In computer science, a theorem-proving method based on match and substitution.

REWORK. The process of remedying an incorrectly manufactured part, sub-assembly or assembly by reprocessing it either in the normal manufacturing system or in a dedicated facility.

ROBOT. (1) An automatic apparatus or device that performs functions ordinarily ascribed to human beings or operates with what appears to be almost human intelligence. (2) A mechanical device that accomplishes what may be thought of as human motions under computer control; an arm with a brain. (3) A reprogrammable multi-functional manipulator designed to move material, parts, tools, or specialized devices through variable programmed motions for the performance of a variety of tasks.

ROLL. The angular displacement around the principal axis of a body, especially its line of motion.

ROUTING. (1) In production, the sequence of operations to be performed in order to produce a part or an assembly. (2) In telecommunications, the assignment of the communications path by which a message can reach its destination.

ROUTING SHEET. A document which lists the sequence of moves between operations or machines, used to control the movement of material in a job shop.

RUNG. A grouping of PC instructions which controls one output. This is represented as one section of a logic ladder diagram.

RUN TIME. The standard time allowed for the processing of one piece of a specific item for a specified machine and operation.

SATURATION. A range within which the output is constant regardless of input.

SCHEDULE. A listing of jobs to be processed through a work center, department, or plant and their respective start dates as well as other related information.

SCHEDULING. The process of setting operation start dates for jobs to allow them to be completed by their due date.

SCRAP. (1) Fragments of stock removed from a part during production. (2) A rejected or discarded part.

SCRAP ALLOWANCE. A factor that expresses the quantity of a particular component that is expected to be scrapped while that component is being built into a given assembly. Also, a factor that expresses the amount of raw material needed in excess of the exact calculated requirement to produce a given quantity of a part.

SCRAP USAGE. The expected average quantity of an item to be scrapped each period.

SELF COMPLIANT ARM FOR ROBOT ASSEMBLY (SCARA) ROBOT. A robot that uses a linear axis joint to achieve vertical displacement and two or more parallel revolute joints (with axes of rotation perpendicular to the horizontal plane) to achieve end-effector positions in a horizontal plane.

SENSOR. A transducer or other device whose input is a physical phenomenon and whose output is a quantitative measure of that physical phenomenon.

SEQUENCE ROBOT. A robot whose motion trajectory follows a preset sequence of positional changes.

SERVOMECHANISM. An automatic control mechanism consisting of a motor driven by a signal that is a function of the difference between commanded position and/or rate and measured actual position and/or rate.

SERVOVALVE. A transducer whose input is a low-energy signal and whose output is a high-energy fluid flow that is proportional to the low energy signal.

SETTLING TIME. The time for a damped oscillatory response to decay within some given limit.

SETUP. The process of making a machine or work cell ready to operate or perform a particular function.

SETUP TIME. Time required to change over a machine, removing tooling and attaching the new tooling to make a particular product.

SHOP FLOOR CONTROL. The system which monitors shop operations, such as product flow, machine status, and order completion.

SHORT-TERM REPEATABILITY. Closeness of agreement of position movements, repeated under the same conditions during a short time interval, to the same location.

SHOULDER. The manipulator arm linkage joint that is attached to the base.

SIMULATION. The representation of certain features of the behavior of a physical or abstract system by the behavior of another system, typically a physical or computer model.

SLEW RATE. (1) The maximum velocity at which a manipulator joint can move; a rate imposed by saturation somewhere in the servo loop controlling that joint (e.g., by a valve's reaching its maximum open setting). (2) The maximum speed at which the tool tip can move in an inertial Cartesian frame.

SOLENOID. A cylindrical coil of wire surrounding a movable core which, when energized, sets up a magnetic feld and draws in the core.

SPAN TIME. Actual time from part design to the completion of the finished product.

SPHERICAL COORDINATE SYSTEM. A coordinate system, two of whose dimensions are angles, the third being a linear distance from the point of origin. These three coordinates specify a point on a sphere.

SPLIT LOT. A production order quantity that has been divided into two or more smaller quantities, usually after the order is in process.

SPRINGBACK. The deflection of a body when external load is removed. May refer to deflection of the end effector of a manipulator arm.

STANDARD. An accepted criterion or an established measure for performance, practice, or design.

STATIC ACCURACY. (1) Deviation from time value when relevant variables are not changing with time. (2) Difference between actual position response and position desired or commanded of an automatic control system as determined in the steady state, i.e., when all transient responses have decayed.

STATIC MODEL. A model, or representation, that ignores the effects of time on the process.

STATION. A physical location where a part normally stops either to have an operation performed on it or to wait for clearance to proceed to the next station.

STEPPING MOTOR. An electric motor whose windings are arranged in such a way that the armature can be made to step in discrete rotational increments (typically 1/200th of a revolution) when a digital pulse is applied to an accompanying "driver" circuit. The armature displacement will stay locked in this angular position independent of applied torque, up to a limit.

STIFFNESS. The risistance to displacement when an object has a force applied to it.

STOP. A mechanical constraint or limit on some motion which can be set up to stop the motion at a desired point.

STRAIN GAUGE. A sensor that, when cemented to elastic materials, measures very small amounts of stretch by the change in its electrical resistance. When used on materials with high modules of elasticity, strain gauges become force sensors.

SUBASSEMBLY. An assembly which is used at a higher level to make up another assembly.

SUPERVISORY CONTROL. A control scheme whereby a person or computer monitors and intermittently re-programs, sets subgoals, or adjusts control parameters of a lower-level automatic controller, while the lower-level controller performs the control task continuously in real time.

SUPINATION. Orientation or motion toward a position with the front, or unprotected side, facing up or exposed.

SYNCHRO. A shaft encoder based upon differential inductive coupling between an energized rotor coil and field coils positioned at different shaft angles.

SYNCHRONOUS. Operating according to an overall timing source, i.e., at regular intervals.

SYSTEM. An organized collection of personnel, machines, and methods required to accomplish a set of specific functions.

SYSTEM DESIGN. The arrangement of a group of functionally distinct devices, components, and computer programs which regularly interact and which are interdependent and meet specific performance levels.

SYSTEMS ANALYSIS. The study of collection of interacting entities, with emphasis on their operation as a whole.

TACHOMETER. A rotational velocity sensor.

TACTILE SENSOR. A sensor that makes physical contact with an object in order to sense it; includes touch sensors, tactile arrays, force sensors, and torque sensors. Tactile sensors are usually constructed from microswitches, strain gauges, or pressure-sensitive conductive elastomers.

TEACH. To guide a manipulator arm through a series of points or in a motion pattern as a basis for subsequen automatic action by the manipulator.

TEACH PENDANT. Cable and keypad device for programming and controlling a machine.

TEARDOWN. The opposite of setup, taking place at the en of an operation, e.g., dismantling of assembly jigs, cleaning of vats or machines, etc.

TELEOPERATOR. A device having sensors and actuators for mobility and/or manipulation, remotely controlled by a human operator. A teleoperator allows an operator to extend his sensor-motor function to remote or hazardous environment.

TEMPLATE MATCHING. Pixel-by-pixel comparison of an image of a sample object with the image of a reference object, usually for purposes of identification, but also applicable to inspection.

THRESHOLDING. The process of quantizing pixel brightness to a small number of different levels (usually two levels, resulting in a binary image). A threshold is a level of brightness at which the quantized image brightness changes.

THROUGHPUT. The number of parts a manufacturing system produces per time unit.

TILT. (1) Orientation of a view, as with a video camera, in elevation. (2) Motion in the elevation direction.

TIMER. (1) In relay-panel hardware, an electromechanical device which can be wired and preset to control the operating interval of other devices. (2) In programmable controllers, a logical analog of the electromechanical device internal to the processor, which is controlled by a user-programmed instruction.

TOOLING. A set of required standard or special tools for production of a particular part, including jigs, fixtures. gauges, cutting tools, etc. Specifically excludes machine tools.

TOOL LIFE. The anticipated life of a cutting tool. It is usually expressed as either the number of pieces the tool is expected to make before it wears out or as the number of hours of use anticipated.

TOOL MODE. In robotics, the use of a coordinate system with its origin at the base of the end effector.

TRACKING. Continuous position control response to a continuously changing input.

TRAJECTORY. The time of intermediate robot arm configurations between initial location and goal.

TRANSDUCER. A device used to convert physical parameters, such as temperature, pressure, weight, etc., into electrical signals.

TRANSFER LINE. A linear network of workstations or machines, possibly separated by buffer storages, in which material flows from one end to the other in a sequence using a materials handling system.

TRANSFER MACHINE. An apparatus or device for grasping a work piece and moving it automatically through stages of a manufacturing process.

TRANSIENT. (1) General temm referring to a value that changes in time. (2) Response of a dynamic system to a transient input such as a step or a pulse.

TRANSLATION. Movement of a body such that all axes remain parallel to what they were, i.e., without rotation.

UNDERSHOOT. The degree to which a system response to a step change in reference input falls short of the desired value.

UNIT LOAD. In material handling, the largest quantity that can be conveniently and economically moved.

UPLOAD. Transfer of files from a local computer/controller to a computer.

VARIANT PROCESS PLANNING. Process planning based on the retrieval and modification of existing plans of similar components.

VELOCITY ERROR. In a servomechanism that operates a manipulator joint, the difference between the rate of change of the actual position of that joint and the rate of change of the commanded position.

VIA POINT. In robot trajectory control, intemmediate locations used to guide the robot path.

VIRTUAL CELL. A software implementation of a manufacturing cell where work stations or support systems arc dynamically assigned or allocated to a cell controller which is responsible for the manufacture of a family of parts.

VOLATILE MEMORY. A memory that loses its information if the power is removed from it.

WAIT TIME. The time that a job spends waiting to be moved or waiting to be worked on in the shop.

WASTE HANDLING SYSTEM. System used to collect, move, store and dispose of process waste.

WINDUP. Colloquial temm describing the twisting of a shaf' under torsional load, so called because the twist usually unwinds, sometimes causing vibration or other negative effects.

WORD LENGTH. The number of bits in a word. In PC literature these are generally only data bits. One PC word equals sixteen data bits.

WORK CENTER. (1) An administrative or accounting subdivision of a department. (2) A specific production facility which may consist of one or more persons or machines.

WORK COORDINATES. The coordinate system with reference to the work-piece jig, or fixture.

WORK ELEMENT. In process planning, a single task to be performed that cannot be subdivided.

WORKING ENVELOPE. The set of points representing the maximum extent or reach of the robot hand or working tool in all directions.

WORK-IN-PROCESS (WIP). Products in various stages of completion throughout the production cycle, including raw material that has been released for initial processing and completely processed material awaiting final inspection and acceptance as finished product.

Work-piece. Any part in any stage of manufacture prior to its becoming a finished part.

Work Station. Manufacturing unit consisting of one or more numerically controlled machine tools serviced by a transport system. In assembly, the space assigned to one worker and his/her supply of materials.

World Coordinates. The coordinate system with reference to the machine pedestal, frame, or the shop floor.

World Mode. In robotics, the use of a coordinate system with its origin at the base of the robot.

Worm Gear. A short screw-like gear that mates to a secondary gear whose axis of rotation is perpendicular to and offset from that of the short screw. When the screw is turned, it drives the gear in rotation.

Wrist. The manipulator arm joint to which a hand or end effector is attached.

Yaw. An angular displacement left or right viewed from along the principal axis of a body having a top side, especially along its line of motion.

BIBLIOGRAPHY

Black, J. Temple, *The Design of the Factory with a Future*, McGraw-Hill, New York, 1991.

Computer-Aided Manufacturing International (CAM-I). *Functional Specifications for Advanced Factory Management System*. March 1979.

Chang, T.C. The advances of computer-aided process planning. Technical Report NBS-GCR-83-441. National Bureau of Standards, 1983.

Fisher, E.W. (ed.) *Robotics and Industrial Engineering: Selected Readings*. Norcross, GA: Industrial Engineering and Management Press. Institute of Industrial Engineers, 1983.

Groover, Mikell P., *Fundamentals of Modern Manufacturing*, Prentice-Hall, Englewood Cliffs, NJ, 1996.

Kimbler, D.L. The industrial robot and automated manufacturing. Microbot, 1986.

Ludema, K.C., Cadell, R. M., and Atkins, A.G., Manufacturing Engineering: Economics and Processes, Prentic-Hall, Englewood Cliffs, NJ, 1987.

Monden, Yasuhiro, *Toyota Production System: An Integrated Approach to Just-In-Time*, IE and Management Press, Norcross, GA, 1993.

Womack, J.P., D.T. Jones and Dan Roos, *The Machine that Changed the World*, HarperPerennial, 1990.

Wysk, R.A. Contributor of "programmable controller" terms.

Yeomans, R.W., Choudry, A. and ten Hagen, P. J. W., *Design Rules for a CIM System*, Elsevier, Science Publishing Co., New York, 1985.

Z94.12
Materials Processing

The definitions developed for this section represent an extensive revision, modification, and addition to the 1982 edition. Many new terms are added to reflect the technological advancements in the materials processing field in general and in the manufacturing automation and computer control fields in particular. Many terms peculiar to the general area of manufacturing have definitions that have evolved through common usage. The compilation of terms in the subsection on manufacturing automation and computer control is by no means exhaustive. The reader is advised to refer to section 17 on Manufacturing Systems for additional terms. For the added terms, the Materials Processing Subcommittee used many already accepted definitions such as those published by technical societies, namely:

Glossary of Terms: Computer Integrated Manufacturing, published by Computer and Automated Systems Association (CASA) of Society of Manufacturing Engineers, 1984.

Volume 1, Machining, and *Volume 2, Forming of the Tool and Manufacturing Engineers Handbook* (TMEH), published by Society of Manufacturing Engineers, 1984.

Metal Cutting Technology, written by E. Orth and J. I. ElGomayel, and published by the Metals Engineering Institute of the American Society of Metals.

Common Words as They Relate to N/C Software, publication of the National Machine Tool Builders' Association, 1972.

The definitions in this section appear in the following subsections: Manufacturing Automation and Computer Control, Forging, Foundry (Casting), Metal Forming, Metal Machining, Plastics, and Welding. Although the work of the earlier subcommittee must be acknowledged, this revised section was the result of the time, effort, expertise, and devotion of the following members of the subcomittee:

Chairman

Professor Joseph I. ElGomayel
Purdue University

Subcommittee

Anthony M. Bratkovich
Engineering Director
National Machine Tool Builders' Association

Ron Fowler
President
Metal Fabricating Institute

Eugene Orth
Engineering Consultant

Other members of the subcommittee who lent their support were:

O. D. Lascoe
Professor Emeritus
Purdue University

Ray Kirschbaum
Mechanical Engineer
Rock Island Army Arsenal

Donald Smith
Director of Engineering
Federal Press Company

Z94.12 Materials Processing
Table of Contents
Manufacturing Automation and Computer Control..12-3
Forging..12-25
Foundry, Casting..12-30
Metal Forming...12-48
Metal Machining..12-60
Plastics..12-68
Welding...12-86

MANUFACTURING AUTOMATION AND COMPUTER CONTROL

ACCURACY. The degree of freedom from error, *i.e.* the degree of conformity to some standard. Accuracy is contrasted with precision. For example: four place numbers are less precise than six place numbers, however, a properly computed four place number might be more accurate than an improperly computed six place number.

ADAPT. A computer-aided NC parts programming language similar to APT, but with fewer capabilities. Developed for small to medium-scale computers and used basically for two-axis contouring.

ADAPTIVE CONTROL (AC). (1) The ability of a control system to change its own parameters in response to a measured change in operating conditions. (2) Machine control units in which feeds and/or speeds are not fixed. The control unit, working from feed back sensors, is able to optimize favorable situations by automatically increasing or decreasing the machining parameters. This secures optimum tool life or surface finish and/ or machining costs or production rates.

ADAPTIVE CONTROL CONSTRAINED (ACC). A control system in which improved machine productivity is obtained through in-process measurement by using limiting values for machine parameters such as torque or spindle deflection.

ADAPTIVE CONTROL OPTIMIZED (ACO). A control system in which optimum machine productivity is obtained through in process measurement and adjustment of operating parameters.

ADDRESS. A character or group of characters that identifies a register, a particular part of storage, or some other data source or destination.

ALGORITHM. A prescribed set of well-defined rules for the solution of a problem in a finite number of steps.

ALPHANUMERIC CODE. A coding system consisting of characters, including numbers, letters, punctuation marks, and such signs as $, @, and #. Also referred to as alphameric code.

AMPLIFIER. Device for controlling power from a source so that more is delivered at the output than is supplied at the input. Source of power may be mechanical, hydraulic, pneumatic, electric, etc.

ANALOG. The use of physical variables, such as distance and rotation, to represent and correspond with numerical variables occurring in a computation. In NC, a system utilizing magnitudes or ratios of electrical voltages to represent physical axis positions.

ANALOG COMPUTER. A continuously measuring computer in which quantities are represented by physical variables. Problem parameters are translated into mechanical or electrical circuits as an analog for the physical phenomenon in question. An analog is used for each variable and produces analogs as output.

ANALOG-TO-DIGITAL-CONVERTER (ADC). Refers to a device which produces a digital output from an input in the analog form of physical motion or electrical voltages.

ANSI. American National Standards Institute, formerly known as ASA and USASI, its purposes include: (1) serving as the national institution for voluntary standardization and certification, (2) furthering voluntary standards, (3) assuring the interests of the public, including consumers, labor, industry, and government, (4) providing the need for standards and certification programs, (5) establishing, promulgating, and administering procedures for recognition and approval of standards as American National Standards, (6) encouraging existing organizations to prepare and submit certification programs for accreditation, (7) cooperating with government agencies, (8) promoting knowledge and use of American National Standards, (9) representing USA interests in international nontreaty standardization and certification organizations, (10) serving as a clearinghouse for information on standards, standardization, and certification in the USA and abroad.

APL (A PROGRAMMING LANGUAGE). A problem solving language designed for use at remote terminals, with special capabilities for handling arrays and for performing mathematical functions.

APT (AUTOMATICALLY PROGRAMMED TOOLS). A computer-assisted program system describing parts illustrated on a design and defining in a sequence of statements, the part geometry, cutter operations, and machine tool capabilities used for turning, point-to-point work, and multiaxis milling.

ARCHITECTURE. Preset, physical, and logical operating characteristics of a control system or control unit.

ARTIFICIAL INTELLIGENCE. Research and study in methods for the development of a machine that can improve its run operations. The development or capability of a machine that can proceed or perform functions that are normally concerned with such human intelligence as learning, adapting, reasoning, self-correction, automatic improvement. In a more restricted sense, the study of techniques for more effective use of digital computers by improved programming techniques.

ASCII (AMERICAN STANDARD CODE FOR INFORMATION INTERCHANGE). A standard data-transmission code that was introduced to achieve compatibility between data devices. It consists of 7 information bits and one parity bit for error-checking purposes, thus allowing 128 code combinations.

ASSEMBLE. The basic element employed when one or more objects are put on or into another object so that they fit or contact each other in a predetermined relation to form a unit.

ASSEMBLY LANGUAGE. (1) A computer-oriented language whose instructions are usually one-to-one correspondence with computer instructions and that may provide facilities such as the use of macroinstructions. (2) A low-level symbolic language, usually machine dependent, that requires an assembler program to translate the symbolic code to the corresponding machine code (contrast with compiler language.)

AUTOMATA THEORY. Relates to the development of theory which relates the study of principles of operations and applications of automatic devices to various behaviorist concepts and theories.

AUTOMATED ASSEMBLY. Assembly by means of operations performed automatically by machines. A computer system may monitor the production and quality levels of the assembly operations.

AUTOMATED PROCESS PLANNING. Creation of process plans, with partial or total computer assistance, for items in a particular family.

AUTOSPOT (AUTOMATIC SYSTEM FOR POSITIONING OF TOOLS). A general-purpose computer program used in preparing instructions for NC positioning and straight-cut systems.

AUTOMATION. The implementation of processes by automatic means; the theory, art or technique of making a process more automatic; the investigation, design, development, and application of methods of rendering processes automatic, self-moving, or self-controlling; the conversion of a procedure, process, or equipment to automatic operation.

AUXILIARY FUNCTION. A function of a machine other than the control of the coordinates of a workpiece or tool. Usually on/off type operations such as starting and stopping a spindle or coolant pump.

AXIS. A general direction relative motion between cutting tool and workpiece. The understanding of axes in rectangular coordinates is the basic keystone to understanding NC.

BATCH PROCESSING. A manufacturing operation in which a designated quantity of material is treated in a series of steps. Also, a method of processing jobs so that each is completed before the next job is initialized.

BATCH PRODUCTION. Non-Continuous processing of unlike parts. Contrast with mass production.

BAUD. (1) A unit of signalling speed equal to the number of discrete conditions or signal events per second. For example, one baud equals one-half dot cycle per second in Morse Code, one bit per second in a train of binary signals, and one 3-bit value per second in a train of signals each of which can assume one of eight different states. (2) In asynchronous transmission, the unit of modulation rate corresponding to one unit interval per second, *i.e.*, if the duration of the unit interval is 20 milliseconds, the modulation rate is 50 baud.

BEGINNER'S ALL-PURPOSE SYMBOLIC INSTRUCTION CODE (BASIC). A procedure-level computer language that is easy to learn and well suited for time-sharing communication via terminals connected with a remotely located computer.

BEHIND THE TAPE READER (BTR). A means of putting data directly into a machine control unit from an external source other than a tape reader.

BCD (BINARY CODED DECIMAL). A system of number representation; that is, an information code in which each decimal digit is represented by 4 binary digits.

BINARY. A numerical system pertaining to characteristics involving a selection or condition in which two possibilities exist.

BINARY DIGIT (BIT). (See BIT.)

BIONICS. A technology attempting to relate the functions, characteristics, and phenomena of living systems to those of hardware systems.

BIT. (1) Bit is an abbreviation for binary digit. Most commonly a unit of information equalling one binary decision, or the designation of one of two possible and equally likely values or states, usually conveyed as 1 or 0, which may also mean "yes" or "no." (2) A single character in a binary number. (3) A single pulse in a group of pulses. (4) A unit of information, capacity of a storage device. The capacity in bits is the logarithm to the base two of the number of possible states of the device.

BLOCK ADDRESS FORMAT. (1) A tape programming method for NC systems in which only instructions that need changing are punched into the tape. (2) A means of identifying words by use of an address specifying the format and meaning of words in the block.

BLOCK DIAGRAM. A simplified schematic drawing setting forth the sequence of operations to be performed for handling a particular application.

BOOLEAN ALGEBRA. A process of reasoning or a deduction system of theorems using symbolic logic and dealing with classes, propositions, yes/no criteria, etc., for variables rather than numeric quantities. Developed by George Boole, this algebra includes operators, such as AND, OR, NOT, EXCEPT, IF...THEN, that permit mathematical calculations.

BOOTSTRAP. (1) A technique or device designed to bring itself into a desired state by means of its own action, *e.g.*, a machine routine whose first few instructions are sufficient to bring the rest of itself into the computer from an input device. (2) To use a bootstrap. (3) That part of a computer program used to establish another version of the computer program.

BREADBOARD. Usually refers to an experimental or rough construction model of a process, device, or construction.

BRIDGE. In a system of measurement, the instruction in which part or all of a bridge circuit is used in measuring one or more electrical quantities. In relation to a fully electronic stringed instrument, the bridge converts the mechanical vibrations produced by the strings into electrical signals.

BUBBLE MEMORY. Such memories are actually tiny cylinders of magnetization whose axes lie perpendicular to the plane of the single-crystal sheet that contains them. Magnetic bubbles arise when two magnetic fields are applied perpendicular to the sheet. A constant field strengthens and fattens the regions of the sheet whose magnetization lies along it. A pulsed field then breaks the strengthened regions into isolated bubbles, which are free to move within the plane of the sheet. Because the presence or absence of bubbles can represent digital information, and because other external fields can manipulate this information, magnetic-bubble devices find uses in data-storage systems.

BUFFER. (1) A "machine" designed to be inserted between forms of storage, usually between internal and external. (2) An input device in which information is assembled from external or secondary storage and stored ready for transfer to internal storage. (3) An output device into which information is copied from internal storage and held for transfer to secondary or external storage. Computation continues while transfers between buffer storage and secondary or internal storage or vice versa takes place. (4) Any device which stores information temporarily during data transfers.

BUFFER STORAGE. (1) A device for storing information for eventual transfer to active storage. It enables the control system to act on stored data without waiting for tape reading. (2) A register used for intermediate storage of data during the transfer of it from or to the computer's accumulators and a peripheral device. (3) A synchronizing element between two forms of storage; computation continues while information is transferred between the buffer storage device and the secondary storage device.

BUG. (1) A program defect or error. Also refers to any circuit fault due to improper design or construction. (2) A mistake or malfunction.

BUS. A conductor, or group of conductors considered as a single entity, which transfers signals or power between elements.

BYTE. A sequence of binary digits operated upon as a single unit. A byte may be comprised of 8, 12, or 16 binary digits, depending upon the system. An IBM developed term used to indicate a specific number of consecutive bits treated as a single entity. A byte is most often considered to consist of eight bits which as a unit can represent one character or two numerals.

CAD. Computer-Aided Design, using computers to aid in designing products. (See COMPUTER-AIDED DESIGN.)

CADAM. The Computer-Graphics Augmented Design

and Manufacturing (CADAM) System is an interactive graphics system for computer-aided design and manufacturing. The system includes a design/drafting package, together with a number of aids to design analysis.

CALIBRATION. The adjustment of a device so that output is within a designated tolerance for specific input values.

CAPACITANCE. The property of a circuit or body that permits it to store an electrical charge equal to the accumulated charge divided by the voltage. The unit of capacitance is a farad.

CAPACITOR. An electric device consisting of conducting surfaces separated by thin layers of insulating material (a dielectric). It introduces capacitance into a circuit, stores electrical energy, blocks the flow of direct current, and permits the flow of alternating current to a certain degree. Also referred to as a condensor.

CAPP (COMPUTER-AIDED PROCESS PLANNING). A prototype software development that provides a data management framework designed to assist the functions of process planning in the manufacture of discrete parts. The system enables a process planner to automatically access standard process plan specification data in an interactive and dynamic manner.

CARD. An information-carrying medium that introduces instructions to computers, either directly or indirectly and often via punched codes.

CARRIER. (1) A particular wave which has constant amplitude and frequency, and a phase which can be modulated by changing amplitude, frequency, or phase. Also, an entity which has the ability to carry an electric charge through a solid. For example, holes and conduction electrons in semiconductors. (2) A continuous frequency capable of being modulated or impressed with a second (information carrying) signal.

CARTESIAN COORDINATE SYSTEM. A system of two or three axes that intersect each other at right angles forming rectangles. Any point within the rectangular space can be identified by the distance and direction from any other point. Also known as rectangular coordinate system.

CASCADE CONTROL. An automatic control system in which the control units, linked in sequence, feed into one another in succession, each regulating the operation of the next in line.

CASSETTE RECORDER. A peripheral device for transferring data to or from a cassette tape.

CASSETTE TAPE. Magnetic tape stored on spools within a standard cartridge — self-contained and operable from a cassette recorder.

CATHODE RAY TUBE (CRT). An electronic vacuum tube in which an electron beam can be focused on a small area of a luminescent screen and varied in position and intensity to form alphanumeric or graphic representations.

C AXIS. An angle defining rotary motion of a machine tool part or slide around the Z axis.

CENTRAL PROCESSING UNIT (CPU). The portion of a computer that is the basic memory or logic. It includes the circuits controlling the interpretation and execution of instructions. Also known as the processor, frame, or main frame.

CHAD. The tiny piece of paper removed when a hole is punched into a card or paper tape.

CHANNEL. (1) That portion of a computer's storage medium which is also accessible to a given reading station. (2) That part of a communication system that connects the message source with the message sink. (3) A path along which signals can be sent, *e.g.*, a data channel, output channel. (4) The portion of a storage medium that is accessible to a given reading or writing station *e.g.*, track, bank. (5) In communication, a means of transmission. Several channels may share common equipment. For example, in frequency multiplexing carrier systems, each channel uses a particular frequency band that is reserved for it.

CHARACTER. One of a set of elements which may be arranged in ordered groups to express information. Each character has two forms: (1) a man-intelligible form, the graphic, including the decimal digits 0-9, and letters A-Z, punctuation marks, and other formatting and control symbols; and (2) its computer intelligible form, the code, consisting of a group of binary bits.

CHIP. (1) A small piece of semiconductor material on which electrical components are formed; an electronic circuit element prior to the addition of terminal connections and prior to being encased. (2) A piece of silicon cut from a slice by scribing or breaking, possibly containing one or more circuits but packaged as a unit. Also known as a die. An electronic circuit element prior to having terminal connections added and prior to being encased for physical protection.

CINTURN II. A programming system customized to the capabilities, ranges, and specification of CINTURN NC Turning and Chucking Centers. CINTURN II Programming utilizes the speed and power of the Fortran computer language to generate part programs written in APT. The CINTURN II programming system is designed to process parts in three simple steps; (1) describe the part by its geometry. (3) describe the roughing operation. (3) describe the finishing operation. Using the corner concept, parts are defined by diameter and length dimensions.

Circuit Breaker. Refers to various devices for opening electric circuits under abnormal operating conditions; *e.g.* excessive current, heat, high ambient radiation level, etc. Also called constant breaker.

Circuit, Integrated (IC). Refers to one of several logic circuits, gates, flip-flops which are etched on single crystals, ceramics or other semiconductor materials and designed to use geometric etching and conductive ink or chemical deposition techniques all within a hermetically sealed chip. Some chips with many resistors and transistors are extremely tiny, others are in effect "sandwiches" of individual chips.

Circuit, Printed. Refers to resistors, capacitors, diodes, transistors and other circuit elements which are mounted on cards and interconnected by conductor deposits. These special cards are treated with light-sensitive emulsion and exposed. The light thus fixes the areas to be retained and an acid bath eats away those portions which are designed to be destroyed. The base is usually a copper clad card.

Circular Interpolation. The control of a cutting tool in a complete circle or arc by a machine control unit which has been given basic statements such as coordinates of center point, radius, direction of travel, and coordinate locations of arc end points.

Clock. (1) The most basic source of synchronizing signals in most electronic equipment, especially computers. (2) That specific device or unit designed to time events. (3) A data communications clock which controls the timing of bits sent in a data stream, and controls the timing of the sampling of bits received in a data stream.

Closed Loop. (1) A circuit in which the output is continuously fed back to its source for constant comparison. Also a group of indefinitely repeated computer instructions. (2) The complete signal path in a control system represented as a group of units connected in such a manner that a signal started at any point follows a closed path and can be traced back to that point.

Closed loop System. A system in which a reference signal from a controller is compared with a position signal generated by a monitoring unit on the machine tool (feedback). The difference is used to adjust the machine tool to reduce the difference to zero.

CL Tape. Abbreviation for either "center line" or "cutter line" tape which is the initial output of a computerized NC program giving the coordinate locations of where the cutting tool center line will travel to machine the workpiece. The program is then post processed to take into account the particular features of the machine tool/control unit combination on which the program will actually be run and the part produced.

COBOL (Common Business-Oriented Language). An English-style programming language oriented to business applications.

Code. (1) A system of organized symbols (bits) representing information in a language that can be understood and handled by a control system. (2) A system of symbols that can be used by machines such as computers. Special external meaning is dictated by the specific arrangement of the symbols.

Command. (1) A pulse, signal or set of signals that initiate a performance. (2) A signal from a machine control unit that initiates one step in a complete program.

Communication Interface Circuit. A communications interface circuit (USART) is a peripheral device programmed by the CPU to operate using virtually any serial data transmission technique currently in use. It will typically have a speed of 4 megabits per second for synchronous operation and 250 kilobits per second for asynchronous operations. (USART refers to Universal Synchronous/Asynchronous Receiver/Transmitter.)

COMPACT II. A universal NC programming system for point-to-point and contouring applications on mills, drills, lathes (including 4 axes), punches, flame cutters and EDM machines.

Computer-Aided Design (CAD). (1) Any computer system or program that supports the design process. Both business and scientific systems are included. (2) The use of computers to assist Engineering Design in developing, producing, and evaluating design, data, and drawings. (For brevity, CAD is also referred to as the organization engaged in computer-aided design.)

Computer-Aided Manufacturing (CAM). (1) The use of computers to aid in the various phases of manufacturing. Numerical Control (NC) is a subset of CAM. (2) The effective utilization of computer technology in the management, control, and operations of the manufacturing facility through either direct or indirect computer interface with the physical and human resources of the company.

Computer Graphics. The process of communicating between a person and a computer in which the computer input and output are pictorial in nature, having the form of charts, drawings, or graphs. Cathode ray tubes, curve tracers, mechanical plotting boards, coordinate digitizers, and light pens are employed in the creation of graphic design.

Computer-Integrated Manufacturing System (CIMS). A multimachine manufacturing complex linked by a material handling system and including features such as toolchangers and load/unload stations. Under the control of a computer, various workpieces are introduced into the system, then randomly and simultaneously transported to the NC machine tools and other processing stations.

Computer Numerical Control (CNC). A self-contained NC system for a single machine tool utilizing a dedicated computer controlled by stored instructions to perform some or all of the basic NC functions. Punched tape and tape readers are not used except possibly as backup in the event of computer failure. Through a direct link to a central processor, the CNC system can become part of a Direct Numerical Control (DNC) system.

Computer Part Programming. The preparation of a manuscript, in an NC computer language, to define the necessary calculations to be performed by the computer.

Computer Program. A detailed set or series of instructions or statements in a form acceptable as input to a computer to achieve a specific result.

Configuration. (1) A group of units that are interconnected and arranged to operate, as a system. (2) The arrangement of software routines or hardware instructions when combined to operate as a system.

Continuous Path Operation. An operation in which the rate and direction of relative movement of machine members are under continuous control so that the machine travels through the designated path at a specified rate without pausing.

Continuous Production. A production system in which the productive units are organized and sequenced according to the steps to produce the product. The routing of the jobs is fixed and set ups are seldom changed.

Contouring. Also known as "continuous path" is the method of NC machining where the control system generates a contour by keeping the cutting tool in constant contact with the workpiece.

Contouring Control System. An NC system that generates a contour by controlling a machine or cutting tool in a path resulting from the coordinated, simultaneous motion of two or more axes.

Control. Measurement of performance or actions and comparison with established standards in order to maintain performance and actions within permissible limits of variance from the standard. May involve taking corrective action to bring performance into line with the plan or standards.

Control System. An arrangement of interconnecting elements which interact to maintain a specific machine condition or modify it in a prescribed manner.

Controller. (1) An element or group of elements that take data proportional to the difference between input and output of a device or system and convert this data into power used to restore agreement between input and output. (2) A module or specific device which operates automatically to regulate a controlled variable or system.

Core. Refers to tiny "doughnuts" or magnetizable metal that can be in either an on or off state and can represent either a binary 1 (on) or binary 0 (off). Commonly called magnetic core and formerly used as the basic type of main memory for many computers.

Conversational Mode. Communication between a human operator and a computer via a keyboard terminal or other input/output device. Questions and responses are elicited from the computer by the operator and vice versa.

Criterion. Refers to a value used for testing, comparing, or judging; *e.g.*, in determining whether a condition is plus or minus, true or false; also, a rule or test for making a decision in a computer or by humans.

Cross Assembler. (1) Refers to a program run on one computer for the purpose of translating instructions for a different computer. (2) Programs are usually assembled by the same assembler or assembly program contained within or used by the processor on which they will be run. Many microprocessor programs, however, are assembled by other computer processors whether they be standard, time-shared, mini or other microcomputers. This process is referred to as cross-assembly, and the programs are not designed for specific microprocessors but are to be used on other computers. They are known as cross-assemblers.

Cross Talk. Generally cross talk occurs when signals on one circuit emerge on another circuit as interference. The circuit which is the source of the signals is known as the disturbing circuit, and that on which the signals are heard is the disturbed circuit.

Cryogenics. The area of technology that uses properties assumed by metals at extremely low temperatures.

Cursor. A visual movable pointer used on a CRT screen to indicate the position at which data entry or editing is to occur.

CUTS II. Computerized part generation system written by Warner and Swasey for their SC line of turning machines.

Cutter Compensation. A method of adjusting for the difference between the actual and programmed size of a tool.

Cutter Location (CL) Data. (1) Information describing the coordinates of the path of the cutter center resulting from a computer program. Common to all machine tool system combinations this information serves as input to the post-processor. (2) The file produced by numerical control systems containing the cutting tool center line at the tip of the tool with an assumed orientation perpendicular to the XY plane unless otherwise indicated. (See CL tape.)

Cutter Offset. (1) The difference between a part surface and the axial center of a cutter or cutter path during a machining operation. (2) An NC feature enabling a machine operator to use an oversized or undersized cutter.

Cutter Path. The cutting path described by the cutter center.

Cutting Speed. The relative velocity, usually expressed in feet per minute, between a cutting tool and the surface of the material from which it is removing stock.

Cybernetics. (1) The field of technology relating to the comparative study of the control and communications of information-handling machines and living organisms. (2) The diverse field encompasses a) integration of communication, control and systems theories; b) development of systems engineering technology; and c) practical applications at both the hardware and software levels. The term was originated by Norbert Wiener, American mathematician (1894-1964) from the Greek word for "steersman." Recent and projected developments in cybernetics are taking place in at least five important areas: technological forecasting and assessment, complex systems modeling, policy analysis, pattern recognition, and artificial intelligence.

Cycle. (1) A sequence of operations repeated regularly. (2) The time necessary for one sequence of operations to occur.

Damping. Refers to a characteristic built into electrical circuits and mechanical systems to prevent rapid or excessive corrections which may lead to instability or oscillatory conditions, *i.e.* connecting a register on the terminals of a pulse transformer to remove natural oscillations or placing a moving element in oil or sluggish grease to prevent mechanical overshoot of the moving parts.

Data. A representation of facts, instructions, concepts, numerical and alphabetical characters, etc., in a manner suitable for communicating, interpreting, and processing by humans or by automatic means such as NC systems.

Data File. A collection of related data records or application data values organized in a specific manner and stored after, and separated from, the user program area.

Data Manipulations. The process of altering and/or exchanging information between storage words through user-programmed instructions to vary application functions. Functions include sorting, merging, input/output, and report generation.

Data Processing. A computer procedure involving one or more operations for collecting data and producing a specified result.

Dead Band. A specific range of values in which the incoming signal can vary without changing the output.

Deadtime. A definite delay between two related actions.

Debugging. The process of finding and eliminating errors.

Decibel. A standard unit expressing a loss or gain in transmission power level. Abbreviated db. The term "DBM" is also used when a power of one milliwatt is the reference level. Db indicates the ratio of power output to power input: $P\ 1\ db = 10 \log 10\ P\ 2$

Diagnostic Routine. A maintenance test of key NC system components, performed by use of a special programmed tape and/or electronic instruments, to discover failure or potential failure of a machine element as well as the location of the failure. Also known as diagnostic check, diagnostic subroutine, diagnostic test, or error detection routine.

Differential Transducer. A type of device which can simultaneously measure two separate stimuli and provide an output proportionate to the difference between them.

Digit. (1) One of the integers in a numbering system such as 0 to 9 in a decimal system. (2) A character or symbol used alone or in combination with other digits to convey a specific numerical quantity.

Digital. Refers to the use of discrete integral numbers in a given base to represent all the quantities that occur in a problem or calculation. It is possible to express in digital form all information stored, transferred, processed, or transmitted by a dual-state condition (*i.e.*, on-off, true-false, and open-closed).

Digital-to-Analog-Converter (DAC). Converts digital signals into a continuous electrical signal suitable for input to an analog computer.

Digital Computer. A computer which processes information represented by combinations of discrete or discontinuous data as compared with an analog computer for continuous data. More specifically, it is a device for performing sequences of arithmetic and logical operations, not only on data but also on its own program. Still more specifically it is a stored program digital computer capable of performing sequences of internally stored instructions, as opposed to calculators, such as card programmed calculators, on which the sequence is impressed manually. Related to machine, data processing.

Digitize. (1) To obtain a digital representation of the value of an analog quantity. (2) To convert scaled, nonmathematical drawings or physical part dimensions to digital data. Refers to a specific device which converts an analog measurement into digital form. *Syn*: quantizer.

DIP (Dual In-Line Package). (1) The most popular IC packaging in the mid-1970's is the plastic, dual-in-line case, using plastic for economic reasons and the dual-in-line package (DIP) configuration for manufacturing efficiently. (2) Chips are enclosed in dual in-line packages which take their names from the double parallel rows of leads which connect them to the circuit board. DIPs are sometimes called "bugs."

Direct Memory Access (DMA). (1) Direct Memory Access, sometimes called data break, is preferred form of data transfer for use with high-speed storage devices such as magnetic disk or tape units. The DMA mechanism transfers data directly between memory and peripheral devices. The CPU is involved only in setting up the transfer; the transfers take place with no processor intervention on a "cycle stealing" basis. The DMA transfer rate is limited only by the bandwidth of the memory and the data transfer characteristics of the device. The device generates a DMA Request when it is ready to transfer data. (2) High-speed data transfer operation in which I/O channel transfers information directly to or from the memory. Also called "data break" or "cycle stealing."

Direct Numerical Control (DNC). The use of a shared computer to program, service, and log a process such as a machine tool cutting operation. Part program data is distributed via data lines to the machine tools.

Discrete. (1) The state of being a separate entity or having unconnected elements. (2) Pertaining to distinct elements or representation of data by means of distinct elements such as characters; pertaining to physical quantities having distinct values only.

Disk Drives. A typical unit consists of a drive, power supply (100-125 v AC, 60 Hz), cooling fan, disk buffer and address select electronics. It is capable of storing over 300,000 words on a flexible disk and several megabytes on a hard disk. Up to 16 disk drives can be controlled by one disk controller.

Disk Storage. A computer memory device capable of storing information magnetically on a disc similar in appearance to a phonograph record.

Documentation. Manuals and other printed materials, such as tables, tape, listings, and diagrams, which provide instructive information regarding the operation, installation, and maintenance of a manufactured product.

Downtime. The time period in which a system or machine tool is not available for use due to failure or routine maintenance. Also known as cumulative lost time.

Dynamic RAM. Data is stored capacitively, and must be recharged (refreshed) periodically (every 2 ms. or so) or it will be lost. (See RANDOM ACCESS MEMORY.)

Dynamic Response. The specific behavior of the output of a device as a function of the input, both with respect to time.

Edit. To modify the form, format, or content of data.

EIA (Electronic Industries Association). A trade association of the electronics industry which formulates technical standards, disseminates marketing data, and maintains contact with government agencies in matters relating to the electronics industry.

Electrode. A conductor through which an electric current enters or leaves, establishing contact with a nonmetallic part of a circuit.

Electronic Data Processing (EDP). Processing of data by means of equipment such as a digital computer that is electronic in nature.

Electronic Industries Association. (See EIA.)

End of Block (EOB). The end of one block of data.

End-of-block Signal. A code that indicates the end of a block of data.

End of Program (EOP). Miscellaneous function signifying the last block of a program and the completion of a workpiece.

End of a Tape (EOT). Miscellaneous function signaling the spindle, coolant, and feed to stop after the completion of all the commands in a block. This function is also used to reset control and/or the machine.

Equation Solver. Usually an analog device used to solve linear simultaneous nondifferential equations, etc.

Erasable Programmable Read Only Memory (EPROM). Most of these ROMs are 256 words by 8 bit electrically programmable and are ideally suited for uses where fast turnaround and pattern experimentation are important. They generally undergo complete programming and functional testing on each bit position prior to shipment insuring 100% programmability. Some types are packaged in a 24 pin dual in-line package with a transparent quartz lid. The transparent quartz lid allows the user to expose the chip to ultraviolet light to erase the bit pattern. A new pattern can then be written into the device. This procedure can be repeated as many times as required. The circuitry of some types is entirely static; no clocks are required. A pin-for-pin metal mask programmed ROM is ideal for large volume production runs of systems; most are fabricated with silicon gate technology. This low threshold technology allows the design and production of higher performance MOS circuits and provides a higher functioning density on a monolithic chip than many conventional MOS technologies.

Error. (1) Refers to any incorrect step, process, or result in a computer or data-processing system. The term also refers to machine malfunctions or "machine errors," and to human mistakes or "human errors." (2) Any discrepancy between a computed, observed, recorded, or measured quantity and the true, specified, or theoretically correct value or condition. Contrast with mistake.

EXAPT. (1) Acronym for Extended Subset of APT, a language process developed in Germany and commonly used for point-to-point or lathe work. (2) The processors select the work sequence and tool and determine the cutting data such as feed rates, spindle speeds, depths of cut and so on. The processor is based on APT.

EZAPT. An APT based NC processor language that is implemented on a microcomputer system with two disk drives. The processor can be used to program all NC machine tools with two-axis circular and three-axis linear and four-axis position (rotary tables and so on) capabilities.

Fabrication. (1) A term used to distinguish production operations for components as opposed to assembly operations. (2) Processing of natural or synthetic materials for desired modification of shape and property.

FAPT. An interactive APT processor with graphic display that provides instant program verification. It simultaneously provides a readout of the commands programmed and generates 2-D contours or simulated 3-D surfaces for visual program check.

Feedback. The signal or data sent back to a commanding unit from a control machine or process for use as input in subsequent operations. When applied to a transmission, feedback is the return of a fraction of the output to the input. In a closed-loop system, it is the part of the system which brings back information about the condition under control.

FEEDBACK CONTROL. Action in which a measured variable is compared to its desired value to bring it closer to the desired value.

FEEDBACK CONTROL LOOP. A closed transmission path which includes an active transducer and consists of a forward path, a feedback path, and one or more mixing points arranged to maintain a prescribed relationship between the loop input and output signals.

FEED RATE. The rate of movement between a machine element and a workpiece in the direction of cutting. Expressed as a unit of distance relative to time; a machine function such as spindle rotation or table stroke.

FEELER. A device used to sense a "two-state condition," such as on-off, go-no-go, open-closed.

FIELD. A set of one or more characters which is treated as a unit of information.

FIRMWARE. (1) A term usually related to microprogramming and those specific software instructions that have been more or less permanently burned into a ROM control block. (2) An extension to a computer's basic command (instruction)- repertoire to create microprograms for a user-oriented instruction set. This extension to the basic instruction set is done in read-only memory and not in software. The read-only memory converts the extended instructions to the basic instructions of the computer.

FIXED-BLOCK FORMAT. An arrangement of data in which the number and sequence of words and characters in successive blocks, as determined by hardware requirements or the programmer, are constant.

FIXED SEQUENTIAL FORMAT. A numerical control format where each word in the format is identified by its position.

FIXED STORAGE. A storage device used to store data that is not changeable by computer instructions, such as magnetic core storage with a lockout feature.

FIXED ZERO. A reference position of the origin of the coordinate system; usually a characteristic of machines with absolute feedback elements.

FIXTURE. A device to hold and locate a workpiece during inspection or production operations. Many fixtures are used on machine tools and often provide means for cutter setting.

FLEXOWRITER. An automatic typewriter incorporating an eight-track tape reader and punch for preparation of punched tape.

FLIP FLOP. (1) A type of circuit having two stable states and usually two input terminals (or signals) corresponding to each of the two states. The circuit remains in either state until the corresponding signal is applied. Also, a similar bistable device with an input which allows it to act as a single-stage binary counter. (2) A bistable device; a device capable of assuming two stable states; a bistable device which may assume a given stable state depending upon the pulse history of one or more input points and having one or more output points. The device is capable of storing a bit of information, controlling gates, etc.; a toggle.

FLOATING ZERO. A characteristic of an NC machine control unit allowing the zero reference point of an axis to be established at any position over the full travel of the machine tool.

FLOPPY DISK. A flexible, magnetic-based disk used to store data input to NC machine control units and computers.

FLOPPY DISK SYSTEMS. A typical floppy disk provides random access program/data storage. Hard-sector formatted, each disk holds over 300,000 data bytes. Because many floppy controllers have all of their intelligence in microcode, some microcontrollers offer features not practical in designs implemented with hard-wired logic. The host-computer driver need only issue a small sequence of commands to write or read data from the disk.

FLOWCHART. A graphical representation of a problem or system, in which interconnected symbols signify operations, data flow, equipment, etc. It is used in defining, analyzing, or solving a problem.

FLUIDICS. The technique of control that uses only a fluid to perform sensing, control, information processing, and actuation functions without moving elements.

FORMAT. The physical arrangement of data on a program tape, and the pattern in which it is organized for presentation.

FORMULA TRANSLATION (FORTAN). Any of a family of universal procedure-oriented languages used to describe numeric processes in such a way that both humans and computers can understand them.

FUNCTION. A specific purpose or characteristic action of an entity, such as a subroutine of a program.

FOURIER ANALYSIS. The determination of the harmonic components of a complex waveform either mathematically or by a waveanalyzer device.

GAIN. Refers to the ratio between the output signal and the input signal of a device.

G CODE. A preparatory numerical code in a program addressed by the letter G indicating a special function or cycle type in an NC system. Also known as G function.

GENESIS. A two axis contouring and simultaneous third axis linear capability processor language for all types of NC/CNC machine tools. Many macros that provide automatic cycles that minimize the programmer's time and eliminates repetitive calculations.

GETURN. General Electric Company's information services division offers GETURN for the programming of lathe parts. It was developed with the cooperation of the TNO Metaalinstituut of the Netherlands, and has been operational in Europe under the name MITURN since 1970.

GNC. Graphical Numerical Control is a part programming system using interactive graphics, mills, lathes, flame cutters, drills, EDM systems and large machining centers. GNC provides effective tape generation by providing graphic displays of the part, the tool path, and the tools themselves.

GRAPHIC INPUT. Input of symbols to NC systems that comes from lines drawn on a cathode ray tube or information obtained from drawings by a scanner.

GRAPHIC PANEL. Master control panel in automation and remote control systems, which shows the relations and functioning of the different parts of the control equipment by means of colored block diagrams.

GROUP TECHNOLOGY. (1) The classification and coding of parts on the basis of similarity of parts. (2) The grouping of parts based on processing similarities so that they can be processed together. (3) The grouping of various machines to produce a family of parts.

HANDSHAKING. (1) A descriptive term often used interchangeably with "buffering" or "interfacing," implying a direct connection or machining of specific units or programs. Some computer terminal programs are called "handshaking" if they greet and assist the new terminal operator to interface with or use the procedures or programs of the system. Other handshaking relates to direct package-to-package connections as regards circuits, programs, or procedures. (2) Exchange of predetermined signals when a connection is established between two data set devices.

HARD COPY. Any visually readable form of data output produced by a computer. For example, a printed listing, punched cards, or paper tape.

HARDWARE. The physical equipment of a system, as opposed to software; the mechanical, electrical, magnetic features of a system that are permanent components. Refers to the metallic or "hard" components of a computer system in contrast to the ``soft'' or programming components. The components of circuits may be active, passive, or both.

HARD-WIRE LOGIC. Refers to logic designs for control or problem solutions that require interconnection of numerous integrated circuits formed or wired for specific purposes and relatively unalterable. A hard-wired diode matrix is hard-wired logic whereas a RAM, ROM, or SPU can be reprogrammed with little difficulty to change the purpose of operation. Hard-wired interconnections are usually completed by soldering or by printed circuits and are thus in contrast to software solutions achieved by programmed microcomputer components.

HARD-WIRED SYSTEM. An NC system with a fixed wired program built in when manufactured and not subject to changes by programming. Changes are possible only through altering the physical components or interconnections.

HEURISTIC METHOD. An exploratory method of problem solving in which various types of solutions that may or may not work are systematically applied and evaluated until a solution is found.

HIERARCHY. A group or series classified and arranged in rank order.

HEXADECIMAL. Refers to whole numbers in positional notation with 16 as the base. Hexadecimal uses 0 through 15, with the first ten represented by 0 through 9 and the last six digits represented by A,B,C,D,E, and F.

HIGH-LEVEL LANGUAGE. Computer language which uses readily understood symbols and command statements. Each statement typically represents a series of computer command statements. Each statement typically represents a series of computer instructions. Examples of high-level languages are BASIC, FORTRAN, and APT.

HOST COMPUTER. (1) The primary or controlling computer in a multiple operation. (2) A computer used to prepare programs for use on another computer or on another data processing system; for example, a computer used to compile, link, edit, or test programs to be used on another system.

HUNTING. Refers to a continuous attempt on the part of an automatically controlled system to seek a desired equilibrium condition. The system usually contains a standard, a method of determining deviation from this standard, and a method of influencing the system such as the difference between the standard and the state.

HYBRID COMPUTER. Various specially-designed computers with both digital and analog characteristics, combining the advantages of analog and digital computer when working as a system. Hybrid computers are being used extensively in simulation of process control systems where it is necessary to have a close representation with the physical world. The hybrid system provides good precision that can be attained with analog computers and greater control than is possible with digital computers, plus the ability to accept input data in either form.

HYSTERESIS. The difference between the response of a unit or system to an increasing signal and the response to a decreasing signal.

IMPULSE. (1) A pulse that begins and ends within so short a time that it may be regarded mathematically as infinitesimal. However, the change in the medium is usually of a finite amount. (2) A change in the intensity or level of some medium, usually over a relatively short period of time; *e.g.*, a shift in electrical potential of a point for a short period of time compared to the time period, *i.e.*, if the voltage level of a point shifts from -10 to 20 volts with respect to ground for a period of two microseconds, one says that the point received a 30-volt, 2-microsecond pulse. (3) A change in the intensity or level of some medium over a relatively short period of time.

INCREMENTAL COORDINATES. Coordinates measured from an origin defined by the preceding value in a sequence of values.

INCREMENTAL SYSTEM. An NC system in which each coordinate or positional dimension, whether input or feedback, is taken from the last position instead of from a common data position, as in an absolute system.

INDEX. A point on a continuous function at which a specific action is desired; or the pulse count of a digital function at which a specific action is desired.

INFORMATION. The knowledge of facts, measurements and requirements necessary for accomplishing useful work.

INFORMATION THEORY. Mathematical analysis of efficiency with which communication channels are employed—the aim being to find the most efficient system of coding for any channel.

INPUT. (1) An adjective referring to a device or collective set of devices used for bringing data into another device. (2) A channel for impressing a state on a device or logic element. (3) Pertaining to a device, process, or channel involved in an input process or to the data or states involved in an input process. In the English language, the adjective "input" may be used in place of "input data," "input signal," "input terminal," etc., when such usage is clear in a given context. (4) Pertaining to a device, process, or channel involved in the insertion of data or states, or to the data or states involved. (5) One, or a sequence of, input states.

INPUT EQUIPMENT. (1) The equipment used for transferring data and instructions into an automatic data processing system. (2) The equipment by which an operator transcribes original data and instructions to a medium that may be used in an automatic data processing system.

INPUT/OUTPUT DEVICE. Equipment such as limit switches, pressure switches, and pushbuttons, used to communicate with a control system.

INSTRUMENT. A device that is capable of converting usable intelligence into electrical signals or a device that is capable of recording, measuring, or controlling.

INSTRUMENTATION. The application of devices for measuring, recording and/or controlling of physical properties and movements.

INTEGRATED CIRCUIT (IC). A combination of passive and active circuit elements that are interconnected and incorporated on or within a continuous substrate.

INTEGRATOR. (1) Any device which integrates a signal

over a period of time. (2) Unit in a computer which performs the mathematical operation of integration, usually with reference to time. (3) A resistor-condensor circuit at the input to the vertical oscillator. (4) A device whose output is proportional to the integral of the input variable with respect to time.

INTERFACE. (1) Refers to instruments, devices, or a concept of common boundary or matching of adjacent components, circuits, equipment, or system elements. An interface enables devices to yield and/or acquire information from one device or program to another. Although the terms adapter, handshake, and buffer have similar meaning, interface is more distinctly a connection to complete an operation. (2) A common boundary—*e.g.*, physical connection between two systems or two devices. (3) Specifications of the interconnection between two systems or units.

INTERFEROMETER. An instrument that uses light interference phenomena to precisely determine wave length, spectral fine structure, indexes of refraction, and small linear displacements.

INTERPOLATION. A function of control enabling data points to be generated between specific coordinate positions to allow simultaneous movement of two or more axes of motion in a defined geometric pattern. For example, in NC, curved sections can be approximated by a series of straight lines or parabolic segments. Also known as linear interpolation.

INTERRUPT. Various interrupts relate to the suspension of normal operations or programming routines of microprocessors and are most often designed to handle sudden requests for service or change. As peripheral devices interface with CPUs, various interrupts occur on frequent bases. Multiple interrupt requests require the processor to delay or prevent further interrupts; to break into a procedure; to modify operations, etc, and after completion of the interrupt task, to resume the operation from the point of interrupt.

INVERTER. (1) A circuit which takes in a positive signal and puts out a negative one, or vice versa. (2) A device that changes AC to DC or vice versa. It frequently is used to change 6 volt or 12 volt direct current to 110 volt alternating current. (3) Arrangement of modulators and filters for inverting speech or music for privacy.

JIG. A device which holds a piece of work in a desired position and guides the tool or tools which perform the necessary operation.

JOB SHOP. A manufacturing enterprise devoted to producing special or custom-made parts or products, usually in small quantities for specific customers.

KINEMATIC COUPLING DESIGN. A design that makes contact at a number of points equal to the number of degrees of freedom that are to be restrained.

LAG. A delay in the response of a system or device.

LAMA-25. A multi-axis contouring processor with line, circle and arc capability for use with CNC machine tools.

LINEAR DISPLACEMENT TRANSDUCER. Commonly referred to as LVDT (Linear Variable Displacement Transformer). An electromechanical device which produces an electrical output proportional to the displacement of a separate moveable core.

LINEAR INTERPOLATION. Refers to a mode of machine tool contouring control which uses the information contained in a block to produce constant velocities proportional to the distance moved in two or more axes simultaneously.

LOAD. (1) The power consumed by a machine or circuit in performing its function. (2) A resistor or other impedance which can replace some circuit element. (3) To fill the internal storage of a computer with information obtained from auxiliary or external storage. (4) The process of reading the beginning of a program into virtual storage and making necessary adjustments and/or modifications to the program so that it may have control transferred to it for the purpose of execution. (5) To take information from auxiliary or external storage and place it into core storage.

LOGGER. (1) Colloquialism for recorder or printout device in a control system. (2) A type of instrument which automatically scans conditions such as pressure, temperature, and humidity and records (logs) the findings on a chart. (3) A device which automatically records physical processes and events, usually with respect to time.

LOGIC. (1) As regards microprocessor, logic is a mathematical treatment of formal logic using a set of symbols to represent quantities and relationships that can be translated into switching circuits or gates. Such gates are logical functions such as, AND, OR, NOT, and others. Each such gate is a switching circuit that has two states, on or off (open or closed). They make possible the application of binary numbers for solving problems. The basic logic functions electronically performed from gate circuits is the foundation of the often complex computing capability. (2) The science dealing with the criteria of formal principles of reasoning and thought. (3) The systematic scheme which defines the interactions of signals in the design of an automatic data processing system. (4) The basic principles and application of truth tables and interconnection between logical elements required for arithmetic computation in an automatic data processing system. Related to symbolic logic.

LTTP. A simplified parts programming language for 2 axis turning applications and part of the Manufacturing Software and Services TOOLPATH processor. LTTP is a subset language and may be interchanged with ADAPT and other features of the TOOLPATH processor in the same part program.

M Function. Similar to a G function except that the m functions control miscellaneous functions of the machine tool such as turning on and off coolant or operating power clamps and so on.

Machine Attention Time. That portion of a machining operation during which the workman performs no physical work yet must watch the progress of the work and be available to make necessary adjustments, initiate subsequent steps or stages of the operation at the proper time, and the like.

Machine Language. (1) A language written in symbols, bits, characters, signs, or a series of bits to convey to a computer instructions or information to be processed. (2) The lowest level language, usually different for every computer type. Machine language is written directly by the binary code and does not require a compiler or assembler program to convert from symbolic code to binary-coded machine instructions.

Machine Tool. A powered machine used to form a part, typically by the action of a tool moving in relation to the workpiece.

Machine Utilization. The percent of time that a machine is running production as opposed to idle time. (cf. idle time, running time.)

Machine Vision (computer vision). Computer perception, based on visual sensory input, to develop a concise description of a scene depicted in an image. The term is used synonymously with image understanding and scene analysis.

Machining Center. A machine tool, usually numerically controlled, that can automatically drill, ream, tap, mill, and bore workpieces. It is often equipped with a system for automatic toolchanging.

Macro. (1) A powerful computer instruction from which a string of micro instructions can be called as a unit. (2) A source language instruction from which many instructions can be generated. Also known as macro instruction, macro program, or macro routine.

Magnetic Disk. A rotating circular plate that is coated or permeated with magnetic material on which information is recorded and stored for subsequent use.

Magnetic Tape. A plastic, metal, or paper tape that is coated or permeated with magnetic material. It is capable of storing data by selective polarization of portions of the surface.

Mainframe. Refers to the basic or main part of the computer, *i.e.*, the arithmetic or logic unit. The central processing unit.

Maintenance. Maintenance and repair of physical plant and materials handling equipment. Any activity intended to eliminate faults or to keep hardware or programs in satisfactory working condition, including tests, measurements, replacements, adjustments and repairs.

Management Information System (MIS). An information feedback system in which data is recorded and processed for use by management personnel in decision making.

Manipulated Variable. That quantity or condition which is altered by the controller and applied to the controlled system.

Manual Control. Machine or process control performed by a person.

Manual Data Input (MDI). A means of manually inserting commands and other data into an NC control.

Manual Part Programming. The preparation of a manuscript in machine control language and format to define a sequence of commands required to accomplish a given task on an NC machine.

Manufacturing. A series of interrelated activities and operations involving the design, material selection, planning, production, quality assurance, management, and marketing of discrete consumer and durable goods.

Manufacturing Process. The series of activities performed upon material to convert it from the raw or semifinished state to a state of further completion and a greater value.

Manuscript. Form used by a part programmer to organize machining instructions. From it a computer program is prepared. Refer also to part program.

Mask. (1) A pattern of characters that controls the retention or elimination of portions of another pattern of characters. (2) A filter.

Mass Production. A method of quantity production in which a high degree of planning, specialization of equipment and labor, and integrated utilization of all productive factors are the outstanding characteristics.

MDCAPT. An APT system that provides enhanced part programming capabilities for machined parts ranging from the most simple to the ultimate in complexity.

Measurement. The process of determining the numerical value of dimensions or factors, determining the quality, extent or degree of the dimensions or factors of a process or workpiece.

Memory. (1) One of the three basic components of a CPU memory that stores information for future use. Storage and memory are interchangeable expressions. Memories accept and hold binary numbers or images. (2) In order to be effective, a computer must be capable of storing the data it is to operate on as well as the program that dictates which operations are to be performed. Not only must that memory unit of a computer store large amounts of information, the memory must be designed to allow rapid access to any particular portion of that information. Speed, size and cost are the critical criteria in any storage unit. Various types are: disc, drum, semi-conductor, magnetic core, charge-coupled devices, bubble domain, etc.

Memory Cycle. (1) A computer operation consisting of reading from and writing into memory. (2) The time required to complete this process.

Metalmats. Metalmats is programmed to run on Radio Shack TRS-80 Model 2 or IBM 370 mainframe equipment. Programs include machining, welding/fabrication, and assembly for medium to large job shops. It establishes time standards and provides a routing chart which includes process, manual, and setup times for almost all metal working operations.

Microcomputer. A general term referring to a complete tiny computing system, consisting of hardware and software, whose main processing blocks are made of semiconductor integrated circuits. In function and structure it is somewhat similar to a minicomputer, with the main difference being price, size, speed of execution and computing power. The hardware of a microcomputer consists of the microprocessing unit (MPU) which is usually assembled on a PC board with memory and auxiliary circuits. Power supplies, control console, and cabinet are separate.

Microcontroller. This can mean a microprogrammed machine, a microprocessor or a microcomputer used in a control operation—that is, to direct or make changes in a process of operation. Microcontroller refers to any device or instrument that controls a process with high resolution, usually over a narrow region.

Microprocessor. The semiconductor central processing unit (CPU) and one of the principal components of the microcomputer. The elements of the microcomputer. The elements of the microprocessor are frequently contained on a single chip or within the same package but sometimes distributed over several separate chips. In a microcomputer with a fixed instruction set, it consists of the arithmetic logic unit and control logic unit. In a microcomputer with a microprogrammed instruction set, it contains an additional control memory unit.

Minicomputer. (1) A small, general-purpose computer which has from 4K to 4mb words of memory and employs words of 8, 12, 16, 18, 24 or 32 bits in its basic configuration. Generally a minicomputer is a mainframe that sells for less than $100,000. Usually it is a parallel binary system with 8, 12, 16, 18, 24 or 36-bit word length incorporating semiconductor or magnetic core memory offering from 4K words to 4mb words of storage and a cycle time of 0.2 to 8 microseconds or less. A bare minicomputer (one without cabinet, consol, and power supplies) consisting of a single PC card can sell for less than $1000 in OEM quantities. (2) These units are characterized by higher performance than microcomputers or programmable calculators, richer instruction sets, higher price and a proliferation of high level languages, operating systems, and networking methodologies.

Mirror Image Programming. A feature of some machine control units that provides for a specific axis by means of a switch. This means that a single program can produce two parts that are mirror images.

Miscellaneous Function. One of a group of special or auxiliary functions of a machine, such as spindle stop coolant control, program stop, and clamp control.

Mnemonic. A combination of letters, numbers, pictures, or words that aids in recalling a memory location or computer operation.

Modem. Refers to a MODulation/DEModulation chip or device that enables computers and terminals to communicate over telephone circuits.

Module. An interchangeable hardware subassembly containing electronic components that can be combined with other interchangeable subassemblies to form a complete unit.

Molecular Manufacturing. The construction of objects to complex, atomic specifications using sequences of chemical reactions directed by nonbiological molecular machinery. It is the use of nanoscale mechanical systems to guide the placement of reactive molecules, building complex structures with atom-by-atom control.

Molecular Nanotechnology. Comprises molecular manu- facturing together with its techniques, its products, and their design and analysis; it describes the field as a whole.

Natural Frequency. The frequency of free oscillation of a system.

NANS. (National Automated Nesting System) is designed to meet the needs of the user of numerically controlled routing and flame cutting machine.

Network. A collection of logic elements connected to perform a specific function.

Neutral Zone. A range of values for which no control action occurs.

NGS. (Numerical Geometry System) is a graphics-based, partly interactive, 3-dimensional system for computer-aided design and manufacture. The system permits the design process to be carried out either by direct entry of numerical data or from existing design systems.

Noise. (1) Any unwanted disturbance within a dynamic electrical or mechanical system. (2) Any unwanted electrical disturbance or spurious signal which modifies the transmitting, indicating, or recording of desired data. (3) In a computer, extra bits or words which have no meaning and must be ignored or removed from the data at the time it is used. (4) Random variations of one or more characteristics of any entity such as voltage, current, or data. (5) A random signal of known statistical properties of amplitude, distribution, and spectral density.

Normally Closed. A designation applied to the contacts of a switch or relay when they are connected so that the circuit will be completed when the switch is not activated or the delay coil is not energized. Symbolized by NC.

Normally Open. (1) A designation applied to the contracts of a switch or relay when they are connected so that the circuit will be broken when the switch is not activated or the relay coil is not energized. Symbolized by NO. (2) Specific pairs of contacts on relays which are closed only when the relay coil is energized.

NUFORMS. Consists of Level I and Level II. Level I is a comprehensive system for the preparation of control data for NC tools utilized in the 2 1/2 axis mode. Level II handles three-dimensional contouring where 3, 4, or 5 axes may move simultaneously and continuously.

Numerical Control (NC). (1) A technique for controlling actions of machine tools and similar equipment by the direct insertion of numerical data at a given point. Data is automatically interpreted. (2) Any system of control plus controlled equipment which accepts commands, data, and instructions in symbolic form as an input and converts this information into a physical output and into physical values such as dimensions or quantities.

Numerical Control, Direct (DNC). A system connecting a set of numerically controlled machines to a common memory for part program or machine program storage, with provision for on-demand distribution of data for the machines. Direct numerical control systems typically have additional provisions for collection, display, or editing of part programs, operator instructions, or data related to the numerical control process.

Octal. A numbering system basic to computer operation. A positional notation system using 8 as a base, instead of 2, as in binary, or 10, as in decimal, etc.

Offset. A displacement in the axial direction of a tool which is the difference between the length established by the programmer and the actual tool length.

ONLINE. Operation of peripheral equipment that is under direct control of a central processor.

OPEN LOOP. (1) Refers to a control system in which there is no self-correcting action for misses of the desired operational condition, as there is in a closed loop system. (2) A family of automatic control units, one of which may be a computer, linked together manually by operator action.

OPEN-LOOP SYSTEM. A control system that is incapable of comparing output with input for control purposes, that is, no feedback is obtainable.

OPERATING SYSTEM. (1) A basic group of programs with operation under control of a data processing monitor program. (2) An integrated collection of service routines for supervising the sequencing and processing of programs by a computer. Operating systems may perform debugging, input-output, machine accounting, compilation and storage assignment tasks.

OPTIMIZE. The rearrangement of instructions or data in NC or computer applications to obtain the best set of operating conditions.

OSHA. Acronym for Occupational Safety and Health Act. A federal law which specifies the requirements an employer must follow in order to guard against illness and injury.

OUTPUT. (1) Printed or recorded data resulting from computer source programs. (2) Data transferred from internal storage to output devices or external storage.

OVERLOAD. A load greater than that which a device is designed to handle, possibly resulting in hardware problems.

OVERSHOOT. (1) Extent to which a servo system carries the controlled variables past their final equilibrium position. (2) For a step change in signal amplitude, undershoot and overshoot are the maximum transient signal excursions outside range from initial to final mean amplitude levels.

PAPER TAPE. A continuous strip of paper in which holes can be punched to represent data.

PARABOLIC INTERPOLATION. A high order of interpolation producing contoured shapes by having the cutting tool travel through parabolas or portions of parabolas.

PARITY. A means of testing the accuracy of binary numbers used in transmitted, recorded, or received data. A self-checking code is used in which the total number of 1's or 0's is always even or odd.

PARITY CHECK. A computer checking method in which the total number of binary 1's (or 0's) is always even or always odd. It is either an even-parity or odd number of bit patterns to signify a character; thus, all characters are added, modulo 2, and the sum checked against a single, previously computed parity digit, *i.e.*, a check which tests whether the number of ones in a word is odd or even. Synonymous with odd-even check and related to check, redundant and to check, forbidden combination.

PART PROGRAM. A complete set of data and instructions written in source language for computer processing or written in machine language for manual programming for the manufacturing of parts on an NC machine.

PARTS EXPLOSION. A list or drawing of all parts used in all the subassemblies of an assembly or product.

PERIPHERAL EQUIPMENT. Auxiliary machines and storage devices which may be placed under control of a central computer and used on or off line to provide a system with outside communication; for example, tape readers, high-speed printers, CRT's magnetic tape feeds, and magnetic drums or disks.

PHASE SHIFT. (1) A time difference between the input and output signals of a system. (2) A change in the phase of a periodic quantity. (3) A change in time relationship of one part of a signal waveform with another, with no change in the basic form of the signal. The degree of change varies with frequency as a signal passes through a channel.

PHOTOCELL (PHOTOELECTRIC CELL). A solid-state photosensitive electron device in which use is made of the variation of the current-voltage characteristics as a function of incident radiation.

PHOTOSENSOR. A photo-sensitive electric device incorporated in an electric circuit and used for controlling mechanical devices; also referred to as an electric eye or photocell.

PLOTTER. (1) A device which will draw a facsimile of coded data input, such as the cutter path of an NC program. (2) A visual display or board on which a dependent variable can be drawn automatically as a function of one or more variables.

Point-to-point Control System. An NC system which controls motion only to move from one point to another without exercising path control during the transition from one end point to the next.

Polar Coordinates. (1) A mathematical system of coordinates for locating a point in a plane by the length of the plane's radius vector and the angle the vector makes with a fixed line. (2) Either of two numbers that locate a point in a plane by its distance from a fixed point on a line and the angle this line makes with a fixed line.

Polling. (1) Refers to an important multiprocessing method used to identify the source of interrupt requests. When several interrupts occur simultaneously, the control program makes the decision as to the one which will be serviced first. (2) Refers to a technique by which each of the terminals sharing a communications line is periodically interrogated to determine whether it requires servicing. The multiplexor or control station sends a poll which, in effect, asks the terminal selected "do you have anything to transmit." (3) A flexible, systematic, centrally-controlled method of permitting terminals on a multi-terminal line to transmit without contending for the line. The computer contacts terminals according to the order specified by the user, and each terminal contacted is invited to send messages. (4) Refers to centrally controlled method of calling a number of points to permit them to transmit information.

Position Analog Unit (PAU). The unit which feeds analog information about the position of a machine slide to the servo amplifier for comparison with positional input information.

Positioning/Contouring System. An NC system that is able to contour in two axes, without buffer storage, and position in a third axis for operations such as drilling, tapping, and boring.

Position Sensor. A device used in measuring a position and converting the measurement into a form which facilitates transmission.

Postprocessor. 1. A computer program that takes a generalized or centerline output and adapts it to the particular machine control unit/machine tool combination that will machine the part. 2. Refers to a set of computer instructions which transform tool centerline data into machine motion commands using the proper tape code and format required by a specific machine control system. Instructions such as feed rate calculations, spindle speed calculations, and auxiliary function commands may be included.

Precision. The degree of exactness with which a quantity is stated. Contrasted with accuracy; for example, a quantity expressed with 10 decimal digits of precision may have only one digit of accuracy.

Preset Tool. A cutting tool placed in a holder so that a predetermined geometrical relationship exists with a gage point.

Printed Circuit. (1) A circuit in which interconnecting wires have been replaced by conductive strips printed, etched, etc., onto an insulating board. It may also include similarly formed components on the baseboard. (2) Refers to resistors, capacitors, diodes, transistors and other circuit elements which are mounted on cards and interconnected by conductor deposits. These special cards are treated with light-sensitive emulsion and exposed. The light thus fixes the areas to be retained and an acid bath eats away those portions which are designed to be destroyed. The base is usually a copper clad card.

Printed Circuit Board. A board on which a predetermined conductive pattern, which may or may not include printed components, has been formed.

Printer. An output device that prints or types characters from parallel or serial entry.

Printout. A printed output of a system giving all data that has been processed by a program.

Process Control. (1) Automatic control of industrial processes in which continuous material or energy is produced. (2) Pertaining to systems whose purpose is to provide automation of continuous operations. This is contrasted with numerical control, which provides automation of discrete operations.

Production. (1) The manufacturing of goods. (2) The act of changing the shape, composition or combination of materials, parts, or subassemblies to increase their value. (3) The quantity of goods produced.

Productivity. The quality or state of being productive. As applied to manufacturing, it means the measured output of goods from a productive facility relative to some standard, norm, or potential maximum.

Program. (1) The plan or procedure for the operation of a machine or process especially when the machine is a computer. (2) A set of instructions arranged in proper sequence to cause the desired operations.

PROGRAMMABLE CONTROLLER (PC). A solid-state control system which has a user programmable memory for storage of instructions to implement specific functions such as: I/O control logic, timing, counting, arithmetic, and data manipulation. A PC consists of central processor, input/output interface memory, and programming device which typically uses relay-equipment symbols. A PC is purposely designed as an industrial control system which can perform functions equivalent to a relay panel or wired solid state logic control system.

PROM. (1) Programmable read-only memory is generally any type which is not recorded during its fabrication but which requires a physical operation to program it. Some PROMs can be erased and reprogrammed through special physical processes. (2) A semiconductor diode array which is programmed by fusing or burning out diode junctions.

PROPORTIONAL CONTROL ACTION. Refers to designed control action in which there is a continuous linear relation between the output and the input. Such a condition applies when both the output and the input are within their normal operating ranges and when an operation is at a frequency below a limiting value.

PROTOCOL. A former agreement between two communicating devices. It defines how data is formatted what the control signals mean, how error checking is performed, and the order and priority of various types of messages.

PUNCH TAPE. An input/output medium in the form of punched holes along a continuous strip of nonmagnetic tape. The tape is used to record and store data and/or programs and/or job control statements (*Syn*: perforated tape).

PUNCHED CARD. (1) A card of constant size and shape on which information is represented by holes in specific positions. (2) A piece of lightweight cardboard on which information is represented by holes punched in specific positions.

QUADRANT. Any of the four quarters of the rectangular or Cartesian coordinate dimensioning system.

RAM (RANDOM ACCESS MEMORY). This type memory is random because it provides access to any storage location point in the memory immediately by means of vertical and horizontal co-ordinates. Information may be "written" in or "read" out in the same very fast procedure.

RANDOM ACCESS. (1) Access to a computer storage under conditions whereby there is no rule for predetermining the position from where the next item of information is to be obtained. (2) Describes the process of obtaining information from or placing information into a storage system where the time required for such access is independent of the location of the information most recently obtained or placed in storage. (3) Describing a device in which random access can be achieved without effective penalty in time.

RANDOM TOOL SELECTION. A feature allowing the next tool to be loaded from any position in an automatic toolchanger rather than from the next location in the changer.

RAPID TRAVERSE. Tool movement at a maximum feed rate from one cutting operation to another.

READ-ONLY MEMORY (ROM). Digital storage device that can be read from but cannot be written into by the computer. The ROM is used to store the microprogram or a fixed program depending upon the microprogrammability of the CPU. The microprogram provides the translation from the higher-level user commands, such as ADD, SUBTR, etc., down to a series of detailed control codes recognizable by the microprocessor for execution. The size of the ROM varies according to user requirements within the maximum allowed capacity dictated by the addressing capability of the microprocessor.

READ OUT. The presentation of output data by means of visual displays, punched tape, etc.

REAL TIME. The ability of a computer to function and control a process as the process occurs. (1) In solving a problem, a speed sufficient to give an answer within the actual time the problem must be solved. (2) Pertaining to the actual time during which a physical process transpires. (3) Pertaining to the performance of a computation during the actual time that the related physical process transpires in order that results of the computation can be used in guiding the physical process.

RECTANGLAR COORDINATES. A system of two or three mutually perpendicular axes along which any point can be located in terms of distance and direction from any other point.

Reed Switch. Refers to a special switching device which consists of magnetic contactors which are sealed into a glass tube. The contactors are actuated by the magnetic field of an external solenoid, electromagnet, or a permanent magnet.

Relay. An electromagnetic switching device in which contacts are opened and/or closed by variations in the conditions of one electric circuit and thereby affect the operation of these devices in the same or other electric circuits.

Remote Control. Any system of control performed from a distance. The control signal may be conveyed by intervening wires, sound (ultrasonics), light, or radio.

Repeatability. The closeness of agreement among a number of consecutive measurements of a constant signal, approached from the same direction.

Resolution. (1) The smallest incremental step in separating a measurement into its constituent parts. In a digital system, resolution is one count in its least significant digit. (2) The ratio of maximum to minimum readings of a measuring system. (3) The process of separating the parts which compose a mixed body.

Resolver. (1) A means for resolving a vector into two mutually perpendicular components. (2) A transformer, the coupling between primary and secondary of which can be varied. (3) A small section with a faster access than the remainder of the magnetic-drum memory in a computer. (4) A device which separates or breaks up a quantity, particularly a vector, into constituent parts or elements; *e.g.*, the mutually perpendicular components of a plane vector. (5) A small section storage, particularly in drum, tape, or disk storage units that has much faster access than the remainder of the storage.

Response Time. (1) The time (usually expressed in cycles of the power frequency) required for the output voltage of a magnetic amplifier to reach 63% of its final average value in response to a step-function change of signal voltage. (2) The time required for the pointer of an instrument to come to apparent rest in its new position after the measured quantity abruptly changes to a new constant value. (3) The amount of time elapsed between generation of an inquiry at a data communications terminal and receipt of a response at that same terminal. Response time, thus defined includes: transmission time to the computer, processing time at the computer, including access time to obtain any file records needed to answer the inquiry, and transmission time back to the terminal.

Retrofit. Modification of a machine originally operated by manual or tracer control to one that operates by NC controls.

Robot. A robot is a reprogrammable, multifunctional manipulator designed to move material, parts, tools, or specialized devices through variable programmed motions for the performance of a variety of tasks.

Routine. A series of computer instructions which performs a specific application function.

RS-232C. Electronic Industries Association (EIA) standard for data communications, RS-232 type C. Data is provided at various rates, 8 data bits per character. Refers to the interface between a modem and the associated data terminal equipment. It is standardized by EIA Standard RS 232. For voice-band modems the interface leads are single leads with a common ground return. Polar type signals are specified with a minimum amplitude of +3 V at the terminating end. The maximum allowable voltage is +25V. A ground potential difference between equipment of up to 2V is allowed for by specifying that the driver sources must provide a +5V signal. The terminating impedance is required to be in the 3000-7000 ohm range. The drivers typically provide voltages in the range +6V to +10V with a source impedance of a few hundred ohms. A negative polarity indicates the binary state "1", marking, or an OFF control state. The positive polarity indicates the binary state "0" spacing, or an ON control state.

Scanner. Refers to an instruction which automatically samples or interrogates the state of various processes, files, conditions, or physical states and initiates action in accordance with the information obtained.

Scrap Allowance. The factor that expresses the quantity of a particular component that is expected to be scrapped while that component is being built into a given assembly. Also a factor that expresses the amount of raw material needed in excess of the exact calculated requirement to produce a given quantity of a part. The factor, dependent upon the type of assembly or part, is carried in the product structure segment and is used to increase the requirements as the component requirements are exploded.

Sensitivity. (1) Ratio of the response of a measuring device, to the magnitude of the measured quantity. It may be expressed directly in divisions per volt, milliradians per microampere, etc., or indirectly by stating a property from which sensitivity can be computed (*e.g.*, ohms per volt for a stated deflection). (2) The degree of response of an instrument of control unit to a change in the incoming signal.

SENSOR. A transducer or other device whose input is a quantitative measurement of an external physical phenomenon and whose output can be monitored by a computer or other control system.

SEQUENTIAL LOGIC. Refers to a circuit arrangement in which the output state is determined by the previous states of the input.

SERVOMECHANISM. 1. A power device for affecting machine motion. It embodies a closed-loop system in which the controlled variable is mechanical position and velocity. (2) Closed-cycle system in which a small input power controls a much larger output power, *e.g.*, movement of a gun turret may be accurately controlled by movement of a small knob or wheel.

SET POINT. In a feedback control loop, the point which determines the desired value of the quantity being controlled.

SHOP FLOOR CONTROL. A system for utilizing data from the shop floor as well as data processing files to maintain and communicate status information on shop orders and work centers. The major subfunctions of shop floor control are: (a) Assigning priority to each shop order. (b) Maintaining WIP quantity information for MRP. (c) Conveying shop order status information to the office. (d) Providing actual output data for capacity control purposes. (*cf.* Closed-loop MRP)

SIGNAL. (1) A visible, audible, or other conveyor of information. (2) The intelligence, message, or effect to be conveyed over a communication system. (3) A signal wave. (4) The physical embodiment of a message.

SOFTWARE. All of the program manuscripts, tapes, decks of cards, methods sheets, flow charts, and other programming documentation associated with computers and numerical control systems. Its counterpart is the physical hardware comprising the computer or NC system.

SOFTWIRED. A system in which a computer generates control logic, as determined by a software program.

SOLID STATE. Pertaining to an electrical circuit having no moving parts, relays, vacuum tubes, or gaseous tube components.

SPECIFICATION. A clear, complete, and accurate statement of the technical requirements descriptive of a material, an item, or a service, and of the procedure to be followed to determine if the requirements are met.

STAND-ALONE SYSTEM. A complete operational system that does not require support from other devices or systems.

STANDARD. An acceptance criterion or an established measure for performance, practice, or design.

STATIC BEHAVIOR. The behavior of a control system or an individual unit under fixed conditions (as contrasted to dynamic behavior, under changing conditions).

STEADY STATE. A condition in which only negligible change is evident with time.

STEP 7. A turning processor language that develops machining data from the descriptions of the rough stock and the finished workpiece.

STEPPING MOTOR. A bidirectional, permanent magnet motor which turns through one angular increment for each pulse applied to it. Refers to one in which rotation occurs in a series of discrete steps controlled electromagnetically by individual (digit) input signals.

STORAGE. Pertaining to a device into which data can be entered, in which they can be held, and from which they can be retrieved at a later time.

STRAIGHT-CUT SYSTEM. A system which has feed rate control only along the axes and controlled cutting action that occurs only along a path parallel to the linear (or circular) machine ways.

SUBASSEMBLY. Two or more parts joined together to form a unit which is only a part of a complete machine, structure, or other article.

SUBROUTINE. A portion of an NC program, stored in memory and capable of being called up to accomplish a particular operation. It reverts to the master routine upon completion.

SURFACE. A geometric shape used for controlling the location of a tool in space. As a cutter is directed along a path, it is guided by two surfaces from the programmer's viewpoint. One is called the part surface. Generally, the bottom of the cutter moves along the part surface while the side of the cutter is guided by the drive surface. A third surface, the check surface, is used to check or halt the movement of the tool in its progress along the DS-PS pair.

System. An organized collection of interdependent and interactive personnel, machines and methods combined to accomplish a set of specific functions as a larger unit having the capabilities of all the separate units.

Switch. A mechanical or electrical device that completes or breaks the path of the current or sends it over a different path.

Tab. A nonprinting spacing action on a tape preparation device whose code separates groups of characters in a tab sequential format.

Tab Sequential Format. A means of identifying a word by the number of tab characters that precede it in a block. The first character of all words is a tab character. Words are presented in a specific order, but all characters in a word, except the tab character, may be omitted when the command represented by the word is not desired.

Tape Leader. The front or lead end of a tape.

Terminal. A point in a system or communication network at which data can either enter or leave.

Time Sharing. A mode of operation that provides for the interleaving of two or more independent processes on one functional unit. The interleaved use of time on a computing system that enables two or more users to execute computer programs concurrently. Although the computer actually services each user in sequence, the high speed of the computer makes it appear that the users are all handled simultaneously.

Tool Offset. (1) An incremental displacement correction for tool position parallel to a controlled axis. (2) The ability to reset tool position manually to compensate for tool wear, finish cuts, and tool exchange.

Tooling. A set of required standard or special tools for production of a particular part, including jigs, fixtures, gauges, cutting tools, etc. Specifically excludes machine tools.

Toolpath. A generalized NC processor operable on IBM 4300 and 370 computers. Both DOS (12K partition) and OS (200K region) versions of TOOLPATH are available. TOOLPATH contains a complete ADAPT language plus comprehensive two- or three-axis, point-to-point and contouring machining operations.

TPTURN. A generalized 2 axis turning post-processor written in FORTRAN IV and compatible with the Manufacturing Software and Services TOOLPATH NC processor or APT/ADAPT processor.

Transient. A phenomenon caused in a system by a sudden change in conditions, and which persists for a relatively short time after the change.

Turning Center. A lathe-type NC machine tool capable of automatically boring, turning outer and inner diameters, threading, and facing parts. It is often equipped with a system for automatically changing or indexing cutting tools.

Turn Key System. An NC or computer system installed by a supplier who has total responsibility for building, installing, and testing the system.

UCC-APT. A version of APT developed by University Computing Company that contains special programming features: all fortran functions, fortran DO loops, logical IF, expanded solid and plane geometry capabilities, bounded FMILL, formatted printing, rough turning, turning and threading subsets.

Undercut. A cut shorter than the programmed cut resulting after a command change in direction.

UNIAPT. The minicomputer implementation and extension of the APT part programming language for small and medium size computers.

Verify. To check, usually by automatic means, one typing or recording of data against another to minimize the number of errors in the data transcription.

Word Address Format. An NC tape format in which each word in a block is identified by one or more preceding characters.

X-axis. The axis of motion that is horizontal and parallel to the workholding surface.

Y-axis. The axis of motion that is perpendicular to the X and Z axes.

Zero Offset. A characteristic of an NC machine which permits the zero point on an axis to be shifted readily over a specified range. A feature of an NC unit which permits the zero point on an axis to be located anywhere within a specific range while actually retaining information of the "permanent zero."

FORGING (See also METAL FORMING.)

Air-lift Hammer. A type of gravity drop hammer where the ram is raised for each stroke by an air cylinder. Since length of strike may be controlled, ram velocity and thus energy delivered to the workpiece may be varied.

Anisotropy. The characteristic of exhibiting different values of a property in different directions with respect to a fixed reference system in the material.

Anvil (base). Extremely large, heavy block of metal which supports entire structure of conventional gravity or steam driven forging hammers. Also, the block of metal on which hand (or smith) forgings are made.

Bark. The decarburized layer just beneath the scale produced by heating steel in an oxidizing atmosphere.

Batch-type Furnace. A furnace for heating materials in which the loading and unloading is done through a single door or slot.

Bender. Term denoting a die impression, tool, or mechanical device designed to bend forging stock to conform to the general configuration of die impressions subsequently to be used.

Billet. A semi-finished, cogged, hot-rolled, or continuous-cast metal product of uniform section, usually rectangular with radiused corners. Billets are relatively larger than bars.

Blank. A piece of stock (also called a "slug" or "multiple") from which a forging is to be made.

Blister. A defect caused by gas bubbles either on the surface or beneath the surface of the metal.

Blocker (blocking impression). The impression in the dies (often one of a series of impressions in a single die set) which imparts to the forging an intermediate shape, preparatory to forging of the final shape.

Blocker Dies. Blocker dies are characterized by generous contours, large radii, draft angles of 7° or more, and liberal finish allowances.

Blocker-type Forging. A forging which approximates the general shape of the final part with relatively generous finish allowance and radii. Such forgings are sometimes specified to reduce die casts where only a small number of forgings are desired and the cost of machining each part to its final shape is not excessive.

Blocking. A forging operation often used to impart an intermediate shape in the finishing impression of the dies. Blocking can ensure proper "working" of the material and contribute to great die life.

Blow. The impact or force delivered by one workstroke of the forging equipment.

Blowhole. A cavity produced by gas evolved during solidification of metal.

Board Hammer. A type of gravity drop hammer where wood boards attached to the ram are raised vertically by action of contra-rotating rolls, then released. Energy for forging is obtained by the mass and velocity of the freely falling ram and the attached upper die.

Bolster. The plate secured to the bed of a press for locating and supporting the die assembly.

Boss. A relatively short protrusion or projection on the surface of a forging often cylindrical in shape.

Burnt. Permanently damaged metal caused by heating conditions producing incipient melting or intergranular oxidation.

Burst. An internal discontinuity caused by improper forging.

Buster (pre-blocking impression). A type of die impression sometimes used to combine preliminary forging operations such as edging and fullering with the blocking operation to eliminate blows.

Camber. Deviation from edge straitness, usually referring to the greatest deviation of side edge from a straight line. Sometimes used to indicate crown on flat rolls.

Check. Crack in a die impression, generally due to forging pressure and/or excessive die temperature. Die blocks too hard for the depth of the die impression have a tendency to check or develop cracks in impression corners.

Closed-die Forging. (See IMPRESSION DIE FORGING.)

Close-tolerance Design. A forging designed with commercially recommended draft radii and finish allowances, but with dimensional tolerances of less than one-half the commercial tolerances recommended for otherwise similar parts. Often little or no machining is required after forging.

Cogging. The process of forging ingots to produce blooms or billets.

Coining. The process of applying necessary pressure to all or some portion of a forging's surface in order to obtain closer tolerances, smoother surfaces, or to eliminate draft. Coining may be done while forgings are hot or cold and is usually performed on surfaces parallel to the parting line of the forging.

Cold Heading. Working metal at room temperature in such a manner that the cross-sectional area of a portion or all of the stock is increased.

Cold Inspection. A visual (usually final) inspection of the forgings for visible defects, dimensions, weight, and surface condition at room temperature. The term may also be used to describe certain nondestructive tests such as a magnetic particle, dye penetrant, and sonic inspection.

Cold Lap. A flaw caused when a workpiece fails to fill the die cavity during first forging. A seam is formed as subsequent dies force metal over this gap to leave a seam on the workpiece surface.

Cold Trimming. Removing flash or excess metal from the forging in a trimming press when the forging is at room temperature.

Cold Working. Permanent plastic deformation of a metal at a temperature below its recrystallization point — low enough to produce strain hardening.

Compressive Strength. The maximum stress that a material subjected to compression can withstand when loaded without deformation or fracture.

Core Forging. The process of displacing metal with a punch to fill a die cavity.

Counterblow Forging Equipment. A category of forging equipment wherein two opposed rams are activated simultaneously, striking repeated blows on the workpiece of a midway point. Action may be vertical, as in the case of counterblow forging hammers, or horizontal as with the "Impacter."

Counterlock. A jog in mating surfaces of dies to prevent lateral die shifting from side thrust developed in forging irregular shaped pieces.

Creep. Time-dependent strain occurring under stress. The resistance to creep, or creep strength, decreases with increasing temperature.

Critical (temperatures). Temperatures at which phase changes take place in metals.

Cross Forging. Preliminary working of forging stock in flat dies so that the principal increase in dimension is in the transverse direction with respect to the original axis of the ingot.

Die Block. A block (usually) of heat-treated steel into which desired impressions are machined or sunk and from which closed-die forgings are produced on hammers or presses. Die blocks are usually used in pairs with part of the impression in one of the blocks and the balance of the impression in the other.

Die Forging. (1) Compression in a closed impression die. (2) A product of such an operation.

Die Lubricant. A compound sprayed, swabbed or otherwise applied on die surfaces of forgings during forging to reduce friction between the forging and the dies. Lubricants may also ease release of forgings from the dies and provide thermal insulation.

Die Match. The condition where dies, after having been set up in the forging equipment, are in proper alignment relative to each other.

Dies (die blocks). The metal blocks into which forging impressions are machined and from which forgings are produced.

Die Shift. A condition requiring correction where, after dies have been set up in the forging equipment, displacement of a point in one die from the corresponding point in the opposite die occurs in a direction parallel to the fundamental parting line of the dies.

Directional Properties. Anisotropic values. Physical or mechanical properties varying with the relation to a specific direction, resulting from structural fibering and preferred orientation.

Dowel. A metal insert placed between mating surfaces of the die shank and die holder in the forging equipment to assure lengthwise die match.

Draft. The amount of taper on the sides of the forging

necessary for removal of the workpiece from the dies. Also the corresponding taper on the side walls of the die impressions.

Draft Angle. The angle of taper, expressed in degrees, given to the sides of the forging and the side walls of the die impression.

Drawing. A forging operation in which the cross section of forging stock is reduced and the stock lengthened between flat or simple contour dies. (See FULLER.)

Drop Forging. A forging produced by hammering metal in a drop hammer between dies containing impressions designed to produce the desired shape. (See IMPRESSION DIE FORGING.)

Drop Hammer. A term generally applied to forging hammers wherein energy for forging is provided by gravity, steam, or compressed air. (See AIR-LIFT HAMMER, BOARD HAMMER, STEAM HAMMER.)

Edger (EDGING IMPRESSION). The portion of the die impression which distributes metal during forging into areas where it is most needed to facilitate filling the cavities of subsequent impressions to be used in the forging sequence. (See FULLER.)

Extrusion. The process of forcing metal to flow through a die orifice in the same direction in which energy is being applied (forward extrusion); or in the reverse direction (backward extrusion) in which case the metal usually follows the contour of the punch or moving forming tool. The extrusion principle is used in many impression die forging applications.

Extrusion Forging. (l) Forcing metal into or through a die opening by restricting flow in other directions. (2) A part made by the operation.

Fillet. The concave intersection of two surfaces. In forging, the desired radius at the concave intersection of two surfaces is usually specified.

Finish Allowance. The amount of excess metal surrounding the intended final shape. Sometimes called clean-up allowance, forging envelope, or machining allowance.

Flakes. Short, discontinuous, internal fissures in ferrous metals attributed to stresses caused by localized transformation and decreased solubility of hydrogen during cooling after hot working.

Flash. Necessary metal in excess of that required to completely fill the finishing impression of the dies. Flash extends out from the body of the forging as a thin plate at the line where the dies meet and is subsequently removed by trimming. Cooling faster than the body of the component during forging, flash can serve to restrict metal flow at the line where dies meet, thus assuring complete filling of the finishing impression.

Flash Gutter. An additional cavity machined along the parting line of the die cavity to receive the excess metal as it flows out of the die cavity through the flash gap.

Flash Land. Configuration in the finishing impression of the dies designed either to restrict or to encourage growth of flash at the parting line, whichever may be required in a particular instance to ensure complete filling of the finishing impression.

Flat Dies Forging (OPEN-DIE FORGING). Forging worked between flat or simple contour dies by repeated strokes and manipulation of the workpiece. Also known as "hand" or "smith" forging.

Flow Lines. Patterns in a forging resulting from the elongation of nonhomogeneous constituents and the grain structure of the material in the direction of working during forging; usually revealed by macroetching. (See GRAIN FLOW.)

Flow Stress. (l) The shear stress required to cause plastic deformation of solid metals. (2) The uniaxial true stress required to cause flow at a particular value of strain.

Forgeability. Term used to describe the relative ability of material to deform without rupture.

Forging. The product of work on metal formed to a desired shape by impact or pressure in hammers, forging machines (upsetters), presses, rolls, and related forming equipment Forging hammers, counterblow equipment, and high-energy-rate forging machines impart impact to the workpiece, while most other types of forging equipment impart squeeze, but the majority of metals are made more plastic for forging by heating.

Forging Machine (UPSETTER OR HEADER). A type of forging equipment, related to the mechanical press, in which the main forming energy is applied horizontally to the work-piece which is gripped and held by prior action of the dies.

Forging Rolls. Rolling mills that forge comparatively

uniform shapes by using variable radii around the circumference of rolls that rotate in the opposite direction from those ordinarily used for rolling.

Forging Strains. Differential strains that result from forging or from cooling from the forging temperature, and that are accompanied by residual stresses.

Forging Stresses. Elastic stresses induced by forging or cooling from the forging temperature; sometimes erroneously referred to as forging strains.

Fuller (fuller impression). Portion of the dies which is used in hammer forging primarily to reduce the cross section and lengthen a portion of the forging stock. The fullering impression is often used in conjunction with an edger (or edging impression).

Gate (sprue). A portion of the die which has been removed by machining to permit a connection between multiple impressions or between an impression and the bar of stock.

Gathering. An operation which increases the cross section of part of the stock above its original size.

Gathering Stock. Any operation whereby the cross section of a portion of the forging stock is increased above its original size.

Grain Flow. Fiber-like lines appearing on polished and etched sections of forgings which are caused by orientation of the constituents of the metal in the direction of working during forging. Grain flow produced by proper die design can improve required mechanical properties of forgings.

Grain Size. The average size of the crystals or grains in a metal as measured against an accepted standard.

Gravity Hammer. A class of forging hammer wherein energy for forging is obtained by the mass and velocity of a freely falling ram and the attached upper die. Examples: board hammers and air-lift hammers.

Gutter. A slight depression machined around the periphery of an impression in the die which allows space for the excess metal (flash during forging).

Hammer Forging. Shaping of metal by impact between dies in one of several types of equipment known as forging hammers. (See air-lift hammer, board hammer, counterblow forging equipment, steam hammer.)

Heat (forging). Amount of forging stock placed in a batch-type furnace at one time.

Heat of Metal. The quantity of material manufactured from one melt at the metal producer's facility. Metal from a single heat is extremely uniform in chemical analysis.

High-energy-rate Forging (high velocity or high speed forging). The process of producing forgings on equipment capable of extremely high ram velocities resulting from the sudden release of a compressed gas against a free piston.

Hot Stamp. Impressing markings in a forging while the forging is in the heated, plastic condition.

Hot Trim. Removing flash or excess metal from the forging in a trimming press while the forging is in the heated state.

Hot Working. The mechanical working of metal at a temperature above its recrystallization point — a temperature high enough to prevent strain hardening.

Hydraulic Hammer. A gravity drop forging hammer which uses hydraulic pressure to lift the hammer between strokes.

Impression. A cavity machined into a forging die to produce a desired configuration in the workpiece during forging.

Impression Die Forging. A forging that is formed to the required shape and size by machined impressions in specially prepared dies which exert 3-dimensional control on the workpiece.

Insert. A component which is removable from a die. An insert can be used to fill a cavity or to replace a portion of the die with a material which gives better service.

Ironing. (1) A press operation used to obtain a more exact alignment of the various parts of a forging, or to obtain a better surface condition. (2) An operation to increase the length of a tube by reduction of wall thickness and outside diameter. (See coining) (swaging.)

Isothermal Forging. A forging operation performed on a workpiece during which the temperature remains constant and uniform. Generally used when aluminum, nickel, or titanium is being forged.

LAP. A surface irregularity appearing as a seam, caused by the folding over of hot metal, fins, or sharp corners and by subsequent rolling or forging (but not welding) of these into the surface.

LOCKS. Changes in the plane of the mating faces of the dies. Locks aid in holding die alignment during forging by counteracting lateral thrust which is present to an extent dependent on the shape of the workpiece.

MANDREL FORGING. (See RING ROLLING.)

MANIPULATOR. A mechanical device for handling an ingot or billet during forging.

MATCH. A condition in which a point in one die-half is aligned properly with the corresponding point in the opposite die-half within specified tolerance.

MATCHED EDGES (MATCH LINES). Two edges of the die face which are machined exactly at 90° to each other, and from which all dimensions are taken in laying out the die impress and aligning the dies in the forging equipment.

MATCHING DRAFT. When unsymmetrical ribs and side walls meet at the parting line it is standard practice to provide greater draft on the shallower die to make the forging's surface meet at the parting line. This is called matching draft.

MECHANICAL WORKING. Subjecting metal to pressure, exerted by rolls, hammers, or presses, in order to change the metal's shape or physical properties.

NATURAL DRAFT. After the parting line has been established and a machining allowance is provided, a shape may have what is called natural draft.

NO-DRAFT FORGING. A forging with extremely close tolerances and little or no draft, requiring a minimum of machining to produce the final part. Mechanical properties can be enhanced by closer control of grain flow and retention of surface material in the final component.

OPEN-DIE FORGING. Hot mechanical forming of metals between flat or shaped dies where metal flow is not completely restricted. Also known as hand or smith forging.

PARTING LINE. The line along the surface of a forging where the dies meet, or the line along the corresponding edge of the die impression.

PARTING PLANE. The plane which includes the fundamental parting line of the dies; the dividing plane between dies.

PLATTER. The entire workpiece upon which the forging equipment performs work, including the flash, sprue, tonghold, and as many forgings as are made at one time.

PRESS FORGING. The shaping of metal between dies by mechanical or hydraulic pressure. Usually this is accomplished with a single workstroke of the press for each die station.

PROOF. Any reproduction of a die impression in any material, frequently a lead or plaster cast.

PUNCH. (1) The movable die in a trimming press or forging machine. (2) A tool used in punching holes in metal.

PUSHER FURNACE. A continuous type furnace where stock to be heated is charged at one end, carried through one or more heating zones, and discharged at the opposite end.

QUANTITY TOLERANCE. Allowable variation of quantity to be shipped on a purchase order. This tolerance is properly agreed to by forging producer and purchaser when order is placed.

RAM. The moving part of a forging hammer, forging machine, or press, to which one of the tools is fastened.

RESTRIKING. Striking a trimmed forging in order to align or size its several components or sections. The operation can be performed hot or cold.

RING ROLLING. The process of shaping weldless rings from pierced discs or thick-walled, ring-shaped blanks, between rolls which control wall thickness, ring diameter, height, and contour.

ROLLER (ROLLING IMPRESSION). The portion of a forging die where cross sections are altered by hammering or pressing while the workpiece is being rotated.

ROLL FORGING. The process of shaping stock between power driven rolls bearing contoured dies. The workpiece is introduced from the delivery side of the rolls, and is reinserted for each succeeding pass. Usually used for pre-forming, roll forging is often employed to reduce thickness and increase length of stock.

ROTARY FURNACE. A circular furnace constructed so that the hearth and workpieces rotate around the furnace's axis during heating.

Saddling (Mandrel Forging). The process of rolling and forging a pierced disc of stock over a mandrel in order to produce a weldless ring.

Scale. The oxide film that is formed on forgings, or other heated metal, by chemical action of the surface metal with the oxygen in the air.

Scale Pit. A surface depression formed on the forging operation.

Seam. A crack or inclusion on the surface of forging stock which may carry through forging and appear on the finished product.

Semi-finisher (semi-finishing impression). An impression in the forging die which only approximates the finish dimensions of the forging. Semi-finishers are often used to extend die life of the finishing impression, to assure proper control of grain flow during forging, and to assist in obtaining desired tolerances.

Shank. The portion of the die or tool by which it is held in position in the forging unit.

Shoe. A holder used as a support for the stationary portions of trimming and forming dies; sometimes termed sow block.

Shrinkage. The contraction of metal during cooling after forging. Die impressions are made oversize according to precise shrinkage scales to allow forgings to shrink to design dimensions and tolerances.

Shrink Scale. A measuring scale or rule, used in die layout, on which graduations are expanded to compensate for thermal contraction (shrinkage) of the forging during cooling.

Sizing. A process employed to control precisely a diameter of rings or tubular components.

Slug. (l) Metal removed when punching a hole in a forging. Also termed "punch out'." (2) Forging stock for one workpiece cut to length. (See BLANK.)

Snag Grinding (snagging). The process of removing portions of forgings not desired in the finished product, by grinding.

Sow Block. Metal die holder employed in a forging hammer to protect the hammer anvil from shock and wear. Also called anvil cap or shoe.

Steam Hammer. A type of drop hammer where the ram is raised for each stroke by a double-action steam cylinder and the energy delivered to the workpiece is supplied by the velocity and weight of the ram and attached upper die driven downward by steam pressure. Energy delivered during each stroke may be varied.

Swage (swedge). Operation of reducing or changing the cross sectional area by revolving the stock under rapid impact blows.

Tonghold. The portion of the stock by which the operator grips the stock with tongs during forging.

Trimming. The process of removing flash or excess metal from a forging.

Upsetting. Working metal in such a manner that the cross-sectional area of a portion of all of the stock is increased.

Vent. A small hole in a punch or die which permits the passage of air or gas. Venting prevents trapping air that interferes with forming of a vacuum, which interferes with stripping.

FOUNDRY, CASTING

Abrasion Resistance. Degree of resistance of a material to abrasion or wear. White iron is extremely abrasion resistant because of its large amount of combined carbon in the form of iron carbide, Fe_3C.

Abrasives. Hard granular materials for grinding, polishing, blasting either bonded together in the form of wheels or bricks or bonded to paper or cloth belts or discs by glue or resins; or used loose and propelled by centrifugal force from a wheel or an entrained in an air stream. Metallic shot, sand, or grit are used in the latter application. Natural abrasives include emery, garnet, and silica; synthetic abrasives are corundum and silicon carbide.

Accelerator. A material which increases the reaction rate of organic binders. It usually acts as a catalyst in the reaction.

Acid. A term applied to refractories and minerals or slags which have a high silica content. Most frequently used with reference to the characteristic reactions between a slag and the refractory lining of a cupola or furnace used for ferrous alloys.

ADDITION AGENT. Any material added to a charge of molten metal in the bath or ladle to bring the alloy to the required specification.

ADDITIVE. Any material added to a molding sand to promote casting "peel", to improve flowability or, to reduce expansion, *e.g.*, sea coal, pitch, cereal graphite, plumbage.

AERATION. The fluffing of a molding sand just before dropping it into the flask.

A.F.S. (AMERICAN FOUNDRYMEN'S SOCIETY). This abbreviation will be found in connection with various tests, numbers, etc., among the foundry definitions.

AIR BELT. Chamber surrounding the cupola at the tuyeres, to equalize volume and pressure of blast and deliver it to the tuyeres. Commonly called a windbox.

AIR DRIED. Surface drying of cores in open air before baking in an oven; also applied to molds which air-dry when left open, which may cause crumbling or crushing when the mold is closed; a core or mold dried in air, without application of heat.

AIR-DRIED STRENGTH. Tenacity (compressive, shear, tensile, or transverse) of a sand mixture after being air dried at room temperature.

AIR FURNACE. A reverbatory type of furnace which is fired by fuel burning at one end of the hearth; the metal is melted by the hot gases which pass over the charge and move toward the stack at the other end. Heat is reflected from the parabolic roof and side walls.

AIR QUENCHING (NORMALIZING). Cooling of alloys in air from above the transformation range.

ALLOY. A metallic mixture composed of two or more chemical elements of which at least one must be a metal. Usually an alloy has properties which are different from its components.

ANNEALING. A heat treating process in which castings are heated in a furnace above the transformation temperature and then are allowed to cool to room temperature before removing from the furnace. A full annealed casting has few residual stresses and is in its softest condition.

APRON FEEDER. A continuous pan conveyor, carried by two strands of chain, used for carrying castings or feeding sand at a controlled rate from a storage hopper.

AS CAST. Referring to a casting which has received no heat treatment other than normal gate and riser removal and sand blasting.

BACK DRAFT. A reverse taper which prevents removal of a pattern from the mold.

BACKING SAND. Reconditioned sand used for ramming main part of mold after pattern has been covered with facing sand.

BAGHOUSE. Large chamber for holding bags used in filtration of gases from a furnace to recover metal oxides and other solids suspended in the gases; one form of dust collector.

BAKED CORE. A core which has been heated through sufficient time and temperature to produce the desired physical properties attainable from its oxidizing or thermal-setting binders.

BAKED PERMEABILITY. The property of a molded mass of sand which when baked at a temperature above 230°F (110°C) and when cooled to room temperature permits passage through it of the gases resulting when molten metal is poured in a mold.

BAKED STRENGTH. Compressive, shear, tensile, or transverse strength of a molded sand mixture when baked at a temperature above 230°F (110°C) and then cooled to room temperature.

BANK SAND. Sedimentary deposits, usually containing less than five per cent clay, occurring in banks or pits, used in core making and in synthetic molding sands.

BARS (CLEATS). Ribs of metal or wood placed across the cope portion of a flask to help support the sand in the cope flask.

BASE PERMEABILITY. That physical property which permits gases to pass through packed dry sand grains containing no clay or other bonding substance.

BASIN. The enlarged mouth of the sprue into which the molten metal is first poured.

BED OR BED CHARGE. Initial charge of fuel in cupola upon which the melting is started.

BED COKE. First layer of coke placed in the cupola. Also the coke used as the foundation in constructing a large mold in flask or pit.

Bench Molder. A craftsman who makes molds for smaller type castings, working at the molder's bench only.

Bentonite. A colloidal clay derived from volcanic ash and employed as a binder in connection with synthetic sands, or added to ordinary natural (clay-bonded) sands where strength is required; western or sodium bentonite is found in Wyoming and South Dakota; also in the south central states which have quantities of southern or calcium.

Binder. Artificially added bond (usually used to indicate other than clay) to foundry sand, such as cereal, pitch, resin, oil, sulphite by-product, etc.

Blacking. Carbonaceous materials such as plumbage, graphite or powdered carbon, usually mixed with a binder and frequently carried in suspension in water or other liquid; used as a thin facing applied to surfaces of molds or cores to improve casting finish.

Blast Furnace. In ferrous metallurgy, a shaft furnace supplied with an air blast (usually hot) and used for producing pig iron by smelting iron ore in a continuous operation. The raw materials (iron ore, coke, and limestone) are charged at the top, and the molten pig iron and slag which collect at the bottom, are tapped out at intervals. In nonferrous metallurgy, a shaft type of vertical furnace is used for smelting coarse copper, lead, and tin ores; similar to the type used for smelting iron, but smaller.

Blast Pressure. Pressure of air in blast pipe or wind belt of cupola, depending on location of indicating instrument; usually given in ounces per square inch or inches of water.

Bleeder. A defect wherein a casting lacks completeness due to molten metal draining or leading out of some part of the mold cavity after pouring has stopped.

Blowhole. Irregular shaped cavities with smooth walls produced in a casting when gas, entrapped while the mold is being filled, or evolved during solidification of the metal, fails to escape and is held in pockets.

Bob. A riser or feeder, usually blind, to provide molten metal to the casting during solidification, thereby preventing shrinkage cavities.

Bond, Bonding Substance, or Bonding Agent. Any material other than water, which when added to foundry sands, imparts bond strength.

Bond Clay. Any clay suitable for use as a bonding material.

Bond Strength. Property of a foundry sand by virtue of which it offers resistance to deformation.

Boss. A projection of circular cross-section on a casting. Usually intended for thickening a section so that it can be drilled and tapped.

Bot. A mass of clay used to stop the flow of metal from the taphole of the cupola.

Break-off Core. A thin core connecting the riser and casting which, while not impeding the flow of metal, serves as a notch to assist in riser removal.

Buckle. An indentation in a casting, resulting from expansion of the sand. May be termed the start of an expansion defect.

Bumper. Machine for ramming sand in a flask by repeated jarring or jolting.

Burned Sand. Sand from which the binder or bond has been wholly or partially lost by contact with molten metal.

Butt-off. Operation performed at times to supplement ramming by jolting, either hand or air rammer.

Carbon Dioxide Process (CO_2 process). A process for hardening cores or molds in which carbon dioxide gas is blown for a few seconds through a sand bonded by sodium silicate. The sand mass instantly becomes rigid because of the formation of a silica gel.

Carbon Steel. Steel that owes its properties chiefly to the presence of carbon, without substantial amounts of alloying elements; also termed "ordinary steel," "straight carbon steel," "plain carbon steel," etc.

Casting. (*noun*) Metal object cast to the required shape by pouring or otherwise injecting liquid metal into a mold, as distinct from one shaped by a mechanical process.

Casting. (*verb*) Act of pouring metal into a mold.

Casting, Centrifugal. A process of filling molds by (1) pouring the metal into a sand or permanent mold that is revolving about either its horizontal or its vertical axis; or (2) pouring the metal into a mold that is subsequently revolved before solidification of the metal is complete.

Casting, Die. (*verb*) The injection of molten (primarily nonferrous) metal into a metal mold under pressure.

Casting, Machine. (*verb*) Process of casting by machine.

Casting, Open Sand. (*noun*) Casting poured into an uncovered mold.

Casting, Permanent Mold. A casting produced in a reuseable metal or refractory mold.

Casting, Plaster. A casting made in a plaster mold.

Casting, Precision. A casting of high dimensional accuracy produced by such processes as investment casting, plaster mold.

Casting, Sand. A casting produced in a mold made of green sand, dried sand, or a core sand.

Casting Yield. The weight of casting or castings divided by the total weight of metal poured into the mold, expressed as a percent.

Cast Iron. Essentially an alloy of iron, carbon, and silicon in which the carbon is present in excess of the amount which can be retained in solid solution. It is austenite at the eutectic temperature. When cast iron contains a specially added element or elements in amounts sufficient to produce a measurable modification of the physical properties of the section under consideration, it is called alloy cast iron. Silicon, manganese, sulphur, and phosphorus, as normally obtained from raw materials, are not considered as alloy addition.

Cast Iron, White. Cast iron in which substantially all the carbon is present in the form of iron carbide. Such a material has a white fracture.

Cavity. (1) The impression in a mold produced by withdrawal of the pattern and to be filled by the casting metal. (2) A hollow or sunken space, or a void in the interior of a casting.

Cement Sand. A synthetic sand that is bonded with Portland cement.

Ceramic Molding. A molding process that uses a ceramic shell or mold made by alternately dipping a pattern in dipcoat slurry and stuccoing with coarse ceramic particles until the shell of desired thickness is obtained.

Cereal Binder. A binder used in core mixtures and molding sands, derived principally from corn flour.

Chaplets. Metal supports or spacers used in molds to maintain cores, or parts of the mold which are not self-supporting in their proper positions during the casting process.

Charge. A given weight of metal or fuel introduced into the cupola or furnace.

Check. A minute crack in the surface of a casting caused by unequal expansion or contraction during cooling.

Cheek. Intermediate section of a flask that is inserted between cope and drag to decrease the difficulty of molding unusual shapes or to fill a need for more than one parting line.

Chill. (1) Addition of solid metal to molten metal in ladle to reduce temperature before pouring. (2) Depth to which chilled structure penetrates a casting.

Chill Block. A cast iron test block in which the depth of chill, as determined by fracture, is used as an estimate of the cast iron's quality.

Chills. Metal inserts in molds or cores at the surface of a casting or within the mold to hasten solidification of heavy sections and cause the casting to cool at a uniform rate.

Choke. A restriction in the gating system for the purpose of keeping dirt, or slag from entering the casting proper.

Chromite. A naturally occurring mineral which is a solid solution of chromium and iron oxides, used as a molding sand for cores and molds.

Clamp-off. An indention in the casting surface due to displacement of sand in the mold.

Clay. An earthy or stony mineral aggregate consisting essentially of hydrous silicates of alumina, plastic when sufficiently pulverized and wetted, rigid when dry, and vitreous when fired at a sufficiently high temperature. Clay minerals most commonly used in the foundry are montmorillonites and kaolinites.

Clay Substance (A.F.S. clay). That portion of foundry sand which, when suspended in water, fails to settle 1 inch per minute and which consists of particles less than 20 microns (0.02 mm or 0.0008 inches) in diameter.

Clay-wash. Thin emulsion of clay and water for coating gaggers and the inside of flasks. Also used as grout.

Coke. A porous, gray infusible product resulting from the dry distillation of bituminous coal, petroleum, or coal tar pitch, which drives off most of the volatile matter. Used as a fuel in cupola melting.

Cold Box Process. A two part organic resin binder system, mixed in conventional mixers and blown into shell or solid core shapes at room temperature. A vapor mixed with air is blown through the sand, resulting in instant curing, stripping, setting and immediate pouring of metal around it.

Cold Setting Binder. Term used to describe any binder that will harden the core sufficiently at room temperature so that the core can be removed from its box without distortion. Commonly used in reference to oil-oxygen types of binders.

Cold Setting Process. Any of several systems for bonding mold or core aggregates by means of organic binders, relying on the use of catalysts rather than heat for polymerization (curing).

Cold Shut. A casting defect caused by imperfect fusing of molten metal coming together from opposite directions in a mold, or due to folding of the surface.

Collapsible Sprue. A sprue pattern of flexible material or of spring tube design for use in the squeeze molding of match plate patterns. Frequently a pouring cup is incorporated in the design.

Colliodal Clay. Finely divided clay of the montmorillonite (bentonite), kaolinite (fire clay) or illite class; prepared for use as sand binders.

Colloidal Material. Finely divided materials which are less than 0.5 micron (0.000015 in.) such as albumin, glue, starch, gelatin, and bentronite.

Columnar Structure. A course structure of parallel columns of elongated grains which is caused by a highly directional solidification pattern which results from steep thermal gradients across the mold metal interface.

Combined Carbon. Carbon in iron and steel which is combined chemically with other elements; not in the free state as graphitic or temper carbon.

Combined Water. That water in mineral matter which is chemically combined and drives off only at temperatures above 110°C (230°F).

Compressive Strength, Sand. Maximum stress in compression which an AFS Standard compacted sand specimen is able to withstand without significant dimensional deformation.

Continuous Casting. A process for forming a bar of constant cross section directly from molten metal by gradually withdrawing the bar from the die as the metal flowing into the die solidifies.

Conveyor, Belt. A continuously moving belt used in an automated or semi-automated foundry to move material from one station to another.

Conveyor, Pallet. A material handling conveyor that holds one or more molds per section and transports them from the molding station, through pouring to shake out.

Cope. Upper or topmost section of a flask, mold, or pattern.

Cope, False. Temporary cope used only in forming the parting and therefore not a part of the finished mold.

Coping Out. The extension of sand of the cope downward into the drag, where it takes an impression of a pattern.

Core. A preformed sand aggregate inserted in a mold to shape the interior or that part of a casting which cannot be shaped by the pattern.

Core Assembly. Putting together a complex core made of a number of sections.

Core Binder. Any material used to hold the grains of core sand together.

Core Blow. A gas pocket in a casting adjacent to a cored cavity and caused by entrapped gases from the core.

Core Box. Wood, metal, or plastic structure, the cavity of which has the shape of the desired core which is to be made therein.

Core Break-Off. A core designed to produce a sharp break line in the gate for removal of the feeder.

Core Collapsibility. The rate of disintegration of a core at elevated temperatures.

CORE DRIERS. Supports used to hold cores in shape while being baked; constructed from metal or sand for conventional baking, or from plastic material for use with dielectric core baking equipment.

CORE, DROP. A type of core used in forming comparatively small openings occurring above or below the parting; the seat portion is so shaped that the core is easily dropped into place.

CORE FILLER. Material used to replace sand in the interior of large cores - coke, cinder, saw dust, etc. Usually added to aid collapsibility to reduce weight and to save core binders.

CORE, GREEN SAND. (1) A core formed from the molding sand and generally an integral part of the pattern and mold. (2) A core made of unbaked molding sand. (See CORE SAND.)

COREMAKER. A craftsman skilled in the production of cores for foundry use.

CORE OIL. Linseed base or other oil used as a binder for baked cores.

CORE OVENS. Low-temperature ovens used for baking cores. Maximum operation temperature is 600°F.

CORE PASTE. A prepared adhesive for joining sections of baked or cured cores.

CORE PRINT. Projections attached to a pattern in order to form recesses in the mold at points where cores are to be supported.

CORE, SAG. A decrease in the height of a core, usually accompanied with an increase in width, as a result of insufficient green strength of the sand to support its own weight.

CORE SAND. Sand for making cores to which a binding material has been added to obtain good cohesion and porosity after curing. (See CORE, GREEN SAND.)

CORE SHIFT. A variation from specified dimensions of a cored section because of a change in position of the core or misalignment of cores in assembling.

CORE SHOOTER. A device using low air pressure to fluidize a core sand mix. The air is released quickly so that it forces the mix into a core box.

CORE, STRAINER. A baked sand or a refractory disc with holes of a uniform size through its thickness. Used to control the discharge of metal from pouring basins or to regulate the flow of metal in gating systems of molds; also to prevent entrance of dross or slag into the mold cavity.

CORE VENTS. (1) A wax product, round or oval in cross section, used to form the vent passage in a core. Also refers to a metal screen or slotted piece used to form the vent passage in the core box employed in a core blowing machine. (2) Holes made in a core to facilitate the escape of gases.

CORE WASH. A suspension of fine clay or graphite applied to cores by brushing, dipping or spraying to improve the cast surface of the cored portion of the casting.

CRACK, HOT TEAR. A rupture occurring in a casting at or just below the solidifying temperature by a pulling apart of the soft metal; caused by thermal contraction stresses.

CRACKING STRIP. A fin of metal molded on the surface of a casting to prevent cracking.

CRITICAL TEMPERATURE. Temperature at which metal changes phase. In usual iron alloys, the temperature at which alpha iron transforms to gamma iron or vice versa. Actually, a temperature range for cast irons.

CRONING PROCESS (C PROCESS). Now called shell molding or shell core practice. The process was named after its German inventor Johannes Croning. (See SHELL MOLDING.)

CRUCIBLE. A ceramic pot or receptical made of material such as graphite or silicon carbide, with relatively high thermal conductivity, bonded with clay or carbon, and used in melting metals; sometimes applied to pots made of cast iron, steel, or wrought steel. The zone in the cupola between the bottom and the tuyere is also known as the crucible zone. The name is derived from the cross (crux) with which ancient alchemists adorned it.

CRUCIBLE FURNACE. A furnace fired with coke, oil, gas, or electricity in which metals are melted in a refractory crucible.

CUPOLA. A cylindrical shaft furnace lined with refractories for melting metal in direct contact with the fuel by forcing air under pressure through openings near its base.

Cupola Drop. The sand bottom, bed, and unmelted charges dropped from the cupola at the end of a heat.

Curing Time (no bake). That period of time needed before a sand mass reaches maximum hardness and rigidity.

Cutoff Machine, Abrasive. A device which uses a thin, bonded abrasive wheel rotating at high speed to cut off gates and risers from castings.

Cuts. Defects in a casting resulting from erosion of the sand by metal flowing over the mold or cored surface.

Cyclone (centrifugal collector). In air pollution control, a controlled descending vortex created to spiral objectionable gases and dust to the bottom of a collector cone.

Decarburization. Loss of carbon from the surface of a ferrous alloy as a result of heating in a medium, usually including oxygen, which reacts with carbon.

Deformation Test. An AFS sand property test which is determined on an instrument such as the Dietest Universal Sand Strength Testing Machine. The test determines the amount of deformation in thousandths of an inch that occurs before a sand specimen ruptures.

Degasser. A material for removing gases from molten metals and alloys (usually nonferrous metals).

Degree of Ramming. A measure of the amount of compaction which has occurred in a molding operation. (See RAMMING.)

DeLavaud Process. A centrifugal process employed chiefly for making cast iron pipe.

Deoxidation. Removal of dissolved oxygen from molten ferrous alloys, usually accomplished by adding materials with a high affinity for oxygen, the oxides of which are either gaseous or which readily form slags.

Dewaxing. The process of melting out the expendable wax pattern from an investment mold by heating, usually at temperatures less than 121°C (250°F).

Diatomaceous Earth. A hydrous form of silica which is soft, light in weight and consists mainly of microscopic shells of diatoms or other marine organisms. Widely used for furnace insulation.

Die casting. A high production casting process in which the molten metal (usually a nonferrous alloy) is forced under greater than atmospheric pressure into a water cooled metal mold cavity.

Dielectric Oven. A rapid curing high frequency electric oven used to cure cores which are bonded with non-conducting materials.

Dip Coat. In solid and shell mold investment casting, a fine ceramic coating applied as a slurry to the pattern to produce maximum surface smoothness, followed by a cheaper investment material.

Direct-arc Furnace. An electric arc furnace in which the metal being melted is one of the poles.

Directional Solidification. The solidification of molten metal in a casting in such a manner that feed metal is always available for that portion that is just solidifying.

Dispersed Shrinkage. Small shrinkage cavities dispersed through a casting, which are not necessarily a cause for rejection.

Distribution, Sand Grain. Variation or uniformity in the particle size which comprise sand aggregate when properly screened by U.S. Standard Sieves.

Downcomer. In air pollution control, a pipe for conducting gases down into a conditioner for subsequent cleaning.

Drag. Lower or bottom section of a mold or pattern. Originally called a nowel.

Draw. (*noun*) A term sometimes used to denote a shrink appearing on the surface of a casting or a riser.

Draw. (*verb*) To remove the cope or drag from a pattern plate or in the pattern from the cope or drag.

Draw Bar. A bar used for lifting the pattern from the sand mold.

Draw Plate. A plate attached to a pattern to facilitate drawing of the pattern from the mold.

Dried Sand. Sand which has been dried by a mechanical drier prior to its use in core making.

Drop. A casting defect caused by dropping of sand from the cope or other overhanging section, which results in a casting defect known as a drop after the mold has been poured.

Dross. Metal oxides in or on the surface of molten metal.

Dry Permeability. The property of a molded mass of bonded or unbonded sand, dried at (104° to 110°C) (220° to 230°F) and cooled to room temperature, that allows passage of gases resulting during pouring of molten metal into a mold.

Dry Sand Mold. A mold from which the moisture content has been removed by suitably heating the mold prior to pouring molten metal therein.

Dry Strength. The maximum strength of a molded sand specimen that has been thoroughly dried at 220°-230°F (104°-110°C) and cooled to room temperature. Also known as dry bond strength.

Ductility. The property permitting permanent deformation without rupture in a material by stress in tension.

Duplexing. A method of producing molten metal of desired analysis, the metal being melted in one furnace and refined in a second.

Durvill Process. A casting process that involves a rigid attachment of the mold in an inverted position above the crucible. The metal is poured by tilting the entire assembly, causing the metal to flow along a connecting laader and down the side of the mold.

Dust. Small solid particles created by breaking up larger particles by any process.

Ejector Marks. Marks left on castings by ejector pins.

Electric Furnace. A furnace for industrial purposes, either melting or heat-treating, in which the heat source is an electric current.

Erosion Scab. A casting defect which occurs when the molten metal has been agitated, boiled, or has partially eroded the sand in the mold, leaving a solid mass of sand and metal at that particular spot.

Exothermic Riser Sleeve. A riser sleeve which is made of a material which liberates heat upon contact with molten alloy. Riser toppings are frequently composed of similar materials.

Expansion Scab. Rough thin layer of metal partially separated from the body of the casting by a thin layer of sand, and held in place by a thin vein of metal, usually resulting in an indentation in the casting, caused by spalling of the mold face.

Expendable Pattern. A pattern that is destroyed in making a casting. It is usually made of wax or foamed plastic.

Facing, Facing Material. Coating material applied to the surface of a mold to protect the sand from the heat of the molten metal; also to impart smooth surface to casting.

Feeder, Feeder Head. A reservoir of molten metal to compensate for the contraction of metal as it solidifies. Molten metal flowing from the feed head, also known as a riser, prevents voids in casting.

Feeder, Sand. A device for discharging a uniform thickness or volume of sand onto a belt or other conveying equipment to maintain uniform delivery.

Feeding. Pourning additional molten metal into a freshly poured mold to compensate for volume shrinkage while the casting is solidifying. Also the continuous supply of molten metal, as from a riser, to the solidifying metal in the casting. Also refers to keeping risers open by manipulation of feeding rods.

Ferritic Steels. Steels in which ferrite is the predominate phase. These steels are magnetic.

Ferritic Matrix. In iron castings, the term refers to cast irons that have a large percentage of ferrite in the microstructures.

Fettle. A British term that refers to the process of removing all runners and risers and cleaning off adhering sand from the casting surface.

Fillet. Concave corner piece usually used at the intersection of right-angle surfaces (that would otherwise meet at an angle) on pattens and core boxes. A struck fillet is one that is dressed to shape in place, usually of wax. A planted fillet is one made separately and affixed in place. Fillets used at recentrant angles in cast shapes lessen the danger of cracks and aviod "fillet shrinkages."

Fin. A thin projection of metal from the casting, formed as a result of imperfect mold or core joints.

Fineness, Sand. The extent of subdivision of a foundry sand, as determined by the AFS fineness test.

Fines. A term the meaning of which varies with the type of foundry or the type of work. It refers to those sand grain sizes substantially smaller than the predominating grain size.

FINISH ALLOWANCE. Amount of stock left on the surface of a casting for machine finish.

FIRECLAY. A clay with a high fusion temperature.

FIRE SAND. A refractory sand which resists high temperatures.

FLASK. Metal or wood frame without top and without fixed bottom used to hold the sand of which a mold is formed; usually consists of two parts, cope and drag.

FLASKLESS MOLDING. Sand is blown into a slip flask and then it is squeezed so firmly around the pattern that it is difficult to strip the mold without a special stripping station. Once made the mold is so strong that no flasks are needed. The making time is about 5 seconds per cycle. With a shuttle, twice that speed can be achieved.

FLOWABILITY. The property of a foundry sand mixture which enables it to fill pattern recesses and move in any direction against pattern surfaces under pressure.

FLUIDITY. The ability of molten metal to flow readily as measured by the length of standard spiral casting.

FOUNDING. The science of melting and casting of metals into useful objects to serve the needs of man and industry.

FOUNDRY. A building, establishment, or works where metal castings are produced.

FOUNDRY FACING. Material, usually carbonaceous, applied to the surface of a sand mold to prevent the molten metal from penetrating and reacting with the molding sand.

FOUNDRYMAN. Craftsman employed in the production of metal castings.

FOUNDRY RETURNS. Metal in the form of gates, sprues, runners, risers, and scrapped castings returned to the furnace for remelting.

FREEZING RANGE. That range of temperature between liquidus and solidus temperatures in which molten and solid constituents coexist.

FRONT SLAGGING. A process wherein both slag and molten metal flow out through the taphole. The slag is then skimmed off the surface of the molten metal.

GAGGERS. Metal pieces of irregular shape used to reinforce and support sand in deep pockets of molds.

GAS HOLES. Rounded cavities, either spherical, flattened, or elongated, in a casting, caused by the generation and/or accumulation of gas or entrapped air during solidification of the casting.

GATE. End of the runner in a mold where molten metal enters the casting or mold cavity; sometimes applied to entire assembly of connected channels, to the pattern parts which form them, or to the metal which fills them, and sometimes is restricted to mean the first or main channel.

GATE, RUNNER. A horizontal channel for running metal into the mold cavity.

GATE, SLOT. A gate used on vertical cylindrical castings in which the down sprue and casting are connected over a large part or all of the height of the casting.

GATE STRAINER. A gate designed to prevent slag and dirt from entering the mold and also to control the rate at which metal enters the mold cavity.

GATING SYSTEM. The complete assembly of sprues, runners, gates and individual casting cavities in the mold. Term also applies to similar portions of master patterns, pattern die, patterns, investment mold, and the finished casting.

GOOSENECK. Spout connecting a metal pot or chamber with a nozzle or sprue hole in the die and containing a passage through which molten metal is forced on its way to the die. It is the metal-injection mechanism in a hot-chamber type of die-casting machine.

GRAIN FINENESS NUMBER (AFS). Approximately the number of mesh per inch of that sieve which would just pass the sample if its grains were of a uniform size. That is, the average of the sizes of grains in the sample. It is approximately proportional to the surface area per unit weight of sand, exclusive of clay.

GRAY CAST IRON. Cast iron which contains a relatively large percentage of its carbon in the form of graphite, and substantially all of the remainder of the carbon in the form of eutectoid carbide. Such material has a gray fracture.

GREEN CASTING. A casting in the as-cast unheat-treated, or unaged condition.

Green Permeability. The ability of a molded body of sand in a tempered condition to permit passage of gases through its mass.

Green Sand. A naturally-bonded sand or compounded molding sand mixture which has been tempered with water for use while still in the damp or wet condition.

Green Strength. Strength of a tempered sand mixture at room temperature.

Heap Sand. Also referred to as system sand or unit sand; usually regarded as sand "heaped" on the foundry floor after it has been reclaimed and placed in "heaps" for reuse; chiefly used as "backing sand."

Heat. A stated tonnage of metal obtained from a period of continuous melting in a cupola or furnace; or the melting period required to obtain that tonnage.

Heat Checking. Formation of fine cracks in a die surface due to alternate heating and cooling. These cracks are reflected in the surface of the casting.

Heel. Metal left in ladle after pouring has been completed, metal kept in channel type induction furnaces during stand-by periods.

High Pressure Molding. A mold which has been compacted by hydraulic squeezing at a pressure greater than 100 pounds per square inch on the pattern plate. The resulting mold must have a hardness greater than 85 on the AFS B scale meter.

Hot Blast. Blast which has been heated prior to entering into the combustion reaction.

Hot Box Process. A furan resin based core-making process similar to shell core-making. Cores are solid throughout unless a mandrel is used.

Hot-chamber Machines. Die-casting machines which have plunger or injection system in continuous contact with molten metal.

Hot Spots. Localized areas of a mold or casting where high temperatures are reached and maintained for a period of time.

Hot Strength. Strength of sand mix as determined at any temperature above room temperature.

Hot Tear. Surface discontinuity of fracture caused by either external loads or internal stresses or a combination of both acting on a casting during solidification, and subsequent contraction at temperatures near the solidus.

Ilite. A mineral, typically $KAl_3Si_3O_{10}(OH)_2$ found in many clays; large clay deposits of ilite are found in Illinois and Michigan.

Impregnation. The treatment of castings with a sealing medium to stop pressure leaks, such as soaking under pressure with or without prior evacuation but usually in heated baths. Media include sodium silicate solutions, drying oils or styrenes, plastics or propietary compounds.

Inclusions. Particles of impurities (usually oxides, sulphides, silicates and such) that are held mechanically, or are formed during solidification or by subsequent reaction within the solid metal.

Inconel. Any of a series of oxidation resistant alloys usually containing up to 80% or more nickel, 8% to 14% chromium and 5% or so of iron.

Indirect-arc Furnace. An electric-arc furnace in which the metal bath is not one of the poles of the arc.

Induction Furnace. A melting furnace which uses the heat developed by an electrical induction coil.

Ingot. A mass of metal cast in a heavy cast iron permanent mold. The ingot is cast in a convenient size and shape for remelting or hot working operations.

Injection. The process of forcing molten metal into a die.

Inoculant. Materials which when added to molten metal modify the structure, and thereby change the physical and mechanical properties to a degree not explained on the basis of the change in composition resulting from their use.

Insulating Pads and Sleeves. Insulating material such as gypsum, diatomaceous earth, etc., used to lower the rate of solidification. As riser sleeves and topping they keep the riser liquid thus increasing feeding efficiency.

Internal Shrinkage. A void or network of voids within a casting caused by improper feeding of that section during solidification.

Inverse Chill. The condition in a casting section where the interior is mottled while the outer sections are gray iron.

Investing. The process of pouring the investment slurry into a flask surrounding the pattern to form the mold.

Investment. A flowable mixture of a graded refractory filler, a binder and liquid vehicle which when poured around the patterns conforms to their shape and subsequently set hard to form the investment mold.

Investment Casting. The process of casting metal into an investment mold.

Iron, Hard or White. Irons possessing white fractures because substantially all of the carbon is in the combined form. Irons to be malleabilized are cast white, as are many abrasion-resistant irons.

Iron, Malleable. A mixture of iron and carbon, including smaller amounts of silicon, manganese, phosphorus, and sulphur, which after being cast (white iron, carbon in combined form as carbides) is converted structurally by heat-treatment into a matrix of ferrite containing nodules of temper carbon.

Iron, Pearlitic. A cast iron (gray ductile or malleable) having a more or less pearlitic matrix.

Iron Oxide. A core sand additive to increase the high temperature and penetration resistance of core sand mixes. It contains about 85% of pulverized ore.

Iserine. A black sand which consists mainly of magnetic iron ore, but which also contains a considerable amount of titanium.

Izod Test. A pendulum type impact test, in which the specimen is supported at one end as a cantilever beam so the energy required to break the specimen is a measure of impact strength.

Jacket, Mold. Wooden or metal form which is slipped over a mold made in a snap or slip flask, to support the four sides of the mold during pouring; maintains alignment of cope and drag halves of mold. After pouring jackets and mold, weights are shifted to another row of molds.

Jet Scrubber. In air pollution control, a high velocity water jet directed into the throat of a venture section of a cupola to separate out particulates.

Jobbing Foundry. A foundry engaged in the manufacture of numerous types of castings not intended for use in its own product. Usually refers to a foundry making castings for many other companies.

Jolt Squeeze Molding. Compaction of a sand mold which weighs 75 pounds or less by first using jolt action to compact the sand around the pattern and then using a squeezing action to complete the molding operation.

Kaolin. Fire clay which is the purest form of china clay which consists of a silicate of aluminum. The name is derived from a hill in China.

Knockout Pins (ejector pins). Pins of a small diameter affixed to die casting or permanent mold dies to eject the casting upon opening the die. Also used with shell pattern plates to strip the shell after curing.

Ladle. Metal receptacle frequently lined with refractories used for transporting and pouring molten metal. Types include hand, bull, crane, bottom-pour, holding, teapot, trolley, shank, lip-pour, buggy, truck, mixing reservoir.

Leakers. Foundry term for castings which fail to meet liquid or gas pressure tests.

Lifters. Metal tool for removing loose sand from cope or drag before closing mold. Also, a device which is attached to the cope, to hold the sand together when the cope is lifted.

Lining. Inside refractory layer or firebrick, clay, sand, or other material in a furnace or ladle.

Lining, Monolithic. A lining made without the customary layers and joints of a brick wall. Usually made by tamping or casting refractory material into place, drying and then burning in places on the job.

Linseed Oil. The most common of all drying oils and used in the foundry principally as a base for core oil. Produced by applying pressure to the seeds of flax, the liquid constituent being linseed oil.

Loam. A mixture of sand, silt, and clay particles in about equal portions; *i.e.* roughly 50% sand and 50% silt and sand.

Lost Wax Process. A pattern casting process in which a wax or thermoplastic pattern is used. The pattern is invested in a refractory slurry; after the mold is dry, the pattern is melted or burned out of the mold cavity.

Malleable Iron. (See iron, malleable.)

Master pattern. A pattern embodying a contraction allowance in its construction, used for making castings to be employed as patterns in production work.

METALLOID. An element intermediate between metals and nonmetals possessing both metallic and non-metallic properties, as arsenic.

METALLOSTATIC PRESSURE. The pressure developed within a molten metal while it is still liquid. The metal head can exert considerable lifting force on a cope particularly if the cope is deep and the metal is poured rapidly.

MELTING LOSS. Loss of metal in charge during the operation of melting, usually due to oxidation or volatilization.

MELTING POINT. The temperature at which a metal begins to liquify. Pure metals, eutectics and some intermediate phases, melt at a constant temperature. Alloys generally melt over a range of temperature.

MELTING RATIO. The proportion of the weight of metal to the weight of fuel used in melting.

METAL PENETRATION. A casting surface defect which appears as if the metal had filled the voids between the sand grains without displacing them.

MICROHARDNESS. The hardness of microconstituents of a material as determined by using a diamond indenter and a special machine equipped with a microscope.

MICROSTRUCTURE. The structure of polished and etched specimens of casting sections as revealed by a microscope at more than 90 diameters.

MISRUN. Casting not fully formed resulting from metal poured so cold that it is solidified before filling the mold completely.

MODIFICATION. A process in which the eutectic temperature, structure and composition of aluminum-silicon alloys are apparently altered by the addition of small amounts of a third element, such as sodium. A similar effect can be achieved by chill casting.

MOISTURE CONTENT. Amount of water contained in a substance that can be driven off by heating at 104°-110°C (220°-230°F).

MOISTURE TELLER. A patented apparatus for the rapid determination of the moisture content of molding sands using the reaction between calcium carbide and water to produce a gas pressure which is then measured.

MOLD. The form, made of sand, metal or any other investment material, which contains the cavity into which molten metal is poured to produce a casting of definite shape and outline.

MOLD COATING. Coating on sand molds to prevent metal penetration and to improve casting finish.

MOLD CAVITY. In a mold, the hole which, when filled with metal becomes the casting. Gates and risers are not considered part of the mold cavity.

MOLDABILITY. Ability of sand to flow into a flask and around a pattern, measured by the amount of sand falling through an inclined screen or slot.

MOLDING, MACHINE. May refer to squeezer or jolt-squeezer machines on which one operator makes the entire mold, or to similar or larger machines including jolt-squeeze-strippers, and jolt and jolt-rollover pattern draw machines on which the cope and drag halves of molds are made.

MOLDING SANDS. Sands containing over 5 per cent natural clay, usually between 8 and 20 per cent.

MOTTLED CAST IRON. Cast iron which consists of a mixture of variable proportions of gray cast iron and white cast iron; such material has a mottled fracture.

MULLER. A type of foundry sand mixing machine.

MULLING. Process of mixing sand and clay particles either by compressing with a heavy roller in preparation for molding or by forcing the sand mixture to flow between wheels mounted on a vertical axis and the side walls of the container. Also, the process of mixing sands with a rubbing action as well as stirring.

MULTIPLE-CAVITY DIE. A die having more than one duplicate impression.

MULTIPLE MOLD. A composite mold made up of stacked sections, each of which produces a complete gate of castings and poured from a central down-gate.

NATURAL SAND. One derived from a rock, in which the grains separate along their natural boundaries. This includes unconsolidated sand, or a soft sandstone where little pressure is required to separate the individual grains.

NI-HARD. Hard white cast iron containing 4% Ni and 2% Cr.

No Bake Binder. A synthetic liquid resin sand binder that hardens completely at room temperatures without baking. Used in the cold setting process.

Nodular Iron. Iron of a normally gray cast iron type that has been suitably treated with a nodularizing agent so that all or the major portion of its graphitic carbon has a nodular or spherulitic form as cast. It is malleable as cast.

Nondestructive Testing (nondestructive inspection). Testing or inspection that does not destroy the object being tested. These include radiography, magnetic particle testing, dye penetrant testing and ultrasonic testing.

Nozzle. Outlet end of a gooseneck or the tubular fitting which joins the gooseneck to the sprue hold.

Oil Core or Mold. A core or mold in which the sand is held together by an oil binder.

Oil-oxygen Binder (cold setting, air-setting binders). A synthetic auto-oxidizing liquid, oil-based binder that partially hardens at room temperature, using an oxygen releasing agent. Baking is needed to complete the curing cycle.

Olivine Sand. $(Mg_2 Fe)_2 SiO_4$, a naturally occurring mineral composed of fosterite and fayalite, crushed and used as a molding sand.

Open Hearth Furnace. A furnace for melting metal, in which the bath is heated by the combustion of hot gases over the surface of the metal and by radiation from the roof.

Optical Pyrometer. A temperature measuring device through which the operator sights the heated object and compares its incandescence with that of an electrically heated filament whose brightness can be regulated to match the intensity of the source. Proper calibration and practice are needed to read consistent temperatures to +-30°C.

Orifice Plate. In a cupola, a device used to measure the volume of air delivered by the blower.

Panoramic Analyzer. An instrument for detecting the level of sounds in a foundry to assure compliance with OSHA standards. The output is either on an oscilloscope or chart.

Parting Compound. A material dusted or sprayed on patterns or mold halves to prevent adherence of sand and to promote easy separation of cope and drag parting surfaces when cope is lifted from drag.

Pattern. A form of wood metal or other materials around which molding material is placed to make a mold for casting metals.

Pattern, Investment Molding. A reproduction with an expendable material of the object to be cast and usually formed in a pattern die.

Patternmaker's Shrinkage. Shrinkage allowance made on all patterns to compensate for the change in dimensions as the solidified casting cools in the mold from freezing temperature to room temperature. Patterns are made larger by the amount of shrinkage characteristic of the particular metal in the casting and the amount of resulting contraction to be encountered. Rules or scales are available for use. (See SHRINKAGE, PATTERNMAKER'S)

Pattern, Master. Pattern constructed with double shrinkage allowance from which production patterns are made.

Pearlitic Malleable. The product obtained by a heat treatment of white cast iron which converts some of the combined carbon into graphite nodules but which leaves a significant amount of combined carbon in the product. Pearlitic malleable may also be obtained through the use of additional alloys which retard the breakdown of pearlite.

Peen. (l) Flat-pointed end of a rammer used in ramming sand into a mold. (2) Process of repairing slight leaks in casting by repeated impacts of a blunt tool or peening hammer. (3) Peening action obtained by impact of metal shot.

Pencil Core. A core projecting to the center of a blind riser to admit atmospheric pressure to force out feed metal.

Pencil Gate. Gating directly into the mold cavity through the cope by means of one or more small vertical gates connecting the pouring basin and mold cavity.

Penetration, Metal. Condition where molten metal has penetrated into the face of a sand mold beyond the midpoint of the first layer of sand grains, resulting in a mixture of metal and sand adhering to the cast surface.

Perlite. A highly siliceous volcanic rock which can be expanded into a porous mass of particles by heating. Perlite can be used as an insulation in foundry sand mixes, or as a riser topping, or as sleeves in steel casting.

Permanent Mold. A metal mold of two or more parts that is used repeatedly for the production of many castings of the same form. Liquid metal is poured in by gravity. Not an ingot mold.

Permeability. As applied to sand molds, permeability means the property of the sand which permits passage of gases. The magnetic permeability of a substance is the ratio of the magnetic induction of the substance to the intensity of the magnetizing field to which it is subjected.

pH. A symbol denoting the negative logarithm of the concentration of the hydrogen ion in gram atoms per liter. Used to express both acidity and alkalinity. pH = log 1/H per liter. An important factor in foundry sand control where a pH = 7 is neutral, values less than 7 are acidic and greater than 7 are basic. (The symbol is derived from the French "pouvoir hydrogène" or "hydrogen power".)

Phenolic Resin. A resin made by polmerization of a phenol with an aldehyde. Used as a binder for cores and sand molds.

Pig Iron. Cast iron generally produced by the reduction of iron ore in the blast furnace; also the over-iron in the foundry poured into pig molds.

Pinhole Porosity. Very small holes scattered through a casting, possibly caused by microshrinkage or gas evolution during solidification.

Pit. A sharp depression in the surface of metal.

Plaster Molds. With plaster as the mold material, liquid plaster slurry is poured over the pattern, allowed to harden and the pattern is then removed. The hardened mold is then heated to drive off the moisture. Smooth finishes and close tolerances are the advantages. Aluminum and copper-base alloys are cast by this process.

Plaster of Paris. A semi-hydrated form of calcium sulphate made by sintering gypsum at 120°-130°C (248°-266°F).

Plumbago. Graphite in powdered form. Plumbago crucibles are made from this graphite plus clay.

Porosity (sand). The ratio of volume of the pores or voids to volume of the entire mass, usually expressed as a percentage. Generally the volume of pores is obtained by determining the amount of water or gas absorbed by the mass, and therefore does not include the volume of sealed pores. The result so obtained is "apparent" rather than "true" porosity. (Not synonymous with permeability.)

Port. Opening through which molten metal enters the injection cylinder of a plunger machine or is ladled into the injection cylinder of a cold-chamber machine.

Pouring. Transfer of molten metal from furnace to ladle, ladle to ladle, or ladle to molds.

Preformed Ceramic Core. A preformed refractory aggregate inserted in a wax or plastic pattern or shape (the interior of that part of a casting which cannot be shaped by the pattern). Sometimes the wax is injected around the preformed core.

Production Foundry. A foundry engaged in manufacture of castings in large quantities, usually highly mechanized to minimize manual labor. May be either jobbing or captive foundry.

Quartz. A form of silica which occurs in hexagonal crystals. It is the most common of all solid minerals.

Radiant Heat. Heat which is transmitted by electromagnetic waves.

Ramming. The operation of packing sand around the pattern in a flask to form a mold.

Rapping. Knocking or jarring the pattern to loosen it from the sand in the mold before withdrawing the pattern.

Release Agent (parting agent). A material such as silicone, stearate, oil or wax for lubricating a die, pattern or core box to facilitate easy removal of a casting, mold or core.

Remelt. Sprues, gates, riser, and defective castings returned directly to the crucible furnace.

Residual Stress. (See stress, residual.)

Respirator. A filtering device which covers the nose and mouth and prevents inhalation of dust or fumes; should have the U.S. Bureau of Mines certificate of approval for the specific contaminant being filtered out.

Ribs. Sections joining parts of a casting to impart greater rigidity.

Riddle. A screening device, manually or mechanically operated, for removing coarse particles and contamination from molding sand.

Riser. A reservoir of molten metal provided to compensate for the internal contraction of the casting as it solidifies.

Rolling Over. The operation of reversing the position of a flask in which the drag part of the pattern has been rammed with the parting surface downward.

Runner. A channel through which molten metal or slag is passed from one receptacle to another, in a mold, the portion of the gate assembly that connects the downgate or sprue with the casting ingate or riser. The term also applies to similar portions of master patterns, pattern dies, patterns, investment molds, and finished castings.

Runout. A casting defect caused by incomplete filling of the mold due to molten metal draining or leaking out of some part of the mold cavity during pouring; escape of molten metal from a furnace, mold, or melting crucible.

Sand. A loose, granular material resulting from the disintegration of rock. The name sand refers to the size of grain and not to mineral composition. Diameter of the individual grains can vary from approximately 6 to 270 mesh. Most foundry sands are made up principally of the mineral quartz (silica). The reason for this is that it is plentiful, refractory and cheap.

bank sand. Sedimentary deposits, usually containing less than 5% clay.

dune sand. Wind-blown deposits of sand found near large bodies of water.

molding sands. Sands which contain over 5% natural clay, usually between 10% and 20%.

silica sand. Although most foundry sands contain a high percentage of silica, the term silica sand is generally reserved for those that show a minimum of 95% silica content. Many high grade silica sands will analyze better than 99% pure silica.

miscellaneous sands. These include zircon, olivine, $CaCO_3$, black sands (lava grains), titanium minerals, etc.

Sand Blast. Sand driven by a blast of compressed air (or steam); used to clean castings; to cut, polish or decorate glass or other hard substances; for cleaning building fronts, etc.

Sand Control. Procedure whereby various properties of foundry sand, such as fineness, permeability, green strength, moisture content, etc., are adjusted to obtain castings free from blows, scabs, veins, and similar defects.

Sand Inclusions. Sand which has loosened from the mold and become entrapped in the molten metal.

Sand Reclamation. Processing of used foundry sand by thermal or hydraulic methods, so that it may be used in place of new sand without substantially changing current foundry sand practice.

Sand Spun Process. A centrifugal casting technique for coating the mold with a thin layer of unbonded refractory parting material. Particularly useful for short lengths of cylindrical shapes.

Sand Toughness Number. The product of deformation, times green compressive strength, times 1000. It is an indication of the workability of the sand mixture, and is usually expressed STN.

Scaling (scale). Surface oxidation, partially adherent layers of corrosion products, left on metals by heating or casting in oxidizing atmospheres.

Scrap. Metal for remelting; includes scrapped machinery, rail, or structural steel and rejected castings.

Screen. A sieve or riddle with openings of a definite size. Sieves are used to separate the different fractions of a sand distribution or to remove agglomerated particles or lumps.

Seam. A surface defect on a casting related to but of lesser degree than a cold shut; a ridge on the surface of a casting caused by a crack in the mold face.

Segregation. (1) A sand problem which occurs during transportation in bulk or in sand silos or hoppers. It is difficult to reblend the various fractions into a homogeneous mix. (2) A casting defect which is evidenced by the concentration of alloying elements in specific regions usually as a result of the primary crystallization of one phase with the subsequent concentration of other elements in the remaining liquid.

Selfcuring Binder. A material used in core making that

sets up to form cured cores without heat or additional assistance.

SEPARATOR. A mechanical device which separates or grades materials into constituent parts, used in the foundry to remove fines from system sands and dust from the air.

SHAKE-OUT. The operation of removing castings from the mold. A mechanical unit for separating the molding material from the solidified metal casting.

SHAW PROCESS. A precision casting process which uses ceramic molds in five stages and using conventional metal patterns. The patented mold material microcrazes to produce sufficient permeability to pour castings weighing up to 1500 pounds (682 kg).

SHELL CORE PROCESS. Resin-coated sand is blown into a heated core box. The sand against the box hardens. The balance of the sand is drained out to make a hollow core.

SHELL MOLDING. A process for forming a mold from resin-bonded sand mixtures brought in contact with pre-heated (300°-500°F; 150°-260°C) metal patterns, resulting in a firm shell with cavity corresponding to the outline of the pattern. Also called Croning process.

SHOT BLASTING. A casting cleaning process which uses a metal abrasive (grit or shot) propelled by centrifugal or pneumatic force.

SHRINKAGE. Change in size as the metal passes from the fluid to the solid state in the mold.

SHRINKAGE, PATTERNMAKER'S. A scale divided in excess of standard measurement to allow for the difference in size between the casting and the corresponding mold cavity. Used by patternmakers to avoid calculations for shrinkage.(See PATTERNMAKER'S SHRINKAGE)

SIEVE ANALYSES. Determination of the partial size distribution of a molding sand expressed in terms of the weight retained on each of a series of standard sieves stacked according to the decreasing mesh size.

SILICA. Silicon dioxide, the prime ingredient of sharp sand and acid refractories.

SILICA BRICK. Refractory material of Ganister bonded with hydrated lime and fired at high temperature.

SILICA FLOUR. Material commonly produced by pulverizing pure grains of quartz sand in large ball mills. It is available in several mesh sizes, generally from 80 to 325 mesh. In analysis, it is 99% pure silica, and is commonly used in foundry practice for improving finish and increasing hot strength. It is also used in production of core and mold washes.

SILT. Very fine particles that pass a No. 270 mesh sieve, but which are not plastic or sticky when wet. (Between -50 and + 5 microns in size.)

SINTERING. The bonding of adjacent surfaces of particles or a compact by heating to a suitable temperature and cooling.

SLAG. A nonmetallic covering which forms on the molten metal as a result of the flux action in combining impurities contained in the original charge, some ash from the fuel, and any silica and clay eroded from the refractory lining. It is skimmed off prior to tapping the heat.

SLICK, SLICKER, SMOOTHER. A tool used for mending and smoothing the surfaces of a mold after withdrawal of pattern and before closing mold.

SLIP CASTING. In ceramics, the pouring of slip (water suspension of finely ground clay) into a plaster of Paris mold. After hardening, it is dried and fired.

SLURRY. A term loosely applied to any clay-like dispersion. It may be used to wash ladles or other refractory linings to impart a smoother surface, as a bonding addition to molding sand; as a thin loam over specially made molds; or as a mixture to fill in joints or cracks of a core, etc.

SLUSH CASTING. A casting made from an alloy that has a low melting point and freezes within a wide range of temperatures. The metal is poured into the mold, and brought into contact with all surfaces so as to form a shell of frozen metal; then the excess metal is poured out. Castings that consist of completely enclosed shells may be made by using a definite quantity of metal and a closed mold.

SNAGGING. The process of rough cleaning castings by grinding.

SOLIDIFICATION. The physical process of change from a liquid to a solid state.

SOLIDIFICATION SHRINKAGE. The decrease in size accompanying the freezing of a molten metal.

Sonic Testing. Using sound waves above audible frequency via a supersonic reflectoscope to measure the time sound waves take as they return from opposite sides of the casting. Defects return the waves in more, or less time.

Spark Test. A method of determining the approximate composition of steel by producing sparks on a grinding wheel.

Spiegeleisen. An alloy of iron and manganese used in basic and acid open hearth steel-making practices. Also can be used in cupola charges. A high manganese pig iron containing 15% to 20% Mn and 4.5% to 6.5% carbon.

Spiral Fluidity Test. A quantitative test for determining the fluidity of an alloy by pouring molten metal into a mold with a long narrow spiral channel. The fluidity index is determined by the length of flow which the alloy achieved.

Splash Core. A core or tile placed in a mold to prevent erosion of the mold at places where metal impinges with more than normal force. Splash cores are commonly used at the bottom of large rammed pouring basins, at the bottom of long downsprues, or at the ingates of large molds.

Sprue. The vertical channel connecting the pouring basin with the skimming gate, if any, and the runner to the mold cavity, all of which together may be called the gate. In top-poured castings the sprue may also act as riser. Sometimes used as a generic term to cover all gates, risers, etc. returned to the melting unit for re-melting. Also applies to similar portions of master patterns, pattern dies, patterns, investment molds, and the finished castings.

Squeeze Board. A board used on the cope half of the mold to permit squeezing it on a jolt squeeze machine.

Standard Pattern. A pattern of high-grade material and workmanship in daily use or used at frequent intervals. A pattern used as a master to make or check production patterns.

Sticker. A lump on the surface of a casting caused by a portion of the mold face sticking to the pattern. Also, a forming tool used in molding.

Streamline Flow. Steady flow of liquid metal with a minimum of changes in direction and subsequent eddying. True streamlined flow cannot be achieved in metal casting but disruptive turbulence can be eliminated.

Stress, Residual. Those stresses set up in a metal as a result of non-uniform plastic deformation or the unequal cooling of a casting.

Styrofoam Pattern. Expendable pattern of foamed plastic, especially polystyrene used in producing castings by the full mold process. Especially useful for large automotive body dies or for art sculpture castings.

Superheat. Any increment of temperature above the melting point of a metal or alloy.

Sweep. A template cut to the profile of the desired mold shape, which when revolved around a stake or spindle, produces that mold.

Swell. A casting defect consisting of an increase in metal section due to the displacement of sand by metal pressure.

Swing Frame Grinder. A device for grinding large castings which must remain stationary. The grinder is suspended by a hoist because of its weight.

Synthetic Molding Sand. Any sand compounded from selected individual materials which, when mixed together, produce a mixture of the proper physical and mechanical properties from which to make foundry molds.

Talc. Hydrated magnesium silicate; it is soft, greasy to the touch and very pliable; widely used in parting compounds, core compounds, and in core paste; often called red talc or yellow talc.

Temper. (*noun*) The moisture content of a sand at which any certain physical test value is obtained, *i.e.*, temper with respect to green compressive strength, permeability, retained compressive strength, etc.

Temper. (*verb*) Mixing sand with sufficient water or other liquid to develop desired molding properties.

Temper Carbon. The free or graphitic carbon that precipitates from solution, usually in the form of rounded or equiaxed nodules in the structure, during the graphitizing or malleabilizing of white cast iron.

Thixotrophy. In sand testing, the proportion of the volumes of liquid and solid matters which is required, one minute after shaking, for the suspension to just flow out of an 8-mm diameter test tube, when the test tube is changed from a vertical to a horizontal position.

Tumbling. Cleaning castings by rotating them in a container in the presence of cleaning materials.

Tuyere. An opening in the cupola shell and refractory lining through which the air blast is forced.

Ultrasonic Cleaning. Immersion cleaning aided by ultrasonic waves which cause microagitation.

Unkilled Steel. A mild steel insufficiently deoxidized so that it evolves gas and blowholes during solidification.

Urea Formaldehyde Resin. A thermosetting product of condensation from urea or thiourea and formaldehyde, soluble in water and used as a sand binder in core and mold compounds.

"V" Process. A Japanese patented process to produce molds using only unbonded sand. The vented pattern plate is covered with a heated plastic film and a mild vacuum draws the film close to the pattern, the flask is located on the plate and loosely bonded sand is added to fill the flask. The plate is rapped to achieve compaction and a second sheet of plastic is placed over the top of the cope. Next a vacuum is drawn through a special port in the flask side. Now the vacuum on the pattern plate is released and the mold is stripped and ready to be assembled to a drag. The mold is poured and the vacuum is released. Clean up and recycling of the sand are simple routines.

Vacuum Casting. A casting process in which metal is melted and poured under very low atmospheric pressure; a form of permanent mold casting where the mold is inserted into liquid metal, vacuum is applied and metal drawn up into the cavity.

Veining. A defect on the surface of a casting appearing as veins or wrinkles and associated with excessive thermal movement of the sand, especially core sands.

Vent. A small opening or passage in a mold or core to facilitate escape of gases when the mold is poured.

Vent Wax. Wire or rod shapes of wax placed during making of the core. During baking the wax is melted, leaving an opening through which gases may escape.

Vibrator. A device, operated by compressed air or electricity, for loosening and withdrawing patterns from a mold, or for vibrating a hopper or chute to promote the flow of material from the hopper or chute.

Virgin Metal (primary metal). Metal extracted directly from the ore; not previously used.

Viscosity. The resistance of a fluid substance to a quantitative measure flow. A characteristic for an individual substance at a given temperature.

Void. A shrinkage cavity produced in castings during solidification.

Warpage. Deformation other than contraction that develops in a casting between solidification and room temperature; also, distortion occurring during annealing, stress relieving, and high-temperature service.

Wash. A casting defect resulting from erosion of sand by metal flowing over the mold or cored surfaces. They appear as rough spots and excess metal on the casting surface. Also called cuts.

Washburn Core. A thin core which constricts the riser at the point of attachment to the casting. The thin core heats quickly and promotes feeding of the casting. Riser removal cost is minimized.

Water Glass. Sodium silicate, a viscous liquid which, when mixed with powdered fireclay, forms a refractory cement; used in CO_2 molding.

Wax Pattern. (1) A precise duplicate, allowing for shrinkage, of the casting and required gates, usually formed by pouring or injecting molten wax into a die or mold. (2) Wax molded around the parts to be welded by a thermit welding process.

Wind Box. The chamber surrounding a cupola through which air is conducted under pressure to the tuyeres.

X-ray. Form of radiant energy with wavelength shorter than that of visible light and with the ability to penetrate materials that absorb or reflect ordinary light. X-rays are usually produced by bombarding a metallic target with electrons in a high vacuum. In nuclear reactions it is customary to refer to photons originating in the nucleus as gamma rays and to those originating in the extranuclear part of the atom as x-rays.

Yield. Comparison of casting weight to total weight of metal poured into mold.

YIELD STRENGTH. The stress at which a material exhibits a specified limiting permanent strain.

YOUNG'S MODULUS (E). The Modulus of Elasticity for pure tension with no other stress activity; it is, in most metals, practically the same in compression. Denoted by E it is the tensile (or compressive) stress per unit of linear strain or tensile stress intensity/tensile strain.

ZIRCON. The mineral zircon silicate, $ZrSiO_4$, a very high melting point acid refractory material used as a molding material in steel foundries.

ZIRCONIA ZrO_2. An acid refractory up to 2,500°C (4,532°F) having good thermal shock resistance and low electrical resistivity.

ZIRCON SAND. A very refractory sand of unusual fineness, low thermal expansion and high thermal conductivity, exceedingly useful and efficient in foundry practice. It consists principally of zirconium silicate and is widely used for improving finish and preventing "burning-in" of sand on the casting surface.

METAL FORMING (See also: FORGING.)

ADJUSTABLE BED. The bed or table of a gap-frame press, such as a horn press, which is bolted to the vertical front surface of the press. It is supported and adjusted (up and down) by means of a screw or screws usually operated by hand. This term also refers to the bed of a large straight-side press mounted and guided in the press frame and provided with a suitable mechanism, usually power operated, for varying the die space shut height. Adjustable-bed presses are also referred to as knee-type presses.

ADJUSTABLE STROKE. The capability of varying the length of a stroke on a press.

ADJUSTMENT SLIDE. The distance that a press slide position can be altered to change the shutheight of the die space. The adjustment may be by hand or by power mechanism.

AGE HARDENING. A process of aging that increases hardness and strength and ordinarily decreases ductility. Age hardening usually follows rapid cooling or cold working.

AIR-HARDENING STEEL. An alloy steel that is hardened by cooling in air from a temperature higher than the transformation range. Also called self-hardening steel.

ALLOY. A substance that has metallic properties and is composed of two or more chemical elements of which at least one is a metal.

ALPHA IRON. The form of iron that is stable below 910°C, (1,970°F) and characterized by a body-centered cubic crystal structure.

ANNEALING. A heat treatment designed to effect: softening of a cold worked structure by recrystallization, or grain growth, or both; softening of an age hardened alloy by causing a nearly complete precipitation of the second phase in relatively coarse form; softening of certain age hardenable alloys by dissolving the second phase and cooling rapidly enough to obtain a supersaturated solution (this usage is generally applied to nickel-base and copper-base alloys the treatment should be called more precisely a solution heat treatment); relief of residual stress and change in properties induced during the forming process.

AUSTEMPERING. A trade name for a patented heat treating process that consists of quenching a ferrous alloy from a temperature above the transformation range in a medium with a heat abstraction rate sufficiently high to prevent the formation of high-temperature transformation products, and maintaining the alloy, until transformation is complete, at a temperature below that of pearlite formation and above that of martensite formation.

AUSTENITE. A solid solution in which gamma iron is the solvent. Characterized by a face-centered cubic crystal structure.

AUTOMATIC PRESS STOP. A machine-generated signal for stopping the action of a press, usually after a complete cycle, by disengaging the clutch mechanism and engaging the brake mechanism.

BACK-GAGE. A surface on two or more supports located behind the shear that can be positioned accurately either manually or automatically to control part size.

BED, PRESS. The stationary part of the press serving as a table to which is affixed the bolster, or sometimes, the lower die directly.

BEAD. A narrow ridge is a sheet metal workpiece or part, commonly formed for reinforcement.

BENDING. The straining of material, usually flat sheet or strip metal, by moving it around a straight axis which lies in the neutral plane. Metal flow takes place within the plastic range of the metal, so that the bent part retains a permanent set after removal of the applied stress. The cross section of the bend inward from the neutral plane is in compression; the rest of the bend is in tension.

BENDING BRAKE OR PRESS BRAKE. A form of open-frame, single-action press comparatively wide between the housings, with bed designed for holding long, narrow forming edges or dies. It is used for bending and forming strips and plates, as well as sheets (made into boxes, panels, roof decks, etc.).

BENDING DIES. Dies used in presses for bending sheet metal or wire parts into various shapes. The work is done by the punch pushing the stock into cavities or depressions of similar shape in the die or by auxiliary attachments operated by the descending punch.

BENDING ROLLS. Staggered rolls (usually three) adjusted to put the desired curvature in a plate or used for coiling and uncoiling strip or wire.

BENDING STRESS. A stress involving both tensile and compressive forces which are not uniformly distributed. Its maximum value depends on the amount of flexure that a given application can accommodate. Resistance to bending may be called "stiffness." It is a function of the modulus of elasticity and, for any metal, is not affected by alloying or heat treatment.

BEND RADIUS. The radius corresponding to the curvature of a bent specimen or bent area of a formed part, and measured on the inside of a bend.

BEND TESTS. Various tests used to determine the ductility of a sheet or plate that is subjected to bending. These tests may include determination of the minimum radius or diameter required to make a satisfactory bend and the number of repeated bends that the material can withstand without failure when it is bent through a given angle and over a definite radius.

BLADE. A replaceable tool having one or more cutting edges.

BLANK. A pressed presintered or fully sintered compact, usually in the unfinished condition and requiring cutting, machining, or some other operation to produce the final shape.

BLANKHOLDER. The tool that prevents the rim of a sheet metal blank from wrinkling while it is being deep drawn.

BLANKHOLDER SLIDE. The outer slide of a multiple-action press. It is usually operated by toggles or cams.

BLANKING. Shearing out a piece of sheet metal in preparation for deep drawing.

BLANKING DIE. A die used for shearing or cutting blanks usually from sheets or strips. The single blanking die used for producing one blank at each stroke of the press is the simplest of all dies, consisting essentially of punch, die block, and stripper.

BLUE BRITTLENESS. Reduced ductility occurring as a result of strain aging, when certain ferrous alloys are worked between 300° and 700°F. This phenomenon may be observed at the working temperature or subsequently at lower temperatures.

BOARD HAMMER. A type of drop hammer in which boards attached to the ram are lifted between rollers.

BOLSTER PLATE. A plate attached to the top of the press bed for locating and supporting the die assembly. It usually has holes or T-slots for attaching the lower die or die shoe. Moving bolster plates are self powered for transferring dies in and out of the press for die setting. Also called rolling bolsters, they may be integral with or mounted to a carriage. They are not to be confused with sliding bolsters, the purpose of which is moving the lower die in and out of the press for workpiece feeding.

BOTTOMING BENDING. Press brake bending process in which the upper die (punch) enters the lower die and coins or sets the material to eliminate springback.

BOW. The tendency of material being sheared to curl downward during shearing, particularly when shearing long narrow strips.

BRAKE. A piece of equipment used for bending sheet; also called a "bar folder." If operated manually, it is called a "hand brake," if power driven it is called a "press brake."

BRALE. A diamond penetrator, conical in shape, used with a Rockwell hardness test for hard metals.

BUCKLING. A bulge, bend, kink, or other wavy condition of the workpiece caused by compressive stress.

BUFF. A polishing wheel usually consisting of a large number of treated or untreated muslin disks sewed together.

BULGING. The process of increasing the diameter of a cylindrical shell (usually to a spherical shape) or of expanding the outer walls of any shell or box shape whose walls were previously straight.

BULL BLOCK. A power-driven reel for drawing heavy-gage wire through a die.

BULLDOZER. Slow-acting horizontal mechanical press with large bed used for bending, straightening, etc. The work, which is done between dies, may be performed either hot or cold. The machine is closely allied to a forging machine.

BURNISHING. (1) Plastic smearing such as may occur on metallic surfaces during buffing. (2) The result of the moveable blade rubbing against the edge of the sheared material due to a blade clearance that is adjusted too tight.

BURNT. A term applied to a metal permanently damaged by having been heated to a temperature close to the melting point.

CAMBER. A slight convexity or rounding of sheet, strip, or plate as might appear along the edge.

CAPACITY, PRESS. The rated force that a press is designed to exert at a predetermined distance above the bottom of the stroke of the slide.

CARBONITRIDING. A process in which a ferrous alloy is case hardened by first being heated in a gaseous atmosphere of such composition that the alloy absorbs carbon and nitrogen simultaneously, and then being cooled at a rate that will produce desired properties.

CARBON STEEL. Steel that owes its properties chiefly to the presence of carbon without substantial amounts of other alloying elements; also termed "ordinary steel," "straight carbon steel," "plain carbon steel."

CARBURIZING. A process that introduces carbon into a solid ferrous alloy by heating the metal in contact with a carbonaceous material — solid, liquid, or gas — to a temperature above the transformation range and holding at that temperature.

CASE HARDENING. A process of hardening a ferrous alloy so that the surface layer or case is made substantially harder than the interior or core.

CHARPY TEST. A pendulum type of impact test in which a specimen, supported at both ends as a simple beam, is broken by the impact of the falling pendulum.

CIRCLE-GRID ANALYSIS. The analysis of deformed circles to determine the severity with which a sheet metal blank has been stretched.

CLOSED DIES. Forging dies in which the compressive force is applied to the whole surface of the forging. In open dies, there is no constraint to lateral flow in some directions.

CLOSED HEIGHT. (See SHUT HEIGHT.)

CLUTCH. An assembly connecting the flywheel to the crankshaft, directly or through a gear train; when engaged, it imparts motion to the mechanical power press brake ram.

COINING. A process of impressing images or characters of the die and punch onto a plane metal surface.

COLD WORK. Plastic deformation at such temperatures and rates that substantial increases occur in the strength and hardness of the metal. Visible structural changes include changes in grain shape and, in some instances, mechanical twinning or banding.

COLD WORKING. Deforming a metal plastically at such a temperature and rate that strain hardening occurs. The upper limit of temperature for this process is the recrystallization temperature.

COMPRESSIVE ULTIMATE STRENGTH. The maximum stress that a brittle material can withstand without fracturing when subjected to compression.

COMPRESSIVE YIELD STRENGTH. The maximum stress that a metal subjected to compression can withstand without a predefined amount of deformation.

CONTINUOUS OPERATION. Uninterrupted multiple strokes of the crosshead without intervening stops at the end of individual strokes.

CREEP. The flow or plastic deformation of metals that are held for long periods of time at stresses lower than the yield strength. Creep effect is particularly important when the temperature of stressing approaches the metal's recrystallization temperature.

CRIMPING. A forming operation used to set down, or close in, a seam.

CROWN. The upper part (head) of a press frame. On hydraulic presses, the crown usually contains the cylinder; on mechanical presses, the crown contains the drive mechanism.

Cupping. The breaking of wire with a cup fracture accompanied by very little reduction of area; observed during cold drawing. Also the forming of sheet into cuplike objects such as shells, by deep drawing.

Curling. Rounding the edge of sheet metal into a closed or partly closed loop.

Cushion, Die. An accessory for a press that provides a resistive force with motion required for some operations, such as blankholding, drawing, or redrawing, maintaining uniform pressure on a workpiece, and knocking out or stripping. Also called pads or jacks. Usually mounted in/under the press bed, they are also used in/on the slide.

Decarburization. The loss of carbon from the surface of a ferrous alloy as a result of heating in a medium that reacts with the carbon.

Deep Drawing. Characterized by production of a parallel-wall cup from a flat blank. The blank may be circular, rectangular, or of a more complex shape. The blank is drawn into the die cavity by action of a punch. Deformation is restricted to the flange areas of the blank. No deformation occurs under the bottom of the punch—the area of the blank that was originally within the die opening. As the punch forms the cut, the amount of material in the flange decreases. Also called cup drawing or radial drawing.

Deflection. The amount of the deviation from a straight line or plane when a force is applied to a press member. Generally used to specify allowable bending of bed, slide, or frame at rated capacity with load of predetermined distribution.

Deformation Limit. In drawing, the limit of deformation is reached when the load required to deform the flange becomes greater than the load-carrying capacity of the cut wall. The deformation limit (limiting drawing ratio, LDR) is defined as the ratio of the maximum blank diameter that can be drawn into a cup without failure to the diameter of the punch.

Die. Generic term used to denote the entire press tooling used to cut or form material. This word is also used to denote just the female half of the press tool.

Die-enclosure Guard. An enclosure attached to the die shoe or stripper, or both, in a fixed position.

Die Pad. A movable plate or pad in a female die, usually for part ejection by mechanical means, springs, or fluid cushions.

Die Set. The assembly of the upper and lower die shoes (punch and dieholders), which usually includes the guide pins, guide pin bushings, and heel blocks. This assembly, which takes many forms, shapes, and sizes is frequently purchased as a commercially available unit.

Diesetter. An individual who places dies in or removes dies from presses, and who makes the necessary adjustments to cause the tooling to function properly and safely.

Die Sinking. Forming or machining a depressed pattern in a die.

Die Shoes. The upper and lower plates or castings which make up a die set (punch and dieholder). Also a plate or block upon which a dieholder is mounted, functioning primarily as a base for the complete die assembly. It is bolted or clamped to the bolster plate or the face of the press side.

Dimpling. A process of cupping sheet so as to permit the use of rivets that have countersunk heads. Dimpling differs from counters in that it is a cupping operation and no metal is cut away after the rivet holes have been drilled.

Direct Extrusion. Extrusion through a die placed at the opposite end of the billet from the ram.

Dishing. The act of forming a large-radiused concave surface in a part.

Distortion. Any deviation from a desired contour or shape.

Double Seaming. The process of joining two edges of metal, each edge being flanged, curled, and crimped within the other.

Draft. (1) The amount of taper in the side walls of die impressions. (2) The taper given to the sides of a pattern to enable it to be withdrawn easily from the mold.

Draw Bead. An insert or riblike projection on the draw ring or hold-down surfaces that aids in controlling the rate of metal flow during draw operations. Draw beads are especially useful in controlling the rate of metal flow in irregular-shaped stampings. (See also BLANKHOLDER.)

Drawing. In general terms, drawing describes the operations used to produce cups, cones, boxes and shell-like parts. The sheet metal being worked wraps around the punch as it descends into the die cavity. Essentially, the metal is drawn or pulled from the edges into the cavity. Shallow drawing applies when the depth of the part is less than one-half of the part radius. Deepdrawn parts are deeper than one-half the part radius. the inner edge of the metal is drawn by the punch.

DRAWABILITY. (1) A quantitative measure of the maximum possible reduction in a drawing process. (2) Reduction in diameter from a blank to a deep-drawn shell of maximum depth.

DUCTILITY. The property that permits permanent deformation before fracture by stress in tension.

DWELL. A portion of the press cycle during which the movement of a member is zero or at least insignificant. Usually refers to the interval when the blankholder in a drawing operation is holding the blank while the punch is making the draw.

EAR. A wavy projection formed in the course of deep drawing as a result of directional properties or anisotropy of the sheet.

EARING. The formation of ears or scalloped edges around the top of a drawn shell, resulting from directional differences in the plastic-working properties of rolled metal with, across, and at angles to the direction of rolling.

EJECTOR. A mechanism for removing work or material from between the dies.

ELASTIC LIMIT. The maximum stress a metal can withstand without exhibiting a permanent deformation upon complete release of the stress. Since the elastic limit may be determined only by successively loading and unloading a test specimen, it is more practical to determine the stress at which Hooke's law (deformation is proportional to stress) no longer holds. It must be remembered that repeated loads which produce any degree of permanent deformation also produce strain-hardening effects in most metals, which in turn, increase the elastic range for load applications after the initial one. The point above which the ratio of stress to strain is no longer constant is called the "proportional limit." It is customary to accept the value of this point as the equivalent of the so-called "elastic limit."

ELONGATION. The amount of permanent extension in the vicinity of the fracture in the tension test; usually expressed as a percentage of the original gage length, such as 25% in 2 inches.

EMBOSSING. A process for producing raised or sunken designs in sheet material by means of male and female dies. Common examples are letters, ornamental pictures and stiffening ribs.

ENDURANCE LIMIT. The maximum stress that a metal can withstand without failure during a specified large number of cycles of stress. If the term is employed without qualification, the cycles of stress are usually such that they produce complete reversal of flexural stress.

ENERGY CURVE. A graphical representation to show available flywheel energy as a function of stroke rate on a variable-speed press.

ENGINEERING STRESS. The load per unit area necessary to elongate a specimen. Computation is based on original cross-sectional area.

EXTRUSION. Shaping metal into a chosen continuous form by forcing it through a die of appropriate shape.

FEEDS. Various devices that move stock or workpieces to, in, or from a die.

FERRITE. A solid solution in which alpha iron is the solvent, and which is characterized by a body-centered cubic crystal structure.

FINISHING TEMPERATURE. The temperature at which hot mechanical working of metals is completed.

FLANGE. A projecting rim or edge of a part, usually narrow and of approximately constant width for stiffening or fastening.

FLATTENING. Removing irregularities of a metal surface by a variety of methods, such as stretching, rolling, and/or roller leveling of sheet and strip.

FLYWHEEL. A heavy, rotating wheel, attached to a shaft, whose principal purpose is to store kinetic energy during the nonworking portion of the press cycle and to release energy during the working portion of the press cycle.

FOOT CONTROL. The foot-operated control mechanism (other than mechanical foot pedal) designed to control the movement of the ram on mechanical, hydraulic, or special-purpose power press brakes.

FORMING. In the context of "Sheet Metal Formability," the term forming covers all operations required to form a flat sheet into a part. These operations include deep drawing, stretching, bending, buckling, etc.

FORMING LIMIT DIAGRAM (FLD). A diagram describing the limits that sheet metal can be stretched under different conditions.

FRAME. The main structure of a press.

FRONT-GAGE. Same as back-gage except that the locating fingers or surfaces are in front of the shear bed blade.

FULL-REVOLUTION CLUTCH. A type of clutch that, when tripped, cannot be disengaged until the crosshead has completed a full cycle.

GAGE, BACK-GAGE. A bar or fingers (located behind the press brake) which can be positioned accurately and quickly so that a sheet inserted into the press brake for bending is positioned to make a bend at the desired point.

GALLING. The friction-induced roughness of two metal surfaces in direct sliding contact.

GAP FRAME. Frame with a cutout to allow slitting or notching.

GIBS. Guides or shoes which ensure the proper parallelism, squareness, and sliding fit between press components such as the slide and frame. They are usually adjustable to compensate for wear and to establish operating clearance.

GOOSENECK PUNCH. A punch that permits making deep, narrow channels, because its shape permits the flange to bend beyond the centerline of the ram. The upper die is relieved on one side past the centerline, to provide clearance for previously formed blanks.

GUARD. A barrier that physically prevents entry of the operator's hands or fingers into a point of operation. (Explanation: In ANSI Standard B11.1 and OSHA requirements, the use of the word guard is reserved exclusively for referring to physical barriers or enclosures designed for safeguarding at the point of operation.) In contrast, a device may be electronic in nature or of a restraining type.

GUERIN FORMING. A method of forming in which the metal is forced to conform to the shape of a male die by the application of a hydrostatic force of confined rubber.

HAMMERING. Beating metal sheet into a desired shape either over a form or on a high-speed mechanical hammer, in which the sheet is moved between a small curved hammer and a similar anvil to produce the required dishing or thinning.

HARDNESS. Defined in terms of the method of measurement. (1) Usually the resistance to indentation. (2) Stiffness or temper of wrought products.

HEAT TREATMENT. A combination of heating and cooling operations, timed and applied to a metal or alloy in the solid state in a way that will produce desired properties.

HEMMING. A bend of 180° made in two steps: First, a sharp-angle bend is made; next, the bend is closed by means of a flat punch and a die.

HOLE FLANGING. Turning up or drawing out a flange or rim around a hole (usually round) in the bottom or side of a shell or in a flat plate. It is essentially the reverse of necking.

HOOKER PROCESS. A process of cold extrusion in a crank press, employed in making small arms cartridge cases and small thin-walled seamless tubes.

HOT FORMING. Working operations such as bending and drawing sheet and plate, forging, pressing, and heading, performed on metal heated to a temperature above room temperature.

HOT PRESSING. The simultaneous forming and heating of a compact.

HOT WORKING. Plastic deformation of metal at such a temperature and rate that strain hardening does not occur. The lower limit of temperature for this process is the recrystallization temperature.

HYDRAULIC PRESS BRAKE. A press brake with the ram actuated directly by hydraulic cylinder.

HYDRAULIC SHEAR. A shear with its crosshead actuated by hydraulic cylinders.

INDIRECT EXTRUSION (INVERTED). An extrusion process in which the metal is forced back inside a hollow ram that pushes the die.

INGOT. A casting intended for subsequent rolling or forging.

INTERNAL FRICTION. Ability of a metal to transform vibratory energy into heat. Internal friction generally refers to low stress levels of vibration; damping has a broader connotation, since it may refer to stresses approaching or exceeding the yield strength.

IRONING. Thinning the walls of deep drawn articles by reducing the clearance between punch and die.

KILLED STEEL. Steel deoxidized with a strong deoxidizing agent such as silicon or aluminum in order to reduce the oxygen content to a minimum so that no reaction occurs between carbon and oxygen during solidification.

KNOCKOUT. A mechanism for releasing workpieces from a die; also called ejector, kickout, or liftout. Crossbars, cams, springs, or air cushions are commonly used to actuate slide knockouts.

LANCING. Slitting and forming a pocket-shaped opening in a part.

LEAF BRAKE. A press brake on which bending action is produced manually by a "leaf" operated by two long handles, or by powered means. (Also called box and pan, or finger brake.)

LIGHTENING HOLE. A hole punched in a part to save weight.

LINE DIES. A sequence of stamping dies, all of which perform operations on a part with manual attendance. As an example, a part blanked out of coil stock may advance through a forming die, a piercing die, a trim die, and/or other dies.

LOAD, PRESS. Amount of force exerted in a given operation.

MECHANICAL PRESS BRAKE. A press brake utilizing a mechanical drive consisting of a motor, flywheel, crankshaft, clutch, and eccentric to generate vertical motion.

MECHANICAL SHEAR. A shear with its crosshead driven by an eccentric which is engaged by a flywheel-clutch combination.

MANNESMANN PROCESS. A process used for piercing tube billets in making seamless tubing. The billet is rotated between two heavy rolls mounted at an angle, and is forced over a fixed mandrel.

MARTEMPERING. The process of quenching an austenitized ferrous alloy in a medium at a temperature in the upper portion of the temperature range of martensite formation, or slightly above that range, and holding in the medium until the temperature through the alloy is substantially uniform. The alloy is then allowed to cool in air through the temperature range of martensite formation.

MECHANICAL WORKING. Subjecting metal to pressure exerted by rolls, dies, presses, or hammers, to change its form or to affect the structure and consequently the mechanical and physical properties.

MODULUS OF ELASTICITY. The ratio of stress to strain; corresponds to slope of elastic portion of stress-strain curve in mechanical testing. The stress is divided by the unit elongation. The tensile or compressive elastic modulus is called "Young's modulus"; the torsional elastic modulus is known as the "shear modulus" or "modulus of rigidity."

MOTION DIAGRAM. A graph or curve which shows the motion of a slide relative to the motion of the driving member, such as the rotation of a crank. It may also show slide velocity and/or acceleration. For a multiple-action press, the motion diagram shows the relative motion and position between the slides.

NATURAL AGING. Spontaneous aging of a supersaturated solid solution at room temperature.

NECKING DOWN. Reduction in area concentrated at the subsequent fracture when a ductile metal is tested in tension.

NECKING FAILURE. The failure of a formed part by thinning abruptly in a narrow localized area. An extreme case of necking failure is splitting.

NITROGEN DIE CYLINDERS. Commercially available gas-charged cylinders manufactured specifically for die applications. These cylinders are used in place of springs or die cushions in applications in which high initial pressure is required, usually in draw, form, and cam dies. Uniformity of pressure is attained by linking cylinders to an accumulator.

NORMALIZING. A process in which a ferrous alloy is heated to a suitable temperature above the transformation range and is subsequently cooled in still air at room temperature, usually done for stress relieving.

NOSING. The act of forming a curved portion, with reduced diameters, at the end of a tubular part.

Notching. The cutting out of various shapes from the edge of a strip, blank, or part.

Overaging. Aging under conditions of time and temperature greater than those required to obtain maximum strength.

Overbending. Allowance for spring-back when bending metal to a desired angle.

Overheated. A term applied when, after exposure to an excessively high temperature, a metal develops an undesirable coarse grain structure but is not permanently damaged.

Overload Relief Device. A mechanism designed to relieve overloads to structural members of the press and/or tooling. The devices can be mechanical, hydropneumatic, or hydraulic and can be located in the slide, connections, bed, or tie rods. These devices have limited strokes, often less than 1" (25.4 mm) of travel.

Pad. The general term used for that part of a die which delivers holding pressure to the metal being worked.

Part-revolution Clutch. A type of clutch that may be engaged or disengaged during any part of the cycle.

Pearlite. The lamellar aggregate of ferrite and carbide. Note: It is recommended that this word be reserved for the microstructures consisting of thin plates or lamellae — that is, those that may have a pearly luster in white light. The lamellae can be very thin and resolvable only with, the best microscopic equipment and technique.

Perforating. The punching of many holes, usually identical and arranged in a regular pattern, in a sheet, workpiece blank, or previously formed part. The holes are usually round, but may be any shape. The operation is also called multiple punching. (See PIERCING.)

Piercing. The general term for cutting (shearing or punching) openings, such as holes and slots in sheet material, plate, or parts. This operation is similar to blanking; the difference being that the slug or piece produced by piercing is scrap, whereas the blank produced by blanking is the useful part.

Pilot. Bullet-nosed component used in dies to maintain correct position of advancing strip. As the strip advances through a sequence of operations, the pilot entry into prepunched holes ensures precise registration of the part at each station of the strip.

Pinch Point. Any point, other than the point of operation, at which it is possible for a part of the human body to be caught between the moving parts of a press or auxiliary equipment, between moving and stationary parts of a press or auxiliary equipment, or between the material and moving part or parts of the press or auxiliary equipment. (Explanation: The expression pinch point, as used in the standard, refers only to parts of the machine or parts associated with it that create a hazard. The expression is not used to describe hazards caused by the tooling at the point of operation, since these hazards are a different problem and require different treatment.)

Plastic Anisotropy. Directional difference in mechanical properties relative to rolling direction applied in producing the sheet metal.

Plastic Deformation. Permanent distortion of a material under the action of applied stresses.

Plastic Flow. The phenomenon that takes place when metals or other substances are stretched or compressed permanently without rupture.

Plasticity. That property or characteristic which permits substances to undergo permanent change in shape without rupturing. Practically, it is the property possessed by useful materials that permits stretching or compressing them into useful shapes.

Platen. The sliding member, slide, or ram of a hydraulic press.

Pneumatic Toggle Links. Special main links of a toggle press which are equipped with pneumatic cushions and a linkage to give air pressure controlled flexibility. These links compensate for variations in material thickness under the blankholder and also can be adjusted to exert different pressures at different corners of the blankholder.

Point of Operation. The area of the press in which material is actually positioned and work is being performed during any process, such as shearing, punching, forming, or drawing.

Poisson's Ratio. The ratio of the lateral expansion to the longitudinal contraction under a compressive load, or the ratio of the lateral contraction to the longitudinal expansion under a tensile load, provided the elastic limit is not exceeded.

Preformed Part. A partially formed part which will be subjected to one or more subsequent forming operations.

PRESS. A machine tool having a stationary bed and a slide or ram that has reciprocating motion at right angles to the bed surface, the slide being guided in the frame of the machine.

PRESS BRAKE. (See BENDING BRAKE.)

PRESS FORGING. The forging process in which metal stock is formed between dies, usually by hydraulic pressure. Press forging is an operation that employs a single, slow stroke.

PRESS FORMING. Any forming operation performed with tooling by means of a mechanical or hydraulic press.

PROGRAMMABLE GAGING. Back-gaging in which the gage bar is driven by an encoder signal or servomotor that can be programmed by the operator.

PROGRESSION. The constant dimension between adjacent stations in a progressive die. As such, it is the precise distance the strip must advance between successive cycles of the press. Accuracy of the progression is guaranteed by piloting and by the feed unit.

PROGESSIVE DIE. A die with two or more stations arranged in line for performing two or more operations on a part, one operation usually being performed at each station.

PUNCH. The moveable part that forces the metal into the die in equipment for sheet drawing, blanking, coining, embossing and the like.

PUNCHING. Shearing holes in sheet metal with punch and die.

QUENCH HARDENING. A process of hardening a ferrous alloy of suitable composition by heating within or above the transformation range and cooling at a rate sufficient to increase the hardness substantially.

QUENCHING. A process of rapid cooling from an elevated temperature by contact with liquids, gases, or solids.

RABBIT EAR. Recess in die corner to allow for wrinkling or folding of the blank.

RAM. The powered, movable portion of the power press brake structure, with die-attachment surface, which imparts the pressing load through male dies onto the piece part and against the stationary portion of the press brake bed. (See SLIDE.)

RECRYSTALLIZATION. A process whereby the distorted grain structure of cold worked metals is replaced by a new, strain-free grain structure during annealing above a specific minimum temperature.

REDRAWING. The second and successive deep drawing operations in which the cuplike shells are deepened and reduced in cross-sectional dimensions (sometimes in wall thickness, by ironing). Redrawing is done in both single-action dies and in double-action dies.

REDUCTION IN AREA. The difference between the original cross-sectional area and the smallest area at the point of rupture, usually stated as a percentage of the original area.

REPEAT. An unintended or unexpected successive stroke of the press resulting from a malfunction.

RESILENCE. The amount of energy stored in a unit volume of metal as a result of applied loads.

RESTRIKING. Cold striking of a finished forging to remove distortions after it has been quenched.

RIB. A long, V-shaped or radiused indentation used to strengthen large panels.

ROLLER STRAIGHTENING. A process involving a series of staggered rolls of small diameter between which rod, tubing, and shapes are passed for the purpose of straightening. The process consists of a series of bending operations.

ROLL FORMING. (1) An operation used in forming sheet. Strips of sheet are passed between rolls of definite settings that bend the sheet progressively into structural members of various contours, sometimes called "molded sections." (2) A process of coiling sheet into open cylinders.

ROLL THREADING. Applying a thread to a bolt or screw by rolling the piece between two grooved die plates, one of which is in motion, or between rotating grooved circular rolls.

ROTARY BLADE. A shearing tool whose cutting edge makes a complete revolution about a fixed axis.

ROTARY SHEAR. A cutting machine with sharpened circular blades or disk-like cutters used for trimming edges and slitting sheet and foil.

ROTARY SWAGER. A machine used for pointing wire rod and tubing and for initial working of moderately brittle metals and alloys in which a pair of tapered dies is used. The dies are rotated and are pressed intermittently against the work material by an eccentric action.

Rubber Pad Forming. A forming operation for shallow parts in which a confined, pliable, rubber pad attached to the press slide is forced by hydraulic pressure to become a mating die for a punch or group of punches that have been placed on the press bed or baseplate. A process developed in the aircraft industry for limited production of a large number of diversified parts. This is the Guerin process. Although this process was used originally on aluminum and magnesium alloys it is now used on light gages of mild and stainless steel.

Scaling. Surface oxidation caused on metals by heating in air or in other oxidizing atmospheres.

Seam. (1) The fold or ridge formed at the juncture of two pieces of sheet material. (2) An extended, narrow defect on the metal surface, resulting from a blow hole or inclusion which has been stretched during processing.

Seaming. The process of joining two edges of sheet material to produce a seam. Machines that do this work are referred to as seaming machines or seamers.

Seizing. Welding of metal from the workpiece to a die member under the combined action of pressure and sliding friction.

Shaving. Trimming heavy-gauge blanks to remove uneven sheared edges.

Shear. (1) A type of cutting operation in which the metal object (sheet, wire, rod, or such) is cut by means of a moving blade and fixed edge or by a pair of moving blades that may be either flat or curved. (2) A type of deformation in which parallel planes in metal crystals slide so as to retain their parallel relation to one another, resulting in block movement.

Shear Strength. The maximum stress that a metal can withstand before fracturing when the load is applied parallel to the plane of stress; contrasted with tensile or compressive force, which is applied perpendicular to the plane of stress. Under shear stress, adjacent planes of metal tend to slide over each other.

Sheet. Any material or piece of uniform thickness and of considerable length and breadth as compared to its thickness is called a sheet or plate. In reference to metal, such pieces under 1/4" thick are called sheets and those 1/4" thick and over are called plates. Occasionally, the limiting thickness for steel to be designated as "sheet steel" is number ten Manufacturer's Standard Gage for sheet steel, which is 0.1345" (3.42 mm) thick.

Shut Height. The distance from the top of the bed to the bottom of the slide of a vertical press, with stroke down and adjustment up. On moving bolster presses, the shutheight is measured from the top of the bolster (when the bolster is integral with the carriage) or the top of the carriage (when the bolster is separate).

Single-acting Hammer. A forging hammer in which the head is raised by a steam cylinder and piston and the blow is delivered by the free fall of the head.

Single-action Press. A forming press that operates with a single function, such as moving a punch into a die with no simultaneous action or holding down the blank or ejecting the formed work.

Single-stroke Mechanism. A mechanical arrangement used on a full-revolution clutch to limit the travel of the slide to one complete stroke at each engagement of the clutch.

Sizing. A final pressing of a sintered compact to secure the desired size.

Skelp. A plate of steel or wrought iron from which pipe or tubing is made by rolling the skelp into shape longitudinally and welding the edges together.

Slide. The main reciprocating member of a press, guided in the press frame, to which the punch or upper die is fastened. Sometimes called the ram. The inner slide of a double-action press is called the plunger or punch-holder slide; the outer slide of a double-action press is called the blankholder slide; the third slide of a trip-action press is called the lower slide; and the slide of a hydraulic press is often called the platen.

Slitting. Cutting or shearing along single lines either to cut strips from a sheet or to cut along lines of a given length or contour in a sheet or workpiece.

Slug. Small pieces of material (usually scrap) which are produced in punching holes in sheet material.

Spinning. The procedure of making sheet metal disks into hollow shapes by pressing a tool against a rotating form (spinning chuck).

SPRINGBACK. An indicator of elastic stresses, frequently measured as the increase in diameter of a curved strip after removing it from the mandrel about which it was held. The measurement is employed as an indicator of the extent of recovery or relief of residual stresses that has been achieved by the transformation of elastic strain to plastic strain during heating or stress relieving.

STAMP. (1) The act of impressing by pressure (sink in) lettering or designs in the surface of sheet material or parts. (2) The act of forming or drawing by pressworking. (3) The general term used to denote all pressworking.

STAMPING. A process used to cut lines of letters, figures and decorations on smooth metal surface. The impact of a punch with comparatively sharp projecting outlines impresses the characters into the surface of the metal.

STEEL RULE DIE. A metal cutting die employing a thin strip of steel (printer's rule) formed to the outline of a part and a thin steel punch mounted to a suitable die set.

STOP. A device for positioning stock or dies in a die.

STOP CONTROL. An operator control designed to immediately deactivate the clutch control and activate the brake to stop slide motion. (Explanation: The term Emergency Stop is sometimes used to refer to this control, even though its use is most commonly not on an emergency basis. Sometimes, the control causes motor shutdown, in addition to clutch disengagement. Also, quite commonly, a top stop control is used to stop continuous stroking at top of stroke or at another predetermined point in stroke. A top stop control action is delayed, after actuation of the operating means, to cause stopping at the predetermined point in stroke.)

STRAIN. A measure of the change in size or shape of a body, due to force, in reference to its original size or shape. Tensile or compressive strain is the change, due to force, per unit of length in an original linear dimension, in the direction of the force.

STRAIN HARDENING. An increase in hardness and strength caused by plastic deformation at temperatures lower than the recrystallization range.

STRESS. The intensity of force within a body which resists a change in shape. It is measured in pounds per square inch or pascals. Stress is normally calculated on the basis of the original cross-sectional dimensions. The three kinds of stresses are tensile, compressive, and shearing.

STRESS RELIEVING. A process of reducing residual stresses in a metal object by heating to a suitable temperature and holding for a sufficient time. This treatment may be applied to relieve stresses induced by casting, quenching, normalizing, machining, cold working, or welding.

STRETCH FORMING. A process of forming panels and cowls of large curvature by stretching sheet over a form of the desired shape. This method is more rapid than hammering and beating.

STRETCHING. Stretching is defined as an extension of the surface of the sheet in all directions. In stretching, the flange of the flat blank is securely clamped. Deformation is restricted to the area initially within the die. The stretching limit is the onset of metal failure.

STRIPPER. A plate designed to surround the piercing punch steels or the piercing punches. Its purpose, quite literally, is to strip the sheet metal stock from the punching members during the withdrawal cycle. Strippers are also employed for the purpose of guiding small precision punches in close toleration dies, to guide scrap away from dies, and to assist in the cutting action. Strippers are made in two types, fixed and movable.

STROKE. The distance between the terminal points of the reciprocating motion of a press slide.

STROKES PER MINUTE (SPM). The specified continuous running speed of a press. It is not the number of permissible single trippings of a press and consequently does not measure the possible production per minute, except when a press is run continuously. The number of single trippings per minute varies with different types and makes of clutches as well as with the dexterity of the operator.

TEMPER BRITTLENESS. Brittleness that results when certain steels are held within, or are cooled slowly through a certain range of temperatures below the transformation range. The brittleness is revealed by notched-bar impact tests at room temperature or lower temperatures.

TEMPERING. A process of reheating quench-hardened or normalized steel to a temperature below the transformation range, and then cooling at any rate desired mainly to reduce brittleness.

Tensile Strength. The maximum tensile stress that a material is capable of withstanding without breaking under a gradually and uniformly applied load. Its value is obtained by dividing the maximum load observed during tensile straining by the specimen cross-sectional area before straining. Other terms that are commonly used are ultimate tensile strength, and less accurately, breaking strength.

Throat (Gap) Depth. The distance from the slide centerline to the frame of a gap-frame press.

Tie Rods. Steel rods, threaded at both ends for nuts, used to prestress straight-side press frames. They are also used to reduce deflection in gap-frame presses, but require careful installation.

Toggle Joint. A connecting mechanism consisting of two links freely pinned together at one end and connected by free pins to other press parts at their other or outer ends.

Tonnage. (See CAPACITY, PRESS.)

Torsional Strength. The maximum stress that a metal can withstand before fracture when subjected to a torque or twisting force. Stress in torsion involves shearing stress, which is not uniformly distributed.

Toughness. As determined by static tests, toughness is considered to be the work per unit volume required to fracture a metal. It is equal to the total area under the stress-strain curve, represents the total energy-absorbing capacity, and includes both elastic and plastic deformation. Toughness in practice is more often considered to be resistance to shock or impact, which is a dynamic property.

Trimming. A secondary cutting or shearing operation on previously formed, drawn, or forged parts in which the surplus metal or irregular outline or edge is sheared off to form the desired shape and size.

Trip (or Tripping). Activation of the clutch to run the press.

Tripping Mechanism. Any auxiliary mechanism, manually, mechanically, or automatically operated, which engages and disengages the clutch for starting and stopping the press.

Twist. The tendency of material (strip) that is being sheared off to curve about a central longitudinal axis.

Ultimate Strength. (See TENSILE STRENGTH.)

Unit Stress. (See STRESS.)

Unitized Tooling. A type of die in which the upper and lower members are incorporated into a self-contained unit, which is arranged in such a way that it holds the die members in alignment. This type of tooling includes continental, subpress, pushthrough, pancake, or short-run dies.

Upsetting. (1) A metal working operation similar to forging. (2) The process of axial flow under axial compression of metal, as in forming heads on rivets by flattening the end of wire.

Vent. A small hole in a punch or die for admitting air to avoid suction holding or for relieving pockets of trapped air that would prevent proper die closure or action.

Wiper Forming (wiping). Method of curving sections and tubing over a form block or die in which this form block is rotated relative to a wiper block or slide block.

Wire Rod. A semifinished product from which wire is made. It is generally of circular cross section approximately 1/4 in, in diameter.

Work Hardness. Hardness developed in metal as a result of cold working.

Wrinkling. The wavy condition that appears on some formed, comparatively thin walled, metal parts due to buckling caused by unbalanced compressive stresses in the drawing or forming operation. This condition often takes place in the flange of a deep-drawn part.

Yield Point. In mild or medium-carbon steel, the stress at which a marked increase in deformation occurs without an increase in load; also called proportional limit. In other steels and in nonferrous metals this phenomenon is not observed. Refer also to yield strength.

Yield Strength. The stress at which a material exhibits a specified permanent plastic yield or set; a limiting deviation from proportionality of stress to strain. An offset of 0.2% is used for many metals, such as aluminium-based and magnesium-based alloys, while 0.5% total elongation under load is frequently used for copper alloys. Also called proof stress.

METAL MACHINING

ABRASIVE MACHINING. The basic process in which chips are removed by very small cutting edges that are integral parts of abrasive particles. Grinding, honing, belt sanding, lapping fall under this category.

ABRASIVES. Hard material that can cut or abrade other substances; they may be natural such as diamond, quartz, sand, garnet, or man made such as silicon carbide (carborundum) or aluminum oxides (alundum).

AISI STEEL DESIGNATION. A system of classifying steel and steel alloys, expanded from the original SAE system. (See SAE STEELS.)

ALLOWANCE. The least intentional difference or the maximum interference between two mating parts.

ALUMINUM OXIDE, AL_2O_3. Fine-grains of this oxide are bonded together and formed into cutting inserts (also known as ceramic or oxides).

AUTOMATIC BAR. Chucking and screw machines, different versions of turning machines for high production of parts from relatively small bar stock. (See SCREW MACHINES.)

AXIAL RAKE ANGLE. Applies to the angular (not helical or spiral) cutting faces: a) The angle of a plane containing the cutting face, or tangent to the cutting face at a given point, and the tool axis. b) The angle between the tool face, on angular teeth, perpendicular to a tangent between the cutting edge and the cutter axis, measured in an axial direction.

AXIAL RELIEF ANGLE. The angle between the relieved surface, or flank of the tooth, and a plane perpendicular to a line between the axis of the cutter and the cutting edge, measured in an axial direction.

BASIC SIZE. The theoretically perfect size from which variation is permitted.

BELT SANDING. A common operation to obtain smooth surfaces. The workpiece is held against a moving, abrasive belt until the desired degree of finish is obtained.

BILATERAL TOLERANCE. The form in which variation is permitted in both directions from the basic size. (See TOLERANCE.)

BORING. The enlargement or relocation of an existing hole or section of a hole by means of a cutting tool.

BRAZED TIP TOOL. One in which the cutting edge of the tool is a different material than the shank to which it has been brazed.

BRINELL HARDNESS TEST. A standardized method of measuring hardness using a steel ball pressed into the workpiece by a standard load. The diameter of the spherical indentation is used to quantify the hardness of the tested material.

BROACHES. A broach is a tool with multiple cutting edges which is passed over an existing surface or through an existing hole to form a specific shape. Each tooth removes a small amount of material until the last tooth generates the final shape and size.

BUILT-UP EDGE. The welding of small segments of the chips from the material being machined to the face of the cutting tool.

CEMENTED CARBIDES. Also known as carbides, are basically composed of carbides with a binder. They are the most commonly used cutting tool materials.

CENTERLESS GRINDING. The generation of a finished cylindrical outside diameter by rotating the workpiece between the grinding wheel and the regulating wheel while constrained by the work rest.

CENTER MOUNTED GRINDING. The generation of finished external cylindrical surfaces on workpieces mounted between centers.

CERAMIC. (See ALUMINUM OXIDE.)

CERMETS. Combinations of metals and ceramics, bonded together in the same manner in which powder metallurgy parts are produced. A typical cermet cutting tool insert is 70% Al_2O_3 and 30% TiC.

CHAMFER ANGLE. The degree of angle wanted for a given chamfer. A chamfer eliminates a sharp corner on the cutting edge.

CHIP. In metal cutting, the metal is removed from the workpiece in the form of chips.

CHIPBREAKER. Means of abrupt diversion of removed metal at the point of cutting causing breakage of a potential string into chips of manageable size.

CHISEL EDGE ANGLE (WEB ANGLE). Both of these terms refer to the angle of the section of the drill between the two cutting edges at the drill point.

CHUCKING INTERNAL GRINDING. The generation of finished internal surfaces on workpieces mounted in chucking devices that provide rotation of the workpiece.

CLEARANCE FIT. A fit between mating parts having limits of size so prescribed that a clearance always results in assembly.

COOLANTS. Usually fluids (sometimes compressed air or paste) to act as temperature reduction agents in machining operations.

COUNTERBORING. Special drills are used to enlarge the top part of a hole.

COUNTERSINKING. Special drills are used to produce a conical shape on the top part of a hole.

CRATER WEAR. Wear on the top (rake) face of the cutting tool which is mainly due to the sliding of the chip.

CREEP GRINDING. Grinding operation utilizing slow feed rates but heavy depth of cut.

CRUSH FORM GRINDING. The grinding of irregular cross sections by a grinding wheel whose face has been formed by feeding a crusher roll against the grinding wheel while rotating at a considerably slower speed than those employed in grinding.

CUBIC BORON NITRIDE (CBN). A super abrasive hard crystal produced by a high-pressure high temperature process similar to that used to make synthetic diamonds. CBN crystals are used most commonly in super abrasive wheels for grinding. The crystals are also compacted to produce polycrystalline cutting tools in the form of inserts. "Borozon" is a G. E. trade name.

CUTTING FLUIDS. Coolants and different liquids used in the cutting operation mainly to reduce the cutting temperature and friction. (See COOLANTS.)

CUTTING FORCE. The force acting normal on the tool during a cutting operation. The force caused by the resistance of the workpiece to being cut.

CUTTING RATIO. The ratio of the chip thickness prior to entering the shear zone and its thickness after passing through the shear zone.

CUTTING TOOL ANGLES. Different angles on the cutting edge of the tool to facilitate the cutting and the sliding of the chip and reduce friction; (cutting edge angles, rake angles, relief angles, clearance angles).

DEBURRING. Removing burrs left by machining, shearing, and casting operations. Deburring is usually done by grinding.

DEPTH OF CUT. The thickness of the workpiece to be removed by the tool.

DIELECTRIC (IN EDM). Fluid used to cool and flush particles as they are machined and form a dielectric barrier between the electrode and the workpiece.

DIVIDING HEAD, UNIVERSAL. (See UNIVERSAL DIVIDING HEAD.)

DRESSING (OF GRINDING WHEELS). Manually or automatically sharpening and/or trueing of grinding wheels, usually with diamond tool.

DULL TOOL. Any cutting tool or grinding wheel which needs resharpening or dressing.

ECONOMIC TOOL LIFE. The period of time, usually given in minutes, during which the tool performs the function required, at cutting conditions that result in least cost per piece.

ELECTROCHEMICAL GRINDING (ECG). Also known as electrolytic grinding. A special form of electrochemical machining (refer to ECM). It employs the combined actions of electrochemical attack and abrasion to rapidly remove material from electrically conductive workpieces (anode) by employing a rotating grinding wheel (cathode).

ELECTRON BEAM MACHINING (EBM). This process uses electrical energy to generate thermal energy for removing material. A pulsating stream of high-speed electrons produced by a generator is focused by electrostatic and electromagnetic fields to concentrate energy on a very small area of work. As the electrons impinge on the work, their kinetic energy is transformed into thermal energy and melts or evaporates the material locally.

ELECTRIC DISCHARGE MACHINING (EDM). A nontraditional method of removing metals by a series of rapidly recurring electrical discharges between an electrode (the tool) and the workpiece in the presence of a dielectric fluid. Minute particles of metal or chips are removed by melting and vaporization, and are washed from the gap by the dielectric fluid which is continuously flushed between the tool and the workpiece.

Electrochemical Machining (ECM). A method of removing metal without the use of mechanical or thermal energy. Electric energy is combined with a chemical to form a reaction of reverse plating. Direct current at relatively high amperage and low voltage is continuously passed between the anode workpiece and cathode tool (electrode) through a conductive electrode.

End Milling. The production of a flat or slotted surface when employing a shank-mounted cutter.

Engine Lathe. The most common turning machine tool, usually manually controlled. Its main components are the bed, carriage, headstock, and tailstock.

Error Budgeting. In precision engineering it is the analytical and experimental procedures to identify and quantify error sources, and to determine their combinational effect on the machining accuracy.

Error Compensation. Procedures to obtain the machining reference and tool dimension offsets and use them to reposition the tool for reducing machining error.

Facing. The production of a flat surface when the cutting tool moves perpendicular to the axis of rotation of the workpiece.

Face Milling. The production of a flat or slotted surface when employing a shank-mounted cutter.

Feed, Feed Rate. The relative movement between the tool and the workpiece as the tool transverses to perform the machining operation.

Finishing. All kinds of machining operations to produce fine surface finish and usually produce accurate dimensions.

Fit. The relationship existing between two mating parts with respect to the amount of clearance or interference which is present when they are assembled.

Fixture. A device for holding the workpiece.

Flank Wear. The wear on the side or flank of the cutting tool which is mainly due to the continuous contact between the tool and the workpiece. It is usually measured by the height of the relatively uniform area of wear.

Flaws. The irregularities of any sort that occur at relatively infrequent intervals.

Flexible Manufacturing System (FMS). An assembly of several programmable machine tools interconnected by work manipulators and controlled by a host computer.

Flying Cutting. Milling operations where the milling cutter has one cutting edge.

Form Milling. The production of a nonlinear surface when employing an arbor-mounted cutter.

Gaging. Measuring process to determine whether the dimensions or characteristic is larger or smaller than the established standard or range of acceptability.

Gear Hobbing. Profile milling of gears by using a worm-like milling cutter referred to as a hob.

Glazed Wheel. An abrasive wheel with a cutting surface too smooth (or dull) to grind efficiently.

Grade (in abrasive wheels). The strength of bonding of a grinding wheel.

Grinding. A precision finishing machining process which uses an abrasive tool rotating at a high rate of speed.

Grinding Ratio. The ratio of volume of metal removed by grinding over the volume of wheel wear. It is a measure of the effectiveness of the grinding operation.

Grit (in abrasives). Identifies the size of abrasive grains in numbers where 10 is very coarse and 500 is very fine.

Gun Drilling. A term used when drilling deep holes. The drilling tool has oil holes for using high pressure coolants.

Hard Turning. A turning process using very hard cutting tools, such as CBN or ceramics, to turn directly hardened materials. This process now can turn steel with hardness up to 65 Rc, and has a competitive edge over grinding. Generally speaking, it produces better surface integrity and requires much less specific energy than does grinding.

Headstock (in lathes). The headstock of the lathe is mounted at the left-hand end of the machine. It is the unit which turns the workpiece. It contains the spindle and series of gears or pulleys by which the spindle is rotated at various speeds.

Helix Angle (drills). The angle between a line tangent to the land of the flute and the axis of rotation at their point of intersection.

Helix Angle (milling cutters). The angle between a line tangent to a point on the helical tooth and the axis of the cutter measured at their point of intersection.

Heterodyne Interferometer. A device using two frequency laser beams to generate fringe patterns for determining machine accuracy such as linear displacement, angular rotation, straitness, squareness, and parallelism. Resolution up to one hundredth of the laser beam wavelength is achievable.

High Efficiency Machining Range. The range of cutting speed which is bounded by the speed for minimum machining cost and the speed for maximum production rate.

High Speed Machining. The term "high speed" is relative and is compared with the speed of conventional machining processes. As a general guide, an approximate range of speeds may be defined as follows:

High Speed	-	2,000 - 6,000 ft./min.
Very High Speed	-	6,000 - 60,000 ft./min.
Ultrahigh Speed	-	Greater than 18,000 ft./min.

High speed machining should be considered basically for situations in which cutting time is a significant portion of floor time of the operation. The workpiece materials used include aluminum alloys.

High Speed Steel (HSS). Conventional material for cutting tools. It is mainly made of high carbon steel alloyed mainly with tungsten and molybdnum.

Honing. Rotational metal removal process using combined rotational and reciprocating motion of bar-shaped abrasive tools mainly for internal cylinders to improve finish and dimensional accuracy.

Horizontal Surface Grinding. The generation of finished flat surfaces by mounting workpieces on a traversing work table below a grinding wheel that rotates about a horizontal axis.

Inserts. Cutting tools made of carbide or ceramics are manufactured in the form of inserts which are usually mechanically held in the tool or cutter. They are available in a large variety of materials and shapes and are also called indexable inserts or throw-away inserts. They have multiple cutting edges and generally are not reground.

Interference Fit. A fit between mating parts having limits of size so prescribed that an interference always results in assembly.

Internal Broaching. The production of internal surfaces other than the origination of holes by employing a tool of increasing size teeth that is actuated by pulling or pushing.

Jig. A device for holding the workpiece and guiding the tool.

Jig Borers. Very precise vertical type boring or drilling machine designed for use in making jigs, fixtures and precision parts.

Knoop Hardness Test. A microhardness test which measures hardness in Knoop hardness numbers (HK). It uses an elongated pyramid like diamond indentor and light load.

Knurling. The production of a pattern-roughened surface by the feed of a free rotating tool.

Lapping. Slow-speed abrasive finishing process used on flat or cylindrical surfaces. The lap is usually made of cast iron, copper or cloth. The abrasive particles are either embedded in the lap or they may be carried through a fluid. Very smooth surface finish can be obtained by this process.

Laser Beam Cutting. Cutting of metal using the laser device as source of energy.

Lathes. Machine tools used mainly for turning, boring, and facing operations. There is a large variety: engine lathe, turret lathe, Swiss machines, bar machines.

Lay. The direction of the predominate surface pattern.

Limits. The extreme permissible values of a dimension.

Lips. The cutting edges of a two-flute drill extending from the chisel edge to the periphery.

Loaded Grinding Wheel. A condition where the porosities on the surface of the wheel become filled or clogged with chips during grinding. This condition hampers the performance of the wheel.

LUBRICANTS IN MACHINING. (See CUTTING FLUIDS.)

MACHINABILITY. The qualitative property of a work material, subjected to the machining operation, that reflects on the ease or difficulty with which the operation takes place. It is usually quantified by tool wear, surface finish, and cutting forces.

MACHINE TOOL ERROR. The difference between the actual postion and the intended position of the tool tip. This error includes the kinematics errors of a cool machine and thermal errors due to changes in temperature distributions in the machine tool structure.

MACHINE TOOL ERROR COMPENSATION. An approach to correct machine tool errors by compensating the estimated errors based on mathematical models and in-process measurements, which can be temperature measurements, measurements of the errors of specific referencing points, etc. A joint study by Purdue University and the National Institute of Standard and Technology published in 1985 indicated the potential improvement in machine tool accuracy can be 10 times. This level of improvement has been subsequently confirmed by many other studies and applications.

MACHINING CELL. An assembly of one or more programmable machine tools served by a single work manipulator.

MACHINING CENTER. A machine tool that is equipped with an automatic tool changer and performs operations on different surfaces of a workpiece. It is usually a computer numerical control machine.

METAL REMOVAL RATE (MRR). The volume of metal removed in one minute of cutting time.

METROLOGY FRAME. A separate stationary reference frame whose position the machine tool's axes are measured with respect to. Metrology frame should not be freed of the effects due to dynamic or static loads and thermal distortions.

MILLING. An important machining operation to produce flat surfaces, slots, and other configurations. Milling cutters have multiple cutting edges.

NOMINAL SIZE. The dimensional designation of size which is used for the purpose of general identification.

NONTRADITIONAL MACHINING PROCESSES. A multitude of metal removal operations which do not rely on mechanical energy for machining; *e.g.* electrodischarge machining, electro-chemical milling.

NOSE RADIUS OF THE CUTTING EDGE OF THE TOOL. It is the radius generated at the intersection of the leading and trailing edges of the tool measured in the plane of the tool face.

NUMERICAL CONTROL. A method of controlling the movements of machine tools by numbers (numerical). It is defined as a system in which actions are controlled by the direct insertion of numerical data at some point; the system must automatically interpret this data.

OBLIQUE CUTTING. An occasionally used term to define the three dimensional cutting operation as opposed to orthogonal cutting.

OFFHAND GRINDING. Grinding work that is held in the operator's hand, also known as freehand grinding.

OPTIMUM CUTTING SPEED. The speed to optimize a machining objective such as minimum machining cost or minimum production time.

ORTHOGONAL CUTTING. Two dimensional cutting operations where all the mechanics of cutting are modelled in one plane.

OUT-OF-ROUND. Having some points on the profile (internal or external cylinders) not equidistant from the common center.

OXIDES. These tool materials consist primarily of fine-grained aluminum oxide (Al_2O_3) particles which are bonded together. They are mainly supplied in indexable inserts and possess high abrasion resistance and hot hardness. They usually are referred to as ceramic tools.

OXYFUEL GAS CUTTING. Instead of shearing, many metals are cut by oxyfuel gas cutting. The metal is merely melted by means of the flame of the oxyfuel gas torch. Acetylene is the most commonly used gas.

PITCH IN GEARS. Distance between corresponding points on equally spaced and adjacent teeth.

PITCH IN THREADS. The pitch of a thread having uniform spacing is the distance, measured parallel to its axis, between corresponding points on adjacent thread forms in the same axial plane and on the same side of the axis.

PLASMA ARC-CUTTING. An arc-cutting process wherein severing of the metal is obtained by melting a localized area with a constricted arc and removing the molten material with a high velocity jet of hot, ionized gas issuing from the orifice.

POINT, DRILL. The cutting end of the drill, made up of the ends of the lands, the web, and the lips. In form, it resembles a cone, but it departs from a true cone to furnish clearance behind the cutting lips.

POWDER METALLURGY. A process wherein fine metal powders are blended, pressed into a desired shape (compacted), and then heated (sintered) in a controlled atmosphere at a temperature below the melting point of the major constituent for sufficient time to bond the contacting surfaces of the particles and establish desired properties. The process, commonly designated as P/M.

PROFILE, SURFACE. This describes the contour of any specified cross section of the surface. It discloses waviness, roughness, and flaws.

PROFILOMETER, SURFACE. An electronic instrument which measures the average surface roughness in microinches, either the root mean square (RMS) or the arithmetic average (AA).

RAKE ANGLE. The angle between the face of the tool or cutter and a plane through the cutting edge. It is the angle on the rake face of the tool where the chip slides.

REAMERS. Multiedged cutting tools, mainly used for two purposes: to bring holes to a more exact size, and to improve the finish of an existing hole by machining a small amount from its surface. No special machines are built especially for reaming. It is usually done on drilling or milling machines. The principal types of reamers are: hand, machine, shell, expansion, and adjustable reamers.

REAMING. The machining operation where reamers are used to improve on a drilled hole (See REAMERS.)

RESIDUAL STRESSES. Stresses that remain within a workpiece after it has been worked on (machined, formed, or heat treated) and all external forces have been removed.

REVERSAL PRINCIPLE. A technique which obtains measurement twice with opposite setups in order to eliminate the error due to the arbor or the instrument.

REVOLUTIONS PER MINUTE (RPM). The number of rotations of the tool or workpiece per minute.

ROCKWELL HARDNESS TESTER. Hardness measuring method in which the hardness value of the test piece is a function of the indentation by an indentor under static load.

ROOT MEAN SQUARE (RMS). A mathematical measure of the average roughness of a surface.

ROUGHING. A machining operation intended for heavy metal removal. Usually followed by finish machining to obtain the required dimension and surface finish.

ROUGHNESS. Surface finish characterized by sharp, closely spaced high and low spots.

SAE STEELS. A system of classification of steels and steel alloys, by their chemistry. It was developed by the Society of Automotive Engineers (SAE). It was later adopted and expanded by the American Iron and Steel Institute (AISI). Both plain-carbon and low-alloy steels are identified by a four-digit number. The first number indicates the major alloying elements and the second, a subgrouping of the major alloy system. The last two digits indicate the approximate carbon content of the metal.

SAWING. A basic machining process in which chips are produced by a succession of small cutting edges or teeth arranged in a narrow line on a saw blade. Saw blades are made in three basic configurations: hacksaw, band saw, and circular saw.

SCLEROSCOPE. An instrument for determining the relative hardness of materials by a drop-and-rebound method.

SCREW MACHINES. Small automatic turret lathes designed for bar stock and usually equipped with an automatic rod-feeding mechanism. They are single or multiple spindle machines. The Brown and Sharpe or the Swiss Automatic are two common types.

SELECTIVE ASSEMBLY. An assembly procedure where the parts are segregated into groups in order to affect particular matings of these component parts.

SHANK. The part of the drill or cutter by which it is held and driven.

SHAPING. A basic machining process where the tool is reciprocated, cutting only during the forward stroke, and the feed is intermittent between strokes.

Shaving, Gear. The most commonly used method for gear finishing. The gear is run at high speed in contact with the shaving tool which is a hardened accurately ground gear that contains a number of peripheral gashes or grooves, thus forming a series of sharp cutting edges on each tooth. The gear and shaving cutter are run in mesh. The cutter removes very fine chips from the gear producing very accurate tooth profile.

Shear, in Metal Cutting. Chips are formed by a localized shear process which takes place over a very narrow region. The shear plane is the area over which the chip is removed from the parent metal. The shear plane is at an angle defined as the (shear angle) from the newly formed surface.

Shot Blasting. A method of abrasive cleaning to remove foreign materials from the formed workpiece. It employs blasting some type of abrasive, usually sand or steel grit or shot, which is impelled under high pressure against the surface to be cleaned.

Silicon Carbide. An abrasive made in the electric furnace from coke and silica SiC.

Single-point Tool. A tool with one cutting edge, usually used in turning operations.

Slotting. The production of a straight-sided groove in the work piece using a radial cutting tool action.

Snag Grinding. The nonprecision removal of unwanted material, such as sprues and gates in foundry operations by means of a grinding operation.

Spade Drills. Usually made of a separate two edged drilling bit held in a solid supporting shank.

Supporting Shank. Widely used for making large holes.

Spark Discharge Machining. (See ELECTRIC DISCHARGE MACHINING.)

Spot Facing. a drilling operation which is done to provide a smooth bearing area on an otherwise rough surface at the opening of a hole and normal to its axis. Counterboring tools are usually used for spot facing.

Standard Hole Tolerancing Practice. The method of allocating dimensions and tolerances in which the minimal hole size is the nominal diameter. Tolerances and allowances are then applied to the shaft as determined from the class of fit.

Standard Shaft Tolerancing Practice. The method of allocating dimensions and tolerances in which the maximum shaft size is the nominal diameter. Classes of fit is determined by varying the hole diameter.

Steady Rest. A device used to support long and slender workpieces during turning or grinding to reduce their deflection due to the force exerted by the cutting action.

Straddle Milling. The production of more than one flat surface on the workpiece while employing more than one milling cutter.

Strain Hardening. Many metals possess a unique property in that after undergoing some deformation, the metal possesses greater resistance to further plastic flow. In essence, metals become stronger when plastically deformed, a phenomenon known as strain hardening or work hardening.

Super Finishing. A variation of the honing process which employs very light, controlled pressure with rapid short strokes and copious amounts of lubricant coolant flooded over the work surface to produce very fine surface finishes.

Surface Feet Per Minute (SFPM). The linear speed of the rotating tool or workpiece. Derived by multiplying the circumference in feet by revolutions per minute.

Surface Grinding. Grinding a plane surface.

Surface Profile. (See PROFILE, SURFACE.)

Surface Roughness. (See ROUGHNESS.)

Swiss Automatic. (See SCREW MACHINES.)

Tailstock. The movable part on the other end of the lathe away from the headstock. It includes the center which is used to hold the workpieces between centers.

Tang. The flattened end of a taper shank of a rotating tool. It is intended to fit into the driving slot in the socket of the machine arbor.

Tapping. The production of screw threads in a hole.

Taps. The cutting tool for tapping operations.

Taylor's Tool Life Equation. A mathematical model developed by F. W. Taylor in 1907 relating the life of the tool to the cutting speed.

Thread Manufacturing. Several methods and machines are used to produce external threads, such as turning, rolling, and milling.

Throw-away-inserts. (See INSERTS.)

Tolerance. The total amount by which a specific dimension is permitted to vary. The tolerance is the difference between the maximum and minimum limits.

Tool and Cutter Grinder. A grinding machine used mainly to sharpen worn tools and cutters.

Tool-chip Interface. The contact surface between the chip and the tool face (on the rake face) during a cutting operation.

Tool Geometry. The dimensional characteristics of the cutting edge such as the cutting edge angles, nose radius, chamfer,.etc.

Tooling. All the equipment used for fixturing workpieces and holding the tools and cutters in metal cutting operations; *e.g.* chucks, holders, face plates, jigs.

Tool Life. The period of time, usually given in minutes, during which the tool performs its function before it is worn-out and needs to be resharpened or replaced.

Toolmaker's Microscope. An optical microscope with adjustable vernier table used mainly to study the cutting edges of tools and cutters.

Tool Materials. The materials from which tools and cutters are made such as high speed steel, carbides, ceramics.

Toolroom Lathe. Precision engine lathe built to closer tolerances for machining of parts with high degree of accuracy.

Tool Wear. The progression of deterioration of the cutting edge of the tool leading to tool failure.

Transfer Machine. Highly automated machines for large volume production of a very small variety of parts (sometime called Detroit lines).

Transition Fit. A fit between mating parts having limits of size so prescribed as to partially or wholly overlap, so that either a clearance or interference may result in assembly.

Trepanning. A drilling operation to produce large holes using a thin, cylindrical cutter and maintaining a core from the drilled hole.

Tribology. The subjects of friction, wear, and lubrication as applied to surface structure of component parts.

Trueing. In grinding, the term applied to the methods of remaking the wheel into a true circle usually by dressing.

Tungsten Carbide (WC). The most commonly used nonferrous carbide material for cutting tools. They are usually made in the form of throw-away inserts.

Turning. The production of a curvilinear surface when the cutting tool moves in a direction parallel to the axis of rotation of the workpiece. Turning is usually done on turning machines: engine lathe, turret lathe, screw machine, etc.

Turret Lathes. Lathes which carry out multiple cutting operations, such as turning, boring, drilling, thread cutting, and facing. Various cutting tools, usually up to six (because of the hexagonal turret) are installed on the turret which is located usually on the other end of the machine away from the headstock.

Twist Drill. The most commonly used drilling tool, usually with two cutting edges and two flutes.

Ultraprecision Machining. Machining process that produces surface finish in the tens of nanometer $(10^{-9}m)$ range, and form accuracies in the µm range or better. The process now uses cutting tools made exclusively of a single crystal diamond, hence the process is also called diamond turning. Applications are seen in manufacturing optical mirrors, computer memory disks, and drum for photo-copying machines. The workpiece materials used to date include copper allouys, aluminum alloys, silver, gold, electroless nickel and plastics (acrylics).

Ultrasonic Machining. The removal of material by high frequency vibration of a tool that imparts energy to abrasive grains introduced between the tool and the workpiece by a slurry. The abrasive crystals then impart hammer blows to the workpiece in removing minute particles of unwanted material.

Unilateral Tolerance. The form in which variation is permitted in only one direction from the basic size. (See TOLERANCE, BILATERAL TOLERANCE.)

Unit Power Consumption. For each different material

to be machined, a measure of the power required is the unit horsepower (uhp), also called specific power consumption (shp). It is defined as the horsepower required to remove one cubic inch of the material in one minute.

UNIVERSAL DIVIDING HEAD. An accessory for milling machines. It is a fixture that rotates the workpiece to specified angles between individual machining steps. Typical uses are in milling parts with polygonal surfaces and in machining gear teeth.

UNIVERSAL GRINDING MACHINE. A machine on which cylindrical, taper, internal, or face grinding can be done as required in toolrooms and machine shops.

VELOCITY, CUTTING SPEED. (See SURFACE FEET PER MINUTE, REVOLUTIONS PER MINUTE.)

VERTICAL MACHINE TOOLS. Machine tools where the power driven spindle which usually carries the cutting tool is vertical e.g. vertical grinding, boring, milling machines and vertical turret lathes.

VISES. Different kinds of work-holding devices usually having one moving jaw and one stationary jaw.

VITRIFIED BOND WHEELS. Grinding wheels where the abrasives are bound together with vitrified ceramic, usually clay and flux.

WATER-JET MACHINING. High pressure jets of water used in various processes such as cutting, and deburring. Water pressure of 1,000 psi at the nozzle is common.

WAVINESS. This consists of recurrent widespread irregularities in the form of waves on which roughness is superimposed.

WEAR, TOOL. (See TOOL WEAR.)

WHEEL, GRINDING. A straight or formed abrasive tool which is rotated to remove metal during the grinding process.

WIRE FEED MACHINE. A numerically controlled EDM machine that uses copper or brass wire for electrode. Cutting much like the band saw, the wire cuts any electrical conductive material.

WORK HOLDING DEVICES. The tooling which is needed to hold workpieces either stationary or moving to perform machining operations.

WORM. A gear with one or more teeth in the form of screw thread.

WORM GEAR. A gear mating with a worm.

PLASTICS

A-STAGE. An early stage in the reaction of certain thermosetting resins, in which the material is still soluble in certain liquids and fusible. Sometimes referred to as resol. (See C-STAGE.)

ABS. (See ACRYLONITRILE BUTADIENE STYRENE.)

ABSORPTION. (1) The penetration into the mass of one substance by another (2) The process whereby energy is dissipated within a specimen placed in a field of radiation energy. Since processes other than absorption occur, e.g., scattering, in which only a fraction of the energy removed from a beam is retained in the specimen. The total amount of energy removed from a well-collimated beam will be greater than the amount actually dissipated within the sample.

ACETAL RESIN. The molecular structure of this polymer is that of a linear acetal, consisting of unbranched polyoxymethylene chains.

ACID-ACCEPTOR. A compound which acts as a stabilizer by chemically combining with acid which may be initially present in minute quantities in a plastic, or which may be formed by the decomposition of the resin.

ACRYLIC PLASTICS. Plastics based on resins made by the polymerization of acrylic monomers, such as ethyl acrylate and methyl methacrylate.

ACRYLIC RESIN. A synthetic resin prepared from acrylic acid or from a derivative of acrylic acid.

ACRYLONITRILE BUTADIENE STYRENE (ABS). Acrylonitrile and styrene liquids and butadiene gas are polymerized in a variety of ratios to produce the family of ABS resins.

ACTIVATION. The process of inducing radioactivity in a specimen by bombardment with neutrons and other types of radiation.

ADDITIVES. Products that are combined with resins and polymers as extenders or modifiers to alter the properties of the base polymer.

ADHEREND. A body which is held to another body by an adhesive.

ADHESION. The state in which two surfaces are held together by interfacial forces which may consist of valence forces or interlocking action or both. (See ADHESION, MECHANICAL; ADHESION, SPECIFIC.)

ADHESION, MECHANICAL. Adhesion between surfaces in which the adhesive holds the parts together by interlocking action. (See ADHESION, SPECIFIC.)

ADHESION, SPECIFIC. Adhesion between surfaces which are held together by valence forces of the same type as those which give rise to cohesion. (See ADHESION, MECHANICAL.)

ADHESIVE. A substance capable of holding materials together by surface attachment. Adhesive is the general term and includes among others cement, glue, mucilage, and paste. All of these terms are loosely used interchangeably. Various descriptive adjectives are applied to the term adhesive to indicate certain characteristics, as follows: physical form, *e.g.* liquid adhesive; tape adhesive chemical type *e.g.* silicate adhesive; resin adhesive materials bonded, *e.g.*, paper adhesive; metal-plastic adhesive; can-label adhesive conditions for use, *e.g.* hot-setting adhesive.

ADHESIVE, ASSEMBLY. An adhesive which can be used in bonding parts together, such as in the manufacture of a boat, airplane, furniture, and the like. The term assembly adhesive is commonly used in the wood industry to distinguish such adhesives (formerly called joint glues) from those used in making plywood (sometimes called veneer glues). It is applied to adhesives used in fabricating finished structures of goods, or subassemblies thereof as differentiated from adhesives used in the production of sheet materials for sale as such, for example, plywood or laminates.

ADSORPTION. A concentration of a substance at a surface or interface of another substance.

AGING. (1) The effect of exposure of plastics to an environment for an interval of time. (2) The process of exposing plastics to an environment for an interval of time. Aging, according to the first definition, may be very great or not, depending on the environmental conditions, such as radiating, moisture, temperature, surrounding gases, etc., and the period of exposure.

AIR-ASSIST FORMING. A method of thermoforming in which air flow or air pressure is used to partially preform the plastics sheet immediately prior to the final pulldown onto the mold using vacuum.

AIR-SLIP FORMING. A variation of snap-back forming in which the male mold is enclosed in a box in such a manner that when the mold moves forward toward the hot plastics, air is trapped between the mold and the plastics sheet. As the mold advances, the plastic is kept away from it by the air cushion until the full travel is attained, at which point a vacuum is applied, removing the air cushion and forming the part against the plug.

AIR VENT. Small outlet usually a groove, to prevent entrapment of gases.

ALKYD PLASTICS. Plastics based on resins composed principally of saturated polymeric esters, in which the recurring ester groups are an integral part of the main polymer chain, and in which ester groups occur in any cross-links that may be present between chains. (POLYESTER PLASTICS.)

ALL-VENEER CONSTRUCTION. Plywood without lumber cores, more frequently multiple for strength requirements, often 7-ply or 9-ply, to equal the thickness of conventional member-core plywood. The maximum thickness of any single sheet of veneer seldom exceeds 1/4".

ALLOY. Composite material produced by blending polymers or copolymers with other polymers or elastomers under controlled conditions.

ALLYL DIGLYCOL CARBONATE (ADC). A crystal clear thermosetting plastic with outstanding scratch resistance; used for goggles, etc.

ALLYL PLASTICS. Plastics based on resins made by additional polymerization of monomers containing ally groups, such as diallyl phthalate.

AMBIENT TEMPERATURE. Temperature of the medium surrounding an object.

AMINO PLASTICS. Plastics based on resins made by the condensation of amines such as urea and melamine, with aldehydes.

AMORPHOUS POLYMERS. Polymers that have randomly oriented molecular structure with no definite order or regularity.

ANCHORAGE. Part of the insert that is molded inside of

the plastic and held fast by the shrinkage of the plastic.

Aniline Formaldehyde Resins. Member of the aminoplastics family made by the condensation of formaldehyde and aniline in an acid solution. These resins are thermoplastic and have high dielectric strength.

Anneal (1) To heat a molded plastic article to a predetermined temperature and slowly cool it to relieve stresses. (2) To heat steel to a predetermined temperature above the critical range and slowly cool it, to relieve stresses and reduce hardness.

Antioxidant. A substance that prevents or slows down oxidation of plastics material that is exposed to air.

Antistatic Agents. Materials and treatments used during or after the molding process to minimize static electricity in plastics materials.

Assembly. The collection of and placing together in proper order of the layers of veneer, lumber, and/or other materials with the adhesive ready to be pressed and bonded into a single unit.

Assembly Time. In adhesives, refers to the elapsed time after the adhesive is spread and until the pressure becomes effective.

Atactic. A random arrangement of the unsymmetrical group with respect to the carbon-carbon backbone plane.

Autoclave. A closed vessel capable of withstanding high temperatures and pressures. Useful for conducting chemical reactions.

Back-pressure-relief Port. An opening from an extrusion die for escape of excess material.

Backing Plate. A plate which backs up the cavity blocks, guide pins, bushings, etc. (Sometimes called support plate.)

Back Taper. Reverse draft used in mold to prevent molded article from drawing freely. (See UNDERCUT.)

Bag Molding. A method of applying pressure during bonding or molding in which a flexible cover exerts pressure on the material being molded, through the application of air pressure or the drawing of a vacuum.

Bakelite. The proprietary name for phenolic and other plastic materials produced by the Union Carbide Corp.

Blanket. Veneers which have all been laid up on a flat table. The complete assembly is placed on or in the mold all at one time; useful only on simple curved surfaces to be molded by the flexible-bag process. Also used to denote a form of bag made of rubber in which the edges are sealed against the mold by clamps.

Bleeding. (1) The flow of color from one region into an adjoining region. (2) In the manufacture of plywood, the escape of a portion of the steam-air mixture during cooking, to permit mixing of the steam and air and to maintain uniform temperature at all levels in the autoclave, when molding with flexible pressure.

Blind Hole. Hole that is not drilled entirely through.

Blind-hole Partial Thread. Thread counterbored from the front for terminal or other assembly fit.

Blister. (1) Undesirable rounded elevation of the surface of a plastic, whose boundaries may be indefinitely outlined, somewhat resembling in shape a blister on the human skin. A blister may burst and become flattened. (2) An elevation of the surface of an adherend, somewhat resembling in shape a blister on the human skin; its boundaries may be indefinitely outlined and it may have burst and become flattened. A blister consists of trapped air, water, or solvent, and can be caused by insufficient adhesive, inadequate curing time, temperature, or pressure.

Blocking. An adhesion between touching layers of plastics, such as that which may develop under pressure during storage or use.

Bloom. A visible exudation or efflorescence on the surface of a plastic. Bloom can be caused by lubricant, plasticizer, etc.

Bolster. Spacer or filler in a mold.

Bond. The attachment at an interface between an adhesive and an adherent. (2) To attach materials together by adhesives.

Bubble. Internal void or a trapped globule of air or other gas. (See JOINT.)

Bulk Density. The density of a molding material in loose form (granular, nodular, etc.) expressed as a ratio of weight to volume (*e.g.* g/cm^3 or b/ft^3).

Bulk Factor. The ratio of the volume of the loose molding compound to the volume of the same quantity in a molded solid piece. Also the ratio of the density of the solid plastic object to the apparent density of the loose molding powder.

Burned. Showing evidence of thermal decomposition through some discoloration, distortion, or destruction of the surface of the plastic.

Butadiene. A gas, insoluble in water but soluble in alcohol and ether, obtained from the cracking of petroleum and other methods. Butadiene is widely used in forming copolymers with styrene and other monomeric substances.

Butadiene Styrene. A thermoplastic polymer used for film and sheet.

C-stage. The final stage in the reactions of a thermosetting resin in which the material is relatively insoluble and infusible. Thermosetting resins in fully cured plastics are in this stage.

Carburize. To increase carbon content of surface of any steel

Cast. (1) The act of forming a plastics object by pouring a fluid monomer-polymer into an open mold, where it finishes polymerizing. (2) The act of forming plastics film and sheet by pouring liquid resin onto a moving belt or by precipitation in a chemical bath.

Cast Film. A film made by depositing a layer of liquid plastic onto a surface and stabilizing this form by evaporation of solvent, by fusing after deposition, or by allowing a melt to cool. Cast films are usually made from solutions or dispersions.

Catalyst. A substance which when added in minor proportion accelerates a chemical reaction.

Caul. In plywood manufacture, a sheet the size of the platens used in hot pressing. Aluminum sheet approximately 1/16 in. thick is generally used. Plywood assemblies are inserted between a pair of cauls, to facilitate loading the press, and to protect plywood faces from contact with the steel plates of the hot press. At one time plywood cauls, 1/16" to 1/4" thick were used; they were replaced by aluminum for quicker heating and better durability.

Caul, Canvas-covered Plywood. A special type of plywood caul used in progressive gluing, to control heat and moisture. Also used for two-plying fragile veneer faces in hot presses.

Caul, Plywood. Used in cold-pressing with conventional adhesives, to assure undamaged faces and to prevent transmission of defects to adjacent assemblies. Usually 1/4" to 3/8" thick, with waxed surfaces to prevent adhesion.

Cavity. That portion of the mold which forms the outer surface of the molded article.

Cell. A single cavity formed by gaseous displacement in a plastic material. (See CELLULAR PLASTIC.)

Cellular Plastic. A plastic whose density is decreased substantially by the presence of numerous cells disposed throughout its mass. (See CELL, FOAMED PLASTICS.)

Cellular Striation. A layer of cells within a cellular plastic which differs greatly from the characteristic cell structure of the material.

Cellulosic Plastics. Plastics based on cellulose compounds, such as esters (cellulose acetate) and ethers (ethyl cellulose).

Cement. (See ADHESIVE, BOND.)

Centipoise. A unit of viscosity, conveniently and approximately defined as the viscosity of water at room temperature. The following table of approximate viscosities at room temperature may be useful for rough comparisons: liquid viscosity in centipoises water 1 kerosene 10 motor oil SAE 10 100 castor oil; glycerine 1000 corn syrup 10000 molasses 100000. Note: Under SL, centipoise has been replaced by pascal-second (Pa-s) where 1 centipoise = 0.001 Pa-s.

Centrifugal Casting. A method of forming thermoplastic resins in which the granular resin is placed in a rotatable container, heated to a molten temperature, and rotated to force the liquid resin to conform to the shape of the container.

Chalking-dry. Chalk-like appearance or deposit on the surface of a plastic. (See HAZE, BLOOM.)

Channel. (See PORT.)

Chase. The main body of the mold which contains the molding cavity or cavities, or cores, the mold pins, the guide pins (or the bushings), etc.

Chase Ring. A ring used in hobbing to restrain the blank against spreading during the sinking of the hob.

Chemically Formed Plastic. A cellular plastic whose structure is produced by gases generated from the chemical interaction of its constituents.

Clamping Plate. A mold plate fitted to the mold and used to fasten the mold to the machine.

Clamp Irons. In plywood manufacture, the pressure-maintenance equipment, which includes the "I" beams or double channel irons, together with clamp screws or turnbuckle rods, to hold bales under pressure after cold gluing.

Closed-cell Foam. A cellular plastic in which there is a predominance of non-interconnecting cells.

Coextrusion. A process for extruding two or more materials in a single film or sheet. Two identical polymers can be laminated by coextrusion. Two or more extruders are used, and the extrudate is usually moved through a common die.

Cohesion. The forces holding a single substance together.

Cold Flow. (See CREEP.)

Cold Molding. The shaping of an unheated compound in a mold under pressure, followed by heating the article to cure it.

Cold Pressing. A bonding operation in which an assembly is subjected to pressure without the application of heat.

Cold Slug. The first material to enter an injection mold, so called because in passing through the sprue orifice it is cooled below effective molding temperature.

Cold-slug Well. Space provided directly opposite the sprue opening of the injection mold to trap the cold slug.

Commodity Plastics. The term commodity identifies a group of plastics that are characterized by high-volume usage and general availability from sources other than the prime resin suppliers. Polyethylene, polystyrene, and polyvinylchloride are examples of commodity plastics.

Compregnated Wood. A consolidation of the term compressed-impregnated wood, referring usually to an assembly of layers of veneer impregnated with a liquid resin and bonded under very high pressures. More commonly, but not always, the veneer layers have parallel grain, *i.e.*, laminated wood construction.

Compression Molding. A technique for thermoset molding in which the molding compound is placed in the open mold cavity, the mold is closed, and heat and pressure are applied until the material is cured.

Compression Ratio. In an extruder screw, compression ratio is the ratio of volume available in the first flight at the hopper to the last flight at the end of the screw.

Condensation. A chemical reaction in which two or more molecules combine, with the separation of water or some other simple substance. If a polymer is formed, the process is called polycondensation. (See POLYMERIZATION.)

Consistency. The resistance of a material to flow or permanent deformation when shearing stresses are applied to it. The term is generally used with materials whose deformations are not proportional to applied stresses. Viscosity is generally considered to be a similar internal friction that results in flow in proportion to the stress applied. (See VISCOSITY, VISCOSITY COEFFICIENT.)

Continuous Tube Process. A blow-molding process that uses a continuous extrusion of tubing to feed into the blow molds.

Cooling Fixture. Equipment used to hold molded article while cooling after removal from the mold; it may be channeled to be cooled with water.

Copolymer. (See HOMOPOLYMER.)

Copolymerization. (See POLYMERIZATION.)

Core. That portion of the mold that forms the required inner surfaces of the molded article.

Cores, Centers. A term usually applied to the central layer of plywood, which in lumber-core construction is the principle strength factor. It is also applied when the central layer is a veneer. The terms are sometimes used in the Pacific Northwest to designate the layer that is spread with adhesive, which agrees with the above in 3-ply construction but is inconsistent when applied to 5-ply. Under this latter usage, center is used to indicate the middle ply. Core may refer also to the remaining part of the log that is too small to be cut into rotary veneer on a lathe.

CRAZING. Development of fine cracks on the surface of a plastic, sometimes extending into the body of the material.

CREEP. The dimensional change with time of a plastic under load, following the instantaneous elastic or rapid deformation. It is the permanent deformation resulting from prolonged application of a stress below the elastic limit. Creep at room temperature is sometimes referred to as cold flow.

CROSSBANDING. The transverse veneer layers that distinguish plywood from laminated wood. Their presence counteracts the tendency of wood to split, as well as to shrink and swell. In standard 5-ply construction, it is the layer between the face and the core and between the back and the core, sometimes called face crossing and back crossing, respectively.

CROSS-LINKING. Applied to polymer molecules, the setting-up of chemical links between the chains. The thermoset plastics are cross-linked at the molecular level.

CRYSTALLINITY. A state of molecular structure in some resins which denotes uniformity and compactness of the molecular chains that form the polymer.

CULL. Material remaining in the transfer pot after the mold has been filled.

CURE. The act of changing the physical properties of a material by chemical reaction, which may be condensation or addition-type polymerization, or vulcanization; accompanied by heat and catalysts, with or without pressure. For room temperature curing systems, heat is generated by an exothermic reaction.

DEGRADATION. A deleterious change in the chemical structure of a plastic. (See DETERIORATION.)

DELAMINATION. The separation of the layers in a laminate.

DETERIORATION. A permanent change in the physical properties of a plastic evidenced by impairment of these properties.

DIAPHRAGM GATE. Gate used in molding annular or tubular articles.

DIE-ADAPTOR. That part of an extrusion die which holds the die block

DIE BLOCK. That part of an extrusion die which holds the forming bushing and core.

DIE BODY. That part of an extrusion die used to separate and form material

DIELECTRIC HEATING. (See HIGH-FREQUENCY HEATING.)

DIFFUSION. The movement of a material in the body of a plastic.

DILATANT. A dilatant fluid, or inverted pseudoplastic, is one whose apparent viscosity increases instantaneously with increasing rate of shear; i.e. the act of stirring creates instantly an increase in resistance to stirring.

DILUENT. In an organosol, a liquid component which has little or no solvating action on the resin, its purpose being to modify the action of the dispersant.

DISC GATE. (See DIAPHRAGM GATE.)

DISHED. Showing a symmetrical distortion of a flat or curved section of a plastic object so that, as normally viewed, it appears concave, or more concave than intended. (See WARP.)

DISPERSANT. In an organosol, a liquid component which has a solvating or peptizing action on the resins, so as to aid in dispersing and suspending it.

DISPERSION. A heterogeneous system in which a finely divided material is distributed in a matrix of another material. In plastics technology, a dispersion is usually the distribution of a finely divided solid in a liquid or solid. For example: pigments or fillers in molded plastics, plastisols, or organosols. (See PLASTISOL, ORGANOSOL.)

DOCTOR ROLL; DOCTOR BAR. A device for regulating the amount of liquid material on the rollers of a spreader.

DOMED. Showing a symmetrical distortion of a flat or curved section of a plastic object, so that, as normally viewed, it appears more convex than intended. (See WARP.)

DOPING. Coating the mold or mandrel with a substance which will prevent the molded plywood part from sticking to it and will facilitate easy removal.

DOWEL. Pin used to maintain alignment between two or more parts of a mold.

DRAFT. The degree of the taper of the side wall or the angle of clearance designed to facilitate removal of molded article from the mold.

Drape-assist Frame. In sheet thermoforming, a frame (made of wires or bars) shaped to the peripheries of the depressed areas of the mold and suspended above the sheet that is to be formed.

Drape Forming. A method of forming a thermoplastic sheet in which the sheet is clamped into a movable frame, heated, and draped over the high points of a male mold. Vacuum or air pressure is then applied to complete the forming operation.

Dry Coloring. The method commonly used for coloring plastics by tumble blending uncolored particles of the plastics material with selected dyes and pigments.

Dry Spot. Area of incomplete surface film on laminated plastics; in laminated glass, an area over which the interlayer and the glass are not bonded.

Duplicate Cavity-plate. Removal plate that retains cavities, used where two-plate operation is necessary for loading inserts, etc.

Durometer Hardness. The hardness of a material as measured by the Shore Durometer.

Elasticity. That property of material by virtue of which it tends to recover its original size and shape after deformation. If the strain is proportional to the applied stress, the material is said to exhibit Hookean or ideal elasticity.

Elastomer. A material which, at room temperature, stretches under low stress to at least twice its length and snaps back to the original length upon release of the stress. The term elastomer is commonly applied to synthetic rubber polymers.

Engineering Plastics. Thermoset and thermoplastic materials whose characteristics and properties enable them to withstand mechanical loads (tension, impact, flexure, vibration, friction, etc.) combined with temperature changes. The reliability and predictability of engineering plastics makes them suitable for application in structural and load-bearing product design elements. Polycarbonate, ABS, acetal, and nylon are among the widely used engineering plastics.

Envenomation. The process by which the surface of a plastic close to or in contact with another surface is deteriorated. Softening, discoloration, mottling, crazing, or other effects may occur.

Epoxy Plastics. Plastics based on resins made by the reaction of epoxides or oxiranes with other materials such as amines, alcohols, phenols, carboxylic acids, acid anhydrides, and unsaturated compounds.

Ethylene Plastics. Plastics based on resins made by the polymerization of ethylene or copolymerization of ethylene with other unsaturated compounds.

Ethylene Tetrafluorethylene Copolymer (ETFE). A moldable variety of fluorocarbon (fluoroplastic) polytetrafluorethylene (PTFE), for example, teflon (TFE).

Exotherm; Exthermic Heat. Heat generated by a chemical reaction; in polymers a particular problem since the heat must be dissipated.

Expandable Plastic. A plastic which can be made cellular by thermal, chemical, or mechanical means.

Expanded Plastics. (See FOAMED PLASTIC.)

Extender. A substance, generally having some adhesive action, added to an adhesive to reduce the amount of the primary binder required per unit area. (See BINDER, FILLER.)

Extraction. The transfer of a material from a plastic to a liquid in contact with it.

Extrusion. A method whereby heated or unheated plastic forced through a shaping orifice becomes one continuously formed piece.

Extrusion, Autothermal. A method of extrusion in which the conversion of the drive energy, through friction, is the sole source of heat.

Fabricating; Fabrication. The manufacture of plastic products from molded parts, rods, tubes, sheeting, extrusions, or other forms by appropriate operations such as punching, cutting, drilling, and taping. Fabrication includes fastening plastic parts together or to other parts by mechanical devices, adhesives, heat-sealing, or other means.

Face. In plywood manufacture, the veneer on the exposed surface of the plywood.

False Body. The deceptively high apparent viscosity of a pseudoplastic fluid at a low rate of shear.

Family Mold. A multicavity mold containing individual cavities of different sizes or designs.

FILLER. An inert substance added to plastics resin to reduce cost and improve physical properties. The filler particles usually are small, in comparison to those used in reinforcements, but there is some overlap between the two additives.

FILLER-SPECKS. Visible specks of a filler, such as woodflour or asbestos, which stand out in color contrast against the background.

FILM. (1) Sheeting having a nominal thickness not greater than 0.010 in. (2) In adhesives, a thin, dry sheet of paper impregnated with a phenol-formaldehyde resin adhesive. Film sometimes is used to refer to a liquid coating of adhesive.

FIN. That part of the flash which remains attached to the molded article.

FINISHING. The removal of defects or development of desired surface characteristics on plastic products.

FISHEYE. Small globular mass which has not blended completely into the surrounding material; particularly evident in a transparent or translucent material.

FLAKE. The dry, unplasticized base of cellulose plastics.

FLASH. That portion of the charge which flows from or is extruded from the mold cavity during the molding.

FLASH RIDGE. That part of a flash mold along which the excess material escapes until the mold is closed.

FLAT GRAIN. Grain produced in approximately a tangential direction, or plain-cs veneers.

FLOATING CHASE. Mold member, free to move vertically, which fits over a lower plug or cavity, and into which an upper plug telescopes.

FLOCK. Short fibers of cotton and other materials, used as fillers for molding materials.

FLOW MARK. A visible mark resulting from the solidification of a pattern of flow.

FLUOROCARBONS. The family of plastics including polytetrafluoroethylene (PTFE), polychlorotrifluoroethylene (PCTFE), polyvinylidence and fluorinated ethylene propylene (FEP).

FLUXING TEMPERATURE. (See FUSION TEMPERATURE.)

FOAM. (See CELLULAR PLASTIC.)

FOAMED PLASTICS. Resins in sponge form.

FOAM MOLDING. A molding process whereby heat-softened plastics containing a foaming (blowing) agent are injection molded into a cavity where they harden to produce a product that has a solid skin contiguous with a foam core. (See STRUCTURAL FOAM, SANDWICH MOLDING.)

FORCE, FORCE PLUG. The force is the member which transmits the pressure from the press unit to the top of the molding charge. In positive molds, there is an extension to or protuberance beneath the force which telescopes within the molding cavity (or loading chamber) and which is termed a force plug (compression molding). (When used in reference to injection molding, See CORE.)

FORCE PLATE. Plate for holding force plug or plugs in compression molding.

FORMING. A process in which the shape of plastic pieces such as sheets, rods, or tubes, is changed to a desired configuration. The use of the term forming in plastics technology does not include such operations as molding, casting, or extrusion, in which shapes or articles are made from molding materials or liquids.

FURANE PLASTICS. Plastics based on resins in which the furane ring is an integral part of the polymer chain, made by the polymerization or polycondensation of furfual, furfuryl alcohol, or other compounds containing a furane ring, or by the reaction of these furane compounds with not more than an equal weight of other compounds. Furane is also spelled furan.

FUSION. In vinyl dispersions the heating of a dispersion to produce a homogeneous mixture. There is an apparent mutual solvation of the resin and plasticizer.

FUSION TEMPERATURE. In vinyl dispersions the temperature at which fusion occurs; also called fluxing temperature.

GAMMA TRANSITION; GLASSY TRANSITION. The change in an amorphous polymer or in an amorphous region of a partially crystalline polymer from (or to) a viscous or rubbery condition to (or from) a hard and relatively brittle one. This transition generally occurs over a relatively narrow temperature region and is similar to the solidification of a liquid to a glassy state; it is not like a first order phase transformation where discontinuities occur. Not only do hardness and brittleness undergo rapid changes in this temperature region, but other properties such as thermal expansibility and specific heat also change rapidly. This phenomenon has been called second-order transition, glass transition, rubber transition, and rubbery transition. The word transformation has also been used instead of transition.

GAMMA-TRANSITION TEMPERATURE; GLASSY-TRANSITION TEMPERATURE. The temperature region in which the gamma or glassy transition occurs. The measured value of gamma- or glassy-transition temperature depends to some extent on the details of the method of test. (See GAMMA TRANSITION.)

GATE. In injection and transfer molding, the gate is the orifice through which the melt enters the cavity.

GEL. (1) A semisolid system consisting of a network of solid aggregates, in which liquid is held. (2) The initial jelly-like solid phase which develops during the formation of a resin from a liquid. Both types of gel have very low strengths and do not flow like a liquid. They are soft and flexible, and will rupture under their own weight unless supported externally.

GELATION. (1) Formation of a gel. (2) In vinyl dispersions, formation of gel in the early stages of fusion.

GEL POINT. The stage at which a liquid begins to exhibit pseudoelastic properties. This stage may be conveniently observed from the inflection point on a viscosity-time plot. (See GEL.)

GLASS. An inorganic product of fusion which has cooled to a rigid condition without crystallizing. Glass is typically hard and relatively brittle, and has a conchoidal fracture.

GLASS FINISH. A material applied to the surface of a glass reinforcement to improve its effect upon the physical properties of the reinforced plastic.

GLASS TRANSITION. (See GAMMA TRANSITION.)

GLASS TRANSITION TEMPERATURE. A reversible change that occurs in an amorphous polymer when it reaches a certain temperature range in which the material undergoes a transition from a hard, brittle, glassy state to a flexible condition. At the glass transition temperature, the polymer chains become free to rotate and to slide past each other. Above the so-called "glass temperature," the material acts like a viscous liquid; below the glass temperature, it behaves like a solid.

GLUE. Originally, a hard gelatin obtained from hides, tendons, cartilage, bones, etc., of animals. Also an adhesive prepared from this substance by heating with water. Through general use the term is now synonymous with the term adhesive. (See ADHESIVE, SIZING.)

GLUE JOINT. That part of an aggregated product which comprises the adhesive (or glue) and the parts in contact therewith.

GRANULAR STRUCTURE. Apparent incomplete fusion of, and at least partial retention of, their original form by the particles from which a plastic is formed.

GRID. Channel shaped mold-supporting members.

GRINDING-TYPE RESIN. A vinyl resin which requires grinding to effect dispersion in plastisols or organosols.

GUIDE PIN. Pin which assures alignment of the mold halves.

GUIDE-PIN BUSHING. A guiding bushing through which the leader pin moves.

GUM. Any of a class of colloidal substances, exuded by or prepared from plants sticky when moist, composed of complex carbohydrates and organic acids, which are soluble or swell in water. The term gum is sometimes used loosely to denote various materials that exhibit gummy characteristics under certain conditions, for example, gum balata, gum benzion, and gum asphaltum. Gums are included by some in the category of natural resins. (See ADHESIVE, GLUE, RESIN.)

HALOCARBON PLASTICS. Plastics based on resins made by the polymerization of monomers composed only of carbon and a halogen or halogens.

HARDENER. A substance or mixture of substances added to an adhesive to promote or control the curing reaction by taking part in it. The term is also used to designate a substance added to control the degree of hardness of the cured film. (See CATALYST.)

HAZE. Indefinite cloudy appearance within or on the surface of a plastic, not describable by the terms chalking, bloom.

HEAD BLOCK; RETAINER BOARD. A thick (3 to 5 in.) large piece of laminated lumber, usually with veneer crossing, used for bottom and top of a bale of plywood, during pressing and clamping.

Heater-adapter. That part of an extrusion die around which heating medium is held.

Heat Forming. (See THERMOFORMING.)

High-density Plywood. Plywood of special construction, made at high specific pressure, usually 500 psi and up. With the increase in pressure comes a corresponding increase in density, or specific gravity.

High-frequency Heating. The heating of materials by dielectric loss in a high-frequency electrostatic field. The material is exposed between electrodes, and by absorption of energy from the electrical field is heated quickly and uniformly throughout.

Hob. A master model used to sink the shape of a mold into a soft steel block.

Hobbing. A process of forming a mold by forcing a hob of the shape desired into a soft steel blank.

Hold-down Groove. A small groove cut into the side wall of the molding surface to assist in holding the molded article in that member while the mold opens.

Homopolymer/Copolymer. These terms are used frequently to differentiate between single and multiple monomers. For example, among acetal resins, Delrin is a homopolymer, whereas Celcon is a copolymer. Polypropylene is available in both molecular forms.

Hot-runner Mold. A mold in which the runners are insulated from the chilled cavities and are kept hot so the material can be used again.

Hot-short. Inelastic, nonstretchable, and easily broken by tension when hot.

Hub. (See HOB.)

Hubbing. (See HOBBING.)

Hydrocarbon Plastics. Plastics based on resins composed of carbon and hydrogen alone.

Inhibitor. A substance which prevents or retards a chemical reaction.

Injection Blow Molding. A blow molding process in which the parison to be blown is formed by injection molding and then blow molded as a secondary operation.

Injection Mold. A mold into which a plasticated material is introduced from an exterior heating cylinder.

Injection Molding. A molding procedure whereby heat-softened thermoplastic or thermoset material is forced from a plasticating barrel into a relatively cool mold cavity for hardening.

Injection Pressure. The pressure on the face of the injection ram and the pressure at which molding material is injected into the mold.

Insert. An object molded or cast into a plastic article for a definite purpose.

Insert, Eyelet-type. Insert having a section which protrudes from the material and is used for spinning over in assembly.

Insert, Floating of. When pressure is applied on the mold and the material softens, it flows upwards. If the insert is loose on the retaining pin, some of the material flows under the insert or into the anchorage points and carries or floats the insert off the retaining pin.

Insert, Flow of Material Into. If a threaded open-hole insert is being molded, the retaining pin does not prevent the material from flowing into the insert unless the retaining pin is threaded, which would increase the cost of molding.

Insert, Open Hole. One having a hole drilled completely through it.

Insert, Protruding. One having a part which protrudes from the molded material.

Insert, Rivet. One having a protruding part which is riveted in assembly.

Insert, Through-type. One which is exposed on both sides of the molded article.

Inventory. In injection molding or extrusion, the amount of plastic contained in the heating cylinder or barrel.

Irradiation. The subjection of a material to radiant energy for the purpose of producing a desired effect or of determining the effect of the radiant energy on the material.

Isocyanate Plastics. Plastics based on resins made by the condensation of organic isocyanates with other compounds. (URETHANE PLASTICS.)

Isocyanate Resins. Most applications for this resin are based on its combination with polyols (polyesters, polyethers, etc.). During the reaction, the ingredients join through formation of the urethane linkage; hence, this technology is generally known as urethane chemistry.

Isotactic. Pertaining to a type of polymeric molecular structure containing sequences of regularly spaced asymmetric atoms arranged in like configuration in the main polymerchain. Materials containing isotactic molecules may exist in high crystalline form, because of the high degree of order that may be imparted to such structure.

Joint. The location at which two adherends are held together with a layer of adhesive. (See BOND.)

Joint, Butt. A type of edge joint in which the edge faces of the two adherends are at right angles to the other faces of the adherends.

Joint, Edge. A joint made by bonding the edge faces of two adherends with adhesives.

Joint, Lap. A joint made by placing one adherend partly over another and bonding together the overlapped portions.

Joint, Scarf. A joint made by cutting away similar angular segments of two adherends and bonding the adherends with the cut areas fitted together.

Joint, Starved. A joint which has an insufficient amount of adhesive to produce a satisfactory bond. This condition may result from too thin a spread to fill the gap between the adherends, excessive penetration of the adhesive into the adherend, too short an assembly time, or the use of excessive pressure.

Knit Line. (See WELD LINE.)

Knockout. Any part or mechanism of a mold used to eject the molded article.

Knockout Pin. (See EJECTOR PIN.)

Laminate. A product made by bonding together two or more layers of material or materials.

Laminated, Cross. Pertaining to a laminate in which some of the layers of materials are oriented at right angles to the remaining layers with respect to the grain or strongest direction in tension. Balanced construction of the laminations about the center line of the thickness of the laminate is normally assumed. (See LAMINATED, PARALLEL.)

Laminated, Parallel. Pertaining to a laminate in which all the layers of material are oriented approximately parallel with respect to the grain or strongest direction in tension.

Laminated Wood. An assembly of wood layers in which the wood grain or fillets of the adjacent layers are parallel, contrasted with plywood, which is characterized by cross layers or crossing, usually alternated with the parallel face, core, and back layers.

Land. (1) The portion of a mold which provides the separation or cutoff of the flash from the molded article. (2) In the screw of an extruder, the bearing surface along the top of the flights. (3) In an extrusion die, the surface parallel to the flow of material.

Landed Force. Force with shoulder which seats on land in landed positive mold.

Landed Plunger. (See LANDED FORCE.)

Latch. Device used to hold two members of a mold together.

L/D Ratio. A term used to define an extrusion screw. Denotes the ratio of screw length to diameter.

Lignin Plastics. Plastics based on resins made by the treatment of lignin with heat or by reaction with chemicals or with not more than an equal weight of synthetic resins.

Loading Board. A device for holding preforms spaced to correspond with the positions of multiple cavities of a compression mold and for dropping the preforms simultaneously into the cavities.

Loading Space. Space provided in a compression mold or in the pot used with a transfer mold to accommodate the molding material before it is compressed.

Locating Ring. A ring which serves to align the nozzle of an injection cylinder with the entrance of the sprue bushing and the mold to the machine platen.

Lumber-core Construction. A type of plywood in which the center layer or core is lumber, usually edge-glued together from narrow strips and seldom less than 3/8 in thick.

LYOPHILIC. In vinyl dispersions, having affinity for the dispersing medium.

LYOPHOBIC. In vinyl dispersions, not having affinity for the dispersing medium.

MAT. A randomly distributed felt of glass fibers used in reinforced plastics lay-up molding.

MATCHED METAL MOLDING. Method of molding reinforced plastics between two close-fitting metal molds mounted in a hydraulic press (similar to compression molding).

MECHANICALLY FOAMED PLASTIC. A cellular plastic whose structure is produced by physically incorporated gases.

MELAMINE FORMALDEHYDE. A synthetic resin derived from the reaction of melamine with formaldehyde or its polymers.

MELAMINE PLASTICS. Plastics based on resins made by the condensation of melamine and aldehydes.

METASTABLE. Describing an unstable condition of a plastic evidenced by changes of physical properties not caused by changes in composition or in environment. Metastable refers, for example, to the temporarily more flexible condition of some plastics after molding. No physical tests should be made while the plastic is in a metastable condition, unless the data regarding this condition are desired.

METHACRYLONITRILE. A vinyl nitrile compound that is similar to ABS.

MIGRATION. The transfer of a material from a plastic to another solid in contact with it.

MOLD BASE. The assembly of all parts making up an injection mold, other than the cavity, cores, and pins.

MOLDING, BAG. A method of molding or laminating which involves the application of fluid pressure, usually by means of air, steam, water, or vacuum, to a flexible material which transmits the pressure to the material being molded or bonded.

MOLDING, BLOW. A method of forming objects from plastic masses by inflating with compressed gas.

MOLDING, COMPRESSION. A method of forming objects from plastics by placing the material in a confining mold cavity and by applying pressure, and usually heat.

MOLDING, CONTACT-PRESSURE. A method of molding or laminating, in which the pressure used is only slightly more than that necessary to hold the materials together during the molding operation. This pressure is usually less than 10 psi (68.948 kPa).

MOLDING, HIGH PRESSURE. Molding or laminating in which the pressure used is greater than 200 psi (1378.96 kPa).

MOLDING, INJECTION. A method of forming plastic objects from granular or powdered plastics by fusing the plastic in a chamber with heat and pressure and then forcing part of the mass into a cooler chamber where it solidifies. This method is commonly used to form objects from thermoplastics.

MOLDING, LOW-PRESSURE. Molding or laminating in which the pressure used is less than 200 psi (1378.96 kPa).

MOLDING TRANSFER. A method of forming molded objects from granular, powdered or preformed plastics by fusing the plastic in a chamber by heat, and then forcing essentially the whole mass into a hot chamber, where it solidifies. This method is commonly used to form objects from the thermosetting plastics.

MOLD INSERT (REMOVABLE). Part of a mold cavity of force which forms undercut or raised portions of a molded article.

MONOMER. A relatively simple compound that can react to form a polymer.

MOUNTING PLATE. (See CLAMPING PLATE.)

MOVABLE PLATEN. The large back platen of an injection molding machine to which the back half of the mold is secured.

NEWTONIAN LIQUID. A liquid in which the rate of flow is directly proportional to the force applied. The viscosity is independent of the rate of shear, and there is no yield value in Newtonian flow.

NONRIGID PLASTIC. A plastic which has a stiffness or apparent modulus of elasticity of not over 10,000 psi (68.948 MPa) at 23°C, when determined in accordance with the Standard Method of Test for Stiffness in Flexure of Plastics (ASTM Designation: D 747).

NOVOLAK. A phenolic-aldehydic resin which, unless a source of methylene groups is added, remains permanently thermoplastic. (See RESINOID, THERMOPLASTIC.)

Nozzle. Restricted orifice at the end of the heating cylinder of an injection or transfer machine.

Nylon The generic name for all synthetic fiber-forming polyamides. These polyamides can be formed into monofilaments and yarns characterized by a high degree of toughness, and elasticity, as well as a high melting point and good resistance to chemicals, but poor resistance to water absorption and penetration.

Nylon Plastics. Plastics based on a resin composed principally of a long-chain synthetic polymeric amide which has recurring amide groups as an integral part of the main polymer chain.

Open-cell Foam. A cellular plastic in which there is a predominance of interconnected cells.

Orange Peel. Uneven surface somewhat resembling an orange peel.

Organosol. A suspension of a finely-divided resin in a volatile organic liquid. The resin does not dissolve appreciably in the organic liquid at room temperature, but does at elevated temperatures. The liquid evaporates at the elevated temperature and the residue on cooling is a homogenous plastic mass. Plasticizers may be dissolved in the volatile liquid.

Overflow Groove. Small groove used in molds to allow material to flow freely to prevent weldlines and low density, and to dispose of excess material.

Parallels. Spaces or supports used under the top and/or bottom halves of the mold to prevent deflection, or to provide space for an ejector mechanism.

Parison. The hollow plastics tube from which a container, toy, etc., is blow molded.

Part. In its proper literal meaning, a component of an assembly. However, the word is widely misused to designate any individual manufactured article, even when (like a cup, a comb, a doll) it is complete in itself, not part of anything.

Permanence. Resistance of a plastic to appreciable changes in characteristics with time and environment.

Phenolic Plastics. Plastics based on resins made by the condensation of phenols such as phenol and cresol, with aldehydes.

Phenolic Resin. A synthetic resin produced by the condensation of an aromatic alcohol with an aldehyde. Phenolic resins are used for thermosetting molding materials and laminated sheets.

Phthalate Esters. A main group of plasticizers, produced by the direct action of alcohol on phthalic anhydride.

Plastic. The adjective plastic indicates that the noun modified is made of, consists of, or pertains to plastic. The above definition may be used as a separate meaning to the definitions contained in the dictionary for the adjective plastic.

Plastic. (*noun*) A material that contains as an essential ingredient an organic substance of large molecular weight, is solid in its finished state, and at some stage in its manufacture or its processing into finished articles, can be shaped by flow.

Plasticate. To soften a material and make it moldable by the use of a plasticizer.

Plasticity. A property of plastics or adhesives which allows the material to be deformed continuously and permanently without rupture upon the application of a force that exceeds the yield value of the material.

Plasticize. To soften by addition of plasticizer.

Plasticizer. Chemical agent (elastomer or plastics) added to plastics compositions to make them softer, more flexible, and more workable.

Plastic, Rigid. A plastic which has a stiffness or apparent static modulus of elasticity greater than 100,000 psi (689.48 MPa) at 23°C, when determined in accordance with the Standard Method of Test for Stiffness in Flexure of Plastics (ASTM Designation: D 747).

Plastics. (*noun*) A generic term for the industry and its products. This term is properly used only as a plural word. The plastics products include polymeric substances, natural or synthetic, and exclude rubber materials.

Plastic, Semirigid. A plastic which has a stiffness or apparent modulus of elasticity of between 10,000 and 100,000 psi (68.948 and 689.48 MPa) at 23°C, when determined in accordance with the Standard Method of Test for Stiffness in Flexure of Plastics (ASTM Designation: D 747).

Plastic Welding. The joining of two or more pieces of

plastic by fusion of the material in the pieces at adjoining or nearby areas, either with or without the addition of plastic from another source.

PLASTICIZER. A material incorporated in a plastic or an adhesive to increase its workability and its flexibility or distensibility. The addition of the plasticizer may lower the melt viscosity, the temperature of the glassy transition, or the elastic modulus of the plastic.

PLASTIFY. (See PLASTICATE.)

PLASTIGEL. A plastisol exhibiting gel-like flow properties; one having an effective yield value.

PLASTISOL. A suspension of finely-divided resin in a plasticizer. The resin does not dissolve appreciably in the plasticizer at room temperature, but does at elevated temperature. On cooling, a homogeneous plastic mass (plasticized resin) results.

PLATFORM BLOWING. A special technique for blowing large parts by use of a moveable table to support the material.

PLUG-AND-RING. Method of sheet forming in which a plug, functioning as a male mold, is forced into a heated plastic sheet that is held in place by a clamping ring.

PLUG FORMING. A thermoforming process in which a plug or male mold is used to partially preform the part before forming is completed by the use of vacuum or pressure. Also called plug assist.

PLUNGER. (See FORCE, POT PLUNGER.)

PLYWOOD. A cross-bonded assembly made of layers of veneer or veneer in combination with a lumber core or plies joined with an adhesive. Two types of plywood are recognized, namely (1) veneer plywood and (2) lumber-core plywood.

POISE. A unit of viscosity in which the shearing stress is expressed in dynes per square centimeter to produce a velocity gradient of one centimeter per second per centimeter. Under SI, viscosity is measured in pascal-seconds, where 1 poise = 0. 100 Pa.s. (See VISCOSITY, CENTIPOISE.)

POLYAMIDE PLASTICS. (See NYLON PLASTICS.)

POLYCONDENSATION. (See CONDENSATION.)

POLYESTER. A resin formed by the reaction between a dibasic acid and a dihydroxy alcohol, both organic. Polyesters modified with fatty acids are called alkyds.

POLYETHYLENE. A thermoplastic material comprised of polymers of ethylene. Normally it is a crystalline, translucent, tough, waxy solid that is unaffected by water and a large variety of chemicals.

POLYMER. A high-molecular-weight organic compound, natural or synthetic, whose structure can be represented by repeated small units (mers). Examples include polyethylene, cellulose, and rubber. Synthetic polymers are formed by addition or condensation polymerization of monomers. If two or more monomers are involved, a copolymer is obtained. Some polymers are elastomer, others are plastics.

POLYMERIZATION. A chemical reaction in which the molecules of a monomer are linked together to form large molecules whose molecular weight is a multiple of that of the original monomer.

POLYPROPYLENE. Tough, lightweight, rigid, crystalline plastics made by the polymerization of high-purity propylene gas in the presence of an organo-metallic catalyst at relatively low pressures and temperatures.

POLYSTYRENE. A water-white thermoplastic produced by the polymerization of styrene (vinyl benzene). The electrical insulating properties of polystyrene are very good and the material is relatively unaffected by moisture, but it generally is brittle.

POLYURETHANE RESINS. A family of resins produced by reacting diisocyanate with organic compounds containing two or more active hydroxyl units to form polymers having free isocyanate groups. These groups, under the influence of heat or catalysts react with each other, or with water, glycols, etc., to form thermosetting or thermoplastic materials.

POLYVINYL ACETATE. A resin prepared by the polymerization of vinyl acetate alone.

POLYVINYL ALCOHOL. A polymer prepared by the hydrolysis of polyvinyl esters.

POLYVINYL CHLORIDE. A resin prepared by the polymerization of vinyl acetate alone.

POLYVINYL CHLORIDE-ACETATE. Copolymer of vinyl chloride and vinyl acetate.

PORT. Inlet or outlet of oil, water, or steam channel.

POSTFORMING. A method of shaping substantially cured thermoset plastic sheets or other forms by heating and stressing into other configurations.

Pot. Chamber to hold and heat molding material for a transfer mold.

Pot Plunger. A plunger used to force softened molding material into the closed cavity of a transfer mold.

Pot-retainer. Plate channeled for heat and used to hold pot of transfer mold.

Preform. A coherent block of granular or fibrous plastic molding compound, or of fibrous material with or without resin. A preform is made by compressing the material sufficiently to produce a coherent block for convenience in handling.

Prepolymer. A chemical structure intermediate between that of the monomer or monomers and the final polymer or resin.

Pressure Pad. Hardened steel reinforcements distributed around a mold to help the land to absorb the final closing pressure.

Primer. A coating applied to a surface, prior to the application of an adhesive or lacquer, enamel, or the like, to improve the performance of the bond.

Progressive Gluing. A method of curing a resin adhesive in successive steps or stages by application of heat and pressure, between the platens of a hot press. Used only for a wood of larger area than the press platens.

Pseudoplastic. A pseudoplastic fluid is one whose apparent viscosity or consistency decreases instantaneously with increase in rate of shear; *i.e.*, an initial relatively high resistance to stirring decreasing abruptly as the rate of stirring is increased.

Reciprocating Screw. An extruder system in which the screw when rotating is pushed backward by the molten polymer which collects in front of the screw. When sufficient material has been collected, the screw moves forward and forces the material at high velocity through the head and the nozzle into a mold.

Reinforced Plastics. Plastics with some strength properties greatly superior to those of the base resin, resulting from the presence of high strength fillers embedded in the composition. The reinforcing fillers are usually fibers, fabrics, or mats made of fibers. The plastic laminates are the most common and strongest type of reinforced plastics. (See FILLER.)

Reinforcement. A strong inert material that is bound into plastics to improve the strength, stiffness, and impact resistance. Reinforcements are usually long fibers of sisal, cotton, glass, etc., in woven or nonwoven form.

Resin. Any of a class of solid or semisolid organic products of natural or synthetic origin. Resins generally are of high molecular weight and have no definite melting point. Most resins are polymers.

Resin, Liquid. An organic polymeric liquid which, when converted to its final state for use, becomes a solid.

Resinoid. (See NOVOLAK.)

Retarder. (See INHIBITOR.)

Ring Gate. Annular opening for entrance of material into cavity of injection or transfer mold.

Rotational Molding (or casting). A method used to make hollow products from powdered plastics, plastisols, and lattices. Powdered plastics or plastisol is charged into a hollow mold that can be rotated in one plane or in two planes. Heat is applied to melt, fuse, or cure the polymer. Cooling is usually required before removing the casting.

Roving. A form of fibrous glass in which spun strands (filaments) are woven into a tubular rope. Chopped roving is commonly used in preforming.

Rubber. An elastomer capable of rapid recovery after being stretched to at least twice its length at temperatures from 0 to 150°F(-18 to 66°C); specifically, Hevea or natural rubber, which is the standard of comparison for elastomers.

Runner. In an injection or transfer mold, the channel, usually circular, that connects the sprue with the gate to the cavity.

Sandwich Molding. A molding process in which two different materials are injected consecutively into a mold cavity to produce products having surfaces of one plastic with desirable characteristics and a core of another material with its desired characteristics.

Saran Plastics. Plastics based on resins made by the polymerization of binylidene chloride or copolymerization of binylidene chloride with not more than an equal weight of other unsaturated compounds.

Sealing Diameter. Portion of the insert which is free of knurl and is allowed to enter into the mold to prevent the flow of material.

Secondary Gluing. The process of gluing together wood and plywood parts in assembling wood products, such as aircraft. Contrasted with primary gluing, when veneers are glued into plywood. (See ASSEMBLY: IN PLASTICS SECTION 12-70.)

Set. To convert a liquid resin or adhesive into a solid state by curing, by evaporation of solvent or suspending medium, or by gelling.

Sheet. A piece of plastic sheeting produced as an individual piece rather than in a continuous length, or cut as individual pieces from a continuous length. (See SHEETING, FILM.)

Sheeting. A form of plastic in which the thickness is very small in proportion to length and width, and in which the plastic is present as a continuous phase throughout, with or without filler. (See FILM.)

Sheet Molding Compound (SMC). A combination of polyester resin, filler, and reinforcement rolled into a sheet form. Thermoplastic resin often is added to obtain a surface with desired characteristics.

Shelf Life. The length of time over which a product will remain fit for use during storage under specific conditions (especially temperature).

Shim. In the manufacture of plywood, a long narrow patch glued into the panel, or into the lumber core.

Silicone. One of a family of polymeric materials in which the recurring chemical group contains silicon and oxygen atoms as links in the main chain. Silicons are derived from silica (sand) and methyl chloride.

Silicone Plastics. Plastics based on resins in which the main polymer chain consists of alternating silicon and oxygen atoms, with carbon-containing side groups.

Sizing. The process of applying a material on a surface in order to fill pores and thus reduce the absorption of the subsequently applied adhesive or coating, or to otherwise modify the surface properties of the substrate to improve adhesion. Also, the material used for this purpose. The latter is sometimes called size.

Sleeve Ejector. Busing-type knockout.

Sliding Plate. (See DUPLICATE CAVITY-PLATE.)

Slip Joint. The method of laying up veneers in flexible-bag molding, wherein the edges are beveled and allowed to overlap part or all of the scarfed area.

Softening Range. The range of temperatures in which a plastic changes from a rigid to a soft state. Actual values will depend on the method of test. Sometimes erroneously referred to as softening point.

Solid-piled. Sometimes called dead-piled or bulked-down. Plywood fresh from clamps or hot press is piled on a solid, flat base, without stickers, and weighted down while reaching normal temperature and moisture content.

Solvation. The process of swelling, gelling, or solution of a resin by a solvent or plasticizer as a result of mutual attraction.

Solvency. Solvent action, or strength of solvent action

Split Cavity. Cavity made in section.

Split-cavity Blocks. Blocks which when assembled contain a cavity for molding articles having undercuts.

Spray-up. The term for a number of techniques in which a spray gun is used as the processing tool.

Spread. The quantity of adhesive per unit joint area applied to an adherend. It is preferably expressed in pounds of liquid or solid adhesive per thousand square feet of joint area. Single spread refers to application of adhesive to only one adherend of a joint. Double spread refers to application of adhesive to both adherends of a joint.

Sprue. (1) The primary feed channel that runs from the outer face of an injection or transfer mold to the mold gate in a single-cavity mold or to the runners in a multiple-cavity mold. (2) The piece formed in a primary feed channel or sprue.

Sprue Bushing. In an injection mold, a hardened steel insert which contains the tapered sprue hole and has a suitable seat for making close contact with the nozzle of the injection cylinder.

Sprue-puller. A pin having a Z-shaped slot undercut in its end, by means of which it serves to pull the sprue out of the sprue bushing.

Stir-in Resin. A vinyl resin which does not require grinding to effect dispersion in a plastisol or an organosol.

Stop Buttons. Multiple island limiting travel of ejector mechanism when returned to molding position.

Stress-crack. External or internal crack in a plastic caused by tensile stresses greater than that of its short-time mechanical strength. The development of such cracks is frequently accelerated by the environment to which the plastic is exposed. The stresses which cause cracking may be present internally or externally or may be combinations of these stresses. The appearance of a network of fine cracks is called crazing.

Stretch Forming. A plastics sheet forming technique in which the heated thermoplastic sheet is stretched over a mold and then cooled.

Stripper Plate. A plate which strips the molded article from mold pin, force or cores.

Stripping Fork. Tool, usually of brass or laminated sheet, used to remove articles from the mold (also called comb).

Structural Foam. This product has a rigid cellular core and a solid integral skin.

Styrene Acrylonitrile (SAN). A thermoplastic copolymer with good stiffness, along with good resistance to chemicals, scratching, and stress cracking.

Styrene Plastics. Plastics based on resins made by the polymerization of styrene or copolymerization of styrene with other unsaturated compounds.

Styrene Rubber Plastics. Plastics consisting of at least 50 percent of a styrene plastic combined with rubbers and other compounding ingredients.

Support Post or Pillar. Post used to resist deflection under pressure.

Syndiotactic. Alternating placement of the group on either side of the chain with respect to the carbon-carbon backbone plane.

Synersis. The contraction of a gel. This is usually evidenced by the separation of a liquid. (See GEL.)

Tack. Stickiness of an adhesive. This property is measured as the pull resistance to effect division without failure or deformation occurring in the adherend surroundings or at the interface while the adhesive still exhibits viscous or plastic flow. The measured value may vary with time, temperature, film thickness, etc.

Temper. To reheat after hardening to some temperature below the critical temperature, followed by air cooling to obtain desired mechanical properties and to relieve hardening strains.

Thermally Foamed Plastics. Cellular plastics produced by applying heat to effect gaseous decompositon or volatilization of a constituent.

Thermoelasticity. Rubber-like elasticity exhibited by a rigid plastic and resulting from an increase of temperature.

Thermoforming. The processes for forming thermoplastic sheet by heating the sheet and using air, vacuum, or mechanical methods to form it onto the surface contour of a mold.

Thermoplastics. Polymers that are capable of being repeatedly softened by heating and hardened by cooling.

Thermoset. A material that will undergo or has undergone a chemical reaction by the action of heat, catalysts, or ultraviolet light and will achieve or has achieved a relatively infusible state.

Thermosetting. Capable of being changed into a substantially infusible or insoluble product when cured by application of heat or chemical means. Thermosetting polymers form bridges between molecule chains referred to as cross-links.

Thixotropic. A thixotropic fluid is one whose apparent viscosity decreases with time, to some constant value at any constant rate of shear; *i.e.*, its apparent viscosity can be gradually decreased to that limit by stirring. When stirring is discontinued, the apparent viscosity increases gradually back to the original value.

Transfer Molding. A method of molding thermosetting materials, in which the plastic is softened by heat and pressure in a transfer chamber, then forced by high pressure through sprues, runners, and gates into a closed mold for final curing.

Undercut. Any depression on the mold which prevents or hinders withdrawal of the molded article.

Urea or Urea Formaldehyde Resin. A synthetic resin derived from the reaction of urea (carbamide) with formaldehyde or its polymers.

UREA PLASTICS. Plastics based on resins made by the condensation of urea and aldehydes.

URETHANE. (See ISOCYANATE RESINS.)

VACUUM FORMING. A thermoforming method of sheet forming in which the plastics sheet is clamped in a stationary frame, heated, and drawn down by vacuum into a mold.

VACUUM METALIZING. A process in which surfaces are thinly coated with metal by exposing them to the vapor of metal that has been evaporated under vacuum.

VENEER. A thin sheet or layer of wood, sliced, rotary-cut, half-round or sawn from a log, block or flitch. There is no sawkerf waste in a knife-cut veneer. Veneer is the raw material from which plywood and laminated wood are assembled. Thicknesses may range from 1/100 in. to 1/4 in. and are seldom greater.

VINYL ACETATE PLASTICS. Plastics based on resins made by the polymerization of vinyl acetate or copolymerization of vinyl acetate with no more than an equal weight of other saturated compounds.

VINYL ALCOHOL PLASTICS. Plastics based on resins made by the hydrolysis of polyvinyl esters or copolymers of vinyl esters.

VINYL CHLORIDE PLASTICS. Plastics based on resins made by the polymerization of vinyl chloride with not more than equal weights of other unsaturated compounds.

VINYLIDENE PLASTICS. (See SARAN PLASTICS.)

VINYL PLASTICS. Plastics based on resins made from vinyl monomers except those specifically covered by other classification, such as acrylic and styrene plastics. Typical vinyl plastics are polyvinyl chloride, polyvinyl acetate, polyvinyl alcohol and polyvinyl butyral, and copolymers of vinyl monomers with unsaturated compounds.

VISCOELASTICITY. The characteristic that causes plastics materials to respond to stress as though they are a combination of elastic solids and viscous fluids. This property is exhibited in varying degrees by all plastics. This term is not applicable to elastomers.

VISCOSITY. The property of resistance to flow exhibited within the body of a material. This property can be expressed in terms of relationship between applied shearing stress and resulting rate of strain in shear. Viscosity is usually taken to mean Newtonian viscosity, in which case the ratio of shearing stress to the rate of shearing strain is constant. In Newtonian viscosity, the ratio of shearing stress to the rate of shearing strain is constant. In non-Newtonian behavior, which is the usual case with plastic materials, the ratio varies with the shearing stress. Such ratios are often called the apparent viscosities at the corresponding shearing stress. (See VISCOSITY COEFFICIENT.)

VISCOSITY COEFFICIENT. The shearing stress necessary to induce a unit velocity flow gradient in a material. In actual measurement, the viscosity coefficient of a material is obtained from the ratio of shearing stress to shearing rate. This assumes the rates to be constant and independent of the shearing stress, a condition which is satisfied only by Newtonian fluids. Consequently, in all other cases, values obtained are apparent and represent one point on the flow curve. In the metric system, the viscosity coefficient is expressed in poises. Under SI, the unit is pascal-seconds.

WARP, WARPAGE. Dimensional distortion in a plastic object after molding or other fabrications. (See DISHED, DOMED.)

WEATHERING. The exposure of plastics outdoors.

WEATHERING, ARTIFICIAL. The exposure of plastics to cyclic laboratory conditions comprising high and low temperatures, high and low relative humidities, and ultraviolet radiant energy, with or without direct water spray, in an attempt to produce changes in their properties similar to those observed on long-time continuous exposure outdoors. The laboratory exposure conditions are usually intensified beyond those encountered in actual outdoor exposure in an attempt to achieve an accelerated effect.

WEB. A textile fabric, paper, or a thin metal sheet of continuous length handled in roll form as contrasted with the same material cut into sheets.

WELD LINE. In a molded article, a zone of defective appearance and/or strength resulting from defective union of two fronts of plastic flowing together within the mold. Sometimes called weld mark.

WINDOW. A defect in a thermoplastic film, sheet, or molding, caused by the incomplete plasticizing of a piece of the material during processing. It appears as a globule in an otherwise blended mass.

WORKING LIFE. The period during which a compound, after mixing with catalyst, solvent, or other compounding ingredients, remains suitable for its intended use.

YIELD VALUE. The force which must be applied to a plastic to initiate flow also called yield stress.

WELDING

AIR CARBON-CUTTING. An arc-cutting process wherein the severing of metals is effected by melting with the heat of an arc between an electrode and the base metal. An air stream is used to facilitate cutting.

ARC BLOW. The deflection of an electric arc from its normal path because of magnetic forces.

ARC-TIME. The length of time the arc is maintained in making an arc weld.

ARC VOLTAGE. The voltage across the welding arc.

ARC WELDING. A group of welding processes wherein coalescence is produced by heating with an arc or arcs, with or without the application of pressure and with or without the use of filler metal.

AUTOMATIC WELDING. Welding with equipment which performs the entire welding operation without constant observation and adjustment of the controls by an operator. The equipment may or may not perform the loading and unloading of the work.

BACKHAND WELDING. A welding technique wherein the welding torch or gun is directed opposite to the progress of welding.

BACKING. Material (metal, weld metal, asbestos, carbon, granular flux, gas, etc.) backing up the joint during welding.

BARE ELECTRODE. A filler-metal electrode, used in arc welding, consisting of a metal wire with no coating other than that incidental to its manufacture or preservation.

BASE METAL. The metal to be welded, soldered, or cut.

BEVEL ANGLE. The angle formed between the prepared edge of a member and a plane perpendicular to the surface of the member.

BRAZE WELDING. A method of welding whereby a weld is made using a filler metal, having a liquidus above 800°F (427°C) and below the solidus of the base metals. The filler metal is not distributed in the joint by capillary attraction.

BUILDUP SEQUENCE. The order in which the weld beads of a multiple-pass weld are deposited with respect to the cross-section of the joint.

BUTT JOINT. A joint between two members lying approximately in the same place.

CARBON ARC-CUTTING. An arc-cutting process wherein the severing of metals is effected by melting with the heat of an arc between a carbon electrode and the base metal.

CARBON ARC-WELDING. An arc-welding process wherein coalescence is produced by heating with an arc between a carbon electrode and the work and no shielding is used. Pressure may or may not be used and filler metal may or may not be used.

COLD WELDING. A solid state welding process wherein coalescence is produced by the external application of mechanical force alone.

COOL TIME. The time interval between successive heat times in multiple-impulse welding or in the making of seam welds by resistance welding.

CORNER JOINT. A joint between two members located approximately at right angles to each other in the form of an L.

COVERED ELECTRODE. A filler-metal electrode, used in arc welding, consisting of a metal core wire with a relatively thick covering which provides protection for the molten metal from the atmosphere, improves the properties of the weld metal and stabilizes the arc.

COVER GLASS. A clear transparent material used in goggles, hand shields, and helmets to protect the filter lens from spattering material.

CRATER. In arc welding, a depression at the termination of a weld bead or in the weld pool beneath the electrode.

CRATER CRACK. A crack in the crater of a weld bead.

CUTTING TORCH. A device used in oxygen cutting for controlling and directing the gases used for preheating and the oxygen used for cutting the metal.

CYLINDER. A portable cylindrical container used for transportation and storage of a compressed gas.

DEPOSITED METAL. Filler metal that has been added during a welding operation.

Deposition Efficiency. The ratio of the weight of deposited metal to the net weight of electrodes consumed, exclusive of stubs.

Deposition Rate. The weight of metal deposited in a unit of time.

Depth of Fusion. The distance that fusion extends into the base metal or previous pass from the surface melted during welding.

Dip Brazing. A brazing process in which the heat required is furnished by a molten chemical or metal bath. When a molten chemical bath is used, the bath may act as a flux. When a molten metal bath is used, the bath provides the filler metal.

Edge Joint. A joint between the edges of two or more parallel or nearly parallel members.

Edge Preparation. The contour prepared on the edge of a member of welding.

Electrode Holder. A device used for mechanically holding the electrode and conducting current to it.

Electron Beam Welding. A welding process wherein coalescence is produced by the heat obtained from a concentrated beam composed primarily of high velocity electrons impinging upon the surfaces to be joined.

Electronic Heat Control. A device for adjusting the heating value (rms value) of the current in making a resistance weld by controlling the ignition or firing of the tubes in an electronic contractor. The current is initiated each half-cycle at an adjustable time with respect to the zero point on the voltage wave.

Electroslag Welding. A welding process wherein coalescence is produced by molten slag which melts the filler metal and the surfaces of the work to be welded. The weld pool is shielded by this slag which moves along the full cross section of the joint as welding progresses. The conductive slag is maintained molten by its resistance to electric current passing between the electrode and the work.

Explosive Welding. A process where an explosive material, usually in the form of a sheet, is placed on the top of two layers of metal and detonated progressively. A compressive stress wave, on the order of thousands of megapascals, progresses across the surface of the plates, so that a small open angle is formed between the two colliding surfaces. Surface films are liquefied or scarfed off the surfaces and jetted out of the interface, leaving clean surfaces which coalesce under the high pressure. The result is a cold weld having a wavy configuration at the interface. The progress is used primarily for bonding sheets of corrosion resistant metals to heavier plates of base metals (cladding).

Face of Weld. The exposed surface of a weld on the side from which welding is done.

Faying Surface. That surface of a member which is in contact or close proximity with another member to which it is to be joined.

Filler Lens. A filler, usually colored glass, used in goggles, helmets and handshields to exclude harmful light rays.

Filler Metal. The metal to be added in making a welded, brazed, or soldered joint.

Fillet Weld. A weld of approximately triangular cross section joining two surfaces approximately at right angles to each other in a lap joint.

Flash. The molten metal which is expelled, or which is squeezed out by the application of pressure, and solidifies around the weld.

Flashback. A recession of the flame into or back of the mixing chamber of the torch.

Flashing Time. The time during which the flashing action is taking place in flash welding.

Flash Welding. A resistance-welding process wherein coalescence is produced, simultaneously over the entire area of abutting surfaces, by the heat obtained from resistance to electric current between the two surfaces, and by the application of pressure after heating is substantially completed. Flashing and upsetting are accompanied by expulsion of metal from the joint.

Flat Position. The position of welding wherein welding is performed from the upper side of the joint and the face of the weld is approximately horizontal.

Flux. Material used to prevent, dissolve or facilitate removal of oxides and other undesirable substances.

Flux Cored Arc-welding. An arc-welding process wherein coalescence is produced by heating with an arc, between a continuous filler metal (consumable) electrode and the work. Shielding is obtained from a flux contained within the electrode. Additional shielding may or may not be obtained from an externally supplied gas or gas mixture.

FOREHAND WELDING. A welding technique wherein the welding torch or gun is directed toward the progress of welding.

FORGE WELDING. A solid state welding process wherein coalescence is produced by heating and by applying pressure or blows sufficient to cause permanent deformation at the interface.

FRICTION WELDING. A solid state welding process wherein coalescence is produced by the heat obtained from mechanically induced sliding motion between rubbing surfaces. The work parts are held together under pressure.

FURNACE BRAZING. A brazing process in which the heat required is obtained from a furnace.

FUSION. The melting together of filler metal and base metal, or of base metal only, which results in coalescence.

GAS METAL ARC-WELDING. An arc-welding process wherein coalescence is produced by heating with an arc drawn between a metal stud or similar part, and the other work part, until the surfaces to be joined are properly heated, when they are brought together under pressure. Shielding is obtained from an inert gas such as helium or argon.

GAS TUNGSTEN ARC-WELDING. An arc-welding process wherein the coalescence is produced by heating with an arc between single tungsten (nonconsumable) electrode and the work. Shielding is obtained from a gas or gas mixture. Pressure may or may not be used and filler metal may or may not be used. (This process has sometimes been called TIG welding.)

GAS WELDING. A group of welding processes wherein coalescence is produced by heating with a gas flame or flames, with or without the application of pressure, and with or without the use of filler metal.

GROOVE ANGLE. The total included angle of the groove between parts to be joined by a groove weld.

HAND SHIELD. A protective device, used in arc welding, for shielding the face and neck. A hand shield is equipped with a suitable filter lens and is designed to be held by hand.

HEAT-AFFECTED ZONE. That portion of the base metal which has not been melted, but whose mechanical properties or microstructure have been altered by the heat of welding, brazing, soldering, or cutting.

HEAT TIME. The time that current occurs during any one impulse in multiple-impulse welding or when making welds by resistance welding.

HELMET. A protective device, used in arc welding, for shielding the face and neck. A helmet is equipped with a suitable filter lens and is designed to be worn on the head.

HOLD TIME. The time during which force is applied at the point of welding after the last impulse of current ceases in resistance welding.

INDUCTION BRAZING. A brazing process in which the heat required is obtained from the resistance of the work to induced electric current.

INDUCTION WELDING. A welding process wherein coalescence is produced by the heat obtained from resistance of the work to induced electric current, with or without the application of pressure.

INTERMITTENT WELD. A weld wherein the continuity of the weld is broken by recurring unwelded spaces.

INTERPASS TEMPERATURE. In a multiple-pass weld, the temperature (minimum or maximum as specified) of the deposited weld metal before the next pass is started.

JOINT DESIGN. The joint geometry together with the required dimensions of the welded joint.

JOINT GEOMETRY. The shape and dimensions of a joint in cross section prior to welding.

JOINT PENETRATION. The minimum depth a groove or flange weld extends from its face into a joint, exclusive of reinforcement.

KERF. The space from which metal has been removed by a cutting process.

LAP JOINT. A joint between two overlapping members.

LASER BEAM CUTTING. Laser beam cutting uses the intense heat from a laser beam to melt and/or evaporate the material being cut. Any known material can be cut by this process. For some nonmetallic materials the mechanism is purely evaporation, but for many metals a gas may be supplied, either inert to blow away the molten metal and provide a smooth, clean kerf, or oxygen to speed the process through oxidation. The temperature achieved may be in excess of 11,093°C (20,000°F), and cutting speeds of the order of 25.4 meters (1,000 inches) per minute are not uncommon in nonmetals and 508 mm (20 inches) per minute in tough steels.

LASER BEAM WELDING. The heat source in laser beam welding is a focused laser beam, usually providing power intensities in excess of 10 kilowatts per square centimeter, but with low heat input -0.1 to 10 joules. The high-intensity beam produces a very thin column of vaporized metal, extending into the base metal. The column of vaporized metal is surrounded by a liquid pool, which moves along as welding progresses, resulting in welds having depth-to-width ratios greater than 4:1. Laser beam welds are most effective for simple fusion welds without filler metal, but filler metal can be added.

LAYER. A stratum of weld metal, consisting of one or more weld beads.

LEAD ANGLE. The angle that the electrode makes in advance of a line perpendicular to the weld axis at the point of welding, taken in longitudinal plane.

LEG OF A FILLET WELD. The distance from the root of the joint to the toe of the fillet weld.

LOCAL PREHEATING. Preheating a specific portion of a structure.

LONGITUDINAL RESISTANCE-SEAM WELDING. The making of a resistance-seam weld in a direction essentially parallel to the throat depth of a resistance-welding machine.

MACHINE WELDING. Welding with equipment which performs the welding operation under the constant observation and control of an operator. The equipment may or may not perform the loading and unloading of the work.

MANIFOLD. A multiple header for interconnection of gas or fluid sources with distribution points.

MANUAL WELDING. Welding wherein the entire welding operation is performed and controlled by hand.

MELTING RATE. The weight or length of electrode melted in a unit of time.

NEUTRAL FLAME. A gas flame wherein the portion used is neither oxidizing nor reducing.

NUGGET. The weld metal joining the parts in spot, seam, or projection welds.

NUGGET SIZE. The diameter or width of the nugget measured in the plane of interface between the pieces joined.

OPEN-CIRCUIT VOLTAGE. The voltage between the output terminals of the welding machine when no current is in the welding circuit.

OVERHEAD POSITION. The position of welding wherein welding is performed from the underside of a joint.

OVERLAP. Protrusion of weld metal beyond the toe or root of the weld.

OXIDIZING FLAME. A flame having an oxidizing effect (excess oxygen).

OXYACETYLENE WELDING. A gas-welding process wherein coalescence is produced by heating with a gas flame or flames obtained from the combustion of acetylene with oxygen, with or without the application of pressure and with or without the use of filler metal.

OXYGEN LANCE. A length of pipe used to convey oxygen to the point of cutting in oxygen-lance cutting.

PASS. A single longitudinal progression of a welding operation along a joint or weld deposit. The result of a pass is a weld bead.

PERCUSSION WELDING. A resistance-welding process wherein coalescence is produced simultaneously over the entire abutting surfaces by the heat obtained from an arc produced by a rapid discharge of electrical energy with pressure percussively applied during or immediately following the electrical discharge.

PLASMA ARC-WELDING. An arc-welding process wherein coalescence is produced by heating with a constructed arc between an electrode and the work piece (transferred arc) or the electrode and the constricting nozzle (non-transferred arc). Shielding is obtained from the hot, ionized gas issuing from the orifice which may be supplemented by an auxiliary source of shielding gas. Shielding gas may be an inert gas or a mixture of gases. Pressure may or may not be used, and filler metal may or may not be supplied.

Platen. A member with a substantially flat surface to which dies, fixtures, backups, or electrode holders are attached, and which transmits the electrode force or upsetting force in a resistance-welding machine.

Platen Force. The force available at the movable platen to cause upsetting in flash or upset welding. This force may be dynamic, theoretical, or static.

Plug Weld. A circular weld made through a hole in one member of a lap or tee joint joining that member to the other. The walls of the hole may or may not be parallel and the hole may be partially or completely filled with weld metal. (A fillet-welded hole or a spot weld should not be construed as conforming to this definition.)

Porosity. Gas pockets or voids in metal.

Postheat Current. The current through the welding circuit during postheat time in resistance welding.

Postheating. The application of heat to an assembly after welding, brazing, soldering, or cutting operations.

Preheating. The application of heat to the base metal immediately before welding, brazing, soldering, or cutting.

Preheat Temperature. The temperature specified that the base metal must attain in the welding, brazing, soldering, or cutting area immediately before these operations are performed.

Pressure Gas Welding. Gas-welding process wherein coalescence is produced, simultaneously, over the entire area of abutting surfaces, by heating with gas flames obtained from the combustion of a fuel gas with oxygen and by the application of pressure, without the use of filler metal.

Procedure Qualification. The demonstration that welds made by a specific procedure can meet prescribed standards.

Projection Welding. A resistance-welding process wherein coalescence is produced by the heat obtained from resistance to electric current through the work parts held together under pressure by electrodes. The resulting welds are localized at predetermined points, by projections, embossments, or intersections.

Quench Time. The time from the end of weld time to the beginning of temper time in resistance welding.

Reducing Flame. A gas flame having a reducing effect. (Excess fuel gas.)

Regulator. A device for controlling the delivery of gas at some substantially constant pressure regardless of variation in the higher pressure at the source.

Resistance Brazing. A brazing process in which the heat required is obtained from the resistance to electric current in a circuit of which the work is a part.

Resistance Seam-welding. A resistance-welding process wherein coalescence at the faying surfaces is produced by the heat obtained from resistance to electric current through the work parts held together under pressure by electrodes. The resulting weld is a series of overlapping resistance-spot welds made progressively along a joint by rotating the electrodes.

Resistance Spot-welding. A resistance-welding process wherein coalescence at the faying surfaces is produced in one spot by the heat obtained from the resistance to electric current through the work parts held together under pressure by electrodes. The size and shape of the individually-formed welds are limited primarily by the size and contour of the electrodes.

Reverse Polarity. The arrangement of direct current arc-welding leads wherein the work is the negative pole and the electrode is the positive pole of the welding arc.

Root Crack. A crack in the weld or heat-affected zone occurring at the root of a weld.

Root Face. That portion of the groove face adjacent to the root of the joint.

Root Opening. The separation between the members to be joined, at the root of the joint.

Semiautomatic Arc Welding. Arc welding with equipment which controls only the filler metal feed. The advance of the welding is manually controlled.

Shielded Metal Arc-welding. An arc-welding process wherein coalescence is produced by heating with an arc between a covered metal electrode and the work. Shielding is obtained from decomposition of the electrode covering. Pressure is not used and filler metal is obtained from the electrode.

Slag Inclusion. Nonmetallic solid material entrapped in weld metal or between weld metal and base metal.

Soldering. A group of joining processes wherein coalescence is produced by heating to a suitable temperature and by using a filler metal having a liquidus not exceeding 800°F (427°C) and below the solidus of the base metals.

Solid State Welding. A group of welding processes wherein coalescence is produced essentially at temperatures below the melting point of the base metals being joined, without the addition of a brazing filler metal. Pressure is not used and filler metal is obtained from the electrode.

Spatter. In arc and gas welding, the metal particles expelled during welding and which do not form a part of the weld.

Spotweld. A weld made between or upon overlapping members wherein coalescence may start and occur on the faying surfaces or may have proceeded from the surface of one member. The weld cross section (plan view) is approximately circular.

Squeeze Time. The time interval between the initial application of the electrode force on the work and the first application of current in making spot and seam welds by resistance welding and in projection or upset welding.

Straight Polarity. The arrangement of direct current arc-welding leads wherein the work is the positive pole and the electrode is the negative pole of the welding arc.

Stress-relief Heat Treatment. Uniform heating of a structure or portion thereof to a sufficient temperature, below the critical range, to relieve the major portion of the residual stresses, followed by uniform cooling. *Note*: Terms normalizing, annealing, etc., are misnomers for this application.

Stud Welding. An arc-welding process wherein coalescence is produced by heating with an arc drawn between a metal stud, or similar part, and the other work part until the surfaces to be joined are properly heated, when they are brought together under pressure. Partial shielding may be obtained by the use of a ceramic ferrule surrounding the stud. Shielding gas or flux may or may not be used.

Submerged Arc Welding. An arc-welding process wherein coalescence is produced by heating with an arc or arcs between a bare metal electrode or electrodes and the work. The arc is shielded by a blanket of granular, fusible material on the work. Pressure is not used and filler metal is obtained from the electrode and sometimes from a supplementary welding rod.

Surfacing. The deposition of filler metal on a metal surface to obtain desired properties or dimensions.

Tack Weld. A weld made to hold parts of a weldment in proper alignment until the final welds are made.

Tee Joint. A joint between two members located approximately at right angles to each other in the form of a T.

Temper Time. That part of the postweld interval following quench time to the beginning of hold time in resistance welding.

Thermit Welding. A group of welding processes wherein coalescence is produced by heating with super heated liquid metal and slag resulting from a chemical reaction between a metal oxide and aluminum, with or without the application of pressure. Filler metal, when used, is obtained from the liquid metal.

Throat of a Fillet Weld. Theoretical - the distance from the beginning of the root of the joint perpendicular to the hypotenuse of the largest right-triangle that can be inscribed within the fillet-weld cross section. Actual - the shortest distance from the root of a fillet weld to its face.

Toe of Weld. The junction between the face of a weld and the base metal.

Torch Brazing. A brazing process in which the heat required is furnished by a gas flame.

Tungsten Electrode. A nonfiller-metal electrode used in arc welding, made principally of tungsten.

Undercut. A groove melted into the base metal adjacent to the toe or root of a weld and left unfilled by weld metal.

Ultrasonic Welding. Ultrasonic welding is a solid-state process wherein coalescence is produced by localized application of very high frequency (10,000 to 200,000 cps) vibratory energy to the workpieces as they are held together under pressure. It uses an ultrasonic transducer which is coupled to a force-sensitive system that contains a welding tip on one end. The pieces to be welded are placed between this tip and a reflecting anvil, thereby concentrating the vibratory energy within the work. Stationary tips for spot welds or rotating disks for seam welds can be used.

Upset. The localized increase in volume in the region of a weld, resulting from the application of pressure.

Upset Welding. A resistance-welding process wherein coalescence is produced, simultaneously over the entire area of abutting surfaces or progressively along a joint, by the heat obtained from resistance to electric current through the area of contact of those surfaces. Pressure is applied before heating is started and is maintained throughout the heating period.

Vertical Position. The position of welding wherein the axis of the weld is approximately vertical.

Weave Bead. A type of weld bead made with transverse oscillation.

Weld. A localized coalescence of metal wherein coalescence is produced either by heating to suitable temperatures, with or without the application of pressure, or by the application of pressure alone, and with or without the use of filler metal. The filler metal either has a melting point approximately the same as the base metals or has a melting point below that of the base metals but above 800°F (427°C).

Weldability. The capacity of a metal to be welded under the fabrication conditions imposed into a specific, suitably designed structure and to perform satisfactorily in the intended service.

Weld Bead. A weld deposit resulting from a pass.

Welder. One who is capable of performing a manual or semiautomatic welding operation. (Sometimes erroneously used to denote a welding machine.).

Welder Certification. Certification in writing that a welder has produced welds meeting prescribed standards.

Welder Qualification. The demonstration of a welder's ability to produce welds meeting prescribed standards.

Weld Gauge. A device designed for checking the shape and size of welds.

Welding Current. The current in the welding circuit during the making of a weld. In resistance welding, the current used during a preweld or postweld interval is excluded.

Welding Goggles. Goggles with tinted lenses, used during welding, brazing, or oxygen cutting, which protect the eyes from harmful radiation and flying particles.

Welding Machine. Equipment used to perform the welding operation. For example, spot-welding machine, arc-welding machine, seam-welding machine, etc.

Welding Operation. (Sometimes erroneously used to denote a welding machine.)

Welding Procedure. The detailed methods and practices including all joint welding procedures involved in the production of a weldment.

Welding Rod. A form of filler metal used for welding or brazing wherein the filler metal does not conduct the electrical current.

Welding Transformer. A transformer used for supplying current for welding.

Weldment. An assembly whose component parts are joined by welding.

Weld Metal. That portion of a weld which has been melted during welding.

Z94.13
Occupational Health and Safety

The terms and their definitions in this standard have primarily resulted from the merging of Z94.15 - Safety and Z94.16 - Occupational Health and Medicine. The resulting compilation has resulted in the elimination of considerable duplication between Z94.15 and Z94.16. In addition, approximately 1/3 of the definitions have been revised/updated and a number of new terms have been added to the standard based on the recommendations of subcommittee members. A current bibliography can be found at the end of this standard, with citations limited to 1980 or later editions.

Recognition is given to the following subcommittee members who contributed their expertise in recommending revisions and additions to the terms in this standard.

Chairperson

John Talty, MS, PE, DEE
Office of Extramural Coordination
    & Special Projects
National Institute for Occupational Safety
    and Health

Subcommittee

Robin Baker, MPH
Director
Labor Occupational Health Program
University of California, Berkeley

Lawrence Gingerich, MS, CIH, CSP
President
Protective Environmental, Inc.

William Milroy, MD, PhD
Plant Medical Director
Caterpillar, Inc.

Timothy Pizatella, MS
Division of Safety Research
National Institute for Occupational Safety
    and Health

Shan Tsai, PhD, FACE
Epidemiologist/Biostatistician
Shell Oil Co.

A-SCALE. A filtering system that has characteristics which roughly match the response characteristic of the human ear at low sound levels (below 55 dB SPL, but frequently used to gauge levels to 85 dB). A-scale measurements are often referred to as dB(A).

ABATE. To eliminate a hazard to comply with a regulatory standard that is being violated, e.g., to correct deficiencies identified during an inspection conducted under the authority of the Occupational Safety and Health Act.

ABNORMAL USE. Using tools and equipment for a purpose other than that for which they were intended to be used, and one which may not reasonably have been foreseen.

ABNORMALITY. A defect. (See INHERITED ABNORMALITY.)

ABRASION. An area of body surface denuded of skin or mucous membrane due to an abnormal or unusual mechanical process. A scraping away of a portion of the skin surface epithelium.

ABRIDGEMENT OF DAMAGES. The right of the court to reduce damages in certain cases.

ABSOLUTE RISK. The probability that a person with stated characteristics will develop a given illness or injury.

ABSORPTION. Penetration of a chemical substance, a pathogen, or radiant energy through the skin or mucous membrane.

ACCELERATION ILLUSIONS. Incorrect perceptions of apparent motion resulting from acceleration-induced stimulation of the semicircular canals. This can be a significant cause of disorientation in aircraft pilots. Three specifically described illusions of frequent occurrence are:

ciriolis – A change of head position during a turning maneuver resulting in illusory motion of the aircraft.

oculogyral – A sensation of spinning in the opposite direction following cessation of a spinning or rolling maneuver.

oculogravic – A perceived tilt in the visual vertical during linear acceleration.

ACCELERATION SYNDROME (G-FORCE SYNDROME). Alterations in physiologic function and/or perceptual-motor performance resulting from the forces imposed on the body by changes in velocity of direction of movement. The most significant of the symptoms are related to the circulatory status of the central nervous system and progress from loss of peripheral vision through blackout to unconsciousness during increases in acceleration.

ACCIDENT. That occurrence in a sequence of events that may produce unintended injury, illness, death, and/or property damage. (Note: The term often implies that the event was not preventable. From a health perspective, use of this term is discouraged since occupational injuries and illnesses should generally be considered preventable.)

ACCIDENT AND SICKNESS BENEFITS (NON-OCCUPATIONAL). Periodic payments to workers who are absent from work due to off-the- job disabilities through accident or sickness.

ACCIDENT CAUSE (CAUSAL FACTORS). One or more factors associated with an accident or a potential accident. Causal factors may be identified as time sequenced events and/or may be categorized as being related to human and/or environmental (e.g., equipment, machinery, atmospheric contaminant, temperature, etc.) influences and their interactions.

ACCIDENT COSTS. Monetary losses associated with an accident.

ACCIDENT EXPERIENCE. One or more indices describing accident performance according to various units of measurement (e.g., disabling injury frequency rate, number of lost-time accidents, disabling injury severity rate, number of first-aid cases, or dollar loss). A summary statement describing accident performance.

ACCIDENT HAZARD. A situation present in an environment or connected with a job procedure or process which has the potential for producing an accident.

ACCIDENT LOCATION. The exact position of the key event that produced the accident.

ACCIDENT POTENTIAL. A behavior(s) or condition(s) having a probability of producing an accident.

ACCIDENT PREVENTION. The application of countermeasures designed to reduce accidents or accident potential within system or organization. Programs directed toward accident avoidance. The reduction or elimination of behaviors and conditions having an accident potential.

Accident Probability. The likelihood of a worker, operation, or item of equipment becoming involved in an accident. The probability of a set of unsafe conditions and/or unsafe acts producing an accident.

Accident Proneness. A discredited theory of accident causation. Originally used to attribute the cause of accidents to personality traits of individuals or groups of workers.

Accident Rate. Accident experience in relation to a base unit of measure (e.g., number of disabling injuries per 1,000,000 employee-hours exposure, number of accidents per 1,000,000 miles traveled, total number of accidents per 100,000 employee-days worked, number of accidents per 100 employees, etc.).

Accident Records. Reports and other recorded information concerning employee accident experience.

Accident Reporting. Collecting information for, and/or preparing and submitting to a designated individual or agency, an official report of an accident.

Accident Statistics. Descriptive or inferential data which provide information about accident occurrences.

Accident Susceptibility. (See ACCIDENT PRONENESS.)

Accident Type. A description of the occurrences directly related to the source of injury classification and explaining how that source produced the injury. Accident type answers the question:
How did the injured person come in contact with the object, substance, or exposure named as the source of injury, or during what personal movement did the bodily motion occur?

Accident-free. A record of no accidents, sometimes of specified types, relating to an operation, activity, or worker performance during a specified time period.

Acclimatization. The state of successful physiologic adaptation of an individual to new climatic conditions.

Accuracy (validity). The closeness with which a measurement approaches the true or actual value.

Acoustic Trauma. Hearing loss caused by sudden loud noise in one ear, or by sudden blow to the head. In most cases, hearing loss is temporary, although there may be some permanent loss.

Acquired Immune Deficiency Syndrome (AIDS). An acquired defect in immune system function that reduces the affected person's resistance to certain types of infections and cancers. The disease is caused by infection with the human immunodeficiency virus (HIV).

Act of God. An accident causal factor generally interpreted as being beyond human control (e.g., lightning, flood, tornado, earthquake, etc.). Although the natural phenomenon itself may be beyond control, its effect can be controlled in many instances. Thus, the act of God classification should not be accepted as an excuse for not taking proper precautions or properly designing structures to reduce the severity of a so-called act of God occurrence.

Actinic Keratoconjunctivitis. (See WELDER'S FLASHBURN.)

Activity Sampling. A measurement technique for evaluating potential accident-producing behavior. It involves the observation of worker and organizational behavior at random intervals and the instantaneous classification of these behaviors according to whether they are safe of unsafe.

Actual Exposure Hours. Employee-hours of exposure taken from payroll or time clock records, wherever possible, and including only actual straight time worked (in hours) and actual overtime hours worked.

Acuity. The sharpness of a sense. Pertaining to the sensitivity of hearing, vision, smell, or touch.

Acute Exposure. Severe, usually critical, often dangerous exposure in which relatively rapid changes are occurring. An acute exposure normally runs a comparatively short course and its effects can be easier to reverse in contrast with a chronic exposure. Exceptions include exposure to electrical energy and radiation.

Acute Mountain Sickness. A disorder occurring in individuals exposed to high altitudes (about 10,000 ft.) for relatively long periods (24 hrs. or more). It presents as malaise, headache, and vomiting attributed to cerebral and pulmonary edema. It is probably related to both hypoxia and decreased atmospheric pressure. It may be prevented by proper acclimatization.

Acute Radiation Syndrome. The organic illness which follows exposure to relatively large acute doses of ionizing radiation. It presents as one of three separately identified dose dependent syndromes:

central nervous system syndrome – Follows doses of 2000 rem [20 sievert (Sv)] or above producing relatively prompt shock and coma, uniformly fatal within hours to a day.

gastrointestinal syndrome – Follows doses of 500 to 2000 rem (5-20 Sv) producing nausea, vomiting, and diarrhea within hours to days and death usually occurring within a week.

hematopoietic syndrome – Follows doses of 100 to 500 rem (1-5 Sv) producing early nausea and vomiting followed by a lag period of several weeks after which infection and hemorrhage occur due to hematopoietic suppression. At doses of 400 to 500 rem (4-5 Sv) death will occur in about 50% of the cases within 30 days. With adequate therapy the majority of affected individuals in this category can recover.

ADJUSTER. An individual who determines the amount of loss suffered in an insurance claim.

ADSORPTION. The adhesion of molecules of a gas, liquid, or dissolved material to a surface.

AEROSOLS. Liquid droplets or solid particles dispersed in air, that are of fine enough particle size to remain so dispersed for a period of time.

AGE-ADJUSTED DEATH RATE (AGE-STANDARDIZED DEATH RATE). The mortality rate of a population calculated for comparative purposes in such a way that allowance is made for the age distribution of the population. When mortality rates for two or more populations are adjusted in the same manner, they may be compared to each other as though the populations had the same age distributions.

AGE-SPECIFIC DEATH RATE. The number of deaths occurring in a defined age group, per unit population in that age group, over a stated period of time.

AGENCY (AGENT). The principal object, substance, or premises, such as tool, machine, or equipment, involved in an accident. The term is used either to designate the object most directly causing an accident, or the object inflicting injury or property damage.

AGENCY PART. The specific hazardous part of the agency that contributed to an accident occurrence or injury.

AIR CONTAMINANTS. Airborne particulates, gases, and vapors that are capable of causing illness or injury to workers upon ingestion, inhalation, or skin or mucous membrane contact.

AIR SAMPLING. Determining quantities and types of atmospheric contaminants by measuring and evaluating contaminants in a known volume of air.

AIRBORNE NOISE. A condition when sound waves are being carried by the atmosphere.

ALLEGED VIOLATION. A written violation issued by a regulatory agency.

ALLERGIC ALVEOLITIS. A granulomatous pneumonitis caused by inhalation of a specific allergen. Examples are: sequoias, farmer's lung, bagassosis, suberosis, and maple bark disease.

ALLERGY. A hypersensitive state acquired through exposure to a particular allergen, re-exposure bringing to light an altered capacity to react.

AMBIENT. Environmental surrounding, background.

AMBIENT NOISE. Total noise present in an environment. Usually a composite of sounds from many sources, including characteristic and background noise.

AMERICAN COLLEGE OF OCCUPATIONAL AND ENVIRONMENTAL MEDICINE (ACOEM). An international medical specialty society whose members promote worker and environmental health through preventive services, clinical practice, research and teaching.

AMERICAN CONFERENCE OF GOVERNMENTAL INDUSTRIAL HYGIENISTS (ACGIH). Professional society of persons employed by official governmental units responsible for full-time programs of industrial hygiene, educators, and others conducting research in industrial hygiene. Functions mainly as a medium for the exchange of ideas and the promotion of standards and techniques in industrial health.

AMERICAN INDUSTRIAL HYGIENE ASSOCIATION (AIHA). Professional society of industrial hygienists. To promote the study and control of environmental factors affecting the health and well-being of workers. Accredits industrial hygiene laboratories.

AMERICAN SOCIETY OF SAFETY ENGINEERS (ASSE). Professional society of safety engineers, safety directors and others concerned with accident prevention and safety programs. Sponsors training, including safety education seminars. Bestows awards; compiles statistics; maintains job placement service.

ANALYSIS, ACCIDENT. A separating or breaking up of an

accident into its parts so as to determine their nature, proportion, function, or relationship. A statement of the results of this process. Commonly involves the identification and evaluation of the unsafe conditions and unsafe acts that are associated with an accident. The dissection of an accident and its results to determine the basic elements and the resultant or potential damage to persons and/or things.

ANALYSIS, JOB HAZARD. (See JOB HAZARD ANALYSIS.)

ANECHOIC ROOM. A room in which all of the sound emanating from a source is essentially absorbed at the walls. Hence, there are no reflections and the spatial sound radiation pattern of a source may be determined. An anechoic room may be described as echoless or acoustically dead. The term may also refer to an electromagnetically echoless room.

ANODIZE. To coat a metal electrolytically with a protective film.

ANTAGONISM. The interaction of two or more physiologically active agents producing an effect which is less than the effect produced using equivalent doses of each agent presented individually.

ANTHRACOSILICOSIS. Pneumoconiosis caused by breathing air containing dust that has free silica as one of its components and that is generated in the various processes in mining and preparing anthracite (hard) coal and to a lesser degree bituminous coal.

APPARENT VIOLATION. An item observed by the compliance officer and mentioned in the closing conference as a violation of federal OSHA standards.

APPROVAL AGENCY. An agency designated or authorized to sanction, consent to, confirm, certify, or accept as good or satisfactory for a particular purpose or use, a method, procedure, practice, tool, or item of equipment or machinery.

APPROVED (METHOD, EQUIPMENT, ETC.). A method, equipment, procedure, practice, tool, etc., which is sanctioned, consented to, confirmed, or accepted as good or satisfactory for a particular purpose or use by a person or organization authorized to make such a judgment.

ARSON. The burning of buildings and/or property with malicious or criminal design, generally utilizing highly flammable materials or explosives to spread the fire quickly, or deliberately placed obstructions to impede fire fighting. This action may be by the owner and/or others and is by law a crime subject to federal jurisdiction.

ARTIFICIAL RADIATION. Man made radioactivity produced by particle bombardment or electromagnetic irradiation (e.g., x-rays) as opposed to natural radioactivity.

ASBESTOS. Naturally-occurring, fibrous minerals that include the serpentine mineral chrysotile and fine amphibole minerals but excludes fibrous forms of other minerals. It is incombustible and is used as thermal insulation. The inhalation of the fibers can cause asbestosis, lung cancer, and mesothelioma.

ASBESTOSIS. A form of lung disease (pneumoconiosis) caused by inhaling fibers of asbestos and marked by interstitial fibrosis of the lung varying in extent from minor involvement of the basal areas to extensive scarring.

ASPHYXIATION. Suffocation resulting from being deprived of oxygen. Simple asphyxiants act mechanically by excluding oxygen from the lungs when breathed in high concentrations (examples: nitrogen, hydrogen, carbon dioxide). Chemical asphyxiants act through chemical action preventing oxygen from reaching the tissue, or else prevent the tissue from using it even though the blood is well oxygenated (e.g., carbon monoxide, hydrogen cyanide, aniline).

ASSIGNABLE CAUSE. A cause factor designated as having a relationship to an accident. A causal factor or several causal factors assigned to an accident based on a predetermined causal classification system.

ASSIGNED RISK. Many states have unsatisfied judgment or financial responsibility laws which make the purchase of insurance mandatory. Some motorists cannot buy insurance for some reason, such as poor accident experience. To make it possible for them to be insured, there are assigned risk plans in which such risks are insured. These risks are rotated among the subscribing companies in proportion to the amount of automobile liability insurance each writes in the state. All companies writing this class of insurance are required to participate in this activity. A comparable system operates in some states with respect to worker's compensation.

ASSOCIATION (CAUSAL). (See CAUSAL ASSOCIATION.)

ASSOCIATION (NONCAUSAL). (See NONCAUSAL ASSOCIATION.)

Assumption of Risk. The legal theory that a person who is aware of a danger and its extent and knowingly exposes himself to it assumes all risks and cannot recover damages, even though he is injured through no fault of his own.

Asthma. A disease characterized by increased responsiveness of the trachea and bronchi to various stimuli, manifested by difficulty in breathing caused by generalized narrowing of the airways.

Atmosphere, Explosive. Atmosphere containing a mixture of vapor gas or particulate matter which is within the explosive or flammable range.

Attack Rate. The number of new cases of a disease or an injury per unit population occurring over a defined period of time.

Attention. Focusing on a task all of the personal abilities (both mental and physical) necessary for the safe accomplishment of that task. The human characteristic of giving sufficient heed or observing and perceiving with sufficient care to help prevent accidents. A readiness to respond to stimuli in a safe manner.

Attenuation. Reduction of intensity; to render less virulent or harmful; to reduce in strength.

Attenuation (sound). The reduction, expressed in decibels, of sound intensity to an observer due to either the distance from the source of noise or due to a barrier or acoustically treated material.

Attributable Risk in the Exposed. The rate of disease or injury in those exposed to a given causal agent that can be attributed to the given causal agent.

Attributable Risk in the Population. The rate of disease or injury in the total population that can be attributed to a given causal agent.

Attributable Risk Percent in the Exposed. The percentage of the rate of disease or injury in those exposed to a given causal agent that can be attributed to the given causal agent.

Attributable Risk Percent in the Population. The percentage of the rate of disease or injury in the total population that can be attributed to a given causal agent.

Audible Range. The frequency range over which normal ears hear. Approximately 20 Hz, through 20,000 Hz. Above the range of 20,000 Hz, the term ultrasonic is used. Below 20 Hz the term subsonic is used.

Audiogram. A record of hearing level measured at several different frequencies, usually 500 to 6,000 Hz. The audiogram may be presented graphically or numerically. Hearing level is shown as a function of frequency.

Audiologist. A person trained in the specialized problems of hearing and deafness.

Audiometer. A signal generator or instrument for measuring objectively the sensitivity of hearing in decibels referred to audiometric zero.

Average Days Charged Per Disabling Injury. This measure expresses the relationship between the total days charged to a disabling injury and the total number of disabling injuries as defined by the ANSI Z16 Standard. The average may be calculated by use of the following formula:

Average days charged per disabling injury =

$$\frac{Total\ days\ charged}{Total\ number\ of\ disabling\ injuries}$$

The following alternate formula may also be used to compute this measure:

Average days charged per disabling injury =

$$\frac{Standard\ Injury\ Severity\ Rate}{Standard\ Injury\ Frequency\ Rate}$$

Avulsions. A tearing away of a portion of the skin or other body structure.

B-Scale. A filtering system that has characteristics which roughly match the response characteristics of the human ear at sound levels between 55 and 85 dB. B-scale measurements are often referred to as dB(B).

Background Noise. The total of all sources of interference in a system used for the production, detection, measurement or recording of a signal, independent of the presence of the signal.

Background Radiation. Natural radiation in the environment including cosmic rays and radiation from naturally radioactive elements; it may also mean radiation from sources other than the one directly under consideration.

BAND PRESSURE LEVEL. Band pressure level of a sound for a specified frequency band is the sound pressure level for the sound contained within the restricted band. The reference pressure must be specified.

BAROTRAUMA. Tissue damage resulting from expansion or contraction in gas-filled spaces within or adjacent to the body. As external pressure varies with ascent or descent, gas volume within these spaces varies resulting in tissue distortion, edema, and hemorrhage. It is the most common occupational disease of divers.

BARRIER GUARD. Protection for operators and other individuals from hazard points on machinery and equipment.

adjustable barrier guard – An enclosure attached to the frame of the machinery or equipment with front and side sections which can be adjusted.

fixed barrier guard – A point of operation enclosure attached to the machine or equipment.

gate or movable barrier guard – A device designed to enclose the point of operation completely before the clutch can be engaged.

interlocked barrier guard – An enclosure attached to the machinery or equipment frame and interlocked with the power switch so that the operating cycle cannot be started unless the guard is in its proper position.

BEAT KNEE. Bursitis of the knee joints due to friction or vibration common in mining.

BELDING-HATCH INDEX. Estimate of the body heat stress of an average man (standard man) for various degrees of activity; also relates to his sweating capacity.

BENDS. (See DECOMPRESSION SICKNESS.)

BENIGN. The property of a neoplasm, of non-invasiveness; i.e., the neoplasm remains localized.

BERYLLIOSIS. Disease due to inhalation of beryllium or beryllium compounds. Acute berylliosis is characterized by bronchitis and pneumonia. The chronic disease which may follow is a systemic disease with pathologic changes not only in the lung, but in numerous other body tissues including the liver, spleen, kidneys, heart, skin, and bone.

BIAS. A consistent tendency for an estimate to overestimate or underestimate the true population value.

BIAS (RECALL). Consistent error due to differences in accuracy or completeness of recall to memory of prior events or experiences.

BIAS (RESPONSE). Consistent error due to differences in characteristics between those who volunteer to participate in a study and those who do not.

BIOENGINEERING. The application of engineering knowledge to the fields of medicine and biology.

BIOHAZARD. A combination of the words biological and hazard. A risk to humans presented by organisms or products of organisms.

BIOLOGICAL AGENTS. Cause of acute and chronic infections, parasitism, and toxic and allergic reactions to plant and animal agents. Infections may be caused by bacteria, viruses, rickettsia, chlamydia, or fungi. These agents act on or within the body to produce disease or infection.

BIOLOGIC HALF-LIFE. The time required for a given species, organ, or tissue to eliminate half of a substance which it takes in.

BIOLOGICAL EFFECTS. Reactions of the body due to exposure to physical, chemical, or biological agents.

BIOLOGICAL EFFECTS OF RADIATION. Any alteration in a biological system resulting from exposure to radiation. Effects may or may not be hazardous or detrimental to the system.

BIOLOGICAL MONITORING. The direct quantitative analysis of expired air, body fluids or tissue for the presence of the hazardous agent or its metabolites and/or evidence of biologic impairment quantified by the use of physiologic, psychometric, biochemical tests.

BIOMECHANICS. The application of mechanical laws to living structures, specifically to the locomotor systems of the human body.

BIOMECHANICS, OCCUPATIONAL. Concerned with the mechanical properties of human tissue, particularly the response of tissue to mechanical stress. A major focus is the prevention of overexertion disorders of the lower back and upper extremities.

BIOMETRY. The application of statistical methods to the

study of numerical data based on biological or biomedical observation and phenomena.

BIRTH COHORT ANALYSIS. The observation over time of persons born within a particular period of time, such as those born during a one-year or a five-year period.

BLIND EXPERIMENT. Experiment in which either the experimenter or the subject(s) does not know the group to which the subject(s) have been allocated. The intent of blind procedures is to eliminate the biases and prejudices of the subject(s) or of the investigator.

BOARD OF CERTIFIED SAFETY PROFESSIONALS (BCSP). An autonomous board established to certify the professional and technical competence of individual practitioners in the safety field.

BODILY INJURY (BI). Injury to a human being, as opposed to injury to property.

BOILER CODES. Standards prescribing requirements for the design, construction, testing, and installation of boilers and unfired pressure vessels. (e.g., American Society of Mechanical Engineers, etc.)

BRIGHTNESS. The light intensity of a surface in a given direction per unit emulsive area as projected on a plane of the same direction. (See ILLUMINATION.)

BUILDING CODE. An assembly of regulations which set forth the standards to which buildings must be constructed.

BURSA. A sac or sac-like structure containing a fluid and situated at places in tissues at which friction would otherwise develop. In industry the bursae of interest are at the shoulder, the elbow, and the anterior surface of the knee.

BURSITIS. Inflammation of a bursa, sometimes with development of calcium deposits. Causes of bursitis are multiple, but inflammation of the shoulder bursa may be associated with repetitive or unusual motion. Bursitis of the elbow or knee may be due to prolonged pressure or blunt trauma.

BYSSINOSIS. A pulmonary disease occurring among cotton textile workers and preparers of flax and soft hemp, due to inhalation of cotton or flax dust. The acute form is characterized by tightness of the chest, wheezing, and cough on return to work after a brief absence. The chronic form, occurring after years of exposure is marked by permanent dyspnea.

C-SCALE. A filtering system that has characteristics which roughly match the response characteristics of the human ear at sound levels above 85 dB. C-scale readings may be referred to as dB(C).

CANCER. Any malignant neoplasm.

CANISTER (AIR-PURIFYING). A container filled with sorbents and catalysts that remove gases and vapors from air drawn through the unit. The canister may also contain an aerosol (particulate) filter to remove solid or liquid particles.

CAPTURE VELOCITY. The air velocity at any point in front of a ventilation hood or at the hood opening necessary to overcome opposing air currents and capture the contaminated air at that point and cause it to flow into the hood.

CARCINOGEN. Any agent capable of inducing or promoting malignant neoplastic changes in living organisms.

CARCINOMA. A malignant neoplasm arising from epithelial cells.

CARDING. The process of unbinding or untangling wool, cotton, etc.

CARPAL TUNNEL. A narrow tunnel structure at the base of the palm as it joins the wrist through which pass flexor tendons to thumb and fingers, blood vessels, and the median nerve.

CARPAL TUNNEL SYNDROME. Impairment resulting from entrapment or compression of the median nerve at the carpal tunnel. Involves a complex of symptoms including pain and numbness in the palm and first three digits. In the workplace it is associated with repetitive or unusual motion in which the median nerve becomes compressed in the carpal tunnel.

CASE. In epidemiology, a person in the population or study group identified as having the particular disease, health disorder, or condition under investigation.

CASE-CONTROL STUDY. A type of epidemiologic study in which persons with a given disease (the cases) and persons without the given disease (the controls) are selected; the proportions of cases and controls with certain background characteristics or exposure to possible risks are then determined and compared.

CASE FATALITY RATE. The number of persons dying of a disease divided by the number of persons who contracted the disease.

CASUAL CONNECTION. An act, agency or force occurring without design or without being foreseen or expected which is a concurrent or contributing factor to the injury, but is not usually the proximate cause thereof.

CASUALTY INSURANCE. Insurance written by companies licensed under the casualty sections of state insurance laws as distinguished from that written under the fire or marine or life insurance sections. This type of coverage is concerned principally with insurance against loss due to accident or other mishap.

CAT. Acronym for computerized axial tomography scan. (See SCAN.)

CATASTROPHE. A loss of extraordinarily large dimensions in terms of injury, illness, death, damage, and destruction.

CAUSAL ASSOCIATION. A statistical association between the occurrence of a factor and a disease or injury, in which available evidence indicates that the statistically associated factor increases the probability of occurrence of the disease or injury and that its removal decreases the probability of occurrence.

CAUSAL FACTORS (OF AN ACCIDENT). A combination of simultaneous or sequential circumstances directly or indirectly contributing to an accident. Modified to identify several kinds of causes such as direct, early, mediate, proximate, distal, etc.

CAVEAT EMPTOR. Latin for let the buyer beware. The rule of law that the purchaser buys at his own risk concerning quality and condition.

CAVEAT VENDITOR. Latin for let the seller beware. The doctrine of law that a seller, if he wishes to absolve himself of future responsibility, must make a specific agreement with the purchaser to that effect.

CERTIFIED SAFETY PROFESSIONAL (CSP). An individual who has been certified by the Board of Certified Safety Professionals as having achieved professional competence in the safety field.

CHAIN OF CAUSATION. The original force is responsible for every subsequent force which it puts in motion, and for the final result.

CHELATING AGENT. Any compound which will inactivate a metallic ion with the formation of an inner ring structure in the molecule, the metal ion becoming a member of the ring. The original ion, thus chelated, is effectively trapped and unable to exert its usual effect.

CHLORACNE. An acneform eruption attributed to chlorinated hydrocarbon exposures. Principal agents are chlornaphthalenes, chlordiphenyls, and chlordiphenyl oxides. Exposure to polychlorinated biphenyls (PCBs) can cause this skin abnormality.

CHRONIC BRONCHITIS. A disorder characterized by excessive mucus secretion in the bronchi manifested by a chronic or recurrent productive cough (arbitrarily for a minimum of three months per year and for at least two successive years) in persons whom other causes of productive cough have been excluded.

CHRONIC VIOLATOR. The chronic or persistent violator is the individual who repeatedly violates established statutes and ordinances as in the case of OSHA regulations, or in the case of company safety rules.

CIRCUMSTANCES (OF AN ACCIDENT). The set of conditions which surround the accident or led to it.

CITATION. Issued by the representative of the Assistant Secretary of Labor, the OSHA Area Director, which alleges conditions which violate specific maritime, construction, or general industry standards.

CLAIM. (1) A broad, comprehensive term whose meanings include, but are not limited to: cause of suit action; judgment; right; a demand for compensation and/or for payment of medical expenses. (2) The amount which a policy holder believes he has coming from an insurance company as the result of some occurrence insured against.

CLAIMANT. One who claims or asserts a right, demand, or claim.

CLUSTER SAMPLING. A procedure in which clusters (e.g., city blocks) rather than individual units are first selected from population and then observations are made on all individual units (e.g., households) in each cluster.

CODES. Rules and standards which have been adopted by a governmental agency as mandatory regulations having the force and effect of law. Also used to describe a body of standards.

COHORT STUDY (PROSPECTIVE). A type of epidemiologic study in which persons initially free from the disease or injury under study are selected and their background characteristics and their exposure to possible risks ascertained; these individuals are then followed through time and the proportions who develop the disease or injury among those with and without certain characteristics and among those exposed and not exposed to the risks are determined and compared.

COHORT STUDY (RETROSPECTIVE). A type of epidemiologic study in which background characteristics and exposure to possible risks at some time in the past are determined in a group of individuals who were at that time free of the disease or injury under study; persons having developed the disease or injury since the time the characteristics or exposures were measured are ascertained, usually through existing records. Comparisons are then made of the proportions having developed the disease or injury among those who did and did not have the characteristics and among those who were and were not exposed to the risk.

COLLISION DIAGRAM. A diagram of an intersection or section of roadway where an accident occurred. The diagram shows the manner of collision, the resting positions of vehicles and other items after the collision occurred by the use of designated symbols.

COMBUSTIBLE GAS METER. Instrument which contains sensitive Wheatstone Bridge. If air containing solvent or combustibles is drawn over one leg of bridge and unbalances it, an appropriate meter indicates change. Signal change is proportional to concentration of contaminant.

COMPARISON (OR CONTROL) GROUP. In a case-control epidemiological study, a group of individuals free of the disease under study who are believed to reflect the characteristics of the population from which the persons with the disease (the cases) were drawn.

COMPENSABLE INJURY. An occupational injury/illness resulting in sufficient disability to require the payment of compensation as prescribed by law. A work injury/illness for which compensation benefits are payable to the worker or beneficiary under worker compensation laws.

COMPENSATION. Indemnity paid to an employee for disability sustained in an occupational accident.

COMPLIANCE. A measure of the ease with which a hollow viscus (lung, urinary bladder) may be distended.

COMPLY. To act in accordance with regulatory standards (e.g., the Occupational Safety and Health Standards); to follow the rules and regulations published in the Code of Federal Regulations.

CONFIDENCE INTERVAL. A range of values for a variable of interest, e.g., a rate, constructed so that this range has a specified probability of including the true value of the variable.

CONFINED SPACE. A space that by design has limited openings for entry and exit, unfavorable natural ventilation, contains or produces dangerous air contaminants, and is not intended for continuous employee occupancy. Confined spaces include storage tanks, compartments of ships, process vessels, pits, silos, vats, degreasers, reaction vessels, boilers, ventilation and other exhaust ducts, sewers, tunnels, underground utility vaults, and pipelines.

CONFLAGRATION. A fire extending over a considerable area, and destroying numbers of buildings and/or substantial amounts of property.

CONFOUNDING VARIABLE. A variable that is associated with both the independent and dependent variables which would not otherwise exist.

CONGENITAL ABNORMALITY. Any defect in the size or function of an organism existing at birth. A congenital abnormality is distinguished from an inborn abnormality in that the change may be hereditary or due to some applied stimulus during the total period of gestation; inborn on the other hand always implies a change secondary to a hereditary influence.

CONSENSUS STANDARD. A standard developed according to a consensus agreement or general opinion among representatives of various interested or affected organizations and individuals.

CONSTRAINT. A restriction or a compelling force affecting freedom or action. Forcing into a holding within close bounds. An operational condition which may necessitate work performance in a less than ideal, safe environment (e.g., building construction) and that therefore requires the provision of special safeguards.

CONTACT DERMATITIS (ALLERGIC TYPE). An inflammation of the skin due to contact with a sensitizing material. Certain substances may not cause a dermatitis on first contact, but may so alter the skin that it becomes inflamed after a second contact 5-10 days later. The first contact is said to have sensitized the skin. The second and subsequent contacts give rise to an allergic reaction often with a minimal exposure.

CONTACT DERMATITIS (IRRITANT TYPE). Inflammation from a material which would cause inflammation in the skin of most people if applied in sufficient concentration over a sufficient length of time. Most industrial dermatitis is due to contact with direct irritants.

CONTAINMENT. Restricting the spreading of fire or toxic or hazardous material.

CONTEST. To object to an alleged violation of regulatory standards. An example would be disputing a violation alleged by an OSHA Area Director, thus placing the dispute before the review commission.

CONTRACTUAL LIABILITY. Liability as set forth by agreements between people as distinguished from liability imposed by law (legal liability).

CONTRIBUTORY NEGLIGENCE. The act or omission amounting to want of ordinary care on part of complaining party, which concurring with defendant☐s negligence is proximate cause of injury. This is different from assumption of risk, which exists where none of fault for injury rests with plaintiff, but where plaintiff assumes consequences of injury occurring through fault of defendant, third person, or fault of no one.

CONTROL GROUP. (See COMPARISON GROUP.)

CONTROL TECHNOLOGY. Engineering measures and techniques designed as a system to eliminate, or reduce to acceptable levels, exposure to harmful agents in the workplace. Includes engineering controls, monitoring, personal protective equipment, and work practices.

CONTUSION. Injury to the body due to a blunt external force from an object, or a fall or bump. Usually accompanied by swelling and black and blue mark due to rupture of veins. Severity may vary from a small bruise to severe underlying damage to bones, vessels and nerves. Severe contusions may be accompanied by lacerations as well.

CORRELATION COEFFICIENT. A measure of the degree of association found between two characteristics in a series of observations on the assumption that the relationship between the two characteristics is adequately described by a straight line. A positive coefficient indicates that as one variable increases in value the other also tends to increase, while a negative coefficient indicates that as one variable increases in value, the other tends to decrease. Its value must be between +1 and -1; either +1 or -1 denotes complete dependence of one characteristic on the other, and 0 denotes no association whatsoever between them.

COST EFFECTIVENESS. In this method, the cost of system changes made to increase safety are compared with either the decreased costs of fewer serious failures, or with the increased effectiveness of the system to perform its task to determine the relative value of these changes.

COVERAGE (INSURANCE). An insured risk or liability. That which is insured, as specified in the insurance policy.

CRASH SAFETY. A system characteristic that allows the system occupants to survive the impact of a crash and to evacuate the system after potentially survivable accidents.

CRASHWORTHINESS. The capacity of a vehicle to act as a protective container and energy absorber during impact conditions.

CRIMINAL NEGLIGENCE. Involving or relating to a legal crime due to failure to use a reasonable amount of care when such failure results in injury, illness, or death to another.

CRITICAL FUNCTION. An operation or activity which is essential to the continuing survival of a system. Those functions which have a major impact on system performance and safety.

CRITICAL INCIDENT TECHNIQUE. A set of procedures for collecting direct observations of human behavior in such a way as to facilitate their potential usefulness in solving practical problems and developing broad psychological principles. The critical incident technique outlines procedures for collecting observed incidents having special significance and meeting systematically defined criteria. A randomly selected sample of critical incidents should permit an inference to be made concerning the existence of similar incidents within the population from which the sample was taken.

CRITICAL ORGAN. That organ or tissue for which radiation injury will be of greatest detriment to health and, therefore, the limiting organ for that particular circumstance. Criticality of an organ may be based on specific radiosensitivity, localized radiation, selective uptake of specific radioisotopes, or the importance of the organ for health.

CROSS-SECTIONAL STUDY. A type of epidemiologic study in which the presence or absence of background characteristics or exposure to possible risks and also the presence or absence of the disease under study are measured at one point in time in the individuals being studied. Prevalence rates among those with and without the characteristics or risks are then compared.

CUMULATIVE TRAUMA DISORDER (CTD). A musculoskeletal injury that arises gradually as a result of repeated microtrauma. CTDs are characterized by injuries to the tendons, nerves, or neurovascular system. Examples of CTDs include tendinitis, tenosynovitis, carpal tunnel syndrome, thoracic outlet syndrome, and Raynaud's phenomenon (white finger disease).

DAMAGE. (1) Loss in value, usefulness, etc., to property or things. Harm causing any material loss. (2) Severity of injury or the physical, functional, or monetary loss that could result if hazard is not controlled.

DAMAGE RISK CRITERION. The suggested base line of noise tolerance, which if not exceeded, should result in no hearing loss due to noise. A damage risk criterion may include in its statement a specification of such factors as time of exposure, noise intensity, noise frequency, amount of hearing loss that is considered significant, percentage of the population to be protected, and method of measuring the noise level.

DAMAGES. Compensation which may be recovered in the courts by a person who has suffered loss, damage, or injury, whether to person or property, through the unlawful act, omission, or negligence of another.

compensatory damages – That amount that will compensate the injured party for injury sustained, and nothing more, to make good or replace the loss.

punitive damages – Damages assessed as a punishment to the wrongdoer, or as an example to others, for outrageous conduct or gross, wanton negligence.

DANGER. A general term denoting liability or potential of injury, illness, damage, loss, or pain.

DANGER ZONE. A physical area or location within which a danger exists.

DANGEROUS. Attended with risk; hazardous; unsafe. Something that if in normal use, danger or injury can be anticipated by the user. Something without adequate protection.

imminently dangerous – Something, by reason of defective construction, that causes an impending or threatening, dangerous situation which could be expected to cause death or serious injury to persons in the immediate future unless corrective measures are taken.

inherently dangerous – Something which is usually dangerous even in its normal or nondefective state, such as explosives or poisons, and requires special precautions and warnings so as to prevent injury.

DAYS OF DISABILITY. Total full calendar days on which an injured person was unable to work as a result of injury. The total does not include the day of the injury or the day of return to work. (See LOST WORKDAYS.)

DE MINIS VIOLATION. Violation of a regulatory standard that does not involve an immediate or direct relationship to the safety or health of an employee.

DEAF. A term used to describe a person who has lost hearing before the speech patterns were established.

DEAFENED. Refers to a person who has lost the ability to hear after normal speech patterns were established.

DEATH (ACCIDENTAL). An injury which terminates fatally and is causally related to an accident. Death resulting from work injuries is assigned a time charge of 6,000 days each according to ANSI 16 standard.

DEATH CERTIFICATE. A vital record signed by a licensed physician that includes cause of death, decedent☐s name, sex, date of birth, date of death, place of residence and of death and usually occupation.

DECIBEL (dB). A unit used to express the ratio of two amounts of electric or acoustic signal power. The decibel is equal to 10 times the logarithm of the signal power ratio as expressed by the following equation:

$$n(dB) = 10 \log_{10} [(P1)/(P2)]$$

DECOMPRESSION SICKNESS (BENDS, CAISSON DISEASE). A condition caused by the formation and growth of bubbles in the blood or tissue resulting from a state of supersaturation with gas. This occurs when the sum of partial pressures of gases dissolved in a tissue exceeds the ambient pressure. It occurs in divers and compressed air workers on return from hyperbaric pressures to surface pressure or in aviators going from surface pressure to hypobaric pressures at altitude. Several specific clinical syndromes are described:

Serious Symptom or Type II
    Cerebral
    Spinal Cord
    Vestibular (the staggers)
    Pulmonary (the chokes)

Simple or Type I
  Pain-only bends
  Skin bends (the niggles)

DECONTAMINATION. Removal of a polluting or harmful substance from air, water, earth surface, etc. For example, the process of removing hazardous chemical contamination from objects or areas.

DECONTAMINATION (RADIATION). The removal of radioactive material from a location where it is not desired. In regard to personnel it would include both removal of external contamination by washing and removal of internal contamination by the use of chelating agents or similar methods.

DEFECT. (1) Known or unknown unsafe or unwanted physical condition of material and/or equipment due to an inherent or created weakness that may lead to an accident. (2) Anything that exceeds specifications or standards.

DEFECTIVE. Lacking in some particular way which is essential to the completeness or security of the object.

DEFLAGRATION. An exothermic reaction which propagates from the burning gases to the unreacted material by conduction, convection, and radiation. The combustion zone progresses through the material at a rate that is less than the velocity of sound in the unreacted material.

DEGENERATIVE JOINT DISEASE. (See OSTEOARTHRITIS.)

DEGREES OF NEGLIGENCE. Ordinary negligence is based upon the fact that one ought to have known the results of unsafe acts. Gross negligence rests on the assumption that one knew the results of acts but was recklessly or wantonly indifferent to the results. All negligence below that called gross or ordinary by the courts is slight negligence.

DENSITY (OF TRAFFIC). The number of vehicles occupying a unit length of the moving lanes of a roadway at a given instant. Usually expressed in vehicles per mile.

DEPENDENT VARIABLE. A variable that is considered to take on values at least in part as a result of the particular value of the independent variable.

DEPOSITION. The testimony of a witness taken upon interrogatories, not in open court, under oath, in writing and duly authenticated and intended to be used as evidence in court. (See INTERROGATORIES.)

DEPTH PERCEPTION (BINOCULAR). The ability to judge distances of nearby objects by the use of both eyes.

DERMATITIS. Inflammation or irritation of the skin. Industrial dermatitis is an occupational skin disease. There are two general types of skin reaction: primary irritation dermatitis and sensitization dermatitis. (See IRRITANT.)

DESIGN (SAFETY). The planning of environments, structures, and equipment, and the establishment of procedures for performing tasks, so that human exposure to injury or illness potential will be reduced or eliminated. In product safety, design of the product for safe use.

DESTRUCTIVE TEST. A procedure for quality testing whereby the material being tested is destroyed in order to obtain the desired measurements.

DETECTOR TUBE. A glass tube containing specific chemicals which have been impregnated on inert material granules and which will change color when chemicals in air are drawn through the tube.

DETONATION. An exothermic reaction that is characterized by the presence of a shock wave in the material that establishes and maintains the reaction. A distinctive feature is that the reaction zone propagates at a rate greater than sound velocity in the unreacted material.

DIRECT CAUSE. Unsafe behaviors or unsafe conditions which contribute sequentially or concurrently in a chain of events leading to an accident.

DIRECT DAMAGE. Damage caused by the direct action of a peril as distinguished from damage done contingently.

DIRECT INJURY COSTS. The sum of compensation payments and medical expenses for an injury.

DISABILITY. Any injury or illness, temporary or permanent, which prevents a person from carrying on usual activity. (See PERMANENT DISABILITY, PERMANENT PARTIAL DISABILITY, PERMANENT TOTAL DISABILITY.)

DISABLING INJURY. ANSI Standard Z16: An injury which prevents a person from performing a regularly established job for one full day (24 hours) beyond the day of the accident.

DISABLING INJURY FREQUENCY RATE. The number of disabling (lost time) injuries per million employee-hours of exposure:

$$DIFR = \frac{Disabling\ Injuries \times 1,000,000}{Employee\text{-}hours\ of\ exposure}$$

(See INCIDENCE RATE.)

DISABLING INJURY INDEX. An index computed by multiplying the disabling injury frequency rate by the disabling injury severity rate and dividing the product by 1,000:

$$DII = \frac{DIFR \times DISR}{1,000}$$

This measure reflects both frequency and severity, yielding a combined index of total disabling injury (ANSI Z16). (See INCIDENCE RATE.)

Disabling Injury Severity Rate. The total number of days charged per million employee-hours of exposure:

$$DISR = \frac{Total\ days\ charged \times 1,000,000}{Employee\text{-}hours\ of\ exposure}$$

(See INCIDENCE RATE.)

DISASTER CONTROL. Advanced planning and established procedures for handling emergency situations.

DISC, INTERVERTEBRAL. A soft tissue structure between the bodies of the vertebrae. The central portion (nucleus) is a soft, pulpy material surrounded by an annular interplacement of tough, fibrous tissue.

DISC, RUPTURED. A condition in which the central portion (nucleus) of the intervertebral disc protrudes or herniates through the annular fibrous tissue; it frequently presses on a nearby nerve. The medical term for this condition is herniation of the nucleus pulposus. Other terms commonly used are slipped disc, ruptured disc, prolapsed disc.

DISCLAIMER. The seller may insert in his contract or agreement a statement that he does not warrant at all, or that he warrants only against specified consequences or costs. Disclaimers do not release the manufacturer or defendant from liability for negligence, nor are disclaimers a defense to statutory violations.

DISEASE, CLASSIFICATION OF. The grouping of persons with similar sign and symptoms and diagnostic test results into a disease entity which permits them to be distinguished from people with other disease entities; also, the arrangement of these disease entities into groups with common characteristics.

DISFIGUREMENT. A blemish, defect, or deformity which harms the appearance or attractiveness of the human body or a physical structure.

DISTAL CAUSE. A behavioral act or condition involved simultaneously or sequentially in the causal factors leading to an accident, but separated from the accident location by time and/or space.

DOSE (RADIATION). The amount of radiation delivered to a specified area or the whole body. A dose meter, or dosimeter is an instrument that measures radiation dose. Dose rate is the dose delivered per unit of time. The term dose or dosage is also used generally to express the amount of energy or substance absorbed in a unit volume by an organ or individual.

DOSE (TOXICOLOGY). The amount of exposure to a bioactive chemical agent that is received by an organism or individual.

DOSE-RESPONSE RELATIONSHIP. A relationship in which a change in amount, intensity, or duration of exposure is associated with a change, either an increase or a decrease, in risk of a specific outcome.

DROWNING. Death from acute asphyxia while submerged, whether or not liquid has entered the lungs.

DRY-CHEMICAL EXTINGUISHER. An extinguisher containing a chemical which extinguishes fire by interrupting the chain reaction wherein the chemicals used prevent the union of free radical particles in the combustion process so that combustion does not continue when the flame front is completely covered with the agent.

DRY-POWDER EXTINGUISHER. A fire extinguisher designed for use on combustible metals fires, such as sodium, titanium, uranium, zirconium, lithium, magnesium, and sodium-potassium alloys.

DUMMY VARIABLE. In regression analysis, a variable that is not continuously distributed, but which has two or more distinct levels; values are assigned to these levels, usually according to their presence or absence. For instance, the value 1 might be assigned to females and the value 0 to males.

Dusts. Small solid particles generated by the breaking up of larger particles by processes such as crushing, grinding, drilling, explosions, etc. Dust particles already in existence in a mixture of materials may escape into the air through such operations as shoveling, conveying, screening, sweeping, etc. Dust is a term used in industry to describe airborne sold particles that range in size from 0.1 to 25 microns (1 micron = 1/10,000 cm = 0.001 mm = 1/25,000 in.). [Editor's Note: Although industry may still use the micron, it is a term outside the International System, and CIPM (International Committee of Weights and Measures) considers it preferable to avoid. Like such terms as fermi, caloric, stere, etc., the micron is a nonacceptable term in American Society of Mechanical Engineers' publications, and it should be converted to its SI counterpart, viz, 0.000 001m.]

Dysbaric Osteonecrosis (aseptic bone necrosis). Lesions occurring in the bones, especially in juxta-articular areas, of divers and caisson workers. Small areas of bone die probably as a result of vascular occlusion by gas bubbles. Once detected, the lesions are essentially irreversible. No accepted method of therapy currently exists.

Ear Protectors. Plugs, muffs, or helmets designed to keep excessive noise from the ear to preserve hearing acuity.

Early Cause. An act on the part of some person or organization or a condition that causes or permits approximate or immediate cause to exist (also referred to as distal cause).

Ecological Correlation. A correlation in which the units studied are populations rather than individuals. Correlations found in this manner may not hold true for the individual members of these populations.

Eczema. A skin disease or disorder. A nonspecific term for any type of dermatitis.

Edema. Excessive accumulation of body fluid in the tissue space of the body, producing a swelling of body tissues.

Efficient Cause. The cause which originates and sets in motion a chain of causation through other causes to the result. The cause of injury is attached to legal liability. This term is not as popular in use as proximate cause.

Electrical Safety. Engineering measures, including devices, used to prevent worker exposure to hazards caused by electricity.

Electrostatic Precipitator. An air cleaning device that involves the following steps: electrical charging of suspended particulate matter; collection of charged particles on a grounded surface; and removal of particulates from the collecting surface by mechanical vibration or flushing with liquid.

Emergency Alarm. A warning device, usually visual or auditory, which indicates the existence of an emergency situation requiring immediate action.

Emergency Procedure. A plan for action in case of emergency.

Emergency Shut-off. A switch placed in a convenient position for cutting off the supply of electricity to a piece of equipment or to a building, in case of emergency.

Emergency Stop. A switch and/or a mechanical device installed in an elevator, or other similar material-handling equipment, by means of which the power to the operating motor can be cut off in case of an emergency.

Emission Control. Engineering measures, including devices, used to prevent worker exposure to contaminants that are released within the workplace. The term also refers to measures used on internal combustion engines, exhaust stacks, and other emission sources that are used to protect the general public.

Emphysema. An enlargement of the air spaces distal to the terminal nonrespiratory bronchioles, with destruction of alveolar walls.

Employee-hours. The total number of hours worked by all employees of an industrial organization or an industry. An employee-hour is the equivalent of one person working for one hour.

Employer's Liability. Legal liability imposed on an employer making the employer responsible for paying damages to an employee injured by the employer's negligence. Generally replaced by Worker's Compensation, which pays the employee whether the employer has been negligent or not.

Enclosed Space. (See CONFINED SPACE.)

Endemic. Usual level of disease occurrence in a population.

ENGINEERING CONTROLS. Basic methods used to prevent worker exposure to harmful chemical, physical, or biological agents by means of material substitution, equipment isolation, and material removal. Examples include: process design and modification, equipment design and enclosure, barrier guards, and ventilation.

EPICONDYLES (OF HUMERUS). The bony prominences at the medial and lateral side of the elbow. The extensor muscles of the wrist and digits are attached to the lateral epicondyle and the flexors to the medial epicondyle.

EPICONDYLITIS (TENNIS ELBOW). An irritation at the common muscle attachment at the epicondyles, almost always at the lateral epicondyle. The cause may be a contusion of the epicondyle area or repetitive motion involving a rotary movement at the elbow. This condition is common in tennis players who serve too vigorously for the first time in spring. Symptoms are tenderness over the external epicondyle, pain in the lateral aspects of the elbow on grasping objects.

EPIDEMIC. Unusually high level of disease or injury occurrence in a population in light of past experience.

EPIDEMIOLOGY. The study of the factors determining and influencing the frequency and distribution of disease, injury, and other health-related events and their causes in a defined human population for the purpose of establishing programs to prevent and control their development and spread.

EQUIVALENT FORM. Any of two or more forms of a test that are closely parallel with respect to the nature of the content and the difficulty of the items included and that will yield very similar average scores and measures of variability for a given group.

ERGONOMICS/HUMAN FACTORS. Field of investigation dealing with interactions between workers and the total working environment. Human factors engineering is the application of information about human characteristics to the design of systems and environments so that people can work safely.

ETIOLOGY. The study of causes of disease or injury, both direct and predisposing, and of their mode of operation.

EXHAUST (GENERAL). Diluting the general room atmosphere with outdoor air fast enough to keep the concentration of toxic vapor in the room air within safe limits. Also known as general ventilation, or dilution ventilation.

EXHAUST (LOCAL). A local exhaust system is used to collect air contaminants at the source, as contrasted with general ventilation which allows the contaminant to spread throughout the workroom, later to be diluted by exhausting quantities of air from the room. Local exhaust may be achieved using an enclosure, a receiving hood, or an exterior hood.

EXHAUST VENTILATION. The removal of air or other gas from any work space, usually by mechanical means.

EXPERIENCE. Data describing past events, e.g., accident experience refers to the frequency number or severity of accidents that have occurred within a specified time period.

EXPERIENCE RATING (MERIT RATING). Process of basing tax rates or insurance premiums on the employer's own record, as in worker's compensation, unemployment insurance, and commercially insured health and insurance programs, so that the employer may benefit from a good record.

EXPERT EVIDENCE. Testimony given in relation to some scientific, technical, or professional matter by an expert witness. An expert can reason out, infer, or conclude from hypothetically stated facts, and can offer opinions involving the subject matter under consideration. It is up to the jury to give whatever credence it wants to the testimony.

EXPERT TESTIMONY. The opinion of a witness skilled in a particular art, trade, or profession or possessed of special knowledge derived from education or experience not within the range of common experience, education, or knowledge.

EXPERT WITNESS. A person possessing particular knowledge, wisdom, skill or information acquired by study, investigation, observation, practice, or experience, regarding the subject matter under consideration and not likely to be possessed by ordinary or inexperienced persons. It is up to the judge to determine if the person is qualified as an expert and if the testimony is pertinent to, or would shed light on the case.

EXPLOSIMETER. An instrument used to determine whether an atmosphere has sufficient gas and oxygen in mixture to be explosive.

EXPLOSION. A rapid increase of pressure in a confined space followed by its sudden release due to rupture of the container (vessel, structure, etc.). The increase in pressure is generally caused by an exothermic chemical reaction or over- pressurization of a system.

Explosion Venting. A means provided for the release of high pressures caused by explosions.

Explosion-proof. An electrical apparatus so designed that an explosion of flammable gas inside the enclosure will not ignite flammable gas outside. (See FLAME-PROOF.)

Explosive Decompression. A sudden rapid decrease in barometric pressure. This may occur from loss of integrity of a pressurized aircraft cabin or recompression chamber or in a diver who blows up, surfacing rapidly from depth due to a loss of buoyancy control. It may result in decompression sickness or pulmonary hyperinflation and air embolism.

Explosive Limits. The minimum (lower) and maximum (upper) concentration of vapor or gas in air or oxygen below or above which explosion or propagation of flame does not occur in the presence of a source of ignition. The explosive or flammable limits are usually expressed in terms of percentage by volume of vapor or gas in air. The difference between the lower and upper flammable (explosive) limits is the "range", expressed in terms of percentage by volume of vapor or gas in air. (See LOWER EXPLOSIVE LIMIT, UPPER EXPLOSIVE LIMIT.)

Explosive Mixture. A mixture of flammable vapor or gas and air within the lower and upper limits of the explosive range.

Explosives. Any chemical compound or mechanical mixture that is used or intended for the purpose of producing an explosion. Contains any oxidizing and combustive units or other ingredients in such proportions, quantities, or packing that an ignition by fire, by friction, by concussion, by percussion, or by detonation of any part of the compound or mixture may cause such a sudden generation of highly heated gases that the resultant gaseous pressures are capable of producing destructive effects on contiguous objects or of destroying life or limb.

Exposure. The quantity of time involved, the level (quantity), and the nature (quality) of involvement with certain types of environments possessing various degrees or types of hazards. The amount of time a worker has been exposed to certain types of job hazards.

Exposure (casualty). Proximity to a condition which may produce injury, death, or damage from dusts, chemicals, high pressure, explosives, etc.

Exposure Hours (employee-hours). Total number of employee- hours worked by all employees including those in operating, production, maintenance, transportation, clerical, administrative, sales, and other activities (See ANSI Z-16).

Extinguisher. (See FIRE EXTINGUISHERS.)

Extinguishing Agent. Material or substance which performs a fire extinguishing function.

Eye Protection. A device which safeguards the eye in an eye-hazard environment. The devices include safety glasses, chemical splash goggles, face shields, etc.

Face Shield. A protective device designed to prevent hazardous substances, dust particles, sharp objects, and other materials from contacting the face.

Facility Layout. The act or process of laying out, or planning in detail, to show the arrangement of equipment and other elements of a facility to establish a safe working environment.

Factor. In mental measure a hypothetical trait, ability, or component of ability, that underlies and influences performance on two or more tests and hence carries scores on the tests to be correlated. The term factor strictly refers to a theoretical variable, derived by a process of factor analysis, from a table of intercorrelations among tests; but it is also commonly used to denote the psychological interpretation given to the variable, i.e., the mental trait assumed to be represented by the variable as verbal ability, numerical ability, etc.

Factor Analysis. Any of the several methods of analyzing the intercorrelations among a set of variables such as test scores. Factor analysis attempts to account for the interrelationships in terms of some underlying factors preferably fewer in number than the original variables and it reveals how much of the variation in each of the original measures arises from or is associated with each of the hypothetical factors. Factor analysis has contributed to our understanding of the organization or components of intelligence, aptitudes, and personality and it has pointed the way to the development of purer tests of the several components.

Factor of Safety (FS). The ratio of ultimate strength of a material or structure to the allowable stress.

Fail Operational Fail Safe. A system characteristic that permits continued operation on occurrence of a failure while remaining acceptably safe. A second type of failure results in the system remaining safe, but non-operational.

Fail Operational. A characteristic design that permits continued operation in spite of the occurrence of a discrete failure.

Fail Safe. Design of a product or equipment, in such a manner that, when it fails or becomes inoperative it will do so in a safe position or condition.

Failure. An inability to perform an intended function.

Failure, Dependent. (See FAILURE, SECONDARY.)

Failure, Independent. (See FAILURE, PRIMARY.)

Failure, Primary. The failure which is responsible for a system malfunction.

Failure, Secondary. A failure which occurs as the consequence of another failure (also dependent failure).

Failure Analysis. The logical systematic examination of an item to identify and analyze the cause, mode, and consequence of a real failure.

Failure Assessment. The process by which the cause, effect, responsibility and cost of any reported problem in the system is determined and reported.

Failure Critical. A failure which could result in major injury or fatality to people or which could result in major damage to any system or loss of a critical function.

Failure Management. Decisions, policies and planning which identify and eliminate or control potential failures and implement corrective or control procedures following real failures.

Failure Mechanism. The physics or chemistry of the failure event, i.e., the cause of the failure.

Failure Mode and Effect Analysis (FMEA). A method of analysis used in system safety. The failure or malfunction of each component is identified, along with the mode of failure (e.g., switch jammed on). The effects of the failure are traced though the system, and the ultimate effect on the task performance is evaluated.

Failure Mode, Effect and Criticality Analysis (FMECA). An extension of an FMEA in which each effect is assigned a criticality index which reflects both the probability of the occurrence of the effect and the seriousness of the effect in terms of loss in performance and/or safety.

Failure Rate. The number of failures of an item per unit time (cycles, hours, miles, events, etc., as applicable for the item).

Fainting. A loss of consciousness as a result of a diminished supply of blood to the brain. Technically called syncope.

False Negative Rate. The proportion of persons with a disease for whom the screening test for that disease is negative.

False Positive Rate. The proportion of persons without a disease for whom the screening test for that disease is positive.

Fatal Accident. An accident resulting in the death of one or more persons.

Fatality. A death resulting from an accident.

Fatigue. The physical and/or mental responses to an activity which show themselves in a diminished capacity for work.

Fault Hazard Analysis. The analysis of hazards or hazard potential situations using fault tree methodology.

Fault Tree Analysis. A method of analysis used in system safety. An undesired event is selected and all possible factors that can contribute to the event are diagrammed in sequence in the form of a tree. The branches of the tree are continued until independent events are reached. Probabilities are determined for the independent events and after simplifying the tree, both the probability of the undesired event and the most likely chain of events leading up to it are computed.

Fibrosis. A reparative or reactive process characterized by the deposition of connective tissue fibers in and around the site of injury.

Final Position (after an accident). The place where objects or persons involved in the accident finally come to rest without application of power. This is the position before anything is moved to help the injured or remove vehicles or equipment.

Fire. Rapid oxidation with the evolution of heat and light.

Fire Alarm. A device or system (visual, auditory, local or transmitted to other locations, etc.) which signals the presence of a fire to occupants and those who will provide assistance.

Fire Classifications. (See FIRE EXTINGUISHERS.)

Fire Doors. Doors rated and tested for resistance to various degrees of fire exposure and utilized to prevent the spread of fire through horizontal and vertical openings. The doors must remain closed normally or be closed automatically in the presence of fire. The degree of resistance required is determined by the type of occupancy, the anticipated fire exposure, and the resistance of the structure in which it is installed.

Fire Extinguishers. Devices having characteristics essential to extinguish flame. Fire extinguishers may contain either liquid or dry chemicals, or gases (water, dry chemicals, carbon dioxide, etc.). They are tested and rated to indicate their ability to handle specific classes and sizes of fires.

class A extinguishers – For ordinary combustibles, such as wood, paper, and textiles, where quenching-cooling effect is required.

class B extinguishers – For flammable liquid and gas fires, such as oil, gasoline, paint, and grease, where oxygen exclusion or flame interruption effect is essential.

class C extinguishers – For fires involving energized electrical wiring and equipment where the nonconductive property of the extinguishing agent is of prime importance.

class D extinguishers – For fires in combustible metals such as magnesium, potassium, powdered aluminum, zinc, sodium, titanium, zirconium, and lithium.

Fire Prevention. Measures or actions specifically directed toward preventing the inception of fires, and minimizing the severity of fires should they occur.

Fire (flame) Proof. Material incapable of burning. The term fire-proof is false. No material is immune to the effects of a fire possessing sufficient intensity and duration. It is commonly, a though erroneously, used synonymously with the term fire resistive. Use of the term is discouraged since it is misleading.

Fire Protection. In its broadest interpretation it embraces all measures in the prevention, detection, and extinguishment of fire; relating to the safeguarding of human life and the preservation of property. In a strict interpretation, it refers to the methods of providing for fire control or fire extinguishment.

Fire Protection Engineering. The field of engineering concerned with the safeguarding of life and property against loss from fire, explosion, and related hazards. It is concerned with integrated programs involving the design and use of structures, equipment, processes and systems, including the areas of prevention, detection and alarm, and fire control and extinguishment, and gives consideration to functional, economic, and operational factors.

Fire Resistive. Refers to properties of materials or designs to resist the effects of any fire to which the material or structure may be expected to be subjected. A building constructed of fire resistive materials can withstand a burnout of its contents without subsequent structural collapse. Fire resistive implies a higher degree of a fire resistance than noncombustible.

Fire Retardant. In general denotes a substantially lower degree of fire resistance than fire resistive. The term is frequently used to refer to materials or structures which are combustible but have been subjected to treatments or surface coverages to prevent or retard ignition or the spread of fire.

Fire Wall. A fire resistant wall designed to prevent the horizontal spread of fire into adjacent areas, generally self-supporting and designed to maintain its integrity if the structure on either side completely collapses. If a wood roof is involved, the wall must extend through and above the roof.

First Aid. The emergency care of a person who is injured or ill, to prevent death or further injury, to relieve pain, and to counteract shock, until medical aid can be obtained.

First Aid Injury. An injury requiring first aid treatment only.

Fit for Purpose Intended. When something sold is useless, or so imperfect that the buyer would not have bought it had he known, the seller has not complied with the statutory warranty against hidden defects and the article cannot be considered fit for the purpose intended, unless the warranty has been specifically waived.

Fitness (or fit). Satisfactoriness, suitability or appropriateness to fulfill the need or use.

Flame. The visible heat rays which appear when the ignition of a material is reached. Hydrogen is one of the exceptions since the heat rays are not visible.

Flame (flash) Arrester. Devices utilized on vents for flammable liquid or gas tanks, storage containers, cans, gas lines or flammable liquid pipelines to prevent flashback (movement of flame) through the line or into the container when a flammable or explosive mixture is ignited.

Flame Propagation (spread). The spread of flame throughout a combustible vapor area which may be in a container or across a surface, independently of the ignition source. Generally used in connection with the capability and rate of such movement.

Flameproof. (See fire proof.)

Flammable. Any substance that is easily ignited, burns intensely, or has a rapid rate of flame spread. The substance may be in the form of an aerosol, a gas, a liquid, or a solid. Flammable and inflammable are identical in meaning; however, the prefix in indicates negative in many words and can cause confusion. Flammable, therefore, is the preferred term.

Flammable Limits. (See explosive limits.)

Flammable Liquids. Any liquid having a flash point below 100°F (37.8°C), except any mixture having components with flash points of 100°F (37.8°C) or higher, the total of which makeup 99 percent or more of the total volume of the mixture. Flammable liquids shall be known as Class I liquids. Class I liquids are divided into three classes as follows: (i) Class IA shall include liquids having flash points below 73°F (22.8°C) and having a boiling point below 100°F (37.8°C); (ii) Class IB shall include liquids having flash points below 73°F (22.8°C) and having a boiling point at or above 100°F (37.8°C); (iii) Class IC shall include liquids having flash points at or above 73°F (22.8°C) and having a boiling point below 100°F (37.8°C).

Flammable Vapor. A concentration, by volume, of vapors in air from a flammable liquid within the lower and upper flammable limits.

Flash Burn. Injury or destruction of body tissue caused by exposure to a flash or sudden release of intense radiant heat.

Flash Point. The lowest temperature of a liquid at which it gives off sufficient vapors to form an ignitible mixture with the air near the surface of the liquid or within the vessel used. The flash point can be determined by the open cup or the closed cup method. The latter is commonly used to determine the classification of liquids which flash in the ordinary temperature range.

Flux. Usually refers to a substance used to clean surfaces and promote fusion in soldering. However, fluxes of various chemical nature are used in the smelting of ores in the ceramic industry, in assaying silver and gold ores, and in other endeavors. The most common fluxes are silica, various silicates, lime, sodium and potassium carbonate and litharge and red lead in the ceramic industry.

Foam. A fluid aggregate of gas- or air-filled bubbles formed by chemical or mechanical means that will float on the surface of flammable liquids or flow over solid surfaces. The foam functions to blanket and extinguish fires and/or to prevent ignition of the material.

Folliculitis. Inflammation of a hair follicle or follicles.

Follow-up Study. (See cohort study.)

Foot Candle. A measure of illuminance produced on a surface all points of which are 1 foot from a directionally uniform point of 1 candela. That illuminance is 1 lumen/ft2 or 1 footcandle. (Editor's Note: Under SI, illuminance is measured in terms of lux units, where

$$1 \text{ lx} = 1 \text{ lm/m}^2 \times cd \times sr$$

where, m = meter; cd = candela; sr = steradian for solid angle. To convert: 1 footcandle = 1.076 391 lx.)

Forced Expiratory Flow ($FEF^{200\text{-}1200}$). The average rate of flow for a specified portion of the forced expiratory volume, usually between 200 and 1200 ml.

Forced Expiratory Volume (qualified by subscript indicating the time interval in seconds, FEV_T). Volume of gas exhaled over a given time interval with expiration as forceful as possible.

Forced Expiratory Volume Percentage Expired ($FEV_T\%$). Forced expiratory volume expressed as a percentage of the forced vital capacity.

$$(FEV_t/FVC) \times 100$$

Forced Mid-expiratory Flow (FEF$_{25-75\%}$). The average rate of flow during the middle half of the forced expiratory volume.

Forced Vital Capacity (FVC). The vital capacity performed with maximum inspiration and expiration as forceful and rapid as possible.

Foreseeability. The legal theory that a person may be held liable for actions that result in injury or damage only where the person was able to foresee dangers and risks that could reasonably be anticipated.

Four-to-one Ratio. An arbitrary ratio frequently used in the comparison of the indirect costs of an accident to the direct costs. Generally considered obsolete since no fixed ratio exists among various types of exposures.

Frequency (in cycles per second, or cps, Hertz, or Hz). The time rate of repetition of a periodic phenomenon. The frequency is the reciprocal of the period. It defines pitch or the highness or lowness of sound.

Frostbite. A freezing injury of the skin due to exposure to extreme cold. It may be recognized by whitening of the skin and loss of sensation. It is treated by rapid thawing. Deep frostbite with freezing of tissues deep to the skin generally results in dry gangrene.

Fume. Solid particles generated by condensation from the gaseous state, generally after volatilization from molten metals. A fume is formed when a volatilized solid, such as metal, condenses in cool air. The solid particles that make up a fume are extremely fine, usually less than 1 μm 1.122(0.001mm). In most cases, the hot material reacts with the air to form an oxide.

Functional Disease. Disease in which some change in function of the body or its parts occurs without changes in structure. Usually refers to psychiatric diseases.

G-force Syndrome. (See ACCELERATION SYNDROME.)

Gamma (γ) Radiation, Ray. Electromagnetic radiation of short wave length omitted from the nucleus of an atom. Gamma radiation is more penetrating than alpha or beta particles. Gamma rays are very similar to high energy X-rays.

Gas Mask. A face covering connected to its own purifying device which filters harmful gases from the air so uncontaminated air may be inhaled. Gas masks do not add oxygen to air and cannot be used where there is oxygen deficiency. (See RESPIRATOR.)

Gases. Normally formless fluids which occupy the space or enclosure and which can be changed to the liquid or solid state only by the combined effect of increased pressure and decreased temperature.

Genetic Drift. Random fluctuation of gene frequencies in a small population.

Genetic Effects of Radiation. Any changes in the genetic material following irradiation which are hereditarily transmitted and expressed in the progeny of the irradiated subject.

Genetic Mutation. A permanent, genetically transmitted variation in which there is an alteration in the molecular arrangement of the inherited DNA in the chromosomes of the germ cells. These changes may occur spontaneously or be induced by the application of an external stimulus. (Note: DNA= Deoxyribonucleic Acid = a nucleic acid originally isolated from fish sperm and thymus gland, but later found in all living cells; on hydrolysis it yields adenine, guanine, cytosine, thymine, deoxyribose, and phosphoric acid. It is the carrier of genetic information for all organisms except RNA viruses. The Watson-Crick Helix is a double helix, each chain of which contains information completely specifying the other chain, representing a structural formulation of the mechanism by which the genetic information of the DNA reproduces itself.)

Geometric Mean. A measure of central tendency frequently used in industrial hygiene sampling data. It is calculated by taking the logarithms of the values, calculating their arithmetic mean, then converted back by taking the antilogarithm.

Glare. The sensation produced by luminance within the visual field that is sufficiently greater than the luminance to which eyes are adapted to cause annoyance, discomfort, or loss in visual performance or ability.

Glassblowers' Cataract. Posterior polar lenticular opacities occurring in occupations entailing long exposures to intense heat and glare. It is probably due to focusing of infrared radiation at the posterior pole of the lens.

Goggles. Large spectacles or glasses, especially those fitted with side guards to protect the eyes against dust, impacting objects, strong light, sparks, or other harmful environmental influences.

Gray (Gy). SI unit (Gy) of absorbed radiation dose equal to one joule/kilogram or 100 RAD.

Ground Fault Circuit Interrupter. A fast-acting circuit breaker that is sensitive to very low levels of current leakage to ground. The interrupter is designed to limit the electric shock to a current and time duration value below that which can produce serious injury.

Grounding. The procedure used to carry an electrical charge to ground through a conductive path. A typical ground may be connected directly to a conductive pipe or to a grounding bus and ground rod.

Guard. An enclosure which prevents entry into the point of operation of a machine or renders contact harmless with any substance or object.

Guard, Fixed Barrier. A point-of-operation enclosure attached to a machine frame by fasteners.

Guard, Interlocking Barrier. An enclosure attached to the frame of a machine and interlocked so that the machine cycle cannot be started normally unless the guard, including its hinged or movable sections, are in position. In some situations, movement of the guards will interrupt the machine cycle.

Guardrail. A device consisting of posts and rail members, or of wall sections erected to mark points of major hazard, and to prevent individuals from coming in contact with the hazard.

Habitual Violator. An individual whose record, during a given time period, shows reports of repeated violations of laws or regulations. In traffic safety, any driver whose record during a consecutive 12-month period shows reports of more than three convictions for traffic violations or more than five times the average number of convictions for all drivers in the state, whichever is greater.

Hard Metal Disease. A condition of fibrosis of the lungs thought to be due to inhalation of dust of tungsten carbide and cobalt used as a binder.

Hazard. That dangerous condition, potential or inherent, which can bring about an interruption or interference with the expected orderly progress of an activity.

Hazard Analysis. An analysis performed to identify hazardous conditions for the purpose of their elimination or control.

Hazard Classification. A designation of relative accident potential based on probability of accident occurrence.

Hazard Control. That function in an organization directed toward the recognition, evaluation, and reduction or elimination of the destructive effects of hazards emanating from human acts of commission and omission and from the physical and environmental aspects of the workplace.

Hazard Level. A qualitative measure of hazards stated in relative terms. (DOD)

Category I. Negligible–Will not result in personal injury or system damage.

Category II. Marginal–Can be counteracted or controlled so that no injury to personnel or major system damage will be sustained.

Category III. Critical–Will cause personal injury or major system damage or both.

Category IV. Catastrophic–Will cause death to personnel.

Hazard Pay. Extra payments to workers in dangerous occupations or while engaged in work where the chances of injury are greater than normal.

Hazard Recognition. The perception of a hazardous condition.

Hazardous Condition. The physical condition or circumstance which is causally related to accident occurrence. The hazardous condition is related directly to both the accident type and the agency of the accident.

Hazardous Materials. Any substance or compound that has the capability of producing adverse health and safety effects on humans.

Health and Insurance Plan. A program of providing financial protection to the worker and family against death, illness, accidents, and other risks, in which the costs may be borne in whole or in part by the employer. One or more of the following major benefits may be provided for the worker and, frequently, dependents: life insurance, accidental death and dismemberment benefits, accident, and sickness.

Health Physicist. A health professional trained in radiation physics and concerned with problems of radiation damage and protection.

Health Risk Appraisal. A generic term applied to methods for describing an individual's chances of becoming ill or dying from selected causes. It is used to indicate risks to health and safety that are influenced by individual's lifestyle behaviors.

HEALTHY WORKER EFFECT. A phenomenon observed in studies of occupational diseases. Workers usually exhibit lower overall mortality rates than the general population, due to the fact that the severely ill and disabled are not employed.

HEARING CONSERVATION. Preventing or minimizing noise-induced hearing loss through the use of hearing protection devices and the control of noise through engineering methods.

HEARING LEVEL (HEARING LOSS). The deviation in decibels of an individual's threshold from the zero reference of the audiometer. Formerly called hearing loss.

HEAT CRAMPS (MINERS' OR STOKERS' CRAMPS). Painful spasms of the voluntary muscles due to salt depletion. These occur in healthy, heat acclimatized individuals and are due to excessive sweating without salt replacement rather than directly to high temperature exposure. They are relieved with fluid and salt replacement.

HEAT EXHAUSTION (HEAT PROSTRATION, HEAT COLLAPSE). A state of peripheral vascular collapse in an unacclimatized individual attributable to exposure to a high temperature environment. Treatment consists of removal to a cool environment, rest, and salt and water replacement. It is prevented by controlling heat exposures, adequate acclimatization, and maintaining adequate salt and water intake.

HEAT PYREXIA (HEAT STROKE). A very serious and often fatal condition resulting from breakdown of thermoregulatory mechanisms during exposure to high temperature environments. It is characterized by extremely high deep-body temperature and an absence of sweating. Treatment consists of rapid cooling in an ice bath.

HEMATOMA. A collection of blood, usually clotted, in an organ, space, or tissue, resulting from a break in the wall of a blood vessel. Frequently occurs with a contusing wound.

HERTZ. The frequency measured in cycles per second (cps). 1 Hz = 1 cps.

HIGH FREQUENCY HEARING LOSS. A hearing deficit starting at 2000 Hz and higher.

HIGH PRESSURE NERVOUS SYNDROME. A disorder occurring at extremely deep depths (in excess of 600 feet or about 20 atmospheres pressure) characterized by tremor, nausea, and decreased psychomotor ability. It appears to be due to both direct hydrostatic pressure and to compression rate. It is counteracted to some extent by addition of a narcotic agent to the breathing mixture. It usually resolves completely with decrease in pressure.

HOISTWAY. A vertical (or in the case of deep ore mines, sloped) passageway designed to enclose and provide support for an elevator, platform, or other lifting device.

HOISTWAY-DOOR INTERLOCK. A hoistway-door interlock is a device, on an elevator shaftway, the purpose of which is, first, to prevent the operation of the elevator machine in a direction to move the car away from a landing unless the hoistway door at that landing at which the car is stopping or is at rest is locked in the closed position, and second, to prevent the opening of the hoistway door from the landing side except by special key, unless the car is at rest within the landing zone, or is coasting through the landing zone with its operating device in the stop position.

HOLD HARMLESS AGREEMENT. A contract under which the legal liability of one party for damages is assumed by another party to the contract. The principal in a large construction project will frequently demand hold harmless agreements from all subcontractors in respect to claims made against the principal arising out of the subcontractors' negligence. The principal often stipulates the purchase of a liability policy by the subcontractor to support the hold harmless agreement.

HOSPITALIZATION BENEFITS. A plan that provides workers, and in many cases their dependents, with hospital room and board or cash allowances toward the cost of such care for a specified number of days, plus the full cost of specified services. Usually part of a more inclusive health and insurance program.

HOST FACTORS. Personal characteristics of the individual. These may be grouped, as Thorndike has done in an analysis of the literature, into a) sensory abilities, b) psycho-motor abilities, c) cognitive and intellectual abilities, d) personal and emotional adjustments (including physiological adjustment), and e) biographical facts.

HOUSEKEEPING. Cleanliness, neatness, and orderliness of a work area.

HYPERBARIC OXYGEN THERAPY. A treatment modality using pure oxygen at high pressure in a recompression chamber. It is useful in the treatment of diving casualties but is also valuable in treatment of medical diseases such as carbon monoxide intoxication and gas gangrene. Hyperbaric oxygen cannot be used at pressures exceeding 3 atmospheres.

HYPERCAPNIA. An increased carbon dioxide level in the blood. It results from carbon dioxide accumulation due to inadequate respiratory exchange or through breathing elevated concentrations. Although not acutely toxic below relatively high concentrations (5%), it is implicated as a synergist in most other high pressure gas-related casualties.

HYPERSENSITIZATION. An antigen-antibody mediated response characterized by an immediate vascular exudative reaction in previously sensitized tissue following exposure to a specific existing agent.

HYPERSUSCEPTIBILITY. A markedly greater quantitative response to a physiologically active agent qualitatively identical to its usual response.

HYPOTHERMIA. Lowered core temperature. Drop in core temperature to 35°C. It produces shivering and discomfort sufficient to adversely affect performance; at about 25°C hypothermia is ordinarily fatal. Cold water immersion produces hypothermia very rapidly whereas exposure in cold air environments is tolerable for much longer periods. Treatment for hypothermia is rapid rewarming in a warm bath.

HYPOTHESIS. A conjecture subject to verification or proof.

ICD. International Classification of Diseases, Injuries, and Causes of Death.

IDLH (IMMEDIATELY DANGEROUS TO LIFE AND HEALTH). The air concentration of a substance at which exposure would result in death or serious injury to a person.

ILLUMINATION. The amount of light flux a surface receives per unit area. May be expressed in lumens per square foot or in footcandles. The rate at which a source emits light energy, evaluated in terms of its visual effect, is spoken of as light flux, and is expressed in lumens.

IMMERSION FOOT (TRENCH FOOT). A painful condition of the feet resulting from relatively long wet exposures at temperatures above freezing. Nerve, muscle, and blood vessel injury occurs as a result of cooling. Prevention and treatment consist of proper foot care including regular drying of the feet, shoes, and socks.

IMMINENT DANGER. An impending or threatening dangerous situation which could be expected to cause death or serious injury to persons in the immediate future unless corrective measures are taken.

IMPERVIOUS. A characteristic of a material that does not allow another substance to pass through it or penetrate it.

IMPINGER. Glass device for drawing contaminated air through a liquid to entrap particles, solvents, and aerosols for later analysis.

IMPLOSION. A rapid expenditure of energy producing an inward burst, opposed to explosion.

INCENDIARY. A substance causing or designed to cause fires. Also, a person who willfully destroys property by fire.

INCIDENCE RATE. The number of injuries, illnesses, or lost workdays related to a common exposure base of 100 full-time workers. The common exposure base enables one to make accurate interindustry comparisons, trend analysis over time, or comparisons among firms regardless of size. This rate is calculated as:

IR = (N/EH) x 200,000

where:

N = number of injuries and/or illnesses or lost work days

EH = total hours worked by all employees during calendar year

200,000 = base for 100 full-time equivalent workers (working 40 hours per week, 50 weeks per year).

INCIDENCE RATE (EPIDEMIOLOGY). The number of new cases, per unit population, occurring during a stated time period.

INCIDENT. An unforeseen event or occurrence which does not result in injury or property damage.

INCUBATION TIME. The elapsed time between exposure to an environmental agent or microorganism and the appearance of biologic alteration.

Independent Variable. A variable that can either be set to a desired value by the investigator or observed as it naturally occurs.

Index Case. The first case in a family or other defined group to come to the attention of the investigator.

Indirect Costs. Monetary losses resulting from an accident other than medical costs and worker's compensation payments. (See ACCIDENT COSTS.)

Indirect Damage. Loss resulting from a hazardous condition or incident but not caused directly thereby.

Industrial Accident. (See OCCUPATIONAL INJURY.)

Industrial Dermatitis. (See DERMATITIS.)

Industrial Disease. (See OCCUPATIONAL DISEASE.)

Industrial Hygiene. Industrial hygiene is that science and art devoted to the anticipation, recognition, evaluation, and control of those environmental factors or stresses arising in or from the workplace which may cause sickness, impaired health and well-being, or significant discomfort and inefficiency among workers or among the citizens of the community.

Industrial Hygienist. A person with the training and ability to: a) recognize the environmental factors and stresses associated with work and work operations to understand their effect on workers and their well-being; b) evaluate, on the basis of experience and with the aid of quantitative measurement techniques, the magnitude of these stresses in terms of ability to impair worker health and well-being; and c) prescribe methods to eliminate, control, or reduce such stresses when necessary to alleviate their effects.

Industrial Medicine. (See OCCUPATIONAL MEDICINE.)

Industrial Safety. (See OCCUPATIONAL SAFETY.)

Industrial Ventilation. An integral part of a system to condition air which may be used in combination with heating, cooling, and humidifying. When used alone, it may be used to remove contaminated air from a work space and for heat control, and includes a supply system and an exhaust system. A well-designed supply system will consist of an air inlet section, filters, heating and/or cooling eqipment, and registers/grilles for air distribution within the work space. The exhaust system may include a general exhaust system and a local exhaust system. (See EXHAUST, GENERAL; EXHAUST, LOCAL.)

Inert Gas Narcosis (nitrogen narcosis, rapture of the depths). The toxic effect of the carrier or inert gas in a breathing mixture at increased pressures. It is characterized by euphoria, impaired coordination, and diminished cognitive function. Breathing air, nitrogen begins to become toxic at about 4 atmospheres absolute pressure and becomes disabling at pressures exceeding 8 atmospheres absolute. It resolves completely with decreasing pressure.

Infant Mortality Rate. The number of deaths of persons under 1 year of age occurring during a stated period of time divided by the total number of live births occurring during that period of time. This rate is often quoted as a useful indicator of the level of health in a community or nation.

Inflammable. A general term once used to describe combustible gases, liquids or solids. Now obsolete. (See FLAMMABLE.)

Inflammation. The reaction of living tissues to injury, whether by infection or trauma. The affected part becoming red, hot, painful, and swollen due to hyperaemia, exudation of fluid, etc.

Ingestion. The act or process of taking in or putting items (food, drugs, etc.) into the body. With regard to certain cells, the act of engulfing or taking up bacteria and other foreign matter.

Inhalation. The act of breathing in, or taking into the lungs, a substance in the form of a gas, vapor, fume, mist, or dust.

Inherited Abnormality. A defective character or quality transmitted from parent to offspring in the genetic material, that may continue to be passed to successive generations.

Inhibitor. A substance or an agent that slows or prevents a chemical or organic reaction, or a material used to prevent or retard rust or corrosion.

Injury. Physical harm or damage to a person resulting in the marring of appearance, personal discomfort, infection, and/or bodily hurt or impairment. Note: The definition of this word is frequently determined by the government agency or other organization using it.

Injury (occupational). Any acute hurt, harm or impairment to a worker that arises out of, or in the course of, employment, and is due to an external cause. The government agency which has jurisdiction determines the exact wording of the term.

INRUNNING NIP POINT. A rotating mechanism that can seize and wind up loose clothing, belts, hair, body parts, etc. It exists when two or more shafts or rolls rotate parallel to one another in opposite directions. It also can occur between a rotating shaft and a fixed surface.

INSURABLE INTEREST. Refers to the relationship of the party of whom the insurance coverage is written to the peril being insured. The party must either own or have a financial stake in the subject that may suffer the damage or loss which would involve out-of-pocket expense to that party if no insurance indemnity was available.

INSURANCE. The making of a legal and enforceable contract between one party (called the insurer or underwriter) with another (called the insured) whereby in consideration of a sum of money (called the premium) the insurer agrees to pay an agreed amount of money to the insured if and when the latter may suffer some loss or may be injured by some event, the happening of which is described in the contract of insurance (which is usually a policy).

INSURED. The person who has purchased a policy of insurance and is protected by it.

INSURED COSTS. Accident losses, which are covered by worker's compensation, medical, or other insurance programs. They comprise the insured element of the total accident cost.

INTENSITY LEVEL (SOUND). The sound-energy flux density level. In decibels of sound, intensity level equals 10 times the logarithm to the base 10 of the ratio of the intensity of this sound to the reference intensity. (See DECIBEL.)

INTERLOCK. A device which interacts with another device or mechanism to govern succeeding operations. For example, an interlocked machine guard will prevent the machine from operating unless the guard is in its proper place. An interlock on an elevator door will prevent the car from moving unless the door is properly closed.

INTERROGATORIES. A set of formal written questions presented to a witness, usually before the trial, who must give written answers under oath.

INTERSTITIAL. (1) Pertaining to the small spaces between cells or structures; (2) occupying the interstices of a tissue or organ; (3) designating connective tissue occupying spaces between the functional units of an organ or a structure.

INTERVENING ACT. The negligent act of a third person to break the chain of causation in such a manner that the injury was not a probable consequence of the original wrongdoer's negligence.

INTERVENING AGENCY. An intermediate agency that disrupts the connection between the negligent act and the injury in such a way to supersede the original act and stand alone as the cause of injury.

INTERVENING CAUSE. An independent cause which breaks the connection between the negligent act and the injury such that it is the immediate cause of injury.

INTERVENING FORCE. One which actively produces harm after the defendant's negligent act has been committed. The word intervening is used in a time sense to include later events, and not conditions or forces already in operation at the time of defendant's conduct. To relieve the defendant from liability for the original negligent act, it must be determined whether the intervening force was the superseding cause of injury, rather than an extraordinary cause or a cause which might reasonably have been expected.

INTERVENING LOSSES. The intrinsic value for the trouble, disadvantage, or deprivation incurred by damage to property or injury to people. Losses in addition to the material or physical losses.

INTERVENING NEGLIGENCE. The negligence of an intervening person or agency between the original negligence of the wrongdoer and the injury.

INTERVENTION STUDY. A type of epidemiologic study in which the investigator assigns, usually at random, individuals to be exposed or not to be exposed to a given factor, such as a treatment or a vaccine. These individuals are then followed through time and the proportion developing or dying from a disease among those exposed to the factor is compared to the proportion developing or dying from the disease among those not exposed, the exposure having been determined by initial assignment.

INTOXICATION. Pertains, in medicine, to poisoning as by a drug, serum, alcohol, or any poison. Also drunkenness or the acute condition produced by over-indulgence of alcohol.

Inversion. Phenomenon of a layer of cool air trapped by a layer of warmer air above it so that the bottom layer cannot rise.

Irradiation. Exposure to radiation.

Irritant. Any external stimulus which produces active responses in a living organism. A primary irritant is one that has been found to produce an irritating effect at the area of the skin contact. Although they affect everyone, they do not produce the same degree of irritation.

Job Hazard Analysis. (See JOB SAFETY ANALYSIS.)

Job Safety Analysis. The breaking down into its component parts of any method or procedure to determine the hazards connected therewith and the requirements or qualifications of those who are to perform it. A method for studying a job in order to: (a) identify hazards or potential accidents associated with each step or task, and (b) develop solutions that will eliminate, nullify, or prevent such hazards or accidents.

Job Safety Training. Training associated with or emphasizing the safety aspects of a job and the hazards of tasks and their interrelationships within a job.

Judgment. The allowance or disallowance of a claim.

Jurisdiction. The power of a court to hear and decide a cause of action, having power over both the parties of the action and the subject matter thereof.

Just Cause. Good or fair reasons for discipline. This term is commonly used in agreement provisions safeguarding workers from unjustified discharge or other punishment.

Key Event (of an accident). That event (or events) in a series of events in an accident that determined the exact time, place, type, and extent of consequence of the accident.

Kickback. A term used to describe the reaction of a piece of material (usually wood) when being cut by a circular saw which is rotating in the opposite direction from the material as it is fed into the saw. Since the saw blade is rotating at a high rate of speed the force that can be exerted on the material is substantial and can cause serious injury to the operator.

Laceration. A term applied to cleanly cut incised wounds as well as jagged, irregular, blunt breaks or tears through the skin. Severity extends from small cuts which can be taped together to severe wounds with damage to underlying structures. The term is also applied to wounds of the mucous membranes and the surface of the eyes.

Laser. An acronym for Light Amplification by Stimulated Emission of Radiation.

Laser Burn. A punctate tissue injury caused by a beam of coherent light. Laser beams are particularly hazardous to the eye, where they may cause burns of the cornea, lens, or retina with consequent effects on visual acuity. A laser burn of the retinal fovea may destroy central vision sufficiently to produce legal blindness.

Late Effects of Radiation. Effects which occur with a long delay time following sublethal doses of radiation or following relatively low dose rates carried over a long period of time. Examples are carcinogenesis and premature aging.

Lead Intoxication. A result of lead absorption, occurring from inhalation of lead dust or fumes or from swallowing lead dust.

Leave of Absence. Generally, excused time to be absent from work or duty, usually for an extended time, without loss of job or seniority.

Legal Liability. Liability imposed by law as opposed to liability arising from an agreement or contract.

Lesion. Injury, damage, or abnormal change to body tissue or organ, especially one that leads to impairment or loss of function of the part involved, or, even if it produces no impairment of function, expresses a symptom or sign of disease.

Level (noise). The logarithm of the ratio of the measured quantity to a reference quantity of the same kind. The base of the logarithms, the reference quantity, and kind of level must be specified.

Liability Insurance. Insurance which obligates the insurance company to pay any liability for which the insured may be covered and, also at the expense of the company, to defend any damage suits brought against it to enforce such liability, thus protecting it from liability and expense of litigation growing out of claims in which it is involved. Insurance which agrees to reimburse the policyholder for sums that may be required to pay others as the result of negligence.

LIFE EXPECTANCY. The average number of years an individual is expected to live if current mortality trends continue to apply. It is a hypothetical measure and indicator of current health and mortality conditions.

LIFE INSURANCE PLAN. Group term insurance coverage for employees, paid for in whole or in part by the employer, providing a lump-sum payment to a worker's beneficiary in the event of death. (See HEALTH AND INSURANCE PLAN.)

LIFE TABLE. A statistical method used to summarize the pattern of mortality and survival in population.

LIFELINE. A rope designed and utilized to provide fall protection that must be secured above the point of operation to an anchorage or structural member. It must also be capable of supporting a minimum specified dead weight.

LIMIT SWITCH. A switch fitted to electric lifts, traveling cranes, etc., in order to cut off the power supply if the liftcar or moving carriage travels beyond a certain specified limit.

LOAD LIMIT. The upper weight limit capable of safe support by a vehicle or floor. The designer of floors to meet load limit requirements should consult ANSI/ASCE Standard 7-95, Minimum Design Loads for Buildings and Other Structures.

LOAD WEIGHT OR ALLOWABLE LOAD. Refers to highest weight of load to be carried by a vehicle safely.

LOCAL EFFECTS OF RADIATION. Effects limited to a specific organ or system due to specific irradiation of that organ or increased sensitivity of the organ.

LOCKOUT-TAGOUT. A program or procedure that prevents injury by eliminating unintentional operation or release of energy within machinery or processes during set-up, start-up, cleaning and clearing jams, or maintenance repairs.

LONGITUDINAL STUDY. (See COHORT STUDY.)

LOSS. In insurance it means the amount the insurer is required to pay because of a judgment that requires the company to pay by virtue of the terms of the insurance contract. Also refers to the overall financial result of some operation, as opposed to profit.

LOSS CONTROL. A program designed to minimize accident-based financial losses. The concept of total loss control is based on studies of near misses (noninjury accidents) and on detailed analysis of both indirect and direct accident costs. Property damage as well as injurious and potentially injurious accidents are included in the analysis.

LOSS PREVENTION. A before-the-loss program designed to identify and correct potential accident problems before they result in actual financial loss or injury.

LOSS RATIO (INSURANCE). A fraction calculated by dividing the amount of losses by the amount of premiums. Expressed as a percentage of the premiums. Various bases are used in calculating the loss ratio, e.g., earned premium loss ratio, written premium loss ratio, etc.

LOSS RESERVE. An estimate of the amount an insurer expects to pay for losses incurred but not yet due for payment.

LOST TIME ACCIDENT (LT). (See LOST TIME INJURY/ILLNESS.)

LOST TIME INJURY/ILLNESS. A work injury/illness which results in death or disability and in which the injured person is unable to report for duty on the next regularly scheduled shift. (See INJURY.)

LOST WORKDAY CASES. Cases which involve days away from work or days of restricted work activity, or both.

LOST WORKDAYS. The number of workdays (consecutive or not), beyond the day of injury or onset of illness, the employee was away from work or limited to restricted work activity because of an occupational injury or illness. (See DAYS OF DISABILITY.)

LOUDNESS. The intensity attribute of an auditory sensation, in terms of which sounds may be ordered on a scale extending from soft to loud. Loudness depends primarily upon the sound pressure of the stimulus, but it also depends upon the frequency and wave form of the stimulus.

LOUDNESS LEVEL. A subjective method for rating loudness in which a 1000 Hz tone is varied in intensity until it is judged by listeners to be equally as loud as a given sound sample. The loudness level in phons is taken as the sound pressure level, in decibels, of the 1000 Hz tone.

LOWER EXPLOSIVE LIMIT (LEL). The minimum concentration of combustible gas or vapor in air of flammable liquids or gases below which propagation of flame does not occur on contact with a source of ignition.

Lumen. Under the common system, it was defined as the flux on one square foot of a sphere, one foot in radius, with a light source of one candle at the center that radiates uniformly in all directions. Under SI, luminous flux is measured in lumen units (symbols: lm) and has as its formula cd x sr which are the SI Base Units of candela and steradian for solid angle. Thus, the lumen is the luminous flux emitted in a solid angle of one lumen uniformly distributed in a solid angle of one steradian by a point source having a uniform intensity of one candela. (See LUX.)

Luminaire. A complete lighting unit consisting of a lamp or lamps together with parts designed to distribute the light, position and protect the lamps and connect the lamps to the power supply.

Luminescent. Emitting light not due to high temperatures, usually caused by excitation by rays of a shorter wavelength.

Lung Volumes.

inspiratory capacity (IC) – Maximal volume inspired from resting expiratory level.

expiratory reserve volume (ERV) – Maximal volume expired from resting expiratory level.

vital capacity (VC) – Maximal volume expelled by complete expiration after a maximal inspiration.

functional residual capacity (FRC) – Volume of gas in the lungs at the resting expiratory level.

residual volume (RV) – Volume of gas in the lungs after maximal expiration.

total lung capacity (TLC) – Volume of gas in the lung after a maximal inspiration; that is, the VC + RV.

tidal volume (TV) – Volume of gas inspired or expired during each breathing cycle.

Lux. The illuminance produced by a luminous flux of one lumen uniformly distributed over a surface of one square meter.

Machine Guarding. The installation of equipment or devices on machines to eliminate hazards created by operation of the machines.

Magnaflux. A test in which particles of iron are applied to the surfaces of a magnetized specimen; the particle pattern indicates surface or near-surface flaws or irregularities.

Maintenance, Preventative. A system of scheduled overhaul and replacement of key parts to forestall breakdown.

Maintenance, Routine and Breakdown. The function of preserving facilities and equipment by making needed repairs.

Major Injury. An injury where there is loss of time to the injured person and a medical expense.

Major Medical Expense Benefit (catastrophe insurance). Plan designed to insure workers against the costly medical expenses resulting from catastrophic or prolonged illness or injury. If the benefit supplements the benefit payable by a basic health insurance plan (hospital, medical, or surgical), it is called a supplementary plan; otherwise, it is called a comprehensive plan.

Makeup Air. Clean, tempered outdoor air supplied to a workspace to create a comfortable environment.

Malformation. A deformity. (See congenital abnormality.)

Malignancy. The property of a neoplasm, marginal invasion, relentless growth, and distal spread with a lethal effect.

Malignant Neoplasm. (See cancer.)

Malpractice. Misconduct or lack of proper professional skill on the part of any professional, such as a doctor, dentist, attorney, or engineer.

Manipulative Dexterity. The degree of skill a person has in using his wrists and fingers for fine tasks.

Manual Control Switch. An auxiliary device for manual operation of an automatic controller. Any control switch that is actuated by a person.

Manual Dexterity. The degree of use and facility a person has in using the hands. Includes gross arm-hand movements as well as fine wrist-finger skills.

Manual Rate. The compensation insurance premium in dollars per hundred dollars of payroll, for a specified classification of operation or risk as listed in an official manual for a given state or jurisdiction.

Margin of Safety. A border, edge, or limit beyond which a particular behavior, condition, or situation becomes hazardous or unsafe.

Mask. A protective covering for the face or head, such as a wire screen, a metal shield, a respirator, or a gas mask.

Masking (noise). The stimulation of one ear of a subject by controlled noise to prevent hearing with that ear the tone or signal given to the other ear. This procedure is used where there is at least 15 to 20 dB difference in the two ears.

Matching Frequency (stratified matching). In a case-control epidemiological study, an attempt to make cases and controls according to certain variables and sampling such that the proportion of controls in each subgroup is the same as the proportion of cases in each subgroup. In a cohort study, the population not exposed to the risk under study would be subdivided and sampled such that the proportion of nonexposed in each subgroup was the same as the proportion of exposed in each subgroup.

Matching. In an epidemiological case-control study, the selection of controls such that the control group has the same distribution of certain variables as the case group; in a cohort study, the selection of those not exposed to a given factor such that they have the same distribution of certain variables as those exposed to the given factor.

Matching Individual. In a case-control epidemiological study, an attempt to make cases and controls comparable by selecting for each case a control who matches the case with respect to specific criteria.

Maximum Permissible Body Burden (MPBB). That amount of an internally deposited radioisotope which will result in a maximum permissible dose to the individual or a critical organ.

Maximum Permissible Concentration (MPC). That amount of radioactive material in air, water, or food which will result in accumulation of a maximum permissible body burden when assimilated at a standard rate for that medium.

Maximum Permissible Dose (MPD). A dose established by competent authority as the highest allowable dose which is not expected to cause injury to a person during his lifetime. A wide safety margin below the dose at which detrimental effects are known to occur is normally provided in establishing MPD'S.

Maximum Voluntary Ventilation (MVV). Volume of air which a subject can breathe with voluntary maximal effort for a given time.

Measurement. The assigning of numbers to observations in such a way that the numbers are amenable to analysis according to certain rules.

Mechanical Hazards. Unsafe conditions involving machinery, equipment, tools, etc.

Median Lethal Dose (LD-50). The acute dose of radiation required to kill 50% of the population within a specified time, usually 30 days.

Medical Benefits. Plans which provide employees, and in most cases their dependents, with specified medical care (other than that connected with surgery) or a cash allowance toward the cost of doctors' visits. Generally part of a health insurance program.

Medical Monitoring. (See BIOLOGICAL MONITORING.)

Medical Only (MO). An injury requiring medical attention only, without loss of time from work. In insurance terminology it describes an injury where medical costs only are paid, even when some loss of work time has been incurred.

Medical Radiation. Radiation from diagnostic or therapeutic radiological procedures, e.g., x-rays, radioactive tracers, cobalt therapy.

Medical Surveillance. Systematic collection, analysis, and interpretation of medical data. In the workplace, surveillance is necessary to establish that a disease condition exists and that the particular disease manifestation may result from exposure to a specific harmful agent.

Membrane Filter. A thin, pliable material of varying porosity designed to be placed in filter holder and have air contaminants drawn across it entrapping particles for gravimetric or instrumental/chemical analysis.

Meta-analysis. The process of using statistical methods to combine the results of different studies.

Metal Fume Fever. An acute illness characterized by the sudden onset of chills and fever several hours after exposure to metal fumes. It is most common in brass foundries and in the smelting of zinc. The condition is thought to be due to inhalation of metal oxides. Zinc, copper, iron, and magnesium have been documented as causing the syndrome, but other metals may behave in a similar manner under certain circumstances.

METASTATIC TUMOR. A secondary malignant growth that originates from a primary tumor and is separated from the primary site.

MICROWAVE BIOEFFECTS. Alteration in biological systems resulting from exposure to microwave radiation. Two broad types of bioeffects are generally described:

Thermal effects – Effects produced by the heating effect of absorbed microwave energy. These effects are fairly well documented and most are accompanied by a detectable rise in tissue temperature. Some effects in which a temperature rise is not detected by available instrumentation are termed microthermal.

Nonthermal effects – Effects which are thought to be due to a mechanism other than heating. Certain experimentally produced field effects fall in this category.

MICROWAVE CATARACTS. Posterior subcapsular lenticular opacities attributed to exposure to microwave radiation. Such opacities have been experimentally produced and several cases involving human exposures have been reported. The occurrence of microwave cataracts in occupationally exposed personnel in normal industrial circumstances has not been demonstrated.

MICROWAVE DOSIMETRY. Measurement of the amount of microwave energy to which a system has been exposed. The term microwave is usually used to cover radiations in the radio and radar frequencies (kilohertz through gigahertz). Absorbed doses cannot currently be measured and can only be roughly estimated from measurements of temperature increase. Time integrating instruments are also lacking so that exposures are usually expressed in terms of power flux density (watts per square centimeter) or field strength (volts per meter).

MICROWAVE HEARING EFFECT. The auditory sensation evoked in people exposed to low level pulsed microwave radiation. It is perceived as an audible buzz or click. It was long thought to demonstrate a direct, nonthermal effect but recent evidence indicates that it may be due to rapid transient thermal expansion of fluid.

MILEAGE DEATH RATE. The number of deaths from vehicle accidents per 100,000,000 miles of vehicle travel in the area for which the rate is computed.

MINOR INJURY. An injury where no lost time or major medical costs are involved.

MISTS. Suspended liquid droplets generated by condensation from the gaseous to the liquid state or by breaking up a liquid into a dispersed state, such as by splashing, foaming, or atomizing. Mist is formed when a finely divided liquid is suspended in the atmosphere.

MOLECULAR WEIGHT. Weight (mass) of a molecule based on the sum of the atomic weights of the atoms that make up the molecule.

MONITORING. Testing to determine if the parameters being measured are within acceptable limits. This includes environmental and medical (biological) monitoring in the workplace.

MONITORING (RADIATION). The periodic or continuous determination of the amount of ionizing radiation or radioactive contamination. Area monitoring is the routine monitoring of the level of radiation or of radioactive contamination of any particular area, building, room, or equipment. Personnel monitoring is the monitoring of any part of an individual, breath, or excretions, or any part of the clothing.

MORBIDITY. A diseased condition or state; the incidence or prevalence of a disease or of all diseases in a population.

MORBIDITY RATE. The number of cases of a specific diseased condition or state occurring per unit population over a stated period of time.

MORTALITY. Death; the incidence or prevalence of death in a population.

MORTALITY RATE. The number of deaths occurring per unit population over a stated period of time.

MOTOR VEHICLE COLLISION. Any accident involving a motor vehicle in motion that results in death, injury, or property damage. However, motion of the motor vehicle is not required in a collision between a railroad train or another motor vehicle.

MOTOR VEHICLE NONTRAFFIC ACCIDENT. Any motor vehicle accident which occurs entirely in any place other than a trafficway.

MUSCULOSKELETAL INJURY. Acute and chronic injury to muscles, tendons, ligaments, nerves, joints, bones, and supporting vasculature.

MUTAGEN. Any agent capable of producing genetic or somatic mutation. (See GENETIC MUTATION, SOMATIC MUTATION.)

NATIONAL FIRE PROTECTION ASSOCIATION (NFPA). Membership drawn from the fire service, business and industry, health care, educational and other institutions, and individuals in the fields of insurance, government, architecture, and engineering. Develops standards intended to minimize the possibility and effects of fire and explosion; conducts fire safety education programs for the public.

NATIONAL INSTITUTE FOR OCCUPATIONAL SAFETY AND HEALTH (NIOSH). Federal agency in the Department of Health and Human Services (DHHS) responsible for research in identifying occupational safety and health hazards and developing means of preventing these hazards, and for conducting educational programs to provide an adequate professional workforce for prevention.

NATIONAL SAFETY COUNCIL (NSC). Voluntary nongovernmental organization. Promotes accident reduction by providing a forum for the exchange of safety and health ideas, techniques, and experiences and the discussion of accident prevention methods. Offers background courses at Safety Training Institute and home study courses for supervisors.

NATURE OF INJURY. The type, or classification of the hurt, harm, or impairment received or inflicted.

NATURE OF WORK. A description of the type or classification of the work actually being done by a person in connection with the work operation. Usually described in reports of occupational injuries.

NEAR ACCIDENT (NEAR MISS). A term used synonymously with non-injury accident. Also applies to an accident having a potential for property damage, but where no property damage was incurred.

NEAR DROWNING. Submersion with asphyxia in which the victim survives the acute episode. Liquid may or may not have entered the lungs. Many near-drowning victims succumb to pulmonary edema hours to days following the acute episode.

NEGLIGENCE. The lack of reasonable conduct or care, characterized by accidental or thoughtlessness, which a prudent person would ordinarily exhibit. There need not be a legal duty.

actionable negligence – The breach or non-performance of a legal duty, through neglect or carelessness, which results in damage or injury to another.

comparative negligence – Where negligence by both the plaintiff and the defendant is concurrent and contributes to the injury. Plaintiff's damages are diminished proportionately, provided fault is less than the defendant's, and that even by exercising ordinary care plaintiff could not have avoided the consequences of defendant's negligence.

contributory negligence – Conduct by the injured person which should have known involved an unreasonable risk. Inattentiveness or carelessness when using an article known to be defective or hazardous, or disregard of warnings or instructions issued by manufacturers and sellers usually constitutes contributory negligence.

degrees of negligence – "Ordinary" negligence is based upon the fact that one ought to have known the results of unsafe acts, while "gross" negligence rests on the assumption that one knew the results of acts but was recklessly or wantonly indifferent to the results. All negligence below that called "gross" or "ordinary" by the courts is "slight" negligence.

NEOPLASM. A local autonomous new growth having no useful function.

NEUROPATHY (PERIPHERAL). Any disease of the peripheral nerves. Principal types are: a) *demyelinating* – a peripheral neuropathy in which the prominent pathologic alteration is an inflammatory destruction of the myelin sheath (*example:* Guillian Barre syndrome); b) Vascular – a peripheral neuropathy secondary to small vessel occlusion. Destruction involves both the axon and myelin sheath (*example:* diabetes mellitus mononeuropathy multiplex); c) *Axonal* – a peripheral neuropathy in which the primary pathologic alteration is the retro-grade destruction of the axon (*examples:* toxic neuropathy caused by exposure to arsenic, n-hexane, acrylamide, methyl N-butyl ketone, triorthocresyl phosphate, carbon disulfide).

NICKEL ITCH. A dermatitis due to sensitization to nickel and an allergic reaction on contact with nickel or nickel compounds. It is seen in nickel-platers and in persons wearing jewelry containing this metal.

Nip Point. The point of intersection or contact of two opposed rotating circular surfaces, or a plane and a rotating circular surface.

Nitrogen Narcosis. (See INERT GAS NARCOSIS.)

Noise. Any undesired sounds usually resulting in an objectionable or irritating sensation.

Noise Control. Engineering measures aimed at: the reduction of noise at the source; precluding the propagation, amplification, and reverberation of noise; and isolating workers. The term is meaningful only when noise control components and the points of observation are fully specified.

Noise Level. Sound level. For air-borne sound, unless otherwise specified, noise level is the weighted sound pressure level called sound level. The weighting must be indicated.

Noise Reduction. A decrease of the sound pressure level at a specified observation point, which is attributable to a designated structure. Noise reduction is also used to designate the differences in sound pressure levels existing at two different locations at a single time, when the designated structures are in position.

Noise-induced Hearing Loss. The slowly progressive inner ear hearing loss that results from exposure to continuous noise over a long period of time as contrasted with acoustic trauma or immediate physical injury to the hearing function.

Nonauditory Effects of Noise. Refers to stress, fatigue, health, work efficiency, and performance effects of loud noise that is continuous.

Noncausal Association. A statistical association between the occurrence of a factor and a disease in which the factor is not a cause of the disease.

Noncombustible. A material or substance that will not burn readily or quickly. Noncombustible implies a lower degree of fire resistance than fire resistive. (See FIRE RESISTIVE.)

Nondestructive Testing. A test to determine the characteristics or properties of a material or substance that does not involve its destruction or deterioration (e.g., x-ray examination, ultra-high frequency sound, etc.)

Nondisabling Injury. An occupational injury which does not result in death, permanent total disability, permanent partial disability, or temporary total disability.

Nonfatal Injury Accident. An accident in which at least one person is injured and no injury terminates fatally.

Nonflammable. A material or substance that will not burn readily or quickly.

Noninjury Accident. Any accident in which there is no personal injury or from which no personal injury results.

Not Otherwise Classified. A general category of items such as might appear in an accident causal classification system to permit the grouping of relatively infrequent dissimilar items.

Nuisance Dust. Generally innocuous dust not recognized as the direct cause of a serious pathological condition.

Occupancy. The use of a building or other structure. The contents of a building or other structure.

Occupational Disease. A disease arising out of, and in the course of employment - resulting from exposure to harmful chemical, biological, or physical agents.

Occupational Illness. Any abnormal physical condition or disorder, other than one resulting from an occupational injury, caused by exposure to environmental factors associated with employment. It includes acute and chronic illness or disease which may be caused by inhalation, absorption, ingestion, or direct contact. Because occupational illnesses are rarely attributable to a specific incident they should be reported in the year in which the illness was first diagnosed and reported to the employer.

Occupational Injury. An acute injury arising out of, and in the course of, employment -resulting from exposure to traumatizing physical or chemical agents in the workplace. Examples include amputations, fractures, eye loss, lacerations, and traumatic deaths.

Occupational Injury and Illness Records, OSHA. A recording of each reportable occupational injury (including fatality) and illness required by every employer covered by the National System for Uniform Recording and Reporting of Occupational Injury and Illness.

Occupational Injury or Illness, Reportable, OSHA. Any disability or permanent impairment to an employee which results from any exposure in the work environment that either: (1) results in death; or (2) prevents the employee from performing normal assignment during the next regular or subsequent work day or shift; or (3) not causing death or loss of time, (a) results in transfer to another job or termination of employment, or (b) requires medical treatment other than first aid, or (c) results in loss of consciousness, or (d) is diagnosed as an occupational illness, or (e) results in restriction of work or motion.

OCCUPATIONAL MEDICINE. The promotion and maintenance of the highest degree of physical, mental, and social well-being of workers in all occupations. The prevention among workers of departures from health caused by their working conditions. The protection of workers in their employment from risks resulting from factors adverse to health. The placing and maintenance of the worker in an occupational environment adapted to his physiological and psychological characteristics.

OCCUPATIONAL RADIATION. Radiation to which an individual is exposed in the course of employment. For instance, x-ray technicians, radioisotope technicians, uranium miners, nuclear reactor technicians may be occupationally exposed to radiation.

OCCUPATIONAL SAFETY. The prevention of personnel and environmental accidents in work-related environments or situations.

OCCUPATIONAL SAFETY AND HEALTH ADMINISTRATION (OSHA). Federal agency in the Department of Labor (DOL) responsible for standard-setting and regulating/enforcing workplace codes, rules, and laws.

OCCUPATIONAL SAFETY AND HEALTH CODES AND STANDARDS. Rules of procedure designed to secure uniformity and protection of life and property having the force of law in certain jurisdictions. Examples include OSHA Standards (29 CFR 1910) and the NFPA National Fire Protection Codes.

OCCUPATIONAL SKIN DISEASES OR DISORDERS. Contact dermatitis, eczema, or rash caused by primary irritants and sensitizers or poisonous plants; oil acne; chronic ulcers; or inflammations, etc., arising out of, or during the course of, employment.

OCCURRENCE. An incident often classified as relatively major or minor. In insurance, distinguished from accident by the fact that it is apparent or foreseen as occurring if certain activities take place.

OCTAVE BAND. A range of frequency where that highest frequency of the band is double the lowest frequency of the band. The band is usually specified by the center frequency.

ODDS RATIO. In an epidemiological study, the odds of affected individuals among those exposed to a given factor divided by the odds of affected individuals among those not exposed to the factor. (For a rare disease, the odds ratio is approximately equal to the relative risk.)

OFF-THE-JOB SAFETY. Accident prevention activities or programs associated with non-job-related activities.

OIL ACNE. A skin condition which affects those surfaces in contact with insoluble oils or oil-soaked clothing. Comedones, raised papules, or an infection about the hair follicles (oil boils) occur.

OLD-AGE SURVIVORS AND DISABILITY INSURANCE (OASDI) BENEFITS. Retirement income and survivors' and disability payments available to eligible workers covered by federal social security legislation.

ONCOLOGY. The study of neoplasia.

ORGANIC DISEASE. Disease in which some change in the structure of body tissue could either be visualized or positively inferred from indirect evidence.

OSH ACT. Occupational Safety and Health Act of 1970.

OSTEOARTHRITIS. Chronic multiple degenerative joint disease.

OTITIS EXTERNA (SWIMMER'S EAR). An inflammatory condition of the external ear canal. It occurs in swimmers and divers and in long-term compression chamber operators, in whom it is one of the most regularly occurring disorders. It is a result of excess moisture in the ear canal resulting in tissue maceration and infection.

OTOSCLEROSIS. A condition caused by a growth of bony tissue about the foot plate of the stapes and the oval window of the inner ear. It results in a gradual loss of hearing.

OXYGEN DEFICIENCY. Designates an atmosphere having less than the percentage of oxygen found in normal air. Normally, air contains approximately 21 percent oxygen at sea level. When the oxygen concentration in air is reduced to approximately 16 percent, many individuals become dizzy, experience a buzzing in the ears, and have a rapid heart beat.

OXYGEN TOXICITY. A disorder associated with increased partial pressures of oxygen. There are two types of oxygen toxicity:

high pressure – Breathing 100% oxygen at pressures greater than 3 ATA may result in acute toxicity producing convulsions.

low pressure – Breathing 100% oxygen at 1 ATA for extended periods (24 hours or greater) may result in pulmonary dysfunction and pulmonary edema.

P VALUE. A statement of probability that the differences observed have occurred by chance under the null hypothesis.

PAID ABSENCE ALLOWANCE. Payment for lost working time available to workers for various types of leave not otherwise compensated for, e.g., excused personal leave.

PARTICULATE. A particle of solid or liquid matter.

PATCH TEST. A test in which a diluted sample of a material suspected of causing an allergic dermatitis is applied to the patient's skin (usually the back) with an adhesive patch and removed in 48 hours. If the patient is sensitive, a small area of dermatitis will appear at the point of contact of the material.

PEDESTRIAN. Any person who at the time of an accident is not in a motor vehicle or non-motor vehicle. Persons on other vehicles such as coaster wagons, child's tricycle, roller skates, etc., are considered pedestrians. A person hitched onto a vehicle is considered a pedestrian unless entirely in or on the vehicle.

PERCEPTION. (1) Awareness of objects or other data through the medium of the senses. The meaning or interpretation given to stimuli received through the senses. (2) Insight or intuition relative to information introduced through sensory reception.

PERIPHERAL NEUROPATHY. (See NEUROPATHY, PERIPHERAL).

PERMANENT DISABILITY. A permanent impairment, includes any degree of impairment from an amputation of a part of finger or a permanent impairment of vision to seriously and permanently nonreversible, nonfatal injuries.

PERMANENT IMPAIRMENT. (See PERMANENT DISABILITY, PERMANENT PARTIAL DISABLILITY, PERMANENT TOTAL DISABILITY.)

PERMANENT PARTIAL DISABILITY. Any injury other than death or permanent total disability which results in the loss, or complete loss of use, of any member or part of a member of the body, or any permanent impairment of functions of the body or part thereof, regardless of any pre-existing disability of the injured member or impaired body function. These cases are used in computing ANSI Standard Z16 injury rates whether or not time is lost.

PERMANENT TOTAL DISABILITY. Any injury other than death which permanently and totally incapacitates an employee from following any gainful occupation, or which results in the loss, or the complete loss of use, of any of the following in one accident: a) both eyes, b) one eye and one hand, or arm, or leg, or foot, c) any two of the following not on the same limb: hand, arm, foot, or leg. Definition can vary from state to state.

PERMISSIBLE EXPOSURE LIMIT (PEL). An occupational exposure limit established by OSHA's regulatory authority. It may be expressed as a time-weighted average (TWA) concentration limit or as a maximum concentration exposure limit. Ceiling concentrations may also be specified.

PERSON-YEARS. A unit of measurement combining persons and time, used as denominator in mortality rates of cohort study. In a cohort study, each subject contributes only as many years of observation to the population at risk as the subject is actually observed; if the subject leaves after one year, one person-year is contributed; if after ten, ten person-years.

PERSONAL FACTOR (UNSAFE). The mental or bodily characteristic which permitted or occasioned an act which contributed to an accidental occurrence.

PERSONAL PROTECTIVE EQUIPMENT. Clothing or devices worn or used by the worker to protect against exposure to hazardous agents in the workplace. Respirators, gloves, and ear protectors are examples.

PESTICIDES. General term for that group of chemicals used to control or kill such pests as rats, insects, fungi, bacteria, weeds, etc., that prey on humans or agricultural products. Among these are insecticides, herbicides, fungicides, rodenticides, miticides, fumigants, and repellents.

Phon. The unit of loudness level.

Physical Factors (unsafe). Environmental factors conducive to accident occurrence. Physical or environmental hazards.

Physiology. The science and study of the functions or actions of living organisms.

Pinch Point. Any point at which it is possible to be caught between the moving parts of a machine or between moving and stationary parts of a machine or other structure or between the material and the moving parts of a machine.

Pitch (hearing). That attribute of auditory sensation in terms of which sounds may be ordered on a scale extending from low to high. Pitch depends primarily upon the frequency of the sound stimulus, but it also depends upon the sound pressure and wave form of the stimulus.

Pneumoconiosis. A condition characterized by permanent deposition of substantial amounts of particulate matter in the lungs, usually of occupational or environmental origin. The condition is also characterized by the tissue reaction to the presence of the particulate matter.

Pneumonitis. An inflammation of the lung.

Point of Operation. The place of contact between the primary functional part of any machine or tool and the material or substances in the production, or process, in which the machine or tool is being used. Generally considered to be the point of greatest danger to the operator.

Point of Perception. The time and place at which the individual actually first perceived"that is, saw, heard, smelled or felt"the hazard, that is, the unusual or unexpected movement or condition that could be taken as a sign of the accident about to occur.

Point of Possible Perception. The place and time at which the unusual or unexpected movement or condition could have been perceived by a normal person. This point always comes at or before the point of (actual) perception.

Potential Hazard. A situation, thing, or event having latent characteristics conducive to an accident occurrence.

Potential Years of Life Lost. A measure of the relative impact of various diseases and injuries on a defined population. Potential years of life lost due to a particular cause is the sum of the years that these persons in a population would have lived had they experienced normal life expectancy.

Power Level (noise). The power level, in decibels, is 10 times the logarithm to the base 10, of the ratio of a given power to a reference power. The reference power must be indicated.

Precautions. Actions taken in advance to reduce the probability of an accident.

Preemployment Examination. A medical examination of a job applicant that is job related, consistent with business necessity, and treated in a confidential manner. In the preoffer phase, the employer may not subject the applicant to a medical exam. In the postoffer phase, the employer may require a medical exam to consider physical and psychological criteria that are relevant to the performance of the contemplated work activity.

Preponderance of Evidence. A greater weight of evidence (facts, information from all sources, past experience) to support one judgment or decision versus the opposite. A phrase frequently used to describe the basis upon which a decision is made, such as determining if an injury is occupational or non-occupational.

Presbycusis. The hearing loss due to age. It is believed to be the degeneration of the nerve cells due to the ordinary wear and tear of the aging process.

Prevalence Rate. The number of new and existing cases, per unit population, occurring during a stated period of time.

Prevention (of accident). The science and the art representing control of worker performance, machine performance and environment to eliminate failures or losses. Prevention connotes correction of conditions as well as elimination or isolation of hazards.

Probability Sampling. Sampling in which each sampling unit has a known non-zero probability of being included in the sample.

Product Liability. The liability a merchant or a manufacturer may incur as the result of some defect in the product sold or manufactured, or the liability a contractor might incur after job completion from improperly performed work. The latter part of product liability is called completed operations.

Property Damage Accident (PD). An accident wherein damage to or destruction of any property is the immediate and direct result. It does not include accidents resulting in loss of human life or personal injury.

Proportionate Mortality Ratios (PMR). A ratio of two proportionate mortality rates. In occupational study, the PMR is a tool for estimating cause-specific risks when the available data consist only of deaths without knowledge of the population characteristics. It can lead to erroneous conclusions if used to compare the mortality experience of populations with different distributions of causes of death.

Prospective Study. (See COHORT STUDY) (RETROSPECTIVE STUDY.)

Protective Clothing. Clothing worn to protect the worker from exposure to or contact with harmful chemical agents or physical agents such as heat.

Protective Coating. A thin layer of metal or organic material as paint applied to a surface primarily to protect it from oxidation, weathering, and corrosion.

Protective Equipment (personal). Equipment to protect the worker from exposure to harmful substances. Such equipment includes safety glasses, face shields, leggings, hard hats, safety shoes, etc.

Protective Hand Cream. A cream designed to protect the hands and other parts of the skin from exposure to harmful substances.

Proximate Cause. The cause which directly produces the effect without the intervention of any other cause. Cause nearest to the effect in time or space.

Prudent. Cautious, careful, attentive, discrete, circumspect, and sensible as applied to action or conduct.

Psychogenic Deafness. That originating in or produced by the mental reaction of an individual to physical or social environment. It is sometimes called functional deafness or feigned deafness.

Psychological Evaluation. Studies of the relationships between accidents and emotional, or intellectual functions as distinct from sensory and sensorimotor factors. Intelligence tests and personality tests are examples of these measurement techniques, as are time sampling, activity analysis, and the observation of critical incidents.

Psychophysical Measurement. Measurements of reaction times and various kinds of sensorimotor and psychomotor coordinations. (Laboratory experiments dealing with vision under conditions of low illumination illustrate this latter group.)

Psychophysical, Characteristics. A combination of mental and physical qualities of humans, such as visual acuity, reaction time, hearing, depth perception, peripheral vision, manipulative dexterity, color vision, etc.

Psychosocial Evaluation. The evaluation of socially conditioned behavior and reactions in relation to the social environment. The measurement of attitudes, studying the effect of supervisory practices, and the use of indices of the adequacy of social adjustment, illustrate methods used in this approach to accident investigation.

Psychosocial Factors. Social influences that are related to or affect psychological factors of human behavior.

Pulmonary Hyperinflation Syndrome (burst lung). Pulmonary barotrauma of ascent. It occurs during decreases in pressure from overdistension and rupture of the lungs by expanding gases. It is a result of breath-holding or air-trapping during ascent. The clinical presentations are pneumothorax, mediastinal or subcutaneous emphysema, and air embolism.

Pure Tone. A pure tone has a unique pitch and is characterized by a sinusoidal variation in sound pressure with time. The frequency spectrum of a pure tone shows a single line at a discrete frequency.

Quality Factor (QF). The factor by which absorbed dose in rads (q.v.) is multiplied to obtain effective or equivalent dose in rem. Sometimes used interchangeably with relative biological effectiveness (RBE), a term which should be limited to research applications. Typical QFs are: x-ray = 1, beta = 1, alpha = 20, protons = 10, fast neutrons = 10.

RAD. Radiation absorbed dose. Under the common system it is the basic unit of absorbed dose of any type radiation equivalent to absorption of 100 ergs per gram of any material. However, under SI, the unit is the gray (Gy). The gray is the energy imparted by ionizing radiation to a mass of matter corresponding to one joule per kilogram or 100 RAD.

Radiation. The emission and propagation of energy through space or a material medium in the form of electromagnetic waves (gamma or x-rays) or particles (alpha and beta). Electromagnetic radiation may be ionizing or nonionizing; all particulate radiation is ionizing.

Radiation Control. Engineering measures, including devices, used to prevent worker exposure to harmful levels of ionizing and nonionizing radiation.

Radiation Dosimetry. Measurement of the amount of radiation delivered to or absorbed at a specific place. Personnel dosimetry is accomplished with such devices as the film badge, thermoluminescent dosimeter, or pocket ionization chamber. In this way continuous recording of cumulative radiation dose can be maintained.

Radiation Monitoring. Measurement of the amount of radiation present in a given area. Radiation exposures can be limited by monitoring and limiting access and stay time in areas in which high levels of radiation are present.

Radioactive Contamination. The deposition or presence of radioactive material in a place where it is not desired and may be harmful. In regard to personnel:

external contamination – The presence of radio active material on the skin.

internal contamination – The presence of radioactive material within the body due to ingestion, inhalation, or absorption.

Random Noise. A sound or electrical wave whose instantaneous amplitudes occur, as a function of time according to a normal (Gaussian) distribution curve. Random noise is an oscillation whose instantaneous magnitude is not specified for any given instant of time. The instantaneous magnitudes of a random noise are specified only by probability functions giving the fraction of the total time that the magnitude, or some sequence of the magnitudes, lies within a specific range.

Random Sampling. A procedure in which each sampling unit in the population has an equal chance of being included in the sample.

Rate. An expression of the speed or frequency with which a certain event or circumstance occurs in relation to a certain period of time, a specific population, or some other fixed standard (e.g., mortality rate, pulse rate, etc.).

Raynaud's Phenomenon (dead hand or white hand syndrome). A vascular disorder of the extremities, usually the hand, frequently resulting from the use of vibrating tools, particularly pneumatic tools. Intermittent cyanosis or pallor of the part, coldness, numbness, and paresthesias are typical. Once the disorder develops it generally does not improve even with cessation of work with vibrating tools.

Recommended Exposure Limit (REL). The NIOSH REL is the recommended maximum allowable airborne concentration of a substance which is not expected to harm workers. It may be expressed as a time-weighted average (TWA) concentration, short-term exposure limit, or a ceiling value.

Recompression Therapy. Treatment of decompression sickness or air embolism by returning the victim to a hyperbaric environment, usually in a recompression chamber. Standard tables for treatment of diving casualties are available, the most frequently used being those developed by the U.S. Navy.

Regression Coefficient. In popular terms, it is the amount of change that will on the average take place in one characteristic when the other characteristic changes by one unit. It is the coefficient of an independent variable in a regression equation.

Relative Risk. In an epidemiological study, the rate of disease among those exposed to a given factor divided by the rate of disease among those not exposed to the given factor. (For a rare disease, the relative risk may be approximated by the odds ratio.)

Reliability. (See REPRODUCIBILITY.)

REM- Roentgen equivalent man. Unit of dose of any type radiation which produces in humans an effect equivalent to exposure to one rad or roentgen of x- or gamma rays. In the SI system the equivalent unit is the Sievert (Sv) = 100 REM.

Repetitive Strain Injury. (See CUMULATIVE TRAUMA DISORDER.)

Replacement Air. The volume of outdoor air that is delivered to a building in a controlled manner to assist in control of contaminants and to replace air being exhausted.

Reproducibility (reliability, precision). The extent to which a measurement produces the same results when repeatedly applied to the same situation.

Research Design. The procedures and methods, predetermined by an investigator, to be adhered to in conducting a research project.

Respirator. A protective device for the human respiratory system designed to protect the wearer from inhalation of harmful air contaminants. There are two types of respiratory protective devices: a) air purifiers, which remove the contaminants from the air by filtering or chemical absorption before inhalation, and b) air suppliers, which provide clean air from an outside source or breathing air from a tank.

Respiratory Diseases. Disease conditions due to toxic agents in the respiratory tract, e.g., pneumonitis, bronchitis, pharyngitis, rhinitis, or acute congestion due to chemicals, dusts, gases, fumes, or infectious agents.

Respiratory Irritants. Irritants affecting the respiratory tract, e.g., dusts, vapors, gases.

Respiratory Protective Equipment. Protective devices for the human respiratory system designed to protect workers from over-exposure by inhalation of air contaminants (air-purifying and air-supplied) and oxygen deficiency (air-supplied). (See RESPIRATOR.)

Respiratory System. A system of the body consisting of (in descending order)-the nose, mouth, nasal passages, nasal pharynx, pharynx, larynx, trachea, bronchi, bronchioles, air sacs (alveoli) of the lungs, and muscles of respiration.

Rest Period (COFFEE BREAK, BREAK TIME). Brief interruption in the workday during which the worker rests or takes refreshments without loss of pay.

Retrograde Shock Amnesia. The loss or impairment of memory due to shock. It can result from shock due to head injuries, severe loss of blood, and other conditions.

Retrospective Study. (See CASE-CONTROL STUDY.)

Reverberation. The persistence of echoing of previously generated sound caused by reflection of acoustic waves from the surfaces of enclosed spaces.

Right-of-way. The right of one vehicle or pedestrian to proceed in a lawful manner in preference to another vehicle or pedestrian approaching under such circumstances of direction, speed, and proximity as to give rise to danger of collision unless one grants precedence to the other.

Riscident. Comes from one of the same root words as accident, and means risk due to fortuitous change. The study of riscidents has been approached in two ways: (a) the analysis of accident records, identifying the maneuvers of the various vehicles or other agents involved, and (b) by recording near accidents.

Riscutant. Refers to the behavior of a person which tends to bring about involvement in riscident or critical situations.

Risk. A probability that an event will occur, e.g., that an individual will become ill, injured, or will die within a stated period of time.

Risk (INSURANCE). The chance of physical or personal loss; the amount of possible loss to the insuring company. Also used in place of insured or prospect. Hazard, danger, peril. A company protected by insurance. A subjective evaluation of relative failure potential.

Risk Factor. A characteristic or agent whose presence increases the probability of occurrence of a disease or injury.

Roentgen (R). The common unit of exposure to x- or gamma rays equivalent to the absorption of 83 ergs per gram of air. Under SI, it is equal to 0.000 258 coulombs per kilogram of air (c/kg).

Routes of Entry. The pathway by which material may gain access to the body including inhalation, ingestion, and skin contact.

Safe. Relatively free from danger, injury, or damage or from the risk of danger.

Safe Workplace. One in which the likelihood of all identifiable undesired events are maintained at an acceptable level.

Safety. The art of performing any activity in the most accident-free manner. Relatively free from hazard.

Safety (CUT-OUT). An overload protective device in an electric circuit.

Safety Belt. A life belt worn by telephone line worker, window washers, construction worker, etc., attached to a secure object (telephone pole, window sill, anchor point, etc.) to prevent injury due to falling. A seat or torso belt securing a passenger in an automobile or airplane to provide body protection during a collision, sudden stop, air turbulence, etc.

SAFETY COUPLING. A friction coupling adjusted to slip at a predetermined torque, to protect the rest of the system from overload.

SAFETY EDUCATION (TRAINING). The transmission of knowledge, skills, attitudes, motivations, etc., concerning the safety requirements of operations, processes, environments, etc., to workers, supervisors, managers, and others.

SAFETY ENGINEERING. Safety engineering is concerned with the planning, development, improvement, coordination, and evaluation of the safety component of integrated systems of individuals, materials, equipment, and environments to achieve optimum safety effectiveness in terms of both protection of people and protection of property.

SAFETY FACTOR. (See FACTOR OF SAFETY.)

SAFETY GLASS. Impact resistant and shatterproof glass used as eye protection and for automobile windows, large architectural windows and doors. Also includes heat-treated glass that breaks into granules instead of sharp-edged strands.

SAFETY HELMETS. Rigid headgear of varying materials designed to protect the head, not only from impact, but from flying particles and electric shock, or any combination of the three. Safety helmets should meet the requirements of ANSI Standard Z89.1, Protective Headware for Industrial Workers.

SAFETY LOCK. A lock which can be opened only by its own key. Often used to lock out the electrical energized sources used in equipment or machinery operation.

SAFETY PROFESSIONAL. An individual who, by virtue of specialized knowledge and skill and/or educational accomplishments has achieved professional status in the safety field, may also have been awarded or earned the status of Certified Safety Professional by the Board of Certified Safety Professionals.

SAFETY RULE. A rule prescribing safeguarding requirements, personal protective equipment, or safe behavior on the job.

SAFETY RULES. Codes of conduct to avoid injury and damage.

SAMPLE. A number of sampling units from the population of units eligible to be included.

SAMPLING (GENERAL). The collection of samples of contaminants in air in the general work environment for a specified length of time.

SAMPLING (GRAB). The collection of samples of contaminants in air in a work environment for small time periods (usually 5 min. or less).

SAMPLING (PERSONNEL). The collection of samples of contaminants in air for the duration of the specific part of an operation by attaching an instrument or device to the worker.

SAMPLING (STATISTICAL). A procedure in which primary sampling units (e.g., municipalities) are first selected from a population, and then secondary units (e.g., city blocks) are sampled from within each chosen primary unit. This may be extended so that tertiary units (e.g., households) or further units (individuals) are selected within the secondary units.

SAMPLING UNIT. The basic unit (e.g., person, household, etc.) around which a sampling procedure is planned.

SARCOMA. A malignant neoplasm arising from connective tissue cells.

SCAN. A shortened form of scintiscan - variously designated according to the organ under examination (e.g., brain scan, lung scan, etc.). Scanning refers to the visual examination of a small area in some detail.

SCHEDULE RATING. A type of rating assigned under the Industrial Compensation Rating Schedule, as approved by the insurance commissioner, by which the Basic Manual Rate is modified to fit the physical conditions related to guarding of machines. It is also affected by the compliance of a safety organization to prescribe insurance standards. Not used in most states any longer.

SCHEDULED CHARGE. The specific charge (in days) assigned to a permanent partial, permanent total, or fatal injury.

SCREENING TEST. A test or series of tests to which an individual submits to determine whether enough evidence of a disease exists to warrant a further diagnostic examination.

SECONDARY ATTACK RATE. The number of new cases of disease which occur over a relatively short period of time within households in which there is a first or primary case of the disease per unit susceptible population in households with primary cases.

SECULAR TRENDS. Changes in incidence rates, mortality rates, and other indicators of disease or injury frequency that occur gradually over relatively long periods of time.

SELF INSURANCE. A term used to describe the assumption of one's own financial risk.

SELF-CONTAINED BREATHING APPARATUS (SCBA). A respiratory protection device that consists of a supply of respirable air, oxygen, or oxygen-generating material worn by the worker.

SENSITIVITY. The extent to which a test identifies as positive all individuals who have a given disease.

SERIOUS INJURY. The classification for a work injury which includes: (1) all disabling work injuries, (2) nondisabling injuries in the following categories: (a) eye injuries from work-produced objects, corrosive materials, radiation, burns, etc., requiring treatment by a physician, (b) fractures, (c) any work injury that requires hospitalization for observation, (d) loss of consciousness (work related), and (e) any other work injury (such as abrasion, physical or chemical burn, contusion, laceration, or puncture wound) which requires: 1) treatment by a medical doctor, or 2) restriction of work, or motion or assignment to another regularly established job.

SERIOUS INJURY FREQUENCY RATE. The number of serious injuries, as defined in ANSI Z16 per 1,000,000 employee-hours of exposure. When serious injury frequency rate is used, it should be clearly identified as serious injury frequency rate, to avoid confusion with other frequency rates. This rate relates serious injuries, as defined, to the employee-hours worked during the period and expresses the number of such injuries in terms of million-hour units by use of the following formula:

SIFR =

$$\frac{\textit{Number of serious injuries} \times 1{,}000{,}000}{\textit{Employee-hours of exposure}}$$

SERIOUS VIOLATION. Any violation in which there is a substantial probability that death or serious physical harm could result from the violative condition. (OSHAct.)

SEVERITY RATE. The total days charged for work injuries as defined in ANSI Z16 per 1,000,000 employee-hours exposure. Days charged include actual calendar days of disability resulting from temporary total injuries and scheduled charges for deaths and permanent disabilities. These latter charges are based on 6,000 days for a death or permanent total disability, with proportionately fewer days for permanent partial disabilities for varying degrees of seriousness. (See standard disabling injury severity rate.)

SR =

$$\frac{\textit{Total days charged for work injuries} \times 1{,}000{,}000}{\textit{Employee-hours exposure}}$$

SHAVER'S DISEASE. A condition of pulmonary fibrosis and emphysema attributed to inhalation of fumes consisting of alumina, silica, iron and other substances fused at 2000°C.

SHOCK. (1) The rapid fall in blood pressure especially following injury, operation, or the administration of anesthesia. (2) A sudden disturbance of mental equilibrium.

SHOP RULES (WORKING RULES). Either regulations established by an employer dealing with day-to-day conduct in the plant operations, safety, hygiene, records, etc., or working rules set forth in collective bargaining agreements and in some union constitutions.

SHORT TERM EXPOSURE LIMITS (STEL). A 15 minute time-weighted average exposure which should not be exceeded at any time during a work shift.

SICK LEAVE. Period of time during which a worker may be absent without loss of job or seniority if unable to work because of illness or accident. A paid sick leave plan provides for full or partial pay for such absence, usually up to a stipulated maximum. Sick leave plans differ from accident and sickness benefits, principally in that the former cover shorter periods of absence, usually provide higher pay, and are uninsured.

SIDEROSIS. A pneumoconiosis resulting from chronic exposure to iron oxide and silicon dioxide.

SIEVERT (Sv). The SI unit of radiation dose equal to the absorption of one gray (Gy), 1 Sv = 100 rem.

SILICATES. Compounds made up of silicon, oxygen, and one or more metals with or without hydrogen. These dusts cause nonspecific dust reactions, but generally do not interfere with pulmonary function.

Silicones. A unique group of compounds made by molecular combinations of the element silicon or certain of its compounds with organic chemicals. Produced in variety of forms, including silicone fluids, resins, and rubber. Silicones have special properties such as water repellency, wide temperature resistance, high durability, and high dielectric strength.

Silicosis. A fibrotic disease of the lungs caused by the chronic inhalation of dust containing silicon dioxide.

Single Point Failure. A failure of a subunit which by itself will cause a failure of the system or equipment.

Sintering. Process of making coherent powder of earthy substances by heating but without melting.

Skin Contamination. Irritations and infections to the skin such as dermatitis and poison ivy. These contaminations are classified as work injuries if they arise out of and in the course of employment.

Slot Velocity. Linear flow rate of contaminated air through the openings in a slot-type hood.

Smoke. An air suspension (aerosol) of particles, often originating from combustion or sublimation. Carbon or soot particles less than 0.1μm in size result from the incomplete combustion of carbonaceous material such as coal or oil. Smoke generally contains droplets as well as dry particles. Tobacco, for instance, produces a wet smoke composed of minute tarry droplets.

Somatic Mutation. An alteration in the molecular arrangement of the inherited DNA of the chromosomes of the general body cells (as opposed to the germ cells) occurring spontaneously or by some abnormal form of external stimulus.

Sonarography. Ultrasonic scan (q.v.) that provides a two-dimensional image corresponding to clear sections of acoustic interfaces in tissue.

Sone. A unit used in judging the loudness of sounds. One sone is defined as the level, at 1000 Hz, that is 40 dB above the subjects' threshold of hearing. The loudness of a given sound is rated by the listener as some multiple of the sone.

Sonometer. An appliance for testing acuteness in hearing.

Sorbent. A material which removes gases and vapors from air passed through a canister or cartridge.

Sound. An oscillation in pressure, stress, particle displacement, particle velocity, etc., which is propagated in an elastic material, in a medium with internal forces (e.g., elastic, viscous), or the super-position of such propagated oscillations. Sound is also the sensation produced through the organs of hearing, usually by vibrations transmitted in a material medium, commonly air.

Sound Absorption. The change of sound energy into some other form, usually heat, in passing through a medium or on striking a surface. In addition, sound absorption is the property possessed by materials and objects, including air, of absorbing sound energy.

Sound Analyzer. A device for measuring the band-pressure level or pressure-spectrum level of a sound as a function of frequency.

Sound Intensity. The average rate at which sound energy is transmitted through a unit area perpendicular to a specified point.

Sound-level Meter. An instrument for the direct measurement of sound pressure level. Sound-level meters often are made with various filtering networks that measure the sound directly on A, B, C, etc., scales. Sound-level meters may also incorporate octave-band filters for measuring sound directly in octave bands.

Sound Pressure Level. The weighted sound pressure in decibels. The level, in decibels, of sound is 20 times the logarithm to the base 10 of the ratio of the pressure of this sound to the referenced pressure. The reference pressure must be explicitly stated. (See DECIBEL.)

Spasm. (1) Sudden, violent, involuntary contraction of muscle(s). (2) Sudden but transitory constriction of a passage, canal, or orifice.

Specific Gravity. The ratio of the density of a material to the density of some standard material, such as water at a specified temperature, or (for gases) air at standard pressure and temperature. Abbreviated sp. gr. Also known as relative density.

Specificity. The extent to which a test identifies as negative all individuals who are free of a given disease.

Speech Perception Test. A measurement of hearing acuity by the administration of a carefully controlled list of words. The identification of correct responses is evaluated in terms of norms established by the average performance of normal listeners.

SPLASH-PROOF GOGGLES. Eye protection constructed of noncorrosive material that fits snugly against the face, and has indirect ventilation ports.

SPONTANEOUS IGNITION. Ignition resulting from a chemical reaction in which there is a slow generation of heat from oxidation of organic compounds until the combustion or ignition temperature of the material (fuel) is reached. This condition is reached only where there is sufficient air for oxidation but not enough ventilation to carry away the heat as fast as it is generated.

SPRAIN. A joint injury in which some of the fibers of a supporting ligament are ruptured, but the continuity of the ligament remains intact.

SPRINKLER SYSTEM. A combination of water discharge devices (sprinklers), distribution piping to supply water to the discharge devices, or more sources of water under pressure, water flow controlling devices (valves), and actuating devices (temperature, rate of rise, smoke or other type device). The system automatically delivers and discharges water in the fire area.

SQUEEZE. This occurs in diving during increases in pressure from the failure or inability to equalize pressure in gas-filled spaces. It occurs most commonly in the middle ear (aerotitis media) due to eustachian tube dysfunction or the nasal sinuses (aerosinusitis) due to blockage of the ostia. Squeeze may also occur in carious or poorly restored teeth, in the lung in breath-hold diving, or in the mask, suit, or helmet.

STANDARD DISABLING INJURY FREQUENCY RATE. (See DISABLING INJURY FREQUENCY RATE.)

STANDARD MAN. A theoretical physically fit man of standard (average) height, weight, dimensions, and other parameters (blood composition, percentage of water, mass of salivary glands, to name a few); used in studies of how heat or ionizing radiation affects humans.

STANDARDIZED MORTALITY RATIO (SMR). The ratio of the number of deaths observed in the study population to the number of deaths that would be expected if the study population had the same specific rates as the standard population, multiplied by 100.

STANNOSIS. A condition due to inhalation of tin or tin oxide dust. It is characterized by the appearance of nodules in the lungs on chest x-ray, but autopsy studies have shown no evidence of fibrosis. The condition at present is thought to be benign.

STENOSING TENOSYNOVITIS (TRIGGER OR SNAPPING FINGER). A localized inflammation of the tendon sheath with localized swelling. The swollen tendon snaps back and forth through the narrowed sheath during flexion and extension. The condition frequently occurs at the base of the thumb or fingers in workers who repeatedly grasp hard or push with the distal palm, causing points of pressure over the tendon sheaths.

STRAIN. Overstretching or overexertion of some part of the body musculature.

STRATIFIED SAMPLING. A procedure in which the population is divided into strata, or groups of units having certain characteristics in common, and a sample of units drawn from each stratum.

SUBROGATION. The legal process by which a company endeavors to recover from a third party the amount paid to an insured under an insurance policy when such third party may have been responsible for the occurrence causing the loss.

SUPPLIED-AIR SUIT. A one- or two-piece suit that is impermeable to most particulate and gaseous contaminants and is provided with an adequate supply of respirable air.

SURGICAL BENEFITS. Plans which provide employees, and in many cases their dependents, with specified surgical care or a cash allowance toward the cost of such care, usually in accordance with a schedule of surgeon's fees. Generally part of a health and insurance program (See HEALTH AND INSURANCE PLAN.)

SURVEILLANCE. Ongoing scrutiny, generally using methods distinguished by their practicability, uniformity, and frequently their rapidity, rather than by complete accuracy. Its main purpose is to detect changes in order to initiate investigative or control measures.

SURVEY. Inspection and comprehensive study or examination of an organization, environment or activity for insurance and/or accident prevention purposes.

SURVIVORS' BENEFITS (TRANSITION BENEFITS, BRIDGE BENEFITS, WIDOWS' ALLOWANCE). Payments to dependents of employees who die prior to retirement, financed in whole or in part by the employer. May be in the form of payments for a fixed period (e.g. 24 months) supplementing regular life insurance benefits, a benefit for life out of a pension program, a lump-sum payment, etc.

Synergism. The interaction of two or more physiologically active agents producing an effect which is equal to (additive) or greater than (potentiation) the effect produced using equivalent doses of each agent presented individually.

System Safety. The application of operating, technical and management techniques and principles to the safety aspects of a system throughout its life to reduce hazards to the lowest level possible through the most effective use of available resources.

System Safety Analysis. The safety analysis of a complex process by means of a diagram or model that provides a comprehensive, overall view of the process, including its principal elements and the ways in which they are interrelated. There are four principal methods of analysis: failure mode and effect, fault tree, THERP, and cost-effectiveness. Each has a number of variations and more than one may be combined in a single analysis. (See FAILURE MODE AND EFFECT ANALYSIS, FAULT TREE ANALYSIS, THERP.)

System Safety Engineering. The application of scientific and engineering principles during the design, development, manufacture, and operation of a system to meet or exceed established safety goals.

Systematic Sampling. A procedure in which the selected sampling units are spaced regularly throughout the population; that is, every nth unit is selected.

Temporary Threshold Shift (TTS). The hearing loss suffered as the result of noise exposure, all or part of which is recovered during an arbitrary period of time when one is removed from the noise. It accounts for the necessity of checking hearing acuity at least 16 hours after a noise exposure.

Temporary Total Disability. The classification for any injury which does not result in death or permanent total or permanent partial disability, but which renders the injured person unable to perform a full day's work. This means that the injured employee cannot perform all the duties of a regularly established job which is open and available; or unable to perform such duties during the entire time interval corresponding to the hours of the regular shift on any one or more days including weekends, holidays, and other days off, or plant shutdown, subsequent to the date of the injury.

Tennis Elbow. (See EPICONDYLITIS.)

Tenosynovitis. Inflammation of a tendon sheath. Causes are varied, but commonly seen in industry associated with repetitive motion in an awkward or strained position. Commonly occurs about the wrist, the extensor tendons of the thumb, or in the fingers.

Teratogen. Any agent or substance capable of producing a permanent alteration in the structure and/or function of cells, tissue and organs in the developing embryo or fetus when a pregnant female is exposed to that substance.

Theory. A system of assumptions, accepted principles and rules of procedure devised to analyze, predict, or otherwise explain a specified set of phenomena.

THERP. This is an acronym for Technique for Human Error Prediction, a means for quantitatively evaluating the contribution of human error to the degradation of product quality. It is often combined with failure mode and effect and fault free methods to identify failures of the human components in a system.

Threshold (hearing). The point at which a person just begins to notice the tone is becoming audible. The level where the first effects occur.

Threshold Dose. The minimum dose of radiation which will produce a detectable biological effect.

Threshold Limit Values (TLVs®). Exposure guidelines that have been established by the American Conference of Governmental Hygienists (ACGIH). TLVs represent conditions under which it is believed that nearly all workers may be repeatedly exposed day after day without adverse health effects. The ACGIH cautions that individuals may experience health problems at the specified levels, and that there may be harmful effects from the interactions of the specified exposures with exposures to other agents, such as tobacco smoke or other chemicals. The guidelines are intended only as recommendations for the control of contaminants in the professional practice of industrial hygiene.

Threshold Limit Value-Ceiling (TLV-C). The ceiling exposure limit-the concentration that should not be exceeded, even instantaneously.

Threshold Limit Value-Time-Weighted Average (TLV-TWA). The allowable time-weighted average concentration of an air contaminant for a normal 8-hour workday or a 40-hour workweek.

Time-weighted Average (TWA). An average over a given (working) period of a person's exposure as generally determined by integrative monitoring techniques.

Tinnitus. A ringing sound in the ears.

Toeboards. A guard commonly installed around flywheels and other equipment in open pits and on overhead catwalks. The installation of toeboards should conform to the ANSI Standard A-1264.1, Safety Requirements for Workplace Floor and Wall Openings, Stairs and Railings. Toeboards should be at least 4 inches high and should be made of wood, metal, or metal grille not exceeding 1-in. mesh. Toeboards at flywheels should be placed as close to the edge of the pit as possible. Wood toeboards for permanent installations should be of 1" x 4" stock or heavier.

Tolerance. The ability of a living organism to resist the usually anticipated stress. The upper and lower limits of permissible variability. The limits of permissible inaccuracy above and below design specifications.

Tolerance (drug). The progressive diminution of susceptibility to the effects of a drug resulting from its continued administration.

Tongs, Safety. A device for feeding objects to and removing them from a danger area.

Total Loss. Loss of all the insured property. Also a loss involving the maximum amount for which a policy is liable.

Total Reaction Distance (motor vehicle). The distance traveled between the point at which the driver perceives that braking evasive action is required and the point at which the contact is made with the braking controls.

Total Reaction Time (motor vehicle). The time required for a vehicle to move the total reaction distance.

Total Stopping Distance (motor vehicle). The distance in which the vehicle comes to rest after the driver discovers a hazard which requires stopping. Includes driver reaction time, brake reaction time, and braking time.

Toxemia. Poisoning by way of the blood stream.

Toxic. Of, pertaining to, or caused by poison. Poisonous, harmful.

Toxic Substance. A substance that demonstrates the potential to induce cancer, to produce short- and long-term disease or bodily injury, to affect health adversely, to produce acute discomfort, or to endanger life of human or animal resulting from exposure via the respiratory tract, skin, eye, mouth or other routes in quantities which are reasonable for experimental animals or which have been reported to have produced toxic effects in humans.

Toxicant. A substance that causes a degenerative alteration in any anatomic, physiologic, or biochemical system of a formed organism.

Toxicity. A relative property of a chemical agent with reference to a harmful effect on some biologic mechanism and the condition under which this effect occurs.

Toxicology. That branch of medical science which deals with the nature and effects of poisons.

Toxicology, Industrial. Study of the nature and action of toxic agents that may cause health impairment to workers.

Traffic Accident. Any accident involving one or more motor vehicles in motion on a roadway. A traffic accident may involve more than one unit if each unit comes in contact with some other unit involved while part or either is in contact with the road or sidewalk.

Transmission Loss (sound). The ratio, expressed in decibels, of the sound energy incident on a structure to the sound energy that is transmitted. The term is applied both to building structures (walls, floors, etc.) and to air passages (muffler, ducts, etc.).

Trauma. An injury, wound, or shock resulting from an accident or outside force.

Tremor. Involuntary shaking, trembling, or quivering.

Trigger Finger. (See STENOSING TENOSYNOVITIS.)

TSCA. Toxic Substances Control Act.

Tumor. Any neoplasm.

Tunnel Vision. Inability to see toward the sides. A narrow field of vision.

Two-hand Controls. Tripping devices for a machine which require simultaneous application of both hands to operate the control, so that the hands of the operator are kept out of the point-of-operation area while the machine is operating.

Type of Accident. The classification of an accident according to the manner of contact of the injured person with the agency. Involves movement of object, material, or person and association with an agency; such as fall, struck by, struck against, etc.

Under the Influence of Alcohol. Sufficient intoxicating liquor in the body system to cause impairment of faculties.

Underwriter. An insurance company employee who accepts or rejects risks on behalf of the insurance company. More broadly, anyone who makes insurance contracts.

Unsafe Act. A departure from an accepted, normal, or correct procedure or practice which has in the past actually produced injury or property damage or has the potential for producing such loss in the future; an unnecessary exposure to a hazard; or conduct reducing the degree of safety normally present. Not every unsafe act produces an injury or loss but, by definition, all unsafe acts have the potential for producing future accident injuries or losses. An unsafe act may be an act of commission (doing something which is unsafe) or an act of omission (failing to do something that should have been done).

Unsafe Condition. Any physical state which deviates from that which is acceptable, normal, or correct in terms of its past production or potential future production of personal injury and/or damage to property or things; any physical state which results in a reduction in the degree of safety normally present. It should be noted that accidents are invariably preceded by unsafe acts and/or unsafe conditions. Thus, unsafe acts and/or unsafe conditions are essential to the existence or occurrence of an accident.

Upper Explosive Limit (UEL). The maximum proportion of vapor or gas in air above which propagation of flame does not occur. The upper limit of the flammable or explosive range.

Vapor. The gaseous form of substances which are normally in the solid or liquid state (at room temperature and pressure). The vapor can be changed back to the solid or liquid states either by increasing the pressure or decreasing the temperature alone.

Variable. A quantity capable of assuming two or more different values.

Ventilation. One of the principal methods used for the engineering control of occupational health hazards. The process causes fresh air to circulate to replace contaminated air that is simultaneously removed.

Vertigo. Dizziness due to vestibular dysfunction. In true vestibular vertigo the victim experiences a sensation of rotary motion of either himself or the environment. Vertigo may be experienced in aviators as a function of acceleration. In divers, vertigo may occur as a result of caloric stimulation or unequal middle ear pressurization (alternobaric vertigo). Permanent vestibular injury and vertigo may result from vestibular decompression sickness or round window rupture.

Vesicant. Any substance that produces blisters.

Vibration. An oscillation wherein the quantity is a parameter that defines the motion of a mechanical system.

Vital Statistics Systematically tabulated information concerning births, marriages, separations, divorces, and deaths based on registrations of these vital events.

Waiting Period. Certain policies which insure against loss of use or against disability provide that they will not begin payment until the interruption or the disability has equalled a specified time, which is called the waiting period.

Wavelength. The distance between a point of a given phase of one wave and a point of the same phase of an adjacent wave.

Welders' Flashburn (actinic keratoconjunctivitis). An inflammatory condition of the corneal and conjunctival epithelium resulting from exposure to intense ultraviolet radiation as from a welder's arc. It is characterized by corneal edema, pain, a gritty sensation, tearing, and photophobia occurring several hours after exposure. It is generally self-limited, resolving spontaneously within 24 to 48 hours and is fully preventable by the placement of shields around welding operations, by the training of welders, and the proper use of welding helmets.

Welding and Cutting Permit. Authorization for the use of open flame and spark-producing devices in areas of combustible and/or hazardous materials, where their use is normally prohibited. Frequently extended to the use of open flame devices and/or high heat-producing devices.

WHITE NOISE. Noise of wide frequency range in which the amplitude of the noise is essentially the same in all frequency bands.

WILLFUL MISCONDUCT. Deliberate failure to comply with statutory regulations.

WILLFUL VIOLATION. An intentional and knowing violation of the OSHAct.

WORK INJURY. Any injury suffered by a person which arises out of and in the course of employment.

WORKER'S COMPENSATION. A quasijudicial system where employers have a statutory obligation to pay medical expenses, disability, and other benefits for work-related injuries or illness. Administrative aspects and extent of coverage of worker's compensation laws vary from state to state.

WORKER'S COMPENSATION INSURANCE. A form of insurance imposed on employers by legislative statute which insures medical disability and other coverage for work-related injuries or illness. Depending on the state, an employer may be self-insured or pay premiums to a carrier or the state to administer the program.

WORKING SURFACE. Any surface or plane on which an employee walks or works. (An agency factor used in classifying accidents.)

WOUND. A break in the continuity of the skin and other body tissues.

BIBLIOGRAPHY

Amdur, M.O., Doull, J., and Klassen, C.D. Casarett and Doul's Toxicology: The Basic Science of Poisons. New York: Pergamon Press, 1991.

Clayton, G.D. and F.C. (eds.). Patty's Industrial Hygiene and Toxicology. New York: Wiley; Vol. 1, 1991; Vols. 2 and 3, 1993.

Grimaldi, J.V. and Simmonds, R.H. Safety Management. Homewood, IL:R.D. Irwin, 1988.

Hammer, W. Occupational Safety Management and Engineering. Englewood Cliffs, NJ: Prentice Hall, 1989.

Hathaway, G.T., et al. Proctor and Hughes' Chemical Hazards of the Workplace. New York: Van Nostrand Reinhold, 1991.

Konzen, R.B., Vernon, R.J., and Cameron, D.B. Glossary of terms common to the industrial hygiene and safety engineering professions. Unpublished report of U.S. DHHS (NIOSH), 1980.

Last, J.M. (ed.). A Dictionary of Epidemiology. New York: Oxford U.Press, 1988.

Last, J.M. Maxcy-Rosenau-Last Public Health & Preventive Medicine. East Norwalk, CT: Appleton & Lange, 1992.

Parker, S.P. McGraw-Hill Dictionary of Scientific & Technical Terms. New York:McGraw-Hill, 1993.

Parmeggiana, L. (ed.). Encyclopedia of Occupational Health and Safety.Geneva: International Labour Office, 1983.

Plog, B.A. (ed.). Fundamentals of Industrial Hygiene. Chicago: National Safety Council, 1988.

Talty, J.T. (ed.). Industrial Hygiene Engineering: Recognition, Measurement, Evaluation & Control. Park Ridge, NJ: Noyes, 1989.

Zenz, C., Dickerson, O.B., and Horvath, E.P. Occupational Medicine: Principles and Practical Applications. St. Louis: Mosby-Year Book, 1993.

Accident Prevention Manual for Business & Industry. Chicago: National Safety Council, Vols. 1 and 2, 1992.

Accident Prevention Manual. Chicago: National Safety Council, Vol. 3, 1994.

Best's Safety Directory. Oldwick, NJ: A.M. Best, 1993.

Dorland's Illustrated Medical Dictionary. Philadelphia: W.B. Saunders,1988.

Industrial Ventilation - A Manual of Recommended Practice. Cincinnati,OH: American Conference of Governmental Industrial Hygienists, 1992.

1993-1994 Threshold Limit Values for Chemical Substances and Physical Agents and Biological Exposure Indices. Cincinnati, OH: American Conference of Governmental Industrial Hygienists, 1993.

Radiation Protection in Occupational Health: Manual for Occupational Physicians. Vienna: International Atomic Energy Agency, 1987.

Accredited Standards Committee Z16. (1998). Draft American National Standard for Occupational Safety and Health Incident Surveillance, BSR Z16.5. Itasca, IL: National Safety Council.

Z94.14
Operations & Inventory Planning & Control

The Industrial Engineering terminology related to production control is rife with acronyms. Every time production control personnel turn around, there is a new system with a new acronym. The proliferation of acronyms occurred with the advent of computers, especially personal computers, and the associated software. The industry is now constantly creating new production control software, creating a new name, a new term, a new acronym, and marketing the software as a new production control concept. Therefore it is impossible and probably inappropriate to document all the new acronyms or terms appearing in the market place. The acronyms and terms presented herein are believed to be generic, not tied to specific software.

Production control is one of the manufacturing system functions that evolved from the plant floor. Many production control personnel assumed those positions without formal training. As such, each manufacturing facility created its own terms or definitions of terms to suit its own needs. Hence, there are many different and acceptable definitions for many terms. Many of the terms presented herein have multiple definitions. The given definition(s) are intended to be as general as possible while still pointing out the key technical points.

As are all industrial engineering specialties, production control is constantly evolving. We have seen the eras of capacity planning, shop floor control, MRP, and JIT as well as many other technologies. Interestingly, these eras of a specific technology do not disappear. Rather they are simply added to, providing more and more capabilities for the production control person. As such, the number of terms defined have increased substantially. In addition, the definitions in many cases have changed or been expanded to accommodate the new technologies.

Where is production planning and control technology going? Many will argue that it is becoming much more data and information systems oriented. With computer proliferation on the production floor, real time data are a reality. The capability of knowing where everything is and what everyone is doing now is possible if not a reality. Unfortunately the result is a data avalanche which in many instances is unmanageable. Therefore, it can be argued that production planning and control is the gathering of the right information at the right time to make decisions right now!

In addition, many production planning and control people will argue that companies are now seeing their function for what it really is, the nervous system running the entire facility. With that as a concept, many companies are providing the production control people much more comprehensive authority and forcing non-production control people to learn more about production control. Hence, there is an ever growing need for non-production control and production control people to know and utilize a common set of accepted terminology.

Those responsible for this terminology are:

Chairman

Timothy J. Greene, Ph.D.
Professor and Department Head
Department of Industrial Engineering and
 Management, Oklahoma State University

Subcommittee

Jim Robbins
Allied Signal Inc.

Marty Muscattelo
Litton PolyScientific Corp.

Colin L. Moodie, Ph.D. Professor of Industrial Engineering
School of Industrial Engineering
Purdue University

Contributors

The following people or organizations provided substantial assistance in compiling these definitions:

William L. Berry, Ph.D.
University of North Carolina

John R. Dougherty
J.R. Dougherty Consulting

J.P. (Spike) Kelleher
Westinghouse, BOM Data Management

American Production and Inventory Control Society
Computer-Aided Manufacturing International, Inc.
Dow Corning
IBM Corporation
General Systems Division
Martin Marietta

ABC Classification. Classification of the items in an inventory in decreasing order of annual dollar volume. This array is then split into three or more classes, called A, B, and C, etc. Class A contains the items with the highest annual dollar volume and receives the most attention. The medium Class B receives less attention, and Class C, which contains the low-dollar volume items, is controlled routinely. The ABC principle is that effort saved through relaxed controls on low-value items will be applied to reduce inventories of high-value items. Syn: distribution by value.

ABC Inventory Control. An inventory control approach based on ABC classification.

Accessory. A choice or feature offered to a customer for customizing the end product. In many companies, this term means that the choice does not have to be specified prior to shipment but could, in fact, be added at a later date. In other companies, however, this choice must be made prior to shipment. (See ATTACHMENT, OPTION.)

Action Message. An output of an MRP system that identifies the need for and the type of action to be taken to correct a current or a potential material coverage problem. Examples of action messages are Release Order, Reschedule In, Reschedule Out, Cancel, etc. *Syn*: exception message.

Aggregate Inventory. The sum of the inventory levels for individual items. For example, the aggregate finished goods inventory would be made up of one half the sum of all the lot sizes plus the sum of all of the safety stocks plus any anticipation inventory plus transportation inventory.

Allocation. (1) In an MRP system, an allocated item is one for which a picking order has been released to the stock room but not yet sent out of the stock room. It is an uncashed stock room requisition. (2) A process used to distribute material in short supply. (See RESERVATION.)

Anticipated Delay Report. A regular report, normally issued by both manufacturing and purchasing to the material planning function, regarding jobs or purchase orders which will not be completed on time, why not, and when they will be completed. This is one essential ingredient of a closed-loop MRP system.

Anticipation Inventory. Additional inventory above what is in-process to cover projected trends of increasing sales, planned sales promotion programs, seasonal fluctuations, plant shutdowns and vacations. (See BASE SERIES.)

Artificial Intelligence. A computer-based system that simulates rational human thought and logic utilizing knowledge data.

Assemble-to-order Product. End items assembled after receipt of a customer order where options or other subassemblies are stocked prior to order arrival. (See MAKE-TO-ORDER PRODUCT.)

Assembly Lead Time. (See LEAD TIME.)

Assembly Line. A manual or automated serial facility where the product is progressively and repetitively manufactured. An assembly line process can be divided into elemental tasks, each with a specified time requirement per unit of product and a sequence relationship with the other tasks. (See LINE BALANCING.)

Attachment. A choice or feature offered to customers for customizing the end product. In many companies, this term means that the choice, although not mandatory, must be selected prior to the final assembly schedule. In other companies, however, the choice need not be made at that time. (See ACCESSORY, OPTION.)

Automatic Relief. (See POST-DEDUCT INVENTORY TRANSACTION PROCESSING, PRE-DEDUCT INVENTORY TRANSACTION PROCESSING.)

Automatic Rescheduling. Allowing the computer to automatically change due dates on scheduled receipts when it detects that due dates and required dates are out of phase.

Available Inventory. The on-hand balance of an item minus outstanding allocations and "usual" quantities held for quality problems.

Available To Promise. The portion of a company's inventory or planned production uncommitted to customer's orders. This figure is frequently calculated from the master production schedule and is maintained as a tool for order promising.

Back Flush. A method of recording issues from inventory by exploding the bill of material for the next higher assembly and deducting the result from the on-hand inventory record. This can be done when the next higher assembly is released or completed.

Backlog. All of the customer orders booked, i.e., orders taken but not yet shipped. Sometimes referred to as "open orders" or the "order board."

BACKORDER. An unfilled customer order or commitment treated as an immediate or past due demand against an item whose inventory is insufficient to satisfy the demand. (See STOCKOUT.)

BACKWARD SCHEDULING. A technique in which the schedule is computed starting with the due date for the order and working backward to determine the required start date. This can generate negative times, thereby identifying where time must be made up. (See FORWARD SCHEDULING.)

BASE SERIES. A standard series of demand-over-time observations used in forecasting seasonal items. This series of factors is usually based upon the relative level of demand during the corresponding period of previous years. The average value of the base series over a twelve-month period will be 1.0. The base series is superimposed upon the average demand trend for the item. *Syn*: base index.

BATCH PROCESS. A manufacturing approach in which product or products are manufactured repetitively, but in specific sized batches or lots.

BILL OF CAPACITY. (See PRODUCT LOAD PROFILE.)

BILL OF LABOR. (See PRODUCT LOAD PROFILE.)

BILL OF MATERIAL (BOM). A listing of all the subassemblies, parts, raw materials, components, bulk products, etc. that go into a parent assembly showing the quantity of each required. There are a variety of BOM formats including single level, indented, modular (planning), transient, matrix, costed, addended, pseudo, etc. *Syn*: assembly list.

BILL OF MATERIAL PROCESSOR. Computer applications supplied by many software manufacturers for maintaining, updating, and retrieving bill of material information.

BILL OF MATERIAL STRUCTURING. The process of organizing bills of material to perform specific functions such as scheduling, inventorying and assemblying. (See PLANNING BILL OF MATERIAL, TRANSIENT BILL OF MATERIAL.)

BILL OF RESOURCES. (See PRODUCT LOAD PROFILE.)

BIN LOCATION FILE. A file that specifically identifies the physical location where each item in inventory is stored. Often the file also maintains quantity information by location.

BIN RESERVE SYSTEM. (See TWO BIN SYSTEM.)

BLANKET ORDER. A long-term commitment to a vendor for material against which short-term releases will be generated to satisfy requirements.

BLENDING. The process of physically mixing two or more lots of material to produce a homogeneous lot. Blends normally receive new identification and require retesting.

BLOW-THROUGH. (See TRANSIENT BILL OF MATERIAL.)

BOOK INVENTORY. An accounting definition of inventory units or value obtained from perpetual inventory records rather than from actual count.

BOM. (See BILL OF MATERIAL.)

BOTTLENECK. A facility, function, department, etc., that impedes production. For example, a machine or work center where jobs arrive at a faster rate than they leave.

BRANCH WAREHOUSE DEMAND. (See WAREHOUSE DEMAND.)

BUCKETED SYSTEM. An MRP system in which all time-phased data are displayed in accumulated time periods or "buckets." If the period of accumulation is one week, then the system would be said to have weekly buckets. (See BUCKETLESS SYSTEM, TIME BUCKET, HORIZONTAL DISPLAY, VERTICAL DISPLAY.)

BUCKETLESS SYSTEM. An MRP system in which all time-phased data are received, stored, processed and reported by specific dates and not in time buckets. (See BUCKETED SYSTEM, TIME BUCKET, HORIZONTAL DISPLAY, VERTICAL DISPLAY.)

BULK ISSUE. The issuance or transfer of parts or materials from their primary storage location to a secondary point-of-use location or directly to work-in-process inventory.

BUSINESS PLAN. A statement of income projections, costs, and profits usually accompanied by a budget, projected balance sheet, and cash flow statement. It is usually stated in terms of dollars only. The business plan and the production plan, although frequently stated in different terms, should be in agreement with each other. (See MANUFACTURING RESOURCE PLANNING.)

Buyer. An individual whose functions may include vendor selection, negotiation, order placement, vendor follow-up, measurement and control of vendor performance, value analysis, evaluation of new materials and processes, etc. In some companies, the functions of order placement and vendor follow-up are handled by the vendor scheduler. (See VENDOR SCHEDULER.)

Buyer/Planner. (See VENDOR SCHEDULER.)

Buying Capacity. (See CAPACITY BUYING.)

By-product. A material of value produced as residual of or incidental to the production process. Ratio of by-product to primary product is usually fixed. By-products may be recycled, sold as is or used for other purposes. *Syn*: co-product.

Cancellation Charges. A fee charged by a seller to cover his costs associated with a customer's cancellation of an order. If the seller has started any engineering work, purchased raw materials, or started any manufacturing operations, these changes would also be included in the cancellation charge.

Capacity. The time available for work at work centers expressed in machine-hours (minutes, etc.) or in man-hours (minutes, etc.).

Capacity Buying. A purchasing practice whereby a company commits to a vendor for a given amount of his capacity per unit of time. Subsequently, schedules for individual items are given to the vendor in quantities to match the committed level of capacity. (See VENDOR SCHEDULING.)

Capacity Control. The process of measuring production output and comparing it with the capacity requirements plan, determining if the variance exceeds pre-established limits, and taking corrective action to get back on plan if the limits are exceeded. (See INPUT/OUTPUT CONTROL, CLOSED LOOP MRP.)

Capacity Planning. (See CAPACITY REQUIREMENTS PLANNING.)

Capacity Requirements Planning (CRP). The function of establishing, measuring, and adjusting limits or levels of capacity. In this context, it is the process of determining how much labor and machine resources are required to accomplish the tasks of production. Open shop orders and planned orders in the MRP system are input to CRP, which "translates" these orders into hours of work by work center by time period. (See RESOURCE REQUIREMENTS PLANNING, INFINITE LOADING, CLOSED-LOOP MRP.)

Capacity Smoothing. (See LOAD LEVELING.)

Carrying Cost. Cost of carrying inventory, usually defined as a percent of the dollar value of inventory per unit of time (generally one year). Depends mainly on cost of capital invested as well as the costs of maintaining the inventory such as tax, insurance, obsolescence, spoilage, and space occupied. (See ECONOMIC ORDER QUANTITY.)

Cellular Manufacturing. Developing the manufacturing flow around the processing of a product so that operators are trained on all processes of a particular product. It is bringing the processes together to build the entire product as opposed to setting up an assembly line or using a job shop layout.

Centralized Dispatching. Organization of the dispatching function into one central location. This often involves the use of data collection devices for communication between the centralized dispatching function, which usually reports to the production control department, and the shop manufacturing departments. (See CONTROL CENTER, DECENTRALIZED DISPATCHING.)

Change Order. (See ENGINEERING CHANGE ORDER.)

Changeover Costs. The sum of the setup cost and the tear-down cost for a manufacturing operation. *Syns*: turnaround costs, shut-down costs, start-up costs.

Classification. A systematic and orderly analysis of items, grouping like things together by their common features and subdividing them by their special features. (See CODING.)

Closed Loop MRP. A system built around MRP including the additional planning functions of production planning, master production scheduling, and capacity requirements planning. The term "closed loop" implies that there is feedback from the execution functions to update the planning. (See MANUFACTURING RESOURCE PLANNING.)

Coding. The assignment of alphanumeric identification symbols to classified descriptions. (See CLASSIFICATION.)

Commodity Buying. Grouping like parts or materials under one buyer's control for the procurement of all requirements.

COMMON PARTS BILL OF MATERIAL. A type of planning bill which groups all common components for a product or family of products into one bill of material. (See PLANNING BILL OF MATERIAL, MODULAR BILL OF MATERIAL, SUPER BILL OF MATERIAL.)

COMPETITIVE BIDDING. The offer of estimates by firms or individuals competing for a contract, privilege, or right to supply specified services or merchandise.

COMPONENT. A raw material, ingredient, part, or subassembly that goes into a higher level assembly, compound or other item.

CONCURRENT ENGINEERING. Designing the product simultaneously with the processing and inspection procedures.

CONFIGURATION CONTROL. Ensuring that the product being built and shipped corresponds to the product ordered and designed. This means that the correct features, customer options and engineering changes have been incorporated.

CONFIRMING ORDER. A purchase order issued to a vendor listing the goods or services and terms of an order placed verbally, or otherwise, in advance of the issuance of the usual purchase document.

CONSIGNED MATERIAL. Component parts, usually as raw material, provided by the customer to a supplier for use in filling an order for the next level part number. This is used where the customer needs to control quality, schedule, or proprietary process issues.

CONSIGNED STOCKS. Inventories generally of finished products which are in the possession of customers, dealers, agents, etc., but which remain the property of the manufacturer by agreement with those in possession.

CONTRACT PEGGING. (See FULL PEGGING.)

CONTROL CENTER. The place at which the dispatching is done in a centralized dispatching operation. (See CENTRALIZED DISPATCHING.)

COSTED BILL OF MATERIAL. A bill of material that extends the quantity of every component by the component costs.

CPIM. Certified in Production and Inventory Management. A certification awarded by The American Production and Inventory Control Society. CFPIM denotes the highest level of certification, the fellow level.

CRITICAL PATH METHOD (CPM). A network planning technique used for planning and controlling the activities in a project. By showing each of these activities and their associated time, the "critical path" can be determined. The critical path identifies those elements that actually constrain the total time for the project. (cf. PERT.)

CRITICAL RATIO. A dispatching rule which calculates a priority index number by dividing the time remaining until due date by the expected time required to finish the job. From the critical ratios, it is possible to determine the orders that are behind (less than 1.0), the orders that are ahead (greater than 1.0), the orders that are on schedule (equal 1.0), the orders that should be processed next, and whether processing rates should be increased. (See DISPATCHING RULE.)

CRP. (See CAPACITY REQUIREMENTS PLANNING.)

CUMULATIVE LEAD TIME. The longest length of time required to accomplish the activity in question. *Syns*: aggregate lead time, stacked lead time, composite lead time, critical path lead time.

CUSTOMER ORDER. An order for a particular product or a number of products from a customer. Often referred to as an "actual demand" to distinguish it from a forecasted demand.

CUSTOMER SERVICE. Delivery of product to the customer at the time which the cutomer or corporate policy specifies. (See PERCENT OF FILL.)

CUSTOMER SERVICE RATIO. A measure of delivery performance usually in the form of a percentage comparing the units or dollars shipped on time to the units or dollars shipped late.

CYCLE COUNTING. A physical inventory procedure in which inventory is counted on a periodic schedule other than once a year. A cycle count may be taken when an item reaches its reorder point, when new stock is received, or on a regular basis usually more frequently for high-value, fast-moving items than for low-value, slow-moving items. Most effective cycle counting systems require the counting of a certain number of items every work day.

CYCLE STOCK. One of the two main components of any item inventory, the other being safety stock. The cycle stock is the most active component depleting gradually and replenished cyclically. (See LOT SIZE, SAFETY STOCK.)

Cycle Time. The time the product is at each work station on an assembly or production line.

Decentralized Dispatching. The organization of the dispatching function into individual departmental dispatchers. (See CENTRALIZED DISPATCHING.)

Delay. The time a product spends waiting to be moved or processed.

Delivery Cycle. The actual time from the receipt of the customer order to the shipment of the product.

Delivery Policy. The company's delivery cycle goal. The policy is sometimes stated as "our quoted delivery time."

Demand. A need for a particular product or component in a given time period coming from any multiple sources. Demand can be created by customer orders; interplant, interplant, and warehouse requirements; and by predictive forecasting methods. (See DEPENDENT DEMAND, INDEPENDENT DEMAND.)

Demand Flow Manufacturing. A means of obtaining material on a just-on-time basis. A pull system versus a push system. A simple means of managing inventory usually without the need of a MRP system.

Demand Management. The task of recognizing and managing all of the demands for products to insure that the master scheduler is aware of them. It encompasses activities including forecasting, order entry, order promising, branch warehouse requirements, interplant orders, and service parts requirements. (See MASTER PRODUCTION SCHEDULE.)

Demonstrated Capacity. Capacity calculated from actual performance data, usually number of items produced times standard hours per item. *Syn*: actual capacity.

Dependent Demand. Demand directly related to or derived from the demand for other items or end products. Such demands are therefore calculated and should not be forecast. A given inventory item may have both dependent and independent demand at any given time. (See INDEPENDENT DEMAND.)

Direct-deduct Inventory Transaction Processing. A method of bookkeeping which decreases the book (computer) inventory of an item as material is issued from stock, and increases the book inventory as material is received into stock. The book record is updated concurrently with the movement of material out of or into stock. As a result, the book record represents what is physically in stock. (See PRE-DEDUCT INVENTORY TRANSACTION PROCESSING, POST-DEDUCT INVENTORY TRANSACTION PROCESSING.)

Disbursement. The issuance of raw material or components from a store room.

Discrete Order Quantity. (See LOT-FOR-LOT.)

Dispatching. Selecting, sequencing, and assigning available jobs to individual work stations and/or workers. Dispatching is also associated with tooling and materials handling devices. (See CENTRALIZED DISPATCHING, DECENTRALIZED DISPATCHING, EXPEDITING, SHOP PLANNING, CLOSED LOOP MRP.)

Dispatching Rule. The logic of assigning jobs priorities to work centers and/or workers. (See CRITICAL RATIO, EARLIEST DUE DATE, FIRST-IN-FIRST-OUT, SHORTEST PROCESSING TIME.)

Distribution Requirements Planning (DRP). Determining the needs to replenish inventory at branch warehouses. Frequently a time-phased order point approach is used where the planned orders at the branch warehouse level are "exploded" via MRP logic to become gross requirements on the supplying source. In the case of multi-level distribution networks, this explosion process can continue down through the various levels of master warehouse, factory warehouse, etc., and become input to the master production schedule. Demand on the supplying source(s) is recognized as dependent. (See TIME-PHASED ORDER POINT, PHYSICAL DISTRIBUTION, PULL DISTRIBUTION SYSTEM.)

Distribution Resource Planning. The extension of distribution requirements planning into the planning of the key resources contained in a distribution system: warehouse space, manpower, money, trucks and freight cars, etc. (See DISTRIBUTION REQUIREMENTS PLANNING.)

Dock to Flow. The process and elapsed time required to make a part from first contact on the receiving dock to being available for immediate use on the factory flow. Dock to flow implies a conscious effort to avoid moving a part into and our of a parts stockroom prior to use.

Dock to Stock. The process and elapsed time required to move a part from first contact on the receiving dock to being fully available in the stock room. Material handling, purchasing, and quality systems and procedures are all involved.

Downtime. Time when the machines in the plant are not producing because they are down for repairs or other reasons.

Drop Shipment. A distribution arrangement in which the seller serves as a selling agent by collecting orders but does not maintain inventory. The orders are sent to the manufacturer which ships directly to the customer.

DRP. (See DISTRIBUTION REQUIREMENTS PLANNING.)

Due Date. The date when purchased material or production material is due to be available for use. Also, often, the date when a customer order has been requested or promised to be shipped.

Dynamic Lot Sizing. (See LEAST TOTAL COST, LEAST UNIT COST, PERIOD ORDER QUANTITY, FIXED ORDER QUANTITY.)

Earliest Due Date (EDD). A priority rule that sequences the jobs in queue according to their due date. Earliness. If a job is finished before its due date, the difference between its completion date and due date.

Economic Order Quantity (EOQ). A type of fixed order quantity which determines the amount of product to be purchased or manufactured at one time in order to minimize the total cost involved, including the ordering costs and carrying costs. The general economic order quantity equation is: where EOQ is the quantity to be ordered, S is the annual sales, A is the ordering cost, r is the carrying cost, and u is the unit cost. (See CARRYING COST, ORDERING COST.)

EDD. (See EARLIEST DUE DATE.)

Effective Date. The date on which a component or an operation is introduced or severed from a bill of material or a routing sheet. The effective dates are used in the explosion process to create demands for the correct material or labor. Normally, bill of material systems provide for an effective "start date" and "stop date." Control may also be by serial number rather than date.

Effectivity. (See EFFECTIVE DATE.)

Efficiency. The relationship between the planned resource requirements, such as labor or machine time, for a task(s) and the actual resource time charged to the task(s).

End Product. The finished product which is shipped from that facility. *Syn*: end item.

Engineering Change Order. A document issued by engineering to modify the bill of material and/or graphic drawing previously released to manufacturing that modifies the configuration of an end product or component. (See CONFIGURATION CONTROL.)

Exception Message. (See ACTION MESSAGE.)

Exception Reports. Reports that list or flag only those items that deviate from plan.

Expediting. The rushing or chasing of production orders which are needed in less than the normal lead time. (See DISPATCHING.)

Explosion. The function of utilizing the parent assembly bill of material to determine the quantity needed of every subassembly or component.

Exponential Smoothing. A type of weighted average forecasting technique in which past observations are geometrically discounted according to their age. The heaviest weight is assigned to the most recent datum. The smoothing is termed "exponential" because data points are weighted in accordance with an exponential function of their age. The technique makes use of a smoothing constant to apply to the difference between the most recent forecast and the critical sales datum, which eliminates the necessity of carrying historical sales data. The approach can be used for data whether or not they exhibit trend or seasonal patterns.

Extrinsic Forecast. A forecast based on a correlated leading indicator. For example, estimating window sales based on housing starts. Extrinsic forecasts tend to be more useful for large aggregations such as total company sales than for individual product sales. (See INTRINSIC FORECAST.)

Family. (See PRODUCT FAMILY.)

FCFS. (See FIRST-COME-FIRST-SERVED.)

Feature. (See ACCESSORY, ATTACHMENT, OPTION.)

FIFO. (See FIRST-IN-FIRST-OUT.)

Final Assembly Schedule (FAS). A schedule of end items needed to replenish finished goods inventory or to satisfy make-to-order demand. For make-to-order products, the FAS is prepared after receipt of a customer order; is constrained by the availability of material and capacity, and schedules the operations required to complete the product from the level where it is stocked (or master scheduled) to the end item level. Also referred to as the "finishing schedule" as it may include other operations than simply the final assembly. *Syn*: blending schedule, pack-out schedule.

FINISHED PRODUCTS INVENTORIES. Inventories on which all manufacturing operations, including final test, have been completed. These may be either finished parts, like renewal parts, or finished assemblies which have been authorized for transfer to the finished stock account.

FINISHING LEAD TIME. The time that is necessary to finish manufacturing a product after receipt of a customer order. An input to the FAS. The finishing lead time should be equal to or less than the company's goal for shipping its product after receipt of a customer order.

FINITE LOADING. Conceptually the term means putting no more work into a factory than the factory can be expected to execute. The specific term usually refers to a technique that involves automatic shop priority revision in order to level load operation. (See INFINITE LOADING.)

FIRM PLANNED ORDER (FPO). A planned order that is frozen in quantity and time. The computer is not allowed to automatically change it; this is the responsibility of the planner in charge of the corresponding item. This technique can aid planners working with MRP systems to respond to material and capacity problems by firming up selected planned orders. Additionally, firm planned orders are the normal method of stating the master production schedule.

FIRST-COME-FIRST-SERVED (FCFS). A priority rule for sequencing jobs such that the first job into the facility receives first priority.

FIRST-IN-FIRST-OUT (FIFO). A priority rule for sequencing jobs such that the first job into the queue is the first job processed.

FIXED ORDER QUANTITY. A lot sizing technique in MRP that will always cause a planned order to be generated for a predetermined fixed quantity or multiples thereof if net requirements for the period exceed the fixed order quantity. (See ECONOMIC ORDER QUANTITY, LOT-FOR-LOT, PERIOD ORDER QUANTITY.)

FLOAT. The amount of work-in-process inventory between two manufacturing operations, especially in repetitive manufacturing.

FLOORSTOCK. Stock of inexpensive production parts held in the factory from which production workers can draw without requisitions.

FLOW SHOP. A manufacturing facility in which the plant layout arranges work stations according to the order in which operations should be performed to optimize cost, lead time, and quality performance. *Syn*: product shop.

FLUCTUATION INVENTORIES. Inventories that are carried as a cushion to protect against forecast error and stockouts. (See SAFETY STOCK.)

FOCUS FORECASTING. A system that allows the user to simulate the effectiveness of numerous forecasting techniques, thereby being able to select the most effective one.

FORECAST. An objective extrapolation of past data to the future. A forecast is analytical versus a prediction which is subjective incorporating management's anticipation of changes and new factors influencing demand.

FORECAST ERROR. The difference between actual demand and forecast demand, typically stated as an absolute value.

FORECAST HORIZON. The period of time into the future for which a forecast is prepared.

FORECAST INTERVAL. The increments of time into which the forecast is divided, sometimes referred to as time buckets. Syn: forecast period.

FORMULATION. (See BILL OF MATERIAL.)

FORWARD SCHEDULING. A scheduling technique where the scheduler proceeds from a known order start date and computes the completion date usually proceeding from the first operation to the last. (See BACKWARD SCHEDULING.)

FULL PEGGING. The ability of a system to automatically trace requirements for a given component all the way up to the end item or contract number.

GANTT CHART. The earliest and best known type of control chart especially designed to show graphically the relationship between planned performance and actual performance. Used especially for machine loading, where one horizontal line represents capacity and another represents load against the capacity. Also used for following job progress where one horizontal line represents the production schedule and another represents the actual progress of the job against the schedule in time. Named after its originator, H. L. Gantt, scientific management pioneer.

Gateway Work Center. A starting work center.

Gross Requirements. The total of independent and dependent demand for a part or an assembly for a specific time period prior to the netting of on-hand inventory and scheduled receipts.

Group Classification Code. A part or material classification technique designating characteristics by hierarchical code groups. For example, classification may denote function, type of material, size, shape, etc. (See GROUP TECHNOLOGY, CODING, CLASSIFICATION.)

Group Technology. The bringing together and organizing (grouping) of common concepts, principles, problems, and tasks (technology) in order to take advantage of their similarities.

Hedge. (1) In master production scheduling, a quantity of stock used to protect against uncertainty in demand. The hedge is similar to safety stock, except that a hedge has the dimension of timing as well as amount. (2) In purchasing, any purchase or sale transaction having as its purpose the elimination of the negative aspects of price fluctuations.

Hierarchical BOM. (See PRODUCT STRUCTURE)

Horizontal Display. A method of displaying output from an MRP system where requirements, scheduled receipts, projected balance, etc., are displayed horizontally, i.e., across the page. Horizontal displays are difficult to use in conjunction with bucketless systems. (See VERTICAL DISPLAY, BUCKETLESS SYSTEM.)

Implementation. The act of installing a system into operation. It concludes the system project with the exception of appropriate follow up or post-installation review.

Indented Bill of Material. A form of multilevel bill of material. It exhibits the highest level sub-assemblies closest to the left margin. All the components going into these sub-assemblies are shown indented to the right of the margin, and all subsequent levels of components are indented farther to the right. If a component is used in more than one sub-assembly within a given product structure, it will appear more than once, under every sub-assembly in which it is used.

Independent Demand. Demand for a finished good or a component unrelated to the demand for other items. Demand for finished goods, parts required for destructive testing and service parts requirements are some examples of independent demand. (See DEPENDENT DEMAND.)

Indirect Labor. Work required to support production without being directly related to a specific product.

Infinite Loading. A loading method for CRP where work is assigned to work centers in specific time periods regardless of the capacity available to perform this work. Though infinite loading is not usually practiced today, it can be used for capacity planning. Infinite implies that a load can be put into a factory regardless of its capability to be processed. The concept obscures the fact that it is necessary to generate capacity requirements and compare these with available capacity before trying to adjust requirements to capacity. (See CAPACITY REQUIREMENTS PLANNING, FINITE LOADING.)

Ingredient. (See COMPONENT.)

Input. Work entering a production facility.

Input/Output Control. Controls production capacity at the work centers. It monitors actual work entering and leaving each work center and compares it against planned inputs and outputs. The purpose is to reduce backlog and thus lead time. (See CAPACITY CONTROL, CLOSED-LOOP MRP.)

Interactive Scheduling. Computer scheduling where the process is either automatic or manually interrupted to allow the scheduler the opportunity to review and/or change the schedule.

Intermittent Production. A production system in which the productive units are organized according to function. The jobs pass through the functional departments in lots and each lot may have a different routing.

Interoperation Time. The time between thecompletion of an order at one work center and its start at the next. (See QUEUE TIME.)

Interplant Demand. Material to be shipped to another plant or division within the corporation. Although it is not a customer order, it is usually handled by the master production scheduling system in a similar manner. (See DEMAND MANAGEMENT.)

INTRANSIT LEAD TIME. The time lag between the date of shipment (at supplier shipping point) and the date of receipt (at the customer's dock). Normally customers' orders specify the date by which goods should be at their dock. Consequently this date should be offset by intransit lead time for establishing a ship date for the supplier.

INTRINSIC FORECAST. A forecast made based on past history that is usually internal to the company, such as a forecast made from a moving average. Syn: statistical forecast. (See EXTRINSIC FORECAST.)

INVENTORY. Items that are in a stock point or work-in-process and which serve to decouple successive operations in the process of manufacturing a product and distributing it to the consumer. Inventories may consist of raw materials, work-in-process, or finished goods.

INVENTORY CONTROL. The activities and techniques of maintaining the stock of items at desired levels, whether they be raw materials, work-in-process, or finished goods. The objective is to minimize inventory while meeting all demands. (See INVENTORY MANAGEMENT.)

INVENTORY MANAGEMENT. The branch of business management concerned with the planning and control of inventories. (See INVENTORY CONTROL.)

INVENTORY POLICY. A definite statement of management philosophy concerning inventories.

INVENTORY SHRINKAGE. Losses resulting from scrap, deterioration, pilferage, etc.

INVENTORY TURNOVER. The number of times that an inventory "turns over" or cycles during the year. One way to compute inventory turnover is to divide the average inventory level into the annual cost of sales. *Syn:* turns.

INVENTORY VALUATION. The value of the inventory at either its cost or its market value. Because inventory value can change with time, some recognition must be made of the age distribution of inventory. Therefore, the cost value of inventory, under accounting practice, is usually computed by first-in-first-out (FIFO), last-in-first-out (LIFO), or a standard cost system to establish the cost of goods sold.

INVENTORY WRITE-OFF. Deducting inventory dollars from the financial statement because the inventory is no longer saleable or because of shrinkage, i.e., the value of the physical inventory is less than its book value. Item. Any individual manufactured or purchased part or assembly. (See END PRODUCT, SUBASSEMBLY, COMPONENT, RAW MATERIALS.)

ITEM NUMBER. A number that serves to uniquely identify a product, component, or raw material. Syns: part number, product number, stock code.

JIT. (See JUST-IN-TIME.)

JOB ORDER. (See MANUFACTURING ORDER.)

JOB SHOP. A process oriented layout used for low volume, batch, or customized products each of which requires a different set or sequence of tasks. *Syn*: process shop, function shop.

JUST-IN-TIME (JIT). An inventory control policy that requires the material to arrive at the next work station precisely when it is needed. (See KANBAN.)

KANBAN. An order point scheduling approach, which uses fixed lot sizes of materials in standard containers with the cards attached to each. Material reorder is triggered at the last minute, when the lot of material is moved to the point of use.

KITTING. The process of removing components of an assembly from the stockroom and sending them to the assembly floor as a kit of parts. (See PICKING.)

LABOR TICKET. A form used to record labor allocated to specific jobs or production operations.

LAP-PHASING. (See OVERLAPPED SCHEDULE.)

LAST-IN-FIRST-OUT (LIFO). A sequencing procedure where the last item in to the queue is the first item out of the queue.

LEAD TIME. A span of time required to perform an activity. In a production and inventory control context, the activity in question is normally the procurement of materials and/or products either from an outside supplier or from one's own manufacturing facility. The individual components of any given lead time can include queue time, move or transportation time, receiving and inspection time, processing time, etc. (See MANUFACTURING LEAD TIME, PURCHASING LEAD TIME.)

Lead Time Offset. A term used in MRP where a planned order receipt in one time period will require the release of that order in some earlier time period based on the lead time. The difference between the order's due date and release date is the lead time offset.

Least Total Cost. A dynamic lot-sizing technique that calculates the order quantity by comparing the carrying cost and the set-up (or ordering) costs for various lot sizes and selects the lot where these are most nearly equal.

Least Unit Cost. A dynamic lot-sizing technique that adds ordering cost and inventory carrying cost for each trial lot size and divides by the lot size, picking the lot size with the lowest unit cost.

Level. A relative position in which a part or assembly is used within a product structure signified by a code. Normally the end items are assigned level "0" and the components/sub-assemblies going into it level "1" and so on. MRP explosion process starts from level "0" and proceeds downward one level at a time.

Leveling. A capacity planning concept where work is balanced between work centers and across time.

LIFO. (See LAST-IN-FIRST-OUT.)

Limiting Operation. The operation having the least capacity. In a series of operations with no alternative routings, the capacity of the total system can be no greater than the limiting operation. As long as this limiting condition exists, the total system can be effectively scheduled by simply scheduling the limiting operation. *Syn:* bottleneck.

Line Balancing. Assignment of elemental tasks to work stations so as to minimize the number of work stations and to minimize the total amount of unassigned time at all stations. Line balancing can also mean a technique for determining the product mix that provides a fairly consistent flow of work at the planned line rate. (See ASSEMBLY LINE.)

Line Item. One item on a customer order, regardless of quantity.

Load. (1) The amount of scheduled work ahead of a manufacturing facility, usually expressed in terms of hours or work units of production. (2) To assign work to the capacity available at particular work stations.

Load Leveling. Spreading orders out over time or rescheduling operations so that the amount of work to be done in the time periods is distributed evenly. (See FINITE LOADING.)

Load Profile. A display of future capacity requirements based on planned and released orders for a single part or all parts over a given time span. Syn: load projection.

Logistics. In an industrial context, the art and science of obtaining and distributing material and product. In a military sense (where it has greater usage), its meaning can also include the transportation of personnel.

Long-range Resource Planning. A planning activity for long-term capacity decisions such as level-loading based on the production plan, and other available data beyond the time horizon for the production plan. This activity is to plan long term capacity needs out to the time period necessary to acquire gross capacity additions (such as a major factory expansion.)

Long Term Agreements (LTA's). A strategic supplier management approach where long-term financial terms, prices, quality and ever improvement requirements are negotiated with a supplier, sometimes by a central purchasing organization. Individual buyers then release orders against the LTA where only quantity and schedule need to be established.

Lot-for-lot. A lot sizing technique in MRP which generates planned orders in quantities equal to the net requirements in each period. *Syn:* discrete order quantity.

Lot Number. A unique identification assigned to a homogenous quantity of material. *Syns:* batch number, mix number.

Lot Size. The amount of a particular item that is ordered from the facility or vendor. *Syns:* order quantity, batch quantity.

Lot Size Inventory. Inventories which are maintained whenever quantity price discounts, shipping costs, or set-up costs, etc. make it more economical to purchase or produce in larger lots than are needed for immediate purposes.

Low Level Code. Identifies the lowest level in any bill of material at which a particular component may appear. Net requirements for a given component are not calculated until all the gross requirements have been calculated down to that level. Low level codes are normally calculated and maintained automatically by the computer software. (See LEVEL.)

LUMPY DEMAND. A demand pattern with large fluctuations from one time period to another. *Syns*: discontinuous demand.

M-DAY CALENDAR. (See MANUFACTURING CALENDER.)

MACHINE LOADING. The accumulation by workstation, machine, or machine group of the hours generated from the scheduling of operations for released orders by time period. Machine loading differs from capacity planning in that it does not use the planned orders from MRP but operates solely from scheduled receipts. (See CAPACITY REQUIREMENTS PLANNING.)

MACHINE UTILIZATION. The percent of the available time that the machine is being operated.

MAJOR SET-UP. The machine set-up and related activities required when one or more items within a group of items is ordered. (See MINOR SET-UP TIME.)

MAKE-OR-BUY DECISION. The act of deciding whether to produce an item in-house or buy it from an outside vendor.

MAKE-TO-ORDER PRODUCT. An end item is finished after receipt of a customer order. Frequently, long lead time components are planned prior to the order arriving in order to reduce the delivery time to the customer. (See ASSEMBLE-TO-ORDER PRODUCT.)

MAKE-TO-STOCK PRODUCT. End items shipped from finished goods, "off the shelf," and, therefore, finished prior to a customer order arriving.

MANUAL RESCHEDULING. The most common method of rescheduling open orders (scheduled receipts). Under this method the MRP system provides information on the part numbers and order numbers that need to be rescheduled. Due dates and/or order quantity changes required are then analyzed and changed by material planners or other authorized persons. (See AUTOMATIC RESCHEDULING.)

MANUFACTURING CALENDAR. A system where only the working days are numbered so that the component and work order scheduling may be done based on the actual number of work days available. *Syn*: M-day calendar.

MANUFACTURING LEAD TIME. The total time required to manufacture an item. Included are order, preparation, queue, set-up, run, move time, inspection, and put-away times.

MANUFACTURING ORDER. A document or group of documents conveying authority for the manufacture of specified parts or products in specified quantities.

MANUFACTURING RESOURCE PLANNING (MRP II). A method for the effective planning of all the resources of a manufacturing company. Ideally, it addresses operational planning in units, financial planning in dollars, and has a simulation capability to answer "what if" questions. It is made up of a variety of functions, each linked together: Business Planning, Production Planning, Master Production Scheduling, Material Requirements Planning, Capacity Requirements Planning and the execution systems for capacity and priority decisions. Outputs from these systems would be integrated with financial reports such as the business plan, purchase commitment report, shipping budget, inventory projections in dollars, etc. Manufacturing resource planning is a direct outgrowth and extension of MRP. (See CLOSED LOOP MRP.)

MARKET HEDGE. (See HEDGE.)

MASTER PRODUCTION SCHEDULE (MPS). A statement of what the company expects to manufacture by item. It is the anticipated build schedule for those items assigned to the master scheduler. The master scheduler maintains this schedule and, in turn, it becomes a set of planning numbers which is an input to MRP. It represents what the company plans to produce expressed in specific configurations, quantities, and dates. The MPS should not be confused with a sales forecast which represents a demand statement. The master production schedule must take forecast plus other important considerations (backlog, availability of material, availability of capacity, management policy and goals, etc.) into account prior to determining the best manufacturing strategy. (See CLOSED LOOP MRP.) *Syn*: master schedule.

MASTER SCHEDULE. (See MASTER PRODUCTION SCHEDULE.)

MASTER SCHEDULE ITEM. A part selected to be planned by the master scheduler. This item is critical in terms of its impact on lower level components and/or resources such as skilled labor, key machines, dollars, etc. A master schedule item may be an end item, a component, a pseudo number or a planning bill of material.

Master Scheduler. The job title of the person who manages the master production schedule. The person should have substantial product and shop knowledge because master scheduling impacts facility performance.

Material Requirements Planning (MRP). A system which uses bills of material, inventory and open order data, and master production schedule information to calculate requirements for materials. It makes recommendations to release replenishment orders to insure availability of materials. Further, since it is time-phased, it makes recommendations to reschedule open orders when due dates and need dates are not in phase. Originally seen as merely a better way to order inventory, today it is thought of primarily as a scheduling technique, i.e., a method for establishing and maintaining valid due dates on orders. (See CLOSED LOOP MRP, MANUFACTURING RESOURCE PLANNING.)

Material Review Board (MRB). An organization within a company, often a standing committee, which has the job of determining disposition of items which have questionable quality or other attributes.

Materials Handling Time. The time necessary to move material from one work center to the next work center. This includes waiting for the material handling equipment and actual movement time.

Materials Management. The grouping of management functions related to the complete cycle of material flow including purchasing, planning and controlling of work-in-process warehousing, shipping, and distributing finished product. Differs from materials control in that the latter term is associated with the internal control of production materials.

Matrix Bill of Material. A chart made up from the bills of material for a number of products in the same or similar families. It is arranged in a matrix with parts in columns and assemblies in rows (or vice versa) so that requirements for common components can be summarized conveniently.

Minimum Order Quantity. An order quantity modifier, applied after the lot size has been calculated, that increases the order quantity to a pre-established minimum.

Min-max System. A type of order point replenishment system used on a fixed-interval, periodic-review basis. The "min" is the order point, and the "max" is the "order-up-to" inventory level. The order quantity is variable, and is the "max" minus available and on order inventory when the latter two are below the "min."

Minor Set-up Time. The incremental preparation activities required when processing other than the first item within a group of items. These are the machine adjustments and related activities required for each item within the group. (See MAJOR SET-UP.)

Modular Bill of Material. A type of planning bill which is arranged in product modules or options. Often used in companies where the product has many optional features, e.g., automobiles. (See PLANNING BILL OF MATERIAL, COMMON PARTS BILL OF MATERIAL, SUPER BILL OF MATERIAL, OPTION.)

Move Ticket. A document used in dispatching to authorize or record movement of a job from one work center to another. It may also be used to report other information such as the active quantity or the material storage location. MPS. (See MASTER PRODUCTION SCHEDULE.)

MRP. (See MATERIAL REQUIREMENTS PLANNING.)

MRP II. (See MANUFACTURING RESOURCE PLANNING.)

Multi-level Bill of Material. Shows all the components that are directly used in an assembly together with the quantity required of every component. If a component is a sub-assembly, all the components of the sub-assembly will also be exhibited in the multi-level bill.

Net Change MRP. A method where the material requirements plan is continually retained in the computer. Whenever there is a change in requirements, open order or inventory status, or engineering usage, a partial explosion is made only for those parts affected by the change. Net change systems may be continuous, transaction oriented, or periodic (often daily). (See REGENERATION MRP, REQUIREMENTS ALTERATION.)

Net Requirements. In MRP, the net requirements for a part or an assembly are derived as a result of netting gross requirements against inventory on hand and the scheduled receipts. Net requirements that are lot sized and offset for lead time become planned orders.

Netting. The process of calculating net requirements.

On Hand. The balance shown in perpetual inventory records as being physically present at the stocking locations.

ON ORDER. The quantity of material represented by the total of all outstanding replenishment orders. The on order balance increases when a new order is released, and it decreases when material is received to fill an order, or when an order is cancelled. (See ON HAND, OPEN ORDER.)

OPEN ORDER. (1) An active manufacturing order or purchase order. (2) An unfilled customer order. *Syn*: scheduled receipt.

OPERATION REPORTING. The recording and reporting of every manufacturing (shop order) operation occurrence on an operation-to-operation basis.

OPERATION SHEET. A form providing information regarding part routing, operation times, tooling, etc. *Syn*: route sheet.

OPERATION START DATE. The date when an operation should be started based upon the work remaining and the available time remaining to complete the job.

OPTION. A choice or feature offered for customizing the end product. Option means either a mandatory choice (the customer must select from one of the available choices) or a voluntary choice (the customer may select an add-on item). (See ACCESSORY, ATTACHMENT.)

ORDER. A catch-all term which may refer to such diverse items as a purchase order, shop order, customer order, planned order, etc.

ORDER BOM. The version of a bill-of-material captured at point of order release in an MRP system. Prior to order release, changes to the BOM are automatically controlled by the system.

ORDER ENTRY. The process of accepting and translating a customer order into terms used by the manufacturer.

ORDERING COST. Includes costs related to preparing, issuing, following, and receiving orders. Also includes cost related to physical handling of goods, inspection, and machine set-up if the order is being manufactured. (See CARRYING COST.) Syn: acquisition cost.

ORDER MULTIPLE. An order quantity modifier applied after the lot size has been calculated to round the order quantity to a predetermined multiple.

ORDER POINT. (1) The inventory level at which action is taken to replenish a material. The order point is normally calculated as forecasted usage during the replenishment lead time plus safety stock. (2) A specific time interval where inventory levels are reviewed for replenishment. (See TIME-PHASED ORDER POINT.) *Syns*: reorder point, trigger level.

ORDER POLICY. (See LOT SIZE INVENTORY.)

ORDER PROMISING. The process of making a delivery commitment, i.e., answering the question "when can you ship?" For make-to-order products this usually involves checking of uncommitted material, available capacity and planned leadtime. *Syns*: order dating, customer order promising.

ORDER QUANTITY MODIFIERS. Special considerations (i.e., price breaks, order multiples) by which an order quantity is adjusted after being calculated according to a lot sizing rule. (See MINIMUM ORDER QUANTITY, ORDER MULTIPLE.)

OUTPUT. Work being completed by production facility.

OUTSIDE SHOP. The vendor. Used to convey the idea of an extension of the inside shop.

OVERLAPPED SCHEDULE. The concept of parts in a lot being scheduled concurrently on two or more successive work centers. *Syns*: lap-phasing, telescoping.

PAPERLESS PURCHASING. A purchasing operation which does not employ purchase requisitions or hard copy purchase orders. In actual practice a small amount of "paperwork" usually remains, normally in the form of the vendor schedule. (See JUST-IN-TIME, VENDOR SCHEDULER.)

PARETO'S LAW. A concept developed by Vilfredo Pareto, an Italian economist, stating that a small percentage of a group accounts for the largest fraction of the effort, value, etc. For example, twenty percent of the inventory items comprise eighty percent of the inventory value. (See ABC CLASSIFICATION.) *Syn:* 80/20 rule.

PART NUMBER. A number which serves to uniquely identify a component, product, or raw material. *Syns:* stock code, product coding.

PART-PERIOD. The relative cost of holding one part for one period. Used in part-period inventory models.

PEGGING. In MRP displaying for a given item the details of parent items gross requirements and/or allocations.

PERCENT OF FILL. A measure of the effectiveness with

which the inventory management system responds to actual demand. For example, the percent of customer orders filled off the shelf can be measured in either units or dollars. (See STOCKOUT PERCENTAGE.)

PERIODIC INVENTORY SYSTEM. A system in which the quantity in storage is reviewed at a fixed time interval. The size of the replenishment order depends upon the number of units in stock at that time, the expected demands and lead time.

PERIOD ORDER QUANTITY. A lot sizing technique under which the lot size will be equal to the net requirements for a given number of future periods (e.g., weeks). (See FIXED ORDER QUANTITY, LOT-FOR-LOT.) *Syns*: days supply, weeks supply.

PERPETUAL INVENTORY SYSTEM. An inventory record-keeping system where each transaction, whether in or out, is recorded and a new balance is computed. (See physical inventory.)

PHANTOM BILL OF MATERIAL. (See TRANSIENT BILL OF MATERIAL.)

PHYSICAL DISTRIBUTION. The combination of activities associated with moving material, usually finished products, from the manufacturer to the customer. In many cases, this movement is made through one or more levels of field warehouses. (See DISTRIBUTION REQUIREMENTS PLANNING.)

PHYSICAL INVENTORY. (1) The actual material held in stock. (2) Determining inventory quantity by actual count. Physical inventories can be taken on a continuous, periodic, or annual basis. (See CYCLE COUNTING.)

PICKING. The process of withdrawing from stock either components for assembly or finished goods to be shipped to a customer (See KITTING.)

PICKING LIST. A document that is used by operating personnel to pick manufacturing or shipping orders.

PICK-TO-ORDER. An order for a customer which is satisfied from current stores. Manufacturing may not be involved.

PIECE-PARTS. Individual items in inventory at the simplest level of manufacturing.

PLANNED ISSUE RECEIPT. A transaction that updates the on-hand balance and the related allocation or open order.

PLANNED ORDER. A suggested order quantity and due date created by MRP processing when it encounters net requirements. In a MRP system planned orders are created by the computer, exist only within the computer, and may be changed or deleted by the computer during subsequent processing if conditions change. Planned orders at one level will be exploded into gross requirements for components at the next lower level. Planned orders also serve as input to capacity requirements planning, along with released orders, to show the total capacity requirements in future time periods.

PLANNING BILL OF MATERIAL. An artificial grouping of items used to facilitate master scheduling and/or material planning. (See COMMON PARTS BILL OF MATERIAL, OPTION, SUPER BILL OF MATERIAL.)

PLANNED QUEUE MANAGEMENT. A technique in on-line MRP systems where planner priorities are specified based on messages in electronic queues. Examples are: Panic-work first, Reschedule out-work second, etc. Planner performance can also be measured based on success in working electronic queues in a specified time. This measures the planner's quantity output.

PLANNING HORIZON. The span of time from the present to some future date for which material plans are generated. This must cover at least the cumulative purchasing and manufacturing lead time, and usually is quite a bit longer to facilitate MRP II planning.

POINT-OF-USE STORAGE. Keeping inventory in specified locations on a plant floor near the operation where it is to be used. Syns: departmental stocks, floor stocks.

POST-DEDUCT INVENTORY TRANSACTION PROCESSING. A method of doing inventory bookkeeping in which the book (computer) inventory of components is reduced only after completion of activity on their upper-level parent or assembly. This approach has the disadvantage of a built-in differential between the book record and what is physically in stock. (See DIRECT-DEDUCT INVENTORY TRANSACTION PROCESSING, PRE-DEDUCT INVENTORY TRANSACTION PROCESSING.)

PRE-DEDUCT INVENTORY TRANSACTION PROCESSING. A method of doing bookkeeping in which the book (computer) inventory of components is reduced prior to issue, at the time a scheduled receipt for their parent or assembly is created. This approach has the disadvantage of a built-in differential between the book record and what is physically in stock. The advantage is that inventory can be committed to a given assembly. (See

DIRECT-DEDUCT INVENTORY TRANSACTION PROCESSING, POST-DEDUCT INVENTORY TRANSACTION PROCESSING.)

PREDICTION. An intuitive estimate of demand taking into account changes and new factors as opposed to a forecast which is an objective projection of the past into the future. (See FORECAST.)

PRE-EXPEDITING. The function of following up on open orders before the scheduled delivery date to ensure the timely delivery of materials in the specified quantity.

PRIORITY. The relative importance of jobs or work stations, i.e. which jobs should be worked on and when. (See SEQUENCING, SCHEDULING.)

PROCESS MANUFACTURING. Production which adds value by mixing, separating, forming and/or chemical reactions. It may be done in either batch or continuous mode.

PROCESSING TIME. The sum of all operation times necessary to complete a component or an assembly.

PRODUCT. Any commodity produced for sale.

PRODUCT FAMILY. A group of products with similar characteristics, often used in production planning.

PRODUCTION CONTROL. The function of directing or regulating the movement of goods through the entire manufacturing cycle from the requisitioning of raw materials to the delivery of the finished product. (See INVENTORY CONTROL.)

PRODUCTION PLAN. The agreed-upon strategy that comes from the production planning function. (See PRODUCTION PLANNING.)

PRODUCTION PLANNING. The function of setting the overall level of manufacturing output. Its prime purpose is to establish production rates that will achieve management's objective in terms of raising or lowering inventories or backlogs, while usually attempting to keep the production force relatively stable. The production plan is usually stated in broad terms (e.g., product groupings, families of products). It must extend through a planning horizon sufficient to plan the labor, equipment, facilities, material, and finances required to accomplish the production plan. Various units of measure are used by different companies to express the plan such as standard hours, tonnage, labor operators, units, pieces, dollars, etc. As this plan affects all company functions, it is normally prepared with information from marketing, manufacturing, engineering, finance, materials, etc. In turn, the production plan becomes management's authorization for the Master Scheduler to convert it into a more detailed plan. (See BUSINESS PLAN, CLOSED LOOP MRP.) *Syn*: production program.

PRODUCTION RATES. The quantity of production usually expressed in units per time, i.e., parts per hour, tons per day, etc. (See PRODUCTION PLANNING.) *Syn*: production levels.

PRODUCTION SCHEDULE. A plan which authorizes the factory to manufacture a certain quantity of a specific item. Usually initiated by the production planning department.

PRODUCTIVITY. Refers to a relative measure of output per labor or machine hour.

PRODUCT LOAD PROFILE. A statement of the resources required to manufacture one unit of a selected item. Often used to predict the impact of the item scheduled in the master production schedule on these resources. *Syn*: bill of labor, bill of resources, resource profile, bill of capacity.

PRODUCT MIX. The combination of individual product types and the volume produced that make up the total production volume. Changes in the product mix can mean drastic changes in the manufacturing requirements for certain types of labor and material.

PRODUCT STRUCTURE. A graphical representation of the bill of materials. (See BILL OF MATERIAL.)

PROGRAM EVALUATION AND REVIEW TECHNIQUE (PERT). A project planning technique similar to the Critical Path Method, which additionally includes obtaining a pessimistic, most likely, and optimistic time for each activity from which the most likely completion time for the project along the critical path is computed. (See CRITICAL PATH METHOD.)

PROJECTED AVAILABLE BALANCE. The inventory balance projected out into the future. It is the running sum of on-hand inventory minus requirements plus scheduled receipts.

PROJECTION. Estimation based on past data. (See FORECAST.) *Syn*: extrapolation.

PSEUDO BILL OF MATERIAL. (See PLANNING BILL OF MATERIAL.)

PULL DISTRIBUTION SYSTEM. A system for replenishing field warehouse inventories wherein replenishment decisions are made at the field warehouse itself, not at the central supply warehouse or plant. (See PUSH DISTRIBUTION SYSTEM.)

PURCHASE ORDER. The purchaser's document used to formalize a purchase transaction with a vendor. A purchase order, when given to a vendor, should contain statements of the quantity, description, and price of the goods or services ordered; agreed terms as to payment, discounts, date of performance, transportation terms, and all other agreements pertinent to the purchase and its execution by the vendor.

PURCHASE REQUISITION. A document conveying authority to the procurement department to purchase specified materials in specified quantities within a specified time.

PURCHASING CAPACITY. The act of buying capacity or machine time from a vendor. This allows a company to use and schedule the capacity of the machine or a part of the capacity of the machine as if it were in their own shop.

PURCHASING LEAD TIME. The total time required to obtain a purchased item. Included are times associated with the following: procurement, vendor, production, transportation, receiving, inspection, and put away.

PUSH DISTRIBUTION SYSTEM. A system for replenishing field warehouse inventories wherein replenishment decision making is centralized, usually at the manufacturing site or central supply facility. (See DISTRIBUTION REQUIREMENTS PLANNING, PULL DISTRIBUTION SYSTEM.)

QUEUE. A waiting line, for example, the jobs at a given work center waiting to be processed. As queues increase, so do average queue time and work-in-process inventory.

QUEUE TIME. The amount of time a job waits at a work center before set-up or work is performed on the job. Queue time is one element of total manufacturing lead time.

RAW MATERIALS. Material that the facility receives but has not performed any process on.

RECEIVING. This function includes the physical receipt of material, the inspection of the shipment for conformance with the purchase order (quantity and damage), identification and delivery to destination, and preparing receiving reports.

RECONCILING INVENTORY. Comparing the physical inventory figures with the perpetual inventory record and making any necessary corrections.

REGENERATION MRP. An approach where the master production schedule is totally re-exploded down through all bills of material at least once per time period, i.e. week, to maintain valid priorities. New requirements and planned orders are completely "regenerated" at that time. (See NET CHANGE MRP, REQUIREMENTS ALTERATION.)

RELEASE. The authorization to produce or ship material which has already been ordered. (See BLANKET ORDER.)

RELEASED ORDER. (See OPEN ORDER.)

REORDER POINT. (See ORDER POINT.)

REORDER QUANTITY. In a fixed reorder system, the quantity which should be ordered each time the available stock (on hand plus on order) falls below the order point. However, in a variable reorder system the amount ordered from time period to time period varies. (See ECONOMIC ORDER QUANTITY, LOT SIZE.) *Syn*: replenishment order quantity.

REPAIR PARTS DEMAND. (See SERVICE PARTS DEMAND.)

REPETITIVE MANUFACTURING. Production of discrete units, planned and executed via a schedule, usually at relatively high speeds and volumes. Material tends to move in a sequential flow.

REPLENISHMENT LEAD TIME. The total period of time that elapses from the moment it is determined that a product is to be reordered until the order is available for use.

REQUIREMENTS ALTERATION. Processing a revised master production schedule in order to review the impact of the changes. Not to be confused with net change, which, in addition to changes to the master production schedule, also processes changes to inventory balances, bill of material, etc. Typically used in MRP systems. (See NET CHANGE MRP, REGENERATION MRP.) *Syn*: alteration planning.

REQUIREMENTS EXPLOSION. A method of calculating future demand for an item. Future production quantities are multiplied by the quantity in the bill of material. The results represent future demand. (See DEPENDENT DEMAND, GROSS REQUIREMENTS, MATERIAL REQUIREMENTS PLANNING.)

RESCHEDULING. The process of changing order or operation due dates, usually as a result of changing priorities.

RESCHEDULING ASSUMPTION. A fundamental piece of MRP logic which assumes that existing open orders can be rescheduled into nearer time periods far more easily than new orders for the same product can be released and received. As a result, planned order receipts are not created until all scheduled receipts have been applied to cover gross requirements.

RESERVATION. The process of designating stock for a specific customer order. (See ALLOCATION.)

RESOURCE PLANNING. (See LONG-RANGE RESOURCE PLANNING.)

RESOURCE PROFILE. (See PRODUCT LOAD PROFILE.)

RESOURCE REQUIREMENTS PLANNING. The process of converting the production plan and/or the master production schedule into the demand on key resources. Product load profiles or bills of resources can be used to accomplish this. The purpose is to evaluate the plan prior to attempting implementation. (See CLOSED-LOOP MRP.) Syns: rough-cut capacity planning, rough-cut resource planning.

REWORK LEAD TIME. The time required to rework material in-house or at a supplier.

REWORK ORDER. A manufacturing order to rework and salvage defective parts or products. *Syns*: repair order, spoiled work order.

ROUGH-CUT CAPACITY PLANNING. (See RESOURCE REQUIREMENTS PLANNING.)

ROUGH-CUT RESOURCE PLANNING. (See RESOURCE REQUIREMENTS PLANNING.)

RUN TIME. The standard hours allowed to perform an operation on one item. The actual time taken to produce one piece may vary from the standard but the latter is used for loading purposes and is adjusted to actual by dividing by the appropriate work center efficiency factor.

SAFETY CAPACITY. The planning for or reserving of excess manpower and equipment above known requirements for unexpected demand. This reserve capacity is in lieu of safety stock. Syn: reserved capacity procurement.

SAFETY STOCK. (1) A quantity of stock planned to be in inventory to protect against fluctuations in demand and/or supply. (2) The average amount of stock on hand when a replenishment quantity is received. (3) In the context of Master Production Scheduling, additional inventory and/or capacity planned as protection primarily against forecast errors and/or short term changes in the backlog. This investment is often under the control of the master scheduler in terms of where it should be planned. Sometimes referred to as "overplanning" or a "market hedge." (See HEDGE.)

SAFETY TIME. The difference between the requirement date and the planned in-stock date. (See SAFETY STOCK.)

SCHEDULE. A listing of jobs to be processed through a work center, department, or plant, and their respective start and completion dates as well as other related information.

SCHEDULED RECEIPTS. (1) A planned order arrival. (2) Within MRP, open production orders and open purchase orders are considered as "scheduled receipts" on their due date and will be treated as part of available inventory during the netting process for the time period in question. Scheduled receipt dates and/or quantities are not normally altered automatically by the MRP system. Further, scheduled receipts are not exploded into requirements for components as MRP logic assumes that all components required for the manufacture of the item in question have been either allocated or issued to the shop floor. (See PLANNED ORDER, FIRM PLANNED ORDER.)

SCHEDULE VARIANCE MANAGEMENT. A technique used in on-line MRP systems to evaluate the quality of planner output. The resulting actual schedule is compared to the system-suggested ideal schedule. Variances are highlighted and summarized on an exception basis for management action.

SCHEDULING. Establishing the order and timing for performing a task. There are various levels of scheduling within a manufacturing company. For example: master production scheduling, requirements scheduling and material shop floor scheduling on time. (See MASTER PRODUCTION SCHEDULE, DISPATCHING, PRIORITY.)

SCHEDULING RULES. Basic rules that are defined ahead of time for consistent use in scheduling jobs.

SEQUENCING. Establishing an order (priority) in which a series of tasks or jobs are to be performed.

SERVICE PARTS. Parts used for the repair and/or maintenance of an assembled product. Typically they are ordered and shipped at a date later than the shipment of the product itself.

SERVICE PARTS DEMAND. The need for a component to be sold by itself, as opposed to being used in production to make a higher level product. *Syn*: repair parts demand.

SET-UP LEAD TIME. The time in hours or days needed to prepare before a manufacturing process can start. Set-up lead time may include run and inspection time for the first piece.

SET-UP TIME. (1) Time required to adjust a machine work center, or line and to attach the proper tooling to make a particular product. (2) Time required for removing tooling, attaching tooling, adjusting, cleaning, warm-up, etc. *Syn*: start-up time.

SHOP FLOOR CONTROL. A system for utilizing data from the shop floor and data processing files to maintain and communicate status information on shop orders and work centers. The major sub-functions of shop floor control are: assigning priority to each shop order, maintaining work-in-process quantity information for MRP, conveying shop order status information to the office, and providing actual output data for capacity control purposes. (See CLOSED LOOP MRP.)

SHOP PACKET. A manufacturing order that travels with the job and includes a group of documents (possibly coded) such as routings, time tickets, etc.

SHOP PLANNING. Coordinating material handling, material availability, and set-up and tool availability so that a job can be processed on a particular machine. Shop planning is often part of the dispatching function although dispatching does not necessarily include shop planning. (See DISPATCHING, CLOSED LOOP MRP.)

SHORTEST PROCESSING TIME (SPT). A sequencing rule where the job requiring the shortest amount of processing time is produced first. *Syn*: shortest operation time.

SIMULATION. A computer method for modeling the behavior of a system used in what-if analysis and as a decision tool. Simulation can also be either physical or theoretical.

SINGLE-LEVEL BILL OF MATERIAL. Shows only those components that are directly used in an upper-level item. It does not show any relationships more than one level down.

SINGLE-LEVEL WHERE USED. Lists each assembly in which a component is directly used and in what quantity. This information is usually made available through the technique known as "implosion." SKU. (See STOCK KEEPING UNIT.)

SLACK. A priority rule for sequencing jobs based on slack time. (See SLACK TIME.)

SLACK TIME. The difference in calendar time between the scheduled due date for a job and the estimated completion date. If a job is to be completed ahead of schedule, it is said to have slack time; if it is likely to be completed behind schedule, it is said to have negative slack time. Slack time can be used to calculate job priorities using methods such as the critical ratio. In the Critical Path Method, total slack is the amount of time a job may be delayed in starting without necessarily delaying the project completion time. Free slack is the amount of time a job may be delayed in starting without delaying the start of any other job in the project. (See CRITICAL RATIO.)

SOURCE DELEGATED INSPECTION. An approach to procured quantity parts that delegates final part inspection to the supplier. The customer validates the supplier's capabilities by approving the supplier's quality system rather than inspecting each part or samples of a lot.

SOURCE INSPECTION. Inspection at the source of supply (e.g., the vendor) instead of inspection following items receipt.

SPLIT LOT. A manufacturing order quantity that has been divided into two or more smaller quantities usually after the order is in process. Lots are sometimes split so that a portion of the lot can be moved through manufacturing faster. This portion is called the send-ahead.

SPT. (See SHORTEST PROCESSING TIME.)

STAGING. Pulling of the material required for an order from inventory before the material is needed. This action is taken as a protection from inaccurate inventory records, but leads to increased problems in inventory records and availability.

START TIME. The calendar time a job starts being manufactured on a machine or in the facility.

STOCK KEEPING UNIT (SKU). Represents an item at a particular location. For example, if a product is stocked at several locations, each combination of product and its stocking location is a different SKU.

STOCKLESS PRODUCTION. (See JUST-IN-TIME.)

STOCKOUT. The lack of materials or components which are needed to be on hand. (See BACKORDER.)

STOCKOUT PERCENTAGE. A measure of the effectiveness with which the inventory management system responds to actual demand. The stockout percent can be a measurement of total stockouts to total line item orders or line items incurring stockouts to total line items in the system. (See CUSTOMER SERVICE RATIO.)

STOCK STATUS. A periodic report showing the inventory on hand and possibly the inventory on order and sales history.

SUBASSEMBLY. An assembly which is used as a component at a higher level in another assembly. (See COMPONENT.) Syn: intermediate.

SUBSTITUTION. The use of a non-primary product or component, normally when the primary item is not available.

SUMMARIZED BILL OF MATERIAL. A form of multi-level bill of material which lists all the parts and their quantities required in a given product structure. Unlike the indented bill of material, it does not list the levels of assembly and lists a component only once showing the total quantity used.

SUPER BILL OF MATERIAL. A type of planning bill, located at the top level in the structure, which ties together various modular bills (and possibly a common parts bill) to define an entire product or product family. The "quantity per" relationship of the super bill to modules represents the forecasted percentage popularity of each module. The master scheduled quantities of the super bill explode to create requirements for the modules which also are master scheduled. (See PLANNING BILL OF MATERIAL, MODULAR BILL OF MATERIAL, COMMON PARTS BILL OF MATERIAL.)

SUPPLIER. (See VENDOR)

SUPPLIER DELIVERY PERFORMANCE. Delivery performance by a supplier can be measured in two ways: (1) On-time performance - parts delivered on-time against a demand line giving credit for partial quantity shipments. This is generally used for interval measurement, i.e. purchasing performance to the assembly line. (2) Fill rate performance - parts delivered on time and to the complete quantity requested either as a result of 0% or 100%. Fill rate is usually used for performance to an external customer.

SUPPLIER PARTNERSHIP. A strategic supplier management approach whereby the number of suppliers to a company is reduced and parts are bought from a smaller member of selected supplier partners. Long term agreements are used to develop a framework between the customer and supplier. This creates a mutually beneficial team approach to the procurement relationship, leading to reduced costs and lead times.

TIME BUCKET. A predefined period of time used for production planning. For MRP, the number of days summarized into one columnar display. A weekly time bucket would contain all of the relevant planning data for an entire week. (See BUCKETLESS SYSTEMS.)

TIME FENCE. A policy or guideline established to note when various restrictions or changes in operating procedures may take place.

TIME-PHASED ORDER POINT (TPOP). An application of MRP for independent demand items, gross requirements coming from a forecast not via explosion. This technique can be used to plan warehouse inventories as well as planning for service (repair) parts since MRP logic can readily handle items with dependent demand, independent demand or a combination of both. (See DISTRIBUTION REQUIREMENTS PLANNING.)

TIME PHASING. The staggering of production of an assembly's components such that all components are available at the correct time for sub- and final assembly.

TIME STANDARD. A preset, known amount of time allowed for performing an operation.

TOP-DOWN PLANNING. An organizational approach to MRP planning in which the individual scheduling the top level assemble also schedules all lower-level components, regardless of commodity. This approach ensures schedule continuity up and down the BOM structure.

TPOP. (See TIME-PHASED ORDER POINT.)

TRANSIENT BILL OF MATERIAL. A bill of material coding and structuring technique used primarily for transient (non-stocked) sub-assemblies. For the transient sub-assembly item, lead time is set to zero and lot-sizing is lot-for-lot. This permits MRP logic to drive requirements straight through the transient item to its components, but retains its ability to net against any occasional inventories of the sub-assembly. This technique also facilitates the use of common bills of material for engineering and manufacturing. *Syns*: phantom bill of material, blow through.

TRANSIT INVENTORY. Inventory that is being transferred from one site to another.

TRANSIT TIME. A standard time allowance for the physical movement of items from one operation to the next. (See MATERIALS HANDLING TIME.)

TRANSPORTATION INVENTORY. Inventories that exist because material must be moved. For example, if it takes two weeks to replenish a branch warehouse, transportation of two weeks of sales will normally be in transit.

TRAVELING PURCHASE REQUISITION. A purchase requisition designed for repetitive use. After a purchase order has been prepared for the goods requisitioned, the form is returned to the originator who holds it until a repurchase of the goods is required. The name is derived from the repetitive travel between the originating and purchasing departments. *Syn*: traveling requisition.

TRAVEL REQUISITION. (See TRAVELING PURCHASE REQUISITION.)

TURNS. (See INVENTORY TURNOVER.)

TWO BIN SYSTEM. A type of fixed order system in which inventory is carried in two bins. A replenishment quantity is ordered when the first bin is empty. When the material is received, the reserve bin is refilled and the excess is put into the working bin. This term is also used loosely to describe any fixed order system even when physical "bins" do not exist.

TWO-LEVEL MPS. A master production scheduling approach wherein a super bill of material is master scheduled along with selected key options, features and attachments.

UNIT OF MEASURE. The unit in which quantitative data regarding an item is expressed. For example: each, pounds, gallons, feet, etc. An issue or receipt transaction which updates the quantity on hand, but for which no order or allocation exists in the data base.

UNPLANNED ISSUE/RECEIPT. An issue or receipt transaction which updates the quantity on hand, but for which no order or allocation exists in the data base.

VENDOR. Entity external to the company who supplies material or services. *Syn*: supplier.

VENDOR LEAD TIME. The time that normally elapses between the time an order is placed with a supplier and the shipment of the material.

VENDOR SCHEDULER. An individual whose main responsibility is insuring vendors conform to the schedule. By using vendor scheduler approach, the buyer (purchasing agent) is then freed from day-to-day order placement and expediting and thus has the time to do cost reduction, negotiation, vendor selection, alternate sourcing, etc.

VERTICAL DISPLAY. A method of displaying or printing output from an MRP system where requirements, scheduled receipts, projected balance, etc. are displayed vertically. Vertical displays are often used in conjunction with bucketless systems. (See HORIZONTAL DISPLAY, BUCKETLESS SYSTEM.)

WAIT TIME. (1) The time a job waits for an available work center or materials handling device. (2) The time a machine stands idle waiting for jobs or maintenance. (See QUEUE TIME.)

WAREHOUSE DEMAND. Demand for an item to replenish a branch warehouse. *Syn*: branch warehouse demand.

WHAT IF ANALYSIS. The process of evaluating alternate strategies considering the consequences of changes to forecasts, manufacturing plans, inventory levels, etc. (See SIMULATION.)

WHERE-USED BOM. (See EXPLOSION.)

WIP. (See WORK-IN-PROCESS.)

Work Center. A specific production entity consisting of one or more people and/or machines considered as one unit for purposes of capacity requirements planning and detailed scheduling.

Work-in-Process (WIP). Product in various stages of completion throughout the plant including raw material that has been released for initial processing and completely processed material awaiting final inspection and acceptance as finished product. Many accounting systems also include semi-finished stock and components in this category. Syn: in-process inventory.

Work Order. (See MANUFACTURING ORDER.)

Yield Rate. The amount of good or acceptable material available after the completion of a process. Usually given as a percentage of the initial amount to the final usable amount.

Zero Inventory. (See JUST-IN-TIME.)

Z94.15
Organization Planning And Theory

The new editors added to the excellent list of terms that previous editors had compiled. There have been some changes which have been added to the list and in addition, some of the new terms of quality management have been added because of the effect of TQM (Total Quality Management) on organizational structure and culture.

Approximately 60 definitions have been added to the list to bring it more up-to-date and to make it more comprehensive. Hopefully the more than 50 years experience will contribute to the needs of the users of the terms.

Chairpersons

Dr. Anita L. Callahan, Ph.D.
Dr. Paul E. Givens, Ph.D.

Industrial & Management Systems Department
College of Engineering
University of South Florida

ADHOCRACY. A structure that is flexible, adaptive, and responsive; organized around unique problems to be solved by groups of relative strangers with diverse professional skills.

ADMINISTRATION. (1) Usually synonymous with the term *management*. However, it sometimes refers to the portion of management exclusive of establishing goals and policies. (2) That group of people who perform the process of administration.

ATTRIBUTION THEORY. When individuals observe behaviors, they attempt to determine whether it is internally or externally caused.

AUTHORITARIANISM. The belief that here should be status and power differences among people in organizations.

AUTHORITY. The right to exercise power and to extend jurisdiction over others for the attainment of performance. (1) The "legitimate"' right to direct or influence the performance of others under the condition of applying rewards and penalties. Authority lies in the position in an organization, not in the man. This authority derived from institutionalized power should be distinguished from other authority concepts. (2) Authority by subordinate acceptance—authority is a function of the degree to which subordinates accept decisions and direction and is derived from the group of subordinates. (3) Authority of person—authority is derived from superior ability or knowledge or charismatic qualities. (4) Authority by legal decree—authority invested by law to enforce statutes. (See DELEGATION, RESPONSIBILITY.)

AUTOCRATIC LEADER. One who tends to control in an absolute manner through the use of personal domination and application of coercive measures. (Compare to participative management.)

AUTONOMOUS WORK-GROUP DESIGN. The design of work around autonomous work teams to incorporate job rotation and/or job enrichment for a greater sense of task meaningfulness.

BEHAVIOR. Those minimum sequences of actions or movements by an individual to which meaning can be assigned. The mode of conduct that a person exhibits.

BEHAVIORAL ETHIC. The set of central cultural values that constitutes a society's expectations about human purposes and its driving force for (or against) social change.

BEHAVIORAL MODIFICATION. A technique developed from the work of B.F. Skinner that provides for a systematic coupling of desired behavior and desired rewards to achieve desired outcomes through scheduling reinforcements.

BEHAVIORAL THEORIES OF LEADERSHIP. Theories proposing that specific behaviors differentiate leaders from non-leaders.

BIOLOGICAL SCHOOL OF JOB DESIGN. Job design efforts that focus on improving the comfort and physical well-being of the employee.

BEHAVIORAL SCIENCE. The currently popular phrase for the various disciplines which study human behavior. As such, all of the traditional social sciences are included. Some argue that in the singular the phrase implies a spurious unity among these various disciplines.

BOUNDED RATIONALITY. Individuals make decisions by constructing simplified models that extract the essential features from problems without capturing all their complexity.

BRAINSTORMING. A idea-generation process that specifically encourages any and all alternatives, while withholding any criticism of those alternatives.

BOARD OF DIRECTORS. A group of people that oversees the governance and management of an organization or corporation. Among its functions are those of trusteeship for constituent groups and clients, determination of policies, overall objectives, selection and/or approval of major executives, review of performance, trusteeship of assets, and distribution of earnings. *Syn*: Board of Trustees in non-profit making organizations.

BUREAUCRACY. A large, formal organization characterized by emphasis on form, hierarchy of levels, specialization of labor, established rules and standards of conduct, records and data keeping, and professionalization of administration.

CASE METHOD. A teaching technique which presents the learner with accounts of real events and challenges him or her to interpret them. It is popular in many business schools but less common among organization trainers because of the skills needed on the part of the instructor. It is one technique among many for increasing the involvement of the learners with the subject matter and with each other.

CENTRALIZATION. (1) The process of consolidating authority and decision making within a single office or person. (2) The act of bringing together, physically or geographically operations or organizational units related by nature or function to form a central grouping.

CHAIN OF COMMAND. The prescribed line connecting the hierarchy of offices or persons through which authority and responsibility flow. (See SCALAR CHAIN.)

CHANGE AGENT. Individual (internal or external) engaged by the client organization to help the organization change.

CHANGE INTERVENTION. A planned action to make things different.

CHARISMATIC LEADERSHIP. Followers make attributions of heroic or extraordinary leadership abilities when they observe certain behaviors.

CHIEF EXECUTIVE OFFICER (C.E.O.). The executive actually responsible for all activities of the firm, his title (in addition to being C.E.O.) can be President and/or Chairman.

CHIEF OPERATING OFFICER (C.O.O.). The person in charge of all operations (day-to-day activities) of a firm. The position reports to the C.E.O. and the holder of the position usually has the additional title of President or Executive Vice President or Senior Vice President.

CLASSICAL ORGANIZATION THEORY. A collection of theories of organization and management that were formulated during the first third of the twentieth century and that were derived from personal experience, observation, and descriptive analysis of organizations. Included within the broad range of classical theory are scientific management pioneered by Frederick W. Taylor, bureaucratic theory developed by Max Weber, and administrative management theory, developed by Henri Fayol. Central features of classical theory are prescriptions for managerial actions and emphasis upon formal organization structure.

COMMITTEE. A group of people or a form of organization established to achieve one or more of the following: a) interchange ideas and information, b) obtain facts and ideas and synthesize into a report or recommendation, c) obtain meeting of minds or a consensus.

COMMUNICATION. The transfer of information and understanding from one point or person to another person. The basic elements in the process of communication are an information source, encoding, transmission, reception, and decoding.

COMMUNICATION CHANNEL. A pathway or route to transmit information and understanding.

COMMUNICATION NETWORK. The system of senders, receivers, and channels for the transmission of information and understanding.

COMPETITIVE BENCHMARKING. Comparing and rating an organization's practices, processes, and products against the world's best, best-in-class, or the competition. Comparisons are not confined to the same industry.

COMPUTER MODELING. A complex computer program that simulates the work environment.

CONFLICT. Interpersonal and social forces, ideas, behaviors, and motivations which are acting in opposition to one another, often involving strife or contest.

CONSENSUS. A collective opinion, or general agreement, that a decision has been accepted by the group. This does not necessarily imply unanimity.

CONSULTATIVE MANAGEMENT. That form of supervision which utilizes full interchange and sharing of ideas between superiors and subordinates, often taking place in group meetings. In consultative supervision the superior retains the power of final decision. This is distinguished from democratic supervision where decision is by group consensus and from autocratic supervision where decision is by superior domination.

CONTROL. Measurement of performance or actions and comparison with established standards in order to maintain performance and actions within permissible limits of variance from the standard. May involve taking corrective action to bring performance into line with the plan or standards.

CONTINGENCY APPROACH. Recognition that few universal principles can be applied in all situations and that the best thing to do depends upon the specific variables operative in each unique situation.

CONTINUOUS PROCESS IMPROVEMENT. Principle used by W. Edwards Deming to look at improvement of product and service. Searching unceasingly for ever-higher levels of quality by isolating sources of defects called KAIZEN in Japan where the goal is zero defects (Crosby)

CRITERION. The standard or rule by which a judgment of the effectiveness of a course of action, or of performance, can be made. (See CONTROL.)

CSI. Continuous Systems Improvement - The concept that to improve the performance of an organization, the systems of the organization should continually be improved thereby keeping the organization viable and wholesome for future production.

CULTURE. The pattern of shared beliefs and values that give members of an organization rules of behavior or accepted norms for conducting operational business.

CYBERNETICS. (1) The field of control and communication theory in general, without specific restriction to any area of application or investigation. (2) The behavior and design of mechanisms, organisms, and/or organizations that receive and generate information and respond to it in order to attain a desired result.

DECENTRALIZATION, AUTHORITY. Placing the authority and decision-making power as close as possible to the organizational level at which the work is done.

DECENTRALIZATION, PHYSICAL. The geographical dispersing of facilities and activities. The management control can either be centralized or decentralized.

DECISION MAKING. The response to a need or stimulus by means of acquiring and organizing information, processing this information to define the major problem and to yield alternative courses of action, and selecting one course of action from among the alternatives. It requires the preparation of a detailed implementation plan and the identification of any "new problems" which may arise.

DECISION RULE. A predetermined criterion which may be used by a decision-maker to select from alternative courses of action.

DECODING. To operate on a signal in order to extract the message and present it in usable form at its destination.

DELEGATION. To grant or confer responsibility and commensurate authority from one executive or organizational unit to another in order to accomplish a particular assignment; often used in the sense of a superior in the organization delegating to subordinates.

DELPHI TECHNIQUE. A group decision method in which individual members, acting separately, pool their judgement in a systematic and independent fashion.

DEPARTMENT. An organizational unit established to operate in, and be responsible for, a specified activity or a physical or functional area.

DEPARTMENTALIZATION (DEPARTMENTATION). The division of an organization into formal groups according to criteria such as function, process, equipment, products, territory, or customer.

DISJUNCTIVE SOCIALIZATION. A socialization process whereby a new recruit does not have a guide on which to model organizational behavior.

DIVISION. One of the groupings into which activities are separated; organizational structure which includes more than two departments within its jurisdiction. Also, a corporate unit which is generally responsible for producing, marketing, and servicing a closely related line of products.

DIVISION OF WORK. The separation of labor tasks into less complex subtasks. This may be to use simpler skills or to make use of special skills. (See SPECIALIZATION.)

DUAL-CAREER COUPLES. A situation where both the husband and the wife have distinct careers outside the home.

DYSFUNCTIONAL CONFLICT. Conflict situations, usually between groups in the same organization, that hinder the achievement of group and organizational goals.

EFFECTIVENESS. Achievement of goal.

EFFICIENCY. The ratio of effective output to the input required to achieve it.

EMPLOYEE. A person who provides services for an organization and receives compensation and benefits for such services in accordance with general rules established for members of the organization, and who is considered a member of the organization for the period of time during which he serves in the sense that he acknowledges the right of the employer to direct and control his services. Excluded is a self-employed individual, *i.e.*, an individual who carries on a business endeavor as a proprietor or partner, or who renders services as an independent contractor.

ENACTED ENVIRONMENT. The unique environment created by an organization through its own process of selectively responding more to some external elements than to others.

ENCODING. To operate on a message from the source of transmission and convert it to signals which the communication channel will accept and which can be decoded.

ENCOUNTER GROUP. A small group of persons meeting with the intention of (1) seeking deeper and more personal relations among members and (2) learning more about themselves and others from the attempt to develop these relations. (See SENSITIVITY TRAINING; T-GROUP.)

ENGINEERING SCHOOL OF JOB DESIGN. A mechanistic approach that focuses on efficiency.

ENTRY PROCESS. Jargon phrase for the highly complex set of enabling conditions by which a *consultant* begins to exert influence. It is regarded as a highly important set of actions separate from the main work the consultant intends, although they are naturally closely related.

EQUITY. Third of four principles of productivity gainsharing (PGS); the concept of a reasonable balance in the distribution of the "gain'" (financial benefits) between the parties (normally the participants and the company) through the design of the productivity measurement system and the related payment system.

EQUITY THEORY. Individuals compare their job inputs and outcomes with those of others and then respond so as to eliminate any inequities.

ERGONOMICS SCHOOL OF JOB DESIGN. Seeks to increase system reliability by developing equipment and jobs that are safe, simple, reliable, and that minimize mental requirements on the worker.

EXECUTIVE. (1) An employee—a) whose primary duty consists of the management of the enterprise in which he is employed or of a recognized department thereof; b) who customarily directs the work of two or more employees therein; c) who has the authority to hire or fire other employees, or whose suggestions and recommendations as to hiring, firing, advancement promotion, demotion or other change in status of other employees will be given particular weight; d) who customarily and regularly exercises discretionary power; e) who does not devote more than 20 percent of his time to activities other than those described in a) through d); f) whose salary and other compensation reflects the intellectual and discretionary content of the position. (Adapted from Explanatory Bulletin, Regulations Part 541, 1956, defining the terms of Section 13a of the Fair Labor Standards Act.) (2) A manager, usually of a top or middle management level.

EXECUTIVE COMMITTEE. A committee consisting of top officers who are on the board of directors and outside board members (usually where the Board only meets quarterly) that assist and act to aid the C.E.O. in making very major policy, planning, and operating decisions between board meetings. In other firms the Executive Committee, if it exists, is composed of a small group of executives reporting to the C.E.O. and assisting the C.E.O. in decision-making.

EXPECTANCY THEORY. The strength of a tendency to act in a certain way depends on the strength of an expectation that the act will be followed by a given outcome and on the attractiveness of that outcome to the individual.

EXPERIENTIAL. A term for a kind of learning process in which the content of what is to be learned is experienced as directly as possible, in contrast to being read about in a book or talked about in lecture and discussion. The term applies to a wide variety of training techniques. It is often used in the phrase, "experiential level," in contrast to *cognitive level*.

FEEDBACK. Information (data) extracted from a process or situation and used in controlling (directly) or in planning or modifying immediate or future inputs (actions or decisions) into the process or situation. Feedback can be either positive or negative, although the field of cybernetics is based upon negative feedback. Negative feedback indicates that the system is deviating from a prescribed course of action and should readjust to a new steady state.

FORECASTING. The process of predicting or projecting the future for the purpose of reducing uncertainties. Specific forecasting techniques include Delphi, trend extrapolation, dynamic modeling, scenario writing, mapping, and the use of relevance or perspective trees, to name some of the more important ones. Among forecasting dimensions often considered are those pertaining to economic, legal, social, political, ecological, and technological conditions.

FOREMAN. An individual at the lowest rank of the managerial hierarchy. He bears general responsibility for the performance of his organizational unit, guides and directs the work of non-supervisory employees, provides face-to-face leadership, and handles such personnel responsibilities as performance appraisal, discipline, instruction, and grievance processing. (See SUPERVISION; MANAGEMENT, LOWER.)

FORMAL (LEADER, ORGANIZATION, SYSTEM). A term introduced originally in the Hawthorne studies to designate the set of organizational relationships that were explicitly established in policy and procedure (*i.e.*, the "formal organization"). Now the term "formal" has been prefixed onto many types of organizational phenomena, but the reference to what is established in policy and procedure remains. "Formal leader," the designated leader of a group, whether he or she has the most influence in it or not, is one of the most common phrases derived from this term. (See INFORMAL GROUP.)

FUNCTION. Usually an organizational sub-goal. May be defined in terms of end to be achieved or behavior required to achieve ends. Example: administrative function, protection function, manufacturing function.

FUNCTIONAL AUTHORITY. The right to exercise control as assigned to a particular organizational unit or person. Conventionally used to designate specific staff authority exercised over line organizational units.

FUNCTIONAL CONFLICT. Conflict situations, usually between groups in different organizations, that serve to stimulate innovations and production.

FUNCTIONAL ORGANIZATION. A form of organization structure in which activities are grouped according to the purpose, service, or utility they possess.

FUNDAMENTAL ATTRIBUTION ERROR. The tendency to underestimate the influence of external factors and overestimate the influence of internal factors when making judgements about the behavior of others.

GENERAL STAFF. A form of organizational unit, most common in the military and in government, composed of a staff group usually assigned to a high level office and principally concerned with developing policies and plans.

GOAL CONFLICT. Conflict that exists when the achievement of one group's goals is perceived as preventing or detracting from another group's goal attainment.

GROUP. Two or more people who share common values differentiating them from other people and regularly interact with each other in striving for a common goal.

GROUP COHESIVENESS. The tendency of a group of people to maintain themselves as a group; the attractiveness which the group holds for its members; the unity and solidarity of the group. It is the total field of forces acting on members to remain in the group.

GROUP DYNAMICS. The behavior and relationships of people derived from their interactions within a group.

GROUP EXECUTIVE. Term usually used in large companies where the Group Executive heads a group of companies and/or plants or operating locations (organizations). This arrangement is frequently used in large manufacturing and/or service organizations. Each component is headed by an executive who has a title such as President (of a wholly or partially owned subsidiary corporation) or Manager. The Group Executive also usually has an additional title such as Senior Vice President, Corporate Vice President, or Division president of the parent company.

GROUP LEADER. (See SUPERVISOR, WORKING.)

GROUP MAINTENANCE. Those behaviors exhibited by members of a group which are *functional* for holding the group together, increasing members' liking for each other, and differentiating the group from its environment. When one member is very prominent in this process, he or she is said to be playing a group-maintenance role.

GROUPTHINK. A situation occurring in highly cohesive groups whereby the desire to agree is so great that it tends to dominate concern for realistic appraisal of alternative courses of action.

HAWTHORNE EFFECT. The novelty or interest in a changed situation which leads, at least initially, to positive results. In the original Hawthorne Studies, this effect was revealed in terms of increased performance resulting from the workers' knowledge that they were being observed with interest by behavioral science researchers, that they were treated as being important, and that their inputs were being taken into consideration. More recent analyses of this 1920's data suggest the interactions in this study of lighting levels and employee productivity are much more complicated.

HIERARCHY. A ranking or ordering of those in authority.

HOMEOSTATIC SYSTEM. Any cybernetic system so arranged as to maintain one particular state or to maintain equilibrium among its component parts.

HUMAN RESOURCE ACCOUNTING. A relatively recent development in management which refers to expanding the accounting procedures of organizations in order to take into consideration the *human* assets along with capital assets. This represents a shift from thinking of organizational members (employees) as costs to regarding them as assets.

HUMAN RESOURCE PLANNING. The process that helps to provide adequate human resources to achieve future organizational objectives. It includes forecasting future needs for employees of various types, accompanying these needs with the present work force, and determining the numbers of types of employees to be recruited or phased out of the organization's employments group.

HYGIENE FACTORS. Those factors - such as company policy and administration, supervision, and salary - that, when present in a job, placate workers. When these factors are present, people will not be dissatisfied.

IDENTIFICATION. The achievement of a general attitude within a company whereby employees recognize the common goals they have with the company.

INCENTIVE. A stimulus which induces action. Any inducement, material or nonmaterial, which impels or encourages a person to behave in a certain way.

INCENTIVE PLAN. A designed program for rewarding individuals or groups for achieving and exceeding performance goals.

INFORMATION. (1) The content of any meaningful communication. (2) In information theory, information refers to a quantitative measure of the amount of order and certainty which exists in a dynamic information system, without reference to the semantic content of the symbols involved.

INFORMAL GROUP. A group that emerges through the efforts of individuals to satisfy personal needs not provided for by the formal organization.

INFORMATION THEORY. A study of the quantity of information contained in a message, and also the capacity of a communications channel to transmit information.

INPUT. That which is put in; in the organizational context usually taken to mean any resource allocation to a system.

INPUT-OUTPUT ANALYSIS. A matrix which provides a quantitative framework for the description of an economic unit. Basic input-output analysis is a unique set of input-output ratios for each production and distribution process. If the ratios of inputs per unit of output are known for all production processes, and if the total production of each end product of the economy, or of that section being studied is known, it is possible to compute precisely the production levels required at every intermediate stage to supply the total sum of end products. Further, it is possible to determine the effect at every point in the production process of a specified change in the volume and mix of end products. Also forms the basis for most measures of organizational productivity; *i.e.,* the ratio of a measure of output to a specific input (or to a group of inputs).

INSTITUTIONAL/TRANSFORMATIVE LEADERSHIP. A social influence approach to guiding and directing organizations in which the leader personifies of transforms values and visions and inspires others without having to communicate directly.

INTEGRATION. How the differentiated parts of a system are coordinated to ensure their contribution to the organization as a whole.

INTERACTION. Virtually any behavior resulting from interpersonal relationships. In human relations it includes all forms of communication, verbal and nonverbal, conscious and unconscious. Interactions speak louder than words.

INTERDISCIPLINARY NATURE. Organizational behavior borrows its core concepts from the three major behavioral-science disciplines and applies principles from the social sciences as well.

INTERFACE. Jargon from the aerospace industry and systems engineering, used both as a noun and a verb to describe one or more *interactions*. It is commonly used in describing work-related interactions among groups.

INTERVENTION. A planned process of introducing change within a group or organizational, initiated or facilitated by a change agent with the help of its members.

JOB. (1) The combination of tasks, duties and responsibilities assigned to an individual employee and usually considered his or her normal or regular assignment. (2) The contents of a work order.

JOB ANALYSIS. Determination of the characteristics of a job through detailed observation and evaluation of the activities, facilities required, conditions of work, and the qualifications needed in a worker.

JOB CLASSIFICATION. The grouping of jobs on the basis of the nature of the functions performed or level of pay, or on the basis of job evaluation, historic groupings, collective bargaining, or arbitrary determination.

JOB CONTENT. The nature of work itself experienced within a job, including the degree to which the tasks are challenging, interesting, and stimulating of personal esteem and growth.

JOB CONTEXT. Working conditions that surround a job, such as the physical environment and facilities, pay and benefits, and relationships with coworkers and managers.

JOB DEPTH. The degree of responsibility and autonomy (for work scheduling, decision prerogatives, and performance control) that is granted to an individual position.

JOB DESCRIPTION. A summary of the essential activities involved in the performance of a job that is abstracted from a job analysis and used in the classifying of jobs and the selection of employees to fill them.

JOB DESIGN. The specification of job content, work methods, and relationships with other jobs in order to satisfy technological and organizational requirements plus the human concerns of the job holders. Techniques used in job design include industrial engineering techniques, human factor analysis, rotation of tasks within a work unit, job enlargement, job enrichment, and creation of self-managed work teams.

JOB ENLARGEMENT. A phrase invented by Charles Walker of Yale University to describe the process by which jobs are redesigned to make them more complex and, presumably, more interesting. The aim is to make the task more of a "natural whole" to the man performing it. Job enlargement has been applied most commonly to so-called blue collar jobs.

JOB ENRICHMENT. The process (or outcome) of designing a job so that it has both considerable responsibility and variety.

JOB EVALUATION. The systematic appraisal of the value of an individual job in an organization in relation to other jobs. Jobs are commonly graded according to differences in skill, effort, physical demands, and working conditions. It is an aid in setting the money worth of jobs for pay purposes.

JOB SCOPE. The degree of variety of number of different tasks a worker is permitted or expected to perform.

JOB SPECIFICATION. A form of job description listing the mental and physical qualifications and special skills required in an individual worker for a given job in order to facilitate the selection and placement of employees.

JOB STANDARDIZATION. The procedure of specifying a standard practice or a standard method for a job.

LAB. A shorthand term for any of a wide variety of programs which derive from the so-called "laboratory method of training," an approach that is primarily *experiential*. The term "lab" has become a suffix for many specific types of training designs; *e.g.,* "conflict lab," "personal growth lab," etc.

LEADERSHIP. Interpersonal influence, exercised in a situation and directed, through the communication process, toward the attainment of a specified goal or goals.

LINE. (1) In the organizational sense line activities are those that contribute most directly to the accomplishment of the organization's primary objective. (2) An authority and responsibility relationship between a superior and subordinate extending vertically in an uninterrupted train from the top to the bottom position in an organization.

MANAGEMENT. (1) The process of utilizing material and human resources to accomplish designated objectives. It involves the activities of planning, organizing, directing, coordinating, and controlling. (2) That group of people who perform the functions described above.

MANAGEMENT AUDIT. The process of evaluating how effectively management has operated the organization. Typical of criteria often employed in this audit are production efficiency, earnings, utilization of assets, management and executive talent, fairness to stockholders, and ROI.

MANAGEMENT BY COMMITTEE. The committee is sometimes used when the subjects or problems are broad or involve many functions. A committee is then formed to manage the responsibility area, usually the members are the managers of the involved functions. Typical examples where this management form is used are patents and new products.

MANAGEMENT BY EXCEPTION. A style of management in which the manager establishes a systematic pattern of acceptable operations thus freeing himself from routine occurrences in order to devote his talents to the more difficult problems which are exceptions to the routine.

MANAGEMENT BY OBJECTIVES. A management strategy developed by Odiorne which makes the establishment and communication of organization objectives the central function of a manager. It is based on the assumption that supervision and leadership will work best under conditions in which both superiors and subordinates have prior "contracts" (*i.e.,* agreements about directions, priorities, and objectives); called MBO.

MANAGEMENT (MANAGER) DEVELOPMENT. The application of planned efforts to assist in supplying, maintaining, and improving managers at, or intended for, the middle and top organizational levels in order that they can more efficiently attain the objectives of the enterprise.

MANAGEMENT FOR QUALITY. The translation of customer focus and quality values into implementation plans for all levels of management and supervision.

MANAGEMENT INFORMATION SYSTEM. An organization's structured system of information inputs covering both the internal and external environments and an associated assimilation, storage and analysis system that provides at the time and place required the information outputs (in an easily understandable manner) necessary to manage the organization efficiently and effectively—and only those outputs unless the system is interrogated and requested to deliver additional data and/or analyses. Usually, in the modern sense, a computer is integral with the system; but this is not a requirement.

MANAGEMENT, LOWER. The first level of supervision which is directly in charge of a group of employees. In very large organizations, having many levels of management, lower management could include the first two levels above the nonsupervisory employee level. The foreman, general foreman, and supervisory positions comprise lower management.

MANAGEMENT, MIDDLE. That broad group of managers and administrators which is located below the top policy-making management level and above the level of supervision. It includes both line and staff personnel.

MANAGEMENT, TOP. The ultimate level of authority consisting of those directors and principal administrative officers of the company who are responsible for the determination of broad policies, procedures, objectives, and goals. It implements these policies and procedures through its continuous function of defining organization, authority, responsibilities, staffing and by coordinating, integrating, measuring and controlling the organization. (See CHIEF EXECUTIVE OFFICER.)

MANAGER. One who plans work and organizes and directs people toward the accomplishment of organizational objectives. (2) One who is responsible for the planning and employment of resources toward the achievement of organizational objectives even though he may not supervise people.

MANAGERIAL GRID. A method of analyzing leadership and managerial styles that was formulated by R.R. Blake and J.S. Mouton. These styles of leadership are classified according to the manager's "concern for people" as one broad dimension and "concern for production" as the other broad dimension. The styles are plotted upon a grid with each dimension scaled from 1 to 9. Blake and Mouton identified five principal styles: 1,1—impoverished; 9,1—task; 1,9—country club; 5,5—middle-of-the-road; 9,9—team management.

MANAGERIAL STYLE. The way in which a manager relates to others in his work place. This includes such characteristics as consideration for people, concern for production, objectivity, work pace, and degree of involvement of subordinates in decision-making.

MATRIX MANAGEMENT. A management system whereby an employee (usually a professional or semi-professional) has two bosses, but each boss is responsible for a different aspect of the work being performed. For example, project managers can be responsible for outputs and budgets but their manpower resides in functional components whose managers retain some responsibility for the functionally oriented activities performed by their personnel.

MEANS-ENDS ANALYSIS. A method of organization planning and program planning that involves a) starting with the general goal to be achieved, b) discovering a set of means for accomplishing this goal, and c) taking each of these means, in turn, as a new subgoal and discovering a set of more detailed means for achieving it and so on to the point where a particular means can be carried out by existing programs of action.

MODEL. A simplification of some phenomenon for purposes of study and understanding. The concrete embodiment of a theory. Behaving in an idealized way so that others might learn or change their behavior through identifying with and adopting those behaviors displayed.

MORALE. The total satisfactions derived from the job, the work group, one's supervisor, the organization, and the general environment. It pertains to the general feeling of well-being, satisfaction, and happiness of people.

MOTIVATION. A willingness to expend energy to achieve a goal or a reward.

MOTIVATION-HYGIENE THEORY. Intrinsic factors are related to job satisfaction, while extrinsic factors are associated with dissatisfaction.

MOTIVE. That which is within the individual, rather than without, that incites him to action.

NEED HIERARCHY. A particular theory about the operation of needs in the organism, introduced by Abraham Maslow. The major assertion is that classes of needs are arranged in a hierarchy, with the most basic biological needs at the bottom and the more variable psychological needs near the top. The theory says that "higher" needs cannot be activated until "lower" needs are relatively satisfied. The theory has led to a variety of shorthand phrases for describing "where a person is" in the need hierarchy at a given time; *e.g.,* "esteem level," "social level," "security level," etc. This particular theory also was the basis for McGregor's *Theory X–Theory Y* formulation.

NETWORK. An increasingly common term for talking about a large set of individuals who are in regular touch by virtue of some common set of interests. They need not belong to the same organization. To say that a person is "in the network" is to say not only that he or she is in communication, but also that his or her values and behavior are somehow *organic* in relation to the others. It is a modern variant of the older phrases, to be a "member of the club" or to be "one of the boys (girls)."

NOMINAL GROUP TECHNIQUE. A group decision method in which individual members meet face to face to pool their judgements in a systematic but independent fashion.

NORMS. Acceptable standards of behavior within a group that are shared by the group's members.

OBJECTIVE. A desired end result, condition, or goal which forms a basis for managerial decision-making. (See MEANS-ENDS ANALYSIS.)

OFFICE OF PRESIDENT (OF CHAIRMAN). An arrangement in which the office of the chief executive consists of more than one executive. This organizational device is sometimes used in very large corporations and typically consists of three to five senior executives who perform collaboratively to make top level decisions and plans. One of these is the chief executive who is accountable to the board of directors.

ORGANIZATION. (1) The classification or groupings of the activities of an enterprise for the purpose of administering them. Division of work to be done into defined tasks along with the assignment of these tasks to individuals or groups of individuals qualified for their efficient accomplishment. (2) Determining the necessary activities and positions within an enterprise, department, or group, arranging them into the best functional relationships, clearly defining the authority, responsibilities and duties of each, and assigning them to individuals so that the available effort can be effectively and systematically applied and coordinated.

ORGANIZATIONAL BEHAVIOR (OB). A field of knowledge and inquiry concerned with the systematic study of organizations; their origins, growth, and effect upon individual members, constituent groups, and other organizations. Organizational behavior is equally concerned with individual behavior, group processes, attitudes and motivations, communication, and the effect of these upon organizational performance.

ORGANIZATIONAL CLIMATE. The multidimensional set of properties of the work environment, perceived directly or indirectly by employees, and assumed to be a major force in influencing their attitudes, motivation, and behavior. The perceived quality and configuration of organizational climate is usually assessed along dimensions such as structure, responsibility, team spirit, standards, warmth-support, rewards, organizational clarity, etc.

ORGANIZATIONAL CULTURE. A pattern of beliefs, values, and assumptions learned and transmitted by organizational members as an effective way of life for adapting to external forces an coping with internal problems.

ORGANIZATIONAL DESIGN. The structural pattern of an organizational which includes the rationale for grouping specialized tasks, locating decision centers, facilitating coordination, and other provisions.

ORGANIZATION CHART. A diagram or graphic representation of an organization which shows, to varying degrees, functions, responsibilities, people, authority and relationships among these. May be used for recording the promotability of individual managers and for orderly planning for succession.

ORGANIZATION CULTURE. The rules, jargon, prejudices, customs, and other traditions that clarify acceptable and unacceptable behavior in an organization.

ORGANIZATION DEVELOPMENT (OD). The planned, organization-wide process of change designed to improve organization effectiveness and adaptation (to changing environmental demands). This is accomplished through planned interventions by an internal or external agent (change agent) using theory and techniques of applied behavioral science. Although the planned interventions or strategies depend on the circumstances and diagnosis, the focus has usually been on the attitudes, norms, values, relations, and organizational climate, rather than on the goals, structure, and technology of the organization. Lately, however, more and more attention has been given to the interplay of structure, technology, and people variables. Techniques employed by OD change agents include sensitivity training, managerial grid applications, goal-setting sessions, team building, confrontation meetings, and interpersonal peacemaking, to name some of the more important ones. Note that OD is much more comprehensive than management development, which is focused only on a particular manager or group of managers in order to change individual managerial behavior. (See CHANGE AGENT.)

ORGANIZATION, FORMAL. A planned and established structure of relationships among people, designed to achieve specified objectives. Thus one may speak of the formal organization of a company, government agency, school or hospital.

ORGANIZATION FOR QUALITY. Structuring organizational activities to effectively serve the accomplishment of the company's customer, quality, innovation, and cycle time objectives.

ORGANIZATION, INFORMAL. The type of organization structure which comprises the authority, responsibility, and communicative and associative relationships among functions, physical factors, and personnel that are supplemental to the "formal" organization structure and may be "for," "against," or "neutral" with regard to the achievement of organizational objectives. It develops spontaneously.

ORGANIZATION PLANNING. The process of planning and designing the organization structure and providing for appropriate interrelationships among people and constituent units. Often involves analysis of goals, grouping of activities, making audits, planning organizational changes, and preparation of policy manuals. May involve manpower planning, training and development of management personnel.

ORGANIZATION STRUCTURE. A framework within which the dynamics of human relations forces take form and come into interaction. A plan under which the totality is subdivided, and job duties, personal relations and lines of authority are specified.

ORGANIZATION THEORY. That branch of study concerned with identification and analysis of the strategic variables and relationships of structure and process involving groups of people engaged in coordinated, purposeful activities over meaningful spans of time.

ORGANIZING. The process of determining the work to be done, grouping work into appropriate units, and defining the desired relationships among people so that the entire body can pursue identified goals.

PARTICIPATION. The process by which people contribute ideas toward the solution of problems affecting the organization and their jobs. The actions by which managers involve their subordinates in the decision making process. It is a second principle for productivity gainsharing and generally provides a formal structure for such "involvement."

PARTICIPATIVE MANAGEMENT. A philosophy and system of management in which employees at one or many hierarchical levels of an organization share in setting goals, making decisions, and solving problems. The degree of employee influence in the decision-making process may range from simply advice-giving to full authority for decisions. Various organizational arrangements may be used such as advisory committees, labor-management committees, workers' councils, self-managed work teams, and employee representatives on governing boards.

PERSONAL/HUMAN RESOURCE MANAGEMENT (P/HRM). A function performed in organizations which facilitates the most effective use of people (employees) to achieve organizations and individual goals. Terms used interchangeably with P/HRM include personnel, human resource management, and employee development.

PERSONNEL MANAGEMENT. The function of acquiring, developing, motivating, and maintaining a competent work force so that the objectives of the organization are properly achieved and so that the members of this work force obtain satisfaction from their participation in their organization.

PLAN. A predetermined course of action over a specified period of time which represents a projected response to an anticipated environment in order to accomplish a specific set of adaptive objectives.

PLANNING. Planning is the process whereby an individual or an organization identifies opportunities, needs, strategies, objectives and policies that are used to guide and manage the organization through future periods. All planning consists of a) accumulation of information, b) sorting and relating bits of information and beliefs, c) establishing premises, d) forecasting future conditions, e) establishing needs, f) identifying opportunities, g) establishing objectives and policies, h) structuring alternative courses of action, i) ranking or selecting total systems of action which will achieve the best balance of ultimate (future) and immediate objectives, j) establishing criteria and means for measuring adherence to the selected program of action and k) so managing the organization to achieve the objectives. Some kinds of planning include short and long range planning, product planning, financial planning, etc.

POLICY. The verbal, written, or implied general plans of action that guide the members of the organization in the conduct of its operation, and incorporating in them broad premise and limitations within which further planning activities take place.

POLITICS. The network of human behaviors and interactions by which social power is acquired, transferred, and exercised.

POWER. The ability to influence the behavior of other persons by any or all of the following means: a) coercion—the application or threat of use of physical sanctions, or use of force to control satisfactions of basic human wants such as food, shelter, comfort and the like, b) control of material rewards such as pay and benefits, c) use of symbolic rewards and penalties via leadership, communication, allocation of prestige symbols, ritual, persuasion, and suggestion. (Adapted from Amitai Etzioni, *A Comparative Analysis of Complex Organizations*, the Free Press of Glencoe, Inc., 1961, p. 4-5).

PRODUCTIVITY. The quantitative and qualitative result of the input of all resources. The most widely used productivity measure is one-dimensional (one measure of input and one measure of output) that defines it as *productivity*—is output per labor input (*e.g.,* number of trees planted per employee hour, etc.). A broader and more modern view involves the relationship of an overall measure of output to the sum of two or more input factors; *i.e.,* Labor, Materials, Capital, Energy, etc.

PRODUCTIVITY GAINSHARING. A motivational process based on developing a method of measuring organizational productivity changes from an historical base period, converting the "gain" to monetary terms, and "sharing" it between the employees and the company. It is based on four principles: Identification, Participation, Equity, and Competent Management.

PROFESSION. A profession is a type of occupation whose work, values, and members ideally conform to the following criteria: a) It requires advanced, specialized formal education and training as distinguished from general academic education or an apprenticeship. b) Professional work requires the consistent exercise of discretion, judgment, and personal responsibility. c) It is based upon a deep and organized body of knowledge. Efforts are continually made to expand the knowledge through research. d) The profession, if advanced and regulated by a national level association of its members which helps establish standards for entry into the profession, establishes standards of ethical practice by members of the profession, and applies sanctions where these standards are violated. e) The members of the profession ideally maintain a social consciousness and sense of trusteeship toward their clients, employers, and the general public. They seek to maintain and update their knowledge and skills in keeping with advances in their field of work.

PROJECT MANAGEMENT. An organizational form that is generally superimposed upon a traditional functional organizational structure. Integrated teams of specialists work under the coordination and direction of a project manager to accomplish a project of limited duration. The project manager coordinates and manages across functional and organizational lines to complete a specific project or program. (See MATRIX MANAGEMENT.)

PROJECT TEAM. (See TASK FORCE.)

PSYCHOLOGICAL SCHOOL OF JOB DESIGN. Redesigning jobs to increase employee satisfaction and motivation.

QUALITY (CONTROL) CIRCLE. A small group of employees drawn from the same work area who meet regularly and voluntarily to identify, solve, and, on occasion, implement solutions to work related problems.

QUALITY FUNCTION DEPLOYMENT (QFD). A system that pays special attention to customer wants and integrates these into the marketing, design, manufacturing, and service processes. Activities that do not contribute to customer wants are considered wasteful.

QUALITY OF WORK LIFE (QWL). (1) A multifaceted concept whereby the work environment is meaningful to employees. Components include autonomy, recognition, belonging, process and development, and external rewards. (2) The extent to which working in an organization enables members to feel that they are masters of their work environment and that they and their work are important

REFERENCE GROUP. A group with which an individual identifies whether or not he or she is an actual member.

RESPONSIBILITY. Responsibility originates when one accepts the assignment to perform assigned duties and activities. The acceptance creates a liability for which the assignee is held answerable for and to their assignor. It constitutes an obligation or accountability for performance.

ROLE. The concept of role can be used in three distinct ways: a) the demands of a given social position as determined by the pressures of the social group on the individual, b) an individual's own conception of the part someone in his position is supposed to play in an organization, c) the actions of individuals in terms of their relevance or implications for a social structure and its norms. It refers to the ways members of a position act with or without conscious design, in accord with or in violation of organizational norms.

SATISFICING. Human behavior and decision-making based upon discovery and selection of satisfactory alternatives rather than optimal alternatives; *i.e.,* optimizing.

SCALAR CHAIN. The chain of direct authority relationships from superior to subordinate and the grading of duties according to degrees of authority and responsibility. Equivalent to chain of command.

SCIENTIFIC MANAGEMENT. A body of literature developed in the early 1900s concerned with incentives, selection, training, and the design of jobs to eliminate time and motion waste.

SEMANTICS. The science dealing with the relationship between signs and symbols (including words), their meaning, and human behavior.

SENSITIVITY TRAINING. The collection of methods for improving the individual's sensitivity to him or herself and others. Although a large number of variations exist, the common ingredients seem to be: (1) the guidance of a trained person or persons; (2) intense *interpersonal* experience by the trainee; (3) a relatively protected environment, free from ordinary pressures and distractions. The *T-group* is the classical, but not the only, means of achieving these three conditions.

SKILL TRAINING. Training that is concerned more with improving effectiveness than learning concepts. It is often used in a mildly negative way to explain that some particular type of training program is "not merely" skill training.

SOCIAL PSYCHOLOGY. A behavioral - science hybrid that integrates the disciplines of psychology and sociology to study how and why individuals behave as they do in groups.

SOCIAL RESPONSIBILITY. The obligation an organization has as a member of society, especially in the areas of ecology, consumerism, and equal employment opportunity.

SOCIAL SYSTEM. The network of task relationships within an organization — of people working with other people to exchange information, make decision and promote purposeful behavior.

Sociometry. The technique of analysis and decision-making for action in small groups whereby the individuals are identified, paired, or clustered according to their expressed preferences for others on such criteria as liking, disliking, leadership, efficiency, and frequency of communication.

Socio-technical System. This concept refers to the same concrete phenomena as *social system* but by the inclusion of the term "technical" emphasizes the physical realities of the system and the *technologies* it employs to do work.

Span of Control. The number of subordinates that a given person supervises.

Specialization. Also referred to as division of work. This concerns dividing or grouping all the work to be done into small homogeneous packages which may be performed by individuals, groups, or departments.

Staff. (1) An organizational unit that serves in an auxiliary and facilitative role in relation to line and operating units or executives. Typical functions of staff are data gathering, preparing plans, giving advice and recommendations, and rendering specialized service. Types of staff are personal or assistant to, general, specialized, and service. (2) Those persons who occupy a staff role in organizations.

Staff, Specialized. A type of staff that provides specialized or technical information, counsel, and service to operating units of the organization. Generally serves the entire organization and often deals in specialized staff work usually possess specialized or technical training and skills.

Staff, Service. A type of staff that provides essential services to the organization and its members. These services are primarily of a physical, maintenance, protection, or welfare nature.

Stakeholders. Those individuals and groups who have a direct interest in the performance of an organization, such as customers, suppliers, shareholders, and employees.

Status. Relative ranking of a person or group in terms of duties, rights, privileges, or prestige. Relative social rank.

Status Congruency. The condition whereby individual group members are ranked high or low in status on all relevant factors.

Strategic Quality Planning. Development of strategic and operational plans that incorporate quality as product or service differentiation and the load bearing structure of the planning process.

Supervision. The function of leading, coordinating, and directing the work of others to accomplish designated objectives. Those people who perform this function; usually applied to those occupying positions of leadership in the lower strata of an organization such as foremen, general foremen, and office supervisors.

Supervisor, Acting (or manager). A temporary managerial position that may or may not become permanent.

Supervisor, Working. A non-supervisory employee who devotes a portion of his time to assigning work to other employees, job instruction, clearing troubles in the work, and checking progress of the work. *Syn*: group leader.

Synergy. Originally a term for the combined and cooperative operations of the bodily organs. Now it is a jargon for any process in which more is accomplished by cooperation than could be done by separate efforts.

System. A system is characterized by: a) a set of components of subsystems linked by information channels, b) engagement in coordinated, goal-directed activity, c) information flow as the basis for control, d) a set of subgoals associated with the individual subsystems or components, e) an external environment which influences the system. A system is said to be an open system if it reacts to its environment and is a closed system if it does not. It is an adaptive system if it reacts to environmental changes in a way that is favorable toward achieving the system goals.

Systems 1, 2, 3, 4. These are four patterns or models of management and organizational climate developed by Rensis Likert. System 1 is designated as exploitive-authoritative, system 2 is benevolent-authoritative (paternalistic), system 3 is consultative, and system 4 is participative. The climate of a given organization is measured by a questionnaire covering such elements as leadership, communication, interaction-influence, and control processes.

Systems Analysis. A method of problem-solving that encompasses the identification, study, and evaluation of interdependent parts and their attributes, functioning as an ongoing process, and constituting an organic whole. For organization theory this involves the analysis of the functioning and integration of systems of people and physical entities.

Systems Theory. A framework for viewing an organization as a system of interrelated parts that when combined are greater than the parts individually.

TASK. A duty, an elementary component of a job or an operation to be performed. Every job normally consists of a number of tasks or basic work requirements.

TASK FORCE. A temporary unit consisting of resources drawn from different parts of the organization and assigned to execute a specific mission. *Syn*: project team.

TEAM BUILDING. A set of activities (interventions) designed to improve a group's effectiveness as a team by enabling the members to examine their working relationships while working on group tasks.

THEORY X. A set of assumptions about the nature of people in organizations that was first formulated by Douglas McGregor. Those holding Theory X views believe that man is basically lazy, lacks ambition, and is indifferent to the needs of the organization. Therefore, the task of management is to employ pressure, coercion, and tight control in order to achieve organizational objectives.

THEORY Y. A set of assumptions about the nature of people in organizations that was first formulated by Douglas McGregor. Those holding Theory Y views believe that people, under proper conditions, seek responsibility and like to work. They have the capacity to exercise imagination and initiative in solving organizational problems. Therefore, management should emphasize self-direction and self-control and the integration of individual and organizational goals.

THEORY Z. A statement concerning the use of Japanese style consensus management initially identified by William Ouchi.

TOTAL QUALITY MANAGEMENT. The application of quality principles for the integration of all functions and processes of the organization. The ultimate goal is customer satisfaction.

TRAINING. A learning experience that seeks a relatively permanent change in an individual that will improve his or her ability to perform on the job.

TSI (TOTAL SYSTEMS IMPROVEMENT). The concept that quality is one of the subsystems of the overall management system and therefore, looking only at Total Quality Management assumes all other systems are part of TQM.

TYPE A BEHAVIOR. Aggressive involvement in a chronic, incessant struggle to achieve more and more in less and less time and if necessary, against the opposing efforts of other things or other people.

TYPE B BEHAVIOR. Rarely harried by the desire to obtain a wildly increasing number of things or participate in an endlessly growing series of events in an ever-decreasing amount of time.

UNITY OF COMMAND. The concept that each subordinate in an organization should report to and be subject to the authority of only one supervisor. Often expressed as a principle of organization.

VALUE ADDED. This is one measure of general output. The basic calculation of Value Added is Net Sales (or Value of Production) less Outside Purchases (and Expenses). It is used in the labor productivity measure which is the basis for the Rucker Plan and, in Great Britain, the Bentley PAR (Performance, Achievement, Recognition) plan. (See PRODUCTIVITY GAINSHARING.)

Z94.16
Quality Assurance and Reliability

Most definitions were taken from existing Terminology standards with appropriate references.

Russell G. Heikes
Professor
Georgia Institute of Technology

Harrison Wadsworth, Jr.
Professor Emeritus
Georgia Institute of Technology

ACCELERATED TEST. Test in which the applied stress level is chosen to exceed that stated in the reference conditions in order to shorten the time required to observe the stress response of the item, or to magnify the response in a given duration. To be valid, an accelerated test must not alter the basic modes and/or mechanisms of failure. [4: 191-14-0]

ACCEPTABLE PROCESS LEVEL (APL). Process level which forms the outer boundary of the zone of process levels which are acceptable. [2: 3.4.10]

ACCEPTABLE QUALITY LEVEL (AQL). When a continuing series of lots is considered, a quality level which for the purposes of sampling inspection is the limit of a satisfactory process average.
Note: The value of the AQL selected is usually dependent upon physical and economic constraints such as the natural process limits (which determine the tolerances that can be set for various technical characteristics) and the costs of inspection balanced against the costs of failure in service. [2: 2.7.1]

ACCEPTANCE CONTROL CHART. A graphical method for the dual purposes of evaluating a process in terms of: (a) whether or not it can be expected to satisfy product or service requirements for the characteristic(s) being measured, and (b) whether or not it is in a "state of statistical control" with respect to within-sample or sub-group variability.
Notes: (1) For variables data, this will require one chart for averages and another for ranges or standard deviations. (2) The emphasis of the acceptance control chart is that the process usually does not need to remain in control about some single standard process level, but as long as the within-sub-group variability remains in control it can run at any level within a zone of process levels acceptable in terms of the process requirements. Some assignable causes will create shifts in the process level which are small when compared with requirements and it is uneconomical to control them too tightly. The issue of narrowing the zone around the target usually involves a different set of problems and actions from problems of process instability within sub-groups. [2: 3.3.16]

ACCEPTANCE CRITERIA. Specification criteria for acceptance of individual product or service characteristics.
Note: Sometimes, as in acceptance sampling, the term "acceptance criteria" is used for a set of several characteristics rather than an individual characteristic. [2: 1.1.10]

ACCEPTANCE INSPECTION. Inspection to determine whether an item or lot delivered or offered for delivery is acceptable. [2: 1.2.3]

ACCEPTANCE NUMBER, (AC). In sampling inspection by attributes, the largest number of nonconformities or nonconforming items found in the sample that permits the acceptance of the lot, as given in the sampling plan. [2: 2.3.10]

ACCEPTANCE SAMPLING. Sampling inspection in which decisions are made to accept or not to accept a lot (or other grouping of product, material, or service) based on the results of a sample or samples selected from that lot. (See SAMPLING INSPECTION.)
Notes: (1) The alternative to acceptance is often termed "rejection" for the purpose of the definition. However, in practice, the alternative may take some form other than outright rejection. (2) In lot-by-lot sampling, acceptance and rejection relate to individual lots. In continuous sampling, acceptance and rejection relate to individual units or to blocks of consecutive units, depending on the stated procedure. [2: 2.3.1]

ACCURACY. The closeness of agreement between a test result and the accepted reference value.
Note: The term accuracy, when applied to a set of test results, involves a combination of random components and a common systematic error or bias component. [1: 3.11]

ACTION LIMITS. ACTION CONTROL LIMITS (UPPER AND/ OR LOWER): In a control chart, the limits above which (upper limit) or below which (lower limit) or the limits outside which the statistic under consideration lies when action should be taken. [2: 3.4.3]

ARITHMETIC MEAN; AVERAGE. The sum of values divided by the number of values.
Notes: (1) The term "mean" is used generally when referring to a population parameter and the term "average" when referring to the result of a calculation on the data obtained in a sample. (2) The average of a simple random sample taken from a population is an unbiased estimator of the mean of this population. However, other estimators such as the geometric or harmonic mean, or the median or mode are sometimes used. [1: 2.26]

ARITHMETIC WEIGHTED MEAN. Weighted Average: The sum of the products of each value and its weight, divided by the sum of the weights where weights are non-negative coefficients assigned to each of the values. [1: 2.27]

ASSIGNABLE CAUSE. A factor (usually systematic) that can be detected and identified as contributing to a change in a quality characteristic or process level.
Notes: (1) Assignable causes are sometimes referred to as special causes of variation. (2) Many small causes of change are assignable, but it may be uneconomic to consider or control them. In that case they should be treated as chance causes. [2: 3.1.8]

AVAILABILITY (ACHIEVED). The probability that an item when used and maintained under stated conditions in an ideal support environment will be in a satisfactory state at any given time.

It may be expressed as:
$$A_a(t) = \frac{MTBM}{MTBM + \overline{M}}$$
where

MTBM = mean time between maintenance – is a function of both preventive and corrective maintenance requirements

$\overline{M}$ = mean active maintenance downtime – a value which is a function of the repair times associated with corrective and preventive maintenance.

AVAILABILITY (INTRINSIC). The probability that an item is in satisfactory state at a stated instant of time (t) when it is used and maintained under stated conditions. It may be expressed as:
$$A_i(t) = \frac{MTBF}{MTBF + MTTR}$$

MTBF = mean time between failures

MTTR = mean time to repair (excludes preventive maintenance downtime, supply down-time, and administrative down time).

AVAILABILITY (OPERATION). The probability that an item is in a satisfactory state at a stated instant of time(t) when it is used and maintained under stated conditions in actual support environment. It may be expressed as:
$$A_o(t) = \frac{MTBM}{MTBM + MDT}$$
where,

MTBM = mean time between maintenance

MDT = mean downtime (includes preventive maintenance, supply and administrative downtime).

AVAILABILITY (PERFORMANCE). The ability of an item to be in a state to perform a required function under given conditions at a stated instant of time or over a given time interval. [4: 191-02-05]

AVERAGE AMOUNT OF INSPECTION. In a given sampling scheme, the number of items expected to be inspected per lot in order to reach a decision for a certain average batch quality level.
Note: This is an average over switching rules, etc., unlike average sample number. It does not include inspecting all items in non-accepted lots as average total inspected requires.
(See AVERAGE SAMPLE NUMBER, AVERAGE TOTAL INSPECTED.) [2: 2.7.7]

AVERAGE CHART; $\overline{X}$ OR $\overline{Y}$ CHART. A control chart for evaluation the sub-group differences in terms of the sub-group averages. [2: 3.3.4]

AVERAGE OUTGOING QUALITY (AOQ). The expected average quality level of outgoing product for a given value of incoming product quality.
Note: In practical cases different definitions of AOQ may be used depending on whether or not nonconforming items removed in the 100% inspection of non-accepted lots are replaced by conforming items. [2: 2.7.4]

AVERAGE OUTGOING QUALITY LIMIT (AOQL). Maximum AOQ over all possible values of incoming product quality level for a given acceptance sampling plan and rectification of all non-accepted lots. [2: 2.7.5]

AVERAGE RUN LENGTH (ARL). (1) sample sense: The average number of times that a process will have been sampled and evaluated before a shift in process level is signalled. (2) item sense: The average number of items that will have been produced before a shift in level is signalled.
Note: A long ARL is desirable for a process located at its specified level (so as to minimize calling for unneeded investigation or corrective action); a short ARL is desirable for a process shifted to some undersirable level (so that corrective action will be called for promptly). ARL curves are used to describe the relative quickness in detecting level shifts of various control chart systems. [2: 3.3.15]

AVERAGE RANGE; MEAN RANGE. The arithmetic mean of the ranges of a set of samples of the same size. [1: 2.31]

Average Sample Number (ASN). Average sample size: Average number of sample units inspected per lot in reaching decisions to accept or not to accept when using a given sampling plan.
Note: ASN is dependent on the actual quality level of the submitted lots.
(See AVERAGE AMOUNT OF INSPECTION.) [2: 2.7.6]

Average Total Inspected (ATI). Average number of items inspected per lot including inspection of all items in non-accepted lots.
Note: Applicable when the procedure calls for 100% inspection of non-accepted lots.
[2: 2.7.8]

Between-Lot (or Between-Batch) Variation. Dispersion of the lot or batch averages of the observations or test results over several lots or batches.
NOTE: The between-batch variation will include a component of within-batch variation which can be reduced by increasing the within-batch sample size.
[2: 3.1.4]

Bias. The difference between the expectation of the test results and an accepted reference value.
Note: Bias is the total systematic error as contrasted to random error. There may be one or more systematic error components contributing to the bias. A larger systematic difference from the accepted reference value is reflected by a larger bias value. [1: 3.13]

Bias Of Estimator. The difference between the expected value of the estimator and the parameter that is to be estimated. [2: 2.54]

Binomial Distribution. The probability distribution of a discrete random variable X that can take any integer value from 0 to n, such that

$$P_r[X = x] = \binom{n}{x} p^x (1-p)^{n-x}$$

with

$x=0,1,2, ...n$; and

parameters $n = 1,2, ...$ and $0<p<1$
where

$$\binom{n}{x} = \frac{n!}{x!(n-x)!}$$

[1: 1.49]

Blemish. An imperfection that occurs with a severity sufficient to cause awareness but that should not cause any real impairment with respect to intended normal, or reasonably foreseeable, usage requirements.

Bulk Sampling. Sampling of materials in lots within which sampling units are not initially readily distinguishable. *Examples*: The sampling of a large bulk of coal for ash content or calorific value, or tobacco for moisture content. [2: 2.1.9]

Capability. The ability of an item to meet a service demand of given quantitative characteristics under given interval conditions. [4: 191-02-04]

Chain Sampling Inspection. Sampling inspection in which the criteria for acceptance of the current lot are governed by the sampling results of that lot and those of a specified number of the preceding consecutive lots. [2: 2.4.8]

Characteristic. A property which helps to identify or differentiate between items of a given population. [2: 1.51]

Chance Causes. Factors, generally many in number but each of relatively small importance, contributing to variation, which have not necessarily been identified.
Note: Chance causes are sometimes referred to as common causes of variation.[2: 3.1.9]

Chi-Squared Distribution; χ^2 Distribution. The probability distribution of a continuous random variable that can take any value from 0 to $+\infty$, the probability density function of which is

$$f(\chi^2;v) = \frac{(\chi^2)^{(v/2)-1}}{2^{v/2}\Gamma(v/2)} \exp\left(-\frac{\chi^2}{2}\right)$$

where

$\chi^2 > 0$ with parameter $v = 1, 2,...$;

Γ is the gamma function.

Notes: (1) The sum of the squares of n independent standardized normal variables is a χ^2 random variable with paramerer n; n is then called degrees of freedom. (2) The probability distribution of the random variable $x^2/2$ is a gamma distribution with parameter $m = v/2$.
[1: 1.39]

Coefficient Of Variation. For a non-negative characteristic the ratio of the standard deviation to the average.
Notes: (1) The ratio may be expressed as a percent-

age. (2) The term "relative standard deviation" is sometimes used as an alternative to "coefficient of variation", but this use is not recommended. [1: 2.35]

Confidence Coefficient; Confidence Level. The value $(1-\alpha)$ of the probability associated with a confidence interval or a statistical coverage interval.
Note: $(1 - \alpha)$ is often expressed as a percentage.
[1: 2.59]

Confidence Interval. An interval calculated from sample data and distribution parameters with a specified probability or confidence. To say that [a,b] is a $1 - \alpha$ confidence interval for the population parameter θ means that the a priori probability that the random interval [A,B] will contain θ is $1 - \alpha$. Another interpretation would be that $(1 - \alpha)(100)$ percent of such intervals calculated by different random samples of the same size would contain θ in the long run.

Confidence Limit. Each of the limits, T_1 and T_2, of the two-sided confidence interval or the limit T of the one-sided confidence interval. [1: 2.60]

Conformity. Fulfilment of specified requirements.
Note: The above definition is valid for the purposes of quality standards. The term "conformity" is defined differently in ISO/IEC Guide 2. [3: 2.9]

Consumer's Risk (CR). For a given sampling plan, the probability of acceptance of a lot or process when the quality level (e.g. fraction nonconforming) has a value stated by the plan as unsatisfactory [e.g. a limiting quality level (LQL)]. [2: 2.6.4]

Consumer's Risk Quality (CRQ). A lot or process quality level which in the sampling plan corresponds to a specified consumer's risk.
Notes: (1) The type of operating characteristic curve should be specified. (2) A special case of this is "limiting quality level" when the operating characteristic curve is of type B.
[2: 2.6.6.]

Consumer's Risk Point (CRP). A point on the operating characteristic curve corresponding to a predetermined and usually low probability of acceptance.
Notes: (1) This probability of acceptance is called the "consumer's risk" and the corresponding lot quality determined by the CRP for the risk is called the 'consumer's risk quality (CRQ)'. (2) The type of operating characteristic curve should be specified.
[2: 2.6.5]

Control Chart. A chart, with upper and/or lower control limits, on which values of some statistical measure for a series of samples or sub-groups are plotted, usually in time or sample number order. The chart frequently shows a central line to assist detection of a trend of plotted values toward either control limit.
Note: In some control charts, the control limits are based on the within-sample or within-sub-group data plotted on the chart; in others, the control limits are based on adopted standard or specified values applicable to the statistical measures plotted on the chart. [2: 3.3.1]

Control Chart Factor. A factor, usually varying with sample size, to convert specified statistics or parameters into a central line value or control limit appropriate to the control chart. [2: 3.4.6]

Control Limits: Shewhart control limits (upper and/or lower). In a control chart, the limit below which (upper limit) or above which (lower limit) or the limits between which the statistic under consideration lies with a very high probability when the process is under control.
[2: 3.4.1]

Continuous Sampling Inspection. Sampling inspection intended for application to a continuous flow of individual items of product that (a)involves acceptance or non-acceptance on an item-by-item basis, and (b)uses alternate periods of 100% inspection and sampling depending on the quality of the observed product. [2: 2.4.5]

Corrective Action. Action taken to eliminate the causes of an existing nonconformity, defect or other undesirable situation in order to prevent recurrence
Notes: (1) The corrective actions may involve changes, such as in procedures and systems, to achieve quality improvement at any stage of the quality loop. (2)There is a distinction between "correction" and "corrective action": "correction" refers to repair, rework or adjustment and relates to the disposition of an existing nonconformity; "corrective action" relates to the elimination of the causes of a non-conformity. [3: 4.14]

Correlation. The relationship between two or several random variables within a distribution of two or more random variables.
Note: Most statistical measures of correlation measure only the degree of linear relationship. [1: 1.13]

Correlation Coefficient. The ratio of the covariance of two characteristics to the product of their standard deviations.

$$r_{xy} = \frac{S_{xy}}{S_x S_y} = \frac{\sum_i (x_i - \bar{x})(y_i - \bar{y})}{\sqrt{\sum_i (x_i - \bar{x})^2 \sum_i (y_i - \bar{y})^2}}$$

where S_{xy} is the covariance of X and Y
S_x and S_y are the standard deviations of X and Y respectively.

Notes: (1) This coefficient is frequently used as a numerical expression for the linear interdependence between X and Y in the series of paired observations. Whenever possible, the scatter diagram should also be examined to verify linearity. (2) The value of rxy will always lie between -1 and +1. When the correlation coefficient equals one of these limits, this means that there exists an exact linear relationship between X and Y in the series of paired observations. (3) This correlation coefficient is for measured characteristics; other correlation coefficients such as Spearman's or Kendall's coefficients are used for ranked data. [1: 2.41]

COUNT CHART; C CHART. A control chart for evaluating the process in terms of the count of nonconformities of a given classification occurring in the sample. [2: 3.3.5]

COUNT PER UNIT CHART (U-CHART). A control chart for evaluating the process in terms of the average count of nonconformities of a given classification per unit occurring within a sample. [2: 3.3.6]

COVARIANCE. The sum of the products of the deviations of x and y from their respective averages divided by one less than the number of observed pairs:

$$s_{xy} = \frac{1}{n-1} \sum_i (x_i - \bar{x})(y_i - \bar{y})$$

where n is the number of observed pairs.
Note: The sample covariance is an unbiased estimator of the population covariance.
[1: 2.40]

CURTAILED INSPECTION. A sampling procedure which contains a provision for stopping inspection when it becomes apparent that adequate data have been collected for a decision. [2: 2.5.7]

CUMULATIVE SUM CHART; CUSUM CHART. A control chart on which the plotted value is the cumulative sum of deviations of successive sample statistics from a target value. When a process change is made, the sum is returned to zero. The ordinate of each plotted point represents the algebraic sum of the previous ordinate and the most recent deviation from the target.
Note: Cusum charts are generally interpreted by masks superimposed on the chart, a signal occurring if the path of the cusum intersects or touches the boundary of the mask. [3.3.12]

CRITICAL VALUE. The limiting value of the critical region. [1: 2.72]

CRITICAL REGION. The set of values of the test statistic for which the null hypothesis is rejected.
Note: Critical regions are determined in such a way that if the null hypothesis is true, the probability of this null hypothesis being rejected should be not more than a given value a which is in general low (for example 5% or 1%). [1: 2.71]

DEFECT. Nonfulfilment of an intended usage requirement or reasonable expectation, including one concerned with safety. [3: 2.11]

DEFECTIVE ITEM; DEFECTIVE UNIT. An item (unit) with one or more defects.
Note: In some cases, a series of imperfections or nonconformities may accumulate to make an item (unit) defective. [2: 1.5.9]

DEGREE OF FREEDOM. In general, the number of terms in a sum minus the number of constraints on the terms of the sum. [1: 2.85]

DEMERIT CHART; QUALITY SCORE CHART. A control chart for evaluating the process in terms of a dermerit (or quality score), e.g. a weighted sum of counts of variously classified nonconformities. [2: 3.3.11]

DEPENDABILITY. The collective term used to describe the availability performance and its influencing factors: reliability performance, maintainability performance and maintenance support performance.
Note: Dependability is used only for general descriptions in non-quantitative terms. [4: 191-02-03]

DERATING. The reduction of the stress ratio on an item. [5: 9.1.11]

DISTRIBUTION FUNCTION. A function giving, for every value x, the probability that the random variable X be less than or equal to x:
$F(x) = P_r [X < x]$. [1: 1.4]

DOUBLE SAMPLING INSPECTION. Sampling inspection in which the inspection of the first sample of size n_1 leads to a decision to accept a lot, not to accept it, or to inspect a second sample of size n_2 before making a decision of acceptance or non-acceptance of the lot. The decisions are made according to defined rules. [2: 2.4.2]

ERROR OF THE FIRST KIND. The error committed in rejecting the null hypothesis (because the statistic takes a value within the critical region), when the null hypothesis is true.
Note: This is often referred to as a type 1 error.
[1: 2.75]

ERROR OF THE SECOND KIND. The error of not rejecting (accepting) the null hypothesis (because the value of the statistic falls outside the critical region), when the null hypothesis is not true.
Note: This is often referred to as a type II error.
[1: 2.77]

ESTIMATION. The operation of assigning, from the observations in a sample, numerical values to the parameters of a distribution chosen as the statistical model of the population from which this sample is taken.
Note: A result of this operation may be expressed as a single value or as an interval estimate.
[1: 2.49]

ESTIMATOR. A statistic used to estimate a population parameter. [1: 2.50]

***F*-DISTRIBUTION.** The probability distribution of a continuous random variable, which can take any value from 0 to $+\infty$, the probability density function of which is

$$f(F; v_1, v_2) =$$

$$\frac{\Gamma[(v_1+v_2)/2]}{\Gamma(v_1/2)\Gamma(v_2/2)}(v_1)^{v_1/2}(v_2)^{v_2/2}\frac{F^{(v_1/2)-1}}{(v_1 F + v_2)^{(v_1+v_2)/2}}$$

where

$F > 0$ with parameters $v1, v2 = 1, 2, ...$;

Γ is the gamma function.
Note: This is the distribution of the quotient of two independent x^2 distributed random variables, each one divided by its number of degrees of freedom. The number of degrees of freedom of the χ^2 random variables of the numerator v_1 and of the denominator v_2 are, in this order, the numbers of degrees of freedom of the *F*-distributed random variable. [1: 1.41]

FAILURE. The termination of the ability of any item to perform its required function. [4:191-04-01]

FAILURE ANALYSIS. The logical, systematic examination of a failed item to identify and analyze the failure mechanism, the failure cause, and the consequences of failure. [4: 191-16-12]

FAILURE, COMPLETE. Failure which results in the complete inability of an item to perform all required functions. [4: 191-04-20]

FAILURE CRITERIA. Rules for failure relevancy such as specified limits for the acceptability of an item.

FAILURE, DEGRADATION. Failures which is both a gradual failure and a partial failure. [4: 191-04-22]

FAILURE, DEPENDENT. One which is caused by the failure of an associated item(s). Not independent. [7]

FAILURE, GRADUAL. A failure due to a gradual change with time of given characteristics of an item.
Note: A gradual failure may be anticipated by prior examination or monitoring and can sometimes be avoided by preventive maintenance. [4: 191-04-11]

FAILURE, INDEPENDENT. One which occurs without being related to the failure of associated items. Not dependent. [7]

FAILURE, INHERENT WEAKNESS. Failures attributable to weakness inherent in the item itself when subjected to stresses within the stated capabilities of that item.

FAILURE MECHANISM. The physical, chemical or other process which results in a failure. [4: 191-04-18]

FAILURE, MISUSE. A failure due to the application of stresses during use which exceed the stated capabilities of the item. [4: 191-04-04]

FAILURE MODE. The effect by which a failure is observed; for example, an open or short circuit condition, or a gain change.

FAILURE, PARTIAL. A failure which results in the inability of an item to perform some, but not all required functions. [4: 191-04-21]

FAILURE, RANDOM. Any failure whose occurrence is unpredictable in an absolute sense but which is predictable only in a probabilistic or statistical sense [7]

Failure Rate. The number of failures of an item per unit measure of life (cycles, time, miles, events, etc., as applicable for the item). [7]

Failure Rate Acceleration Factor. The ratio of the failure rate under accelerated testing conditions to the failure rate under stated reference test conditions. [4:191-14-11]

Failure Rate, Assessed. The failure rate of an item determined as a limiting value or values of the confidence interval with a stated confidence level, based on the same data as the observed failure rate of nominally identical items.

Failure Rate, Extrapolated. Extension by a defined extrapolation or interpolation of the observed or assessed failure rate for durations and/or conditions different from those applying to the conditions of that observed or assessed failure rate.

Failure Rate, Observed. The ratio of the total number of failures in a sample to the total cumulation observed time on that sample. The observed failure rate is to be associated with particular, and stated time intervals (or summation of intervals) in the life of the items, and with stated conditions.

Failure Rate, Predicted. For the stated conditions of use and the design considerations of an item, the failure rate computed from the observed, assessed or extrapolated failure rates of its parts.

Failure, Secondary. Failure of an item caused either directly or indirectly by the failure of another item.

Failure, Sudden. Failures that could not be anticipated by prior examination. [4: 191-04-10]

Failure, Wear-out. A failure whose probability of occurrence increases with the passage of time, as a result of processes inherent in the item. [4: 191-04-09]

Goodness Of Fit Of A Distribution. A measure of the agreement between an observed distribution and either a theoretical distribution specified a priori or one fitted to the observations. [1: 2.63]

Grade. An indicator of category or rank related to features or characteristics that cover different sets of needs for products or services intended for the same functional use. [2: 1.1.3]

Histogram. A graphical representation of the frequency distribution of a quantitative characteristic, consisting of a set a contiguous rectangles, each with a base equal to the class width and an area proportional to the class frequency.[1: 2.17]

Hypergeometric Distribution. A discrete probability distribution with probability function

$$P_r[X = X] = \frac{\binom{M}{x}\binom{N-M}{n-x}}{\binom{N}{n}}$$

where

x = max. $(0, M - N + n)$, max. $(0, M - N + n) + 1, ...,$ min. (M, n)

parameters $N = 1, 2, ...$

$M = 0, 1, 2, ..., N$

$n = 1, 2, ..., N$

and

$$\binom{M}{x} = \frac{M!}{x!(M - x)!}$$

etc.

Note: This distribution arises as the probability distribution of the number of successes in a sample of size n, taken without replacement from a population of size N containing M successes. [1: 1.52].

Independence. Two random variable X and Y are independent if, and only if, their distribution functions are related by

$F(x, y) = F(x, \infty) \cdot F(\infty, y) = G(x) \cdot H(y)$

Where $F(x,\infty) = G(x)$ and $F(\infty, y) = H(y)$ are the marginal distribution functions of X and Y, respectively for all pairs (x,Y).

Notes: (1) For continuous independent random variables, their probability density functions if they exist are related by

$f(x, y) = g(x) \cdot h(y)$

where $g(x)$ and $h(y)$ are the marginal density functions of X and Y, respectively, for all pairs (x,y).

For discrete independent random variables, their probabilities are related by

$P_r(X=x_i, Y=y_j) = P_r(X=x_i) \cdot P_r(Y=y_j)$

for each pair (x_i, y_j). (2) Two events are independent if the probability that both occur is equal to the product of the probabilities of the two events.[1: 1.11]

INDEPENDENT TRIALS. The successive trials of an event are said to be independent if the probability of outcome of any trial is independent of the outcome of the others. The expression is usually confined to cases where the probability is the same for all trials. [6]

INDIFFERENCE QUALITY LEVEL (IQL). The quality level which in the sampling plan corresponds to 0.5 probability of acceptance when a continuing series of lots is considered. [2: 2.6.14].

INDIRECT INSPECTION. An acceptance inspection where a lot is accepted or rejected after examining and verifying the inspection system of the supplier and examining the results it provides; examination of samples from the submitted lot is thus omitted. [2: 1.2.8].

INDIVIDUAL OBSERVATIONS CHART; ORIGINAL DATA CHART. A control chart for evaluating the process level in terms of the individual observations in the sample. [2: 3.3.9].

INHERENT PROCESS VARIABILITY. Variability that is inherent in a process or its products when operating in a state of statistical control.
Notes: (1) Inherent process variability when determined from an individual process (one machine or line, one group of operators, and one batch of material) is usually smaller than when it is determined from an overal process (many machines or lines, groups of operators, and batches of material). The disparity between these two conditions may be due to a number of causes which could be assigned, but are not feasible to control tightly in a routine operation. (2) When the standard deviation is used it is sometimes denoted by σ_i.[2: 3.2.1].

IN-PROCESS INSPECTION. Inspection which is performed during the manufacturing or repair cycle in an effort to prevent nonconformities from occuring and to inspect the characteristics and attributes as early as expedient.

INSPECTION. Activities such as measuring, examining, testing and gauging one or more characteristics of a product or service, and comparing with specified requirements to determine conformity.[2: 1.2.1].

INSPECTION (100%). Inspection of every item of product or service, i.e. the whole (as contrasted with any form of sampling inspection).[2: 1.2.5].

INSPECTION LEVEL. An index of the relative amount of inspection of a sampling scheme, chosen in advance and relating the size of samples to the lot size.
Notes: (1) A lower (higher) level can be selected if past experience shows that a less (more) discriminating operating characteristic curve will be appropriate. (2) This term should not be confused with severity of sampling, which concerns switching rules that operate automatically.[2: 2.5.1].

INSPECTION LOT. A definite quantity of some product, material or service, collected together and submitted for examination.
Note: An inspection lot may consist of several batches or parts of batches. [2: 1.3.5].

INSTANTANEOUS FAILURE RATE (HAZARD). The limit, if it exists, of the ratio of the conditional probability that the instant of time, T, of a failure of an item falls within a given time interval, $(t + \Delta t)$ and the length of this interval, Δt, when Δt tends to zero, given that the item is in an up state at the beginning of the time interval. [4: 191-02-02]

LIMITING QUALITY (LQ). When a lot is considered in isolation, a quality level which for the purposes of sampling inspection is limited to a low probability of acceptance.
Notes: (1) For a particular sampling system, the probability of acceptance will lie within a defined range. (2) Limiting quality is an unsatisfactory term but it is widely used. Isolated lot limiting quality level is a more satisfactory name for the concept. [2: 2.7.3].

LIMITING QUALITY LEVEL (LQL). When a continuing series of lots is considered, a quality level which for the purposes of sampling inspection is the limit of an unsatisfactory process average. [2: 2.7.2].

LOT-BY-LOT INSPECTION. Inspection of product submitted in a series of lots. [2: 1.2.4].

MAINTAINABILITY. The probability that a given active maintenance action, for an item under given conditions of use can be carried out within a stated time interval, when the maintenance is performed under stated conditions and using stated procedures and resources. [4: 191-13-01]

MEAN. The expected value of a random variable. [1: 1.18].

MEAN DEVIATION. The arithmetic mean of the deviations from an origin when all deviations are given a posi-

tive sign.
Note: Generally, the chosen origin is the arithmetic mean, although the mean deviation is minimized by taking the median as the origin. [1: 2.32].

MEAN LIFE. The arithmetic mean of the times-to-failure of the units of a given item.

MEAN-MAINTENANCE-TIME. The total preventive and corrective maintenance time divided by the total number of preventive and corrective maintenance actions during a specified period of time. [7]

MEAN SERVICE RATE. The expectation of the number of services completed in one time unit, given that service is going throughout the entire time unit.

MEAN TIME BETWEEN FAILURES. The expectation of the time between failures. [4: 191-12-08]

MEAN TIME TO FAILURE. The expectation of the time to failure. [4: 191-12-07]

MEAN TIME TO REPAIR (MTTR). The expectation of the time to restoration. [4: 191-13-08]

MEDIAN. If n values are arranged in non-decreasing order of magnitude and numbered 1 to n, the median of these n values is the $[(n + 1)/2]^{th}$ and value, if n is odd. If n is even, the median lies between the $(n/2)^{th}$ and the $[(n/2) + 1]^{th}$ values and is not defined uniquely. Unless otherwise specified, it may be taken to be the arithmetic mean of these two values.[1: 2.28].

METHOD OF ATTRIBUTES. Noting the presence (or absence) of some characteristic or attribute in each of the items in the group under consideration, and counting how many items do (or do not) possess the attribute, or how many such events occur in the item, group, or area.
Note: One of the most common attribute measures for acceptance sampling is the percentage of nonconforming items. [2: 1.52]

MID-RANGE. The arithmetic mean of the largest and the smallest observed value of a quantitative characteristic. [1: 2.29].

MODE. The value(s) of a random variable such that the probability mass (discrete random variable) or the probability density (continuous random variable) has a local maximum for this value (or these values). Note: If there is one mode, the probability distribution of the random variable is said to be "unimodal"; if there is more than one mode the probability distribution is said to be "multimodal" (bimodal if there are two modes.) [1: 1.17].

MOVING AVERAGE. An unweighted average of the latest n observations where the current observations has replaced the oldest of the previous n observations. [2: 3.3.18].

MOVING AVERAGE CONTROL CHART. A control chart for evaluating process level in terms of an arithmetic average of the latest n observations in which the current observation has replaced the oldest of the latest $n + 1$ observations. [2: 3.3.18].

MOVING RANGE CONTROL CHART. A control chart for evaluating the variability within a process in terms of the range of the latest n observations in which the current observation has replaced the oldest of the latest $n + 1$ observations.[2: 3.3.20].

MULTIPLE SAMPLING INSPECTION. Sampling inspection in which, after each sample has been inspected, a decision is made, based upon defined rules, to accept the lot, not to accept it, or to take another sample. The decision rules are based on the cumulative evidence of all the samples from that lot.
Note: For most multiple sampling plans, the largest number of samples that can be taken is specified with an "accept" or "not accept" decision being forced at that point. [2: 2.4.3]

MULTIVARIATE CONTROL CHART. A control chart for evaluating a process in terms of the levels of two or more characteristics. [2: 3.3.23].

NATURAL PROCESS LIMITS. Limits for a characteristic which include a stated fraction of the individuals in a population.
Notes: (1) If limits are set at $\pm 3\sigma$ about the process average for a normal (Gaussian) distribution, they will include 99.7% of the units produced when the process is in a state of statistical control. Other limits for nonmal distributions may be set using tables of the distribution function. for other, non-normal distributions, limits which will include a specified percentage of the units produced should be set by other methods. (2) In many cases several machines making the same product feed into one process. The natural process limits should then include both the natural process limits for a single machine and some measure of the difference between machine averages. In these circumstances it may not be possible to estimate reliably the percentage of units produced which will fall

within the limits. (3) Natural process limits will not ordinarily be the dimensional limits shown on an engineering drawing; they are mostly used to compare the natural capability of the process to tolerance limits. [2: 3.2.4].

NOMINAL VALUE. Value of a characteristic designated in a given design specification or drawing.
Notes: (1) This may be the target value or dimension from which variations are permitted within a specified tolerance zone. (2) In English the term "rated value" is sometimes used with this meaning and "nominal value" is used with the meaning 'a suitable approximation'. [2: 1.4.2].

NONCONFORMING ITEM; NONCONFORMING UNIT. An item (unit) with one or more nonconformities. [2: 1.5.7].

NONCONFORMITY. A departure of a quality characteristic from its intended level or state that occurs with a severity sufficient to cause an associated product not to meet a specification requirement. [2: 1.5.6].

NORMAL DISTRIBUTION; LAPLACE-GAUSS DISTRIBUTION. The probability distribution of a continuous random variable X, the probability density function of which is

$$f(x) = \frac{1}{\sigma\sqrt{2\pi}} \exp\left[-\frac{1}{2}\left(\frac{x-\mu}{\sigma}\right)^2\right]$$

for $-\infty < x < +\infty$

Note: μ is the expectation and s is the standard deviation of the normal distribution. [1: 1.37].

NULL HYPOTHESIS AND ALTERNATIVE HYPOTHESIS. Statements about one or more parameters, or about a distribution, which are to be tested by means of a statistical test.
Note: The null hypothesis (H_0) relates to the statement being tested, whereas the alternative hypothesis (H_1) relates to the statement to be accepted when the null hypothesis is rejected.
Examples:
a) Test of the hypothesis that the expectation μ of a random variable X in a population is not less than a given value, μ_0:
H_0 ($\mu > \mu_0$) and H_1 ($\mu < \mu_0$).

b) Test of the hypothesis that the proportions of nonconforming items in two lots, p_1 and p_2, have the same value (unspecified):
H_0 ($p_1 = p_2$) and H_1 ($p_1 \neq p_2$)

c) Test of the hypothesis that a random variable X is distributed normally (with unspecified parameters). Alternative hypothesis: the distribution is not normal.[1: 2.66].

NUMBER OF NONCONFORMING ITEMS CHART. A control chart for evaluating the process in terms of the total number of items (areas of opportunity) in a sample in which a nonconformity of a given classification occurs. [2: 3.3.10].

OBSERVED VALUE. The value of a characteristic obtained as the result of a single observation. [1: 2.6].

ONE-SIDED CONFIDENCE INTERVAL. When **T** is a function of the observed values such that, θ being a population parameter to be estimated, the probability P_r(T $>\theta$) [or the probability P_r(T $<\theta$) is at least equal to (1 - α) is a fixed number, positive and less than 1], the interval from the smallest possible value of θ up to T (or the interval from T up to the largest possible value of θ) is a one-sided (1 - α) confidence interval for θ. *Note*: The limit T of the confidence interval is a statisticand as such will generally assume different values from sample to sample.[1: 2.58].

ONE-SIDED TEST. A test in which the statistic used is one-dimensional and the critical region is the set of values less than a critical value (or the set of values greater than a critical value). [1:2.73]

OPERATING CHARACTERISTIC CURVE (OC CURVE). (1) Type A: A curve showing, for a given sampling plan, the probability that an acceptability criterion is satisfied as a function of the lot quality level. (2) Type B: A curve showing, for a given sampling plan, the probability of accepting a lot, as a function of the quality level of the process from which the lots come; as used for some types of plans, a curve showing the percentage of lots, or product items, that may be expected to be accepted as a function of the process quality level. (3) Type C: A curve showing, for a continuous sampling plan, the long-run percentage of product accepted during the sampling phase as a function of the quality level of the process.[2.6.1]

ORDER STATISTIC. When the observations in a sample are arranged in non-decreasing order of magnitude, each of these ordered observations is a value of a random variable, known as an order statistic. More generally, any statistic based on order statistics in this narrower sense is also called an order statistic.
Example: The *k*th value in the non-decreasing sequence of observations $x_{[k]}$ is the value of the random

variable $X_{[k]}$ called the "kth order statistic". In a sample of size n, the smallest observation $x_{[1]}$ and the largest observation $x_{[n]}$ are the values of the random variables $X_{[1]}$ and $X_{[n]}$, the first and nth order statistic, respectively. The range, $x_{[n]} - x_{[1]}$, is the value of the order statistic $X_{[n]} - X_{[1]}$.
[2: 2.46].

OUTLIERS. Observations in a sample, so far separated in value from the remainder as to suggest that they may be from a different population, or the result of an error in measurement. [1: 2.64]

PARAMETER. A quantity used in describing the probability distribution of a random variable. [1: 1.12]

PARETO CURVE. An empirical relationship describing the number of persons, y, whose income is x, first advanced by Pareto (1879) in the form

$$y = Ax^{-(1+\alpha)}, \quad 0 \leq x \leq \infty$$

The expression is now used to denote any frequency distribution of this form whether related to incomes or not. The variable x may be measured from some arbitrary value, not necessarily zero.
The coefficient A in the expression for the Pareto curve is generally referred to as the "Pareto Index." If affords evidence of the concentration of incomes, or, more generally, of the concentration of variate values in distributions of the Pareto type. [5]

PERCENT CHART. A control chart for evaluating the process in terms of the percent of the total number of units (areas of opportunity) in a sample in which a nonconformity of a given classification occurs.
[2: 3.3.8].

POWER OF A TEST. The probability of not committing an error of the second kind.
Note: Thus, it is the probability, usually designated (1-ß), of rejecting the null hypothesis when this hypothesis is false. [1: 2.79].

PRECISION. The closeness of agreement between independent test results obtained under stipulated conditions.
Notes: (1) Precision depends only on the distribution of random errors and does not relate to the true value or the specified value. (2) The measure of precision usually is expressed in terms of imprecision and computer as a standard deviation of the test results. Less precision is reflected by a larger standard deviation. (3) "Independent test results" means results obtained in a manner not influenced by any previous result on the same or similar test object. Quantitative measures of precision depend critically on the stipulated conditions. Repeatability and reproducibility conditions are particular sets of extreme stipulated conditions.
[1: 3.14]

PREVENTIVE ACTION. Action taken to eliminate the causes of a potential nonconformity defect or other undesirable situation in order to prevent occurrence. [3: 4.13]

PROBABILITY. A basic concept which may be taken either as expressing in some way a "degree of belief", or as the limiting frequency in an infinite random series. Both approaches lead to much the same calculus of probabilities. [1: 1.1].

PROBABILITY DISTRIBUTION (OF A RANDOM VARIABLE). A function giving the probability that a random variable takes any given value or belongs to a given set of values.
Note: The probability on the whole set of values of the random variable equals 1.[1: 1.3]

PROCEDURE. Specified way to perform an activity.
Notes: (1) In many cases, procedures are documented. (2) When a procedure is documented, the term "written procedure" or "documented procedure" is frequently used. [3: 1.3]

PROCESS. Method of operation in any particular stage of any element, group of elements, or total aspect of production or service.
Note: It is necessary to distinguish between an individual and an overall process. [2: 1.1.1]

PROCESS CAPABILITY. A statistical measure of inherent process variability for a given characteristic.
Notes: (1)Standard measures of process capability have not achieved consensus at the present time. some examples are: (a)the standard deviation (s) or the range, or a multiple thereof based upon the inherent variability; (b)a composite value of the component due to inherent variability and a component due to small assignable causes; (c)a composite value of a multiple of the standard deviation of inherent variability based upon an individual process (which can be denoted by σ), plus an acceptable small range allowed for shifts due to assignable causes. (2) When using the term 'process capability,' it is essential to state which measure is being used (in appropriate cases si, or st may be specified). [2: 3.2.3]

PROCESS CAPABILITY INDEX (PCI). Value of the tolerance specified for the characteristic divided by the process capability.

Notes: (1) When the process capability is defined as 6s
$$PCI_{6\sigma} = (U - L)/6\sigma$$
where U and L are, respectively, the upper and lower tolerance limits specified. $PCI_{6\sigma}$ is sometimes called C_p, but to avoid confusion it is best to restrict C_p when used without qualification to the case where 6σ is used in the definition oc C_p. Similarly,
$$PCI_{\sigma} = (U - L)/\sigma$$
(2) When using the term 'process capability', it is essential to state which measure is being used. (3) The index PCI is often used to classify a process to show the degree to which it can satisfy the tolerance specified.

(a) Low relative process capability:
$$PCI_{\sigma} < 6 \text{ or } PCI_{6\sigma} < 1$$
(difficult to satisfy tolerance).

(b) Medium relative process capability:
$$6 < PCI_{\sigma} < 8 \text{ or } 1 < PCI_{6\sigma} < 1.33.$$

(c) High relative process capability:
$$PCI_{\sigma} > 8 \text{ or } PCI_{6\sigma} > 1.33$$
(not difficult to satisfy tolerance).

To avoid confusion it is best to restrict this classification when used without other qualification to the case which uses σ_t or $6\sigma t$. [2: 3.2.6]

Process In Control; Stable Process. A process in which each of the quality measures (e.g. the average and variability or fraction nonconforming or average number of nonconformities of the product or service) is in a state of statistical control.
Note: Statistical control is usually assessed by using control charts. [2: 3.1.6]

Process Inspection. Inspection of a process by examination of the process itself or of the product characteristics at the appropriate stage(s) of the process. [1: 1.2.2]

Process Quality Control; Process Control. That part of quality control that is concerned with maintaining the product, process or service characteristics, within specified limits. [2: 1.1.6]

Producer's Risk (PR). For a given sampling plan, the probability of non-acceptance of a lot when the lot or process quality level (e.g. the fraction nonconforming) has a value stated by the plan as acceptable [e.g. acceptable quality level (AQL)].[2: 2.6.7]

Producer's Risk Quality (PRQ). A lot or process quality level which in the sampling plan corresponds to a specified producer's risk.
Notes: (1) The type of operating characteristic curve should be specified. (2) A special case of this is 'acceptable quality level' when the operating characteristic curve is type B.[2: 2.6.9]

Producer's Risk Point (PRP). A point on the operating characteristic curve corresponding to the producer's risk.
Note: The type of operating characteristic curve should be specified. [2: 2.6.8]

(Production) Batch. A definite quantity of some commodity or service produced at one time under conditions that are presumed uniform.
Note: The circumstances under which the conditions can be presumed uniform cannot be generally stated; for example, a change in the material or tool used or an interruption in the manufacturing process can give rise to different conditions. [2: 1.3.4]

Proportion Chart; Fraction Chart. A control chart for evaluating the process in terms of the proportion (or fraction) of the total number of units (areas of opportunity) in a sample in which a nonconformity of a given classification occurs. [2: 3.3.7]

Quality. The totality of features and characteristics of a product or service that bear on its ability to satisfy stated or implied needs. [2: 1.1.2]

Quality Assurance. All those planned and systematic actions necessary to provide adequate confidence that a product, process, or service will satisfy given requirements for quality. [2: 1.1.4]

Quality Audit. A systematic and independent examination to determine whether quality activities and related results comply with planned arrangements and whether these arrangements are implemented effectively and are suitable to achieve objectives. [3: 4.9]

Quality Control. Operational techniques and activities that are used to fulfil requirements for quality. [2: 1.1.5]

Quality Level. Any relative quality measure obtained by comparing observed values with the relevant requirements.
Notes: (1) Usually a numerical value indicating either the degree of conformity or nonconformity, especially for specification or sampling inspection purposes. (2) Where possible, a more precise term should

be used, e.g. 'proportion conforming', 'fraction nonconforming', 'acceptable quality level'. [2: 1.1.8]

QUALITY MANAGEMENT. all activities of the overall management function that determine the quality policy, objectives and responsibilities, and implement them by means such as quality planning, quality control, quality assurance and quality improvement within the quality system. [3: 3.2]

QUALITY SYSTEM. The organizational structure, responsibilities, procedures, processes, and resources for implementing quality management., [3: 3.6].

RANDOM VARIABLE. A variable that may take any of the values of a specified set of values and with which is associated a probability distribution. [1: 1.2]

RANDOMIZATION. The process by which a set of items are set into a random order.
Note: If, from a population consisting of the natural numbers 1 to n, numbers are drawn at random (i.e. in such a way that all numbers have the same chance of being drawn), one by one successibely, without replacement, until the population is exhausted, the numbers are said to be drawn' in random order".
If these n numbers have been associated in advance with n distinct objects or n distinct treatments which are then rearranged in the order in which the numbers are drawn, the order of the objects or treatments is said to be randomized. [1: 2.91]

RANGE. The difference between the largest and the smallest observed value of a quantitative characteristic. [1: 2.30]

RANGE CHART; R CHART. A control chart for evaluating the variability within a process in terms of the range of the sub-group. [2: 3.3.13]

RATIONAL SUB-GROUP. In an ordered sequence, one of the sub-groups within which variations may be considered to be due to non-assignable chance causes only, but between which there may be variations due to assignable causes the presence of which is considered both possible and important to detect. [2: 1.3.11]

RECTIFYING INSPECTION. Removal or replacement of nonconforming items during inspection of all the items (or of some specified number of items) in a lot or batch which was not accepted by acceptance sampling. [2: 1.2.7]

REDUNDANCY. In an item, the existence of more than one means for performing a required function. [4: 191-15-01]

REDUNDANCY, ACTIVE. That redundancy wherein all means for performing a required function are intended to operate simultaneously. [4: 191-15-02]

REDUNDANCY, STANDBY. That redundancy wherein a part of the means for performing a required function is intended to operate while the remaining parts of the means are inoperative until needed.[4: 191-15-03]

RELIABILITY (OF AN ITEM, EXPRESSED NUMERICALLY). The probability that an item will perform a required function under stated conditions for a stated period of time.
Note: This definition is used when defining the characteristic intended by such modified terms as "assessed reliability" and "predicted reliability."
[4: 191-12-01]

RELIABILITY (GENERAL DEFINITION). Ability of an item to perform a required function under stated conditions for a stated period of time. [4: 191-02-06]

RELIABILITY, HUMAN PERFORMANCE. The probability that man will accomplish all required human functions under specified conditions. [7]

RELIABILITY, INHERENT. The potential reliability of an item present in its design.

REJECTABLE PROCESS LEVEL (RPL). A process level which forms the inner boundary of the zone of rejectable processes. [2: 3.4.11]

REJECTION NUMBER, RE; NON-ACCEPTANCE NUMBER. In sampling inspection by attributes, the smallest number of nonconformities or nonconforming items found in the sample that requires that the lot be not accepted, as given in the sampling plan. [2: 2.3.11]

REPEATABILITY. Precision under repeatability conditions. [1: 3.15]

REPEATABILITY CONDITIONS. Conditions where independent test results are obtained with the same method on identical test items in the same laboratory by the same operator using the same equipment within short intervals of time.[1: 3.16]

REPEATABILITY STANDARD DEVIATION. The standard deviation of test results obtained under repeatability conditions.

Notes: (1) It is a measure of the dispension of the distribution of test results under repeatability conditions. (2) Similarly "repeatability variance" and "repeatability coefficient of variation" could be defined and used as measures of the dispension of test results under repeatability conditions. [1: 3.17]

REPETITION. A term denoting the execution of a statistical inquiry several times using the same method on the same population under the same conditions. [1: 2.89]

REPLICATION. During the course of an experiment or survey, replication is the determination a a value more than once.
Note: Replication should be distinguished from repetition by the fact that replication denotes determinations carried out at different places and/or times defined in the plan or design. The successive determinations, including the first, are called replicates. [1: 2.90]

REPRODUCIBILITY. Precision under reproducibility conditions. [1: 3.20]

REPRODUCIBILITY CONDITIONS. Conditions where test results are obtained with the same method on identical test items in different laboratories with different operators using different equipment. [1: 3.21]

REPRODUCIBILITY STANDARD DEVIATION. The standard deviation of test results obtained under reproducibility conditions.
Notes: (1) It is a measure of the dispension of the distribution of test results under reproducibility conditions. (2) Similarly "reproducibility variance" and "reproducibility coefficient of variation" could be defined and used as measures of the dispension of test results under reproducibility conditions. [1: 3.22]

RUN. In a series of observations of a qualitative characteristic, the occurrence of an uninterrupted series of the same attribute is called a "run". In a series of observations of a quantitative characteristic, a consecutive set of values which are monotonically increasing or decreasing is said to provide a run "up" or "down" respectively. [1: 2.48]

SAMPLING. The process of drawing or constituting a sample. [2: 2.1.2]

SAMPLING ERROR. That part of the estimation error which is due to the fact that only a sample of size less than the population size is observed. [1: 2.53]

SAMPLING FRAME. A list, compiled for sampling purposes, which designates the items of a population to be considered in a study. [1: 2.4]

SAMPLING INSPECTION. The inspection of products or services using samples (as distinct from 100% inspection).[2: 2.2.1]

SAMPLING INTERVAL. In periodic systematic sampling, an interval at the end of which a sample is taken. [2: 2.1.8]

SAMPLING PLAN. A specific plan which states the sample size(s) to be used and the associated criteria for accepting the lot.
Notes: (1) A criterion is, for example, that the number of nonconforming items is less than or equal to the acceptance number. (2) The sampling plan does not contain the rules on how to take the sample. [2: 2.3.3]

SAMPLING PROCEDURE. Operational requirements and/or instructions relating to the use of a particular sampling plan; i.e. the planned method of selection, withdrawal and preparation of sample(s) from a lot to yield knowledge of the characteristic(s) of the lot. [2: 2.3.2]

SAMPLING SCHEME. A combination of sampling plans with rules for changing from one plan to another.
Note: Some schemes have switching rules for automatic change to tightened inspection plans or reduced inspection plans or change to 100% inspection. [2: 2.3.4]

SAMPLING SYSTEM. A collection of sampling schemes, each with its own rules for changing plans, together with criteria by which appropriate schemes may be chosen. [2: 2.3.5]

SAMPLING UNIT. (1)One of the individual items into which a population is divided. (2) A quantity of product, material or service forming a cohesive entity and taken from one place at one time to form part of a sample.
Notes: (1) A sampling unit may contain more than one item to be tested, e.g., a packet of cigarettes, but one test result or observation will be obtained from it. (2) The unit of a product may be a single item, a pair or a set of items, or it may be a specified quantity of material, such as a length of round brass rod, a volume of paint, or a weight of coal. It need not be the same as the unit of purchase, supply, production, or shipment. [2: 1.3.3].

Sampling With Replacement. Sampling in which each sampling unit taken and observed is returned to the population before the next sampling unit is taken.
Note: In this case the same sampling unit may appear several times in the sample. [1: 4.6]

Sampling Without Replacement. Sampling in which sampling units are taken from the population once only or successively without being returned to the population.[1: 4.7]

Sample. One or more sampling units taken from a population and intended to provide information on the population. [2: 2.1.1]

Sample Size. Number of sampling units in a sample. [2: 2.1.3]

Sample Standard Deviation Chart; S Chart. A control chart for evaluating the variability within a process in terms of the sample standard deviation, s, of the sub-group. [2: 3.3.14]

Screening Inspection. 100% inspection of material or items of a product, with rejection of all items or portions found nonconforming.
Notes: (1) Screening inspection might only be concerned with one particular kind of nonconformity. (2) Screening may be carried out for the purpose of removing nonconforming items from a lot or batch which was not accepted. [2: 1.2.6]

Sequential Sampling Inspection. Sampling inspection in which, after each item has been inspected, the decision is made according to a defined rule based on the cumulative evidence of all the items from that lot inspected so far: to accept the lot, not to accept it, or to inspect another item.
Note: The total number of items to be inspected is not fixed in advance but a maximum number is often agreed upon. [2: 2.4.4]

Servicing. The replenishment of consumables needed to keep an item in operating condition, but not including any other preventive maintenance or any corrective maintenance. [7]

Severity Of Sampling. The degree of discrimination within a sampling scheme for changing from a normal to a reduced (tightened) sampling plan if the quality of the submitted product or service improves (deteriorates).
Note: This term should not be confused with inspection level, which is independent of switching rules. [2: 2.5.2]

Shewhart Control Chart. A control chart to show if a process is in statistical control.
Note: It may be a chart using attributes (e.g. proportion nonconforming) for evaluating a process, or it may be a chart using variables (e.g. average and range) for evaluating a process. [2: 3.3.3]

Significance Level (Of A Test). The given value which is the upper limit of the type 1 error probability.
Note: The significance level is usually designated as α. [1: 2.70]

Significant Result (At The Chosen Significance Level α). The result of a statistical test which leads to the rejection of the null hypothesis. If the hypothesis is not rejected, the result is not significant.
Note: When the test result is called statistically significant, this means that the result is out of the range of values that are likely from random effects if the null hypothesis is true. It does not necessarily mean that this has physical or economic importance.[1: 2.84]

Significance Testing. Statistical appraisal of the outcomes of sampling to note whether or not, at a certain level of risk, the results represent real effects or chance fluctuations of sampling and measurement.

Simple Random Sample. A sample of n items taken from a population of N items in such a way that all possible combinations of n items have the same probability of being taken. [2: 2.1.4]

Single Sampling Inspection. Sampling inspection in which the decision, according to a defined rule, to accept or not to accept a lot is based on the inspection results obtained from a single sample of predetermined size n. [2: 2.4.1]

Skip-Lot Sampling Inspection. Sampling inspection in which some lots in a series are accepted without inspection, when the sampling results for a stated number of immediately preceding lots meet stated criteria. [2: 2.4.9]

Specification. Document stating requirements.
Notes: (1) A qualifier should be used to indicate the type of specification, such as product specification, test specification. (2) A specification should refer to, or include, drawings, patterns or other relevant documents and indicate the means and the criteria whereby conformity can be stating. [3: 3.14]

Standard Deviation (Of A Random Variable, or Of A Probability Distribution). The positive square root of the variance of the random variable. [1: 1.23]

STANDARD DEVIATION (OF A SAMPLE). The positive square root of the variance of the sample
Note: The sample standard deviation is a biased estimator of the population standard deviation. [1: 2.34]

STANDARD ERROR. The standard deviation of an estimator. [1: 2.56]

STANDARDIZED NORMAL DISTRIBUTION; STANDARDIZED LAPLACE-GAUSS DISTRIBUTION. The probability distribution of the standardized normal random variable U, the probability density function of which is

$$f(u) = \frac{1}{\sqrt{2\pi}} \exp\left(-\frac{u^2}{2}\right)$$

for $-\infty < u < +\infty$, [1: 1.38]

STATE OF STATISTICAL CONTROL. A state in which the variations among the observed sampling results can be attributed to a system of chance causes that does not appear to change with time.
Note: Such a system of chance causes will generally behave as though the results are simple random samples from the same population. [2: 3.1.5]

STATISTIC. A function of the sample random variables.
Note: A statistic, as a function of random variables, is also a random variable and as such it assumes different values from sample to sample. The value of the statistic obtained by using the observed values in this function may be used in a statistical test or as an estimate of a population parameter such as a mean or a standard deviation. [1: 2.45]

STATISTICAL QUALITY CONTROL. That part of quality control in which statistical techniques are used.
Notes: (1) These techniques include the use of frequency distributions, measures of central tendency and dispersion, control charts, acceptance sampling, regression analysis, tests of significance, etc. (2) When statistical quality control is used to control the operation of a process rather than to control the quality of materials supplied, the term 'statistical process control' is often used. [2: 1.1.7]

STATISTICAL TEST. A statistical procedure to decide whether a null hypothesis should be rejected in favour of the alternative hypothesis or not rejected.
Notes: (1) The decision on the null hypothesis is taken based on the value(s) of an appropriate test statistic or statistics. Since the test statistic is a random variable, there is some risk of error when the decision is taken. (2) Generally, a test assumes a priori that certain assumptions are fulfilled (for example, assumption of independence of the observations, assumption of normality, etc). [1: 2.65]

STEP STRESS TEST. A test consisting of several increasing stress levels applied sequentially for periods of equal time duration to an item. During each period a state stress level is applied and the stress level is increased from one step to the next. [4: 191-14-08]

STRATIFICATION. The division of a population into mutually exclusive and exhaustive subpopulations (called strata), which are thought to be more homogeneous with respect to the characteristics investigated than the total population.[1: 4.13]

STRATIFIED RANDOM SAMPLING. The process of selecting a simple random sample from each of the population strata. [1: 4.14]

STRATUM. A group of units from a population, sub-population, usually defined by relevant population characteristics. [1: 4.13]

SUB-GROUP (MEASUREMENT SENSE). One of the sets of groups of observations obtained by subdividing a larger group of observations. [2: 1.3.10]

SUB-GROUP (OBJECT SENSE). One of the sets of items or quantities of material obtained by subdividing a larger group of items or quantities of material.[2: 1.3.9]

SURVEILLANCE. Continual monitoring and verification of the status of an entity and analysis of records to ensure that specified requirements are being fulfilled. [3: 4.7]

SURVIVABILITY. The measure of the degree to which an item will withstand hostile man-made environment and not suffer abortive impairment of its ability to accomplish its designated mission. [7]

SWITCHING RULES. Instructions within a sampling scheme for changing from one sampling plan to another of greater or less severity (e.g. normal, reduced, or tightened inspection or discontinuance of acceptance), based on demonstrated quality level history. [2: 2.5.3].

***t*-DISTRIBUTION; STUDENT'S DISTRIBUTION.** The probability distribution of a continuous random variable, the probability density function of which is

$$f(t;v) = \frac{1}{\sqrt{\pi v}} \left(\frac{\Gamma[(v+1)/2]}{\Gamma(v/2)} \right) \left(\frac{1}{(1+t^2/v)^{(v+1)/2}} \right)$$

where
- < t <+ with parameter v = 1,2, ...;
Γ is the gamma function.

Note: The quotient of two independent random variables, the numerator of which is a standardized normal variable, and the denominator of which is the positive square root of the quotient of a χ^2 random variable and its number of degrees of freedom v, is a Student's distribution with v degrees of freedom.[1:1.40]

TIME (AS USED IN RELIABILITY DEFINITIONS). Refers to any duration of observations of the considered items - either in actual operation or in storage, readiness, etc., but excludes downtime due to a failure.
Note: In definitions where "time" is used, this parameter may be replaced by distance, cycles, or other measures of life as may be appropriate. This refers to terms such as acceleration factor, wear-out failure, failure, rate, mean life, mean-time-between-failures, mean-time-to-failure reliability, and useful life.

TIME, CHECKOUT. That part of active corrective maintenance time during which function check out is performed. [4: 191-08-13]

TIME, CLEANUP. That element of maintenance time during which the item is enclosed and extraneous material not required for operation is removed. [7]

TIME DELAY. That element of downtime during which no maintenance is being accomplished on the item because of either supply delay or administrative reasons. [7]

TIME, DOWN (DOWNTIME). The time interval during which an item is in a down state. [4: 191-09-08]

TIME, FAULT CORRECTION. That part of active corrective maintenance time during which fault correction is performed. [4: 191-08-11]

TIME, FAULT LOCATION. That part of active corrective maintenance time during which fault localization is performed. [4:191-08-15]

TIME, INACTIVE. That time during which an item is in reserve (in the inactive inventory). [7]

TIME, ITEM OBTAINMENT. That element of maintenance time during which the needed item or items are being obtained from designated organization stock rooms. [7]

TIME, MISSION. That element of uptime during which the item is performing its designated mission. [7]

TIME, MODIFICATION. The time necessary to introduce any specific change(s) to an item to improve its characteriustics or to add new ones. [7]

TIME, PREPARATION. That element of maintenance time needed to obtain the necessary test equipment and maintenance manuals, and set up the necessary equipment to initiate fault location. [7]

TIME-SERIES. A time-series is a set of ordered observations on a quantitative characteristic of an individual or collective phenomenon taken at different points of time.

TOLERANCE. Difference between the upper and the lower tolerance limits. [2: 1.4.4]

TOLERANCE LIMITS, LMITING VALUES, SPECIFICATION LIMITS. Specified values of the characteristic giving upper and/or lower bounds of the permissible value.
Notes: (1) This term should not be confused with natural process limits or tolerance interval. (2) Tolerance limits may be set on the basis of natural process limits. [2: 1.4.3]

TWO-SIDED CONFIDENCE INTERVAL. When T_1 and T_2 are two functions of the observed values such that, θ being a population parameter to be estimated, the probability $P_r(T_1 \leq \theta \leq T_2)$ is at least equal to $(1 - \alpha)$ [where $(1 - \alpha)$ is a fixed number, positive and less than 1], the interval between T_1 and T_2 is a two-sided $(1 - \alpha)$ confidence interval for θ.
Notes: (1) The limits T_1 and T_2 of the confidence interval are statistics and as such will generally assume different values from sample to sample. (2) In a long series of samples, the relative frequency of cases where the true value of the population parameter θ is covered by the confidence interval is greater than or equal to $(1 - \alpha)$.[1: 2.57].

TWO-SIDED TEST. A test in which the statistic used is one-dimensional and in which the critical region is the set of values less than a first critical value and the set of values greater than a second critical value.
Note: The choice between a one-sided test and a two-sided test is determined by the alternative hypothesis. [1: 2.74]

Type 1 Error Probability. The probability of committing the error of the first kind.
Notes: (1) It is always less than or equal to the significance level of the test. (2) Sometimes called a "type 1 risk". [1: 2.76]

Type II Error Probability. The probability of committing an error of the second kind.
Notes: (1) Its value, usually designated ß, depends on the real situation and can only be calculated if the alternative hypothesis is adequately specified. (2) Sometimes called a "type II risk".[1: 2.78]

Unbiased Estimator. An estimator with zero bias. [1: 2.55]

Universe. The totality of the test of items, units, or measurements, etc., real or conceptual, that is under consideration.

Validation. Confirmation by examination and provision of objective evidence that the particular requirements for a specific intended use are fulfilled. [3: 2.18].

Variables Measurement. Measuring and recording the numerical magnitude of a characteristic for each of the items in the group under consideration; this involves reference to a continuous scale of some kind. [2: 1.5.3]

Variance (Of A Random Variable Or Of A Probability Distribution). The expectation of the square of the centered random variable:

$$\sigma^2 = V(X) = E\left[X - E(X)^2\right]$$

[1: 1.22]

Variance (Of A Sample). A measure of dispersion, which is the sum of the squared deviations of observations from their average divided by one less than the number of observations.
Example:
For n observations $x_1, x_2, ..., x_n$ with average

$$\bar{x} = \frac{1}{n}\sum_1 (x_1 - \bar{x})^2$$

the variance is

$$s^2 = \frac{1}{n-1}\sum_i (x_1 - \bar{x})^2$$

Notes: (1) The sample variance is an unbiased estimator of the population variance. (2) The variance is n/(n - 1) times the central moment of order 2. [1: 2.33]

Variate. In contradistinction to a variable a variate is a quantity which may take any of the values of a specified set with a specified relative frequency or probability. The variate is therefore often known as a random variable. It is to be regarded as defined, not merely by a set of permissible values like an ordinary mathematical variable, but by an associated frequency (probability) function expressing how often those values appear in the situation under discussion. [1: 1.2].

Verification. Confirmation by examination and provision of objective evidence that specified requirements have been fulfilled. [3: 2.17]

Verification Sampling Inspection. Sampling inspection for ascertaining whether the producer's sampling procedures are in accordance with his/her declared sampling scheme.
Note: This is often called an audit of the producer's sampling procedures. [2: 2.4.10]

Warning Limits (Upper And /Or Lower). In a Shewhart control chart, the limit below which (upper limit) or above which (lower limit) or the limits between which the statistic under consideration lies with a high probability when the process is under control.
Notes: (1) When the value of the statistic computed from a sample is outside the warning limits but inside the action limits, increased supervision of the process is generally necessary and rules may be made for action in particular processes. (2) At the warning limits, attention is called to the possibility of out-of-control conditions, but further action is not necessarily required. (3) The warning limits will always be within the action limits. [2: 3.4.4]

Wear-Out Failure Period. That final period, if any, in the life of an item during which the instantaneous failure intensity for a repaired item or the instantaneous failure rate for a non-repaired item is considerably higher then that of the preceding period. [4: 191-10-10]

Within-Lot (Or Within-Batch) Variation. Dispersion of observatiions or test results obtained within a lot or batch.
Note: The within-batch variation may be estimated from a single batch or by pooling the estimates for several batches, as appropriate. [2: 3.1.3]

BIBLIOGRAPHY

[1] ANSI/ISO/ASQC A3534-1, 1993, *Statistics Vocabulary and Symbols - Part 1: Probability and Statistical Terms*, Milwaukee, WI: American Society for Quality Control.

[2] ANSI/ISO/ASQC A3534-2, 1993, *Statistics - Vocabulary and Symbols - Part 2: Statistical Quality Control*, Milwaukee, WI: American Society for Quality Control.

[3] ANSI/ISO/ASQC A8402, 1994, *Quality Management and Quality Assurance - Vocabilary*, Milwaukee, WI: American Society for Quality Control.

[4] International Electrotechnical Commission, Chapter 191, *Dependability and Quality of Service*, 1990.

[5] International Electrotechnical Commission, IEC Publication 271, *List of Basic Terms, Definitions and Related Mathematics for Reliability*, 1991.

[6] Kendall, M. G., and Buckland, W. R. *Dictionary of Statistical Terms*. (2nd ed.) New York: Hafner Publishing Company, 1960.

[7] United States Department of Defense, Mil-Std.-721B, 25 August 1966, *Definitions of Effectiveness Terms for Reliability, Maintainability, Human Factors, and Safety*, Washington, DC: United States Government Printing Office, 1966.

Z94.17 Work Design and Measurement

The definitions in this section are essentially an update of those in the 1989 volume. Work measurement concepts and practices have changed little in the past ten years. Use of the techniques is on a plateau. As industry and commercial organizations mechanize their operations, less attention is given to individual workplace productivity measurement.

Increased use is being made of techniques to measure groups and even entire organizations. As traditional measurement practices are converted to measuring groups, the terminology employed requires few changes to accommodate macromeasurement concepts. Many of the traditional terms are used in organization-wide measurement programs.

With the increased encouragement of employees by industry and commercial firms to become involved with management to raise productivity and reduce costs, there is less need for management to rely on work measurement and so-called fair day's work standards. As human resources concepts improve relations between employees and management, there is less reliance on work measurement to improve productivity. Traditional work measurement will continue to be used in planning and managing operations, and in incentive programs.

Chairman
Mitchell Fein, P.E., C.M.C.
Mitchell Fein, Inc.

Subcommittee

Marvin E. Mundel, Ph.D., P.E.

Alan V. Owen
Woods Gordon

John L. Zalusky
AFL-CIO Research Department

Jack Niles, P.E., C.M.C.
Rath & Strong

Arnold J. Ellenson, P.E., C.M.C.
Mitchell Fein, Inc.

Thomas H. Lennon
Mitchell Fein, Inc.

Joseph E. May
H.B. Maynard and Company, Inc.

ABNORMAL READING. (See ABNORMAL TIME.)

ABNORMAL TIME. A time value which is outside of statistical or policy variance limits. *Syn*: abnormal reading.

ACCEPTABLE PRODUCTIVITY LEVEL (APL). The work pace established by management, or jointly by management and labor, at a level considered satisfactory; it is established at a given relationship to the motivated productivity level. (See NORMAL PERFORMANCE.)

ACCUMULATIVE TIMING. A multiple (usually three) stopwatch technique for time study in which a mechanical linkage pressed at successive cycle breakpoints instantaneously stops, starts, and resets the individual watches so that, respectively, they may be: read for recording the latest element time, timing the element currently being observed, and ready to time the next element.

ACTIVITY SAMPLING. (See WORK SAMPLING.)

ACTUAL HOURS. (See ACTUAL TIME.)

ACTUAL TIME. The unadjusted time for the accomplishment of a defined task or task element as obtained by a timing device. *Syn*: observed time.

ALLOWANCE. (1) Work measurement: a time value or percentage of time by which the normal time is increased, or the amount of non-productive time applied to compensate for justifiable causes or policy requirements. The normal time plus allowances equal the standard time. Usually includes irregular elements, incentive opportunity on machine controlled time, minor unavoidable delays, rest time to overcome fatigue, and time for personal needs. (2) Dimensional: the minimum clearance or maximum interference distance between two interpenetrated objects. (See TRAINING ALLOWANCE.)

ALLOWED HOURS. (See STANDARD HOUR, ALLOWED TIME.)

ALLOWED TIME. A normal time value increased by an appropriate allowance(s). (See STANDARD TIME.)

ARBITRATOR. An impartial third party to whom disputing parties submit their differences for a decision (award). An adhoc arbitrator is selected to act in a specific case or a limited group of cases. A permanent arbitrator is selected to serve for the life of the contract or a stipulated term.

ASSIGNABLE CAUSE. A source of variation in a process which can be isolated, especially when its significantly larger magnitude or different origin readily distinguishes it from random causes of variation.

AUXILIARY PROCESS TIME. The time required for essential supplementary process operations which assure the continuity and completion of the principal process operations. Such auxiliary operations as deburring, straightening, cleaning, and finishing usually result in relatively minor changes in the appearance of physical characteristics of the workpiece in comparison with the effects due to such principal operations as cutting, forming, welding, casting, and assembly.

AVAILABLE MACHINE TIME. The portion of a time cycle during which a machine could be performing useful work.

AVAILABLE PROCESS TIME. The portion of a time cycle during which a process agent or system could be acting usefully on the product.

AVERAGE CYCLE TIME. (1) The sum of observed or actual work times, divided by the number of such cycle observations. Abnormal times should be accounted for and usually should be prorated into cycle time. (2) The sum of the average elemental times. (See AVERAGE ELEMENT TIME.)

AVERAGE ELEMENT TIME. The sum of a series of observed or actual element times, divided by the number of such element observations. Abnormal times should be accounted for and usually should be prorated into cycle time. (See AVERAGE CYCLE TIME.)

AVERAGE TIME. (See AVERAGE CYCLE TIME, AVERAGE ELEMENT TIME.)

AVOIDABLE DELAY. A time delay not allowed in standard time calculations because it is unnecessary and is due to factors under worker control and responsibility; should be accounted for.

BALANCE. (1) The act of distributing the work elements between the two hands performing an operation or between the different operations in a process to achieve essentially equal performance times among them. (2) The state of approximately equal working time distribution among the various components of an operation or process, *e.g.*, the stations on an assembly line.

BALANCED MOTION PATTERN. (1) The sequence of concurrent arm and hand movements over symmetrical paths that produce approximately equal momentum between the arms in directions which facilitate muscular equilibrium. (2) A series of movements with both hands involving negligible delay or idle time for either hand while the other is working.

BALANCING DELAY. (1) The idle time of one hand in an

operation due to imperfect balancing. (2) The idle time of one or more operations in a series due to imperfect balancing. (See BALANCE.) *Syn*: balance delay.

BALLISTIC MOVEMENT. A motion of a body extremity with relatively simple muscle action that is rapid, smooth, and flowing from start to finish. Such action results when a protagonistic muscle group causes motion in the intended direction to attain a peak force and velocity, which retards to zero as an antagonistic muscle group changes the direction or causes motion to cease. This sequential muscle action contrasts with that producing nonballistic motion, in which the protagonistic and antagonistic muscle groups act concurrently to precisely control the force and velocity of movement. Historically, the term is associated with the similarity of the motion plot to projectile trajectory plots.

BANK. (See FLOAT.)

BARGAINING. A method of formulating goals to match agreements for the exchange of goods or services with other management organizations. The process by which persons or groups with partly conflicting and partly harmonious interests try to agree on a procedure for dividing available resources. Bargaining is likely to occur when each group controls resources desired by the other and a rage of agreements can be made that may benefit both groups.

BASE PERIOD. The period of time established as the base from which to measure productivity changes in other periods of time.

BASE PRODUCTIVITY FACTOR (BPF). Is the ratio of total actual hours worked divided by the standard hours produced in the time period measured.

BASIC DIVISION OF WORK. (See THERBLIG.)

BASIC ELEMENT. (See ELEMENTAL MOTION.)

BASIC MOTION. A human motion closely related to primary physiological and/or bio-mechanical performance capabilities of the body or its members (*e.g.*, a therblig or other standard motion defined within a predetermined time system). *Compare*: elemental motion.

BASIC MOTION TIMES (BMT). A proprietary predetermined time standards system.

BENCHMARK. A standard of measurement with enough characteristics common to the individual units of a population to facilitate economical comparison of attributes for units selected from a sample. Benchmarks may be used for job evaluation, performance rating, establishing operational standards, standard data development, cost estimating, and other purposes. (See BENCHMARK JOB, KEY JOB.)

BENCHMARK JOB. A job with enough characteristics common to other jobs, judged acceptable as a gauge for those other jobs without their direct measurement for time standards, job evaluations, or other purposes. (See BENCHMARK.)

BIOMECHANICS. The application of mechanical principles, such as levers and forces, to the analysis of body part structure and movement. This includes studies of range, strength, endurance, and speed of movements, and mechanical responses to such physical forces as acceleration and vibration.

BRAINSTORMING. A problem solving conference technique in which participants announce suggestions in rapid sequence. The brainstorming objective is to uncover a large number of possible problem solution, useful as well as useless. A technique for enhancing group creativity that encourages group members to generate as many novel ideas as possible on a given topic and no one may criticize or evaluate ideas while meeting is in session.

BREAKPOINT. A point in a work cycle readily distinguished by sight and/or sound which is selected as the boundary between two elements for time recording or element definition in motion study. *Syns*: reading point, endpoint.

BUILDING BLOCK. An approach used in standard data development of creating fixed groups or modules of work elements which may be added together to obtain time values for elements and entire operations.

CAMERA STUDY. (See MICROMOTION STUDY, MEMOMOTION STUDY.)

CHANGEOVER TIME. The time required to modify or replace an existing facility or workplace, usually including both teardown time for the existing condition and setup of the new condition. (See SETUP, TEARDOWN.)

CHECK STUDY. A partial or complete review of a job or operation to evaluate the validity of a standard time.

CHRONOCYCLEGRAPH TECHNIQUE. A modification of the cyclegraph technique to permit the computation of motion velocities and accelerations from the spacing of light signals on the film whose shapes indicate the direction of movement, produced by pulsing the lights on and off at regular time intervals. (See CYCLEGRAPH TECHNIQUE.)

CHRONOLOGICAL STUDY. A detailed study and recording of a sequence of events in the order of occurrences. (See PRODUCTION STUDY.)

CODING. (1) Translation of a data processing machine program from descriptive, symbolic, or diagram form into machine language (code) or into an explicit symbolic language that may be translated directly into machine language by means of an assembly program or compiler. (2) Referring numbers to a convenient origin and/or scale for ease of computation. (3) Assigning a numerical and/or alphabetical symbol or group of symbols to a class or variable to achieve consistent identification, location, or interpretation. A desirable property of such symbology is the mnemonic, or memory-jogging, characteristic to promote efficient association of the code meaning.

COMBINED MOTIONS. Two or more elemental motions performed during the same interval by the same body member. Example: regrasping an object while moving it to its destination. Simultaneous motions are performed at the same time by different body members.

COMBINED WORK. (1) The production of a person working with one or more machines in which the output is controlled by the operator. In calculating a standard, the machine portion of the work cycle is not taken into account. (2) The total accomplishment of a group of workers.

CONCURRENT ENGINEERING. Is a systematic approach to the integrated, concurrent design of products and their related processes, including manufacturing and support to cause the developers, from the outset, to consider all elements of the product life cycle from conception through disposal, including quality, cost, schedule, and user requirements.

CONSISTENCY. (1) The absence of noticeable or significant variation in behavioral or numerical data as, for example, in the work pace or method used by a worker. (2) Uniformity or agreement, within stated limits, between repetitive occurrences of an event or a numerical value.

CONSTANT ELEMENT. A job or task element which occurs without significant variation in work content and/or performance time. May be used to describe elements within a given operation or elements common to different operations; an element which occurs in different jobs in which other elements are varied, but the one in question occurs in the same time.

CONTINUOUS METHOD. (See CONTINUOUS TIMING.)

CONTINUOUS READING. (See CONTINUOUS TIMING.)

CONTINUOUS TIMING. A stopwatch technique in which the watch runs continuously throughout the study and readings are made of the cumulative time at the end of each element. Individual element times are then obtained by subtraction. *Syns*: continuous method, continuous reading.

CONTROL SYSTEM. A system that has as its primary function the collection and analysis of feedback from a given set of functions for the purpose of controlling the functions. Control may be implemented by monitoring and/or systematically modifying parameters or policies used in those functions, or by preparing control reports that initiate useful action with respect to significant deviations and exceptions.

COVERAGE, ACTUAL. The number of jobs, the number of personnel, or the total hours which are covered by standards during the reporting period. Commonly expressed as a percentage of the total number of jobs, personnel, or hours. *Syn*: standard coverage.

COVERAGE, POTENTIAL. An estimate of the number of jobs, the number of personnel, or the total hours which can be covered by time standards during the reporting period. Commonly expressed as a percentage of the total number of jobs, personnel, or hours.

COVERAGE, STANDARD. The percent of operations that are performed against time standards. (See COVERAGE, ACTUAL.)

CYCLE. The complete sequence of activities, operations, and machine or process times required to complete one segment, unit, or batch of work. (See MOTION CYCLE, WORK CYCLE.)

CYCLEGRAM. Similar to cyclegraph except that the film being exposed is moved slowly to one side so that retraced motions are shown side by side on the film rather than superimposed on earlier motions.

CYCLEGRAPH. The pattern developed by the cyclegraph technique. (See CYCLEGRAPH TECHNIQUE.)

CYCLEGRAPH TECHNIQUE. The use of small lights on the hands or other body members to indicate their motion patterns. The lights are recorded by a still camera in a darkened room with an exposure time equal to at least one motion cycle.

Cycle Timing. Timing a complete work cycle as a single time value rather than timing the individual elements of the cycle. (See DIFFERENTIAL TIMING.)

Cyclic Element. An element of an operation or process that occurs in every cycle of the operation or process.

Cyclic Timing. (See CYCLE TIMING.)

Day Work. Work for which pay is based on time worked as contrasted to pay based on performance.

Decimal-hour Stopwatch. A timing device with two hands whose movement may be started, stopped, or reset to zero by depressing control buttons on the perimeter of the watch case. A small dial is calibrated in hundredths of an hour and a large dial is calibrated in ten-thousandths of an hour. Thus, the time interval may be read in decimal hours to four decimal places.

Decimal-minute Stopwatch. A timing device with two hands whose movement may be started, stopped, or reset to zero by depressing control buttons on the perimeter of the watch case. A small dial is calibrated in whole minutes and a large dial is calibrated in hundredths of a minute. Thus, the time interval may be read in decimal minutes to two decimal places.

Delay. A pause or interruption in the scheduled work activity of the employee, machine, or product flow. (See AVOIDABLE DELAY, UNAVOIDABLE DELAY, INHERENT DELAY.) *Syns*: interruption, stoppage.

Delay Allowance. (1) A time increment to allow for contingencies and minor delays beyond the control of the operator. May be included in a time standard as a percentage or as a discrete time value. (2) A separate credit (in time or money) to compensate the operator on incentive for a specific instance of delay not covered by the piece rate or standard. (See UNAVOIDABLE DELAY ALLOWANCE.)

Delay Time. A period during which an employee is idle due to breakdown of equipment, lack of tools or materials, or any other factor beyond personal control. (See DOWNTIME, IDLETIME.)

Diagnostic Study. A brief investigation or cursory methods study of an operation, process, group, or individual to discover causes of operational difficulties or problems for which more detailed remedial studies may be feasible. An appropriate work measurement technique may be used to evaluate alternatives or to locate major areas requiring improvement. *Syn*: survey.

Did-Take Standards. Are established time standards based on historical data without the use of performance rating or benchmark references. Did-Take standards are based on actual time taken to perform given tasks. (See SHOULD-TAKE STANDARDS, NORMAL PERFORMANCE, BASE PRODUCTIVITY FACTOR).

Differential Timing. The time study technique used to obtain the time value of an element of extremely short duration. It consists of: (1) obtaining cycle values, first including and then excluding the element for which the time is required, and obtaining the required element time by subtraction; (2) timing the element by combining it with preceding and/or following elements in successive cycles and then obtaining the time for the short element by subtraction.

Direct Labor. (1) Work which is readily chargeable to or identifiable with a specific product. (2) Work performed on a product or service that advances the product or service towards its completion or objectives.

Direct Labor Standard. A standard time set on a direct labor operation. (See DIRECT LABOR.)

Discontinuous Timing. (See REPETITIVE TIMING.)

Division of Labor. The separation of jobs or tasks into less complex jobs or tasks usually to allow use of workers possessing less skill than that required by the overall job or task, or to make use of special skills. *Syn*: division of work.

Downgrade. (1) The lowering of a particular job in scope, authority, responsibility, degree of difficulty, etc., with a possible reduction in wage or salary. (2) Dilution of skills required for the task. (See DIVISION OF LABOR.)

Downtime. A period of time during which an operation is halted or delayed due to the lack of materials, a machinery breakdown, or similar occurrences.

Drop Delivery. (1) The movement of a component or object to some location where it is merely let go rather than being placed where it is to go. (2) Provisions in a workplace for disposal of objects by dropping.

Earned Hours. The time in standard hours credited to a worker or a group of workers as a result of their completion of a given task or group of tasks; usually calculated by summing the multiplication of applicable standard times and the completed work units.

EFFECTIVENESS. (1) The ratio of earned hours to actual hours spent on prescribed tasks. When earned hours equal actual hours, the effectiveness equals 100%. (2) The ratio of standard, estimated, or budgeted performance to actual performance expressed as a percentage. (3) The performance or output received from an approach or program. Ideally it is a quantitative measure which can be used to evaluate the level of performance in relation to some standard, set of criteria, or end objective. (See EFFICIENCY, LABOR.) (4) The extent to which the objectives are met. The statement of objectives must be substantive and quantitative or else it is a statement whose accomplishment cannot be evaluated. In the private sector this is affected, independently of productivity, by the competition, the price indexes, the state of the economy, the tax structure, and the state-of-the-art. In the public sector it is affected, again independent of productivity by, among other factors, public acceptance, public cooperation, competing objectives, legal restraints, resources made available, and the state-of-the-art.

EFFICIENCY, LABOR. (1) The ratio of standard performance time to actual performance time, usually expressed as a percentage. (2) The ratio of actual performance numbers (*e.g.*, the number of pieces) to standard performance numbers, usually expressed as a percentage. (See PRODUCTIVITY.)

EFFORT CONTROLLED CYCLE. (See MANUALLY-CONTROLLED WORK.)

EFFORT, PHYSICAL. The amount of muscle work performed on a job, often referred to as the physical work load. It is often defined by the number of objects handled per shift, their weight, the distance they are transported, and how long the task is performed.

EFFORT RATING. (See PERFORMANCE RATING.)

ELAPSED TIME. (1) The actual time taken by a worker or machine to complete a task, an operation, or an element of an operation. (2) The total time interval from the beginning to the end of a time study. (See ACTUAL TIME.)

ELEMENT. A subdivision of the work cycle composed of one or a sequence of several basic motions and/or machine or process activities which is distinct, describable, and measurable. (See MANUAL ELEMENT, MACHINE-CONTROLLED TIME.)

ELEMENTAL MOTION. Individual manual motions or simple motion combinations used to describe the sensory-motor activity in an operation. Generally refers to the more basic and elementary therbligs. An attempt often is made to define these precisely with associated time values. Typical elemental motions are: reach, move, assemble, pre-position, turn.

ELEMENTAL STANDARD DATA. (1) Standard data. (2) A time value of an individual element in a standard data system.

ELEMENT BREAKDOWN. (1) The separation of a work cycle into two or more elements. (2) A listing of work elements with individual descriptions and/or calculations for each.

ELEMENT TIME. The time to perform a given element. May refer to the observed (raw), average, selected, normal, or standard time.

ENDPOINT. (See BREAKPOINT.)

ENGINEERED PERFORMANCE STANDARD. (See STANDARD TIME.)

ENGINEERED STANDARD. (See STANDARD TIME.)

ENGINEERED TIME STANDARD. (See STANDARD TIME.)

ERGONOMETRICS. (See ERGONOMICS, WORK MEASUREMENT.)

ERGONOMICS. The study of the design of work in relation to the physiological and psychological capabilities of people. One of several terms used to define similar fields of interest; others are human engineering, human factors, and human factors engineering. Ergonomics has been used predominantly outside of the U.S.A. The aim of the discipline is the evaluation and design of facilities, environments, jobs, training methods, and equipment to match the capabilities of users and workers, and thereby to reduce the potential for fatigue, error, or unsafe acts. (See WORK DESIGN, METHODS ENGINEERING, MOTION ANALYSIS, MOTION ECONOMY.)

ESTIMATED TIME. An element or operation time that has been predicted on the basis of such information as may be available without detailed study.

EXCESS WORK ALLOWANCE. An allowance to compensate for work required to be performed on an operation or job in addition to that specified in the standard method. Sometimes applied as a separate grant of time or as a grant of money in a piece-work system. (See ALLOWANCE.)

EXPECTED ATTAINMENT. (See FAIR DAY'S WORK.)

EXPECTED WORK PACE. (1) The work pace necessary for an operator to maintain in order to achieve a specified level of earnings under an incentive system. (2) The work pace required to meet non-incentive production standards.

EXTERNAL ELEMENT. (See EXTERNAL WORK.)

EXTERNAL TIME (JUST-IN-TIME). When setting up machines, an event that can take place before the end of the previous run of parts or after the new run has started; external time does not add time to the setup.

EXTERNAL WORK. Any element of an operation which must be performed by the operator while the machine or process is not in operation and which results in a loss of potential machine or process operating time. The term "external" implies that the element occurs outside the machine or process cycle.

FAIR DAY'S WORK. The amount of work which is expected daily from an employee. May be established solely by management or through mutual agreement. May or may not be established through the use of various measurement techniques. (See NORMAL PERFORMANCE.) *Syn*: expected attainment.

FATIGUE. A psychological and physiological process that reduces the performance capacity and motivation of a worker. The magnitude of the effect varies with the stress induced on the worker resulting from the nature of the work, the work environment, and the physical taxing of the worker.

FATIGUE ALLOWANCE. Time included in the production standard to allow for rest to overcome the effects of fatigue. May be applied either as a percentage of the leveled, normal, or adjusted time or as a stated number of nonproductive minutes per hour. (See ALLOWANCE, STANDARD TIME.)

FILM ANALYSIS. A systematic, detailed analysis of work from a motion picture film or video tape. Usually related to micromotion or memomotion study, or from a video tape.

FILM ANALYSIS CHART. For recording a film analysis. Generally records each successive elemental motion, element, or operation, the beginning and ending clock time (if a clock is included in the picture) or frame number, and its descriptive symbol. (See SIMO CHART, THERBLIG CHART.) *Syn*: film analysis record.

FILM ANALYSIS RECORD. (See FILM ANALYSIS CHART.)

FILMS, RATING. Motion picture films and video tapes containing a consistent or random sequence of work scenes being done at varying performance levels, used to train work measurement analysts in identifying different performance levels. May also be used to attempt to standardize the concept of normal performance such as in card dealing, walking, or typical shop operations.

FIRST PIECE TIME. The allowed time to produce the first piece in an order of several pieces. Intended to compensate for delays resulting from unfamiliarity with the work method, or for extra first piece work such as setup or for inspection and adjustment in setup.

FIXTURE. A device used to position and hold materials which are being worked upon or assembled.

FLOAT. (1) The amount of material in a system or process, at a given point in time, that is not being directly employed or worked upon. (2) The total cushion or slack in a network planning system. *Syn*: bank.

FLOW ANALYSIS. Detailed examination of the progressive travel, either of personnel or material, from place to place and/or from operation to operation.

FLOWCHART. Tabular material, standardized symbols, and explanations depicting the predetermined route of either personnel or material, from place to place and/or from operation to operation in the manufacturing or processing sequence of events.

FLOW DIAGRAM. A representation of the location of activities or operations and the flow of materials between activities on a pictorial layout of a process. Usually used with a flow process chart.

FLOW LINE. (1) The direction taken either by personnel or material as they progress through the manufacturing or processing sequence of events. (2) The path along which personnel or material travel in progressing through the plant. *Syn*: line of flow.

FLOW PATH. The route taken and/or space occupied by the personnel, material, subassembly, or assembly as these progress through the manufacturing process.

FLOW PROCESS CHART. A graphic, symbolic representation of the work performed or to be performed on a product as it passes through some or all of the stages of a process. Typically, the information included in the chart is quantity, distance moved, type of work done (by symbol with explanation), and equipment used. Work times may also be included. Flow process chart symbols generally used are:

○ ASME* Standard Symbol

   operation: A subdivision of a process that changes

or modifies a part, material or product, and is done essentially at one workplace location.

A specialized application for paperwork uses two standard symbols:

◎ creation of a record or set of papers

⊘ addition of information to a record or set of papers

➡ *transportation (move)*: Change in location of a person, part, material, or product from one workplace to another.

☐ *inspection*: Comparison of observed quality or quantity of product with a quality or quantity standard.

▽ *storage*: Keeping a product, material, or part protected against unauthorized removal.

D *delay*: An event which occurs when an object or person waits for the next planned action.

◘ *combined activity*: Adjustment during testing, *e.g.*, would combine the separate operation and inspection symbols.

*See ASME Standard 101
Syns: flow chart, production process chart, product analysis chart.

FOREIGN ELEMENT. An element with a random, usually unpredictable, frequency of occurrence, not part of a normal method, usually accounted for by eliminating the element or by predicting the occurrence frequency and allowing elemental time prorated into the operation time.

FORM PROCESS CHART. A graphic, symbolic representation of the process flow of paperwork forms. Similar to a flow process chart except that the item of interest is one or more forms. A form process chart may show organizations, operations, movements, temporary and controlled storages, inspection or verification, disposal of all forms charted, as well as the source and type of information transmitted between forms. Flow process chart symbols may be adapted to reflect the form processing activity. *Syns*: information process analysis chart, functional forms analysis, forms analysis chart.

FRAME COUNTER. A mechanical or electronic counter which can be used to determine the number of frames that have passed a predetermined point in a motion picture. The frame counter may be attached to any device for showing or viewing motion pictures or video recordings.

FREQUENCY. (1) The number of times a specified value occurs within a sample of several measurements of the same dimension or characteristics on several similar items. (2) In work measurement, the number of times an element occurs during an operation cycle.

FUMBLE. An unintentional human activity referred to as a sensory-motor error that may or may not be avoidable depending upon the working environment or the skill of the operator.

GAINSHARING. The term used to denote any plan for sharing productivity gains with employees, including small group and individual incentives and other pay for performance practices.

GAINSHARING PLANS. Group, department or company-wide bonus systems designed to reward all eligible members for improved performance. Gains are shared with all employees in the unit according to a predetermined formula or target (several measures may be included in the formula). Examples of common gain sharing plans are Scanlon, Rucker, IMPROSHARE®; may include cash profit sharing.

GANG CHART. (See MULTIPLE ACTIVITY PROCESS CHART.)

GANTT CHART. A graphic representation on a time scale of the current relationship between actual and planned performance.

GILBRETH BASIC ELEMENT. (See THERBLIG.)

GRAVITY FEED. The principle of using the force of gravity to convey materials from one location to another by having them fall, slide or roll downward and laterally.

GRIEVANCE. A formal complaint or expressed dissatisfaction by an employee relating to a job, pay, or other employment factors. Methods in how it is handled depends upon policies for handling grievances in a union or non union plant.

GRIEVANCE PROCEDURE. It is the procedure that is defined and handled as specified steps.

GROUP. Used in connection with other words referring to a type of activity. Refers to more than one person working together to accomplish a task.

GUARANTEE. A minimum rate of pay under incentives which is paid regardless of lower productivity, normally it is a time rate for a job upon which compensation for work is calculated.

Halo Effect. Distortion of performance rating factors and/or time value to fit previously established data.

Hand Time. The time required to perform a manual element. (See MANUAL ELEMENT.) *Syn*: manual time.

Handling Time. The time required to move parts or materials to or from an operation or work area.

High Task. Performance of an average experienced operator working at an incentive pace, over an eight-hour day under safe conditions, without undue or cumulative fatigue. Often stated as a percentage above normal performance. (See NORMAL EFFORT, LOW TASK.) *Syns*: incentive pace, motivated productivity level.

Historical Time. (See STANDARD TIME, STATISTICAL.)

Human Engineering. (See HUMAN FACTORS ENGINEERING.)

Human Factors Engineering. A merging of the branches of engineering and the behavioral sciences which are concerned principally with the human component in the design and operation of human-machine systems. Based on a fundamental knowledge and study of human physical and mental abilities and emotional characteristics. (See ERGONOMICS, WORK DESIGN, METHODS ENGINEERING, MOTION ANALYSIS, MOTION ECONOMY)

Human Performance Times (HPT). A predetermined time standards system.

Idle Machine Time. (See MACHINE IDLE TIME.)

Idle Time. Time during which a worker is not working. (See AVOIDABLE DELAY, UNAVOIDABLE DELAY, WAITING TIME.)

IMPROSHARE®. A gainsharing plan in which productivity is measured as output against input in hours, by discrete product.

Incentive Pace. (See HIGH TASK, MOTIVATED PRODUCTIVITY LEVEL.)

Incentive Plan. A program to reward individuals or groups for achieving and exceeding performance standards or goals.

Incentive Operators. (1) Employees whose pay is determined all or in part by the quantity and/or quality of output. (2) Employees working under a wage incentive plan.

Incidental Element. (See IRREGULAR ELEMENT.)

Indirect Labor. Labor which does not add to the value of a product but which must be performed to support its manufacture. May not be readily identifiable with a specific product or service. (See INDIRECT LABOR STANDARD.)

Indirect Labor Standard. (1) An established standard of time for labor performed while rendering services necessary to production, the cost of which cannot be assessed against any part, product, or group of parts or products accurately or without undue effort and expense. (2) A standard time for indirect labor. (See INDIRECT LABOR, STANDARD TIME.)

Industrial Engineer. A person qualified to practice industrial engineering as defined by the Institute of Industrial Engineers.

Industrial Engineering. Concerned with the design, improvement, and installation of integrated systems of people, materials, information, equipment, and energy. It draws upon specialized knowledge and skill in the mathematical, physical, and social sciences together with the principles and methods of engineering analysis and design, to specify, predict, and evaluate the results to be obtained from such systems. (Official definition of the Institute of Industrial Engineers.)

Inherent Delay. (See DELAY TIME.)

Instruction Card or Sheet. A written description provided to a worker which states the standard method or standard practice to be used.

Interference Allowance. An allowance to compensate for interference time. (See INTERFERENCE TIME.)

Interference Time. Idle machine time resulting from the inability of a machine operator, when assigned to two or more semi-automatic machines, to serve one or more of them when they require service. (See MACHINE INTERFERENCE.)

Intermittent Element. (See IRREGULAR ELEMENT, REGULAR ELEMENT.)

Internal Element. (See INTERNAL WORK.)

Internal Time (Just-In-Time). When setting up machines, an event that takes place while the machine is stopped, an event that cannot be made external. (In work measurement) refers to time available to operator for work while machine works.

Internal Work. Manual work performed by an operator

while the machine or process is operating automatically. *Syns*: fill up work, inside work.

IRREGULAR ELEMENT. An element which occurs randomly and can be statistically determined.

JIG. (1) A mechanical device used to guide a cutting tool along a predetermined path when in contact with the material or workpiece supported in the device. (2) A device used to hold parts in position (*e.g.*, welding jig, airframe jig).

JOB. (1) The combination of tasks, duties, and responsibilities assigned to an employee and usually considered as a normal or regular assignment. (2) The contents of a work order.

JOB ANALYSIS. Determination of the requirements of a job through detailed observation and evaluation of the work performed, facilities required, conditions of work, and the qualifications required of a worker. Syn: job study.

JOB BREAKDOWN. The systematic division of an operation into elements, or the results of such an analysis. *Syn*: operation breakdown.

JOB CHARACTERISTIC. (See JOB FACTOR.)

JOB DESIGN. (See WORK DESIGN.)

JOB FACTOR. (1) An element characteristic of a job which provides a basis for selecting and training workers and establishing the wage range for the job. Such characteristics include mental and physical requirements, responsibilities, hazards, and other working conditions. (2) A predetermined element of a job evaluation plan against which jobs are compared. *Syn*: job characteristic.

JOB SKILL. The manual and mental proficiency required to perform a given task. *Syn*: skill.

JOB STANDARDIZATION. The procedure of specifying a standard practice or a standard method for a job.

JOB STUDY. (See JOB ANALYSIS.)

JUST-IN-TIME (JIT). A philosophy concerning the operation of the business that teaches that only activities that add value to a product or service are beneficial; all other activities are viewed as waste. JIT is designed to reduce inventory and work in process by grouping machines and operations into work cells by product, rather than by similar machine groupings.

KEY JOB. A job that is considered representative of similar jobs in the same plant, company, industry, or labor market and hence may be used for comparing the descriptions of other jobs with the key job for job evaluation and job classification purposes. May also be used as an aid in establishing wages for other jobs.

KYMOGRAPH. An electronic time study device used to measure extremely short work time intervals. Consists of a system of transducers (principally microswitches and photoelectric cells) that are activated by an operator performing a job, and a tape puller that records the impulses as a function of time.

LABOR COST. That part of a firm's total costs attributable to wages, salaries, supplementary benefits, and other employment costs.

LABOR-SAVING RATIO. The labor cost saved by an improved method divided by the unit labor cost of the original method.

LABOR STANDARD. (See DIRECT LABOR STANDARD, INDIRECT LABOR STANDARD.)

LEARNER'S ALLOWANCE. (See TRAINING ALLOWANCE.)

LEARNING CURVE. A plot of productive output or unit work times of an individual or group as a function of time or output per unit time; used to predict the learning rate in starting up a new job or project. A learning curve is usually exponential and flattens out with time. (See START-UP CURVE, PROGRESS CURVE.) *Syn*: learner's curve.

LEVELED ELEMENT TIME. (See NORMAL ELEMENT TIME.)

LEVELED TIME. (See NORMAL TIME.)

LEVELING. (See PERFORMANCE RATING.)

LINE OF FLOW. (See FLOW LINE.)

LOOSE STANDARD. A standard time greater than that required by a qualified worker with normal skill and effort, following a prescribed method and utilizing allowances for delays, personal needs, and rest.

LOST TIME. (See WAITING TIME.)

LOW TASK. A term used to indicate that performance rating or production standards are based on daywork levels as contrasted to high task or incentive work performance. Sometimes taken to mean a level of performance below the level expected under measured daywork conditions.

Machine Assignment. (1) The equipment assigned to an operator in the performance of a job. (2) Equipment designed to perform jobs as in production scheduling.

Machine Attention Time. Time during which a machine operator must observe the machine's functioning and be available for immediate servicing, while not actually operating or servicing the machine. *Syn*: service time.

Machine Capability. Qualitative and quantitative measures of acceptable output from a given piece of power equipment.

Machine-controlled Time. The time portion of an operation cycle required by a machine to complete the machine portion of the work cycle. The operator does not control this portion of the cycle time, whether or not attending the machine. *Syns*: independent machine, machine-controlled time allowance, allowance for machine-controlled time.

Machine Element. (See MACHINE-CONTROLLED TIME.)

Machine Hour. A unit for measuring the availability or utilization of machines. It is equivalent to one machine working for 60 minutes, two machines working for 30 minutes, or an equivalent combination of machines and working time.

Machine Idle Time. (1) Time during which a machine is idle during a work cycle awaiting the completion of manual work. (2) Interference time.

Machine Interference. The occurrence of conflicting demands for service by two or more units of equipment.

Machine Load. (1) The planned usage of a unit of equipment during a specified interval of time. (2) The percentage of maximum load at which the machine is actually used.

Machine Pacing. Machine or mechanical control over the rate at which the work progresses, as opposed to pacing by the worker(s). (See MACHINE-CONTROLLED TIME.)

Machine Time. (See MACHINE-CONTROLLED TIME.)

Machine Time Allowance. (See MACHINE-CONTROLLED TIME, ALLOWANCE.)

Macroelement. An element of a work cycle long enough to permit observation and timing by a stopwatch. (See MICROELEMENT.)

Maintenance. Preventive and/or correctional activities to insure that facilities and equipment are functionally capable of expected operation. As a result of these activities, equipment should be in good operating condition (clean, free from hazards, etc.) within specified limitations such as those imposed by age and prior use.

Man-hour. A unit of measure representing one person working for one hour. The combination of "n" people working for "h" hours produces nh man-hours. Frequent qualifications to the definition include: (1) designation of work effort as normal effort; (2) designation of time spent as actual hours. (See MAN-MINUTE.)

Man-machine Chart. (See MULTIPLE ACTIVITY PROCESS CHART.)

Man-minute. A unit for measuring work. It is equivalent to one person working at normal pace for one minute, two people working at normal pace for thirty seconds, or an equivalent combination of people working at normal pace for a period of time. (See MAN-HOUR.)

Man-process Chart. A graphic, symbolic representation of the work steps or activities performed or to be performed by a person. Typically, the information included on the chart is the distance the person moves and type of work done (by symbol with description). Equipment used and work times may also be included.

Manual Element. A distinct, describable, and measurable subdivision of a work cycle or operation performed by hand or with the use of tools, and one that is not controlled by process or machine.

Manual Time. The time required to perform a manual element. (See MANUAL ELEMENT.) *Syn*: hand time.

Manually-controlled Work. A work cycle consisting completely of manual elements or where the manual time controls the pace at which the work progresses. *Syn*: effort-controlled cycle.

Manufacturing Cell. A grouping of equipment to form a self-contained autonomous unit to produce parts or products of similar geometry and specifications; equipment may range from traditional machine tools through computer numerical control machines and robots.

Marstochron. An electric motor driven paper-tape puller used to record motion or work element times. An observer visually detects the endpoints of successive motions or elements and presses one or both of two keys that record these endpoints as successive marks along a time base on the tape. *Syns*: chronograph, marstograph.

Maximum Working Area. That portion of the working area that is easily accessible to the hands of an operator, with arms fully extended, who is in the normal working position with trunk erect and stationary.

Mean Time. (See AVERAGE TIME.)

Measured Day Work. (1) Work performed for a set hourly nonincentive wage where performance is compared to established production standards (most frequent use). (2) An incentive plan wherein the hourly wage is adjusted up or down and is guaranteed for a fixed future period (usually a quarter) according to the average performance in the prior period (infrequently used).

Measured Work. A term used to describe work, operations, cycles, etc., on which a standard has been set using time study or another standard setting technique.

Mechanization. The act or process of using power-driven machinery to perform specific operations or functions usually with the intent of improving productivity and/or quality of the work performed.

Median Time. That time which is greater than or equal to half of the observed times, excluding abnormal times. It is also less than or equal to the other half of the observed times.

Memomotion Study. A work measurement and methods analysis technique using a motion picture camera that records events at less than normal camera speed, *e.g.*, 50, 60, or 100 frames per minute; incorporate video equipment similarly. Same results are obtained by changing the tape transport speed or the frequency of analytical recording. Used for the analysis of long events, group activities, or processes that do not move rapidly. *(Syns*: Camera Study, Time-Lapse Photography).

Mental Work. Work done principally by the mind: logical decision-making, such as sorting, classifying, or inspecting (monitoring): recalling (memory): calculation, such as performing mathematical or verbal operations and inductive policy or hypothesis formulation. The complexity may vary from elementary mental reactions to highly involved judgments based on a large number of variable factors.

Merit Rating. A formalized system of appraising employee performance according to an established group of factors. Usually the appraisal is annual or semiannual and is by immediate supervision. Factors frequently considered include quality of work, quantity of work, reliability, adaptability, initiative, and attitude. *Syns*: performance appraisal, employee rating, personnel rating.

Method. (1) The procedure or sequence of motions by workers and/or machines used to accomplish a given operation or work task. (2) The sequence of operations and/or processes used to produce a given product or accomplish a given job. (3) A specific combination of layout and working conditions; materials, equipment, and tools; and motion patterns involved in accomplishing a given operation or task.

Methods Analysis. That part of methods engineering normally involving an examination and analysis of an operation or a work cycle broken down into its constituent parts for the purpose of improvement, elimination of unnecessary steps, and/or establishing and recording in detail a proposed method of performance.

Methods Engineering. That aspect of industrial engineering concerned with the analysis and design of work methods and systems, including technological selection of operations or processes, specification of equipment type and location, design of manual and worker-machine tasks. May include the design of controls to insure proper levels of output, inventory, quality, and cost. (See WORK DESIGN, MOTION ANALYSIS, MOTION ECONOMY, METHODS ANALYSIS.)

Methods Study. A systematic examination of existing methods with the purpose of developing new or improved methods, tooling, or procedures.

Methods Time Measurement (MTM). A proprietary predetermined time standards system.

Microchronometer. A large-faced electric clock with rapidly moving hands used in micromotion studies (within the camera's view) to indicate the passage of time. The clock usually measures to the nearest wink, or 0.0005 minutes. *Syn*: wink counter.

Microelement. An element of work too short in time to allow it to be observed with the unaided eye. (See ELEMENTAL MOTION.)

Micromotion Study. A work measurement or methods analysis technique using a motion picture camera or video equipment to record events at normal (960 frames per minute) or faster than normal camera speed. Used for the analysis of short, highly detailed movements that are too rapid for satisfactory visual observation. The camera may be driven so as to act as a timing device for the measurement of motions or elements, or there may be a timing device such as a microchronometer in the camera's field of view. *(Syn:* camera study).

MINIMUM TIME. The shortest actual time recorded during a time study for each element of work.

MODAL TIME. The actual time value for an element or operation that occurs more often than any other time value.

MOST®. (MAYNARD OPERATION SEQUENCE TECHNIQUE.) A proprietary predetermined time standards system.

MOTION ANALYSIS. The study of the basic divisions of work involved in the performance of a given operation for the purpose of eliminating all useless motions and arranging the remaining motions in the best sequence for performing the operation. (See PRINCIPLES OF MOTION ECONOMY.)

MOTION CYCLE. The complete sequence of motions and activities required to do one unit of work or to perform an operation once. (See CYCLE.)

MOTION ECONOMY. (See PRINCIPLES OF MOTION ECONOMY, MOTION ANALYSIS.)

MOTION STUDY. (See MOTION ANALYSIS.)

MOTIVATED PRODUCTIVITY LEVEL (MPL). The work pace of a motivated worker possessing sufficient skill to do the job, physically fit to do the job, after adjustment to it, and working at an incentive pace that can be maintained day after day without harmful effect on a safe job. MPL can be used as a base for measuring work performance by establishing normal performance at a specified work level from MPL. For example, if 100% is designated as MPL, and MPL could be 130% with respect to APL, then in this case APL would be 77% and MPL would be 100%. Or, APL can be 100% and MPL could be 130% (See HIGH TASK, INCENTIVE PACE, NORMAL PERFORMANCE, ACCEPTABLE PRODUCTIVITY LEVEL).

MULTIPLE ACTIVITY OPERATION CHART. (See MULTIPLE ACTIVITY PROCESS CHART.)

MULTIPLE ACTIVITY PROCESS CHART. A chart of the coordinated synchronous or simultaneous activities of a work system of one or more machines and/or one or more workers. Each machine and/or worker is shown in a separate, parallel column indicating their activities as related to the rest of the work system. Examples: multiworker process chart, Gantt chart, multiworker-machine process chart, worker-machine process chart, worker-multimachine process chart.

Syns: multiple activity operation chart, multiple activity chart.

MULTIPLE WATCH TIMING. (See ACCUMULATIVE TIMING.)

NONBALLISTIC MOTIONS. (See BALLISTIC MOVEMENT.)

NONCYCLIC ELEMENT. An element of an operation or process that does not occur every cycle of the operation or process, but its frequency of occurrence in the operation or process is specified by the method.

NONREPETITIVE. (1) Generally an operation or process that is performed for only one or a few cycles before it has to be changed significantly to adapt to new requirements. (2) Odd-job production. (3) An operation that does not have a predictable order of elements. (4) An occasional and/or varying element, operation, or job.

NORMAL EFFORT. The effort required in manual work to produce normal performance. (See NORMAL PERFORMANCE.)

NORMAL ELEMENT TIME. (1) The selected (average, modal, or other) element time adjusted by performance rating to obtain the time required by an average qualified worker to perform a single element of an operation while working at a normal pace. (2) The time resulting from application of a predetermined time system to a prescribed work method.

NORMAL PACE. The manual pace required to produce normal performance. (See NORMAL PERFORMANCE.)

NORMAL PERFORMANCE. (1) The work output of a qualified employee considered acceptable in relation to standards and/or pay levels, which result from agreement, with or without measurement, by management or between management and the workers or their representatives. (2) An acceptable amount of work produced by a qualified employee following a prescribed method under standard safe conditions with an effort that does not incur cumulative fatigue from day to day. (3) The base performance level above which incentive bonus is paid. (See FAIR DAY'S WORK, SHOULD-TAKE.)

NORMAL TASK. (See NORMAL PERFORMANCE.)

NORMAL TIME. The time required by a qualified worker to perform a task at a normal pace to complete an element, cycle, or operation using a prescribed method. (See NORMAL PERFORMANCE.) *Syns:* base time, leveled time.

NORMAL WORKING AREA. (1) The area at the workplace which is bounded by the arc drawn by the worker's fingertips moving in the horizontal plane, with the elbow as a pivot, when the worker is standing or seated in the normal working position with the arm close to the body hanging in a stationary position: the section where the right and left hands overlap in front of the worker constitutes the normal working area for the two hands. (2) In a vertical plane, the space on the surface of the imaginary sphere which would be generated by rotating about the worker's body as an axis, the arc traced by the worker's fingertips of the right or left hand when the forearm is moved vertically about the elbow as a pivot. (3) The space within reach of a worker's fingertips as they develop arcs of revolutions, the elbows acting as a pivot when the worker is standing or sitting in the normal working position with the upper arm hanging from the shoulder close to the body in a stationary position.

NORMS. A word frequently employed with different meanings, depending on how it is used. Most frequently it refers to the central value from a population distribution as with normal effort, normal element time, normal pace, normal performance, normal time or normal working area. It is the most frequently used word in work measurement definitions where it usually refers to the central measure of a pertinent distribution. However, with the term *normal operator*, it means one reasonably expectable from the usual distribution.

NUMBERING SYSTEM. (1) A plan for the assignment of numeric keys to items or cases included within a given classification. (2) Means of identifying individual or group arrangements.

NUMERICAL CONTROL. A system of controlling a unit, usually a machine tool, whereby either a binary or decimal digit system is programmed to carry out machining operations through electronic circuits and related activating mechanisms.

OBJECTIVE RATING. A form of performance rating where all jobs are rated against a single standard base. Difficulty adjustment factors are added in terms of objectively describable features of the work. (See PERFORMANCE RATING.)

OBSERVATION. (1) In time study, the act of noting and recording the elapsed time taken by a worker performing an operation or an element of an operation. (2) In motion study, the act of noting and recording the motions used by a worker to perform an operation or an element of an operation. (3) In work sampling, the act of noting and recording what a worker is doing or what is happening in an operation at a specific instant.

OBSERVED TIME. (See ACTUAL TIME.)

OCCURRENCE (FREQUENCY). (1) The number of times an event takes place, usually in a specific time period. (2) The number of times an element occurs per cycle.

ONE BEST WAY. The concept that for every job there is an optimal work method that can be developed and specified. A concept originated by Frank and Lillian Gilbreth.

OPERATION. (1) A job or task, consisting of one or more work elements, usually done essentially in one location. (2) The performance of any planned work or method associated with an individual, machine, process, department, or inspection. (3) One or more elements which involve one of the following: the intentional changing of an object in any of its physical or chemical characteristics; the assembly or disassembly of parts or objects; the preparation of an object for another operation, transportation, inspection, or storage; planning, calculating, or the giving or receiving of information.

OPERATION BREAKDOWN. (See JOB BREAKDOWN.)

OPERATION CHART. (See RIGHT- AND LEFT-HAND CHART.)

OPERATION PROCESS CHART. A graphic, symbolic representation of the act of producing a product or providing a service, showing operations and inspections performed or to be performed with their sequential relationships and materials used. Operation and inspection time required and location may be included.

OPERATIONS ANALYSIS. A study of an operation or scenes of operations involving people, equipment, and processes for the purpose of investigating the effectiveness of specific operations or groups so that improvements can be developed which will raise productivity, reduce costs, improve quality, reduce accident hazards, and attain other desired objectives.

OPERATION TIME CHART. (See OPERATOR PROCESS CHART.)

OPERATOR PROCESS CHART. An operation process chart describing the work done by one operator. (See RIGHT- AND LEFT-HAND CHART.) *Syn*: operation time chart.

OPERATOR PRODUCTIVITY. The ratio of standard time or other performance standard to the actual time or other performance measure for the same task. When this ratio is equal to 1.00 (100%) the operator is meeting standard output. (See PERFORMANCE INDEX.) *Syn*: operator performance.

OUTLIER. In a group of data, a value which is so far removed from the rest of the distribution that its presence cannot be reasonably explained by the random combination of chance causes. (See ABNORMAL TIME.)

OUTPUT. A measure of the value of completed work units, usually expressed as the employee-hours to complete the work units; may be expressed as the money value. The ratio of output to input is a measure of performance.

OUTSIDE WORK. (See EXTERNAL WORK.)

PACE RATING. (See PERFORMANCE RATING.)

PARETO'S LAW. Sometimes called the law of the trivial many and the critical few. A principle which states that, in most activities, a small fraction (commonly estimated at 20%) of the total activity creates the major portion (commonly estimated at 80%) of the work, cost, profit, or other measure of importance. *Syn*: rule of 80-20 (q.v.).

PERFORMANCE. A measure of how much of a goal is achieved. May be called "worker performance"; how well the worker's work time compared to the standard time; "group performance" would be similar. May also be called "productivity performance"; how well the productivity achieved compared to the productivity goal set; "effectiveness performance" as a measure of what was accomplished compared to the goal set for results.

PERFORMANCE EVALUATION. A critical and objective appraisal of performance measurement data and related information to obtain an accurate picture of the overall status of a specific area or persons to ascertain exceptional accomplishments, identify shortcomings and their causative factors, and develop meaningful recommendations.

PERFORMANCE INDEX. The ratio of a performance standard established for a certain quantity of work to the performance actually achieved. When this ratio is equal to 1.00 (100%) the worker or group is meeting standard performance. (See OPERATOR PRODUCTIVITY.)

PERFORMANCE INDICATOR. A significant quantitative measure of performance which provides the best perspective of total management effort being applied in an area.

PERFORMANCE MEASUREMENT. The assessment of accomplishments in terms of historical or objective standards or criteria. (See PERFORMANCE EVALUATION.)

PERFORMANCE RATING. (1) A process whereby an analyst evaluates observed operator performance in terms of a concept of normal performance expressed as benchmarks using APL/MPL concepts. (2) The performance rating factor. *(Syns*: leveling, pace rating, effort rating, objective rating, normal performance).

PERFORMANCE RATING FACTOR. The number (usually a percentage) representing the performance rating.

PERFORMANCE RATING SCALE. A numerical scale of performance which may or may not include defined benchmarks. For example, normal performance might be expressed as 100% or 60 minutes per hour. The 100% scale is the most common scale used.

PERFORMANCE RATIO. (See PERFORMANCE INDEX.)

PERFORMANCE SAMPLING. A technique for determining the performance rating factor to be applied to an operator or a group of operators determined by short randomly spaced observations of the performance.

PERFORMANCE STANDARD. A criterion or benchmark to which actual performance is compared.

PERSONAL ALLOWANCE. An allowance to provide time for the personal needs of the worker during the workday. (See ALLOWANCE.) *Syn:* personal time.

PERSONAL TIME. (See PERSONAL ALLOWANCE.)

PREDETERMINED MOTION TIME SYSTEM. (See PREDETERMINED TIME SYSTEM.)

PREDETERMINED TIME. (See PREDETERMINED TIME SYSTEM.)

PREDETERMINED TIME SYSTEM. An organized body of information, procedures, techniques, and motion times employed in the study and evaluation of manual work elements. The system is expressed in terms of the motions used, their general and specific nature, the conditions under which they occur, and their previously determined performance times. *Syn*: predetermined motion time system.

PRINCIPLES OF MOTION ECONOMY. A general listing of common sense steps and procedures to simplify and improve the effectiveness of manual work.

PROCESS. (1) A planned series of actions or operations (*e.g.*, mechanical, electrical, chemical, inspection, test) which advances a material or procedure from one stage of completion to another. (2) A planned and controlled treatment that subjects materials or procedures to the influence of one or more types of energy (*e.g.*, human, mechanical, electrical, chemical, thermal) for the time required to bring about the desired reactions or results.

PROCESS CHART. A graphic, symbolic representation of the specific steps in a processing activity. (See FLOW PROCESS CHART, OPERATION PROCESS CHART, MAN-PROCESS CHART, FLOWCHART, MULTIPLE ACTIVITY PROCESS CHART, OPERATOR PROCESS CHART.)

PROCESS CHART SYMBOLS. Graphical symbols or signs used on process charts to depict the type of events that occur during a process. (See FLOW PROCESS CHART.)

PROCESS DESIGN. The act of prescribing the production process to produce a product as designed. This may include specifying the equipment, tools, fixtures, machines, and the like required: the methods to be used: the personnel necessary; and the estimated or allowed times. (See METHODS ANALYSIS, PROCESS.)

PROCESS ENGINEER. An individual qualified by education, training, and/or experience to prescribe efficient production processes to safely produce a product as designed and who specializes in this work. This work includes specifying all the equipment, tools, fixtures, human job elements, and the like that are to be used and, often, the estimated cost of producing the product by the prescribed process. (See PROCESS, PROCESS DESIGN.)

PROCESSING. The carrying out of a production process. (See PROCESS.)

PROCESS PLANNING. A procedure for determining the operations or actions necessary to transform material from one state to another.

PROCESS SHEET. A sketch, diagram or listing of the operations in the sequential order necessary to accomplish the desired result (such as transforming material from one state to another).

PROCESS TIME. (1) Time required to complete the machine or process-controlled portion of a work cycle. (2) Time required to complete an entire process.

PRODUCTION STANDARD. (See STANDARD TIME.)

PRODUCTION STUDY. (1) A detailed analysis of a job, operation, process, or group of activities using the techniques of methods engineering and work measurement with the objective of improvement. (2) An extended time study to determine delay allowances or verify other major variables—sometimes called an eight-hour study.

PRODUCTIVE LABOR. (See DIRECT LABOR.)

PRODUCTIVE TIME. Time in which effective work is done in an operation or process, as opposed to nonproductive or idle time.

PRODUCTIVITY. (1) The ratio of output to total inputs. (2) The ratio of actual production to standard production, applicable to either an individual worker or a group of workers.

PRODUCTIVITY COMPANY LEVEL. Ratio of outputs to inputs relative to the whole of a company or part of a company independently serving the open economic market, compared to a similar ratio (using the same factors) for a previous period for the same company or part of a company.

PRODUCTIVITY, GROUP. Ratio of outputs to inputs relative to a generalized work area or function, *e.g.*, materials handling, purchasing, warehousing, etc., compared to a similar ratio (using the same factors) for a previous period of the same work group.

PRODUCTIVITY INDEX. (See PRODUCTIVITY.)

PRODUCTIVITY, NATIONAL. An estimate of the total "value" of outputs of a country, compared to the totality of inputs, all compared to a similar ratio for some base period. The word "estimate" is used because it is most unlikely that the final value will be estimated from a sample of firms; it is unlikely that we will have universal reporting. The word "value" is in quotation marks because the basis of value is not a scientific measurement; we must assign the unit of value.

PRODUCTIVITY, WORKPLACE. Ratio of outputs to inputs relative to a single workplace, compared to a similar ratio (using the same factors) for a previous period at the same workplace.

PROFIT SHARING PLAN. An arrangement under which an organization pays or makes available to employees, subject to eligibility rules, in addition to regular remuneration, special current or deferred sums of money based on the profits of the organization.

PROGRESS CHART. A graphical representation of the status or extent of completion of work in process. (See GANTT CHART.)

PROGRESS CURVE. A plot of work accomplished or productive output versus time. May be accompanied by a plot of expected or planned output for comparison purposes. *Syn*: learning curve

QUALIFIED OPERATOR. A worker who, by virtue of training, skill, and experience, is able to perform a task within acceptable quality and time limits.

QUALITY CIRCLE. A group of employees who volunteer to create improvements in company operations; reward is satisfaction, usually no monetary rewards are earned.

RANDOM ELEMENT. (See FOREIGN ELEMENT.)

RANDOM SAMPLE. A sample selected in such a way that each element of the population being sampled has an equal chance of being selected.

RATE. (1) Hourly wage rate. (2) To evaluate the observed performance of a task in comparison with some concept of normal performance. (3) The quantity of output produced per unit of time. (4) The quantity of output produced expressed as a percent of either capacity or normal output. (5) Piece rate. (See PERFORMANCE RATING.)

RATE CHANGE. (1) Rate refers to production in time; an increase in a production or time standard made because of a revision in product design, quality requirements, production methods, materials, or conditions. (2) Rate refers to amounts of money; an increase or decrease in money paid per unit of time or unit of output.

RATE CUTTING. The arbitrary reduction of a standard time or incentive pay rate. Not considered good practice.

RATED AVERAGE ELEMENT TIME. (See NORMAL ELEMENT TIME.)

RATE SETTING. (1) The establishment of pay per unit for incentive work. (2) The establishment of a standard time. *Syn*: rate determination.

RATING. (See PERFORMANCE RATING.)

RATIO-DELAY STUDY. (See WORK SAMPLING.)

RAW TIME. (See ACTUAL TIME.)

READING POINT. (See BREAKPOINT.)

REASONABLE EXPECTANCY (RE). Usually used in short interval scheduling, the time standards issued with work assignments representing normal performance. RE work loads are set for short term periods. Standards may be estimated or established through work measurement.

REENGINEERING. Is the rethinking and unconstrained re design of business processes to achieve improvements in measures of performance, such as costs, quality, service, organization structures and other company aspects.

REGULAR ELEMENT. An element of an operation or process that occurs either every cycle of the operation or process, or occurs frequently and in a fixed pattern with the cycles of that operation or process. For example, once every third cycle or four cycles out of five.

RELAXATION ALLOWANCE. (See FATIGUE ALLOWANCE, PERSONAL ALLOWANCE.)

REPETITIVE ELEMENT. (See REGULAR ELEMENT.)

REPETITIVE TIMING. A stopwatch technique where a time value is read and recorded at each breakpoint and the watch is instantaneously reset to zero to begin timing the next element. *Syn*: snapback timing.

REST ALLOWANCE. (See FATIGUE ALLOWANCE, PERSONAL ALLOWANCE.)

RESTRICTED ELEMENT. (See RESTRICTED WORK.)

RESTRICTED WORK. Manual or human-machine work for which the pace or speed of work is not completely under the control of the worker. (See MACHINE-CONTROLLED TIME.)

REWORK. (1) The process of correcting a defect or deficiency in a product or part. (2) Units of product requiring correction.

RIGHT- AND LEFT-HAND CHART. A chart on which the motions made by one hand in relation to those made by the other hand are recorded using standard process chart symbols or basic therblig abbreviations or symbols. (See OPERATOR PROCESS CHART.)

RUCKER® PLAN. A proprietary gain sharing plan that measures economic gains as a ratio of dollar payroll to value added, established as sales dollars minus material costs.

RULE OF 80-20. (See PARETO'S LAW.)

RUNOUT TIME. Time required by machine tools after cutting time is completed before the tool and material are completely free of interference so that the next sequence of operations can proceed.

SCANLON PLAN. A gain sharing plan that measures economic gains as a ratio of dollar payroll to sales dollars. The generic plan is in the public domain.

SELECT(ED) ELEMENT TIME. (See SELECT(ED) TIME.)

SELECT(ED) TIME. The time which is chosen by simple observation or by statistical means as being representative of the actual time values (prior to applying a performance rating factor) obtained from the observation of an element or operation. (See AVERAGE CYCLE TIME, AVERAGE ELEMENT TIME.)

SEQUENCING. Specifying the order of performance of tasks so that available production facilities are utilized in an optimal manner.

SETUP. Preparation of a workplace or a machine for a specific work method, activity, or process. Includes installation of all necessary hand tools, jigs, fixtures, and other tools or equipment in the location and condition for proper performance of the work.

SHORT INTERVAL SCHEDULING (SIS). A technique of dispatching batches of work, usually at one hour intervals, accompanied by time standards. Workers are closely monitored to determine reasons for not completing work in an assigned time interval. Standards are called "reasonable expectancies."

SHOULD-TAKE, DID-TAKE TIME STANDARDS. Normal performance usually involves setting standards through performance rating or predetermined time standards, creating time values that an average qualified worker "should take" to perform a task. "Did-take" standards are established based on historical data without the use of performance rating or benchmark references. (See NORMAL PERFORMANCE.)

SIMO CHART (SIMULTANEOUS MOTION CHART). A chart for displaying two-handed work with motion symbols plotted vertically against time. The therblig or motion abbreviation and a brief description are shown for each activity. In addition, individual time values and body member detail may be shown. (See RIGHT- AND LEFT-HAND CHART.)

SIMPLIFIED PRACTICE. (1) The practices or operations resulting from a work simplification or methods study. (2) A description of the work method of a job, specified in somewhat less detail than in a standard practice.

SIMULTANEOUS MOTIONS. Two or more elemental motions performed during the same time interval by different body members.

SKILL. (See JOB SKILL.)

SNAPBACK TIMING. (See REPETITIVE TIMING.)

SPEED RATING. (See PERFORMANCE RATING.)

STANDARD. (1) An established norm for the measure of quantity, weight, extent, value, quality, or time. (2) Standard time.

STANDARD ALLOWANCE. An allowance calculated, arbitrarily set, or negotiated to provide in advance for specified conditions. (See ALLOWANCE.)

STANDARD DATA. A structured collection of normal time values for work elements codified in tabular or graphic form. The data are used as a basis for determining time standards on work similar to that from which the data were collected without making additional time studies. (See SYNTHETIC DATA.)

STANDARD ELEMENT TIME. A standard time for an individual work element. (See STANDARD TIME.)

STANDARD HOUR. The quantity of output required of an operator to meet exactly the production quota for one hour. The production quota is normally based on a standard time. Also used to refer to an hour of less than 60 minutes when allowances are expressed as nonproductive minutes. (See ALLOWED HOURS.)

STANDARD OUTPUT. The reciprocal of standard time expressed in appropriate units (*e.g.*, dozens of units per hour, tons per day, or hundreds of barrels per week).

STANDARD PERFORMANCE. The performance of a person or group achieving standard output.

STANDARD PRACTICE. A description of a work method wherein all of the significant variables of the method have been specified in detail. Usually follows a specified format. (See METHOD.) *Syns*: standard method, written standard practice.

STANDARD SYSTEM. The common name for a codified set of time-motion data, often covering both general and proprietary data sets, considered the usual practice for a given plant or location and thus regarded as authoritative. (See PREDETERMINED TIME SYSTEM.)

STANDARD TIME. A unit time value for the accomplishment of a work task as determined by the proper application of appropriate work measurement techniques by qualified personnel. Generally established by applying appropriate allowances to normal time. Standard time and normal time are identical when nonproductive time is granted in lieu of allowances. (See NORMAL PERFORMANCE.) *Syns*: direct labor standard, engineered performance standard, engineered standard, output standard, production standard, time standard.

STANDARD TIME DATA. (See STANDARD DATA.)

STANDARD TIME, STATISTICAL. A standard time developed from statistical analysis of past performance time data. *Syn:* historical time.

STANDARDS AUDIT. A work measurement study or sequence of studies intended to test the correctness of existing standard times and methods. By means of periodic sampling of work times, an attempt is made to detect significant changes.

STANDBY. A category of time in which the worker is not actively engaged in producing a unit of output but is available to take appropriate action when needed. Standby is recognized only when it is essential to the task and when no other work can be done during the standby period. (See DELAY.)

STANDBY TIME. The time expended in standby status, *e.g.*, the time spent by workers in awaiting equipment, labor crews, or work assignment; or due to failure of utilities, inclement weather, and other similar occurrences.

START-UP CURVE. A learning curve applied to a job or process to adjust for work times longer than standard, or average, as a result of the introduction of a new job or new worker(s). (See LEARNING CURVE.)

STATIC WORK. Work performed by the hands or arms where no significant motion occurs, *e.g.*, holding.

STATISTICAL TIME. (See STANDARD TIME, STATISTICAL.)

STOPWATCH. A portable timing device that can be started or stopped at will by the user to register continuous and/or elapsed time. (See DECIMAL-HOUR STOPWATCH, DECIMAL-MINUTE STOPWATCH.)

SUBTRACTED TIME. The difference between successive stopwatch readings when using a continuous timing technique. Usually represents the time for one element.

SYNCHRONIZATION ALLOWANCE. (See INTERFERENCE ALLOWANCE.)

SYNTHETIC DATA. (1) Work measurement time values not obtained from direct measurement of the work to which they are applied. Generally represent values for task elements that are sufficiently basic as to occur in several jobs, obtained from measuring task elements in similar jobs or from predetermined time systems. (2) Any production data not measured directly from but applicable to a given situation. (See STANDARD DATA, PREDETERMINED TIME SYSTEM.)

SYNTHETIC TIME STANDARD. A standard time determined from synthetic data.

SYSTEM. A set of interrelated parts that operate as a whole in pursuit of common goals; is characterized by: a) a set of components of subsystems linked by information channels, b) engaged in coordinated, goal-directed activity, c) information flow as the basis for control, d) a set of subgoals associated with the individual subsystems or components, e) an external environment which influences the system. A system is said to be an open system if it reacts to its environment and is a closed system if it does not. It is an adaptive system if it reacts to environmental changes in a way that is favorable toward achieving the system goals.

TASK. (See JOB.)

TEARDOWN. All work items required between the end of one operation or job and the start of setup for the next operation or job, both jobs requiring the same machinery or facilities.

TEMPORARY RATE. (1) A production or time standard based on a temporary standard. (2) Wage incentive pay rate based on a temporary standard such as for while awaiting study or other causes for delay.

TEMPORARY STANDARD. An approximate standard time intended to apply for a limited time to account for some unusual job condition or while awaiting restudy of the task to which it applies.

THERBLIG. A short manual work segment used to describe the sensory-motor activities or other basic elements of an operation. Developed by Frank and Lillian Gilbreth, therbligs form a basic language for methods description and, in modified form, for elemental motion time data. The original seventeen are: search, select, grasp, transport empty, transport loaded, hold, release load, position, preposition, inspect, assemble, disassemble, use, unavoidable delay, avoidable delay, plan, rest for overcoming fatigue. *Syn*: Gilbreth basic element, basic division of accomplishment, fundamental motion, basic motion, basic element. (Therblig spelled backwards is Gilbreth.)

Therblig Chart. An operation chart with the suboperations broken down into individual motions, and all motions designated with their appropriate therblig symbols. *Syns*: right- and left-hand chart, simo chart.

Thruput. The quantity of acceptable product processed through a manufacturing line or machine in a specified time (such as an eight-hour shift.)

Tight Rate. (See TIGHT STANDARD.)

Tight Standard. A standard time less than that required by a qualified workman with normal skill and effort following a prescribed method and including allowances for delays, personal needs, and rest. *Syn*: tight rate.

Time Allowance. (See ALLOWANCE.)

Time Formula. A formula for determining the normal time or standard time of a task as a function or one or more variables in the task. Included are coefficients for the variables so that insertion of the variable values allows direct time computation.

Time Standard. (See STANDARD TIME.)

Time Study. A work measurement technique consisting of careful time measurement of the task with a time measuring instrument, adjusted for any observed variance from normal effort or pace and to allow adequate time for such items as foreign elements, unavoidable or machine delays, rest to overcome fatigue, and personal needs. Learning or progress effects may also be considered. If the task is of sufficient length, it is normally broken down into short, relatively homogenous work elements, each of which is treated separately as well as in combination with the rest.

Time Study Observation Sheet. A form for the systematic, detailed recording of element time values, and irregular occurrences observed during a time study. Generally, space is also provided for entering other pertinent information and for computation of standard times from the data. *Syns*: time study computation sheet, time study form.

TMU. (Time Measurement Unit.) Unit of measure for a number of work measurement systems. 100,000 TMU's equal one hour.

Tolerance. (1) A permissible variation in a characteristic of a product or process, usually shown on a drawing or specification. (2) In work measurement, the permissible variation of a time value for an operation or other work unit.

Tolerance Limits. (1) The upper and lower extreme values permitted by the tolerance. (2) In work measurement, the limits between which a specified operation time value or other work unit will be expected to vary.

Training Allowance. (1) An allowance to compensate for an untrained worker's learning effect. (2) An adjustment to an incentive pay piece rate for the same reason. (3) An adjustment to earnings of an individual or group for the same reason. Usually involves adjusting actual earnings under an incentive plan to some guaranteed percentage, or to prior average earnings. (See ALLOWANCE.) *Syn*: learner's allowance.

Travel Chart. A table giving distances travelled between points in a manufacturing facility. Values may be adjusted to reflect weight, value, or some other factor depending on circumstances.

Travel Time. Time required to move material, equipment, personnel, or information from one work or storage area to another.

Unavoidable Delay. A delay which is outside the control or responsibility of the worker.

Unavoidable Delay Allowance. An allowance intended to provide time for expected unavoidable delays in a task. (See ALLOWANCE, UNAVOIDABLE DELAY.)

Unrestricted Element. An operation element that is completely under the control of the worker. Syn: manual element.

Unrestricted Job. A job that is completely under the control of the worker.

Uptime. Time when an operator or a machine is working.

Value Analysis. Review of product costs to evaluate contribution to product value. May include phases of work design, methods engineering, and motion economy to reduce manufacturing costs. May include phases of work simplification and brainstorming to evaluate use of alternate materials, components, or work specifications.

Value Engineering. (See VALUE ANALYSIS.)

Variable Element. (1) An element whose normal time varies significantly from cycle to cycle as a function of one or more job variables. (2) An element common to two different jobs and whose time varies due to differences between the jobs.

Video Tape Analysis. (See FILM ANALYSIS RECORD).

Wage Incentive Plan. A method of compensation based on pay by performance in which improved performance earns extra pay for the employee. Standards may be time values or dollars per piece. Plans usually provide a minimum pay level for work below normal or standard. Plans are usually for individual employees; small group incentives will include groups working together or on assembly lines. These plans are distinguished from broad based plans such as gain sharing, profit sharing, executive bonus plans, and other pay for performance plans.

Waiting Time. The time when an operator or a machine waits for service, parts, inspection, instructions and for other causes.

Wink. One division on the microchronometer equal to 1/2,000 (.0005) minute.

Wink Counter. (See MICHROCHRONOMETER.)

Work Cell. A group of machines arranged to process a family of parts in a way that minimizes material handling and storage between operations; can handle a lot size of one piece, and integrated in such a way that the cell produces a finished product.

Work Cycle. (1) A pattern or sequence of tasks, operations, and/or processes. (2) A pattern of manual motions, elements, operations, and/or activities that is repeated without significant variation each time a unit of work is completed. (See MOTION CYCLE.)

Work Design. The design of work systems. System components include people, machines, materials, sequence, and the appropriate working facilities. The process technology and the human characteristics are considered. Individual areas of study may include analysis and simplification of manual motion components: design of jigs, fixtures, and tooling; human-machine analysis and design; or the analysis of gang or crew work. *Syns*: ergonomics, job design, methods engineering, methods study, motion study, operation analysis, work simplification, motion economy.

Work Factor (WF). A proprietary predetermined time standards system.

Working Area. That portion of the workplace within which an operator moves about in the normal job performance.

Working Conditions. Generally refers to surrounding conditions or factors such as noise, temperature, air pollution, presence of chemical odors, health effecting factors, and so on.

Work Measurement. A generic term used to refer to the setting of a time standard by a recognized industrial engineering technique, such as time study, standard data, work sampling, or predetermined motion time systems. *Syn*. ergonometrics.

Workplace Layout. The manner in which all of the items necessary to perform a work task, as specified by the standard method, are arranged.

Work Sampling. An application of random sampling techniques to the study of work activities so that the proportions of time devoted to different elements of work can be estimated with a given degree of statistical validity.

Work Simplification. A management philosophy of planned improvement using any or all of the tools and techniques of industrial engineering in an atmosphere of creative participation which enables employees to achieve individual goals through the achievement of organizational goals. (See WORK DESIGN.)

Work Station. (See WORKPLACE.)

Work Study. The techniques of methods study and work measurement employed to ensure the best possible use of human and material resources in calling out a specific activity.

Work Task. A specific quantity of work, set of duties or responsibilities, or job function assigned to one or more persons.

Work Unit. An amount of work, or the results of an amount of work, that it is convenient to treat as an integer (an each) when examining work from a quantitative point of view.

Work Unit Analysis. A hierarchical delineation of the objectives and types of outputs of an organization, and the subparts of these outputs, in work-unit terms.

Written Standard Practice. (See STANDARD PRACTICE.)

INDEX

5 (five) competitive forces model, 10-12
5th (fifth) lumbar vertebra, 2-36. *See also* herniated disk; lumbosacral angle
5th to 95th percentile, design criteria, 2-6
80-20 rule. *See* Pareto's law

A

a-scale
A-stage
abandon, 6-3
abandonment, 6-3
abate, 13-2
ABC classification, 14-3. *See also* Pareto's law
ABC inventory control, 14-3
abdominal wall, 2-29
abduction, 2-29, 9-2. *See also* adduction
abduction, angle of, 2-30
abductors, 2-29
abilities test, physical, 6-42
ability, 6-3, 10-2
ability to pay, 6-3
abnormal, 13-2
abnormality, 13-2
abnormality, congenital, 13-10
abnormality, inherited, 13-25
abnormal reading. *See* abnormal time
abnormal time, 17-2. *See also* outlier(s)
abrasion, 13-2
abrasion resistance, 12-30
abrasive machining, 12-60
abrasives, 12-30, 12-60
abridgement of damages, 13-2
ABS (acrylonitrile butadiene styrene), 12-68
absence, 6-3
absence allowance, paid, 6-39, 13-35
absenteeism, 6-3
absentee rate, 6-3
absolute, 11-2
absolute maximum, 1-2
absolute minimum, 1-2
absolute risk, 13-2
absolute threshold, 9-2
absorption, 12-68, 13-2
absorption, sound, 13-42

Ac (acceptance number), 16-2
AC (adaptive control), 11-2, 12-3
ACC (adaptive control constrained), 12-3
accelerated test, 1-2, 16-2
acceleration, 4-2
acceleration, angular, 2-30
acceleration factor, 1-2
acceleration illusions, 13-2
acceleration resistance, 8-2
acceleration signature. *See* signature
acceleration syndrome, 13-2
accelerator, 12-30
acceptable process level (APL), 16-2
acceptable productivity level (APL), 17-2. *See also* MPL (motivated productivity level); normal performance
acceptable quality (AQL), 10-2, 16-2
acceptance control chart, 16-2
acceptance criteria, 16-2
acceptance final (partial), 4-2
acceptance inspection, 16-2
acceptance number (Ac), 16-2
acceptance sampling, 1-2, 10-2, 16-2
acceptance theory of authority, 10-2
accessibility score, 9-2
accession, 6-3
accessory, 8-2, 14-3. *See also* attachment; option
access, random, 1-2
access time, 3-2
access to the work, 4-2
accident, 6-3, 8-2, 9-2, 13-2
accidental death and dismemberment benefits, 6-3. *See also* worker's compensation
accident analysis, 13-4
accident and sickness benefits, 6-3, 13-2. *See also* health and insurance plan; sick leave; temporary disability insurance; worker's compensation
accident, causal factors, 13-9
accident, cause, 13-2
accident, circumstances of an, 13-9
accident costs, 13-2. *See also* indirect cost
accident experience, 13-2
accident, fatal, 13-18
accident, final position after, 13-18
accident, fire, 8-13

accident, first aid case, 6-21
accident-free, 13-3
accident, frequency rate of, 6-21
accident hazard, 13-2
accident, industrial. *See* occupational injury
accident, key event of, 13-27
accident location, 13-2
accident, lost time (LT), 13-28
accident, marine, 8-20
accident, motor vehicle, 8-21
accident, motor vehicle nontraffic, 13-31
accident, near, 13-32
accident, nonfatal injury, 13-33
accident, noninjury, 13-33
accident, PD (property damage), 13-37
accident, potential, 13-2
accident, prevention, 13-2, 13-36
accident, probability, 13-3
accident, proneness, 6-3, 13-3
accident rate, 13-3
accident rate, frequency, 6-21
accident, records, 13-3
accident, reporting, 13-3
accident, severity rate of, 6-51
accident statistics, 13-3
accident susceptibility, 13-3
accident, traffic, 13-45
accident type, 13-3, 13-46
acclimatization, 13-3
accommodation, 9-2, 10-2
accommodation, active, 11-2
accommodation, multivariate, 2-5, 2-6
accommodation, passive, 11-13
accordion roller conveyor, 8-2. *See also* roller conveyor
accountability, 4-2, 6-3, 10-2
account code structure, 4-2
accounting life, 7-6
accounting prices, 1-2
account number, 4-2
accounts. *See* work item
accounts, accrual, 6-3
accounts payable, 4-2. *See also* taxes payable
accounts receivable, 4-2
accrual accounts, 6-3
accrual of benefits, 6-3
accumulating conveyor, minimum pressure, 8-21
accumulative timing, 17-2
accumulator conveyor, 8-2
accuracy, 1-2, 3-2, 9-2, 11-2, 12-3, 13-3, 16-2

accuracy, dynamic, 11-7
acetal resin, 12-68
achievement, 6-3
achievement-oriented, 10-2
acid, 12-30
acid-acceptor, 12-68
acne, oil, 13-34
ACO (adaptive control optimized), 12-3
ACOEM (American College of Occupational and Environmental Medicine), 13-4
acoustic impedance, specific, 2-49
acoustic scattering, 9-2
acoustic stimulus, 2-29, 9-2
acoustic trauma, 9-2, 13-3
acquired immune deficiency syndrome (AIDS), 13-3
acquired-needs theory, 10-2
acquisition, 10-2
acromion, 2-29
across the board increase, 6-3
acrylic plastics, 12-68
acrylic resin, 12-68
acrylonitrile butadiene styrene (ABS), 12-68
actinic keratoconjunctivitis, 13-46. *See also* welder's flashburn
act, intervening, 13-26
action control limits, 16-2
action limits, 16-2
action message, 14-3
action plan, 10-2
action potential, 2-29
action, preventive, 16-12
action research, 10-2
activating receptors, 2-29
activation, 12-68
active accommodation, 11-2
active employees, 6-3
active listening, 10-2
active redundancy, 1-21, 16-14
activity, 1-2, 4-2, 10-2. *See also* work item
activity analysis, 9-2
activity analysis problem, 1-2
activity code, 4-2
activity, completed, 4-7
activity description, 4-2
activity duration, 4-2. *See also* duration
activity identifier, 4-2
activity level, 1-2
activity plateau, 6-3
activity redundancy, 16-14
activity sampling, 9-2, 13-3, 17-2. *See also* work sampling

This index is arranged in letter-by-letter format.

activity splitting, 4-2
activity times, 4-2
activity total slack, 4-2
activity value, 4-35
activity vector, 1-2
act(s) of God, 4-2, 13-3
actual completion date, 4-2
actual cost, 4-2
actual cost of work performed, 4-2
actual damages, 4-11
actual dollars, 7-6. *See also* constant dollars; deflating by a price index; inflating by a price index
actual exposure hours, 13-3
actual finish date, 4-2
actual hours, 17-2
actual-hours-worked, 6-3
actual impossibility, 4-18
actual start date, 4-3
actual time, 17-2. *See also* elapsed time
actuary, 6-3
actuator, 11-2
acuity, 9-2, 13-3
acute exposure, 13-3
acute mountain sickness, 13-3
acute radiation syndrome, 13-3
ADA (Americans with Disabilities Act), 6-5
A/D (analog-to-digital) converter, 11-2, 12-3
ADAPT, 12-3
adaptation, general, 9-2
adaptation, sensory, 9-2
adaptive control, 9-2
adaptive control (AC), 11-2, 12-3
adaptive control constrained (ACC), 12-3
adaptive control optimized (ACO), 12-3
adaptive mode, 10-2
ADC (allyl diglycol carbonate), 12-69
ADC (analog-to-digital converter), 11-2, 12-3
addenda, 4-3
adder. *See* additive, hourly
addition agent, 12-31
additive, 12-31
additive, hourly, 6-3
additives, 12-68
add-on. *See* additive, hourly
address, 3-2, 8-2, 12-3
adduction, 2-29, 9-2. *See also* abduction
adductor, 2-29
ADEA (Age Discrimination and Employment Act), 6-4
adherend, 12-68
adhesion, 12-68
adhesion, mechanical, 12-69
adhesion, specific, 12-69
adhesive, 12-69. *See also* glue; gum
adhesive assembly, 12-69
ad hoc committee, 10-2
adhocracy, 10-2, 15-2
adjacent extreme point methods, 1-2
adjourning, 10-2
adjustable barrier guard, 13-7
adjustable bed, 12-48
adjustable stroke, 12-48
adjuster, 13-4
adjustment, equitable, 4-15
adjustment slide, 12-48
adjustment time, 1-26

ADM (arrow diagramming method), 4-4
ADM (ending node of network), 4-14
administration, 6-4, 15-2
administrative expense, 4-3
administrative labor, 6-4
administrative management, 10-2
administrative protections, 10-2
administrative services only (ASO), 6-4
administrative time, 1-26
administrator, 6-4
admissible basis, 1-2. *See also* basis
adopters, early, 5-6
adoption process, 5-2
ADP (automatic data processing), 3-2, 3-3
adsorption, 12-69, 13-4
advance notice, 6-4. *See also* pay-in-lieu-of notice
advance on wages, 6-4
adverse impact, 10-2
advertising, 5-2, 10-2
advertising, comparative, 5-4
advertising, corrective, 5-4
advertising, distributive lag effect of, 5-6
advertising evaluation, 5-2. *See also* market research
advertising, institutional, 5-7
advertising, point-of-purchase, 5-11
advertising, product, 5-11
advertising, subliminal, 5-15
advisory arbitration, 6-5
advisory light, 9-2
advisory system, 9-2
AEM (automated electrified monorail), 8-3
aeration, 12-31
aerobic metabolism, 2-29. *See also* metabolism
aerosols, 13-4
afferent, 2-29. *See also* efferent
affirmative action, 6-4, 10-2
affirmative action plan, 10-2
affordance, 9-2
AFL-CIO (American Federation of Labor and Congress of Industrial Organizations), 6-4
A.F.S. (American Foundrymen's Society), 12-31
A.F.S. clay, 12-33
AFS (grain fineness number), 12-38
afterimage, 9-2
age-adjusted death rate, 13-4
Age Discrimination and Employment Act (ADEA), 6-4
age hardening, 12-48
agency, 13-4
agency, intervening, 13-26
agency part, 13-4
agency shop, 6-4. *See also* exclusive bargaining rights; union shop
agent, 4-3, 13-4
agent, bargaining, 6-7
age-specific death rate, 13-4
age-standardized death rate, 13-4
aggregate arm-tool, 2-30
aggregate inventory, 14-3
aggregate limb-load, 2-41
aggregate production planning, 10-2
agile (manufacturing), 11-2
aging, 12-69
aging, natural, 12-54

agonist. *See* prime movers
agreement, 4-3, 6-4. *See also* collective bargaining
agreement, bargaining, 6-7
agreement effective date, 4-14
agreement, form, 6-21
agreement, master, 6-34. *See also* multi-employer bargaining; multiplant bargaining
agreement, open-end, 6-38
agreement, standard, 6-53
AGV (automated guided vehicle), 11-2
AGV (automated guided vehicle) routing, 8-2
AGVS (automatic guided vehicle system), 8-3
AI (articulation index), 9-3
AI (artificial intelligence), 3-2, 9-3, 10-3, 12-3, 14-3
AIDS (acquired immune deficiency syndrome), 13-3
AIHA (American Industrial Hygiene Association), 13-4
AIM (Automatic Identification Manufacturers, Inc.), 8-2
air-assist forming, 12-69
air belt, 12-31
airbill, 8-2
airborne noise, 13-4
air carbon-cutting, 12-86
air cargo, 8-2
air contaminants, 13-4
aircraft accident, 8-2
air dried, 12-31
air-dried strength, 12-31
air-float chain conveyor, 8-2
air freight, 8-2
air freight forwarder, 8-2
air furnace, 12-31
air-hardening steel, 12-48
air-lift hammer, 12-25. *See also* drop hammer; hammer forging
air mail, 8-2
air parcel post, 8-2
air quenching (normalizing), 12-31
air sampling, 13-4
air-setting binder, 12-42
air-slip forming, 12-69
air transport, 8-3
air transport, city terminal service, 8-7
air transport, declared value, 8-10
air transport, deferred rate, 8-10
air transport, deferred service, 8-10
air transport, economy service. *See* deferred service
air transport, gateway city, 8-16
air transport, general commodity rate, 8-16
air transport, pick-up and delivery service, 8-25
air transport, restricted articles, 8-28
air vent, 12-69
airway, 2-29
air waybill, 8-2
airway resistance, 2-29
AISI steel designation, 12-60. *See also* SAE steels
aisle, 8-2
aisle captive AS/RS, 8-2
aisle to aisle S/R device, 8-2
aisle transfer car, 8-2
Al_2O_3, 12-60

This index is arranged in letter-by-letter format.

alarm, 9-2
alcoholism program, 6-4
alcohol, under the influence of, 13-46
alert time, 1-26
algebra, boolean, 12-5
ALGOL (algorithmic language), 3-2. *See also* language
algorithm, 1-2, 11-2, 12-3
algorithmic language (ALGOL), 3-2
algorithms, composite, 1-6
alias, 3-2
alkalosis, 2-29
alkyd plastics, 12-69
alleged violation, 13-4
allergic alveolitis, 13-4
allergy, 13-4
allocate, 11-2
allocation, 14-3. *See also* reservation
all or none law, 2-29
allowable load, 13-28
allowance, 4-3, 5-2, 6-4, 12-60, 17-2
allowance, excess work, 17-6
allowance, fatigue, 6-20, 17-7
allowance, promotional, 5-12
allowance, standard, 6-53, 17-18
allowance, subsistence, 6-56
allowance, training, 17-20
allowance, unavoidable delay, 17-20
allowed hours, 17-2. *See also* standard hour
allowed time, 6-4, 17-2. *See also* standard time
alloy, 12-31, 12-48, 12-69
all-veneer construction, 12-69
allyl diglycol carbonate (ADC), 12-69
allyl plastics, 12-69
alpha, 9-3
alpha iron, 12-48
alphanumeric, 3-2
alphanumeric code, 12-3
alphanumeric display, 9-3
alpha testing, 9-3
alternate capacity, 8-2
alternate optima, 1-2
alternate routing, 11-2
alternate work center, 11-2
alternative contingent, 7-6
alternative, economic, 7-6
alternative hypothesis, 16-11
alternative, independent, 7-6
alternative, mutually exclusive, 7-6
alternative work schedules, 10-3
aluminum oxide, 12-60
alveolitis, allergic, 13-4
amalgamated craft union, 6-5
ambient, 13-4
ambient noise, 13-4
ambient temperature, 12-69
ambiguity, 4-3
amendment, 4-3

American Arbitration Association, 6-5
American College of Occupational and Environmental Medicine (ACOEM), 13-4
American Foundrymen's Society (A.F.S.), 12-31
American Industrial Hygiene Association (AIHA), 13-4
American National Standard Code for Information Interchange (ANSII), 12-4
American National Standards Institute (ANSI), 12-3
American Society of Safety Engineers (ASEE), 13-4
Americans with Disabilities Act (ADA), 6-5
amino plastics, 12-69
amnesia, retrograde shock, 13-39
amoral management, 10-3
amorphous polymers, 12-69
amortization, 4-3, 7-6. *See also* capital recovery
amplifier, 12-3
anaerobic metabolism, 2-29. *See also* respiration
analog, 3-2, 12-3
analog computer, 12-3
analog data, 11-2
analog display, 9-3
analog-to-digital converter (A/D, ADC), 11-2, 12-3
analogue, mechanical, 2-42
analysis, 4-3
analysis, accident, 13-4
analysis, job hazard. *See* hazard analysis
analysis, multivariate, 2-5
analysis of variance, 1-2
analyst, 3-2
analytic workplace design, 2-29. *See also* rational workplace design
anatomical position, 2-29
anatomical reference point, 2-30
anatomy, 2-30
anatomy, applied, 2-30
anatomy, functional, 2-37
anatomy of function, 2-30
anchorage, 12-69
anchoring and adjustment, 10-3
and, 3-2
andon, 11-2
anechoic room, 9-3, 13-5
angle, chamfer, 12-60
angle of abduction, 2-30
angle, viewing, 9-25
angular acceleration, 2-30
angular matrix, 1-2
angular velocity, 2-30
aniline formaldehyde resins, 12-70
animated-panel display, 9-3
anisotropic, 2-30
anisotropy, 12-25

anisotropy, plastic, 12-55
anneal, 12-70
annealing, 12-31, 12-48
annual bonus, 6-5
annual cost, 7-6
annual earnings, 6-5
annual equivalent, 7-6
annual improvement factor (productivity), 6-5. *See also* deferred wage increases
annually recurring cost, 4-3
annual wage or employment guarantee, 6-5. *See also* guaranteed wage plan; wage advance plan
annual worth, 7-6
annuity, 4-3, 6-5, 7-6. *See also* pension plan
annuity factor, 7-6. *See also* capital recovery factor
annuity fund, 7-6
annuity fund factor, 7-6. *See also* present worth factor(s); uniform gradient series
annuity plan, group, 6-41. *See also* pension plan
anodize, 13-5
anoxia, 2-30, 2-39, 9-3. *See also* respiration
ANSI (American National Standards Institute), 12-3
antagonism, 13-5
antagonist, 2-30
anterior, 2-23
anthracosilicosis, 13-5
anthropometric tables, 2-30
anthropometry, 2-30, 9-3
anthropometry, civilian survey results, miscellaneous, 2-5t
anthropometry, correlated survey selection, 2-4
anthropometry, design principles, 2-3, 2-4
anthropometry, design requirements, general, 2-4
anthropometry, design safety considerations, 2-5
anthropometry, further reading, 2-24–2-27
anthropometry, historical overview, 2-2, 2-3
anthropometry, introduction, 2-2
anthropomorphic, 2-30
anticipated delay report, 14-3
anticipation inventory, 14-3
anticipatory breach, 4-3
antifreeloader argument, 10-3
antioxidant, 12-70
anti-racketeering law, 6-5
antistatic agents, 12-70
anti-strikebreaker law, 6-5. *See also* strikebreaker
anvil (base), 12-25
anxiety, 6-5
AOQ (average outgoing quality), 16-3
AOQL (average outgoing quality limit), 16-3

APL (acceptable process level), 16-2
APL (acceptable productivity level), 17-2. *See also* normal performance
APL (A Programming Language), 12-3
aponeurosis, 2-30
apparent motion, 9-3
apparent violation, 13-5
appendix, 6-5
applicant, 6-5
application, 3-2, 9-3
application blank, 10-3
application form, 6-5
application for payment, 4-3
application program, 11-2
applications generator, 3-2
application software packages, 10-3
apportion, 7-6
appraisal, 4-35
apprentice, 6-5. *See also* learner
apprentice rate, 6-5
approval agency, 13-5
approve, 4-3
approved, 13-5
approved industrial truck, 8-2
A Programming Language (APL), 12-3
apron, 8-2
apron conveyor, 8-2. *See also* slope conveyor
apron feeder, 12-31
APT (automated programmed tool), 11-2, 12-3
aptitude, 6-5
aptitude test, 6-5
AQL (acceptable quality level), 10-2, 16-2
arbitration, 4-3
arbitration, advisory, 6-5
arbitration, binding, 6-8
arbitration, board of, 6-8
arbitration, compulsory, 6-5
arbitration, voluntary, 6-5
arbitration, wage, 6-63
arbitrator, 6-6, 17-2. *See also* impartial chairman
arbitrator, permanent, 6-42
arc, 1-3
arc blow, 12-86
arc-cutting, carbon, 12-86
architecture, 11-2, 12-3
arc-time, 12-86
arc voltage, 12-86
arc-welding, 12-86
arc-welding, carbon, 12-86
arc-welding, flux cored, 12-87
arc-welding, gas metal, 12-88
arc-welding, gas tungsten, 12-88
arc-welding, plasma, 12-89
arc-welding, semiautomatic, 12-90
arc-welding, shielded metal, 12-90
arc-welding, submerged, 12-91
area differential, 6-6, 8-2. *See also* intercity differential

This index is arranged in letter-by-letter format.

area sampling, 5-2
area wage survey, 6-6. *See also* community wage survey
areaway, 8-2
argument, 3-2
arithmetic
   average, 1-3
   mean, 1-3, 16-2
   operator, 3-2
   weighted mean, 16-2
ARL (average run length), 16-3
arm, 11-2
arm conveyor, 8-3. *See also* lowering conveyor
arm length, 2-11
arm reach, 2-8
arm reach contours, functional, 2-8
arm to column connector, 8-3
arm-tool aggregate, 2-30. *See also* limb-load aggregate
array, 3-2. *See also* programmable logic array
arrival rate, constant, 1-3
arrival rate distribution, constant, 1-3
arrival rate, mean, 1-3
arrival time, constant, 1-3
arrow, 4-3
arrow diagram, 4-3
arrow diagramming method (ADM), 4-4
arson, 13-5
arthritis, 2-30. *See also* joint
arthritis, traumatic, 2-52
articulation, 2-30. *See also* joint
articulation index (AI), 9-3
artificial basis, 1-3
artificial horizon, 9-3
artificial intelligence (AI), 3-2, 9-3, 10-3, 12-3, 14-3
artificial radiation, 13-5
artificial reality. *See* virtual environment
artificial variables, 1-3
artificial vector, 1-3
asbestos, 13-5
asbestosis, 13-5
as-built schedule, 4-4
A-scale, 13-2
as cast, 12-31
ascending ramus, 2-23
ASCII (American National Standard Code for Information Interchange), 3-2, 12-4
ASEE (American Society of Safety Engineers), 13-4
aseptic bone necrosis, 13-15
ASME type LP gas cylinder, 8-3
ASN (average sample number), 16-4. *See also* average amount of inspection
ASO (administrative services only), 6-4
aspect ratio, 8-3
asphyxiation, 13-5
aspiration, level of, 6-32

AS/RS (automated storage and retrieval system), 8-3, 11-2
assemble, 12-4
assembler, 3-2
assemble-to-order product, 14-3. *See also* make-to-order product
assembly, 11-2, 12-4, 12-70
assembly, automated, 11-2, 12-4
assembly language, 3-2. *See also* language
assembly line, 14-3. *See also* line balancing
assembly line balancing problem, 1-3
assembly robot, 11-2
assembly, selective, 12-65
assembly service, 8-3
assembly time, 12-70
assessed failure rate, 1-10
assessed value, 4-4
assessment, 6-6
assessment center, 10-3
asset management ratios, 10-3
assets, fixed, 7-10
assignable cause, 13-5, 16-3, 17-2
assigned risk, 13-5
assignment problem, 1-3
association agreement, 6-6. *See also* employers' association; multi-employer bargaining
association, causal, 13-9
association, noncausal, 13-33
assumption of risk, 13-6
a-stage, 12-68
asthenia, 2-30
asthma, 13-6
asynchronous, 11-2
asynchronous operation, 3-2
atactic, 12-70
ATI (average total inspected), 16-4. *See also* average amount of inspection
atmosphere, explosive, 13-6
atrophy, 2-31
attachment, 14-3. *See also* accessory; option
attack rate, 13-6
attack rate, secondary, 13-40
attainment, expected. *See* fair day's work
attendance bonus, 6-6
attention, 6-6, 9-3, 13-6
attention, divided, 9-3
attention, selective, 9-3
attenuation, 13-6
attic, 8-3
attitude, 6-6
attitude survey, 6-6
attributable risk
   in the exposed, 13-6
   percent in the exposed, 13-6
   percent in the population, 13-6
   in the population, 13-6
attribute, 3-2
attributes inspection, 1-3
attribution theory, 15-2
attrition arrangement, 6-6

audible range, 13-6
audiogram, 13-6
audiogravic illusion, 9-3
audiogyral illusion, 9-3
audiologist, 13-6
audiometer, 9-3, 13-6
audit, 11-2
audit, external, 10-11
audit, functional, 10-3
audit, internal, 10-16
auditory after-effect, 9-3
auditory fatigue, 9-3
auditory flutter, 9-3
auditory fusion, 9-3
auditory lateralization, 9-3
auditory localization, 9-3
auditory stimulus, 9-2
augmented matrix, 1-3
austempering, 12-48
austentite, 12-48
authoritarianism, 15-2
authority, 6-6, 10-3, 15-2
authority, functional, 15-5
authorization card, 6-6. *See also* card check; check-off
authorized work, 4-4
autoclave, 12-70
autocratic, 10-3
autocratic leader, 15-2
autokinetic effect, 9-4
automata theory, 12-4
automated assembly, 11-2, 12-4
automated electrified monorail, 8-3
automated guided vehicle (AGV), 11-2
automated process planning, 11-2, 12-4
automated programmed tool (APT), 11-2, 12-3
automated storage and retrieval system (AS/RS), 8-3, 11-2
automatic, 11-2
automatic bar, 12-60. *See also* screw machines
automatic coupler, 8-3
automatic data processing (ADP), 3-2, 3-3
automatic dispenser, 8-3
automatic guided vehicle system (AGVS), 8-3
Automatic Identification Manufacturers, Inc. (AIM), 8-2
automatic increases, 6-6
automaticity, 9-4
automatic press stop, 12-48
automatic progression. *See* automatic increases
automatic relief. *See* post-deduct inventory transaction processing; pre-deduct inventory transaction processing
automatic rescheduling, 14-3. *See also* manual rescheduling
automatic retirement, 6-6. *See also* compulsory retirement
automatic selling, 5-2

automatic system for positioning of tools (AUTOSPOT), 12-4
automatic wage adjustment, 6-6
automatic wage progression, 6-6
automatic welding, 12-86
automation, 6-6, 9-4, 11-3, 12-4
autonomation, 11-3
autonomous work group, 10-3
autonomous work-group design, 15-2
autonomy, 10-3
AUTOSPOT (automatic system for positioning of tools), 12-4
auxiliary function, 12-4
auxiliary operation, 3-3
auxiliary process time, 17-2
auxiliary storage, 3-3
availability, 1-3, 9-4, 10-3, 11-3
availability, achieved, 1-3, 16-3
availability, intrinsic, 1-3
availability, measure of, 9-4
availability, operation, 1-4, 16-3
availability, performance, 16-3
available, 11-3
available inventory, 14-3
available machine time, 17-2
available motions inventory, 2-31. *See also* demand motions inventory
available process time, 17-2
available stock, 5-2
available to promise, 14-3
average, 16-2. *See also* arithmetic, average
average amount of inspection, 16-3
average chart, 16-3
average cost, 5-2
average cycle time, 17-2. *See also* select(ed) time
average days charged per disabling injury, 13-6
average earned rate, 6-6
average earnings, 6-6
average element time, 17-2. *See also* select(ed) time
average fixed cost, 5-2
average hourly earnings exclusive of overtime payments, 6-7. *See also* gross average hourly earnings
average incentive earnings, 6-7
average inventory, 5-2
average man concept, 9-4
average outgoing quality (AOQ), 16-3
average range, 16-3
average revenue, 5-2
average run length (ARL), 16-3
average sample number (ASN), 16-4
average straight-time hourly earnings, 6-7
average time, 6-7, 17-2
average total inspected (ATI), 16-4
average variable cost, 5-2

This index is arranged in letter-by-letter format.

avoidable delay(s), 6-7, 17-2. *See also* delay; idle time
avoidance, 10-3
avulsions, 13-6
award, 6-7
award a job, 6-7
award, notice of, 4-23
axial rake angle, 12-60
axial relief angle, 12-60
axis, 12-4
axis, C, 12-6
axis control, coordinated, 11-5
axis of rotation, 2-31
axis of thrust, 2-31

B

backbone (computer), 3-3
backcharge, 4-4
back cross brace, 8-3
back diagonal brace, 8-3
back draft, 12-31
back flush, 14-3
back-gage, 12-48, 12-53
background noise, 13-6
background processing, 3-3
background radiation, 13-6
backhand welding, 12-86
backhaul, 5-2
backing, 12-86
backing plate, 12-70
backing sand, 12-31
backlash, 9-4, 11-3
backlog, 14-3
backorder, 10-3, 14-4. *See also* stockout
back pay, 6-7
back post ties, 8-3
back-pressure-relief port, 12-70
back taper, 12-70. *See also* undercut
back-to-back tie, rigid, 8-29
back-to-work movement, 6-7
backup, 3-3, 4-4
backward chaining, 9-4
backward pass, 4-4
backward scheduling, 11-3, 14-4. *See also* forward scheduling
bag conveyor, double helical, 8-3
baghouse, 12-31
bag molding, 12-70
bait pricing, 5-2
baked core, 12-31
baked permeability, 12-31
baked strength, 12-31
bakelite, 12-70
balance, 17-2
balanced motion pattern, 17-2
balance of payments, 10-3
balance of trade, 10-3
balance sheet, 10-3
balance sheet budget, 10-3
balancing, 11-3
balancing delay, 17-2

balcony, 8-3
ballistic movement, 17-3
bandpass, 9-4
band pressure level, 13-7
bandwidth, 6-7, 9-4
bang-bang control, 9-4, 11-3
bang-bang-off control, 11-3
bank, 6-7. *See also* float
bankruptcy, 10-3
bank sand, 12-31
bar, 8-3
bar, automatic, 12-60. *See also* screw machines
bar code, 3-3, 8-3
bar code density, 8-3
bar code label, 8-3
bar code reader, 8-3
bar code, vertical, 8-36
bare electrode, 12-86
bargaining, 10-3, 17-3
bargaining agent, 6-7
bargaining agreement, 6-7. *See also* agreement
bargaining, collective, 6-11. *See also* agreement
bargaining committees, continuous, 6-13. *See also* human relations committees
bargaining, crisis, 6-14. *See also* interim committees
bargaining, industry-wide, 6-27
bargaining, multi-employer, 6-36. *See also* association agreement; master agreement
bargaining, multiplant, 6-36
bargaining pattern, 6-40
bargaining productivity, 6-45
bargaining rights, 6-7
bargaining rights, exclusive, 6-19. *See also* agency shop
bargaining unit, 6-7. *See also* representation election
bar graph, 9-4
bark, 12-25
bar length, 8-3
barotrauma, 13-7
barrier guard, 13-7
barrier guard, adjustable, 13-7
barrier guard, fixed, 13-7
barrier guard, gate, 13-7
barrier guard, interlocked, 13-7
barrier guard, movable, 13-7
bars, 12-31
BARS (behaviorally anchored rating scales), 10-4
bar width, 8-3
base, 11-3
base, cantilever rack, 8-3
base date. *See* base time
based contract price, proposed, 4-27
base division of work. *See* therblig
base guide rail, 8-4
base inventory level, 5-2
basement, 8-4
base metal, 12-86
base pay, 6-7

base period, 4-4, 17-3
base permeability, 12-31
base point for escalation, 4-4
base points, 6-7
base, portable metal stacking rack, 8-4
base productivity factor (BPF), 17-3. *See also* did-take standards
base rate, 6-7, 9-4
base rate, guaranteed, 6-23
base rate insensitivity, 9-4
base salary, 6-7
base series, 14-4
base time, 4-4
base wage rate, 6-7
BASIC (beginner's all-purpose symbolic instruction code), 3-3, 12-4
basic element. *See* elemental motion
basic feasible solution, 1-4
basic grasp, 2-31
basic motion, 17-3
basic motion times (BMT), 17-3
basic piece rate, 6-8
basic size, 12-60
basic solution, 1-4
basic variables, 1-4
basin, 12-31
basis, 1-4
basis inverse, 1-4
basis matrix. *See* basis
basis variable, 1-4
basis vector, 1-4
batch
 manufacturing, 11-3
 picking, 8-4
 process, 14-4
 processing, 3-3, 8-4, 10-3, 12-4
 production, 11-3, 12-4, 16-13
batch-type furnace, 12-25
battery-electric truck, 8-4
battery limit, 4-4, 8-4. *See also* offsites
baud, 3-3, 12-4
bay, 8-4
Bayesian statistics, 7-6
bay marker, 8-4
BCD (binary-coded decimal notation), 3-3, 12-4
BCG growth-share matrix, 10-3
BCSP (Board of Certified Safety Professionals), 13-8
BCWP (budget cost of work performed), 4-6
BCWS (budget cost of work scheduled), 4-6
bead, 12-48
beam clamp, 8-4
beams, 8-4
bearing brace, 8-4
bearing plate, 8-4
beat knee, 13-7
beats, 9-4
bed, 12-31
bed, adjustable, 12-48

bed charge, 12-31
bed coke, 12-31
bed press, 12-48
begin date, expected, 4-16
beginner rate, 6-8
beginning event, 4-4
beginning network event, 4-4
beginning node of a network, 4-4
beginning spurt, 9-4
behavior, 15-2
behavioral
 displacement, 10-4
 ethic, 15-2
 modification, 15-2
 science, 10-4, 15-2
 sciences, 6-8
 theories of leadership, 15-2
 viewpoint, 10-4
behaviorally anchored rating scales (BARS), 10-4
behavior modification, 10-4
behind the tape reader (BTR), 12-4
belding-hatch index, 13-7
belongingness needs, 10-4
belt conveyor, 8-4. *See also* slope conveyor
belt conveyor, closed, 8-4
belt conveyor, flat, 8-4
belt conveyor, magnetic, 8-4
belt conveyor, multiple cord, 8-4
belt conveyor, multiple ribbon, 8-4
belt conveyor, portable, 8-5
belt conveyor, troughed, 8-5
belt feeder, 8-5. *See also* conveyor type feeder; feeder
belt, safety, 8-30, 13-39
belt sanding, 12-60
benchmark, 2-31, 6-8, 17-3. *See also* key job
benchmark evaluation method, 6-8
benchmark indexes, 4-4
benchmarking, competitive, 15-3
benchmark job, 6-8, 17-3
bench molder, 12-32
benchwork job, 6-8
bender, 12-25
bending, 12-48
bending brake, 12-49
bending dies, 12-49
bending rolls, 12-49
bending stress, 12-49
bend radius, 12-49
bends, 13-12. *See also* decompression sickness
bend tests, 12-49
beneficial occupancy, 4-4. *See also* substantial completion
benefit-cost, 7-6
benefit-cost ratio and method, 7-6
benefit limitations, 6-8
benefits, 6-8, 10-4
benefits, accidental death and dismemberment, 6-3
benefits, accident and sickness, 6-3, 13-2. *See also* health and insurance plan; sick leave; tem-

This index is arranged in letter-by-letter format.

porary disability insurance; worker's compensation
benefits, accrual of, 6-3
benefits, bridge, 13-43
benefits, cafeteria, 6-9
benefits, cafeteria style, 10-4
benefit segmentation, 5-2
benefits, employee, 6-17
benefits, fringe, 4-17
benefits, hospitalization, 6-25, 13-23
benefits, maintenance of, 6-33
benefits, major medical expense, 6-33, 13-29
benefits, maternity, 6-34
benefits, medical, 6-34, 13-30. See also health and insurance plan; health center
benefits, old age, survivors and disability insurance (OADSI), 6-38
benefits, out-of-work, 6-39
benefits plan, non-contributory, 6-37
benefits, short workweek, 6-52
benefits, supplemental unemployment benefit plan (SUB), 6-56
benefits, surgical, 6-57, 13-43. See also health and insurance plan
benefits, survivors', 6-57
benefits, transition, 13-43
benefits, widows' allowance, 13-43
benign, 13-7
bentonite, 12-32
bereavement pay. See funeral leave pay
berylliosis, 13-7
beta, 9-4
beta testing, 9-4
between-batch variation, 16-4
between-lot variation, 16-4
bevel angle, 12-86
BFOQ (bonafide occupational qualification), 6-8
bias, 1-4, 13-7, 16-4
bias of estimator, 16-4
bias recall, 13-7
bias response, 13-7
BI (bodily injury), 13-8
bicanthic diameter, 2-8
bicipital, 2-31
bid, 4-4
bid bond, 4-4, 4-5
bidder, 4-4
bidding, competitive, 1-4, 5-4, 14-6
bidding documents, 4-5
bidding requirements, 4-5
bidirectional path, 8-5
bidirectional read, 8-5
bidirectional symbol, 8-5
bid security, 4-4
bid shopping, 4-5
bifurcation, 2-31
bilateral modification, 4-22
bilateral tolerance, 12-60. See also tolerance; unilateral tolerance
billet, 12-25
bill of capacity. See product load profile
bill of labor. See product load profile
bill of lading, 8-5
bill of lading, order, 8-22
bill of lading, straight, 8-33
bill of material (BOM), 10-4, 14-4. See also product structure
bill of material (BOM), common parts, 14-6
bill of material (BOM), hierarchical. See product structure
bill of material (BOM), indented, 14-10
bill of material (BOM), matrix, 14-14
bill of material (BOM), modular, 14-14. See also option
bill of material (BOM), multi-level, 14-14
bill of material (BOM), order, 14-15
bill of material (BOM), planning, 14-16. See also option
bill of material (BOM), processor, 14-4
bill of material (BOM), single-level, 14-20
bill of material (BOM), structuring, 14-4
bill of material (BOM), summarized, 14-21
bill of material (BOM), super, 14-21
bill of material (BOM), transient, 14-22
bill of materials (BOM), indented, 14-10
bill of material structuring. See planning bill of material; transient bill of material
bill of resources. See product load profile
binary, 12-4
binary-coded decimal notation (BCD), 3-3, 12-4
binary digit (bit), 3-3, 9-4, 11-3, 12-5
binary search, 3-3
binder, 8-5, 12-32
binder, cereal, 12-33
binder, selfcuring, 12-44
binding arbitration, 6-8
bin location file, 14-4
binocular parallax, 9-4. See also parallax
binomial distribution, 16-4
bin reserve system. See two bin system
bin storage, 8-5
biocontrol system, 2-31
bioengineering, 13-7
biohazard, 13-7
biological agents, 13-7
biological effects, 13-7
biological effects of radiation, 13-7
biological monitoring, 13-7
biological school of job design, 15-2
biologic half-life, 13-7
biomechanical environment, internal, 2-39. See also external mechanical environment
biomechanical hypothesis, 2-31
biomechanical profile, 2-31. See also electromyography
biomechanics, 2-31, 9-4, 13-7, 17-3
biomechanics, occupational, 13-7
biometry, 13-7
bionic device, 2-31
bionics, 9-4, 12-4
biotaxis, 2-31, 2-43
birth-and-death process, 1-4
birth cohort analysis, 13-8
birth process, 1-4. See also branching process
bit (binary digit), 3-3, 11-3, 12-5
bit mapping, 3-3
bitrochal seat, 2-31
black box, 4-5
blacking, 12-32
blade, 12-49
blank, 11-3, 12-25, 12-49
blanket, 12-70
blanket bond, 4-5
blanket order, 14-4. See also release
blanket routing, 11-3
blank holder, 12-49. See also draw bead
blankholder slide, 12-49
blanking, 12-49
blanking die, 12-49
blast furnace, 12-32
blast pressure, 12-32
bleeder, 12-32
bleeding, 12-70
blemish, 1-4, 16-4
blending, 14-4
blending conveyor. See mixing conveyor, paddle type
blind experiment, 13-8
blind hole, 12-70
blind-hole partial thread, 12-70
blind-positioning reactions, 9-4
blind tests, 5-2
blister, 12-25, 12-70
block, 8-5
block address format, 12-5
block box, 1-4
block column, 8-5
block diagram, 12-5
blocked space (air transport), 8-5
blocker, 12-25
blocker dies, 12-25
blocker-type forging, 12-25
blocking, 12-25, 12-70
blocking impression, 12-25
block leg, 8-5
block pivot, 1-4, 3-3, 3-4
block stacking, 8-5
block-triangular-matrix, 1-4
blood flow, lack of, 2-40
blood pressure, 2-31
bloom, 12-70. See also chalking-dry
blow, 12-25
blowhole, 12-25, 12-32
blow moldings, injection, 12-77
blow-through. See transient bill of material
BLS (Bureau of Labor Statistics), 4-5, 6-9
BLS (Bureau of Labor Statistics) periodicals, list of, 4-5
blue brittleness, 12-49
blue-collar workers, 6-8. See also pink collar worker; white-collar worker
BMT (basic motion times), 17-3
BNA (Bureau of National Affairs), 6-9
board hammer, 12-25, 12-49. See also drop hammer; hammer forging
board of arbitration, 6-8
board of certified safety professionals (BCSP), 13-8
board of directors, 15-2
board of fact-finding, 6-20
board of inquiry, 6-8
bob, 12-32
bodily injury (BI), 13-8
body breadth, maximum, 2-11
body depth, maximum, 2-11
body fat estimating, skinfold measurement for, 9-20
body-load aggregate, 2-32
bogey, 6-8
bogus, 6-8
boiler codes, 13-8
bolster, 12-25, 12-70
bolster plate, 12-49
BOM (bill of material), 10-4, 14-4. See also product structure
BOM (bill of material), common parts, 14-6
BOM (bill of material), hierarchical. See product structure
BOM (bill of material), indented, 14-10
BOM (bill of material), matrix, 14-14
BOM (bill of material), modular, 14-14
BOM (bill of material), multi-level, 14-14
BOM (bill of material), order, 14-15
BOM (bill of material), planning, 14-16
BOM (bill of material), processor, 14-4
BOM (bill of material), pseudo. See planning bill of material
BOM (bill of material), single-level, 14-20

This index is arranged in letter-by-letter format.

BOM (bill of material), structuring, 14-4
BOM (bill of material), summarized, 14-21
BOM (bill of material), super, 14-21
BOM (bill of material), transient, 14-22
bonafide occupational qualification (BFOQ), 6-8
bond, bid, 4-5
bond, blanket, 4-5
bond clay, 12-32
bonding agent, 12-32
bonding surface, 12-32
bond, payment, 4-5
bond performance, 4-5
bond(s), 4-5, 12-32, 12-70. *See also* joint
bond strength, 12-32
bone necrosis, aseptic, 13-15
bones. *See also* palmar arch; ulnar nerve
aseptic necrosis of, 13-15
   cheek, 2-24
   hamate, 2-37
   metacarpal, 2-42
   navicular, 2-43
   scaphoid, 2-43
   sesamoid, 2-48, 2-49
   tubular, 2-53
bonus, 6-8
bonus, annual, 6-5. *See also* equivalent uniform annual cost
bonus, attendance, 6-6
bonus, determinant, 6-8
bonus earnings, 6-8
bonus, group, 6-23
bonus, long term, 6-32
bonus, non-production, 6-37
bonus-penalty, 4-5
bonus-penalty contracts, 4-9
bonus plan, 6-8
bonus, production, 6-44
bonus restriction, 6-8
bonus, step, 6-54
book inventory, 14-4
book member. *See* union member
book value, 4-5, 7-6
book value, depreciated, 4-12
boolean algebra, 12-5
boolean operation, 3-4
boom, 8-5
boom conveyor, 8-5. *See also* stacker conveyor
booster conveyor, 8-5
bootleg wages, 6-9. *See also* kickback
bootstrap, 3-4, 12-5
boring, 12-60
boss, 12-25, 12-32
bot, 12-32
bottleneck, 11-3, 14-4
bottom discharge bucket conveyor, 8-5
bottoming bending, 12-49
bottom out, 6-9

bottom-up budgeting, 10-4
boundary spanning, 10-4
bounded rationality, 10-4, 15-2
bounded variable, 1-5
bounded variable problem, 1-5
bow, 12-49
boycott, 6-9. *See also* hot-cargo clause
BPF (base productivity factor), 17-3
BPM (cardiac rate), 2-32
brace, 8-5
brace, back cross, 8-3
brace, back diagonal, 8-3
brace, bearing, 8-4
brace, top horizontal, 8-34
brachialis muscle, 2-32
brainstorming, 6-9, 10-4, 15-2, 17-3
brake, 12-49
brake, parking, 8-25
brale, 12-49
branch and bound, 1-5
branch house, 5-2
branching process, 1-5. *See also* birth process
branch office, 5-3
branch store, 5-3
branch warehouse demand. *See* warehouse demand
brand, 5-3
brand insistence, 5-3
brand management, 5-3
brand recall, 5-3
brand recognition, 5-3
brands, private, 5-11
brand switching, 5-3
brazed tip tool, 12-60
braze welding, 12-86
brazing, induction, 12-88
brazing, resistance, 12-90
breach of contract, 4-5
bread-and-butter unionism. *See* business unionism
breadboard, 12-5
breadth, ear, 2-9
break-even analysis, 5-3, 10-4
break-even chart, 4-5, 7-6
break-even point, 4-5, 7-6
break-in allowance, 6-9
break in service, 6-9
break-off core, 12-32
breakout schedule, 4-5
breakpoint, 17-3
break time, 13-39. *See also* rest period
breathing. *See* pulmonary ventilation; respiration
bridge, 12-5
bridge benefits, 13-43. *See also* survivors' benefits
bridge crane. *See* crane; fixed crane; overhead crane
bridge plate, 8-5
bridge rail, 8-5
brief, 6-9
brightness, 9-4, 13-8. *See also* illumination

brightness contrast, 9-4. *See also* foot candle
brightness control, 9-5
brightness, surround, 9-22
Brinell hardness test, 12-60
broaches, 12-60
broaching, internal, 12-63
broker, 5-3
bronchitis, chronic, 13-9
B-scale, 13-6
BTPS conditions, 9-5. *See also* STPD condition
BTR (behind the tape reader), 12-4
bubble, 12-70. *See also* joint
bubble memory, 12-5
bubble sort, 3-4
bucket conveyor, gravity discharge, 8-16
bucketed system, 14-4. *See also* horizontal display; time bucket; vertical display
bucket elevator, 8-5. *See also* marine leg
bucket elevator, double leg, 8-11
bucket elevator, gravity discharge, 8-16
bucket elevator, internal discharge, 8-18
bucket elevator, pivoted, 8-6
bucket elevator, positive discharge, 8-27
bucketless system, 14-4. *See also* horizontal display; vertical display
bucket loader, 8-5. *See also* portable conveyor
bucket, time, 14-21
buckle, 12-32
buckling, 12-49
budget, 4-6
budget, balance sheet, 10-3
budget, capital expenditures, 10-5
budget, cash, 10-5
budget cost of work performed (BCWP), 4-6
budget cost of work scheduled (BCWS), 4-6
budget estimate, 4-15
budget, expense, 10-11
budget, financial, 10-12
budgeting, 10-4
budgeting, bottom-up, 10-4
budgeting, capital, 4-6, 7-7
budgeting, error, 12-62
budgeting, top-down, 10-31
budgeting, zero-base (ZBB), 10-33
budget, merit, 6-35
budget, operating, 10-22
budget, profit, 10-24
budget, revenue, 10-26
budget, salary, 6-49
budget, sales, 5-13, 10-27
buff, 12-49
buffer, 11-3, 12-5

buffering, 10-4
buffer stock, 5-3
buffer storage, 12-5
bug, 12-5. *See also* union label
building, 8-5
building block, 17-3
building code, 13-8
buildup sequence, 12-86
built-up edge, 12-60
bulging, 12-50
bulk density, 12-70
bulk factor, 12-71
bulk issue, 14-4
bulk material, 4-6
bulk sampling, 16-4
bulk stock, 8-5
bulk storage, 8-5
bull block, 12-50
bulldozer, 12-50
bumper, 12-32
bumper pad, 8-6
bumping, 6-9
burden, 4-6
burden of proof, 4-6
bureaucracy, 15-2
bureaucratic control, 10-4
bureaucratic management, 10-4
Bureau of Labor Statistics (BLS), 4-5, 6-9
Bureau of National Affairs (BNA), 6-9
burned, 12-71
burned sand, 12-32
burn-in, 1-5
burnishing, 12-50
burnout, job, 6-27
burnt, 12-25, 12-50
bursa, 13-8
bursitis, 13-8
burst, 12-25
burst lung, 13-37
bus, 3-4, 12-5
business agent, 6-9
business, international, 10-16
business-level strategy, 10-4
business plan, 10-4, 14-4. *See also* manufacturing resource planning (MRP II); production planning
business planning, 4-6
business unionism, 6-9
buster (pre-blocking impression), 12-25
butadiene, 12-71
butadiene styrene, 12-71
butt joint, 12-86
buttock depth, 2-9
buttock-knee length, 2-9
buttock-leg length, 2-9
buttock-popliteal length, 2-9
butt-off, 12-32
buyer, 14-5. *See also* vendor scheduler
buyer's market, 5-3
buying capacity, 14-5
buying power, 5-12
buy-out, 6-9

This index is arranged in letter-by-letter format.

bylaws, 6-9
by-product, 14-5
byssinosis, 13-8
byte, 3-4, 12-5

C

C axis
C chart
C-scale
C-stage
cab, 8-6
cab controlled, 8-6
cable drive, 11-3
cable-screw conveyor, 8-6
cableway, 8-6
cableway, slack line, 8-6
cache memory, 3-4
CADAM (Computer-Graphics Augmented Design and Manufacturing), 12-5
CAD/CAM (computer-aided design/computer-aided manufacturing), 11-4
CAD (computer-aided design), 10-6, 11-4, 12-5, 12-7
CAE (computer-aided engineering), 11-5
cafeteria benefits, 6-9
cafeteria style benefit plan, 10-4
caged storage, 8-6
CAI (computer-aided instruction), 10-6
Caisson disease, 13-12
calendar, 4-6
calendar range, 4-6
calendar start date, 4-6
calendar unit, 4-6
calibration, 11-3, 12-6
calibration time, 1-26
call, 3-4
callback pay, 6-9
call-in pay, 6-9. *See also* reporting pay
CAMAC, 11-3
camber, 12-25, 12-50
camber (girder), 8-6
CAM (computer-aided manufacturing), 10-6, 11-5, 12-8
camera, matrix-array, 11-11
camera study. *See* memomotion study; micromotion study
cancellation charges, 14-5
cancelled rate, 6-9
cancer, 13-8
candela (CD), 9-5
candlepower, 9-5
canister, air-purifying, 13-8
canopy, 8-6
cantilever rack, 8-6
cantilever rack base, 8-3
cantilever truck, 8-6
capability, 1-5, 16-4
capacitance, 12-6

capacitated transportation problem, 1-5
capacitor, 12-6
capacity, 8-6, 11-3, 14-5
capacity, alternate, 8-2
capacity argument, 10-4
capacity buying, 14-5. *See also* vendor scheduler
capacity cargo container, 8-6
capacity control, 14-5. *See also* closed loop MRP; input/output control
capacity factor, 4-6, 7-6. *See also* demand factor; load factor
capacity loading, 11-3
capacity planning, 10-5, 11-3, 14-5
capacity planning, rough-cut, 14-19
capacity, press, 12-50
capacity requirements, 11-3
capacity requirements planning (CRP), 10-5, 14-5. *See also* infinite loading
capacity, safety, 14-19
capacity smoothing. *See* load leveling
capital, 7-7. *See also* investment
capital budgeting, 4-6, 7-7
capital, direct. *See* direct cost
capital expenditures budget, 10-5
capital, fixed, 4-6
capital, indirect. *See* indirect cost(s)
capitalist economy, 10-5
capitalized asset, 7-7
capitalized cost, 7-7. *See also* perpetual endowment
capital, operating, 4-6
capital recovery, 7-7. *See also* amortization; depletion; depreciation
capital recovery factor, 7-7. *See also* annuity factor
capital recovery with return, 7-7
capital, sustaining, 4-6
capital tooling, 11-4
capital, total, 4-6
capital, venture, 4-6
capital, working, 4-6
capitulum of humerus, 2-32. *See also* radiohumeral joint
CAPP (computer-aided process planning), 11-5, 12-6
capped rate, 6-10
captive container, 8-6
capture velocity, 13-8
carbides, cemented, 12-60
carbon arc-cutting, 12-86
carbon arc-welding, 12-86
carbon, combined, 12-34
carbon dioxide, 2-32
carbon dioxide process, 12-32
carbonitriding, 12-50
carbon steel, 12-32, 12-50
carburize, 12-71
carburizing, 12-50

carcinogen, 13-8
carcinoma, 13-8
card, 12-6
card, authorization, 6-6
card check, 6-10
cardiac rate, 2-32
cardinal change, 4-7
carding, 13-8
career, 10-5
career management, 10-5
cargo, 8-6
cargo container
    capacity, 8-6
    displacement, 8-6
    height, 8-6
    length, 8-6
    width, 8-6
cargo, measurement, 8-21
cariation, seasonal, 4-31
carousel conveyor, 8-6
carousel, horizontal, 8-6
carousel, vertical, 8-6
carpal tunnel, 2-32, 13-8
carpal tunnel syndrome (CTS), 2-32, 13-8
carriage, 8-6
carrier, 8-6, 12-6
carrier head, 8-6
carriers
    common, 5-4
    contract, 5-4
carrying cost, 10-5, 14-5. *See also* ordering cost
Cartesian coordinate system, 11-4, 12-6
cartilaginous plates, 2-32
car type conveyor, 8-6. *See also* floor conveyor
cascade control, 12-6
case, 13-8
case-control study, 13-8
case fatality rate, 13-8
case hardening, 12-50
case method, 15-2
cash-and-carry wholesaler, 5-3
cash budget, 10-5
cash costs, 4-7
cash cows, 5-3
cash discount, 5-3
cash flow, 4-7, 7-7
cash flow diagram, 7-7
cash flow table, 7-7
cash, maximum out-of-pocket, 4-22
cash return, percent of total capital, 4-7
cassette recorder, 12-6
cassette tape, 12-6
cast, 12-71
cast film, 12-71
casting, 12-32
casting, centrifugal, 12-32, 12-71
casting, continuous, 12-34
casting, die, 12-33
casting, machine, 12-33
casting, open sand, 12-33
casting, permanent mold, 12-33

casting, plaster, 12-33
casting, precision, 12-33
casting, rotational, 12-82
casting, sand, 12-33
casting, slip, 12-45
casting yield, 12-33
cast iron, 12-33
cast iron, gray, 12-38
cast iron, mottled, 12-41
cast iron, white, 12-33
casual connection, 13-9
casualty insurance, 13-9
casual workers, 6-10
CAT, 13-9. *See also* scan
catalyst, 12-71. *See also* hardener
catalyst costs, 4-13
catastrophe, 13-9
catastrophe insurance, 6-10, 13-29. *See also* major medical expense benefit
catastrophic failure, 1-10
category killer, 5-3
caterer problem, 1-5
cathode ray tube (CRT), 12-6
cathode ray tube (CRT) display, 3-4
catwalk, 8-6
caul, 12-71
caul, canvas-covered plywood, 12-71
caul, plywood, 12-71
causal association, 13-9
causal factors, accident, 13-9
causal model, 10-11
causation, 4-7
causation, chain of, 13-9
cause
    early, 13-15
    intervening, 13-26
caveat emptor, 13-9
caveat venditor, 13-9
cavity, 12-33, 12-71
C axis, 12-6
CBISs (computer-based information systems), 10-6
CBN (cubic boron nitride), 12-61
C chart, 16-6
CD (candela), 9-5
ceiling effect, 9-5
ceiling tie, 8-7
cell, 3-4, 11-4, 12-71
cell control, 11-4
cell manning, 6-10
cellular layout, 8-7
cellular manufacturing, 11-4, 14-5
cellular plastic, 12-71. *See also* foamed plastics
cellular striation, 12-71
cellulosic plastics, 12-71
cemented carbides, 12-60
cement sand, 12-33
census, 5-3
center, 11-4
center control, 11-4
center control truck, 8-7
centerless grinding, 12-60
center mounted grinding, 12-60

This index is arranged in letter-by-letter format.

center of gravity, 2-32
center of mass, 2-32
center of rotation, 2-32
centipoise, 12-71
central hearing loss, 9-5
centralization, 10-5, 15-2
centralized dispatching, 14-5. *See also* control center; decentralized dispatching; dispatching
central labor council, 6-10
central processing unit (CPU), 3-5, 10-5, 12-6
centrifugal casting, 12-71
centrifugal collector, 12-36
C.E.O. (chief executive officer), 15-3. *See also* management, top
ceramic, 12-60
ceramic core, preformed, 12-43
ceramic molding, 12-33
cereal binder, 12-33
cerebellum, 2-32, 2-33
cerebrum, 2-33
ceremonial, 10-5
cermets, 12-60
certification, 6-10
Certified Safety Professional (CSP), 13-9
chad, 12-6
chain conveyor, 8-7. *See also* tow conveyor
chain conveyor, drag, 8-12
chain conveyor, rolling, 8-29
chain conveyor, sliding, 8-32
chain drive, 11-4
chain elevator. *See* bucket elevator
chain feeder. *See* conveyor type feeder
chain (in a graph), 1-5
chain index, 4-7
chaining, backward, 9-4
chain, kinematic, 2-40
chain of causation, 13-9
chain of command, 10-5, 15-3
chain sampling inspection, 16-4
chain, scalar, 15-12
chain stores, 5-3
chain store system, 5-3
chairman, impartial, 6-25. *See also* arbitrator
chairman, office of, 15-9
chalking-dry, 12-71. *See also* bloom; haze
challenger, 7-7. *See also* MAPI method
chamfer angle, 12-60
chance causes, 1-5, 16-4
chance constrained programming, 1-5
change, 4-7, 10-5
change agent, 10-5, 15-3. *See also* organizational development (OD)
change, cardinal, 4-7
change, classification, 6-10
change, constructive, 4-7
change, creeping method, 6-14
changed conditions. *See* differing site conditions
change, general wage, 6-22
change in scope, 4-7
change in sequence, 4-7
change intervention, 15-3
change order, 4-7, 11-4
change order, engineering, 14-8. *See also* configuration control
change order, proposed, 4-28
changeover costs, 14-5
changeover time, 6-10, 17-3. *See also* setup; teardown
change, planned, 10-23
change, rate, 6-46, 17-17
change, reactive, 10-26
change, salary level, 6-50
change, unilateral. *See* modification, unilateral
channel, 3-4, 12-6. *See also* port
channel alignment, 5-3
channel capacity, 9-5
channels, multiple, 1-5
chaplets, 12-33
chapter. *See* local union
character, 3-4, 8-7, 12-6
character density, 8-7
character height, 9-5
characteristic, 1-5, 16-4
characteristic root, 1-5
character reading, 8-7
character set, 8-7
character width, 9-5
charade, 6-10
charge, 6-10, 12-33
charged and dry, 8-7
charged and wet, 8-7
charge, scheduled, 13-40
charisma, 10-5
charismatic leadership, 15-3
charpy test, 12-50
chart, average, 16-3
chart, C, 16-6
charter, 6-10
chart of accounts. *See* code of accounts
chase, 12-71
chase ring, 12-72
check, 12-25, 12-33
check character, 8-7
check digit, 3-4, 8-7
check-off, 6-10. *See also* authorization card; union dues
checkout, 1-5
checkout time, 1-26, 16-18
check study, 17-3
cheek, 12-33
cheekbone, 2-24
cheek breadth, 2-9
chelating agent, 13-9
chemical engineering plant cost index, 4-7
chemically formed plastic, 12-72
chemicals costs, 4-13
chemoreceptor, 9-5
chemotaxis, 2-33
chernoff display, 9-5
chest breadth, 2-9
chest depth, 2-9
chief executive officer (C.E.O.), 15-3. *See also* management, top
chief operating officer (C.O.O.), 15-3
chill block, 12-33
chill/chills, 12-33
chip, 12-6, 12-60
chipbreaker, 12-60
chisel edge, 12-61
chi-squared distribution, 16-4
chloracne, 13-9
choice reaction time, 9-5
choke, 12-33
chromite, 12-33
chronic bronchitis, 13-9
chronic violator, 13-9
chronocyclegraph, 2-33. *See also* signature
chronocyclegraph technique, 17-3. *See also* cyclegraph technique
chronological study, 17-4. *See also* production study
chucking internal grinding, 12-61
chute conveyor, 8-7
chute, pneumatic, 8-26
CICHME (College-Industry Council on Material Handling Education), 8-7
CIM (computer-integrated manufacturing), 10-6, 11-5, 12-8
CINTURN II, 12-7
circle-grid analysis, 12-50
circuit breaker, 12-7
circuit, integrated (IC), 12-7, 12-14
circuit, printed, 12-7
circular interpolation, 12-7
circumstances, of an accident, 13-9
citation, 13-9
city terminal service (air transport), 8-7
Civil Rights Act, 6-10. *See also* Fair Employment Practice Laws
Civil Service Reform Act, 6-10
claim, 4-7, 13-9
claimant, 13-9
clamp, 8-7
clamp arms, 8-7
clamp, beam, 8-4
clamping plate, 12-72
clamp irons, 12-72
clamp-off, 12-33
clan control, 10-5
class-based storage policy, 8-7
classical organizational theory, 15-3
classical viewpoint, 10-5
classification, 6-10, 11-4, 14-5. *See also* coding
classification act employees, 6-10. *See also* wage board employees
classification and coding system, 8-7
classification change, 6-10
classification code, group, 14-10. *See also* coding; group technology
classification index, 6-10
classification method of job evaluation, 6-10
classification, primary, 4-26
class of positions, 6-10
class rate, 6-11
clavicle, 2-33
clay, 12-33
clay, colloidal, 12-34
clay substance, 12-33
clay-wash, 12-34
CL (cutter location) data, 12-9
cleanup time, 1-26, 6-11, 16-18
clearance fit, 12-61
cleats, 12-31
clerical payroll costs, 4-13
click, 9-5
clients, 10-5
client-server, 3-4
clock, 3-4, 12-7
clock, master, 3-10
clock-off, 6-11
clock-out, 6-11
clock rate, 11-4
closed belt conveyor, 8-4
closed-cell foam, 12-72
closed circuit conveyor, 8-7
closed-die forging. *See* impression die forging
closed dies, 12-50
closed height. *See* shut height
closed loop, 9-5, 12-7
closed loop control, 11-4
closed loop MRP, 14-5. *See also* capacity control; dispatching; input/output control; MPS (master production schedule); MRP (materials requirements planning); MRP II (manufacturing resource planning); production planning; shop floor control; shop planning
closed loop system, 11-4, 12-7
closed shop, 6-11
closed system, 10-5
closed union, 6-11. *See also* open union
closet, 8-7
close-tolerance design, 12-25
clothes changing time, 6-11. *See also* cleanup time
clothing allowance, 6-11
clo unit, 9-5
CL tape, 12-7
cluster sampling, 1-5, 13-9
clutch, 12-50
CMM (coordinate measuring machine), 11-5
CNC (computer numerical control), 11-5, 12-8
COBOL (common business oriented language), 3-4, 12-7
code, commodity, 8-8. *See also* SITC

This index is arranged in letter-by-letter format.

code, continuous, 8-8
code dating, 5-4
code, discrete, 8-11
code, group classification, 14-10
code of accounts, 4-7
code reader, 8-7
code(s), 3-4, 8-7, 12-7, 13-9
code scanner, 8-7
Codes of Ethical Practices, 6-11
codetermination, 6-11
coding, 11-4, 14-5, 17-4. *See also* classification
coefficient cost, 1-7
coefficient matrix, 1-5
coefficient of loss of service, 1-5
coefficient of variation, 16-5
coercive power, 10-5
coextrusion, 12-72
coffee break, 13-39. *See also* rest period
cogging, 12-26
cognition, 9-5
cognitive engineering, 9-5
cognitive resources, 10-5
cognitive resource theory, 10-5
cognitive theories, 10-5
cohesion, 12-72
cohesiveness, group, 10-14, 15-6
cohort study, prospective, 13-9
cohort study, retrospective, 13-10
COI (cube-per-order index), 8-10
coining, 12-26, 12-50. *See also* ironing
coke, 12-34
COLA (cost of living adjustment), 6-11, 6-13. *See also* consumer price index (CPI); escalator clause
cold box process, 12-34
cold flow. *See* creep
cold heading, 12-26
cold inspection, 12-26
cold lap, 12-26
cold molding, 12-72
cold pressing, 12-72
cold setting binder, 12-34, 12-42
cold setting process, 12-34
cold shut, 12-34
cold slug, 12-72
cold-slug well, 12-72
cold trimming, 12-26
cold welding, 12-86
cold work, 12-50
cold working, 12-26, 12-50
collaboration, 10-5
collapsible container, 8-8
collapsible sprue, 12-34
collective bargaining, 6-11. *See also* agreement
collectors, 8-8
collector shoe, 8-8
collector wheel, 8-8
College-Industry Council on Material Handling (CICHME), 8-7
collision, 3-4
collision avoidance rules, 8-8

collision diagram, 13-10
collision, motor vehicle, 13-31
colloidal clay, 12-34
colloidal material, 12-34
coloring, dry, 12-74
column, 8-8
columnar structure, 12-34
column spacing, 8-8
column vector, 1-6
combined activity (symbol), 17-8
combined carbon, 12-34
combined motions, 17-4
combined water, 12-34
combined work, 17-4
combustible gas meter, 13-10
COM (computer output microfilming), 3-4
command, 9-5, 12-7
command, chain of, 10-5, 15-3
command cycle, dual, 8-12
command cycle, single, 8-31
command group, 10-5
command language, 9-6, 11-4
command line interface, 9-5
commission earnings, 6-11. *See also* drawing account
commission merchant, 5-4, 6-11
commitments, 4-7
committee, 15-3
commodities, equivalent sets, 4-15
commodities, seasonal, 4-31
commodity, 4-7
commodity buying, 14-5
commodity code, 8-8. *See also* SITC
commodity plastics, 12-72
commodity rate, 8-8
commodity rate, general, 8-16
common carriers, 5-4, 8-8
common costs, 7-7
common labor, 6-11
common labor rate, 6-11
common parts, 11-4
common parts bill of material (BOM), 14-6. *See also* bill of material (BOM)
communication, 6-11, 10-5, 15-3
communication channel(s), 10-6, 15-3
communication, informal, 10-15
communication interface circuit, 12-7
communication network, 10-6, 15-3
communication, nonverbal, 10-21
communication, one-way, 10-21
communications link, 11-4
communication, two-way, 10-32
communication, upward, 10-32
community wage survey, 6-11. *See also* area wage survey
COMPACT II, 12-7
company bargaining. *See* multi-plant bargaining
company union, 6-11
comparable rate, 6-11
comparable wages, 6-12

compa-ratio, 6-12
comparative advertising, 5-4
comparison group, 13-10
compartment (battery), 8-8
compatibility, 9-6
compensable delay, 4-11
compensable injury, 6-12, 13-10
compensation, 6-12, 10-6, 11-4, 13-10
compensation, cutter, 12-9
compensation, direct, 6-15
compensation, executive, 6-19
compensation, indirect, 6-26
compensation, sales, 6-50
compensation, supplemental, 6-56
compensation, total, 6-59
compensation, total cash, 6-59
compensatory display, 9-6
compensatory time off, 6-12
competition, 10-6
competitive advantage, 10-6
competitive advantage of nations, 10-6
competitive benchmarking, 15-3
competitive bidding, 1-4, 5-4, 14-6
competitive wages, 6-12
competitors, 10-6
compile, 3-4, 3-5
compiler, 3-5
complacency, 10-6
complaint, 6-12
complementary slackness theorem, 1-6
completed activity, 4-7
complete failure, 1-10
completion date, actual, 4-2. *See also* actual finish date
completion date, earliest expected, 4-14
completion date, required, 4-29
completion date, scheduled, 4-30
completion, mechanical, 4-22
compliance, 11-4, 13-10
comply, 13-10
component, 11-4, 14-6. *See also* inventory write-off; subassembly
components, major, 4-21
composite algorithms, 1-6
composite component, 11-4
composite price index, 4-7
composite routing, 11-4
compound amount, 7-7
compound amount factor(s), 7-7
compounding, continuous, 7-7
compounding, discrete, 7-7
compounding period, 7-7
compound interest, 7-7
compound interest factors, functional forms of, 7-3–7-5
compregnated wood, 12-72
compressed workweek, 10-6
compression, 6-12
compression molding, 12-72, 12-79
compression ratio, 12-72
compressive strength, 12-26

compressive strength, sand, 12-34
compressive ultimate strength, 12-50
compressive yield strength, 12-50
compromise, 10-6
compulsory arbitration, 6-5
compulsory retirement, 6-12. *See also* automatic retirement
computed path control, 11-4
computer-aided design (CAD), 10-6, 11-4, 12-5, 12-7
computer-aided design/computer-aided manufacturing (CAD/CAM), 11-4
computer-aided engineering (CAE), 11-5
computer-aided instruction (CAI), 9-6, 10-6
computer-aided manufacturing (CAM), 10-6, 11-5, 12-8
computer-aided manufacturing, integrated (ICAM), 11-9
computer-aided process planning (CAPP), 11-5, 12-6
computer, analog, 12-3
computer backbone, 3-3
computer-based information systems (CBISs), 10-6
computer, digital, 3-6
computer graphics, 9-6, 12-8
Computer-Graphics Augmented Design and Manufacturing (CADAM), 12-5
computer hardware, 3-7
computer hardware, mainframe, 3-10
computer hardware, microcomputer, 3-11
computer hardware, modem, 3-11
computer hardware, mouse, 3-11
computer-integrated manufacturing (CIM), 10-6, 11-5
computer languages
 ADAPT, 12-3
 ALGOL (algorithmic language), 3-2
 APL (A Programming Language), 12-3
 BASIC (beginner's all-purpose symbolic instruction code), 3-3, 12-4
 COBOL (common business oriented language), 3-4, 12-7
 FORTRAN (formula translation), 3-7, 12-13
 JCL (job control language), 3-9
 natural, 3-11
 PL/I, 3-13
 procedure-oriented, 3-13
 source, 3-16
 symbolic language, 3-16
computer modeling, 15-3
computer network, 11-5
computer numerical control (CNC), 11-5, 12-8
computer output microfilming (COM), 3-4

This index is arranged in letter-by-letter format.

computer part programming, 12-8
computer program, 12-8. *See also* software
computer supported cooperative work (CSCW), 9-6
computer virus, 10-6
computer vision, 12-16
concave function, 1-6
concentrated marketing, 5-4
concentration, 10-6
concept testing, 5-4
concept trainers, 9-6
conceptual schedule, 4-7, 4-8
conceptual skills, 10-6
concession bargaining, 6-12
conciliation. *See* mediation
concurrent control, 10-6
concurrent delay, 4-11
concurrent engineering, 10-6, 11-5, 14-6, 17-4
concurrent operation, 3-5
condensation, 12-72. *See also* polymerization
conditioned reflex, 2-33. *See also* simple reflex
conductive deafness, 2-33
conductive hearing loss, 9-6. *See also* perception deafness
conductors, 8-8
conductors, enclosed, 8-8
cone, 1-6
cones, 9-6
confidence coefficient, 16-5
confidence interval, 1-6, 13-10, 16-5
confidence interval, one-sided, 16-11
confidence interval, two-sided, 16-18
confidence level, 1-6, 16-5
confidence limit, 1-6, 16-5
configuration, 12-8
configuration control, 11-5, 14-6. *See also* engineering change order
configuration management, 11-5
confined contract price, proposed, 4-28
confined space, 13-10
confirming order, 14-6
conflagration, 13-10
conflict, 6-12, 10-6, 15-3
conflict, dysfunctional, 15-4
conflict, functional, 15-5
conflict in plans and specifications, 4-8
conformity, 6-12, 16-5
confounding variable, 13-10
congenital abnormality, 13-10. *See also* malformation
conjoing measurement, 5-4
connecting carrier, 8-8
connector (battery), 8-8
connector lock, 8-8
consensus, 15-3
consensus standard, 13-10
consent of surety, 4-8

consideration, 10-6
consideration, individualized, 10-15
consigned material, 14-6
consigned stocks, 14-6
consistency, 12-72, 17-4. *See also* viscosity; viscosity coefficient
consistent mapping, 9-6
console, 3-5
consolidated metropolitan statistical areas (CSMAs), 5-4
consolidating area, 8-8
consolidating point, 8-8
consonance, 9-6
constant arrival rate distribution, 1-3
constant arrival time, 1-3
constant basket, 4-8
constant basket price index, 4-8
constant dollars, 4-8, 7-7. *See also* actual dollars; deflating by a price index; inflating by a price index
constant element, 17-4
constant failure period, 1-6
constant failure rate period, 1-6
constant service rate distribution. *See* constant arrival rate distribution
constant service time. *See* constant arrival rate distribution
constant sharing plan, 6-12
constant total cost plan, 6-12
constant unit labor cost plan, 6-12
constant utility price index, 4-8
constant vector, 1-6
constraint date, 4-8
constraint matrix, 1-6
constraint, nonegativity, 1-17
constraint, nonlinear, 1-17
constraint qualification, 1-6
constraint(s), 1-6, 10-6, 13-10. *See also* restraint
construction, all-veneer, 12-69
construction cost, 4-8
construction, lumber-core, 12-78
construction management, 4-8
construction, phased, 4-25
constructive change, 4-7
consultation, 6-12
consultative management, 15-3
consumables, 4-8
consumer behavior, 5-4
consumer goods, 5-4
consumerism, 5-4
consumer panels, 5-4
consumer price index (CPI), 4-8, 6-13. *See also* cost of living adjustment (COLA); escalator clause
consumer product safety commission (CPSC), 5-4
consumer's cooperative, 5-4
consumer's risk (CR), 1-6, 16-5
consumer's risk point (CRP), 16-5
consumer's risk quality (CRQ), 16-5

consumer surplus, 5-4
consumer, ultimate, 5-15
contact dermatitis
    allergic type, 13-10
    irritant type, 13-10
contact grasp, 2-33
contact sensor, 11-5
container, 8-8
container, collapsible, 8-8
container corner fittings, 8-8
container, demountable cargo, 8-10
container, expendable, 8-13
container load, 8-8
container marking, 8-8
container, reusable, 8-28
container, rigid, 8-29
container, two-way, 8-35
containment, 13-11
contention, 3-5
contest, 13-11
contingency, 4-8
contingency analysis, 9-6
contingency approach, 15-3
contingency planning, 10-7
contingency theory, 10-7
contingent, alternative, 7-6
continuous bargaining committees, 6-13. *See also* crisis bargaining; human relations committees
continuous casting, 12-34
continuous code, 8-8
continuous method, 17-4
continuous operation, 6-13, 12-50. *See also* round-the-clock operations
continuous path control, 11-5
continuous path operation, 12-8
continuous process, 6-13, 8-8
continuous process improvement, 10-7, 15-3
continuous production, 12-8
continuous reading, 17-4
continuous sampling inspection, 16-5
continuous systems improvement (CSI), 15-3
continuous timing, 17-4
continuous tube process, 12-72
contouring, 12-8
contouring control system, 12-8
contract bar, 6-13
contract, breach of, 4-5
contract carrier(s), 5-4, 8-8
contract completion date, 4-8
contract date, 4-8
contract documents, 4-8
contract, exclusive dealing, 5-6
contract, fixed price, 4-9, 4-16. *See also* lump-sum
contractile tissue, 2-33
contracting-out, 6-13. *See also* subcontracting
contraction, eccentric, 2-34
contract, long term, 6-32
contractor, 4-8

contract pegging. *See* full pegging
contract price, 4-8
contract price (confined), proposed, 4-28
contract, "read as whole", 4-8
contract(s), 4-9. *See also* agreement
contracts, bonus-penalty, 4-9
contracts, cost plus, 4-9
contracts, cost plus fixed fee, 4-9
contracts, cost plus fixed sum, 4-9
contracts, cost plus percentage burden and fee, 4-9
contracts, cost plus percentage fee, 4-9
contracts, fixed price, 4-9
contracts, guaranteed maximum (target price), 4-9
contracts, lump sum, 4-9
contracts, negotiating, 10-20
contracts, unit price, 4-9
contract time, 4-8
contract time, date for commencement, 4-11
contractual liability, 13-11
contract wage payment, 6-13
contract work breakdown structure (CWBS), 4-36
contrast, 9-6
contribution plan, defined, 6-15
contributory negligence, 13-11
contributory pension plan, 6-13. *See also* non-contributory pension plan
control, 4-9, 9-6, 12-8, 15-3. *See also* criterion
control bars, 8-8
control center, 14-6. *See also* centralized dispatching
control chart, 1-6, 16-5
control chart factor, 16-5
control chart, multivariate, 16-10
control-display ratio, 9-6
control-force curve, 9-6
control group, 13-11
control layout, 9-6
controlled system, 9-6
controlled velocity roller conveyor, 8-8. *See also* roller conveyor
controller, integrated, 9-11
controller, magnetic, 8-20
controller(s), 8-8, 11-5, 12-8
control limits, 16-5
controlling, 10-7
control point, 6-13
control, single speed, 8-8
control system, 9-6, 10-7, 12-8, 17-4
control technology, 13-11
control, variable speed, 8-9
control, voltage, 8-9
control, wage and salary, 6-13
contusion, 13-11
convenience goods, 5-4
convergence, 9-6
convergent thinking, 10-7
conversational mode, 12-8

This index is arranged in letter-by-letter format.

convex combination, 1-6
convex cone, 1-6
convex function, 1-6
convex hull, 1-7
convex polyhedral cone, 1-7
convex polyhedron, 1-7
convex programming, 1-7
convex set, 1-7
conveyor, 8-9
conveyor, accordion roller, 8-2
conveyor, accumulator, 8-2
conveyor, air-float chain, 8-2
conveyor, apron, 8-2
conveyor, arm, 8-3
conveyor, bag, double helical, 8-3
conveyor, belt, 8-4, 12-34
conveyor, belt, closed, 8-4
conveyor, belt, flat, 8-4
conveyor, belt, magnetic, 8-4
conveyor, belt, multiple cord, 8-4
conveyor, belt, multiple ribbon, 8-4
conveyor, belt, portable, 8-5
conveyor, belt, troughed, 8-5
conveyor, boom, 8-5
conveyor, booster, 8-5
conveyor, bottom discharge bucket, 8-5
conveyor, bucket, 8-5. *See also* bucket elevator; positive discharge bucket elevator
conveyor, cable-screw, 8-6
conveyor, carousel, 8-6
conveyor, car type, 8-6
conveyor, chain, 8-7
conveyor, chute, 8-7
conveyor, closed belt, 8-4
conveyor, closed circuit, 8-7
conveyor, controlled velocity roller, 8-8
conveyor, cross bar, 8-9
conveyor, curved belt, 8-10
conveyor, double leg en masse, 8-12
conveyor, drag chain, 8-12
conveyor, el, 8-12
conveyor, elevating, 8-13
conveyor-elevator, 8-9
conveyor, enclosed track trolley, 8-13
conveyor, en masse, 8-13
conveyor, extendable, 8-13
conveyor, fixture, 8-14
conveyor, flat top, 8-14
conveyor, flight, 8-14
conveyor, floor, 8-14
conveyor, flume, 8-15
conveyor, gravity discharge bucket, 8-16
conveyor, hugger belt, 8-17
conveyor, hydraulic, 8-17
conveyor, inclined reciprocating, 8-17
conveyor, internal ribbon, 8-18
conveyor, inverted power and free, 8-18
conveyor, lowering, 8-20

conveyor, magnetic-belt, 8-4
conveyor, metering, 8-21
conveyor, minimum pressure accumulating, 8-21
conveyor, mixing, paddle type, 8-21
conveyor, mixing, screw type, 8-21
conveyor, multiple cord belt, 8-4
conveyor, multiple ribbon belt, 8-4
conveyor, multiple strand, 8-22
conveyor, natural frequency vibrating, 8-22
conveyor, oscillating, 8-23
conveyor, over-and-under, 8-23
conveyor, overhead. *See* power-and-free conveyor; trolley conveyor
conveyor, overhead trolley. *See* trolley conveyor
conveyor, paddle. *See* mixing conveyor, paddle type
conveyor, pallet, 12-34
conveyor, pallet type, 8-25
conveyor, pan, 8-25
conveyor, passenger, 8-25
conveyor, pin type slat, 8-26
conveyor, pivoted bucket, 8-26
conveyor, plain chain. *See* sliding chain conveyor
conveyor, pneumatic, 8-26
conveyor, pocket, 8-26
conveyor, pocket belt, 8-26
conveyor, portable, 8-26
conveyor, portable drag, 8-26
conveyor, power, 8-27
conveyor, power-and-free, 8-27
conveyor, power driven roller. *See* roller conveyor, live
conveyor, pusher bar, 8-27
conveyor, pusher chain, 8-27
conveyor, reciprocating beam, 8-28
conveyor, retarding, 8-28
conveyor, ribbon flight screw, 8-28
conveyor, roller, 8-29
conveyor, roller, herringbone, 8-29
conveyor, roller, hydrostatic, 8-29
conveyor, roller, live, 8-29
conveyor, roller, shock absorbing, 8-29
conveyor, roller, skewed, 8-29
conveyor, roller, speed-up, 8-29
conveyor, roller, spool type, 8-29
conveyor, roller, spring mounted, 8-29
conveyor, roller, troughed, 8-29
conveyor, rolling chain, 8-29
conveyor, rope and button, 8-29
conveyor, screw, 8-30
conveyor screw, multiple flight, 8-21
conveyor screw, rotating casing, 8-30
conveyor screw, vertical, 8-30
conveyor, self-feeding, 8-30

conveyor, self-feeding portable, 8-30
conveyor, shaker. *See* oscillating conveyor
conveyor, shuttle, 8-31
conveyor, side-pull en masse, 8-31
conveyor, side pusher, 8-31
conveyor, skate-wheel, 8-31
conveyor, slat, 8-32
conveyor, sliding chain, 8-32
conveyor, slope, 8-32
conveyor, sorting, 8-32
conveyor, spindle, 8-32
conveyor, stacker, 8-32
conveyor, steel belt, 8-33
conveyor, suspended tray, 8-33
conveyor, telescoping, 8-34
conveyor, tow, 8-34
conveyor, trolley, 8-35
conveyor, twist, 8-35
conveyor type feeder, 8-9. *See also* belt feeder
conveyor, vertical belt, 8-36
conveyor, vertical chain, opposed shelf type, 8-36
conveyor, vertical reciprocating, 8-36
conveyor, vibrating, 8-37
conveyor, vibrating, balanced, 8-37
conveyor, vibrating, natural frequency, 8-37
conveyor, walking-beam, 8-37
conveyor, wheel, 8-37
conveyor, wicket, 8-37
convolution, 1-7
C.O.O. (chief operating officer), 15-3
coolants, 12-61. *See also* cutting fluids
cooling fixture, 12-72
cooling-off period, 6-13. *See also* national emergency dispute
cool time, 12-86
cooperative marketing, 5-4
cooperative processing, 3-5
co-opting, 10-7
coordinated axis control, 11-5
coordinate measuring machine (CMM), 11-5
cope, 12-34
cope, false, 12-34
coping out, 12-34
copolymer. *See* homopolymer
copolymerization. *See* polymerization
copper-harper-scale, 9-6
copy, 3-5
core, 12-8, 12-34, 12-72. *See also* force
core assembly, 12-34
core binder, 12-34
core blow, 12-34
core box, 12-34
core break-off, 12-34
core collapsibility, 12-34
core driers, 12-35

core, drop, 12-35
core filler, 12-35
core forging, 12-26
core, green sand, 12-35
coremaker, 12-35
core oil, 12-35
core ovens, 12-35
core paste, 12-35
core print, 12-35
core, sag, 12-35
core sand, 12-35
cores, centers, 12-72
core shift, 12-35
core shooter, 12-35
core, strainer, 12-35
core vents, 12-35
core wash, 12-35
coriolis effect, 9-6
corner joint, 12-86
corner marker, 8-9
coronal, 2-23
coronal plane, 2-33. *See also* midsagittal plane; transverse plane
coronoid fossa, 2-33
corporate culture, 10-7
corporate-level strategy, 10-7
corporate philanthropy, 10-7
corporate social responsibility, 10-7
corporate social responsiveness, 10-7
correction period, 4-9
corrective action, 16-5
corrective advertising, 5-4
correlation, 1-7, 6-13, 16-5
correlation coefficient, 13-11, 16-5
correlative kinesiology, 2-33
corridor principle, 10-7
corrosion, 8-9
corrosion prevention, 8-9
corrosion preventive, 8-9
cost, 4-9
cost accounting, 4-9
cost accounting, marketing, 5-9
cost analysis, 4-9
cost analysis, marketing, 5-9
cost and schedule control systems criteria (C/SCSC), 4-10
cost, annual, 7-6
cost, annually recurring, 4-3
cost approach, 4-10
cost, average, 5-2
cost, average fixed, 5-2
cost, average variable, 5-2
cost-benefit, 7-6
cost catalyst, 4-13
cost category, 4-10
cost center, standard, 10-29
cost, chemicals, 4-13
cost, clerical payroll, 4-13
cost coefficient, 1-7
cost, common, 7-7
cost, construction, 4-8
cost control, 4-10
cost, development, 4-12, 7-9
cost deviation, 4-12

This index is arranged in letter-by-letter format.

cost, direct, 4-12, 4-13, 7-9, 11-6. *See also* marginal cost; traceable cost
costed bill of material, 14-6
cost effectiveness, 13-11
cost effectiveness analysis, 7-8
cost engineer, 4-10
cost estimate, 4-15
cost estimation, 4-10
cost, field, 4-16
cost, first, 4-16
cost, fixed, 4-16, 5-6, 7-10, 8-14
cost function, 1-7
cost, functional replacement, 4-17
cost, holding, 10-5
cost, home office, 4-18
cost, idle equipment, 4-18
cost, impact, 4-18
cost, incremental, 4-18
cost index, 4-10. *See also* price index
cost, indirect, 4-18, 7-11, 11-9, 13-25. *See also* accident costs
cost, initial. *See* first cost
cost, input material, 4-13
cost, insured, 13-26
cost, item, 10-16
cost leadership strategy, 10-7
cost, least total, 14-12
cost, least unit, 14-12
cost, local, 4-21
cost, maintenance, 4-13
cost, maintenance and repair, 4-21
cost, manufacturing, 4-21
cost, marginal, 4-21, 5-8, 7-12
cost, material, 4-22
cost of capital, 4-10, 7-8. *See also* required return
cost of living adjustment (COLA), 6-11, 6-13. *See also* consumer price index (CPI); escalator clause
cost of living allowance, 6-13
cost of living index, 4-10
cost of lost business advantage, 4-10
cost of ownership, 4-10
cost of quality, 4-10
cost of quality conformance, 4-10
cost of quality nonconformance, 4-10
cost of work performed, actual, 4-2
cost, operating, 4-24
cost, planned, 4-25
cost-plus, 5-4
cost plus contract, 4-9
cost plus fixed sum contract, 4-9
cost plus percentage burden and fee contract, 4-9
cost range, 1-7
cost row, 1-7
cost, startup, 4-32
cost, sunk, 4-33
cost, traceable, 7-14
cost, unit, 4-35
cost value, 4-17

cost, variable, 4-36, 5-16
Council of Economic Advisors, 6-13. *See also* guideposts
count chart, 16-6
counter, 11-5
counteradvertising, 5-4
counterbalanced front sideloader truck, 8-9
counterbalanced truck, 8-9
counterblow forging equipment, 12-26. *See also* hammer forging
counterboring, 12-61
counterlock, 12-26
countersinking, 12-61
count per unit chart, 16-6
coupler, automatic, 8-3
coupler height, 8-9
coupling design, kinematic, 12-15
coupling equations, 1-7
coupling, safety, 13-40
coupon, 5-5
covariance, 16-6
coverage
 actual, 17-4
 insurance, 13-11
 potential, 17-4
 standard, 17-4
covered electrode, 12-86
cover glass, 12-86
covert lifting task, 2-33
CPI (consumer price index), 6-13. *See also* cost of living adjustment (COLA); escalator clause
CPIM (certified in production and inventory management), 14-6
CPM (critical path method), 5-5, 14-6. *See also* PERT (program evaluation and review technique)
c process, 12-35. *See also* shell molding
CPSC (consumer product safety commission), 5-4
CPU (central processing unit), 3-5, 10-5, 12-6
crack, hot tear, 12-35
cracking strip, 12-35
craft, 6-13
craft union, 6-13. *See also* horizontal union; skilled trades
crane, 8-9
crane attachment, 8-9
crane bar, portable metal stacking rack. *See* tiering guide
crane, bridge. *See* crane; fixed crane; overhead crane
crane bridge (top-running), 8-9
crane, crawler, 8-9
crane, derrick, 8-11
crane, double girder gauge, 8-16
crane drive, individual, 8-9
crane, fixed, 8-14
crane, gantry, 8-16
crane, girder, 8-16
crane, jib, 8-18
crane, locomotive, 8-19
crane, overhead, 8-23

crane, portable, 8-26
crane, semi-gantry, 8-30
crane, top-running, 8-34
crane, tower, 8-34
crane, traveling, 8-35
crane, truck, 8-35
crane, under-running, 8-35
crane, wall, 8-37
crash safety, 13-11
crashworthiness, 13-11
crate, 8-9
crater, 12-86
crater crack, 12-86
crater wear, 12-61
crawler crane, 8-9
crazing, 12-73
CR (consumer's risk), 16-5
creativity, 10-7
credited service, 6-14
credit, line of, 4-20
credit union, 6-14
credit union deductions, 6-14
creep, 6-14, 12-26, 12-50, 12-73
creep grinding, 12-61
creeping method changes, 6-14
crew station, 9-6
criminal negligence, 13-11
crimping, 12-50
crisis bargaining, 6-14. *See also* continuous bargaining committees; interim committees
crisis problem, 10-7
criteria, 4-10
criterion, 12-8, 15-3. *See also* control
critical activity, 4-10
critical flicker frequency, 9-6
critical function, 13-11
critical incidents, 6-14
critical incident technique, 9-6, 13-11
critical items, 11-5
criticality, 4-10
critical organ, 13-11
critical path, 4-10, 10-7. *See also* network analysis
critical path method (CPM), 4-10, 5-5, 14-6. *See also* PERT (program evaluation and review technique)
critical path scheduling, 5-5. *See also* network analysis
critical ratio, 11-5, 14-6. *See also* dispatching rule; slack
critical ratio scheduling, 11-5
critical region, 16-6
critical temperature, 12-26, 12-35
critical value, 16-6
critical work center, 11-5
croning process, 12-35. *See also* shell molding
cross aisle, 8-9
cross assembler, 12-8
crossbanding, 12-73
cross bar conveyor, 8-9
cross-docking, 8-9
cross-linking, 12-73

cross-sectional study, 13-11
cross stacking, 8-10
cross talk, 12-9
cross tie, 8-10
crown, 12-50
CRP (capacity requirements planning), 14-5
CRP (consumer's risk point), 16-5
CRQ (consumer's risk quality), 16-5
CRT (cathode ray tube), 12-6
CRT (cathode ray tube) display, 3-4
crucible, 12-35
crucible furnace, 12-35
crude materials, 4-10, 4-11
crush form grinding, 12-61
cryogenics, 12-9
crystallinity, 12-73
C-scale, 13-8
C/SCSC (cost and schedule control systems criteria), 4-10
CSCW (computer supported cooperative work), 9-6
CSI (continuous systems improvement), 15-3
CSMAs (consolidated metropolitan statistical areas), 5-4
CSP (Certified Safety Professional), 13-9
C-stage, 12-71
CTD (cumulative trauma disorder), 13-12
CTS (carpal tunnel syndrome), 2-32, 13-8. *See also* carpal tunnel
cube, 8-10
cube-per-order index (COI), 8-10
cube per order index storage policy, 8-10
cubic boron nitride (CBN), 12-61
cue, 9-7
cull, 12-73
culture, 15-3
cumulants, 1-7
cumulative distribution function, $F(x)$, 1-7
cumulative lead time, 14-6
cumulative quantity discounts, 5-5
cumulative sum chart, 16-6
cumulative trauma, 2-33. *See also* trauma
cumulative trauma disorder (CTD), 13-12
cupola, 12-35
cupola drop, 12-36
cupping, 12-51
cure, 12-73
curing time (no bake), 12-36
curling, 12-51
current cost accounting, 4-11
current dollars, 4-11. *See also* actual dollars; deflating by a price index
current period (of a given price index), 4-11
cursor, 3-5, 9-7, 12-9
curtailed inspection, 16-6

This index is arranged in letter-by-letter format.

curved belt conveyor, 8-10
cushion, die, 12-51
cushion tire, 8-10
customer divisions, 10-7
customer order, 14-6
customers, 10-7
customer service, 14-6
customer service level, 5-5
customer service ratio, 14-6. *See also* stockout percentage
custom in the industry, 4-11
cusum chart, 16-6
cut, 1-7
cut and paste, 3-5, 9-7
cutaneous, 2-23
cutoff machine, abrasive, 12-36
cutoff rate of return, 7-8. *See also* required return
cut-out, safety, 13-39
cuts, 12-36
CUTS II, 12-9
cutter compensation, 12-9
cutter location (CL) data, 12-9
cutter offset, 12-9
cutter path, 11-5, 12-9
cutting fluids, 12-61. *See also* coolants
cutting force, 12-61
cutting, laser beam, 12-88
cutting, oblique, 12-64
cutting, orthogonal, 12-64
cutting, oxyfuel gas, 12-64
cutting plane, 1-7
cutting, plasma arc, 12-65
cutting ratio, 12-61
cutting speed, 12-9
cutting speed, optimum, 12-64
cutting tool angles, 12-61
cutting torch, 12-86
CWBS (contract work breakdown structure), 4-36
cybernetic control system, 10-7
cybernetics, 3-5, 9-7, 12-9, 15-3
cycle, 12-9, 17-4
cycle counting, 14-6. *See also* physical inventory
cyclegram, 17-4
cyclegraph, 17-4
cyclegraph technique, 17-4. *See also* chronocyclegraph technique
cycle, motion, 17-13
cycle stock, 5-5, 14-6. *See also* lot size; safety stock
cycle time, 5-5, 6-14, 11-5, 14-7
cycle time, average, 17-2. *See also* average element time; select(ed) time
cycle timing, 17-5. *See also* differential timing
cycle, work, 6-65, 9-27, 17-21
cyclic element, 17-5
cyclic timing, 17-5
cycling (linear programming), 1-7
cyclone, 12-36
cylinder, 12-86
cylindrical coordinate system, 11-5

D

DAC (digital to analog converter), 11-6, 12-10
D/A (digital-to-analog) converter, 11-6
daily rate, 6-14
damage, direct, 13-13
damage, indirect, 13-25
damage risk criterion, 13-12
damage(s), 13-12
damages, actual, 4-11
damages, liquidated, 4-11
damages, mitigation of, 4-22
damages, ripple. *See* impact cost
damping, 11-6, 12-9
danger, 13-12
danger, imminent, 13-24
dangerous, 13-12
danger zone, 13-12
dark adaptation, 9-7
data, 3-5, 10-7, 12-9
data, analog, 11-2
data base, 3-5, 10-7
data base management systems (DBMS), 3-5, 10-7
data base manager, 3-5
data chart, original, 16-9
data collection, 11-6
data date (DD), 4-11. *See also* status line
data directory, 3-5
data entry, 9-7, 11-6
data file, 12-9
data glove, 9-7
data hierarchy, 11-6
data logging, 11-6
data manipulations, 12-9
data, primary, 5-11
data, primary classification, 5-11
data processing, 12-9
data processing, automatic (ADP), 3-2, 3-3
data, standard, 6-53
data, synthetic, 17-19
date, effective, 14-8
date for the commencement of the contract time, 4-11
date, imposed, 4-18
date, imposed finish, 4-18
date of the agreement, effective, 4-14
dates, interim, 4-19
day rate, 6-14
days charged per disabling injury, average, 13-6
days of disability, 13-12. *See also* lost workdays
day work, 6-14, 17-5
dazzle, 9-7
dB (decibel), 13-12
DBMS (data base management systems), 3-5
DDC (direct digital control), 11-6
DD (data date), 4-11
dead band, 9-7, 11-6, 12-9

dead hand syndrome, 13-38
deadheading pay, 6-14
dead loads, 8-10
deadman control, 9-7
deadtime, 12-9. *See also* downtime
dead work, 6-14
deaf, 13-12
deafened, 13-12
deafness. *See also* conductive hearing loss
 conductive, 2-33
 perception, 9-15
 psychogenic, 13-37
dealer, 5-5
death, accidental, 13-12
death benefit, 6-14. *See also* life insurance plan
death certificate, 13-12
death rate, 13-4
death rate, age-adjusted, 13-4
death rate, age-specific, 13-4
death rate, age-standardized, 13-4
death rate, mileage, 13-31
debt capital, 10-7
debt management ratios, 10-7
debug, 9-7
debugging, 1-7, 12-9
deburring, 12-61
decarburization, 12-36, 12-51
deceleration, 2-33, 4-11
decentralization, 10-8
decentralization, authority, 15-4
decentralization, physical, 15-4
decentralized dispatching, 14-7. *See also* centralized dispatching; dispatching
decentralized storage, 8-10
decibel (dB), 9-7, 12-10, 13-12. *See also* sound pressure level
deciding to decide, 10-8
decimal-hour stopwatch, 17-5. *See also* stopwatch
decimal-minute stopwatch, 17-5. *See also* stopwatch
decision, 6-14
decision function, 1-7
decision making, 6-14, 9-7, 10-8, 15-4
decision-making models, descriptive, 10-8
decision-making models, normative, 10-21
decision matrix. *See* payoff table
decision rule, 15-4
decisions, nonprogrammed, 10-21
decisions under certainty, 1-7, 7-8
decisions under risk, 1-7, 7-8. *See also* risk analysis
decisions under uncertainty, 1-8, 7-8
decision support-system (DSS), 10-8
decision support-system, integrated (IDSS), 11-9
decision table, 3-5
decision theory, 7-8

decision time, 9-7
decision tree, 7-8, 10-8. *See also* risk analysis
deck, 8-10
deckmember, 8-10
deckmember chamfer, 8-10
deck opening, 8-10
deck, pallet storage rack, 8-10
deck, portable metal stacking rack, 8-10
deck spacer, 8-10
declared value (air transport), 8-10
declining balance depreciation, 4-11, 7-9. *See also* Matheson formula
decoder logic, 8-10
decoding, 10-8, 15-4
decomposition principle, 1-8. *See also* master program
decompression sickness, 13-12
decontamination, 13-13
decontamination, radiation, 13-13
decoupler, 11-6
dedicated storage policy, 8-10
deductible, 6-14. *See also* major medical expense benefits
deep drawing, 12-51
deep lane storage, 8-10
de-escalate, 4-11
defect, 1-8, 4-11, 13-13, 16-6
defective, 4-11
defective item, 16-6
defective specifications, 4-11
defective unit, 16-6
defect, latent, 4-11
defect, patent, 4-11
defender, 7-8
defensive avoidance, 10-8
defensive strategies, 10-8
deferred bonus payment, 6-14
deferred compensation, 6-14
deferred rate (air transport), 8-10
deferred service (air transport), 8-10
deferred wage increases, 6-15. *See also* annual improvement factor (productivity)
Deficit Reduction Act of 1984, 6-14
defined benefit pension plan, 6-15
defined contribution plan, 6-15
definitions lists, 7-2
definitive estimate, 4-15
deflagration, 13-13
deflating by a price index, 7-8. *See also* actual dollars; constant dollars; inflating
deflation, 4-11, 7-8. *See also* inflation
deflection, 12-51
deformation limit, 12-51
deformation test, 12-36
degasser, 12-36
degenerate solution, 1-8
degenerative joint disease. *See* osteoarthritis
degradation, 12-73. *See also* deterioration

This index is arranged in letter-by-letter format.

degradation failure, 1-10
degree of freedom, 11-6, 16-6
degree of ramming, 12-36. *See also* ramming
degrees of negligence, 13-13
dehumidify, 8-10
delamination, 12-73
DeLavaud process, 12-36
delay, 6-15, 14-7, 17-5. *See also* standby; waiting time
delay allowance, 6-15, 17-5
delay allowance, unavoidable, 17-20
delay, avoidable, 6-7. *See also* idle time
delay, balancing, 17-2. *See also* balance
delay, compensable, 4-11
delay, concurrent, 4-11
delay, excusable, 4-11, 4-12
delay, inexcusable, 4-12
delay, nonprejudicial, 4-12
delay, prejudicial, 4-12
delay report, anticipated, 14-3
delay (symbol), 17-8
delay time, 17-5. *See also* downtime; idle time
delay, unavoidable, 17-20
delegation, 10-8, 15-4
delegation and responsibility. *See* authority
deliverable, 4-12
delivery cycle, 11-6, 14-7
delivery policy, 14-7
delivery table, 8-10
Delphi method, 9-7, 10-8
Delphi technique, 15-4
deltoid, 2-33
demand, 14-7
demand curve, 5-5
demand, dependent, 14-7
demand, elastic, 5-6
demand factor, 4-12, 7-8. *See also* capacity factor; load factor
demand flow manufacturing, 14-7
demand, independent, 14-10
demand, interplant, 14-10
demand, lumpy, 14-13
demand management, 14-7. *See also* master production schedule (MPS)
demand motions inventory, 2-34. *See also* available motions inventory
demerit chart, 16-6
de minis violation, 13-12
democracy, industrial, 6-26
democratic, 10-8
demographic index, 4-12
demographics, 5-5
demonstrated capacity, 14-7
demotion, 6-15
demountable box, 8-10
demountable cargo container, 8-10
demurrage, 4-12, 8-11
density, bulk, 12-70
density, of traffic, 13-13

deoxidation, 12-36
department, 15-4
departmentalization, 10-8, 15-4
departmentation, 15-4
department store, 5-5
dependability, 1-8, 16-6
dependent demand, 14-7. *See also* demand; independent demand; requirements explosion
dependent demand inventory, 10-8
dependent failure, 1-10. *See also* failure, secondary
dependent variable, 13-13. *See also* independent variable
depletion, 4-12, 7-8. *See also* capital recovery
depletion allowance, 7-8
deposited metal, 12-86
deposition, 13-13. *See also* interrogatories
deposition efficiency, 12-87
deposition rate, 12-87
depreciated book value, 4-12
depreciation, 4-12, 7-8. *See also* capital recovery
depreciation, accelerated, 7-9
depreciation allowance, 7-9
depreciation basis, 7-9
depreciation, declining balance, 4-11, 7-9
depreciation method, multiple straight-line, 4-23, 7-9
depreciation, multiple straight-line, 7-9
depreciation, sinking fund, 7-9
depreciation, straight-line, 7-9
depreciation, sum-of-years-digits, 7-9
depressed (incentive), 6-15
depressed (standard), 6-15
depth of cut, 12-61
depth of field, 8-11
depth of fusion, 12-87
depth perception, 9-7
depth perception, binocular, 13-13
De Quervain's disease, 2-34
derating, 1-8, 16-6
derivative control, 11-6
dermatitis, 13-13. *See also* irritant
derrick crane, 8-11
describing function, 9-7
descriptive decision-making models, 10-8
desertification, 6-15
desertification election, 6-15
desiccant, 8-11
design, 11-6
designated imperfections, 1-8
design criteria, 5th to 95th percentile, 2-6
design, safety, 13-13
design specification, 4-32
destination inspection, 8-11
destructive test, 13-13
detailed engineering, 4-12
detailed schedule, 4-12
detectability, 9-7

detector tube, 13-13
deterioration, 7-9, 8-11, 12-73. *See also* degradation
deterministic model, 1-8
detonation, 13-13
developed countries, 10-8
development costs, 4-12, 7-9
deviation, 4-12
deviation costs, 4-12
deviation, mean, 16-9
deviation, radial, 2-46
deviation, standard, 1-24, 4-32, 16-17
devil's advocates, 10-8
dewaxing, 12-36
dexterity, manual, 13-29
diagnostic routine, 12-10
diagnostic study, 17-5
dialectical inquiry, 10-8
dialog box, 9-7
diameter, sealing, 12-83
diaphragm gate, 12-73
diaphysis, 2-34
diatomaceous earth, 12-36
dichotic, 9-7
Dictionary of Occupational Titles, 6-15
did-take standards, 17-5. *See also* base productivity factor (BPF); normal performance
did-take time standard, 17-18
die-adaptor, 12-73
die, blanking, 12-49
die block, 12-26, 12-73
die body, 12-73
die casting, 12-36
die-enclosure guard, 12-51
die forging, 12-26
dielectric, 12-61
dielectric heating. *See* high-frequency heating
dielectric oven, 12-36
die lubricant, 12-26
die, multiple-cavity, 12-41
die pad, 12-51
die(s), 12-26, 12-51
diesel-electric truck, 8-11
die set, 12-51
diesetter, 12-51
die shift, 12-26
die shoes, 12-51
die sinking, 12-51
diet problem, 1-8
difference limen, 9-7. *See also* limen
differential, intercity, 6-27
differential marketing, 5-5
differential piece rate, 6-15
differential piecework, 6-15. *See also* multiple piece rate plan
differential price escalation rate, 4-12
differential skill, 6-15
differential time plan, 6-15. *See also* multiple time plan
differential timing, 17-5. *See also* cycle timing

differential transducer, 12-10
differentiation, 10-8
differentiation paradox, 10-8
differentiation strategy, 10-8
differing site conditions, 4-12
diffusion, 5-5, 12-73
digit, 12-10
digital, 11-6, 12-10
digital computer, 3-6, 12-10
digital-to-analog converter (D/A, DAC), 11-6, 12-10
digitize, 3-6, 12-10
dilatant, 12-73
diluent, 12-73
dimensional weight, 8-11
dimensional weight rule, 8-11
dimensions, qualifying, 5-12
dimensions, types of, 2-3
dimpling, 12-51
diophantine programming. *See* integer linear programming
diotic, 9-7
dip brazing, 12-87
dip coat, 12-36
DIP (dual in-line package), 12-10
direct access, 3-6
direct-arc furnace, 12-36
direct capital. *See* direct cost
direct cause, 13-13
direct compensation, 6-15
direct contact, 10-8
direct cost, 4-12, 4-13, 7-9, 11-6. *See also* marginal cost; traceable cost
direct damage, 13-13
direct-deduct inventory transaction processing, 14-7. *See also* post-deduct inventory transaction processing; pre-deduct inventory transaction processing
direct digital control (DDC), 11-6
direct extrusion, 12-51
direct injury costs, 13-13
direct interlock, 10-8
direct investment, 10-9
directional properties, 12-26
directional solidification, 12-36
directive, 10-9
direct labor, 6-15, 17-5. *See also* productive labor
direct labor budget, 6-15
direct labor standard, 17-5
direct manipulation, 9-7
direct marketing, 5-5
direct memory access (DMA), 12-10
direct numerical control (DNC), 11-6, 12-10, 12-18
direct observation, 6-15
directorates, interlocking, 10-16
directors, board of, 15-2
directory. *See* data directory
dirty money, 6-15
disability, 6-15, 13-13
disability, long-term, 6-32
disability pay, 6-15
disability, permanent, 13-35

This index is arranged in letter-by-letter format.

15

disability, permanent and total, 6-42
disability, permanent partial, 13-35
disability, permanent total, 13-35
disability retirement, 6-15
disability, short-term, 6-51
disability, temporary total, 13-44
disabling injury, 13-13
disabling injury frequency rate, 13-13. *See also* incidence rate
disabling injury index, 13-14
disaffiliation, 6-15
disassembly, 11-6
disaster control, 13-14
disbursement, 14-7
disc gate, 12-73
discharge, 6-15
discharged and wet, 8-11
disc, intervertebral, 13-14
disciplinary interview, 6-15. *See also* interview
discipline, 6-15
disclaimer, 13-14
disconnect device, main line, 8-11
discontinuous timing. *See* repetitive timing
discount, cash, 5-3
discounted cash flow, 7-9. *See also* internal rate of return; investor's method; profitability index (PI); rate of return
discount, quantity, 5-12
discount rate. *See* discounted cash flow; interest rate
discounts, cumulative quantity, 5-5
discount, trade, 5-15
discrete, 11-6, 12-10
discrete code, 8-11
discrete manufacturing system, 8-11
discrete order quantity. *See* lot-for-lot
discrete parts manufacturing, 11-6
discrete variable problem. *See* integer linear programming
discretionary expense center, 10-9
discrimination, 6-16
discriminatory analysis, 5-5
disc, ruptured, 13-14
disc, slipped, 2-38
disease classification, 13-14
disease(s)
 Caisson's, 13-12
 degenerative joint. *see* osteoarthritis
 De Quervain's, 2-34
 functional, 13-21
 hard metal, 13-22
 occupational, 13-33
 organic, 13-34
 Raynaud's, 2-47
 respiratory, 13-39
 Shaver's, 13-41
disfigurement, 13-14
dished, 12-73. *See also* warp

dishing, 12-51
disinflation, 4-13
disjunctive socialization, 15-4
disk/diskette, 3-6
disk/diskette, flexible, 3-7
disk/diskette, floppy, 3-7
disk/diskette, magnetic, 3-10, 12-16
disk drive, 3-6, 12-10
disk operating system (DOS), 3-6
disk storage, 12-10
dismissal compensation. *See* severance pay
dispatching, 11-6, 14-7. *See also* closed loop MRP; expediting; scheduling; shop planning
dispatching, centralized, 14-5. *See also* control center
dispatching, decentralized, 14-7
dispatching rule, 14-7. *See also* critical ratio
dispatch rules, 8-11
dispenser, automatic, 8-3
dispersant, 12-73
dispersed shrinkage, 12-36
dispersion, 4-13, 12-73. *See also* organosol; plastisol
displacement, cargo container, 8-6
displacement, lateral, 2-41
displacement, medial, 2-42
display, 9-7
display, analog, 9-3
display, cathode ray tube, 3-4
display, compensatory, 9-6
display, dynamic, 9-7
display, go, no-go, 9-9
display, graphical, 9-9
display, head up (HUD), 9-10
display, helmet mounted (HMD), 9-10
display, integrated, 9-11
display layout, 9-7
display, predictor, 9-16
disposable income, 5-5
dispute, 4-13, 6-16
dispute, jurisdictional, 6-29
disruption, 4-13
dissatisfies, 10-9
distal, 2-34, 11-6
distal cause, 13-14
distinctive competence, 10-9
distortion, 12-51
distributables, 4-13
distributed control, 9-7
distributed decision making, 9-7
distributed numerical control, 11-6
distributed processing, 10-9
distribution, 5-5. *See also* warehouse site selection
distribution, binomial, 16-4
distribution center, 5-6
distribution, dual, 5-6
distribution, exclusive, 5-6
distribution-free method, 1-8, 4-13
distribution function, 16-6
distribution model, 10-27
distribution, normal, 16-11

distribution, physical, 5-11, 14-16
distribution points, 8-11
distribution, probability, 5-11
distribution requirements planning (DRP), 14-7. *See also* time-phased order point (TPOP)
distribution resource planning, 14-7
distribution, sand grain, 12-36
distribution, selective, 5-13, 5-14
distribution service, 8-11
distribution system, pull, 14-18
distribution system, push, 14-18
distribution warehouses, 5-6
distributive lag effect of advertising, 5-6
divergent thinking, 10-9
diversification, 5-6, 10-9
diverter, 8-11
divestiture, 10-9
divided attention, 9-3
dividend, 6-16
dividing head, universal, 12-67
division, 15-4
divisionalized form, 10-9
divisional structure, 10-9
division of labor, 6-16, 17-5. *See also* downgrade
division of work, 15-4. *See also* specialization
division of work, base. *See* therblig
DMA (direct memory access), 12-10
DNC (direct numerical control), 11-6, 12-10, 12-18
dock, 8-11
dockboard, 8-11
dock leveler, 8-11
dock, saw-tooth, 8-30
dock shelter, 8-11
dock to flow, 14-7
dock to stock, 14-7
doctor bar, 12-73
doctor roll, 12-73
documentation, 12-10
dollars, actual, 7-6. *See also* deflating by a price index; inflating by a price index
dollars, constant, 4-8, 7-7, 7-9. *See also* deflating by a price index; inflating by a price index
dollars, current, 7-9
dollars, real, 7-9
dollars, then current, 7-9
dolly, 8-11
domain shifts, 10-9
domed, 12-73. *See also* warp
doping, 12-73
Doppler effect, 2-34
dorsiflexion, 2-34
DOS (disk operating system), 3-6
dose, radiation, 13-14
dose-response relationship, 13-14
dose, toxicology, 13-14
dot matrix printer. *See* matrix printer

double arm, 8-11
double back pay, 6-16
double click, 9-7
double deep storage, 8-11
double dipping, 6-16
double leg bucket elevator, 8-11. *See also* bucket elevator
double leg en masse conveyor, 8-12
double precision, 3-6
double sampling inspection, 16-6
double seaming, 12-51
double time, 6-16
dowel, 12-26, 12-73
downcomer, 12-36
downgrade, 6-16. *See also* division of labor
downgrading, 6-16
download, 11-6
downsizing, 10-9
downtime, 1-26, 6-16, 11-6, 12-10, 14-8, 16-18, 17-5. *See also* delay time; work design and measurement
downtime, machine, 6-33
downward communication, 10-9
draft, 12-26, 12-51, 12-73
draft angle, 12-27
draft, matching, 12-29
draft, natural, 12-29
drag, 12-36
drag chain conveyor, 8-12. *See also* conveyors
drag conveyor, portable, 8-6
dragging, 11-6
drape-assist frame, 12-74
drape forming, 12-74
draw, 12-36
drawability, 12-52
draw bar, 12-36
drawbar pull, maximum, 8-12
drawbar pull, rated normal, 8-12
draw bead, 12-51. *See also* blank holder
drawing, 12-27, 12-51
drawing account, 6-16. *See also* commission earnings
drawings, plans, 4-13
draw plate, 12-36
dressing (grinding wheels), 12-61
dried sand, 12-36
drift, 11-7
drill point, 12-65
drill, twist, 12-67
drive, chain, 11-4
drive girder, 8-12
drive in rack, 8-12
drive thru rack, 8-12
driving forces, 10-9
driving head, 8-12
drop, 12-36
drop delivery, 17-5
drop forging, 12-27. *See also* impression die forging
drop hammer, 12-27. *See also* air-lift hammer; board hammer; steam hammer

This index is arranged in letter-by-letter format.

drop section, 8-12
drop shipment, 14-8
drop shipper, 5-6
dross, 12-37
drowning, 13-14
drowning, near, 13-32
DRP (distribution requirements planning), 14-7
drum grooves, 8-12
drum handler, 8-12
drum support, 8-12
dry-chemical extinguisher, 13-14
dry coloring, 12-74
dry permeability, 12-37
dry-powder extinguisher, 13-14
dry sand mold, 12-37
dry spot, 12-74
dry strength, 12-37
DSS (decision support-system), 10-8
dual-career couples, 15-4
dual command cycle, 8-12
dual distribution, 5-6
dual in-line package (DIP), 12-10
duality theorem for linear programming, 1-9
dual linear program problems, 1-8
dual pay, 6-16
dual problem. *See* dual linear program problems
dual simplex algorithm, 1-9
dual unionism, 6-16
ductility, 12-37, 12-52
due date, 11-7, 14-8
dues, 6-16
dull tool, 12-61
dummy activity, 4-14
dummy battery, 8-12
dummy start activity, 4-14
dummy variable, 13-14
dumper, 8-12
dunnage, 8-12
duplex, 11-7
duplexing, 12-37
duplex transmission, 3-6
duplicate cavity-plate, 12-74
durable goods, 4-14
duration, activity, 4-2
duration, expected, 4-16
duration, original, 4-24
durometer hardness, 12-74
durvill process, 12-37
dust, 12-37
dust, nuisance, 13-33
dusts, 13-15
duty cycle, 11-7
dwell, 12-52
dynamic accuracy, 11-7
dynamic display, 9-7
dynamic lot sizing. *See* fixed order quantity; least total cost; least unit cost; period order quantity
dynamic lowering, 8-12
dynamic model, 11-7
dynamic moment, 2-34
dynamic programming, 1-9

dynamic RAM, 12-11. *See also* RAM (random-access memory)
dynamic response, 12-11
dynamics, group, 15-6
dynamic system, 9-8
dynamic visual acuity, 9-8. *See also* visual acuity
dynamic work, 2-34. *See also* isometric work
dynamometer, 2-34, 9-8
dysbaric osteonecrosis, 13-15
dysfunctional conflict, 15-4

E

EAP (employee assistance programs), 6-17, 10-10
ear, 12-52
ear breadth, 2-9
earcon, 9-8
ear height, 2-9
earing, 12-52
earliest due date (EDD), 14-8
earliest expected completion date, 4-14
early adopters, 5-6
early cause, 13-15
early event time (EV), 4-14
early failure period, 1-9
early finish time (EF), 4-14
early retirement age, 6-16
early start time (ES), 4-14
early work schedule, 4-14
earned hours, 6-16, 17-5
earned rate, 6-16
earned rate, average, 6-6
earned value, 4-14
earned value concept, 4-14
earned value reports, 4-14
earning power of money, 7-10
earnings, 6-16. *See also* straight time
earnings, annual, 6-5
earnings, average, 6-6
earnings, average hourly exclusive of overtime payments, 6-7
earnings, average straight-time hourly, 6-7
earnings efficiency, 6-16
earnings, expected (level), 6-19
earnings, interim, 6-27
earnings, potential, 6-43
earnings, spendable, 6-53. *See also* take-home pay
earnings value, 4-14, 6-16, 7-10
ear protectors, 13-15
EBCDIC (extended binary-coded decimal interchange code), 3-7
EBM (electron beam machining), 12-61
eccentric contraction, 2-34
echo, 9-8
echography, 2-34. *See also* ultrasonics

ECM (electrochemical machining), 12-62
ecological, 9-8
ecological correlation, 13-15
ecological psychology, 9-8
ecological stress factor, 2-34
ecological validity, 9-8
ecology, occupational, 2-43
econometric models, 10-9
economic alternative, 7-6
economic element, 10-9
economic life, 7-10. *See also* life
Economic Opportunity Act, 6-16
economic order quantity (EOQ), 5-6, 10-9, 14-8. *See also* carrying cost; fixed order quantity; ordering cost; reorder quantity
Economic Recovery Act of 1981 (ERTA), 6-19
economic return, 4-14
economic strikes, 6-17
economic supplements, 6-17
economic tool life, 12-61
economic value, 4-14
economies of scale, 5-6
economy, 4-14, 7-10
economy service (air transport). *See* deferred service
economy, socialist, 10-28
ectocanthus, 2-23
eczema, 13-15
EDD (earliest due date), 14-8
edema, 13-15
edge joint, 12-78, 12-87
edgemember, 8-12
edge preparation, 12-87
edger, 12-27. *See also* fuller
edges, matched, 12-29
edging impression, 12-27
EDI (electronic data interchange), 3-6
edit, 3-6, 9-8, 12-11
editor program, 3-6
EDM (electric discharge machining), 12-61
EDP (electronic data processing), 10-9, 12-11
educational assistance, 6-17
education leave, paid (PEL), 6-41
education, safety, 6-49, 13-40
EEOC (Equal Employment Opportunity Commission), 6-18
EF (early finish time), 4-14
effective date, 14-8
effective date of the agreement, 4-14
effective interest. *See* interest rate, effective
effectiveness, 7-10, 10-9, 15-4, 17-6
effective temperature, 2-34. *See also* thermal environment
effectivity, 14-8
efferent, 2-34. *See also* afferent
efficiency, 6-17, 10-9, 14-8, 15-4
efficiency, labor, 17-6. *See also* productivity

efficient cause, 13-15
effort, 6-17
effort, normal, 17-13. *See also* normal performance
effort-performance expectancy, 10-9
effort, physical, 17-6
effort rating, 6-17. *See also* leveling; normalize; performance rating
EIA (Electronic Industries Association), 12-11
ejector marks, 12-37, 12-52
ejector pins, 12-40
elapsed time, 17-6. *See also* actual time
elapsed time, expected, 4-16
elastic demand, 5-6
elasticity, 12-74
elasticity, modulus of, 12-54
elastic limit, 12-52
elastic limit of tissues, 2-34
elastic supply, 5-6
elastomer, 12-74
elbow height, 2-9, 9-17
elbow rest height, sitting, 2-9
elbow-to-elbow breadth, 2-9
el conveyor, 8-12. *See also* roller conveyor; wheel conveyor
electrically baffled, 8-12
electrical safety, 13-15
electrical-silence, 2-34
electric discharge machining (EDM), 12-61
electric furnace, 12-37
electric tractor, 8-12
electric truck, 8-12
electrified monorail, automatic, 8-3
electrocardiograph, 2-34
electrochemical grinding, 12-61
electrochemical machining (ECM), 12-62
electrode, 2-34, 12-11
electrode, bare, 12-86
electrode holder, 12-87
electromyographic kinesiology, 2-35
electromyography, 2-35. *See also* biomechanical profile
electron beam machining (EBM), 12-61
electron beam welding, 12-87
electronic data interchange (EDI), 3-6, 5-6
electronic data processing (EDP), 10-9, 12-11
electronic heat control, 12-87
Electronic Industries Association (EIA), 12-11
electronic mail, 3-6. *See also* e-mail
electronic mail system, 10-9. *See also* e-mail
electronic monitoring, 10-10
electrophysiological apparatus, 2-35

This index is arranged in letter-by-letter format.

electroslag welding, 12-87
electrostatic plotter, 3-6
electrostatic precipitator, 13-15
element, 6-17, 17-6. *See also* machine-controlled time; manual element
elemental breakdown, 17-6
elemental motion, 17-6. *See also* microelement
elemental standard data, 17-6
elementary commodity groups, 4-14
elementary matrix, 1-9
element, basic. *See* elemental motion
element, external, 17-7
element, internal. *See* internal work
element, international, 10-16
element, irregular, 17-10
element time, 6-17, 17-6
element time, normal, 17-13
elevating conveyor, 8-13
elevator, bucket, 8-5. *See also* bucket conveyor; bucket loader; marine leg
elevator, bucket pivoted, 8-6
elevator, chain. *See* bucket elevator
elevator, double leg bucket, 8-11
elevator, gravity discharge bucket, 8-16
elevator, internal discharge bucket, 8-18
elevator pivoted bucket, 8-6
elevator, positive discharge bucket, 8-27
elimination method, 1-9
ELISA test, 6-17
elongation, 12-52
e-mail, 3-6. *See also* electronic mail system
embossing, 12-52
emergency alarm, 13-15
Emergency Board, 6-17
emergency exit, 8-13
emergency procedure, 13-15
emergency shut-off, 13-15
emergency stop, 13-15
emergent feature, 9-8
emission control, 13-15
emphysema, 13-15
empirical data, 11-7
empirical workplace design, 2-35. *See also* rational workplace design
employee, 6-17, 15-4
employee assistance programs (EAP), 6-17, 10-10
employee benefits, 6-17
employee buy-out, 6-17
employee-centered, 10-10
employee classification, 6-17
employee contributions, 6-17
employee-hours, 13-15, 13-17
employee involvement programs, 6-17

employee involvement teams, 10-10
employee pacing, 6-17
employee perquisites, 6-17
Employee Polygraph Protection Act, 6-17
Employee Retirement Income Security Act (ERISA), 6-19
employees, active, 6-3
employees, classification act, 6-10. *See also* wage board employees
employee services, 6-17. *See also* perquisite
employees, exempt, 6-19
employees, experienced, 6-19
employees, ghost, 6-22
employees, non-exempt, 6-37
employees, part-time, 6-40
employees, protected, 6-45
employees, regular, 6-47
employees, temporary, 6-58. *See also* probationary period
Employee Stock Ownership Plan (ESOP), 6-17
employees, wage board, 6-63
employer, 6-18. *See also* management
employers' association, 6-18. *See also* association agreement
employer's liability, 13-15
employment at will, 10-10
employment contract, 6-18
employment director, 6-18
employment, guaranteed, 6-23
employment manager, 6-18
employment, seasonal, 6-50
employment security. *See* job security
employment test, 10-10
empty field myopia, 9-8, 9-21
empty pallet stacker, 8-13
enacted environment, 15-4
enclosed dock, 8-13
enclosed space. *See* confined space
enclosed track trolley conveyor, 8-13. *See also* trolley conveyor
encode, 3-6
encoded area, 8-13
encoder, 11-7
encoding, 10-10, 15-4
encounter group, 15-4. *See also* sensitivity training
end control truck, 8-13
end effector, 11-7
endemic, 13-15
end frames, portable metal stacking rack, 8-13
ending event, 4-14
ending node of network (ADM), 4-14
end item, 11-7
end milling, 12-62
end network event, 4-14
endochthon, 2-35. *See also* exochthon

end of a tape (EOT), 12-11
end of block (EOB), 12-11
end-of-block signal, 12-11
end of program (EOP), 12-11
endowment, 4-15, 7-10
endowment method, 7-10
endowment, perpetual, 7-13
endpoint. *See* breakpoint
end-point control, 11-7
end-point rigidity, 11-7
end product, 14-8. *See also* inventory write-off
end stop, 8-13
end tie, 8-13
end trucks, 8-13
endurance limit, 12-52
end user, 10-10
end-user computing, 10-10
energy curve, 12-52
energy expenditure, 9-8
engineer, cost, 4-10
engineered performance standard. *See* standard time
engineered standard. *See* standard time
engineered time standard. *See* standard time
engineer, in contracts, 4-15
engineer, industrial, 17-9
engineering, bioengineering, 13-7
engineering change order, 14-8. *See also* configuration control
engineering, cognitive, 9-5
engineering, computer-aided (CAE), 11-5
engineering, concurrent, 10-6, 11-5, 14-6, 17-4
engineering controls, 13-16
engineering, detailed, 4-12
engineering economy, 7-10
engineering, environmental, 2-35
engineering, fire protection, 13-19
engineering, human, 2-38, 9-10, 17-9. *See also* ergonomics
engineering, human factors, 17-9. *See also* ergonomics; motion analysis; motion economy; work design
engineering, methods, 6-35, 17-12. *See also* methods analysis; motion analysis; motion economy; motion study; work design
engineering plastics, 12-74
engineering, preliminary, 4-26
engineering psychology, 9-8. *See also* ergonomics
engineering, safety, 6-49, 13-40
engineering school of job design, 15-4
engineering stress, 12-52
engineering, system, 9-22
engineering, system safety, 13-44
engineering, value, 4-35, 17-20
engineer, methods, 6-35
engineer, process, 17-16. *See also* process; process design
engineer, resident, 4-29

engine lathe, 12-62
enlightened self-interest argument, 10-10
en masse conveyor, 8-13
enterprise integration, 10-10
enterprise resource planning (ERP), 11-7
entrance rate, 6-18. *See also* minimum plant rate; probationary rate
entrepreneur, 10-10
entrepreneurial mode, 10-10
entrepreneurial team, 10-10
entrepreneurship, 10-10
entropy, negative, 10-20
entry level, 6-18
entry process, 15-4
entry rate, 6-18
enumeration, implicit, 1-12
envenomation, 12-74
environmental complexity, 10-10
environmental dynamism, 10-10
environmental engineering, 2-35
environmental inputs, 2-35
environmental munificence, 10-10
environmental stress factor, 2-35
environmental uncertainty, 10-10
environment, external, 10-11
environment, external mechanical, 2-36. *See also* Gilbrethian variables
environment, input variables of external, 2-39
environment, internal, 10-16
environment, internal (bio)mechanical, 2-39
environment, thermal, 2-51
EOB (end of block), 12-11
EOP (end of program), 12-11
EOQ (economic order quantity), 5-6, 10-9, 14-8. *See also* carrying cost; ordering cost
EOT (end of a tape), 12-11
epicondyle, 2-35
epicondyles of humerus, 13-16
epicondylitis, 2-35, 13-16
epidemic, 13-16
epidemiology, 13-16
epiphysis, 2-35
epoxy plastics, 12-74
EPROM (erasable programmable read-only memory), 3-6, 12-11
Equal Employment Opportunity Commission (EEOC), 6-18
Equal Pay Act of 1963, 6-18
equal pay for comparable work, 6-18
equal pay for equal work, 6-18
equation, 1-9
equation, nonlinear, 1-17
equation solver, 12-11
equilibrium, 2-35
equipment cost, rental (leased), 4-28
equipment, peripheral, 11-13, 12-19
equitable adjustment, 4-15

This index is arranged in letter-by-letter format.

equity, 6-18, 15-5
equity capital, 10-10
equity funding, 6-18
equity, individual, 6-26
equity, internal, 6-27
equity theory, 10-10, 15-5
equivalence operation, 3-6
equivalent form, 13-16
equivalent sets of commodities, 4-15
equivalent uniform annual cost (EUAC). *See* annual cost
equivalent workers, 6-19
erasable programmable read-only memory (EPROM), 3-6, 12-11
erector spinae muscles, 2-35. *See also* sacrospinalis muscle
ergometer, 9-8
ergonometrics. *See* ergonomics; work measurement
ergonomic analysis, 2-35
ergonomics, 2-35, 9-8, 11-7, 13-16, 17-6. *See also* engineering psychology; human engineering; human factors; human factors engineering; methods engineering; motion analysis; motion economy; work design
ergonomics school of job design, 15-5
Erg theory, 10-10
ERISA (Employee Retirement Income Security Act), 6-19
erosion, rate, 6-46
erosion scab, 12-37
ERP (enterprise resource planning), 11-7
error, 4-15, 12-11
error budgeting, 12-62
error compensation, 12-62
error, constant, 9-8
error mean-square, 1-9
error, mode, 9-14
error of the first kind, 16-7
error of the second kind, 16-7
error, position, 11-13
error probability, type I, 16-19
error probability, type II, 16-19
errors and omissions, 4-15
error, standard, 16-17
error sum of squares, 1-9
error, variable, 9-8
ERTA (Economic Recovery Act of 1981), 6-19. *See also* IRA (individual retirement account); tax shelter
ERV (expiratory reserve volume), 13-29. *See also* lung volumes
escalation, 4-15
escalation, nonrational, 10-21
escalation situations, 10-10
escalator, 8-13
escalator clause, 4-15, 6-19. *See also* consumer price index (CPI); cost of living adjustment (COLA)
escape clause, 6-19

ES (early start time), 4-14
ES (expert system), 10-11
ESOP (Employee Stock Ownership Plan), 6-17
ESS (executive support system), 10-11
esteem needs, 10-11
esteem value, 4-17
estimate, 7-10
estimate, budget, 4-15
estimate, cost, 4-15
estimate, definitive, 4-15
estimated time, 17-6
estimate, order of magnitude, 4-15
estimate-to-complete, 4-15
estimation, 1-9, 16-7
estimator, 16-7
estimator, unbiased, 16-19
ETFE (ethylene tetrafluoroethylene copolymer), 12-74
Ethical Practices, Codes of, 6-11
ethics, 10-11
ethnocentric orientation, 10-11
ethylene plastics, 12-74
ethylene tetrafluoroethylene copolymer (ETFE), 12-74
etiology, 2-35, 9-8, 13-16
EUAC (equivalent uniform annual cost). *See* annual cost
Euler diagram. *See* Venn diagram
evaluated rate, 6-19
evaluation advertising, 5-2. *See also* market research
evaluation, promotion, 5-12
evalucomp, 6-19
EV (early event time), 4-14
event, 1-9, 4-15, 10-11. *See also* node
event, beginning, 4-4
event, beginning network, 4-4
event, intermediate, 4-19
event name, 4-15
event number, 4-15
event slack, 4-15
event time, 4-15
event time, early (EV), 4-14
event time, scheduled, 4-30
evidence, expert, 13-16
evidence, preponderance of, 13-36
evolutionary operation, 1-9
EXAPT, 12-11
exception message. *See* action message
excess work allowance, 17-6. *See also* allowance
exchange rate, 7-10, 10-11
exchange value, 4-17
exclusive bargaining rights, 6-19. *See also* agency shop
exclusive dealing contract, 5-6
exclusive distribution, 5-6
exclusive outlet selling, 5-6
excusable delay, 4-11, 4-12
execute, 3-7
executive, 15-5
executive board, 6-19
executive committee, 15-5

executive compensation, 6-19
executive compensation plan, 6-19
executive, group, 15-6
executive support system (ESS), 10-11
exempt, 4-16
exempt employee, 6-19
exempt job, 6-19
exhaust, general, 13-16. *See also* industrial ventilation
exhaust, local, 13-16. *See also* industrial ventilation
exhaust, ventilation, 13-16
existence needs, 10-11
exit, 3-7
exochthon, 2-35. *See also* endochthon
exotherm, 12-74
exothermic heat, 12-74
exothermic riser sleeve, 12-37
expandable plastic, 12-74
expanded plastics. *See* foamed plastics
expansion, 4-16
expansion scab, 12-37
expatriates, 10-11
expectancy theory, 10-11, 15-5
expected attainment. *See* fair day's work
expected begin date, 4-16
expected duration, 4-16
expected earnings (level), 6-19
expected elapsed time, 4-16
expected value, 10-11
expected work pace, 17-6
expected yield, 7-10
expediting, 11-7, 14-8. *See also* dispatching
expendable container, 8-13
expendable pattern, 12-37
expense, 4-16. *See also* plant overhead
expense account, 6-19
expense, administrative, 4-3
expense budget, 10-11
experience, 13-16
experience curve. *See* learning curve
experienced employee, 6-19
experience rating, 6-19, 13-16
experiential, 15-5
expert, 9-8
expert evidence, 13-16
expert power, 10-11
expert system (ES), 3-7, 9-8, 10-11, 11-7
expert testimony, 13-16
expert witness, 6-20, 13-16
expiration date, 6-20
expiratory flow, forced, 13-20
expiratory reserve volume (ERV), 13-29. *See also* lung volumes
expiratory volume, forced, 13-20
explanatory model, 10-11
explosimeter, 13-16
explosion, 11-7, 13-16, 14-8
explosion-proof, 13-17. *See also*

flame proof
explosion venting, 13-17
explosive atmosphere, 13-6
explosive decompression, 13-17
explosive limits, 13-17. *See also* lower explosive limit (LEL); upper explosive limit (UEL)
explosive mixture, 13-17
explosives, 13-17
explosive welding, 12-87
exponential service time, 1-9
exponential smoothing, 14-8
exporting, 10-11
exposure, 13-17
exposure, acute, 13-3
exposure, casualty, 13-17
exposure hours, 13-17
exposure hours, actual, 13-3
expropriation, 10-11
extendable conveyor, 8-13. *See also* telescoping conveyor
extended binary-coded decimal interchange code (EBCDIC), 3-7
extended leave plan, 6-20
extended vacation plan, 6-20. *See also* paid vacations
extender, 12-74
extension, 2-35, 11-7
extensor muscle, 2-35. *See also* flexor muscle
extensor retinaculum, 2-35
extensor tendon, 2-36. *See also* insertion; tendon
external audit, 10-11
external element, 17-7
external environment, 10-11
external mechanical environment, 2-36. *See also* Gilbrethian variables; internal (bio)mechanical environment
external rate of return, 7-10. *See also* internal rate of return; rate of return
external sensor, 11-7
external time, 17-7
extinction, 10-11
extinguisher, dry-chemical, 13-14
extinguisher, dry-powder, 13-14
extinguisher, fire, 13-19
extinguishing agent, 13-17
extracanthic diameter, 2-10
extraction, 12-74
extrapolated failure rate, 1-10, 16-8
extrapolation, 4-16
extremal problem. *See* master program
extreme point, 1-9
extrinsic, 2-36
extrinsic forecast, 14-8. *See also* intrinsic forecast
extrinsic rewards, 10-11
extrusion, 12-27, 12-52, 12-74
extrusion, autothermal, 12-74
extrusion, direct, 12-51
extrusion forging, 12-27

extrusion, indirect, 12-53
extrusion, inverted, 12-53
eye contours, 2-8
eye height, sitting, 2-10
eye height, standing, 2-10
eye protection, 13-17
eye scanning, 2-36
EZAPT, 12-11

F

f distribution
fabricating, 12-74
fabrication, 11-7, 12-11, 12-74
face, 12-74
face milling, 12-62
face of weld, 12-87
face shield, 13-17
face validity, 9-8
facial length, 2-10
facilities, 10-11
facility, 8-13
facility layout, 13-17
facing, 12-37, 12-62
facing material, 12-37
fact-finding board, 6-20. *See also* board of inquiry
factor, 5-6, 13-17
factor analysis, 1-9, 13-17
factor comparison, 6-20
factoring, 5-6
factor of safety (FS), 13-17
factor scales, 6-20
factor weight, 6-20
factory, 11-7
factory control, 11-7
factory expense. *See* plant overhead
factory mutual listing (FM), 8-13
fail operational, 13-18
fail operational fail safe, 13-18
fail safe, 9-8, 13-18
failure, 1-9, 13-18, 16-7
failure analysis, 1-10, 13-18, 16-7
failure assessment, 13-18
failure, catastrophic, 1-10
failure, complete, 1-10, 16-7
failure criteria, 1-10, 16-7
failure, critical, 13-18
failure, degradation, 1-10, 16-7
failure, dependent, 1-10, 16-7
failure, gradual, 1-10, 16-7
failure, independent, 1-10, 16-7
failure, inherent weakness, 1-10, 16-7
failure management, 13-18
failure mechanism, 1-10, 16-7
failure, mechanism, 13-18
failure, misuse, 1-10, 16-7
failure mode, 16-7
failure, mode, 1-10
failure, mode and effect analysis (FMEA), 13-18
failure, mode, effect and critical analysis (FMECA), 13-18

failure, partial, 1-10, 16-7
failure period, constant, 1-6
failure period, early, 1-9
failure, primary, 13-18
failure, random, 1-10, 16-7
failure rate, 1-10, 13-18, 16-8
failure rate acceleration factor, 1-10, 16-8
failure rate, assessed, 1-10, 16-8
failure rate, extrapolated, 1-10, 16-8
failure rate, instantaneous (HAZARD), 16-9
failure rate, observed, 1-10, 16-8
failure rate period, constant, 1-6
failure rate, predicted, 1-10, 16-8
failure, secondary, 1-10, 13-18, 16-8
failure, single point, 13-42
failure, sudden, 1-10, 16-8
failure, wear-out, 1-10, 16-8
fainting, 13-18
fair day's pay, 6-20
fair day's work, 6-20, 17-7. *See also* normal performance
Fair Employment Practice Laws, 6-20. *See also* Civil Rights Act
fair game, 1-10
Fair Labor Standards Act (FLSA), 6-20
fair rate of return, 7-10
fair trade, 5-6
fair value, 4-16
false body, 12-74
false negative rate, 13-18
false positive rate, 13-18
family. *See* product family
Family and Medical Leave Act (FMLA), 6-20
family mold, 12-74
FAPT, 12-11
Farkas' lemma, 1-10
farming out. *See* contracting-out
FAS (final assembly schedule), 14-8
fastener, 8-13
fatal accident, 13-18
fatality, 13-18
fathom, 1-11
fatigue, 2-36, 6-20, 9-8, 13-18, 17-7
fatigue allowance, 6-20, 17-7. *See also* allowance
fatigue, auditory, 9-3
fault, 9-8
fault correction time, 1-26
fault hazard analysis, 13-18
fault localization, 9-8
fault location time, 1-26
fault tree analysis, 9-8, 13-18. *See also* system safety analysis
"Favored Nations" Clause, 6-20
faying surface, 12-87
FCFS (first-come-first-served), 14-9
FCL (full container load), 8-16
F-distribution, 16-7

feasible basis, 1-11
feasible solution, 1-11
feasible solution, basic, 1-4
featherbedding, 6-20
feature. *See* accessory; attachment; option
Federal Labor Relations Authority (FLRA), 6-20
Federal Mediation and Conciliation Service (FMCS), 6-20. *See also* mediation; mediator
federation, 6-21
feedback, 9-8, 10-11, 12-11, 15-5
feedback control, 10-11, 12-12
feedback control loop, 12-12
feedback data, 11-7
feedback, sensory, 9-20
feedback system, 2-36, 4-16. *See also* homeostasis
feeder, 8-13, 12-37. *See also* belt feeder
feeder and catchers table, 8-13
feeder, apron, 12-31
feeder, chain. *See* conveyor type feeder
feeder head, 12-37
feeder, sand, 12-37
feedforward control, 10-11
feeding, 12-37
feed rate, 12-12, 12-62
feed(s), 12-52, 12-62
feeler, 12-12
femur, 2-36
ferrite, 12-52
ferritic matrix, 12-37
ferritic steels, 12-37
fettle, 12-37
FF (free float), 4-17
fibrosis, 13-18
fibrositis, traumatic, 2-52
fibula, 2-36
fidelity, 9-8
field, 12-12
field cost, 4-16, 8-13
Fielder's contingency model, 10-12
field labor overhead, 4-16
field of view, 9-9
field order, 4-16
field separator, 8-13
field study, 9-9
field supervision, 4-16
FIFO (first-in-first-out), 3-7, 4-16, 5-6, 14-9. *See also* LIFO
fifth (5th) lumbar vertebra, 2-36. *See also* herniated disk; lumbosacral angle
fifth (5th) to 95th percentile, design criteria, 2-6
file, 3-7
filler, 12-75. *See also* reinforced plastics
filler lens, 12-87
filler metal, 12-87
filler-specks, 12-75
fillet, 12-27, 12-37

fillet weld, 12-87
film, 12-75. *See also* sheet; sheeting
film analysis, 17-7. *See also* allowance; standard time
film analysis chart, 17-7. *See also* Simo chart; Therblig chart
film analysis record, 17-7
film, cast, 12-71
film, micromotion, 2-43
film, rating, 17-7
filter, membrane, 13-30
fin, 12-37, 12-75
final assembly schedule (FAS), 14-8
final average pay (pension benefit formula), 6-21
final pay (pension benefit formula), 6-21
final position after an accident, 13-18
financial budgets, 10-12
financial incentive plan, 6-21. *See also* productivity gainsharing (PGS)
financial life. *See* venture life
financial statement, 10-12
fineness, sand, 12-37
fines, 12-37
finger length, 2-10
finish allowance, 12-27, 12-38
finish date, actual, 4-2. *See also* actual completion date
finish date, imposed, 4-18
finish date, project, 4-27
finish date, projected, 4-27
finished goods, 4-16
finished-goods inventory, 10-12
finished products inventories, 14-9
finish-go-home basis of pay, 6-21
finishing, 12-62, 12-75
finishing lead time, 14-9
finishing, super, 12-66
finishing temperature, 12-52
finish schedule, level (SF), 4-20
finish time, early (EF), 4-14
finite and infinite games, 1-11
finite loading, 14-9. *See also* infinite loading; load leveling
fink, 6-21. *See also* strikebreaker
fire, 13-19
fire accident, 8-13
fire alarm, 13-19
fire classifications, 13-19
fireclay, 12-38
fire doors, 13-19
fire extinguishers, 13-19
fire prevention, 13-19
fire proof, 13-19
fire protection, 13-19
fire protection engineering, 13-19
fire resistive, 13-19
fire retardant, 13-19
fire sand, 12-38
fire wall, 13-19
firm planned order (FPO), 14-9. *See also* scheduled receipts

This index is arranged in letter-by-letter format.

firmware, 12-12
first aid, 13-19
first aid case accidents, 6-21
first aid injury, 13-19
first-come-first-served (FCFS), 14-9
first cost, 4-16, 7-10
first event number, 4-16
first floor, 8-13
first-in-first-out (FIFO), 3-7, 4-16, 5-6, 14-9. *See also* LIFO
first line manager, 10-12
first line supervisor, 10-12
first piece time, 17-7
first read rate, 8-13
fisheye, 12-75
fit, 12-62, 13-20
fit for purpose intended, 13-19
fitness, 13-20
five competitive forces model, 10-12
fixed assets, 7-10
fixed barrier guard, 13-7
fixed benefit retirement plan, 6-21
fixed-block format, 12-12
fixed capital, 4-6
fixed charge problem, 1-11
fixed cost, average, 5-2
fixed costs, 4-16, 5-6, 7-10
fixed crane, 8-14. *See also* bridge crane; crane; job crane; portable crane; tower crane; traveling crane
fixed-interval schedule of reinforcement, 10-12
fixed linkage mechanism, 2-36
fixed order interval systems, 5-7
fixed order quantity, 14-9. *See also* economic order quantity (EOQ); lot-for-lot; period order quantity
fixed platform truck, 8-14
fixed position layout, 8-14, 10-12
fixed price contract(s), 4-9, 4-16. *See also* lump-sum
fixed-ratio schedule of reinforcement, 10-12
fixed sequential format, 12-12
fixed shift, 6-21. *See also* rotating shift; shift; split shift
fixed-stop robot, 11-7
fixed storage, 12-12
fixed zero, 12-12
fixture, 11-7, 12-12, 12-62, 17-7. *See also* jig
fixture conveyor, 8-14
flag, 3-7
flagged rate. *See* red circle rate
flake, 12-75
flakes, 12-27
flame, 13-20
flame arrester, 13-20
flame, neutral, 12-89
flame proof, 13-19. *See also* explosion-proof; fire proof
flame propagation, 13-20
flame spread, 13-20

flammable, 13-20. *See also* inflammable
flammable limits. *See* explosive limits
flammable liquids, 13-20
flammable vapor, 13-20
flange, 12-52
flank wear, 12-62
flash, 12-27, 12-75, 12-87
flash arrester, 13-20
flashback, 12-87
flash blindness, 9-9
flash burn, 13-20
flash gutter, 12-27
flashing time, 12-87
flash land, 12-27
flash point, 13-20
flash ridge, 12-75
flash welding, 12-87
flask, 12-38
flaskless molding, 12-38
flat belt conveyor, 8-4
flat benefit retirement plan, 6-21
flat dies forging, 12-27
flat grain, 12-75
flat increase, 6-21
flat position, 12-87
flat rate, 6-21
flat structure, 10-12
flattening, 12-52
flat top conveyor, 8-14
flat wire conveyor belt, 8-14
flaws, 12-62
FLD (forming limit diagram), 12-53
flexible benefits program. *See* cafeteria benefits
flexible disk, 3-7. *See also* floppy disk
flexible manufacturing cell (FMC), 11-7
flexible manufacturing system (FMS), 10-12, 11-8, 12-62
flexible workstation, 11-8
flexion, 2-36, 11-8
flexor muscle, 2-36
flexor retinaculum, 2-37
flexowriter, 12-12
flextime, 6-21, 10-12
flicker, 9-9
flicker fusion, 9-9. *See also* stroboscope
flight, 8-14
flight conveyor, 8-14. *See also* slope conveyor
flight deck, 9-9
flight simulator, 9-9
flight trainer, 9-9
flip flop, 12-12
float, 4-16, 14-9, 17-7
floating chase, 12-75
floating-point representation system, 3-7
floating zero, 12-12
flock, 12-75
floor, 8-14
floor controlled units, 8-14

floor conveyor, 8-14. *See also* car type conveyor; pallet type conveyor; slat conveyor
floor loading rating, 8-14
floor stocks, 11-8, 14-9
floor-to-floor time, 11-8
floor under wages. *See* minimum wage
floppy disk, 12-12. *See also* flexible disk
floppy disk systems, 12-12
flowability, 12-38
flow analysis, 8-14, 17-7
flowchart, 3-7, 8-14, 12-12, 17-7. *See also* process chart
flow diagram, 8-14, 17-7
flow line(s), 11-8, 12-27, 17-7. *See also* line of flow
flow mark, 12-75
flow path, 8-14, 17-7
flow process chart, 17-7. *See also* process chart; process chart symbols
flow progress chart, 8-14
flow rack, 8-15
flow shop/flowshop, 8-15, 11-8, 14-9
flow stress, 12-27
FLRA (Federal Labor Relations Authority), 6-20
FLSA (Fair Labor Standards Act), 6-20
fluctuation inventories, 14-9. *See also* safety stock
fluid coupling, 8-15
fluidics, 12-12
fluidity, 12-38
flume conveyor, 8-15
fluorocarbons, 12-75
flush dock, 8-15
flutter, auditory, 9-3
flux, 12-87, 13-20
flux cored arc-welding, 12-87
fluxing temperature, 12-75
flybar (flying by auditory reference), 9-9
flying cutting, 12-62
flywheel, 12-52
FMC (flexible manufacturing cell), 11-7
FMCS (Federal Mediation and Conciliation Service), 6-20
FMEA (failure mode and effect analysis), 13-18
FMECA (failure mode, effect and critical analysis), 13-18
FM (factory mutual listing), 8-13
FMLA (Family and Medical Leave Act), 6-20
FMS (flexible manufacturing system), 10-12, 11-8, 12-62
foam, 13-20. *See also* cellular plastic
foam, closed-cell, 12-72
foamed plastics, 12-75. *See also* cellular plastic
foam molding, 12-75. *See also*

sandwich molding
foam, open cell, 12-80
F.O.B. origin, 5-7
focus forecasting, 14-9
focus strategy, 10-12
folding container, 8-15
folliculitis, 13-20
follow-up study. *See* cohort study
font, 3-7
font, multiple, 8-21
foot breadth, 2-10
foot candle, 9-9, 13-20. *See also* brightness contrast; foot-lambert; reflectance; visual acuity
foot control, 12-52
foot-lambert, 9-9. *See also* foot candle; reflectance
foot length, 2-10
force, 12-75. *See also* core
forced expiratory flow, 13-20
forced expiratory volume, 13-20
forced expiratory volume percentage expired, 13-20
forced mid-expiratory flow, 13-21
forced vital capacity (FVC), 13-21
force-field analysis, 10-12
force, ground reaction, 2-37
force, intervening, 13-26
force, joint reaction, 2-40
force joystick, 9-9
force, moment of, 2-43. *See also* torque
force-paced, 9-9
force plate, 12-75
force platform (plate). *See* reactance platform
force plug, 12-75. *See also* core
force sensor, 11-8
force-time, 2-37
forearm-hand length, 2-10
forecast, 4-16, 14-9. *See also* prediction; projection
forecast error, 14-9
forecast, extrinsic, 14-8
forecast horizon, 14-9
forecasting, 15-5
forecasting, judgemental, 10-17
forecasting, qualitative, 10-25
forecasting, quantitative, 10-25
forecasting, social, 10-28
forecasting, technological, 10-31
forecast interval, 14-9
forecast, intrinsic, 14-11
forecast, product, 5-12
forecast, sales, 4-30, 5-13
forehand welding, 12-88
foreign element, 17-8
foreign service premium, 6-21
foreman, 15-5. *See also* management, lower; supervision
forgeability, 12-27
forge welding, 12-88
forging, 12-27
forging, blocker-type, 12-25
forging, closed-die. *See* impression die forging
forging, core, 12-26

This index is arranged in letter-by-letter format.

forging, die, 12-26
forging, drop, 12-27
forging, extrusion, 12-27
forging, flat dies, 12-27
forging, hammer, 12-28
forging, heat, 12-28
forging, high-energy-rate, 12-28
forging, high speed, 12-28
forging, high velocity, 12-28
forging, impression die, 12-28
forging, isothermal, 12-28
forging machine (upsetter or header), 12-27
forging, mandrel, 12-30. See also ring rolling
forging, no-draft, 12-29
forging, open die, 12-27, 12-29
forging, press, 12-29
forging, roll, 12-29
forging rolls, 12-27
forging strains, 12-28
forging stresses, 12-28
fork adapter, 8-15
fork adjuster, 8-15
fork height, 8-15
forklift opening, 8-15
forklift truck, 8-15
fork opening, 8-15
fork pockets, 8-15
forks, 8-15
fork spacing, 8-15
fork spacing, maximum inside, 8-15
fork spacing, maximum outside, 8-15
fork spacing, minimum inside, 8-15
fork spacing, minimum outside, 8-15
forks, safety, 8-15
fork truck with tilting mast, 8-20
form agreement, 6-21. See also standard agreement
formal communication, 10-12
formaldehyde, melamine, 12-79
formal group, 10-12
formalization, 10-12
formal leader, 15-5. See also informal group
formal organization, 15-5. See also informal group
formal system, 15-5. See also informal group
format, 3-7, 12-12
forming, 10-12, 12-53, 12-75
forming limit diagram (FLD), 12-53
form milling, 12-62
form process chart, 17-8
formulation. See bill of material (BOM)
forseeability, 13-21
FORTRAN (formula translation), 3-7, 12-13
forward chaining, 9-9
forwarding area, 8-15
forward pass, 4-16, 4-17

forward pricing, 4-26
forward scheduling, 11-8, 14-9. See also backward scheduling
founding, 12-38
foundry, 12-38
foundry facing, 12-38
foundryman, 12-38
foundry returns, 12-38
Fourier analysis, 12-13
four-to-one-ratio, 13-21
four-way container, 8-15
FPO (firm planned order), 14-9
fractile, 4-17
fractional programming, 1-11
fraction chart, 16-13
fragile, 8-15
frag net, 4-17. See also subnet
frame, 1-11, 12-53
frame buffer, 11-8
frame, cantilever rack, 8-15
frame counter, 17-8
framing, 10-12
franchise, 5-7, 10-12
franchisee, 10-12
Frankfort plane, 2-10, 2-23
FRC (functional residual capacity), 13-29. See also lung volumes
free float (FF), 4-17
free haul, 4-17
free lift, 8-15
free riders, 6-21, 10-13
free standing rack, 8-15
free standing rack structure, 8-15
freezing range, 12-38
freight, 8-15
freight forwarder, 5-7, 8-15
frequency, 2-37, 13-21, 17-8
frequency distribution, 5-7
frequency, matching, 13-30
frequency, natural, 12-18
frequency rate of accidents, 6-21
friction, internal, 12-53
friction welding, 12-88
friendship group, 10-13
fringe benefit cost, 4-13
fringe benefits, 4-17, 6-21
frontal, 2-23
front end, 6-21
front-gage, 12-53
front slagging, 12-38
front to back members, 8-16
front to back members, bolted, 8-16
front to back members, drop-in-place, 8-16
front to back members, welded, 8-16
frostbite, 13-21
frustration regression principle, 10-13
FS (factor of safety), 13-17
full container load (FCL), 8-16
full crew law, 6-21
fuller, 12-28. See also edger
full-function wholesaler, 5-7
full pegging, 14-9

full-revolution clutch, 12-53
full-time earnings, 6-21
full-time worker rate, 6-21
fumble, 17-8
fume, 13-21
function, 3-7, 4-17, 12-13, 15-5
functional anatomy, 2-37
functional arm reach contours, 2-8
functional audit, 10-13
functional authority, 15-5
functional conflict, 15-5
functional disease, 13-21
functional group, 10-13
functional leg reach contours, 2-8
functional-level strategy, 10-13
function allocation, 9-9
functional managers, 10-13
functional organization, 15-5
functional replacement cost, 4-17
functional residual capacity (FRC), 13-29. See also lung volumes
functional system, 4-17
functional use area, 4-17
functional worth, 4-17
functional worth, cost value, 4-17
functional worth, esteem value, 4-17
functional worth, exchange value, 4-17
functional worth, use value, 4-17
function, cost, 1-7
function, nonlinear, 1-17
fund, 6-21. See also trustee
fundamental attribution error, 15-6
funeral leave pay, 6-22
furane plastics, 12-75
furnace, air, 12-31
furnace, basic, 12-25
furnace, batch-type, 12-25
furnace, blast, 12-32
furnace brazing, 12-88
furnace, crucible, 12-35
furnace, direct-arc, 12-36
furnace, electric, 12-37
furnace, indirect-arc, 12-39
furnace, induction, 12-39
furnace, open hearth, 12-42
furnace, pusher, 12-29
furnace, rotary, 12-29
fusion, 12-75, 12-88
fusion, auditory, 9-3
fusion temperature, 12-75
future worth, 7-10
futurists, 10-13
fuzzy logic control, 11-8
FVC (forced vital capacity), 13-21

G

G code
G-force
g-tolerance
gage, 12-53

gaggers, 12-38
gaging, 12-62
gain, 9-9, 12-13
gainsharing, 6-22, 10-13, 17-8. See also improshare; profit sharing; Rucker plan
gainsharing plans, 17-8
gainsharing, productivity (PGS), 6-45, 15-11. See also financial incentive plan; Rucker plan; Scanlon plan; sharing plan; value added
gait analysis, 2-37
galling, 12-53
gambler's ruin, 1-11
gamble, standard, 1-24
game(s)
    finite and infinite, 1-11
    matrix, 1-15, 1-28
    multiperson, 1-17
    n-person, 1-17
    saddlepoint of a, 1-22
    value of a, 1-27
    zero-sum two-person (matrix), 1-28
game theory, 1-11, 10-13
Gamma ray, 13-21
gamma transition, 12-75
gamma-transition temperature, 12-76
Gamma (Y) radiation, 13-21
gang chart. See multiple activity process chart
ganged controls, 9-9
gangrene, 2-37
gantry crane, 8-16. See also traveling cranes
gantry robot, 11-8
Gantt chart, 10-13, 14-9, 17-8. See also progress chart
gap depth, 12-59
gap frame, 12-53
garbage-can model, 10-13
garnishment of wages, 6-22
gas-electric truck, 8-16
gases, 13-21
gas holes, 12-38
gas mask, 13-21. See also respirator
gas metal arc-welding, 12-88
gas meter, combustible, 13-10
gas tungsten arc-welding, 12-88
gas welding, 12-88
gate, 12-28, 12-38, 12-76
gate barrier guard, 13-7
gate, runner, 12-38
gate, slot, 12-38
gate strainer, 12-38
gateway city (air transport), 8-16
gateway work center, 14-10
gathering, 12-28
gathering stock, 12-28
gating system, 12-38
gauge of double girder cranes, 8-16
gauger table, 8-16
Gaussian elimination, 1-11

This index is arranged in letter-by-letter format.

G code, 12-13
gear hobbing, 12-62
gear shaving, 12-66
GE business screen, 10-13
gel, 12-76. *See also* gel point; synersis
gelation, 12-76
gel point, 12-76. *See also* gel
general adaptation, 9-2
general commodity rate (air transport), 8-16
general managers, 10-13
general merchandise stores, 5-7
general overhead, 4-17. *See also* overhead
general purpose index, 4-17
general requirements, 4-17
general staff, 15-6
general terms and conditions, 4-17
general wage changes, 6-22
generative process planning, 11-8
genesis, 12-13
genetic drift, 13-21
genetic effects of radiation, 13-21
genetic mutation, 13-21. *See also* mutagen
geocentric organization, 10-13
geographic differentials, 6-22
geographic divisions, 10-13
geometric mean, 13-21
geometric solution, 1-11
GETURN, 12-13
G-force syndrome, 13-2
ghost employees, 6-22
gibs, 12-53
Gilbreth basic element. *See* therblig
Gilbrethian "systems concept", 2-37
Gilbrethian variables, 2-37. *See also* external mechanical environment
girder crane, 8-16
giveaways, 6-22
givebacks, 6-22
given year, 4-17
glabella, 2-23
glare, 9-9, 13-21
glassblower's cataract, 13-21
glass cockpit, 9-9, 12-76
glass finish, 12-76
glass, safety, 13-40
glass transition, 12-76
glass transition temperature, 12-76
glassy transition, 12-75
glassy-transition temperature, 12-76. *See also* gamma transition
glazed wheel, 12-62
glenoid cavity, 2-37
global, 3-7
globalization, 10-13
global optimum, 1-11
glue, 12-76. *See also* adhesive; gum; sizing
glue joint, 12-76
gluing, secondary, 12-83
gluteus maximus, 2-37

GNC, 12-13
GNP (gross national product), 4-17
goal, 10-13
goal commitment, 10-13
goal conflict, 15-6
goal incongruence, 10-13
goal programming, 1-11
goals, operational, 10-22
God, acts of, 4-2, 13-3
goggles, 13-21
goggles, splash-proof, 13-43
going-concern value, 7-10
going rate, 6-22. *See also* prevailing rate
going wages, 6-22
golden handcuffs, 6-22
golden parachute, 6-22. *See also* worker parachutes
GOMS, 9-9
goniometer, 2-37
goniometry, reflex, 2-47
go, no-go display, 9-9
gonveal, 6-22
goodness of fit of a distribution, 16-8
goods
  convenience, 5-4
  durable, 4-14
  industrial, 5-7
  nondurable, 4-23
  shopping, 5-14
  specialty, 5-14
good-will value, 7-10
goon, 6-22
gooseneck, 12-38
gooseneck punch, 12-53
government agencies, 10-13
government, hand of, 10-14
grade, 16-8
grade (abrasive wheels), 12-62
grade clearance, 8-16
grade description. *See* job description; position description
grade resistance, 8-16
gradient factors, 7-11. *See also* uniform gradient series
gradient methods, 1-12
gradient of a function, 1-12
gradual failure, 1-10
grain fineness number (AFS), 12-38
grain flow, 12-28
grain size, 12-28
grandfathering. *See* red circle rate
grand strategy, 10-14
granular structure, 12-76
grapevine, 10-14
graph, bar, 9-4
graphic, 3-7
graphical display, 9-9
graphical user interface (GUI), 9-10
graphic input, 12-13
graphic panel, 12-13
graphic rating scales, 10-14
graph (linear), 1-12

graph, signal flow, 1-24
grasp, basic, 2-31
grasp, pinch, 2-45
grasp reflex, 2-37. *See also* prehensile
grasp, tripodal, 2-52, 2-53
grasp, wraparound, 2-54
graveyard shift, 6-22. *See also* shift
gravity, center of, 2-32. *See also* center of mass
gravity discharge bucket conveyor, 8-16. *See also* bucket conveyor
gravity discharge bucket elevator, 8-16
gravity feed, 17-8
gravity hammer, 12-28
gravity hopper, 8-16
gray cast iron, 12-38
gray (Gy), 13-21
grayout, 9-10
green casting, 12-38
green circled rate, 6-22
green permeability, 12-39
green sand, 12-39
green strength, 12-39
grey-scale picture, 11-8
grid, 12-76
grievance, 6-22, 17-8
grievance committee. *See* shop committee
grievance procedure, 6-22, 17-8
grievance steps, 6-22
grinding, 12-62
grinding, centerless, 12-60
grinding, center mounted, 12-60
grinding, chucking internal, 12-61
grinding, creep, 12-61
grinding, crush form, 12-61
grinding, electrochemical, 12-61
grinding, horizontal surface, 12-63
grinding machine, universal, 12-68
grinding, offhand, 12-64
grinding ratio, 12-62
grinding, snag, 12-30, 12-66
grinding, surface, 12-66
grinding-type resin, 12-76
grinding wheel, loaded, 12-63
gripper, 11-8
grip, power, 2-45
grit (in abrasives), 12-62
groove angle, 12-88
grooving, 2-37
gross area, 4-17
gross average hourly earnings, 6-22. *See also* average hourly earnings exclusive of overtime payments
gross average weekly earnings, 6-23
gross national product (GNP), 4-17
gross profit, 4-26
gross requirements, 14-10. *See also* requirements explosion
gross weight, 8-16
ground, 9-10

ground fault circuit interrupter, 13-22
grounding, 13-22
ground reaction force, 2-37
group, 10-14, 15-6, 17-8
group annuity plan, 6-23, 6-41
group annuity plan, endowment. *See* pension plan
group bonus, 6-23
group classification code, 14-10. *See also* classification; coding; group technology
group cohesiveness, 10-14, 15-6
group dynamics, 15-6
group executive, 15-6
group, formal, 10-12
group, functional, 10-13
group incentive plan, 6-23. *See also* incentive wage system
group, informal, 10-15, 15-7
group jobs, 6-23
group leader. *See* supervisor, working
group maintenance, 15-6
group maintenance roles, 10-14
group payment, 6-23
group piecework plan, 6-23
group task roles, 10-14
group technology, 10-14, 11-8, 12-13, 14-10. *See also* group classification code
group technology code (GT code), 11-8
groupthink, 10-14, 15-6
growth needs, 10-14
growth-need strength, 10-14
growth strategies, 10-14
GT (group technology) code, 11-8
g-tolerance, 9-10
guarantee, 6-23, 17-8
guaranteed annual wage, 6-23
guaranteed base rate, 6-23
guaranteed employment, 6-23
guaranteed hourly rate, 6-23
guaranteed maximum (target price) contracts, 4-9
guaranteed rate, 6-23
guaranteed standard, 6-23
guaranteed time, 6-23
guaranteed wage plan, 6-23. *See also* annual wage or employment guarantee; wage advance plan
guaranteed wage rate, 6-23
guarantee or trial rate, 6-23. *See also* temporary rate
guard, 12-53, 13-22
guard, fixed barrier, 13-22
guard, interlocking barrier, 13-22
guardrail, 13-22
guerin forming, 12-53
guided rail (drive-in, drive-thru racks), 8-16
guided vehicle, automated (AGV), 11-2
guided vehicle system, automatic (AGVS), 8-3

This index is arranged in letter-by-letter format.

guideline, 4-17
guideline method of job evaluation, 6-23
guideline method of job pricing, 6-23
guide pin, 12-76
guide-pin bushing, 12-76
guideposts, 6-23. *See also* Council of Economic Advisors
guide rail, base, 8-4
guide schedule, monthly, 4-22
GUI (graphical user interface), 9-10
gum, 12-76. *See also* adhesive; glue; resin
gun drilling, 12-62
gutter, 12-28
Gy (gray), 13-21

habitual violator, 13-22
hackers, 10-14
half-duplex, 3-7
half-life, biologic, 13-7
half-space, 1-12
halfword, 3-7
halocarbon plastics, 12-76
halo effect, 6-23, 10-14, 17-9
halo error, 6-23
hamate bone, 2-37. *See also* palmar arch; ulnar nerve
hammer, air-lift, 12-25
hammer, board, 12-25, 12-49
hammer, drop, 12-27
hammer forging, 12-28. *See also* counterblow forging equipment
hammer, gravity, 12-28
hammer, hydraulic, 12-28
hammering, 12-53
hammer, single-acting, 12-57
hammer, steam, 12-30
hammock, 4-17
hamstrings, 2-38
hand, 8-16, 11-8
hand breadth at metacarpal, 2-10
handicapped worker rate, 6-24. *See also* substandard rate
handicapped workers, 6-24
hand length, 2-10
handling time, 17-9
hand of government, 10-14
hand of management, 10-14
handshaking, 12-13
hand shield, 12-88
hand thickness at metacarpal III, 2-10
hand time, 17-9. *See also* manual element
hand truck, 8-16
hanger, 4-17
hanger rods, 8-16
hard copy, 3-7, 12-13
hardener, 12-76. *See also* catalyst

hard metal disease, 13-22
hardness, 12-53
hardship allowance premium, 6-24
hard turning, 12-62
hardware, 3-7, 10-14, 12-13
hard-wired system, 12-13
hard-wire logic, 12-13
harmonic drive, 11-8
harvest, 10-14
hash total, 3-8
haul distance, 4-17
Hawthorne effect, 9-10, 10-14, 15-6
Hawthorne studies, 10-14
hay system, 6-24
hazard analysis, 13-22
hazard classification, 13-22
hazard control, 13-22
hazard function, 1-12
HAZARD (instantaneous failure rate), 16-9
hazard level, 13-22
hazard, mechanical, 13-30
hazardous condition, 13-22
hazardous material, 8-16, 13-22
hazardous operation, 8-16
hazard pay, 6-24, 13-22. *See also* high time
hazard, potential, 13-36
hazard recognition, 13-22
hazard(s), 6-24, 13-22
haze, 12-76. *See also* chalking-dry
HCI (human-computer interaction), 9-10
head block, 12-76
head breadth, 2-10
head contours, 2-8
head length, 2-10
head length, maximum, 2-11
head of radius, 2-38
head scanning, 9-10
headstock (lathes), 12-62
head up display (HUD), 9-10
health and insurance plan, 6-24, 13-22. *See also* accident and sickness benefits; hospitalization benefits; life insurance plan; major medical expense benefits; medical benefits; surgical benefits
health center, 6-24. *See also* medical benefits
health maintenance organization (HMO), 6-24
health physicist, 13-22
health risk appraisal, 13-22
healthy worker effect, 13-23
heap sand, 12-39
hearing, 6-24
hearing conservation, 13-23
hearing level, 13-23
hearing loss, 13-23
hearing loss, central, 9-5
hearing loss, conductive, 9-6. *See also* perception deafness
hearing loss, high frequency, 13-23

heart rate, 9-10
heat, 12-39
heat acclimatization, 2-38
heat-affected zone, 12-88
heat checking, 12-39
heat collapse, 13-23
heat cramps, 13-23
heater-adapter, 12-77
heat exhaustion, 13-23
heat, exothermic, 12-74
heat forging, 12-28
heat forming. *See* thermoforming
heating, dielectric. *See* high-frequency heating
heating, high-frequency, 12-77
heat of metal, 12-28
heat prostration, 13-23
heat pyrexia, 13-23
heat, radiant, 12-43
heat stress, 9-10
heat stress index, 9-10
heat stroke, 2-38, 13-23
heat time, 12-88
heat treatment, 12-53
hedge, 4-17, 14-10
heel, 12-39
heel, cantilever rack, 8-16
height above deck, portable metal stacking rack, 8-16
height, cargo container, 8-6
height, ear, 2-9
height of scan, 8-16
helium neon laser, 8-16
helix angle (drills), 12-63
helix angle (milling cutters), 12-63
helmet, 12-88
helmet mounted display (HMD), 9-10
helmets, safety, 13-40
help system, 9-10
hematoma, 13-23
hemming, 12-53
hemodynamics, peripheral, 2-44
herniated disk, 2-38. *See also* fifth lumbar vertebra
hertz, 13-23
heterodyne interferometer, 12-63
heuristic, 9-10, 11-8
heuristic method, 3-8, 12-13
hexadecimal, 12-13
hierarchical bill of material. *See* product structure
hierarchical control, 11-8
hierarchical menu system, 9-10
hierarchical model, 3-8
hierarchy, 12-13, 15-6
hierarchy of needs theory, 10-14
high-density plywood, 12-77
high efficiency machining range, 12-63
high-energy-rate forging, 12-28
highest and best use, 4-18
high frequency hearing loss, 13-23
high-frequency heating, 12-77
high-level language, 3-8, 12-14. *See also* language

high lift platform truck, 8-17
high lift truck, 8-17
highlight, 9-10
high pressure molding, 12-39
high pressure nervous syndrome, 13-23
high rise AS/RS, 8-17
high speed forging, 12-28
high speed machining, 12-63
high speed steel (HSS), 12-63
high task, 6-24, 17-9. *See also* incentive pace; low task; MPL (motivated productivity level); normal performance
high time, 6-24. *See also* hazard pay
high velocity forging, 12-28
hinged feeder, 8-17
hinge joint, 2-38
hip breadth, sitting, 2-11
hip breadth, standing, 2-11
hiring, 6-24
hiring hall, 6-24
hiring, preferential, 6-43
hiring rate. *See* entrance rate
histogram, 16-8
historical time. *See* standard time
hit-the-bricks, 6-24
HMD (helmet mounted display), 9-10
HMO (health maintenance organization), 6-24
hob, 12-77
hobbing, 12-77
hoist, 8-17
hoistway, 13-23
hoistway-door interlock, 13-23
holdback pay, 6-24
hold-down groove, 12-77
hold harmless agreement, 13-23
holding cost, 10-5
holding time, 1-12, 4-18
hold time, 12-88
hole flanging, 12-53
holiday pay, 6-24
holidays, paid, 6-39
holidays, unpaid, 6-61
home leave, 6-24
home office cost, 4-18
homeostasis, 2-38. *See also* feedback system
homeostatic system, 15-6
homework, 6-24
homogenous transform, 11-8
homologous motion, 2-38
homopolymer/copolymer, 12-77
honeycombing, 8-17
honing, 12-63
hook approach, 8-17
hooker process, 12-53
horizontal bay clearance, 8-17
horizontal channel integration, 5-7
horizontal communication, 10-14
horizontal coordination, 10-14
horizontal display, 14-10. *See also* bucketed system; bucketless system; vertical display
horizontal rail clearance, 8-17

This index is arranged in letter-by-letter format.

horizontal surface grinding, 12-63
horizontal union, 6-24. *See also* craft union
hospitalization benefits, 6-25, 13-23. *See also* health and insurance plan
host, 11-8
host computer, 12-14
host factors, 13-23
hot blast, 12-39
hot box process, 12-39
hot-cargo clause, 6-25. *See also* boycott; struck work or goods
hot-chamber machines, 12-39
hot forming, 12-53
hot pressing, 12-53
hot-runner mold, 12-77
hot-short, 12-77
hot spots, 12-39
hot stamp, 12-28
hot strength, 12-39
hot tear, 12-39
hot trim, 12-28
hot working, 12-28, 12-53
hour, allowed, 17-2
hourly additive, 6-3
hourly earnings exclusive of overtime payments, average, 6-7. *See also* average straight-time hourly earnings; gross average hourly earnings
hourly rate, 6-25
hourly rate, guaranteed, 6-23
hourly rate wage plan, 6-25
hour plan, standard, 6-54
hours, actual, 17-2
hours produced, standard, 6-54
hour, standard, 6-53, 17-18
housekeeping, 13-23
house organ(s), 6-25
housing allowance, 6-25
HPT (human performance times), 17-9
HRM (human resource management), 10-14
hub, 12-77
hubbing, 12-77
hubrid structure, 10-15
HUD (head up display), 9-10
hue, 9-10
hugger belt conveyor, 8-17
human capital theory, 6-25
human-computer interaction (HCI), 9-10
human engineering, 2-38, 9-10, 17-9. *See also* ergonomics
human error, 9-10
human factors, 9-10, 11-8, 13-16. *See also* ergonomics
human factors engineering, 17-9. *See also* ergonomics; methods engineering; motion analysis; motion economy; work design
human-machine system, 9-11
human operator, 9-10
human performance reliability, 1-22

human performance times (HPT), 17-9
human relations, 6-25
human relations committees, 6-25. *See also* continuous bargaining committees; interim committees
human resource accounting, 15-6
human resource management (HRM), 10-14
human resource planning, 10-15, 15-6
human resources department, 6-25
human resources programs, 6-25
human skills, 10-15
human transfer function, 9-10
humeral rotation, 2-38
humerus, 2-38
hundred percent incentive, 6-25
Hungarian method, 1-12
hunting, 12-14
hurdle rate, 7-8
hybrid computer, 12-14
hydraulic conveyor, 8-17
hydraulic hammer, 12-28
hydraulic motor, 11-9
hydraulic press brake, 12-53
hydraulic shear, 12-53
hydrocarbon plastics, 12-77
hygiene factors, 10-15, 15-6
hygiene, industrial, 2-39, 13-25
hygienist, industrial, 13-25
hyperbaric oxygen therapy, 13-24
hypercapnia, 13-24
hypergeometric distribution, 16-8
hyperplane, 1-12
hypersensitization, 13-24
hypertest, 9-11
hyperventilation, 2-39
hypothenar, 2-23
hypothenar eminence, 2-23
hypothermia, 13-24
hypothesis, 13-24
hypothesis, alternative, 16-11
hypoxia, 2-39, 9-11
hysteresis, 11-9, 12-14

I

ICAM (integrated computer-aided manufacturing), 11-9
ICC type LP gas cylinder, 8-17
ICD, 13-24
ICHCA (International Cargo Handling Coordination Association), 8-17
IC (inspiratory capacity), 13-29. *See also* lung volumes
IC (integrated circuit), 12-7, 12-14
icon, 3-8, 9-11
idea champion, 10-15
ideal index, 4-18
identification, 15-6
identity matrix, 1-12

idiopathic, 2-39
idle equipment cost, 4-18
idle machine time, 6-33
idler sheave, 8-17
idle time, 3-8, 6-25, 11-9, 17-9. *See also* avoidable delay(s); delay time; unavoidable delay; waiting time
idle time, machine, 6-33
IDLH (immediately dangerous to life and health), 13-24
IDSS (integrated decision support system), 11-9
if-and-only-if operation, 3-8
if-then operation, 3-8
IISS (integrated information support system), 11-9
iliac crest, 2-39
ilite, 12-39
illegal character, 3-8
illumination, 13-24. *See also* brightness
illusion, audiogravic, 9-3
illusion, audiogyral, 9-3
imbalanced load, carousel, 8-17
immediately dangerous to life and health (IDLH), 13-24
immersion foot, 13-24
imminent danger, 13-24
immoral management, 10-15
impact, adverse, 10-2
impact cost, 4-18
impact factor, 8-17
impact printer, 3-8
impairment, 2-39
impartial chairman, 6-25. *See also* arbitrator
impasse. *See* impartial chairman
impedance, specific acoustic, 2-49
imperfection, 4-18
impervious, 13-24
impinger, 13-24
implementation, 14-10
implicit enumeration, 1-12
implicit prices, 1-12
implosion, 13-24
import quota, 10-15
imposed date, 4-18
imposed finish date, 4-18
impossibility, 4-18
impossibility, actual, 4-18
impracticability, 4-18
impregnation, 12-39
impression, 12-28
impression die forging, 12-28. *See also* drop forging
improshare, 6-25, 17-9. *See also* gainsharing
improvement approach to workplace design, 2-39. *See also* rational workplace design; workplace layout
improvement factor, 6-25
improvement factor, annual, 6-5. *See also* deferred wage increases
impulse, 9-11, 12-14
impulse items, 5-7

impulse noise, 9-11
imputation (of price movement), 4-18
inactive time, 1-26
in bond cargo, 8-17
incendiary, 13-24
incentive, 6-25, 15-6
incentive earnings, 6-25. *See also* incentive pay
incentive earnings, average, 6-7
incentive operators, 17-9
incentive opportunity, 6-25
incentive pace, 6-26. *See also* high task; motivated productivity level; MPL (motivated productivity level)
incentive pay, 6-26. *See also* production bonus
incentive performance, 6-26
incentive plan, 6-26, 15-6, 17-9
incentive plan, group, 6-23
incentive plan, non-financial, 6-37
incentive plan, piecework, 6-26
incentive plan, standard hour, 6-26
incentive rate, 6-26. *See also* piece rate
incentive, supervisory, 6-56
incentive wage system, 6-26. *See also* piecework
incidence rate, 13-24. *See also* disabling injury frequency rate
incidence rate, epidemiology, 13-24
incident, 13-24
incidental element. *See* irregular element
incident rate, 6-26
inclined reciprocating conveyor, 8-17. *See also* lowering conveyor; vertical reciprocating conveyor
inclusions, 12-39
inclusions, sand, 12-44
income, 4-18, 6-26. *See also* profit
income, disposable, 5-5
income policy, 6-26
income security plans, 6-26
income statement, 10-15
inconel, 12-39
increase, promotion, 6-45
increases, automatic, 6-6. *See also* automatic progression
increases, lump sum, 6-33
incremental coordinates, 12-14
incremental costs, 7-11. *See also* marginal cost
incremental costs (benefits), 4-18
incrementalist approach, 10-15
incremental model, 10-15
incremental system, 12-14
incremental transfer effectiveness, 9-11
increment cost, 7-11
incubation time, 13-24
incubator, 10-15
incumbent, 6-26
indented bill of material (BOM), 14-10

This index is arranged in letter-by-letter format.

independence, 16-8
independent, alternative, 7-6
independent demand, 14-10. See also demand; dependent demand
independent demand inventory, 10-15
independent equations, 1-12
independent failure, 1-10
independent trials, 1-12, 16-9
independent union, 6-26
independent variable, 1-12, 13-25
index, 3-8, 12-14. See also price index
index, articulation (AI), 9-3
index, belding-hatch, 13-7
index, benchmark, 4-4
index case, 13-25
index, chain, 4-7
index, chemical engineering plant cost, 4-7
index, classification, 6-10
index, cost, 4-10
index, cost of living, 4-10
index, cube-per-order (COI), 8-10
index, demographic, 4-12
index, disabling injury, 13-14
index, general purpose, 4-17
index, heat stress, 9-10
index, ideal, 4-18
indexing, 6-26, 8-17
index of work tolerance, 2-39
index, performance, 17-15
index, process capability (PCI), 16-12
index, productivity, 17-16
index, profitability (PI), 4-26, 7-13
index, speech articulation, 9-21
index, sub-, 4-33
index, urban wage rate, 6-61
indicator, 9-11
indifference quality level (IQL), 16-9
indirect-arc furnace, 12-39
indirect capital. See indirect cost(s)
indirect compensation, 6-26
indirect cost(s), 4-18, 7-11, 11-9, 13-25. See also accident costs
indirect damage, 13-25
indirect extrusion, 12-53
indirect inspection, 16-9
indirect interlock, 10-15
indirect labor, 6-26, 14-10, 17-9
indirect labor standard, 17-9
indirect materials, 11-9
individual equity, 6-26
individualism-collectivism, 10-15
individualized consideration, 10-15
individual job rate structure, 6-26
individual, matching, 13-30
individual observation chart, 16-9
individual price index, 4-18
individual rate, 6-26
individual retirement account (IRA), 6-26

individual wage determination, 6-26. See also maturity comparisons
induction brazing, 12-88
induction furnace, 12-39
induction motor, 11-9
induction welding, 12-88
industrial accident. See occupational injury
industrial democracy, 6-26
industrial dermatitis. See dermatitis
industrial disease. See occupational disease
industrial engineer, 17-9
industrial goods, 5-7
industrial hygiene, 2-39, 13-25
industrial hygienist, 13-25
industrial medicine. See occupational medicine
industrial relations, 6-26
industrial relations center, 6-26
industrial safety. See occupational safety
industrial tow tractor, 8-17
industrial truck, 8-17
industrial union, 6-27
industrial ventilation, 13-25. See also exhaust, general; exhaust, local
industry-wide bargaining, 6-27
inefficiency, 4-18
inequality, 1-12
inequality relation, 1-12
inequity, 6-27
inequity adjustment, 6-27
inert gas narcosis, 13-25
inertia, moment of, 2-43
inervertebral discs, 2-39
inexcusable delay, 4-12
infant mortality rate, 13-25
infeasible basis, 1-12
inferior, 2-23
infinite loading, 14-10. See also capacity requirements planning; finite loading
inflammable, 13-25. See also flammable
inflammation, 13-25
inflating by a price index, 7-11. See also actual dollars; constant dollars; deflating by a price index
inflation, 4-18, 6-27, 7-11. See also deflation
informal communication, 10-15
informal group, 10-15, 15-7. See also formal group
informal leader, 10-15
information, 3-8, 9-11, 10-15, 12-14, 15-6
informational picketing, 6-27, 6-43
information center, 10-15
information power, 10-15
information-processing approach, social, 10-28

information sharing, 5-7
information support system, integrated (IISS), 9-11
information system, 10-15
information systems marketing, 5-9
information theory, 9-11, 12-14, 15-7
infraorbitale, 2-23
infrared photography, 2-39. See also thermograph
infrastructure, 10-16
ingestion, 13-25
ingot, 12-39, 12-53
ingredient. See component
ingress, 9-11
inhalation, 13-25
inherent delay. See delay; delay time
inherent process variability, 16-9
inherent reliability, 1-22, 16-14
inherent weakness failure, 1-10
inherited abnormality, 13-25. See also abnormality
inhibitor, 12-77, 13-25
initial basis, 1-12
initial cost. See first cost
initial feasible basis, 1-13
initialize, 3-8
initial program loader (IPL), 3-8
initial solution, 1-13
initiating structure, 10-16
initiation fee, 6-27
injection, 12-39
injection blow molding, 12-77
injection mold, 12-77
injection molding, 12-77
injection pressure, 12-77
injunction, 6-27
injury, 9-11, 13-25
injury, compensable, 6-12, 13-10
injury frequency rate, serious, 13-41
injury, lost time, 13-28
injury, major, 13-29
injury, musculoskeletal injury, 13-32
injury, nature of, 13-32
injury, nondisabling, 13-33
injury, occupational, 13-25
injury, personal, 8-25
injury, serious, 13-41
ink jet printer, 3-8
in-line package (DIP), dual, 12-10
innovation, 10-16
inoculant, 12-39
I-node (ADM), 4-18
in-place value, 4-18
in-process inspection, 16-9
in-process inventory, 11-9
in-progress activity, 4-18
input, 3-8, 9-11, 12-14, 14-10, 15-7
input and output (I/O) point, 8-17
input device, 9-11
input equipment, 12-14
input material costs, 4-13

input-output analysis, 4-18, 4-19, 15-7
input-output coefficient, 1-13
input/output control, 14-10. See also capacity control; closed loop MRP
input/output device, 12-14
input-output (I/O), 3-8
input/output (I/O), 11-9
input/output module, 11-9
inputs, 10-16
input variables of external environment, 2-39
inquiry, board of, 6-8. See also fact-finding board
inrunning nip point, 13-26
insert, 3-8, 12-28, 12-77
insert, eyelet-type, 12-77
insert, floating of, 12-77
insert, flow of materials into, 12-77
insertion, 2-39. See also extensor tendon; origin (muscle)
insert, open hole, 12-77
insert, protruding, 12-77
insert, rivet, 12-77
inserts, 12-63
insert, through-type, 12-77
inspection, 1-13, 16-9
inspection, average amount of, 16-3. See also average sample number (ASN); average total inspected (ATI)
inspection, chain sampling, 16-4
inspection, curtailed, 16-6
inspection, double-sampling, 16-6
inspection, indirect, 16-9
inspection, in-process, 16-9
inspection level, 16-9
inspection lot, 16-9
inspection, lot-by-lot, 16-9
inspection, nondestructive, 12-42
inspection, screening, 16-16
inspection, single sampling, 16-16
inspection, skip-lot sampling, 16-16
inspection, symbol, 17-8
inspiratory capacity (IC), 13-29. See also lung volumes
instantaneous failure rate (HAZARD), 16-9
institutional advertising, 5-7
institutional power, 10-16
institutional/transformative leadership, 15-7
instruction, 3-8
instruction card, 17-9
instruction set, 3-8
instruction sheet, 17-9
instrument, 12-14
instrumentation, 12-14
insulating pads and sleeves, 12-39
insurable interest, 13-26
insurance, 13-26
insurance, catastrophe, 6-10, 13-29. See also major medical expense benefit

This index is arranged in letter-by-letter format.

insurance, key man, 6-30
insurance, strike, 6-55
insured, 13-26
insured costs, 13-26
intangibles, 4-19, 7-11. See also irreducibles
integer form, 1-13
integer linear programming, 1-13
integral control, 11-9
integrated circuit (IC), 12-7, 12-14
integrated computer-aided manufacturing (ICAM), 11-9
integrated controller, 9-11
integrated decision support system (IDSS), 11-9
integrated display, 9-11
integrated information support system (IISS), 11-9
integrated salary structure, 6-27
integrated surface myogram, 2-39
integration, 10-16, 15-7
integration, horizontal channel, 5-7
integrator, 12-14
intellectual stimulation, 10-16
intelligent robot, 11-9
intelligent system, 11-9
intensity level (sound), 13-26
interaction, 15-7
interactive, 9-11, 11-9
interactive scheduling, 14-10
interarrival time, 1-13
interchangeable item, 1-13
intercharacter gap, 8-17
intercity differential, 6-27. See also area differential
interdependence, reciprocal, 10-26
interdependence, sequential, 10-27
interdependence, technological, 10-31
interdisciplinary nature, 15-7
interest, 7-11. See also time value of money
interest, compound, 7-7
interest factors, functional forms of compound, 7-3–7-5
interest group, 10-16
interest rate, 7-11
interest rate, effective, 7-11
interest rate, market, 7-11
interest rate, nominal, 7-11
interest rate-nominal. See simple interest
interest rate, real, 7-11
interest, simple, 7-14
interface, 3-8, 9-11, 11-9, 12-15, 15-7
interface activity, 4-19
interface node, 4-19
interference, 4-19
interference allowance, 17-9
interference fit, 12-63
interference, machine, 17-11
interference time, 17-9
interferometer, 12-15
interim committees, 6-13. See also crisis bargaining; human relations committees

interim dates, 4-19
interim earnings, 6-27
interindustry (input-output) analysis, 1-13
interleaved bar code, 8-17
interline, 8-17
interlock, 9-11, 11-9, 13-26
interlocked barrier guard, 13-7
interlock, indirect, 10-15
interlocking directorates, 10-16
interlocking mechanism, 8-17
interlocks, electrical, 8-18
interlocks, mechanical, 8-18
intermediate events, 4-19
intermediate materials, 4-19
intermediate node, 4-19
intermediate package, 8-18
intermittent element. See irregular element; regular element
intermittent production, 11-9, 14-10
intermittent weld, 12-88
internal audit, 10-16
internal (bio)mechanical environment, 2-39. See also external mechanical environment
internal broaching, 12-63
internal combustion engine truck, 8-18
internal combustion tractor, 8-18
internal discharge bucket elevator, 8-18. See also bucket elevator
internal element. See internal work
internal environment, 10-16
internal equity, 6-27
internal friction, 12-53
internal rate of return, 7-11, 7-13. See also discounted cash flow; external rate of return; investor's method; rate of return
internal ribbon conveyor, 8-18
internal sensor, 11-9
internal shrinkage, 12-39
internal sort, 3-9
internal time, 17-9
internal work, 17-9
international business, 10-16
International Cargo Handling Coordination Association (ICHCA), 8-17
international element, 10-16
international management, 10-16
interoperability, 3-9
interoperation time, 14-10. See also queue time
interpass temperature, 12-88
interphalangeal joints, 2-39
interplant demand, 14-10
interpolation, 12-15
interpreter, 3-9
interpupillary, 2-11, 2-23
interrogatories, 13-26. See also deposition
interrupt, 3-9, 11-9, 12-15
interscapulae, 2-23

interstitial, 13-26
intervening act, 13-26
intervening agency, 13-26
intervening cause, 13-26
intervening force, 13-26
intervening losses, 13-26
intervening negligence, 13-26
intervening study, 13-26
intervention(s), 10-16, 15-7
interview, 6-27
interview, disciplinary, 6-15
interview, selection, 10-27
intoxication, 13-26
intransit lead time, 14-11
intrapreneurs, 10-16
intrapreneurship, 10-16
intrinsic availability, 1-3
intrinsic forecast, 14-11. See also extrinsic forecast
intrinsic muscle, 2-39
intrinsic rewards, 6-27, 10-16
inventory, 4-19, 8-18, 10-16, 12-77, 14-11
inventory, active, 1-13
inventory, aggregate, 14-3
inventory, anticipation, 14-3
inventory, available, 14-3
inventory, available motions, 2-31
inventory, average, 5-2
inventory, book, 14-4
inventory control, 14-11. See also production control
inventory, demand motions, 2-34
inventory, dependent demand, 10-8
inventory, finished-goods, 10-12
inventory, inactive, 1-13
inventory, independent demand, 10-15
inventory, in-process, 16-9
inventory level, base, 5-2
inventory, lot size, 14-12
inventory management, 14-11
inventory models, 10-16
inventory, motions, 2-43
inventory, physical, 14-16. See also cycle counting
inventory policy, 14-11
inventory, raw materials, 10-26
inventory, reaction, 2-47
inventory, reconciling, 14-18
inventory, shrinkage, 14-11
inventory, skills, 6-52, 10-28
inventory system
 periodic, 14-16
 perpetual, 14-16
inventory, transit, 14-22
inventory, transportation, 14-22
inventory turnover, 14-11
inventory, valuation, 14-11
inventory write-off, 14-11. See also component; end product; raw materials; subassembly
inverse, basic, 1-4
inverse chill, 12-39
inverse of a matrix, 1-13
inversion, 1-13, 13-27

inverted extrusion, 12-53
inverted power and free conveyor, 8-18
inverter, 12-15
investing, 12-40
investment, 4-19, 7-11, 12-40. See also capital
investment casting, 12-40
investment center, 10-16
investment cost, 4-19
investment, direct, 10-9
investment, prudent, 4-28
investor's method, 7-11. See also discounted cash flow; internal rate of return; profitability index (PI); rate of return
invisible hand, 10-16
I/O (input-output), 3-8, 11-9
IPL (initial program loader), 3-8
IQL (indifference quality level), 16-9
IRA (individual retirement account), 6-26. See also ERTA (Economic Recovery Act of 1981)
iron
 alpha, 12-48
 hard, 12-40
 malleable, 12-40
 nodular, 12-42
 pearlitic, 12-40
 white, 12-40
ironing, 12-28, 12-54. See also coining
Iron Law of Responsibility, 10-16
iron oxide, 12-40
irradiation, 12-77, 13-27
irreducibles, 7-11. See also intangibles
irregular element, 17-10
irritant, 13-27. See also dermatitis
ischemia, 2-40. See also respiration
ischial tuberosity, 2-40
ischium, 2-40
iserine, 12-40
isocyanate plastics, 12-77
isocyanate resins, 12-78
isoinertial, 2-40
isokinetic, 2-40
isometric work, 2-40. See also dynamic work; negative work; positive work
isotactic, 12-78
ISO (The International Organization for Standardization), 3-9
isothermal forging, 12-28
issue receipt, planned, 14-16
issues management, 10-16
itch, nickel, 13-32
item, 4-19
item cost, 10-16
item, interchangeable, 1-13
item number, 14-11
item obtainment time, 1-26
item, replaceable, 1-13
item, substitute, 1-13

This index is arranged in letter-by-letter format.

item-to-person AS/RS, 8-18
item, unit, 1-13
iteration, 1-13
iterative operation, 3-9
izod test, 12-40

J

jacket mold, 12-40
Japanese management, 10-16
jaw breadth, 2-11
jaw height, total, 2-11
JCL (job control language), 3-9
jet scrubber, 12-40
jib crane, 8-18. *See also* crane
jig, 11-10, 12-15, 12-63, 17-10.
    See also fixture
jig borers, 12-63
JIT (job instruction training), 6-28
JIT (just-in-time), 11-10, 14-11,
    17-7, 17-9, 17-10. *See also*
    paperless purchasing
JIT (just-in-time) inventory control, 10-17
J-Node (ADM), 4-19. *See also*
    successor event
job, 3-9, 6-27, 15-7, 17-10. *See
    also* task
job analysis, 6-27, 9-11, 10-16,
    15-7, 17-10
job assignment, 11-10
job banks, 6-27
job, benchmark, 6-8, 17-3
jobber, 5-7
jobbing foundry, 12-40
job breakdown, 6-27, 17-10
job burnout, 6-27
job-centered, 10-17
job characteristic, 6-27, 17-10
job characteristics model, 10-17
job class, 6-28
job classification, 6-28, 6-30, 15-7
job cluster, 6-28
job combination, 6-28
job comparison, 6-28
job content, 6-28, 15-7
job content evaluation method, 6-28
job context, 15-7
job control language (JCL), 3-9.
    See also language
job crane. *See* fixed crane
job cycle, 6-28
job definition, 6-28
job depth, 10-17, 15-7
job description, 6-28, 10-17, 15-7
job design, 10-17, 15-7
job design, biological school of,
    15-2
job design, engineering school of,
    15-5
job design, ergonomics school of,
    15-5
job design, psychological school
    of, 15-12

job enlargement, 6-28, 9-11,
    10-17, 15-7
job enrichment, 6-28, 10-17, 15-7
job evaluation, 6-28, 10-17, 15-8.
    See also labor grade
job evaluation, classification
    method of, 6-10
job evaluation committee, 6-28
job evaluation, guideline method,
    6-23
job evaluation manual, 6-28
job evaluation, non-quantitative,
    6-37
job evaluation point method, 6-28
job evaluation, quantitative, 6-46
job evaluation, ranking method of,
    6-46
job factor, 6-28, 17-10
job family, 6-28
job feedback, 6-28
job freezing, 6-28
job grade, 6-30. *See also* pay
    grade
job, group, 6-23
job hazard analysis, 13-27. *See*
    hazard analysis
job instruction training (JIT), 6-28
job, key, 17-10. *See also* benchmark
job levels, 6-29
job matching, 6-29
job order. *See* manufacturing
    order
job overhead, 4-19
job posting, 6-29, 10-17
job pricing, 6-29
job pricing, guideline method,
    6-23
job profile, 6-29
job ranking, 6-29
job rate, 6-29. *See also* minimum
    job rate; standard rate
job rate structure, individual, 6-26
job rating. *See* job evaluation
job relatedness, 6-29
job, restricted, 6-48
job rotation, 10-17
job safety analysis, 13-27
job safety training, 13-27
job scope, 6-29, 10-17, 15-8
job security, 6-29
job sharing, 6-29
job shop, 8-18, 11-10, 12-15,
    14-11
job simplification, 10-17
job skill, 17-10. *See also* skill
job specification, 6-29, 10-17, 15-8
job standardization, 15-8, 17-10
job status, 11-10
job study, 17-10
job title, 6-29
joining, 3-9
joint, 2-40, 11-10, 12-78. *See also*
    arthritis; articulation; bond;
    bubble; tennis elbow
joint action process. *See* union-
    management cooperation

joint and survivor option, 6-29
joint bargaining, 6-29
joint board, 6-29
joint, butt, 12-78, 12-86
joint, corner, 12-86
joint council, 6-29
joint design, 12-88
joint, edge, 12-78, 12-87
joint geometry, 12-88
joint, hinge, 2-38
joint, interphalangeal, 2-39
joint, lap, 12-78, 12-88
joint lockout. *See* lockout
joint, lumbosacral, 2-41
joint mode, 11-10
joint penetration, 12-88
joint, pivot, 2-45
joint, radiocarpal, 2-46. *See also*
    radius
joint, radiohumeral, 2-46. *See also*
    capitulum of humerus; radial
    deviation; radius
joint, radio-ulnar, 2-46. *See also*
    radial deviation; tennis elbow
joint rate, 8-18
joint rate setting, 6-29
joint reaction force, 2-40
joint, scarf, 12-78
joint, slip, 12-83
joint socket, 2-37
joint space, 11-10
joint, starved, 12-78
joint venture, 10-17
jolt squeeze molding, 12-40
journeyman, 6-29
journeyman rate, 6-29
joy stick/ joystick, 3-9, 9-12, 11-10
judgemental forecasting, 10-17
judgemental sampling, 4-19
judgment, 13-27
jurisdiction, 6-29, 13-27
jurisdictional dispute, 6-29
jurisdictional strike, 6-29
jury-duty pay, 6-29
jury of executive opinion, 10-17
just cause, 6-30, 13-27
justification, 11-10
just-in-time (JIT), 5-7, 11-10, 14-
    11, 17-7, 17-9, 17-10. *See also*
    kanban; paperless purchasing
just-in-time (JIT) inventory control, 10-17
just noticeable difference, 9-12

K

K, 3-9
kaizen, 11-10
kanban, 10-17, 11-10, 14-11. *See
    also* just-in-time (JIT)
kaolin, 12-40
Karmarkar's algorithm, 1-13
Karuch-Kuhn-Tucker conditions,
    1-13

Keough Retirement Plan, 6-30
kerf, 12-88
key, 3-9
key activity, 4-19
keyboard, 9-12
keyclick, 9-12
key event of an accident, 13-27
key job, 17-10. *See also* benchmark; benchmark job
key man insurance, 6-30
keypad, 9-12
keyword, 3-9
kick-back, 6-30, 13-27. *See also*
    bootleg wages
killed steel, 12-54
kinematic chain, 2-40
kinematic coupling design, 12-15
kinematics, 2-40
kinesic behavior, 10-17
kinesiology, 2-40
kinesiology, correlative, 2-33
kinesiology, electromyographic,
    2-35
kinesiometer, 2-40
kinesthesis, 9-12
kinetics, 2-40
kitting, 14-11. *See also* picking
knapsack problem, 1-13
knee height, sitting, 2-11
knee switch, 2-40
knee-to-knee breadth, sitting, 2-11
knit line. *See* weld line
knockdown container. *See* collapsible container
knockout, 12-54, 12-78
knockout pin/pins, 12-40. *See also*
    ejector pin
knoop hardness test, 12-63
knowledge-based system. *See*
    expert system
knowledge of results, 9-12
knuckle height, 2-11
knurling, 12-63
kymograph, 17-10

L

lab, 15-8
label, 3-9, 5-7
labor, 6-30
labor, administrative, 6-4
labor area, 6-30
laboratory, 8-18
labor budget, direct, 6-15
labor burden, 4-19
labor class, 6-30
labor, common, 6-11
labor cost, 17-10
labor cost, manual, 4-19, 6-30
labor cost, non-manual, 4-19
labor, direct, 6-15, 17-5
labor dispute. *See* dispute
labor, division of, 6-16, 17-5. *See
    also* downgrade

This index is arranged in letter-by-letter format.

labor economics, 6-30
labor factor, 4-19
labor force, 6-30
labor force participation rate, 6-30
labor grade, 6-30. *See also* job evaluation
labor, indirect, 6-26, 14-10, 17-9
labor injunction. *See* injunction
labor-management committee, 6-30
labor-management relations, 10-17
Labor Management Relations Act, 6-30. *See also* National Labor Relations Act; National Labor Relations Board (NLRB); unfair labor practice
Labor-Management Reporting and Disclosure Act of 1959, 6-30
labor market area, 6-31
labor market, relevant, 6-47
labor mobility, 6-31
labor movement, 6-31
labor, non-productive, 6-37
labor, productive, 6-44
labor productivity, 6-31
labor rate, common, 6-11
labor relations, 6-31
labor-saving ratio, 6-31, 17-10
labor, semi-skilled, 6-50
labor, skilled, 6-52
labor standard, direct, 17-5. *See also* direct labor
labor standard, indirect, 17-9
labor supply, 6-31, 10-17
labor supply curve, 6-31
labor ticket, 14-11
labor turnover, 6-31
laceration, 13-27
lactic acid, 2-40
ladder diagram, 11-10
laddering, 4-19
ladle, 12-40
lag, 4-19, 12-15
laggards, 5-7
LaGrange multipliers, 1-13
lag relationship, 4-19, 4-20
laissez-faire, 10-18
LAMA-25, 12-15
Lambert surface, 2-41
laminate, 12-78
laminated, cross, 12-78
laminated, parallel, 12-78
laminated wood, 12-78
lancing, 12-54
land, 12-78
landed force, 12-78
landed plunger, 12-78
Landolt ring, 9-12
Landrum-Griffin Act, 6-30
LAN gateway, 3-9
language, 3-9. *See also* computer languages
language, ALGOL (algorithmic language), 3-2
language, APL (A Programming Language), 12-3

language, assembly, 3-2
language, BASIC (beginner's all-purpose symbolic instruction code), 12-4
language, COBOL (common business oriented language), 3-4
language, command, 9-6, 11-4
language, high-level, 3-8, 12-14
language, job control language (JCL), 3-9
language, machine, 12-16
language, natural, 3-11
language, object, 10-21
language, procedural, 3-13
language, procedure-oriented, 3-13
language, programming, 3-13
language, source, 3-16
language, symbolic, 3-16
LAN (local area network), 3-10, 10-18, 11-11
lanyard, 8-18
lap, 12-29
lap joint, 12-88
LaPlace-Gauss distribution, 16-11
Laplace transform, 1-14
lap-phasing. *See* overlapped schedule
lapping, 12-63
LA prospective, 10-17
large-batch, 10-18
laser, 13-27
laser beam cutting, 12-63, 12-88
laser beam welding, 12-89
laser burn, 13-27
laser printer, 3-9, 3-10
Laspwyres-type price index (strict appellation), 4-20
last-in-first-out (LIFO), 3-10, 4-20, 5-7, 14-11. *See also* FIFO
latch, 12-78
late effects of radiation, 13-27
late finish (LF), 4-20
latent condition, 4-20
latent defect, 4-11
lateral, 2-23, 9-12
lateral displacement, 2-41. *See also* medial displacement
lateral relations, 10-18
lateral transfer, 2-41
late start, 4-20
latest event time (LET), 4-20
latest revised estimate, 4-20
lathes, 12-63
latisssimus dorsi, 2-41
law(s)
  all or none, 2-29
  anti-racketeering, 6-5
  anti-strikebreaker, 6-5. *see also* strikebreaker
  of effect, 10-18
  Fair Employment Practice Laws, 6-20. *see also* Civil Rights Act
  full crew law, 6-21
  Iron Law of Responsibility, 10-16

Pareto's law, 14-15, 17-15. *see also* ABC classification
  prevailing wage law, 6-44
  and regulations, 4-20
  Right-To-Work Law, 6-48. *see also* Section 14 (B)
  wage-hour law. *see* Fair Labor Standards Act
  Weber-Fechner law, 9-26
  Weber's law, 9-26
lay, 12-63
layer, 12-89
layoff, 6-31
layoff allowance. *See* severance pay
layoff costs/benefits, 6-31
layoff pay, 6-31. *See also* severance pay
layout, 8-18
layout, fixed position, 8-14, 10-12
layout, office, 8-22
layout, single-row, 8-31
LCC (life-cycle cost) method, 4-20
LCL (less-than-carload), 8-18
LD-50 (median lethal dose), 13-30
LDCs (less developed countries), 10-18
L/D ratio, 12-78
lead, 4-20
lead angle, 12-89
leader, autocratic, 15-2
leader, formal, 15-5. *See also* informal group
leader, group. *See* supervisor, working
leader, informal, 10-15
leadership, 10-18, 15-8
leadership, charismatic, 15-3
leadership, romance of, 10-27
leadership, substitutes for, 10-29
leader, transactional, 10-31
leader, transformational, 10-31
leading, 10-18
leading indicators, 10-18
lead intoxication, 13-27
lead screw, 11-10
lead storage battery, 8-18
lead time, 14-11
lead time, cumulative, 14-6
lead time, finishing, 14-9
lead time, intransit, 14-11
lead time, manufacturing, 14-13
lead time offset, 14-12
lead time, order, 5-10
lead time, purchasing, 14-18
lead time, replenishment, 14-18
lead time, rework, 14-19
lead time, set-up, 14-20
lead time, vendor, 14-22
leaf brake, 12-54
leakers, 12-40
lean production, 11-10
learner, 6-31. *See also* apprentice
learner rates, 6-31
learner's allowance, 6-31. *See also* training allowance

learner's certificates, 6-31
learning, 9-12
learning control, 11-10
learning curve, 4-20, 9-12, 17-10. *See also* progress curve; start-up curve
learning curve, manufacturing progress function (MPF), 6-32
learning theory, social, 10-28
learning, transfer of, 9-24
learning, trial-and-error, 9-12
learning, whole vs. part, 9-12
leaseback, 7-11
least total cost, 14-12
least unit cost, 14-12
leave, military, 6-35
leave of absence, 6-32, 13-27
leave, personal, 6-42
leave plan, extended, 6-20
left-hand side, 1-14
legal liability, 13-27
legally required benefits, 6-32
legal plan, prepaid, 6-32
legal-political element, 10-18
legitimate power, 10-18
leg of a fillet weld, 12-89
leg reach contours, functional, 2-8
LEL (lower explosive limit), 13-28
lesion, 2-41, 13-27
less developed countries (LDCs), 10-18
less-than-carload (LCL), 8-18
less-than-truckload, 8-18
lethal dose, median (LD-50), 13-30
LET (latest event time), 4-20
letter of credit, 4-20
level, 8-18, 14-12. *See also* low level code
leveled element time. *See* normal element time
leveled time. *See* normal time
level finish schedule (SF), 4-20
level float, 4-20
level income option. *See* social security adjustment option
leveling, 6-32, 14-12. *See also* effort rating; normalize; performance rating
levelized fixed-charge rate, 4-20
level (noise), 13-27
level of aspiration, 6-32
level of effort (LOE), 4-20
level production, 11-10
level start schedule (SS), 4-20
leverage ESOP, 6-18
leverage (trading on equity), 4-20
lexicographic ordering, 1-14
LF (late finish), 4-20
liability, contractual, 13-11
liability insurance, 13-27
liaison role, 10-18
library, 3-10
licensing, 10-18
life, 4-20, 7-11. *See also* study period

This index is arranged in letter-by-letter format.

life-cycle cost, 7-12
life-cycle cost (LCC) method, 4-20
life cycle, product, 5-11
life cycle(s), 10-18. *See also* life; study period
life, economic, 7-10
life expectancy, 13-28
life insurance plan, 6-32, 13-28. *See also* death benefit; health and insurance plan
lifeline, 8-18, 13-28
life, mean, 1-15, 16-10
life style, 5-7
life support, 9-12
life table, 13-28
life, venture, 4-36
life, working, 12-85
LIFO (last-in-first-out), 3-10, 4-20, 5-7, 14-11
lift, 8-18
lifters, 12-40
lifting magnet, 8-18
lifting task, 2-41
lifting task, overt, 2-44
lifting torque, 2-41
lift-out links, 8-18
lift section, 8-18
lift speed, 8-18
lift table, 8-19
lift truck, 8-19
ligament, 2-41
ligament, transverse, 2-52
ligament, transverse dorsal, 2-35, 2-36
light adaptation, 9-12
lightening hole, 12-54
light flux, 9-12
light pen, 3-10, 8-19, 9-12
light task, 2-41. *See also* metabolic cost
lignin plastics, 12-78
Likert scale, 9-12
limb-load aggregate, 2-41. *See also* arm-tool aggregate
limen, 9-12
limen, difference, 9-7
limit determination, multivariate, 2-6
limited-line store, 5-8
limiting operation, 14-12
limiting quality level (LQL), 16-9
limiting quality (LQ), 16-9
limiting values, 16-18
limits, 12-63
limits, action, 16-2
limits, action control, 16-2
limit switch, 8-19, 11-10, 13-28
limit verification, multivariate, 2-6
line, 15-8
linear array camera, 11-10
linear combination, 1-14
linear constraint, 1-14
linear displacement transducer, 12-15
linear equation, 1-14
linear estimator, 1-14

linear function, 1-14
linear independence, 1-14
linear inequality, 1-14
linear interpolation, 12-15
linearity, 11-10
linearly dependent, 1-14
linear model, 1-14
linear programming (LP), 1-14, 3-10, 10-18
linear programming (LP), duality theorem, 1-9
linear programming (LP) problem, 1-14
linear programming (LP), upper bounded problem, 1-27
linear program problems, dual, 1-8
line, assembly, 14-3
line authority, 10-18
line balancing, 11-10, 14-12
line dies, 12-54
line item, 8-19, 14-12
line layout, 8-19
line of credit, 4-20
line of flow, 8-19. *See also* flow line
line of progression (LOP), 6-32. *See also* progression
line position, 10-18
line printer, 3-10
lines, match, 12-29
line synchronization, 11-10
lining, 12-40
lining, monolithic, 12-40
linkage. *See* linkage mechanism, fixed
linkage editor, 3-10
linkage mechanism, fixed, 2-36
link analysis, 9-12
linking pin, 10-18
linking procedure, 4-20. *See also* splicing technique
linseed oil, 12-40
lips, 12-63
liquefied petroleum gas (LP gas), 8-19
liquidated damages, 4-11
liquidation, 10-18
liquidity ratios, 10-18
liquids, flammable, 13-20
list, 3-10
listening, active, 10-2
list price, 5-8
live load, 8-19
living document, 6-32
living wage, 6-32
load, 3-10, 9-12, 12-15, 14-12
load, allowable, 13-28
load axle, 8-19
load backrest extension, 8-19
load bar, 8-19
load block, 8-19
load brake, mechanical, 8-21
load capacity, 11-10
load carrying flange, 8-19
load center, 8-19
load centering device, 8-19

load deflection, 11-10
loaded, 8-19
loaded grinding wheel, 12-63
load factor, 4-21, 7-12. *See also* capacity factor; demand factor
load height, 8-19
loading board, 12-78
loading platform, 8-19
loading, self, 8-30
loading space, 12-78
load length, 8-19
load leveling, 4-21, 11-11, 14-12. *See also* finite loading
load limit, 13-28
load moment, 8-19
load moment constant, 8-19
load overhang, 8-19
load, press, 12-54
load profile, 14-12
load, sensory, 9-12
load weight, 13-28
load width, 8-19
loafing, social, 10-28
loam, 12-40
local area network (LAN), 3-10, 10-18, 11-11
local cost, 4-21
local effects of radiation, 13-28
localization, auditory, 9-12
local optimum. *See* global optimum
local preheating, 12-89
local union, 6-32
locating ring, 12-78
locating surfaces, 11-11
location factor, 4-21
locator system, 8-19
lockout, 6-32. *See also* work stoppage
lockout, joint. *See* lockout
lockout-tagout, 13-28
locks, 12-29
lock, safety, 13-40
locomotion, 2-41
locomotive crane, 8-19. *See also* portable crane
locomotor system, 2-41
LOE (level of effort), 4-20
loft, 8-19
logger, 12-15
logic, 12-15
logical office, 10-18
logical record, 3-10
logical restraint, 4-21
logic, preferential, 4-26
logic, sequential, 12-23
logistics, 14-12
longevity pay, 6-32
longitudinal resistance-seam welding, 12-89
longitudinal study. *See* cohort study
long-range resource planning, 14-12
long term agreements (LTA's), 14-12
long term bonus, 6-32

long term contract, 6-32
long-term disability, 6-32
loop, 3-10, 4-21
loop (in a graph), 1-14
looping table, 8-20
loose rate, 6-32. *See also* rate erosion
loose standard, 17-10. *See also* loose rate
LOP (line of progression), 6-32. *See also* progression
lordosis, 2-41
lordotactic, 2-41
loss, 13-28
loss control, 13-28
losses, intervening, 13-26
loss leader, 5-8
loss of productivity/efficiency. *See* inefficiency
loss of service, coefficient of, 1-5
loss, prevention, 13-28
loss ratio (insurance), 13-28
loss reserve, 13-28
loss, total, 13-45
lost time. *See* waiting time
lost time illness, 13-28. *See also* injury
lost time injury, 13-28. *See also* injury
lost time (LT) accident, 13-28
lost wax process, 12-40
lost workday cases, 13-28
lost workdays, 13-28. *See also* days of disability
lot batch, 4-21
lot-by-lot inspection, 16-9
lot-for-lot, 14-12. *See also* fixed order quantity; period order quantity
lot number, 14-12
lot size, 4-21, 11-11, 14-12. *See also* cycle stock; reorder quantity
lot size inventory, 14-12
lot sizing, dynamic. *See* fixed order quantity; least total cost; least unit cost; period order quantity
lot tolerance percent defective (LTPD), 1-14
loudness, 9-13, 13-28
loudness contour, 9-13
loudness level, 9-13, 13-28
lower arm length, 2-11
lower explosive limit (LEL), 13-28. *See also* explosive limits
lowering conveyor, 8-20. *See also* arm conveyor; inclined reciprocating conveyor; suspended tray conveyor; vertical reciprocating conveyor
lowering, dynamic, 8-12
low level code, 14-12. *See also* level
low-lift pallet truck, 8-20
low-lift truck, 8-20
low solid tire, 8-20

This index is arranged in letter-by-letter format.

low task, 6-33, 17-10. *See also* high task
LPC orientation, 10-18
LP gas (liquefied petroleum), 8-19
LP (linear programming), 1-14, 3-10, 10-18
LQ (limiting quality), 16-9
LQL (limiting quality level), 16-9
LTA's (long term agreements), 14-12
LT (lost time) accident, 13-28
LTTP, 12-16
lubricant, die, 12-26
lubricants in machining, 12-64
lumbar spine, 2-41
lumbar vertebrae, 2-41
lumbar vertebra, fifth, 2-36. *See also* herniated disk; lumbosacral angle
lumber-core construction, 12-78
lumbosacral angle, 2-41. *See also* fifth lumbar vertebra
lumbosacral joint, 2-41
lumen, 9-13, 13-29. *See also* lux
lumen of the transverse canal, 2-42
luminaire, 13-29
luminescent, 13-29
luminous flux, 9-13
lump-sum, 4-21. *See also* fixed price contract(s)
lump sum award, 6-33
lump sum contracts, 4-9
lump sum increase, 6-33
lumpy demand, 14-13
lunch period, paid, 6-39
lung capacity, total (TLC), 13-29
lung volumes, 13-29
lux, 13-29. *See also* lumen
lyophilic, 12-79
lyophobic, 12-79

M

machine assignment, 17-11
machine-assignment problem, 1-15
machine attention time, 12-16, 17-11
machine bureaucracy, 10-18
machine capability, 17-11
machine center, 11-11. *See also* work center
machine code, 3-10
machine-controlled time, 6-33, 17-11. *See also* element; restricted work
machine down time, 6-33
machined surface, 11-11
machine element, 17-11
machine guarding, 13-29
machine hour, 6-33, 17-11
machine idle time, 6-33, 17-11
machine interference, 17-11. *See also* interference time

machine language, 12-16
machine layout, 8-20
machine load, 17-11
machine loading, 14-13
machine pacing, 6-33, 17-11
machine time, 17-11
machine time allowance, 17-11
machine time, available, 17-2
machine tool, 11-11, 12-16
machine tool error, 12-64
machine tool error compensation, 12-64
machine utilization, 12-16, 14-13
machine vision, 12-16
machine welding, 12-89
machining, abrasive, 12-60
machining cell, 12-64
machining center, 12-16, 12-64
machining, high speed, 12-63
machining process, 11-11
machining processes, nontraditional, 12-64
machining process, non-traditional, 12-64
machining range, high efficiency, 12-63
machining, ultraprecision, 12-67
machining, ultrasonic, 12-67
macro, 9-13, 12-16. *See also* macroinstruction
macroelement, 17-11. *See also* microelement
macroinstruction, 3-10
magazine, 8-20
magnaflux, 13-29
magnetic belt conveyor, 8-4
magnetic controller, 8-20
magnetic disk, 3-10, 12-16
magnetic storage, 3-10
magnetic tape, 12-16
mail order house (retail), 5-8
main aisle, 8-20
main floor, 8-20
mainframe, 3-10, 12-16
main menu, 9-13
maintainability, 1-15, 16-9
maintenance, 1-15, 11-11, 12-16, 17-11
maintenance and repair cost, 4-21
maintenance, breakdown, 13-29
maintenance, corrective, 1-15
maintenance costs, 4-13
maintenance, group, 15-6
maintenance of benefits, 6-33
maintenance-of-membership clauses, 6-33. *See also* union security
maintenance, preventive, 1-15, 13-29
maintenance roles, group, 10-14
maintenance, routine, 13-29
major components, 4-21
major injury, 13-29
major medical expense benefits, 6-33, 13-29. *See also* catastrophe insurance; deductible; health and insurance plan

major milestone, 4-21
major set-up, 14-13. *See also* minor set-up
major system acquisition projects, 4-21
make-or-buy decision, 14-13
make-to-order product, 14-13. *See also* assemble-to-order product
make-to-stock product, 14-13
makeup air, 13-29
make-up pay, 6-33
make-work. *See* featherbedding
malformation, 13-29. *See also* congenital abnormality
malignancy, 13-29
malignant neoplasm. *See* cancer
malleable iron, 12-40
malpractice, 13-29
management, 6-33, 10-18, 15-8. *See also* employer
management, administrative, 10-2
management, amoral, 10-3
management audit, 15-8
management, brand, 5-3
management, bureaucratic, 10-4
management by committee, 15-8
management by exception, 10-19, 15-8
management by objectives (MBO), 6-33, 10-19, 15-8
management by wandering around (MBWA), 10-19
management, career, 10-5
management, configuration, 11-5
management, construction, 4-8
management, consultative, 15-3
management, demand, 14-7
management development, 15-8
management, failure, 13-18
management for quality, 15-8
management, hand of, 10-14
management, hand of, 10-14
management, human resource (HRM), 10-14
management, immoral, 10-15
management information system (MIS), 3-10, 10-19, 12-16, 15-8
management, international, 10-16
management, inventory, 14-11
management, issues, 10-16
management, Japanese, 10-16
management, lower, 15-8. *See also* foreman
management, middle, 15-8
management, moral, 10-20
management of technology (MOT), 10-19
management, operations, 10-22
management, participative, 15-11
management, personal/human resource (P/HRM), 15-11
management, personnel, 15-11
management, planned queue, 14-16
management planning, strategic, 5-15
management prerogatives, 6-33

management, product, 5-11
management, project, 4-27, 5-12, 15-12
management, quality, 16-14
management, schedule variance, 14-19
management science, 10-19
management, scientific, 10-27, 15-12
management, span of, 10-28
management, strategic, 10-29
management, supply chain, 5-15
management, top, 15-8. *See also* chief executive officer (C.E.O.)
management, total quality, 15-14
management, traffic, 5-15
manager, 15-9
manager, acting, 15-13
manager development, 15-8
manager, first line, 10-12
manager, functional, 10-13
manager, general, 10-13
managerial ethics, 10-19
managerial grid, 15-9
managerial integrator, 10-19
managerial style, 15-9
manager, sales, 5-13
manager, top, 10-31
mandatory bargaining subjects, 6-33
man-days of strike idleness, 6-33
mandible, 2-23
mandrel forging, 12-30. *See also* ring rolling
man-environment interface, 2-42
man-equipment task system, 2-42
man-function, 1-15
man-hour, 6-33, 17-11. *See also* man-minute
man-hour output, 6-34
manifest, 8-20
manifold, 12-89
manipulated variable, 12-16
manipulation, 2-42
manipulation, direct, 9-7
manipulative dexterity, 13-29
manipulative skill, 2-42
manipulator, 11-11, 12-29
manlift, 8-20
man-machine chart. *See* multiple activity process chart
man-minute, 6-34, 17-11. *See also* man-hour
mannesmann process, 12-54
manning table, 6-34. *See also* skills inventory
man-on-board, 8-20
manpower, 6-34
man-process chart, 17-11. *See also* process chart
man, standard, 13-43
man-task system, 2-42
manual, 9-13
manual control, 12-16
manual control switch, 13-29
manual data input (MDI), 12-16
manual dexterity, 13-29

This index is arranged in letter-by-letter format.

manual element, 17-11. *See also* element; hand time
manual labor cost, 4-19. *See also* labor cost, non-manual
manually-controlled work, 17-11
manual part programming, 12-16
manual rate, 13-29
manual rescheduling, 14-13. *See also* automatic rescheduling
manual time, 17-11
manual welding, 12-89
manual workers. *See* blue-collar workers
manufacturer's agent, 5-8
manufacturing, 11-11, 12-16
manufacturing automation protocol (MAP), 11-11
manufacturing calendar, 14-13
manufacturing cell, 17-11
manufacturing, cellular, 11-4, 14-5
manufacturing cost, 4-21. *See also* operating costs
manufacturing lead time, 14-13. *See also* lead time
manufacturing, molecular, 12-18
manufacturing order, 14-13
manufacturing planning, 11-11
manufacturing process, 12-17
manufacturing progress function learning curve, 6-32
manufacturing, repetitive, 14-18
manufacturing resource planning (MRP II), 10-19, 14-13. *See also* business plan; closed loop MRP
manufacturing system, 11-11
manufacturing system, discrete, 8-11
manuscript, 12-17
MAPI method, 7-12. *See also* challenger
MAP (manufacturing automation protocol), 11-11
mapping, perceptual, 5-10
marginal analysis, 4-21
marginal cost, 4-21, 5-8, 7-12. *See also* direct cost; incremental cost; traceable cost
marginal cost (benefit), 4-21
marginal productivity theory of wages, 6-34
marginal revenue, 5-8
margin of safety, 13-30
marine accident, 8-20
marine leg, 8-20. *See also* bucket elevator
market, 5-8
market comparison study, 6-34
market control, 10-19
market delineation, 5-8
market growth, 5-8
market hedge. *See* hedge
marketing, 5-8
marketing channels, 5-9
marketing, concentrated, 5-4
marketing concept, 5-9

marketing, cooperative, 5-4
marketing cost accounting, 5-9
marketing cost analysis, 5-9
marketing, differential, 5-5
marketing, direct, 5-5
marketing function, 5-9
marketing information systems, 5-9
marketing management, 5-9
marketing mix, 5-9
marketing model, 5-9
marketing, multilevel, 5-10
marketing planning, 5-9
marketing policy, 5-9
marketing, producers' cooperative, 5-11
marketing research, 5-9
marketing, undifferentiated, 5-15, 5-16
market potential, 5-8
market pricing, 6-34
market rates, 6-34
market research. *See* advertising evaluation
market segmentation, 5-8
market share, 5-8
market testing, 5-8
market, total, 5-8
market, total available, 5-15
market value, 4-21
mark sensing, 3-10, 8-20
mark up, 4-22, 5-10
marstochron, 17-11
martempering, 12-54
masculinity-femininity, 10-19
mask, 3-10, 12-17, 13-30
masking, 9-13
masking (noise), 13-30
mass, center of, 2-32. *See also* center of gravity
mass production, 8-20, 10-18, 11-11, 12-17
mast, 8-20
master agreement, 6-34. *See also* multi-employer bargaining; multiplant bargaining
master clock, 3-10. *See also* clock
mastercraft, 6-34
master pattern, 12-40
master production schedule (MPS), 10-19, 14-13. *See also* demand management; scheduling
master program, 1-15. *See also* decomposition principle
master schedule, 14-13
master schedule item, 14-13
master scheduler, 14-14
master-slave manipulator, 11-11
master switch, crane, 8-20
mast, nontelescoping, 8-22
mast vertical, 8-20
mat, 12-79
match, 12-29
matched edges, 12-29
matched metal molding, 12-79
matching, 13-30

matching draft, 12-29
matching frequency, 13-30
matching individual, 13-30
match lines, 12-29
material control, 11-11
material cost, 4-22
material difference, 4-22
material flow network, 8-20
material handling, 11-11
material handling assignment, 8-20
Material Handling Institute of America (MHIA), 8-21
material handling system selection, 8-20
material, hazardous, 8-16, 13-22
material, indirect, 11-9
material, intermediate, 4-19
material review board (MRB), 14-14
materials handling equipment, 8-20
materials handling time, 14-14. *See also* transit time
materials management, 14-14
materials planning, 11-11
materials requirements planning (MRP), 10-19, 14-14. *See also* requirements explosion
maternity benefits, 6-34
maternity leave, 6-34
mathematical model, 3-10, 11-11
mathematical programming, 1-15
Matheson formula, 7-12. *See also* declining balance depreciation
matrix, 1-15, 3-10, 3-11
matrix, angular, 1-2
matrix array camera, 11-11
matrix, augmented, 1-3
matrix, basic. *See* basis
matrix bill of material, 14-14
matrix, coefficient, 1-5
matrix, constraint, 1-6
matrix element, 1-15
matrix game, 1-15, 1-28
matrix, identity, 1-12
matrix management, 15-9. *See also* project management
matrix, nonsingular, 1-18
matrix printer, 3-11
matrix, rank of, 1-21
matrix, singular, 1-24
matrix structure, 10-19
matrix, technology, 1-26
matrix, transformation, 1-27
matte feeder, 8-20
maturity comparisons, 6-34. *See also* individual wage determination
maturity curve, 6-34
max-flow min-cut theorem, 1-15
maximal network flow problem, 1-15
maximax criterion, 7-12
maximum, absolute, 1-2
maximum body breadth, 2-11
maximum body depth, 2-11

maximum fork height, 8-21
maximum head length, 2-11
maximum lift, 8-21
maximum mouth breadth, 2-11
maximum out-of-pocket cash, 4-22
maximum permissible body burden (MPBB), 13-30
maximum permissible concentration (MPC), 13-30
maximum permissible dose (MPD), 13-30
maximum voluntary ventilation (MVV), 13-30
maximum working area, 17-12
Mayard operation sequence technique (MOST), 17-13
MBO (management by objectives), 6-33, 10-19, 15-8
MBWA (management by wandering around), 10-19
M-day calendar, 14-13
MDCAPT, 12-17
MDI (manual data input), 12-16
mealtime. *See* paid lunch period
mean, 1-15, 16-9
mean, arithmetic weighted, 16-2
mean arrival rate, 1-3
mean deviation, 16-9
mean, geometric, 13-21
mean life, 1-15, 16-10
mean-maintenance-time, 1-15, 16-10
mean range, 16-3
means-end analysis, 15-9. *See also* objective
mean service rate, 1-15, 16-10
mean time. *See* average time
mean time between failures (MTBF), 1-15, 3-11, 9-13, 11-12, 16-10
mean time to failure, 1-15, 16-10
mean time to repair (MTTR), 1-16, 11-12, 16-10
measured day work/daywork, 6-34, 17-12
measured work, 17-12
measurement, 12-17, 13-30
measurement cargo, 8-21
mechanical adhesion, 12-69. *See also* adhesion
mechanical analogue, 2-42
mechanical completion, 4-22
mechanical environment, external, 2-36
mechanical hazards, 13-30
mechanical load brake, 8-21
mechanically foamed plastic, 12-79
mechanical press brake, 12-54
mechanical shear, 12-54
mechanical working, 12-29, 12-54
mechanism, interlocking, 8-17
mechanistic characteristics, 10-19
mechanization, 11-12, 17-12
mechanoreceptor, 2-42
mechanotaxis, 2-42

This index is arranged in letter-by-letter format.

Med/Arb (mediation/arbitration), 6-34
medial, 2-23
medial displacement, 2-42. *See also* lateral displacement
medial plane, 9-13
medial popliteal nerve, 2-42
median, 1-16, 16-10
median lethal dose (LD-50), 13-30
median nerve, 2-42
median time, 17-12
mediation, 6-34. *See also* Federal Mediation and Conciliation Service
mediation/arbitration (Med/Arb), 6-34
mediation, preventive, 6-44
mediation, wage, 6-63
mediator, 6-34. *See also* Federal Mediation and Conciliation Service
medical benefits, 6-34, 13-30. *See also* health and insurance plan; health center
medical monitoring. *See* biological monitoring
medical only (MO), 13-30
medical radiation, 13-30
medical surveillance, 13-30
medicine, industrial. *See* occupational medicine
medium, 10-19
mega-environment, 10-19
melamine formaldehyde, 12-79
melamine plastics, 12-79
melting loss, 12-41
melting point, 12-41
melting rate, 12-89
melting ratio, 12-41
member, 6-35
member-in-good-standing. *See* union member
member, union, 6-60
membrane filter, 13-30
memomotion study, 17-12
memorandum tariff, 8-21
memory, 3-11, 9-13, 12-17
memory, bubble, 12-5
memory, cache, 3-4
memory cycle, 12-17
memory, erasable programmable read-only (EPROM), 3-6
memory, programmable read-only (PROM), 3-13
memory, random-access (RAM), 3-14
memory, read-only (ROM), 3-14
memory, reprogrammable read-only, 3-14
mental model, 9-13
mental work, 17-12
mental workload, 9-13
menton, 2-23
mentor, 10-19
menu, 3-11, 9-13
menu, breadth, 9-13

menu, depth, 9-13
menu navigation, 9-13
menu system, 9-13
merchandise stores, general, 5-7
merchandising, 5-10
merge, 3-11
merger, 10-19
merging feeder, 8-21
merit budget, 6-35
merit increase, 6-35. *See also* wage review
merit pay, 6-35
merit pool, 6-35
merit progression, 6-35
merit rating, 6-35, 13-16, 17-12
merit rating increase, 6-35
merit shop. *See* open shop
merry-go-round conveyor. *See* carousel conveyor
message, 10-19
meta-analysis, 13-30
metabolic cost, 2-42. *See also* light task
metabolic rate, 2-42
metabolic reserves, 9-13
metabolism, 2-42, 9-13. *See also* respiration; thermodynamic efficiency
metabolism, aerobic, 2-29
metabolism, anaerobic, 2-29
metabolism, resting, 2-48
metabolism, work, 2-54
metacarpal, 2-23
metacarpal bones, 2-42
metal, base, 12-86
metal fume fever, 13-30
metalloid, 12-41
metallostatic pressure, 12-41
metallurgy, powder, 12-65
metalmats, 12-17
metal penetration, 12-41
metal, primary, 12-47
metal removal rate (MRR), 12-64
metastable, 12-79
metastatic tumor, 13-31
metering conveyor, 8-21
methacrylonitrile, 12-79
method, 17-12. *See also* standard practice
method of
 adjustment, 9-13
 attributes, 16-10
 constant stimuli, 9-13
 equal sense distances, 9-13
 limits, 9-14
 magnitude estimation, 9-14
 magnitude production, 9-14
 paired comparisons, 9-14
 performance, 4-22
 rank order, 9-14
 ratio production, 9-14
 single stimuli, 9-14
methods analysis, 6-35, 17-12. *See also* process design
methods engineer, 6-35
methods engineering, 6-35, 17-12. *See also* ergonomics; human factors engineering; motion analysis; motion economy; motion study; work design
methods study, 6-35, 17-12. *See also* motion study
methods time measurement (MTM), 6-36, 17-12
me-too clause, 6-35
metrology frame, 12-64
mets, 9-14
mezzanine, 8-21
M function, 12-16
MHIA (The Material Handling Institute of America), 8-21
microbiotaxis, 2-43
microchronometer, 17-12
microcomputer, 3-11, 11-12, 12-17
microcontroller, 12-17
microelement, 17-12. *See also* elemental motion; macroelement
microfiche, 3-11
microhardness, 12-41
micro load AS/RS, 8-21
micromotion film, 2-43
micromotion study, 17-12
microprocessor, 3-11, 11-12, 12-17
microstructure, 12-41
microwave
 bioeffects, 13-31
 cataracts, 13-31
 dosimetry, 13-31
 hearing effect, 13-31
middle managers, 10-20
midpoint, 6-35
midpoint progression, 6-35
midpoint separation, 6-35
mid-range, 16-10
mid-sagittal, 2-23
mid-sagittal plane, 2-43. *See also* coronal plane; transverse plane
migration, 12-79
migratory workers, 6-35
mileage death rate, 13-31
milestone flag, 4-22
milestone level, 4-22
milestone, major, 4-21, 4-22
milestone report, 4-22
milestone schedule, 4-22
military leave, 6-35
milling, 12-64
miners' cramps, 13-23
minicomputer, 12-17
mini-load, 8-21
minimal cost flow problem, 1-16
minimax criterion, 7-12
minimax principle, 1-16
minimax regret criterion, 7-12
minimax theorem, 1-16
minimin criterion, 7-12
minimum, absolute, 1-2
minimum attractive rate of return, 7-12. *See also* required return
minimum cost life. *See* economic life
minimum funding, 6-35

minimum guarantee. *See* guarantee
minimum job rate, 6-35. *See also* job rate
minimum manning, 6-35
minimum order quantity, 14-14. *See also* order quantity modifiers
minimum plant rate, 6-35. *See also* entrance rate
minimum pressure accumulating conveyor, 8-21. *See also* accumulator conveyor
minimum rate, 6-36
minimum separable acuity, 9-14
minimum time, 6-36, 17-13
minimum underclearance, 8-21
minimum wage, 6-36
min-max system, 14-14
minor injury, 13-31
minor set-up time, 14-14. *See also* major set-up
mirror image programming, 12-17
miscellaneous function, 12-18
MIS (management information system), 3-10, 10-19, 12-16, 15-8
misrepresentation, 4-22
misrun, 12-41
mission, 1-16, 10-20
mission statement, 10-20
mission time, 1-26
mists, 13-31
misuse failure, 1-10
mitigation of damages, 4-22
mixed-integer programming, 1-16
mixed strategy, 1-16
mixing conveyor
 paddle type, 8-21
 screw type, 8-21
MNC (multinational corporation), 10-20
mnemonic, 9-14, 12-18
mnemonic symbol, 3-11
mobile rack, 8-21
modal time, 17-13
mode, 1-16, 9-14, 10-20, 16-10
mode error, 9-14
mode failure, 1-10
model, 1-16, 11-12, 15-9
model, dynamic, 11-7
modeling, 10-20
model pricing, 4-22
models, nonrational, 10-21
modem, 3-11, 12-18
modification, 12-41
modification, bilateral, 4-22
modification time, 1-26
modification, unilateral, 4-22
modified union shop. *See* union shop
modular bill of material (BOM), 14-14. *See also* common parts bill of material; option; planning bill of material; super bill of material
modular drawer, 8-21
module, 12-18
modulus of elasticity, 12-54

This index is arranged in letter-by-letter format.

moisture content, 12-41
moisture teller, 12-41
mold, 12-41
moldability, 12-41
mold base, 12-79
mold cavity, 12-41
mold coating, 12-41
molding
    bag, 12-70, 12-79
    blow, 12-79
    ceramic, 12-33
    cold, 12-72
    compression, 12-72, 12-79
    contact-pressure, 12-79
    flaskless, 12-38
    foam, 12-75
    high pressure, 12-39, 12-79
    injection, 12-77, 12-79
    injection blow, 12-77
    jolt squeeze, 12-40
    low-pressure, 12-79
    machine, 12-41
    matched metal, 12-79
    rotational, 12-82
    sands, 12-41
    sandwich, 12-82. *see also* foam molding
    shell, 12-45
    transfer, 12-79, 12-84
mold insert (removable), 12-79
mold, multiple, 12-41
mold(s)
    dry sand, 12-37
    family, 12-74
    hot-runner, 12-77
    injection, 12-77
    jacket, 12-40
    oil, 12-42
    permanent, 12-43
    plaster, 12-43
molecular manufacturing, 12-18
molecular nanotechnology, 12-18
molecular weight, 13-31
MO (medical only), 13-30
moment concept, 2-43
moment, dynamic, 2-34
moment of force, 2-43. *See also* torque
moment of inertia, 2-43
money, dirty, 6-15
money-purchase plan. *See* pension plan
money wages, 6-36
monitor, 3-11, 9-14
monitoring, 4-22, 13-31
monitoring, biological, 13-7
monitoring methods, 10-20
monitoring, radiation, 13-31
monitorship, 6-36. *See also* trusteeship
monomer, 12-79
monorail, 8-21
monotone, 1-16
monotonic increasing function, 1-17
Monte Carlo method, 1-17
monthly guide schedule, 4-22

monthly labor review, 6-36
month-to-month price index, 4-22
moonlighting, 6-36. *See also* sunlighting
morale, 15-9
moral management, 10-20
morbidity, 13-31
morbidity rate, 13-31
morphological analysis, 10-20
mortality, 13-31
mortality rate, 13-31
mortality rate, infant, 13-25
most likely time estimate, 4-22
MOST (Maynard operation sequence technique), 17-13
motion analysis, 17-13. *See also* ergonomics; human factors engineering; methods engineering; motion study; principles of motion economy
motion and time study, 9-14
motion, apparent, 9-3
motion, basic, 17-3
motion chart, simultaneous, 17-18
motion cycle, 17-13. *See also* cycle; work cycle
motion diagram, 12-54
motion economy. *See also* ergonomics; human factors engineering; methods engineering; motion analysis
motion economy, principles of, 17-15
motion parallax, 9-14
motion pattern, balanced, 17-2
motions, combined, 17-4
motions inventory. *See* available motions inventory; demand motions inventory
motions pathway, 2-43. *See also* predetermined motion time system
motions, simultaneous, 17-18
motion study, 6-36. *See also* methods engineering; methods study; motion analysis
motion time system, predetermined, 2-45
motivated productivity level (MPL), 17-13. *See also* acceptable productivity level (APL); high task; incentive pace; normal performance
motivation, 10-20, 15-9
motivational research, 5-10
motivators, 10-20
motive, 15-9
motive-hygiene theory, 15-9
MOT (management of technology), 10-19
motor brake, 8-21
motor, hydraulic, 11-9
motor, induction, 11-9
motorized hand/rider truck, 8-21
motorized hand truck, 8-21
motor nerve, 2-43. *See also* synapse

motor skill, 9-14
motor, stepping, 11-17, 12-23
motor vehicle accident, 8-21
motor vehicle collision, 13-31
motor vehicle nontraffic accident, 13-31
mottled cast iron, 12-41
mountain sickness, acute, 13-3
mouse, 3-11, 9-14
mouth breadth, maximum, 2-11
movable barrier guard, 13-7
movable platen, 12-79
move, 3-11
move ticket, 14-14
move time, 11-12
moving average, 1-17, 4-22, 16-10
moving average control chart, 16-10
moving beam bar code reader, 8-21
moving range control chart, 16-10
MPBB (maximum permissible body burden), 13-30
MPC (maximum permissible concentration), 13-30
MPD (maximum permissible dose), 13-30
MPF (learning curve, manufacturing progress function), 6-32
MPL (motivated productivity level), 17-13. *See also* acceptable productivity level (APL); high task; incentive pace; normal performance
MPS (master production schedule), 10-19, 14-13. *See also* closed loop MRP
MRB (material review board), 14-14
MRP (materials requirements planning), 10-19, 14-14. *See also* closed loop MRP
MRP II (manufacturing resource planning), 10-19, 14-13. *See also* business plan; closed loop MRP
MRR (metal removal rate), 12-64
MTBF (mean time between failures), 1-15, 3-11, 9-13, 11-12, 16-10
MTM (methods time measurement), 6-36, 17-12
MTTR (mean time to repair), 1-6, 11-12, 16-10
muller, 12-41
mulling, 12-41
multicommodity network problem, 1-17
multicriteria optimization, 1-17
multidimensional scaling, 5-10
multi-employer bargaining, 6-36. *See also* association agreement; master agreement
multifocal strategy, 10-20
multi-level bill of material (BOM), 14-14
multilevel marketing, 5-10

multimedia, 9-14
multinational corporation (MNC), 10-20
multiperson game, 1-17
multiplant bargaining, 6-36. *See also* master agreement
multiple activity operation chart, 17-13
multiple activity process chart, 17-13. *See also* process chart
multiple-cavity die, 12-41
multiple control systems, 10-20
multiple cord belt conveyor, 8-4
multiple finish network, 4-23
multiple flight conveyor screw, 8-21
multiple font, 8-21
multiple mold, 12-41
multiple piece rate plan, 6-36. *See also* differential piecework
multiple rates of return, 7-12
multiple regression, 1-17
multiple ribbon belt conveyor, 8-4
multiple roots, 7-12
multiple sampling inspection, 16-10
multiple screw feeder, 8-22
multiple start network, 4-23
multiple straight-line depreciation method, 4-23, 7-9
multiple strand conveyor, 8-22
multiple time plan, 6-36. *See also* differential time plan
multiple watch timing. *See* accumulative timing
multiplexer, 3-11
multiplexing, 11-12
multiprocessing, 3-11
multiprogramming, 3-11
multi-row layout, 8-22
multitasking, 3-11, 9-14
multivariate accommodation, 2-5, 2-6
multivariate analysis, 2-5
multivariate control chart, 16-10
multivariate limit determination, 2-6
multivariate limit verification, 2-6
muscle, brachialis, 2-32
muscle, erector spinae, 2-35
muscle, extensor, 2-35. *See also* extensor tendon
muscle, flexor, 2-36
muscle, intrinsic, 2-39
muscle, sacrospinalis, 2-48. *See also* erector spinae muscles
muscle, sternocleidomastoid, 2-50
muscle, trapezious, 2-52
musculoskeletal injury, 13-32
musculoskeletal system, 2-43
mutagen, 13-32. *See also* somatic mutation
mutation, genetic, 13-21
mutually exclusive alternative, 7-6
MVV (maximum voluntary ventilation), 13-30
myogram, integrated surface, 2-39

This index is arranged in letter-by-letter format.

myography. See electromyography
myopia, empty field, 9-8
myopia, space, 9-21

N

NACH (need for achievement), 10-20
NAFF (need for affiliation), 10-20
nanotechnology, molecular, 12-18
NANS (National Automated Nesting System), 12-18
narcosis, inert gas, 13-25
narcosis, nitrogen gas, 13-25
narrow aisle truck, 8-22
nasal breadth, 2-11
nasal height, 2-12
nasal septum, 2-23
Nash-Harsonyi bargaining model, 1-17
National Automated Nesting System (NANS), 12-18
national brand, 5-10
national emergency dispute, 6-36. See also cooling-off period
National Fire Protection Association (NFPA), 13-32
National Institute of Occupational Safety and Health (NIOSH), 6-37, 13-32
National Labor Relations Act, 6-37. See also Labor Management Relations Act
National Labor Relations Board (NLRB), 6-37. See also Labor Management Relations Act
National Mediation Board, 6-37
National Railroad Adjustment Board, 6-37
national responsiveness strategy, 10-20
National Safety Council (NSC), 13-32
National Union, 6-37
natural aging, 12-54
natural draft, 12-29
natural frequency, 12-18
natural frequency vibrating conveyor, 8-22
natural language, 3-11. See also language
natural process limits, 16-10
natural sand, 12-41
natural selection model. See population ecology model
nature of injury, 13-32
nature of work, 13-32
navicular bone, 2-43
NC (noise criterion) curves, 9-15
NC (numerical control), 3-12, 11-12, 12-18, 12-64, 17-14
near accident, 13-32
near drowning, 13-32
near miss, 13-32

near-optimum solution, 1-17
necking down, 12-54
necking failure, 12-54
necrosis, aseptic bone, 13-15
need for achievement (NACH), 10-20
need for affiliation (NAFF), 10-20
need for power (NPOW), 10-20
need hierarchy, 15-9
needs analysis, 10-20
negative definite quadratic form, 1-17
negative entropy, 10-20
negative reinforcement, 10-20
negative synergy, 10-20
negative transfer, 9-14
negative work, 2-43. See also isometric work; positive work
negligence, 4-23, 13-32
negligence, actionable, 13-32
negligence, comparative, 13-32
negligence, contributory, 13-32
negligence, degrees of, 13-32
negligence, intervening, 13-26
negotiating contracts, 10-20
negotiation. See collective bargaining
neoplasm, 13-32
nerve, medial popliteal, 2-42
nerve, median, 2-42
nerve, motor, 2-43. See also synapse
nerve, phrenic, 2-45
nerve, radial, 2-46
nerve, sensory, 9-20. See also synapse
nerve, spinal, 2-49
nerve, ulnar, 2-53. See also hamate bone
nervous syndrome, high pressure, 13-23
nesting target, portable metal stacking rack, 8-22
net area, 4-23
net benefits (savings), 4-23
net change MRP, 14-14. See also regeneration MRP; requirements alteration
net load capacity, 11-12
net profit, 4-23, 4-26
net profit, percent of sales, 4-23
net purchases (concept of), 4-23
net requirements, 14-14
netting, 14-14
net weight, 8-22
network, 3-11, 4-23, 10-20, 12-18, 15-9
network analysis, 4-23, 5-10. See also critical path; critical path scheduling
network diagram, 10-21
network diagram, standard, 4-31
network event, beginning, 4-4
network flow problem, 1-15
network model, 3-11
network, multiple finish, 4-23
network, multiple start, 4-23

network planning, 4-23, 5-10
network, summary, 4-33
neuromuscular, 9-15
neuropathy, 2-43
neuropathy, peripheral, 13-32
neutral flame, 12-89
neutralizers, 10-21
neutral zone, 12-18
newly industrialized countries (NICs), 10-21
Newtonian liquid, 12-79
new venture, 10-21
new venture teams, 10-21
new venture units, 10-21
NFPA (National Fire Protection Association), 13-32
NGS (numerical geometry system), 12-18
NGT (nominal group technique), 10-21, 15-9
nickel-iron storage battery, 8-22
nickel itch, 13-32
NICs (newly industrialized countries), 10-21
night shift. See shift
night vision, 9-15
Ni-hard, 12-41
NIOSH (National Institute of Occupational Safety and Health), 6-37, 13-32
nip point, 13-33
nitrogen die cylinders, 12-54
nitrogen gas narcosis, 13-25
nitrogen narcosis. See inert gas narcosis
NLRB (National Labor Relations Board), 6-37. See also Labor Management Relations Act
no bake binder, 12-42
node, 1-17, 3-11, 4-23. See also event
node, interface, 4-19
node, intermediate, 4-19
node of a network, beginning, 4-4
node of network, ending (ADM), 4-14
no-draft forging, 12-29
nodular iron, 12-42
noise, 9-15, 10-21, 11-12, 12-18, 13-33
noise, airborne, 13-4
noise, ambient, 13-4
noise, background, 13-6
noise control, 13-33
noise criterion (NC) curves, 9-15
noise-induced hearing loss, 13-33
noise level, 13-33
noise, nonauditory effects of, 13-33
noise, power level, 13-36
noise, random, 13-38
noise reduction, 13-33
nominal group technique (NGT), 10-21, 15-9
nominal interest. See interest rate-nominal
nominal size, 12-64

nominal value, 16-11
nominal wages. See money wages
non-acceptance number, 16-14
non-additive percentiles, 2-5, 2-6
nonauditory effects of noise, 13-33
nonballistic motions. See ballistic movement
nonbasic variable, 1-17
noncash, 4-23
noncausal association, 13-33
noncombustible, 13-33
nonconforming item, 16-11
nonconforming unit, 16-11
nonconformity, 1-17, 16-11
non-contributory benefits plan, 6-37
non-contributory pension plan, 6-37. See also contributory pension plan
noncrisis problem, 10-21
noncybernetic control system, 10-21
noncyclic element, 17-13
nondegeneracy assumption, 1-17
nondegenerate feasible solution, 1-17
nondestructive inspection, 12-42
nondestructive testing, 12-42, 13-33
nondisabling injury, 13-33
nondurable goods, 4-23
nonegativity constraint, 1-17
nonexempt, 4-23
non-exempt employees, 6-37
nonfatal injury accident, 13-33
non-financial incentive (plan), 6-37
non-financial rewards, 6-37
nonflammable, 13-33
nonimpact printer, 3-11
noninjury accident, 13-33
nonlinear constraint, 1-17
nonlinear equation, 1-17
nonlinear function, 1-17
nonlinear programming, 1-18, 3-12
non-metric scaling, 5-10
non-parametric statistics, 5-10
non-performance pay, 6-37
nonperishables, 8-22
non-positive tactile stimuli, 2-43
nonprejudicial delay, 4-12
non-production bonus, 6-37. See also bonus plan; production bonus
non-productive labor, 6-37
nonprogrammed decisions, 10-21
nonqualified pension plans, 6-37
non-quantitative job evaluation, 6-37
nonrational escalation, 10-21
nonrational models, 10-21
non-read, 8-22
nonrepaired unit, 1-27
nonrepetitive, 17-13
nonrigid plastic, 12-79

This index is arranged in letter-by-letter format.

nonrigid tie, 8-22
nonsingular matrix, 1-18
nontelescoping mast, 8-22
nontraditional machining processes, 12-64
nonverbal communication, 10-21
nonwork unit, 4-23
no-read, 8-22
normal and proper usage, 8-22
normal distribution, 16-11
normal effort, 17-13
normal element time, 17-13
normalize, 6-37. *See also* effort rating; leveling
normalizing, 12-54
normally closed, 12-18
normally open, 12-18
normal operation, 6-37. *See also* work design and measurement
normal pace, 17-13
normal performance, 6-38, 17-13. *See also* APL (acceptable productivity level); did-take standards; fair day's work; high task; MPL (motivated productivity level); should-take time standard; should-take, did-take time standards; standard time
normal retirement. *See* retirement
normal retirement age, 6-38
normal task, 17-13
normal time, 17-13
normal working area, 17-14
normal work rate. *See* normal performance
normative decision-making models, 10-21
norming, 10-21
norms, 10-21, 15-9, 17-14
northwest corner rule, 1-18
nose bridge height, 2-12
nose radius of the cutting edge of the tool, 12-64
nosing, 12-54
no-strike, no-lockout clause. *See* normal performance
notched stringer, 8-22
notching, 12-55
notch, semilunar, 2-48
not-for-profit organization, 10-21
notice, advance, 6-4
notice of award, 4-23
notice, pay-in-lieu-of, 6-40
notice to proceed, 4-23
not otherwise classified, 13-33
novolak, 12-79. *See also* resinoid; thermoplastics
nozzle, 12-42, 12-80
n-person game, 1-17
NPOW (need for power), 10-20
NSC (National Safety Council), 13-32
NT-hard problems, 1-18
NUFORMS, 12-18
nugget, 12-89
nugget size, 12-89
nuisance dust, 13-33

null hypothesis, 16-11
numbering system, 17-14
number, non-acceptance, 16-14
number of nonconforming items chart, 16-11
number, summary, 4-33
numerical, 3-12
numerical control (NC), 3-12, 11-12, 12-18, 12-64, 17-14
numerical display, 9-15
numerical geometry system (NGS), 12-18
nylon, 12-80
nylon plastics, 12-80
nystagmogram, 2-43
nystagmus, 9-15

OADSI (old age, survivors and disability insurance benefits), 6-38, 13-34
OALH (overall lowered height), 8-23
OAS (office automation system), 10-21
object code, 3-12
objective, 15-9. *See also* means-end analysis
objective event, 4-23
objective function, 1-18, 10-21
objective function element, 1-18
objective, product, 5-12
objective rating, 17-14. *See also* performance rating
objective value, 1-18
object language, 10-21
oblique cutting, 12-64
OB (organizational behavior), 15-10
observation, 17-14
observation chart, individual, 16-9
observation, direct, 6-15
observed failure rate, 1-10
observed time. *See* actual time
observed value, 1-18, 16-11
observer, standard, 9-21
obsolescence, 4-23, 7-12
obsolescence, planned, 5-11
obstacle, 8-22
occiput, 2-23
occupancy, 13-33
occupancy, beneficial, 4-4. *See* substantial completion. *See also* substantial completion
occupational disease, 13-33
occupational ecology, 2-43
occupational illness, 13-33
occupational injury, 13-33
occupational injury and illness records (OSHA), 13-33
occupational injury or illness, reportable (OSHA), 13-33
occupational medicine, 2-44, 13-34

occupational radiation, 13-34
occupational safety, 13-34
Occupational Safety and Health Administration (OSHA), 13-34
occupational safety and health codes and standards, 13-34
occupational skin diseases, 13-34
occupational skin disorders, 13-34
occurrence, 13-34
occurrence (frequency), 17-14
OC curve, 16-11
OCR (optical character recognition), 3-12, 11-12
octal, 12-18
octave band, 13-34
oculogram, 2-44
odds ratio, 13-34
OD (organizational development), 6-38, 10-22, 15-10
offhand grinding, 12-64
office, 8-22
office automation system (OAS), 10-21
office layout, 8-22. *See also* layout
office of chairman, 15-9
office of president, 15-9
off-line/offline, 3-12, 11-12
offset, 12-18
offset, cutter, 12-9
offsites, 4-23. *See also* battery limit
off-the-job safety, 13-34
oil acne, 13-34
oil core, 12-42
oil, linseed, 12-40
oil mold, 12-42
oil-oxygen binder, 12-42
old age, survivors and disability insurance benefits (OADSI), 6-38, 13-34
olecranon process, 2-44
oligopoly, 5-10
olivine sand, 12-42
ombudsperson, 10-21
omission, 4-24
oncology, 13-34
one best way, 17-14
one-sided confidence interval, 16-11
one-sided test, 16-11
one-way communication, 10-21
on hand, 14-14
online, 3-12, 11-12, 12-19
on-line processing, 10-21
on order, 14-15
on-stream factor, 4-24
open-cell foam, 12-80
open-circuit voltage, 12-89
open dating, 5-10
open-die forging, 12-27, 12-29
open dock, 8-22
open-end agreement, 6-38
open hearth furnace, 12-42
open loop, 9-15, 12-19
open-loop control, 11-12
open-loop system, 12-19

open order, 14-15
open pay system, 6-38
open shop, 3-12, 4-24, 6-38. *See also* union security
open system, 10-22
open union, 6-38. *See also* closed union
operable, 1-18
operating budget, 10-22
operating capital, 4-6
operating characteristic curve, 16-11
operating costs, 4-13, 4-24. *See also* manufacturing cost
operating profit, 4-26
operating supplies costs, 4-13
operating system, 3-12, 12-19
operation, 6-38, 11-12, 17-14
operational, 1-18
operational control, 10-22
operational goals, 10-22
operational plans, 10-22
operation analysis chart, 6-38
operation breakdown. *See* job breakdown
operation chart. *See* right- and left-hand chart
operation, continuous, 6-13, 12-50. *See also* continuous process; round-the-clock operations
operation, hazardous, 8-16
operation, normal, 6-37. *See also* work design and measurement
operation process chart, 17-14. *See also* process chart
operation reporting, 14-15
operations analysis, 17-14
operation sheet, 14-15
operations management, 10-22
operations research, 1-18, 9-15, 10-22
operation start date, 14-15
operation (symbol), 17-7
operation time chart, 17-14
operator, 9-15
operator inputs, 9-15
operator outputs, 9-15
operator overload, 9-15
operator process chart, 17-14. *See also* process chart; right- and left-hand chart
operator productivity, 17-14. *See also* performance index
operator, qualified, 6-46, 17-17
opinion leader, 5-10, 6-38
opportunity cost, 1-18, 7-12
opportunity problem, 10-22
optical character reader, 8-22
optical character recognition (OCR), 3-12, 11-12
optical pyrometer, 12-42
optical scanner, 11-12
optima, alternate, 1-2
optimal control, 11-12
optimistic time estimate, 4-24
optimization, 11-12

This index is arranged in letter-by-letter format.

optimize, 1-18, 12-19
optimum cutting speed, 12-64
optimum-location principle, equipment design, 9-15
optimum plant size, 4-24
optimum solution, 1-18
option, 14-15. *See also* accessory; attachment; modular bill of material; planning bill of material
optional feature, 8-22
orange peel, 12-80
orchestrator, 10-22
order, 14-15
order batching, 8-22
order bill of lading, 8-22. *See also* bill of lading
order bill of material (BOM), 14-15
order, blanket, 14-4. *See also* release
order collation, 8-22
order entry, 14-15
ordering cost, 10-22, 14-15. *See also* carrying cost
order lead time, 5-10
order multiple, 14-15
order of magnitude estimate, 4-15
order picker truck, high lift, 8-22
order picking, 8-23
order picking and stacking truck, 8-23
order picking sequence, 8-23
order picking truck, 8-23
order, planned, 14-16. *See also* scheduled receipts
order point, 14-15. *See also* time-phased order point (TPOP)
order policy. *See* lot size inventory
order promising, 14-15
order quantity, discrete. *See* lot-for-lot
order quantity, minimum, 14-14
order quantity modifiers, 14-15
order-statistic(s), 1-18, 16-11
organic characteristics, 10-22
organic disease, 13-34
organization, 10-22, 15-10
organizational behavior (OB), 15-10
organizational chart, 10-22, 15-10
organizational climate, 15-10
organizational codes, 4-24
organizational cultural change, 10-22
organizational culture, 10-22, 15-10
organizational design, 10-22, 15-10
organizational development (OD), 6-38, 10-22, 15-10. *See also* change agent
organizational picketing, 6-43
organizational problems, 10-22
organizational social responsibility, 10-22
organizational social responsiveness, 10-22

organizational stress, 6-38
organizational structure, 10-22
organizational termination, 10-22
organizational theory, classical, 15-3
organization culture, 15-10
organization, formal, 15-10. *See also* informal group
organization for quality, 15-10
organization, functional, 15-5
organization, geocentric, 10-13
organization, informal, 15-10
organization, not-for-profit, 10-21
organization planning, 15-10
organization structure, 15-10
organization theory, 15-11
organizer, 6-38
organizing, 10-22, 15-11
organosol, 12-80. *See also* dispersion
orientation programs, 6-39
orifice plate, 12-42
original data chart, 16-9
original duration, 4-24
original package, 8-23
origin (muscle), 2-44. *See also* insertion
orthoaxis, 2-44
orthocentre, 2-44
orthogonal cutting, 12-64
orthosis, 2-44. *See also* prosthesis
oscillating conveyor, 8-23. *See also* vibrating conveyor
oscillating feeder. *See* conveyor type feeder
oscillograph, 2-44
OSH Act, 13-34
OSHA (Occupational Safety and Health Administration), 12-19, 13-34
ossification, 10-22
osteoarthritis, 13-34
otitis externa, 13-34
otosclerosis, 13-34
outlaw strike, 6-65. *See also* strike; wildcat strike
outlet selling, exclusive, 5-6
outlier(s), 16-12, 17-15. *See also* abnormal time
out-of-line rate, 6-39. *See also* red circle rate
out-of-line request, 6-39
out-of-pocket cash, maximum, 4-22
out-of-round, 12-64
out-of-work benefits, 6-39
output device, 9-15
output per man-hour, 6-39
output(s), 3-12, 10-23, 11-12, 12-19, 14-15, 17-15
output, standard, 17-18
outrigger, 8-23
outside shop, 14-15
oven, dielectric, 12-36
overaging, 12-55
overall depth, 8-23
overall height, 8-23

overall lowered height (OALH), 8-23
overall vehicle height, 8-23
overall width, 8-23
over-and-under conveyor, 8-23
overbending, 12-55
overconfidence, 10-23
overcontrol, 10-23
overflow, 3-12
overflow groove, 12-80
over-haul, 4-24
overhead, 4-24
overhead conveyor. *See* power-and-free conveyor; trolley conveyor
overhead crane, 8-23. *See also* traveling cranes
overhead, general, 4-17
overhead position, 12-89
overhead trolley conveyor. *See* trolley conveyor
overheated, 12-55
overlap, 12-89
overlapped schedule, 14-15
overlay, 3-12
overload, 12-19
overload principle, 2-44
overload relief device, 12-55
overrun (underrun), 4-24
overrun (underrun), projected, 4-27
overshoot, 11-12, 12-19
overtime, 6-39. *See also* straight time
overtime pay, 6-39. *See also* premium rate of pay
overtime premium pay, 6-39. *See also* premium rate of pay
overt lifting task, 2-44
over(under) plan, 4-24
owner, 4-24
oxides, 12-64
oxidizing flame, 12-89
oxyacetylene welding, 12-89
oxyfuel gas cutting, 12-64
oxygen consumption, 9-15
oxygen debt, 2-44
oxygen deficiency, 13-34
oxygen lance, 12-89
oxygen toxicity, 13-35

P

P&D stations (pick-up and delivery), 8-26
Paasche-type price index, 4-24
pace, 6-39
pace, normal, 17-13. *See also* normal performance
pace rating, 6-39, 17-15. *See also* performance rating
pace setter, 6-39
package, intermediate, 8-18
package, original, 8-23

package settlement, 6-39
packaging, 8-23
packet switching, 3-12
pact. *See* agreement
pad, 12-55
padding, 3-12
paddle conveyor. *See* mixing conveyor, paddle type
paddle conveyor screw, 8-23
paddle washer, 8-23
padd mixer. *See* mixing conveyor, paddle type
page printer, 3-12
paging, 3-12
paid absence allowance, 6-39, 13-35
paid education leave (PEL), 6-41
paid holidays, 6-39. *See also* holiday pay; unpaid holidays
paid lunch period, 6-39
paid vacations, 6-39. *See also* extended vacation plan
pain, referred, 2-47
pallet, 8-23, 11-12
pallet, all-way entry, 8-23
pallet, captive, 8-23
pallet dispenser, 8-23
pallet, double deck, 8-23
pallet, double entry. *See* pallet, two-way entry
pallet, double faced. *See* pallet, double deck
pallet, double wing, 8-24
pallet, euro, 8-24
pallet, expendable container, 8-24
pallet, flush, 8-24
pallet frame. *See* pallet storage shelf
pallet, full four-way entry, 8-24
pallet height, 8-24
palletizer, 8-24
pallet jack, 8-24
pallet length, 8-24
palletless loader, 8-24
pallet load detierer, 8-24
pallet loader, 8-24
pallet load tierer, 8-24
pallet, noncaptive, 8-24
pallet, nonreversible, 8-24
pallet, partial four-way entry, 8-24
pallet, reusable, 8-24
pallet, reversible, 8-24
pallet, single deck, 8-24
pallet, single faced, 8-24
pallet, single wing, 8-24
pallet size, 8-24
pallet, slave, 8-32
pallet stacker. *See* empty pallet stacker
pallet, stacking, 8-24
pallet storage rack, 8-24
pallet storage shelf, 8-25
pallet, take-it-or-leave-it, 8-25
pallet-to-post clearance, 8-25
pallet truck, 8-25
pallet truck openings, 8-25
pallet, two-way entry, 8-25

This index is arranged in letter-by-letter format.

pallet type conveyor, 8-25. *See also* floor conveyor
pallet unloader, 8-25
pallet width, 8-25
pallet, wing, 8-25
palmar arch, 2-44. *See also* hamate bone
palpate, 2-23
pan, 11-12
pan conveyor, 8-25
pan feeder. *See* conveyor type feeder
panic, 10-23
panoramic analyzer, 12-42
paperless purchasing, 14-15. *See also* just-in-time (JIT); vendor scheduler
paper locals, 6-39
paper tape, 12-19
parabolic interpolation, 12-19
parachute, golden, 6-22
parachute, pension, 6-41
parachute, tin, 6-59
parachute, worker, 6-65
paralanguage, 10-23
parallax, 9-15. *See also* binocular parallax
parallax, binocular, 9-4. *See also* parallax
parallel computer, 3-12
parallel processing, 3-12
parallels, 12-80
parameter, 1-18, 3-12, 16-12
parameters, definitions and symbols, 7-2
parametric programming, 1-18
parcel post air freight, 8-25
Pareto curve, 1-18, 16-12
Pareto optimality (non-dominated solution), 1-19
Pareto's law, 14-15, 17-15. *See also* ABC classification
parison, 12-80
parity, 12-19
parity bit, 3-12
parity check, 12-19
parking brake, 8-25
part, 11-12, 12-80
part classification, 11-13
part family, 11-13
part handling system (PHS), 11-13
partial-factor productivity, 10-23
partial failure, 1-10
partial utilization, 4-24
participation, 15-11
participation point, 6-39
participative, 10-23
participative management, 15-11
particulate, 13-35
particulotaxis, 2-44
parting agent, 12-43
parting compound, 12-42
parting line, 12-29
parting plane, 12-29
partition, 1-19
part number, 14-15

part orientation, 11-13
part-period, 14-15
part program, 11-13, 12-19
part-revolution clutch, 12-55
parts, common, 11-4
parts explosion, 12-19
parts manufacturing, discrete, 11-6
part task simulation, 9-15
part-time employee, 6-40
part time, permanent, 6-42
part-time worker rate, 6-40
pass, 12-89
passenger conveyor, 8-25
passive accommodation, 11-13
password, 3-13
past practice, 6-40
past service, 6-40
patch test, 13-35
patent defect, 4-11
path, 4-24, 11-13
path control, continuous, 11-5
path float. *See* float
path-goal theory, 10-23
pathocumulus, 2-44
pathological process, 2-44
path operation, continuous, 12-8
pattern, 12-42
pattern bargaining, 6-40
pattern, expendable, 12-37
pattern, investment molding, 12-42
patternmaker's shrinkage, 12-42
pattern, master, 12-40, 12-42
pattern recognition, 3-13, 9-15, 11-13
pattern, standard, 12-46
paulin, 8-25
PAU (position analog unit), 12-20
pay, ability to, 6-3
pay adjustment, 6-40
pay-as-you-go. *See* unfunded plan
pay, back, 6-7
payback method. *See also* simple payback period
payback (payoff) period. *See* payout time
payback period, 7-12
payback period, discounted, 7-12
pay, base, 6-7
pay, bereavement. *See* funeral leave pay
pay, callback, 6-9
pay, call-in, 6-9. *See also* reporting pay
pay compression. *See* compression
pay curve. *See* wage curve
pay, deadheading, 6-14
pay, disability, 6-15
pay, double back, 6-16
pay, dual, 6-16
pay, fair day's, 6-20
pay, funeral leave, 6-22
pay grade, 6-40
pay grade overlap, 6-40
pay grade width. *See* salary ranges

pay, hazard, 6-24, 13-22
pay, holdback, 6-24
pay, holiday, 6-24. *See also* paid holidays
pay, incentive, 6-26
pay-in-lieu-of notice, 6-40. *See also* advance notice
pay, jury-duty, 6-29
pay, layoff, 6-31
payload, 11-13
pay, longevity, 6-32
pay, make-up, 6-33
payment bond, 4-5
payment by result, 6-40
payment certain guarantee. *See* period certain option
payment, group, 6-23
payments, balance of, 10-3
payments, questionable, 10-25
payments, terms of, 4-34
pay, merit, 6-35
pay, non-performance, 6-37
payoff, 1-19, 10-23
payoff function, 1-19
payoff matrix, 1-19
payoff (payback) period. *See* payout time
payoff period. *See* payback period
payoff table, 10-23. *See also* payback period
payout period. *See* payback period
payout time, 4-24
pay, overtime, 6-39
pay, overtime premium, 6-39
pay plan, 6-40
pay, premium rate of, 6-44
pay, promotional, 6-45
pay range. *See* salary ranges
pay, reporting, 6-48
pay, retroactive, 6-48
payroll burden, 4-24
payroll (clerical) costs, 4-13
payroll deduction, 6-40
payroll period, 6-40
payroll tax, 6-40
pay, severance, 6-51
pay, skill-based, 10-28
pay, status, 6-54
pay steps, 6-40
pay structure, 6-40
pay survey, 6-40, 10-23
pay, take-home, 6-57. *See also* spendable earnings
pay tend line, 6-40
pay, total, 6-59
pay, transfer, 6-59
pay, unemployment, 6-60
pay, vacation, 6-61
PBGC (pension benefit guarantee corporation), 6-41
PCI (process capability index), 16-12
PC (programmable controller), 11-14, 12-20
PDM arrow, 4-24
PDM finish to finish relationship, 4-24, 4-25

PDM finish to start relationship, 4-25
PDM (precedence diagram method), 4-25
PDM start to finish relationship, 4-25
PDM start to start relationship, 4-25
PD (property damage) accident, 13-37
pearlite, 12-55
pearlitic malleable, 12-42
pedestal robot, 11-13
pedestrian, 13-35
peen, 12-42
pegging, 14-15
peg point, 6-41
PEL (paid education leave), 6-41
PEL (permissible exposure limit), 13-35
penalty rate, 6-41. *See also* premium rate of pay
pencil core, 12-42
pencil gate, 12-42
penetration, metal, 12-42
pension, 6-41
pension administrator, 6-41
pension benefit guarantee corporation (PBGC), 6-41
pension liability, 6-41
pension parachute, 6-41
pension plan, 6-41. *See also* retirement
pension plan, contributory, 6-13
pension plan, defined benefit, 6-15
pension plan, non-contributory, 6-37
pension plans, nonqualified, 6-37. *See also* annuity; group annuity plan
pension trust fund, 6-41
penthouse, 8-25
per capita tax, 6-41
percentage increase, 6-41
percent chart, 16-12
percent complete, 4-25
percent grade, 8-25
percentiles, non-additive, 2-5, 2-6
percentile threshold, 9-15
percent of fill, 14-15. *See also* stockout percentage
percent on diminishing value. *See* declining balance depreciation
perception, 5-10, 9-15, 10-23, 13-35
perception deafness, 9-15. *See also* conductive hearing loss
perceptual augmentation, 9-15
perceptual defense, 10-23
perceptual mapping, 5-10
perceptual-motor task, 9-16
perceptual overload, 9-15
perceptual skill, 9-16
percussion welding, 12-89
perforating, 12-55. *See also* piercing

This index is arranged in letter-by-letter format.

performance, 6-41, 17-15
performance appraisal, 6-41, 10-23
performance bond, 4-25
performance decrement, 9-16
performance efficiency, 6-41
performance evaluation, 17-15
performance, incentive, 6-26
performance index, 17-15. *See also* operator productivity
performance indicator, 17-15
performance measurement, 17-15
performance measurement baseline, 4-25
performance measures, 9-16
performance, normal, 6-38, 17-13. *See also* APL (acceptable productivity level); did-take standards; fair day's work; high task; MPL (motivated productivity level); normal effort; normal pace; normal time; should-take time standard; should-take, did-take time standards; standard time
performance-outcome expectancy, 10-23
performance rating, 6-41, 17-15. *See also* effort rating; objective rating; pace rating; rate; rating
performance rating factor, 6-41, 17-15
performance rating scale, 6-41, 17-15
performance ratio, 17-15
performance sampling, 17-15
performance specification, 4-32
performance standard, 17-15. *See also* standard performance
performance, standard, 6-54, 17-18
performance standards, 6-41
performing, 10-23
period, base, 4-4, 17-3
period certain option, 6-42
periodic inventory system, 14-16
period order quantity, 14-16. *See also* fixed order quantity; lot-for-lot
peripheral equipment, 11-13, 12-19
peripheral hemodynamics, 2-44
peripheral neuropathy, 13-32
peripheral unit, 3-13
peripheral vision, 9-16, 9-26
perk (perquisite), 6-42
perlite, 12-42
permanence, 12-80
permanent and total disability, 6-42. *See also* disability; disability retirement
permanent arbitrator, 6-42
permanent disability, 13-35. *See also* disability
permanent impairment. *See* permanent disability; permanent partial disability; permanent total disability

permanently feasible set (stochastic programming), 1-19
permanent mold, 12-43
permanent partial disability, 13-35. *See also* disability
permanent part time, 6-42
permanent piece rate, 6-42
permanent total disability, 13-35
permanent umpire. *See* permanent arbitrator
permeability, 12-43
permeability, baked, 12-31
permeability, base, 12-31
permeability, dry, 12-37
permissible exposure limit (PEL), 13-35
permit card, 6-42
perpetual endowment, 7-13. *See also* capitalized cost; endowment
perpetual inventory system, 14-16. *See also* physical inventory
perquisite, 6-42. *See also* employee services; perk
personal allowance, 17-15
personal factor (unsafe), 13-35
personal/human resource management (P/HRM), 15-11
personal injury, 8-25
personal leave, 6-42
personal power, 10-23
personal protective equipment, 8-25, 13-35
personal rate. *See* green circled rate; red circle rate
personal time, 6-42, 17-15
personnel aisle, 8-25
personnel-assignment problem. *See* assignment problem
personnel management, 15-11
personnel relations. *See* labor relations
personnel subsystem (PS), 1-19, 9-16
person-to-item AS/RS, 8-25
person-years, 13-35
PERT (program evaluation and review technique), 10-23, 10-25, 14-17. *See also* critical path method
PERT (project evaluation review technique), 4-25, 14-17
perturbation techniques, 1-19
pessimistic time estimate, 4-25
pesticides, 13-35
PGS (productivity gainsharing), 6-45, 15-11. *See also* sharing plan; value added
pH, 12-43
phalanx, 2-45
phantom bill of material. *See* transient bill of material
phantom stock, 6-42
phantom stock plans, 6-42
p-hard problems, 1-19
pharmacokinesis, 2-45

phased construction, 4-25
phase I, 1-19
phase II, 1-19
phase process, 1-19
phase queuing model, 1-19
phase shift, 12-19
phase-type random variable, 1-19
phenolic plastics, 12-80
phenolic resin, 12-43, 12-80
phon, 9-16, 13-36
photocell, 12-19
photoelectric cell, 12-19
photography, infrared, 2-39. *See also* thermograph
photometry, visual, 9-26
photopic adaptation, 9-16
photopic vision, 9-16
photoreceptors, 9-16
photosensor, 12-19
phrenic nerve, 2-45
P/HRM (personal/human resource management), 15-11
PHS (part handling system), 11-13
phthalate esters, 12-80
physical abilities test, 6-42
physical distribution, 5-11, 14-16. *See also* distribution requirements planning (DRP)
physical factors (unsafe), 13-36
physical inventory, 14-16. *See also* cycle counting; perpetual inventory system
physical progress, 4-25
physical work environment, 6-42
physiological needs, 10-23
physiological optimal alignment, 2-45
physiological response, 2-45
physiology, 2-45, 13-36
pick and place, 11-13
pick and place robot, 11-13
picketing, 6-42
picketing, informational, 6-27, 6-43
picketing, organizational, 6-43
picketing, recognitional, 6-43
picking, 14-16. *See also* kitting
picking list, 14-16
pick list, 8-25
pickstation, 8-25
pick-to-order, 14-16
pick-up and delivery service (air transport), 8-25
pick-up and delivery stations (P&D stations), 8-26
pick-up/drop off point, 8-26
PID (proportional-integral-derivative) control, 11-14
piece parts, 14-16
piece rate, 6-43. *See also* incentive rate
piece rate, basic, 6-8
piece rate, differential, 6-15
piece rate, permanent, 6-42
piece rate plan, 6-43. *See also* piecework plan; piecework rate

piece rate plan, multiple, 6-36. *See also* differential piecework
piece scale, 6-43. *See also* price list
piece time, first, 17-7
piecewise linear approximation, 1-19
piecework, 6-43. *See also* incentive wage system
piecework, differential, 6-15. *See also* multiple piece rate plan
piecework earnings, 6-43
piecework incentive plan, 6-26
piecework plan, 6-43. *See also* piece rate plan
piecework plan, group, 6-23
piecework rate, 6-43
piecework without base guarantee, 6-43
pie chart, 9-16
piercing, 12-55. *See also* perforating
piezoelectric effect, 2-45
piggy-back, 8-26
pig iron, 12-43
pillar, support, 12-84
pilot, 12-55
pilot plant, 8-26
pilot study, 9-16
pinch grasp, 2-45
pinch point, 12-55, 13-36
pinhole porosity, 12-43
pink collar worker, 6-43. *See also* blue-collar workers; white-collar worker
pins, 8-26
pin type slat conveyor, 8-26
PI (profitability index), 4-26
pit, 8-26, 12-43
pitch, 11-13
pitch (hearing), 13-36
pitch in gears, 12-64
pitch in threads, 12-64
pivotal method, 1-20
pivot column, 1-19
pivoted bucket conveyor, 8-26
pivoted bucket elevator, 8-6
pivot element, 1-19
pivot joint, 2-45
pivot row, 1-19
pivot step, 1-20
pixel, 3-13, 9-16
plain chain conveyor. *See* sliding chain conveyor
plan, 4-25, 10-23, 15-11
plane, coronal, 2-33
plane, medial, 9-13
plane, mid-sagittal, 2-43
plane, sagittal, 2-48
planetary drive, 11-13
plane, transverse, 2-52
planned change, 10-23
planned cost, 4-25
planned issue receipt, 14-16
planned obsolescence, 5-11
planned order, 14-16. *See also* scheduled receipts
planned queue management, 14-16

planning, 4-25, 10-23, 15-11
planning bill of material (BOM), 14-16. *See also* bill of material structuring; common parts bill of material; modular bill of material; option; super bill of material
planning department, 8-26
planning horizon, 7-13, 14-16. *See also* study period; utility
planning mode, 10-23
planning package, 4-25
planning staff, 10-23
planning, top-down, 10-31
planograph, 8-26
plans, drawings, 4-13
plans, operational, 10-22
plantar flexion, 2-45
Plant Closing Act (1988), 6-43
plant closings, 10-23
plant cost index, chemical engineering, 4-7
plant layout, 8-26
plant location analysis, 8-26
plant overhead, 4-25. *See also* expense
plant site selection, 5-11
plant size, optimum, 4-24
plasma arc-cutting, 12-65
plasma arc-welding, 12-89
plaster molds, 12-43
plaster of paris, 12-43
plastic, 12-80
plastic, acrylic, 12-68
plastic, alkyd, 12-69
plastic, allyl, 12-69
plastic, amino, 12-69
plastic anisotropy, 12-55
plasticate, 12-80
plastic, cellular, 12-71. *See also* cell; foamed plastics
plastic, cellulosic, 12-71
plastic, chemically formed, 12-72
plastic, commodity, 12-72
plastic deformation, 12-55
plastic, engineering, 12-74
plastic, epoxy, 12-74
plastic, ethylene, 12-74
plastic, expandable, 12-74
plastic, expanded. *See* foamed plastics
plastic flow, 12-55
plastic, foamed, 12-75. *See also* cellular plastic
plastic, furane, 12-75
plastic, halocarbon, 12-76
plastic, hydrocarbon, 12-77
plastic, isocyanate, 12-77
plasticity, 12-55, 12-80
plasticize, 12-80
plasticizer, 12-80, 12-81
plastic, lignin, 12-78
plastic, mechanically foamed, 12-79
plastic, melamine, 12-79
plastic, nonrigid, 12-79
plastic, nylon, 12-80

plastic, phenolic, 12-80
plastic, polyamide. *See* nylon plastics
plastic, pseudoplastic, 12-82
plastic, reinforced, 12-82. *See also* filler
plastic, rigid, 12-80
plastic, saran, 12-82
plastic, semi-rigid, 12-80
plastic, silicone, 12-83
plastic, styrene, 12-84
plastic, styrene rubber, 12-84
plastic, thermally foamed, 12-84
plastic, thermoplastics, 12-84
plastic, urea, 12-85
plastic, vinyl, 12-85
plastic, vinyl acetate, 12-85
plastic, vinyl alcohol, 12-85
plastic, vinyl chloride, 12-85
plastic, vinylidene. *See* saran plastics
plastic welding, 12-80
plastify, 12-81
plastigel, 12-81
plastisol, 12-81. *See also* dispersion
plate, bearing, 8-4
platen, 12-55, 12-90
platen force, 12-90
platform, 8-26
platform blowing, 12-81
platform truck, powered, 8-26
platform truck, unpowered. *See* dolly
platter, 12-29
playback accuracy, 11-13
play for position, 2-45
PL/I, 3-13
plotter, 3-13, 12-19. *See also* printer
plotter, electrostatic, 3-6
plug-and-ring, 12-81
plug date, 4-25
plug forming, 12-81
plug, portable metal stacking rack, 8-26
plug weld, 12-90
plumbago, 12-43
plywood, 12-81
plywood, high-density, 12-77
P.M. (premium money/push money), 6-43, 6-46
PMR (proportionate mortality ratios), 13-37
PNA (project network analysis), 4-27
pneumatic chute, 8-26
pneumatic conveyor, 8-26
pneumatic spout, 8-26
pneumatic toggle links, 12-55
pneumoconiosis, 13-36
pneumonitis, 13-36
pneumotachograph, 2-45
pocket belt conveyor, 8-26
pocket conveyor, 8-26
point, 6-43
point, drill, 12-65

point factor method, 10-24
point method. *See* job evaluation point method
point of operation, 12-55, 13-36
point of perception, 13-36
point of possible perception, 13-36
point-of-purchase advertising, 5-11
point-of-use storage, 14-16
point plan. *See* job evaluation point method
point system. *See* job evaluation point method
point-to-point control, 11-13
point-to-point control system, 12-20
poise, 12-81
Poisson's ratio, 12-55
poka-yoke, 11-13
polar cone, 1-20
polar coordinates, 12-20
polar coordinate system, 11-13
polarity, reverse, 12-90
policy, 10-24, 15-11
political risk, 10-24
politics, 15-11
polling, 3-13, 11-13, 12-20
polyamide plastics. *See* nylon plastics
polycentric orientation, 10-24
polycondensation. *See* condensation
polyester, 12-81
polyethylene, 12-81
polygraph, 2-45
polymer, 12-81
polymerization, 12-81. *See also* condensation
polymers, amorphous, 12-69
polypropylene, 12-81
polystyrene, 12-81
polyurethane resins, 12-81
polyvinyl
  acetate, 12-81
  alcohol, 12-81
  chloride, 12-81
  chloride-acetate, 12-81
pooled interdependence, 10-24
popliteal height, sitting, 2-12, 2-23
population ecology model, 10-24
population stereotype, 9-16
pop up menu, 9-16
porosity, 12-90
porosity (sand), 12-43
port, 12-43, 12-81
portability, 3-13, 6-43
portable belt conveyor, 8-5
portable conveyor, 8-26. *See also* bucket loader; stacker conveyor
portable crane, 8-26. *See also* crane; crawler crane; locomotive crane; truck crane
portable drag conveyor, 8-26. *See also* drag chain conveyor
portable metal stacking rack, 8-26

  base, 8-4
  under clearance, 8-35
  crane bar. *see* tiering guide
  deck, 8-10
  end frames, 8-13
  height above deck, 8-16
  nesting target, 8-22
  plug, 8-26
  removable posts, 8-28
  rigid post, 8-29
  tiering guide, 8-34
portal-to-portal, 6-43
portfolio strategy approach, 10-24
port of entry, 8-27
port-of-origin cargo clearance, 8-27
position, 6-43
position analog unit (PAU), 12-20
position class, 6-43
position control, 11-13
position description. *See* job description
position error, 11-13
position guide, 6-43
positioning, 5-11
positioning/contour system, 12-20
position movement, primary, 9-16
position movement, second corrective, 9-16
position tracking, 9-16. *See* rate tracking
positive definite quadratic form, 1-20
positive discharge bucket elevator, 8-27. *See also* bucket conveyor; bucket elevator
positive synergy, 10-24
positive work, 2-45. *See also* isometric work; negative work
post, 8-27
post-deduct inventory transaction processing, 14-16. *See also* direct-deduct inventory transaction processing; pre-deduct inventory transaction processing
posterior, 2-23
postforming, 12-81
postheat current, 12-90
postheating, 12-90
posting. *See* job posting
postprandial, 2-45
postprocessor, 3-13, 11-13, 12-20
post, support, 12-84
pot, 12-82
potential earnings, 6-43
potential hazard, 13-36
potential years of life lost, 13-36
pot plunger, 12-82
pot-retainer, 12-82
pouring, 12-43
powder metallurgy, 12-65
power, 6-43, 10-24, 15-11
power-and-free conveyor, 8-27
power bars, 8-27
power, coercive, 10-5
power conveyor, 8-27

This index is arranged in letter-by-letter format.

power curve, 8-27
power distance, 10-24
power driven roller conveyor. *See* roller conveyor, live
powered industrial truck, 8-27
power, expert, 10-11
power grip, 2-45
power level (noise), 13-36
power of a test, 16-12
power, personal, 10-23
practice effect, 9-16
precautions, 13-36
precedence diagram method (PDM), 4-25
preceding event. *See* beginning event
precision, 1-20, 3-13, 11-13, 12-20, 13-38, 16-12
precision, single, 3-15
preconstruction CPM, 4-25
predecessor activity, 4-26
predecessor event. *See* beginning event
pre-deduct inventory transaction processing, 14-16. *See also* direct-deduct inventory transaction processing; post-deduct inventory transaction processing
predetermined motion time system, 2-45, 17-15. *See also* motions pathway
predetermined time, 17-15
predetermined time system, 17-15. *See also* standard system; synthetic data
predicted, 1-20
predicted failure rate, 1-10
prediction, 14-17. *See also* forecast
predictor display, 9-16
preemployment examination, 13-36
preemptive priority, 1-20
preemptive service, 1-20
pre-expediting, 14-17
preferential hiring, 6-43
preferential logic, 4-26
preform, 12-82
preformed ceramic core, 12-43
preformed part, 12-55
preheating, 12-90
preheat temperature, 12-90
prehensile, 2-46. *See also* grasp reflex
prehension, 2-46
prejudicial delay, 4-12
preliminary CPM plan, 4-26
preliminary engineering, 4-26
premium, 5-11, 6-43
premium money (P.M.), 6-43
premium overtime pay. *See* overtime pay
premium rate of pay, 6-44. *See also* overtime pay; overtime premium pay; penalty rate; skill differential

prepaid legal plan, 6-32
preparation time, 1-27
prepolymer, 12-82
preponderance of evidence, 13-36
preprinted symbol, 8-27
preprocessor, 3-13
presbycusis, 13-36
presence, 9-16
present value, 7-13
present worth, 7-13
present worth factor(s), 7-13. *See also* annuity fund factor
preservation, 8-27
preservatives, 8-27
preset tool, 12-20
president, office of, 15-9
press, 12-56
press brake, 12-49
press brake, hydraulic, 12-53
press brake, mechanical, 12-54
press forging, 12-29, 12-56
press, single-action, 12-57
press stop, automatic, 12-48
pressure gas welding, 12-90
pressure, injection, 12-77
pressure pad, 12-82
pressure sensor, 2-46
prevailing rate, 6-44. *See also* going rate
prevailing wage law, 6-44
prevalence rate, 13-36
prevention, 4-26
prevention of accident, 13-36
preventive action, 16-12
preventive mediation, 6-44
preview control, 9-17
price, 4-26
price change, pure, 4-28
price cutting, 5-11
price escalation rate, differential, 4-12
price index, 4-10, 4-26. *See also* cost index
price index, composite, 4-7
price index, constant basket, 4-8
price index, constant utility, 4-8
price index, consumer, 4-8
price index, consumer (CPI), 6-13
price index, deflating by a, 7-8. *See also* constant dollars; inflating
price index, individual, 4-18
price index, inflating by a, 7-11
price index, Laspwyres-type, 4-20
price index, master program. *See also* current dollars
price index, month-to-month, 4-22
price index, Paasche-type, 4-24
price index, spot market, 4-32
price index, year-to-year, 4-37
price leader, 5-11
price list, 6-44. *See also* piece scale
price, list, 5-8
price, promotional, 5-12
price, protest, 6-45
price relatives, 4-26

prices, implicit, 1-12
prices, shadow, 1-23
price, transfer, 4-35
pricing, 4-26
pricing, bait, 5-2
pricing, forward, 4-26
pricing, model, 4-22
pricing, retrospective, 4-26
pricing, unit, 5-16
primary classification, 4-26
primary data, 5-11
primary metal, 12-47
prime movers, 2-46
primer, 12-82
principal, 7-13
principles of motion economy, 17-15. *See also* motion analysis
printed circuit, 12-7, 12-20
printed circuit board, 12-20
printer, 3-13, 12-20. *See also* plotter
printer, dot matrix, 3-6
printer, impact, 3-8
printer, ink jet, 3-8
printer, laser, 3-9, 3-10
printer, line, 3-10
printer, matrix, 3-11
printer, nonimpact, 3-11
printer, page, 3-12
printer, serial, 3-15
printout, 12-20
print wheel, 3-13
priority, 14-17. *See also* scheduling; sequencing
priority scheduling rule, 11-13
privacy rights, 6-44
private brands, 5-11
proactive inhibition, 9-17
probability, 1-20, 16-12
probability distribution, 5-11, 16-12
probability sampling, 13-36
probationary period, 6-44. *See also* temporary employee
probationary rate, 6-44. *See also* entrance rate
problem(s)
  activity analysis, 1-2
  assembly line balancing, 1-3
  assignment, 1-3
  bounded variable, 1-5
  capacitated transportation, 1-5
  caterer, 1-5
  crisis, 10-7
  diet, 1-8
  discrete variable. *see* integer linear programming
  dual. *see* dual linear program problems
  dual linear program, 1-8
  extremal. *see* master program
  fixed charge, 1-11
  knapsack, 1-13
  linear programming, 1-14
  machine-assignment, 1-15
  maximal network flow, 1-15
  minimal cost flow, 1-16

multicommodity network, 1-17
network flow, 1-15
noncrisis, 10-21
NT-hard, 1-18
opportunity, 10-22
organizational, 10-22
personnel-assignment. *see* assignment problem
p-hard, 1-19
programming, 1-20
shortest route, 1-24
transportation, 1-27
transshipment, 1-27
traveling-salesman, 1-27
trim, 1-27
upper bounded linear programming, 1-27
problem solving, 9-17
problem space, 9-17
procedural language, 3-13
procedure, 10-24, 16-12
procedure-oriented language, 3-13. *See also* language
procedure qualification, 12-90
process, 1-20, 11-13, 16-12, 17-15
process capability, 16-12
process capability index (PCI), 16-12
process chart, 17-16. *See also* flowchart
process chart, flow, 17-7
process chart, man-, 17-11
process chart, multiple activity, 17-13
process chart, operation, 17-14
process chart, operator, 17-14
process chart symbols, 17-16. *See also* flow process chart
process consultation, 10-24
process control, 3-13, 9-17, 11-14, 12-21, 16-13
process design, 17-16. *See also* methods analysis
process detailing, 11-14
process engineer, 17-16
process improvement, continuous, 10-7, 15-3
process in control, 16-13
processing, 17-16
processing time, 14-17
process inspection, 16-13
process instruction, 11-14
process layout, 8-27, 10-24
process limits, natural, 16-10
process manufacturing, 14-17
processor, 11-14
process plan, 11-14
process planning, 11-14, 17-16
process planning, automated, 11-2, 12-4
process quality control, 16-13
process sequencing, 11-14
process sheet, 17-16
process time, 17-16
process time, auxiliary, 17-2
process time, available, 17-2
process variability, inherent, 16-9

This index is arranged in letter-by-letter format.

procurement, 4-26
producers' cooperative marketing, 5-11
producer's risk point (PRP), 16-13
producer's risk (PR), 1-20, 16-13
producer's risk quality (PRQ), 16-13
product, 14-17
product advertising, 5-11
product divisions, 10-24
product family, 14-17
product forecast, 5-12
product form of inverse (computing form), 1-20
product group, 11-14
production, 6-44, 11-14, 12-20
production batch, 16-13
production bonus, 6-44. *See also* bonus; incentive pay; non-production bonus
production, continuous, 12-8
production control, 11-14, 14-17. *See also* inventory control
production cycle time, 11-14
production foundry, 12-43
production, intermittent, 11-9, 14-10
production levels, 11-14
production monitoring, 11-14
production plan, 11-14
production planning, 14-17. *See also* business plan; closed loop MRP
production planning, aggregate, 10-2
production rate, 6-44
production rates, 11-14, 14-17
production schedule, 14-17
production schedule, master (MPS), 10-19, 14-13. *See also* demand management; scheduling
production standard(s), 6-44. *See also* standard time
production study, 17-16. *See also* chronological study
production workers, 6-44
productive labor, 6-44, 17-16. *See also* direct labor
productive time, 6-45, 17-16
productivity, 4-26, 6-45, 10-24, 12-20, 14-17, 15-11, 17-16. *See also* annual improvement factor; efficiency, labor
productivity bargaining, 6-45
productivity company level, 17-16
productivity gainsharing (PGS), 6-45, 15-11. *See also* financial incentive plan; gainsharing; Rucker plan; Scanlon plan; sharing plan; value added
productivity, group, 17-16
productivity index, 17-16
productivity measurement, 6-45
productivity, national, 17-16
productivity, operator, 17-14. *See also* performance index
productivity, total-factor, 10-31

productivity tour. *See* annual improvement factor
productivity, workplace, 17-16
product layout, 8-27, 10-24
product liability, 13-36
product life cycle, 5-11, 10-24
product line, 5-11
product load profile, 14-17
product management, 5-11
product/market evolution matrix, 10-24
product mix, 5-12, 11-14, 14-17
product objective, 5-12
product potential, 5-12
product recall, 5-12
product structure, 14-17. *See also* bill of material (BOM)
profession, 15-11
professional bureaucracy, 10-24
profile, surface, 12-65
profilometer, surface, 12-65
profit, 4-26. *See also* income
profitability, 4-26
profitability analysis, 4-26
profitability index (PI), 4-26, 7-13. *See also* discounted cash flow; investor's method; rate of return
profitability ratios, 10-24
profit budget, 10-24
profit center, 10-24
profit, gross, 4-26
profit margin. *See* net profit, percent of sales
profit, net, 4-23, 4-26
profit, operating, 4-26
profit range, 1-20
profit sharing, 6-45. *See also* gainsharing
profit sharing plan, 6-45, 17-16
profit sharing trust, 6-45
program, 3-13, 4-27, 10-25, 12-20, 12-21
program evaluation and review technique (PERT), 10-23, 10-25, 14-17. *See also* critical path method
programmable controller (PC), 11-14, 12-20
programmable gaging, 12-56
programmable logic array. *See* array
programmable read-only memory (PROM), 3-13, 12-21
program manager, 4-27
program, master, 1-15. *See also* decomposition principle
programmed decisions, 10-25
programmed tool, automated (APT), 11-2, 12-3
programmer, 3-13
programming, 1-20, 3-13
programming, chance constrained, 1-5
programming, computer part, 12-8
programming, convex, 1-7
programming, diophantine. *See* integer linear programming
programming, dynamic, 1-9
programming, fractional, 1-11
programming, goal, 1-11
programming language, 3-13. *See also* language
programming, manual part, 12-16
programming, mathematical, 1-15
programming, mirror image, 12-17
programming, mixed-integer, 1-16
programming, nonlinear, 1-18, 3-12
programming, parametric, 1-18
programming problems, 1-20
programming, quadratic, 1-21
programming, separable, 1-23
programming, stochastic, 1-25
programming, symmetric parametric, 1-26
program, reusable, 3-14, 3-15
program scan, 11-14
progress, 4-27. *See also* status
progress chart, 17-16. *See also* Gantt chart
progress curve, 17-17. *See also* learning curve
progression, 6-45, 12-56. *See also* line of progression (LOP)
progression, automatic. *See* automatic increases
progression system, 6-45
progressive die, 12-56
progressive gluing, 12-82
progress trend, 4-27
project, 4-27, 10-25
project control, 4-27
project duration, 4-27
projected available balance, 14-17
projected finish date, 4-27
projected start date, 4-27
projected underrun (overrun), 4-27
project evaluation review technique (PERT), 4-25
project finish date (schedule), 4-27
projection, 4-27, 10-25, 14-17. *See also* forecast
projection welding, 12-90
project life. *See* economic life
project management, 4-27, 5-12, 15-12. *See also* matrix management
project manager, 4-27, 10-25
project network analysis (PNA), 4-27
project office, 4-27
project phase, 4-27
project plan, 4-27
project start date (schedule), 4-27
project summary work breakdown structure (PSWBS), 4-27, 4-36
project team. *See* task force
project time, 4-27
promotion, 6-45
promotional allowance, 5-12
promotional pay, 6-45

promotional price, 5-12
promotion cost, 7-13
promotion evaluation, 5-12
promotion increase, 6-45
PROM (programmable read-only memory), 3-13, 12-21
pronation, 2-46, 11-14. *See also* range of forearm pronation; supination
proof, 12-29
proof, burden of, 4-6
property damage, 8-27
property damage (PD) accident, 13-37
proportional control, 11-14
proportional control action, 12-21
proportional-integral-derivative control (PID), 11-14
proportionate mortality ratios (PMR), 13-37
proportion chart, 16-13
proposal schedule, 4-27
proposed base contract price, 4-27
proposed change order, 4-28
proposed confined contract price, 4-28
proprioception, 2-46, 9-17
proprioceptor, 9-17
prospective study. *See* cohort study; retrospective study
prospect theory, 10-25
prosthesis, 2-46. *See also* orthosis
protected employees, 6-45
protections, administrative, 10-2
protective clothing, 13-37
protective coating, 13-37
protective equipment, personal, 13-37
protective hand cream, 13-37
protest price, 6-45. *See also* temporary rate
protocol, 3-13, 11-14, 12-21
prototype, 10-25
prototyping, 10-25
prototyping tools, 9-17
protuberance, 2-23
proxemics, 10-25
proximal, 2-46
proximate case, 13-37
proximity sensor, 11-14
PRP (producer's risk point), 16-13
PR (producer's risk), 16-13
PRQ (producer's risk quality), 16-13
prudent, 13-37
prudent investment, 4-28
pseudo bill of material. *See* planning bill of material
pseudoplastic, 12-82
PS (personnel subsystem), 9-16
PSWBS (project summary work breakdown structure), 4-27, 4-36
psychogenic deafness, 13-37
psychological evaluation, 13-37
psychological school of job design, 15-12

This index is arranged in letter-by-letter format.

psychology, social, 15-12
psychometrics, 9-17
psychomotor ability, 9-17
psychomotor task, 9-17
psychophysical characteristics, 13-37
psychophysical measurement, 13-37
psychophysical methods, 9-17
psychophysical quantity, 9-17
psychophysics, 9-17
psychosocial evaluation, 13-37
psychosocial factors, 13-37
psychosomatic response, 2-46
public affairs department, 10-25
public relations, 5-12, 10-25
pull distribution system, 14-18. *See also* distribution requirements planning (DRP); push distribution system
pull down menu, 9-17
pulling strategy, 5-12
pull system, 11-14
pulmonary hyperinflation syndrome, 13-37
pulmonary ventilation, 2-46
pulse rate ratio, 2-46
punch, 12-29, 12-56
punched card, 12-21
punching, 12-56
punchlist, 4-28
punch tape, 12-21
punishment, 10-25
purchase order, 14-18
purchase requisition, 14-18
purchasing, 10-25
purchasing capacity, 14-18
purchasing lead time, 14-18. *See also* lead time
purchasing power, 5-12, 6-46
pure price change, 4-28
pure strategy, 1-20
pure tone, 13-37
purkinje effect, 9-17
pursuit tracking, 9-17
push-back rack, 8-27
push button station (crane), 8-27
push distribution system, 14-18. *See also* distribution requirements planning (DRP); pull distribution system
push diverter, 8-27
pusher bar conveyor, 8-27
pusher chain conveyor, 8-27
pusher furnace, 12-29
pushing strategy, 5-12
push money (P.M.), 6-43, 6-46
push-pull, 8-27
P value, 13-35
pyramiding, 6-46

Q

QC (quality circle), 10-25, 17-17
QFD (quality function deployment), 15-12
QF (quality factor), 13-37
quadrant, 12-21
quadratic form, 1-20
quadratic form, negative definite, 1-17
quadratic form, semidefinite, 1-17, 1-20
quadratic function, 1-21
quadratic programming, 1-21
qualification submittals, 4-28
qualified operator, 6-46, 17-17
qualified stock option, 6-46
qualifying dimensions, 5-12
qualitative forecasting, 10-25
qualitative research, 5-12
quality, 1-21, 10-25, 11-14, 16-13
quality assurance, 1-21, 16-13
quality audit, 16-13
quality, average outgoing (AOQ), 16-3
quality bonus, 6-46
quality circle (QC), 6-46, 10-25, 17-17
quality control, 1-21, 11-15, 16-13
quality (control) circle, 15-12
quality control, total (TQC), 10-31
quality factor (QF), 13-37
quality function deployment (QFD), 15-12
quality level, 16-13
quality management, 16-14
quality management, total, 15-14
quality of worklife program, 6-46
quality planning, strategic, 15-13
quality score chart, 16-6
quality system, 16-14
quality work life (QWL), 15-12
quantal response, 1-21
quantitative data, 1-21
quantitative display, 9-17
quantitative forecasting, 10-25
quantitative job evaluation, 6-46
quantitative research, 5-12
quantity discount, 5-12
quantity ratio, 4-28
quantity survey, 4-28
quantity surveyor, 4-28
quantity tolerance, 12-29
quartz, 12-43
quench hardening, 12-56
quenching, 12-56
quench time, 12-90
questionable payments, 10-25
queue, 1-28, 8-27, 11-15, 14-18
queue discipline, 1-21
queueing, 11-15
queue management, planned, 14-16
queue time, 14-18. *See also* interoperation time; wait time
queuing model, 10-25
queuing model, phased, 1-19
queuing, multistage, 1-21
queuing theory, 1-21

quickening, 9-17
quickie strike. *See* strike
quiet area, 8-27
quit, 6-46
quota hiring, 6-46
quotas, 6-46
QWL (quality work life), 15-12

R

rabbit ear, 12-56
rack, drive in, 8-12
rack, drive thru, 8-12
rack, single, 8-31
rack supported building structure, 8-27
radial deviation, 2-46. *See also* radio-ulnar joints; radiohumeral joint
radiale height, 9-17
radial nerve, 2-46
radiant heat, 12-43
radiation, 13-38
radiation, absorbed dose (RAD), 13-37
radiation, artificial, 13-5
radiation, background, 13-6
radiation, biological effects of, 13-7
radiation control, 13-38
radiation dosimetry, 13-38
radiation, late effects of, 13-27
radiation, local effects of, 13-28
radiation, medical, 13-30
radiation monitoring, 13-38
radiation syndrome, acute, 13-3
radioactive contamination, 13-38
radiocarpal joint, 2-46. *See also* radius
radio controlled, 8-27
radio frequency (RF) identification, 8-28
radiohumeral joint, 2-46. *See also* capitulum of humerus; radial deviation
radio-ulnar joints, 2-46. *See also* radial deviation; tennis elbow
radius, 2-47. *See also* ulna
RAD (radiation absorbed dose), 13-37
raiding (no-raiding agreement), 6-46
rail, 8-28
rail and support height, 8-28
rail guide, 8-28
rail projection, 8-28
rail span, 8-28
rail support, 8-28
rail support connector, 8-28
rail tie, 8-28
raise, 6-46
rake angle, 12-65
ram, 8-28, 12-29, 12-56. *See also* slide

ramming, 12-43
ramming, degree of, 12-36
ramp, 8-28
RAM (random-access memory), 3-14, 12-21
RAM (random-access memory), dynamic, 12-11
ramus, ascending, 2-23
random, 1-21
random access, 1-2, 3-13, 3-14, 12-21
random-access memory (RAM), 3-14, 12-21
random-access memory (RAM), dynamic, 12-11
random element. *See* foreign element
random failure, 1-10
randomization, 16-14
random noise, 13-38
random sample, 17-17
random sampling, 13-38
random storage policy, 8-28
random tool selection, 12-21
random variable, 1-21, 16-14
random walk, 1-21
range, 1-21, 16-14
range, average, 16-3
range chart, 16-14
range, cost, 1-7
range, mean, 16-3
range of forearm pronation, 2-47. *See also* pronation
rank and file, 6-46
ranking method of job evaluation, 6-46
rank of a matrix, 1-21
rapid transverse, 12-21
rapping, 12-43
RAP (resource allocation process), 4-29
rapture of the depths, 13-25
raster graphics, 3-14
rate, 6-46, 13-38, 17-17. *See also* performance rating
rate, beginner, 6-8
rate change, 6-46, 17-17
rate control, 11-15
rate cutting, 6-46, 17-17
rated average element time. *See* normal element time
rated load (crane), 8-28
rate erosion, 6-46. *See also* loose rate
rate, individual, 6-26
rate of return, 7-13. *See also* discounted cash flow; investor's method; profitability index (PI)
rate of return, cutoff, 7-8. *See also* required return
rate of return, external, 7-10, 7-13
rate of return, fair, 7-10
rate of return, internal, 7-11, 7-13. *See also* discounted cash flow
rate of return, multiple, 7-12
rate range, 6-47
rate setting, 6-47, 17-17

This index is arranged in letter-by-letter format.

rate, single, 6-52
rate tracking, 9-17. *See also* position tracking; tracking
ratification, 6-47
rating, 1-21. *See also* leveling; normalize
rating, effort, 6-17
rating, objective, 17-14
rating, pace, 6-39, 17-15
rating, performance, 6-42, 17-15
rating scale, 6-47
ratio analysis, 10-25
ratio, critical, 11-5, 14-6. *See also* dispatching rule; slack
ratio-delay study. *See* work sampling
rationalization, 10-25
rationalize, 11-15
rational model, 10-25
rational sub-group, 16-14
rational workplace design, 2-47. *See also* analytic workplace design; empirical workplace design; improvement approach to workplace design
rationing, 10-25
raw materials, 14-18. *See also* inventory write-off
raw materials inventory, 10-26
raw time. *See* actual time
Raynaud's disease, 2-47. *See also* scalenus anticus syndrome
Raynaud's phenomenon, 13-38
RCC (remote center compliance), 11-15
R chart, 16-14
reach truck, 8-28
reactance platform (force platform), 2-47
reaction distance, total (motor vehicle), 13-45
reaction inventory, 2-47
reaction time (RT), 9-17
reaction time (RT), choice, 9-5
reaction time (RT), complex, 9-17
reaction time (RT), simple, 9-18
reactive change, 10-26
read, 3-14
reading, continuous, 17-4
reading point. *See* breakpoint
read-only memory (ROM), 3-14, 12-21
read-only memory (ROM), reprogrammable, 3-14
read out, 12-21
read rate, first, 8-13
ready rate, operational (combat), 1-21
real dollars, 7-9. *See also* actual dollars; constant dollars
real estate, 4-28
realistic job preview, 10-26
reality, artificial. *See* virtual environment
real property, 4-28
real time/real-time, 3-14, 9-18, 11-15, 12-21

real wages, 6-47
reamers, 12-65
reaming, 12-65
reasonable expectance (RE), 17-17
reasonableness standard, 4-28
rebasing, 4-28
recall, 6-47
receipts, scheduled, 14-19. *See also* firm planned order (FPO); planned order
receiver-operating-characteristic, 9-18, 10-26
receiving, 8-28, 14-18
receiving area, 8-28
receptor, 9-18
receptors, activating, 2-29
reciprocal interdependence, 10-26
reciprocating beam conveyor, 8-28
reciprocating screw, 12-82
reclamation, sand, 12-44
reclassification, 6-47
recognition, 9-18
recognitional picketing, 6-43
recognition, hazard, 13-22
recognition, union, 6-61
recommended exposure limit (REL), 13-38
recompression therapy, 13-38
reconciling inventory, 14-18
record, 3-14
recorder, cassette, 12-6
record layout, 3-14
record length, 3-14
record-playback robot, 11-15
record size, 3-14
recruitment, 6-47, 10-26
recrystallization, 12-56
rectangular coordinates, 12-21
rectifying inspection, 16-14
rectilinear, 1-21
rectus abdominus, 2-47
red circle rate, 6-47. *See also* grandfathering; out-of-line rate
red-green blindness, 9-18
redout, 9-18
redrawing, 12-56
reducing flame, 12-90
reduction in area, 12-56
reduction in force, 6-31, 6-47
redundancy, 1-21, 9-18, 11-15, 16-14
redundancy, active, 1-21, 16-14
redundancy, standby, 1-21, 16-14
redundant equations, 1-22
reed switch, 12-22
reengineering, 17-17
reentry point, 3-14
reevaluation, 6-47
reference checks, 10-26
reference groups, 5-12, 15-12
referendum, 6-47
referent power, 10-26
referred pain, 2-47
reflectance, 9-18. *See also* foot candle; foot-lambert
reflex goniometry, 2-47. *See also* goniometer

reflex, simple, 2-49. *See also* spasticity; synapse
refractory period, 9-18
refresh, 3-14
regeneration MRP, 14-18. *See also* net change MRP; requirements alteration
regional differential, 6-47
register, 3-14, 11-15
regression coefficient, 13-38
regression line, 6-47
regression models, 10-26
regression, multiple, 1-17
regular element, 17-17
regular employee, 6-47
regular rate, 6-47
regulator, 12-90
rehabilitation, 2-47
rehearsal, 9-18
rehire, 6-47
reinforced plastics, 12-82. *See also* filler
reinforcement, 9-18, 12-82
reinforcement, negative, 10-20
reinforcement, schedules of, 10-27
reinforcement theory, 10-26
reinstatement, 6-47
rejectable process level (RPL), 16-14
rejection number, re, 16-14
relatedness needs, 10-26
relative, 11-15
relative address, 3-14
relative risk, 13-38
relaxation allowance. *See* fatigue allowance; personal allowance
relay, 12-22
release, 14-18. *See also* blanket order
release agent, 12-43
released order. *See* open order
relevant labor market, 6-47
reliability, 1-22, 9-18, 11-15, 13-38, 16-14
reliability, human performance, 1-22, 16-14
reliability, inherent, 1-22, 16-14
reliability, of an item, 16-14
reliability, of an item (expressed numerically), 1-22
reliability, psychological test, 9-18
relief, automatic. *See* post-deduct inventory transaction processing; pre-deduct inventory transaction processing
relief time, 6-47. *See also* rest period
relocatable program, 3-14
relocation reimbursement, 6-48
REL (recommended exposure limit), 13-38
remaining available resources, 4-28
remaining duration, 4-28
remaining float (RF), 4-28
remelt, 12-43

remote center compliance (RCC), 11-15
remote control, 12-22
remote handling, 9-18
remote indicating, 9-18
remote job entry (RJE), 3-14
removable posts, portable metal stacking racks, 8-28
REM (Roentgen equivalent man), 13-38
rental (leased) equipment cost, 4-28
reopening clause, 6-48. *See also* wage reopening
reorder point (ROP), 5-12, 10-26. *See also* order point
reorder quantity, 14-18. *See also* economic order quantity (EOQ); lot size
repair. *See* maintenance, corrective
repairable unit, 1-27
repair parts demand. *See* service parts demand
repeat, 12-56
repeatability, 11-15, 12-22, 16-14
repeatability condition, 16-14
repeatability (measurement). *See* precision
repeatability standard deviation, 16-14
repetition, 16-15
repetitive element, 17-17
repetitive manufacturing, 14-18
repetitive strain injury. *See* cumulative trauma disorder
repetitive timing, 17-17
replaceable item, 1-13
replacement, 4-28
replacement air, 13-38
replacement chart, 10-26
replacement cost, 4-29
replacement cost, functional, 4-17
replacement policy, 7-13
replacement study, 7-13
replacement value, 4-29
replenishment, 8-28
replenishment lead time, 14-18
replication, 16-15
reporting pay, 6-48. *See also* call-in pay
report program generator (RPG), 3-15
representation election, 6-48. *See also* bargaining unit; runoff election
representativeness, 9-18, 10-26
reproducibility, 13-38, 16-15
reproducibility conditions, 16-15
reproducibility standard deviation, 16-15
reproduction cost, 4-29
reprogrammable read-only memory (ROM), 3-14
reprogramming, 4-29
repudiation. *See* anticipatory breach
required completion date, 4-29

This index is arranged in letter-by-letter format.

required return, 4-29, 7-13. *See also* cost of capital; cutoff rate of return; minimum attractive rate of return
requirement, 4-29
requirements alteration, 14-18. *See also* net change MRP; regeneration MRP
requirements explosion, 14-19. *See also* dependent demand; gross requirements; materials requirements planning (MRP)
RE (reasonable expectance), 17-17
resale price maintenance, 5-12
resale value, 4-29
reschedule, 4-29
rescheduling, 14-19
rescheduling assumption, 14-19
rescheduling, automatic, 14-3
rescheduling, manual, 14-13
research design, 5-12, 13-39
research expense, 4-29
research, motivational, 5-10
research, qualitative, 5-12
research, quantitative, 5-12
reservation(s), 10-26, 14-19. *See also* allocation
reserved word, 3-14
reserve stock. *See* safety stock
reset, 3-14
residency period, 6-48
resident, 3-14
resident buyer, 5-12
resident engineer, 4-29
residual rights, 6-48
residual stress(es), 12-46, 12-65
residual volume (RV), 13-29. *See also* lung volumes
resilience, 12-56
resin, 12-82. *See also* gum
resin, acetal, 12-68
resin, acrylic, 12-68
resin, aniline formaldehyde, 12-70
resin, grinding-type, 12-76
resin, isocyanate, 12-78
resin, liquid, 12-82
resinoid. *See* novolak
resin, phenolic, 12-43, 12-80
resin, polyurethane, 12-81
resin, stir-in, 12-84
resin, urea formaldehyde, 12-47, 12-84
resistance brazing, 12-90
resistance seam-welding, 12-90
resistance spot-welding, 12-90
resistors, ballast (crane or carrier), 8-28
resolution, 11-15, 12-22
resolver, 12-22
resolving power, 9-18
resource, 4-29
resource allocation, 3-14
resource allocation process (RAP), 4-29
resource availability date, 4-29
resource availability pool, 4-29

resource code, 4-29
resource dependence, 10-26
resource dependence model, 10-26
resource description, 4-29
resource histogram, 4-29
resource limited scheduling, 4-29
resource planning. *See* long-range resource planning
resource plot, 4-29
resource profile. *See* product load profile
respiration, 2-47. *See also* anaerobic metabolism; anoxia; ischemia
respirator, 12-43, 13-39. *See also* gas mask
respiratory diseases, 13-39
respiratory irritants, 13-39
respiratory protective equipment, 13-39
respiratory quotient, 2-47, 2-48, 9-19
respiratory system, 13-39
response, 1-22, 9-19
response, dynamic, 12-11
response function, 5-13
response time, 3-14, 9-19, 12-22
responsibility, 4-29, 6-48, 10-26, 15-12
responsibility center, 10-26
responsibility, social, 15-12
responsible organization, 4-29
rest allowance. *See* fatigue allowance; personal allowance
resting metabolism, 2-48
rest period, 6-48, 13-39. *See also* break time; coffee break; relief time
restraining forces, 10-26
restraint, 4-29, 4-30
restraint, logical, 4-21
restricted articles (air transport), 8-28
restricted element, 17-17
restricted job, 6-48
restricted work, 17-17. *See also* machine-controlled time
restriction, 1-22
restriking, 12-29, 12-56
restructuring, 10-26
retailer, 5-13
retainer board, 12-76
retarder, 8-28. *See* inhibitor
retarding conveyor, 8-28
retention, 4-30
retina, 9-19
retinal disparity, 9-19
retinal field, 9-19
retinal illuminance, 9-19
retinal rivalry, 9-19
retirement, 6-48. *See also* pension plan; social security act
retirement account, individual (IRA), 6-26
retirement age, early, 6-16
retirement age, normal, 6-38

retirement, automatic, 6-6
retirement, compulsory, 6-12
retirement, normal. *See* retirement
retirement of debt, 4-30, 7-13
retirement plan. *See* pension plan
retirement plan, fixed benefit, 6-21
retirement plan, flat benefit, 6-21
retraining, 6-48
retroactive pay, 6-48
retrofit, 12-22
retrograde shock amnesia, 13-39
retrospective pricing, 4-26
retrospective study. *See* case-control study
return, 3-14
return on investment (ROI), 5-13
reusable container, 8-28
reusable program, 3-14, 3-15
revenue, average, 5-2
revenue budget, 10-26
revenue center, 10-26
revenue, marginal, 5-8
Revenue Reconciliation Act of 1993, 6-48
revenue, sales, 4-30
reverberation, 13-39
reversal principle, 12-65
reverse polarity, 12-90
revised simplex method (computing form), 1-22
revision, 4-30
revitalization, 10-26
revolutions per minute (RPM), 12-65
reward power, 10-26
rewards, extrinsic, 10-11
rewards, intrinsic, 6-27, 10-16
rewards, non-financial, 6-37
rewarehousing, 8-28
rework, 11-15, 17-17
rework lead time, 14-19
rework order, 14-19
RF (radio frequency) identification, 8-28
RF (remaining float), 4-28
rheology, tissue, 2-52
rhodospin, 9-19
rib, 12-56
ribbon belt conveyor. *See* belt conveyor, multiple ribbon
ribbon conveyor, internal. *See* internal ribbon conveyor
ribbon flight, 8-28
ribbon flight conveyor screw, 8-28
ribbon flight screw conveyor, 8-28
ribs, 12-43
riddle, 12-44
rider truck, 8-29
right- and left-hand chart, 17-17. *See also* operator process chart
right angle stack aisle width, 8-29
right-hand side, 1-22
right-hand side element, 1-22
right-hand side range, 1-22
right-of-way, 13-39
rights, bargaining, 6-7

Right-To-Work Law, 6-48. *See also* Section 14 (B)
rigid back-to-back tie, 8-29
rigid container, 8-29
rigid post, portable metal stacking rack, 8-29
ring gate, 12-82
ring rolling, 12-29
riscident, 13-39
riscutant, 13-39
riser, 12-44
riser sleeve, exothermic, 12-37
risk, 4-30, 7-13, 10-26, 13-39
risk, absolute, 13-2
risk analysis, 7-13. *See also* decision tree; decisions under risk
risk, assigned, 13-5
risk, assumption of, 13-6
risk factor, 13-39
risk (insurance), 13-39
rite, 10-27
RJE (remote job entry), 3-14
RMS (root mean square), 12-65
robot, 11-15, 12-22
robot, fixed-stop, 11-7
robot, intelligent, 11-9
robot, sequence, 11-16
Rockwell hardness tester, 12-65
rods, 9-19
roentgen (R), 13-39
ROI (return on investment), 5-13
role, 10-27, 15-12
role-playing, 6-48
roll, 11-15
roller, 12-29
roller conveyor, 8-29. *See also* el conveyor
roller conveyor, accordion, 8-2
roller conveyor, controlled velocity, 8-8
roller conveyor, herringbone, 8-29
roller conveyor, hydrostatic, 8-29
roller conveyor, live, 8-29
roller conveyor, shock absorbing, 8-29
roller conveyor, skewed, 8-29
roller conveyor, speed-up, 8-29
roller conveyor, spool type, 8-29
roller conveyor, spring mounted, 8-29
roller conveyor, troughed, 8-29
roller spiral, 8-29
roller straightening, 12-56
roller table, 8-29
roll forging, 12-29
roll forming, 12-56
rolling. *See* bumping
rolling chain conveyor, 8-29. *See also* chain conveyor
rolling impression, 12-29
rolling over, 12-44
rolling resistance, 8-29
roll threading, 12-56
rollway skid, 8-29
romance of leadership, 10-27
ROM (read-only memory), 3-14, 12-21

This index is arranged in letter-by-letter format.

ROM (read-only memory), reprogrammable, 3-14
room, 8-29
root crack, 12-90
root face, 12-90
root mean square pressure of fundamental speech sounds, 9-19
root mean square (RMS), 12-65
root opening, 12-90
roots, multiple, 7-12
rope and button conveyor, 8-29
ROP (reorder point), 10-26
rotary blade, 12-56
rotary furnace, 12-29
rotary shear, 12-56
rotary swager, 12-56
rotary switch, 8-29. *See also* switch, monorail
rotary table feeder, 8-29
rotary vane feeder, 8-29
rotating shift, 6-49. *See also* fixed shift; shift; split shift; swing shift
rotation, 2-48
rotational casting, 12-82
rotational molding, 12-82
rotation, axis of, 2-31
rotation, center of, 2-32
rotation (zero g), 9-19
rotator, 8-29
rotator cuff, 2-48
rough-cut capacity planning, 14-19
roughing, 12-65
roughness, 12-65
rough terrain truck, 8-29
round-the-clock operations. *See* continuous operation
routes of entry, 13-39
routine, 3-15, 12-22
routing, 11-15
routing, alternate, 11-2
routing, blanket, 11-3
routing model, 10-27
routing sheet, 11-15
roving, 12-82
row, 8-30
row, cost, 1-7
royalties, 4-30
royalty, 6-49
RPG (report program generator), 3-15
RPL (rejectable process level), 16-14
RPM (revolutions per minute, 12-65
R (roentgen), 13-39
RS-232C, 12-22
RT (reaction time), 9-17
rubber, 12-82
rubber pad forming, 12-57
Rucker plan, 6-49, 17-17. *See also* gainsharing; productivity gainsharing (PGS); sharing plan
rule, 10-27
rule of 80-20. *See* Pareto's law
run, 3-15, 16-15

runaway, 8-30
runaway conductors, 8-30
runaway rate, 6-49
runaway shop, 6-49
rung, 11-15
runner(s), 8-30, 12-44, 12-82
running sheave, 8-30
runoff election, 6-49. *See also* representation election
runout, 12-44
runout time, 17-17
run time, 11-15, 14-19
RV (residual volume), 13-29. *See also* lung volumes

saccadic movements, 9-19
sacrospinalis muscle, 2-48. *See also* erector spinae muscles
sacrum, 2-48
saddlepoint, 1-22
saddlepoint of a game, 1-22
saddling, 12-30
SAE steels, 12-65. *See also* AISI steel designation
safe, 13-39
safestock, 5-13
safety, 6-49, 9-19, 13-39
safety belt, 8-30, 13-39
safety capacity, 14-19
safety coupling, 13-40
safety cut-out, 13-39
safety education, 6-49, 13-40
safety engineering, 6-49, 13-40
safety factor. *See* factor of safety
safety glass, 13-40
safety helmets, 13-40
safety, industrial. *See* occupational safety
safety lock, 13-40
safety, margin of, 13-30
safety needs, 10-27
safety professional, 13-40
safety rule(s), 13-40
safety spreaders, 8-16
safety stock, 14-19. *See also* cycle stock; fluctuation inventories
safety time, 14-19
safety training, 13-40
safety work surface, 8-30
safe workplace, 13-39
sagittal, 2-23
sagittal plane, 2-48. *See also* transverse plane
salary, 6-49
salary administration, 6-49
salary and commission. *See* commission earnings
salary, base, 6-7
salary budget, 6-49
salary compression, 6-49
salary control, 6-13
salary curve, 6-49

salary differential, 6-49
salary grade, 6-30
salary guide, 6-49
salary level change, 6-49
salary plus bonus, 6-50
salary plus commission, 6-50
salary profile, 6-50
salary ranges, 6-50
salary rate, 6-50
salary scattergram, 6-50
salary, straight, 6-55
salary structure, 6-50
salary structure, integrated, 6-27
sales, 4-30
sales analysis, 4-30, 5-13
sales budget, 5-13, 10-27
sales compensation, 6-50
sales-force composite, 10-27
sales forecast, 4-30, 5-13
sales manager, 5-13
sales planning, 5-13
sales price, 4-30
sales profile, 4-30
sales promotion, 5-13
sales quota, 5-13
sales research, 4-30
sales revenue, 4-30
sales territory, 5-14
salvage value, 4-30, 7-13
sample, 1-22, 13-40, 16-16
sample, random, 17-17
sample size, 1-23, 16-16
sample standard deviation chart, 16-16
sampling, 5-14, 16-15
sampling, bulk, 16-4
sampling, cluster, 1-5, 13-9
sampling error, 16-15
sampling frame, 16-15
sampling, general, 13-40
sampling, grab, 13-40
sampling inspection, 16-15
sampling inspection, continuous, 16-5
sampling inspection, multiple, 16-10
sampling inspection, sequential, 16-16
sampling interval, 16-15
sampling, judgemental, 4-19
sampling, personnel, 13-40
sampling plan, 16-15
sampling procedure, 16-15
sampling, random, 13-38
sampling scheme, 16-15
sampling, sequential, 1-23
sampling, severity of, 16-16
sampling, statistical, 13-40
sampling system, 16-15
sampling unit, 13-40, 16-15
sampling without replacement, 16-16
sampling with replacement, 16-16
sand, 12-44
sand, backing, 12-31
sand, bank, 12-31
sand blast, 12-44

sand, burned, 12-32
sand casting, 12-33
sand, cement, 12-33
sand control, 12-44
sand, core, 12-35
sand, dried, 12-36
sand feeder, 12-37
sand fineness, 12-37
sand, fire, 12-38
sand grain distribution, 12-36
sand, green, 12-39
sand, heap, 12-39
sand inclusions, 12-44
sanding, belt, 12-60
sand mold, dry, 12-37
sand, natural, 12-41
sand, olivine, 12-42
sand porosity, 12-43
sand reclamation, 12-44
sand spun process, 12-44
sand, synthetic molding, 12-46
sand toughness number, 12-44
sandwich molding, 12-82. *See also* foam molding
sand, zircon, 12-48
SAN (styrene acrylonitrile), 12-84
saphenous veins, 2-48
saran plastics, 12-82
sarcoma, 13-40
satisfaction model, 10-27
satisfaction-progression principle, 10-27
satisficing, 15-12
satisfiers, 10-27
saturation, 9-19, 11-15
savings (net benefits), 4-23
savings plan. *See* thrift plan
sawing, 12-65
saw-tooth dock, 8-30
SBU (strategic business unit), 5-15, 10-29
scab. *See* strikebreaker
scalar chain, 15-12
scale, 12-30, 12-44. *See also* union rate
scalenus anticus syndrome, 2-48. *See also* Raynaud's disease; syndrome
scale pit, 12-30
scales, factor, 6-20
scaling, 12-44, 12-57
scaling, multidimensional, 5-10
scaling, non-metric, 5-10
scan, 3-15, 8-30, 13-40. *See also* CAT
scan area, 8-30
Scanlon plan, 6-50, 17-18. *See also* productivity gainsharing (PGS); sharing plan
scanner, 12-22
scanner, optical, 11-12
scanning curtain, 8-30
scanning, social, 10-28
scaphoid, 2-48
scaphoid bone, 2-43
scaphoid tubercle, 2-48
scapula, 2-23, 2-48

This index is arranged in letter-by-letter format.

SCARA (self compliant arm for robot assembly), 11-16
scattergram, 6-50
scattergram, salary, 6-50
scenarios, 10-27
S chart, 16-16
schedule, 4-30, 11-16, 14-19
schedule, conceptual, 4-7, 4-8
scheduled charge, 13-40
scheduled completion date, 4-30
scheduled date. *See* contract date
schedule, detailed, 4-12
scheduled event time, 4-30
scheduled receipts, 14-19. *See also* firm planned order (FPO); planned order
scheduled variance, 4-30
schedule finish, level (SF), 4-20
schedule item, master, 14-13
schedule, master, 14-13
schedule, master production (MPS), 10-19, 14-13
schedule rating, 13-40
scheduler, master, 14-14
schedules of reinforcement, 10-27
schedule variance, 4-30
schedule variance management, 14-19
scheduling, 4-30, 8-30, 11-16, 14-19. *See also* dispatching; priority
scheduling, backward, 11-3, 14-4
scheduling, critical path, 5-5. *See also* network analysis
scheduling, critical ratio, 11-5
scheduling, forward, 11-8, 14-9
scheduling, interactive, 14-10
scheduling rules, 4-30, 14-20
school of job design, biological, 15-2
school of job design, engineering, 15-4
school of job design, ergonomics, 15-5
school of job design, psychological, 15-12
scientific management, 10-27, 15-12
scintillation, 9-19
scleroscope, 12-65
scoop, 8-30
scope, 4-30
scope change, 4-30
scope, change in, 4-7
scotoma, 9-19
scotopic adaptation, 9-19
scotopic vision, 9-19
scrap, 11-16, 12-44
scrap allowance, 11-16, 12-22
scrap usage, 11-16
screen, 12-44
screen dump, 9-19
screening inspection, 16-16
screening test, 1-23, 13-40
screw conveyor, 8-30
screw conveyor, rotating casing, 8-30

screw conveyor, vertical, 8-30
screw, cut-and-folded flight, 8-30
screw, cut flight, 8-30
screw, double flight, 8-30
screw feeder, multiple, 8-22
screw machines, 12-65. *See also* automatic bar
screw, multiple flight conveyor, 8-21
scroll, 9-19
scrolling, 3-15
SCUBA (self-contained underwater breathing apparatus), 13-41
sealing diameter, 12-83
seam, 12-30, 12-44, 12-57
seaming, 12-57
seam-welding, resistance, 12-90
search, 3-15
search key, 3-15
seasonal cariation, 4-31
seasonal commodities, 4-31
seasonal employment, 6-50
seat reference point (SRP), 9-19
Secchi disk, 9-19
secondary attack rate, 13-40
secondary boycott. *See* boycott
secondary failure, 1-10
secondary float (SF), 4-31
secondary gluing, 12-83
section, 8-30
Section 14 (B), Labor Management Relations Act, 6-50. *See also* right-to-work law
sector, 3-15
secular trend(s), 4-31, 13-41
segregation, 12-44
seizing, 12-57
select(ed) element time, 17-18
select(ed) time, 17-18. *See also* average element time
selection, 6-50, 10-27
selection interview, 10-27
selection model, natural. *See* population ecology model
selective assembly, 12-65
selective attention, 9-3
selective distribution, 5-14
selective rack, 8-30
self-actualized needs, 10-27
self compliant arm for robot assembly (SCARA), 11-16
self-contained underwater breathing apparatus (SCUBA), 13-41
self-control, 10-27
selfcuring binder, 12-44
self-efficacy, 10-27
self-feeding conveyor, 8-30
self-feeding portable conveyor, 8-30
self insurance, 13-41
self loading, 8-30
self-managing team, 10-27
self-oriented roles, 10-27
self-paced, 9-19
self-serving bias, 10-27
selling, 5-14
selling, automatic, 5-2

selling, exclusive outlet, 5-6
selling expense, 4-31
selling price. *See* sales price
selling up, 5-14
sellion, 2-23
semantic blocks, 10-27
semantic net, 10-27
semantics, 15-12
semiautomatic arc welding, 12-90
semidefinite quadratic form, 1-17, 1-20
semi-finisher, 12-30
semi-finishing impression, 12-30
semi-gantry crane, 8-30. *See also* traveling cranes
semilunar notch, 2-48
semi-skilled labor, 6-50
semispinalis capitis, 2-48
sender, 10-27
seniority, 6-50
seniority rights, 6-51
seniority, super, 6-56
sensation, 9-19
sensing time, 9-20
sensitivity, 1-23, 7-13, 12-22, 13-41
sensitivity analysis, 1-23, 4-31, 7-14
sensitivity testing, 1-23
sensitivity training, 6-51, 15-12. *See also* encounter group
sensor, 11-16, 12-23
sensor, external, 11-7
sensor, internal, 11-9
sensory adaptation, 9-2
sensory end organs, 9-20
sensory feedback, 9-20
sensory nerve, 9-20. *See also* synapse
sentimental value, 4-31
separable acuity, minimum, 9-14
separable constraint, 1-23
separable objective function, 1-23
separable programming, 1-23
separation pay. *See* severance pay
separations, 6-51
separator, 12-45
sequence, 3-15
sequence, change in, 4-7
sequence-of-use principle (equipment design), 9-20
sequence robot, 11-16
sequencing, 14-20, 17-18. *See also* priority
sequential access, 3-15
sequential interdependence, 10-27
sequential logic, 12-23
sequential sampling, 1-23
sequential sampling inspection, 16-16
serial, 3-15
serial code, 8-30
serial printer, 3-15
serial sort, 3-15
serious injury, 13-41
serious injury frequency rate, 13-41

serious violation, 13-41
serviceability, 1-23, 4-31
service aisle, 8-30
service bay, 8-4
service fee, 6-51
service, length of, 6-51
service life. *See* life
service parts, 14-20
service parts demand, 14-20
service rate distribution, constant. *See* arrival rate distribution, constant
service rate, mean. *See* mean service rate
services, 5-14
service test model, 9-20
service time, constant. *See* arrival rate distribution, constant
service worth value, 4-31
servicing, 1-23, 4-31, 16-16
servomechanism, 9-20, 11-16, 12-23
servovalve, 11-16
sesarnoid bone, 2-48, 2-49
set, 1-23, 3-15, 12-83
set point, 12-23
settling time, 11-16
setup, 11-16, 17-18. *See also* changeover time
set-up lead time, 14-20
set-up time, 14-20
setup time, 11-16
SEU (subjective expected utility), 9-22
severance pay, 6-51. *See also* layoff pay
severity of sampling, 16-16
severity rate, 13-41
severity rate of accidents, 6-51
sexual harassment, 6-51
SFPM (surface feet per minute), 12-66
SF (schedule finish, level), 4-20
SF (secondary float), 4-31
shade, 9-20
shadow prices, 1-23
shadow stock. *See* phantom stock
shaft, 8-30
shake-out, 12-45
shaker conveyor. *See* oscillating conveyor
shake table, 9-20
shank, 12-30, 12-65
shape coding, 9-20
shapeup, 6-51
shaping, 10-28, 12-65
shared storage policy, 8-30
share-the-work. *See* work sharing
sharing plan, 6-51. *See also* productivity gainsharing (PGS); Rucker plan; Scanlon plan
sharing plan, constant, 6-12
Shaver's disease, 13-41
shaving, 12-57
shaving, gear, 12-66
shaw process, 12-45
shear, 2-49, 12-57

This index is arranged in letter-by-letter format.

shear, hydraulic, 12-53
shear, in metal cutting, 12-66
shear, mechanical, 12-54
shear, rotary, 12-56
shear strength, 12-57
sheet, 12-57, 12-83. *See also* film
sheeting, 12-83. *See also* film
sheet molding compound (SMC), 12-83
shelf, 8-31
shelf connector, 8-31
shelf life, 1-25, 12-83
shell core process, 12-45
shell molding, 12-45. *See also* croning process
shestil, 6-51
Shewhart control chart, 16-16
Shewhart control limits, 16-5
shielded metal arc-welding, 12-90
shift, 6-51. *See also* watch
shift, core, 12-35
shift, die, 12-26
shift differential, 6-51
shift, domain, 10-9
shifter, side, 8-31
shift, fixed, 6-21. *See also* rotating shift; shift; split shift
shift, graveyard, 6-22
shifting base, 4-31
shift, phase, 12-19
shift premium. *See* shift differential
shift, rotating, 6-49
shift, split, 6-53
shift, swing, 6-57
shift, temporary threshold (TTS), 13-44
shift, threshold, 9-23
shim, 12-83
shipper, drop, 5-6
shipping, 8-31
shipping area, 8-31
ship's conveyor elevator, 8-31
shock, 13-41
shoe, 12-30
shop, 8-31
shop, closed, 6-11
shop committee, 6-51
shop drawings, 4-31
shop floor control, 11-16, 12-23, 14-20. *See also* closed loop MRP
shop order number. *See* account number
shop packet, 14-20
shopping center, 5-14
shopping goods, 5-14
shop planning, 14-20. *See also* closed loop MRP; dispatching
shop rules, 13-41. *See also* work rules
shop steward, 6-51
shortest processing time (SPT), 14-20
shortest route problem, 1-24
short interval scheduling (SIS), 17-18
short-term activities. *See* monthly guide schedule

short-term disability, 6-51
short-term exposure limits (STEL), 13-41
short-term income protection, 6-52
short-term memory, 9-20
short-term repeatability, 11-16
short workweek benefit, 6-52
shot blasting, 12-45, 12-66
shoulder, 11-16
shoulder breadth, 2-12
shoulder-elbow distance, 9-20
shoulder elbow length, 2-12
shoulder height, sitting, 2-12
shoulder height, standing, 2-12
should-take, did-take time standards, 17-18. *See also* normal performance
should-take time standard. *See also* did-take standards
shrinkage, 5-14, 12-30, 12-45
shrinkage, internal, 12-39
shrinkage, patternmaker's, 12-42, 12-45
shrinkage, solidification, 12-45
shrink scale, 12-30
shrinkwrap, 8-31
shut-down, 6-52
shutdown point, 4-31
shut height, 12-57
shuttle, 8-31
shuttle conveyor, 8-31
sick leave, 6-52, 13-41. *See also* accident and sickness benefits
SIC (standard industrial classification) code, 4-31, 4-32, 5-14
side loader, 8-31
side loading attachment, 8-31
side-mounted operator compartment, 8-31
side-pull en masse conveyor, 8-31
side pusher conveyor, 8-31
siderosis, 13-41
side shifter, 8-31
side tone, 9-20
sieve analysis, 12-45
sievert (Sv), 13-41
signal, 12-23
signal detection theory, 9-20
signal flow graph, 1-24, 3-15
signature, 2-49. *See also* chronocyclegraph
signature, velocity, 2-53
significance level, of a test, 16-16
significance testing, 1-24, 16-16
significant result, 16-16
significant risks, 6-52
significant variances, 4-31
silica, 12-45
silica brick, 12-45
silica flour, 12-45
silicates, 13-41
silicon carbide, 12-66
silicone, 12-83
silicone plastics, 12-83
silicones, 13-42
silicosis, 13-42

silo, 8-31
SIL (speech interference level), 9-21
silt, 12-45
Simo chart, 17-18. *See also* film analysis chart; right- and left-hand chart
simple interest, 7-14. *See also* interest rate-nominal
simple payback period (SPP), 4-31. *See also* payback method
simple random sample, 16-16
simple reflex, 2-49. *See also* conditioned reflex; spasticity; synapse
simple structure, 10-28
simplex, 1-24
simplex algorithm. *See* simplex method
simplex algorithm, dual, 1-9
simplex method, 1-24
simplex multipliers, 1-24
simplified practice, 17-18
simulation, 1-24, 3-15, 9-20, 10-28, 11-16, 14-20. *See also* what if analysis
simulator, 9-20
simultaneous motion chart, 17-18
simultaneous motions, 17-18
single-acting hammer, 12-57
single-action press, 12-57
single arm, 8-31
single command cycle, 8-31
single company union, 6-52
single deep storage, 8-31
single-level bill of material (BOM), 14-20
single-level where used, 14-20. *See also* stock keeping unit (SKU)
single point failure, 13-42
single-point tool, 12-66
single precision, 3-15
single rack, 8-31
single rate, 6-52
single-row layout, 8-31
single sampling inspection, 16-16
single-stroke mechanism, 12-57
single-use plans, 10-28
singular matrix, 1-24
sink, 1-24
sinking fund, 7-14
sinking fund depreciation, 7-9
sinking fund factor, 7-14
sintering, 12-45, 13-42
SIS (short interval scheduling), 17-18
SITC (Standard International Trade Classification), 8-31, 8-32. *See also* commodity code
sitdown strike. *See* strike
site preparation, 4-31
sitting eye height, 2-10
sitting height, 2-12, 9-20
sitting hip breadth, 2-11
sitting knee height, 2-11
sitting knee-to-knee breadth, 2-11

sitting popliteal height, 2-12
sitting shoulder height, 2-12
sitting thigh clearance height, 2-12
situational leadership theory, 10-28
situation awareness, 9-20
SI units, 9-20
size, basic, 12-60
sizing, 12-30, 12-57, 12-83. *See also* glue
skate-wheel conveyor, 8-31. *See also* wheel conveyor
skeletal configuration, 2-49
skelp, 12-57
skew table, 8-31. *See also* roller conveyor, skewed
skid support, 8-31
skill, 6-52
skill-based pay, 10-28
skill differential, 6-52. *See also* premium rate of pay
skill, differential, 6-15
skilled labor, 6-52
skilled trades, 6-52. *See also* craft union
skill, job, 17-10
skills inventory, 6-52, 10-28. *See also* manning table
skills, technical, 10-31
skill training, 15-12
skill variety, 10-28
skimming price policy, 5-14
skin contamination, 13-42
skinfold measurement for estimating body fat, 9-20
skinfolds, 2-7
skinfold, subscapular, 2-7
skinfold, triceps, 2-7
skip hoist, 8-32
skip-lot sampling inspection, 16-16
SKU (stock keeping unit), 8-32, 14-21
slack, 1-24, 10-28, 14-20. *See also* float; slack time
slack paths, 4-31
slack resources, 10-28
slack time, 4-31, 4-32, 14-20. *See also* critical ratio; slack
slack variable, 1-24
slack vector, 1-24
slag, 12-45
slag inclusion, 12-90
slat conveyor, 8-32. *See also* floor conveyor
slave pallets, 8-32
sleep envelope, 2-8
sleeve ejector, 12-83
slew rate, 11-16
slick, 12-45
slicker, 12-45
slide, 12-57. *See also* ram
slide, adjustment, 12-48
slider bed, 8-32
sliding chain conveyor, 8-32. *See also* chain conveyor; drag chain conveyor

This index is arranged in letter-by-letter format.

sliding plate. *See* duplicate cavity-plate
sliding scale wage plan, 6-52
sliding type switch, 8-32. *See also* switch, monorail
sling, 8-32
slip casting, 12-45
slip joint, 12-83
slipped disc, 2-38
slitting, 12-57
slope conveyor, 8-32. *See also* apron conveyor; belt conveyor; flight conveyor
slotting, 6-52, 12-66
slotting fee, 5-14
slot velocity, 13-42
slowdown, 6-52. *See also* strike
slug, 12-30, 12-57
slurry, 12-45
slush casting, 12-45
small-batch production, 10-32
SMC (sheet molding compound), 12-83
smoke, 13-42
smoother, 12-45
smoothing, 10-28
SMR (standardized mortality ratio), 13-43
SMSA (standard metropolitan statistical area), 6-54
snagging, 12-30, 12-45
snag grinding, 12-30, 12-66
snapback timing. *See* repetitive timing
snapping finger, 13-43
snow-blindness, 9-20
social audit, 10-28
social forecasting, 10-28
social information-processing approach, 10-28
socialist economy, 10-28
socialization, disjunctive, 15-4
social learning theory, 10-28
social loafing, 10-28
social psychology, 15-12
social responsibility, 15-12
social scanning, 10-28
Social Security Act, 1935, 6-52. *See also* retirement
Social Security Adjustment Option, 6-53
Social Security offset, 6-53
social system, 15-12
sociocultural element, 10-28
sociometry, 15-13
sociopsychological conditions, 6-53
sociotaxis, 2-49
socio-technical system, 15-13
softening range, 12-83
software, 3-15, 10-28, 12-23
software programs
  DOS (disk operating system), 3-6
  editor, 3-6
  linkage editor, 3-10
  operating system, 3-12

reusable, 3-14, 3-15
routine, 3-15
source, 3-16
spreadsheet, 3-16
supervisory, 3-16
utility, 3-17
softwired, 12-23
soldering, 10-28, 12-91
solenoid, 11-16
solidification, 12-45
solidification shrinkage, 12-45
solid-piled, 12-83
solid state, 12-23
solid state welding, 12-91
solid tire, 8-32
solution, 1-24
solution, basic, 1-4
solution, basic feasible, 1-4
solution, degenerate, 1-8
solution, feasible, 1-11
solution, geometric, 1-11
solution, initial, 1-13
solution level, 1-24
solution, near-optimum, 1-17
solution, nondegenerate feasible, 1-17
solution, non-dominated, 1-19
solution, optimum, 1-18
solution, starting, 1-24
solution, unbounded, 1-27
solution, unique optimal, 1-27
solvation, 12-83
solvency, 12-83
somatic mutation, 13-42. *See also* mutagen
somatotyping, 2-49, 9-20
sonarography, 13-42
sone, 9-21, 13-42
sonic testing, 12-46
sonography, 2-49. *See also* ultrasonics
sonometer, 13-42
sorbent, 13-42
sort, 3-15
sortation system, 8-32
sorting conveyor, 8-32
sort, internal, 3-9
sort key, 3-15
sound, 13-42
sound absorption, 13-42
sound analyzer, 13-42
sound intensity, 13-42
sound-level meter, 13-42
sound pressure level, 13-42. *See also* decibel
source, 1-24
source delegated inspection, 14-20
source inspection, 14-20
source language, 3-16. *See also* language
source program, 3-16
sow block, 12-30
space, 8-32
space myopia, 9-21
spade drills, 12-66
span, 8-32
span of control, 6-53, 10-28, 15-13

span of management, 10-28
span of reach, 2-49
span time, 11-16
spark discharge machining. *See* electric discharge machining
spark enclosing equipment, 8-32
spark test, 12-46
sparse vector or matrix, 1-24
spasm, 2-49, 13-42
spasticity, 2-49. *See also* simple reflex
spatter, 12-91
specialization, 15-13. *See also* division of work
special permit rate, 6-53
specialty goods, 5-14
specific acoustic impedance, 2-49
specific adhesion, 12-69. *See also* adhesion; adhesion, mechanical
specification, design (prescriptive), 4-32
specification limits, 16-18
specification, performance, 4-32
specification(s), 4-32, 12-23, 16-16
specific gravity, 13-42
specificity, 13-42
spectrum colors, 9-21
specular reflection, 9-21
speech articulation index, 9-21
speech interference level (SIL), 9-21
speech perception test, 13-42
speech recognition, 9-21
speech synthesis, 9-21
speed rating. *See* leveling; performance rating
speedup, 6-53. *See also* stretchout
spell, 6-53
spendable earnings, 6-53. *See also* take-home pay
spherical coordinate system, 11-16
spiegeleisen, 12-46
spike, 2-49
spinal nerve, 2-49
spindle conveyor, 8-32
spine, 2-49
spine layout, 8-32
spinning, 12-57
spinous process of the vertebra, 2-49. *See also* vertebra
spin table, 9-21
spiral fluidity test, 12-46
spirometry, 2-50
splash core, 12-46
splash-proof goggles, 13-43
splicing technique, 4-32. *See also* linking procedure
split cavity, 12-83
split-cavity blocks, 12-83
split lot, 11-16, 14-20
split shift, 6-53. *See also* fixed shift; rotating shift; shift; swing shift
SPM (strokes per minute), 12-58
sponsor, 10-29

spontaneous ignition, 13-43
spooling, 3-16
spot, 8-32
spot, dry, 12-74
spot facing, 12-66
spot market price index, 4-32
spotweld, 12-91
spot-welding, resistance, 12-90
spout, pneumatic, 8-26
SPP (simple payback period), 4-31
sprain, 13-43
spray-up, 12-83
spread, 12-83
spreaders, safety, 8-16
spreadsheet, 3-16
spread-the-work. *See* work sharing
springback, 11-16, 12-58
sprinkler system, 13-43
sprue, 12-28, 12-46, 12-83
sprue bushing, 12-83
sprue, collapsible, 12-34
sprue-puller, 12-83
SPT (shortest processing time), 14-20
spurt, beginning, 9-4
spur track, 8-32
squaring shaft, 8-32
squatting height, 9-21
squeeze, 9-21, 13-43
squeeze board, 12-46
squeeze time, 12-91
SRP (seat reference point), 9-19
SS (start schedule, level), 4-20
stability, 8-32
stability strategy, 10-29
stabilization (zero g), 9-21
stable process, 16-13
stacker conveyor, 8-32. *See also* boom conveyor; portable conveyor
staff, 15-13
staffing, 10-29
staff position, 10-29
staff, service, 15-13
staff, specialized, 15-13
stage of processing, 4-32
staging, 14-21
stair, 8-32
stairway, 8-32
stakeholders, 15-13
stamp, 12-58
stamping, 12-58
stand-alone system, 12-23
standard, 6-53, 11-16, 12-23, 17-18. *See also* work standard
standard agreement, 6-53. *See also* form agreement
standard allowance, 6-53, 17-18. *See also* allowance
standard allowed time, 6-53
standard cost center, 10-29
standard data, 6-53, 17-18. *See also* synthetic data
standard deviation, 1-24, 4-32, 16-16
standard deviation, repeatability, 16-14

This index is arranged in letter-by-letter format.

standard disabling injury frequency rate. *See* disabling injury frequency rate
standard element time, 17-18
standard error, 16-17
standard gamble, 1-24
standard, guaranteed, 6-23
standard hour, 6-53, 17-18. *See also* allowed hours
standard hour plan, 6-54
standard hours produced, 6-54
standard industrial classification (SIC) code, 4-31, 4-32, 5-14
Standard International Trade Classification (SITC), 8-31, 8-32. *See also* commodity code
standardization, 5-14
standardized LaPlace-Gauss distribution, 16-17
standardized mortality ratio (SMR), 13-43
standardized normal distribution, 16-17
standard man, 13-43
standard metropolitan statistical area (SMSA), 6-54
standard network diagram, 4-31
standard observer, 9-21
standard output, 17-18
standard pattern, 12-46
standard performance, 6-54, 17-18
standard practice, 17-18. *See also* method
standard rate, 6-54. *See also* job rate
standards audit, 17-19
standard shaft tolerancing practice, 12-66
standard system, 17-18. *See also* predetermined time system
standard time, 6-54, 17-18. *See also* normal performance; production standard; time; work standard
standard time data, 6-54
standard time, statistical, 17-19
standby, 17-19. *See also* delay
standby redundancy, 1-21, 1-22, 16-14
standby time, 17-19
stand hole tolerancing practice, 12-66
standing committee, 10-29
standing eye height, 2-10
standing hip breadth, 2-11
standing knee height, 9-21
standing plans, 10-29
standing shoulder height, 2-12
stannosis, 13-43
start date, actual, 4-3
start date, project, 4-27
start date, projected, 4-27
starting basis, 1-24
starting event. *See* beginning event
starting rate, 6-54
starting solution, 1-24
start schedule, level (SS), 4-20

start/stop character, 8-32
start time, 14-21
start time, early (ES), 4-14
startup costs, 4-32
start-up curve, 17-19. *See also* learning curve
start-up/startup, 4-32, 10-29
statement, 3-16
state of statistical control, 16-17
state probability in queuing models, 1-24
static accuracy, 11-16
static behavior, 12-23
static display, 9-21
static model, 11-17
static work, 2-50, 17-19
station, 8-33, 11-17
stationary process, 1-25
statistic, 1-25, 16-17
statistical process control, 10-29
statistical quality control, 1-25, 16-17
statistical test, 16-17
statistical time, 17-19
statistic, order, 16-11
statistics, non-parametric, 5-10
stature, 2-12, 9-21
status, 4-32, 5-15, 15-13. *See also* progress
status congruency, 15-13
statusing, 4-32. *See also* updating
status line, 4-32. *See also* data date (DD); time now line
status pay, 6-54
steady rest, 12-66
steady state, 1-25, 12-23
steam hammer, 12-30. *See also* drop hammer; hammer forging
steel
 air-hardening, 12-48
 carbon, 12-32, 12-50
 ferritic, 12-37
 high speed (HSS), 12-63
 killed, 12-54
 SAE, 12-65
 unkilled, 12-47
steel belt conveyor, 8-33
steel rule die, 12-58
STEL (short term exposure limits), 13-41
stenosing tenosynovitis, 13-43
STEP 7, 12-23
step bonus, 6-54
stepping motor, 11-17, 12-23
steppingstone method, 1-25
step rates, 6-54. *See also* automatic progression
step stress test, 1-25, 16-17
stereoscopic acuity, 9-21
stereotyping, 10-29
sternocleidomastoid muscles, 2-50
sternum height, 9-21
sticker, 12-46
stiffness, 2-50, 11-17
stimulation, intellectual, 10-16
stimulus, 9-21
stimulus, acoustic, 2-29, 9-2

stimulus, auditory, 9-2
stimulus generalization, 9-21
stint, 6-54
stir-in resin, 12-84
stochastic, 1-25
stochastic programming, 1-25
stock and bond value, 4-32
stock, available, 5-2
stock bonus, 6-54
stock, bulk, 8-5
stock, cycle, 5-5, 14-6
stock keeping unit (SKU), 8-32, 14-21. *See also* single-level where used
stockless production. *See* just-in-time (JIT)
stock number, 8-33
stock option, 6-54
stock option discount, 6-54
stock option exercise, 6-54
stock option forfeit, 6-54
stock option holding period, 6-54
stock option, non-qualified, 6-54
stock option plans, 6-54
stock option, qualified, 6-46
stock option term, 6-55
stockout, 14-21. *See also* backorder
stockout cost, 10-29
stockout percentage, 14-21. *See also* customer service ratio; percent of fill
stockouts, 5-15
stock purchase plan, 6-55
stock, safety, 14-19. *See also* fluctuation inventories
stock status, 14-21
stock turnover, 5-15
stock warrant, 6-55
stokers' cramps, 13-23
stop, 11-17, 12-58
stop buttons, 12-84
stop control, 12-58
stopping distance, total (motor vehicle), 13-45
stopping tolerance (carousel), 8-33
stopwatch, 17-19
stopwatch, decimal-hour, 17-5
stopwatch, decimal-minute, 17-5
stopwatch waiting time. *See* deadtime
stop work order. *See* suspension of work, directed
storage, 3-16, 12-23
storage and retrieval system, automated (AS/RS), 8-3, 11-2
storage, auxiliary, 3-3
storage bay, 8-4
storage, bin, 8-5
storage, bulk, 8-5
storage, caged, 8-6
storage, decentralized, 8-10
storage, fixed, 12-12
storage life (shelf life), 1-25
storage module, 8-33
storage policy, 8-33

storage policy, class-based, 8-7
storage policy, random, 8-26
storage/retrieval (S/R) machine, 8-33
storage, single deep, 8-31
storage (symbol), 17-8
storehouse, 8-33
store, limited-line, 5-8
stores, chain, 5-3
stores, general merchandise, 5-7
storming, 10-29
story, 8-33, 10-29
STPD (standard temperature and pressure, dry) condition, 9-21. *See also* BTPS conditions
straddle milling, 12-66
straddle truck, 8-33
straight bill of lading, 8-33. *See also* bill of lading
straight chute, 8-33
straight commission, 6-55
straight-cut system, 12-23
straight-in straight-out dock, 8-33
straight-line depreciation, 4-33, 7-9
straight polarity, 12-91
straight salary, 6-55
straight time, 6-55. *See also* earnings; overtime
straight time rate, 6-55
strain, 2-50, 12-58, 13-43
strain gauge, 11-17
strain hardening, 12-58, 12-66
strain propagation, 2-50
strain synthesis, 2-50. *See also* work stress
strain, work, 2-54
strapping, 8-33
strap slot, 8-33
strategic business unit (SBU), 5-15, 10-29
strategic control, 10-29
strategic control points, 10-29
strategic goals, 10-29
strategic management, 10-29
strategic (management) planning, 5-15
strategic plans, 10-29
strategic quality planning, 15-13
strategy formulation, 10-29
strategy implementation, 10-29
strategy (mixed). *See* minimax theorem
strategy, pulling, 5-12
strategy, pushing, 5-12
strategy/strategies, 1-25, 10-29
stratification, 16-17
stratified matching, 13-30
stratified random sampling, 1-25, 16-17
stratified sampling, 13-43
stratum, 1-25, 16-17
streamline flow, 12-46
strength, baked, 12-31
strength considerations in anthropometric design, 2-5
strength, dry, 12-37

This index is arranged in letter-by-letter format.

strength, torsional, 12-59
strength, ultimate. *See* tensile strength
stress, 9-21, 12-58
stress, bending, 12-49
stress-crack, 12-84
stress, engineering, 12-52
stress equivalent, 2-50
stress, organizational, 6-38
stress-relief heat treatment, 12-91
stress relieving, 12-58
stress, residual, 12-46, 12-65
stress test, step, 1-25, 16-17
stress transmittal, 2-50
stretch forming, 12-58, 12-84
stretching, 12-58
stretchout, 6-55. *See also* speedup
stretchwrap, 8-33
striation, cellular, 12-71
strictly concave function, 1-25
strictly convex function. *See* convex function
strictly increasing function, 1-25
strike, 6-55. *See also* slowdown
strike benefits, 6-55
strikebreaker, 6-55. *See also* anti-strikebreaker law; fink
strike deadline, 6-55
strike fund, 6-55
strike, general walkout, 6-55
strike insurance, 6-55
strike, jurisdictional, 6-29
strike notice, 6-55
strike, outlaw, 6-55
strike, quickie, 6-55
strike, sit-down, 6-55
strike, slowdown, 6-55
strike, sympathy, 6-55
strike vote, 6-55
strike, wildcat, 6-55
string, 3-16
stringer, 8-33
stringermember, 8-33
strip mall, 5-15
stripper, 12-58
stripper plate, 12-84
stripping fork, 12-84
stroboscope, 9-21. *See also* flicker fusion
stroke, 12-58
stroke, adjustable, 12-48
strokes per minute (SPM), 12-58
strophospheres, 9-22
struck work or goods, 6-56. *See also* hot-cargo clause
structural bay, 8-4
structural foam, 12-84. *See also* foam molding
structure, simple, 10-28
student's distribution, 16-17
stud welding, 12-91
study, intervening, 13-26
study period, 4-33, 7-14. *See also* life; planning horizon
style development rate, 6-56
stylus, 3-16
styrene acrylonitrile (SAN), 12-84

styrene plastics, 12-84
styrene rubber plastics, 12-84
styrofoam pattern, 12-46
subassembly, 11-17, 12-23, 14-21. *See also* component; inventory write-off
subcontract, 4-33
subcontracting, 6-56. *See also* contracting-out
subcontractor, 4-33
sub-group, measurement sense, 16-17
sub-group, object sense, 16-17
subindex, 4-33
subject, 9-22
subjective expected utility (SEU), 9-22
subliminal advertising, 5-15
submerged arc welding, 12-91
subminimal rate, 6-56
subnasale, 2-23
subnet, 4-33. *See also* frag net
suboptimal, 1-25
suboptimization, 5-15
subrogation, 13-43
subroutine, 3-16, 12-23
subroutine call, 3-16
subscapular skinfold, 2-7
subscript, 3-16
subsistence allowance, 6-56
substandard rate, 6-56. *See also* handicapped worker rate
substantial completion, 4-33. *See also* beneficial occupancy
substitute item, 1-13
substitutes, 10-29
substitutes for leadership, 10-29
substitution, 14-21
SUB (supplemental unemployment benefit plan), 6-56
subsystem, 4-33
subtracted time, 17-19
successor activity, 4-33
successor event, 4-33. *See also* J-Node (ADM)
sudden failure, 1-10
suggestion system, 6-56
summarized bill of material (BOM), 14-21
summary item, 4-33
summary network, 4-33
summary number, 4-33
sum-of-digits-method, 4-33
sunk cost, 4-33, 7-14, 10-30
sunlighting, 6-56. *See also* moonlighting
superannuated rate, 6-56
superannuated workers, 6-56
super bill of material (BOM), 14-21. *See also* common parts bill of material; modular bill of material; planning bill of material
supercraft. *See* mastercraft
super finishing, 12-66
superheat, 12-46
superior knowledge. *See* misrepresentation

supermarket, 5-15
superordinate goals, 10-30
super seniority, 6-56. *See also* seniority
supervision, 6-56, 15-13. *See also* foreman
supervision costs, 4-13
supervisor, 6-56
supervisor, acting, 15-13
supervisor, first line, 10-12
supervisor, working, 15-13
supervisory control, 9-22, 11-17
supervisory incentive, 6-56
supervisory program, 3-16
supination, 2-50, 11-17. *See also* pronation; range of forearm pronation
supplemental compensation, 6-56
supplemental unemployment benefit plan (SUB), 6-56
supplementary benefits. *See* fringe benefits
supplementary conditions, 4-33
supplements to wages and salaries, 6-57
supplied-air suit, 13-43
supplier, 4-33. *See also* vendor
supplier delivery performance, 14-21
supplier partnership, 14-21
suppliers, 10-30
supply chain management, 5-15
supply, elastic, 5-6
supporting hyperplane, 1-25
supporting shank, 12-66
supportive, 10-30
support pillar, 12-84
support post, 12-84
supports, conductors, 8-33
surety, 4-33
surface, 12-23
surface feet per minute (SFPM), 12-66
surface grinding, 12-66
surface profile, 12-65
surface roughness. *See* roughness
surfacing, 12-91
surgical benefits, 6-57, 13-43. *See also* health and insurance plan
surplus variable, 1-25
surround brightness, 9-22
surveillance, 1-25, 13-43, 16-17
surveillance, medical, 13-30
survey, 5-15, 13-43
survey, attitude, 6-6
survivability, 1-25, 16-17
survivor protection, 6-57
survivors' benefits, 6-57, 13-43
survivors' option. *See* joint and survivor option
suspended tray conveyor, 8-33. *See also* lowering conveyor
suspended tray elevator, 8-33
suspended tray lift, 8-33
suspension, 6-57
suspension of work, constructive, 4-33

suspension of work, directed, 4-33
sustaining capital, 4-6
Sv (sievert), 13-41
swager, rotary, 12-56
sweatshop, 6-57
swedge, 12-30
sweep, 12-46
sweetheart agreement, 6-57
sweethearting, 5-15
swell, 12-46
swimmer's ear, 13-34
swing frame grinder, 12-46
swing shift, 6-57. *See also* rotating shift; shift; split shift
swiss automatic. *See* screw machines
switch, 12-24
switching rules, 16-17
switch, monorail, 8-33
switch, rotary, 8-29, 12-56
switch, sliding type, 8-32
SWOT analysis, 10-30
symbol, 3-16, 8-33, 10-30
symbol density, 8-33
symbolic language, 3-16. *See also* language
symbolic processes, 10-30
symbol length, 8-34
symbols lists, 7-2
symmetric parametric programming, 1-26
sympathy strike. *See* strike
synapse, 2-50. *See also* motor nerve; sensory nerve; simple reflex
synchro, 11-17
synchronization allowance. *See* interference allowance
synchronous, 11-17
synchronous transmission, 3-16
syndiotactic, 12-84
syndrome, 2-50
syndrome, acceleration, 13-2
syndrome, acquired immune deficiency (AIDS), 13-3
syndrome, acute radiation, 13-3
syndrome, carpal tunnel (CTS), 2-32, 13-8
syndrome, dead hand, 13-38
syndrome, G-force, 13-2
syndrome, high pressure nervous, 13-23
syndrome, pulmonary hyperinflation, 13-37
syndrome, scalenus anticus, 2-48
syndrome, white hand, 13-38
synectics, 10-30
synergism, 13-44
synergist, 2-50
synergy, 10-30, 15-13
synergy, negative, 10-20
synersis, 12-84. *See also* gel
synovia, 2-51
synovial fluid, 2-50
synovial structures, 2-51
syntax, 3-16

This index is arranged in letter-by-letter format.

synthetic data, 17-19. *See also* predetermined time system; standard data
synthetic molding sand, 12-46
synthetic time standard, 6-57, 17-19
SYSGEN (system generation), 3-16
system, 1-26, 3-16, 4-33, 9-22, 10-30, 11-17, 12-24, 15-13, 17-19
system acquisition projects, major, 4-21
system analysis, 9-22
systematic sampling, 13-44
system design, 11-17
system, dynamic, 9-8
system effectiveness, 1-26
system engineering, 9-22
system, formal, 15-5. *See also* informal group
system, functional, 4-17
system generation (SYSGEN), 3-16
system, intelligent, 11-9
system, linear, 1-26
systems #1, 2, 3, 4, 15-13
system safety, 13-44
system safety analysis, 13-44. *See also* fault tree analysis; THERP
system safety engineering, 13-44
systems analysis, 11-17, 15-13
systems development in life cycle, 10-30
systems improvement, continuous (CSI), 15-3
systems improvement, total (TSI), 15-14
systems studies, 4-34
systems theory, 10-30, 15-13

T

tab, 12-24
table, 3-16
tableau, 1-26
table feeder, rotary, 8-29
table lookup, 3-16
tablet, 3-16
tab sequential format, 12-24
tabular display, 9-22
tachistoscope, 9-22
tachometer, 11-17
tack, 12-84
tack weld, 12-91
tactical control, 10-30
tactical goals, 10-30
tactical plans, 10-30
tactile, 9-22
tactile control, 9-22
tactile sense, 9-22
tactile sensor, 11-17
tactile stimuli, non-positive, 2-43
tactile stimulus, 9-22

Taft-Hartley Act. *See* Labor Management Relations Act
tag line, 8-34
tailstock, 12-66
take-home pay, 6-57. *See also* spendable earnings
take-home wages, 6-57
take-off, 4-34
takeover, 10-30
talc, 12-46
tall structure, 10-30
tang, 12-66
tangential, 2-51
tangibles, 4-34
tape, cassette, 12-6
tape leader, 12-24
tape, magnetic, 12-16
tape, paper, 12-19
tapping, 12-66
taps, 12-66
tare, 8-34
target, 6-57, 9-22
target acquisition, 9-22
target date, 4-34
target discrimination, 9-22
target reporting, 4-34
target return objectives, 5-15
target start date. *See* expected begin date
tariff, 10-30
tariff, air transport, 8-34
tariff, memorandum, 8-21
tariff, motor carrier, 8-34
tariff, railroad, 8-34
task, 3-16, 4-34, 6-57, 9-22, 15-14. *See also* job
task analysis, 9-22
task and bonus plan, 6-57
task coding, 9-22
task, continuous, 9-23
task element, 9-23
task environment, 10-30
task equivalence, 9-23
task force, 10-30, 15-14
task group, 10-30
task identity, 10-30
task load, 2-51
task monitor, 4-34
task, normal, 17-13
task, procedural, 9-23
task roles, group, 10-14
task significance, 10-30
tax credit ESOP, 6-57
Tax Equity and Fiscal Responsibility Act, 1982 (TEFRA), 6-58
taxes payable, 4-34
tax, per capita, 6-41
Tax Reform Act Stock Ownership Plan (TRASOP), 6-57
tax shelter, 6-58. *See also* ERTA
Taylor's tool life equation, 12-66
t-distribution, 16-17
teach, 11-17
teach pendant, 11-17
team, 10-30
team building, 10-30, 15-14

team manning, 6-58
teardown, 11-17, 17-19. *See also* changeover time
technical skills, 10-31
technique for human error prediction (THERP), 13-44. *See also* system safety analysis
technological element, 10-31
technological forecasting, 10-31
technological interdependence, 10-31
technological transfer, 10-31
technological unemployment, 6-58
technology, 10-31
technology code (GT code), group, 11-8
technology, group, 10-14, 11-8, 12-13, 14-10. *See also* group classification code
technology matrix, 1-26
technostructural activities, 10-31
tee joint, 12-91
TEFRA (Tax Equity and Fiscal Responsibility Act, 1982), 6-58
telecommunication, 3-16
telecommuting, 10-31
teleoperation, 9-23
teleoperator, 11-17
telescoping conveyor, 8-34. *See also* extendable conveyor
telescoping mast, 8-34
temper, 12-46, 12-84
temperature, ambient, 12-69
temperature, critical, 12-26, 12-35
temperature, effective, 2-34. *See also* thermal environment
temperature, finishing, 12-52
temperature, fluxing, 12-75
temperature, fusion, 12-75
temperature, gamma-transition, 12-76
temperature, glass transition, 12-76
temperature, glassy-transition, 12-76
temperature, interpass, 12-88
temperature, preheat, 12-90
temper brittleness, 12-58
temper carbon, 12-46
tempering, 12-58
temper time, 12-91
template, 3-17
template matching, 11-17
temporary construction cost, 4-34
temporary disability insurance, 6-58. *See also* accident and sickness benefits; worker's compensation
temporary employee, 6-58. *See also* probationary period
temporary rate, 6-58, 17-19. *See also* guarantee or trial rate; protest price
temporary standard, 6-58, 17-19
temporary threshold shift (TTS), 13-44

temporary total disability, 13-44
tendinitis, 2-51. *See also* tenosynovitis
tendon, 2-51. *See also* tenosynovitis
tennis elbow, 2-32, 2-51, 13-16. *See also* epicondylitis; joint; radio-ulnar joints
tenosynovitis, 2-51, 13-44. *See also* tendinitis; tendon
tenosynovitis, stenosing, 13-41
tensile strength, 12-59
teratogen, 13-44
terminal, 3-17, 12-24
termination, 4-34, 6-58
termination pay. *See* severance pay
terms and conditions, general, 4-17
terms of payment, 4-34
territory assignment, 5-15
territory, sales, 5-14
test, accelerated, 1-2, 16-2
test, alpha, 9-3
test, aptitude, 6-5
test, bend, 12-49
test, beta, 9-4
test, blind, 5-2
test, Brinell hardness, 12-60
test, charpy, 12-50
test, concept, 5-4
test, deformation, 12-36
test, destructive, 13-13
test, employment, 10-10
tester, Rockwell hardness, 12-65
test, hypertest, 9-11
testimony, expert, 13-16
testing, 1-26
testing, nondestructive, 12-42, 13-33
test, izod, 12-40
test, knoop hardness, 12-63
test, market, 5-8
test, one-sided, 16-11
test, patch, 13-35
test, physical abilities, 6-42
test, power of, 16-12
test, screening, 1-23, 13-40
test, sensitivity, 1-23
test, service model, 9-20
test, significance, 1-24, 16-16
test, significance level of a, 16-16
test, sonic, 12-46
test, spark, 12-46
test, speech perception, 13-42
test, spiral fluidity, 12-46
test, statistical, 16-17
test, two-sided, 16-18
test, usability, 9-25
test, validity, 9-25
tethered-electric truck, 8-34
text, 3-17
text processor, 3-17
TF (total float), 4-34
thenar, 2-23
thenar eminence, 2-23

theorems
    complementary slackness, 1-6
    duality theorem for linear programming, 1-9
    max-flow min-cut, 1-15
    minimax, 1-16
theoretical biomechanics, 2-51
theory, 13-44
theory, acquired-needs, 10-2
theory, attribution, 15-2
theory, automata, 12-4
theory, behavioral, 15-2
theory, classical organizational, 15-3
theory, cognitive, 10-5
theory, cognitive resource, 10-5
theory, contingency, 10-7
theory, decision, 7-8
theory, equity, 10-10, 15-5
theory, Erg, 10-10
theory, expectancy, 10-11, 15-5
theory, game, 1-11, 10-13
theory, hierarchy of needs, 10-14
theory, human capital, 6-25
theory, information, 9-11, 12-14, 15-7
theory, marginal productivity of wages, 6-34
theory, motive-hygiene, 15-9
theory, organization, 15-11
theory, path-goal, 10-23
theory, prospect, 10-25
theory, queuing, 1-21
theory, reinforcement, 10-26
theory, signal detection, 9-20
theory, situational leadership, 10-28
theory, social learning, 10-28
theory, systems, 10-30, 15-13
theory, trichromatic, 9-25, 10-2
theory, two-factor, 10-32
Theory X, 6-58, 15-14
Theory Y, 6-58, 15-14
Theory Z, 6-58, 10-31, 15-14
Therblig, 17-19
Therblig chart, 17-20. *See also* film analysis chart
thermal environment, 2-51. *See also* effective temperature
thermally foamed plastics, 12-84
thermal stress, 2-51
thermit welding, 12-91
thermodynamic efficiency, 2-51. *See also* metabolism
thermoelasticity, 12-84
thermoforming, 12-84
thermograph, 2-51. *See also* infrared photography
thermoplastics, 12-84. *See also* novolak
thermoset, 12-84
thermosetting, 12-84
THERP (technique for human error prediction), 13-44. *See also* system safety analysis
thigh clearance height, sitting, 2-12

thinking, convergent, 10-7
thinking, divergent, 10-9
third party claim, 4-34
third-party intervention, 10-31
thixotrophy, 12-46
thixotropic, 12-84
thread manufacturing, 12-67
threshold, absolute, 9-2
threshold, absolute visual, 9-23
threshold contrast, 9-23
threshold, difference, 9-23
threshold dose, 13-44
threshold (hearing), 13-44
thresholding, 11-17
threshold limit values-ceiling (TLV-C), 13-44
threshold limit values-time-weighted average (TLV), 13-44
threshold limit values (TLV), 13-44
threshold, masked differential, 9-23
threshold, of audibility, 9-23
threshold, of detectability, 9-23
threshold, of discomfort, 9-23
threshold, of pain, 9-23
threshold shift, 9-23
threshold, terminal, 9-23
thrift plan, 6-58
throat depth, 12-59
throat of a fillet weld, 12-91
throughput, 3-17, 11-17
through-put capacity, 9-23
throw-away-inserts. *See* inserts
thruput, 17-20
thrust, axis of, 2-31
thumbwheel, 3-17
tibia, 2-52
tidal volume (TV), 13-29. *See also* lung volumes
tied activity, 4-34
tie, nonrigid, 8-22
tier, 8-34
tierer, pallet load. *See* pallet load tierer
tiering, 8-34
tiering guide, portable metal stacking rack, 8-34
tie rods, 12-59
tight rate, 6-58, 17-20
tight standard, 6-58, 17-20
tilt, 8-34, 11-17
timbre, 9-24
time, abnormal, 17-2. *See also* outlier(s)
time, access, 3-2
time, activity, 4-2
time, actual, 17-2
time, adjustment, 1-26
time, administrative, 1-26
time, alert, 1-26
time allowance. *See* allowance
time allowance, machine, 17-11
time, allowed, 6-4, 17-2
time and link analysis, 9-24
time and one-half, 6-58
time, assembly, 12-70

time, auxiliary process, 17-2
time, available machine, 17-2
time, available process, 17-2
time, average, 6-7, 17-2
time, average cycle, 17-2
time, average element, 17-2
time, base, 4-4
time, break, 13-39. *See also* rest period
time bucket, 14-21. *See also* bucketed system; bucketless system
time, calibration, 1-26
time, changeover, 6-10, 17-3. *See also* setup; teardown
time, checkout, 1-26, 16-18
time, cleanup, 1-26, 6-11, 16-18
time, clothes changing, 6-11
time, constant arrival, 1-3
time, constant service. *See* arrival rate distribution, constant
time, contract, 4-8
time, cool, 12-86
time, cycle, 5-5, 6-14, 11-5, 14-7
time, decision, 9-7
time delay, 1-26, 16-18
time, delay, 17-5. *See* downtime; idle time
time, double, 6-16
time, down. *See* downtime
time, early event (EV), 4-14
time, early finish (EF), 4-14
time, early start (ES), 4-14
time, elapsed, 17-6
time element, 17-6
time, element, 6-17
time, estimated, 17-6
time estimate, most likely, 4-22
time estimate, optimistic, 4-24
time estimate, pessimistic, 4-25
time, expected elapsed, 4-16
time, exponential service, 1-9
time extension, 4-34
time, external, 17-7
time, fault correction, 1-26, 16-18
time, fault location, 1-26
time, fault locator, 16-18
time fence, 14-21
time, first piece, 17-7
time, flashing, 12-87
time formula, 17-20
time, guaranteed, 6-23
time, hand, 17-9. *See also* manual element
time, handling, 17-9
time, heat, 12-88
time, high, 6-24. *See also* hazard pay
time, historical. *See* standard time
time, hold, 12-88
time, holding, 1-12, 4-18
time horizon. *See* study period
time, human performance (HPT), 17-9
time, idle, 3-8, 6-25, 11-9, 17-9. *See also* avoidable delay(s); unavoidable delay
time, idle machine. *See* machine idle time

time, inactive, 1-26, 16-18
time, incubation, 13-24
time, in reliability, 16-18
time, interarrival, 1-13
time, interference. *See also* interference allowance; machine interference
time, internal, 17-9
time, interoperation, 14-10. *See* queue time
time, item obtainment, 1-26, 16-18
time, latest event (LET), 4-20
time, lead, 14-11
time, leveled. *See* normal time
time-limited scheduling, 4-34
time-line analysis, 9-24
time, lost. *See* waiting time
time, machine, 17-11
time, machine attention, 17-11
time, machine-controlled, 17-11. *See also* restricted work
time, machine idle, 17-11
time, manual, 17-11. *See also* manual element
time, materials handling, 14-14
time, mean. *See* average time
time, mean-maintenance, 1-15
time measurement unit (TMU), 17-20
time, median, 17-12
time, minimum, 6-36, 17-13
time, minor set-up, 14-14. *See also* major set-up
time, mission, 1-26, 16-18
time, modal, 17-13
time, modification, 1-26, 16-18
time, move, 11-12
time, normal, 17-13. *See also* normal performance
time, normal element, 17-13
time now line, 4-34. *See also* status line
time, observed. *See* actual time
time off, compensatory, 6-12
time of the essence, 4-34
time out, 3-17
time, payout, 4-24
time, personal, 6-42, 17-15
time-phased order point (TPOP), 14-21. *See also* distribution requirements planning (DRP); order point
time phasing, 14-21
time plan, differential, 6-15. *See also* multiple time plan
time plan, multiple, 6-36
time, predetermined, 17-15
time, preparation, 1-27, 16-18
time, process, 17-16
time, processing, 14-17
time, productive, 6-45, 17-16
time, project, 4-27
time, quench, 12-90
time, queue, 14-18
timer, 11-17

This index is arranged in letter-by-letter format.

time rate, 6-59. *See also* time wage
time rate earnings, 6-59
time, reaction (RT), 9-17
time, real, 3-14, 9-18, 11-15, 12-21
time (reliability), 1-26
time, relief, 6-47
time, response, 3-14, 9-19, 12-22
time, run, 11-15, 14-19
time, runout, 17-17
time, safety, 14-19
time scaled CPM, 4-34
time, scheduled event, 4-30
time, select(ed), 17-18
time, select(ed) element, 17-18
time, sensing, 9-20
time-series, 1-27, 16-18
time-series methods, 10-31
time, settling, 11-16
time, set-up/setup, 11-16, 14-20
time sharing/time-sharing, 3-17, 9-24, 12-24
time, slack, 4-31, 4-32, 14-20. *See also* critical ratio; slack
time, span, 11-16
time, squeeze, 12-91
time standard, 6-59, 14-21. *See also* standard time
time, standard, 6-54, 17-2, 17-18. *See also* normal performance; production standard; work standard
time, standard allowed, 6-54
time, standard element, 17-18
time standard, synthetic, 17-19
time, standby, 17-19
time, start, 14-21
time, statistical, 17-19
time, straight, 6-55. *See also* earnings; overtime
time study, 6-59, 17-20
time study observation sheet, 17-20
time, subtracted, 17-19
time, temper, 12-91
time, tool maintenance. *See* allowed time
time, transit, 14-22
time, travel, 6-59, 17-20
time, turnaround, 3-17
time unit. *See* calendar unit
time value of money, 7-14. *See also* interest
time wage, 6-59. *See also* time rate
time, wait, 11-18, 14-22
time, waiting, 1-28, 17-21
time, warm-up, 9-26
time, wash-up, 6-64
time-weighed average (TWA), 13-45
time, work, 6-66
timing, continuous, 17-4
timing, differential, 17-5. *See also* cycle timing
tin handcuffs, 6-59

tinnitus, 13-45
tin parachutes, 6-59
tint, 9-24
tip, 6-59
tip tool, brazed, 12-60
TIPTURN, 12-24
tire, low solid, 8-20
tire, solid, 8-32
tissue rheology, 2-52
TLC (total lung capacity), 13-29. *See also* lung volumes
TLV-C (threshold limit values-ceiling), 13-44
TLV (threshold limit values), 13-44
TLV-TWA (threshold limit values-time-weighted average), 13-44
TMU (time measurement unit), 17-20
toe board/toeboards, 8-34, 13-45
toe, cantilever rack, 8-34
toe of weld, 12-91
toggle joint, 12-59
toggle links, pneumatic, 12-55
tolerance, 12-67, 13-45, 16-18, 17-20
tolerance, bilateral, 12-60
tolerance, drug, 13-45
tolerance limits, 16-18, 17-20
tolerance, specification, 1-27
tolerance, unilateral, 12-67
tonal, 9-24
tonal gap, 9-24
tonal island, 9-24
tonality, 9-24
tone, 9-24
tonghold, 12-30
tongs, safety, 13-45
tongue switch, 8-34
tonnage. *See* capacity, press
tonnage rate, 6-59
tonus, 2-52
tool allowance, 6-59
tool and cutter grinder, 12-67
tool-chip interface, 12-67
tool cutting edge, nose radius, 12-64
tool geometry, 12-67
tooling, 11-18, 12-24, 12-67
tool life, 11-18, 12-67
tool maintenance time. *See* allowed time
toolmaker's microscope, 12-67
tool materials, 12-67
tool mode, 11-18
tool offset, 12-24
toolpath, 12-24
toolroom lathe, 12-67
tool selection, random, 12-21
tool, single-point, 12-66
tool wear, 12-67
top cross braces, 8-34
top diagonal brace, 8-34
top-down budgeting, 10-31
top-down planning, 14-21
top horizontal brace, 8-34
top managers, 10-31

top-running crane, 8-34
top tie, 8-34
torch brazing, 12-91
torque, 2-52. *See also* moment of force
torque converter transmission, 8-34
torque, full load, 8-34
torque, lifting, 2-41
torsional strength, 12-59
total available market, 5-15
total capital, 4-6
total cash compensation, 6-59
total compensation, 6-59
total cost bidding, 4-34
total-factor productivity, 10-31
total float (TF), 4-34
total loss, 13-45
total lung capacity (TLC), 13-29. *See also* lung volumes
total market, 5-8
total pay, 6-59
total quality control (TQC), 10-31
total quality management, 15-14
total reaction distance (motor vehicle), 13-45
total stopping distance (motor vehicle), 13-45
total systems improvement (TSI), 15-14
touch labor, 6-59
touch screen, 9-24
toughness, 12-59
tour of duty. *See* shift
tow conveyor, 8-34. *See also* chain conveyor
tower crane, 8-34. *See also* crane; fixed crane
tow tractor, industrial, 8-17
toxemia, 13-45
toxic, 13-45
toxicant, 13-45
toxicity, 13-45
toxicology, 13-45
toxicology, industrial, 13-45
toxic substance, 13-45
Toxic Substances Control Act (TSCA), 13-45
TPOP (time-phased order point), 14-21
TPS (transaction-processing system), 10-31
TQC (total quality control), 10-31
trace, 3-17
traceable costs, 7-14. *See also* direct cost; marginal cost
trackball, 9-24
tracking, 4-35, 9-24, 11-18. *See also* rate tracking
tracking, rate, 9-17
track openers, 8-18
tractive effort, 8-34
tractive effort, required return, 8-34
tractor, internal combustion, 8-18
tractor, motor driven, 8-34
trade associations, 10-31

trade, balance of, 10-3
trade council. *See* joint board
trade discount, 5-15
trademark, 5-15
tradeoff analysis, 5-4
trade union. *See* union
trading area, 5-15
traffic accident, 13-45
traffic control, 8-35
traffic density, 13-13
traffic management, 5-15
tragion, 2-23
trailing costs. *See* layoff costs/benefits
trainee, 6-59
training, 9-24, 15-14
training allowance, 17-20. *See also* allowance; learner's allowance
training and development, 10-31
training, job instruction (JIT), 6-28
training, job safety, 13-27
training, safety, 13-40
training, sensitivity, 6-51, 15-12
training, skill, 15-12
training, vestibule, 6-62
training, vocational, 6-62
traits, 10-31
trajectory, 11-18
tramway, 8-35
transactional leaders, 10-31
transaction-processing system (TPS), 10-31
transducer, 11-18
transducer, differential, 12-10
transducer, ultrasonic, 2-53
transfer card, 6-59
transfer, lateral, 2-41
transfer lift section, 8-35
transfer line, 11-18
transfer machine, 11-18, 12-67
transfer molding, 12-84
transfer, negative, 9-14
transfer of learning, 9-24
transfer pay, 6-59
transfer price, 4-35
transfer, technological, 10-31
transformational leaders, 10-31
transformation matrix, 1-27
transformation of a matrix, 1-27
transformation processes, 10-31
transient, 11-18, 12-24
transient bill of material (B)M), 14-22. *See also* bill of material structuring
transient state, 1-27
transillumination, 9-24
transit inventory, 14-22
transition benefits, 13-43. *See also* survivors' benefits
transition fit, 12-67
transit time, 14-22. *See also* materials handling time
translation, 11-18
translator, 3-17
transmission, 3-17

This index is arranged in letter-by-letter format.

transmission loss (sound), 13-45
transmittance, 9-24
transportation, 5-15
transportation inventory, 14-22
transportation problem, 1-27
transportation problem, capacitated, 1-5
transportation (symbol), 17-8
transshipment problem, 1-27
transverse, 2-23
transverse dorsal ligament, 2-35, 2-36
transverse ligament, 2-52
transverse plane, 2-52. *See also* coronal plane; mid-sagittal plane; sagittal plane
trapezious muscle, 2-52
TRASOP (Tax Reform Act Stock Ownership Plan), 6-57
trauma, 2-52, 13-45. *See also* cumulative trauma
trauma, acoustic, 9-2, 13-3
trauma, cumulative, 2-33
traumatic arthritis, 2-52
traumatic fibrositis, 2-52
travel chart, 17-20
traveling crane, 8-35. *See* fixed crane. *See also* crane; gantry crane; overhead crane; semi-gantry crane; wall cranes
traveling purchase requisitions, 14-22
traveling-salesman problem, 1-27
travel requisition, 14-22
travel speed, 8-35
travel time, 6-59, 17-20
travel time models, 8-35
tray, 8-35
tray, battery, 8-35
tree, 1-27
tremor, 2-52, 9-24, 13-45
trench foot, 13-24
trepanning, 12-67
trial, 9-24
trials, independent, 1-12, 16-9
trianomaly, 9-24
tribology, 12-67
triceps skinfold, 2-7
trichromatic theory, 9-25
trichromatism, 9-25
trick. *See* shift
trigger finger, 2-52, 13-43
trimming, 12-30, 12-59
trim problem, 1-27
trip, 12-59
triple precision, 3-17
tripodal grasp, 2-52, 2-53
tripping, 12-59
tripping mechanism, 12-59
tritanopia, 9-25
trochlea, 2-52
trolley conveyor, 8-35. *See also* enclosed track trolley conveyor
trolley, top running, 8-35
trolley, underhung. *See* carrier
troughed belt conveyor, 8-5

truck
 approved industrial, 8-2
 battery-electric, 8-4
 cantilever, 8-6
 center control, 8-7
 counterbalanced, 8-9
 counterbalanced front side-loader, 8-9
 diesel-electric, 8-11
 electric, 8-12
 end, 8-13
 end control, 8-13
 fixed platform, 8-14
 forklift, 8-15
 fork truck with tilting mast, 8-20
 gas-electric, 8-16
 hand, 8-16
 high lift, 8-17
 high lift platform, 8-17
 industrial, 8-17
 internal combustion engine, 8-18
 lift, 8-19
 low-lift, 8-20
 low-lift pallet, 8-20
 motorized hand, 8-21
 motorized hand/rider, 8-21
 narrow aisle, 8-22
 order picker, high lift, 8-22
 order picking, 8-23
 order picking and stacking, 8-23
 pallet, 8-25
 platform, 8-26
 powered industrial, 8-27
 reach, 8-28
 rider, 8-29
 rough terrain, 8-29
 straddle, 8-33
 tethered-electric, 8-34
 variable reach lift, 8-36
truck/air/truck service, 8-35
truck bay, 8-4
truck crane, 8-35. *See also* portable crane
truck leveler, 8-35
trueing, 12-67
truncation, 3-17
trunk, 9-25
trustee, 6-59. *See also* fund
trusteeship, 6-59. *See also* monitorship
trust fund. *See* fund
trust fund, pension, 6-41
truth table, 3-17
TSCA (Toxic Substances Control Act), 13-45
TSI (total systems improvement), 15-14
TTS (temporary threshold shift), 13-44
tube process, continuous, 12-72
tuberosity, 2-53
tubular bone, 2-53
tuition payment plan, 6-60
tumbling, 12-47

tumor, 13-45
tumor, metastatic, 13-31
tungsten carbide (WC), 12-67
tungsten electrode, 12-91
tunnel, 8-35
tunnel vision, 13-45
Turing machine, 3-17
turn. *See* shift
turnaround, 10-31
turnaround time, 3-17, 9-25
turning, 12-67
turning center, 8-35, 12-24
turning radius, inside, 8-35
turning radius, outside, 8-35
turn key system, 12-24
turnover, 6-31. *See also* labor turnover
turnover ratio, 4-35
turns. *See* inventory turnover
turntable, 8-35
turret lathes, 12-67
tuyere, 12-47
TV (tidal volume), 13-29. *See also* lung volumes
TWA (time-weighted average), 13-45
twist, 12-59
twist conveyor, 8-35
twist drill, 12-67
two bin system, 14-22
two-factor theory, 10-32
two-hand controls, 13-45
two-level MPS, 14-22
two-sided confidence interval, 16-18
two-sided test, 16-18
two-way communication, 10-32
two-way container, 8-35
type A behavior, 15-14
type B behavior, 15-14
type I error probability, 16-19
type II error probability, 16-19
type of accident, 13-46

U

UCC-APT, 12-24
U chart, 16-6
UEL (upper explosive limit), 13-46
u-line layout, 8-35
ulna, 2-53. *See also* radius
ulnar deviation, 2-53
ulnar nerve, 2-53. *See also* hamate bone
ultimate consumer, 5-15
ultimate strength. *See* tensile strength
ultraprecision machining, 12-67
ultrasonic, 9-25
ultrasonic cleaning, 12-47
ultrasonic machining, 12-67
ultrasonics, 2-53. *See also* echography; sonography

ultrasonic transducer, 2-53
ultrasonic welding, 12-91
UL (underwriter's laboratories) listing, 8-36
umpire, permanent. *See* permanent arbitrator
unaffiliated union. *See* independent union
unavoidable delay, 17-20. *See also* delay; idle time
unavoidable delay allowance, 17-20. *See also* allowance; delay allowance
unbiased estimator, 16-19
unbounded solution, 1-27
unburdening, 9-25
uncertainty, 4-35, 7-14, 10-32
uncertainty avoidance, 10-32
under clearance, portable metal stacking rack, 8-35
undercontrol, 10-32
undercut, 12-24, 12-84, 12-91. *See also* back taper
underdrive, 8-35
undereye height, 2-12
underground facilities, 4-35
underhung crane. *See* traveling cranes
under-running crane, 8-35
underrun (overrun), 4-24
underrun (overrun), projected, 4-27
undershoot, 11-18
under the influence of alcohol, 13-46
undervoltage protection, 8-36
underwriter, 13-46
underwriters' laboratories listing (UL), 8-36
undifferentiated marketing, 5-16
unemployment compensation, 6-60
unemployment insurance, 6-60. *See also* short-term income protection
unemployment pay, 6-60
unemployment, technological, 6-58
unfair labor practice, 6-60. *See also* Labor Management Relations Act
unfair labor practice strike, 6-60
unfair list, 6-60
unfunded plan, 6-60
UNIAPT, 12-24
unidirectional path, 8-36
uniform gradient series, 7-14. *See also* annuity fund factor; gradient factors
unilateral modification, 4-22
unilateral tolerance, 12-67. *See also* bilateral tolerance; tolerance
union, 4-35, 6-60
union agreement. *See* agreement
union, amalgamated craft, 6-5
union, closed, 6-11

This index is arranged in letter-by-letter format.

union, company, 6-11
union contract. *See* agreement
union convention, 6-60
union dues, 6-60. *See also* check-off
union, independent, 6-26
union, industrial, 6-27
unionism, bread-and-butter. *See* business unionism
unionism, business, 6-9
unionism, dual, 6-16
union label (bug), 6-60
union leave, 6-60
union, local, 6-32
union-management cooperation, 6-60
union member, 6-60. *See also* member
union, open, 6-38
union organizer. *See* organizer
union rate, 6-61
union recognition, 6-61
unions, 10-32
union scale. *See* union rate
union security, 6-61. *See also* maintenance-of-membership clauses; open shop
union shop, 6-61. *See also* agency shop
union, single company, 6-52
union steward. *See* shop steward
unique optimal solution, 1-27
unit cost, 4-35
unit item, 1-13
unitized tooling, 12-59
unit load, 8-36, 11-18
unit, nonrepaired, 1-27
unit of issue, 8-36
unit of measure, 14-22
unit of production, 6-61
unit, peripheral, 3-13
unit power consumption, 12-67
unit price contracts, 4-9
unit pricing, 5-16
unit production, 10-32
unit, repairable, 1-27
unit stress. *See* stress
unity of command, 15-14
univariate standard scores, anthropometry, 2-5
universal dividing head, 12-68
universal grinding machine, 12-68
universal product code (UPC), 5-16, 8-36
universe, 1-27, 16-19
univocal, 2-53
unjust enrichment doctrine, 4-35
unkilled steel, 12-47
unlicensed personnel, 6-61
unpaid holidays, 6-61. *See also* paid holidays
unplanned issue/receipt, 14-22
unrestricted element, 17-20
unrestricted job, 6-61, 17-20
unsafe act, 13-46
unsafe condition, 13-46
unusually severe weather, 4-35

UPC (universal product code), 5-16, 8-36
update. *See* statusing
updating, 4-35. *See also* statusing
upgrading, 6-61
upload, 11-18
upper bounded linear programming problem, 1-27
upper explosive limit (UEL), 13-46. *See also* explosive limits
upright, cantilever rack, 8-36
upright frame, 8-36
upset, 12-92
upsetting, 12-30, 12-59
upset welding, 12-92
uptime, 9-25, 17-20
upward communication, 10-32
urban wage rate index, 6-61
urea, 12-84
urea formaldehyde resin, 12-47, 12-84
urea plastics, 12-85
urethane. *See* isocyanate resins
usability, 9-25
usability testing, 9-25
usage rate, 5-16
U.S. Department of Labor, 6-61
use area, functional, 4-17
useful life, 1-27, 4-35
user, 10-32
use value, 4-17
utilities costs, 4-13
utility, 1-27, 7-14. *See also* planning horizon
utility function, 7-14
utilityman, 6-61
utility operator, 6-61
utility price index, constant, 4-8
utility program, 3-17
utility worker, 6-61
utilization parameter, 1-27
utilization, partial, 4-24

vacation pay, 6-61
vacation plan, extended, 6-20
vacations, paid, 6-39
vacuum casting, 12-47
vacuum forming, 12-85
vacuum metalizing, 12-85
valence, 10-32
validation, 3-17, 16-19
validity, 6-61, 10-32, 13-3
validity test, 9-25
valuation charges, 8-36
valuation or appraisal, 4-35, 7-14
value, 9-25
value, activity, 4-35
value added, 15-14. *See also* productivity gainsharing (PGS)
value added by distribution, 4-35, 5-16

value added by marketing, 4-35, 5-16
value-added services (in a warehouse), 8-36
value analysis, 17-20
value, assessed, 4-4
value effective, 4-35
value engineering, 4-35, 17-20
value engineering cost avoidance, 4-35
value engineering cost reduction, 4-35
value engineering job plan, 4-36
value, expected, 10-11
value of a game, 1-27
value of work performed to date, 4-35
value, present, 7-13
vane feeder, rotary, 8-29
vapor, 13-46
vapor, flammable, 13-20
variable, 1-27, 3-17, 13-46
variable, basic, 1-4
variable, bounded, 1-5
variable, confounding, 13-10
variable cost(s), 4-36, 5-16, 7-14
variable, dependent, 13-13
variable, dummy, 13-14
variable element, 17-20
variable error, 9-8
variable, independent, 1-12, 13-25
variable-interval schedule of reinforcement, 10-32
variable-length record, 3-17
variable, manipulated, 12-16
variable, nonbasic, 1-17
variable, phase-type random, 1-19
variable problem, discrete. *See* integer linear programming
variable, random, 1-21, 16-14
variable-ratio schedule of reinforcement, 10-32
variable reach lift truck, 8-36
variables, artificial, 1-3
variable, slack, 1-23
variables measurement, 16-19
variable, surplus, 1-25
variance, 1-28, 4-36
variance, analysis of, 1-2
variance management, schedule, 14-19
variance of a probability distribution, 16-19
variance of a random variable, 16-19
variance of a sample, 16-19
variance, scheduled, 4-30
variant process planning, 11-18
variate, 1-28, 16-19
variation, between-batch, 16-4
variation, between-lot, 16-4
variation, coefficient of, 16-4
variation in estimated quantity, 4-36
varied mapping, 9-25
vasoconstriction, 2-53
vault, 8-36

VC (vital capacity), 13-29. *See also* lung volumes
VEBA (Voluntary Employees' Beneficiary Association), 6-61
vector, 1-28, 3-17
vector, artificial, 1-3
vector, basic, 1-4
vector, constant, 1-6
vector, slack, 1-23
veining, 12-47
veins, saphenous, 2-48
velocity, angular, 2-30
velocity, capture, 13-8
velocity, cutting speed. *See* revolutions per minute; surface feet per minute
velocity error, 11-18
velocity signature, 2-53. *See also* signature
velocity, slot, 13-42
vendor, 14-22
vendor lead time, 14-22
vendor scheduler, 14-22. *See also* buyer; capacity buying; paperless purchasing
veneer, 12-85
Venn diagram, 1-28
vent, 12-30, 12-47, 12-59
ventilation, 13-46
ventilation exhaust, 13-16
ventilation, industrial, 13-25. *See also* exhaust, general; exhaust, local
venture capital, 4-6
venture life, 4-36. *See also* life
venture, new, 10-21
venture team, 10-32
venture teams, new, 10-21
venture units, new, 10-21
venture worth, 4-36
vent wax, 12-47
verbal communication, 10-32
verbal mediation, 9-25
verbal protocol analysis, 9-25
verification, 16-19
verification sampling inspection, 16-19
verify, 3-17, 12-24
vertebra, 2-53
vertebrae, fifth lumbar, 2-36. *See also* herniated disk; lumbosacral angle
vertebrae, lumbar, 2-41
vertebral column, 2-53
vertebra, spinous process of, 2-49
vertex, 1-28, 2-24
vertical bar code, 8-36
vertical belt conveyor, 8-36
vertical chain conveyor, opposed shelf type, 8-36
vertical channel integration, 5-16
vertical clearance, 8-36
vertical communication, 10-32
vertical coordination, 10-32
vertical display, 14-22. *See also* bucketed system; bucketless system; horizontal display

This index is arranged in letter-by-letter format.

vertical event numbering, 4-36
vertical integration, 10-32
vertical lift module, 8-36
vertical loading, 6-62
vertical machine tools, 12-68
vertical marketing system, 5-16
vertical position, 12-92
vertical reciprocating conveyor, 8-36. *See also* inclined reciprocating conveyor; lowering conveyor
vertical union. *See* industrial union
vertigo, 9-25, 13-46
vesicant, 13-46
vested rights, 6-62
vestibule training, 6-62
vesting, 6-62
via point, 11-18
vibrating conveyor, 8-37. *See also* oscillating conveyor
vibrating conveyor, balanced, 8-37
vibrating conveyor, natural frequency, 8-37
vibration, 13-46
vibrator, 12-47
vicarious learning, 10-32
video tape analysis. *See* film analysis record
video teleconferencing, 10-32
viewing angle, 9-25
viewpoint, classical, 10-5
vigilance, 9-25
vinyl acetate plastics, 12-85
vinyl alcohol plastics, 12-85
vinyl chloride plastics, 12-85
vinylidene plastics. *See* saran plastics
vinyl plastics, 12-85
violation
  alleged, 13-4
  apparent, 13-5
  serious, 13-41
violator
  chronic, 13-9
  habitual, 13-22
virgin metal, 12-47
virtual cell, 11-18
virtual environment, 9-25
virtual image, 9-25
virtual storage, 3-17
virus, computer, 10-6
viscoelasticity, 12-85
viscosity, 12-47, 12-85. *See also* consistency; viscosity coefficient
viscosity coefficient, 12-85. *See also* consistency; viscosity
vises, 12-68
visibility, 9-25
vision
  foveal, 9-25
  machine, 12-16
  night, 9-15
  peripheral, 9-16, 9-26
  persistence of, 9-26
  photopic, 9-16

visual acuity, 9-25, 9-26. *See also* foot candle
visual acuity, dynamic, 9-8
visual adaptation, 9-26
visual angle, 9-26
visual field, 9-26
visual field defect, 9-25
visual photometry, 9-26
visual range, 9-26
visual space, 9-26
vital capacity (VC), 13-29. *See also* lung volumes
vital capacity (VC), forced, 13-21
vital statistics, 13-46
vitrified bond wheels, 12-68
Vocational Rehabilitation Act of 1973, 6-62
vocational training, 6-62
void, 12-47
volatile memory, 11-18
volume (tonal), 9-26
voluntary arbitration, 6-5
Voluntary Employees' Beneficiary Association (VEBA), 6-61
V process, 12-47

W

W2 form, 6-62
wage, 6-62, 12-30
wage adjustment, automatic, 6-6
wage advance plan, 6-62. *See also* annual wage or employment guarantee
wage and price control, 6-62
wage and salary administration, 6-62
wage and salary control, 6-13
wage and salary curve, 6-63
wage and salary receipts, 6-62
wage and salary structure, 6-62
wage and salary survey, 6-62
wage arbitration, 6-62
wage assignment, 6-62
wage award, 6-63
wage board employees, 6-63. *See also* classification act employees
wage changes, general, 6-22
wage control, 6-13
wage curve, 6-63
wage deductions, 6-63
wage determination, 6-63
wage determination, individual, 6-26. *See also* maturity comparisons
wage differential, 6-63
wage drift, 6-63
wage escalation. *See* escalator clause
wage, guaranteed annual, 6-23
wage guide. *See* salary guide
wage guidelines, 6-63
wage-hour law. *See* Fair Labor Standards Act

wage incentive, 6-63
wage incentive plan, 6-63, 17-21
wage inequality, 6-63
wage inequity, 6-63
wage law, prevailing, 6-44
wage leaders, 6-63
wage leadership, 6-63
wage level, 6-63
wage, living, 6-32
wage mediation, 6-63
wage, minimum, 6-36
wage movement, 6-63
wage or employment guarantee, annual, 6-5
wage pattern, 6-64
wage payment, contract, 6-13
wage plan, 6-63
wage plan, guaranteed, 6-23. *See also* annual wage or employment guarantee
wage plan, hourly rate, 6-25
wage plan, sliding scale, 6-52
wage policy, 6-64
wage profile. *See* salary profile
wage progression. *See* automatic progression; step rates
wage progression, automatic, 6-6
wage rate, 4-36, 6-64
wage rate, base, 6-7
wage rate bracket, 6-64
wage rate, guaranteed, 6-23
wage rate index, urban, 6-61
wage reopening, 6-64. *See also* reopening clause
wage review, 6-64. *See also* merit increase
wages
  advance on, 6-4
  bootleg, 6-9. *see also* kick-back
  comparable, 6-12
  competitive, 6-12
  deferred wage increases, 6-15. *see also* annual improvement factor (productivity)
  garnishment of, 6-22
  going, 6-22
  marginal productivity theory of, 6-34
  money, 6-36
  real, 6-47
  take-home, 6-57
wages and salaries, 6-62
wages and salaries, supplements to, 6-57
wage structure, 6-64. *See also* salary structure
wage survey. *See* wage and salary survey
wage survey, area, 6-6
wage survey, community, 6-11
wage system, incentive, 6-26
wage, time, 6-59
Wagner Act of 1935, 6-64. *See also* National Labor Relations Act
waist depth, 2-12, 9-26
waist-thumbtip length, 2-12

waiting-in-line models, 10-32
waiting line, 1-28
waiting period, 13-46
waiting time, 1-28, 17-21. *See also* deadtime; delay time; downtime; idle time
waiting time, stopwatch. *See* deadtime
wait time, 11-18, 14-22. *See also* queue time
walkie stacker, 8-37
walking beam conveyor, 8-37
walkout, 6-64
walk, random, 1-21
wall cranes, 8-37. *See also* traveling cranes
wall tie, 8-37
Walsh-Healy Public Contracts Act of 1936, 6-64
wand scanner, 8-37
warehouse demand, 14-22
warehouse layout, 8-37
warehouses, distribution, 5-6
warehouse site selection, 5-16, 8-37. *See also* distribution
warm-up time, 9-26
warning, 9-26
warning limits, 16-19
warp, 12-85. *See also* dished; domed
warpage, 12-47, 12-85. *See also* dished; domed
warranty, 5-16
wash, 12-47
washburn core, 12-47
wash-up time, 6-64
waste bonus, 6-64
waste handling system, 11-18
watch, 6-64. *See also* shift
watchkeeping, 9-26
water, combined, 12-34
water glass, 12-47
water-jet machining, 12-68
water muffler, 8-37
wavelength, 13-46
wave picking, 8-37
waviness, 12-68
wax pattern, 12-47
waybill, 8-37
WBS (work breakdown structure), 4-36
WC (tungsten carbide), 12-67
weak color, 9-26
wear-out failure, 1-10
wear-out failure period, 1-28, 16-19
weathering, artificial, 12-85
weather, unusually severe, 4-35
weave bead, 12-92
web, 12-85
web angle, 12-61
Weber-Fechner law, 9-26
Weber's law, 9-26
weight, dimensional, 8-11
weight, factor, 6-20
weightlessness, 9-26
weights, 4-36

This index is arranged in letter-by-letter format.

weld, 12-92
weldability, 12-92
weld bead, 12-92
welder, 12-92
welder certification, 12-92
welder qualification, 12-92
welder's flashburn, 13-46
weld gauge, 12-92
welding and cutting permit, 13-46
welding, automatic, 12-86
welding, backhand, 12-86
welding, braze, 12-86
welding, cold, 12-86
welding current, 12-92
welding, electron beam, 12-87
welding, electroslag, 12-87
welding, explosive, 12-87
welding, flash, 12-87
welding, forehand, 12-88
welding, forge, 12-88
welding, friction, 12-88
welding, gas, 12-88
welding goggles, 12-92
welding, induction, 12-88
welding, laser beam, 12-89
welding, longitudinal resistance-seam, 12-89
welding machine, 12-92
welding, machine, 12-89
welding, manual, 12-89
welding operation, 12-92
welding operator, 12-92
welding, oxyacetylene, 12-89
welding, percussion, 12-89
welding, plastic, 12-80
welding, pressure gas, 12-90
welding procedure, 12-92
welding, projection, 12-90
welding, resistance seam, 12-90
welding, resistance spot, 12-90
welding rod, 12-92
welding, solid state/solid-state, 12-91
welding, stud, 12-91
welding, thermit, 12-91
welding transformer, 12-92
welding, ultrasonic, 12-91
welding, upset, 12-92
weld, intermittent, 12-88
weld line, 12-85
weldment, 12-92
weld metal, 12-92
weld, toe of, 12-91
welfare plan, 6-64
wellness program, 6-64
WF (work factor), 17-21
what if analysis, 14-22. *See also* simulation
what you see is what you get (WYSIWIG), 9-27
wheel base, 8-37
wheel conveyor, 8-37. *See also* el conveyor; skate-wheel conveyor
wheel, driving, 8-37
wheel, grinding, 12-68
wheel load, 8-37
wheel spiral, 8-37

where-used bill of material. *See* explosion
whipsawing, 6-65
whistle-blower, 10-32
Whistleblower Protection Act of 1989 (WPA), 6-65
white collar union, 6-65
white-collar worker, 6-65. *See also* blue-collar workers; pink collar worker
white hand syndrome, 13-38
white noise, 9-26, 13-47
whiteout, 9-26
wholesaler, 5-16
wholesaler, full-function, 5-7
wholly owned subsidiary, 10-33
wicket conveyor, 8-37
wide area networks, 10-33
widows' allowance, 13-43
wildcat strike, 6-65. *See also* strike
willful misconduct, 13-47
willful violation, 13-47
wind box, 12-47
windchill, 9-27
window, 3-17, 12-85
windup, 11-18
wink, 17-21
wink counter. *See* microchronometer
wiper forming, 12-59
wiping, 12-59
WIP (work-in-process), 11-18, 14-23
wire feed machine, 12-68
wire rod, 12-59
withholding, 6-65
within-batch variation, 16-19
within-lot variation, 16-19
witness, expert, 6-20, 13-16
wood, compregnated, 12-72
word, 3-18
word address format, 12-24
word length, 11-18
word processor. *See* text processor
work, 2-54, 4-36, 6-65, 9-27
work agenda, 10-33
work, authorized, 4-4
work, base division of. *See* therblig
work breakdown structure, contracts (CWBS), 4-36
work breakdown structure (WBS), 4-36
work cell, 17-21
work center, 11-18. *See also* machine center
work center, alternate, 11-2
work, combined, 17-4
work content, 6-65
work coordinates, 11-18
work cycle, 6-65, 9-27, 17-21. *See also* cycle; motion cycle
work design, 17-21. *See also* ergonomics; human factors engineering; methods engineering

work design and measurement. *See* normal operation
work directive change, 4-36
work, division of, 15-4. *See also* specialization
work, dynamic, 2-34
work element, 11-18
work environment, physical, 6-42
worker parachutes, 6-65. *See also* golden parachute
worker participation clause, 6-65
workers, blue-collar, 6-8. *See also* pink collar worker; white-collar worker
worker's compensation, 6-66, 13-47. *See also* accident and sickness benefits; accidental death and dismemberment benefits; temporary disability insurance
worker's compensation insurance, 13-47
workers, equivalent, 6-18
workers, manual. *See* blue-collar workers
workers, migratory, 6-35
workers, productive, 6-44
workers, superannuated, 6-56
work factor (WF), 17-21
work force. *See* labor force
work group, autonomous, 10-3, 15-2
work hardness, 12-59
work holding devices, 12-68
workhour, 4-36
Work Hours Act, 6-65
working area, 17-21
working area, maximum, 17-12
working area, normal, 17-14
working capital, 4-6, 7-14. *See also* capital
working conditions, 6-66, 17-21
working envelope, 11-18
working force. *See* labor force
working life, 12-85
working, mechanical, 12-29, 12-54
working rules, 13-41. *See also* work rules; worker parachutes
working surface, 13-47
work injury, 13-47
work-in-process, 4-36
work-in-process inventory, 10-33
work-in-process (WIP), 11-18, 14-23
work, internal, 17-9
work, isometric, 2-40
work item, 4-36. *See also* activity
workload, 6-66
workload, mental, 9-13
work, measured day, 6-34, 17-12
work measurement, 2-54, 6-65, 17-21
work, mental, 17-12
work metabolism, 2-54
work, nature of, 13-32
work, negative, 2-43

work order. *See* manufacturing order
work pace, 9-27
work pace, expected, 17-6
work package, 4-36
work physiology, 2-54, 9-27
work-piece, 11-19
workplace, 6-66
workplace design
  analytic, 2-29
  empirical, 2-35
  improvement approach, 2-39
  rational, 2-47
workplace layout, 2-54, 6-66, 17-21
workplace, safe, 13-39
work, positive, 2-45
workpower leveling. *See* load leveling
work rate, normal. *See* normal performance
work rest cycle, 9-27
work, restricted, 17-17. *See also* machine-controlled time
work rules, 6-65
work sampling, 17-21. *See also* activity sampling
work sampling study, 6-65
work schedule, early, 4-14
work schedules, alternative, 10-3
work sharing, 6-65
work simplification, 6-65, 17-21
work site, 4-37
workspace, 9-27
work space layout, 9-27
work specialization, 10-33
work standard, 6-66. *See also* standard; standard time
work, static, 2-50
work station, 11-19
work stoppage, 6-66. *See also* lockout
work strain, 2-54. *See also* strain
work stress, 2-54. *See also* strain synthesis
work study, 6-66, 9-27, 17-21
work surface, safety, 8-30
work task, 17-21
work ticket, 6-66
work time, 6-66
work tolerance, 2-54
work tolerance, index of, 2-39
work unit, 4-37, 17-21
work-unit analysis, 17-21
work week, 6-66
workweek, compressed, 10-6
world coordinates, 11-19
world mode, 11-19
worldwide integration strategy, 10-33
worm, 12-68
worm gear, 11-19, 12-68
worth, 4-37
worth, annual, 7-6
worth, functional, 4-17
worth, present, 7-13
wound, 13-47

This index is arranged in letter-by-letter format.

WPA (Whistleblower Protection Act), 1989, 6-65
wraparound, 3-18
wraparound grasp, 2-54
wrinkling, 12-59
wrist, 2-46, 11-19
write, 3-18
written amendment, 4-37
written standard practice. *See* standard practice
WYSIWIG (what you see is what you get), 9-27

zircon sand, 12-48
zone picking, 8-37
zone pricing, 5-16
zooming, 3-18
Z scores, anthropometry, 2-5
Z, Theory, 6-58, 10-31, 15-14
zygomatic arch, 2-24

X-axis, 12-24
X (bar) chart, 16-3
X-ray, 12-47
X, Theory, 6-58, 15-14

Y

yaw, 9-27, 11-19
yaw axis, 9-27
yawing movement, 9-27
Y-axis, 12-24
Y (bar) chart, 16-3
years of life lost, potential, 13-36
year-to-year price index, 4-37
yes-no experiment, 9-27
yield, 7-14, 12-47
yield, expected, 7-10
yield point, 12-59
yield rate, 14-23
yield strength, 12-48, 12-59
yield value, 12-86
Young's modulus (E), 12-48
Y, Theory, 6-58, 15-14

Z

ZBB (zero-base budgeting), 10-33
ZEBRA plan (Zero-balance Reimbursement Account), 6-66
Zero-balance Reimbursement Account (ZEBRA plan), 6-66
zero-base budgeting (ZBB), 10-33
zero defects, 10-33
zerofill, 3-18
zero inventory. *See* just-in-time (JIT)
zero offset, 12-24
zero-sum two-person game (matrix game), 1-28
zircon, 12-48
zirconia ZrO_2, 12-48

This index is arranged in letter-by-letter format.